ENGINEERING MATERIALS

Properties and Selection

Eighth Edition

Kenneth G. Budinski
Technical Director
Bud Labs

Michael K. Budinski
Senior Materials Engineer
General Motors Corporation

Upper Saddle River, New Jersey
Columbus, Ohio

Library of Congress Cataloging-in-Publication Data
Budinski, Kenneth G.
Engineering materials : properties and selection / Kenneth G. Budinski, Michael K. Budinski.— 8th ed.
p. cm.
Includes bibliographical references and index.
1. Materials. I. Budinski, Michael K. II. Title.

TA403.B787 2004
620.1'1—dc22

2004012298

Executive Editor: Debbie Yarnell
Production Editor: Louise N. Sette
Production Supervision: Susan Free, *The GTS Companies*/York, PA Campus
Design Coordinator: Diane Ernsberger
Cover Designer: Jim Hunter
Production Manager: Deidra Schwartz
Marketing Manager: Jimmy Stephens

This book was set in Times Roman by *The GTS Companies*/York, PA Campus. It was printed and bound by R. R. Donnelley & Sons Company. The cover was printed by Phoenix Color Corp.

Pearson Education Ltd.
Pearson Education Singapore Pte. Ltd.
Pearson Education Canada, Ltd.
Pearson Education—Japan
Pearson Education Australia Pty. Limited
Pearson Education North Asia Ltd.
Pearson Educacion de Mexico, S.A. de C.V.
Pearson Education Malaysia Pte. Ltd.

10 9 8 7 6 5 4 3 2 1
ISBN 0-13-183779-6

Dedicated to Robert V. Fister, a co-worker who freely shared his considerable technical knowledge,
and to Emilie,
a future materials scientist

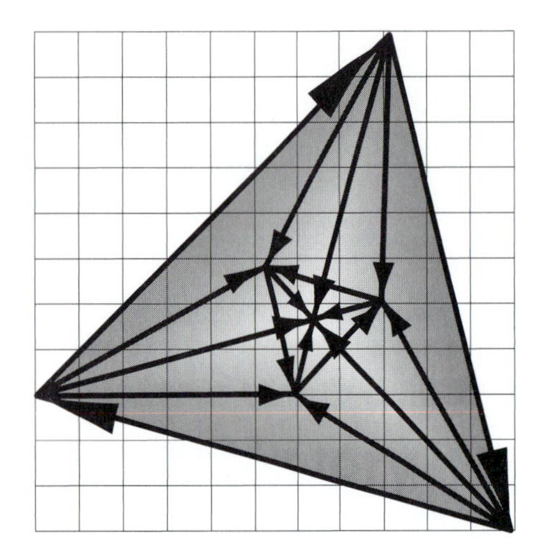

About the Cover

The cover shows a rocket engine from a spacecraft. This engine and many other components on spacecraft would not be possible without the many engineering material innovations that have taken place in the last fifty years. Engineers had to identify materials that could be used to make pumps, cryogenic fluids, etc., that resist erosion from hot gases at 5000°F, that withstand heat stresses and impact, and that are resistant to fatigue and corrosion. Space exploration requires the gamut of engineering materials. Rocket engine structure members are made from super alloys; ceramics are used to resist hot gas erosion; plastics are used for seals and for self-lubricating mechanisms; and composites are used for weight reduction and insulation. Spacecraft are engineering marvels. They demonstrate what can be accomplished through an understanding of the properties of engineering materials.

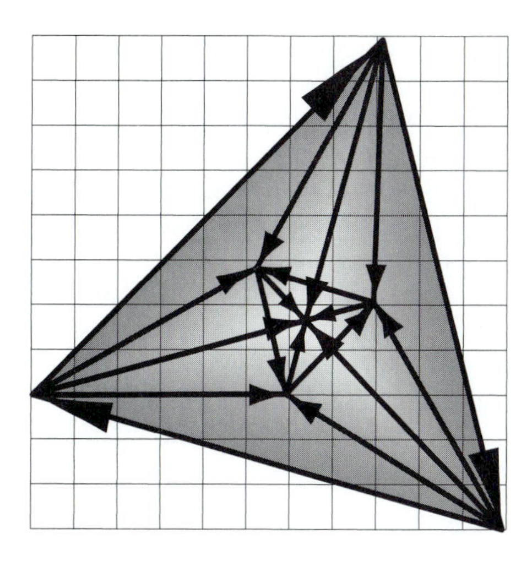

Preface

This is the 25th year that this title has been in print, but it is certainly not a 25-year-old book. Rather, it chronicles what has happened in engineering materials worldwide over that time period. The original purpose, objective, and scope remain unchanged. This book presents the materials information that designers, engineers, and manufacturing staff need to know to do their jobs effectively. The scope of the book has remained materials for machine design, but the list of applicable materials has increased quite a lot since the first edition. This is why each edition has been "fatter" than the previous one; there are more material options that designers need to know about, and we know more about why things happen the way they do in materials engineering. We apologize to the students who have to carry this heavier-than-most textbook.

The biggest change in this edition is the addition of another chapter on plastics. The material that was in Chapter 4 in the last edition is now in two more teachable and digestible chapters. New terms were added to the glossary in Chapter 2. A Rationale was added to each chapter to show students why they need to learn the material in that chapter. The latest technical developments were added to each chapter. In addition, each chapter has new Case Histories of material selection situations, new questions, and additional references. A new chapter on surface engineering (Chapter 19) was written from a blank sheet of paper and reviewed by eminent surface engineers. The reviewers preferred the 7th Edition format, so we kept much from the 7th Edition and added what is happening today in this important field.

Another significant change in this edition is a thorough review of the entire book by three users in universities and reviews of individual chapters by eminent materials engineers. They identified errors, added new theories, and suggested rewrites in some areas. We are quite confident that the material in this edition is as up to date, correct, complete, and relevant as a materials text can be. It reflects the changed repertoire of materials prompted by a global economy. The book is appropriate for international users because we have made a conscious effort to list international standards that apply to material designations and property tests.

In summary, we believe that this book is the best that there is on selection of engineering materials, and we hope that students and teachers agree. We are very grateful to the reviewers who made suggestions for updating and improving the text. Professor Robert Snyder of Rochester Institute of Technology reviewed the entire book and suggested new test questions. We acknowledge the user reviews commissioned by Prentice Hall: William Dalton, Purdue University; David Domermuth, Appalachian State University; and Roger L. Ballerand, Triton College. A special thanks to Professor William Dalton from Purdue University, who performed the most comprehensive review that we have ever received. Peter Blau of Oak Ridge National Labs reviewed the Tribology chapter, Mark Kohler of Eastman Kodak reviewed the Surface Engineering chapter, Mahmoud Abd Elhamed of General Motors reviewed the Corrosion chapter, Keith Newman of General Motors reviewed the Castings and Powder Metals chapter, and Angela Leisner did most of the typing and organizing needed for the book and instructor's manual.

We sincerely hope that this book becomes a "keeper" for students who use it. It contains the net learning of our collective 65 years in the materials engineering business as well as the theories and fundamentals that make materials engineering a science.

Best regards,
Ken Budinski (father)
Michael Budinski (son)

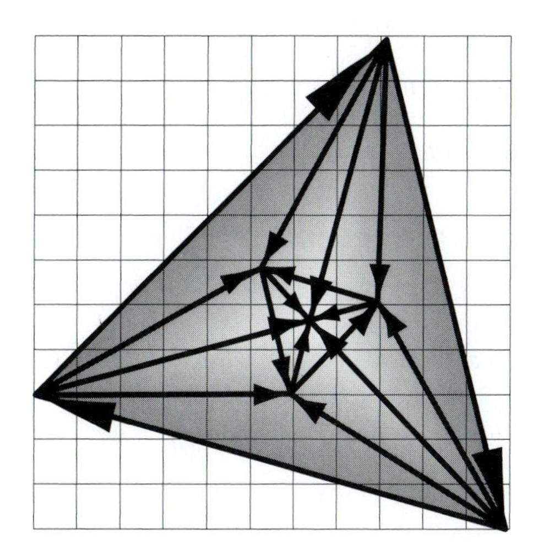

Contents

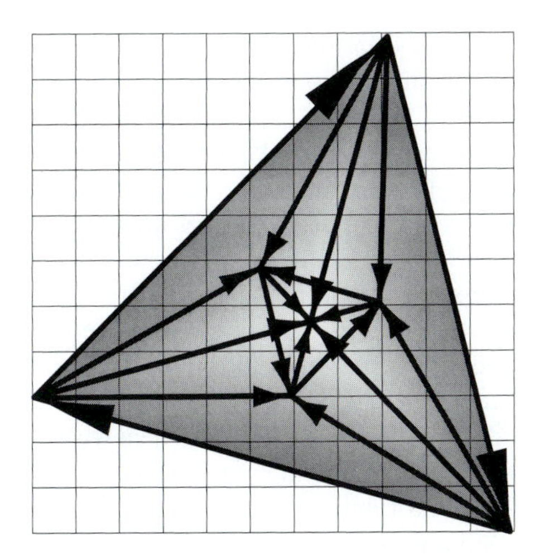

CHAPTER 1

The Structure of Materials

Chapter Goals

1. An understanding of how the elements are the building blocks for engineering materials.
2. A review of basic chemistry, the nature of the atom, and how the elements combine; and the establishment of the language of materials.
3. An understanding of how engineering materials, metals, polymers, ceramics, and composites are related in origin and structural characteristics.

Rationale

The subject of engineering materials is vast; there are thousands of different plastics, thousands of metal alloys, and countless ceramic compounds and composite possibilities that comprise this area of study. However, all of these engineering materials are made from the hundred or so stable elements that have been discovered. In fact, this rather thick book is all about 51 of these elements. That is all that it takes to make the thousands of material options that are available to design engineers. The elements are the basis of engineering materials which is why this text starts with a discussion of elements and the principles of chemistry—the universal language for dealing with our elements and their reactions.

Engineering materials are the materials that make up the products, structures, devices, and mechanisms that we use to maintain life and to improve living conditions. Metals are key to structures and machinery; plastics are invaluable in packaging, medical devices, consumer goods, and even clothing. Ceramics are necessary for the electronics that have become so much a part of our lives. Composites use metals, plastics, and ceramics to make new materials with properties superior to those of the component materials.

The evolution of the human species follows the evolution of engineering materials. At the time of Christ, only a handful of engineering materials were available. Of course, people had stone, wood, and clay to use for structures, but they only had copper, tin, iron, gold, lead, and silver for tool materials. Needless to say, only iron and some copper alloys had utility as tools for shaping useful items that would improve living conditions.

In the past 200 years, scientists have identified all of the stable materials (the *elements*) that exist in our world. We have learned how to combine them to form useful *compounds*, structural materials, tool materials, and electronic devices and we are using them for constructing impressive buildings, for communications, for medical purposes—for all of the things that we want and need. With engineering materials, we can make it all happen.

The overall objective of this book is to present enough theory and property information on engineering materials to give the reader enough knowledge to be able to select materials for designing engineering assignments. The reader will develop a repertoire of materials that he or she can comfortably use to build a bridge, to design plastic parts, to make a tool, to support an integrated circuit, to conduct heat or electricity, or to support a broken bone. And there should be no failures. This book will layout both the good and the bad of every material system discussed, and suggestions will be made on how to deal with the property limitations of each material. Information will be presented that a designer needs for making designs that will last for their intended lifetime.

The theory of material systems will be minimized, but enough will be presented to provide a foundation for selection information. All of the important material systems will be covered: polymers, ceramics, metals, composites, and combinations of these systems. Few machines work well using only polymers or only metals. All material systems should be considered for use. As an introduction to the materials concept, this chapter will review basic chemistry and show how engineering materials are interrelated in concept and properties.

1.1 The Origin of Engineering Materials

Materials engineering is based largely on the pure sciences of chemistry and physics. This text will assume that the reader has a general knowledge of these subjects. Because engineering materials involve many chemical terms, material discussions will be prefaced with a brief review of some of the more important chemical fundamentals and terms.

All materials obey the laws of physics and chemistry in their formation, reactions, and combinations. The smallest part of an element that retains the properties of that element is the *atom*. Atoms are the building blocks for engineering materials. All matter is composed of atoms bonded together in different patterns and with different types of bonds. As shown in Figure 1–1, most substances that we deal with in industry and in everyday life can be categorized as organic or inorganic. *Organic materials* contain the element carbon (and usually hydrogen) as a key part of their structure, and they are usually derived from living things. Petroleum products are organic; crude oil is really the residue of plants that lived millions of years ago, and all plants and animals are organic in nature. *Inorganic materials* are those substances not derived from living things. Sand, rock, water, metals, and inert gases are inorganic materials. Chemistry as a science is usually separated into two fields based on these two criteria. Some chemists specialize in organic chemistry; others specialize in inorganic chemistry. Metallurgists and ceramists deal primarily with inorganic substances. Plastic engineers, on the other hand, deal primarily with organic substances. The field of materials engineering deals with both areas, as does this text.

We shall review the list of basic ingredients that are used to make both organic and inorganic materials, the elements, in order to address engineering materials on a chronological basis. An *element* is a pure substance that cannot be broken down by chemical means to a simpler substance. About 90 elements occur naturally in the earth's crust; some elements are unstable and occur as the result of fission or fusion reactions. Most chemistry texts list 109 elements, but inclusion of laboratory-synthesized elements brings the total number of elements to more than 120.

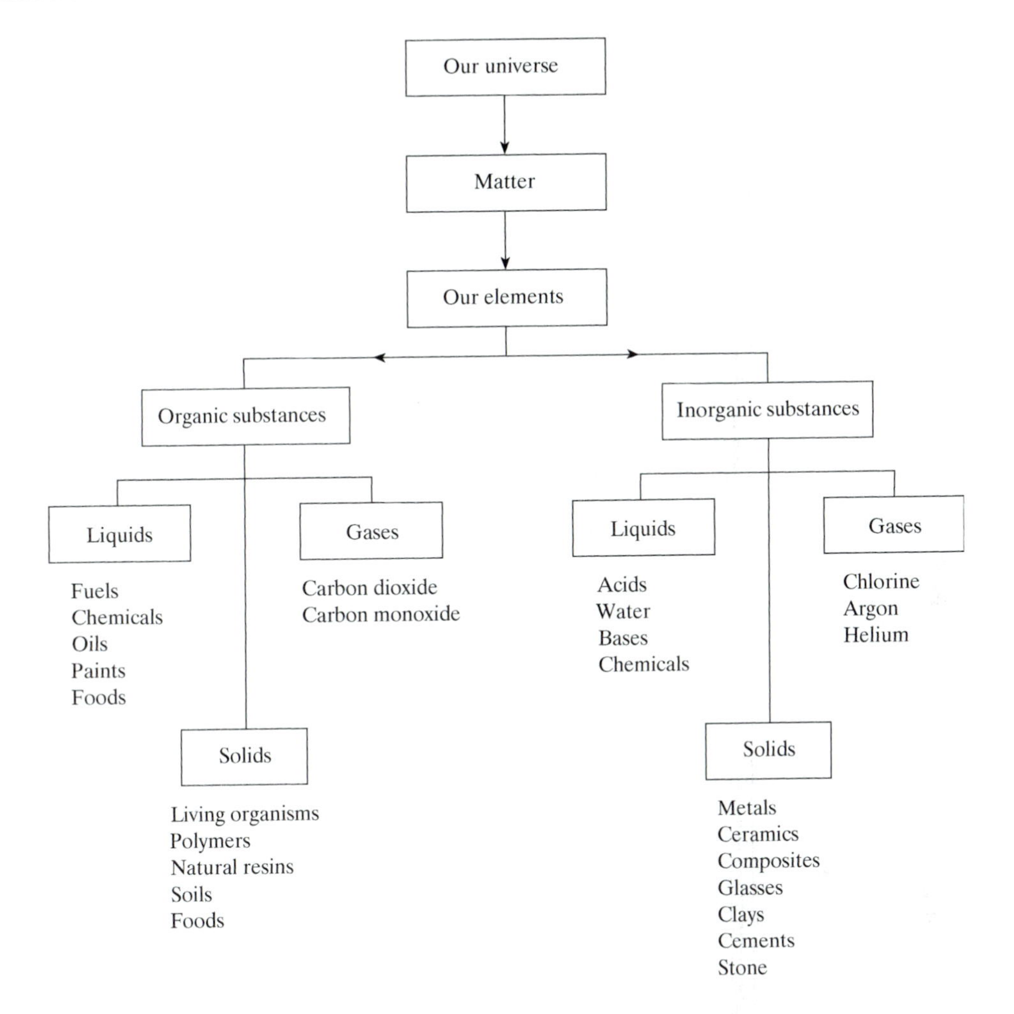

Figure 1–1
The elements are the building blocks for all materials.

Many of these elements have little industrial significance, but it is important in engineering materials to recognize the names and chemical symbols for the more useful elements. Figure 1–2 shows a common version of the periodic table. This table lists elements by atomic number. The element hydrogen was assigned an atomic number of 1, and all of the other elements derive their atomic number from a comparison of their subatomic makeup compared to the element hydrogen. The *atomic number* is really the number of protons in the nucleus of an atom. Atoms are far more complicated than we probably even know, but, current knowledge characterizes atoms as being composed of *protons* (positively charged particles), *neutrons* (neutral particles), and *electrons*, which orbit the *nucleus*, or core, of an atom. For simplicity, atoms are often characterized as a "sun" (nucleus), surrounded by orbiting "planets" (electrons). Electrons have mass. Both neutrons and protons also have mass. It is generally agreed that protons have a nominal mass of

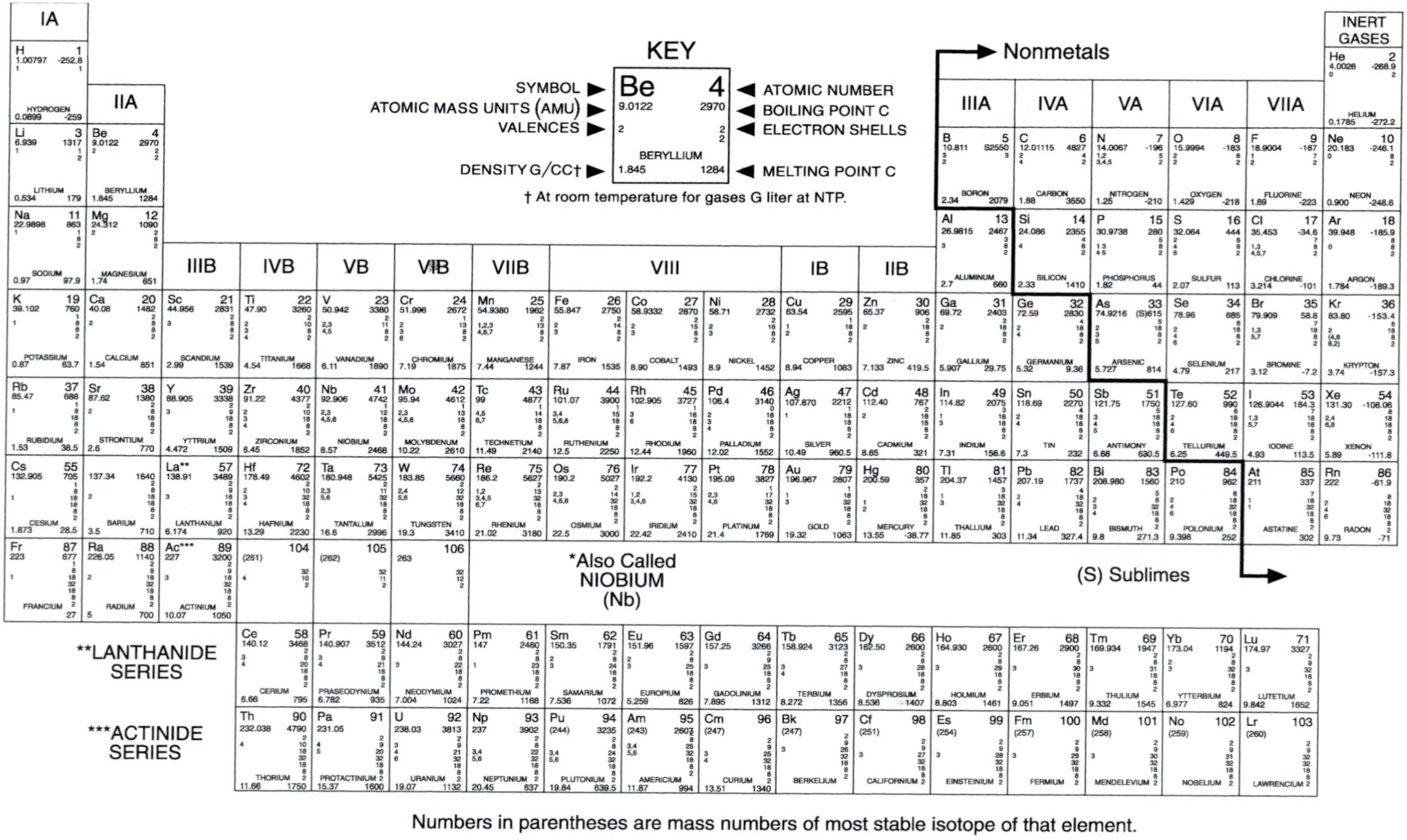

KEY

SYMBOL ▶	Be	4	◀ ATOMIC NUMBER
ATOMIC MASS UNITS (AMU) ▶	9.0122	2970	◀ BOILING POINT C
VALENCES ▶	2	2 2	◀ ELECTRON SHELLS
	BERYLLIUM		
DENSITY G/CC† ▶	1.845	1284	◀ MELTING POINT C

† At room temperature for gases G liter at NTP.

→ Nonmetals

*Also Called NIOBIUM (Nb)

(S) Sublimes

Group	Atomic number	Symbol	Name	AMU	Boiling point C	Melting point C	Density	Valences	Electron shells
IA	1	H	HYDROGEN	1.00797	-252.8	-259	0.0899	1	1
INERT GASES	2	He	HELIUM	4.0026	-268.9	-272.2	0.1785	0	2
IA	3	Li	LITHIUM	6.939	1317	179	0.534	1	1, 2
IIA	4	Be	BERYLLIUM	9.0122	2970	1284	1.845	2	2, 2
IIIA	5	B	BORON	10.811	S2550	2079	2.34	3; 2	3
IVA	6	C	CARBON	12.01115	4827	3550	1.88	2; 4	4, 2
VA	7	N	NITROGEN	14.0067	-196	-210	1.25	1,2; 3,4,5	5, 2
VIA	8	O	OXYGEN	15.9994	-183	-218	1.429	2; 2	6, 2
VIIA	9	F	FLUORINE	18.9004	-187	-223	1.69	1; 2	7, 2
INERT GASES	10	Ne	NEON	20.183	-246.1	-248.6	0.900	0	8, 2
IA	11	Na	SODIUM	22.9898	863	97.9	0.97	1	1, 8, 2
IIA	12	Mg	MAGNESIUM	24.312	1090	651	1.74	2	2, 8, 2
IIIA	13	Al	ALUMINUM	26.9815	2467	660	2.7	3	3, 8, 2
IVA	14	Si	SILICON	24.086	2355	1410	2.33	4	4, 8, 2
VA	15	P	PHOSPHORUS	30.9738	280	44	1.82	1 3; 4 5	5, 8, 2
VIA	16	S	SULFUR	32.064	444	113	2.07	2; 4; 6	6, 8, 2
VIIA	17	Cl	CHLORINE	35.453	-34.6	-101	3.214	1,3; 4,5,7	7, 8, 2
INERT GASES	18	Ar	ARGON	39.948	-185.9	-189.3	1.784	0	8, 8, 2
IA	19	K	POTASSIUM	39.102	760	63.7	0.87	1	1, 8, 8, 2
IIA	20	Ca	CALCIUM	40.08	1482	851	1.54	2	2, 8, 8, 2
IIIB	21	Sc	SCANDIUM	44.956	2831	1539	2.99	3	2, 9, 8, 2
IVB	22	Ti	TITANIUM	47.90	3260	1668	4.54	2; 3; 4	2, 10, 8, 2
VB	23	V	VANADIUM	50.942	3380	1890	6.11	2,3; 4,5	2, 11, 8, 2
VIB	24	Cr	CHROMIUM	51.996	2672	1875	7.19	2; 3; 6	1, 13, 8, 2
VIIB	25	Mn	MANGANESE	54.9380	1962	1244	7.44	1,2,3; 4,6,7	2, 13, 8, 2
VIII	26	Fe	IRON	55.847	2750	1535	7.87	2; 3	2, 14, 8, 2
VIII	27	Co	COBALT	58.9332	2870	1493	8.90	2; 3	2, 15, 8, 2
VIII	28	Ni	NICKEL	58.71	2732	1452	8.9	2; 3	2, 16, 8, 2
IB	29	Cu	COPPER	63.54	2595	1083	8.94	1; 2	1, 18, 8, 2
IIB	30	Zn	ZINC	65.37	906	419.5	7.133	2	2, 18, 8, 2
IIIA	31	Ga	GALLIUM	69.72	2403	29.75	5.907	2	2, 18, 8, 2
IVA	32	Ge	GERMANIUM	72.59	2830	9.36	5.32	2; 4	4, 18, 8, 2
VA	33	As	ARSENIC	74.9216	(S)615	814	5.727	2; 3; 5	5, 18, 8, 2
VIA	34	Se	SELENIUM	78.96	685	217	4.79	2; 4; 6	6, 18, 8, 2
VIIA	35	Br	BROMINE	79.909	58.8	-7.2	3.12	1,3; 5,7	7, 18, 8, 2
INERT GASES	36	Kr	KRYPTON	83.80	-153.4	-157.3	3.74	2; (4,6; 8,2)	8, 18, 8, 2
IA	37	Rb	RUBIDIUM	85.47	688	38.5	1.53	1	1, 8, 18, 8, 2
IIA	38	Sr	STRONTIUM	87.62	1380	770	2.6	2	2, 8, 18, 8, 2
IIIB	39	Y	YTTRIUM	88.905	3338	1509	4.472	3	2, 9, 18, 8, 2
IVB	40	Zr	ZIRCONIUM	91.22	4377	1852	6.45	2; 3; 4	2, 10, 18, 8, 2
VB	41	Nb	NIOBIUM	92.906	4742	2468	8.57	2,3; 4,5,6	1, 12, 18, 8, 2
VIB	42	Mo	MOLYBDENUM	95.94	4612	2610	10.22	2,3; 4,5,6	1, 13, 18, 8, 2
VIIB	43	Tc	TECHNETIUM	99	4877	2140	11.49	4,5; 6,7	1, 14, 18, 8, 2
VIII	44	Ru	RUTHENIUM	101.07	3900	2250	12.5	3,4; 5,6,8	1, 15, 18, 8, 2
VIII	45	Rh	RHODIUM	102.905	3727	1960	12.44	3; 6	1, 16, 18, 8, 2
VIII	46	Pd	PALLADIUM	106.4	3140	1552	12.02	2; 3; 4	0, 18, 18, 8, 2
IB	47	Ag	SILVER	107.870	2212	960.5	10.49	1	1, 18, 18, 8, 2
IIB	48	Cd	CADMIUM	112.40	767	321	8.65	2	2, 18, 18, 8, 2
IIIA	49	In	INDIUM	114.82	2075	156.6	7.31	1; 3	3, 18, 18, 8, 2
IVA	50	Sn	TIN	118.69	2270	232	7.3	2; 4	4, 18, 18, 8, 2
VA	51	Sb	ANTIMONY	121.75	1750	630.5	6.68	3; 4; 5	5, 18, 18, 8, 2
VIA	52	Te	TELLURIUM	127.60	990	449.5	6.25	2; 4; 6	6, 18, 18, 8, 2
VIIA	53	I	IODINE	126.9044	184.3	113.5	4.93	1,3; 5,7	7, 18, 18, 8, 2
INERT GASES	54	Xe	XENON	131.30	-108.06	-111.8	5.89	2,4; 6,8	8, 18, 18, 8, 2
IA	55	Cs	CESIUM	132.905	705	28.5	1.873	1	1, 8, 18, 18, 8, 2
IIA			BARIUM	137.34	1640	710	3.5	2	2, 8, 18, 18, 8, 2
IIIB	57	La**	LANTHANUM	138.91	3489	920	6.174	3	2, 9, 18, 18, 8, 2
IVB	72	Hf	HAFNIUM	178.49	4602	2230	13.29	2; 3; 4	2, 10, 32, 18, 8, 2
VB	73	Ta	TANTALUM	180.948	5425	2996	16.6	2,3; 5,6	2, 11, 32, 18, 8, 2
VIB	74	W	TUNGSTEN	183.85	5660	3410	19.3	2,4; 5,6	2, 12, 32, 18, 8, 2
VIIB	75	Re	RHENIUM	186.2	5627	3180	21.02	1,2; 3,4,5; 6,7	2, 13, 32, 18, 8, 2
VIII	76	Os	OSMIUM	190.2	5027	3000	22.5	2,3; 4,6,8	2, 14, 32, 18, 8, 2
VIII	77	Ir	IRIDIUM	192.2	4130	2410	22.42	1,2; 3,4,6	2, 15, 32, 18, 8, 2
VIII	78	Pt	PLATINUM	195.09	3827	1769	21.4	2,3; 4,6	1, 17, 32, 18, 8, 2
IB	79	Au	GOLD	196.967	2807	1063	19.32	1; 3	1, 18, 32, 18, 8, 2
IIB	80	Hg	MERCURY	200.59	357	-38.77	13.55	1; 2	2, 18, 32, 18, 8, 2
IIIA	81	Tl	THALLIUM	204.37	1457	303	11.85	1; 3	3, 18, 32, 18, 8, 2
IVA	82	Pb	LEAD	207.19	1737	327.4	11.34	2; 4	4, 18, 32, 18, 8, 2
VA	83	Bi	BISMUTH	208.980	1560	271.3	9.8	2; 3; 4	5, 8, 32, 18, 8, 2
VIA	84	Po	POLONIUM	210	962	252	9.398	2; 4; 6	6, 18, 32, 18, 8, 2
VIIA	85	At	ASTATINE	211	337	302		1; 1	7, 18, 32, 18, 8, 2
INERT GASES	86	Rn	RADON	222	-61.9	-71	9.73	2; 4; 6	8, 18, 32, 18, 8, 2
IA	87	Fr	FRANCIUM	223	677	27		1	1, 8, 18, 32, 18, 8, 2
IIA	88	Ra	RADIUM	226.05	1140	700	5	2	2, 8, 18, 32, 18, 8, 2
IIIB	89	Ac***	ACTINIUM	227	3200	1050	10.07	3	2, 9, 18, 32, 18, 8, 2
IVB	104			(261)				4	32, 10, 2
VB	105			(262)					32, 11, 2
VIB	106			263					32, 12, 2

**LANTHANIDE SERIES

Atomic number	Symbol	Name	AMU	Boiling point C	Melting point C	Density	Valences	Electron shells
58	Ce	CERIUM	140.12	3468	795	6.66	3; 4	2, 8, 20, 18, 8, 2
59	Pr	PRASEODYNIUM	140.907	3512	935	6.782	3; 4	2, 8, 21, 18, 8, 2
60	Nd	NEODYMIUM	144.24	3027	1024	7.004	3	2, 8, 22, 18, 8, 2
61	Pm	PROMETHIUM	147	2460	1168	7.22	1	2, 8, 23, 18, 8, 2
62	Sm	SAMARIUM	150.35	1791	1072	7.536	2; 3	2, 8, 24, 18, 8, 2
63	Eu	EUROPIUM	151.96	1597	826	5.259	2; 3	2, 8, 25, 18, 8, 2
64	Gd	GADOLINIUM	157.25	3266	1312	7.895	3	2, 9, 25, 18, 8, 2
65	Tb	TERBIUM	158.924	3123	1356	8.272	3; 4	2, 8, 27, 18, 8, 2
66	Dy	DYSPROSIUM	162.50	2600	1407	8.536	3	2, 8, 28, 18, 8, 2
67	Ho	HOLMIUM	164.930	2600	1461	8.803	3	2, 8, 29, 18, 8, 2
68	Er	ERBIUM	167.26	2900	1497	9.051	3	2, 8, 30, 18, 8, 2
69	Tm	THULIUM	169.934	1947	1545	9.332	3	2, 8, 31, 18, 8, 2
70	Yb	YTTERBIUM	173.04	1194	824	6.977	2; 3	2, 8, 32, 18, 8, 2
71	Lu	LUTETIUM	174.97	3327	1652	9.842	3	2, 9, 32, 18, 8, 2

***ACTINIDE SERIES

Atomic number	Symbol	Name	AMU	Boiling point C	Melting point C	Density	Valences	Electron shells
90	Th	THORIUM	232.038	4790	1750	11.66	4	2, 10, 18, 32, 18, 8, 2
91	Pa	PROTACTINIUM	231.05		1600	15.37	4; 5	2, 9, 20, 32, 18, 8, 2
92	U	URANIUM	238.03	3813	1132	19.07	3; 4; 6	2, 9, 21, 32, 18, 8, 2
93	Np	NEPTUNIUM	237	3902	637	20.45	3,4; 5,6	2, 8, 22, 32, 18, 8, 2
94	Pu	PLUTONIUM	(244)	3235	639.5	19.84	3,4; 5,6	2, 8, 24, 32, 18, 8, 2
95	Am	AMERICIUM	(243)	2607	994	11.87	3,4; 5,6	2, 8, 25, 32, 18, 8, 2
96	Cm	CURIUM	(247)		1340	13.51	3; 4	2, 9, 25, 32, 18, 8, 2
97	Bk	BERKELIUM	(247)				3	2, 9, 26, 32, 18, 8, 2
98	Cf	CALIFORNIUM	(251)				3	2, 9, 27, 32, 18, 8, 2
99	Es	EINSTEINIUM	(254)				3	2, 9, 28, 32, 18, 8, 2
100	Fm	FERMIUM	(257)				3	2, 9, 29, 32, 18, 8, 2
101	Md	MENDELEVIUM	(258)				3	2, 9, 30, 32, 18, 8, 2
102	No	NOBELIUM	(259)					2, 9, 31, 32, 18, 8, 2
103	Lr	LAWRENCIUM	(260)					2, 9, 32, 32, 18, 8, 2

Numbers in parentheses are mass numbers of most stable isotope of that element.

Figure 1–2
Periodic table of the elements

Source: Cabot Corp., Boyertown, PA. Reprinted with permission

1 atomic mass unit (AMU). The neutrons have a slightly larger mass than the protons. Electrons have relatively small mass compared with the protons and neutrons (about 1/1837 the mass of a proton).

Electron "orbits" are not well-defined rings. Quantum mechanics tells us that electrons have both properties of particles and properties similar to those of energy waves. The electronic configuration of an atom is defined by quantum numbers. One cannot say that a particular electron orbits the nucleus of an atom at, for example, a distance of 1 Å from the nucleus. Instead, the position of electrons associated with a particular atom is described by four quantum numbers that essentially state the probability of a particular electron being in a particular relationship with the nucleus of an atom. This concept is illustrated in Figure 1–3.

Quantum numbers and the electron configuration of atoms (Figure 1–4) are used in a variety of ways in engineering materials. For example, the electron configuration of carbon atoms determines molecular bonding characteristics in polymers. In organic chemistry, electron configuration is often related to crystal structure. Electron configurations and available energy levels are extremely important in solid-state physics and electronics. Design engineers may think that they will never use this concept in design engineering, but they may use this material without being aware of it. Advanced analytical techniques that investigate the nature of surface films (XPS and Auger spectroscopy) often analyze 1*s*, 2*s*, 2*p* energy levels to identify surface contaminants, and surface chemical composition. Designers may use these analytical techniques to solve a paint adhesion or welding problem.

Many intricacies are involved in analyzing the nuclear atom. The structure of the atom or of the nucleus of atoms is unimportant in most work in ordinary materials engineering, but it

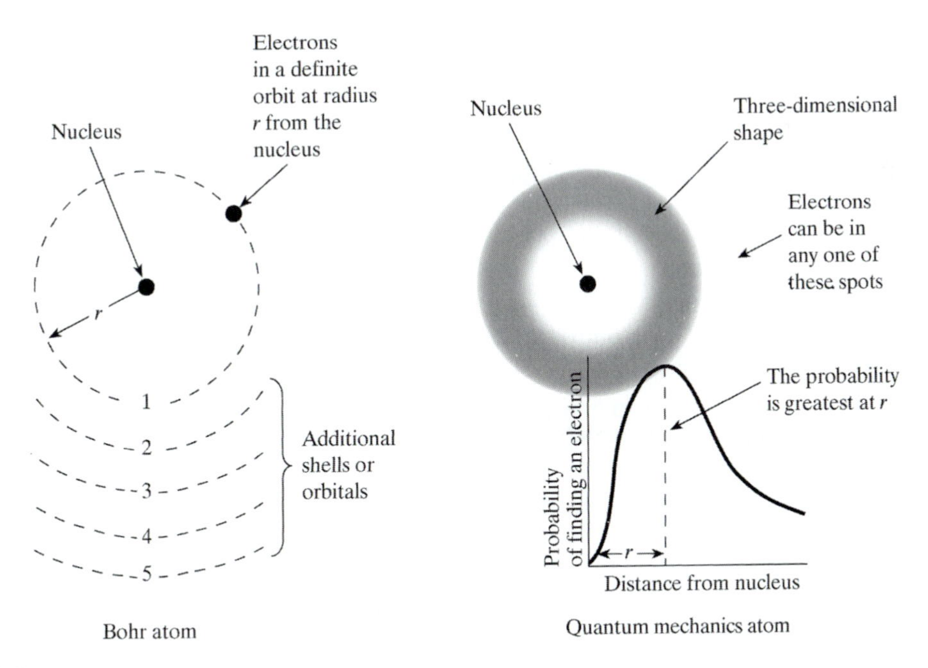

Figure 1–3
The Bohr atom compared with an atom described by quantum mechanics

		Principal quantum number (n) Increasing energy →																			Orbitals
		1	2		3			4				5				6				7	Sub levels
Atomic number	**Element**	Number of electrons in sublevel ($2n^2$)																			**Notation for the electron configuration**
		s	*s*	*p*	*s*	*p*	*d*	*s*	*p*	*d*	*f*	*s*	*p*	*d*	*f*	*s*	*p*	*d*	*f*	*s*	
1	Hydrogen, H	1																			$1s^1$
2	Helium, He	2																			$1s^2$
3	Lithium, Li	2	1																		$1s^2 2s^1$
4	Beryllium, Be	2	2																		$1s^2 2s^2$
5	Boron, B	2	2	1																	$1s^2 2s^2 2p^1$
6	Carbon, C	2	2	2																	$1s^2 2s^2 2p^2$
7	Nitrogen, N	2	2	3																	$1s^2 2s^2 2p^3$
8	Oxygen, O	2	2	4																	$1s^2 2s^2 2p^4$
9	Fluorine, F	2	2	5																	$1s^2 2s^2 2p^5$
10	Neon, Ne	2	2	6																	$1s^2 2s^2 2p^6$
18	Argon, Ar	2	2	6	2	6															$1s^2 2s^2 2p^6 3s^2 3p^6$
26	Iron, Fe	2	2	6	2	6	6	2													(Ar)*$3d^6 4s^2$
54	Xenon, Xe	2	2	6	2	6	6	2	6	10	–	2	6								$1s^2 2s^2 2p^6 3s^2 3p^6 3d^6 4s^2 4p^6 4d^{10} 5s^2 5d^6$
74	Tungsten, W	2	2	6	2	6	10	2	6	10	14	2	6	4	–	2					(Xe)$5d^4 6s^2$
86	Radon, Rn	2	2	6	2	6	10	2	6	10	14	2	6	10	–	2	6				(Xe)$5d^{10} 6s^2 6p^6$

*Write the notation for Ar and add additional notation shown.

Figure 1–4
Electron configuration of some elements. An electron is completely described by four quantum numbers: n, the principal quantum number; l, the angular quantum number (goes from $l = 0$ to $n - l$); the magnetic quantum number (goes from $-l$ to $+l$), and the spin quantum number (goes from $+ \frac{1}{2}$ to $- \frac{1}{2}$). No two electrons have the same four quantum numbers.

can have some application in deducing bonding tendencies between atoms. Several general rules about the electronic configuration of atoms are worthy of note:

1. Electrons associated with an atom occupy orbitals and subshells within orbitals.
2. The exact location of electrons in orbitals is defined by four quantum numbers that refer to the energy of the electron (principal quantum number), the shape of an orbital (angular momentum quantum number), the orientation of an orbital (magnetic quantum number), and the spin of an electron (spin quantum number).
3. No two electrons can have the same four quantum numbers (they cannot be in the same place at the same time). This is the Pauli exclusion principle.
4. When two electrons reside in the same orbitals, their spins must be paired.
5. The maximum number of electrons possible in a given orbital is $2n^2$, where n is the principal quantum number.
6. When atoms interact—for example, to form compounds—electrons go into unoccupied orbitals rather than into a partially occupied orbital.
7. The outermost, or *valence*, electrons largely determine the chemical behavior of elements. The maximum number is eight.
8. In chemical reactions, most elements attempt to attain an electron structure of eight

Figure 1–5
Quantum mechanics oxygen atom compared to the Bohr oxygen atom

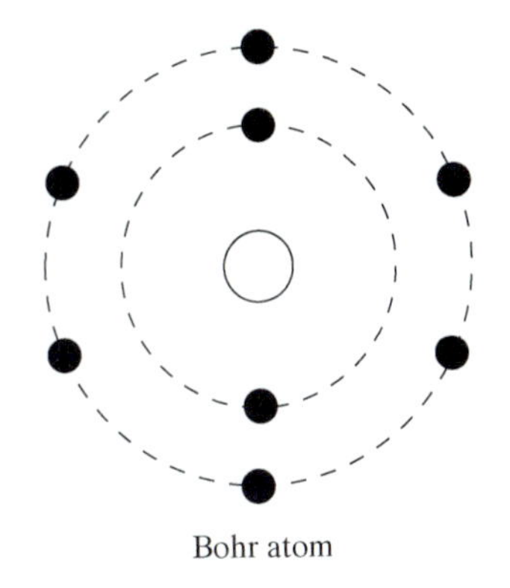
Bohr atom

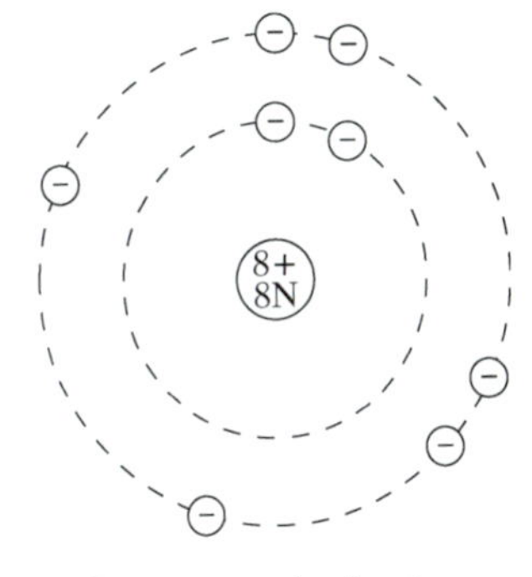

Quantum mechanics atom
$1s^2\ 2s^2\ 2p^4$

electrons in the outermost energy level. This is the most stable configuration.

The term *quantum* is used in physics to describe the amount of energy that is given off when an electron moves from one orbit to a lower orbit. Quantum mechanics and quantum numbers deal with electron configurations and things that happen when these electrons and atomic particles are manipulated in atomic reactions. The difference between the Bohr atom and the quantum mechanics atom is illustrated in Figure 1–5. The Bohr atom model of the element oxygen shows two electrons in the *k* orbital and six electrons in the *l* orbital. The quantum mechanics atom shows that some of the electrons are paired in the two orbitals. The paired electrons are in a sublevel of an orbit. The orbitals are designated by numbers from 1 to 7, and these numbers are called the principal quantum numbers. The sublevels in each orbital (principal quantum number) are identified by letters. The sublevel with the lowest energy state is called the *s* sublevel, the next is the *p* sublevel, and then there is a *d* and an *f* sublevel. If there are two electrons in the *s* sublevel of the second orbital, this is designated as $2s^2$. There is a quantum number designation for each element. For example, the configuration of carbon is $1s^2 2s^2 p^2$; the configuration of the elements using the old Bohr atom notation is shown in the far right column in the boxes of the periodic table (see "electron shells," Figure 1–2).

Quantum numbers and quantum mechanics are of limited significance to the material user, but these designations are sometimes used in chemical analysis techniques. Some instruments that break down compounds into atoms for chemical analysis make identifications by referring to peaks for 2*s* or 2*p* electrons and similar quantum mechanics designations. It is probably sufficient that material users simply be aware that the makeup of atoms is important in determining their reactivity in forming usable compounds, and it is the makeup of atoms that determines their engineering properties. Quantum mechanics is the current system for designating the component parts of atoms, and these atoms are the building blocks for engineering materials.

1.2 The Periodic Table

The properties of elements tend to be a periodic function of their atomic numbers. It is common practice to list the elements in the array shown in Figure 1–2, the *periodic table*. The atomic number of the elements increases horizontally in the table, and the vertical groupings are based on similarities in valence electron configurations and similarities in chemical and physical properties of the elements. The elements in group

IA are called alkali metals; group IIA elements are alkaline-earth metals. The groups listed as transition elements are metals with a particular electron subshell configuration (incomplete subshell). Groups IIIA, IVA, VA, VIA, and VIIA are mostly nonmetals (as shown by the heavy line), and the elements in the last vertical grouping are inert gases. The groups of elements in the separate horizontal blocks, lanthanide series and actinide series, really belong in periods (i.e., rows) 6 and 7, respectively, but to list them in this way would make the table unbalanced in shape. The elements in each series behave the same chemically; thus this deviation is logical as well as practical.

The periodic table was developed in the mid-nineteenth century by chemists who were trying to arrange the elements known at that time by similarities in chemical behavior. A Russian scientist, D. I. Mendeleev, has been accepted as the author of the table, which looks generally like the one in use today. The horizontal rows are the periods. They start from the left, and each element (going right) has one more nuclear charge than the preceding element. These charges are neutralized by an additional electron. The period ends with a noble gas with eight electrons in its valence (outer) shell. There is a periodic variation in atomic configuration based on which electron shell is being filled. It is this periodic variation in electron configuration that leads to periodic property variations. Elements in a particular vertical group all have the same number of electrons in their valence shell (with some exceptions), and this is thought to be the reason why they have the same general chemical behavior. The noble (inert) gases have similarities in properties. The elements in Group VIIA, called *halogens*, have chemical similarities, and so on, for each group.

Elements in the periodic table with an atomic number greater than 92 do not exist in nature; they were produced by nuclear reactions. There is no definite end to the periodic table. Elements with atomic numbers as high as 120 have been identified, but they are relatively unstable, and some are even unnamed.

What is the significance of the periodic table in engineering materials? Foremost, it is the dictionary for the names and chemical symbols of the elements that are the building blocks for all engineering materials. The chemical symbols for the elements are used throughout subsequent discussions of materials and their processing. The family groupings indicate which elements behave similarly. This can sometimes be an aid in selection problems. The atomic weight is the average weight of the common isotopes of a particular element. The atomic weight is an indicator of the density of an element, a physical property that can also enter into selection. The simplified electron structure shown in Figure 1–2 shows the number of electrons in the various orbitals. The number at the bottom of the vertical column indicates the number of electrons in the valence shell. This number and the element grouping provide indicators of how a particular element might combine with other elements.

1.3 Forming Engineering Materials from the Elements

Some of the elements are used as engineering materials in their pure elemental state. Many metals fall into this category; beryllium, titanium, copper, gold, silver, platinum, lead, mercury, and many of the refractory metals (W, Ta, Mo, Hf) are used to make industrial items. Many metals are used in the pure state for electroplating durable goods, tools, and electrical devices: Cr, Ni, Cd, Sn, Zn, Os, Re, Rh. In the nonmetal category, carbon is used in industrial applications for motor brushes and wear parts, and in the cubic form as diamond for tools. The inert gases are other nonmetals that are used in the elemental (*ions* or *molecules*) form for industrial applications for protective atmospheres and the like. Table 1–1 presents some property information on elements that are commonly used in the field of engineering materials.

Table 1–1
Properties of some elements commonly used in engineering materials

Element	Symbol	Melting Point [°F (°C)]	Density[a] (g/cm^3)	Resistivity at 20°C (10^{-6} Ω cm)	Linear Coefficient of of Thermal Expansion[b] (10^{-6} in./in./°F)	Thermal Conductivity at 25°C (W/cm/°C)[c]
Silver	Ag	1761 (960)	10.49	1.59	10.9	4.29
Aluminum	Al	1220 (660)	2.70	2.66	13.1	2.37
Gold	Au	1945 (1062)	19.32	2.44	7.9	3.19
Beryllium	Be	2345 (1285)	1.84	4.20	6.4	2.01
Carbon	C	6740[d] (3726)	2.25	75.00	0.3–2.4	2.1
Calcium	Ca	1564 (851)	1.54	4.60[g]	12.4	2.01
Niobium[e]	Nb	4474 (2467)	8.57	14.60	4.06	0.537
Cerium	Ce	1463 (795)	6.66	75.00[g]	4.44	0.113
Cobalt	Co	2719 (1492)	8.90	5.68[f]	7.66	1.00
Chromium	Cr	3407 (1825)	7.19	12.80	3.4	0.939
Copper	Cu	1981 (1082)	8.94	1.69	9.2	4.01
Iron	Fe	2795 (1535)	7.87	10.70	6.53	0.804
Germanium	Ge	1717 (936)	5.32	60×10^6	3.19	—
Hafnium	Hf	4032 (2222)	13.29	35.5	3.1	0.230
Mercury	Hg	−70 (−38)	13.55	95.78	—	0.083
Iridium	Ir	4370 (2410)	22.42	5.30[f]	3.8	1.47
Lanthanum	La	1688 (892)	6.17	57.00[g]	3.77	0.134
Magnesium	Mg	1204 (651)	1.74	4.46	15.05	1.56
Manganese	Mn	2271 (1244)	7.49	185	12.22	0.078
Molybdenum	Mo	4730 (2610)	10.22	5.78[g]	2.7	1.38
Nickel	Ni	2646 (1452)	8.90	7.8	7.39	0.909
Osmium	Os	5432 (3000)	22.50	9.5	2.6	0.076
Lead	Pb	621 (327)	11.34	22	16.3	0.353
Palladium	Pd	2826 (1552)	12.02	10.3	6.53	0.718
Platinum	Pt	3216 (1768)	21.40	10.58	4.9	0.716
Plutonium	Pu	1183 (639)	19.84	146.45	30.55	0.067
Rhenium	Re	5755 (3179)	21.02	19.14	3.7	0.480
Rhodium	Rh	3560 (1960)	12.44	4.7	4.6	1.50
Silicon	Si	2520 (1382)	2.33	15×10^6[h]	1.6–4.1	1.49
Tin	Sn	449 (232)	7.30	11.5	13	0.668
Tantalum	Ta	5425 (2996)	16.60	13.6[g]	3.6	0.575
Thorium	Th	3182 (1750)	11.66	18[g]	6.9	0.540
Titanium	Ti	3035 (1668)	4.54	42	4.67	0.219
Uranium	U	2070 (1132)	19.07	30	3.8–7.8	0.275
Vanadium	V	3486 (1918)	6.11	24.8	4.6	0.307
Tungsten	W	6170 (3410)	19.30	5.5	2.55	1.73
Yttrium	Y	2748 (1508)	4.47	65	—	0.172
Zinc	Zn	787 (419)	7.13	5.75	22	1.16
Zirconium	Zr	3366 (1852)	6.45	44	5.8	0.227

[a] At 68°F (20°C)
[b] Multiply by 1.8 to convert to cm/cm/°C
[c] Multiply by 55.7 to convert to Btu/ft/°F
[d] Sublimes
[e] Also called columbium (Cb)
[f] At 0°C
[g] At 25°C
[h] At 300°C

A larger percentage of engineering materials utilize the elements in combined forms, in *alloys* (a metal combined with one or more other elements), in *compounds* (chemically combined elements with definite proportions of the component elements), and, to a smaller degree, in *mixtures* (a physical blend of two or more substances). The difference between an alloy and a mixture is that in alloys the elements added to the host metal can be in solid solution. These atoms can take positions in the *crystal structure*. The components in a mixture are not normally in solid solution. A slurry is a mixture of solid particles in a fluid. There is no bonding between the fluid and particles. These combinations of the elements can be solids, liquids, or gases. Our discussions will concentrate on elements combined to make solids. Many solids are compounds formed by chemical reactions. The smallest part of a compound that still retains the properties of that compound is the *molecule*. Usually, molecules contain various atoms with definite ratios of one atom to the other:

H_2O—a water molecule formed by two hydrogen atoms and one oxygen atom.

Gases often form molecules with like atoms:

H_2—a molecule of hydrogen gas composed of two hydrogen atoms.

The subscripts denote the number of atoms in a chemical compound or in a reacting species. When more than one molecule of a substance is involved in a chemical reaction, a prefix is used to indicate this number:

$4H_2O$—four molecules of water.

Chemical reactions are denoted by writing equations using this symbol system:

$H_2O \rightarrow H_2 + O$—a formula for the breakdown of water into its component parts, hydrogen and oxygen.

In actual practice it is known that dissociation of water produces *diatomic* (two of the same atoms in a molecule) hydrogen and oxygen. Thus the equation must be balanced to take this into account:

$$2H_2O \rightarrow 2H_2\uparrow + O_2\uparrow \qquad \text{(gas)}$$

All chemical reactions must be balanced. The number of atoms in the reacting species must equal the number of atoms in the reaction products.

Many laws of chemistry and physics control chemical bonding and the tendencies for bonding. A simple rule that fits most reactions is that atoms tend to combine in such a manner that their outer electron shell (containing valence electrons) will be complete when it contains eight electrons. In the water molecule (Figure 1–6), hydrogen has a valence of 1 and oxygen has a valence of 6. When they combine in a ratio of two hydrogen atoms to one oxygen atom, the component atoms can share electrons and complete their outer shells.

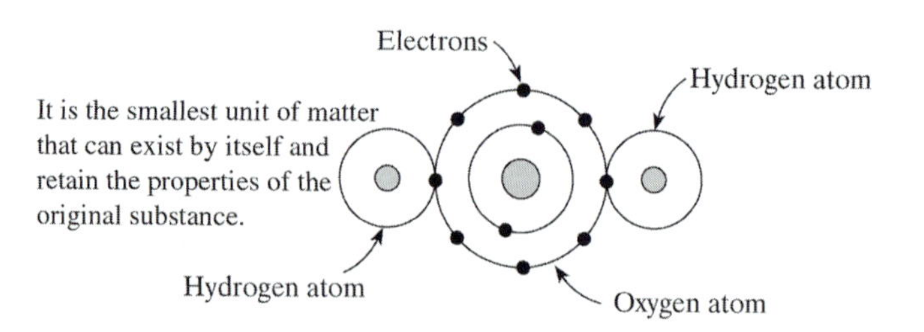
Electrons
Hydrogen atom
It is the smallest unit of matter that can exist by itself and retain the properties of the original substance.
Hydrogen atom
Oxygen atom

Figure 1–6
Water molecule

The valence band of the oxygen atom is now complete. It shares the electrons of two hydrogen atoms. The electron configurations shown in the periodic table can then be used to predict how atoms will combine to form compounds. It is not always this simple, but basically the valence of an atom or a group of atoms (polyatomic) determines its tendency to combine and form new substances.

1.4 The Solid State

We have depicted engineering materials as solids formed from various elements. A solid can be a pure element such as gold; it can be a compound such as sand, a combination of silicon and oxygen (SiO_2); or it can be a combination of molecules. Most living plants are a complex network of cellulose molecules. Some solids are mixtures of the preceding, but within such solids each component retains its identity. Concrete is a composite of cement (a compound) and aggregate (another compound).

Solids are formed when definite bonds exist between component atoms or molecules. In the liquid or gaseous state, atoms or molecules are not bonded to each other. The bonds that hold atoms or molecules together can be very specific and orderly, or they can be less well defined. Solids of the former type have a crystalline structure. Solids that do not have a repetitive three-dimensional pattern of atoms are said to be *amorphous*. The dictionary definition of *amorphous* is "without form." Most metals and inorganic compounds have a crystalline structure, whereas glasses and plastic materials are often amorphous.

When a solid has a crystalline structure, the atoms are arranged in repeating structures called *unit cells*. The cells form a larger three-dimensional array called a *lattice*. The unit cells can be of a dozen or so different types, but some of the more common *crystal* cells are shown in Figure 1–7. They each have a different name. When a crystalline solid starts to form from the molten or gaseous state, these cells will tend to stack in a three-dimensional array, with each cell perfectly aligned, and they will form a crystal. If many crystals are growing in a melt at the same time, the crystals will eventually meet and form

Figure 1–7
Typical crystal structures

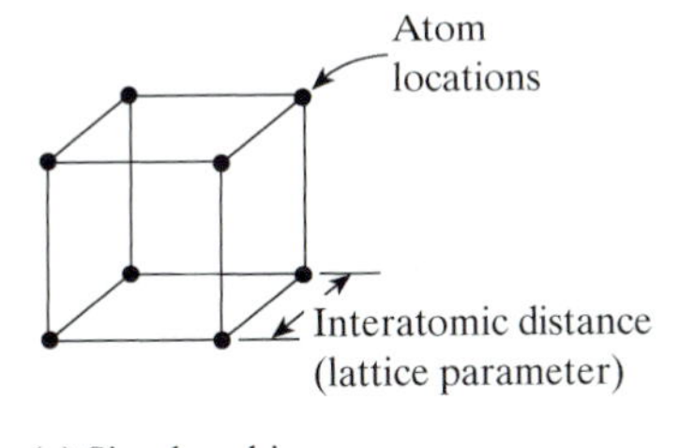

(a) Simple cubic

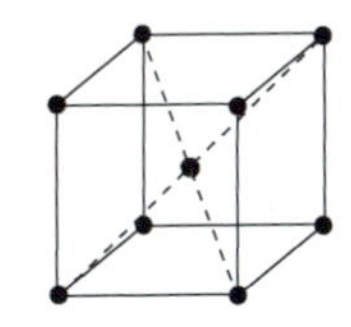

(b) Body-centered cubic

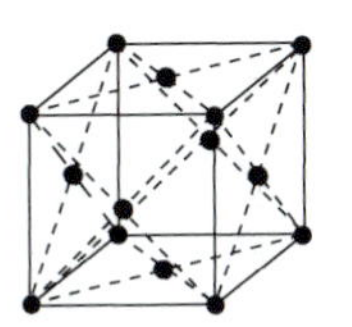

(c) Face-centered cubic

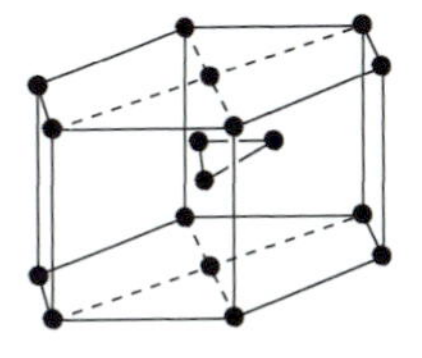

(d) Hexagonal close-packed

grains. The junctions of the crystallites are called *grain boundaries* (Figure 1–8).

The properties of crystalline materials are affected by the type of crystal structure [body-centered cubic (BCC), face-centered cubic (FCC), and so on], the crystal or *grain* size, and the strength of the bonds between atoms.

Amorphous materials are not really as disorganized as the name implies. Crystallinity or the lack of same is measured by electron or *x-ray diffraction* techniques. When a crystalline solid is exposed to a collimated beam of x-rays, the beam is diffracted by the ordered planes of atoms and the crystal cell size and location of atoms can be measured. This is how the materials engineer knows what type of structure a solid has. Amorphous materials do not have a structure ordered enough to allow distinct diffraction patterns. It has been learned, however, that most amorphous materials have short-range order. For example, an amorphous material may consist of long-chain molecules with significant order between molecules making up the molecular chains (short-range order), but there may be little order between the chains (long range). From the property standpoint, amorphous materials have different solidification characteristics than crystalline materials, but this does not necessarily detract from usable engineering properties. Crystallinity can be an important selection factor in plastics. This will be covered in detail in a later chapter.

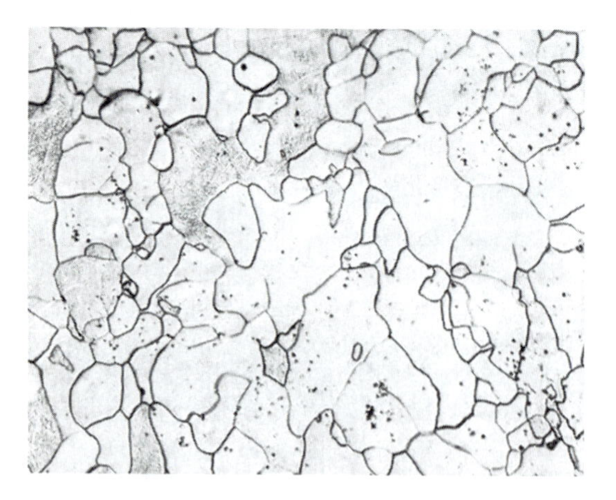

Figure 1–8
Microstructure of pure iron (×100). Dark areas are grain boundaries. Each grain is a crystal.

1.5 The Nature of Metals

In chemistry, a metal is defined as an element with a valence of 1, 2, or 3. However, a metal can best be defined by the nature of the bonds between the atoms that make up the metal crystals. Metals can be defined as solids composed of atoms held together by a matrix of electrons (Figure 1–9). The electrons associated with each individual atom are free to move throughout the volume of the crystal or piece of metal. This is why metals are good conductors of heat and electricity. Current flow requires a flow of electrons. Other properties that distinguish metals from other materials are their malleability (their ability to deform plastically), their opacity (light cannot pass through them), and their ability to be strengthened.

When crystalline solids are subjected to loads, on the atomic scale, there is a tendency to separate the atoms. If the bonds between the atoms are very strong, there is a tendency

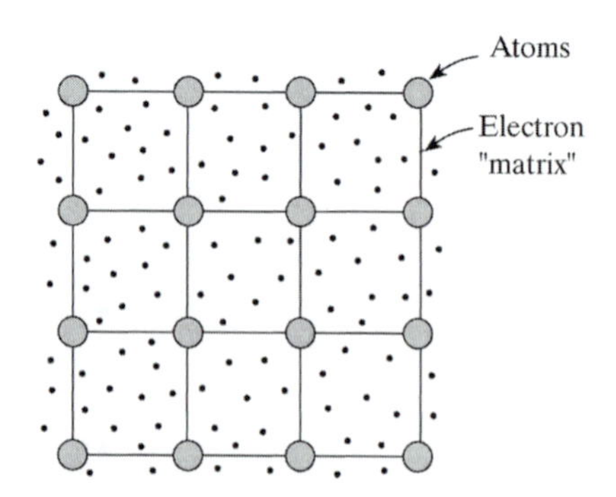

Figure 1–9
Electron matrix of a metal atom

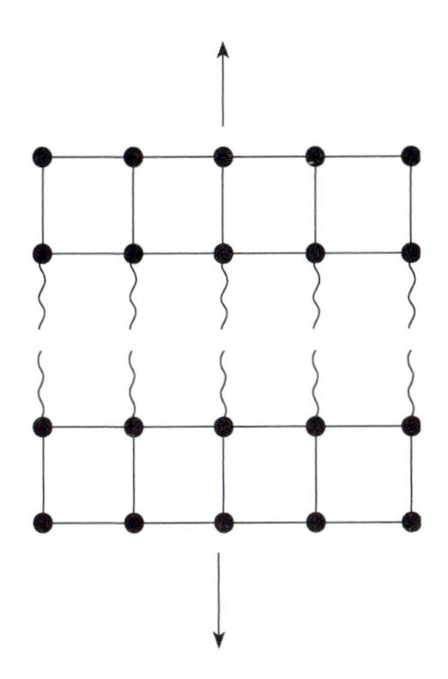

Figure 1–10
Cleavage failure of brittle crystalline materials

to cleave the crystals apart (Figure 1–10). In metals and some other crystalline materials, the interatomic bonds are such that rather than causing cleavage, loading can cause atomic slip.

A *dislocation* is a crystal imperfection characterized by regions of severe atomic misfit where atoms are not properly surrounded by neighboring atoms. When metals are deformed, the atoms making up the crystalline structure of the metal rearrange to accommodate the deformation by various mechanisms. Dislocation motion is a primary mechanism. The simplest type of dislocation, the edge dislocation, is shown schematically in Figure 1–11. The dislocation is the extra half-plane of atoms in Figure 1–11(a). When the dislocation reaches an outside surface of the crystal, it can cause a *slip step* [Figure 1–11(b)]. The extra half-plane of atoms will protrude from a free surface and can be observed on suitably prepared surfaces at high magnification.

If one were to calculate the theoretical strength of a metal, assuming a perfect crystalline structure, he or she would find that it is

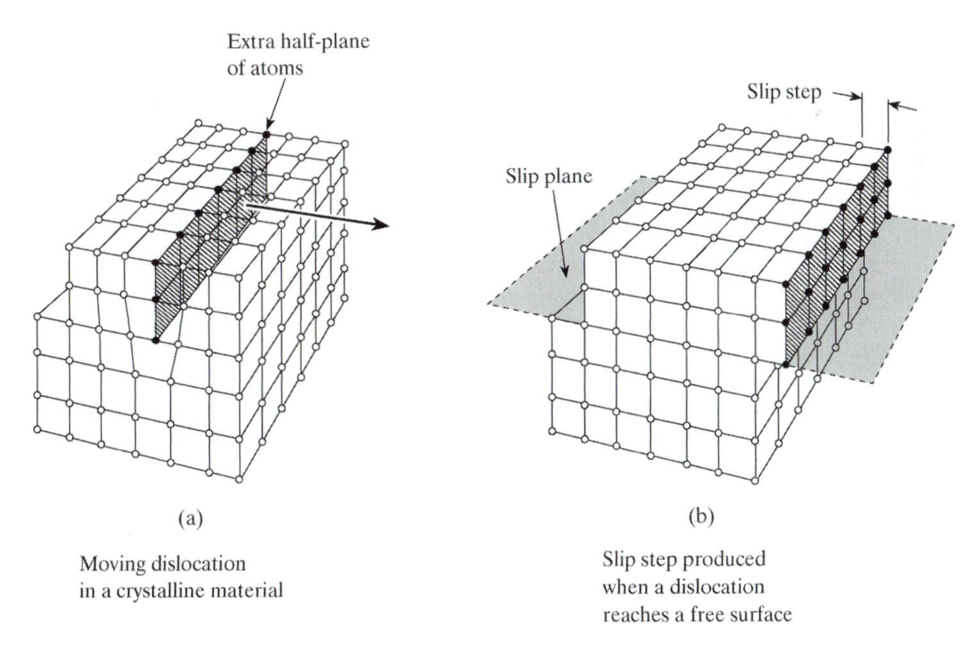

Figure 1–11
Slip produced by movement of an edge dislocation

one to two orders of magnitude higher than the actual strength of the metal. As depicted in Figure 1–12(a), if a shearing stress is applied to a perfect crystal, all of the atomic bonds along the slip plane would have to be broken and re-created in order to deform such a crystalline array. Deforming a crystal in such a manner is analogous to pulling a long length of carpeting across a floor. Clearly, it takes a lot of energy to overcome the frictional forces to drag the carpet in this manner. Similarly, it takes a lot of energy to deform a crystal by breaking and re-forming all of the bonds at once along the slip plane.

If that same crystalline array had dislocations, as shown in Figure 1–12(b), the material may deform under an applied shear stress by only breaking and re-forming one row of bonds at a time as the dislocation moves along the slip plane. Deformation in this manner is analogous to moving a long carpet across a floor by pulling up a "hump" of carpet and then pushing the "hump" along the length of the carpet. It takes much less energy to deform a metal or move a carpet in this manner. As metals plastically deform, dislocation defects in the crystal matrix move around in response to the applied forces, allowing the crystal to deform with minimal energy input. Discussions in a later section will show that impeding the mobility of the dislocations in the material will strengthen the material.

There are many types of dislocations and mechanisms by which dislocations interact. Because dislocations are atomic in size, they can usually be seen and studied only with the use of special microscopic and etching techniques. Their study is an important part of physical metallurgy.

Where do dislocations come from? Dislocations can be produced by crystal mismatch in solidification. They can be introduced by external stresses such as plastic deformation, they can occur by phase transformations that cause atomic mismatch, or they can be caused by the atomic mismatch effects of adding alloy elements. The importance of dislocations to the metal user is that dislocation interactions within a metal are a primary means by which metals are deformed and strengthened. When metals deform by dislocation motion, the more barriers the dislocations meet, the stronger the metal.

Deformation by dislocation motion is one of the characteristics of metals that make them the most useful engineering materials. The metallic bond is such that strains to the crystal lattice are accommodated by dislocation motion. Materials with strong *covalent* bonds or ionic bonds (as in some ceramics) will tend to cleave rather than deform by atomic movement. Many metals can tolerate significant plastic deformation before failing. This is a property that is rather unique to metals. They can be bent and formed into a desired shape. This cannot be said about many plastics, ceramics, and composites or cermets (ceramic/metal blends).

Metals can be strengthened by *solid solution strengthening*. This term means that impurity atoms are added to a pure metal to make an alloy. If the atoms of the alloying element are significantly larger than the atoms of the host metal, these large atoms can impede the motion of dislocations and thus strengthen the metal [Figure 1–13(a)]. Mechanical working strengthens metals by multiplication of dislocations. The dislocations interact with each other and with such things as grain boundaries, and thus movement of individual dislocations becomes difficult and the metal is strengthened [Figure 1–13(b)].

Precipitation hardening is used to strengthen many nonferrous metals. By choosing a suitable alloying element, it is possible by heat-treating techniques to get alloying elements to agglomerate within the metal lattice. The agglomerated alloy element atoms create atomic mismatches and strains that serve as barriers to dislocation motion and thus strengthen the metal [Figure 1–13(c)].

Dispersion strengthening is similar to precipitation hardening in concept: fine, nondeformable particles are added to metals (usually as a dispersion in the molten alloy), and these particles strengthen by inhibiting dislocation motion. Aluminum oxide particles dispersed in an

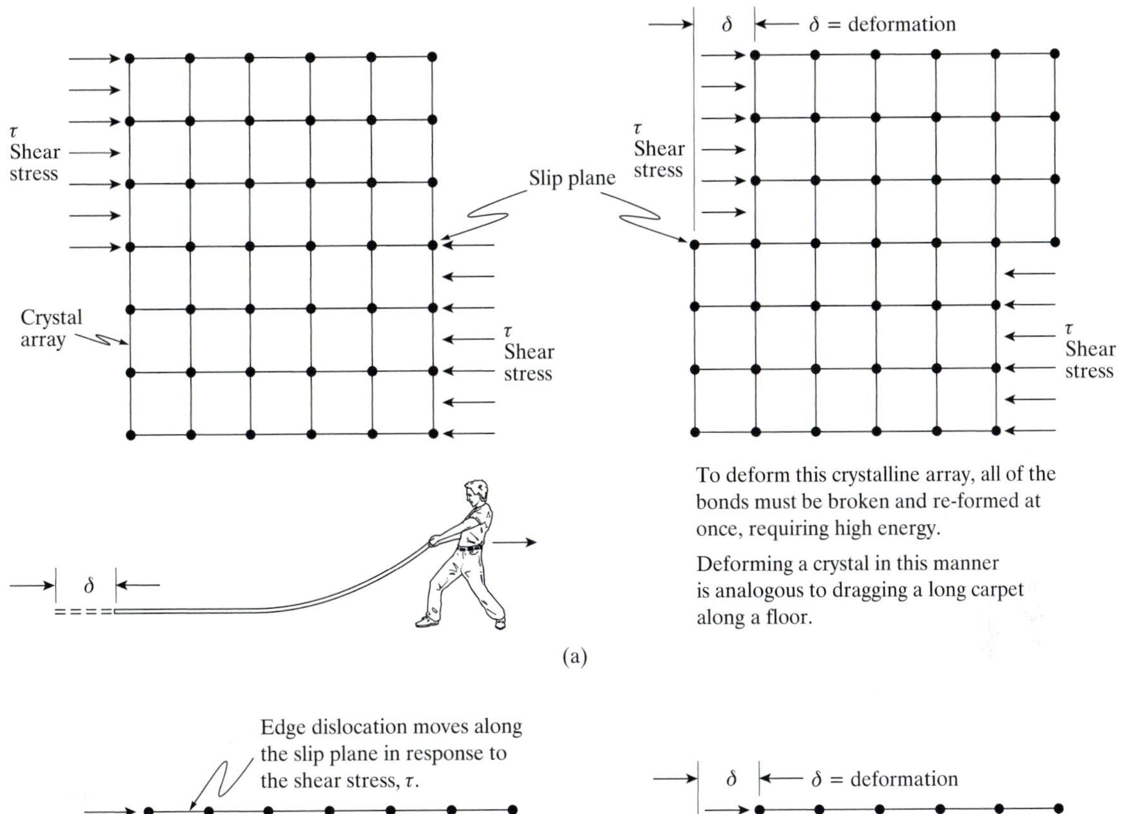
δ
δ = deformation
τ
Shear
stress
Slip plane
τ
Shear
stress
Crystal
array
τ
Shear
stress
τ
Shear
stress
To deform this crystalline array, all of the bonds must be broken and re-formed at once, requiring high energy.
Deforming a crystal in this manner is analogous to dragging a long carpet along a floor.
δ
(a)

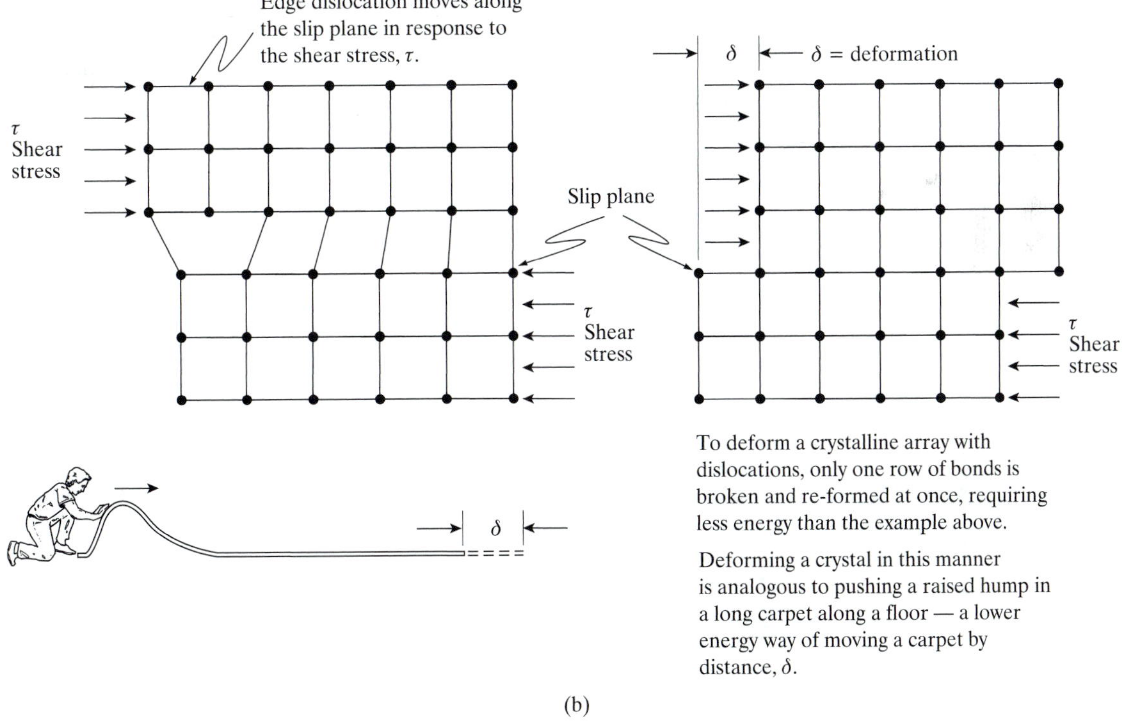
Edge dislocation moves along the slip plane in response to the shear stress, τ.
δ
δ = deformation
τ
Shear
stress
Slip plane
τ
Shear
stress
τ
Shear
stress
To deform a crystalline array with dislocations, only one row of bonds is broken and re-formed at once, requiring less energy than the example above.
Deforming a crystal in this manner is analogous to pushing a raised hump in a long carpet along a floor — a lower energy way of moving a carpet by distance, δ.
δ
(b)

Figure 1–12
Deformation of a crystalline material with dislocations is analogous to moving a carpet by progressive movement of a carpet "wave."

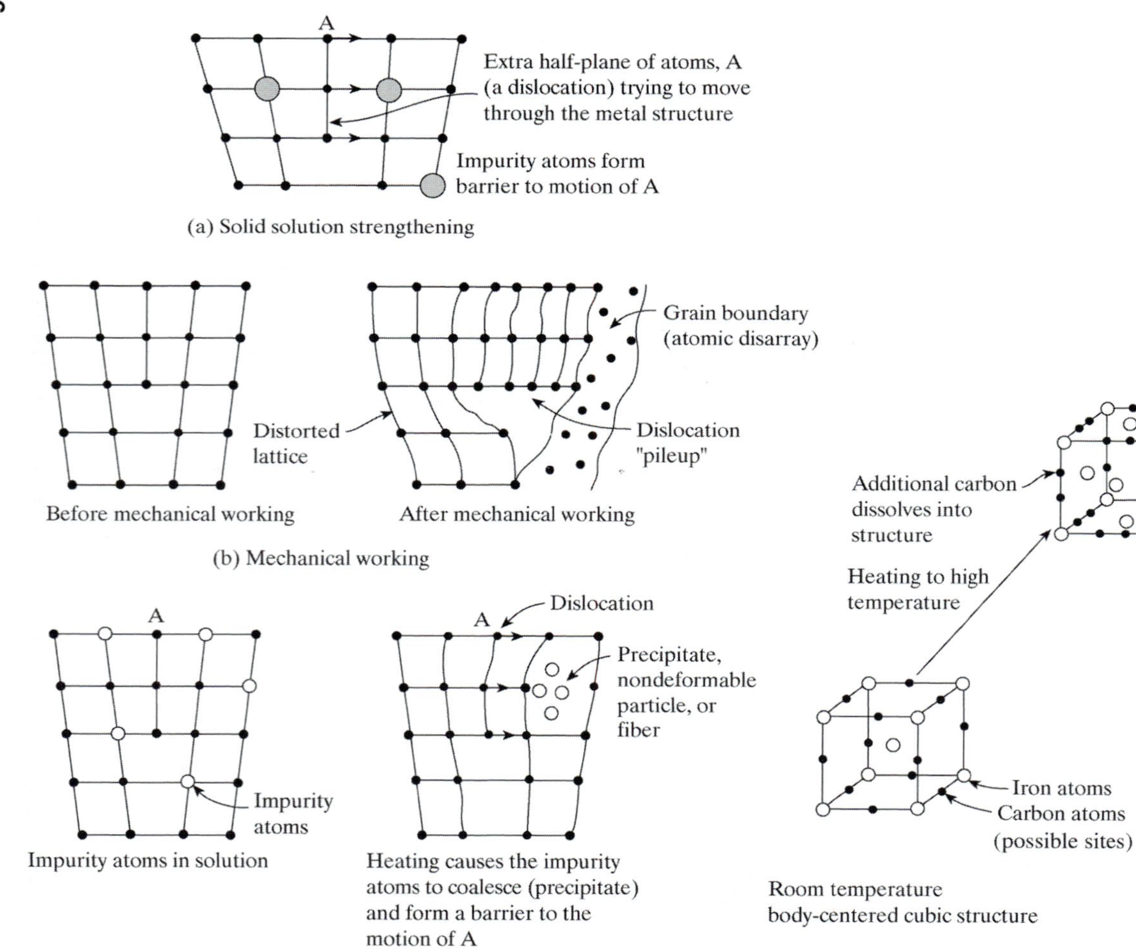
A
Extra half-plane of atoms, A (a dislocation) trying to move through the metal structure
Impurity atoms form barrier to motion of A
(a) Solid solution strengthening
Distorted lattice
Grain boundary (atomic disarray)
Dislocation "pileup"
Before mechanical working
After mechanical working
(b) Mechanical working
A
Impurity atoms
Impurity atoms in solution
A
Dislocation
Precipitate, nondeformable particle, or fiber
Heating causes the impurity atoms to coalesce (precipitate) and form a barrier to the motion of A
(c) Precipitation hardening
Additional carbon dissolves into structure
Face-centered cubic structure
Heating to high temperature
Rapid cooling
Iron atoms
Carbon atoms (possible sites)
Room temperature body-centered cubic structure
Room temperature body-centered tetragonal structure
(d) Quench hardening

Figure 1–13
Strengthening mechanisms in metals

aluminum matrix are an example of a dispersion-strengthened alloy. Dispersion strengthening does not have the commercial significance of precipitation hardening.

Metal matrix composites are metals reinforced by ceramics or other materials, usually in fiber form. The role of the reinforcing material in this class of materials is to strengthen by impeding dislocation motion. Continuous reinforcements such as fibers also help to distribute the strains throughout the structure, and they can increase the metal's stiffness if the modulus of elasticity of the reinforcement is higher than that of the matrix metal. Metal matrix composites are in commercial use in areas where metals by themselves did not have adequate properties. Silicon carbide–reinforced aluminum is used for connecting rods in high-performance automobile engines.

The final and most industrially important strengthening mechanism in metals is *quench hardening* [Figure 1–13(d)]. Quench hardening is a heat treating process used to induce atomic strains into a metal lattice. The strains are produced by quench-induced trapping of solute atoms into the lattice. The trapped atoms actually change the atomic spacing. The distorted lattice and the action of the quenched-in solute atoms impede dislocation motion and thus strengthen the metal. The strengthening effect of some of these processes on alloys of iron and carbon (steels) is illustrated in Figure 1–14.

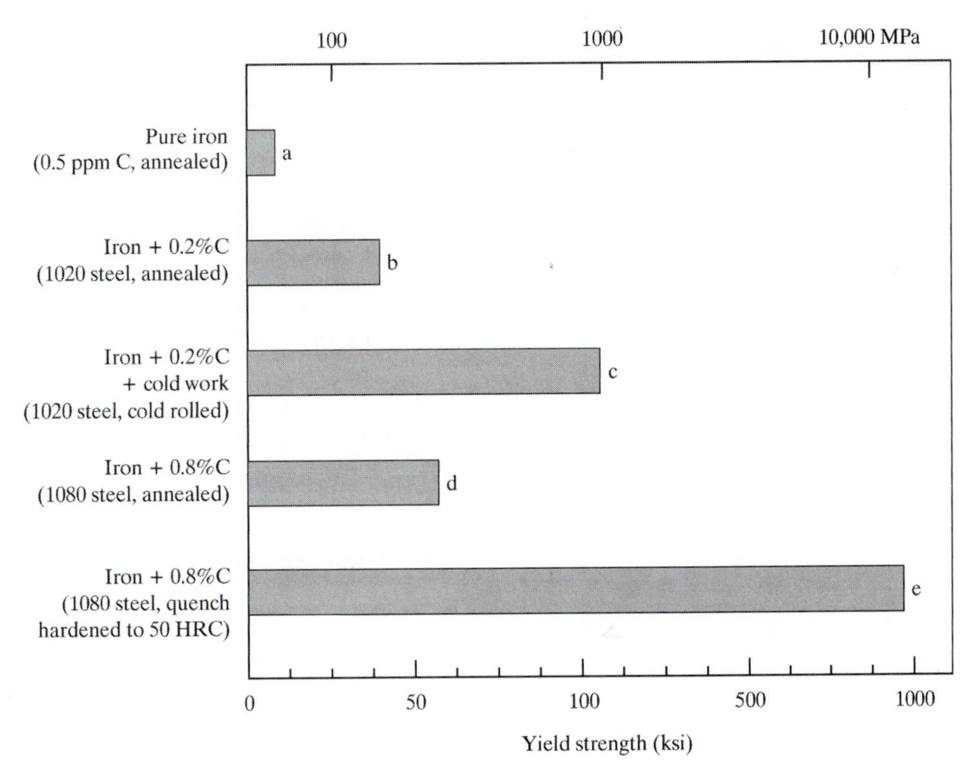

Figure 1–14
Strengthening iron: by alloying with carbon (a → b; b → d); by cold work (b → c); by quench hardening (d → e). Note the significant effect of alloying and quench hardening.

1.6 The Nature of Ceramics

In terms of basic chemistry, a nonmetallic element has a valence of 5, 6, or 7. Elements with a valence of 4 are *metalloids*; sometimes they behave as a metal, sometimes as a nonmetal. Elements with a valence of 8 are inert. They have a low tendency to combine with other elements—for example, inert gases. A *ceramic* can be defined as a combination or compound of one or more metals with a nonmetallic element. What really distinguishes a ceramic from other engineering materials, however, is the nature of the bond between atoms. As opposed to the long-range electron matrix bond in metals, ceramic materials usually have very rigid covalent or ionic bonds between adjacent atoms. As shown in Figure 1–15, the ceramic aluminum oxide is formed by the combination of three oxygen atoms and two aluminum atoms in such a manner that by sharing valence electrons, each atom has eight electrons in its outer shell. This sharing of electrons is called *covalent bonding*.

An *ion* is an atom that has lost or gained an electron. In ionic bonding, valence electrons from one atom are transferred to another atom, and the atoms involved are then held together by the electrostatic attraction between the two oppositely charged ions.

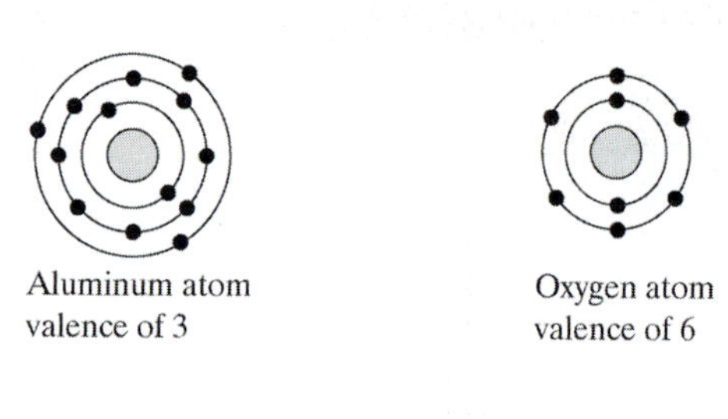

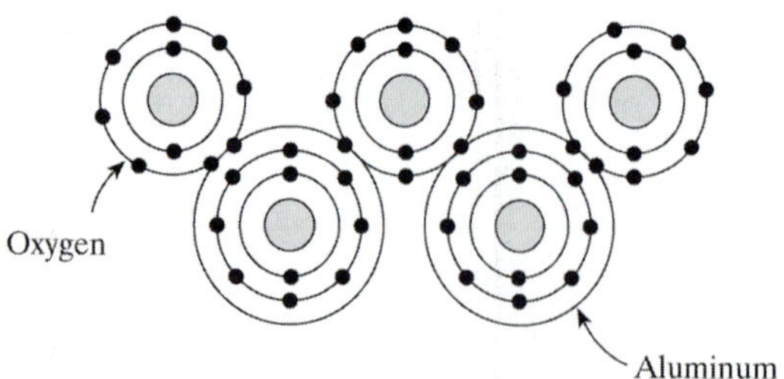

Figure 1–15
Structure of aluminum oxide (Al_2O_3)

Both ionic and covalent bonds involve very strong bonds between neighboring atoms. Thus crystalline ceramics with this type of bond tend to be very brittle. Tensile loading tends to result in crystal cleavage. Deformation by dislocation motion or atomic slip is difficult. Other property manifestations of these strong bonds are high hardness, chemical inertness, and electrical insulation. Ceramics tend to be electrical insulators because the electrons are tied up in bonding and are not free to move throughout the crystal. Ceramics can be strengthened by adding other elements, but the effect is usually not pronounced. They usually cannot be strengthened by cold working or by precipitation hardening. Some ceramics can be strengthened by cold working or by precipitation hardening. Some ceramics can be strengthened by changes in crystal structure; for example, hexagonal boron nitride is very soft and cubic boron nitride is very hard, but such cases are the exception rather than the rule. Fibers and other materials are used to strengthen ceramics. Silicon carbide fibers are added to silicon nitride to improve its characteristics for metal cutting tools. Ceramics are also blended or alloyed to obtain better use properties. Toughened zirconia is added to brittle aluminum oxide to make a ceramic material with better strength and toughness than pure aluminum oxide.

1.7 The Nature of Polymers

The engineering materials known as plastics are more correctly called *polymers*. This term comes from the Greek words *poly*, which means "many," and *meros*, which means "part." Polymers are substances composed of long-chain repeating molecules (mers). In most cases the element carbon forms the backbone of the chain (an organic material). The atoms in the repeating molecule are strongly bonded (usually covalent), and the bonds between molecules may be due to weaker secondary bonds or similar covalent bonds. The common polymer polyethylene is composed of repeating ethylene molecules (C_2H_4). Using the *rule of eight*, the carbon atoms have unsaturated valence bands (carbon has a valence of 4 and hydrogen has a valence of 1). If ethylene molecules attach to each side of the one illustrated in Figure 1–16, the valence bands on the carbon atoms in the center molecule will be satisfied. This is why these materials tend to form long chains: carbon-to-carbon bonds satisfy valence requirements. When the chains grow very long and get tangled, they tend to lose their three-dimensional symmetry, and they appear amorphous when analyzed by x-ray diffraction techniques. Thus the degree of crystallinity in a polymer often depends on chain

Figure 1–16
Structure of polyethylene

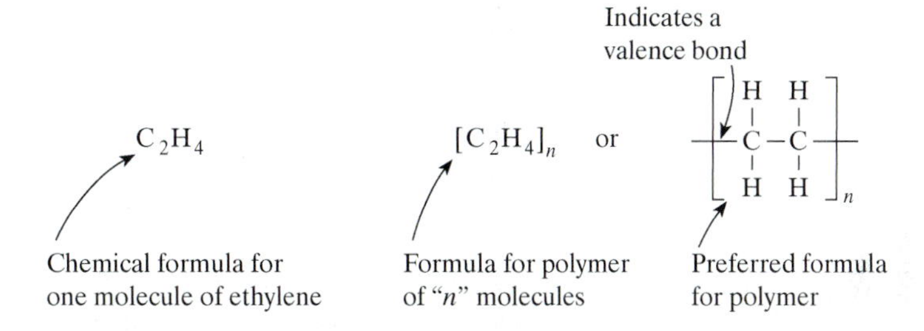

A Single Molecule of Polyethylene

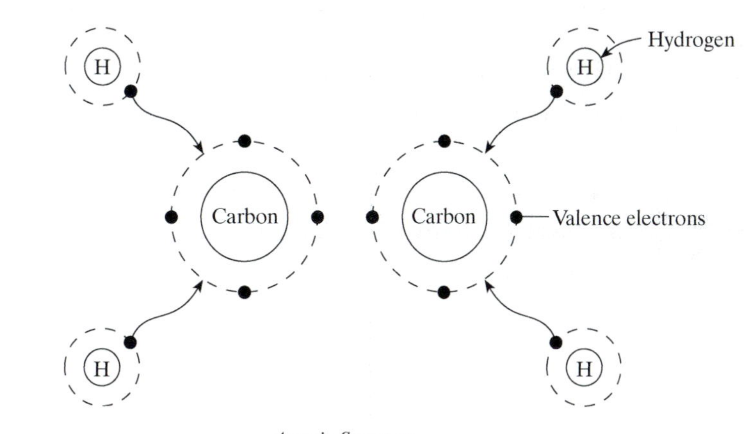

Atomic Structure

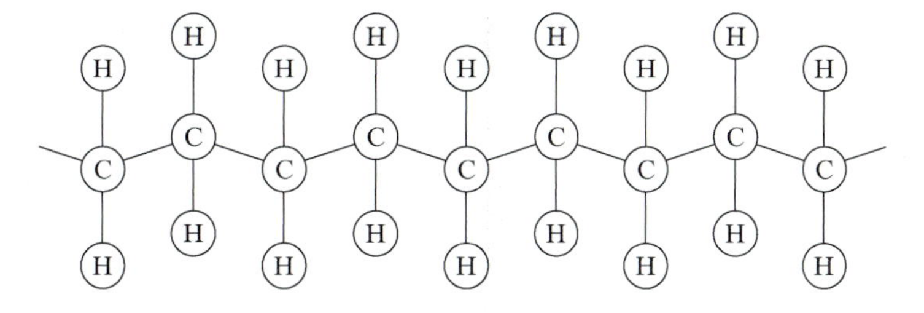

Physical Structure of Polyethylene

alignment. Some polymers have a high degree of crystallinity; some do not.

Long-chain polymers are usually weaker than most ceramics and metals because the molecular chains are bonded to each other only with rather weak electrostatic forces called *van der Waals* bonds. When loaded, the long-chain molecules slip with respect to each other. Strengthening is accomplished by techniques that retard chain movement: fillers, cross-linking of chains, chain branching, and the like.

1.8 The Nature of Composites

A *composite* is a combination of two or more materials that has properties that the component materials do not have by themselves. Nature made the first composites in living things. Wood is a composite of cellulose fibers held together with a glue or matrix of soft lignin. In engineering materials, composites are formed by coatings, internal additives, and laminating. An important metal composite is clad metal. Thermostatic controls are made by roll-bonding a high-expansion alloy such as copper to a low-expansion alloy like steel. When the composite is heated, it will deflect to open electrical contacts. Plywood is a similarly common composite. Because wood is weaker in its transverse direction than in its long direction, the alternating grain in plywood overcomes the transverse deficiency.

At the current time the most important composites are combinations of high-strength, but crack-sensitive, ceramic-type materials and polymers. The most common example of such a system is fiberglass (fiber-reinforced plastic). Glass fibers are very strong, but if notched they fracture readily. When these fibers are encapsulated in a polyester resin matrix, they are protected from damage, and the polyester transfers applied loads to the glass fibers so that their stiffness and strength can be utilized. More advanced components use fibers of graphite and boron. These fibers are very stiff and strong, yet lightweight. The strengthening effect of the reinforcements in composites depends on the orientation of the reinforcements to the direction of the loads.

Besides polyester, suitable composite matrixes are polyimides, epoxies, and even metals such as aluminum and copper.

Composites as a class of engineering materials provide almost unlimited potential for higher strength, stiffness, and corrosion resistance over the "pure" material systems of metals, ceramics, and polymers. Composites are widely used for sports equipment (tennis rackets, golf clubs, skis, snowboards, boats, etc.), in aircraft (spars, radomes, cabin liners, etc.), and in the chemical process industry for piping and tanks. Most gasoline stations in the United States now have fiber-reinforced plastic (FRP) in-ground storage tanks.

Summary

This first chapter may be completely unnecessary for students with a recent exposure to chemistry, but if it has been a while, a review can be helpful. The most important concept that is promoted in this chapter is that metals, plastics, ceramics, and composites, all of our engineering materials, have their origin in the elements. All materials are related by their atomic structure. The atoms of engineering materials all have the same component parts (neutrons, protons, and electrons), and the configuration of these atomic parts in an element determines the properties of that element. When elements combine, it is the nature of the bond between atoms and/or molecules that determines the properties of macroscopic things made from applications of atoms.

Metals are good conductors because of the electron mobility provided by metallic bonding (mobile valence electrons). Plastic properties depend on chains of molecules mostly made up of carbon-to-carbon atomic bonds or on interpenetrating bonds between complex organic molecules. Composites are physical blends of the primary material systems (metals, plastics, and ceramics). They are engineered to optimize the strong points of each system; for instance, strong glass fibers reinforce plastics to make the composite we often recognize as fiberglass. The glass fibers would be useless without the polymer matrix, and vice versa. Some additional concepts to remember are as follows:

- The periodic table is the reference sheet for the elements that can be used to form engineering materials.
- The valence electrons in the outermost shell of atoms determine the ability of those atoms to combine with other atoms.
- Materials can be amorphous or crystalline (or mixtures of both, as in some plastics) in the solid state. Crystallinity, or the lack of it, often determines the use properties of a material.
- Dislocation motion is produced when crystalline solids are strained, and these types of atomic movements are responsible for the malleability of metals.
- Organic materials are based on the element carbon and they come from living matter; everything else is inorganic.
- Some elements (mostly the metals) are used as engineering materials in elemental form. The other engineering materials are made from compounds formed by the elements (plastics, ceramics, some composites).
- The rules of chemistry and physics apply to engineering materials, chemistry in the formation of materials, and physics (quantum mechanics and the like) in the study of atomic reactions and atomic bonding.
- We know quite a bit about why things happen and how to make a wide variety of engineering materials. Future developments in materials will depend on new knowledge of chemistry and atomic structure. We will probably not find any new stable elements; we must become more creative with what we have.
- Metals have a crystalline structure and a bond between atoms characterized as a sea of electrons. This structure produces good electrical and thermal conductivity and malleability.
- Ceramics are crystalline materials, usually compounds formed with strong covalent or ionic bonds between atoms.
- Plastics are materials formed from repeating organic molecules. They can be crystalline, amorphous, flexible, or brittle depending on the nature of the bonds between the polymer molecules.

Critical Concepts

- The elements are the building blocks for all engineering materials.
- Atomic properties and interactions determine the ultimate use characteristics of engineering materials.
- There are atomic differences between metals, ceramics, plastics, and composites.
- Engineering materials are strengthened by "happenings" at the atomic level.

Terms You Should Remember

organic material
inorganic material
element
crystal structure
amorphous
compound
molecule
valence
crystal
lattice
grain

unit cell
x-ray diffraction
polymer
ion
covalent
dislocation
solid solution

Case History

VIEWING ATOMS AND DISLOCATIONS

In this chapter we discuss atoms, quantum mechanics, crystal structures, and dislocations. How do we know these entities really exist? For example, by the way crystalline materials diffract x-rays, we can infer certain information about the dimensions and atomic arrangement within a crystal lattice. However, advances in high-resolution analytical methods now allow us to directly see the basic structure of materials.

The scanning tunneling microscope (STM) allows us to "see" atoms on the surface of a material. The STM is based on the concept of a tunneling current that flows when an infinitesimally sharp probe or tip is placed within 1 nm of the surface of a conductive material. The tip is attached to a piezo electric actuator that controls the distance between the tip and the sample surface. A second pair of actuators allows the probe to scan an area of the sample. The system controls the probe height to maintain a constant tunneling current as the probe scans across the sample creating a surface "topography" map. On very smooth samples analyzed in vacuum under ideal conditions, arrays of atoms on the surface of a sample may be observed. More accurately, you see the electromagnetic field created by the electron clouds surrounding each atom. Highly skilled microscopists can manipulate individual atoms on the surface. Some have drawn their company logo using single atoms.

Dislocations have been studied for many years using transmission electron microscopes (TEM). Small samples are thinned to a few hundred angstroms using an ion or chemical mill. High-energy electrons are focused on and penetrate through the sample. A recording device below the sample records the resultant image, in which actual arrays of dislocations are viewed. Using a tensile stage, scientists can deform metal during imaging with the TEM allowing observation of dislocation interactions.

The inventors of both scanning tunneling microscopy and electron microscopy have been awarded Nobel Prizes for their profound contributions in developing instruments for unlocking the mysteries of the structure and behavior of materials.

Questions*

Section 1.1

1. What is the largest principal quantum number for iron?
2. What is the difference between an element and a compound?
3. How many valence (outer shell) electrons are there in an atom of (a) lithium? (b) beryllium? (c) iron?

Section 1.2

4. What are the heaviest and the lightest stable elements?
5. What is the purpose of the periodic table?
6. Balance the following equation:

$$Zn + HCl \rightarrow ZnCl + H_2$$

7. Why are gases such as neon, argon, helium, and krypton inert?

*Use conversions in Table 1–2.

Table 1–2
Conversion to SI units per ASTM E 380

		SI Unit
Volume		
1 gallon (U.S. liquid) (gal)	$= 3.785 \times 10^{-3}$ m^3	cubic meter
inch^3 (in.^{3})	$= 1.638 \times 10^{-5}$ m^3	cubic meter
liter (l)	$= 1 \times 10^{-3}$ m^3	cubic meter
cubic foot (ft^3)	= 0.0283 m^3	cubic meter
Velocity		
foot/minute (ft/min)	$= 5.080 \times 10^{-3}$ m/s	meter/second
mile/hour (mph)	$= 4.470 \times 10^{-1}$ m/s	meter/second
Pressure/Stress		
kips[a]/inch^2 (ksi)	= 6.894 MPa	megapascal
pounds/inch^2 (psi)	$= 6.894 \times 10^{-3}$ MPa	megapascal
kilogram/millimeter^2 (kg/mm^2)	= 9.8067 MPa	megapascal
Force		
pound (force)	= 4.448 N	newton
kilogram (force)	= 9.806 N	newton
Mass		
pound (lb)	= 0.4536 kg	kilogram
ton (short)	= 907.1847 kg	kilogram
Area		
square foot (ft^2)	= 0.0929 m^2	square meter
square inch (in.^{2})	$= 6.4516 \times 10^{-4}$ m^2	square meter
Length		
inch (in.)	$= 25.4 \times 10^{-3}$ m	meter
foot (ft)	= 0.3048 m	meter
angstrom (Å)	$= 1 \times 10^{-10}$ m	meter
microinch (μin.)	= 0.0254 μm	micrometer
angstrom (Å)	= 0.10 nm	nanometer
micron (μm)	$= 1 \times 10^{-6}$ m	meter
Impact Energy		
foot pound (ft-lb)	= 1.356 J	joule
Temperature		
degree kelvin (K)	= tc[b] + 273.15	kelvin
	$tc = \dfrac{(\text{degrees Fahrenheit} - 32)}{1.8}$	
Heat		
Btu in./h ft^2 °F	$= 1.441 \times 10^{-1}$ W/m · K	watt/meter · kelvin

[a]1 kip = 1000 pounds
[b]tc = temperature Celsius (°C)

8. How many electrons are there in the element aluminum? How many protons?
9. What is the theoretical density of an 80–20 cupronickel alloy (80% Cu, 20% Ni)?
10. State the quantum numbers for the electron configuration of the neutral atom of beryllium.
11. List five elements that are metals and five that are nonmetals.
12. If a neutral atom has an atomic number of 12, how many neutrons does it have in its nucleus?
13. What is the theoretical density of the intermetallic compound $NiAl_2$?

Section 1.4

14. Explain the difference between an amorphous and a crystalline material.
15. Why are metals better conductors than ceramics?

Section 1.5

16. What is a metal?
17. State two methods for hardening metals.

Section 1.6

18. What is a ceramic?
19. Why are some ceramics brittle?
20. How many atoms are there in a molecule of aluminum oxide?

Section 1.7

21. What is a polymer?
22. What is the nature of molecular bonds in polymers?

Section 1.8

23. What is a composite?
24. What is the matrix in a fiberglass (FRP) composite?
25. Identify two consumer items made from composites?
26. What is a metal matrix composite?

To Dig Deeper

Boikers, Robert S., and Edward Edelson. *Chemical Principles*. New York: Harper & Row Publishing Co., 1978.

Dorin, Henry. *Chemistry: The Study of Matter*. Newton, MA: Cebco, Division of Allyn & Bacon, Inc., 1982.

Hall, Nina, Ed. *The New Chemistry*. Cambridge: Cambridge University Press, 2000.

Hertzberg, R. W. *Deformation and Fracture Mechanics of Engineering Materials*, 3rd ed. New York: John Wiley & Sons, 1989.

Horath, Larry. *Fundamentals of Materials Science for Technologists*. Upper Saddle River, NJ: Prentice Hall, Inc., 1995.

Hume-Rothery, William, and G. V. Raynor. *The Structure of Metals and Alloys*. London: Institute of Metals, 1962.

Kaxiras, E., *Atomic Structure of Solids*. New York: Cambridge University Press, 2002.

Lide, David K., Ed. *CRC Handbook of Chemistry and Physics*, 73rd ed. Boca Raton, FL: CRC Press, Inc., 1992.

Pollack, Daniel D. *Physics of Engineering Materials*. Upper Saddle River, NJ: Prentice Hall, Inc., 1990.

Puddephatt, R. J., and P. K. Monaghan. *The Periodic Table of the Elements*, 2nd ed. Oxford: Clarendon Press, 1986.

Read, W. T., Jr. *Dislocations in Crystals*. New York: McGraw-Hill, Inc., 1953.

Shackelford, James F. *Introduction to Materials Science*. New York: Macmillan Publishing Co., 1988.

Van Vlack, L. *Materials for Engineering: Concepts and Applications*. Reading, PA: Addison-Wesley, Inc., 1982.

Winter, M. J. Printable periodic table. www.webelements.com/support/media/pdf/WEB_ELEM.PDF 1998.

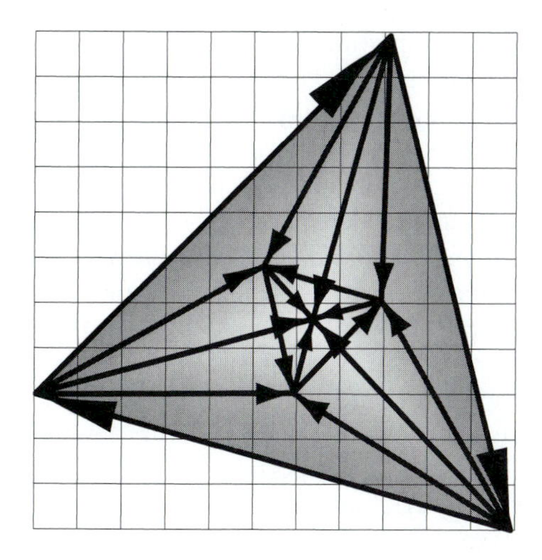

CHAPTER 2

Properties of Materials

Chapter Goals

1. Familiarization with the properties that must be reviewed when making material selections.
2. A knowledge of how properties apply to different material systems.
3. An understanding of the pitfalls to avoid in performing property tests and in using property data.
4. A thorough understanding of the differences among the properties of stiffness, strength, and toughness.

Rationale

Properties determine the usefulness of an engineering material. We all use properties for every decision in life. What we eat for breakfast depends on what properties we want. Some people want something hot, some sweet, others cold. These are "material" properties. We measure them with our senses. In the same way, when a material is selected for design application, which material you choose will depend on its properties. The properties of engineering materials that are important for use in a product can be measured by a plethora of tests. This chapter will discuss the properties that are often key in making material selections, how they are measured, and how to interpret the data from various tests. We want you to be educated consumers *of engineering materials—able to discern both what properties are important in selecting a material for an application and what tests yield appropriate data for your engineering-material decisions.*

Material selection is based on properties. The designer must decide the properties required of a material for a part under design and then weigh the properties of candidate materials. Before we can discuss the relative merits of various material systems, we must establish the vocabulary of properties. It is the purpose of this chapter to define the properties that are important for selection; to show how they apply to the major material systems, polymers, metals, and ceramics; and to show how these properties are used to select materials. Hundreds of properties are measured in laboratories for the purpose of comparing materials. We cannot discuss all these in a single chapter, so we shall concentrate on the more important ones. In some cases, we shall describe measuring techniques.

2.1 The Property Spectrum

When the average person shops for an automobile, he or she establishes selection criteria in several areas—possibly size, appearance, performance, and cost. Certain things are desired in each of these areas, and each automobile will have different characteristics in these areas. The thoughtful car buyer will look at several brands, rate each in various categories, and then make a selection. The goal is usually the car that will provide the best service at an affordable price. Material selection should be approached in this same manner.

The major categories to be considered in material selection are shown in Figure 2–1. *Chemical properties* are material characteristics that relate to the structure of a material and its formation from the elements. These properties are usually measured in a chemical laboratory, and they cannot be determined by visual observation.

Physical properties are characteristics of materials that pertain to the interaction of these materials with various forms of energy and with other forms of matter. In essence, they pertain to the science of physics. They can usually be measured without destroying or changing the material. Color is a physical property; it can be determined by just looking at a substance visually or with an instrument. Density can be determined by weighing and measuring the volume of an object; it is a physical property. The material does not have to be changed or destroyed to measure this property.

Mechanical properties are the characteristics of a material that are displayed when a force is applied to the material. They usually relate to the elastic or plastic behavior of the material, and they often require the destruction of the material for measurement. Hardness is a mechanical property because it is measured by scratching or by application of a force through a small penetrator. This is considered to be destructive because even a scratch or an indentation can destroy a part for some applications. The term *mechanical* is applied to this category of properties because they are usually used to indicate the suitability of a material for use in mechanical applications—parts that carry a load, absorb shock, resist wear, and the like.

Procurement/manufacturing considerations are not listed in property handbooks, and they are not even a legitimate category by most standards. However, the available size, shape, finish, and tolerances on materials are often the most important selection factors. Thus we have established a category of properties relating to the shape of a material and its surface characteristics. Surface roughness is a dimensional property. It is measurable and important for many applications.

Material properties apply to all classes of materials, but certain specific properties may apply to only one particular class of materials. For example, flammability is an important chemical property of plastics, but it is not very important in metals and ceramics. Metals and ceramics can burn or sustain combustion under some conditions, but when a designer selects a steel or ceramic for an application, it is likely that he or she will not even question the flammability rating of the metal or ceramic. In Figure 2–1 we have taken the important classes of engineering materials—metals, plastics, ceramics, and composites—and have tried to list some types of mechanical and chemical properties as well as procurement considerations that are likely to be important in the selection of these materials for a particular application. Many more specific properties could have been listed for each class of materials, but those listed are the ones most likely to be of importance. It is not possible to list the important physical properties for metals, plastics, ceramics, or composites because the physical properties that are important for a particular application are unique to that application, and all physical properties apply to all materials. For example, *ferromagnetism* applies to all materials. A material is either ferromagnetic or it is not.

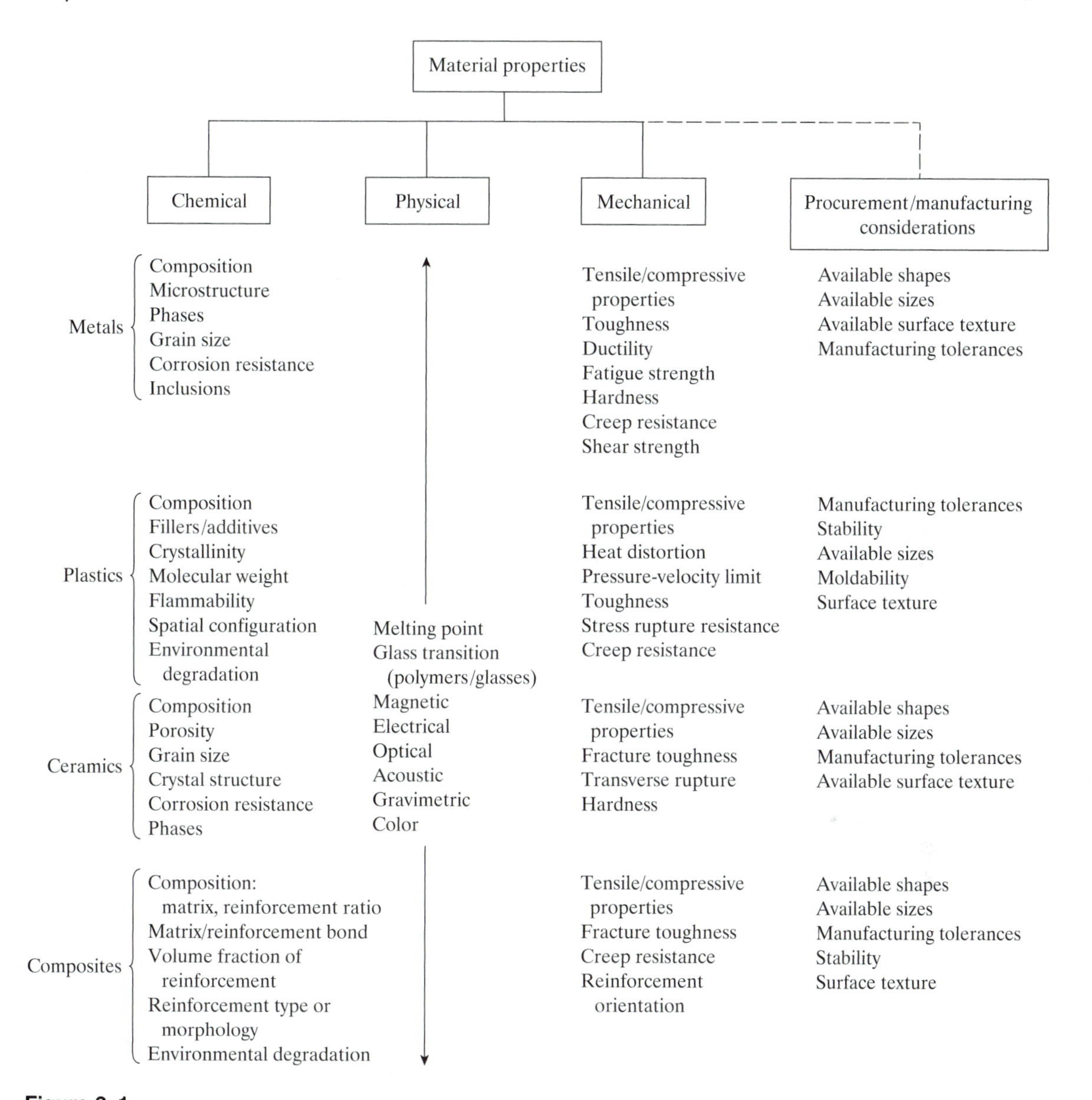

Figure 2–1
Spectrum of material properties and how they apply to various material systems (physical properties apply equally to all systems)

For some applications this is important, for others it is not. All materials have thermal properties such as *thermal expansion* characteristics, *thermal conductivity*, specific heat, latent heat, and so on, but only the application will determine if any of these properties are important selection factors.

We are discussing material properties early in this text because they are used as the basis for selection; they help us to discriminate one material from another. To properly choose a material for an application—our ultimate goal—it is necessary to understand what these properties mean, how they are measured, and how they should be compared in the selection process.

The trend toward global industrial markets that started in the 1990s has urged many industries to seek certification of their quality programs by the International Standards Organization (ISO). Some international customers will deal only with organizations who have demonstrated conformance to ISO 9000 and other quality criteria. Part of the certification process is to document testing procedures. The easiest way to meet test standards of quality organizations is to measure material properties with standard tests. Various international standards organizations have developed standard tests for measuring most material properties:

Standards Organization	Acronym
ASTM International	ASTM
International Standards Organization	ISO
American National Standards Institute	ANSI
European Committee for Standardization	CEN
Deutsches Institut für Normung e.V. (German Institute for Standards)	DIN
British Standards Institution	BSI

The use of these standard tests is advised. Their use makes accreditation to ISO 9000, and the like, much easier. Standard test methods are usually material specific; for example, there are test methods for the chemical analysis of aluminum alloys, copper alloys, steels, and so on. There are so many tests that it is not possible to list them in this text, but there are indexes to ASTM test methods in most libraries. This index is the place to start in the search for standard tests on a particular material.

In the remainder of this chapter we will discuss some of the more widely used chemical, mechanical, physical, and dimensional properties. However, keep in mind that there are many properties that we have not listed, and it is the designer's responsibility to establish the properties that are important for an application and to use these property requirements as the criteria for material selection. As we describe specific properties, we will point out how they should be used in the selection process.

2.2 Chemical Properties

Composition This property can be determined by analytical chemistry techniques. In metals, *composition* usually means the percentage of the various elements that make up the metal, for example, 80% copper, 20% nickel. The composition of a polymer consists of the chemical notation of the monomer with an indication of the chain length (number average molecular weight):

$$\left[\begin{array}{cc} H & H \\ | & | \\ -C & -C- \\ | & | \\ H & H \end{array}\right]_{20,000,\ \text{etc.}}$$

If the polymer is a mixture of polymers, the component homopolymers (polymers of a single monomer species) and their percentages should be stated. If the polymer contains a filler such as glass fiber, this should also be stated with its proportion. The composition of a ceramic is usually the stoichiometric makeup (the quantitative relationship of elements in combination) of the compound or compounds (for example, the composition of the ceramic aluminum oxide would be Al_2O_3).

If a ceramic material consists of a number of different compounds, the stoichiometry of these compounds should be stated along with the volume fraction of each component. Some ceramics contain binders. If this is the case, the nature of the binder should be stated along with its volume fraction. Additional items to specify are phases present, crystal structure, grain size, and porosity; for example,

Material: alpha silicon carbide, 1- to 2-μm grain diameter, porosity <0.01 %

The composition of a composite requires a statement of the details of the matrix and the reinforcement and the volume fraction of each. Often the direction or orientation of the reinforcement is also indicated.

In material selection, composition is a fundamental consideration. The designer should always have some idea of what a material is made from. All too often one hears designers claim that Ripon is a much better steel than Excelsior XL. If the designer took the time to look up the composition, he or she might find out that Ripon and Excelsior XL are identical. The same thing is true of polymers and ceramics. Avoid selecting by trade names. Try to obtain and understand the composition of materials used in design.

Microstructure Studies of the microstructure of metals are highly useful tools of the metallurgist. Microstructure studies indicate grain size, phases present, heat treatment, inclusions, and the like. If, for example, the metallurgist wishes to measure the grain size of a steel, he or she cuts a small piece from the part, mounts it in a potting compound, polishes it to a mirror finish, and then applies a chemical etching reagent. Grain boundaries will etch at a different rate than the grains, thus leaving the grains standing out, and they become visible with a reflected light microscope. Figure 2–2 is a typical metallurgical microscope, and Figure 2–3 is a typical photomicrograph of a metal structure. A similar polish-and-etch procedure is used to reveal the microstructure of ceramics and metal matrix composites. This technique for analyzing materials is referred to as *metallography* for metals and *ceramography* for ceramics. Because polymers are resistant to chemical etches, the microstructure of polymers is often revealed by special dyeing and lighting techniques. Microscopic analysis of polymers is often conducted by

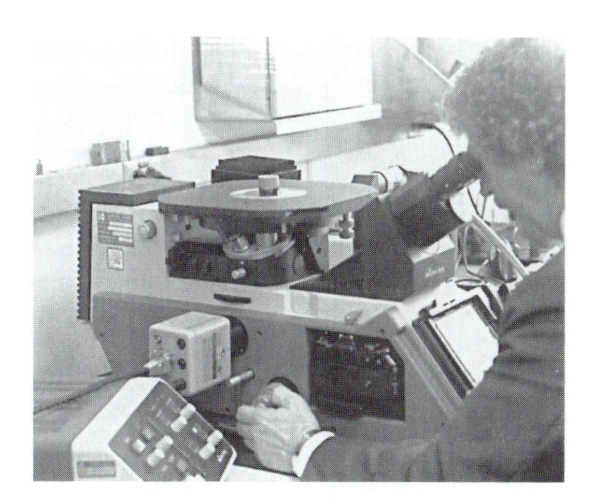

Figure 2–2
Metallurgical microscope

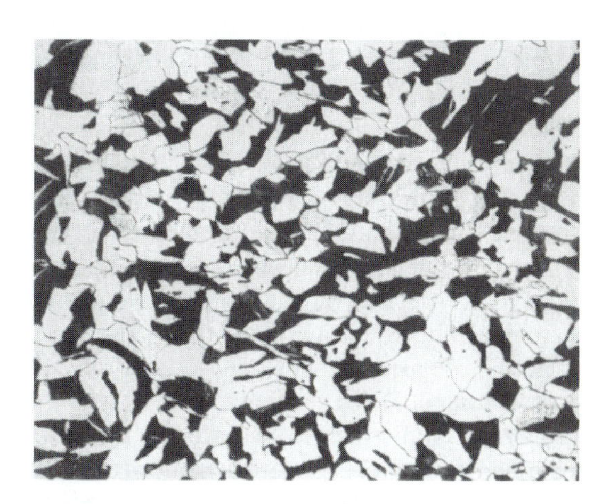

Figure 2–3
Microstructure of a plain carbon steel. Two phases are present. The light regions are ferrite, and the dark regions are pearlite. This microstructure is made visible by metallography. A polished sample of the steel is etched to reveal the grains and phases and then analyzed using a reflected light microscope (nital etch, × 100).

transmitting polarized light through thin wafers or shavings removed from the surface of the sample using a *microtome*. This technique is called *microtomy*.

All through this text we shall be referring to microstructures. This is how samples are made and studied. Microstructures are invaluable to the materials engineer in solving problems and in understanding material responses to treatments. To the designer, microstructures may be indirectly involved in selection. For example, if a designer reads about a supercoating that will solve all wear problems, he or she can have it analyzed with metallographic techniques before committing an expensive part to be processed. Microstructure studies are also an important tool in studying why a part failed in service.

Crystal Structure and Stereospecificity The atomic or molecular structure of metals, polymers, and ceramics is not an easy property to determine. Sophisticated techniques such as x-ray diffraction are required. Structure analysis tells the materials user if the structure is, for example, face-centered cubic or body-centered cubic, and it is also used to study structural changes during processing. To be correct, crystal structure applies only to crystalline materials, but many polymers can have structure with enough order that they are considered crystalline or semicrystalline. The property of stereospecificity refers to the three-dimensional spatial arrangement of elements in polymer molecules. Figure 2–4 shows an isotactic form of polypropylene. Using techniques developed for studying crystal structure, researchers have determined that all of the methyl groups (CH_3) in this form of polypropylene are located on one side of the polymer chain. If these groups were connected at random to the chain, there would be no stereospecificity.

Designers often select plastic materials based on crystal structure. In metals and ceramics it is a property of a material that helps the user understand other, more design-related, characteristics. For example, metals with a body-centered cubic structure often get very brittle at sub-zero temperatures. Metals with a face-centered cubic structure usually do not. In polymers, crystallinity and stereospecificity affect melting characteristics and many mechanical properties. Similarly, many properties of ceramics are affected by the type of crystal structure or lack of it.

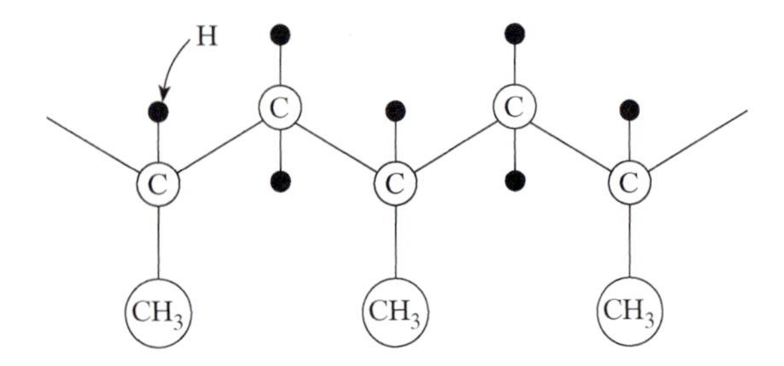

Figure 2–4
Isotactic form of polypropylene

Corrosion Resistance There are many forms of corrosion and many ways of measuring it. Essentially, corrosion is degradation of a material by reaction with its environment. It does not have to involve immersion of a material in a chemical. Many polymers become brittle on exposure to sunlight. The property *environmental degradation* is usually used in place of chemical resistance because of factors such as sunlight's effect. An entire chapter (Chapter 13) is devoted to this subject. At this point it is sufficient to say that this property applies to all materials and is an important selection factor.

2.3 Physical Properties

Use of Physical Properties

There are so many physical properties that it is not possible to even briefly describe all of them and how they are measured. Figure 2–1 is an attempt to categorize them, but this list (thermal, electrical, magnetic, gravimetric, optical, acoustic) is still inadequate. With the advent of nuclear materials in engineering, a whole new group of physical properties has arisen. Physicists measure a material's neutron absorption characteristics and susceptibility to radiation damage, for example.

The designer must be concerned with the thermal properties of materials any time a part is to be used at some temperature other than the temperature at which it was fabricated or if it is expected to perform some heat transfer function. Thermal conductivity is important in many machine applications: heat-sealing heads, heat exchangers, heat sinks, heating platens, die casting, and plastic-molding cavities. This property has strange units: Btu ft/hr/ft^2/°F in the English system and W/m · K (watt/meter-kelvin) in the metric system. Figure 2–5 shows the basic equation for steady-state heat flow. This property is simply K in this equation. Fortunately, techniques for measuring thermal conductivity are such that K values for polymers, ceramics, and metals can be compared for selection purposes.

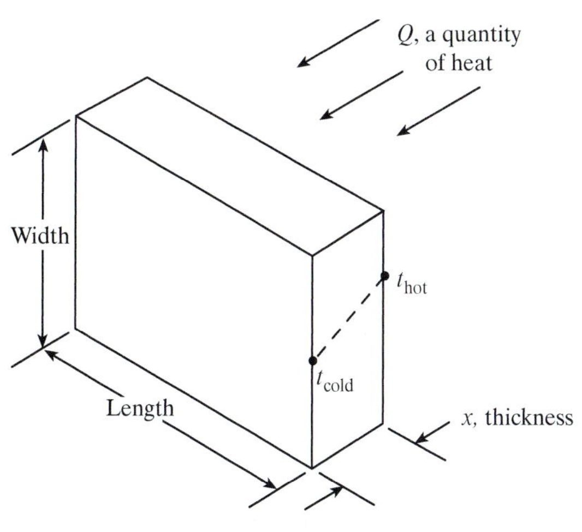

$$Q = KA\frac{\Delta t}{x}$$

where

Q = Heat flow through a material (Btu) [Watt] per unit time
K = Thermal conductivity ($Q \cdot x / A \cdot \Delta t$)
A = Area through which heat will flow ($W \times L$)
Δt = ($t_H - t_C$) temperature differential
x = Thickness

Figure 2–5
Steady-state heat flow through a solid

Thermal expansion is important when dissimilar materials will be fastened and heated, and when materials are locally heated. As an example, a designer complained to a materials engineer that his aluminum heat seal block (a heated plate used to seal plastic wrapping on products) was not dimensionally stable at elevated temperatures. He forgot to consider that he had the aluminum bolted to steel, and aluminum expands twice as much as steel—thus the cause of the distortion. One constant source of confusion regarding this property relates to the question, if you heat a doughnut-shaped material, does the hole in the center get bigger or smaller? Many people argue over what happens, but the fact is that thermal expansion takes place on the volume of material. The hole in the doughnut gets bigger when it is heated, as do the outside diameter (OD) and the thickness. The units for thermal expansion are in./in. °F (inch/inch–degree Fahrenheit) or cm/cm °C (centimeter/centimeter–degree Celsius). Every dimension on the piece must be multiplied by this factor.

One of the most important selection factors for polymers is maximum use temperature. Many polymers lose all useful engineering properties at temperatures above 100°C. However, the upper use temperature of polymers is continually improving. The property of *heat distortion temperature* is more quantitative in that it does show what the material will do under a specified stress. Whenever a polymer is used at elevated temperatures, a thorough study should be made of its response to the specific environmental conditions of the intended service.

In polymers, *water absorption* can be a very important selection factor if close dimensional tolerances need to be maintained on a part. A large number of engineering polymers swell significantly with increases in ambient relative humidity. The moisture-absorption factor should always be as low as possible if polymer part dimensions are critical.

The electrical properties of materials are very important when a design calls for a part to be an electrical conductor or insulator. Electrical resistivity ρ is analogous (inversely) to K in the steady-state conduction equation in Figure 2–5. It determines the rate at which current will flow through a given cross section and through a given length. The unit used for metals is $\mu\Omega$-cm (microohm-centimeter) or Ω-m (ohm-meter) in metric. The resistivity of wires can be measured with an ohmmeter. Resistivity is found by the formula

$$\rho = \frac{AR}{L}$$

where A = cross section of the wire in square centimeters
R = resistance in ohms
L = length of the wire over which the measurement in made

Thus the units come out Ω-cm. There are other ways to measure and report the resistance of a material to the flow of electricity, but resistivity suffices for metals; conductivity is simply the reciprocal of the resistivity. Metals have resistivities in the range of about 1 to 200 Ω-cm, whereas ceramics and other good insulators have values usually greater than 10^{15} Ω-cm.

Ceramics and polymers are frequently used as electrical insulators. Their relative ability to insulate is measured by such things as arc resistance, dielectric strength, and dissipation factor. The measurement techniques for determining these properties will be discussed in Chapter 7.

The use of magnetic properties in material selection can be more complicated than the use of electrical properties. At least 20 of these properties are listed in handbooks (ASTM A 340): permeability, retentivity, hysteresis loss, coercivity, intrinsic induction, and so on. Like some of the electrical properties, effective use of these properties in material selection will require more explanation than can be afforded in this discussion. Of all the magnetic properties of materials, the one that designers will have most occasion to use is simply whether a material will be attracted by a magnetic field.

This property is called *ferromagnetism*, and only five of the stable elements have it: iron, nickel, cobalt, gadolinium, and dysprosium. The latter two are not likely to be encountered in everyday materials usage, but alloys of the other elements make up the important ferromagnetic materials. The property of ferromagnetism is the result of a set of circumstances associated with the magnetic moments of electrons and the configuration of atoms in a material. Many oxides and other ceramic-type compounds can be ferromagnetic, and we shall discuss these in Chapter 8.

All metals and many nonmetals have some sort of response to magnetic fields, and this response is measured by a number of magnetic properties. One common way to rate a material's response to a magnetic field is by measuring the flux density (B) in the material when it is placed in a magnetic field of varying magnetic strength (H), and this field strength is increased and decreased. The slope of the B–H curve is a measure of the magnetic characteristics of a material, and the shape of the curve that results when the material is tested in a cyclic magnetic field is a measure of how the material will behave under such conditions. If a material magnetizes and demagnetizes easily, the B–H curve exhibits a hysteresis loop like that illustrated in Figure 2–6(a). A material that tends to remain magnetized as the field strength is decreased will have a hysteresis loop resembling that illustrated in Figure 2–6(b). A material with this latter type of hysteresis loop would be a good permanent magnet but a poor material for an electrical device like a solenoid core iron that must be easily magnetized and demagnetized. Low-carbon steels with from 0.5% to 5% silicon have a small hysteresis loop,

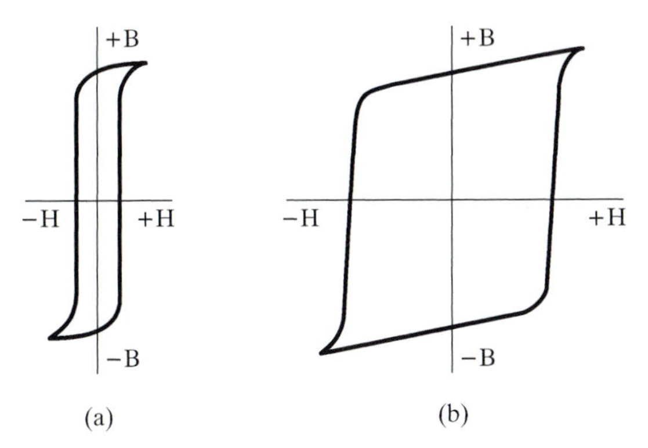

Figure 2–6
Magnetic hysteresis curves. (a) A soft magnetic material (easily magnetized and demagnetized material). (b) A hard magnetic material, one that retains magnetism as in a permanent magnet.

and they are widely used for laminae in electric motors. The area within the hysteresis loop is a measure of the energy lost in a magnetization/demagnetization cycle; in electric motors it is desirable to keep these losses to a minimum because they produce unwanted heating of the motor. The hard magnetic materials that make good permanent magnets include hardened low-alloy steels and specialty iron alloys containing aluminum, nickel, and cobalt (Alnicos).

All steels except those with austenitic structures (FCC) are ferromagnetic at temperatures below the Curie temperature, and all other metals with the exception of some nickel and cobalt alloys are not ferromagnetic. When using the latter alloys, it is advisable to check alloy data sheets on these materials if the property of ferromagnetism is a selection factor ; some nickel and cobalt alloys are not ferromagnetic even at room temperature.

Another category of physical properties listed in Figure 2–1, gravimetric, has to do with the weight or mass of materials. The most important properties relating to material selection are *density* and specific gravity. It is good practice to factor in densities in comparing costs of various materials. Many engineering plastics cost about $5 per pound. It may appear that one of these plastics is too expensive to be used to replace a $1-per-pound steel part. If the plastic has a density of 0.04 pound per cubic inch and the steel has a density of 0.28 pound per cubic inch, the price differential can become small if the part has low volume. If the part requires only a cubic inch of material, the cost of the material in the plastic part will be $0.20 and of that in the steel part $0.28. Density can be an important part of material selection.

Another useful application of density or specific gravity information is to determine porosity. Materials such as ceramics and powder metals that are made by compaction and sintering have varying degrees of porosity. Sometimes property tabulations will show the percent theoretical density or apparent density. These types of data will supply information on porosity. In structural parts, porosity is often undesirable. In powdered metal wear parts, porosity is usually desired for retention of lubrication.

The preceding discussion of physical properties excluded 50 or so properties that could be important for some applications. The optical properties of polymers and ceramics are sometimes important selection factors. Design situations arise where even the velocity of sound in a material (an acoustic property) is an important selection factor. Unfortunately, there are just too many physical properties to be able to cover them all in detail. The limited number of properties that have been discussed are some of the more important ones, and they should provide guidance for most selection problems involving the physical properties of materials.

2.4 Mechanical Properties

Mechanical properties are of foremost importance in selecting materials for structural machine components. Think of any tool, any power transmission device, or any wear member, and

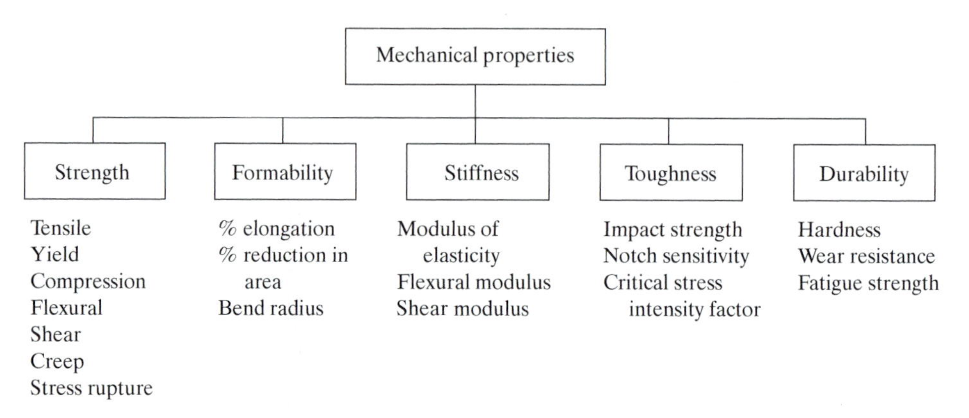

Figure 2–7
Serviceability factors and related mechanical properties

list the properties needed for serviceability. Your list would probably include a number of the factors listed in Figure 2–7: strength, formability, stiffness, toughness, and durability. There are many tests to measure mechanical properties, but three tests supply the most useful information for most applications: tensile, impact, and hardness tests. These tests will be referred to throughout this text, and materials will be compared on data developed from these tests. It is imperative that the designer understand how these tests are run and how to interpret the test information.

Stress

Mechanical components are subjected to forces that pull, push, shear, twist, or bend as depicted in Figure 2–8. A component deforms in response to the forces applied to it. If the forces are relatively low, the material deforms *elastically* and the component returns to its original shape on removal of the forces. However, if the magnitude of the forces is increased, the component may permanently deform. Permanent deformation is called *plastic* deformation.

The external forces applied to a component result in stresses within the material. A *stress* is a measure of the force in a component relative to the cross-sectional area over which the force is applied. For example, the stress applied by a person wearing spike-heeled shoes is high enough to dent a wooden floor. However, if the same person wore flat-heeled loafers, the stress on the floor would be significantly lower—resulting in no dents in the floor. The force did not change; however, the area over which the force was applied was significantly increased when flat-bottomed shoes were worn. Hence stress is a relative measure of the *intensity* of the force.

If a mechanical component such as the tie rod in Figure 2–8 is pulled in tension, a force is transmitted through the tie rod. As shown in Figure 2–9 if an imaginary knife cuts the tie rod in half, one can imagine a distribution of internal forces across the area of the cut surface. These internal forces are called stress. Stress is measured in force per unit area, for example, lb/in.^{2} or N/m^2 (Pascal, Pa). Hence the stress in the main body of the tie rod loaded in tension is simply the applied force divided by the cross-sectional area of the tie rod as shown in the following equation:

$$\sigma_t = \frac{F_t}{A}$$

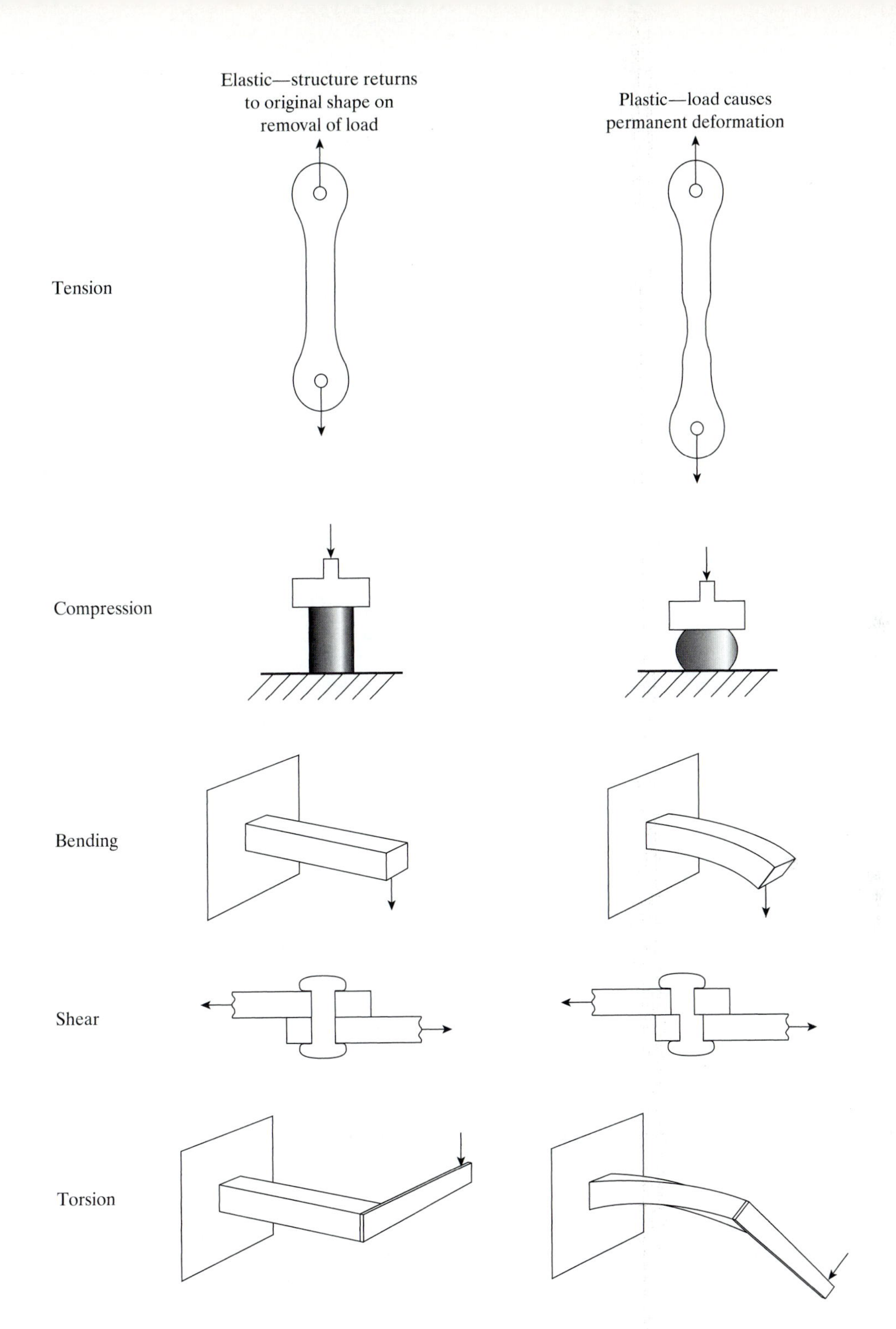
Elastic—structure returns to original shape on removal of load
Plastic—load causes permanent deformation
Tension
Compression
Bending
Shear
Torsion

Figure 2–8
Examples of five basic types of stress

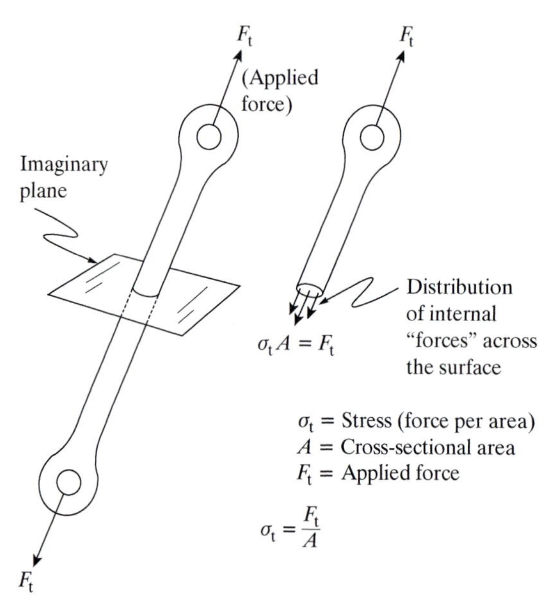

Figure 2–9
Formation of tensile stresses in a mechanical component as the result of an applied load

where σ_t is the stress, F_t is the force, A is the area, and the subscript t denotes that the force and stress are in tension.

Calculating the stress in the tie rod provides a convenient method for selecting tie rod materials. For example, one could make smaller, less expensive test samples out of candidate materials, load the test samples to failure, calculate the failure stress in the test samples, and then compare those failure stresses with the anticipated stress in the tie rod. As long as the failure stress in the candidate material is higher than the stress in the tie rod, the tie rod will not break when the force is applied.

The calculation of stresses is not always simple. Stress calculations can be long and involved. Frequently, finite element modeling, a form of computer modeling, is used to calculate the stresses in a complicated component. There are five basic types of stress as shown in Figure 2–8.

1. *Tensile stress* pulls a member apart.
2. *Compressive stress* crushes, collapses, or buckles a member.
3. *Shear stress* cleaves a member.
4. *Torsional stress* twists a member.
5. *Bending stress* deflects a member.

Depending on the orientation of the forces on a part, the part may contain tensile stresses, compressive stresses, and shear stresses. In many instances combinations of tension, compression, and shear stresses in a part may result in a combined stress that is even higher than any individual stress.

Elastic Modulus

Most engineering materials are considered to be elastic. As such, they basically behave as "springs." That is, they may be distorted up to a certain point and on removal of the load return to their original dimensions. Elasticity in materials was documented by Robert Hooke over 300 years ago. He observed that if a length of wire, say 30 or 40 feet long, was pulled in tension with weights the distance the wire stretched was directly proportional to the weight applied to the wire (see Figure 2–10.) Hence

$$F = K\delta$$

where F is the weight applied to the wire, δ is the distance the wire stretched due to the applied weight, and K is a constant of proportionality often called a spring constant.

On removal of the weight, he also observed that the wire returned to its original length. The tensile stress in the wire, σ, is calculated as

$$\sigma = \frac{F}{A}$$

where F is the weight applied to the wire and A is the cross-sectional area of the wire.

If the change in length of the wire due to the applied weight is divided by the original length of the wire, the *strain* in the wire may be calculated. Strain is a measure of deformation (either

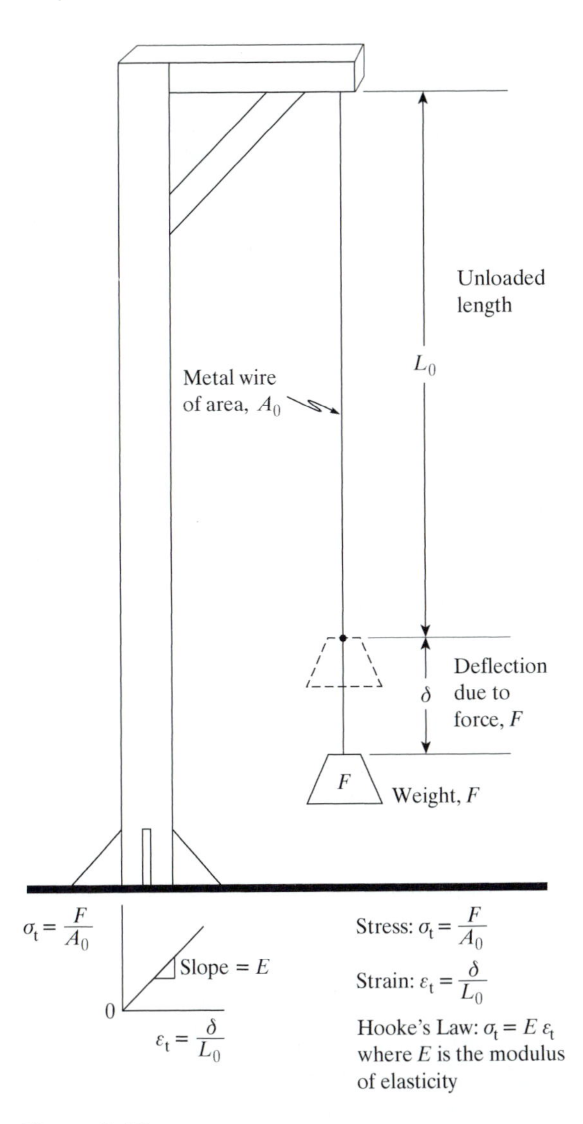

Figure 2–10
Hooke's law

elastic or permanent). Hence, strain in the wire, denoted as ε, may be written as

$$\varepsilon = \frac{\delta}{L}$$

where L is the original length of the wire and δ is the change in length that occurs when the wire is pulled by weight F.

In terms of stress and strain, Hooke's law may be rewritten as

$$\sigma = E\varepsilon$$

where E is a constant of proportionality referred to as *Young's modulus* or the *elastic modulus*. Young's modulus is analogous to a spring constant in that it is a measure of the relative elastic stiffness of a material. Young's modulus may be used in compression as well as in tensile applications.

Example

Figure 2–10 illustrates a 240-in. (6.09 m)–long spring steel wire, 0.0630 in. (1.6 mm) in diameter loaded with 300 lb (136 kg). The length of the wire increases to 240.768 in. (6.116 m) once the load is applied. Calculate the tensile stress, tensile strain, and modulus of elasticity of the wire.

Stress is the force (F) per unit area (A); r-wire radius:

$$\sigma_t = \frac{F}{A} = \frac{F}{\pi r^2} = \frac{300\text{ lb}}{\pi(0.0315\text{ in.})^2}$$
$$= 96{,}239\,\frac{\text{lb}}{\text{in.}^2}\,(663\text{ MPa})$$

Strain is the change in length divided by the original length:

$$\varepsilon_t = \frac{\delta}{L} = \frac{240.768\text{ in.} - 240\text{ in.}}{240\text{ in.}}$$
$$= 0.0032\,\frac{\text{in.}}{\text{in.}}$$

The modulus of elasticity, a measure of the elasticity or spring rate of the material, is calculated from Hooke's law:

$$\sigma_t = E\varepsilon_t$$

On rearrangement,

$$E_t = \frac{\sigma_t}{\varepsilon_t} = \frac{96239\text{ lb/in.}^2}{0.0032\text{ in./in.}}$$
$$= 30{,}100{,}000\text{ lb/in.}^2\,(207.5\text{ GPa})$$

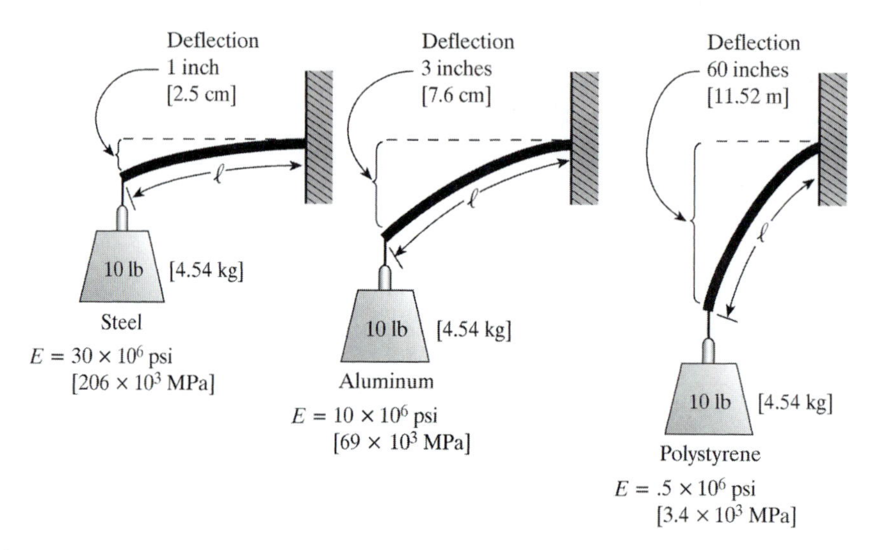

Figure 2–11
Effect of modulus of elasticity on elastic deflection. All beams have the same length and cross section.

The *modulus of elasticity* is of crucial importance in materials selection. It determines the elastic deflection of a structural member under load. Polymers often have very low elastic moduli compared with metals and ceramics (Figure 2–11).

Tensile Test

As we discussed earlier, to select a material for a component with given applied stresses, one must know the strength of candidate materials. Tensile testing measures the strength of a material. A material's response to tensile, compressive, and shear forces may be measured on a universal testing machine—commonly referred to as a tensile tester (see Figure 2–12). These machines can pull on a sample to measure *tensile strength* or push on a sample to measure *compressive strength*. *Shear strength* is measured using a special fixture that applies shear forces to a small pin sample.

Tensile samples are typically cylindrical rods with a reduced diameter in the center or flat plates with a narrow section in the center as shown in Figure 2–13. Prior to testing, the original cross-sectional area of the reduced section of the tensile sample is measured and two fiducial marks (*gage marks*) of known separation distance are placed on the sample. The test sample is mounted in a tensile machine using grippers or jaws. One jaw is fixed to the base of the testing machine, and the other is affixed to a

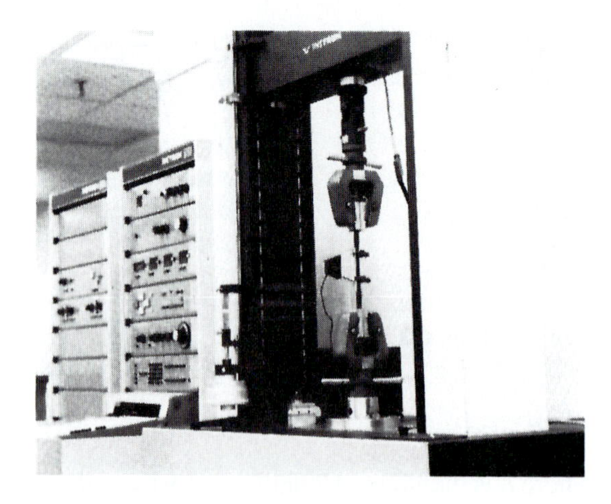

Figure 2–12
Typical universal testing machine

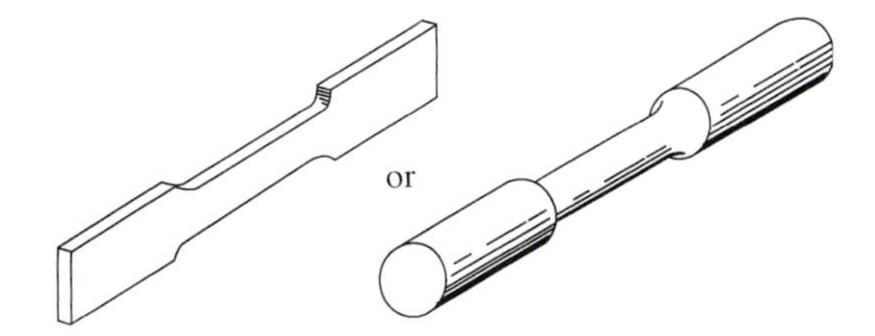

Figure 2–13
Tensile samples

movable crosshead. The crosshead is driven in such a manner as to pull the sample apart. A force transducer, connected to one of the jaws, measures the applied force during the test. The length of the reduced area section of the sample is also measured using an *extensometer* (a special device that clamps on the body of the test sample to convert length change to an electrical measurement).

During the test, the crosshead is moved (by power screws or hydraulics) to create a tensile force in the test sample. The force is measured by the force transducer (a special device that converts force into an electrical measurement), and the tensile stress is calculated by dividing the force by the original cross-sectional area. Simultaneously, the length of the reduced section of the test specimen is also measured by the extensometer. The strain is calculated by dividing the change in length of the sample by the original length. Instrumentation such as a computer then calculates a stress–strain graph as shown in Figure 2–14. During the early stages of the test, the sample will behave elastically. That is, if the crosshead were stopped and returned to its starting position, the sample would return to its original length. As the test continues, the stress in the sample increases, and the length of the sample within the gage region becomes longer. The stress–strain curve is approximately linear, and the slope of that curve is the elastic modulus. Materials that exhibit a linear stress–strain curve in the elastic range are said to be Hookean (after Robert Hooke).

The modulus of elasticity is the slope of the linear portion of the stress–strain curve (Figure 2–14) and may be calculated through the following equation.

$$\text{modulus of elasticity}, E = \frac{\Delta\text{stress}}{\Delta\text{strain}}$$

Where Δstress and Δstrain are the respective increments in stress and strain on the linear portion of the stress–strain curve.

Over a range of stresses, the stress–strain curve begins to deviate from linearity, even though it is still in the elastic range. This transition to nonlinearity occurs at a point referred to as the *proportional limit*. Below this point, the material exhibits Hookean elastic behavior (Figure 2–14). In some instances, the material may still behave elastically for a small increment of strain above the proportional limit although in a nonlinear, non-Hookean manner.

As the stress continues to increase in the sample, the sample continues to elongate until it finally reaches a point where some permanent plastic deformation occurs. If the sample were unloaded at this point, the specimen would be slightly longer than its original length. This permanent increase in length is referred to as plastic deformation. The point at which the deformation of the material transitions from elastic deformation to plastic deformation is referred to as the *elastic limit* or the *yield point*. Because the yield point is difficult to determine precisely, engineers often consider the yield point to occur at an offset strain of 0.2% (for metallic materials) as shown in Figure 2–14. Testing standards, such as ASTM, also designate the stress at 0.2% offset strain as the *yield stress* or *yield strength* of the material. Yield strength is typically measured in psi or Pa. Most published handbook data report metal yield points as the yield strength measured at an offset strain of 0.2%.

As the crosshead in the tensile tester continues to elongate the sample, the tensile force in the sample continues to increase in a nonlinear

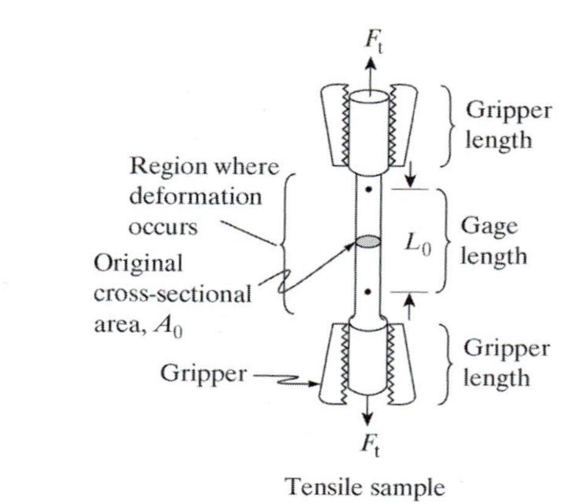

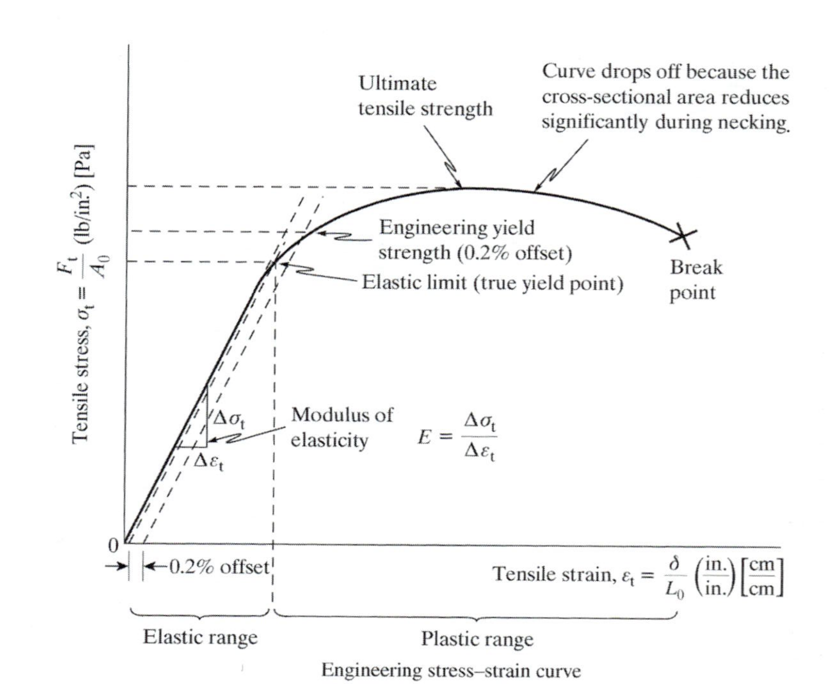

Figure 2–14
Tensile test method and the resultant data

manner as shown in Figure 2–14. Simultaneously, the cross-sectional area of the reduced area section of the test sample begins to decrease as the sample length increases. Initially, this reduction in cross-sectional area occurs uniformly over the reduced section of the sample.

As the tester continues to pull the sample, two competing effects occur. The material continues to harden and gets stronger; however, at the same time the cross-sectional area of the sample begins to rapidly decrease, reducing the load-carrying capacity of the test specimen. The force curve reaches a peak and then begins to drop off with continued extension. The stress calculated at this peak load is called the *ultimate tensile strength* of the material (see Figure 2–14). Before the ultimate tensile stress is reached, the rate of work hardening in the material increases faster than the reduction in cross-sectional area. At the ultimate tensile stress, the rate of cross-sectional area reduction occurs faster than the work hardening rate, so the load-carrying capacity of the specimen decreases and the curve begins to drop off. Typically, at the ultimate tensile stress point, the reduction in cross-sectional area occurs in a pronounced, localized spot on the test sample. This localized reduction in area is known as necking (see Figure 2–15). As the test proceeds, ultimately the sample fractures into two halves.

Once the sample fractures, it is removed from the tensile testing machine and two measurements are made (see Figure 2–15).

- The fractured sample is reassembled as best as possible and the final length of the gage area is measured.
- The final diameter of the necked down portion of the sample is also measured.

These two measurements are used to assess the *ductility* of the material. Ductility, a measure of a material's ability to be stretched or drawn, is typically reported as *percent elongation* or *percent reduction in area*. The percent elongation is a ratio of the increase in length incurred in the sample up to the point of fracture relative to the initial length of the sample. Percent elongation is calculated by the following equation (see Figure 2–15).

$$\text{Percent elongation} = \frac{\text{final length} - \text{initial length}}{\text{initial length}}$$

The higher the percent elongation, the more ductile and formable the material is. When comparing the ductility of materials, always verify that the percent elongation data are based on the same gage length.

Percent reduction in area is measured by comparing the original cross-sectional area in the gage section of the test specimen with the necked-down area of the specimen after fracture (usually done with cylindrical tensile specimens). To measure the final cross section of the sample, the fractured sample is assembled as best as possible and the final diameter in the necked-down region is measured (the initial diameter must also be measured before testing). Percent reduction in area is calculated by the following equation (see Figure 2–15):

$$\text{Percent reduction in area} = \frac{\text{initial area} - \text{final area}}{\text{initial area}}$$

There are essentially two types of stress-strain curves: engineering stress–strain curves and true stress-strain curves. The stress values in *engineering* stress–strain curves are calculated by dividing the force measured during the tensile test by the original cross-sectional area of the specimen. However, in reality, the cross-sectional area of the test specimen actually reduces as the sample is stressed in the tensile tester (see Figure 2–15). In *true* stress–strain curves, the stress is calculated by dividing the force measured during the tensile test by the actual or instantaneous cross-sectional area of the specimen. Similarly, the

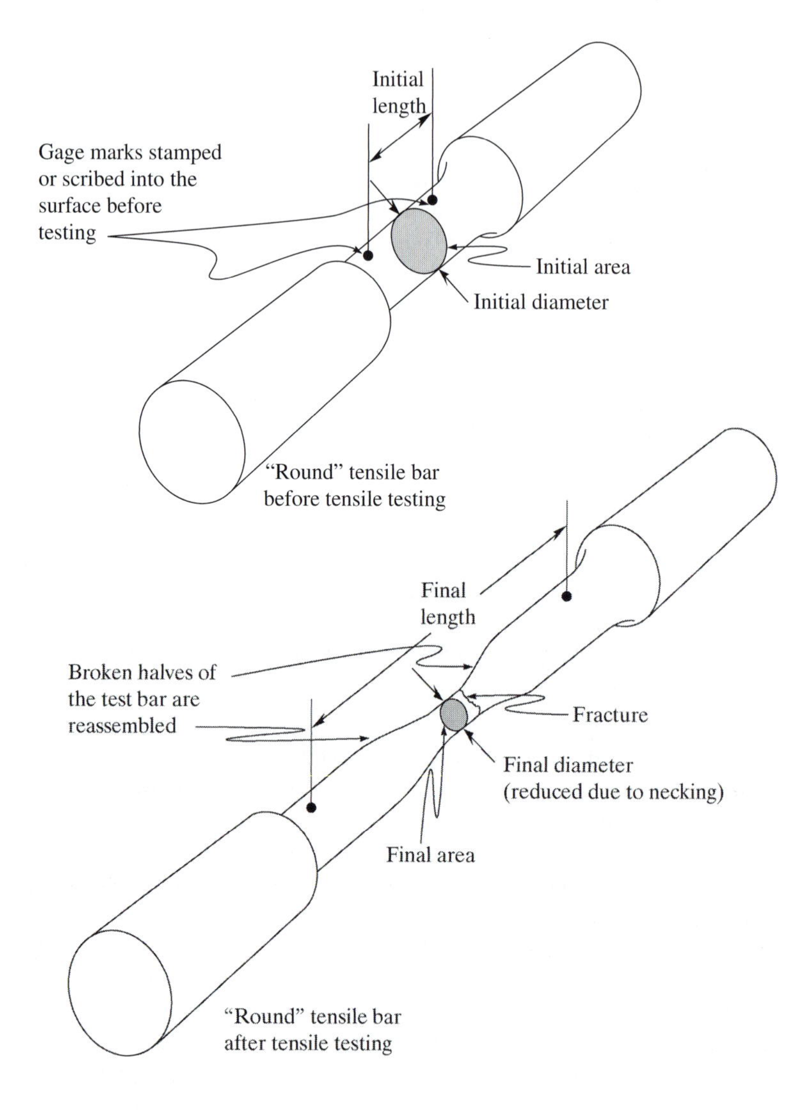

Figure 2–15
Tensile test samples before and after testing, and the measurements required to calculate the percent elongation and the reduction in area

strain is calculated by using an instantaneous gage length rather than the original gage length used for engineering strain calculations. Figure 2–16(a) shows a representation of engineering and true stress–strain curves.

When a material is stressed beyond its elastic limit, it undergoes plastic deformation. As reviewed in Section 1.5, when a metal is plastically deformed, interactions with dislocations in the material's structure can cause the material to become stronger and harder. This phenomenon is known as *work hardening* or *strain hardening*. In true stress–strain testing, an equation may be used to approximate the shape of the plastic region of the stress–strain curve:

$$\sigma_{\text{true}} = K\varepsilon_{\text{true}}^{n}$$

where σ_{true} is the stress, $\varepsilon_{\text{true}}$ is the true strain, K is a strength coefficient, and n is the strain-hardening exponent. The strain-hardening exponent is a parameter that defines a material's

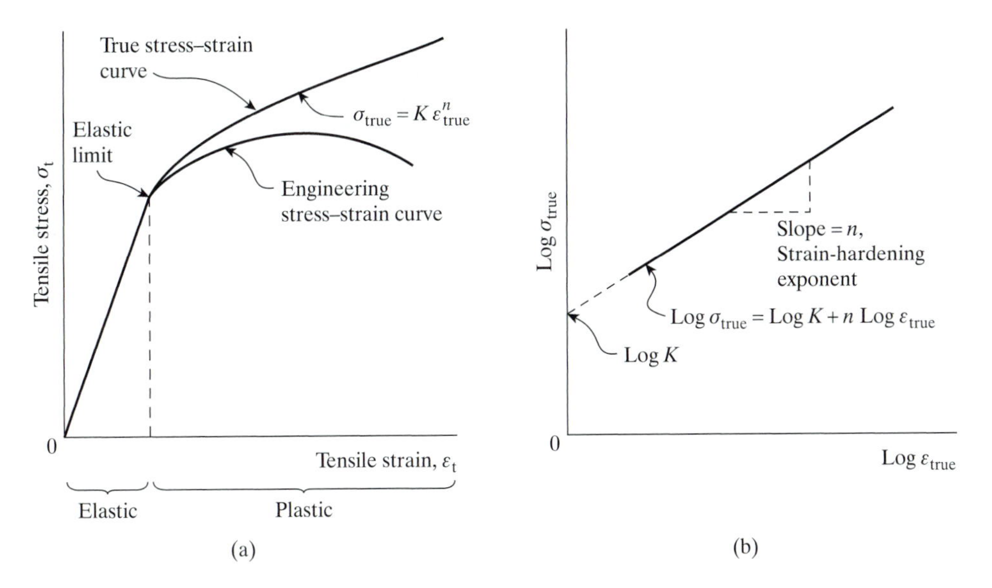

Figure 2–16
Comparison of true and engineering stress–strain curves and the calculation of the strain-hardening exponent from true stress–strain data.

tendency to work harden when plastically deformed. A material with a high strain-hardening exponent will become very strong when plastically deformed whereas the strength of a material with a low strain-hardening exponent does not increase significantly with plastic deformation [see Figure 2–16(a) and (b)].

The strain-hardening exponent is an important parameter for applications such as forming a fender for an automobile. A fender is formed from steel sheet by a process called deep drawing. A flat sheet of steel is placed between a punch and die tool set and the metal is drawn into the die as the punch is pushed into the sheet. To successfully make a deep drawn part such as a fender, the steel should have a low yield strength, high ductility (high percent elongation), and a high strain-hardening exponent. A high strain-hardening exponent for deep drawing may seem counterintuitive; however, it is desirable for the material to become stronger as it is deformed so that it pulls fresh material into the tool as the deep-drawing process progresses.

When a tensile sample such as that shown in Figure 2–13 or Figure 2–15 is stretched during a tensile test, the diameter of the reduced section (or gage section) decreases in order to maintain constant sample volume (see Figure 2–17). The reduction in diameter constitutes a strain that develops in a direction that is perpendicular or transverse to the longitudinal or axial direction of the test specimen. The strain in the axial direction can be related to the transverse strain (e.g., strain due to change in diameter of a round tensile bar) through a constant called *Poisson's ratio*

$$\nu = \left| \frac{\varepsilon_{transverse}}{\varepsilon_{longitudinal}} \right|$$

where ν is Poisson's ratio, $\varepsilon_{longitudinal}$ is the axial strain, and $\varepsilon_{transverse}$ is a strain perpendicular to the axial strain. Poisson's ratio typically ranges from 0.2 to 0.35 for many materials and is only relevant below the proportional limit. It is an important property for stress analysis calculations.

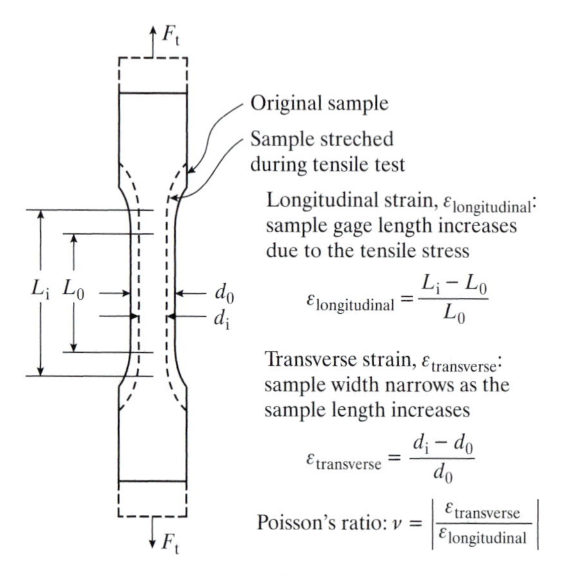
F_t
Original sample
Sample streched during tensile test
Longitudinal strain, $\varepsilon_{\text{longitudinal}}$: sample gage length increases due to the tensile stress
$\varepsilon_{\text{longitudinal}} = \frac{L_i - L_0}{L_0}$
Transverse strain, $\varepsilon_{\text{transverse}}$: sample width narrows as the sample length increases
$\varepsilon_{\text{transverse}} = \frac{d_i - d_0}{d_0}$
Poisson's ratio: $\nu = \left| \frac{\varepsilon_{\text{transverse}}}{\varepsilon_{\text{longitudinal}}} \right|$
L_i L_0
d_0
d_i
F_t

Figure 2–17
Poisson's ratio may be calculated from the stress–strain test by measuring the ratio of transverse strain to axial strain.

With regard to the properties dicussed in this section, most materials have specific methods for conducting mechanical tests as shown in Table 2–1. Typically, published materials property data are reported with reference to these methods.

Significance of Stress–Strain Testing

Stress–strain testing provides basic information on the stiffness, strength, and ductility of a material. Table 2–2 summarizes the basic parameters obtained from tensile testing. The shape of the stress–strain curve can also reveal characteristics of the material as shown in Figure 2–18. Brittle materials, for example, exhibit very little plastic deformation before fracture [Figure 2–18 (c)]. Alternatively, some polymers exhibit significant orientation strengthening and long sample extensions before fracture [Figure 2–18 (d)].

Table 2–1
Some ASTM International tests for mechanical properties. Tests are material specific.

ASTM Test	Description
Tension testing	
A 370	Standard methods for steel products
B 208	Cast copper alloys
B 557	Cast aluminum and magnesium alloys
D 638	Plastics
D 3039	Resin-matrix composites
D 3379	High-modulus single-filament materials
E 8	Tension testing of metallic materials
E 21	Elevated-temperature tests, metallic materials
E 292	Time for rupture, notch tension
E 329	Concrete, steel, and bituminous materials
E 338	Sharp-notch, high-strength sheet materials
E 345	Metallic foil
E 517	Plastic strain ratio for sheet metal
E 607	Sharp-notch, cylindrical specimens
E 646	Tensile strain-hardening exponents
F 606	Fasteners
Compression testing	
A 256	Cast irons
C 773	Ceramic materials
D 695	Plastics
E 9	Metallic materials at room temperature
E 209	Metallic materials at elevated temperature
Elastic constant testing	
C 674	Ceramics
D 790	Plastics, flexural modulus
D 3379	High-modulus single-filament materials
E 111	Metals at room temperature
E 132	Metals, Poisson's ratio
Shear strength	
B 562	Aluminum rivets
D 732	Plastics
D 4255	Composites
E 229	Structural adhesives

Resilience is a property that defines a material's ability to absorb eleastic energy. For example, a spring used for shock absorption should probably have high resilience. The area under the elastic portion of the stress–strain curve

Table 2–2
Material properties obtained from tensile testing

Parameter	Description
Modulus of elasticity	Used to measure the relative stiffness of materials
Yield strength	Design stresses must be lower than the yield strength to ensure that a part does not fail by plastic deformation. Shear strength may be estimated from the yield strength.
Ultimate tensile strength	The ultimate tensile stress is the maximum stress observed in a tensile test. Necking begins when this value is reached.
Ultimate tensile strength/yield strength ratio	The ratio of the ultimate tensile strength to the yield strength provides an indication of the degree of work hardening that has occurred.
Percent elongation	Indication of material ductility and toughness
Percent reduction in area	Indication of material ductility and toughness
General shape of the curve	Area under the curve provides a relative indication of material toughness. Interstitial activity in the material may be observed. Relative levels of work hardening are assessed.

provides an indication of a material's resilience. For linear elastic (Hookean) materials, the following equation applies:

$$\text{Resilience} = \tfrac{1}{2}\sigma_{\text{el}}^2\varepsilon_{\text{el}} = \frac{\sigma_{\text{el}}^2}{2E}$$

where σ_{el} is the stress at the elastic limit, ε_{el} is the strain at the elastic limit, and E is the modulus of elasticity. For applications in which large energy absorption is required, suitable materials have a high elastic limit (or yield point) and a low elastic modulus.

Toughness is another parameter that may be determined from the stress–strain curve. Toughness is defined as the ability of a material to absorb energy before fracturing. A way to assess toughness is to calculate the total area under the stress–strain curve up to the point of fracture. The stress-strain curve in Figure 2–19 shows the measurement of resilience and toughness.

When designing mechanical structures, the design is governed by these strength parameters, accuracy of the strength data, statistical variations of the material from lot to lot or between different suppliers, and anticipated manufacturing variations. Often safety factors are used to account for these unknowns.

In high-performance applications, such as military, defense, or aerospace, high design efficiency (to save weight for example) requires that material properties and the stress state of the components be well understood, because only small safety factors are used. Civil engineering projects often use substantial safety factors in their design to account for unknown conditions.

Shear Properties

A common application of metals in engineering design is in shear loading. Bolts, rivets, and drive keys are loaded in such a manner as to cleave the material in half. The shear strength of a material is the stress at which a shear-loaded member will fail. A shear test can be performed in a tensile machine, with special grips replacing the tensile specimen (Figure 2–20). Double shear is the standard test for metals and ASTM D 732 is a suitable test for plastics.

It is not common to use polymers or ceramics as shear-loaded devices in machines, so their shear properties are seldom reported in handbooks. The application of the property of shear strength in machine design is obvious. It is this property that must be considered on shear-loaded fasteners and the like. Unfortunately, it is often difficult to find good tabulations in the

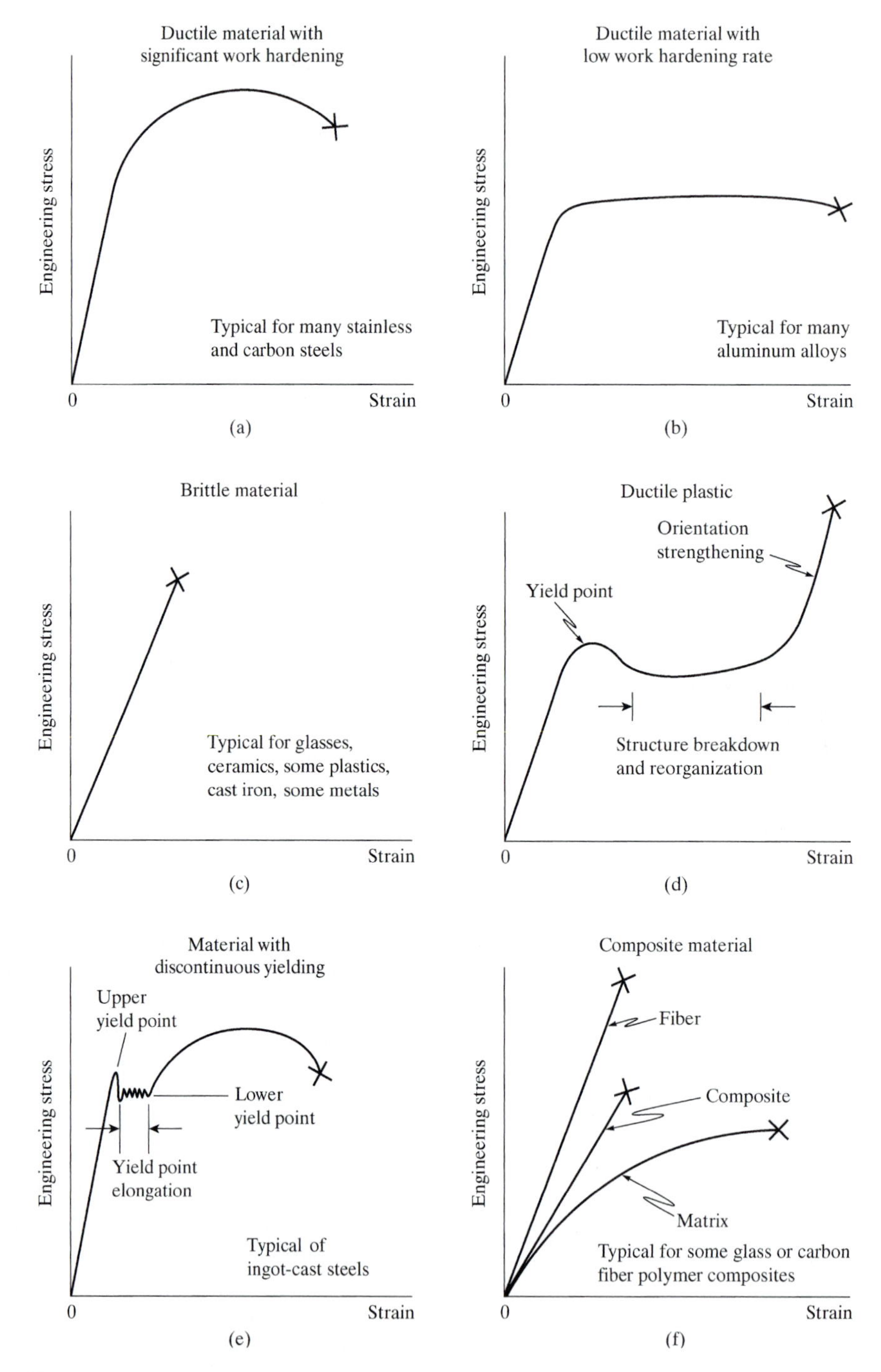

Figure 2–18
Examples of stress–strain response in various materials.

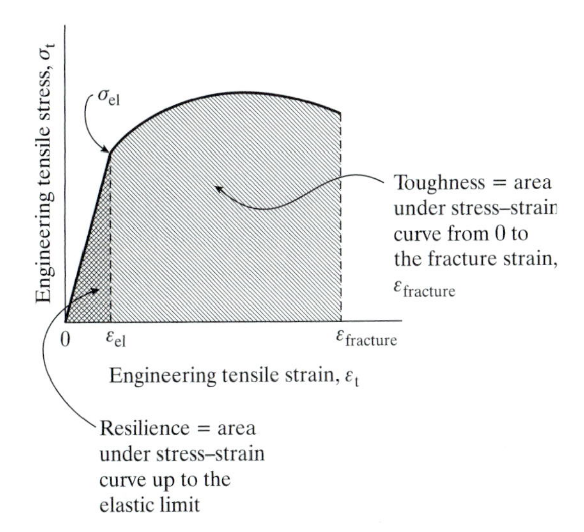
Engineering tensile stress, σ_t
σ_{el}
Toughness = area under stress–strain curve from 0 to the fracture strain, $\varepsilon_{fracture}$
0
ε_{el}
$\varepsilon_{fracture}$
Engineering tensile strain, ε_t
Resilience = area under stress–strain curve up to the elastic limit

Figure 2–19
Resilience and toughness of a material may be estimated from stress–strain data.

literature on shear strength. Based on the Huber– von Mises–Henkey distortion energy theory for ductile material failure, the following relationship is useful

$$\text{Shear yield strength} \approx 57.7\% \text{ of the tensile yield strength}$$

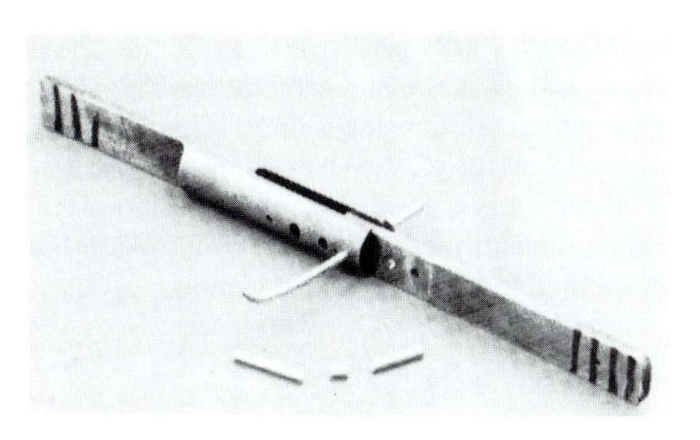

Figure 2–20
Shear test fixture for use in a tensile machine

Materials also behave elastically in shear (at stresses below the elastic limit). For example, torsion bars used for automotive suspensions behave as elastic springs that are energized by twisting (see Figure 2–8). The stresses from torsion loads are shear stresses. Figure 2–21 shows the distortion that develops in a bar that is loaded in torsion. The amount of twisting or angular distortion that occurs due to the application of a shear stress defines the shear strain. Shear stresses are proportionally related to the shear strains through an elastic constant called the *shear modulus*:

$$\sigma_s = G\varepsilon_s$$

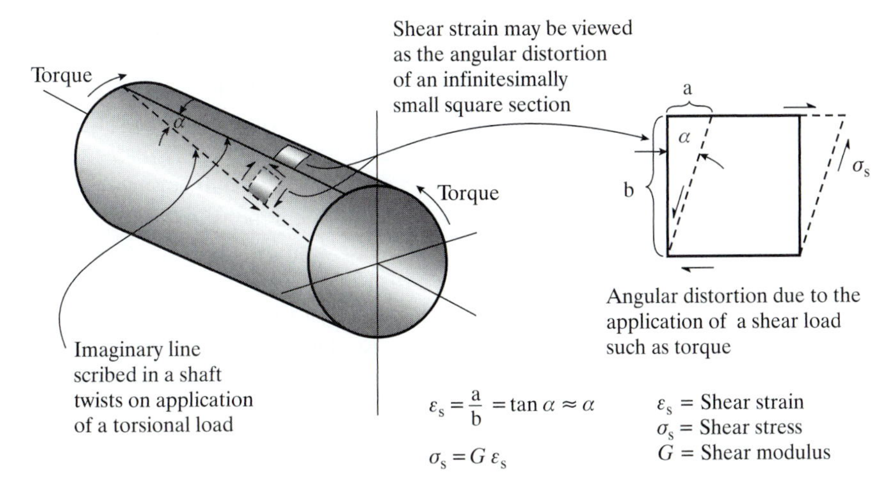
Torque
α
Shear strain may be viewed as the angular distortion of an infinitesimally small square section
Torque
a
b
α
σ_s
Angular distortion due to the application of a shear load such as torque
Imaginary line scribed in a shaft twists on application of a torsional load
$\varepsilon_s = \frac{a}{b} = \tan \alpha \approx \alpha$
$\sigma_s = G\,\varepsilon_s$
ε_s = Shear strain
σ_s = Shear stress
G = Shear modulus

Figure 2–21
Depiction of shear stress and strain in a round bar with torsional loads

where σ_s is the shear stress, ε_s is the shear strain, and G is the shear modulus. The shear stress–strain relationship is directly analogous to Hooke's law for elastic stresses in tension. The shear modulus is an important property for calculating the stiffness or rigidity of a torsion bar or another component loaded in shear.

Hardness Tests

Hardness is probably one of the most used selection factors. The hardness of materials is often equated with wear resistance and durability. This is not a completely accurate concept, but in steels it serves as a measure of abrasion resistance and strength.

There are probably 100 ways of measuring hardness (Table 2–3). In the early days of metallurgy, heat-treated steels were tested for hardness by filing an edge. If it did not file, it was hard. The hardness of ceramics and minerals was and still is measured by scratching the surface with different types of minerals. This is called the *Mohs* hardness test. Most present-day hardness tests consist of pushing a penetrator into the material and measuring the effects. Some of the most commonly used penetrators are shown in Figure 2–22. The loading mechanism varies with the various tests, as does the mechanism for measuring the effect of the indentation. In the *Brinell* hardness test, a 10-mm-diameter ball is pushed into the surface, and an optical measuring device is used to measure the diameter of the resulting indentation. This diameter is then used to calculate a Brinell hardness number (HB).

Rockwell hardness testers (Figure 2–23) use different loads and penetrators, and the depth of the indentation is measured by the machine and converted on a dial to a hardness number. Microhardness testers use much smaller penetration loads than Rockwell scales. Penetration

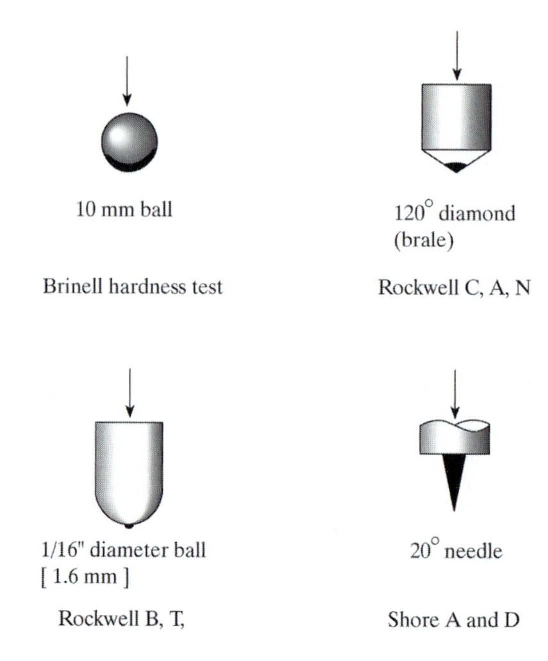

Figure 2–22
Typical penetrators used in hardness tests

Table 2–3
Some ASTM hardness test methods

ASTM Test	Description
A 370	Steel products
E 10	Brinell, metallic materials
B 299	Cemented carbides
B 578	Metallic plating
C 730	Glass
C 748	Carbon/graphite
C 849	Ceramics
D 530	Hard rubber products
D 785	Plastics, by Rockwell Indenter
D 1474	Organic coatings
D 2134	Plastics, by Sward Rocker
D 2240	Plastics/elastomers, by Durometer
D 2583	Plastics, by Barcol Impressor
E 18	Metals, Rockwell
E 92	Vickers
E 110	Portable testers
E 140	Conversion tables
E 384	Microhardness
E 448	Metals, Scleroscope

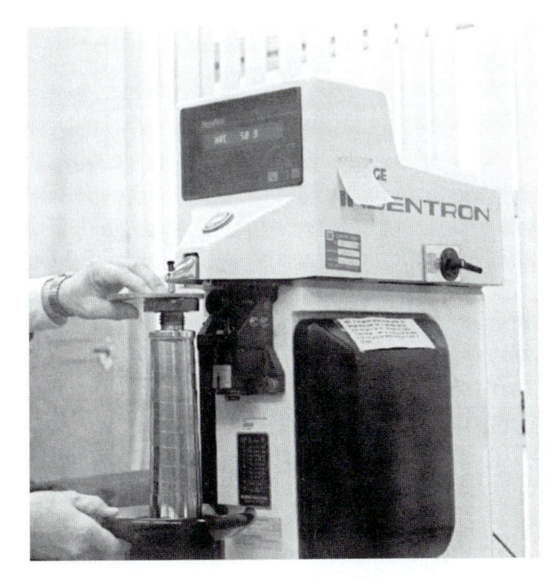

Figure 2–23
Hardness tester

Figure 2–24
Shore Durometer hardness tester. Needle on the bottom is the penetrator (ASTM D 2240).

loads are as low as 10 g on conventional microhardness testers and as low as a few nanonewtons on nanoindentor machines. Indentation measurements are measured with sophisticated optical and electronic techniques. These types of testers are used to measure the hardness of individual microconstituents in a material structure, the hardness of thin film coatings, and soft materials such as plastic films.

A common hardness test for polymers and elastomers (plastics that behave like rubber) is the Shore Durometer test. Hardness is measured by pushing a spring-loaded needle into the material (Figure 2–24).

There are pros and cons to each hardness test, but the hardness tests of most importance in material selection are those shown in Figure 2–25. They differ in penetrator, load, and applicability. Unfortunately, hardness numbers measured on one test cannot always be converted to a comparable hardness measured on another scale. An approximate conversion between some scales is shown in Figure 2–25. As a bare minimum, the designer should become familiar with the Rockwell B, C, and R tests; the Brinell test; and Shore Durometer tests. It is the designer's responsibility to specify the desired hardness on engineering drawings where the material can be hardened by heat treatment or fabrication. Of primary concern are hardenable metals and elastomers. The Shore Durometer hardness of elastomers can vary over a wide range. A typical drawing specification for an elastomer is as follows:

> *Material:* Polyurethane, Durometer 60, Shore A

There are various ways of specifying the hardness of metals; a system used by many industries is shown in Figure 2–26. The hardness of ceramics cannot usually be measured by any of the tests listed in Figure 2–25 except the Knoop (HK) or Vickers (HV). The scale is shown to stop at 1000 HK, but it continues well above this value. Absolute hardness is measured

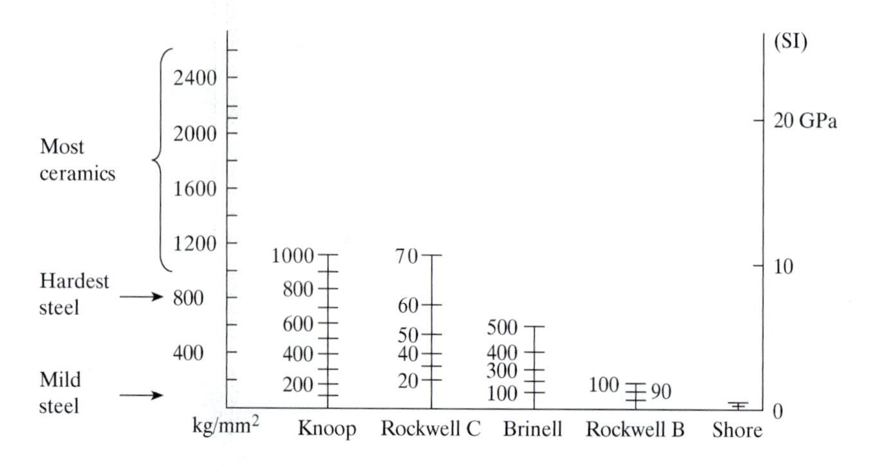

Hardness Test	Indenter	Load	Application
Knoop or Vickers[a]	Diamond	1 g to 2000 g	Microhardness of soft steels to ceramics
Brinell[b]	Ball	500 & 3000 kg	Soft steels & metals to 40 HRC*
Rockwell B[c]	Ball	100 kg	Soft steels & nonferrous metals
Rockwell T[c]	Ball	15, 30 & 45 kg	Thin soft metals
Rockwell N[c]	Diamond	15, 30 & 45 kg	Hard thin sheet metals
Rockwell A[d]	Diamond	50 kg	Cemented carbides
Rockwell R[e]	Ball	10 kg	Polymers
Shore Durometer[f]	Needle	Spring	Elastomers
Rockwell C[c]	Diamond	150 kg	Hardened metals (thick)

*HRC = Hardness Rockwell C
ASTM test methods: [a]=E 364, [b]=E 10, [c]=E 18, [d]=B 294, [e]=D 785, [f]=D 2240

Figure 2–25
Comparison of hardness tests

by a microhardness machine, and the measurement is expressed as a pressure in kilograms/square millimeter (kg/mm^2). These hardness values are obtained by dividing the penetration load by the projected area of the indentation. This way of expressing hardness is common in the scientific community, and it is usually expressed in gigapascals, GPa.

Description of Hardness Code

Specify hardness according to the code described in Figure 2–26. This code is in agreement with the method of designation used by the following standards organizations:

1. ASTM International (ASTM)

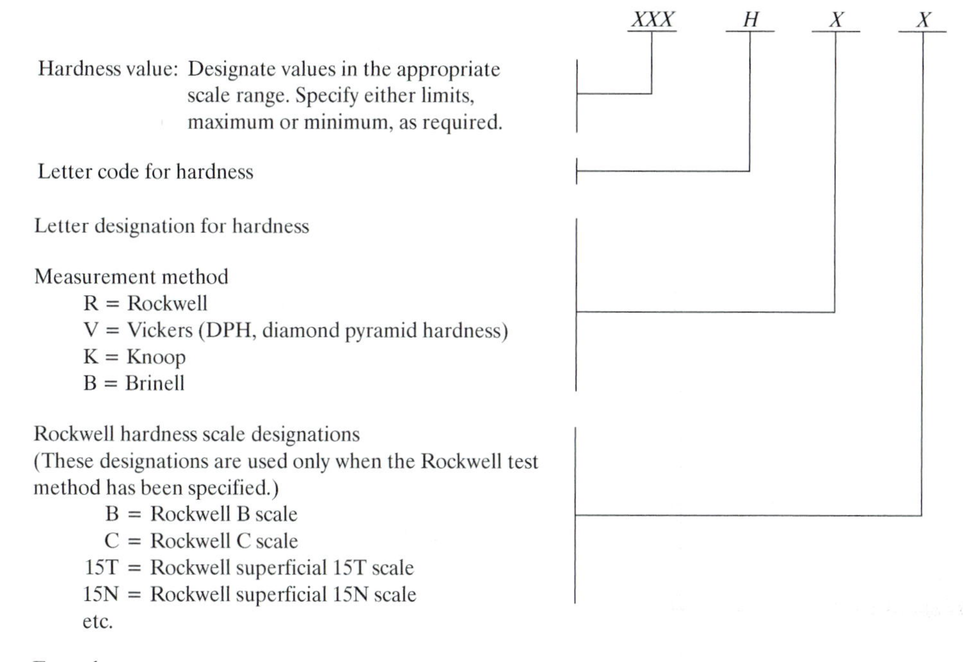

XXX H X X
Hardness value: Designate values in the appropriate scale range. Specify either limits, maximum or minimum, as required.
Letter code for hardness
Letter designation for hardness
Measurement method
R = Rockwell
V = Vickers (DPH, diamond pyramid hardness)
K = Knoop
B = Brinell
Rockwell hardness scale designations
(These designations are used only when the Rockwell test method has been specified.)
B = Rockwell B scale
C = Rockwell C scale
15T = Rockwell superficial 15T scale
15N = Rockwell superficial 15N scale
etc.

Example

1. 50–60 HRC means: a hardness value of 50 to 60 using the Rockwell C scale
2. 85 HR15T max means: a maximum hardness value of 85 using the Rockwell Superficial 15T scale
3. 185–240 HV_{100} means: a hardness value of 185–240 using the Vickers hardness tester and a test load of 100 kilogram-force, use HK for Knoop
4. 200 HB min means: a minimum hardness value of 200 using the Brinell hardness tester

Figure 2–26
Specification of hardness numbers for metals (ASTM E 10, E 384, and E 18). See ASTM E 140 for scale conversions.

2. American National Standards Institute (ANSI)
3. International Standards Organization (ISO)

Impact Tests

Impact strength is used to measure a material's ability to withstand shock loading. The classic definition of impact strength is the energy required to fracture a given volume of material. The units of this property are reported in the literature as foot-pounds (ft-lb) in the English system and joules (J) or joules/cubic centimeter (J/cm^3) in the metric system.

In metals and polymers the impact strength is most commonly measured by a pendulum-type impacting machine. Ceramics and brittle metals such as gray cast iron have negligible toughness, and consequently one will seldom find handbook data covering these materials. The most common types of impact tests used today are illustrated in Figure 2–27.

For most metals the specimen as shown in the illustration has a notch in it to prompt fracture in the desired spot. When the impact data

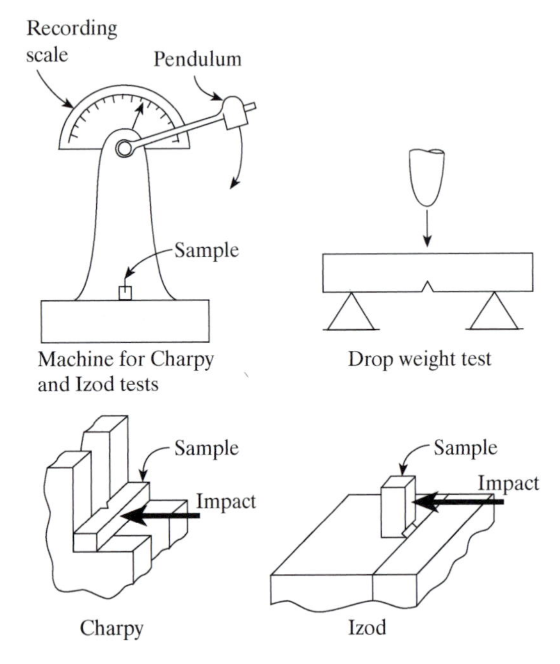

Figure 2–27
Common impact tests (ASTM E 23)

are reported as Charpy Vee, or notched Izod, it means that notched specimens were used. In data that do not indicate a notched specimen, chances are that the material is really quite brittle. Notched impact data cannot be compared with unnotched. The drop-weight type of test is a more recent addition to the toughness testing field. Its big advantage is that it uses relatively large specimens (as large as several hundred square inches in cross section). The use of large specimens was prompted by research reports that indicated that the data obtained on the small specimens used in standard Charpy and Izod tests do not confirm service characteristics on heavy sections. The standard notched and unnotched pendulum test specimens have a cross section only on the order of 0.2 in.^{2} The drop-weight tests (some types are called *dynamic tear*) have the major disadvantage of being quite expensive.

Impact strength can be affected by temperature. This is especially true for carbon steels and other metals with a body-centered cubic (BCC) or hexagonal crystal (HCP) structure. Metals with a face-centered cubic (FCC) structure (such as austenitic stainless steel, copper, and aluminum) strengthen slightly at low temperatures, but there is not a significant lowering of impact strength as can be the case with carbon steels. During World War II many ships developed catastrophic failures in cold North Atlantic waters. It was learned that the cause of the failures was a significant decrease of toughness in the ship steels at low water temperatures. If low temperatures are a possible service environment for a structure, it is advisable to look into the *nil ductility temperature* (NDT). This parameter, a product of Charpy Vee or dynamic tear tests, is defined as the temperature at which the toughness of the material drops below some predetermined value (usually 15 ft-lb; 21 J) for the Charpy Vee test. A typical impact-strength-versus-temperature curve showing the NDT is illustrated in Figure 2–28. If the NDT of a steel is 32°F (0°C), it should not be used for impact-loaded parts that may operate at this temperature or lower.

Impact tests are intended to measure the toughness of materials. A more recent test is a fracture toughness test (ASTM E 399). Essentially the test is conducted by putting a fatigue crack in a test bar and cyclic loading the cracked bar to produce crack propagation. The tougher the material, the more slowly the crack propagates through the test sample. The measured parameter is the critical stress intensity factor. Like impact strength, it is a measure of a material's toughness.

In summary of our discussion of the property of impact strength, it is important to compare data only from identical test methods. If temperature is a design factor, the effect of temperature on impact strength should be investigated. The most common application of impact strength is in comparing the toughness of steels at different hardness levels and in selecting polymeric materials for their relative resistance to shock.

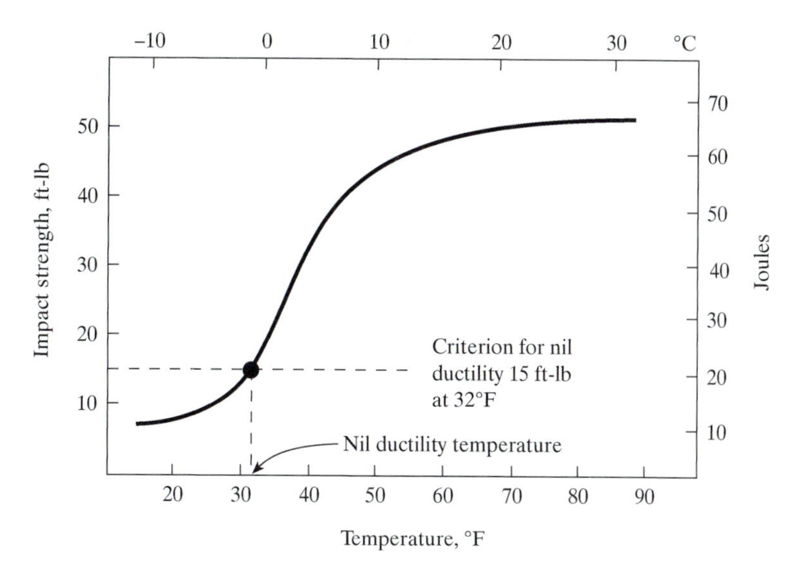

Figure 2–28
Use of impact test data to determine NDT

Long-Term Serviceability

Tensile tests provide information to be used in comparing the relative strengths of engineering materials. However, because this test usually takes only a few minutes to perform, it is not a good measure of how well a material will withstand dynamic loads or loading at elevated temperatures. Additional mechanical properties are used to rate materials for these types of service.

Endurance Limit The endurance limit or *fatigue strength* of a material is obtained by repeatedly loading a specimen at given stress levels until it fails. Any form of loading can be used, and the stress level is usually calculated or measured by strain gages. A bending fatigue setup is illustrated in Figure 2–29. The specimen is loaded until, for example, the maximum stress in the sample is 40 ksi (275 MPa). At this stress level it may fail in 10 cycles. These data are recorded, and the stress level is reduced to maybe 30 ksi (206 MPa). A specimen may break after 1000 stress cycles at this low stress level. This procedure is repeated until a stress level is determined below which failure does not occur. A test duration of 10 million stress cycles is usually considered infinite life. Obviously, this test is expensive. It involves a large number of samples and statistical evaluation. The end result, the endurance limit of a material, is an extremely important design property. This property, rather than allowable static stress, should be used in determining allowable operating stresses in components that are subjected to cyclic loading in service. As an example, the American Institute of Steel Construction recommends a design stress for a 60-ksi tensile strength (ASTM A 36) steel of 22 ksi (152 MPa) for static loading. In cyclic loading situations, the allowable stress is only about 13 ksi (89 MPa).

Cyclic loading significantly reduces the allowable stress that a material can withstand. If handbook data are not available on the endurance limit of a material under consideration for use, a percentage of the tensile strength can

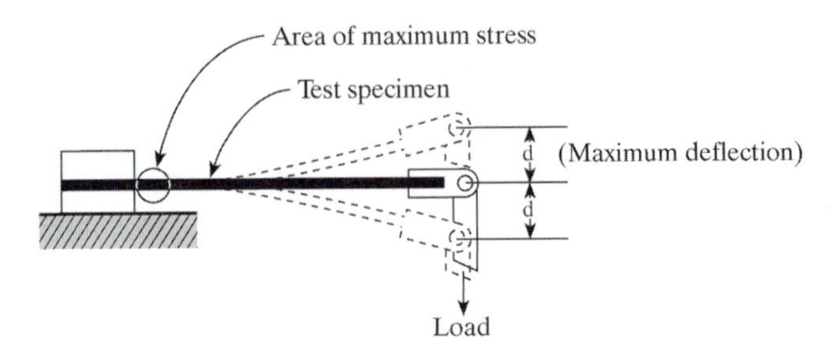

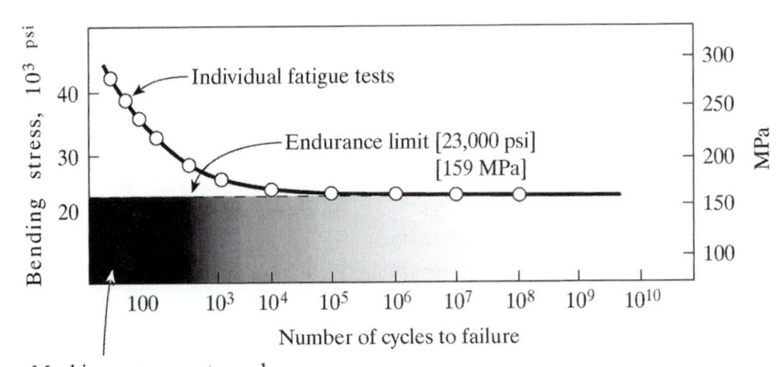

Figure 2–29
Use of an *S–N* (stress–number of cycles) curve to establish fatigue strength

be used. This percentage varies with different material systems, but for steels with a tensile strength less than 200 ksi (and no stress concentrations), the endurance limit can be approximated as 50% of the tensile strength. A value of 30% to 40% of the tensile strength is used for nonferrous metals.

Ceramics are usually not used for cyclic loading, and there are very few engineering data on endurance limits. Polymeric materials and composites are very subject to fatigue, and efforts should be taken to arrive at endurance-limit information if fail-safe design is desired.

Creep This property is used to rate the resistance of a material to plastic deformation under sustained load. For metals, *creep strength* is often expressed as the stress necessary to produce 0.1% strain in 1000 h. In polymers, a percent deformation at a given stress is often used. Creep data must also show the testing temperature. Typical creep testing is illustrated in Figure 2–30. Creep is not too important with most ferrous metals unless the operating temperature is above 800°F (426°C). Creep can be an important selection factor with low-melting-temperature metals and polymers. It is a principal cause of failure of fixtures and hangers in heat treat furnaces. In epoxy-bonded piping systems, the creep strength of the epoxy is often the weak link in the system. Polymeric bearings often develop excessive clearance owing to compressive creep. The solution to these types of problems is to use materials with good creep characteristics and to keep the operating stress below the creep strength.

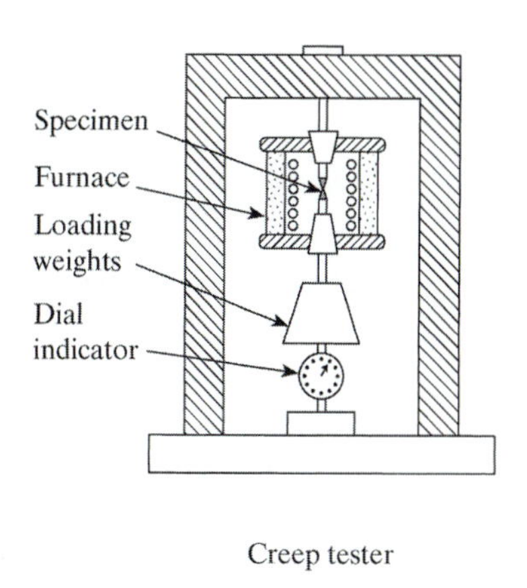

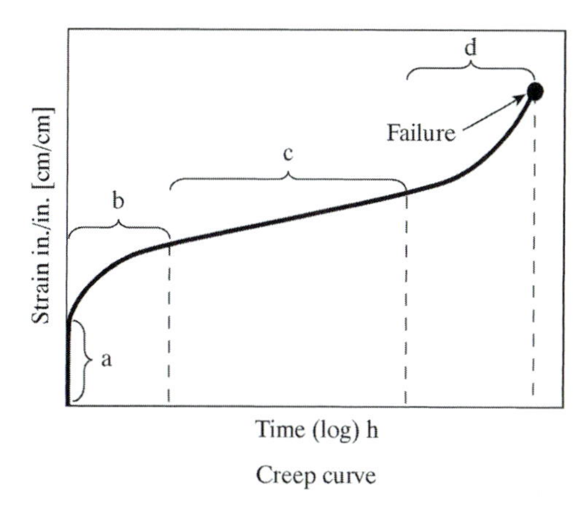

Figure 2–30
Creep testing of metals. In the schematic creep curve, region a is elastic strain, region b is creep at decreasing rate, region c is steady-state creep, and region d is creep at increasing rate (ASTM E 139).

Creep resistance should be considered for any part or structure that is subjected to sustained load in service. Plastic creep data are often expressed in units of percent strain (ASTM D 2990) in a period of time—for example, 10,000 h. These strain data can be used to predict dimensional changes in service.

Stress Rupture This property complements creep data; it shows the stress at which a part will fail under sustained load at elevated temperature. *Stress rupture strength* tests are usually conducted with dead-weight loading of the specimen, and the strain is not reported. A typical stress rupture curve is shown in Figure 2–31.

Plastics are not commonly used at elevated temperature, but when they are, their stress rupture characteristics should be assessed. Stress rupture data are important for metals or ceramics intended for high-temperature service. Reviewing the data in Figure 2–31, note how stress rupture data are used. If a part were to be used at 1000°F (538°C), it would last only 1000 h if the operating stress were 52 ksi (359 MPa). If the operating stress were lowered to 20 ksi (138 MPa), the expected service life would be in

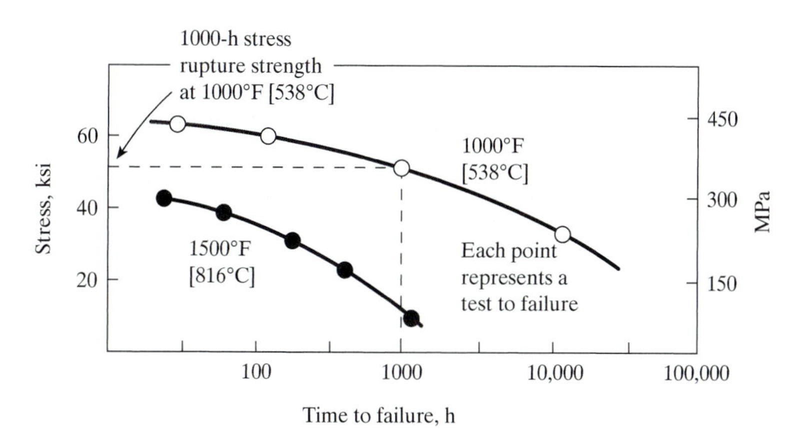

Figure 2–31
Typical stress rupture data (ASTM E 139)

excess of 10,000 h. At 1500°F (816°C), a stress level of 20 ksi would result in failure after less than 1000 h of service. Thus stress rupture data can be a useful tool in selecting materials and operating stresses.

Fracture Mechanics

The critical stress intensity factor derived in fracture mechanics tests can be a useful selection aid in designing for prevention of mechanical failure, and the average designer can use it without getting involved with the mathematical entanglements of hard-core fracture mechanics.

Fracture mechanics is based on analysis of the state of stress at the tip of a crack in a material. The most important material parameter used in fracture mechanics is the critical (c) stress intensity factor, K_c. It is measured in a universal testing machine similar to that used for tensile tests. The most popular test specimens (ASTM E 399) are heavy notched plates. A small fatigue crack is produced at the root of the notch on a separate fatigue tester. The precracked plate is then put in the tensile test machine, and the crack propagation characteristics of the material are measured. The net result of the test is the critical stress intensity factor, K_c.

This term is a function of the type of material, the condition of heat treat, the microstructure, and the residual stress conditions; and on a more micro scale it can depend on grain size, inclusion level, dislocation density, and even atomic bonds. It is an intrinsic property of the particular piece of material that was tested. How a material responds to the presence of the ultimate stress concentration—a fatigue crack—is a true measure of fracture toughness.

Critical stress intensity factors are tabulated in handbooks and in material selection tables. They can also be measured for any material of interest if a laboratory has a tensile test machine and a fatigue cracking machine. They can be used in design to determine the stress that will produce fracture when a part contains a known size crack:

$$S_c = \frac{K_c}{B\sqrt{\pi a}}$$

where S_c = critical stress that will produce fracture

K_c = critical stress intensity factor

B = a dimensionless parameter that is a function of the type of crack, where it is in a member, and the mode of loading of the member

a = crack size

This same equation can be rearranged to allow calculation of the size crack that will produce fracture (a_c) when the part is loaded to a particular stress level S_n:

$$a_c = \frac{1}{\pi}\left(\frac{K_c}{BS_n}\right)^2$$

The time that it takes for a crack to grow to the critical size that will cause fracture could also be calculated using related fracture mechanics concepts.

At the current time, fracture mechanics is used more for failure analysis than for design, but in more progressive industries it is used to calculate allowable operating stresses using the assumption that the material may contain a certain size crack. The crack size assumed is usually the largest defect that could go undetected in their nondestructive testing program (x-ray, ultrasonics, and so on). The uninitiated design engineer would have to invest a considerable amount of study to apply fracture mechanics in strength and fatigue calculations, but without knowing any more than the fact that the critical stress intensity factor is a measure of a material's fracture resistance, the average designer can look at tabulations such as those in Table 2–4 and use these data for material selection. The higher the critical stress intensity factor is, the

Table 2–4
Fracture toughness of various engineering materials

Material		Critical Stress Intensity Factors ksi[a] ($\sqrt{\text{in.}}$)	(MN[b] $m^{-3/2}$)
Metals:	Pure ductile metals (Cu, Ni, Ag, etc.)	90–320	100–350
	HY 130 steel	155–160	170–175
	Low-carbon steel	110–127	120–140
	18% Ni maraging steels	75–100	75–100
	PH stainless steels	60–130	66–143
	Titanium alloys	32–98	35–108
	4340 steel	40–86	44–78
	Medium-carbon steel	36–46	40–50
	Aluminum alloys	20–45	22–50
	Gray cast iron	5–18	6–20
	Cemented carbides	13–15	14–16
	Beryllium	3–5	4
Plastics:	Glass-reinforced epoxy	38–55	42–60
	(ABS)	2.7–3.6	3–4
	Acrylonitrile butadiene styrene		
	Polypropylene, nylon	2.7	3
	High-density polyethylene	1.8	2
	Polystyrene	1.8	2
	Polycarbonate	0.9–2.4	1–2.6
Ceramics:	Aluminum oxide, Al_2O_3	2.7–4.5	3–5
	Silicon nitride, Si_3N_4	3.6–4.5	4–5
	Silicon carbide, SiC	2.7	3
	Magnesium oxide, MgO	2.7	3
	Soda lime glass	0.6–0.7	0.7–0.8

Data from various sources.
[a]ksi, Kips per square inch where a kip is 1000 lb.
[b]MN, Meganewton

more resistant that material is to mechanical breakage. It is recommended that these data be used when evaluating candidate materials for high-performance structural components.

2.5 Manufacturing Considerations

Surface Finish

The surface characteristics of engineering materials often have a significant effect on serviceability and thus cannot be neglected in design. It is the designer's responsibility to specify the nature of the surface on machine components. Currently, more than 20 mathematical parameters are applied to the characterization of a surface, but the most commonly specified parameters are *roughness*, *waviness*, and lay (Figure 2–32). Lay is usually macroscopic and can be measured visually or with a simple loupe. Total surface profile, which is the net of the surface roughness and waviness, is usually measured by profilometer devices that electronically measure surface texture with a stylus not unlike a phonograph needle

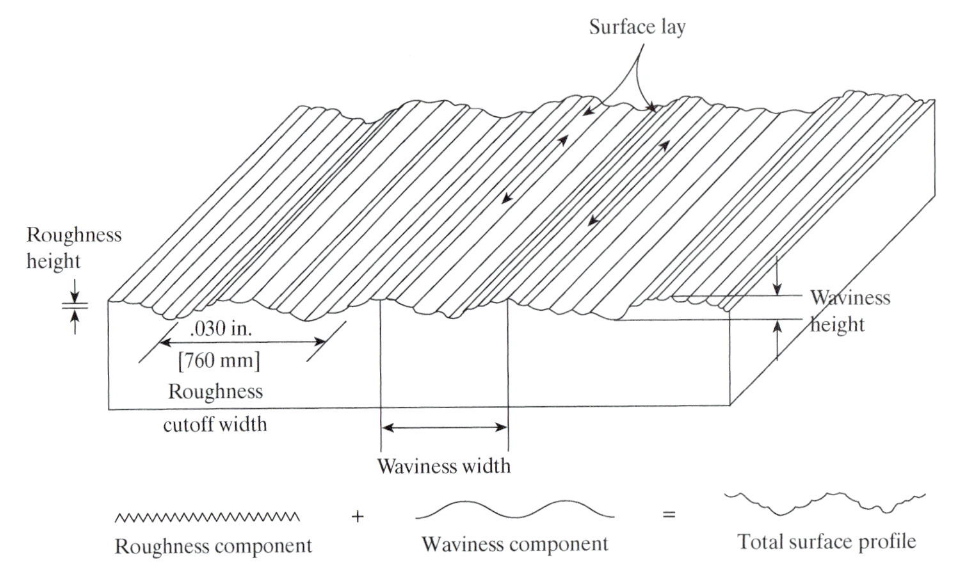

Figure 2–32
Components of surface microtopography

(Figure 2–33). The simplest profilometers yield only surface roughness data. The more sophisticated devices yield contour maps, single-line surface profiles, and roughness average data (Figure 2–34). Some profilometers have been made noncontact; surface feature measurements

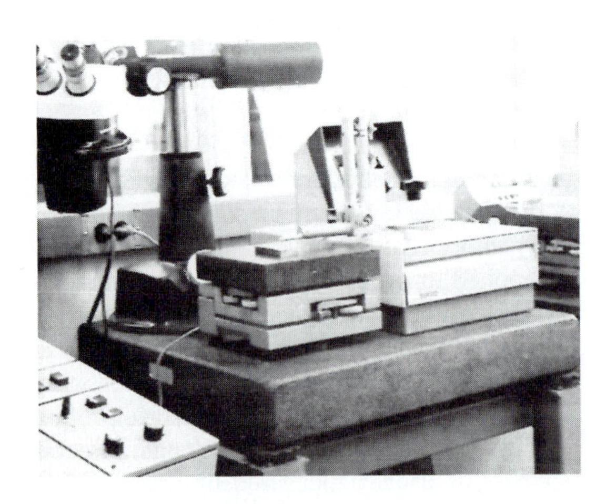

Figure 2–33
Stylus-type profilometer

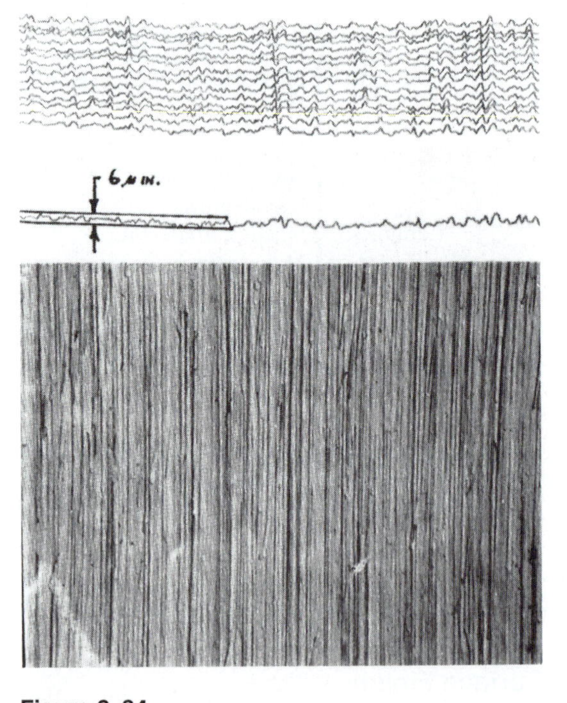

Figure 2–34
Profilometer map of a ground surface (top), single-line trace (middle), and photomicrograph at ×200 (bottom)

are made with laser beams and optical measuring systems. Surface roughness is usually expressed as the arithmetic average (Ra) of the peak-to-valley height of surface asperities in microinches (μin.). The SI units are micrometers (μm). Because a profilometer stylus has a finite radius (usually 2, 5, or 10 μm), it cannot reach the bottom of valleys of surface features; it cannot measure true depth. The Ra roughness is approximately 25% of the true peak-to-valley height on a uniform surface.

Atomic force miscroscopes, which came into widespread use in about 1990, are available with surface styli having radii as small as 100 Å compared to a few micrometers for conventional profilometers. These devices produce truer representations of surface features.

Most profilometers average the surface roughness over a set increment of stylus travel. This is called the *cutoff width*. All the surface peaks in this distance are integrated to yield a single roughness reading.

The parameters of waviness are waviness height and waviness width. Surface lay has no quantitative units, but there are symbols to indicate a desired lay. The American Society of Mechanical Engineers (ASME) has devised a system for describing surface texture on engineering drawings (Figure 2–35).

How does surface finish relate to selection and serviceability? There is an optimum range of surface roughness for parts intended for accurate fits, wear applications, release characteristics, and even nonfunctional surfaces. If a rotating shaft is too rough, it could abrade a soft bearing material. Coarse machine marks cause stress concentrations that can lead to fatigue failures. Surfaces with a finish that is "too good" unnecessarily increase fabrication costs.

Figure 2–36 presents some experience guidelines on finishes to call for on selected machine components. Figure 2–37 shows the surface finish ranges possible with various machining techniques. This illustration points out the inadequacy of using surface descriptors such as grind, turn, and drill. A ground surface can be as rough as 120 μin. (3 μm) Ra or as smooth as 4 μin. (0.1 μm) Ra. Surface finish requirements should always be expressed by using quantitative limits on at least surface roughness. The preferred technique in the United States is the use of the ASME system outlined in Figure 2–35. Some fabrication shops may not have equipment for quantitative measurement of surface texture. If this is the case, it will be very difficult to meet drawing specifications on surface texture. Visual examination is usually the only recourse, but few machinists have eyes sufficiently trained to discriminate different levels of surface roughness or other texture parameters. The photographs in Figure 2–38 show ground steel surfaces magnified seven times. This illustration simulates what a machinist would see using a ×7 loupe to examine his or her work. It is obvious that the machinist may be able to tell if a surface roughness is less than 8 μin. (0.2 μm) Ra because grinding scratches start to disappear, but it would be difficult to tell the difference between a 63- and 32-μin. (1.6- and 0.8-μm) Ra surface. The point is that visual examination of surface finish is a poor substitute for instrument measurement.

Size and Shape Considerations

A primary material selection factor used by designers is material availability in the size and shape required for the part under design. A mechanical-property study may show that type 317 stainless steel is the best material for a support column under design. If the job requires a 10-ft-long, 4-in.-by-6-in. channel and this shape is not available in small quantities from a warehouse, this material cannot be used. Similarly, if a material is required for an accurate machine baseplate, a primary selection factor may be the availability of a material with good flatness tolerances.

Camber, an edge bow in sheet or strip, is important in using sheet and strip materials. If a

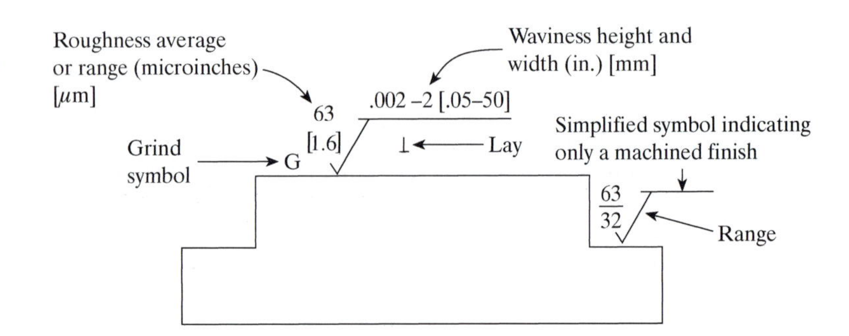

Figure 2–35
Specification of surface texture (American Standard Surface Texture ASME Y14.36M 1996)

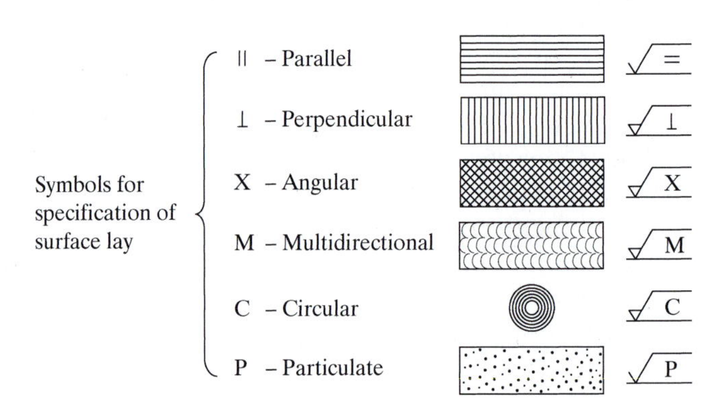

material is available in the desired thickness and overall size but comes in with excessive camber, it may not be usable for the intended application.

Stock tolerances are important if areas of a part are to be used in the as-received condition. Some material shapes are made with a tolerance of plus 1% of the nominal thickness, minus nothing. If a part requires a minimum thickness of, for example, 0.500 in., the thickness tolerances should be investigated on candidate materials for a part under design. If a material with a nominal thickness of 0.500 in. is ordered and it comes in with a thickness of 0.490 in., it may be useless.

Occasionally, a particular material form will lead to dimensional problems. Hot-rolled steels have a loose "flaky" scale, and the surface finish is usually too poor to use without machining. If the designer did not consider this when the material was specified, it may make the purchased material unusable or it may significantly add to machining costs.

Castings may come from the foundry with gouges left from gate removal; sometimes flash or mold pickup are not removed. The designer can control these factors by drawing notation calling for sandblasted, flash-free castings free of surface defects.

Extruded shapes are usually bowed and twisted when they are made. If long lengths are required, the designer should use materials that are available as straightened extrusions.

The preceding are some of the size and shape considerations that are part of the dimensional properties of materials. How should a designer deal with these factors? A checklist on dimensional property requirements should be

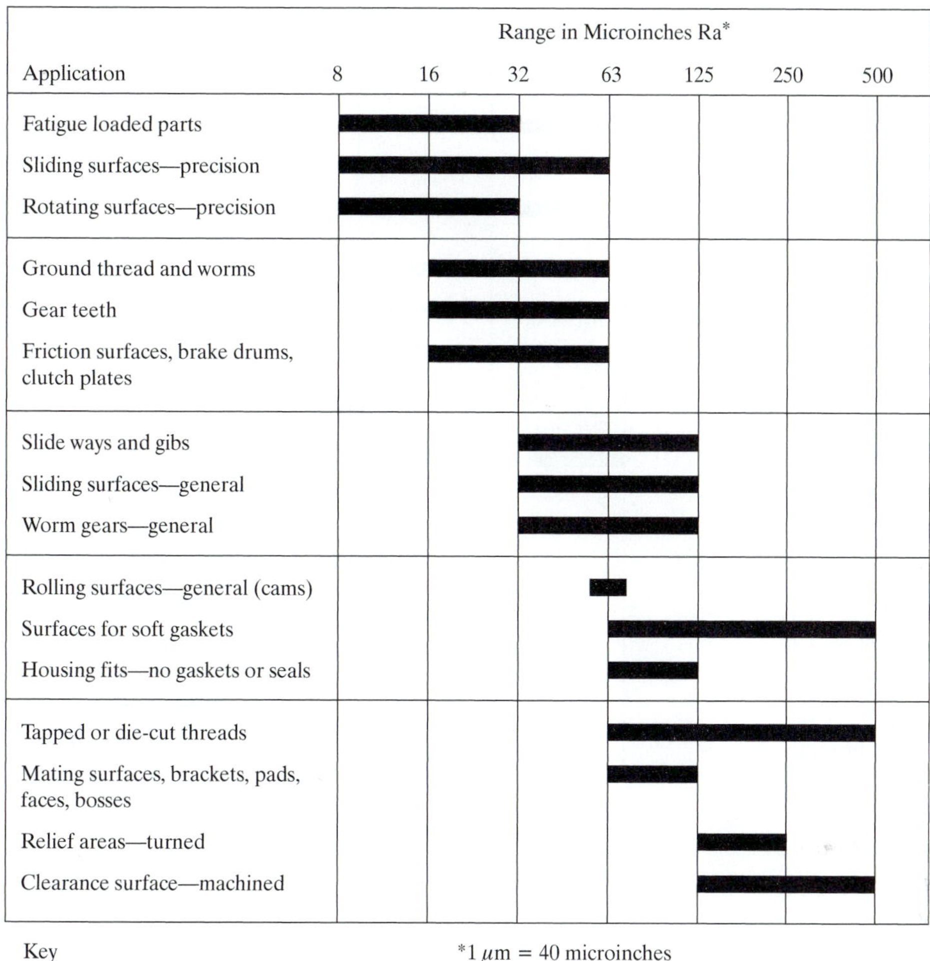

The ranges shown are ordinarily considered desirable for the conditions listed.
Higher or lower values may be obtained.

Figure 2–36
Recommended surface roughness for machine parts

mentally reviewed immediately after the part is designed. The checklist should contain the following factors:

1. Surface texture, lay, roughness, and like requirements
2. Flatness requirements
3. Allowable surface defects
4. Stock dimensional tolerances
5. Camber
6. Surface cleanliness
7. Edge tolerances
8. Bow tolerances

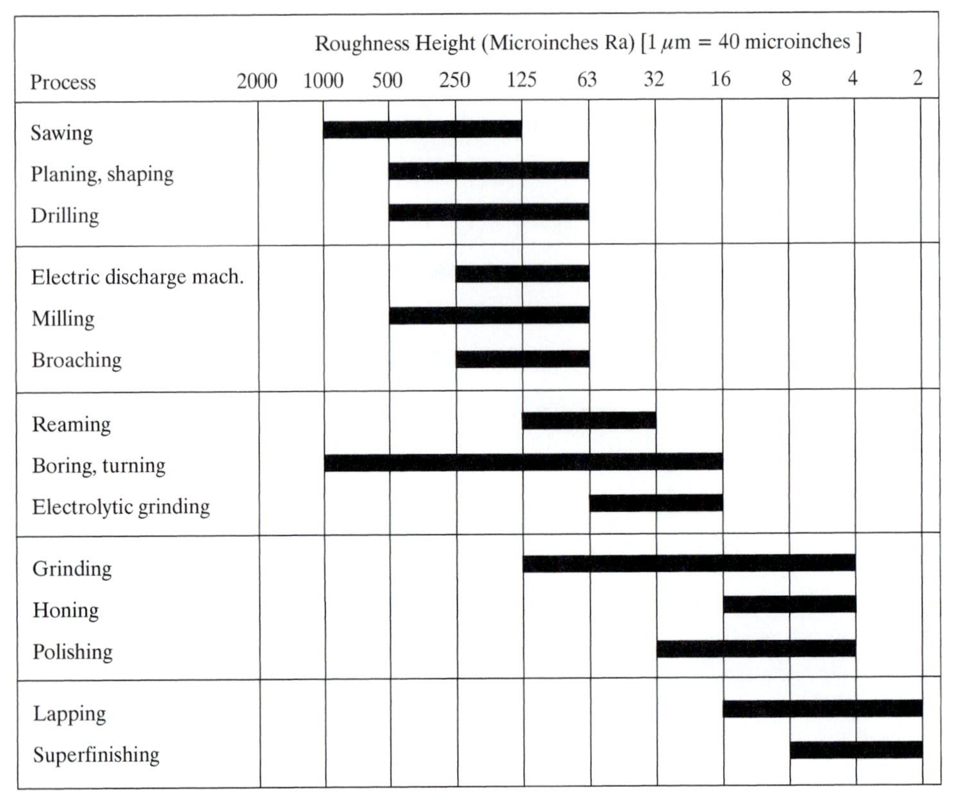

Key

Average application

The ranges shown are typical of the processes listed.
Higher or lower values may be obtained under special conditions.

Figure 2–37
Surface roughness produced by various machining techniques

9. Surface reflectance

10. Should prefinished material be used?

Other factors could be listed, but the foregoing are some of the more important dimensional considerations. The designer should establish which of these factors, if any, will affect part serviceability. If it is clear that some of these factors are important, then steps should be taken to specify dimensional requirements on engineering drawings and purchasing specifications.

2.6 Property Information

All remaining chapters will present additional property information on engineering materials, but undoubtedly this book will not have all of the property information that a designer may need for property comparison. The references listed at the end of the chapter are those frequently used by mechanical designers. In addition to these references, material property databases can be accessed by designers on personal computers (PCs)

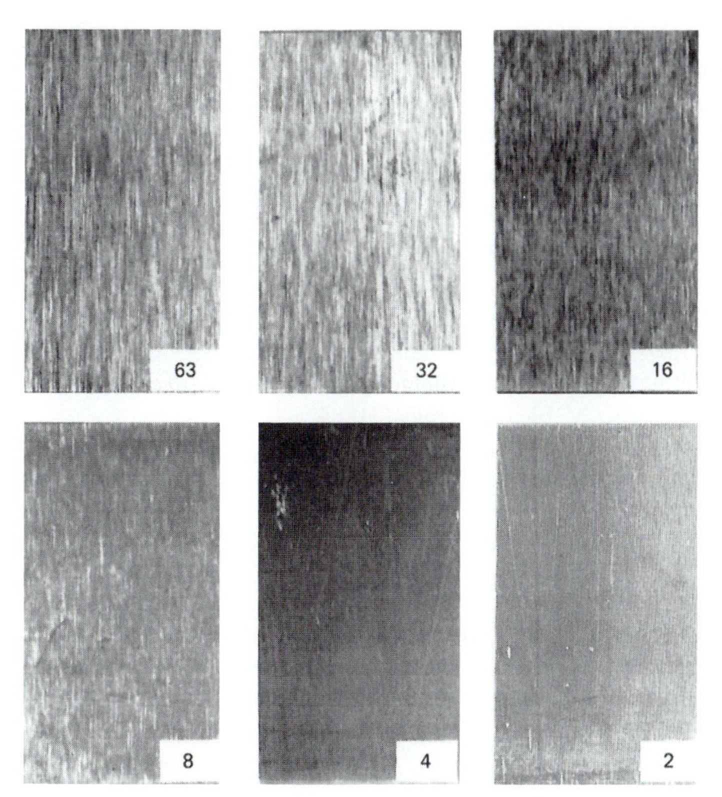

Figure 2–38
Photographs of ground steel surfaces (×7). The number in the lower corner represents the average surface roughness in microinches (divide by 39.37 to get micrometers).

and terminals and through national networks. The PC databases can be purchased as a set of flexible disks or compact disks that are updated periodically, and they contain all sorts of property information on metals, plastics, and ceramics. The large national databases contain a wide array of property information, and they are usually subscribed to by libraries or information centers. It is current practice in the plastic industry to use floppy disks instead of catalogs for material information on plastics. These disks can often be obtained free from plastic suppliers, but suppliers usually include only their materials. This can be addressed by getting disks from a number of companies.

In summary, practicing materials engineering means evaluating numerous properties of many materials. Someday a knowledge-based computer system may do all material selection, but for the foreseeable future we recommend understanding the properties that we have discussed in this chapter and using the references that we cite to obtain property information. List your use requirements and use property information to make the best fit of a material to operational needs. There will be more on this in the chapters on the selection process.

Glossary of Terms

Chemical Properties

Additive: In plastics, a substance added to alter properties such as flammability, UV resistance, etc.

Amorphous: Noncrystalline—absence of long-range order.

Anisotropy: Directional properties in a material (e.g., yield strength aligned with the rolling direction is different from the yield strength transverse to the rolling direction).

Composition: The elemental or chemical components that make up a material, and the relative proportions of these components (units—volume or weight %).

Corrosion resistance: The ability of a material to resist deterioration by chemical or electrochemical reaction with its environment (units—mils/year + others).

Crystallinity: In plastics, the degree to which molecules are arranged in a spacially repeating manner.

Crystal structure: The ordered, repeating arrangement of atoms or molecules in a material (units—HCP, FCC, etc.).

Environmental resistance: In plastics, the ability of a material to be unaffected in a particular environment.

Equiaxed: Grains that are the same size in all directions (not elongated, etc.).

Filler: In plastics, a substance added to alter properties such as density, strength, etc.

Flammability: In plastics, the extent to which a material will support combustion.

Grain size: The average diameter of a grain or crystallite in a material as measured by cross-section microscopy or other technique (units—μm).

Inclusion: A foreign body in the microstructure of a material.

Matrix: In composites, the base material that is reinforced.

Matrix/reinforcement bond: In composites, how effectively or completely a matrix adheres to a reinforcement and the relative strength of that adhesion.

Microstructure: The structure of polished and etched materials as revealed by microscope magnifications greater than ten diameters; structure includes the phases present, the morphology of the phases, and their volume fractions.

Molecular weight: In plastics, average number of repeating molecules in a molecular chain.

Orientation: The alignment of phases, molecules, grains, etc., in a preferred direction in a material.

Passivity: Ability of a material's surface to remain unaffected in an environment that is capable of chemical attack.

Phase: A homogeneous and distinct component of a microstructure.

Porosity: The number or percentage of voids in a solid (units—sometimes % theoretical density).

Reinforcement: In composites, a substance that is added to strengthen a neat material—usually fibers or woven fabric.

Segregation: A gradient in chemical composition within a material (e.g., the surface is purer than the center).

Stereospecificity: A tendency for polymers and molecular materials to form with an ordered, spatial, three-dimensional arrangement of monomer molecules.

Volume fraction: How much of a material is composed of a phase or secondary microconstituent (unit—%, as in % ferrite, or % acrylonitrile, or % carbon fiber).

Physical Properties

Curie point: The temperature at which ferromagnetic materials can no longer be magnetized by outside forces [ASTM E 1033; units—°F (°C)].

Density: The mass of a material per unit volume [units—(kg/m^3)].

Dielectric strength: The highest potential difference (voltage) that an insulating material of given thickness can withstand for a specified time without occurrence of electrical breakdown through its bulk (ASTM D 150).

Electrical conductivity: The ability of a material to allow electron flow when there is a voltage drop or other driving force to make electron flow occur.

Electrical resistivity: The electrical resistance of a material per unit length and cross-sectional area or per unit length and unit weight (ASTM D 257; units—Ω-m).

Emissivity: The ability of a material to absorb/emit infrared radiation [units—A number from 0 to 1 with 1 being a perfect absorber (black body)].

Ferromagnetism: The characteristic that allows a material to be attracted by a magnetic field or permanent magnet.

Glass transition temperature: In plastics, the temperature at which a material stops behaving as a rigid solid and behaves like leather or rubber [units—°F (°C)].

Heat distortion temperature: The temperature at which a polymer under a specified load shows a specified amount of deflection [ASTM D 1637; units—°F (°C)].

Magnetically hard: The ability of a ferromagnetic material to remain magnetized after removal of an externally applied magnetic field—a good permanent magnet.

Magnetically soft: The ability of a ferromagnetic material to quickly and easily return to a nonmagnetized condition after removal of an external magnetic field—a material that is easily magnetized and demagnetized.

Magnetic permeability: The ratio of the magnetic induction (B) of a material to the magnetic field strength (H) that produces the magnetic induction.

Melting point: The point at which a material liquefies on heating or solidifies on cooling. Some materials have a melting range rather than a single melting point [ASTM E 794; units—°F (°C)].

Permeability: The passage of gas, liquid, or solid through a material without physically or chemically affecting it.

Poisson's ratio: The absolute value of the ratio of the transverse strain to the corresponding axial strain in a body subjected to uniaxial stress (ASTM E 132; unitless).

Refractive index: The ratio of the velocity of light in a vacuum to its velocity in another material (unitless).

Solidification shrinkage: The relative dimension change (shrink) that a material undergoes in transforming from a molten to a solid condition [units—in./in. (m/m)].

Specific gravity: The ratio of the mass or weight of a solid or liquid to the mass or weight of an equal volume of water (ASTM D 792; unitless).

Specific heat: The ratio of the amount of heat required to raise the temperature of a unit mass of a substance 1° (Celsius or Fahrenheit) to the heat required to raise the same mass of water 1° [ASTM D 2766; units—BTU/lb/°F (J/kg °C)].

Thermal conductivity: The rate of heat flow per unit time in a homogeneous material under steady-state conditions, per unit area, per unit temperature gradient in a direction perpendicular to area [ASTM C 1095; units—BTU/hr/ft^2/°F/in. (W/in. · °C)].

Thermal expansion (linear coefficient of): The amount that a material elongates when heated. The coefficient is expressed as a unit increase in length per unit rise in temperature within a specified temperature range [ASTM E 831; units—in./in. °F (m/m °C)].

Volume resistivity: The electrical resistance between opposite faces of a cube of material, usually 1 cm on an edge (units—Ω-cm).

Water absorption: The amount of weight gain experienced in a polymer after immersion in water for a specified length of time under a controlled environment (ASTM D 570; units—% in 24 h).

Mechanical Properties

Compressive strength: The maximum compressive stress that a material is capable of withstanding; based on original area (Table 2–1) [units—psi (MPa)].

Compressive yield strength: The stress in compression at which a material exhibits a specified deviation from the proportionality of stress and strain [units—psi (MPa)].

Creep: Time-dependent permanent strain under stress (units—%/time).

Creep strength: The constant nominal stress that will cause a specified quantity of creep in a given time at constant temperature [units—psi (MPa)].

Ductility: The ability of a material to plastically deform without fracture.

Endurance limit: The maximum stress below which a material can theoretically endure an infinite number of stress cycles [units—psi (MPa)].

Fatigue strength: Same as endurance strength.

Flexural modulus: The stiffness of a material determined by a bending test (flex). (Usually the specimen is a simply supported beam loaded in the center.) [units—psi (MPa)].

Flexural strength: The outer fiber stress developed when a material is loaded as a simply supported beam and deflected to a certain value of strain (Table 2–1) [units—psi (MPa)].

Formability: The ease of shaping metals as measured by tests that stretch, bend, draw, or plastically deform in some other way.

Fracture toughness: A numerical parameter that rates a material's resistance to propagation of a preexisting crack or defect [units—ksi $\sqrt{\text{in.}}$ ($\text{MN} \cdot \text{m}^{3/2}$)].

Hardness: The resistance of a material to plastic deformation, usually by indentation [units—many scales: force/unit area of indentation (kg/mm^2)].

Impact strength: The amount of energy required to fracture a given volume of material [units—ft-lb (J)].

Malleability: The ability of a material to plastically deform in compression without fracture.

Modulus of elasticity: The ratio of stress to strain in a material loaded within its elastic range; a measure of rigidity (Table 2–1) [units—psi (GPa)].

Notch sensitivity: The ability of a material to resist fracture at sharp reentrant corners and geometric stress concentrations.

Percent elongation: In tensile testing, the increase in the gage length measured after the specimen fractures within the gage length (units—%).

Percent reduction in area: In tensile testing, the difference, expressed as a percentage of original area, between the original cross-sectional area of a tensile test specimen and the minimum cross-sectional area measured after fracture (units—%).

Proportional limit: In tensile testing, the stress at which a material starts to plastically (rather than elastically) deform [units—psi (MPa)].

Shear strength: The stress required to fracture a shape in a cross-sectional plane that is parallel to the force application. When a bolt or rivet is sheared in service, the fracture plane is parallel to the direction of motion of the plates (or whatever the bolts were holding) that had to move to produce the shear fracture (Table 2–1) [units—psi (MPa)].

Stress rupture strength: The nominal stress at fracture in a tension test at constant load and constant temperature (usually elevated) [units—psi (MPa)].

Tensile strength (ultimate strength): The ratio of the maximum load in a tension test to the original cross-sectional area of the test bar (Table 2–1) [units—psi (MPa)].

Toughness: The energy required to fracture a given volume of material [units—ft-lb/given sample size (J/m^3)].

Yield strength: The stress at which a material exhibits a specified deviation from proportionality of stress and strain (Table 2–1) [units—psi (MPa)].

Dimensional Properties

Camber: Deviation from edge straightness; usually the maximum deviation of an edge from a straight line of given length.

Lay: The direction of a predominating surface pattern, usually after a machine operation.

Out of flat: The deviation of a surface from a flat plane, usually over a macroscopic area.

Roughness: Relatively finely spaced surface irregularities, the height, width, and direction of which establish a definite surface pattern.

Surface finish: The microscopic and macroscopic characteristics that describe a surface.

Waviness: A wavelike variation from a perfect surface; generally much wider in spacing and higher in amplitude than surface roughness.

Summary

The purpose of this chapter is to establish the language of engineering materials: the properties that belong to materials and how these properties are used to select materials. The essence of material selection is weighing properties of one material over another. The important concepts that should be understood about material properties include the following:

- The difference between mechanical and physical properties.
- The most used mechanical and physical properties should be well understood—what they mean and how they are used. Do not confuse elastic modulus with strength or toughness.
- Critical mechanical properties that apply to all materials include tensile strength, yield strength, percent elongation, modulus of elasticity, hardness, and impact strength.
- Critical mechanical properties that apply to plastics include heat distortion temperature and compressive strength.
- One of the most important mechanical properties for ceramics is toughness as measured by resistance to crack propagation (critical stress intensity factor).
- Physical properties can be more important than all other properties, depending on the application. They should be scanned for applicability and importance.
- Knowing the chemical composition of any engineering material is fundamental to understanding that material. It may not be necessary to know the percentage of each element in the material, but the user should know the basic components and what family of materials a material comes from.
- Surface texture is often a critical use property of an engineering material, and material users should understand how to specify and obtain appropriate surface finishes.
- The available shapes and sizes of materials are often a key selection factor (for example, when the time required to obtain a material is crucial).
- The properties of composites depend on the nature of the reinforcement and the matrix; each one is unique. Generic property data on composites may not be applicable.

There are probably other important statements that we could make about the use of property information, but these reflect the situation in machine design. When a designer has the task of filling in the material blank on a drawing, he or she will undoubtedly make a material selection by weighing the properties of several candidate materials. This chapter has defined most of the properties that will be important for most design situations. The remaining chapters will introduce more properties, and all of these properties will be referenced in the index.

Critical Concepts

- There are different categories of material properties. Some can be altered by processing; some cannot.
- Material selection is usually based on comparison of standard properties.
- There are standard tests for measuring most properties.
- There is a significant difference between stiffness, toughness, and strength.

Terms You Should Remember

composition
crystal structure
density
thermal conductivity
thermal expansion
heat distortion temperature
water absorption
ferromagnetism
tensile strength
yield strength
percent reduction in area
percent elongation
modulus of elasticity
compressive strength
flexural strength
shear strength
hardness
impact strength
fatigue strength
creep strength
stress rupture strength
roughness
waviness
camber
microtopography

Case History

Stiffness Driven Designs

The stiffness or rigidity of a structural component is primarily controlled by the elastic and shear modulus of the material and the cross-sectional shape. Making a component from a stronger or harder material may not make the component stiffer unless the elastic or shear modulus of the material is increased. Certain cross-sectional shapes may be more effective for specific loading situations as shown below.

Load	Efficient shape
Torsion	**Hollow tube**
Bending	**I-Beam**
Axial compression	**Square or round tube**
Axial tension	**Shape not critical**

As an example, a windshield wiper design was developed for a new minivan with a particularly large windshield. The linkages or link arms used to actuate the windshield wiper blades needed to be long and slender but also had to sustain compression and bending forces. Additionally, the linkages had to be resistant to buckling yet be low mass to minimize inertia effects (for example, cycling a windshield wiper at its fastest speed could actually rock the entire vehicle). Thin-walled round steel tubes were chosen to optimize buckling resistance while minimizing component weight and cost. Round steel tubing is cost efficient to fabricate by continuously shaping sheet steel into a tube and seam welding. Special ends were cost-effectively crimped onto the tubes. This solution was found to be strong and more cost effective (finished part cost) than a lighter weight material like aluminum.

Questions

Section 2.1

1. Explain the difference between physical properties and mechanical properties and list five of each.
2. Your one-year-old car shows rust spots. What properties are deficient?
3. What kind of a material property is gloss?

Section 2.2

4. Describe how the composition of a tool steel, a plastic, and a ceramic is specified.
5. How do you check the composition of a plastic, metal, or ceramic?
6. How can you tell if a material is crystalline?

Section 2.3

7. Calculate the heat flow per unit area through the wall of a steel furnace with a thickness of 1 in. when the temperature gradient across the one wall is 10 °F. The thermal conductivity of the steel is 26 Btu · ft/h · ft^2 · °F.
8. Calculate the resistance of 100 ft of 12-gage (0.080-in.-diameter) DHP copper wire with a resistivity of 1.7×10^{-8} Ω-cm.
9. Calculate the final size of the bore of a steel bushing, ½-in. internal diameter, ¼-in. wall, and 1 inch long, when the bushing is raised in temperature 100°F. The coefficient of thermal expansion of steel is 6×10^{-6} in./in.°F.
10. Explain the difference between specific gravity and density. What is the weight of a 2-in. cube of a plastic that has a specific gravity of 1.7?
11. What is the Curie point of steel? How is it used?

Section 2.4

12. Is Poisson's ratio a physical property or a mechanical property? Why?
13. Explain how the information from a tensile test can be used in material selection.
14. Calculate the strain that will occur in a ½-in.-diameter steel eyebolt that is 11 in. long when it carries a load of 2000 lb. The modulus of elasticity of steel is 30×10^6 psi.
15. Compare the hardness of cemented carbide (2100 kg/mm^2) and a hardened steel file with a Rockwell hardness of 62 HRC on the same scale.
16. A steel has a tensile strength of 110,000 psi. What is a reasonable estimate of its fatigue strength?
17. When would you use creep strength as a basis for a structural design?
18. Calculate the tensile strength of a ½-in.-diameter bar of aluminum that broke when a 1-ton load was applied in tension.
19. Describe the difference between elastic and plastic deformation and give an example of each.
20. How does hardness relate to toughness?
21. A metal has a hardness of 30 HRC. What is its Brinell hardness?
22. What is the stretch in a 1-in.-diameter steel rod that is carrying a 10,000-psi axial load? Show calculations.
23. What mechanical properties determine the formability of a metal?
24. A surface has a finish made up of grinding scratches with a peak-to-valley height of 20 μin. What is the average roughness, Ra, of this surface?

25. Estimate the relative costs to produce a ground and polished surface (5 μin. Ra) and to produce a planed surface with a roughness of 63 μin. on a steel plate that has finished dimensions of ¾ × 24 × 24 in. Estimate labor hours.
26. Cite an example where surface roughness can affect service life. Explain.
27. A metal tension specimen of initial diameter 0.505 in. fractures at a load of 10,000 lb. Final fracture diameter is 0.376 in. Calculate (a) percent reduction in area, (b) true stress at fracture, and (c) engineering stress at fracture.

References for Property Data

Metals

Alloy Digest. ASM International, Materials Park, OH: 1997.

Aluminum Standards and Data. Washington, DC: Aluminum Association, 1993.

Materials Selector—Materials Engineering. Cleveland, OH: Penton Publishing, 1994.

Metals Handbook, 10th ed. *Volume 1, Properties and Selection; Irons, Steels and High Performance Alloys* (1990); *Volume 2, Properties and Selection: Nonferrous Alloys and Special Purpose Materials*. Materials Park, OH: ASM International, 1991.

Standards Handbook—Cast Copper, Bronze and Copper Alloy Products. New York: Copper Development Association, 1978.

Standards Handbook—Wrought Copper Alloy Mill Products. New York: Copper Development Association, 1973.

Steel Products Manual. Washington, DC: American Iron and Steel Institute. (There is a publication on each type of steel product, i.e., tool steel, sheet steel, tin mill, etc.)

Woldman's Engineering Alloys, 8th ed. Materials Park, OH: ASM International, 1994. (Also available in CD-ROM)

Plastics

Engineering Plastics. Volume 2, Engineered Materials Handbook. Metals Park, OH: ASM International, 1988.

Juran, R., Ed. *Plastics Encyclopedia*, Volume 66, Number 11. New York: McGraw-Hill Book Co., 1993.

Ceramics

Kingery, W. D., D. P. Birnie, and Y. Chaing. *Physical Ceramics: Principles for Ceramics Science and Engineering*. New York: John Wiley & Sons, Inc., 1995.

Kingery, W. D., H. K. Bowen, and D. R. Uhlman. *Introduction to Ceramics*, 2nd ed. New York: John Wiley & Sons, Inc., 1976.

Composites

Composites. Volume 1, Engineered Materials Handbook. Metals Park, OH: ASM International, 1987.

Test Methods

Annual Book of ASTM Standards. West Conshohocken, PA: American Society for Testing and Materials, updated and published annually.

To Dig Deeper

Ashby, M. F., and D. R. H. Jones. *Engineering Materials*, 2nd ed. Oxford: Butterworth-Heinemann, 1996.

Dieter, George E. *Mechanical Metallurgy*. New York: McGraw-Hill Book Co., 1961.

Driscoll, S. D., Ed. *The Basics of Testing Plastics: Mechanical Properties, Flame Exposure, and General Guidelines*. West Conshohocken, PA: ASTM International, 1998.

Flinn, Richard A., and Paul K. Trojan. *Engineering Materials and Their Application*. Boston: Houghton Mifflin Co., 1975.

Schaffer, J. P. *The Science and Design of Engineering Materials*, 2nd ed. New York: McGraw-Hill, 1999.

Shockelford, J. F., and W. Alexander. *Material Science and Engineering Handbook*, Boca Raton, FL: CRC Press LLC, 2000.

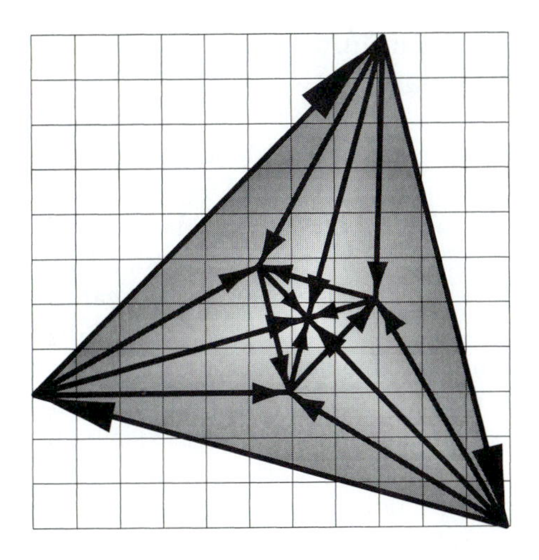

CHAPTER 3

Tribology

Chapter Goals

1. An understanding of tribology and its importance.
2. Knowledge of the fundamentals of friction.
3. A knowledge of the types of wear.
4. A review of the basics of bearings and lubricants.

Rationale

All past editions of this book have contained a significant amount of discussion on friction and wear, but it has been scattered throughout various chapters. In this edition as well as in the previous edition, we moved much of the fundamental discussions to the front of the book, and wear and friction characteristics of specific materials will be discussed in their respective chapters. Tribology *is a term that was coined in England in the 1960s to describe the "science and technology of interacting surfaces in relative motion and related practices" (Dawson, 1979), in other words, friction, wear, lubrication, and subjects such as bearings and hardfacings that are inexplicably part of friction wear and lubrication.*

Tribology is still a largely unknown word in most circles, and we only use it here because we want at least engineering students to know what the word means and what it has to do with engineering. A significant number of universities have courses in tribology, and universities in a number of countries (UK, Japan, Finland, Sweden, etc.) offer baccalaureate and advanced degrees in the subject. The overall objective of this chapter is to provide the reader with knowledge of the basis of tribology and how tribological properties of materials affect material selection.

The importance of *friction*, wear, and *lubrication* in modern life may best be illustrated with the automobile. Friction between the tires and the road is essential for getting the automobile moving, and friction between brake pads and rotors is needed to stop the vehicle. Wear and friction occur in all moving systems. Lubricants are employed to mitigate both, but wear will eventually win and destroy engines and other sliding systems. Wear is the number one cause of death of automobiles. Death may come at 150,000 miles, or even 200,000 miles, but it will come. It is said that rust never sleeps; similarly, friction and wear never stop when surfaces in contact experience relative motion. It is a primary cause of failure in most machines.

Earthquakes are the result of *stick-slip friction* between rock plates that support land masses. The "slip" event causes the earthquake. Erosion, a form of wear, is the cause of shoreline changes as well as natural wonders like the Grand Canyon in the western United States. Friction and wear is very important in the human body. Arthritis is the result of loss of lubricating cartilage in joints. A significant part of tribology research is directed toward prosthetic devices: hips, knees, and so on. Which materials make the best couples? Is stainless steel on ultra-high-molecular-weight polyethylene (UHMWPE) better for a 5-year hip joint and ceramic on UHMWPE better for a 10-year joint couple? Examples of the importance of tribology are everywhere, and that is why we want to discuss this subject up front, before discussing any specific material systems.

The annual cost of wear is conservatively estimated to be 5% of the gross domestic product of any country. Understanding friction, wear, erosion, and lubrication can help reduce that cost. For example, if a rotating system has perfect lubrication, the rotating members are not in contact, friction is very low, and wear does not occur. The evidence of this is the number of electrical generators still in use after 50 or even 100 years. Friction and wear cannot be eliminated, but their effects can be reduced to economically tolerable levels. This chapter will not address the wear properties of different materials, which will appear in the respective chapters. Rather, this chapter will present wear and friction fundamentals by discussions of contact mechanics, tribological terms, tribosystems, abrasion, solid-interaction wear, surface fatigue, erosion, bearings, and lubricants. It will serve as the foundation for subsequent discussions.

3.1 Historical Studies of Friction and Wear

Prehistoric people learned how to use tribological devices to make work easier, improving their quality of life. When they killed a deer or other large animal, they learned that it was easier to drag their kill than carry it. If a deer carcass weighed 100 lb, the force needed to pull it on a sled was only a fraction of that. Wheels lowered that force even more. Early Egyptians used liquid lubrication to help slide stones used in building their monuments. Early Greek scholars wrote about friction but did not establish any models or rules to deal with it. Leonardo da Vinci, in 1495, documented testing devices that were used to study friction (Figure 3–1). These same devices still serve that purpose. Guillaume Amontons, in 1699, wrote that the force to pull an object on a level plane was about one-third of the weight of the object and that force originated from interlocking of asperities on the mating surfaces. He essentially postulated the first laws of friction: The friction force is proportional to the force applied to the object that is being set into motion, and that force is independent of area. Charles A. de Coulomb, in 1785, investigated friction variables and observed that the friction force was independent of velocity.

The present-day definition of friction force is the resisting force tangential to the interface between two bodies when, under the action of an external force, one body moves or tends to move relative to the other—ASTM G 40.

What we now call the coefficient of friction is the "one-third" relationship proposed by Amontons. He is usually credited with the mathematical expression of the first law of friction—that the force of friction is proportional to the applied force and the force of friction is independent of contact area:

$$F = \mu N$$

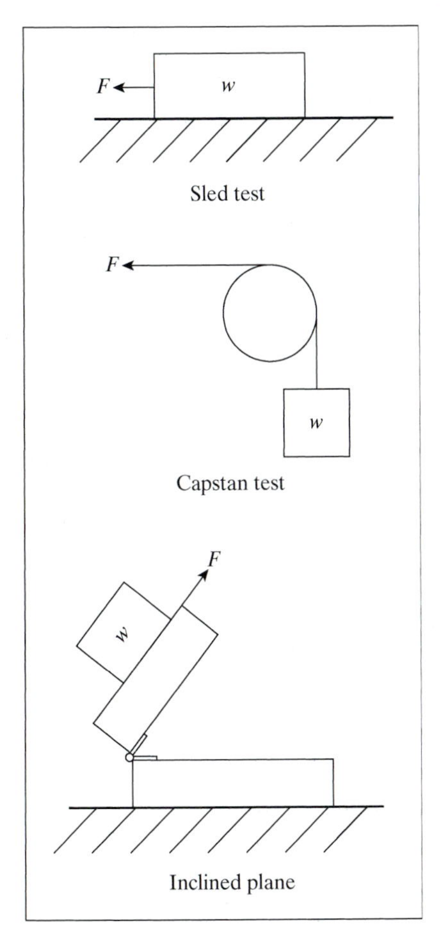

Figure 3–1
Schematic of the devices that Leonardo da Vinci used to study friction

where F = friction force

μ = a proportionality constant (now referred to as the coefficient of friction)

N = the normal or downward force on the object to be moved

Many others contributed to the body of knowledge that we now have on friction; however, we have not progressed to the point where we can "engineer" friction for most practical sliding systems. We mostly measure and predict from lab studies. Friction, as a topic, is broader than solids sliding on other solids. A general definition may be, friction consists of forces on an object resisting motion when motion is attempted. Friction occurs between fluids and the surfaces over which a fluid flows. It occurs in anything that moves, and nobody has learned how to prevent it. Many people have tried to make a perpetual motion device, but none has succeeded. The friction force required to start an object sliding is usually different from that required to sustain motion. That is why the terms *static friction* (starting) and *kinetic friction* (moving) are used. Designers need to make sure that the appropriate value is used in calculations.

Wear is similar to friction in that people have addressed it since prehistoric times, but it has not been eliminated, and we have not identified the laws even as well as friction. The evolution of tools essentially chronicles how man has addressed wear. The first tools were probably wood. They did not last very long. Stones were attached to wood to make hatchets, hammers, knives, and arrows. Stones did not wear as fast as wood. Then they found that some stones, such as flint, lasted longer than other stones. Then metals were discovered. These made tools that lasted longer and worked better than stone. Then it was discovered that metals could be treated to make them harder and more wear resistant. Concurrent with tools were lubricants and special bearing materials that reduced the detrimental effects of rubbing in mechanical devices. In the year 2004, we have many ingenious devices, materials, and lubricants in our arsenal to combat friction, wear and, erosion, and we will discuss these in subsequent chapters.

3.2 Contact Mechanics

When tribologists gather at conferences, the attendees will invariably come from diverse academic backgrounds, but predominately physics, chemistry, mechanical engineering, and material science. The physicists usually concern themselves with the atomic aspects of contacting surfaces; the chemists concern themselves with the formulation of robust organic materials to separate and lubricate surfaces; the mechanical engineers are concerned with design of parts and machines, the mechanics of tribosystems—fluid mechanics, temperature rises, stresses, and other mechanical aspects of the tribosystem. Finally, the materials engineers concentrate on responses of various materials to wear processes. Together, they make up the tribology community and there is considerable synergy as a team charged with reducing the cost of friction and wear.

Contact mechanics is a product of the 1960s focus on tribology in the United Kingdom. It is a form of engineering mechanics concentrating on stresses and deformation at the microscopic areas that form the real areas of contact between solids. Figure 3–2 illustrates the concept

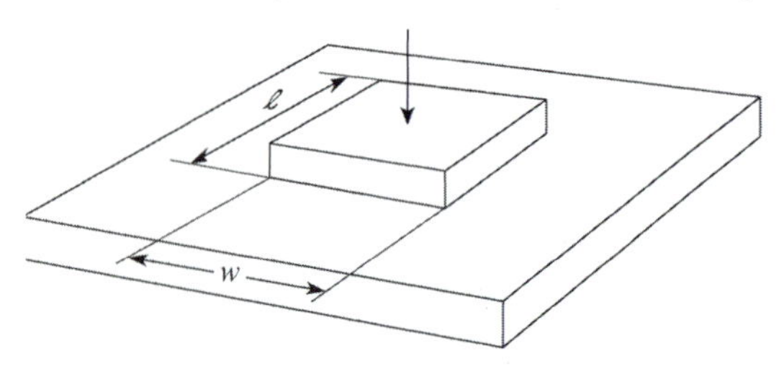

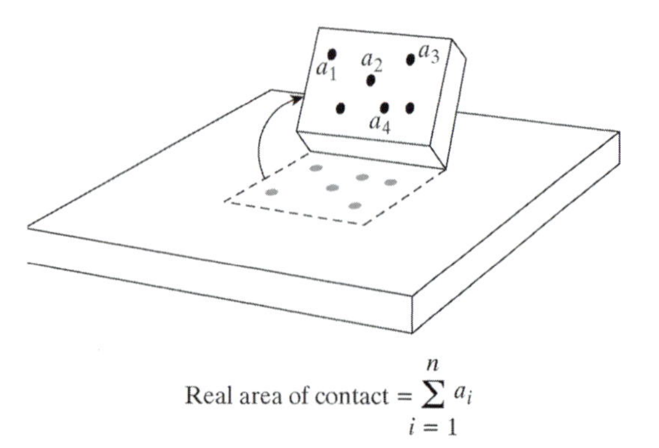

Figure 3–2
Concept of real and apparent area of contact

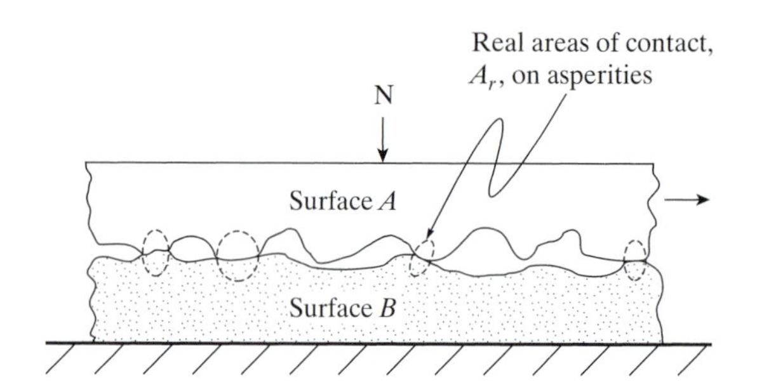

Figure 3–3
When real surfaces contact, the up-features of the surfaces carry the load

of real area of contact and apparent area of contact. Because real surfaces contain errors of form from manufacturing processes when one surface is placed on another, the surface features above the mean asperity height (up-features) will contact each other and produce junctions of various sizes depending on the load. The higher the load, the larger the *contact area*. Contact mechanics assumes that surface texture features such as asperities carry the loads (Figure 3–3) and provides mathematical treatments to relate friction and wear to what happens as asperities on conforming surfaces interact. There are many mathematical models that allow calculation of the *coefficient of friction* for a sliding couple knowing the surface texture of both surfaces and the force pushing them together. However, few practicing engineers rely on these models to solve real wear problems because they do not have demonstrated real-life reliability.

Wear does not always involve contact between conforming surfaces. In fact, as we shall see, some forms of wear occur from the mechanical action of fluids on a solid surface and many wear processes involve concentrated contacts like a grain of abrasive imposed on a surface or a ball or roller rolling on a surface. One of the most useful aspects of contact mechanics is determination of point and line contact stresses with *Hertzian stress* equations. In the nineteenth century, Heinrich Hertz developed models that

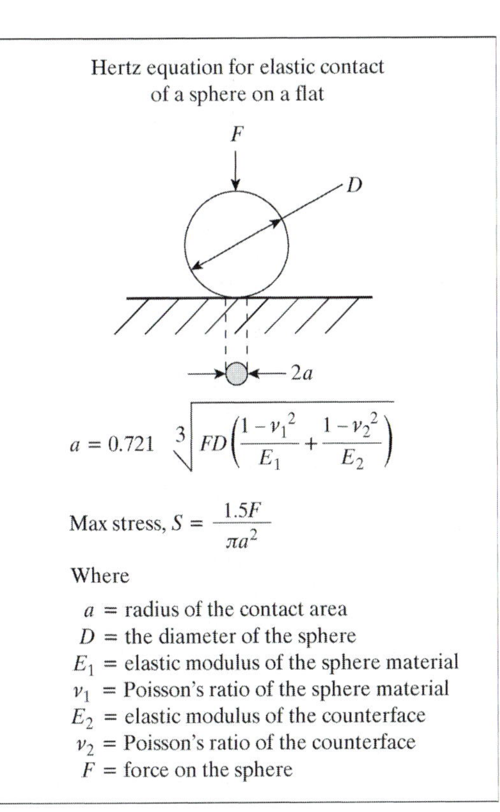

Hertz equation for elastic contact
of a sphere on a flat
F
D
2a
$a = 0.721 \sqrt[3]{FD\left(\frac{1-\nu_1^2}{E_1} + \frac{1-\nu_2^2}{E_2}\right)}$
Max stress, $S = \frac{1.5F}{\pi a^2}$
Where
a = radius of the contact area
D = the diameter of the sphere
E_1 = elastic modulus of the sphere material
ν_1 = Poisson's ratio of the sphere material
E_2 = elastic modulus of the counterface
ν_2 = Poisson's ratio of the counterface
F = force on the sphere

Figure 3–4
Hertz equation for the contact of a sphere on a flat surface under elastic conditions

allow calculation of the elastic stresses for systems, such as a ball on a flat surface, crossed cylinders, or a cylinder on a flat surface. These relationships are still widely used, and they allow approximation of stresses in many types of tribosystems. Figure 3–4 illustrates the classic equation for a ball on a flat surface.

Hertz's stress equations are in countless handbooks, and they provide an insight into the material stresses that play an important role in wear processes. The stress at a point contact is a function of both the geometry of the contact as well as the stiffness (modulus of elasticity) and, to a lesser degree, Poisson's ratio for the contacting materials. The bases of many wear models are the real area of contact of surfaces and the properties of the contacting materials. In summary, contact mechanics is an aid for understanding the mechanical aspects of interacting surfaces, but it is not currently a tool that can be used to solve a problem with rapid wear of a punch press die in blanking silicon steel.

3.3 Friction

What Causes Friction? Some tribologists believe that surface "rugosities" provide inclined planes that must be overcome to produce and sustain motion (Figure 3–5). Others believe that friction originates from atomic forces tending to bond the two bodies together. When two atoms are brought together, there is attraction until the atoms get close enough (a few atomic spacings), and then repulsion occurs. Rubbing conditions can be such that atoms of one surface are in attraction range of the other surface atom at asperities. Still others maintain that surface films play a significant role in friction. All real surfaces are covered with films that can interact. For example, in high-humidity conditions, surfaces tend to stick. Your arms stick to your desk. Windows and doors stick. There can be a meniscus effect when moisture-covered surfaces are brought in contact, and this can contribute to the friction force. Most tribologists believe that

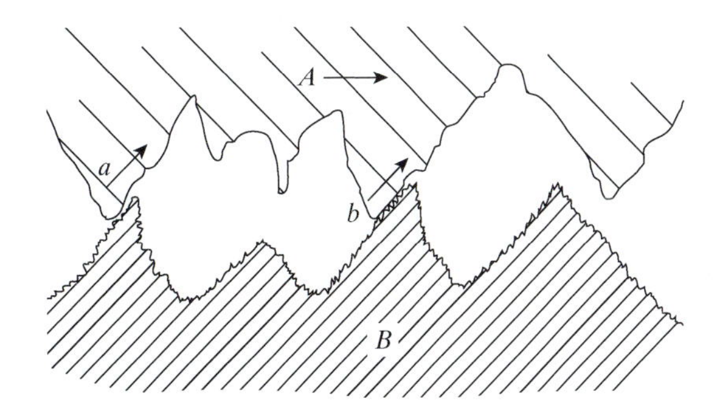

Figure 3–5
The force required to move surface *A* up the rugosities on surface *B*

the presence of third bodies, such as sand particles between conforming bearing surfaces, affects friction.

Finally, real surfaces do not only contain surface asperities, but they also contain errors of form, such as wave forms, machining features, and feature patterns resulting from peculiarities of the machine tool used to generate a surface. If these up-features (asperities or waves that protrude above the mean texture height) interact with the conforming surface, there can be a plowing component in friction. The missing component is the force required to plastically deform interacting surface features. Third bodies between surfaces can act as miniature rolling element bearings. So which of these factors is responsible for the friction force? Probably all of them contribute to the friction force:

$$F = F_a + F_p + F_s + F_n$$

where F = observed friction force
F_a = forces to break adhered junctions
F_p = forces to plow and deform surface features
F_s = forces to shear films between surfaces
F_n = forces due to the nature of the sliding system

The area independence of the friction force is explained in this way: The friction force is equal to the real area of contact (A_r) between surfaces times the shear strength (S) of the junctions that are formed between the two surfaces.

$$F = SA_r$$

It is assumed that the real area of contact is determined by the normal force (N) pressing two bodies together and the penetration hardness (P) of the weaker of the friction couple

$$A_r = \frac{N}{P}$$

Combining these equations yields a relationship for friction coefficient that is a function of the shear stress of junctions and the hardness of the softer member

$$\mu = \frac{F}{N} = \frac{SA_r}{PA_r} = \frac{S}{P}$$

This model explains why the friction coefficient is never 0 (a material always has a shear strength and a hardness) and the ratio is seldom over 1. It is not common to have a material with a shear strength that is many times its hardness, like some rubbers. Similarly, it helps explain why the friction coefficient is not significantly affected by temperature or sliding velocity until conditions are such that there is likely to be thermal softening of the material (low P, low S).

Obviously, friction is affected by the nature of the material, because S and P are different for various materials and films, and contacting surfaces interfere with junction formation so they will affect frictional behavior. There is also a chemical effect influencing the shear strength of junctions. Some metals and plastics (such as copper and PTFE) readily adhere to many other surfaces. This tendency to form adhesive junctions makes some materials have friction characteristics that set them apart from other materials. Rubbers can have very high friction coefficients against most other solids (over 1). The tendency, in part, is due to their high conformability and shear strength to hardness ratio. The real area of contact becomes very large with relatively small contact forces. Second, rubbers are viscoelastic materials. There is time-dependent recovery from strain. Any viscoelastic effects that occur in a rubber or other solid sliding system affect friction. They are delayed energy, and friction is loss of energy at an interface.

Rolling friction is more complicated than we have space to address. For occasional machine design situations, it is probably sufficient to know that rolling friction depends on the stiffness of the roller and counterface. The harder and stiffer the couple, the lower the rolling friction. A wheelbarrow with half-flat rubber tires will require much more energy to move than one with properly inflated tires.

In summary, friction is the sum of the forces impeding motion in sliding. Static friction is the net result of forces resisting motion when motion is attempted. Friction forces are affected by microscopic and macroscopic factors that control adhesion of contacting surfaces and the mechanics of motion. The role of atomic and molecular forces is probably minor in most engineering sliding systems, but as designs evolve into smaller mechanical devices, such as MEMS (micro-electro-mechanical systems), they may become significant. Sliding surfaces in machines are almost always covered with films (absorbed moisture, oil, contaminants, oxides, etc.). The force needed to shear these films to produce motion will be a component of the friction force. The other component that definitely plays a role in the friction force is the plastic deformation that occurs at real areas of contact. Manufactured surfaces are not perfectly flat or smooth. When you put one of these imperfect planes on another, they will contact at "spots." When motion is attempted and during motion, these spots elastically or plastically deform. They carry the forces of the weight of the sliding member as well as the other forces on the sliding member. During sliding, these surfaces deform, and this requires energy. The friction force is the manifestation of energy dissipation required to alter surfaces in relative motion. It is an energy dissipation process. When a solid surface rubs on another solid surface, energy is required to accommodate the system changes that occur at the rubbing interface (deformation, wear, oxide removal, film shear, etc.). This energy must be dissipated. Some goes into the friction force; some goes into heat or some other energy dissipative processes. Friction cannot be eliminated, because when one surface slides on another, "things" happen and these things require energy.

Measurement Many friction tests have been devised over the hundreds of years that it has been studied, but the most commonly used unlubricated tests are those illustrated in Figure 3–6. Friction force recordings are interpreted to yield a starting or breakaway friction force and a force for continuous sliding. The former is used to calculate the static coefficient of friction—the latter, the kinetic coefficient of friction. Typical friction force curves are illustrated in

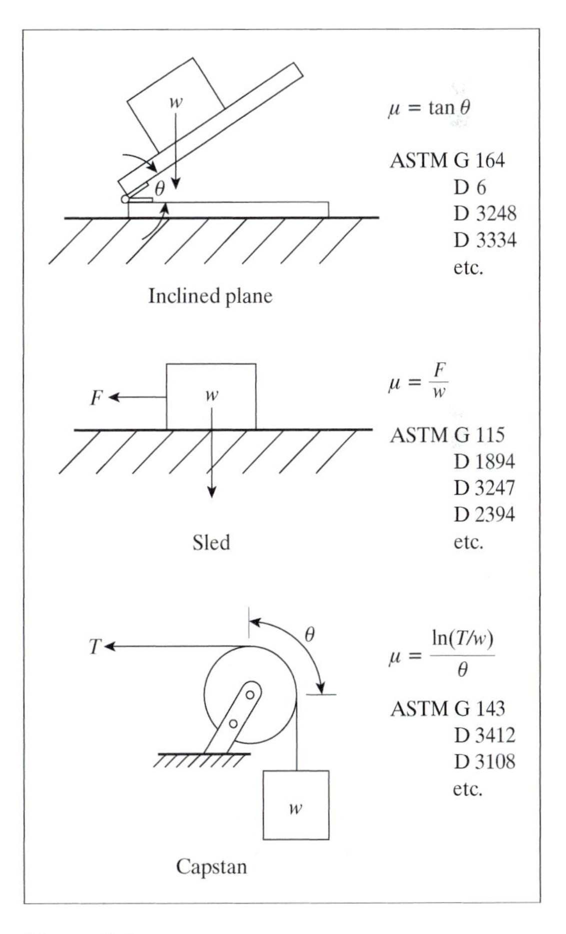

Figure 3–6
Equations for friction force

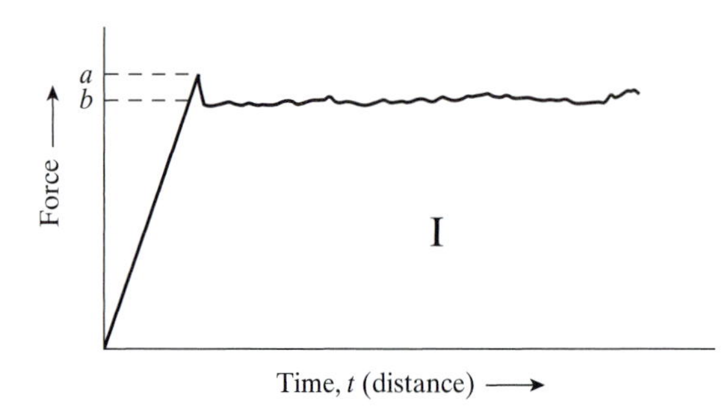

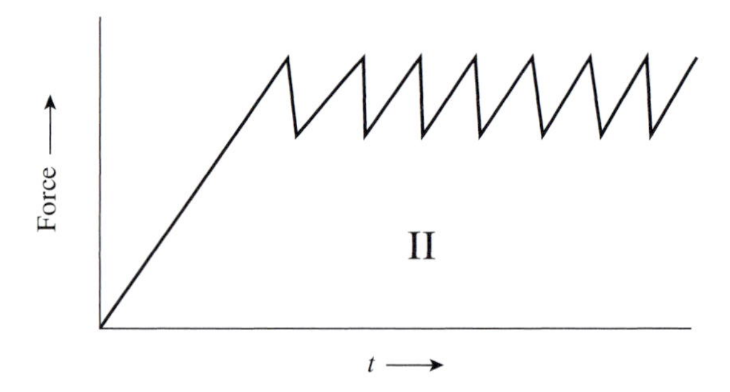

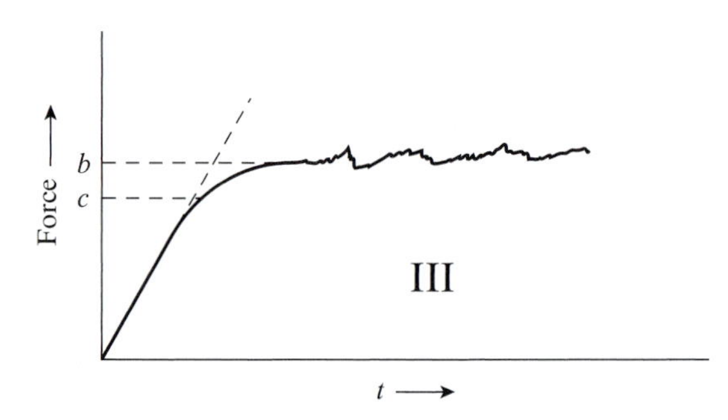

Figure 3–7
Types of friction force recordings that can be encountered

Figure 3–7. Friction is not usually consistent in sliding systems, but a periodic sawtooth type of force response like case II is called stick-slip behavior. The body is at rest, and the pulling device (which is elastic) is moving. At some point, the object breaks away, slides to catch up with the pulling device, and then stops again. The process is repeated. Coefficients are not valid for a system such as this, and results are usually reported as "stick-slip." Whenever a sliding system squeals, stick-slip is probably occurring. The squeal of automobile tires in a panic stop is stick-slip between the tires and the pavement. Case I and case III are the normal results, and forces *a* and *c* are used to calculate the static coefficient of friction, while force *b* is used to calculate kinetic friction. Sometimes, friction is measured at various sliding distances. Friction data from wear tests include the friction effects of wear debris. This should be considered when using these data.

Friction coefficients are needed for many types of engineering calculations and models. Valid data usually come from tests conducted with test systems that simulate the application. Data from the literature do not apply unless they were generated in a tribosystem similar to yours. Remember that a piece of material does not have a coefficient of friction; it is a system property. By definition, the material must be mated with another material, and it makes a difference what that material is.

Finally, lubricated and rolling friction can be very low. As shown in Figure 3–8, well-lubricated surfaces can have friction as low as about 0.15. Most metals and plastics exhibit lower friction when normal loads get very high. This is because surface deformation dominates and the force needed to deform the surface remains constant as the load increases. Temperature and velocity may have a similar affect on some systems, especially a plastic/metal or plastic system. Rubbers tend to have high friction against most other counterfaces because they are so compliant. The friction shear area is very large. Clean metals free of surface films and contaminants can have a friction coefficient near unity. Rolling element bearings and hydrodynamically lubricated systems can have friction coefficients that are less than 0.1.

The decrease in friction coefficient with pressure shown in Figure 3–8 can occur with sliding velocity. Plastic deformation, even melting, occurs at high speeds, and friction becomes low because surfaces can be liquid lubricated if melting occurs. This condition is readily seen with

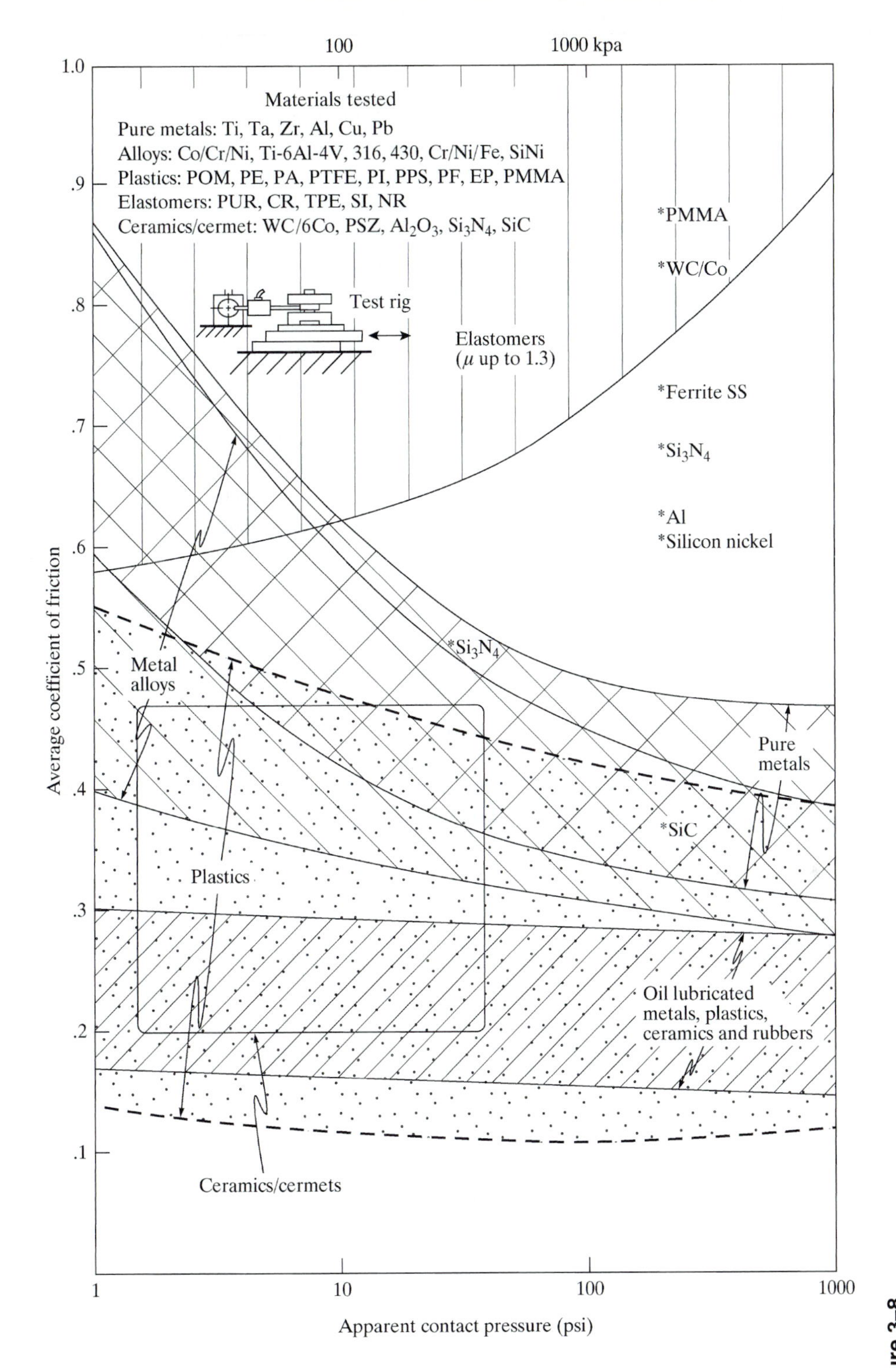

Figure 3–8
Average friction coefficients for various materials in reciprocating motion of an annular ring rider (0.1 in.^{2} area) on a type 316 stainless steel counterface at 20°C 50% relative humidity (RH) at various normal forces. The stroke was 50 mm and the frequency was 0.5 Hertz. The friction force was averaged for eight cycles in each test.

plastic/metal couples at high speeds. Some researchers believe that there is a critical velocity for "normal" friction, equivalent to one-tenth of the velocity of sound in the weaker materials in the sliding couple. Above this velocity, the friction starts to decrease. Surface roughness also plays a role in the decrease of friction. Rough surfaces take at least ten times longer to get into the low-friction regime. Interface deformation is accommodated by asperity deformation. Rough surfaces (abrasive blasted) are much more *galling* resistant than ground surfaces. The transition from normal friction to low friction with increasing contact force occurs when visible plastic deformation or melting occurs on the rubbing surfaces. Surface yielding is occurring. With plastics, this can happen at quite low forces; with metals and hard solids, seizure can occur and motion is stopped. If sufficient force and clearance is available, the surfaces simply keep deforming and the friction force becomes low.

The friction map of Figure 3–8 was developed from reciprocating sliding of a conforming rider (disk) on a flat surface. If the sliding system or counterface were different, the map might look entirely different. It is intended to serve as a rough guide of friction regimes in sanitary sliding systems where one member must be type 316 stainless steel. It also points out that the friction coefficient is not constant for all sliding conditions and that self-lubricated plastics and lubricated metals provide low friction if that is a design requirement. Brake and clutch materials do not produce low friction against metal counterfaces because they have engineered friction properties. Brake materials are often made by adding friction modifiers to a polymer base until a particular friction/pressure/velocity profile is achieved. Finally, friction data from handbooks are probably no better than the rough estimates shown in our friction map.

Measurement of friction in lubricated systems is often performed on actual machinery by measuring the power expended by prime movers. Internal combustion engines are put on a dynamometer, and an energy balance can be calculated to show energy in and energy out. The difference yields information on energy lost to friction. Figure 3–8 shows a low range of friction for liquid-lubricant surfaces, but the friction in liquid-lubricated systems is known to be a function of the nature of the lubricant and the sliding speed. The Stribeck curve (Figure 3–9), which is a

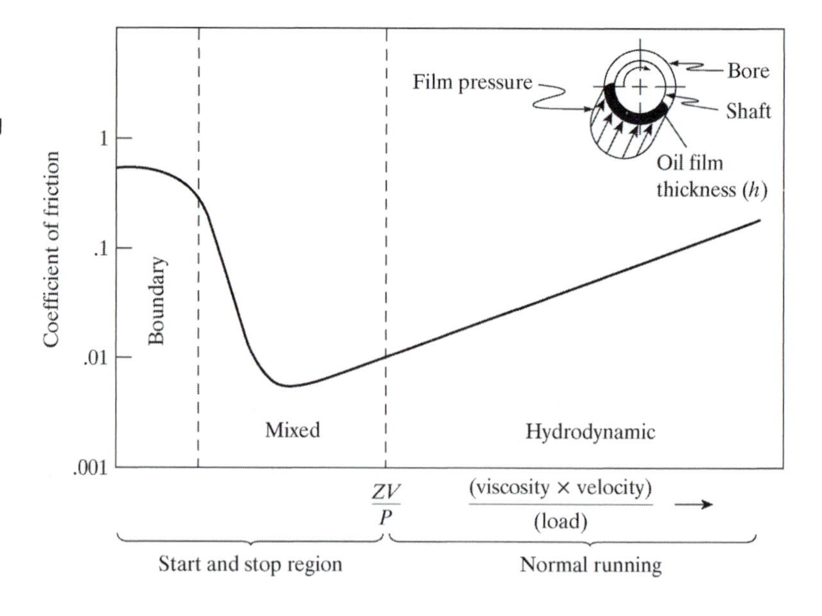

Figure 3–9
Stribeck curve showing how lubricated sliding systems can have friction vary with operating conditions and lubricant properties

classic lubricant relationship, shows that the coefficient of friction varies with the lubricant (the *viscosity*) and operating conditions in a lubricated journal/shaft tribosystem. The significance of this relationship is that you will not achieve the low friction of *hydrodynamic lubrication* unless your operating conditions are suitable to formation of fluid-film separation of the rubbing parts. This is the meaning of hydrodynamic lubrication. The rubbing surfaces are completely separated by a film of some thickness. *Boundary lubrication* means that some solid contacts are present. Wear will be nil when hydrodynamic lubrication is present; wear will occur when boundary lubrication exists. For example, running a motor at 500 rpm may result in boundary lubrication, and it may take 1500 rpm to get hydrodynamic lubrication. Low-speed rotation will produce more wear than high-speed rotation. *Elastohydrodynamic lubrication*, as occurs in a cam and follower, is a special form of hydrodynamic lubrication and the Stribeck curve is normally not applied. In these systems, there is elastic deflection of the cam and follower to produce a larger contact area, thus the term *elastohydrodynamic lubrication*.

The separating oil film that produces hydrodynamic lubrication can have different thicknesses. The thickness must be greater than the peaks and valleys that comprise surface roughness. This is called the *lambda ratio* (film thickness/composite surface roughness of the mating surfaces). The *film thickness* can be calculated knowing the properties of the lubricant (viscosity) and the sliding speed and the shape of the sliding surfaces. Another aspect of lubricating fluids that pertains to friction is the traction force of the fluid. Some tribosystems such as automobile transmissions require oil films to transmit power. Oils specially formulated to do this job are called traction fluids. The oil film is under a shear stress and under a high state of shear it behaves as a "pseudo-solid" to transmit energy. These fluids are formulated to transmit energy rather than be slippery. Essentially, they produce high friction (higher than lubricants).

Finally, the friction of oil and grease films can be measured in the laboratory in rolling element bearings by essentially turning the inner race while holding the outer race with a torque measuring device. Lubricating fluids do have different friction characteristics. This is especially true of greases, which are combinations of an oil and thickeners to make the oil stay where you want it. Some greases and oils can be quite "sticky" under some conditions. Thus, friction is a factor in lubricated systems because different lubricants lead to different system friction coefficients. Friction can even be relatively high, as is the case with traction fluids.

Significance The value of controlled friction is self-evident. We use controlled friction to stop motion with brakes. We use the high friction of tires on pavement to transmit engine torque to forward motion and to transmit brake friction to the tire/pavement tribosystem. Low friction is equally important, but in another way. Friction is a cost—a significant cost. Some estimates indicate that about 20% of the heat energy from burning fuel in an automobile engine is expended in overcoming the friction of the various tribosystems in the engine: the rings/cylinder systems, the valve train, the connecting rod/crank system, the main bearings/crank system, the oil pump, and so on. Lower friction in automobiles could amount to huge savings in the use of fuel and significant reductions in the pollution produced by burning petroleum fuels.

The overall objective of tribology is control of friction and wear. However, friction is different from wear. They often do not correlate. PTFE (Teflon) usually displays low friction in sliding contact with other surfaces, yet it has notoriously poor wear properties. It must be reinforced with glass or other materials when used as a bearing. On the other end of the scale, polyurethane rubber has high friction against most other surfaces and is well known for its abrasion resistance. There are explanations of these phenomena, but the point to remember is not to use friction data alone when selecting materials for wear resistance.

Use of Friction Data in Design Friction information is very important in manufacturing products in web form. Paper, plastic films, sheet metal, fabrics, flooring, and so on are conveyed in continuous form through and on rollers. If the web does not have the proper friction characteristics in contact with the rollers, it may not convey properly, or if it is slipping on the rollers, the surface could be damaged. It is recommended to measure the friction characteristics of the product/roller couple and design web tensions to accommodate that level of friction. This same type of analysis is needed for handling rope, yarn, and other strands, which are also conveyed in long lengths in manufacture.

Designs involving rotating machinery should always include a friction check of each tribosystem that the prime mover drives. You need this information to size the motor. The same is true with linear systems. The use of finite element modeling techniques is widely used in engineering to calculate stresses and motions in products and critical machine components. If you are analyzing the stress on a punch that is perforating a steel strip, you need to input to the computer the friction coefficient of the steel against the punch and die. The model will only be as good as the friction data that you input. The other data for the model are easier to obtain; handbooks contain reliable data on elastic modulus and Poisson's ratio, but, unfortunately, there is no handbook of friction coefficients. Figure 3–8 can only be used for that particular tribosystem; testing under simulated service conditions produces the most valid data. Be wary of supplier data claiming a material has a low coefficient of friction. A material does not have a coefficient of friction. Friction is the energy dissipation in a sliding system. Sometimes high friction is desirable (brakes, clutches, tires, etc.); sometimes low friction is desired in a sliding system (bearings, slides, cams, etc.). Designers should consider system friction as an operating parameter to be dealt with.

3.4 Definition of Wear

Wear is defined by the ASTM G 40 standard on wear and erosion terminology as "damage to a solid surface, generally involving progressive loss of material, due to relative motion between the surface and a contacting substance or substances." The standard defines *erosion* as "progressive loss of original material from a solid surface due to mechanical interaction between that surface and a fluid, a multicomponent fluid, or an impinging liquid or solid particles." Sometimes sliding contact between solids results in displacement of material by plastic deformation rather than removal. This definition of wear still applies because it states that wear is "damage" to a surface and deformation would be included. Material displacement rather than removal is common in plastic-to-metal wear systems. Displaced material can be measured and included in wear volume measurements.

Now that we have defined wear and erosion, we state that components do not just wear or erode. They wear or erode in different ways. There are about 13 forms of wear and 5 forms of erosion that are generally recognized by the tribology community (Figures 3–10 and 3–11).

3.5 Types of Erosion

Solid Particle This type of erosion is damage to a surface caused by impingement of solid particles carried by a gas. This gas is usually air, but it could be any gas. There is a standard test, ASTM G 76, that uses a timed stream of oxide particles directed at a solid surface by an air jet. This form of wear occurs in particle conveying systems such as soot blowers and cyclone separators. It also occurs on fans used in dirty areas, even in jet engine blades. The degree of damage is usually a function of the mass of particles

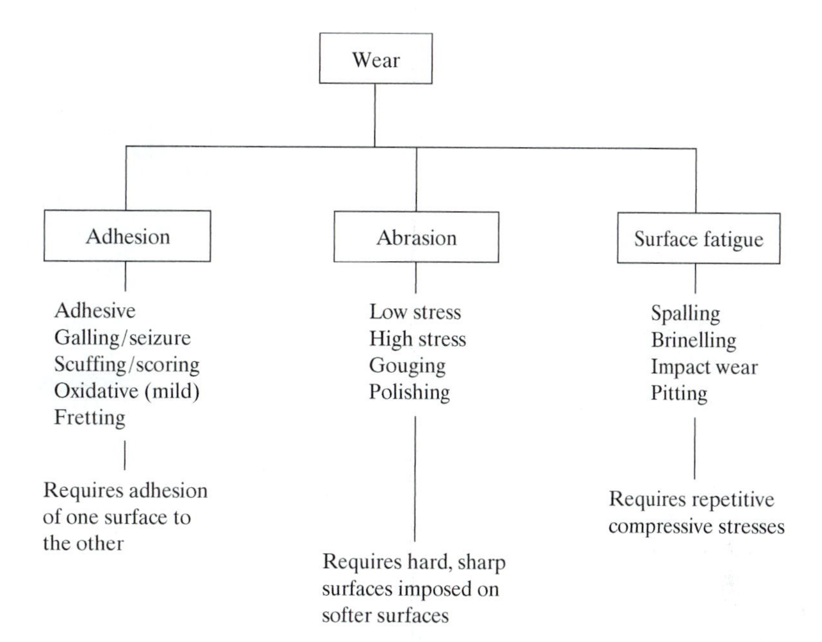

Figure 3–10
Major categories of wear and specific types of wear in each category

impacting a surface (M), their velocity (V), the angle of impingement (r), and the particle sharpness (P).

$$\text{Erosion (particle)} = \alpha M V^2 (r)(P)$$
$$\alpha = \text{constant for the system}$$

The effect of angle depends on the system. Brittle materials show the highest erosion with normal impact (they spall), whereas soft metals show highest erosion with an impingement angle between 20 and 30°. The mechanism of material removal can range from spalling in brittle materials to cutting in ductile materials. Abrasive grains in sandblasting can be rotating when they impact the target surface, and they behave as tiny cutters to form chips. Mostly, material is removed by a fatigue mechanism. A particle forms a crater, repeated strikes hit the extended material around the crater, and the lips eventually fracture (Figure 3–12).

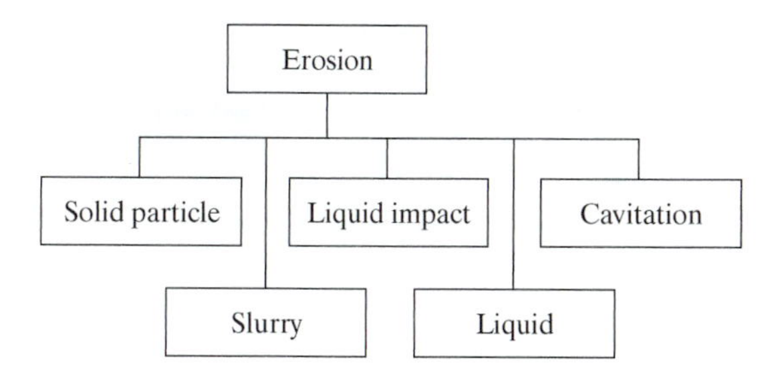

Figure 3–11
Types of erosion

Slurry Erosion This is material removal caused by slurry motion across a solid surface. This form of erosion often occurs in pumps handling drilling and mine waters or in conveying slurries in pipelines (Figure 3–13). Some coal pipelines are hundreds of miles long. The *erosivity* of a particular slurry or the ability of a material to resist erosion by a particular slurry can be tested by the ASTM G 75 Miller number test. This number, which ranges from 1 to 1000, rates the erosivity of a slurry; a Miller number of 1000 is most erosive. This form of wear correlates with the same factors as solid particle erosion, the mass of particles that contact the surface, their velocity, and their shape factor.

Figure 3–12
(a) Schematic of solid particle erosion; (b) erosion of a wearback from a pipe carrying fly ash. Note hole and wavy surface.

The mechanism of removal is similar to that for solid particles; however, slurry systems usually involve mostly parallel flow, and the action of the particles is more like scratching. Very often, slurry erosion involves a corrosion component from the conveying fluid. The Miller number test has a procedure that allows determination of the corrosion component (material dissolution) of the erosion as well as the mechanical action (scratching) component of the erosion.

Liquid Impact This form of erosion is produced by impingement of liquid droplets on a solid surface. This is a special form of erosion that occurs when surfaces are impacted by very high-velocity liquid droplets. Rain erosion of aircraft is

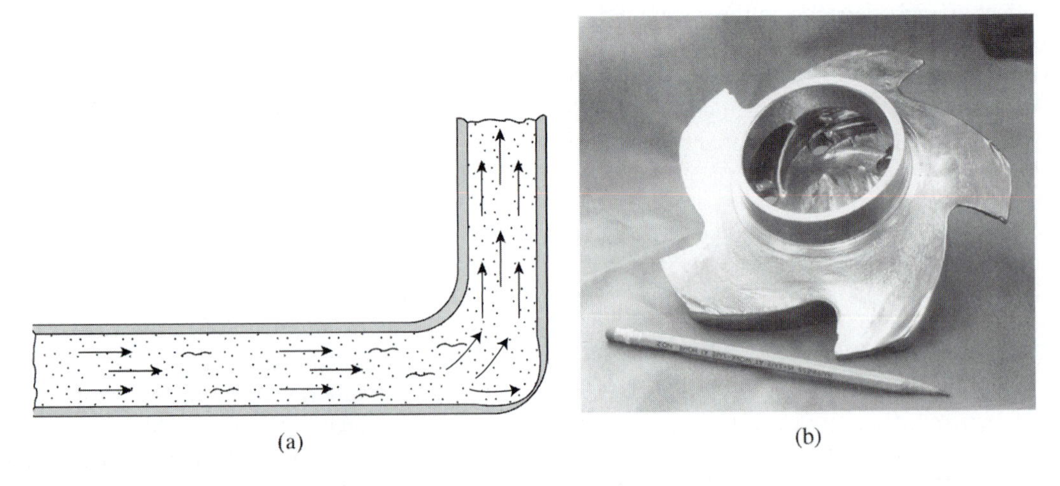

Figure 3–13
(a) Schematic of slurry erosion; (b) pump impeller showing erosion damage from pumping a slurry of silica and water

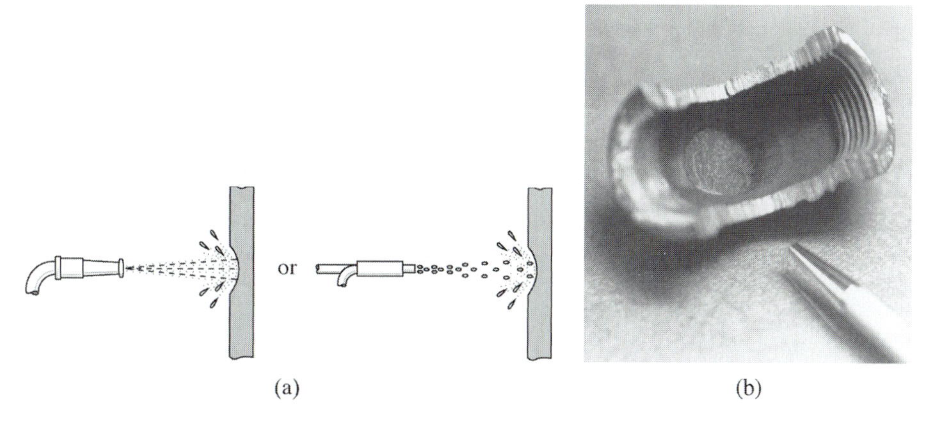

Figure 3–14
(a) Schematic of liquid impingement erosion; (b) pipe elbow perforated by impingement from high-velocity fluid in a pipeline

the classic example. When aircraft enter a rain field at hundreds of miles per hour or even supersonic speeds, the impact with the droplets can be very destructive to the aircraft surfaces contacted. The nose of the aircraft (made from fiber-reinforced plastics) is particularly susceptible, as are windshields and leading edges of spars. In industry, this kind of erosion can be experienced in steam turbines when condensate is present in the steam. Nonmetals are particularly susceptible to damage. The ASTM G 73 test for liquid impingement erosion testing uses a sample on the end of a propeller that can rotate at speeds of hundreds of meters per second. The propeller rotates in an artificial rain field, and the damage to test specimens is measured as a volume loss versus time. Mostly, this is a form of erosion that is restricted to special tribosystems.

Liquid Erosion This form of wear is characterized by removal of material by the action of a liquid impinging or moving along the surface of a solid (Figure 3–14). Liquid erosion is often encountered in piping, especially in pipes made from materials that rely on a passive oxide films for corrosion resistance. The mechanical action of the liquid removes the protective oxide. It reforms and is removed again. This repeated action can perforate pipe elbows when velocity is high. Plastics and ceramics that do not rely on protective films for corrosion resistance are relatively immune to this form of erosion, but metal piping of steel, stainless steel, and copper is particularly prone. There is a critical velocity below which this does not occur. Thus, the normal remedy is to control liquid velocity at a safe level.

Cavitation *Cavitation* is damage to a solid surface by implosion of bubbles near a surface. When a bubble collapses, the surrounding liquid rushes in to fill the void. A jet is formed that can develop stresses that exceed the yield strength of materials. The mechanism of material removed is microscopic fracture to form pits (Figure 3–15). Even noble metals such as gold are susceptible, so it is not always removal of oxides that removes material (gold does not have an oxide film). Cavitation occurs in fluid propulsion devices such as pumps, on ship propellers, in dam spillways—wherever a cavitation field is produced. The wide industrial use of ultrasonic debubblers and cleaners has made this form of erosion a very formidable factor to be dealt with. Tanks often perforate where transducers are attached.

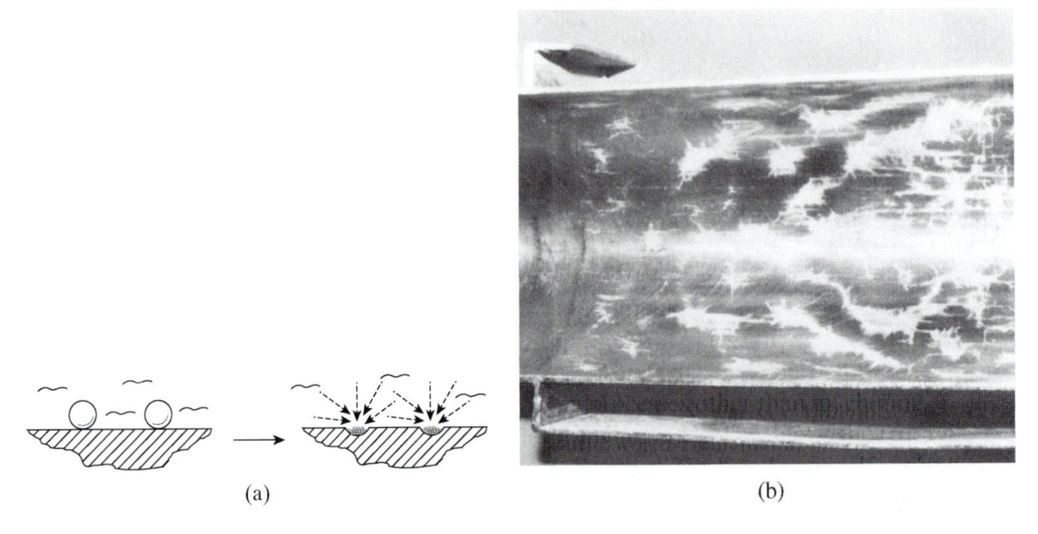

(a) (b)

Figure 3–15
(a) Schematic of cavitation; (b) cavitation on a stainless steel tank. An ultrasonic agitation device was attached to the other side of this section of the tank (diameter = 15 cm).

3.6 Types of Wear

Sliding Contact

Adhesive In our discussion of friction, we showed how part of the friction force between conforming solids is due, in part, to bonding of asperities or high spots on the mating surface. When this bonding becomes significant, particles from one surface can adhere to the other. Wear (removal) has occurred. When the surfaces are relatively clean (no third bodies present), this is called *adhesive wear* (Figure 3–16). Adhesive wear is defined in ASTM G 40 as "wear due to localized bonding between contacting solids leading to material transfer or loss from either contacting surface." All material systems are susceptible, and the factors that control this form of wear are often described in the *Archard equation*:

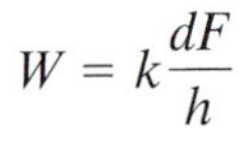

$$W = k\frac{dF}{h}$$

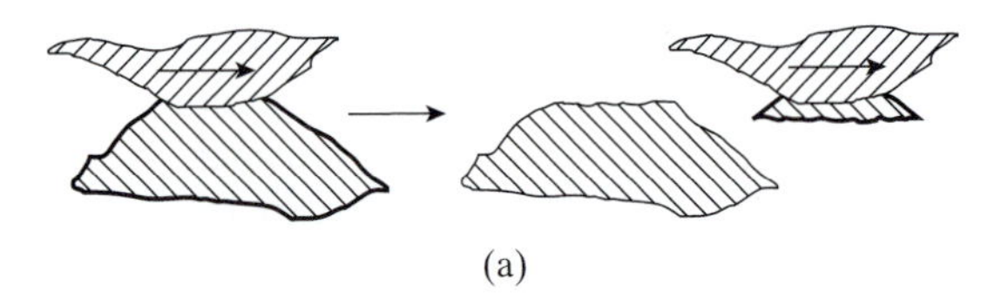

(a)

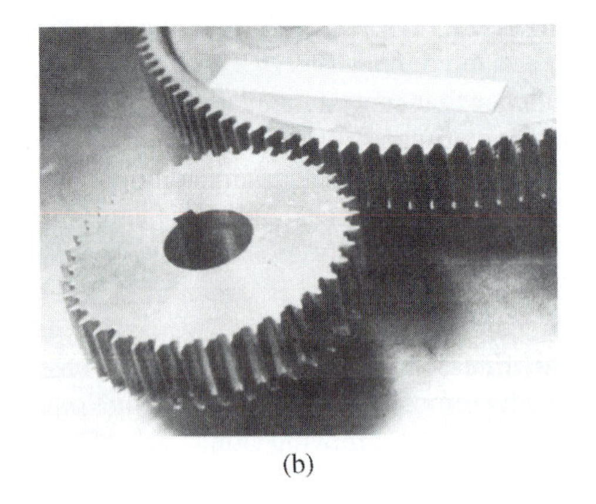

(b)

Figure 3–16
(a) Adhesion of asperities in adhesive wear; (b) metal-to-metal wear on gear teeth (no lubrication)

where W = wear volume (mm^3)
k = a proportionality constant for the system (dimensionless)
d = sliding distance (m)
F = normal force (N)
h = hardness of the softer member (kg/mm^2)

Tribologists often perform wear tests like the block-on-ring test (ASTM G 77) to measure the wear volume, then solve for k use this as a metric for the adhesive wear behavior of various couples. Unfortunately, adhesive wear is not that predictable in laboratory tests. Coefficients of variation can be as high as 50%. Wear coefficients are available in the literature, but it is advisable to be cautious in using them in the Archard equation to calculate system wear rates. Wear is system dependent, and your system may be much different from the one used in the literature to calculate a wear coefficient. Some examples of wear coefficients (from various sources) are

Clean, unlubricated soft metals self-mated	1 to 3 $\times 10^{-3}$
Clean, unlubricated hard metals self-mated	2 to 5 $\times 10^{-5}$
Clean, unlubricated hard metal to plastic	1 to 3 $\times 10^{-6}$
Clean, unlubricated plastic to plastic	1×10^{-3} to 3×10^{-6}
Well-lubricated hard metals self-mated	1 to 3 $\times 10^{-7}$

Many sliding systems experience rapid wear-in or a transition from low wear at first to more rapid wear later on. The Archard equation does not handle these situations well, and wear coefficients are at best only approximations of wear rate.

Galling When adhesive wear becomes severe, it leads to *galling*. This form of wear is characterized by the formation of macroscopic excrescences. Material flows up from the surface (Figure 3–17), and this large up-feature often leads to *seizure* in systems where the sliding

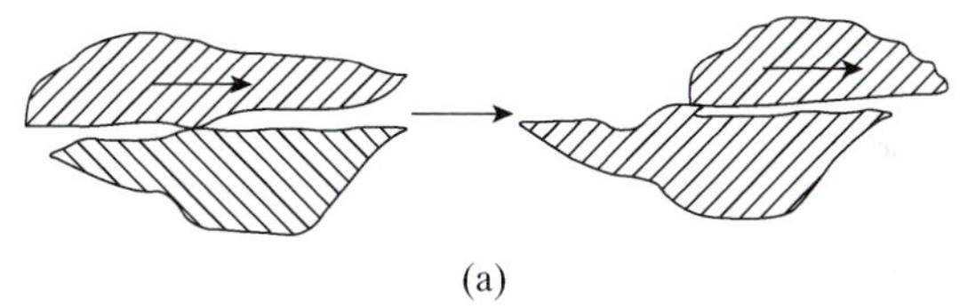

(a)

(b)

Figure 3–17
(a) Schematic of formation of an excrescence in galling; (b) galling damage on the polished conforming surfaces of special nuts after one use

members rub with little clearance. Soft single-phase metals (stainless steels) are particularly prone to galling. There is a standard galling test, ASTM G 98, which rates a couple's galling resistance with a threshold galling stress. A flat-ended round pin is rotated on end against a flat specimen of the same material or another material. The pin is loaded to an arbitrary apparent contact pressure, for example, 1 ksi, and rotated 360°. Both surfaces are examined for severe deformation in the track and excrescences. If none are there, new samples are loaded to a higher pressure. This is repeated until galling occurs. The threshold galling stress is the highest stress that the couple can withstand without galling. A 316 stainless steel couple would have a threshold galling stress of about 1 ksi, while the threshold for a hardened 440C stainless couple may be 20 ksi. This test is a good tool for screening materials for applications like plug valves that are prone to galling.

Scuffing* and *Scoring These are moderate forms of adhesive wear characterized by macroscopic scratches or surface deformation aligned with the direction of motion. If a piston in a cylinder is working properly, only mild wear occurs, the cylinder stays smooth, and its inside diameter just gets larger as wear occurs. When *scoring* (the preferred term) occurs, the surface is roughened. Scoring leads to unacceptable high wear rates. Overloaded or underlubricated gears often score, and the scoring can be measured in gear accuracy. It is a form of wear to be avoided.

Oxidative Wear This is also termed *mild wear*. It is the least severe form of adhesive wear. It starts by adhesion at the real areas of contact, then wear particles form and with repeated rubbing, these particles react with the environment and usually form an oxide. In ferrous systems, the rubbing surfaces appeared to be rusted (Figure 3–18). Almost all household door hinge pins develop oxidative wear because most homeowners neglect to oil them. Hard/hard couples such as hardened steels often display this form of wear. The rubbing surfaces appear to be rusted. They are oxidized, but from rubbing not from contact with water or chemicals.

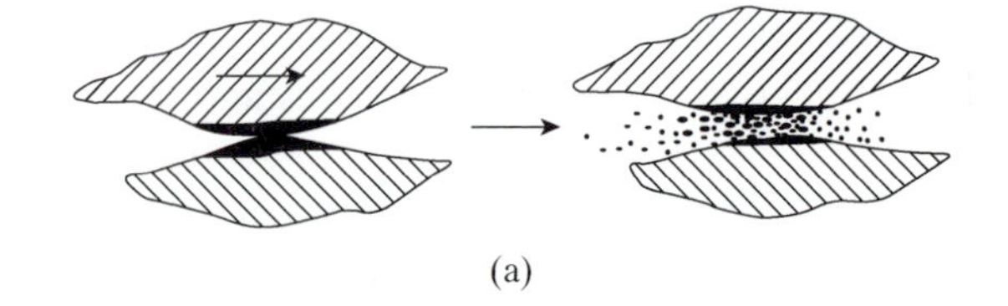

(a)

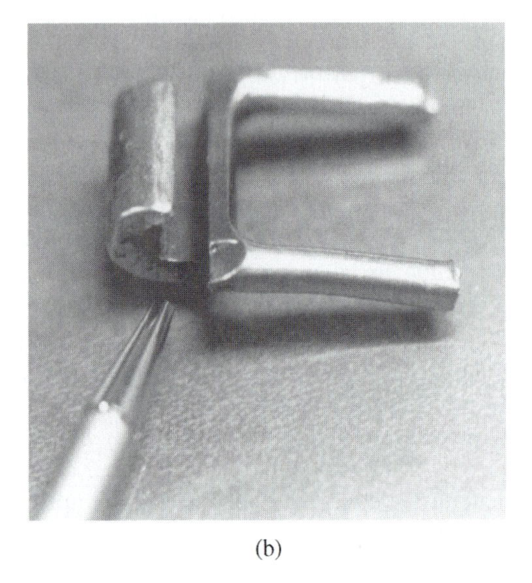

(b)

Figure 3–18
(a) Schematic of oxidative wear; (b) oxidative wear occurred from low-speed moving with a mating chain link (dark area)

Fretting This is oscillatory motion of small amplitude (less than 300 μm). It can produce *fretting wear* or *fretting corrosion*. Fretting wear is characterized by a surface that looks gnarled and pitted (Figure 3–19). Fretting corrosion looks like rust, with pits in the rusty area. With oscillatory rubbing of small amplitude, contacting surfaces locally adhere at asperities or

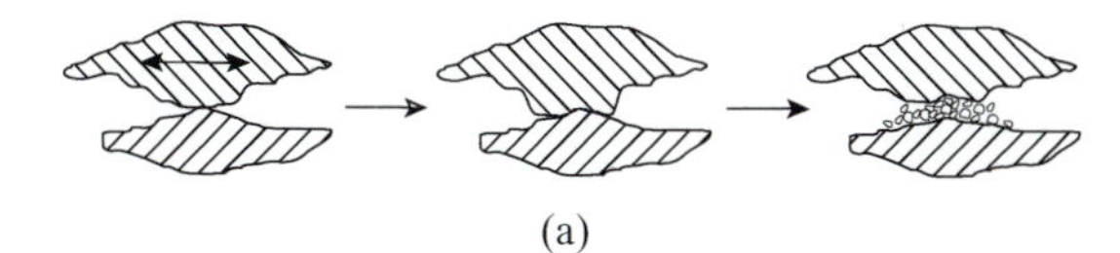

(a)

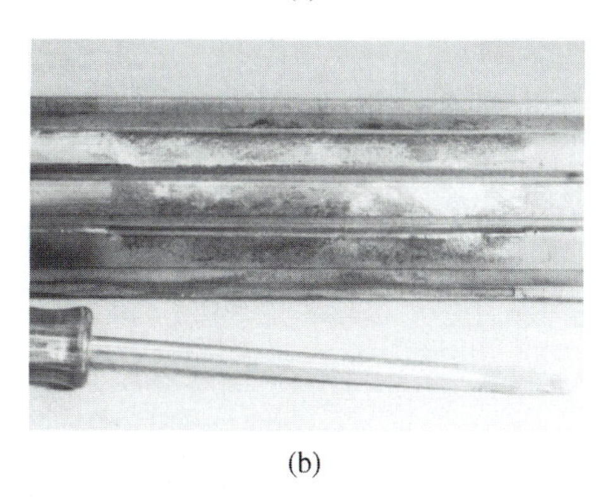

(b)

Figure 3–19
(a) Schematic of asperity interaction in fretting wear; (b) fretting damage on a splined shaft from relative motion of a mating part

up-features, the junction is broken, and, with repeated rubbing, wear particles roll back and forth in between the contacting surfaces. If the repeated rubbing causes the particles to react with the ambient environment, the damage is called fretting corrosion. If the surfaces and particles do not react with the environment, the result is fretting wear. Ferrous metals almost always display fretting corrosion under fretting conditions whereas plastics exhibit fretting wear. They do not react with the environment. There are no universally identified miracle couples that are immune to fretting damage. Most tribologists try to solve fretting problems by preventing the fretting motion.

All of the wear processes in the adhesion category at least start by surface adhesion. Also, it is typified by conforming surfaces in sliding contact. Motion can be unidirectional, rotational, or oscillating. Damage follows the Archard equation. Longer times and higher loads exacerbate the wear. High hardness reduces it.

Abrasion

The definition of *abrasive wear* is unintentional wear produced by a hard/sharp particle or protuberance imposed on and moving on a softer surface. Filing is abrasive wear, but more often, abrasion is produced by particles of hard substances. The hard particles can be fixed—as in sandpaper, pavement, or grinding wheels—or free to roll about between two conforming surfaces. The former is called *two-body abrasion*; the latter is called *three-body abrasion*, with the loose abrasive particles called *third bodies*. Your shoes wear by both two- and three-body abrasion. Rubbing on rough concrete produces two-body abrasion (fixed particles), and dirt particles between your soles and flooring or walkways produce three-body abrasion. Usually, two-body abrasion is more severe.

There are no agreed-to models for abrasive wear, but tribologists have adapted the Archard equation to apply by adding a factor relating to the sharpness of abrasive particles (B):

$$W = \frac{kdF}{h}B$$

$$B = 2\cot\theta/\pi$$

where θ is the included angle of the indenting point of an abrasive particle. The sharper the particle, the greater the wear.

As is the case with adhesive wear processes, the model is not developed to the point where it can be used to calculate wear solutions. It only tells the user what matters. We will now describe the different types of abrasive wear.

Low-Stress Abrasion This is characterized by fine scratches of the surface and is the type of abrasion that might be encountered in mixing dry sand and Portland cement with a hoe to make concrete. The tool is rubbed all over with sand and stones. Coal sliding down a chute produces low-stress abrasion as do cutting paper products and plastics containing glass fillers.

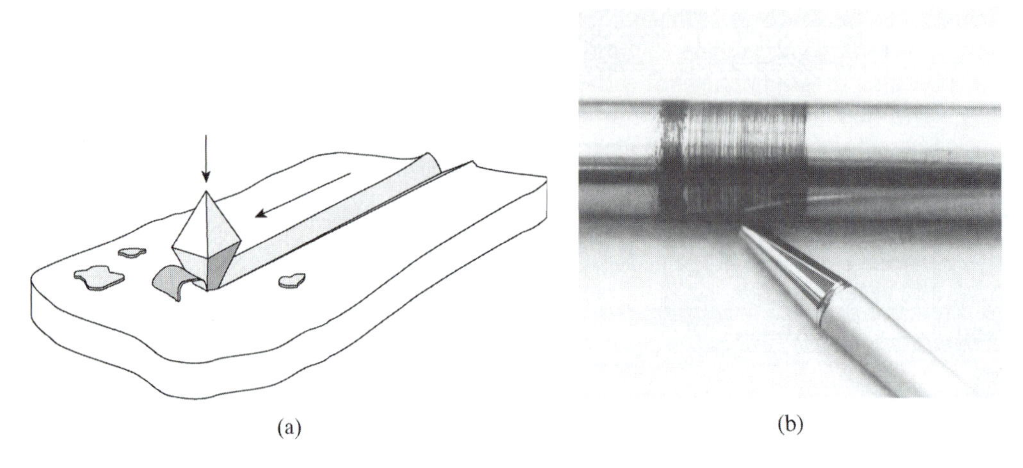

Figure 3–20
(a) Schematic of low-stress abrasion; (b) low-stress abrasion of a shaft from hard contaminants in a plastic bushing

The mechanism of material removed can be chip formation from a fixed grain sliding on a surface, or that grain could just produce a scratch—it plows a furrow, but does not fracture off a chip. Chips eventually are generated from the up-features of the furrows, but the metal removal mechanism is like that of grinding (Figure 3–20).

The ASTM G 65 dry-sand rubber wheel test simulates low-stress abrasion in the lab. A flat test specimen (of any material) is loaded tangentially against a rubber-tired wheel and a special sand is metered into the specimen/wheel interface. The wheel rotates against the stationary specimen for a prescribed number of revolutions. This is three-body abrasion, and wear volume is calculated from the specimen mass loss. This test correlates with many industrial processes such as punch press tooling, where abrasion to tools comes from dirty stock, nonmetallic inclusions, or abrasive work material.

High-Stress Abrasion This is a more severe form of abrasion in which the abrasive substance is imposed on the surface with sufficient stress to cause the abrasive to fracture or crush. Surface grinding is a form of high-stress abrasion. The abrasive grains or the grinding wheel fracture during the process. This kind of abrasion occurs in moving parts on tracks for bulldozers and similar earthmoving equipment. The mechanism of material removal is usually scratching. Furrows are plowed in the material from the sharp edges of the abrasive, and chips are formed. The usual way to simulate this form of abrasion in the laboratory is to rub the test material on a copper or other soft metal lap charged with the abrasive of interest. The classic rig uses a large copper ring with a conforming rider. Figure 3–21 shows high-stress abrasion in a trash grinder.

Gouging Abrasion This is damage to a solid surface characterized by macroscopic plastic deformation from a single impact. The damage is in the form of a gouge groove or deep scratch (Figure 3–22). It is the type of damage produced by dropping a large rock on a rigid metal surface or pressing the rock against the surface with sufficient force to crush it. This type of abrasion occurs in mining equipment and rock crushers and on power shovel teeth and buckets and similar earthmoving equipment. The ASTM test for a material's resistance to this form of wear (ASTM

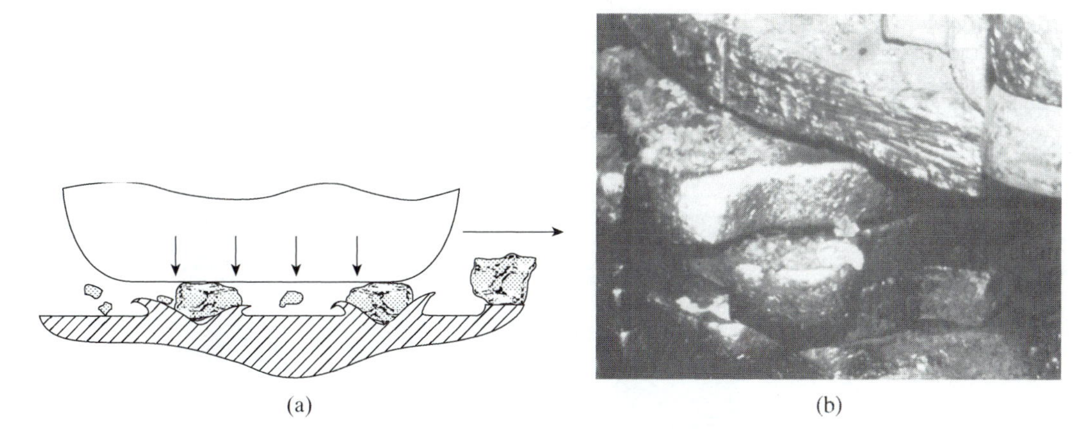

Figure 3–21
(a) Schematic of high-stress abrasion; (b) star wheels on a refuse grinder that have been subjected to high-stress abrasion. Wheels are 2 in. (50 mm) thick.

G 81) is a jaw crusher that crushes rocks from 1 to 2 in. in diameter to pebble size by a plate covered with a reference steel and the test material. Five hundred pounds of rock are crushed and the mass loss is measured on the test plates and reference plates. This is a special form of abrasion, but one that is important economically because it has to do with the cost of obtaining and processing ores, building roadways, and much behind-the-scenes infrastructure work.

Polishing This is produced by fine abrasives, and yet there are no scratches or furrows produced by the polishing abrasive. For example,

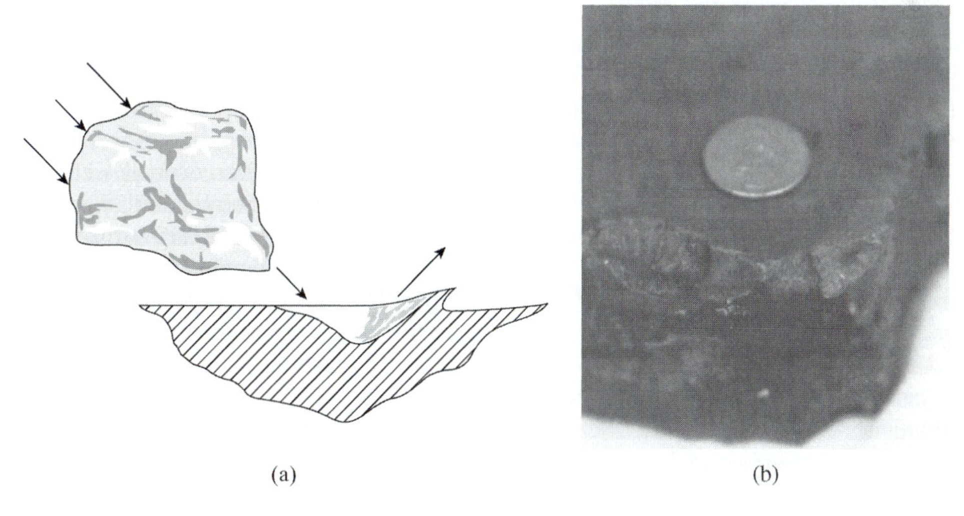

Figure 3–22
(a) Schematic of gouging abrasion; (b) gouging damage caused by grinding of rocks

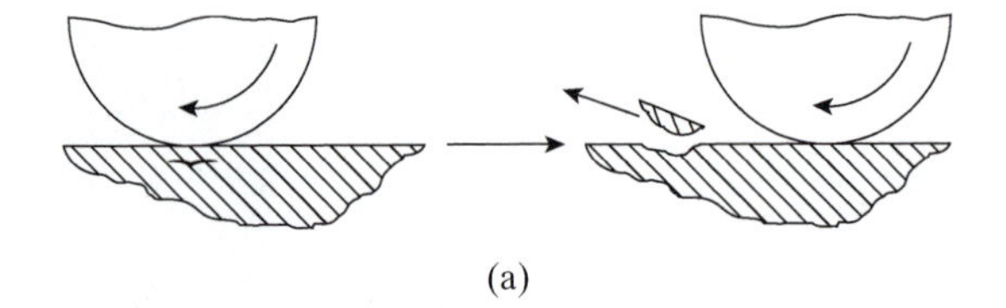

(a)

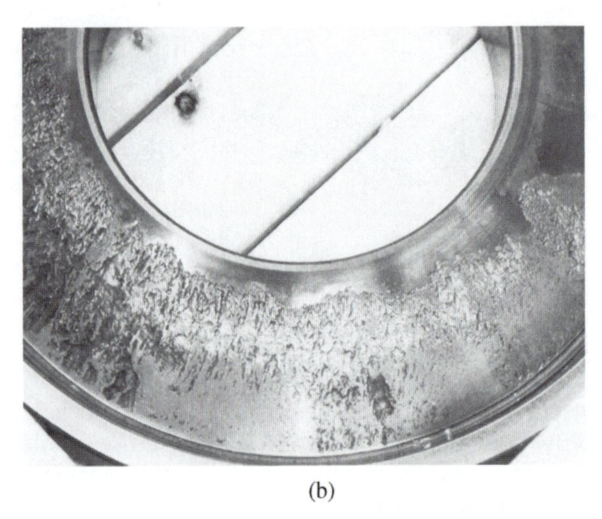

(b)

Figure 3–23
(a) Schematic of pitting due to surface fatigue; (b) pitting of a large roller thrust bearing race due to surface fatigue

polishing of metal is often performed on steel with aluminum oxide or diamond particles with a mean diameter of 1 μm (micrometer). These particles do not scratch the surface, but they remove scratches and polish. The abrasive particles remove oxides, the polishing fluids corrode the surface, and this removes material. This is the mechanism of material removal in planarizing silicon for the manufacture of integrated circuits. It is called *chemical mechanical polishing* because aluminum oxide and other abrasives are used in a liquid that is capable of attacking the material.

Surface Fatigue

Pitting This is one of the most common forms of surface fatigue. It is the formation of cavities when regions of a tribosurface spall. Figure 3–23 shows pitting on the rolling face of a million-pound *thrust bearing* from an extruder. Pits initiate as subsurface cracks, and the cracks propagate to allow ejection of a fragment. Once a metal or hard material fragment contaminates the rolling surface, it is rolled into the surface, other pits form, and shortly the entire bearing is ruined. Fatigue life tests are used to rate the life span that can be expected of rolling element bearings. Pitting also occurs in almost all heavily used railroad tracks.

Impact Wear This form of wear occurs on surfaces subjected to repeated impact. Riveting, anvils, and tools are subject to this type of wear, as are chisels, impact drivers, and hammers of all types. The mechanism of damage is usually plastic flow like that on the end of a well-used cold chisel. Sometimes, a subsurface crack occurs and fragments can spall from the surface. The example in Figure 3–24 is the end of a sledge hammer that failed in subsurface fatigue. The end fell off. Very special hardness profiles are used in high-quality hammers to minimize impact wear and subsurface fatigue.

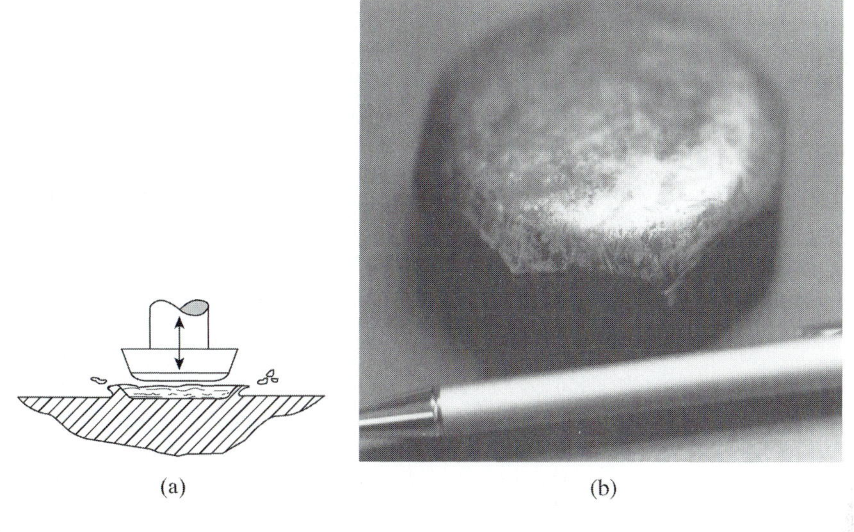

(a) (b)

Figure 3–24
(a) Schematic of impact wear; (b) impact wear on the striking face of a battering tool

Spalling This is the fracture of a portion of the surface of a material that is subjected to repeated stresses—usually compressive, usually Hertzian. *Spalling* can be caused by stresses other than those produced by rolling contact, such as thermal cycling stresses, but in mechanical systems, rolling elements are very susceptible. Figure 3–25 shows chromium plating spalling under ball contacts in a reciprocating tool with a rather short stroke. When a hard brittle coating or hardened layer is applied to a significantly softer substrate, Hertzian loading can

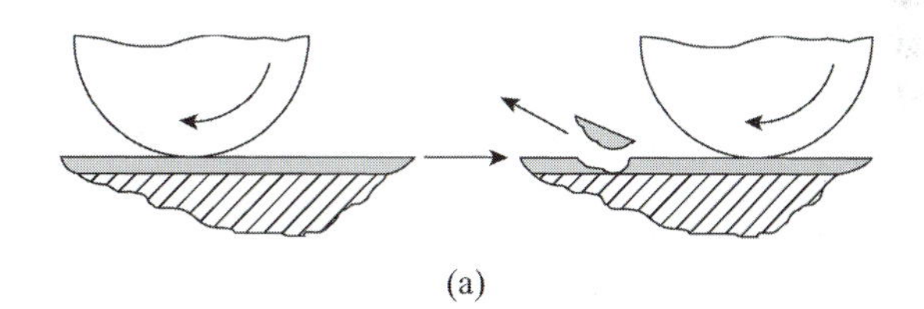

(a)

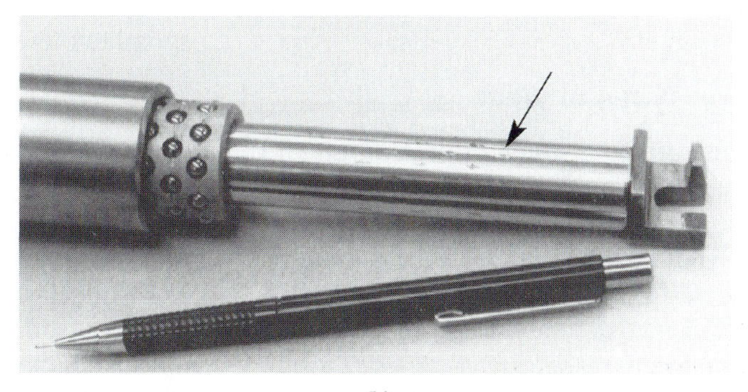

(b)

Figure 3–25
(a) Spalling of a coating from surface fatigue; (b) spalling of plating due to surface fatigue (oscillatory movement of about 5 mm)

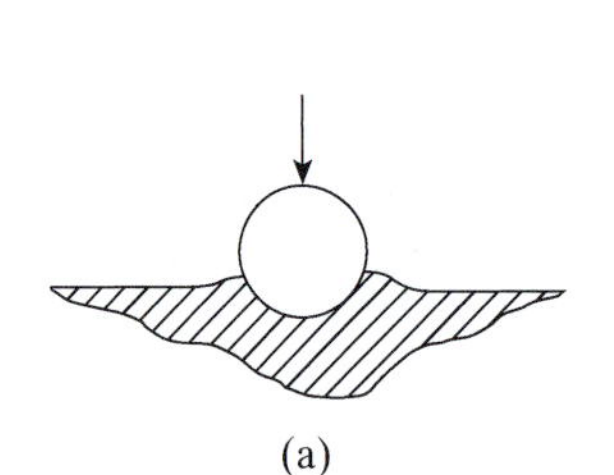
(a)

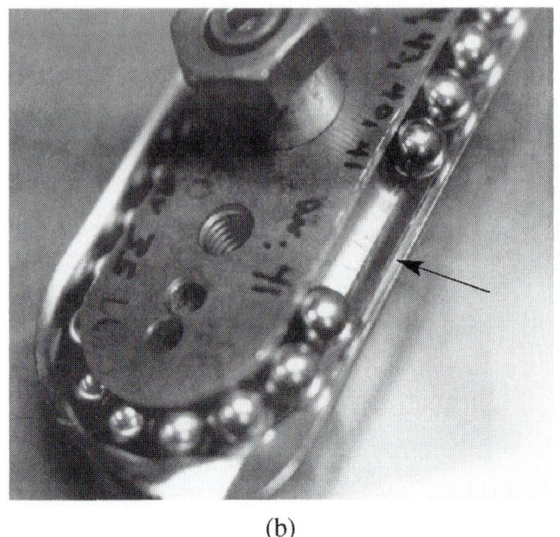
(b)

Figure 3–26
(a) Schematic of brinelling;
(b) brinelling of a bearing race by static overload

cause spalling. The usual solution is to back up thin hard coatings with a surface that is only about 20% softer. Graded coatings are used in physical vapor deposition (PVD) cutting tool coatings for stress accommodation. This is the best way to reduce the risk of spalling of hard surface layers.

Brinelling This is not really a wear process. It is indentation of a surface by static overload of balls or rollers (Figure 3–26). It is common to include it in wear discussions, because it happens on ball and roller bearings and it destroys the bearing in the same way that pitting and spalling destroy rolling element bearings. The solution to *brinelling* is to avoid shock loads on tribosystems that involve Hertzian contacts.

Summary: Types of Wear

We have described 5 kinds of erosion and 13 kinds of wear. It should be apparent at this point that the materials that prevent or reduce these different types of wear are different. A material is not wear resistant. Whatever the material is, it is unlikely that it can survive all 18 processes. Certain materials resist certain types of wear, and we will discuss the wear resistance of specific materials in their respective chapters. The material characteristics that control wear start at the atomic level and progress to the surface texture of a part (Figure 3–27). At the atomic level, adhesion of surfaces can be affected by the nature of contacting atoms. On the nano level, dislocations in both members may control wear characteristics. At the mezzo-scale level, the grain diameter, or the size of the second phase, may play a role in determining wear characteristics. Next, the surface asperities can affect wear characteristics and, at the bulk level, the errors of form on conforming surfaces can affect wear behavior. We will focus on the microstructure level in our wear discussions in later chapters on various types of materials. We will also present data on how the various material systems respond to these different types of wear.

3.7 Bearings

As shown in Figure 3–1, tribology encompasses more than just the material aspects of friction and wear. The mechanical engineering aspect of tribology addresses types of bearings, the fluid mechanics of fluid lubrication, and the selection of other tribocomponents such as gears, clutches, brakes, seals, power transmission equipment, and, since about 1990, prosthetic devices. It is the purpose of this section to briefly discuss one of the most important parts of tribology—the technology of bearings. Bearings come in many forms, and they are the simplest triboelement. Bearings are devices that support and locate rotating or sliding elements with respect to other parts of a mechanism or structure. Conceptually, there are only two types of bearings—flat pad and those that support or slide on a cylindrical element (Figure 3–28). Any type of motion can be accommodated by a combination of flat pad and revolute bearings. The objective of this section is to briefly discuss plain and rolling element bearings, because a major use of engineering materials is making these types of triboelements. Bearings can often be the limiting factor in a design application. For example, jig grinding small-diameter holes to micrometer accuracy requires a tiny wheel, but these wheels must be rotated at a certain speed for them to work. On a 1-mm-diameter hole, the required cutting speed may

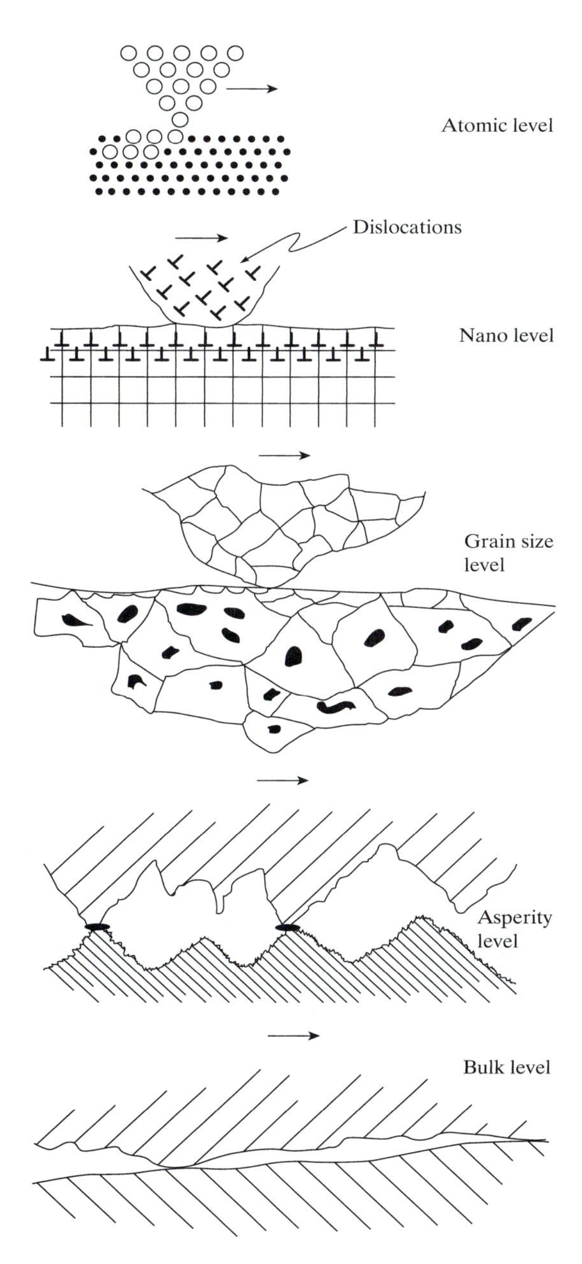

Figure 3–27
Factors that affect wear and friction at different scales

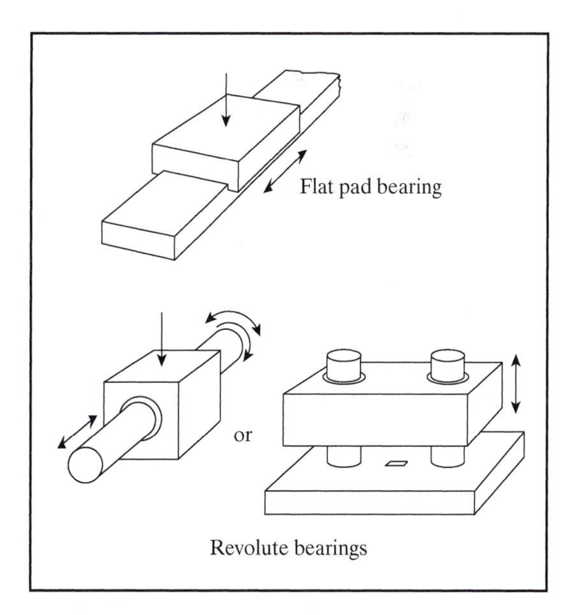

Figure 3–28
Fundamental categories of bearings

translate to a spindle rotation of one million revolutions per minute. To date, we have not uncovered a jig grinder spindle that can meet this requirement. Most present-day bearings cannot accommodate this speed.

This section will briefly describe plain bearings, types, and materials and then rolling element bearings and some factors to consider in their selection and use. The last section in this chapter will briefly discuss types of lubrication for triboelements.

Plain Bearings These are bearings that have conformal contact with the sliding surface. They can have countless shapes, but the common feature is that they are the sliding surface for some geometric shape, and the rubbing occurs within a defined area of contact as opposed to *rolling element bearings*, where the load is carried by balls or rollers that have Hertzian contact with the triboelements involved. Another way to look at it is this: *Plain bearings* are two-body sliding systems (the bearing and what slides against the bearing), and rolling element bearings are three-body tribosystems (stationary surface–ball/roller–moving member).

Plain bearings can operate with hydrodynamic lubrication conditions if the lubrication, bearing design, and operating conditions are right. There can be lubricant film separation of the rubbing surfaces, and wear will only occur at startup and shutdown. Similarly, the sliding member can be supported by pumping a pressurized fluid into the bearing–moving member interface to create a hydrostatic (liquid as the fluid) or air bearing (gas as the separating fluid). The greatest number of plain bearings in use are oil-impregnated porous metal (P/M) or self-lubricating plastic bearings. These are called bushings if they have a cylinder configuration. P/M plain bearings are typically made from steels or bronzes and are pressure and vacuum impregnated with lubricant, usually mineral oil. The oil trapped in the pores wicks into the bearing clearance in operation, and hydrodynamic lubrication is possible. Plastic plain bearings can be any shape, and the lubricating substances can be fluorocarbons or solid-film lubricants such as molybdenum disulfide or graphite. Large flat pad and revolute bearings are often made from soft metals such as lead, tin, and combinations thereof (*babbitts*). The soft metals can embed contaminants that find their way into bearings, and they can tolerate the sliding at startup and shutdown. Under steady-state operation, an oil film keeps the soft metal and moving member separated. We will discuss the plain bearing characteristics of various engineering materials in subsequent chapters.

Plain bearings are the lowest-cost bearings, and they are the best choice for many applications. The competition is rolling element bearings, and plain bearings have the following major differences with rolling elements.

Cost	Lower than rolling element
Precision	Not as good, require a running clearance
Friction	Not as low, rolling friction is usually lower than sliding friction
Speed	Limited to lower speed than rolling element
Load capacity	Can be better than rolling element (large contact area)

If you are designing a low-speed motion device, plain bearings will probably be the cost-effective choice. If you are designing a cooling fan that runs over 3000 rpm, ball bearings are preferred. If you are designing a bearing for an automobile trunk lid, a self-lubricating plastic plain bearing may be the appropriate choice. The designer must weigh the pluses and minuses of each system.

Rolling Contact Bearings Greased precision rolling element bearings are usually the first choice for manufacturing equipment that must run continuously. In 2004 in the United States, it is common to rank the effectiveness of manufacturing equipment by a number that is essentially the number of hours that a machine produces good product (in a day, week, or month) divided by the hours that the machine was available to run. This kind of scrutiny makes it essential that machines run a lot, preferably around the clock. It is not uncommon to schedule only two weeks' shutdown a year in an operation that runs 24 h a day. Greased rolling element bearings are often used in this type of environment, but a number of concerns need to be addressed if these bearings are to last as long as intended.

Bearing Life If a bearing is properly selected, mounted, lubricated, and maintained, it will eventually reach its normal life and fail due to fatigue. For this reason, the life of an individual bearing is defined as the total number of revolutions or hours at a given constant speed that a bearing runs before the first evidence of fatigue develops. The American Bearing Manufacturers Association (ABMA) established a standard method of evaluating load and life ratings of ball (ANSI/ABMA 9, 1990) and roller (ANSI/ABMA 11, 1990) bearings. This standard method is called *rated life* (L_{10}) and should be initially used and specified.

The rated life is the number of revolutions (or hours) at a given constant speed that 90% of an apparently identical group of bearings will survive before any fatigue is evident. The basic load rating (C) is the radial load that a ball bearing can withstand for one million revolutions of the inner ring. Its value depends on bearing type, bearing geometry, accuracy of fabrication, and bearing material. The equivalent load (P) is the constant, stationary radial load, applied to a bearing with rotating inner ring and stationary outer ring, that gives the same life span the bearing will attain under actual load and rotation conditions. The following equation is used to determine the rated life for *ball bearings* and *roller bearings*.

$$L_{10} = \frac{16{,}700}{N}(C/P)^k$$

where L_{10} = rated life, revolutions
C = basic dynamic load rating, lb
P = equivalent radial load, lb
k = constant: 3 for ball bearings, 10/3 for roller bearings
N = rpm

Selection of Proper Bearing Precision The American Bearing Manufacturers Association has established industry standards for bearing tolerances (ANSI/ABMA 20, 1996). The Annular Bearing Engineering Committee (ABEC) functions under the jurisdiction of the American Bearing Manufacturers Association (ABMA), which advances standards to international status. There are five precision grades (ABEC 1, 3, 5, 7, and 9). Each precision grade has increasingly better controlled tolerance on the outside diameter (OD), inside diameter (ID), width, inner race, and outer race run-out. ABEC 1 bearings usually can be used in 80% of the applications; ABEC 7 and 9 are normally specified in extreme precision applications; ABEC 3 and 5 are seldom specified. U.S. bearing manufacturers producing standard bearings must conform to these standard dimensions and tolerances. Other countries have similar standards.

Selection of Bearing Type per Application Rolling contact bearings consist of ball (Figure 3–29) and roller bearings (Figure 3–30). There are over 40 different types of ball bearings, each with its unique operating characteristics and capabilities. Ball bearings typically have higher speed and lower load capabilities than roller bearings. Deep-groove radial ball bearings are the most widely used bearing in industry; they represent approximately 80% of the applications. This bearing type can accommodate 70% of its radial capacity in thrust. Spherical roller and needle bearings are in general the most common type of roller bearings.

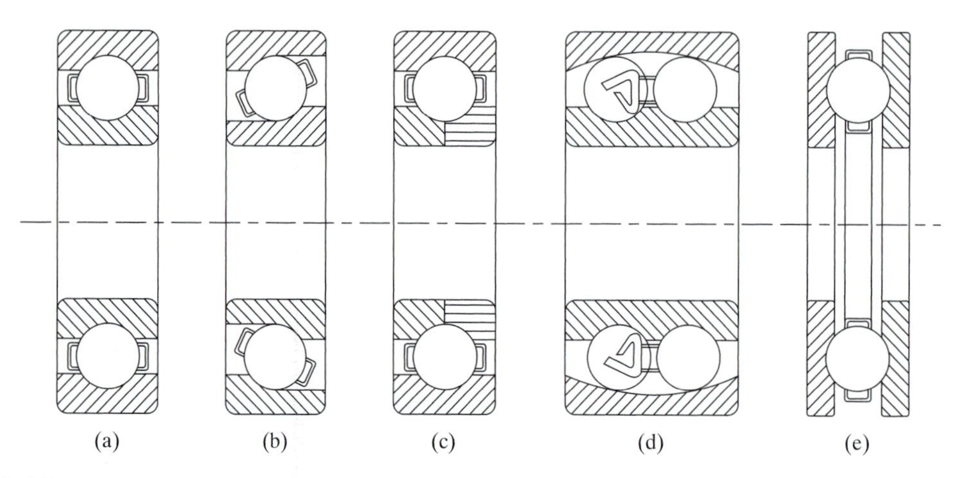

Figure 3–29
Examples of various types of ball bearings: (a) deep groove, single row (for radial loads, limited axial); (b) *angular contact* (takes radial load and axial loads in one direction); (c) split inner ring (special applications); (d) self-aligning double row (tolerates some misalignment); (e) ball thrust bearing (for axial loads only)

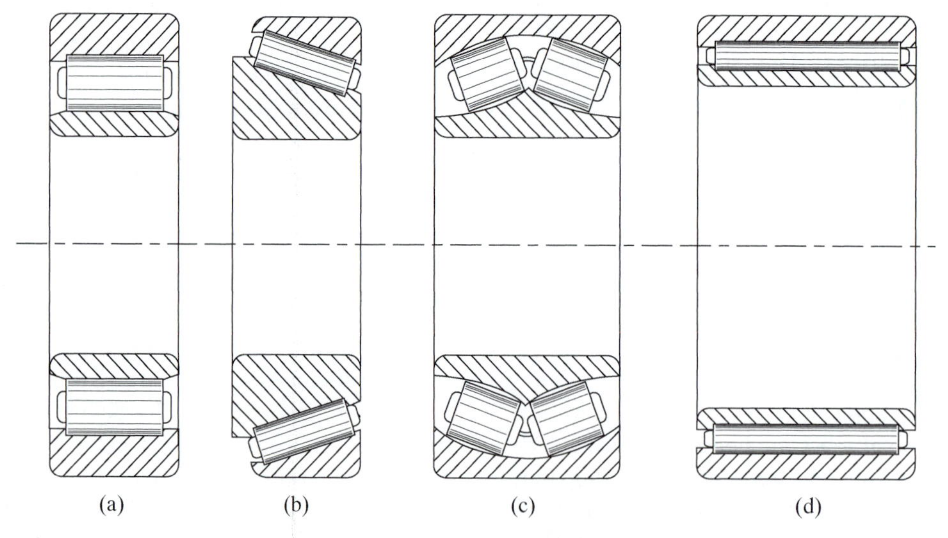

Figure 3–30
Examples of roller bearings: (a) cylindical (accommodates high radial load, no axial loads); (b) tapered (accommodates radial and axial loads); (c) spherical (accommodates misalignment); (d) needle (high radial load capacity for its size)

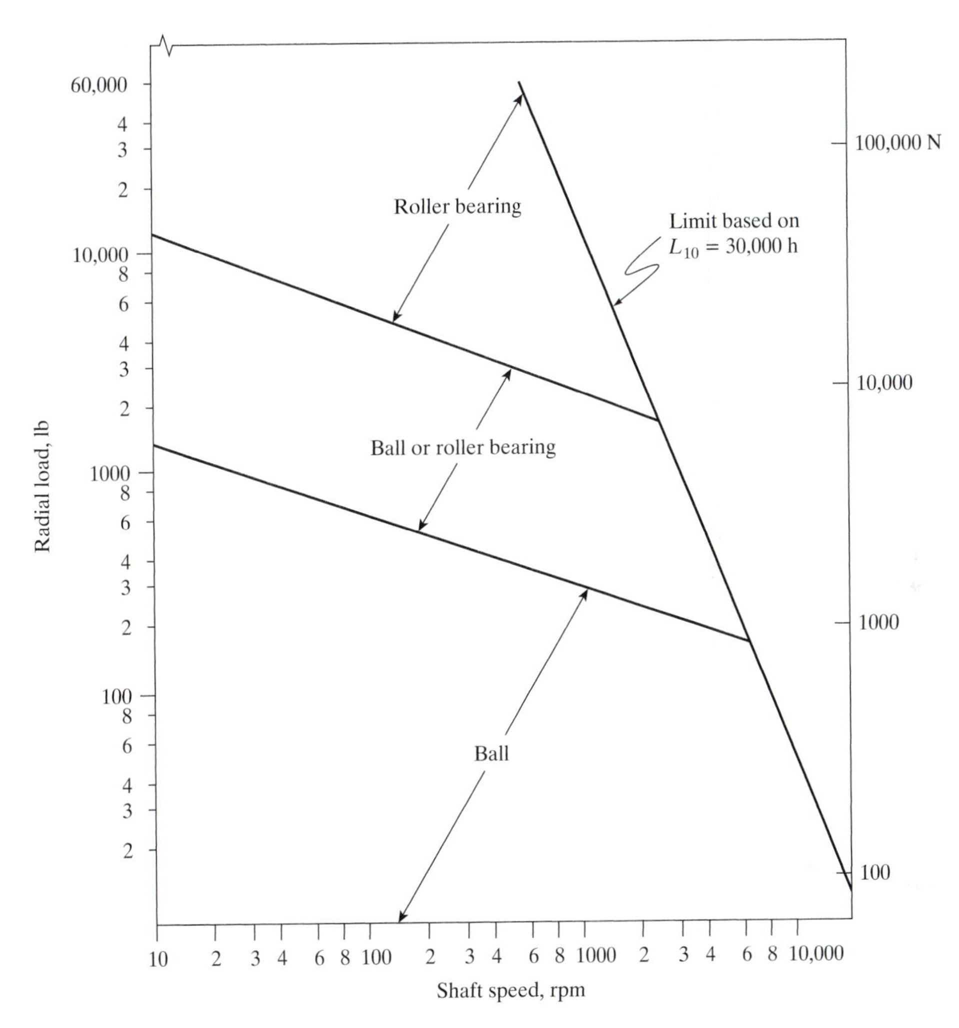

Figure 3–31
Ball versus roller bearing selection

Figure 3–31 can be used to initially determine if a ball or roller bearing is required for an application based on load and speed conditions only. Actual selection of the exact type of bearing (tapered roller, maximum capacity ball, spherical roller, needle bearing, cam follower, etc.) is based on the application and operating requirements.

Required Internal Clearance The operating performance of rolling contact bearings is directly related to the amount of internal clearance in the bearing. Internal clearance is the freedom or looseness of the bearing rings in relation to each other. The amount of internal clearance influences noise, vibration, heat buildup, and fatigue life. The selection of the initial internal clearance of a bearing is mostly affected by press fits and temperature gradients in the system. Internal clearances are arranged into classes as shown here. Class 3 bearings are becoming the standard internal clearance specification.

a. Extra-Loose; ABMA Class 4 Fit

b. Loose; ABMA Class 3 Fit

c. Standard; ABMA Class 0 Fit

d. Tight; ABMA Class 2 Fit

Selection of Cage Materials *Cages* are sometimes called separators or retainers. The purpose of the cage is to equally space the rolling elements (balls or rollers) and prevent contact between them. The cage forms a pocket around each rolling element while still allowing it to rotate freely. The cage also enables the grease to remain in the bearing to provide for effective lubrication.

The style of the cage and the material are a critical part of the rolling contact bearing, especially as it relates to the speed limit. The speed limit of a bearing is based on several factors such as lubricant type, bearing material, and cage material. Following are the most common types of cages specified.

a. Molded nylon cages are the standard for most small ball bearings along with two-piece pressed steel cages.

b. Miniature ball bearings normally use one-piece steel snap-in cages.

c. High-precision bearings (ABEC 7 and 9) are normally made from one-piece phenolic cages.

d. Large roller bearings are normally made from machined solid brass.

Selection of Shaft and Housing Fits and Material Specifications ABMA and each bearing manufacturer recommend tolerances (for shaft and housing fits) for mounting bearings under various applications and conditions. Without the proper mounting, bearings will have reduced life. A loose fit between the shaft and bearing inner ring can lead to relative movement and can cause wear or fretting corrosion. Normally, a loose fit is used on the outer ring (if stationary) to allow for assembly and thermal expansion. Therefore, it is important to consider the following criteria when selecting bearing fits:

a. **Accuracy required.** As the application becomes more precise, bearing tolerances and fits become closer.

b. **Rotational speed.** As the speed increases, mounting must be more accurate and a precision bearing must be used.

c. **Load requirements.** The greater the load or shock, the more rigid the mount must be.

d. **Temperature requirements.** Under high-temperature conditions, clearance for thermal expansion must be considered and fits must be loosened accordingly.

e. **Economics.** Initial cost of construction and ease of maintenance should be considered. The more precision required, the more expensive the manufacture.

f. **Ring rotation.** Which ring is rotating, inner or outer? Generally, the rotating part should have the interference fit.

g. **Materials.** Determine the composition of the housing and shaft materials. Varying strengths and thermal expansion rates make the proper fit different for each material.

Selection of Bearing Material The U.S. bearing industry has used SAE 52100 steel (at 60–62 HRC) as the standard bearing material. This is a high-carbon, chromium steel that is usually manufactured using an induction vacuum-melt process that provides for high dynamic load capacities and increased reliability because of lower inclusion levels. For corrosive environments, AISI 440C (58 HRC) stainless steel is the U.S. standard; however, it has lower dynamic capacity than SAE 52100. Other common materials available include

a. AISI M50 tool steel that is used for high temperatures up to 660°F (315°C)

b. Hybrid bearings using steel rings and silicon nitride ceramic balls, which provide for increased load and speed capacity

Other Bearing Design Factors Are

- Reliability requirements of the system
- Maintenance considerations for relubrication
- Operating torque level requirements
- Selection of proper enclosures (seals/shield, types of materials)
- Quality of bearing procured (all bearings manufacturers are not equal)
- Temperature and speed limits of bearing types
- Misalignment capabilities

3.8 Lubricants

Lubricants by definition are substances that separate rubbing surfaces and readily shear while adhering to the surfaces. As shown in Figure 3–32, there are three major categories of lubricants: oils, greases, and solid-film lubricants. Oils and greases can be made from crude oil or from chemical feedstocks. The former are called *mineral oils*, the latter, *synthetic oils*. *Greases* are a combination of an oil, a *thickener*, and an *additive* package. Most greases are between 80% and 90% oil. The bulk of the remainder is the thickener, which is like a three-dimensional sponge that releases oil when the device is in use and retains the oil when not in use. *Solid lubricants* are usually applied as coatings to metals and are compounded (mixed) into plastics. They develop films between rubbing surfaces, which reduce friction and contact of the rubbing surfaces. Graphite is an ancient solid-film lubricant.

In the twentieth century, an average machine designer did not have to concern himself or herself with lubricant selection. Large companies usually had a lubrication engineer who took care of these matters. After the corporate downsizing and mergers of the 1990s, many designers find themselves faced with specifying lubrication for a tribosystem. Thus, this section is intended to present some guidelines on how to go about this task.

Oil

Oil Selection There are countless oils commercially available. Which should you specify for a machine that you are designing? The short answer is to use the recommendations of the component manufacturer. For example, if you are putting a speed reducer on your machine,

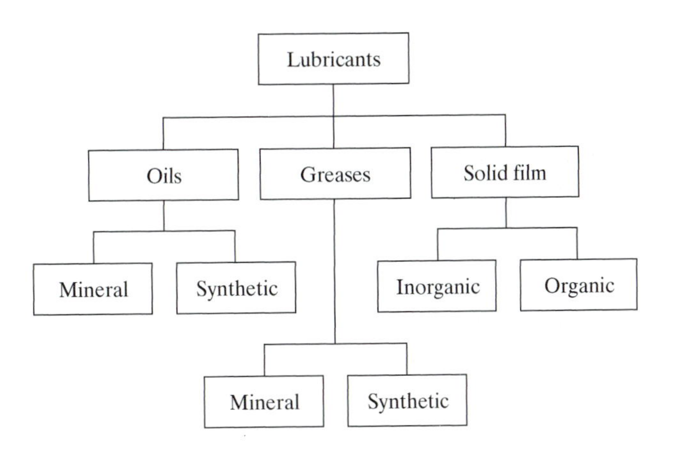

Figure 3–32
Types of lubricants

the manufacturer will likely specify a type of oil. If you have to specify a manufacturer and a grade, you will probably need more information than we can present in this limited discussion, but we will at least give you a start.

Mineral oils are crude oil that is refined to remove certain molecular fractions. Refined crude yields a base oil (paraffinic, naphthenic, or aromatic). Commodity oils are formulated from base oils by adding chemicals to alter the pour point (low-temperature fluidity), viscosity modifiers, and additions that are intended to make that oil superior to the competition. *Viscosity index* (VI) modifiers are important because they control the change in viscosity with heat and, as shown in Figure 3–9, viscosity is a key parameter in film formation. Additives include chemicals to form bonds to the rubbing surface (ZDDP), oxidation and corrosion inhibitors, agents to reduce foaming, friction modifiers, *detergents*, and so on. Additives can sometimes make up as much as 30% of the volume of the oil.

Mineral oils are refined from crude oil extracted from the earth. They contain a broad mix of hydrocarbon molecules, which are separated in refining. Synthetic oils have all the same molecules as the *base stock*. They are manufactured that way in chemical processes. The most common synthetic oils are polyalphaolefins (PAOs). Other synthetic base oils include various esters and silicones. These oils can be 5 to 20 times as expensive as mineral oils, but they have better heat resistance, longer life, less evaporation, and, in some cases, they produce lower friction in sliding systems than mineral oil. The use of these lubricants is often an economic decision. The cost of more frequent oil changes and machine downtime must be weighed against the five times higher cost. Some fleet truck companies claim net fuel savings of several percent with the use of synthetic lubricants. There are some risks. Some synthetic oils attack conventional rubber seals in gear boxes, transmissions, and engines. This needs to be investigated, but PAO is used the most in these areas. It can also be blended with mineral oil. In the year 2004, most industrial users of oils tend to limit synthetic oils to strategic uses such as special gears, compressors, engines, and hydraulic fluids—severe applications.

It is appropriate to use oil for systems that can be splash lubricated or lubricated by a pumping system. Oils remove heat from tribosystems, and greases are poor in this regard. If an application pushes the temperature limits of mineral oils, synthetics are justified. Mineral oils are suitable for most applications. The viscosity index (VI) is the usual measure of an oil's ability to accommodate elevated temperature: the higher the VI, the better the high-temperature ability to provide film separation. Multigrade mineral oils have high VIs. The viscosity needed for an application (and the specific type) is usually recommended by the device manufacturer. Ball bearings may require a viscosity of 13 centistokes (CSt); roller bearings, 21; and thrust bearings, 30. The overall message is to not take the oil for granted if it is a critical application. Probe the matter and satisfy yourself that the issue has been adequately addressed and resolved. Liquid lubrication of plastic bearings must be done judiciously. Lubricants can attack the plastic. We recommend checking with the plastic manufacturer before using oils (or grease) for lubrication. Self-lubricating plastics usually do not need additional lubrication.

Greases

It is the oil in a grease that does the lubricating. Oil makes up 90% of most greases, and the oil can be mineral or synthetic. The reasons for choosing one over the other are the same as mentioned with oils. We have been stressing that rolling element bearings are the preferred bearings for manufacturing machines and strategic equipment. Grease lubrication is usually the lowest-cost way to lubricate a rolling element bearing. There are three main categories of greases based on thickener type:

Maximum Operating Temperature	*Thickener Type*
150°F (121°C)	Lithium
	Sodium
350°F (177°C)	Polyurea
	Lithium complex
	Aluminum complex
>350°F (177°C)	Bentonite clay
	Polytetrafluoroethylene
	Perfluoropolyether

Bearing speed also affects the type of lubricant that is used. For revolute bearings, a *DN* value is commonly used in the United States to rate the "speed limit" of bearings, and the *DN* value is different when lubricated by oil or grease. In fact, it can be used to determine if a bearing should be lubricated with grease or oil. The *DN* value is numerically equal to the product of the bearing bore in millimeters (mm) and the rotational speed in revolutions per minute.

$$\text{DN} = \text{bore (mm)} \times \text{speed (rpm)}$$

Table 3–1 compares the DN limits for greases and oils at a particular temperature.

These data show that bearings have higher DNs with oil rather than grease lubrication. However, greases should be the lubricant of choice for most applications. Grease is the most convenient and the easiest to use, and, with seals, it is possible to lubricate a device for over 20 years. Oil lubrication requires continual checking of oil supply.

Table 3–1
DN limits for greases and oils

Bearing Type	Grease	Oil
Single raw ball	200,000	300,000
Double raw ball	160,000	220,000
Spherical roller	150,000	170,000
Cylindrical roller	170,000	200,000
Tapered roller	150,000	170,000

Solid-Film Lubricants

Solid-film lubricants are useful where oils and greases simply cannot be tolerated (e.g., sanitary systems, vacuum, in liquids, etc.). As shown in Figure 3–32, solid-film lubricants fall into two main categories—organic and inorganic. Polymers are the most important organic solid-film lubricants, with fluoropolymers being the most popular. The list of inorganic solid-film lubricants is long (Table 3–2). Of course, graphite belongs in the list, but theoretically it is organic and belongs with the polymer group.

These materials are applied as coatings from a fraction of a micrometer to 50 μm or more in thickness. Application processes include physical vapor deposition, spraying, burnishing, and impingement.

Each of these solid-film lubricants has its application niche, and selection can be guided by Web site information from suppliers of these lubricants. There are suppliers who will coat your parts or sell you the material for in-house coating. One supplier compares the top three in popularity with limits in apparent pressure:

$$\text{Molybdenum disulfide} = 100{,}000 \text{ psi}$$
$$\text{Graphite} = 40{,}000 \text{ psi}$$
$$\text{PTFE fluorocarbon} = 6{,}000 \text{ psi}$$

These data suggest that molybdenum disulfide is the most robust coating, but it is not always the

Table 3–2
Solid-film lubricants

Molybdenum disulfide	Indium
Tungsten disulfide	Silver
Antimony trioxide	Zinc oxide
Boric acid	Bismuth
Calcium fluoride	Lead oxide

best choice. It rapidly oxidizes at use temperatures over 750°F (400°C), whereas some of the others on the list do not. In summary, solid-film lubricants should be a tool for solving lubrication problems where oils and greases do not work. Selection of a specific type will probably require testing and evaluation in conjunction with supplier recommendations.

Summary

We mentioned a number of tribotests in our descriptions of wear modes. This chapter can be summarized with the statement that friction wear and lubrication models are not developed to the point where they can be used to supply specific answers to specific problems. We can say with relative certainty that increasing velocity and/or load will increase wear, but we cannot calculate the amount without the empirically derived wear coefficient in the Archard equation. In the year 2004, tribology depends heavily on tribotesting. Screening materials, lubricants, and devices in the lab is the current status.

Wear and friction testing is not a problem. The ASTM G 2, D 2, D 20, and various other committees have developed tests that address just about every form of tribology problem, and there are more standard grease oil tests (in ASTM) than we care to go into. Wear tests are necessary for many situations and Table 3–3 lists some common friction and wear tests. The

Table 3–3
Some standard tribotests

Purpose	Description	ASTM No.
Friction Testing Standards		
Web vs. roller	Capstan test rig	G 143
Conforming surfaces	Sled test	D 1894
Waxed surfaces	Paper clip/inclined plane	G 164
Testing and recording data	Lists many different tests	G 115
Erosion Testing Standards		
Cavitation	Vibratory horn immersed in fluid	G 32
Liquid impingement	Samples on a rotor in a rain field	B 73
Solid particle impingement	Sandblast nozzle aimed at surface	G 76
Slurry erosion	Miller number test	G 75
Cavitation	Water jet	G 134
Wear Testing Standards		
Low-stress abrasion	Dry-sand rubber wheel	G 65
Metal to metal	Block on ring	G 77
Gouging abrasion	Crushing rocks with metal plates	G 81
Metal to metal	Crossed cylinder	G 83
Galling	Pin rotation on flat	G 98
Screening mating couples	Pin on disk	G 99
High-stress abrasion	Pin on sandpaper	G 132
Reciprocating sliding	Pin reciprocating on flat	G 133
Plastic to metal	Plastic block on metal ring	G 137

following are some factors to keep in mind in dealing with friction wear and lubrication:

- Friction is a system effect (a material does not have a coefficient of friction).
- Friction data for engineering problems should be obtained from tests designed to simulate the operation under study.
- Always start a wear investigation by determining the wear mode in the system of interest.
- Use wear tests that produce wear that looks like the problem being addressed.
- Wear tests can have high variability—test enough replicates.
- Always lubricate a sliding system if possible.
- Rolling element bearings are usually preferred over plain bearings for high-speed applications.
- Greases are the lowest cost lubricants for rolling element bearings.
- Solid-film lubricants are useful on sanitary systems.
- Synthetic oils and greases may attack rubbers and plastics not attacked by mineral oils. Check compatibility before using them.
- Elevated temperatures (over 120°C) lower lubricant load-carrying capability and may require special lubricants.
- Self-lubricating plastics versus metals is a candidate tribosystem where lubricants cannot be tolerated.

Critical Concepts

- Friction and wear are system effects; they are controlled by what rubs on what and the conditions of rubbing.
- Lubrication always reduces wear and friction, and should be considered in every tribosystem design.
- Wear and friction models are not developed to the point where they can be used like mechanical properties in design calculations.
- Specific oils and greases (and their application) should be specified on critical tribosystems.

Terms You Should Remember

galling
adhesive wear
abrasive wear
surface fatigue
seizure
high-stress abrasion
low-stress abrasion
polishing
fretting
fretting corrosion
fretting wear
scoring
scuffing
gouging abrasion
brinelling
spalling
pitting
friction
coefficient of friction
static friction
kinetic friction
contact area
contact mechanics
Hertzian stress
lubricant
grease
viscosity
lambda ratio
film thickness
boundary lubrication
hydrodynamic lubrication
elastohydrodynamic lubrication
solid-film lubricant
stick-slip friction
mineral oil
synthetic oil
DN
Archard equation
additive
friction modifiers
detergent
viscosity index
thrust bearing
rolling element bearing
ball bearing
roller bearing
plain bearing
angular contact bearing
base stock
thickener
rated L_{10} life

Case History

Tenters Used in Making Plastic Films

Plastic films are everywhere: packaging, movie film, magnetic media, even shrink-wrap for boats. The stronger plastic films get their strength from oriented polymer molecules. Films start by casting plastic on a wheel or metal band, and then they are stretched axially (drafting) and widthwise (tentering) to increase their strength. Drafting is accomplished by overdriving the rollers that pull the plastic web through the machine. Tentering is more complicated. In fact, tenter mechanisms are a tribological nightmare. Clips grab the sides of the web. These clips are attached to diverging chains, like big bicycle chains, that may be hundereds of feet long. The clips have bearings to allow opening and closing and are attached to the chain with more plain bearings, which also have to go around a sprocket at speeds that may be several hundred feet per minute. In addition, the tentering section of the machine is hot, sometimes much hotter than petroleum lubricants can tolerate.

Many plain bearing materials were tested in the laboratory and field to find a material for clip and chain bearings. Candidates ranged from dry-film lubricated metals, to carbon products, to plastics, to ceramics, to composites. Glass/fluorocarbon composites emerged as the winners in the 1990s, but as new self-lubricating plastics are developed, they will probably find their way into screening tests for this application. In 2003, bearing life span was still only about a year, and continual cost reductions will try to improve on this life span.

Questions

Section 3.1

1. What is the origin of friction?
2. How is friction measured?
3. Why is friction independent of area of contact?
4. Compare the contact area for a 0.5-in.-diameter ball under a load of 10 lb on aluminum and on steel. Compare the contact stress also.

Section 3.3

5. Describe boundary, elastohydrodynamic, and hydrodynamic wear.
6. If a bearing has a surface roughness of 1 μm and a lambda ratio of 0.8, what is the film thickness?
7. Explain the role of viscosity and speed in lubricant film thickness.
8. A 15-lb block rests on a horizontal surface. In order to start the block sliding, a horizontal force of 4.5 lb must be applied. What is the coefficient of friction between the block and the surface? At what angle would the surface need to be inclined so that the block would just begin to slide?
9. What happens to the friction of a rolling element bearing at low speed?

Section 3.5

10. What is wear?
11. What is erosion?
12. State five kinds of erosion.
13. Explain fretting wear and corrosion.
14. Describe surface fatigue wear in gears.
15. How does cavitation damage metals?

Section 3.6

16. Describe three important types of bearings.
17. List four advantages of ball bearings over plain bearings.
18. List four advantages of roller bearings over ball bearings.

Section 3.8

19. What is the difference between a mineral oil and synthetic oil?

20. When are solid-film lubricants used?

21. How do you determine the life span of a ball bearing?

22. What is a grease?

23. Explain how the DN number is used.

24. How do you evaluate the cavitation resistance of a material?

To Dig Deeper

Baurewell, F. T. *Bearing Systems—Principles and Practice*. Oxford: Oxford University Press, 1979.

Bayer, R. G. *Wear Analysis for Engineers,* New York: HNB Publishing, 2002.

Bhushan, B. *Modern Tribology Handbook*. Boca Raton, FL: CRC Press LLC, 2000.

Blau, P. J. *Friction Science and Technology*. New York: Marcel Dekker, Inc., 1995.

Booser, E., Ed. *Tribology Data Handbook*, Boca Raton, FL: CRC Press LLC, 1998.

Bowden, F. P., and D. Tabor. *The Friction and Lubrications of Solids*. Oxford: Clarendon Press, 1950.

Bowden, F. P., and D. Tabor, *The Friction and Lubrication of Solids—Part II*. Oxford: Clarendon Press, 1964.

Buckley, D. H. *Surface Effects in Adhesion, Friction, Wear and Lubrication*. Amsterdam: Elsevier, 1981.

Budinski, K. G. *Surface Engineering for Wear Resistance*. Upper Saddle River, NJ: Prentice Hall, 1979.

Dawson, D. *The History of Tribology*. New York: Longman Group, Ltd., 1979.

Glaeser, W. A. *Materials for Tribology*. Amsterdam: Elsevier, 1992.

Hutchings, I. M. *Tribology: Friction and Wear of Engineering Materials*. New York: CRC Press, 1992.

Ikada, Y., and Y. Uyama. *Lubricating Polymer Surfaces*. Lancaster, PA: Technomic Publishing Co., 1993.

Lancaster, J. K. Solid Lubricants. In E. Booser, Ed., *Handbood of Lubrication*, vol. II. New York and Boca Raton, FL: CRC Press, 1984.

Ludema, K. C. *Friction, Wear, Lubrication*. New York: CRC Press, 1996.

Rabinowicz, E. *Friction and Wear of Materials*. New York: John Wiley & Sons, 1965.

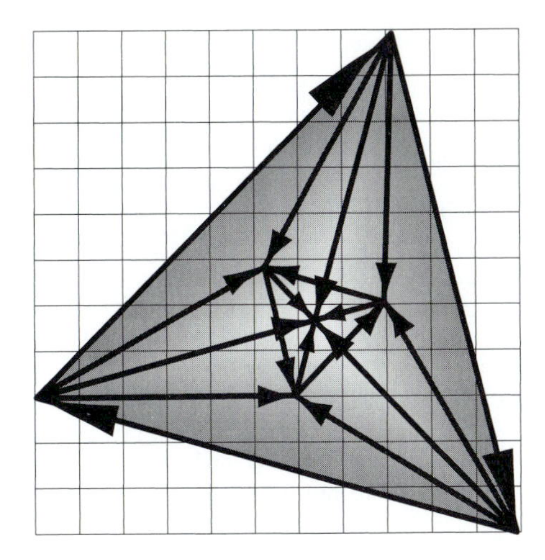

CHAPTER 4

Principles of Polymeric Materials

Chapter Goals

1. An understanding of how polymerized organic materials form engineering materials.
2. An understanding of the chemical makeup of polymeric materials.
3. An understanding of the techniques that are used to strengthen polymers.

Chapter Rationale

The next three chapters will be full of "poly's." There are thousands of commercially available plastics whose chemical names usually start with poly: *polyethylene, polypropylene, polysulfide, poly*-something. *You will remember many of these names when you finish this book, but as some of them fade in your memory with time, this book will remain as a useful reference. This chapter is about where plastics come from, how they form, what holds them together, what makes them stronger, and how they are or are not alike. We will demonstrate how the elements, mostly carbon and hydrogen, play a key role in forming plastics. This chapter will present the fundamentals that you need to remember throughout your technical career. These fundamentals help you to understand polymer families and all of the other details that you will need to properly use and specify the important engineering materials that we know and love as "plastics."*

In Chapter 1, we characterized polymers as long chains of repeating molecules based on the element carbon. Polymers have their origin in nature. The building blocks of animal life—animal proteins—are polymers, as are rosin, shellac, natural rubber, and a host of other familiar substances. Wood is composed of chains of cellulose molecules bonded together by another natural polymer called *lignin*. Natural polymers have been around since the beginning of life.

The polymeric materials that we use today in machines, packaging, appliances, automobiles, and the like are not made from natural polymers. They are manufactured, and they have come into importance as engineering materials only within the past 75 years (Figure 4–1). The first synthetic moldable polymer was cellulose nitrate, or celluloid. It was not widely used, but the next generation of polymers, the phenolics, found wide application as structural and

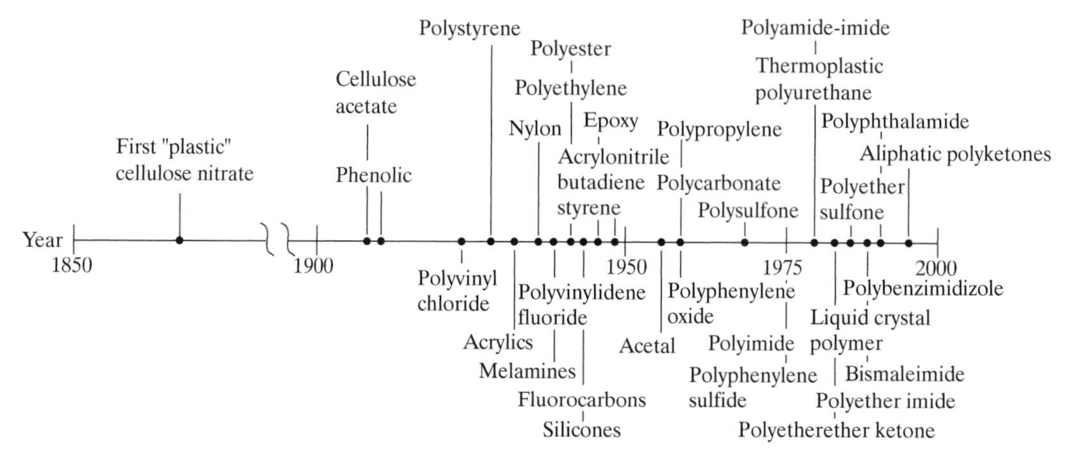

Figure 4–1
Chronological development of important engineering polymers

insulating materials in electrical devices. Light switches are still normally made from phenolic polymers. The era of World War II was characterized by a scarcity of materials, and the evolution of polymers took place at a rapid pace. Nylon replaced silk, vinyls replaced leathers, and so on. In 1979, production of plastics (on a tonnage basis) exceeded steel production, and this situation still exists. Today there are literally thousands of polymers that have utility in design engineering, and it is the purpose of this chapter to lay the foundation in polymer chemistry and properties so that designers can understand the differences among polymer systems. A subsequent chapter will stress selection criteria for engineering polymers for specific applications.

The term *polymer* is used interchangeably with the term *plastic*. Neither term is accurate. *Plastic* means "pliable"; most engineering polymers are not plastic at room temperature. On the other hand, *polymer* can include every sort of material made by *polymerization* with repeating molecules. The ASTM definition (D 883) of a plastic is "a material that contains as an essential ingredient an organic substance of large molecular weight, is solid in its finished state, and, at some stage in its manufacture or in its processing into finished articles, can be shaped by flow." Rather than engage in a lengthy semantics discussion, we have opted to use the term *polymer* in discussing the chemical formation of polymers and the term *plastic* for polymeric materials in finished form. Basically a plastic is an organic material with repeating molecular units that can be formed into usable solid shapes by casting, sintering, or melt processing.

4.1 Polymerization Reactions

Most polymers consist of long chains of repeating molecules. There are basically two ways that the individual molecules attach to each other: (1) The molecules can physically link to each other, like stringing beads on a string; or (2) a new molecule can be attached to another molecule when it is formed by a chemical reaction. The former process is called *addition* polymerization; the latter, *condensation* polymerization. In addition polymerization, the starting material is a monomer in a solution, emulsion, or vapor, or even in bulk, and the polymer resulting from the polymerization has the same repeating unit as the starting monomer. In condensation polymerization, the repeating molecules in the polymer chain are different from the starting materials. Water is commonly a byproduct of the chemical reaction involved—thus the term *condensation*. Some polymeric materials such as epoxies are formed by strong primary chemical bonds called *cross-linking*. The resulting material can be considered a single macromolecule. These concepts of polymer formation are illustrated in Figure 4–2.

Obviously, these reactions are not as simple as indicated by the illustration. A catalyst of some form or an elevated temperature or pressure may be needed to initiate both the addition and the condensation reactions. The length of the polymer chains (molecular weight) can, in many cases, be controlled by processing parameters or by the type of catalyst. This is an important concept to keep in mind, since the length of polymer chains has a profound effect on properties. To the fabricator, the molecular weight is a factor that plays an important role in how easily the polymer molds. To the designer, molecular weight is important in the determination of properties. For example, a high-molecular-weight polyethylene has different properties than a low-molecular-weight polyethylene; a low-molecular-weight synthetic oil has different properties than a high-molecular-weight oil.

In chemistry, a polymer is described by drawing its physical structure of atoms. Polyethylene is described as follows:

$$\left[\begin{array}{cc} H & H \\ | & | \\ -C- & -C- \\ | & | \\ H & H \end{array}\right]_n$$

where H = hydrogen atom

C = carbon atom

n = molecular weight (number of repeating units)

[ ] = monomer symbol

— = electron bond

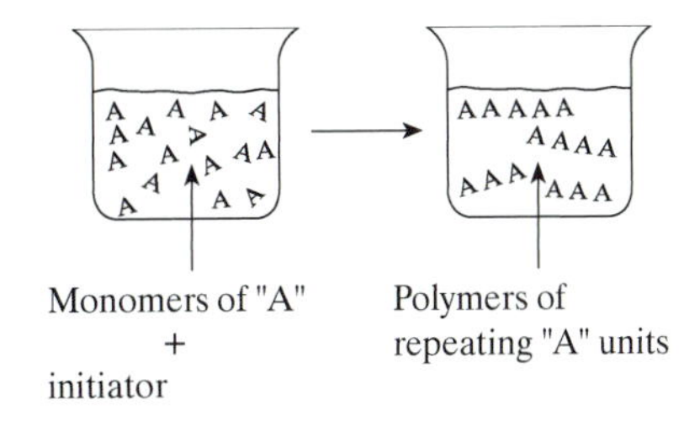

Addition polymerization

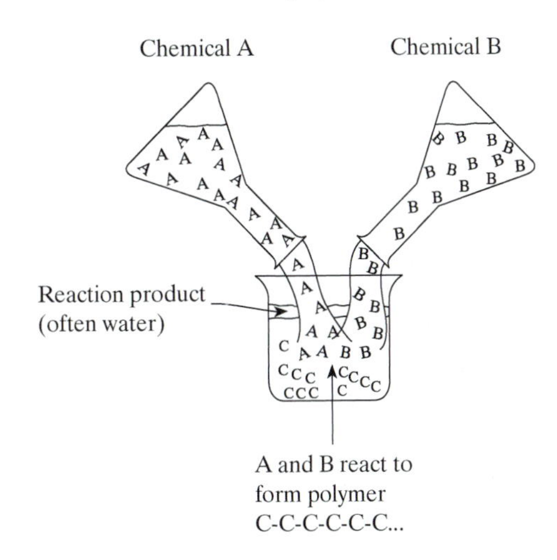

Condensation polymerization

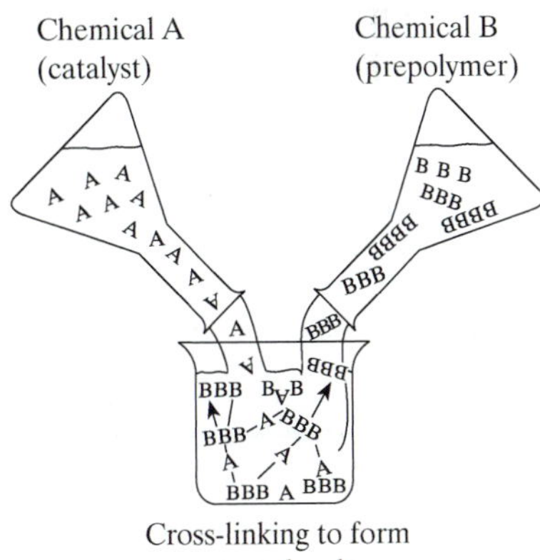

Figure 4–2
Polymerization reactions

The unsaturated bonds on the carbon atoms indicate that molecules can form covalent bonds with like molecules to form a polymer chain. Many polymers can be made from the same carbon atom backbone as that shown in polyethylene. To accomplish this may involve complex processing equipment, but, from the chemistry standpoint, it simply involves a substitution of another element or chemical compound for one or more of the hydrogen atoms. As an example, we can take the polyethylene monomer and substitute a chlorine atom for one of the hydrogens; we would end up with polyvinyl chloride, or just plain vinyl as it has come to be known:

Replace with Cl:

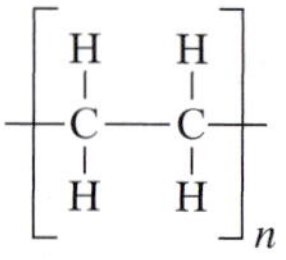

$$\left[-\overset{\displaystyle H}{\underset{\displaystyle H}{\overset{|}{\underset{|}{C}}}}-\overset{\displaystyle H}{\underset{\displaystyle H}{\overset{|}{\underset{|}{C}}}}- \right]_n \qquad \left[-\overset{\displaystyle H}{\underset{\displaystyle H}{\overset{|}{\underset{|}{C}}}}-\overset{\displaystyle Cl}{\underset{\displaystyle H}{\overset{|}{\underset{|}{C}}}}- \right]_{n'}$$

Polyethylene 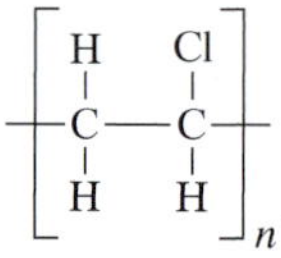 Polyvinyl chloride

Chemical reaction:

$$\underset{\text{Ethylene}}{CH_2{=}CH_2} + \underset{\text{Chlorine gas}}{Cl_2} \xrightarrow[\text{260°C in reactor}]{\text{heat to}}$$

$$\underset{\text{Polyvinyl chloride monomer}}{CH_2{=}CHCl} + \underset{\text{Hydrogen chloride}}{HCl}$$

If instead we replace a hydrogen atom with a more complex substance than an element, it would be termed a *substituent* or *functional group*—usually a group of atoms that acts like a single atom and as a group has a definite valence. The polymer polyvinyl alcohol has an oxygen and a hydrogen atom as a replacement for a hydrogen atom in the ethylene monomer. These oxygen and hydrogen atoms are a substituent, more specifically, a hydroxyl substituent.

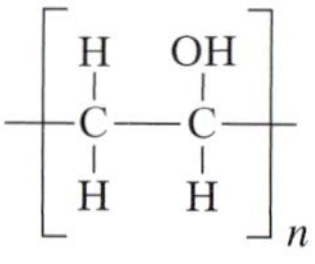

$$\left[-\overset{\displaystyle H}{\underset{\displaystyle H}{\overset{|}{\underset{|}{C}}}}-\overset{\displaystyle OH}{\underset{\displaystyle H}{\overset{|}{\underset{|}{C}}}}- \right]_n$$

Polyvinyl alcohol

Some of the substituents encountered in describing polymer molecules are much more complex, so the group formula is replaced by R. The important point to be illustrated by this discussion is not so much the chemical designations for polymers, but that many polymers are formed by modifying the structure of an existing polymer with the substitution of a new element or substituent group.

How are the molecular weight and chemical structure of a polymer determined? The organic chemist uses numerous tools to determine what atoms or functional groups make up a polymer chain and where they are, but from the standpoint of identifying the substituents and basic polymer family, most analytical laboratories have adopted the use of infrared spectroscopy. In this technique, the polymer is dissolved in a suitable solvent or converted to a film; a source of infrared energy is passed through the solution or film; and the transmitted infrared energy is measured and recorded. With a newer form of this process, *Fourier transform infrared spectroscopy* (FTIRS), thin films and organic coatings on solids can be analyzed by using an IR beam and by recording changes in the reflected beam. Each functional component of the polymer chain will exhibit a characteristic peak that can be identified by comparison with standards of known composition. The amplitude of the peak provides a means of getting a quantitative analysis. A typical IR curve is illustrated in Figure 4–3.

The determination of the exact atomic structure of a polymer is much more involved; in many cases, a new polymer may even be in commercial use before the complete analysis of its structure is made. It should be kept in mind that these polymer chains that we sketch as carbon and

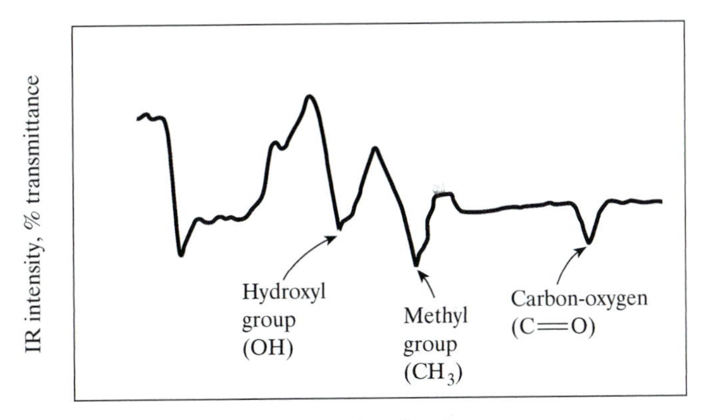

Figure 4–3
Infrared spectroscopy for analysis of polymers

hydrogen "balls" are often complex, three-dimensional substances. The molecular weight of a polymer is just as important as the chemical nature of the polymer. However, not all of the chains that make up a polymeric substance are the same length, so essentially an averaging technique must be employed. One common technique for measuring the molecular weight of thermoplastics is by measuring the volume of molten polymer at a specified temperature that can pass through a given orifice under a controlled head and in a predetermined elapsed interval of time. The volume that passes through the orifice depends on the polymer's viscosity, and the viscosity of a certain type of polymer at a prescribed temperature is proportional to its molecular weight. The net result of this test is the melt flow rate. The lower the melt flow rate, the greater the molecular weight. This property, commonly referred to as melt flow index, is usually presented in polymer data sheets; in systems where polymer strength is important, it can be of value in material selection. In most polymer systems, mechanical properties such as tensile strength and flexural strength increase with the molecular weight; a high-molecular-weight polymer will be stronger than the same polymer with low molecular weight. On the other hand, processibility (molding, forming) generally decreases as molecular weight increases. Thus there is a practical limit on molecular weight.

Before leaving the subject of polymerization reactions, we shall discuss several additional concepts. *Copolymerization*, as the name implies, is a polymer production process that involves forming a polymer chain containing two different monomers. Each monomer by itself is capable of forming a polymer chain, but intentionally it is linked to another monomer; the resulting copolymer contains chain increments of both polymers. For example,

```
  B         B
  |         |
  B         B
  |         |
A–A–A–A–A–A–A–A–A
      |
      B        (graft)
      |
      B
```

```
A–A–A–A–B–B–B–B–B–A–A–A–B–B–B–B
            (block)
```

where A = monomer of A chemistry
B = monomer of B chemistry

The increments of A and B do not necessarily have to be the same, nor does their spacing. Techniques are also available to make graft and block reactions in specific places in a polymer chain. Additional techniques (often catalysts) can be used to design the spatial relationship of grafts; *stereospecificity* can be controlled. Attachments to a primary chain can be alternating on sides of the chain (syndiotactic) or random (atactic). The important aspect of copolymerization is that it affords a way of improving the properties of a polymer. It also provides almost infinite types of plastics. It is very important commercially. Many older polymers have been improved by copolymerization with other polymers.

Akin to copolymerization is the use of plastic *blends* and *alloys*. Blends and alloys are made by physically mixing two or more polymers (also called *homopolymers*). At least 5% of another polymer is necessary to create a blend or an alloy, and a number of microstructures can result. If the homopolymers are miscible, a single-phase blend or alloy will result. If the component polymers are immiscible, a multiphase alloy or blend results. The additive polymers can form spheres in the host matrix; they can form cylinders that are usually oriented in the direction of flow in processing; or they can form lamella (alternating layers, like plywood). If the resultant polymer behaves as a single polymer, it is usually considered an alloy. If the copolymer retains some of the characteristics of the original polymers, it is usually considered a blend.

The technology of alloying and blending is often proprietary. It is not a trivial matter to mix different polymers and have a synergistic effect; but in the 1980s, alloying and blending were important means of improving serviceability and applicability of polymers. The cost of developing a completely new polymer may be as high as several hundred million dollars, and five years of development time may be required. Alloys and blends can cost a small fraction of the cost of developing a new polymer, and the time can be as short as months. Customer acceptance is often better as well. People know the properties of the host polymer, and when they buy the host polymer with something added, they are more prone to expect better properties. Most of the new thermoplastic materials introduced since the 1980s have been alloys and blends.

A–A–A–A–A–A–A

B–B–B–B–B   A–A–A   B–B–B–B

A–A–A–A–A–A–A–A   B–B–B–B

A–A–A–A–A–A–A–A–A

Polymer blend

Once again, the combining of two or more polymers is done to improve some use characteristic. In both copolymerization and plastic alloys, this approach to improving polymers is not limited to adding only one polymer. There are also *terpolymers* (three monomers in a chain) and plastic alloys with several polymer additives.

In summary, polymerization, the process of making long-chain polymer molecules, is achieved by causing identical monomers to add on to each other or by a condensation reaction that involves combining a number of chemicals to form a polymer as well as additional reaction products. Different polymers can be formed by substituting different functional groups or elements in a basic carbon chain. In all polymerization reactions, the molecular weight produced is important because it has a strong influence on mechanical and physical properties. In an effort to obtain improved polymer properties, polymer chemists have also devised ways of copolymerizing two or more different monomers, as well as ways of blending two or more polymers into a plastic alloy.

4.2 Basic Types of Polymers

The subject of polymeric materials may seem overwhelming to the average material user because there are so many different plastics with different properties and structures. How does a person remember the differences and know which is the best to use for a particular application? We will address the latter question in our discussions on selection, but the answer to the former question is, become familiar with the important systems.

Figure 4–4 shows how the field of polymeric materials can be broken into various categories having to do with use characteristics and origin. We will discuss each of these categories, with the exception of the polymers that are part of biological systems such as the human body—enzymes, proteins, and the like. In this chapter we will discuss the group of polymeric materials that users call plastics. Plastics can be divided into two categories relating to their elevated temperature characteristics; plastics are usually either *thermoplastic* or *thermosetting*. All plastics behave like molten materials at some point in their processing; this is part of the definition of a plastic. A thermoplastic material will flow at elevated temperatures (above the *glass transition* temperature or crystalline melting point), and the solidified polymer can be reheated as many times as desired and it will do the same thing. As a matter of interest, most plastic fabricators do recycle scrapped parts as well as the unused portions of moldings. On the other hand, thermosetting polymers, once their shape has been made by casting or by plastic flow at elevated temperature, will no longer melt or flow on reheating. Polymerization has occurred by strong network bonds (cross-linking) produced by catalysis or by the application of heat and pressure, and these strong bonds keep the material from remelting. When reheating is attempted, these materials will char, burn, or in some cases sublime; thermosetting materials cannot be recycled.

Thermoplastic polymers tend to consist of long polymer chains with little breadth, essentially two-dimensional structures. Polymers with this type of chain structure are also referred to as *linear polymers*. Thermosetting polymers have a structure that is characterized by a three-dimensional network of molecules. As we shall see as we discuss fabrication techniques, the property of being thermoplastic or thermosetting has a profound effect on the potential uses of a polymer. There are many low-cost fabrication options with thermoplastic materials, but the thermosetting materials usually require more expensive fabrication processes.

The other major way of classifying polymers—by chemical families—means that different polymers can be made by changing substituent groups on some monomer (for example, substituting chlorine for one hydrogen atom changes polyethylene into polyvinyl chloride). We shall discuss each family of polymers that is important in engineering design in detail, but the families that are illustrated in Figure 4–4 are the most important from the usage standpoint. We list 19 polymer families. There are countless specific plastics commercially available from these families. For example, the plastic that everyone knows as nylon comes from the polyamide family. There are about ten different types of nylon; compounders add fillers and other additives to these, and in the end the plastic user can select from 100 grades of nylon. If we include blends and alloys, the number of options is even greater. However, if the user becomes familiar with the use properties of some of the nylon homopolymers, it will be possible to make intelligent decisions on selecting a grade of nylon and in comparing this polymer family with other families.

About 75% of the two billion pounds of plastics produced in 1994 came from only three basic polymer families: olefins (polymers derived from ethylene), vinyls, and styrenes. *Engineering plastics* are the higher-strength, high-performance plastics. They represent only about 10% of the expected usage, but they are extremely important because these are the polymer families that allow heretofore impossible designs to happen. The usage of thermosetting resins is typically only about 20% of the usage of thermoplastics, but a similar situation exists in their use. About 90% of the usage comes from a few basic families: phenolics, unsaturated polyesters, and ureas. The point of this discussion is that the subject of polymeric materials can be simplified by concentrating on polymer families rather than by trying to memorize plastic trade

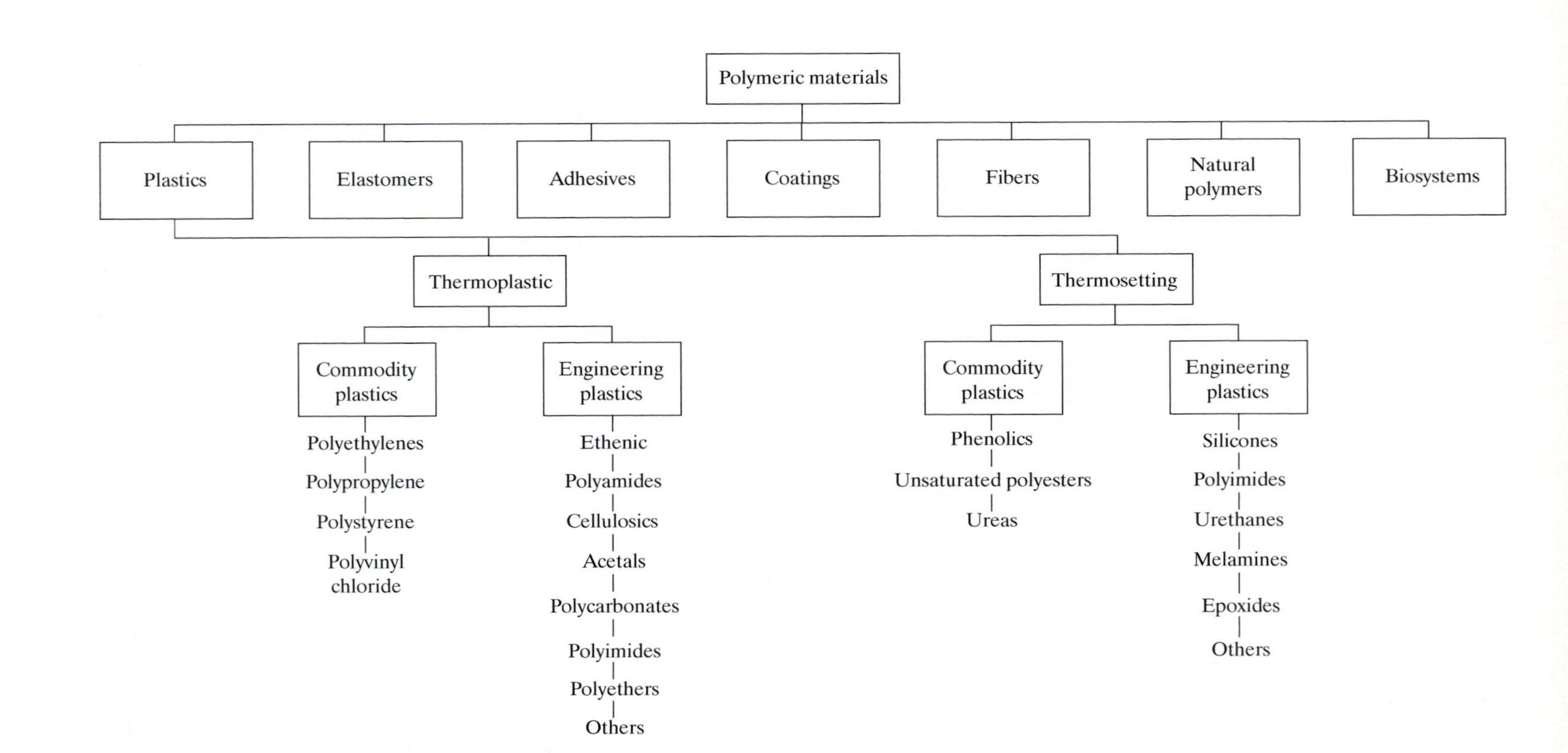

Figure 4–4
Spectrum of polymeric materials and some of the important thermoplastic and thermosetting plastic families. Commodity plastics generally have lower cost, and they are the most used plastics. Engineering plastics include all others.

names; the user must have an understanding of olefins, polyamides, styrenes, phenolics, and other important families.

A final point to be made about polymer families is how to identify specific polymers in these families. The plastics industry is, unfortunately, fraught with product secrecy, and some plastics manufacturers are reluctant to disclose even the basic polymer system that is used in one of their grades. Attempts have been made to develop a generic identification system for specific plastics. The ASTM D 4000 specification is one such system. This specification shows how the user can use an alphanumeric identifier for a specific type of nylon or other plastic. It can be used to show desired fillers and additives, even coloring agents. Unfortunately, this system has not been widely adopted in industry. Specification by trade name still predominates. This situation is not likely to change in the near future, and our recommendation is, where possible, to use generic specifications. Where trade names must be used, make certain that you know the basic polymer and what fillers and additives it contains. Handbooks are available that convert trade names to generic polymer families; use these if necessary.

4.3 Altering Properties of Polymers

With a few exceptions, strengthening polymers is the job of the polymer chemist. One engineering plastic, polyamide-imide, can be increased in strength by a postmolding thermal treatment, but the normal situation is that a plastic will have certain strength characteristics after molding and thereafter nothing can be done to make it stronger. The user can, however, specify fillers and additives that modify strength, and it is the purpose of this discussion to describe the techniques that are used by compounders and polymer manufacturers to control and improve mechanical and other use properties of basic polymer systems. An understanding of these factors will assist the user in selecting plastics that are modified for improved properties.

Linear Polymers

We have already mentioned how the molecular weight of polymers has an effect on polymer properties. In general, increasing the molecular weight of a polymer increases its tensile and compressive strength. Similarly, copolymerization and alloying can improve mechanical properties. In some instances, alloys or composite polymers contain a weaker polymer, which has an adverse effect on mechanical properties but may be used to increase lubricity or reduce friction characteristics. The addition of fluorocarbons to various polymers is an example of an alloying agent used for this purpose. The most common factor responsible for high strengths and rigidity in polymers is bonding between polymer chains. The simplest thermoplastic polymers have a *linear* structure. There is a two-dimensional array of polymer chains, with each chain behaving like a chain. That is, there is little breadth of the chain, but significant length. Such a structure is illustrated in Figure 4–5.

The bond between the polymer chains is due to such things as *van der Waals forces,* hydrogen bonding, or interaction of polar groups, but in linear polymers, the polymer chains are usually flexible to the degree that they intertwine

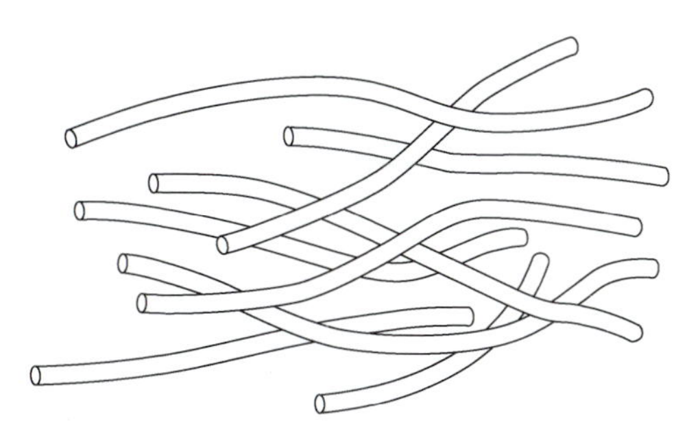

Figure 4–5
Schematic of polymer chains in a linear polymer

and lie in a single plane in space. This type of structure would be analogous to a bowl of spaghetti. Depending on the pendent groups, degree of chain *branching*, or other factors, linear polymers may have either *amorphous* or crystalline structures.

Branched Polymers

If intentionally or by chance in the polymerization reaction, a chain continues to grow concurrently as two chains, it is said to have a *branched* structure. This concept is illustrated in Figure 4–6.

Branching usually causes strengthening and stiffening, since deformation of the polymer requires the movement of chains that are much more entwined than in linear polymers. Special catalysts and processing techniques are used to promote and control branching. Many elastomers or polymeric rubbers have a branched type of structure. This is why they have such resilience and can withstand significant stretch without breaking. The entangled chains stretch when strained but are not pulled apart.

Cross-Linking

If during the polymerization reaction we could get the individual chains to form chemical bonds

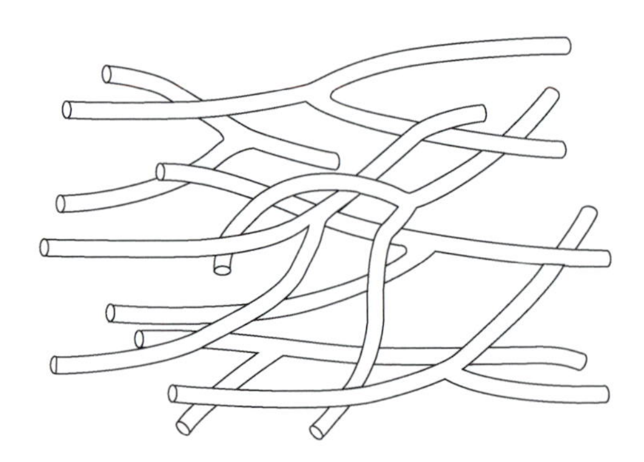

Figure 4–6
Polymer chains in a branched polymer

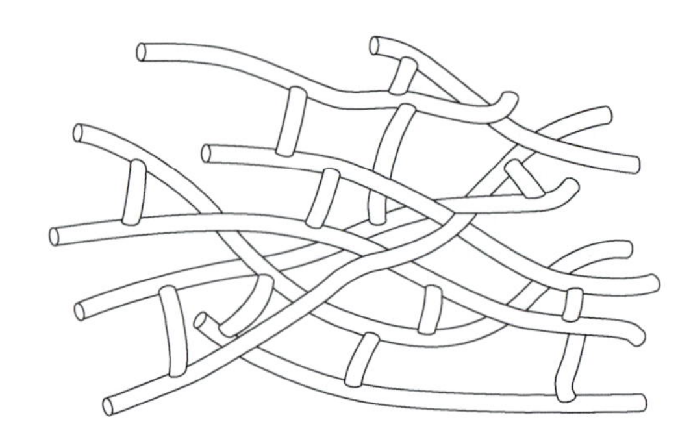

Figure 4–7
Cross-linked polymer

to each other, we would expect the resultant structure to be very strong and rigid. This, in fact, is what happens in most thermosetting polymers. The phenomenon of *cross-linking* involves primary bonds between polymer chains, as illustrated in Figure 4–7.

The polymer formed usually cannot be remelted because the bonds between the chains are too strong. Interactions between polymer chains, such as branching and cross-linking, do not occur to the same extent in different polymers. Some polymers may be slightly branched or highly branched. The same thing is true with cross-linking. The greater the degree of cross-linking, the greater the rigidity of the material, the less soluble it is, and the less it responds to remelting.

Chain Stiffening

A polymer strengthening mechanism significantly different from those already discussed is chain stiffening, caused by large substituent groups on the monomers making up a polymer chain. We have illustrated how linear polymer chains can intertwine and bend around each other. The carbon-to-carbon bonds act as pivot points for chain flexure. Suppose that a polymer has a monomer that is physically large and asymmetrical; the ability of a chain to flex will be

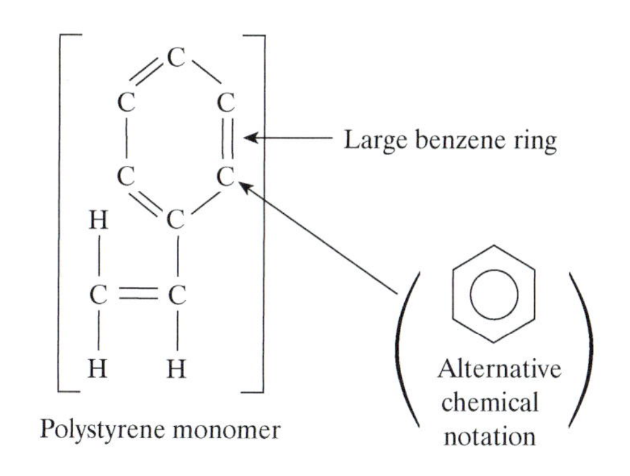

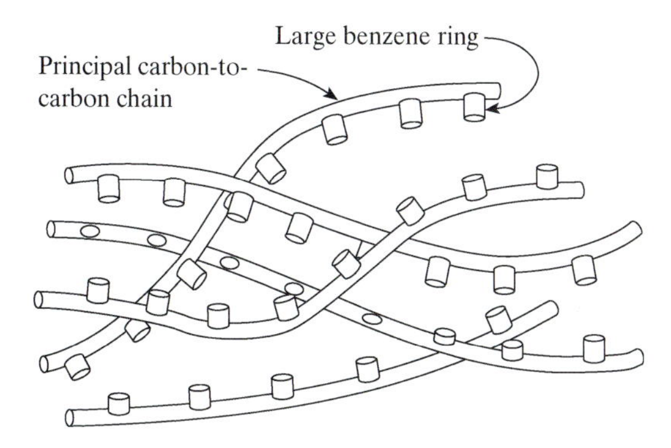

Figure 4–8
Strengthening due to chain stiffening

impaired. Polystyrene, a typical example of such a system, is illustrated in Figure 4–8.

Polystyrene, a popular commodity plastic, which is used for disposable items such as plastic forks, knives, and spoons, is a rigid and relatively brittle thermoplastic, as would be expected by its physical structure. Once again, the basic chain structure is the same as that of soft and ductile polyethylene. The presence of the large benzene ring as an integral part of the polystyrene monomer causes a reduction in chain mobility and thus an increase in rigidity. Substituent groups, like benzene in polystyrene, are generally referred to as *pendent* groups. The relative size of the pendent groups can affect the properties of the thermoplastic. Larger pendent groups typically increase the stiffness and strength. Pendent group size is also referred to as *steric hinderance*.

Aside from the size of the pendent groups on the chain, the strength and stiffness of a polymer may be altered by the location of the pendent groups on the main chain. In linear (*aliphatic*) polymers, the pendent groups along the main chain may be arranged in several basic ways. As shown in Figure 4–9, the methyl groups (CH_3) in polypropylene may all be on one side of the main chain (*isotactic*), may alternate regularly along both sides of the chain (*syndiotactic*), or may be randomly placed anywhere along the chain (*atactic*). The relative location of the pendent groups is called *stereo-regularity*, *stereo-specificity*, or tacticity. Stereoregularity of the pendent groups along the carbon backbone can affect the mechanical properties of the polymer. Polymers with greater degrees of stereoregularity (order) will tend to be crystalline. Isotactic polypropylene is highly crystalline. Syndiotactic polypropylene has lower levels of *crystallinity* and lower strength properties compared with isotactic polypropylene. Atactic forms of polypropylene are amorphous and the properties are rather soft and rubbery. Plastics such as polystyrene and polymethylmethacrylate are typically atactic. However, syndiotactic grades of polystyrene, with improved dimensional stability and creep resistance, are being considered for markets such as compact disks as a replacement for more costly plastics such as polycarbonate.

Structural and Melting Characteristics of Crystalline and Amorphous Thermoplastics

Polymers in the solid state may have either a predominately *amorphous* structure or a *semi-crystalline* structure. As discussed in Chapter 1, crystalline materials have atomic bonds that are regular and repeat in a specific and orderly manner. In general, amorphous materials do not have regular, repeating three-dimensional arrays of atoms. Factors such as chain branching, stereoregularity, polarity, degree of cross-linking, and *steric hindrance* (the relative size of the

Figure 4–9
The relative location of the methyl pendent group in polypropylene greatly affects its properties

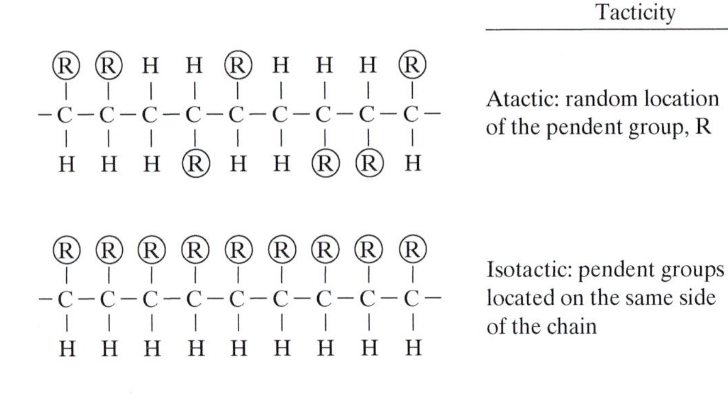
Tacticity
Atactic: random location of the pendent group, R
Isotactic: pendent groups located on the same side of the chain
Syndiotactic: pendent groups occupy alternate locations along the chain
R = H–C–H
Methyl group

pendent groups), all influence the structure of a polymer.

Polymers that crystallize tend to have regular, straight-chain (linear, or *aliphatic*) structures with small pendent groups. Some crystallizable polymers have ring structures in the main chain, but no large pendent groups. As shown in Figure 4–10, when a crystalline polymer cools from the liquid state, the long polymer chains fold up like an accordion to form a regular, repeating structure. This regular structure constitutes a crystal. Similar to many metals, crystalline polymers are actually polycrystalline.

In the solid state, a crystalline polymer is stiff and rigid with a very high "*viscosity*." As shown in Figure 4–11, when a crystalline polymer is heated to the melting point, the viscosity of the polymer significantly and abruptly drops to a much lower level. Similarly, ice, when it melts, readily transforms from a hard and high-viscosity solid to a low-viscosity, flowing liquid.

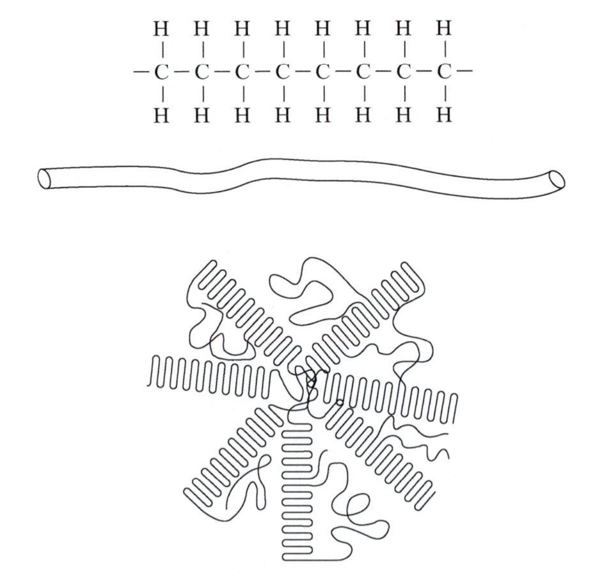

Figure 4–10
Crystalline polyethylene polymer. The straight chain and absence of pendent groups allows the chain to fold into a regular, repeating structure.

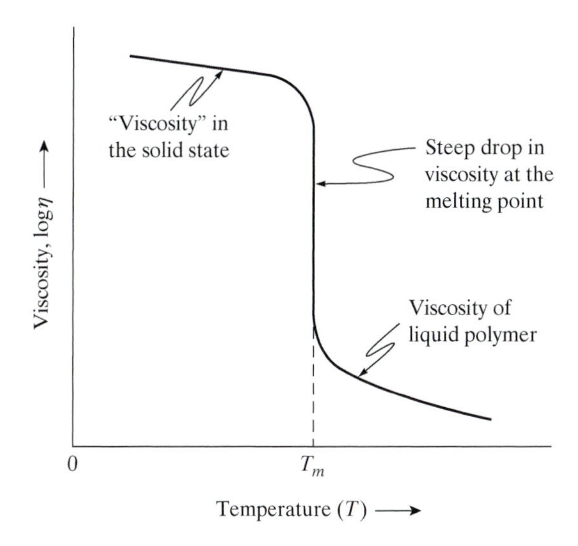

Figure 4–11
Viscosity versus temperature curve for a crystalline thermoplastic. Note the steep viscosity drop at the crystalline melting point.

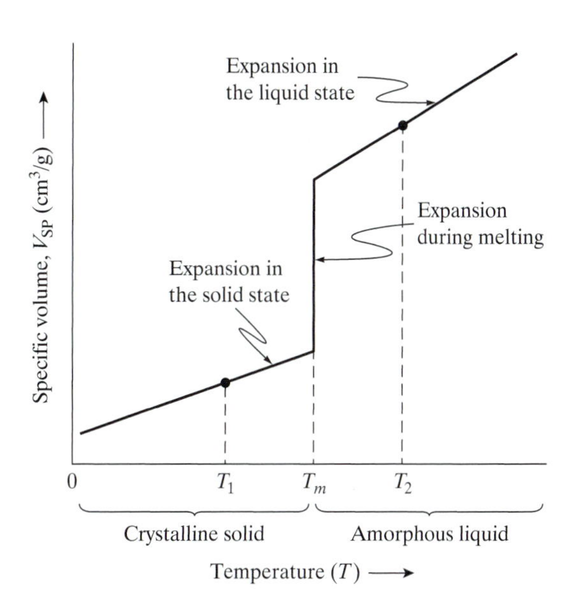

Figure 4–12
Specific volume of a crystalline thermoplastic as a function of temperature. Note the sharp increase in specific volume at the crystalline melting point.

Another means of characterizing the behavior of a polymer is by measuring the *specific volume* in cm^3/g (inverse of density) as a function of temperature. Typically an instrument such as a *dilatometer* is used to make such measurements. As shown in Figure 4–12, when a crystalline polymer sample of known volume is heated from the solid state, it will expand at a certain rate (based on the coefficient of thermal expansion). At the melting point (also known as the crystalline melting point, T_m), the polymer experiences a significant and abrupt increase in specific volume. This volumetric expansion occurs because the tightly compacted crystal structure present in the solid state reaches a point where it breaks down and the polymer transforms into an amorphous liquid. When this transition occurs, the internal volume, or *free volume,* of the polymer structure dramatically increases. With further heating, the liquid polymer expands at a rate proportional with the coefficient of thermal expansion of the liquid polymer.

Below its crystalline melting point the polymer behaves like a rigid solid—as shown in the stress–strain curve in Figure 4–13. However, above the melting point the polymer transforms to a liquid incapable of supporting mechanical stress.

When a polymer's structure has large pendent groups with significant chain branching, or a lack of stereoregularity, the chains become so stiff and bulky that they cannot readily crystallize through chain folding as they cool from a liquid state. As shown in Figure 4–14, when atactic polystyrene cools from a liquid state, the polymer chains condense to form a high-viscosity solid structure. However, the stiffening effect caused by the large benzene-ring appendages prevents the chains from folding into a regular, repeating structure. Instead, the polymer chains will form a rather random structure. Polymers with this structural morphology are called amorphous polymers.

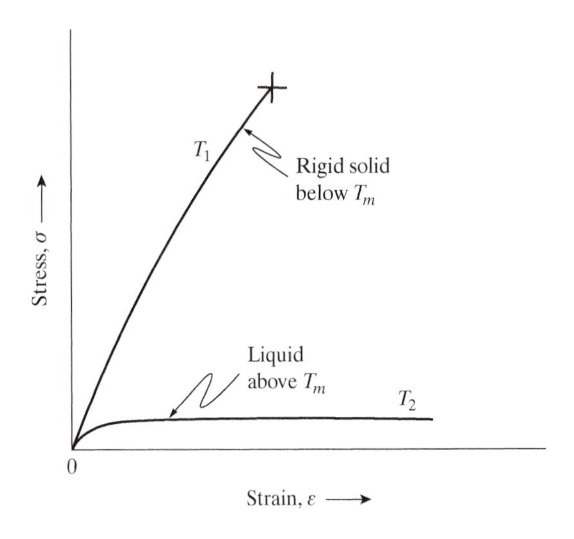

Figure 4–13
Typical stress–strain curves for a crystalline thermoplastic at temperatures, T_1 and T_2, above and below the melting point, respectively. See Figure 4–12 for the location of reference temperatures T_1 and T_2 on the specific volume versus temperature curve.

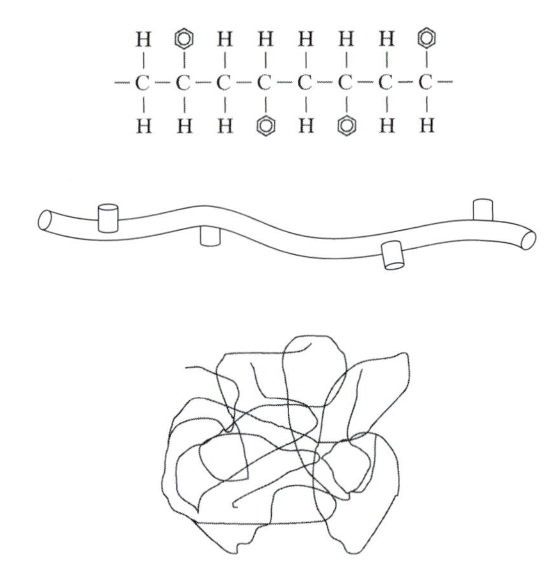

Figure 4–14
Amorphous polystyrene polymer. The bulky atactic pendent groups prohibit the chain from folding into a regular, repeating structure. The resultant structure has no long-range order.

If an amorphous thermoplastic, such as atactic polystyrene, is heated, it initially transforms from a hard, brittle state to a leathery state. With additional heating it turns soft and rather tacky before it develops into a viscous liquid. When heated further, the polymer appears to have "melted." However, there is no sharp transition from a solid to a liquid but rather a gradual change from hard/brittle, to leathery/rubbery, to viscous liquid over a broad temperature range. This change in viscosity as a function of temperature is illustrated in Figure 4–15. Indeed, amorphous thermoplastics do not have a distinct melting point—they "liquefy" over a broad temperature range.

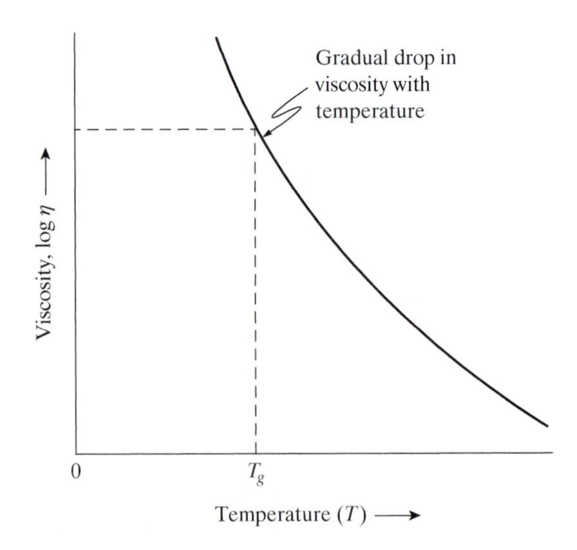

Figure 4–15
Viscosity versus temperature curve for an amorphous thermoplastic. Note the gradual drop in viscosity with increasing temperature. There is no notable viscosity transition at T_g.

When a liquid or molten sample of amorphous polymer cools slowly, the volume of the sample contracts based on the coefficient of thermal expansion of the material (Figure 4–16). The viscosity of the polymer also increases. However, if the viscosity is still relatively low, structural changes (reorientation) can

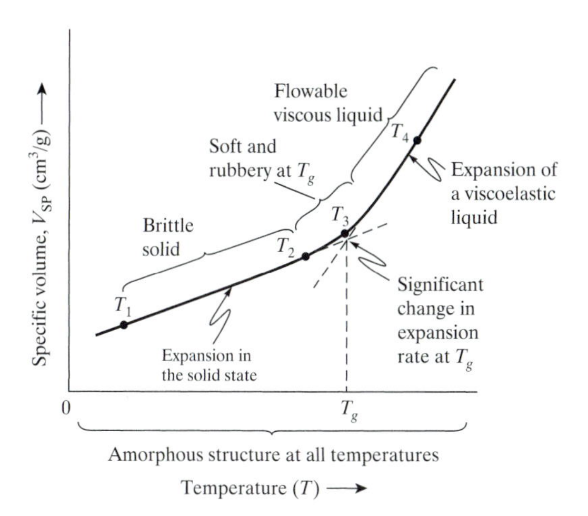

Figure 4–16
Specific volume of an amorphous thermoplastic as a function of temperature. Note the increase in slope of the specific volume curve at the glass transition temperature, T_g. Temperatures T_1–T_4 are reference temperatures discussed in Figure 4–17.

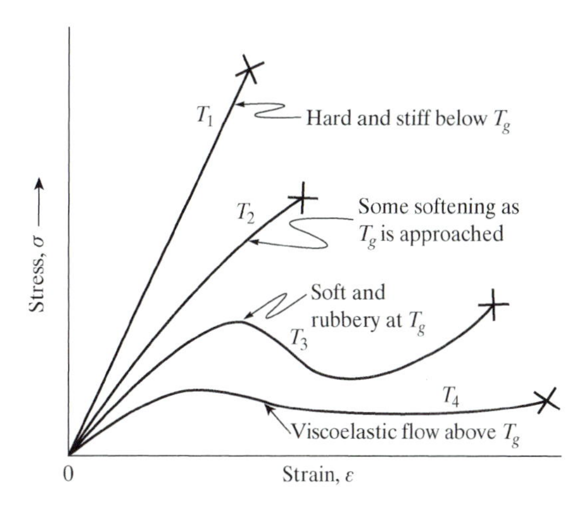

Figure 4–17
Sample stress–strain curves for an amorphous thermoplastic at various temperatures above and below the glass transition temperature. See Figure 4–16 for the location of reference temperatures T_1–T_4 on the specific volume versus temperature curve.

occur at the same rate as the cooling rate. With further cooling, the volume of the polymer continues to reduce and the polymer chains rearrange into a more dense structure as the viscosity of the polymer continues to increase. At some point, the viscosity of the polymer increases at a higher rate than the rate of structural rearrangement. At this transition temperature, the slope of the specific volume versus temperature curve changes (see Figure 4–16). This change in slope occurs at the *glass transition temperature*, T_g. The polymer is resistant to crystallization because the chains, with their rather stiff structures or bulky pendent groups, do not coil into an orderly structure but remain in a random, amorphous state.

The mechanical properties of amorphous polymers significantly change when the polymer is heated near the glass transition temperature (T_g). As shown in Figure 4–17, at temperatures below the T_g, the polymer behaves as a relatively stiff and rigid material. However, as the polymer is heated to temperatures approaching T_g, the mechanical properties degrade and the material transitions to a soft, rubbery, or leathery state. With continued heating above the T_g, the material eventually behaves as a *viscoelastic* liquid, which has elastic and viscous characteristics—like gelatin.

Amorphous polymers tend to be completely amorphous, but crystalline polymers typically have some level of amorphous structure. As shown in Figure 4–18, when the long chains in a crystalline polymer fold into a regular, repeating structure, some amount of amorphous material is present to accommodate mismatch and other defects within the crystals. Figure 4–19 shows the microstructure of a crystalline polypropylene thermoplastic. Most crystalline polymers are actually semicrystalline, where the percentage of crystalline phase can range from 40 to 90%. Semicrystalline polymers exhibit well-defined melting points as well as a T_g (due to the influence of the amorphous phase). The level of crystallinity is controlled by many factors including the cooling rate and

Figure 4–18
Structure of a semicrystalline polymer. Long-range order, produced by the folding of polymer chains, creates crystalline lamellae. The spherical morphology is called a sperulite.

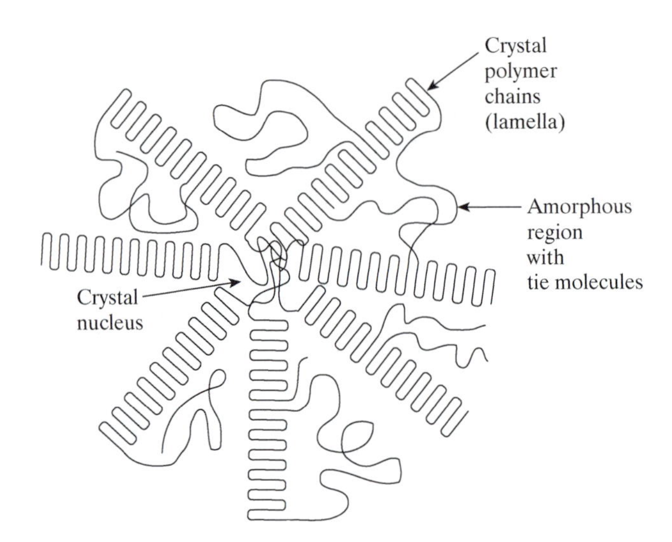

the presence of *additives* such as *nucleating agents*. For example, if a crystalline thermoplastic is quenched rapidly from the liquid state to a temperature below the T_g, the amorphous phase will dominate the structure. However, with time, some of the amorphous phase may convert to a crystalline phase (the material shrinks during this phase change).

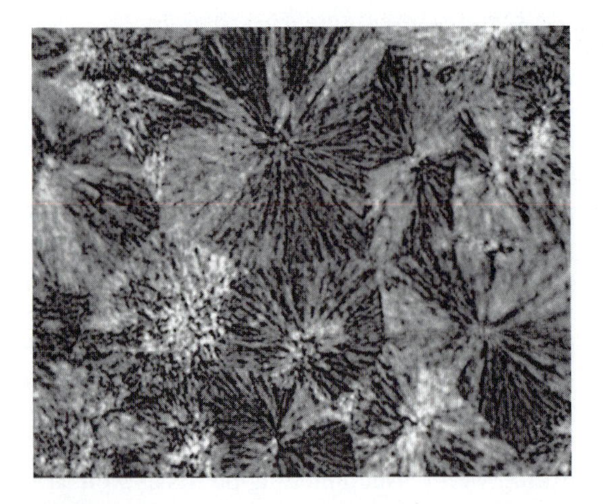

Figure 4–19
Structure of crystalline isotactic polypropylene (×640)

Large ring structures (*aromatic rings*) in the main chain (*backbone*) of a polymer increase the strength and stiffness of the polymer as well as the T_g (see Figure 4–20). For example, the linear (or aliphatic) chains in polyethylene are very flexible. The flexibility of the backbone causes the T_g to be very low (below room temperature). Alternatively, polymers with aromatic ring groups in the main chain, such as polycarbonate, tend to be significantly stiffer and have higher T_g's. Aromatic rings located along the backbone prevent the backbone of the polymer from flexing, twisting, and other forms of segmental motion. Ring groups in the backbone of a polymer stiffen the structure in a manner analogous to increasing the web in an I-beam.

Blending and alloying polymers (copolymers) can also influence the structure and thermal properties of the polymer. For example, when polyethylene and polypropylene (two relatively stiff crystalline polymers) are blended, the resultant polymer is a very soft, rubbery, and amorphous polymer. Copolymerization processes that reduce regularity within the structure tend to inhibit crystallization.

The stereoregularity of a polymer can affect its ability to crystallize. For example, syndiotactic

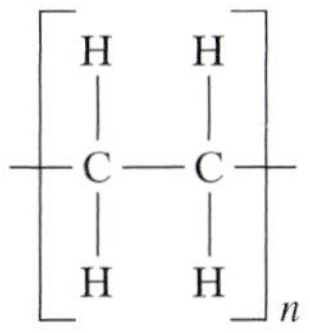

$$\left[-\overset{\displaystyle H}{\underset{\displaystyle H}{\overset{|}{\underset{|}{C}}}} - \overset{\displaystyle H}{\underset{\displaystyle H}{\overset{|}{\underset{|}{C}}}} - \right]_n$$

Polyethylene monomer

- The very flexible chain allows the polyethylene chain to fold into a semicrystalline structure.
- Due to high chain flexibility, the T_g of polyethylene is very low (about –120°C).

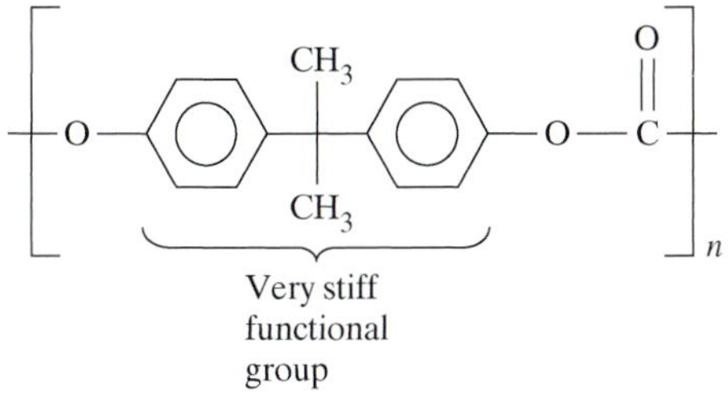

$$\left[-O-C_6H_4-C(CH_3)_2-C_6H_4-O-\overset{\displaystyle O}{\overset{||}{C}}- \right]_n$$

Very stiff functional group

Polycarbonate monomer

- The structure of polycarbonate is amorphous because ring structures in the main chain inhibit crystallization.
- Ring structures in the main chain increase the stiffness of the polymer and the T_g (about 149°C).

Figure 4–20
Effect of aromatic functional groups in the main chain on stiffness, structure, and T_g

and isotactic versions of polypropylene crystallize readily due to the regular placement of the methyl pendent groups. Atactic polypropylene, conversely, forms an amorphous polymer. The random location of pendent groups in atactic polymers affects the polymer chain's ability to fold and pack into regular, crystalline structures. Polystyrene is commonly encountered as an amorphous polymer with atactic benzene-ring pendent groups. However, a newer form of polystyrene with syndiotactic pendent groups crystallizes because of the regular arrangement of the pendent groups.

Most thermosetting polymers have a cross-linked structure that is amorphous. The cross-linking prevents the polymer chains from folding or arranging into a regular, repeating structure. Fully cross-linked polymers do not have a melting point. However, they do exhibit a T_g. For a given polymer material, such as bisphenol A epoxy, for example, a higher T_g, indicates a greater degree of cross-linking. Cross-linking tends to decrease the free volume in the polymer structure, causing the chains to be more closely tied together. This effect, in general, tends to increase the T_g.

Summary: Amorphous Polymers

- Thermoplastics with extensive chain branching, large pendent groups, and low stereoregularity tend to favor an amorphous structure.
- Thermosetting polymers are amorphous because the cross-linking inhibits crystallization.
- Thermoplastic and thermosetting amorphous polymers exhibit glass transition temperatures (T_g).
- The mechanical properties of amorphous polymers significantly degrade near the T_g.
- Amorphous thermoplastics melt or liquefy over an extended temperature range.
- Thermosetting polymers do not melt but will degrade above the T_g.

Summary: Semicrystalline Polymers

- Polymers with long, slender aliphatic chains and lower levels of chain branching tend to crystallize.
- Semicrystalline polymers have a defined melting point (T_m).
- Most crystalline polymers contain some degree of amorphous polymer.
- A T_g may be detected for the amorphous phase present in a crystalline polymer.
- The amorphous phase in a crystalline polymer can have profound effects on the polymer's mechanical properties.

Polymer properties are significantly affected by the degree of crystallinity. Crystallinity tends to increase mechanical properties such as tensile strength and hardness while diminishing ductility, toughness, and elongation. The amorphous phase present in semicrystalline polymers, however, can improve the toughness. Chemical permeability and solubility generally decrease with increasing crystallinity. Crystalline polymers are generally optically opaque or translucent, whereas amorphous polymers have the greatest optical clarity.

Semicrystalline thermoplastics will generally shrink more than amorphous thermoplastics when cooled from the injection molding temperature. Over time, the crystallinity of a semicrystalline plastic part may increase, causing dimensional changes (shrinkage). Amorphous polymers are typically more dimensionally stable.

Additives

Engineering and commodity plastics, as well as other polymeric materials such as paints and adhesives, use special additives to improve their properties and performance. Additives improve mechanical properties, thermal processing, surface characteristics, and chemical properties as well as appearance and aesthetic properties.

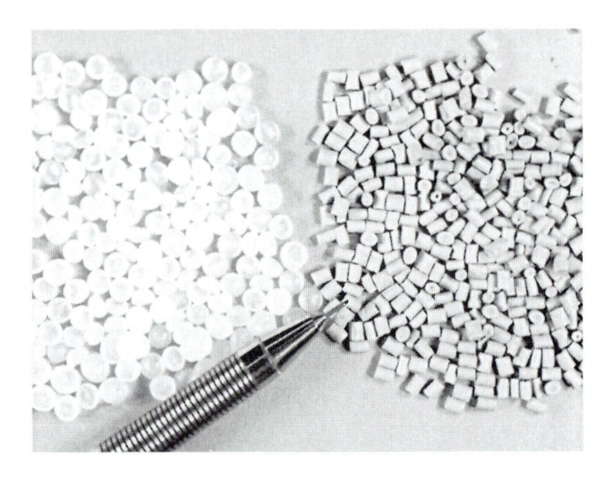

Figure 4–21
Typical forms of thermoplastic beads and pellets

Additives may be incorporated into a polymer in various ways. In thermosetting polymers, for example, most of the additives are added to the resin though special mixing and dispersion processes. Most thermoplastics are sold in the form of small beads or pellets (Figure 4–21). Pellets and beads are made using a pelletizer machine, in which molten polymer is extruded through a plate with many small holes. As the polymer extrudate exits the extrusion die, the molten polymer is simultaneously quenched in a liquid bath and cut into small pellets with a knife. The polymer pellets are removed from the bath, dried, and packaged for use. During the pelletizing process, various additives are typically added to the polymer. Examples of typical polymer additives are listed in Table 4–1.

Fully compounded resins have additives added in the proper proportion through the entire batch of resin. Alternatively, additives are sometimes added to a plastic by dispersing polymer pellets or beads that contain concentrated additives into a base resin (virgin pellets or beads) before molding. Polymer pellets or beads with concentrated additives are called *masterbatch* resins. Color concentrates, for example, are often added as masterbatch additions. Instead of purchasing large quantities of

Table 4–1
Additives and modifiers for polymeric materials (from various sources)

Additive Type	Purpose	Examples
Mechanical property modifiers		
Impact modifiers	Improve fracture toughness and impact strength	Polybutadiene rubber, EVA copolymers, SIS block copolymers
Plasticizers	Improve flexibility and toughness, reduce modulus	Phthalates, adipates, trimellitates, ricinolates, acetoxy stearates, sebecates
Nucleating agents	Control the rate of crystallization	
Reinforcements	Increase strength	Carbon fiber, glass fiber, fabrics (cotton, canvas), mica
Fillers	Control shrinkage or reduces resin cost	wood flour, titanium dioxide, mica, alumina, calcium carbonate, polymer or glass spheres, silicas, titanates, ground nut shells or rice husks, clays, talc
Surface property modifiers		
Internal lubricants	Improve tribological properties, prevent polymer from adhering to processing equipment	Silicone, waxes, stearates, fatty acid amides, fluorodispersions, glycerides, petrolatum
External lubricants	Improve tribological properties	PTFE, silicone, molybdenum disulfide, graphite, waxes, stearates
Slip and antiblocking agents	Reduce the tendency of films and sheets to stick together	Precipitated silica, aluminum silicates, fatty acids, stearates
Antistatic agents	Reduce static charge buildup	Pyrogenic alumina, metal powders, polyether block amides, carbon, alkali salts
Wetting agents	Stabilize dispersions of fillers	
Antifogging agents	Disperse moisture droplets on films	Fatty acid esters of glycerol, sorbitan, fatty alcohols
Chemical property modifiers		
Antioxidants/heat stabilizers	Prevent thermal or oxidative degradation: chain scission, cross-linking, crazing gelation depolymerization	Hindered phenols, mercaptobenzoimidizoles, organophosphites, organotins
UV stabilizers	Mitigate ultraviolet light degradation	Hindered amines, hydroxyphenyl benzotriazoles, benzophenones
Flame retardants, synergists	Suppress flammability	Antimony oxide, borates, reactive bromates, intumescent phosphates
Smoke suppressants	Suppress smoke	Aluminum trihydrate, magnesium hydroxide, zinc oxide

Table 4–1
Additives and modifiers for polymeric materials (from various sources) (*Continued*)

Additive Type	Purpose	Examples
Antistats	Reduce static charge buildup	Pyrogenic alumina, metal powders, polyether block amides, carbon, alkali salts
Biocides, fungicides, preservatives	Reduce microbial activity, mildew, and fungus formation	
Processing modifiers		
Plasticizers	Lower the viscosity of polymer during processing	Phthalates, adipates, trimellitates, ricinolates, acetoxy stearates, sebecates
Lubricants	Applied to pellets to prevent the polymer from sticking to processing equipment or thin films from sticking together	Silicone, waxes, stearates, fatty acid amides, fluorodispersions, glycerides, petrolatum
Thixants	Increase viscosity—used with adhesives and coatings	Fumed silica
Slip agents	Prevent polymer from sticking to itself (e.g., films) and to processing equipment	
Heat stabilizers	Prevent thermal degradation during processing	Organophosphites, organotins
Foaming/blowing agents	Allow polymers to foam	Azodicarbonamide-based tetrazoles, hydrazide
Reactive diluents	Reduce the viscosity of liquid resins	
Promoters, catalysts, and curing agents	Promote or increase the rate of cross-linking in thermosets	Aliphatic amines, aliphatic/aromatic polyamine adducts, organic peroxides, organometallic complexes
Deaerating agents		
Mold release agents (internal)	Aid in part release from tooling	Waxes, silicones, PTFE, stearates, fatty acids
Aesthetic property modifiers		
Colorants, dyes, pigments	Control color, optical opacity	Metal oxides, sulfur compounds, quinacridones, azo/anthaquinone, dyes, carbon, titanium dioxide
Antifogging agents	Improve transparency	Fatty acid esters of glycerol, sorbitan, fatty alcohols
Nucleating agents	Control the rate of crystallization and transparency	
Fragrances (odorants)	Impart fragrance	Fragrance oils
Deodorants	Prevent odors	
Biocides	Prevent mold, fungus, and biological growth	Triclosan-based, oxybisphenoxarsine

resins in various colors, using color concentrates allows a molder to buy large quantities of a naturally colored base resin and then control the color of various parts by adding a color concentrate.

The strength of a polymeric material may be improved by using *fillers* such as glass, carbon, or aramid (Kevlar®) fibers. The effect of glass-fiber reinforcement on the properties of a nylon thermoplastic is shown in Figure 4–22. Mineral fillers such as talc, wood pulp, milled glass, mica, calcium carbonate, and so on may be added to reduce the cost of a polymer resin or to increase dimensional stability and reduce shrinkage. Plant-based fibers can be used in making biodegradable polymers. For example, cellulose-based plastics have been made biodegradable by filling with soybean fibers. The toughness of polymers is typically improved through the incorporation of impact modifiers. High-impact polystyrene, for example, contains small rubber particles (e.g., styrene butadiene rubber) that are grafted to the main polymer chain. The rubber particles arrest crazes and cracks—thereby increasing the toughness and fracture resistance of the material.

Lubricants such as molybdenum disulfide, graphite, silicone compounds, or fluoropolymers (Teflon®) allow a polymer to be self-lubricating for wear application such as bearings. Other lubricants (or slip agents) such as fatty acid esters migrate to the surface to help lubricate a polymer during processing to prevent the polymer from sticking to processing equipment or to itself.

Plasticizers are added to thermoplastics to improve flexibility and toughness. Polyvinyl chloride, for example, is commonly plasticized for applications such as automotive interiors. High-molecular-weight phthalate esters added to PVC increase the polymer's internal free volume, providing more space within the polymer structure for long-range motion of the

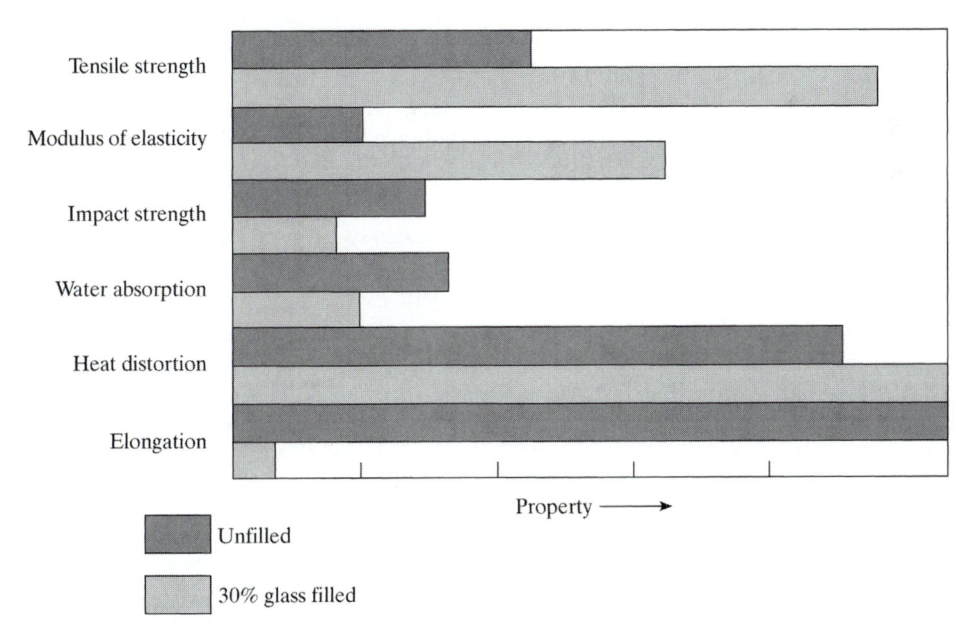

Figure 4–22
Effect of glass-fiber reinforcement on the properties of nylon 6/6

Figure 4–23
General effect of plasticizers on properties

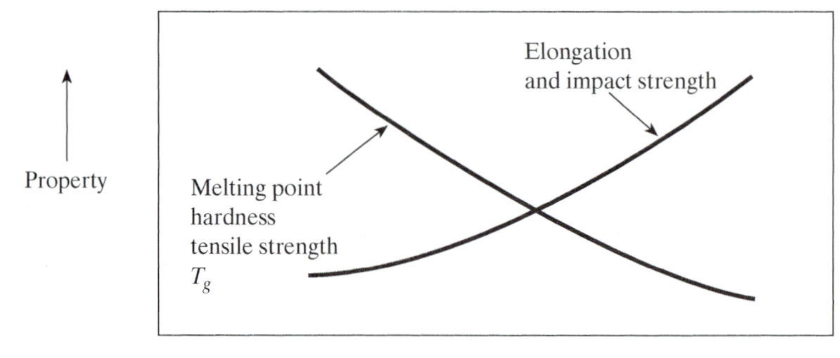

polymer chains. While improving flexibility, plasticizers reduce the T_g, melt viscosity, and tensile strength of a thermoplastic (Figure 4–23). The solubility and chemistry of plasticizers in thermoplastics is important, because plasticizers can migrate to the surface of a polymer, causing greasy films, or may vaporize from the surface, causing the material to become brittle. Certain phthalate esters such as di(2-ethylhexyl) phthalate (DEHP), used in medical devices, food packaging, and children's toys, have come under scrutiny by some regulatory bodies. Some studies suggest that such plasticizers may cause medical problems when small quantities are ingested. In general, the propensity for plasticizers or other polymer additives to leach or bloom on a polymer surface should be fully investigated before a product is commercialized.

Probably the most important additives are stabilizers and antioxidants. As polymers age or are exposed to degradative environments, the structure of the polymer may change. For example, heat during injection molding can cause the polymer chain size to be reduced through chain scission. The polymer may also cross-link, forming gel slugs in the part. Stabilizers and antioxidants halt degradative processes such as chain scission, cross-linking, crazing, hydrolysis, or depolymerization.

Stabilizers and antioxidants also prevent degradation of the polymer during use. Exposure to heat, moisture, ultraviolet light, and chemicals, for example, can degrade a polymer chain by causing brittle cross-linking, depolymerization, crazing, chain scission, or other forms of degradation. Due to aging, the part may also yellow, chalk, or change mechanical properties (e.g., become brittle).

Polymer additives may simulaneously impair other properties while enhancing one property. The art (and science) of polymer formulation is balancing the composition to enhance the beneficial polymer properties while minimizing any deleterious effects. Polymer suppliers strive to develop polymers with the proper balance of additives for a particular application. In 2003, clay nanoparticles ($<$100 nm mean diameter) were being investigated as fillers for plastics to be used as auto body panels. Larger fillers often have a detrimental effect on the surface finish of molded panels. The goal is strengthening without the loss of surface finish. Differences in polymer structure as well as additive packages make it difficult to substitute polymers of a similar family from different suppliers. For example, similar polymers from two different suppliers may not process or perform the same in end use because of differences in the additive package. Significant testing is typically required when selecting or changing polymer suppliers. It is important when selecting and specifying a polymer to review the additive package with the supplier and verify that the proper additives are present for the application at hand.

Blending and Alloying

In our discussion of polymerization processes, the concept of polymer alloys and blends was mentioned. Alloying and blending are techniques that are used to increase strength or to alter other properties. Using these techniques is infinitely cheaper than developing a new plastic. Different types of blends can be produced; if two or more plastics that are to be blended are miscible in each other, it is possible to form a single-phase material, which is usually called an *alloy*. One of the oldest of such alloys is a combination of polyvinyl chloride and acrylic. This material is often used in sheet form for thermoforming and similar applications, and the alloy has some of the flexibility of the vinyl and some of the strength and sunlight resistance of the acrylic component. There is synergism.

If the polymers that are mixed together are immiscible, they will form a two-phase material, and the components with the lower concentration will be a separate phase in the other material. It is common to add rubberlike polymers to rigid polymers to produce a blend with improved flexibility. The problem with immiscible blends is that the separate phases may have different glass transition temperatures, which may produce complications in molding.

The third possibility that occurs in blends is that the polymers making up the blend have partial miscibility. These types of blends consist of more than one phase, and the separate phases are blends of the components. For example, one phase may be matrix rich, with a small amount of the additive polymer. The other phase may be solute rich, with a small amount of the matrix material in solution.

In addition to the problems with differing melting characteristics of the phases, potential problems must be solved with the adhesion of the phases to each other. Achieving a blend of polymers with the desired compatibility requires considerable research and development, but this approach to modifying polymer properties is still much easier and more cost effective than the development of new polymers. For this reason this technique will continue to be widely used in the twenty-first century. Old, well-accepted polymers will be blended to make new families of polymers, and new polymers as they are developed will be blended to generate even more polymer families. Unfortunately, most of these blends and alloys are proprietary to specific polymer suppliers. This is not a problem other than that property information is not as easily available as one would like. Designers must rely on manufacturer's brochures for property information. The best protection against bias is to accept only data from tests that were conducted by a standard test procedure (like those of the ASTM). Disregarding the problems with manufacture and property information, blends and alloys are important parts of polymer strengthening mechanisms.

Interpenetrating Networks

We have mentioned how polymer blends are used to produce synergistic effects, polymer properties that are better than those of the components by themselves. A related system of strengthening that came into use in the mid-1980s is the *interpenetrating network* (IPN): A cross-linked polymer is penetrated throughout with another polymer such that the penetrant becomes the matrix and the network polymer is like a three-dimensional reinforcement of the matrix polymer, or cross-linked polymers can have another polymer cross-linked within the network of the first polymer. A semi-interpenetrating network is compared with a fully interpenetrating network in Figure 4–24. Semi-interpenetrating networks are usually formed by polymerizing a linear polymer within

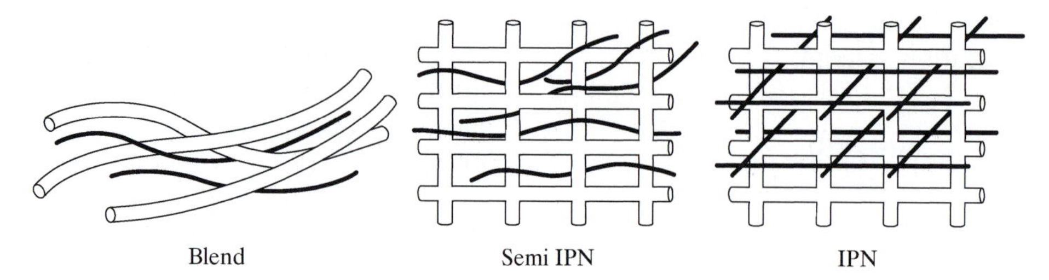

Figure 4–24
Formation of interpenetrating networks formed by interpenetrating cross-linking of different polymers

the network of a cross-linked polymer; a fully interpenetrated network can be formed by cross-linking a second polymer within the network of a previously cross-linked polymer. As we might expect, the rigidity of interpenetrating structures increases mechanical properties and other properties such as chemical resistance. The cross-linked polymer in IPNs is usually a thermosetting material, and full IPNs may require conjoint cross-linking. One such system is polyurethane and isocyanate. A semi-IPN can be made by polymerizing a rubbery thermoplastic like polysulfone within a cross-linked epoxy. This concept can be used even with a single polymer system; for example, cross-linked nylon can be penetrated by linear nylon. These latter types of network polymers can usually be fabricated by injection molding. The thermoset—thermoset networks and the thermoplastic—thermoset networks usually require processing by thermosetting processes such as compression molding. The use of these systems is still in its infancy, but this strengthening technique will undoubtedly produce many new engineering plastics with properties not currently available.

Summary

In this chapter we discussed polymerization, basic polymer systems, altering polymer properties, structure, additives, and alloys. Grounding in the structure–property relationships of polymers will help clarify the various polymer families that are reviewed in Chapter 5. At first, polymer systems and their respective structures can seem complicated. Even the vocabulary is full of strange tongue twisters. However, remembering the points that follow will improve confidence and understanding.

- Plastics from different families have different molecular structures.
- All plastics are made up of repeating units of a single molecule.
- Some plastics are thermoplastic: They can be remelted. Some plastics are thermosetting: They cannot be recycled; they will char or burn, but not remelt.
- Polymers are made by polymerization, a sometimes complicated chemical process.
- Amorphous polymers have a random (spaghetti-like) arrangement of polymer molecules.
- Crystalline polymers have alignment of molecular chains.
- Many thermosetting polymers have chemical bonds between chains, and they form a rigid three-dimensional network of macromolecules.

- Alloys and blends are simple mixtures of two or more plastics; there are no chemical or molecular links between the various polymer molecules (copolymers have bonds between different polymers).
- Cross-linking is one of the most common methods of strengthening polymers, and many thermosets have a cross-linked structure.
- Usually the only user option for strengthening a polymer is to specify the use of fillers such as glass, carbon fibers, and the like.
- Users of polymers should understand the various strengthening techniques that are used by polymer manufacturers: fillers, blending, interpenetrating networks, and crystallinity.
- Users of polymers should understand the basic processes that are used to produce shapes in both thermoplastic and thermosetting plastics. Injection molding is the most frequently used process (there were over 65,000 of these machines in the United States in the 2000s). Many rod and bar shapes for machined parts are made by extrusion.
- The first thing to question about a new plastic is if it is thermoplastic or thermosetting. The fabrication differences are significant.

Critical Concepts

- Plastics are made from repeating molecules of organic compounds.
- Plastics can be melted and remelted (thermoplastic) or melted only once (thermoset).
- Some plastics have an amorphous structure; others are semicrystalline.
- Elastomers/rubbers are differentiated from plastics by their high elongation (at least 200% before breaking).
- The glass transition temperature (T_g) of a thermoplastic is a measure of when it becomes rigid on cooling and it correlates with some mechanical properties.
- Thermosetting plastics do not have a T_m, and they can have temperature resistance superior to that of thermoplastics.

Terms You Should Remember

lignin
plastic
polymer
polymerization
condensation
addition
cross-linking
substituent
copolymerization
stereospecificity
glass transition
terpolymer
thermoplastic
thermosetting
engineering plastics
linear polymers
branching
crystallinity
plasticizer
fillers
interpenetrating network
homopolymer
amorphous

Case History

Microwave Sterilization

A new system sterilizes medical and dental instruments using microwave technology similar to domestic microwave ovens. The tools are placed in a polymer pouch that is then heat sealed. A water-filled pillow is installed in a second vestibule or chamber of the pouch. The pouch is loaded with the instruments, inserted in a special cassette, and then placed in a microwave oven. Microwave energy boils the water in the pillow, the pillow melts at the boiling point of the water, and steam expels into the pouch containing the instruments. The microwave energy continues heating the water to steam (133°C at 350 kPa) for several minutes. The microwave system has a distinct advantage over conventional steam autoclaves because of its bench-top practicality and short sterilization time.

Polypropylene is used for the pouch material because of its strength and resistance to the sterilizing temperature and pressure. The water pillow is made from polyethylene because it has a lower melting point than polypropylene. The polypropylene pouch contains an FDA-approved powdered corrosion inhibitor additive to prevent corrosion of the surgical and dental instruments during sterilization and storage before use. The polypropylene resin used for the pouch is compounded with the corrosion inhibitor additive and then fabricated into a thin film (0.14 mm thick) suitable for manufacturing pouches. The corrosion inhibitor in the polypropylene film is designed to vaporize during the sterilization process, filling the inside of the pouch with a partial pressure of vapor phase corrosion inhibitor—preventing instrument corrosion for many months. This is a high-tech use of commodity plastics.

Questions

Section 4.1

1. Draw the molecular structure of polyvinyl chloride with a molecular weight of 5.
2. Explain the difference between a copolymer and a terpolymer. Cite an example of each.
3. Explain the difference between a polymer blend and a homopolymer.
4. What type of bond is formed between carbon atoms when a polymerization reaction takes place?
5. Name three natural polymer resins and state where they can be used.

Section 4.2

6. What advantages do thermoplastic polymers have over thermosetting polymers, and vice versa?
7. What is the difference in molecular structure between thermoset and thermoplastic polymers?
8. How do you designate a type of polymer on an engineering drawing?
9. What is the difference between a commodity plastic and an engineering plastic?

Section 4.3

10. Explain cross-linking, branching, linear polymers, and chain stiffening.
11. Explain the role of stereoregularity in polymer strengthening.
12. How does a polymer's T_g relate to its strength?
13. Explain how a polymer develops crystallinity and what effect crystallinity has on properties.
14. What is the role of aramid fibers, talc, glass, silicone, and antimony oxide on polymer properties?

15. What is the purpose of alloying? How is it done?
16. Explain how an interpenetrating network strengthens.

To Dig Deeper

Billmeyer, F. W., Jr. *Textbook of Polymer Science*, 3rd ed. New York: John Wiley & Sons, Inc., 1984.

Clegg, D. W., and A. A. Collyer. *The Structure and Properties of Polymeric Materials*. London: The Institute of Materials, 1993.

Ebewele, R. O. *Polymer Science and Technology*. Boca Raton, FL: CRC Press, LLC, 2000.

Hall, C. M. *Polymeric Materials*. New York: John Wiley & Sons, Inc., 1981.

Murray, G. T. *Introduction to Engineering Materials*. New York: Marcel Dekker, Inc., 1993.

Painter, P. C., and M. M. Coleman. *Fundamentals of Polymer Science: An Introductory Text*. Lancaster, PA: Technomic Publishing Company, Inc., 1994.

Seymour, R. B., and C. E. Carraher, Jr. *Polymer Chemistry*. New York: Marcel Dekker, Inc., 1981.

Stevens, M. P. *Polymer Chemistry. An Introduction,* 2nd ed. New York: Oxford University Press, 1990.

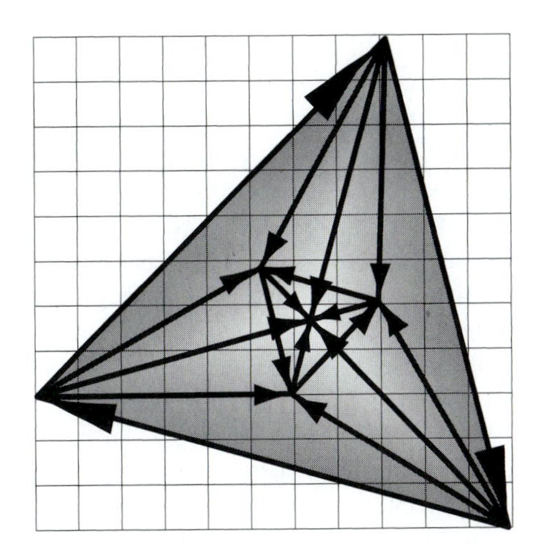

CHAPTER 5

Polymer Families

Chapter Goals

1. An understanding of polymer chemistry and of how chemical similarities create polymer and elastomer families.
2. A feeling for the relative differences between polymer families.
3. A simplification of the field of plastics through knowledge of polymer families.

Rationale

There are "families" in every category of engineering material. Going back in antiquity, people built from stone, wood, and bricks, each material having distinct "families." In the stone category, for instance, builders could select from granites, marbles, sandstones, limestones, flints, etc. Flints were usually the hardest; sandstones were usually the easiest to carve. Builders also recognized how families of woods varied in properties. Pines were light in weight and strong; oaks were durable but more difficult to shape; and willows were too unstable to use for construction. Bricks similarly varied in properties depending on the starting clays and curing. Thus builders knew when and where to use materials based on their knowledge of the family that a specific material came from.

Today we have over 15,000 commercially available plastics, and no designer can be familiar with this many specific materials. However, all of these 15,000 plus plastics come from about 20 polymer families. We propose that a working knowledge of the discriminating differences between these polymer families will help make you proficient in selecting plastics for design engineering assignments.

5.1 Identification of Polymer Families

The mer structures are analogous to the chemical compositions of metals. It is common practice to define a metal alloy by its chemical composition: the percentage of iron, the percentage of chromium, the percentage of nickel, and so on. These composition recipes are published for all standard alloys, and when a person gains a little metallurgical background, he or she can look at an alloy composition and have a good idea of the properties of the alloy. For example, if an alloy contains only iron, silicon, and about 3% carbon, a metallurgist will recognize this material as cast iron. If a material has a composition of 18% chromium, 0.15% carbon, and the remainder iron, a metallurgist will recognize this alloy as a 400 series stainless steel. In this chapter we hope to present enough information that the reader can do this sort of thing with plastics. We would like to give the reader enough background in the chemistry of common plastics that he or she can look at a mer structure for a plastic and have an idea of the family of plastics it belongs to and know something about its properties.

Polymer chemistry is an entire curriculum in some universities, so we will not go into significant detail; rather, we will show the structures of the common plastics and make some statements on the properties of each polymer or family of polymers. The subject of polymer chemistry can be very confusing, but it is not complicated if the basic polymer families are understood. There are thousands of specific plastics, but these thousands of plastics can be put into categories and families, which simplifies things.

As shown in Figure 5–1, the "universe" of plastics can be broken into just two categories: *thermoplastic* and *thermosetting*. Immediately it can be seen (Figure 5–1) that the thermoplastic materials are used much more than the thermosetting materials. The other pie charts show that only a few plastics make up the bulk of both thermoplastic and thermosetting production. About 80% of the production of thermoplastics comes from only four polymers (Figure 5–1): polyethylene (several types), polypropylene, polystyrene, and polyvinyl chloride. A similar situation occurs in thermosetting polymers; the bulk of the usage comes from only four polymer families—polyurethanes, phenolics, unsaturated polyesters, and ureas. Knowing that only eight plastics make up about 80% of the usage of plastics should help to make the subject less intimidating. It also points out the need to become familiar with these eight popular materials.

The widely used plastics are often called *commodity* plastics. The other plastics, which have higher strength, greater environmental resistance, or better physical properties, are often called *engineering* plastics. The implication is that these "fancier" plastics are more suitable for durable goods than commodity plastics. Although commodity plastics are used for throwaway items such as hot drink cups, it is not true that their properties are not good enough for "engineering applications." Rigid PVC, for example, is an excellent material for chemical storage tanks, exhaust systems, and many other chemical plant applications. Probably the more accurate discriminator is cost; commodity plastics are usually cheaper than engineering plastics. Commodity plastics in 2003 generally cost less than $1/lb ($2.2/kg). Engineering plastics come in several price range increments. There are many in the $1.50 to $6/lb ($3.30 to $13.20/kg) category; a significant number have prices in the range of $10 to $30/lb ($22 to $66/kg); and there are a few plastics with prices as high as $600/lb ($1320/kg). This is an important factor in selecting plastics; some are so expensive that they are used only as a last resort.

To summarize, in the remaining sections we will discuss the chemical and property differences between important families of thermoplastic and thermosetting plastics. We will categorize plastic families into thermoplastic commodity plastics, thermoplastic engineering plastics, thermosetting commodity plastics, and thermosetting engineering plastics; we will conclude with a discussion of elastomers. Keep in mind that most of the plastics that we will discuss actually comprise a polymer family. There may be 200 commercially available grades of polycarbonate. Some are reinforced; some contain fillers; some contain additives to reduce UV absorption; some contain pigments; and some are impact modified. All 200 will have the same basic mer and the same polymer molecule, but in the plastic databases of 15,000 or so plastics, each is considered to be a separate "plastic." Try to leave this chapter with a basic image of the mer that is common to a polymer family, and try

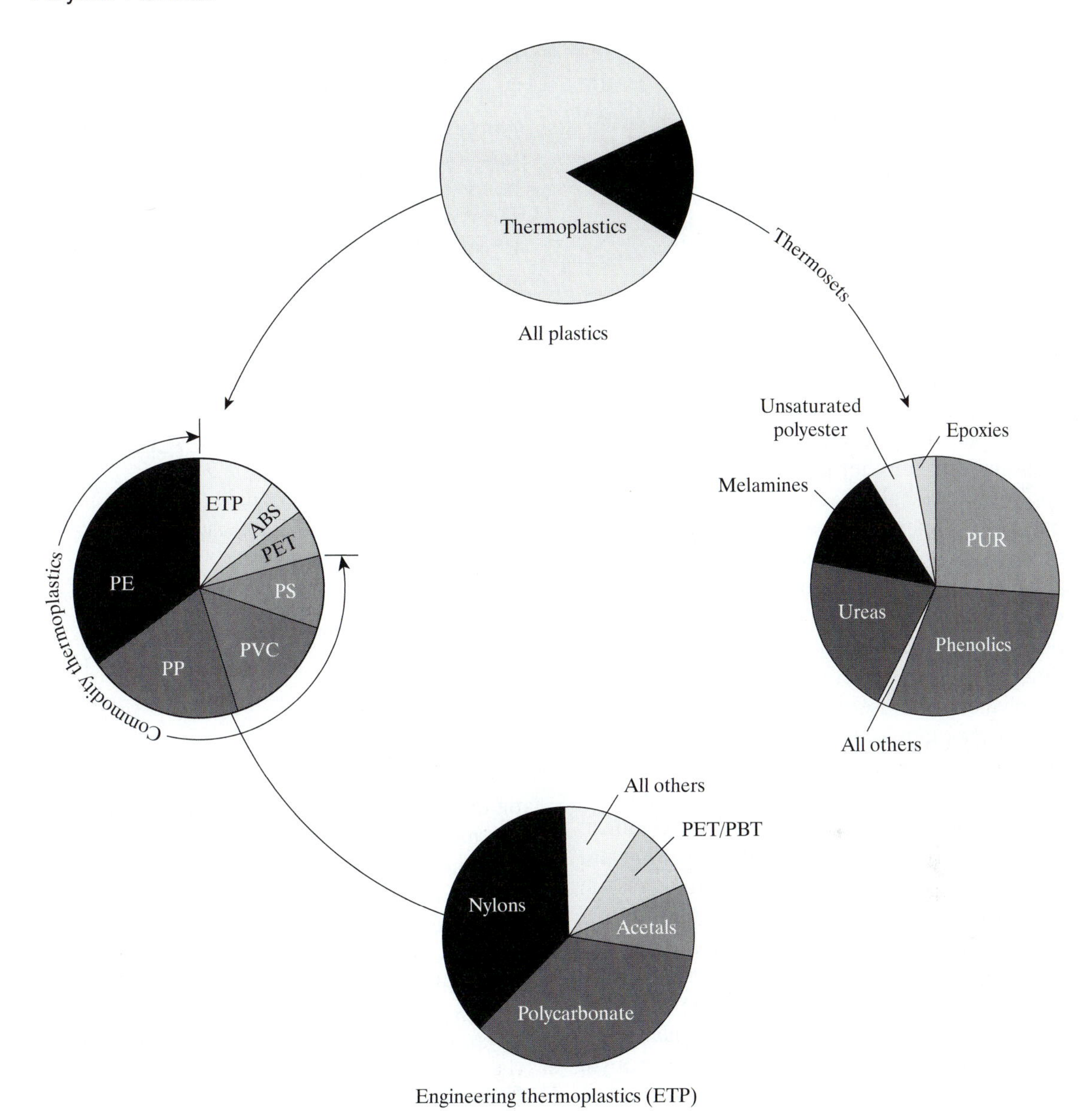

Figure 5–1
Estimated relative usage of thermoplastic and thermosetting plastics in the United States in 2003 (various sources)

to remember the one or two key property factors that set each polymer family apart from the others. Do not try to become familiar with all 15,000 plastics in the databases.

5.2 Thermoplastic Commodity Plastics

In our discussion of strengthening mechanisms, it was shown that the properties of a given polymer could be altered by such techniques as copolymerization, fillers, blending, and orienting. A polymer without additives and without blending with another polymer is called a *homopolymer.* We shall discuss the properties of important polymer families with the polymers in those families in homopolymer form. There are so many options to altering polymers that it would be impossible to discuss, for example, polystyrene and all the commercial modifications; but we shall discuss some of the important copolymers. If a user understands the basic homopolymer family, he or she will have a very good feel for the properties that can be obtained when this material is blended or reinforced.

We shall identify polymers by their chemical name or their accepted acronym. Acronyms are probably the only generic identification system that is widely accepted and used. Homopolymers and some blends, copolymers, and terpolymers are assigned letters that abbreviate their name, and there is no provision in the system to show additives, fillers, and reinforcement (like the ASTM D 4000 system). Where an acronym exists, we place it next to the chemical name:

PS	polystyrene
PA	polyamide
PC	polycarbonate
SAN	styrene acrylonitrile (copolymer)

These acronyms may vary somewhat around the world, but the ones that we use are generally recognized worldwide.

Ethenic Polymers

The term *ethenic* comes from the fact that this family of polymers has the basic monomer structure of ethylene, a carbon-to-carbon double bond ($H_2C{=}CH_2$). As mentioned previously, a series of polymers with greatly different characteristics can be obtained by altering a monomer structure by removing a hydrogen atom and substituting another functional atom or group of atoms (substituent group). This is how a family of ethenic polymers is formed:

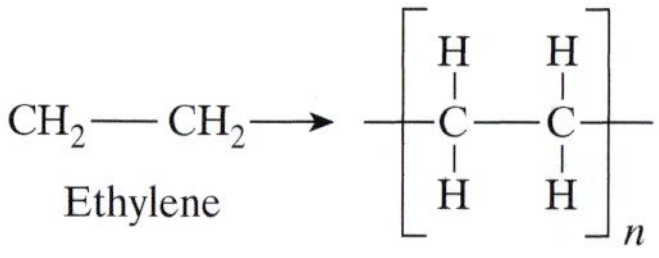

$$\underset{\text{Ethylene}}{CH_2{-}CH_2} \longrightarrow \underset{\text{Polyethylene (PE)}}{\left[\begin{array}{cc} H & H \\ | & | \\ -C- & -C- \\ | & | \\ H & H \end{array}\right]_n}$$

$$\left[\begin{array}{cc} H & H \\ | & | \\ -C- & -C- \\ | & | \\ H & H \end{array}\right]_n + \text{substitution of Cl} \longrightarrow \underset{\text{Polyvinyl chloride (PVC)}}{\left[\begin{array}{cc} H & Cl \\ | & | \\ -C- & -C- \\ | & | \\ H & H \end{array}\right]_{n'}}$$

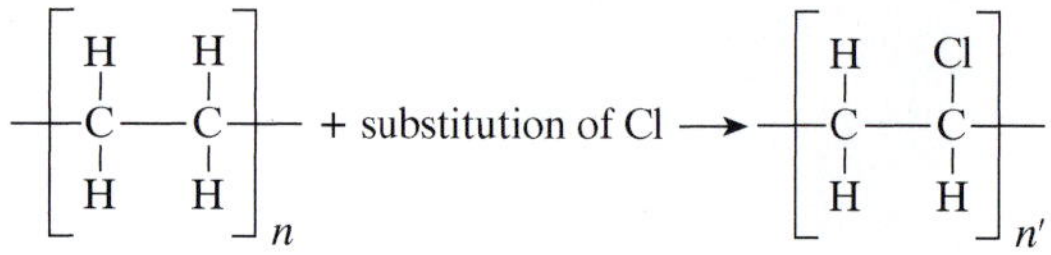

Table 5–1 lists some of the polymers based on the ethylene structure; most of the commodity thermoplastics come from this polymer family. The most important thermoplastic commodity plastics—polyethylenes [both low (LD) and high (HD) density], polypropylene, polystyrene, and PVC—cost less than \$0.50/lb (\$1.10/kg) in 2003. Ironically, some of the most expensive engineering plastics [often more than \$10/lb (\$22/kg)], the fluoropolymers, come from this same family.

Polyethylene *Polyethylene* was introduced on a commercial basis in the 1940s, and its applications and modifications have been increasing ever since. In most producing countries it surpasses all other plastics in quantity produced. The basic monomer is ethylene. The polymers made from the ethylene monomer vary in molecular weight (chain length), crystallinity (chain orientation), and branching characteristics (branches between chains).

Table 5–1
Ethenic polymers

Polymer	Substitution
[a]Polyvinyl chloride	Cl
Polyvinylidene chloride	Cl + Cl
[a]Polypropylene	$[CH_3]$[b]
Polyisobutylene	$[CH_3] + [CH_3]$
Polytetrafluoroethylene	F + F + F + F
Polytrifluorochloroethylene	F + F + F + Cl
Polyvinyl acetate	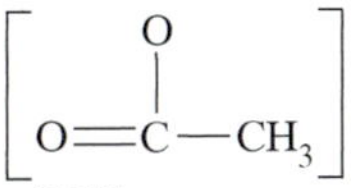 $\left[\mathrm{O{=}\overset{\overset{O}{\vert}}{C}{-}CH_3}\right]$
Polyvinyl alcohol	[OH]
[a]Polystyrene 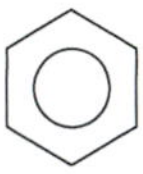	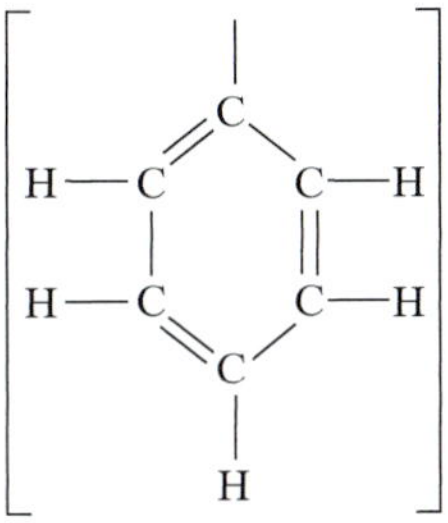 ⌬ or $[\mathrm{C_6H_5}]$ (ring: C with $\mathrm{H{-}C}$, $\mathrm{C{-}H}$, $\mathrm{H{-}C}$, $\mathrm{C{-}H}$, $\mathrm{C{-}H}$)
Polymethyl methacrylate	$[CH_3] + [COOH_3]$
Polyacrylonitrile	[CN]

[a]Commodity plastics
[b]Each bracketed group replaces one hydrogen atom.

Polyethylene is made from petroleum or natural gas feedstocks. The early processes for production involved extremely high pressures (up to 30,000 psi; 207 MPa) and temperatures as high as 400°C. High-pressure processes are still used, and the polymer produced is called low-density polyethylene. The density classification system relating to polyethylenes refers essentially to the normal density in g/cm^3 of a particular resin. Based on density, there are four classifications of polyethylene:

Low density	0.910–0.925 g/cm^3
Medium density	0.926–0.940 g/cm^3
High density	0.941–0.959 g/cm^3
Very high density	0.959 g/cm^3 or higher

High-density polyethylene is made by low-pressure and low-temperature techniques. Polymerization is achieved by sophisticated and often proprietary catalysis techniques. The current trend is to use these latter processes because of their lower energy requirements. The most recent processes produce a polyethylene grade called linear low-density polyethylene (LLDPE). The supermarket bags that have largely replaced paper bags are often made from high-density polyethylene.

The importance of the density ranges that apply to polyethylene is that structure, processibility, and use characteristics vary with density. Low-density grades have a significant degree of branching, which in turn means a lower melting point, lower strength, and lower elasticity. As the density increases, the linearity of the chain orientation increases; the high-density grades are more crystalline. Other properties that discriminate between different grades of polyethylene are melt index and molecular weight distribution. A high melt index indicates greater fluidity at processing temperatures. As molecular weight distribution increases, the variability of chain lengths increases. Higher-strength resins usually have low molecular weight distribution and a low melt index (Figure 5–2).

Low-density grades have good clarity in film form, and they are most often used for packaging films, paper coatings, wire coatings, and similar nonstructural applications. The high-density grades of polyethylene are predominately used for injection molded consumer items, blow molded bottles, and underground piping. Ultra-high-molecular-weight polyethylene (*UHMWPE*) is a linear polyethylene with a molecular weight of 3 to 5 million and is suitable for applications in industrial machines for wear-resistant parts. It has limited moldability; thus it is most often applied in the form of extruded shapes. The limited moldability occurs because UHMWPE does not melt like lower-molecular-weight grades; at 150°C it still behaves like a rubber. The flow temperature of high-density polyethylene is 136°C,

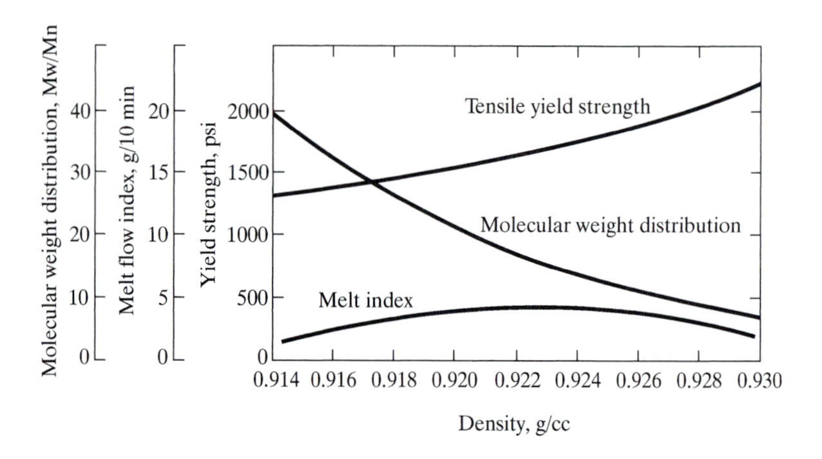

Figure 5–2
Variation of room temperature yield strength of polyethylene with density

and the lower-density grade is molded at temperatures as low as 110°C.

Some injection moldable grades of high-density polyethylene have use properties that approach those of UHMWPE. They may do the job if injection molding is a "must" design requirement. There are cross-linked grades of polyethylene that are used for flexible hot and cold water plumbing and heating systems. These applications require indefinite usable lifetime because they are often buried in structures.

The characteristics that distinguish polyethylenes from other polymer systems are their chemical resistance and flexibility. They are resistant to many acids and bases. They can be used for contact with most organic chemicals and solvents. Their flexibility is midway between the rigid plastics such as polystyrene and the very flexible plastics such as plasticized vinyls. Probably the most significant characteristics that limit polyethylene's applications are its low strength and low heat resistance, compared with engineering polymers, and its mechanical properties, which degrade when exposed to certain environments, such as ultraviolet (UV) radiation from sunlight.

Polypropylene The monomer of *polypropylene* contains a methyl group (CH_3) substituent:

$$\left[-\underset{H}{\overset{H}{\underset{|}{\overset{|}{C}}}}-\underset{CH_3}{\overset{H}{\underset{|}{\overset{|}{C}}}}- \right]_n$$

Polypropylene (PP)

There are fewer grades of polypropylene than of polyethylene, and this material is stereospecific. It is made from propylene gas, and various three-dimensional modifications in the chain structure can be made by catalysis and by special processing techniques. These structural modifications can affect use properties. The following is an illustration of a polypropylene chain with an isotactic structure; all the methyl groups are on one side of the chain:

$$\underset{CH_3}{\overset{H}{C}}-\underset{H}{\overset{H}{C}}-\underset{CH_3}{\overset{H}{C}}-\underset{H}{\overset{H}{C}}-\underset{CH_3}{\overset{H}{C}}-\underset{H}{\overset{H}{C}}-\underset{CH_3}{\overset{H}{C}}$$

Polypropylene is similar to high-density polyethylene, but its mechanical properties make it more suitable for molded parts than polyethylene. It is stiffer, harder, and often stronger than many grades of polyethylene. It has excellent fatigue resistance, and it is often used for molded parts that contain an integral hinge. It also has a higher use temperature than HD polyethylene. The trade-offs for these property benefits are less toughness and more low-temperature brittleness than polyethylene.

The chemical resistance of polypropylene is similar to and somewhat better than that of polyethylene, with the exception of oxidative environments and ultraviolet rays from sunlight. Stabilization is required to make it suitable for outdoor exposure. There is no solvent for polypropylene at room temperature. Thus it is suitable for containers such as molded bottles and chemical tanks.

A unique property of polypropylene is its low density. It is one of the lightest plastics, if not *the* lightest, with a density of 0.9 to 0.915 g/cm^3. This property makes polypropylene suitable for water ski tow ropes; it floats.

The two largest uses for polypropylene are for injection molded parts, such as luggage, battery cases, and tool boxes, and for fibers or filaments. Carpeting and ropes make up the bulk of the fiber and filament production.

Polyallomers The term *polyallomer* is used to describe crystalline copolymers of *olefins* (unsaturated hydrocarbons with the general formula C_nH_{2n}). A polyallomer of commercial significance is a block copolymer of polyethylene and polypropylene:

$$-\left[\begin{array}{ccc} H & & H \\ | & & | \\ C & - & C \\ | & & | \\ H & & CH_3 \end{array}\right]_m -\left[\begin{array}{ccc} H & & H \\ | & & | \\ C & - & C \\ | & & | \\ H & & H \end{array}\right]_n -$$

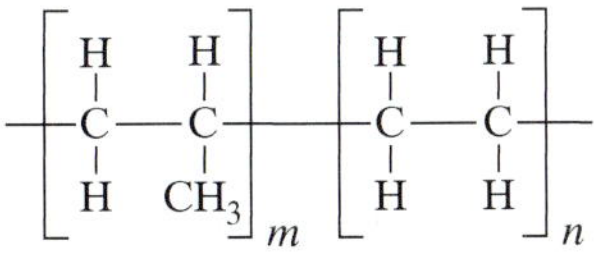

$$m > n$$

Properties can be varied by chain length combinations, but, in general, polyallomers are highly crystalline, with some of the properties of both polyethylene and polypropylene. The ease of molding is similar to polyethylene, yet the high degree of crystallinity provides the excellent flex resistance synonymous with polypropylene. Tensile strength and other tensile properties of polyallomers are usually midway between those of polyethylene and polypropylene. The density is about 0.9 g/cm^3, making it an extremely lightweight plastic.

The applications of polypropylene–polyethylene polyallomer are similar to those of polypropylene; it is used for a wide variety of molded parts. Its principal advantage over the component homopolymers is a better combination of moldability and mechanical properties.

Polybutylene Polybutylene is made from polyisobutylene, a distillation product of crude oil. It is made in many molecular weights to give a wide range of properties, from a sticky liquid to an elastomeric material. Its monomer molecule contains two methyl groups (CH_3) substituted for hydrogen atoms in the basic ethylene molecule:

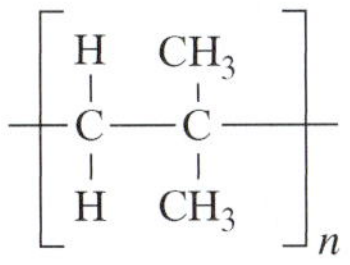

$$-\left[\begin{array}{ccc} H & & CH_3 \\ | & & | \\ C & - & C \\ | & & | \\ H & & CH_3 \end{array}\right]_n -$$

Polybutylene (PB)

The most important application of polybutylene is for making butyl rubber, which is used as a rubber, as an additive to caulking compounds, or as an adhesive. In the sticky liquid form (polyisobutylene), it can be mixed with petroleum oils to make an oil that will not migrate from a surface. This product is commonly sold in automotive stores as an oil additive. In machines, butyl rubbers are useful for gaskets and seals, and as an oil additive it is useful in lubricating shafts for linear bushings.

Polyvinyl Chloride The widely used acronym for *polyvinyl chloride* is PVC. It is one of the most widely used plastics in terms of volume produced. PVC is made by reacting acetylene gas

(C_2H_2) with hydrochloric acid in the presence of a suitable catalyst. The monomer has one chlorine atom substituted for a hydrogen atom:

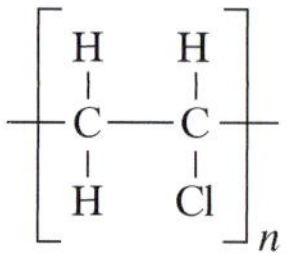

$$\left[-\overset{\displaystyle H}{\underset{\displaystyle H}{\overset{|}{\underset{|}{C}}}} - \overset{\displaystyle H}{\underset{\displaystyle Cl}{\overset{|}{\underset{|}{C}}}} - \right]_n$$

Polyvinyl chloride (PVC)

Plasticized PVC is more commonly recognized as vinyl. In this form it has low strength, and it is used for imitation leather, decorative laminates, upholstery, wall coverings, and myriad consumer products. It has industrial application in the form of flexible tubing and for corrosion-resistant coatings applied to metals. PVC is often highly plasticized or put into a dispersion to be used as a coating. The heavy insulating coatings used on the handles of electrical tools are often plastisol PVC coatings. PVC is also the most widely used insulating coating on wires. PVC is often used over other plastics for these types of applications because it is considered nonflammable; it does not sustain combustion.

Rigid PVC (without plasticizer) has sufficient strength and stiffness to almost qualify as an engineering plastic. It is thermoplastic with good moldability and fabricability. In the extruded form, rigid PVC is widely used for low-cost piping. It is available in all building supply and hardware stores for household plumbing. It is accepted practice to use PVC piping for cold water or chemicals and chlorinated PVC (CPVC) for hot water piping (temperatures up to 180°F, or 82°C). In industry, rigid PVC can be used for guards, ducts, tanks, fume hoods, and similar components that require corrosion resistance. It is very resistant to many acids and bases, but it has poor resistance to some organic solvents and organic chemicals. In this respect, the olefins are superior. The other major limitation of rigid PVC is poor toughness and notch sensitivity. Care should be taken to prevent stress concentrations in structural applications. PVC–plastic alloys are available with improved toughness. Besides alloys, PVC can be copolymerized with a number of other polymers, such as vinyl acetate and acrylonitrile, to get property modifications.

Polyvinyl Acetate The monomer of polyvinyl acetate (PVA) has an acetate substituent replacing a hydrogen atom on an ethylene molecule:

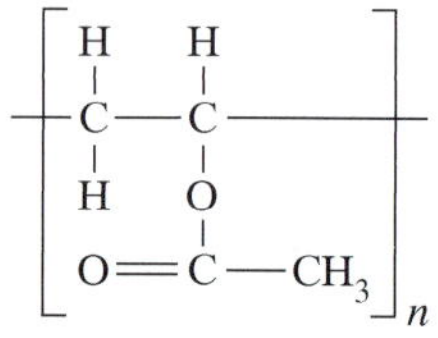

$$\left[-\overset{\displaystyle H}{\underset{\displaystyle H}{\overset{|}{\underset{|}{C}}}} - \overset{\displaystyle H}{\underset{\displaystyle O-\overset{\displaystyle O}{\overset{\|}{C}}-CH_3}{\overset{|}{\underset{|}{C}}}} - \right]_n$$

Polyvinyl acetate (PVA)

Polyvinyl acetates are relatively soft thermoplastics used predominately for coatings and adhesives rather than for molded parts. PVA is a popular polymer for household and industrial latex paints. As an adhesive, it is the familiar wood glue found in most hardware stores. Other uses include textile fibers and textile finish coatings. It is not important as a structural plastic.

Polyvinyl Alcohol Similar to PVA, polyvinyl alcohol (PVAL) is not a particularly important polymer for molded or extruded parts. It is more often used for coatings. PVAL contains one hydroxyl substituent (OH) for a hydrogen in an ethylene molecule.

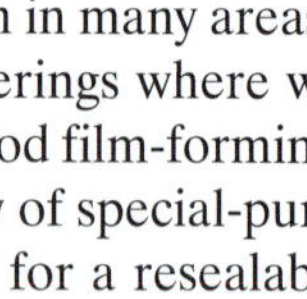

$$\left[-\overset{\displaystyle H}{\underset{\displaystyle H}{\overset{|}{\underset{|}{C}}}} - \overset{\displaystyle H}{\underset{\displaystyle OH}{\overset{|}{\underset{|}{C}}}} - \right]_n$$

Polyvinyl alcohol (PVAL)

Polyvinyl alcohol is water soluble, thus limiting its application in many areas. It is sometimes used for pill coverings where water solubility is desired. It has good film-forming ability, and it is used for a variety of special-purpose coatings. It can also be used for a resealable adhesive such as on envelopes. In industrial applications, it sometimes is used for tubing for resistance to oils and organic solvents. It has excellent oil and grease resistance. The mechanical properties are similar to those of plasticized PVC.

Polyvinylidene Chloride The acronym for *polyvinylidene chloride* is PVDC; it is made from trichloroethane, a common solvent. The monomer of PVDC has two chlorine atoms substituted for hydrogen atoms in the ethylene molecule:

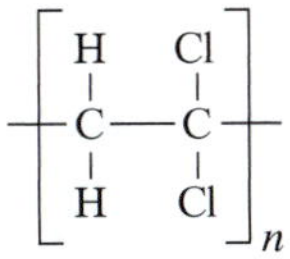

```
 ┌ H   Cl ┐
 |  |   |  |
─┼─C───C─┼─
 |  |   |  |
 └ H   Cl ┘n
```

Polyvinylidene chloride (PVDC)

In domestic applications it is used for packaging (Saran Wrap® and the like). In industrial applications it is used for the same types of applications as PVC. However, it has better strength, temperature resistance, and solvent resistance than PVC.

Polystyrene This plastic accounts for about 20% of all thermoplastics in commercial use. It is low in cost and thus is widely used in throwaway items (plastic tableware, food containers, and the like) and toys. Polystyrene is made from ethylbenzene. A large benzene ring replaces a hydrogen atom on an ethylene molecule:

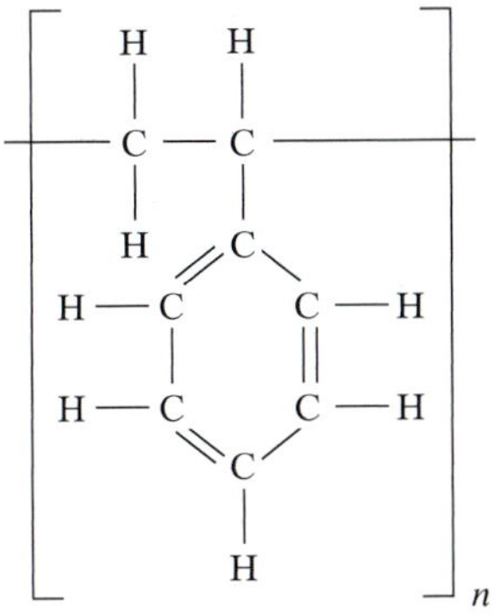

```
 ┌                       ┐
 |    H     H            |
 |    |     |            |
─┼─── C ─── C ───────────┼─
 |    |     |            |
 |    H    //C\          |
 |  H ─ C        C ─ H   |
 |      |        ||      |
 |  H ─ C        C ─ H   |
 |        \\C/           |
 |          |            |
 |          H            |
 └                       ┘n
```

Polystyrene (PS)

Polystyrenes are predominately amorphous and atactic. The benzene rings are random in their location on polymer chains. The presence of a large molecule (benzene) as the substituent on a carbon atom chain causes *chain stiffening.* Deformation by relative chain movement is inhibited by the interaction of benzene rings on adjacent chains. This structural effect is responsible for the inherent brittleness and hardness of this plastic.

Because of its poor impact strength, polystyrene cannot compete with engineering plastics for use on machines. It is used mainly in consumer products.

Polystyrene has excellent moldability: its melting point is only 240°F (115°C), and it can be made into a rigid foam. The foams are widely used for thermal insulation and for flotation devices. Unpigmented polystyrene is clear with very good light transmission characteristics, but it is not widely used for windows because it yellows with repeated UV exposure. It is not corrosion resistant like polyethylene and polypropylene; it is attacked by many solvents.

Summary: Commodity Ethenic Polymers

Most commodity ethenic polymers are linear in structure with a minimum of branching and cross-linking. Most are thermoplastic and have good processibility; the exceptions are some of the fluorocarbons and ultra-high-molecular-weight polyethylene. Four ethenic polymers—polypropylene, polyethylene, polyvinyl chloride, and polystyrene—are so widely used that they make up about 80% of the tonnage of polymers annually used in the United States. All ethenic polymers are based on the ethylene molecule, and most are products of crude oil. The mechanical properties vary from very low strength to strength comparable to that of engineering plastics (Figure 5–3).

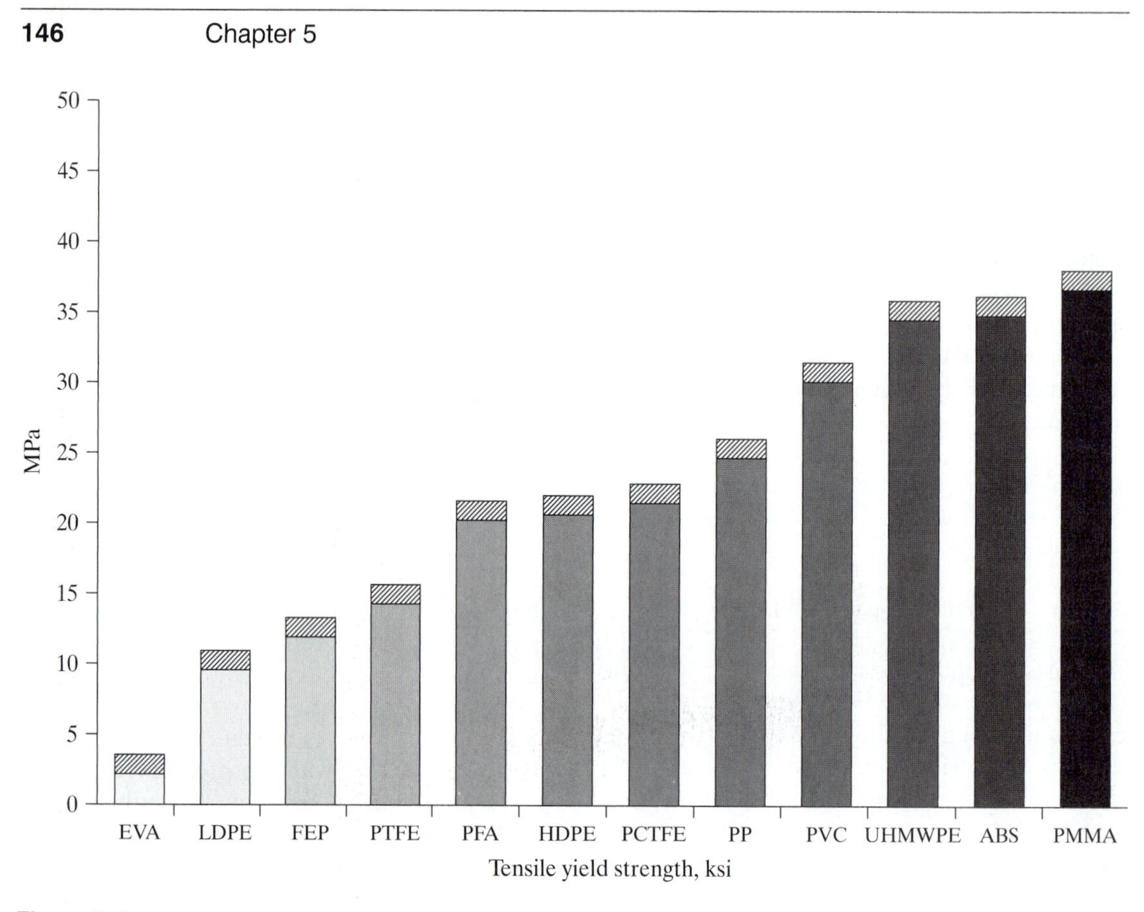

Figure 5–3
Approximate tensile yield strength of ethenic homopolymers

The following are additional characteristics to keep in mind:

- Polyethylenes can have significantly different properties depending on the molecular weight (the length of the molecular chains).
- Polyethylenes are mostly low-strength polymers, but the UHMWPE grades can have strengths comparable to engineering plastics.
- Polypropylene has improved strength over general-purpose polyethylenes.
- Polystyrenes have excellent moldability and are often first choices for recyclable molded items (packaging, and the like); some grades are often the lowest-cost plastic.
- Rigid PVC has excellent resistance to many chemicals and is widely used in plumbing.
- Plasticized PVC is the most widely used plastic for upholstery and for flexible plastic films for a wide variety of applications.

There are so many ethenic polymers that it is difficult to make general statements about their use characteristics, but it is probably safe to say that the ethenic family of polymers is the most important of the families that we shall discuss.

5.3 Thermoplastic Ethenic Engineering Plastics

Polymethyl Pentene

This polymer essentially contains a short carbon atom chain as a substituent on an ethylene molecule:

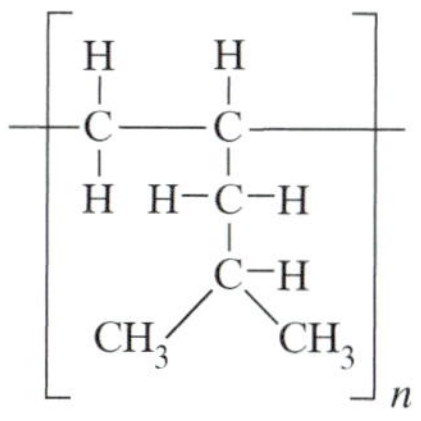

Polymethyl pentene (PMP)

It is thermoplastic, with good optical clarity, and has a use temperature that can be as high as 300°F (149°C). The electrical properties are similar to those of the fluorocarbons. The combination of transparency, high-temperature resistance, and good electrical properties makes this plastic suitable for lighting and appliance components.

Acrylonitrile Butadiene Styrene

Many of the property disadvantages of polystyrene can be overcome by plasticization or copolymerization. High-impact styrenes are made by adding plasticizers. Important copolymers of polystyrene are styrene butadiene and styrene acrylonitrile:

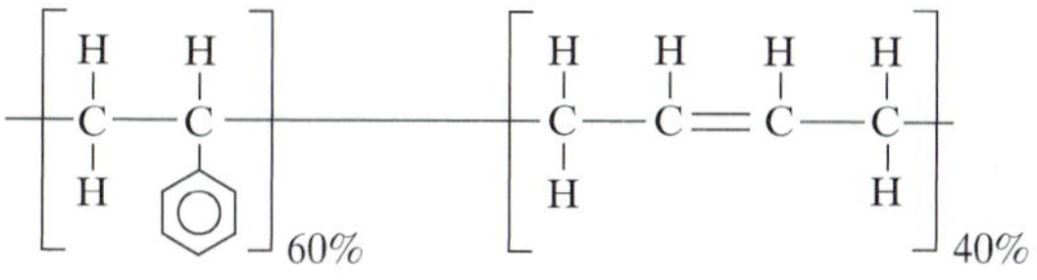

Styrene butadiene (SB)

Styrene butadiene copolymer is often used in latex paints, and it is the basis of an important rubber. Styrene acrylonitrile has improved impact strength, heat resistance, and environmental resistance over the styrene homopolymer. A more important modification of styrene is a combination of the foregoing: a polystyrene terpolymer of acrylonitrile, butadiene, and styrene (*ABS*):

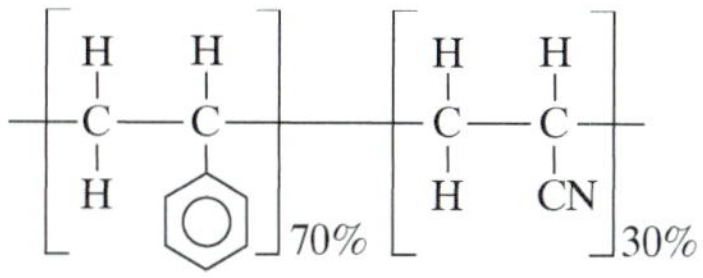

Styrene acrylonitrile (SAN)

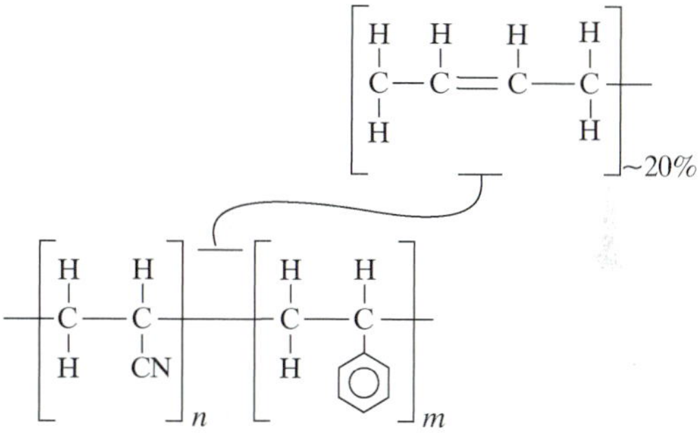

Acrylonitrile butadiene styrene (ABS)

This is a graft terpolymer, meaning that the butadiene chain attaches to the side of the acrylonitrile–styrene chain. ABS is an important thermoplastic for the manufacture of consumer goods. As an example of the general properties of this plastic, most telephones are injection molded from ABS. Almost everyone has dropped a telephone receiver, and the receiver seldom breaks. ABS has excellent toughness and moderately high strength and stiffness. It has good processibility. Most of the thermoplastic forming and molding processes apply; this includes foam processes. Some grades of ABS are even well suited to electroplating.

Polymethyl Methacrylate

This complicated name applies to a polymer that everyone is familiar with. *Polymethyl methacrylate* is the polymer used to make the clear sheet material in unbreakable windows and the low-cost lenses in cameras, flashlights, and safety glasses. It may be more readily recognized by the trade name Plexiglas®. Another name commonly used for polymers based on polymethyl methacrylate is acrylics. Acrylics are esters obtained by reacting an acid such as methylacrylic acid $[CH_2{=}C(CH_3)COOCH_3]$ with an alcohol. The monomer has a methyl group (CH_3) replacing one hydrogen atom on an ethylene molecule and a $COOCH_3$ substituent (from the acid) replacing another hydrogen atom:

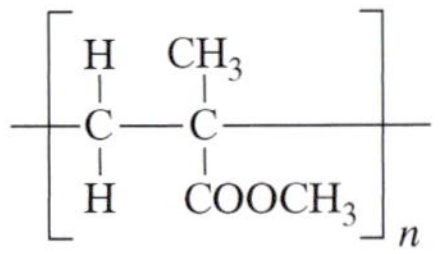

$$\left[\begin{array}{ccc} & H & CH_3 \\ & | & | \\ - & C - & C - \\ & | & | \\ & H & COOCH_3 \end{array}\right]_n$$

Polymethyl methacrylate (PMMA)

The acrylics are rigid and clear, making them useful on machines for guards, sight glasses, and covers. Another major industrial use is in coatings. They are used for clear lacquers on decorative parts, and pigmented they are used for paints. Of general interest, most false teeth, auto light lenses, and watch crystals are made from acrylics. They are readily injection molded.

Fluorocarbons

Fluorocarbons are a family of polymers based on a fluorine atom substitute. The oldest fluorocarbon, *polytetrafluoroethylene*, may be more familiar under the trade name Teflon®. It is made by polymerizing a gas with the same chemical name. The acronym for this polymer is PTFE. The monomer has one fluorine atom replacing each hydrogen atom on an ethylene molecule:

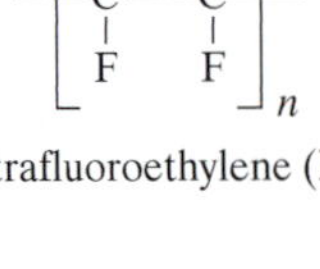

$$\left[\begin{array}{ccc} & F & F \\ & | & | \\ - & C - & C - \\ & | & | \\ & F & F \end{array}\right]_n$$

Polytetrafluoroethylene (PTFE)

PTFE is one of the most chemically inert materials known, so it is used for seals, tubing, and small vessels used to handle very aggressive chemicals. The major limitations of PTFE are that it is expensive, it is not moldable by conventional techniques, and it has low strength and creep resistance. It should not be used for structural components.

The principal fabrication limitation of PTFE is that it does not melt and flow like most thermoplastics (even though it is usually considered to be thermoplastic). Shapes and parts are compacted from powder and sintered with techniques similar to those used for powdered metals. A number of melt processible grades of fluorocarbons have been developed to circumvent this limitation.

Polytrifluorochloroethylene (PTFCE) can be injection molded, and it is used for the same types of applications as PTFE. The sacrifice for moldability is a decrease in maximum use temperature (200°C vs. 280°C) and a decrease in chemical resistance compared with PTFE.

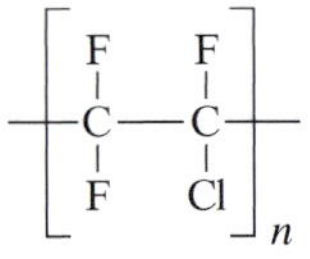

$$\left[\begin{array}{ccc} & F & F \\ & | & | \\ - & C - & C - \\ & | & | \\ & F & Cl \end{array}\right]_n$$

Polytrifluorochloroethylene (PTFCE) [Kel-F®]

Fluorinated ethylene propylene (FEP) is a copolymer of PTFE and hexafluoropropylene that more closely approaches the properties of PTFE than other fluorocarbons. Its uses are mostly in chemical handling.

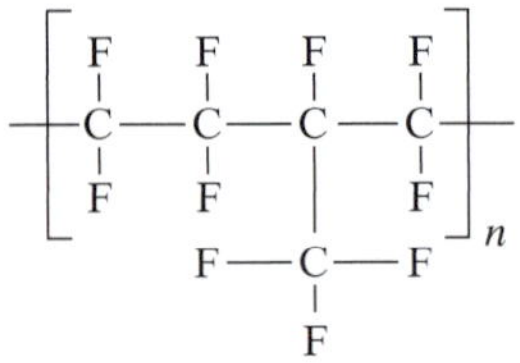

$$\left[\begin{array}{ccccc} & F & F & F & F \\ & | & | & | & | \\ - & C - & C - & C - & C - \\ & | & | & | & | \\ & F & F & | & F \\ & & F - & C & - F \\ & & & | & \\ & & & F & \end{array}\right]_n$$

Fluorinated ethylene propylene (FEP)

Perfluorinated alkoxy is a newer grade of injection moldable fluorocarbon that has an even higher use temperature (260°C) than PTFCE and FEP, and better creep resistance than PTFE. The chemical resistance approaches that of PTFE.

$$\left[-\mathrm{CF_2}-\mathrm{CF_2}-\mathrm{CF(O{-}R)}-\mathrm{CF_2}-\mathrm{CF_2}- \right]_n \quad \mathrm{R}\ (\mathrm{C}_X\mathrm{F}_{2X-1})$$

Perfluorinated alkoxy (PFA)

Polyvinyl fluoride is not as chemical or heat resistant as PTFE, PTFCE, FEP, or PFA, but it has better melt processibility. It can be used for molded parts, but more frequently it is used for wire insulation and for decorative and weather-resistant coatings. Interior wall coverings can be made washable and stain resistant by PVF coatings. PVF films are also laminated to aluminum to make house siding and architectural panels.

$$\left[-\mathrm{CH_2}-\mathrm{CHF}- \right]_n$$

Polyvinyl fluoride (PVF) [Tedlar®]

Vinylidene fluoride is one of the most moldable fluorocarbons. It is used mostly for molded parts requiring chemical resistance. It is not as corrosion resistant as the other fluorocarbons (except PVF), and the maximum use temperature is only 300°F (150°C).

$$\left[-\mathrm{CH_2}-\mathrm{CF_2}- \right]_n$$

Vinylidene fluoride (PVDF) [Kynar®]

Ethylene tetrafluoroethylene is a PTFE ethylene copolymer with excellent moldability. It has lower density, greater toughness, and higher strength than PTFE and FEP. The maximum use temperature, 350°F (180°C), however, is much lower.

$$\left[-\mathrm{CH_2}-\mathrm{CH_2}- \right]_n \left[-\mathrm{CF_2}-\mathrm{CF_2}- \right]_m$$

Ethylene tetrafluoroethylene (ETFE) [Tefzel®]

Ethylene chlorotrifluoroethylene is a copolymer similar to FEP in moldability and uses. It has better environmental resistance and a slightly higher use temperature, 390°F (200° C), than ETFE.

$$\left[-\mathrm{CH_2}-\mathrm{CH_2}- \right]_n \left[-\mathrm{CF_2}-\mathrm{CClF}- \right]_m$$

Ethylene chlorotrifluoroethylene (ECTFE) [Halar®]

There are a few additional commercially available fluorocarbons, but they are essentially modified versions of the foregoing. The modifications include alloying, blends with other polymers, and various reinforcements.

Summary: Ethenic Engineering Plastics

As was the situation with the commodity plastics, the ethenic polymers are the most used. ABS, acrylics, SAN, and the fluorocarbons easily make up more than half of the production of engineering thermoplastics in the United States. Some people do not consider ABS, SAN, and acrylics to be engineering plastics, but without a doubt these plastics are used for engineering applications. In the 1970s a production automobile (the Cord reproduction) had a body made from ABS. PTFE is a relatively weak material, but because of its low friction when mated with steel under heavy loads, it is used for bearing pads to support high-rise buildings and bridges in earthquake zones. Acrylics are used in architectural applications for many structural items, but one of the most taxing applications is for windshields and canopies on jet aircraft.

In summary, the substitution of a functional group on the ethylene molecules has created the lowest-cost commodity plastics as well as some of the plastics most used for engineering applications. The mechanical properties range from intentionally very weak in plasticized vinyl (PVC) to very strong in acrylics. Chemical properties range from extremely chemical-resistant fluorocarbons to plastics that dissolve in water (polyvinyl alcohol, PVAL). The most used ethenic polymers make the lowest-cost plastic components; other ethenic polymers are among the most expensive plastics. This polymer family is probably the most important thermoplastic family, and designers should become familiar with at least the ethenic plastics that we have described.

5.4 Other Engineering Thermoplastics

Polyamides

The polyamides are polymers formed as a condensation product of an acid and an amine. They all contain the characteristic amide group: CO—NH. The basic reaction involved in forming a polyamide is

$$\underset{\text{Acid}}{R^*-C(=O)-OH} + \underset{\text{Amine}}{NH_2R} \rightarrow \underset{\text{Polyamide (PA)}}{[R-C(=O)-NHR]_n} + H_2O$$

*Symbol for any functional group

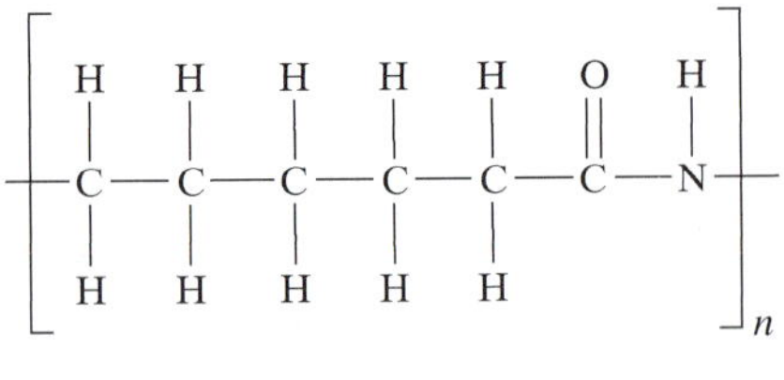

$$\left[-\overset{H}{\underset{H}{C}}-\overset{H}{\underset{H}{C}}-\overset{H}{\underset{H}{C}}-\overset{H}{\underset{H}{C}}-\overset{H}{\underset{H}{C}}-\overset{O}{\overset{\|}{C}}-\overset{H}{N}- \right]_n \quad \text{nylon 6}$$

There are a number of common polyamides; the various types are usually designated as nylon 6, nylon 6/6, nylon 6/10, nylon 6/12, nylon 11, and nylon 12.

These suffixes refer to the number of carbon atoms in each of the reacting substances involved in the condensation polymerization process. They have been maintained as an identification factor. Nylons with a period between the numbers in the suffix are homopolymers; nylons with a slash (/) between the numbers are copolymers; nylon 6/12 is a copolymer of nylons 6 and 12.

The term *polyamide* is the true generic name for this class of polymers, but the term *nylon* been used for so long that it has lost its trade name status and is now an acceptable generic name. Nylon 6 was developed in the 1930s and was brought to prominence in World War II as a replacement for silk for parachutes. Nylons are crystalline thermoplastics with good mechanical properties and good melt processibilty. Nylon was one of the first engineering plastics. It has sufficient mechanical properties to allow its use in structural parts. The mechanical properties of nylon homopolymers are among the highest of all engineering plastics. Continuous use temperatures can be as high as 260°F (120°C), but properties can degrade with time due to oxidation. The biggest disadvantage of nylons as engineering materials is their tendency to absorb moisture. Nylon 6 can absorb up to 10% by weight [equilibrium at 20°C and 100% relative humidity (RH)]. This moisture absorption causes a lowering of tensile strength and stiffness but an increase in toughness. Nylons 6 and 6/6 are very prone to moisture absorption, and nylons 11 and 12 are the most resistant of the group. The moisture absorption characteristics of the other grades are intermediate. The most troublesome aspect of moisture absorption is not the effect on strength (this can be dealt with by design) but the size change in use and in fabrication. A moisture absorption of 10% in a nylon 6 part could result in a dimensional change as high as 0.025 in./in. (0.06 cm/cm). This means that a molded part could change size with variations in the ambient relative humidity. For this reason, nylons are best suited for applications in which size change in use would not be detrimental to function. Examples of such applications are nylon coatings on dishwasher racks, wire covering, extruded hose, appliance wheels, soles for footware, awnings, rope, tire reinforcement, textiles for clothing, and carpeting.

Nylon is widely used for gears, cams, slides, and the like on machinery where size change in service could affect serviceability. In these applications, manufacturers' data on size change must be used in design calculations on operating clearances.

New grades of nylon and new reinforcement systems are being developed to overcome some of the limitations of the 6, 6/6, and 6/10 nylons. Aramid resins (Nomex, Kevlar—see the *Modern Plastics Encyclopedia* for plastic trade name attributions) are polyamides that contain an aromatic functional group in their monomer structure:

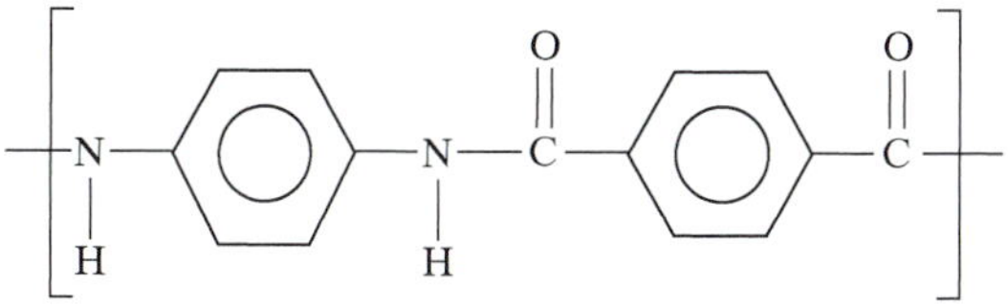

$$\left[-\underset{H}{\underset{|}{N}}-C_6H_4-\underset{H}{\underset{|}{N}}-\overset{O}{\overset{\|}{C}}-C_6H_4-\overset{O}{\overset{\|}{C}}- \right]$$

Aromatic polyamide (aramid) [Kevlar®]

They are produced mainly in the form of fibers and sheet. Some aramids have high tensile strength, and they can withstand continuous use at temperatures as high as 500°F (260°C). Chopped Nomex fibers have been used to replace asbestos as a reinforcement in clutch and brake friction materials. Nomex sheet is used for insulation on electrical devices that see high temperatures. The aramids and other new grades of nylon will ensure that this class of polymers will continue to be competitive in properties with more recently developed polymer systems. It is the major engineering plastic in volume.

Nylons have tensile properties comparable to some of the softer aluminum alloys. Types 6 and 6/6 are the most popular grades. Together they comprise over 90% of the production in the United States. About two-thirds of this production is of type 6/6, and one-third of the production is nylon 6. These two grades have the highest strength. Their impact strength is fair in the dry condition, but when they are saturated with water the impact properties increase substantially. The absorbed moisture acts as a plasticizer. Manufacturers of these materials expect that they will maintain their popularity well into the new century, and this family of plastics should be considered for applications in which mechanical properties are important and moisture-induced size changes will not affect function.

Polyacetals

Polyacetal is different from many of the other polymers in that polymer molecules have a carbon-to-oxygen bond rather than the more typical carbon-to-carbon bond. Such a structure is called a *heterochain* as opposed to a carbon chain. The basic monomer unit is formaldehyde, and the polymer chains have ending groups:

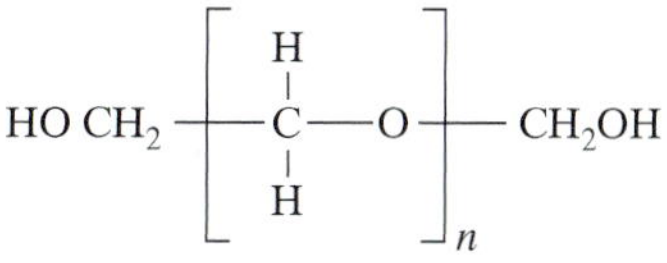

$$HO\,CH_2 \left[-\underset{H}{\underset{|}{\overset{H}{\overset{|}{C}}}}-O- \right]_n CH_2OH$$

Polyoxymethylene (POM)

Polyoxymethylene (POM) is the correct chemical term for this polymer, but acetal is the accepted generic term.

There are basically three types of acetals in wide commercial use: the homopolymer (Delrin 500®), a copolymer (with trioxane) (Celcon®), and PTFE-filled versions (Delrin AF®, Acetron®, etc.). A number of others are

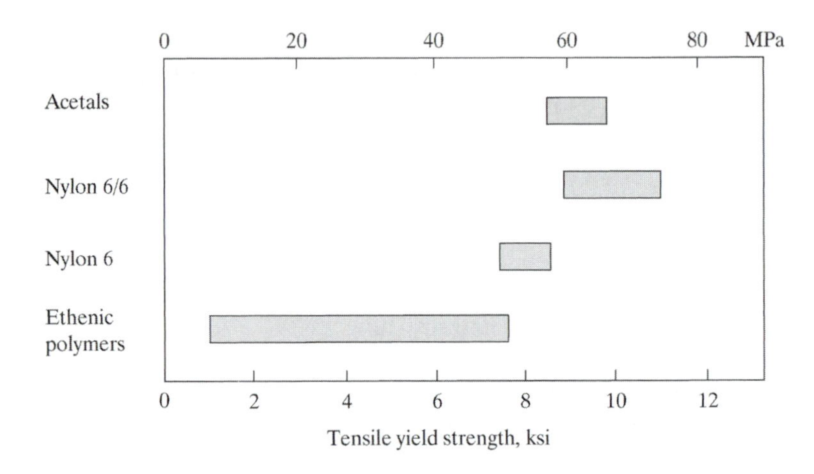

Figure 5–4
Approximate tensile yield strength of nylon and acetal homopolymers

manufactured, but the grade variations remain essentially the same: There is basically an unlubricated grade and a grade with PTFE added to improve tribological properties. It is not common to glass reinforce these polymers, and most of the usage is in the form of one of the three variations that we have listed. All grades are highly crystalline, with mechanical properties similar to those of nylons. As is shown in Figure 5–4, this family of polymers is higher in strength than most of the ethenic polymers, and the strength is slightly lower than that obtainable with the nylons.

The homopolymer has higher strength than the copolymers, but the copolymers are more stable at elevated temperatures; maximum continuous use temperatures are as high as 220°F (104°C), compared with 195°F (90°C) for the homopolymer. One advantage of acetals over nylons is their lower tendency for moisture absorption. They still absorb moisture and change size, but to a lesser degree. This makes acetals more stable in service as cams, gears, and other mechanical devices for which size change is detrimental. The PTFE-filled acetal has even lower moisture absorption tendencies and is even more stable in service. The PTFE addition also allows the use of this polymer for self-lubricating bushings and sliding devices. Unlubricated nylons and acetals make poor bearings.

A disadvantage that limits the use of acetals is high shrinkage in injection molding. It is difficult to mold unfilled grades to accurate part dimensions. Glass filling and foaming are often required to minimize molding shrink.

Thus, there are advantages and disadvantages of acetals, but, in general, they are one of the most important engineering plastics and one of the most frequently used in machine design and in structural components of consumer items.

Cellulosics

Cellulose is a natural substance that makes up a significant portion of all plant life. The cellulose molecule (monomer) is quite complex, and bonding is obtained by oxygen linkages to form a polymer:

OH OH
CH—CH H
C C—O
H CH—O
CH_2OH
n

Most polymers are made from petroleum or natural gas feedstocks. The raw material for the manufacture of *cellulosics* is cotton linters or wood pulp.

The most commonly used cellulose polymers are cellulose acetate (CA), cellulose acetate butyrate (CAB), and cellulose acetate propionate (CAP). The cellulose nitrates are now seldom used because of their flammability. Cellulose acetate is a thermoplastic often used for films and fibers. It is an ester resulting from the reaction of cellulose with acetic acid. Cellulose acetate propionate and cellulose acetate butyrate are mixed esters with improved physical and mechanical properties over the cellulose acetates. They are thermoplastics with good molding characteristics and thus find wide application in injection molded and vacuum formed domestic articles. The familiar roadside signs with fluorescent lights inside are usually thermoformed from cellulose acetate butyrate. One of the most commonly encountered cellulosics is cellulose produced by the viscous process. The familiar name for the end product of this process is rayon. Another common cellulosic is cellophane, which is also a product of the viscous cellulose regeneration process. In general, cellulosics are low-cost thermoplastic polymers with good moldability, desirable colorability, and weatherability. Their application is more often found in domestic goods than in machine and industrial applications. A good consumer application is cigarette filters. They are made from cellulose triacetate.

Most photographic film is made with a cellulose acetate support. Other film applications include safety goggles, sunglasses, screwdriver handles, and ski goggles. This material is selected for goggle applications because it has good mechanical properties, it is not brittle, and cellulosics have excellent resistance to sunlight. Many plastics, possibly most plastics, cannot withstand sunlight. Ultraviolet radiation causes the chemical breakdown of molecular bonds in many plastics, and undesirable things can happen with repeated exposure. Polyethylenes get brittle, many elastomers craze, plasticized vinyls lose their plasticizer, and rigid vinyls can form a powdery surface. Essentially, most plastics are not well suited to outdoor use. This is one of the outstanding properties of the cellulosics. They can withstand UV from sunlight without substantial loss of mechanical properties, color, or transparency.

Thermoplastic Polyesters

Thermoplastic polyesters have been used for about 40 years, predominately in films for packaging and in fibers. Everyone is familiar with polyester clothing and polyester auto tire reinforcement. These are thermoplastic polyesters, usually polyethylene terephthalate (PET):

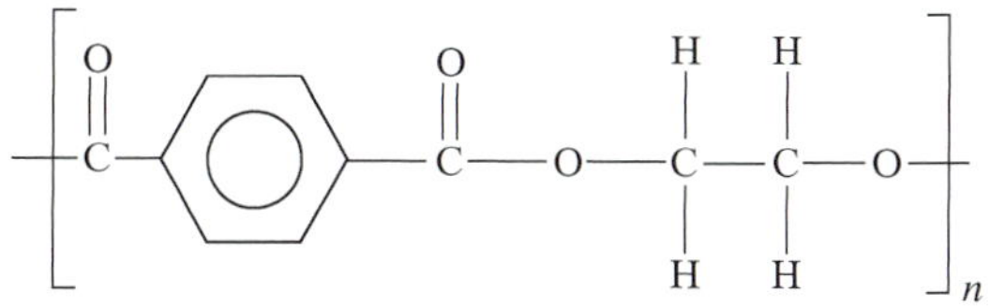

$$\left[-\overset{O}{\overset{\|}{C}}-C_6H_4-\overset{O}{\overset{\|}{C}}-O-CH_2-CH_2-O- \right]_n$$

Polyethylene terephthalate (PET)

PET has been used for plastic liter-size beverage bottles and for engineering plastic applications such as auto parts, gears, and cams. PETG is a glycol-modified PET that is used for similar applications.

The other basic type of thermoplastic polyester that is used for similar engineering applications is polybutylene terephthalate (PBT):

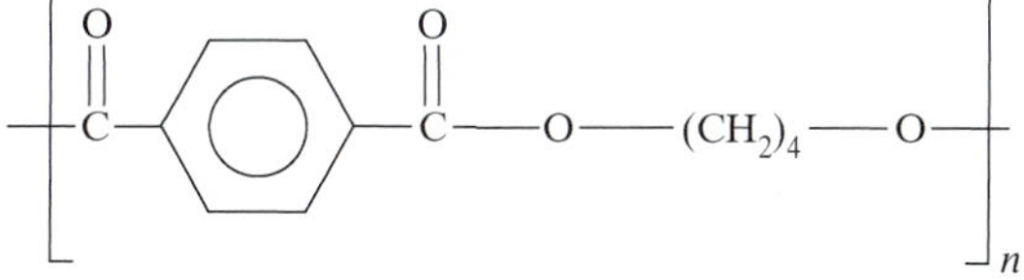

$$\left[-\overset{O}{\overset{\|}{C}}-C_6H_4-\overset{O}{\overset{\|}{C}}-O-(CH_2)_4-O- \right]_n$$

Polybutylene terephthalate (PBT)

PBT has good injection molding characteristics and mechanical properties similar to nylons. It has a use temperature above that of most nylons, and it does not have the moisture absorption problems of nylon. Glass-reinforced PBT has a stiffness similar to that of thermosetting resins while maintaining good toughness. As a class of materials, thermoplastic polyesters have performance characteristics that make them likely candidates for structural applications in appliances, automobiles, and consumer products.

About 45% of the production of PET goes into films for photography and packaging, and in fibers for clothing, tire reinforcement, and the like. A similar percentage of the U.S. production is used in blown beverage bottles. Thus the major usage of this polymer is for nonstructural applications. About 10% of the production is in grades that are made to be injection moldable, and these grades are used for all sorts of mechanical parts. It is common practice to reinforce thermoplastic molding grades of PET and PBT with 15% to 50% glass or minerals. The reinforced grades of PBT and PET are similar in mechanical properties to nylons, and they are used for the same types of applications. Both PET and PBT are crystalline and somewhat notch sensitive. Mechanical parts made from the engineering grades of these materials should have generous radii on reentrant corners.

PET and PBT are closely related polymers in use properties with only subtle differences. PET is slightly stronger and lower in cost. Some grades of PBT have processing advantages over PET grades. In any case, the grades that are used for engineering applications have very similar properties, and those properties are competitive with nylons, but these families of resins do not have the moisture absorption problems that are indigenous to nylons. PBT is widely used in blends: mixed with PET it is General Electric's family of Valox® resins; mixed with polycarbonate it forms Xenoy®; mixed with polyphenylene ether it forms Gemax® resins. Other blends are available, but as a family of plastics, the thermoplastic polyesters are made up of two basic polymer systems, PET and PBT, and there are many grades of each of these polymers for films, bottles, and molding resins. Both polymer systems produce engineering grades with excellent properties for general use in automotive, electrical, and appliance applications. In general, polyesters have higher strengths and use temperatures than commodity plastics like PE and PP, and they compete with nylons and acetals in mechanical properties.

Polycarbonates

A *polycarbonate* is really a polyester because both are esters of carbonic acid and an aromatic bisphenol:

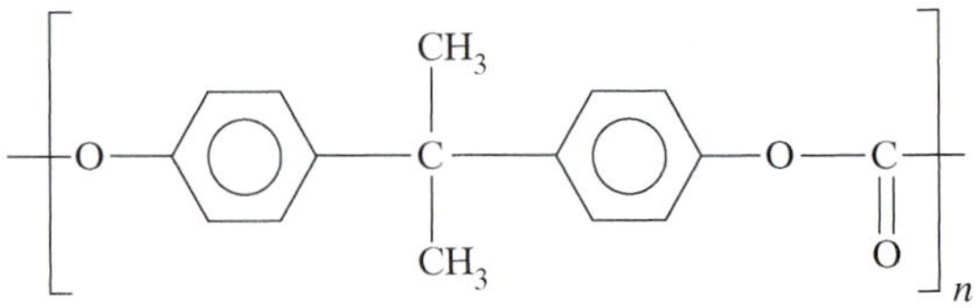

$$\left[-O-C_6H_4-C(CH_3)_2-C_6H_4-O-\overset{\displaystyle O}{\overset{\|}{C}}- \right]_n$$

Polycarbonate (PC)

The polycarbonates are amorphous linear polyesters with excellent moldability and impact strength. The combination of impact strength, temperature resistance, and transparency makes them suitable for the same types of applications as acrylics: guards, sight glasses, and the like. Their cost is usually higher than that of acrylics, so they are used only where the higher impact strength is needed. Polycarbonate is widely used for helmets and face shields. For structural applications, it is similar in properties to the nylons and acetals. The tensile properties of polycarbonates are similar to those of nylons and acetals, but the impact strength can be as much as ten times greater. Thus it is replacing materials such as ABS and nylon for automotive parts such as dashboards, window cranks, small gears, and similar mechanical components.

Polycarbonate is an important engineering thermoplastic for industrial applications as well as for consumer goods such as power tool housings, telephones, cellular phones, and sporting goods.

Each manufacturer of polycarbonate has at least a dozen grades. Some are specially formulated for electrical applications, some are designed for clear applications, and others are designed for mechanical applications. Grades are available with glass or other reinforcement, and some grades are lubricated with PTFE or similar materials for tribological applications. Probably the most familiar application of polycarbonates is for clear sheets (GE's Lexan® and others) and lenses for automotive lights. Polycarbonate competes with acrylics (PMMA) for safety glass types of applications. It is more expensive than acrylics, but it can have up to 16 times the impact strength. A similar situation exists with clear styrenes; polycarbonates cost more, but they are used because of their increased toughness. The molding grades of polycarbonate have similar toughness, with strengths comparable to acetals, nylons, and polyesters. These plastics are also candidates for use outdoors because of their good resistance to UV light. Windows and similar clear panels can tolerate outdoor exposure without degradation in clarity or loss of properties. As molding resins, PC parts have good stability and make accurate molded parts. Principal applications are for automotive trim, lenses and dashboard components, electrical connectors, and the like, and for all sorts of appliance parts. Clear impact-resistant sheet is the most familiar application. It is almost a standard material for construction hard hats, football helmets, and similar applications where moldability, toughness, and good appearance are required. The thickness or section size of polycarbonate parts can affect the apparent toughness of polycarbonate. When thicknesses are greater than about 0.10 in. (2.5 mm), polycarbonate may behave in a brittle manner. Other factors such as residual stress, notches, or annealing may also reduce the toughness of polycarbonate. Polycarbonates are susceptible to environmental stress cracking in certain solvents, and they can craze (develop fine cracks) under sustained load if creep stress limits are exceeded. These factors should be considered in design.

Polyimides

The polyimides came into commercial use as engineering plastics in the 1970s. They have very complex structures characterized by chain stiffening due to the presence of ring structures in the mer:

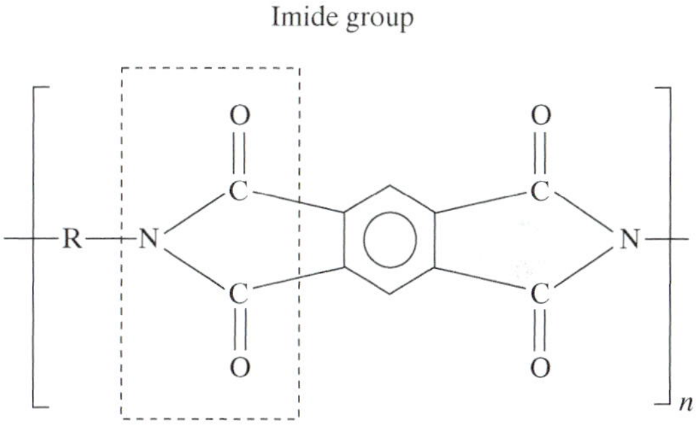

Imide group

Polyimide (PI)

The presence of the imide group in the mer structure makes a *polyimide*. The different types of polyimides depend on the reacting species. The thermoplastic types have a linear structure, and thermosetting types have a cross-linked structure. The usual method for making polyimides is by reacting aromatic diamines and dianhydrides. The common feature of this family of polymers is resistance to elevated temperature. Some of the materials have continuous use temperatures in the range of 500 to 600°F (260 to 315°C). They maintain stability and decent mechanical properties at these temperatures, and adequate data are tabulated on elevated temperature properties to allow the prediction of service at these high temperatures.

Unfortunately, the stable aromatic structure of these materials creates processing problems. The thermoplastic grades of polyimide require very high molding temperatures that often tax the capabilities of conventional plastic molding machines. Some of the polyimides are direct formed to shape at about 800°F (426°C) with techniques that are similar to those used in powder metallurgy. Powder is compacted and subsequently sintered into a shape under pressure in a mold. Sometimes vacuum sintering is used. The powder particles do not melt; the particles develop sufficient plasticity to fill voids, and the particles coalesce to form a full, dense part or standard shape. One commonly used polyimide, Vespel®, is offered in direct formed standard shapes that can be machined to the desired part shape. Individual parts can be direct formed if there is sufficient part volume to justify the compaction tooling.

In addition to good mechanical properties at elevated temperatures, polyimides have excellent mechanical properties. Tensile strength can be higher than 15 ksi (103 MPa); compression strength can be higher than 30 ksi (206 MPa). One reason for the thermal stability of polyimides is their low coefficient of thermal expansion (as low as 28×10^{-6} in./in. °F). This property makes polyimides candidates for precision parts that must function at high temperatures.

Polyimides are available in molding pellets, in direct formed shapes for machining, in direct formed parts, in films, as resins for adhesives, and as a matrix for advanced composites. Two popular additives to the molding grades are graphite and PTFE. These materials are added to grades destined for tribological applications. These lubricious fillers provide the capability for using these materials as self-lubricating bearings. Other applications are for structural parts that must maintain good mechanical strength at temperatures that destroy many other plastics. Films (Kapton®) are used for circuit board types of applications where operating temperatures are too high for the conventional phenolic circuit board materials.

Some polyimides are injection moldable—if this is a design consideration. Polyimides are not very stable in sunlight, and they can be attacked by oxidizing acids. However, probably the most significant limitation of this family of plastics is their high cost. Molded parts can be competitive in cost with other, more conventional plastics, but direct formed standard shapes, such as rod stock or flats, are extremely expensive. Depending on the shapes, it is possible to have a cost of $1000 for a kilogram or so of material. This apparent high cost is often justified when we consider the functional advantages over other types of materials. Polyimides are often the most cost-effective solution to application problems that involve elevated temperature, exceptional electrical properties, or unusual tribological properties.

Polyamide-imide

As implied by the name, *polyamide-imide* polymers have a chemical structure that is a combination of the typical nitrogen bond of a nylon (polyamide) and the ring structure of a polyimide:

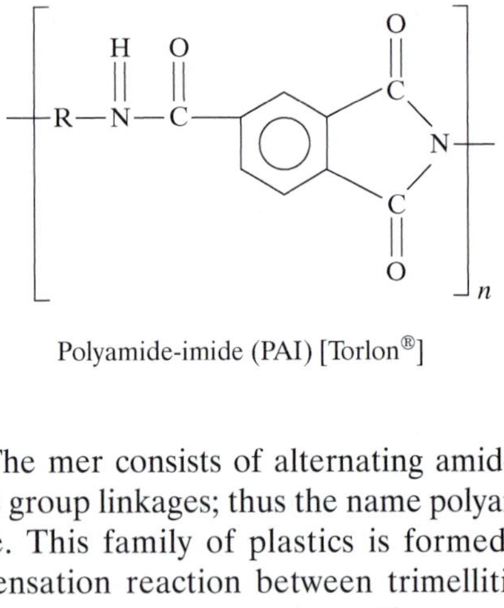

Polyamide-imide (PAI) [Torlon®]

The mer consists of alternating amide and imide group linkages; thus the name polyamide-imide. This family of plastics is formed by a condensation reaction between trimellitic anhydride and various diamines. The commonly used grades are amorphous and thermoplastic. Polyamide-imides are the only commonly used plastics that are strengthened by postmolding heat treating. Postcuring for 1 to 6 days at about 500°F (250°C) can double the tensile strength and percent elongation. Similarly, the heat deflection temperature can rise 50°F (25°C). This is not a phase-change hardening operation like that for metals. What is happening is additional polymerization (chain lengthening). This heat treatment is recommended whenever high strength is a selection criterion.

This family of polymers is the offspring of high-temperature wire varnishes, and commercial availability as an engineering plastic occurred in about 1973. Grades are available with no reinforcement, with glass, with carbon fibers, and with a variety of lubricants. The latter are for tribological applications, and the common lubricants are PTFE (as much as 8%) and graphite (as much as 20%). The lubricated grades are intended to compete with the Vespel types of polyimides that were just discussed.

The distinguishing characteristic of this family of polymers is high strength and a high maximum operating temperature (as high as 525°F or 275°C). One manufacturer of these polymers makes the claim that the unfilled grade has the highest strength of any unreinforced grade of plastic (23,000 psi or 158 MPa). Polyamide-imides have a high modulus of elasticity (as high as 1.5×10^6 psi or 10 GPa); the impact strength is comparable to that of other engineering plastics (1.5 ft-lb/in. or J/m). Equilibrium moisture absorption is less than 2% long term, and in 24 h the absorption rate is about 1%.

These engineering plastics are used for a wide variety of applications. Most involve carrying loads at elevated temperatures or some elevated-temperature electrical application. Some of the underhood plastic parts on automobile engines are made from these materials. In such applications, users capitalize on this PAI's elevated temperature strength. Valves made from PAI replace bronze castings in hot water plumbing systems. Polyamide-imides are not cheap. Resin costs in 2003 were in the range of \$20 to \$30/lb (\$44 to \$66/kg). These are premium engineering plastics. They are lower in cost than the polyimides, but they are still much higher priced than the commodity plastics (<\$1/lb). They should be used when elevated-temperature strength and injection moldability are important selection factors.

Polysulfone

Polysulfone thermoplastics came into commercial use around 1970. The basic monomer contains stable linkages of benzene rings:

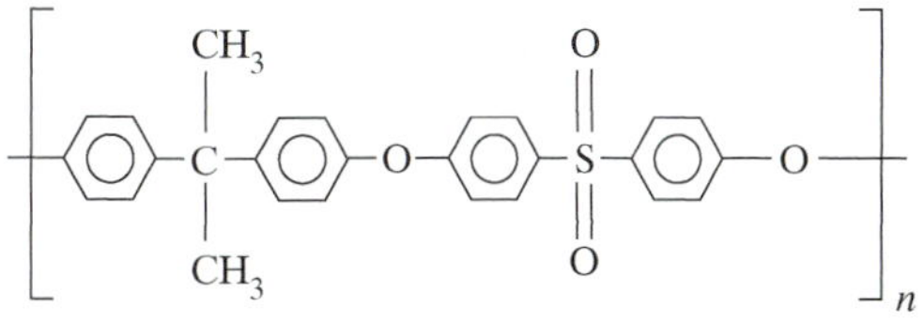

Polysulfone (PSU) [Udel®]

The mechanical properties of molded polysulfones are similar to those of nylons at room temperature, but useful strength and rigidity are maintained to temperatures as high as 350°F (180°C). Ordinary nylons can only be used at temperatures to 250°F (120°C). Polysulfones have good optical clarity and lower moisture absorption than nylons.

Polysulfone is an amorphous, rigid thermoplastic material made from dichlorodiphenylsulfone and bisphenol A. The continuous use temperature, as high as 320°F (160°C), is significantly above that of other clear plastics, such as polystyrene, acrylics (PMMA), and polycarbonate:

Plastic	Heat Distortion Temperature at 264 psi (1.8 MPa)
PS	130°F (54°C)
PMMA	170°F (76°C)
PC	270°F (132°C)
PSU	345°F (173°C)

Polysulfone is higher than the other materials in cost, but it is less expensive than some of the clear engineering plastics such as polyether sulfone.

The property combination of good strength at elevated temperatures and transparency make this material popular for containers and cookware. It is popular for microwave cooking utensils, and in the health field it is used for medical containers and supplies that must withstand repeated sterilization. Polysulfone retains its strength after repeated autoclave cycles, and it is very resistant to degradation by hot water. Certain types of polysulfone have been approved by the U.S. Federal Drug Administration (FDA) for repeated use in contact with food.

The electrical properties of polysulfone are good enough to allow use for circuit boards and other electrical devices that were formerly made from a hard-to-process thermosetting resin. Its favorable high-frequency electrical properties allow its use for microwave service. The elevated temperature properties allow polysulfone an area of application in piping for chemicals at temperatures too high for the lower-cost PVC or CPVC pipes and tubes. Polysulfone is available with fiber reinforcement, but the unfilled grades are more frequently used.

Polyethersulfone

Polyethersulfone became commercially available in the United States in 1973. It is an amorphous, aromatic, high-temperature thermoplastic characterized by *ether*-type R—O—R monomer linkages:

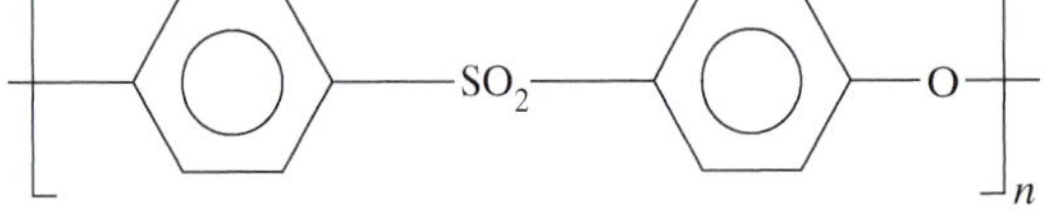

Polyethersulfone (PES)

This plastic is similar to polysulfone in chemical nature and properties. It is amorphous, it is thermoplastic, and it has good mechanical strength and retention of strength at elevated temperature. It is fairly transparent (light transmission 60% to 90%), but it has a slight yellow tint. The cost of PES was in the range of $4 to $5/lb in 2003 ($10/kg). It is more expensive than polysulfone but is supposed to last longer than PSU, justifying the higher cost.

The mechanical properties are similar to those of nylon and other engineering plastics, but it has better creep resistance than many competitive thermoplastics. Use temperatures can be as high as 390°F (200°C); it has one of the highest glass transition temperatures of any of the thermoplastics, 428°F (220°C). This is the reason for the high use temperatures. This high glass transition temperature is probably also responsible for the exceptional creep resistance of this material. The long-term creep characteristics are superior to those of many other engineering thermoplastics, such as nylon, polycarbonate, and acetal.

Another property that distinguishes this material from other engineering plastics is its flame resistance. Flame spread is very low, and smoke emission on burning is lower than that of PC, PSU, PTFE, PS, PVC, and some other thermoplastics. It is even lower than some thermosetting polymers such as some phenolics.

A limiting characteristic of PES is its moisture absorption. It can absorb as much as 2.3% immersed in 100°C water. Degradation of properties by immersion in hot water is very low, reportedly lower than that of polysulfone, but the absorption of water will result in size change. The application areas of this plastic are similar to those of polysulfone and polyphenylene oxide, mostly applications that involve some type of elevated temperature service. PES is used in hot water piping and low-pressure steam sterilizers, where service temperatures can be as high as 280°F (140°C). PES has been used for molded printed circuit boards (to eliminate thermosetting plastics). This application requires good electrical properties as well as a temperature resistance that will allow running these boards through wave soldering machines. In summary, this is a high-temperature, semitransparent engineering-grade thermoplastic for molded parts that must withstand some type of hot water or other form of elevated temperature service. It costs much more than commodity plastics, so its use must be justified by property needs.

Polyphenylene Sulfide

Polyphenylene sulfide came into commercial use in the early 1970s. It is a rigid, crystalline thermoplastic with a stable structure and rigidity provided by the chain stiffening of benzene rings:

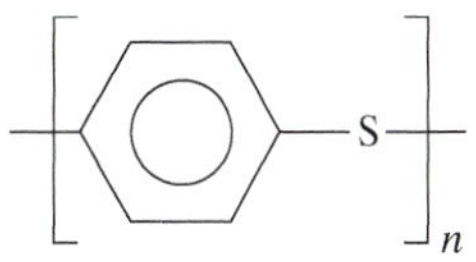

Polyphenylene sulfide (PPS)

Polyphenylene sulfide is made by reacting *p*-dichlorobenzene with sodium sulfide in the presence of a polar solvent. The actual polymerization process is proprietary. The degree of crystallinity depends on molding conditions, but most manufacturers use processing conditions that produce a primarily crystalline polymer. Basically, four different versions of PPS are commercially available: glass filled, mineral filled, PTFE filled, and carbon fiber filled. PPS used to be available without fillers, but the product has been withdrawn and now the most commonly used grade contains 40% glass. The PTFE version is used for tribological applications; the mineral-filled grades are used for electrical applications (they are better than the glass-filled grades), and the carbon fiber grades are intended for applications that require very high strength and stiffness. The normal glass-filled grades cost about $5 to $6/lb ($12 to $14/kg), but the carbon fiber reinforced grades cost more.

PPS competes with PPO, PC, and other engineering plastics when elevated temperature service is anticipated. Continuous use temperatures can be as high as 500°F (260°C). The room temperature mechanical properties are similar to those of nylons, but at elevated temperatures PPS has much better strength retention and creep resistance than many of the competitive thermoplastics. PPS has good chemical resistance. It resists degradation by fuels and lubricants but is not resistant to sunlight (UV). There are no known solvents for this material below about 400°F (204°C). PPS is resistant to attack in water applications, and its resistance to moisture absorption is exceptional (<0.05% at 20°C in 24 h).

These plastics are opaque, and they have low flammability. The low flammability and high-temperature resistance make PPS suitable for automotive applications such as under-the-hood mechanical devices. The 40% glass-filled grades are easily injection molded, and they have strengths and stiffness greater than most other engineering plastics. It is not exactly fair to compare a glass-filled plastic with unfilled homopolymers, but this material is not available (under normal conditions) without a filler. These plastics have great compatibility with fillers, but they have poor compatibility with most other plastics. This latter fact explains why blends and alloys are not normally available.

As a family of material, the polyphenylene sulfides are crystalline thermoplastics that can be used for many applications, but the most suitable applications are those that take advantage of their high strength and elevated temperature capabilities. With glass or carbon fiber reinforcement, they are among the highest-strength thermoplastics. They are often used to replace metals.

Polyetherether Ketone

In the late 1980s, a new family of polymers was developed with the facetious sounding acronym *PEEK*. The acronym stands for polyetherether ketone. This family of high-temperature plastics has a mer structure similar to some of the other high-temperature plastics:

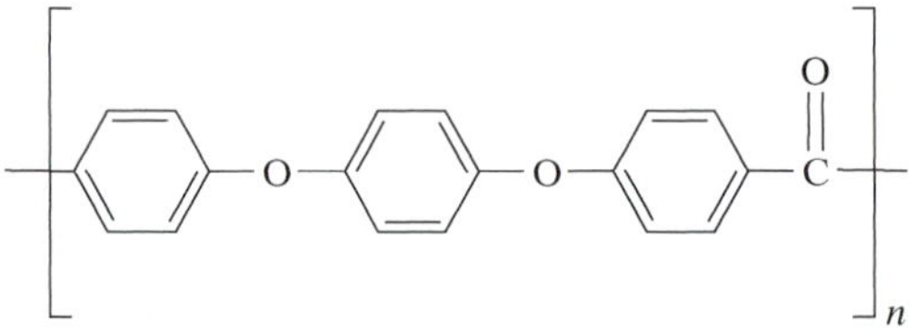

Polyetherether ketone (PEEK) [Victrex®]

Benzene rings are linked with an oxygen bond, forming a typical R—O—R type of ether structure. These plastics are partially crystalline; they are thermoplastic and can be molded in conventional equipment. Molding temperature can be as high as 750°F (400°C). PEEKs have low flammability, and they have one of the lowest smoke emission ratings of any plastic. The mechanical properties are as high as or higher than those of nylons, and they have good dimensional stability at elevated temperature.

The usual types of resins are available: unfilled, glass filled, lubricated with PTFE, and carbon fiber filled. These plastics are fairly new. In 2003 they were still carving out markets, but the applications to date are for things such as mechanical and electrical components that require resistance to elevated temperatures. Continuous use temperatures can be as high as 600°F (315°C). PEEK has excellent resistance to attack by hot water and low-pressure steam. The various grades of PEEK are very resistant to many other chemicals, including many organic solvents. When PEEKs are used as the matrix resin for carbon fiber composites, the resulting composites can have a tensile modulus of elasticity of about 18 million psi (125 GPa), which makes this composite much stiffer than aluminum and comparable in stiffness to titanium and copper alloys. The specific stiffness (modulus/density) is almost three times that of aluminum. These types of mechanical properties make this family of plastics a candidate for aerospace and aircraft types of applications for which weight reduction is very important.

PEEK is not a cheap plastic; in 2003 the high-performance grades could cost as much as $30/lb ($66/kg). As is the case with all of the high-performance engineering plastics, these materials are to be used where the properties can justify the cost. Essentially, the area of application is mostly high-strength mechanical components that require additional features, such as elevated temperature resistance, chemical resistance, or flammability, or all of these.

Liquid Crystal Polymers

Chemically, liquid crystal polymers can have completely different mers; the thread of commonality is that they are crystalline in the liquid or melt state as well as in the solid state, thus the term *liquid crystal*. The high-strength aramid fiber that was discussed in the previous section on polyamides, Kevlar, is a liquid crystal. Chemically, liquid crystals have an intermediate in their structure, hydroxybenzoic acid (HBA), that promotes the liquid crystal state. The following is the mer structure of one type of liquid polymer:

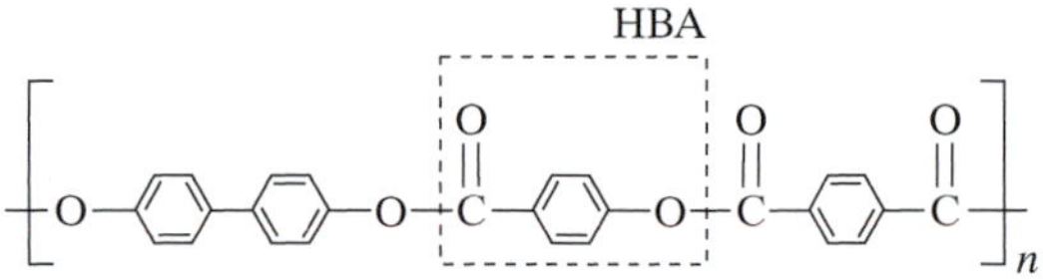

HBA

Liquid crystal polymer (LCP)

There are other possible mer structures. It is the morphology of the polymer that makes it a liquid crystal. Aligned polymer chains form in the melt, and these aligned chains are maintained into the solid condition. One manufacturer of these resins uses the term *self-reinforced plastics* to describe this family of polymers, thus suggesting that the polymer chains that form in the melt are analogous to the fiber reinforcement used in polymer composites.

The properties of liquid crystal polymers (*LCPs*) depend on the specific mer formed by the reacting species, but as a family of materials the characteristics that set them apart from other polymer systems are high strength and high-temperature capability without the molding and processing problems that are characteristic of many other high-temperature thermoplastics. They have good injection molding characteristics; they do not require long curing times in the mold before ejection and have good dimensional stability; the mechanical properties can be higher than those for nylon, and glass reinforcement gives them tensile strengths in excess of 20 ksi (140 MPa). The tensile modulus can be in excess of 20 million psi (138 GPa) with good fracture toughness. The competitive materials are engineering plastics such as PEEK, PPO, and some of the more advanced types of commodity plastics, such as *polyesters*.

The cost of LCPs in the early 2000s was in the range of \$4 to \$7/lb (\$9 to \$15/kg). This puts this material in a favorable position compared to high-performance engineering plastics such as the polyimides, polyphenylene sulfides, and the like. The LCPs are relatively low in cost with premium properties.

Applications of these materials are common in the electronic and electrical area. They have good stability, are easily molded, have low flammability, have good electrical properties at elevated temperature, and have very low coefficients of thermal expansion, which allows for stability in service. Obviously, this mix of properties coupled with excellent strength makes LCPs candidates for many mechanical parts that require something more than is offered by a commodity plastic. New applications are proliferating, and new LCP types are being developed to further reduce costs. Eventually, these plastics may replace many of the glass-filled plastics that are used when high strengths are required. This is a desirable goal, because glass reinforcement produces tool wear in all phases of processing.

Polyphenylene Oxide

The acronym PPO (*polyphenylene oxide*) has been registered as a trademark for General Electric company's Noryl® resins. There are about 30 grades of these. Other manufacturers of these resins make additional grades and blends, including copolymers. The generic name for this family of resins could be polyphenylene ether (PPE), since PPO and at least one of the copolymers have an R—O—R type of structure, which makes them an ether. In the early 1990s, the PPO acronym was more commonly used to denote this family of plastics, and the modified PPO grades actually refer to blends of the above mer with polystyrene.

All the commercially popular grades are thermoplastic materials. There are other blends in addition to PPO/PS; PPO/nylon is one blend that is available for chemical resistance that is an improvement over the other blends. The ratio of polyphenylene ethers to polystyrene in one of the popular blends is 4 to 1 (20% PS). Similar fractions are used for a popular nylon blend. Up to 20% glass can be added for improved mechanical properties. The basic mer is

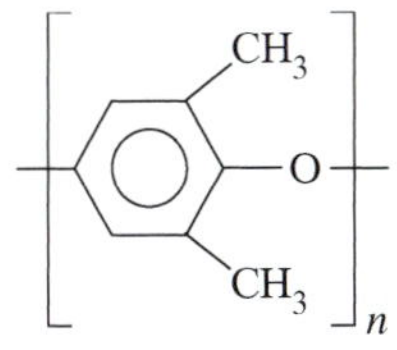

Polyphenylene oxide (PPO)

PPE homopolymer has poor processing characteristics, so it is really never used. The materials from this family of polymers with commercial importance are the blends and copolymers that have been formulated for improved processing characteristics. The homopolymer is thermoplastic, but it has poor melt flow characteristics. The PPO blends and similar blends have strength characteristics similar to those of ABS or some grades of polycarbonate. They are not transparent, so they do not compare with the acrylics and other materials that have this property. These resins are competitive in cost with ABS, polycarbonate, and similar types of thermoplastics. The heat distortion temperature at 264 psi (1.8 MPa) is about 200°F (100°C) for the unfilled grades. They have good electrical properties, but the properties that make this family of plastics stand out from some of the other families are the high impact strength of about 4 ft-lb/in. or 9.6 J/m at room temperature and the low water absorption rate. PPOs have one of the lowest water absorption rates ($<1\%$ at equilibrium). This latter property produces molded parts with very good dimensional stability.

As a family of plastics, the PPO/PPEs are widely used for things such as appliance and computer housings. They are stable, tough, and fairly low in cost. In addition, there are grades that are platable and grades that can be foamed. Foaming these materials further reduces costs (more volume per unit mass). Thus these plastics are aimed at production markets mostly in durable goods. They have better properties than the commodity plastics but cost less than the engineering plastics. As blends and copolymers they have good processibility.

Polyetherimide

Developed by General Electric in 1982, *polyetherimide* is a high-strength thermoplastic polymer composed of imide groups linked by bisphenol A and ether bonds (—O— groups):

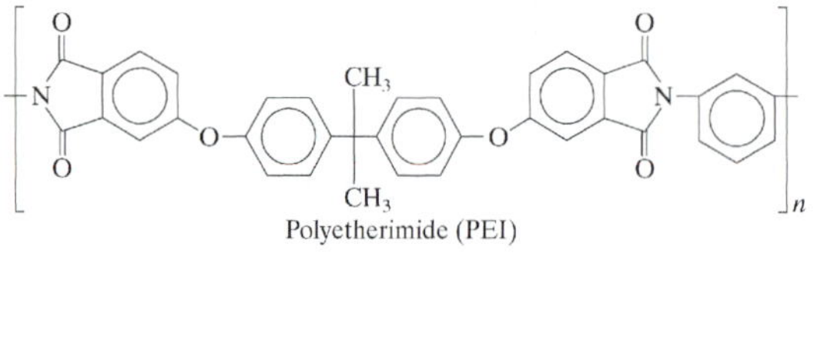

Polyetherimide (PEI)

The ether linkages in the backbone of the polymer allow PEI to be readily melt processed while the imide groups provide stiffness, strength, and high-temperature resistance. Although the structure of PEI is quite regular, the polymer is amorphous. Features of PEI include high strength and modulus at elevated temperatures, inherent flame resistance, low smoke evolution, high T_g (215°C), good hydrolytic stability, excellent melt processibility, and excellent dimensional stability. Natural, unfilled PEI is orange-colored and translucent.

PEI is used for many electrical applications because of its low flammability and excellent electrical characteristics (excellent insulator, high volume resistivity, high arc resistance and dielectric constant). Applications include underhood automotive applications, printed circuit boards, microwave equipment, consumer and industrial electronic components, and medical applications.

In both the glass-reinforced and unreinforced conditions, PEI is one of the strongest thermoplastics available. Even at higher temperatures (190°C) PEI retains much of its room-temperature strength. The modulus of PEI is also quite high relative to other engineering thermoplastics, but it is quite sensitive to notches and stress concentrations. Moisture in molding pellets can lower the toughness of PEI, hence this resin must be thoroughly dried before molding.

Polyphthalamide

Polyphthalamide (PPA), a semiaromatic polyamide, is essentially a nylon polymer with ring structures in the main chain. The aromatic structure of PPA arises from the polymerization of primarily terephthalic acid (with small amounts of isophthalic acid) and an amine such as hexamethylene diamine (along with adipic acid):

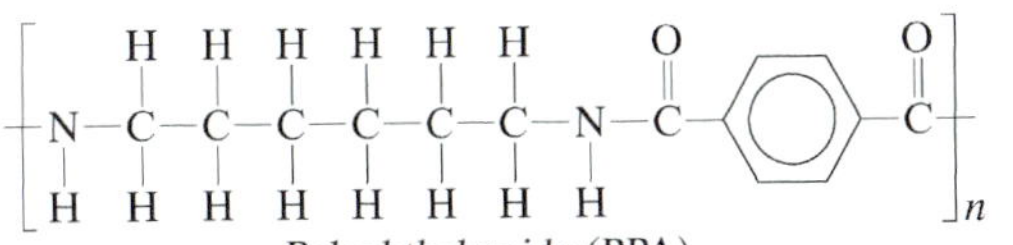

Polyphthalamide (PPA)

Although most injection molding grades of PPA are semicrystalline, amorphous versions do exist for film and coating applications. PPA is somewhat sensitive to moisture during molding and requires dry pellets (<0.15% moisture).

Although PPA has better thermal properties, higher strength, and lower moisture absorption than aliphatic polyamides such as nylon 6/6, it has lower impact strength (although toughened grades exist). The aromatic or ring structures afford PPA improved mechanical properties over engineering plastics such as polycarbonate, nylon, polyester, and acetal while providing a cost advantage over specialty polymers such as liquid crystal polymers, polyphenylene sulfide, and polyetherimide.

PPA is used regularly for underhood automotive application, plumbing fittings, gears, aerospace components, electrical connectors, housings, components in electric motors, and small motors. In electrical applications, PPA can withstand infrared and vapor-phase soldering temperatures. Wear grades of PPA, containing solid lubricants such as PTFE, are also available for bearings and other tribological applications. Mechanical applications use PPA for its high strength, stiffness, and toughness. Glass reinforced resins of PPA can be stronger than similarly reinforced PEI. PPA has a T_g of 127°C. Essentially, PPA is a high-strength, higher-temperature, and more dimensionally stable nylon. Some PPA resins can cost as little as $4–6/lb ($9–13/kg).

Aliphatic Polyketone

A terpolymer of polyethylene, carbon monoxide, and a small amount of polypropylene, aliphatic polyketone is a semicrystalline thermoplastic with the following structure.

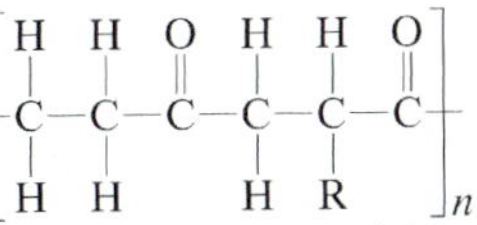

Aliphatic polyketone (PK)

Shell Chemical Company sells aliphatic polyketone under the trade name Carilon. With high chemical and permeation resistances, excellent wear resistance, high creep rupture strength, and excellent toughness, PK is a high-performance material that competes with other ethenic polymers such as polypropylene. Injection molding shrinkage is slightly less than polyacetal. Aliphatic polyketone has a T_m of 220°C and a T_g of 5–20°C. Aliphatic polyketones compete with nylon and polyacetal in tribological applications.

Summary: Thermoplastic Engineering Plastics

In the preceding discussion on families of thermoplastic engineering plastics we tried to show how various engineering plastics are chemically different, and we tried to present enough property information to distinguish one plastic family from the other. Each family of plastics has properties that set it apart from other plastics. These should be kept in mind when selecting materials for design applications. The fluorocarbons, in general, have chemical resistance that is superior to that of many other plastics, the polycarbonates have excellent toughness, the polysulfones are good in hot water, and so on. All of the plastics that we have discussed are considered to be thermoplastics, and this categorization is extremely important. It means that these plastics (with a few exceptions) can be injection molded; this is very important because injection molding is usually the lowest-cost method for converting a polymer into a usable shape. It also means that scrap and trim from these plastics can be remelted for reuse. These materials can be recycled if there is a mechanism to prevent mixing with other plastics.

A number of thermoplastic polymer families were not included in our discussion. There are a number of reasons for omission: the plastics may be proprietary, and there is no information on the chemical makeup; a plastic may be too new to the industry to have carved out a distinguishable area of application; finally, some plastics were left out because the authors have not had personal experience with them. The plastic families that we have listed should take care of most design situations. Whatever the plastic, users should begin evaluation of a plastic by understanding the chemistry of the material, the fillers that are present, the available blends and alloys, and the morphology. The next step is to review the properties to determine if they meet the expected service conditions of a potential application. We shall discuss the selection process further in subsequent chapters. The point we are trying to make here is that users should try to understand the chemistry of the polymers that are under consideration for an application.

5.5 Thermosetting Polymers

About 15% by volume of the plastics manufactured in the United States are thermosetting materials. As was the case with the thermoplastic materials, the bulk of the production is made up of commodity plastics, and the high-tech or high-performance plastics usually make up less than 10% of the market. This being the case, we will concentrate our discussion on the thermosets that make up the bulk of the usage, and we will only briefly describe the lesser-used materials. Figure 5–1 illustrated the relative market shares of the various thermosetting plastics. About two-thirds of the production comes from four families of thermosets: urethanes, *phenolics*, unsaturated polyesters, and ureas.

We will discuss the chemistry and properties of these, but to start our discussion of thermosetting polymers we reemphasize that the difference between these families of plastics and the thermoplastic families is ease of molding. Thermosets cannot be injection molded with the standard equipment that is used for injection molding of thermoplastics. Once they are molded or machined into shape, thermosetting parts cannot be remelted. When they are overheated, they char rather than melt and flow. These materials are used less than the thermoplastic materials mostly because they cannot be injection molded and thermoset fabrication processes are more expensive. From the property standpoint, thermosetting plastics are usually more brittle than thermoplastic plastics; they are generally rigid. Most thermosetting plastics are harder than thermoplastic materials, and they are usually not used without some type of reinforcement or filler.

Some of the rubbers and elastomers that are commercially important can be considered to be thermosetting polymers. We will discuss these in the following section on elastomers. Also, some polymers that we have just discussed as thermoplastic materials are also available in formulations that make them thermosetting. These are special materials, and we will not discuss them in this section. The point to be remembered is that in plastics, as well as in many other materials, there are exceptions to just about every rule. When we say that all phenolics are thermosetting, some company will point out a grade that is thermoplastic. Thus what we say about families applies to most specific grades in a family, but there will always be some variants that violate general statements.

Polyurethanes

Polyurethanes are condensation polymerization reaction products that come from reacting an isocyanate (—NCO) and an alcohol (ROH) with a reactive hydroxyl functional group (OH).

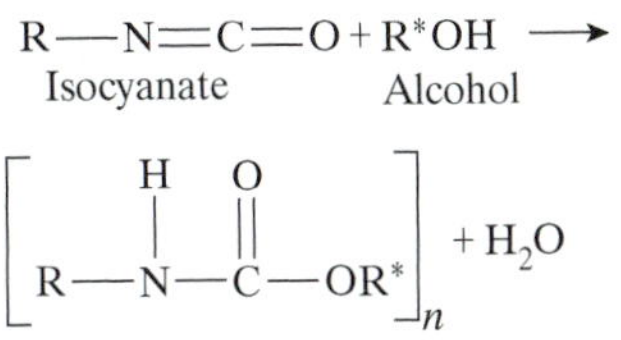

$$\underset{\text{Isocyanate}}{R{-}N{=}C{=}O} + \underset{\text{Alcohol}}{R^{*}OH} \longrightarrow \left[R{-}\overset{\displaystyle H}{\overset{|}{N}}{-}\overset{\displaystyle O}{\overset{\|}{C}}{-}OR^{*} \right]_n + H_2O$$

Polyurethane (PUR)

The alcohols are called "polyols" and they can be a polyester or a polyether polyol. Most polyurethanes are made from polyethers. Different properties are obtained depending on the reacting species and the reaction technique. Polyurethanes can have properties ranging from rigid thermoset, to rubber, to *thermoplastic elastomer*. A serendipitous effect of the condensation reaction is that some formulations produce carbon dioxide as a condensation product. This allows some polyurethanes to "foam themselves." If the foaming reactants are trapped in a closed mold they can form a plastic with a dense skin and foam core—a great weight saver. In fact, foams are the largest use for polyurethanes. Elastomeric foams are used for furniture cushions, bedding, and auto padding and rigid foams are used for thermal insulation. Reaction molded foams are widely used for trucking and automotive application—for fascia, dash pads, even bumpers. Polyurethanes have been used for many years in making furniture that looks and feels like wood, but is not. The building industry in the United States is running out of select lumber to use for moldings and decorative house parts. Molded polyurethanes are filling the bill.

The elastomeric polyurethane foams used for bedding and cushions are made to many different densities, depending on the application. The lower density foams are often made by forming giant foam "buns" by uncontained reaction of the isocyanate and polyol. These large buns are then skived with razor-type knives to make sheets of different thicknesses. Reaction-injection molded (Rimmed) parts are made by pumping the reactants into a closed mold, and the density is a function of the quantity of reactants. As a polymer family, polyurethanes are entrenched in so many applications that it is hard to envision modern life without them. Most of us sleep on polyurethane foam and walk on carpeting supported by polyurethane padding. Our furniture and wood floors are finished with polyurethene varnishes and many parts of our automobiles are held together with polyurethane adhesives. Polyurethanes, in general, have toughness as a noteworthy property along with stiffness, density, and strength that can be engineered to suit any application.

Phenolics

Phenolics are the oldest family of thermosetting plastics, dating back to the 1870s. This family of polymers is based on the presence of a ring structure alcohol, phenol, with the following formula:

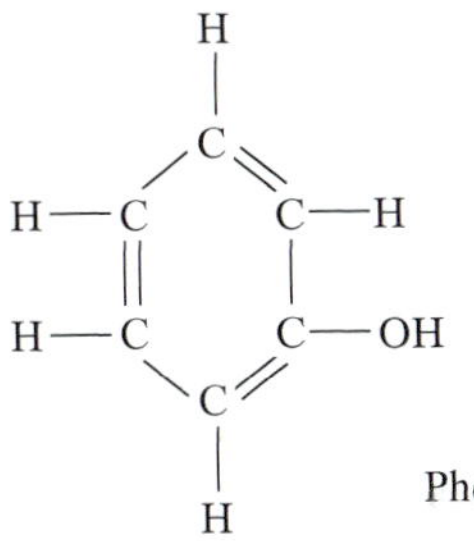

H, H—C, C—H, H—C, C—OH, H (ring of six C atoms)

Phenol

OH (ring symbol)

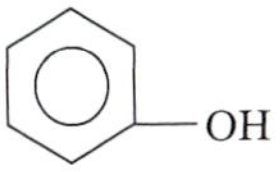

Chemical symbol

This starting material for phenolic-based polymers is obtained by various processes, but mostly it comes from petroleum distillates (benzene and propylene). Phenolic molding and laminating resins formed by the reaction of phenol with formaldehyde (CH_2O) contain the following monomers:

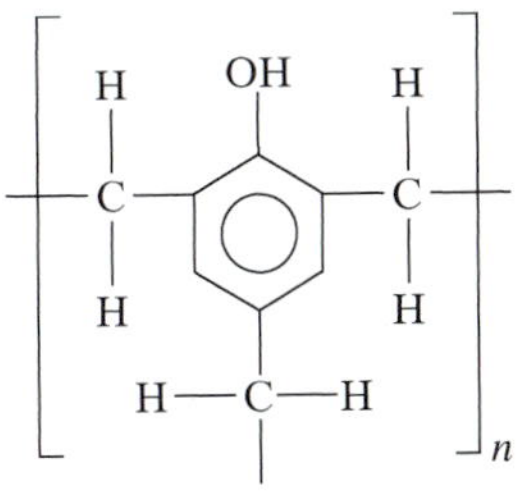

[H, OH, H; —C— (ring) —C—; H, H; H—C—H]$_n$

Phenol-formaldehyde (PF)

These monomers form a rigid network structure, which in turn forms a hard, rigid plastic. Polymerization is accomplished by cross-linking of these mers into a three-dimensional network. The cross-linking reaction requires heat or some other form of energy, but many thermosetting resins can exist in various stages. In the A stage, cross-linking has not started and the resins' components can sit around for some time before any significant cross-linking occurs. This time interval is referred to as *potlife*. As time progresses, cross-linking will gradually occur, and this transition time period is called the B stage. Some thermosets can exist in this form for as long as 24 months at room temperature and possibly longer with refrigeration.

Most thermosetting resins are rubbery and tacky in the B stage, and they are often compounded with reinforcements and fillers in this stage. They are then fully cross-linked to the C stage by heating. Phenolic resins are somewhat different in that they form a brittle solid in the B stage. This material can be ground into powder and used as molding resin, which fully cross-links in the heated molding operation.

The first commercial phenol–formaldehyde polymers were produced in the early part of this century under the trade name Bakelite. This material was filled with woodflour or mineral fillers to obtain desired thermal and mechanical properties, and it was widely used for compression molded electrical parts such as switches, distributor caps, and the like. This family of thermosets continues to be used for these types of applications. Molded phenolics are characterized by good dielectric properties, low moisture absorption, and relatively high use temperatures [about 400°F (204°C)]. The major fault of this family of thermosetting resins is that they are relatively brittle.

Phenolics represent the largest tonnage of thermosetting resins. The high usage comes from applications that are not very visible to the casual observer. About one-third of the usage is for the adhesive in plywood. Other significant applications include adhesives for fiberglass insulation, binders for brake and clutch friction materials, and binders in other technical applications such as foundry cores. Another significant application is for electrical circuit boards. In this application, phenolics are usually reinforced by cotton or glass fabrics. They are compression molded into flat sheets. Another very prominent electrical application is for electrical outlets, switches, boxes, and related household wiring supplies. The properties that make phenolics suitable for these types of applications are that they are good electrical insulators and dielectrics, and, probably more important, when there is a short or some similar incident of high-temperature exposure, they char rather than melt and catch fire; they can withstand significant damage from an arc and still provide decent electrical insulation. This is one property that sets this family of resins apart from competing thermoplastics materials.

In addition to electrical equipment parts, phenolics are frequently compression molded into standard shapes that are subsequently machined into a wide variety of machine components: rolls, gears, cams, platens, and many mechanical devices. Phenolics have excellent compressive strength. Flatwise, they can have strengths as high as 40 ksi (275 MPa). With woven reinforcement, these materials have excellent toughness and can take considerable abuse. From the cost standpoint, they are one of the lowest-cost plastics. In 2003 the cost per pound was in the range of $0.50 to $1. In summary, phenolics are important plastics. They are among the oldest plastics, but they will continue to be used in large quantity for some time because of their favorable electrical and mechanical properties and low cost.

Alkyds

Alkyd is a term derived from al/cohol (al) and a/cid (kyd), and it is used to denote a family of thermosetting polymers that are reaction products of alcohols and acids. They are really polyesters, but it is accepted practice to keep them separate from thermoplastic polyesters such as polyethylene terephthalate (PET) and from thermosetting polyester molding resins (unsaturated) of the type used in fiber-reinforced plastics. A wide variety of acids and alcohols can be used, but phthalic and isophthalic are the more common acids, and glycerine and ethylene glycol are the more common alcohols. Alkyd resins can be used for compression molding, but most alkyd resins end up as paints. The alkyd resins are mixed with oils (such as linseed) and with pigments (such as titanium dioxide). The resin is usually polymerized in a kettle with the oil and is thinned and pigmented to make a paint. Drying of the paint causes oxygen linkages between the alkyd–oil polymer chain to form a durable film. Alkyds are not important engineering plastics, but they are important paint bases.

Melamine Formaldehyde

Melamine formaldehyde is a rigid thermosetting polymer with a network type of structure:

Melamine formaldehyde (MF)

Units similar to the preceding join at the stars and an interpenetrating network is formed. The main importance of melamine resins is in food handling. Many plastic dishes are made from this material. The polymer is easily molded, low in cost, rigid, nontoxic, and easily colored, and has good toughness, abrasion resistance, and temperature resistance as high as 300°F (148°C).

Engineering applications of melamine plastics are largely for electrical devices and for laminates for surfaces.

Urea Formaldehyde

This polymer is similar in use characteristics to melamine formaldehyde. It has a different repeating structure, but it is also a thermosetting network polymer:

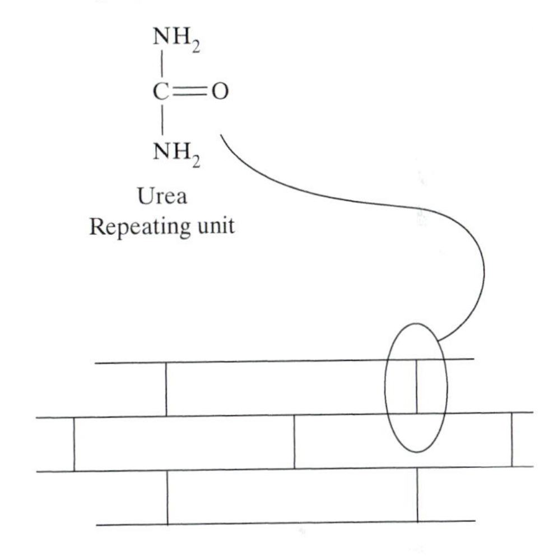

Network structure of urea formaldehyde (UF)

The urea formaldehydes are used for many of the same types of applications as the melamines and phenolics. They are used for electrical devices, circuit breakers, switches, and the like. Technical applications include use as additives to papers (for wet strength of toweling) and for construction adhesives. At the peak of the building surge in the early 1980s, as much as 60% of the usage of urea resins was for the adhesive in particle board. They are often used as the adhesives for large molded building trusses and beams. Urea resin adhesives can be catalyzed so that they will cure at room temperature. This allows structures to be made that are too large to fit in curing ovens. Its use as an engineering plastic is only about 10%.

The ureas have mechanical properties that are similar to those of the melamines, but they do not have the UV resistance of the melamines: They cannot be used outdoors. They are lower in cost than the melamines. This family of plastics is not likely to be used by the average machine designer for miscellaneous machine components. If it is used at all, it may be invisible. It may be an adhesive in some construction form or composite. Ureas are not commonly available from plastic suppliers in standard shapes. In some form applications they offer a cost or property advantage over competing thermoset families (phenolics, melamines, and others).

Unsaturated Polyesters

The term *polyester* is analogous to the term *steel* in metals. Just as there are many types of steels with widely varying properties, so too there are a multitude of polyesters with a significant range of properties. Some of the cellulosics are polyesters, as are a number of other plastics previously discussed. The polyesters referred to in this discussion are those that are by industry convention called polyesters. There are two groups in this category: (1) the thermosetting resins, which are usually typified by a cross-linked structure, and (2) the thermoplastic polyesters, which are highly crystalline in structure with comparatively high melting points.

Esters are the reaction product of the combination of an organic acid and an alcohol:

$$\underset{\text{Acid}}{R-\overset{\overset{\displaystyle O}{\|}}{C}-OH} + \underset{\text{Alcohol}}{R'-\underset{\underset{\displaystyle H}{|}}{\overset{\overset{\displaystyle H}{|}}{C}}-OH} \rightleftharpoons$$

$$\underset{\text{Ester}}{R-\overset{\overset{\displaystyle O}{\|}}{C}-O-\underset{\underset{\displaystyle H}{|}}{\overset{\overset{\displaystyle H}{|}}{C}}-R'} + \underset{\text{Water}}{HOH}$$

In making polyester resins, the water from the esterification is removed to prevent reversal of the reaction. The most important polyesters from the standpoint of engineering materials are the polyester resins used as the resin matrix for reinforced composites.

Polyesters are used worldwide as the matrix material for reinforced composites, or RTP (reinforced thermoset plastics). Everyone is familiar with the use of this material for boats, recreational vehicles, and a wide variety of other structural applications, such as tanks and portable shelters. About six different polyester resins are used for composite applications. Composites can be reinforced with a variety of materials, but the use of glass fibers is the most common. The properties of a composite made with an *unsaturated polyester* (UP) depend on the nature of the ester. When different acids and alcohols are used, they form different mers. Orthophthalic polyester is formed from the reaction of orthophthalic acid with an alcohol. Other polyesters are formed from other reactants.

Unsaturated polyester resins are sold in a liquid form that requires catalyzation to cure. The liquid resin is really a solvent containing relatively low-molecular-weight ($n = 200$ to 20,000) polymers. A common solvent is styrene monomer. Composites are made by catalyzing the resin mixture with a material such as MEK (methyl ethyl ketone) peroxide, saturating woven reinforcement with the resin, and allowing the saturated material to cure. The curing cycle is really the cross-linking reaction, and the low-molecular-weight molecules become part of a giant macromolecule by cross-linking. Usually only a very small amount of catalyst is needed to initiate the curing reaction (a few milliliters per liter). Reaction time depends on temperature and the amount of catalyst, but curing times can be as short as minutes to overnight. If a user wishes to fine-tune an application involving unsaturated polyester resins, the properties of the specific resins should be investigated along with the effect of various amounts of glass reinforcement on these resins.

The general-purpose polyester resin used for boats, fishing poles, autos, and the like usually contains an ester produced by the reaction of ethylene glycol (alcohol) and maleic acid. This ester is *unsaturated;* that is, it is capable of bonding to other polymers because of available bonding sites (at carbon double bonds). The ethylene glycol maleate polyester is reacted with styrene monomer, and the result is essentially a copolymer:

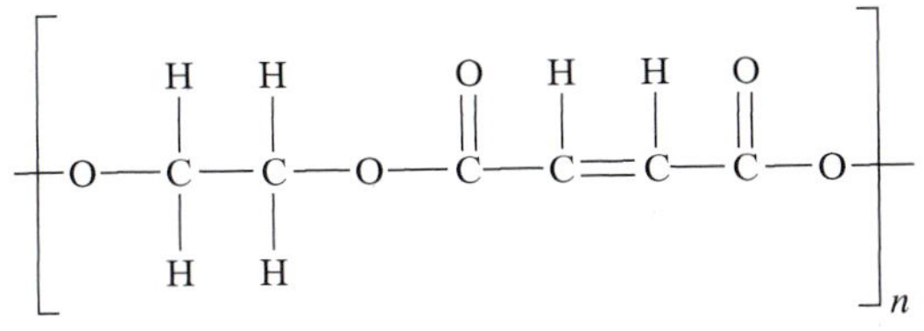

```
[      H   H        O   H   H   O    ]
       |   |        ||  |   |   ||
 --O---C---C---O----C---C===C---C---O--
       |   |
       H   H                         ]n
```

Ethylene glycol maleate polyester

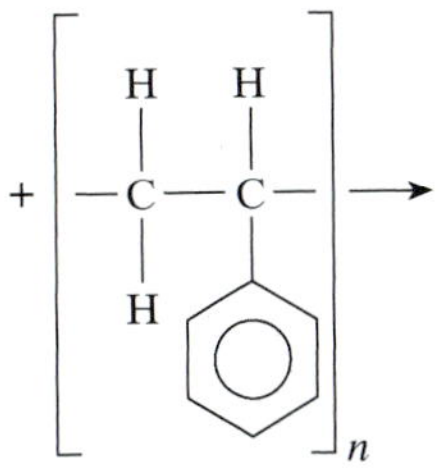

```
    [  H   H  ]
       |   |
 +  [ -C---C- ]  -->
       |   |
       H   C6H5 ]n
```

Styrene

```
                            *
[      H   H        O   H   ~   O    ]
       |   |        ||  |   ~   ||
 --O---C---C---O----C---C---C---C---O--
       |   |            |   |
       H   H            |   H        ]n
                        |
                    H---C---H
                        |
                    H---C---C6H5
                        |
[      H   H        O   |   H        ]
       |   |        ||  |   |
 --O---C---C---O----C---C---C---C---O--
       |   |            |   ~   ||
       H   H            H   ~   O    ]n
                            *
```

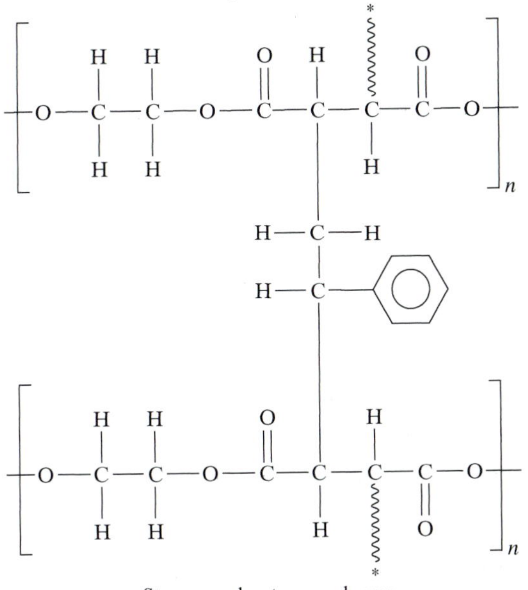

Styrene polyester copolymer
(*sites for additional cross-linking)

There is a bright future for this family of thermosetting plastics. Its use in boats, recreation, and construction will increase, as will its use in transportation. There still is only one production automobile (the Corvette) made in the United States from reinforced polyester, but as fabrication processes become more cost effective, there will probably be increased use in vehicles. The cost of this material varies monthly with market forces, but in 2003 it was priced in the range of \$0.50 to \$2.00/lb in the United States.

Summary: Thermosetting Plastics

The workhorse thermosets are phenolics, ureas, and unsaturated polyesters. They are low in cost like commodity plastics, but they can have properties that match or surpass many of the engineering plastics. Essentially, these plastics are used where thermoplastics simply will not do the job. There are sophisticated plastics like polybenzimidazole (PBI) that have high use temperature [heat distortion temperature = 815°F (435°C)], but they can cost \$300 per pound. Phenolics resist heat better than most thermoplastics, and they may cost less than \$1 per pound. Similarly, unsaturated polyesters (UPs) are normally the resins of choice for "fiberglass" structures because they are low in cost (about \$1/pound in 2003) and easy to use. *Ureas* do the "dirty jobs"—such as making plywood. We will discuss epoxies and some of the other thermosets used in making composites in Chapter 6. We conclude this discussion of thermosets with this advice: Save them for the tough jobs that are troublesome for thermosplastics.

5.6 Elastomers

The term *elastomer* is used to describe a polymer that has rubberlike properties. The derivation of the term comes from *elastic,* the ability of a material to return to its original dimensions when unloaded, and *mer,* the term used in polymer chemistry to refer to a molecule. By definition, a *rubber* is a substance that has at least

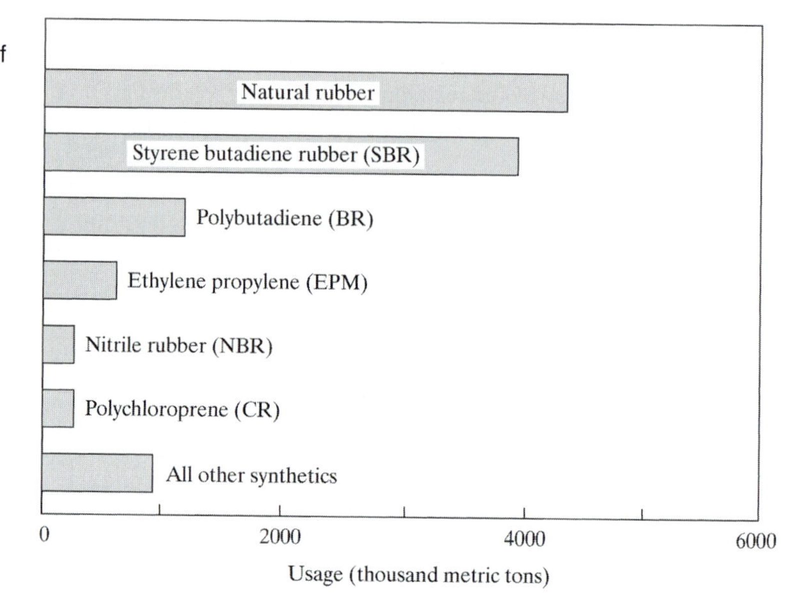

Figure 5–5
Estimated world consumption of rubber (new) in the 1990s. These usage ratios were still typical in 2003.
Source: International Institute of Synthetic Rubber Producers.

200% elongation in a tensile test and is capable of returning rapidly and forcibly to its original dimensions when load is removed. Not all elastomers meet the definition of a rubber, because recovery in some polymers is not rapid. A more usable definition of an elastomer is a polymeric material that has elongation rates greater than 100% and a significant amount of resilience. The term *resilience* refers to a material's ability to recover from elastic deflections. It is measured by the amount of energy lost in deflection. A material with good resilience when dropped from a height onto a rigid surface will rebound to almost its original height.

The relative usage of important rubbers is shown in Figure 5–5. *Silicones* and *polyurethanes* are included in the "all other" category. In the remainder of this section, we shall present a brief description of the other important synthetic rubbers or elastomers.

Natural Rubber

In its crude form, natural rubber is simply the sap from certain trees with the moisture removed by smoking. It has rather poor mechanical properties, it is tacky, and its industrial uses are limited. Natural rubber, like many of the synthetic rubbers, is converted into usable products through compounding with additives and subsequent vulcanization. The additives are often blended by repeatedly running the starting material through a set of smooth rolls or by masticating the crude in a powerful blade-type mixing machine. In this process, coloring agents, curing agents, and inert fillers may be added. Sulfur is used in most natural rubbers to enhance the final step in the manufacturing process: vulcanization. *Vulcanization* involves heating the blended starting material to about 150°C with the crude in a steel mold of the desired shape. The vulcanization process is really a cross-linking reaction that creates the polymer chain bonds that are responsible for elastomeric properties. Sulfur molecules form cross-links between C=C bonds on adjacent chains.

Natural rubbers have good electrical properties, excellent resilience, and tear resistance. They soften with exposure to sunlight and when in contact with certain organic solvents and some strong acids. The tear resistance and low hysteresis (heat buildup on flexure) make natural rubbers prime candidates for shock absorbing parts and for large vehicle tires. Additives and controlled environments are used to circumvent weathering susceptibility.

Polyisoprene

During World War II, sources of supply for natural rubber (Malaysia and elsewhere) were embroiled in the conflict. To meet the need for natural rubber, *polyisoprene* was developed. It has the same chemical structure as natural rubber minus natural impurities:

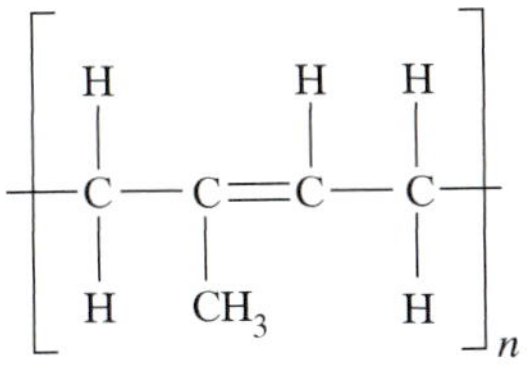

Polyisoprene (IR)

The properties are the same as natural rubber without the processing steps needed to grade natural rubber and to remove impurities. It is still more economical to use natural rubber. Thus synthetic polyisoprene is not as widely used as natural rubber.

Styrene Butadiene Rubber (SBR)

SBR is really a copolymer of polystyrene and polybutadiene:

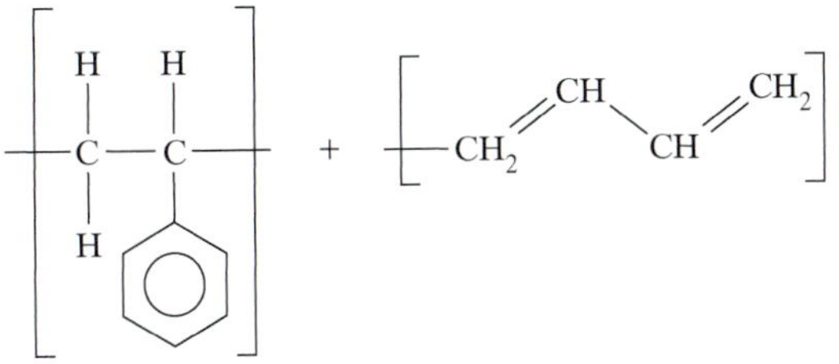

Polystyrene 1 part + Butadiene 3 parts

(SBR)

The butadiene mer can have various configurations (isomers) with the same number of carbon and hydrogen atoms. The significant use of this rubber (commonly called buna S) is largely due to its low cost. It is used for hoses, belts, gaskets, shock mounts, and most automobile tires.

Chloroprene Rubber

Chloroprene rubber is widely recognized by the trade name *neoprene*. It is a carbon chain polymer containing chlorine:

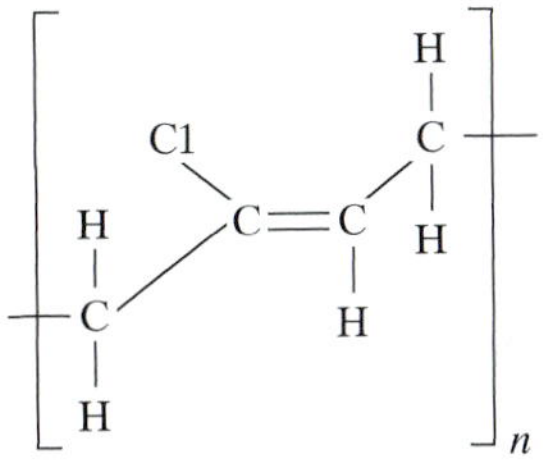

Polychloroprene (CR)

It was developed around 1930 and was the first commercial synthetic rubber. The main advantage of chloroprene over natural rubber is its improved resistance to weather, sunlight, and petroleum oils. It is widely used as a general-purpose industrial rubber for gaskets, shock mounts, seals, and conveyor belts. Chloroprene rubber is often used for automotive hoses and other under-the-hood rubber components. It has excellent flame resistance, but its electrical properties are inferior to those of natural rubber and many other rubbers.

Polybutadiene

This rubber is similar to natural rubber in its properties. It is a straight carbon chain elastomer:

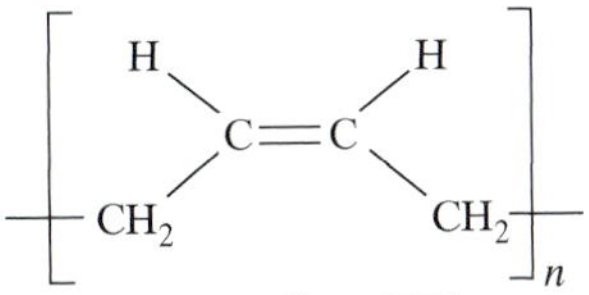

Polybutadiene (BR)

Polybutadiene is more costly to process into shapes than rubbers such as SBR. For this reason, it is most often used as an additive (up to 75%) to improve the hysteresis and tear resistance of other rubbers.

Nitrile Rubber

This rubber is really a copolymer of butadiene and acrylonitrile:

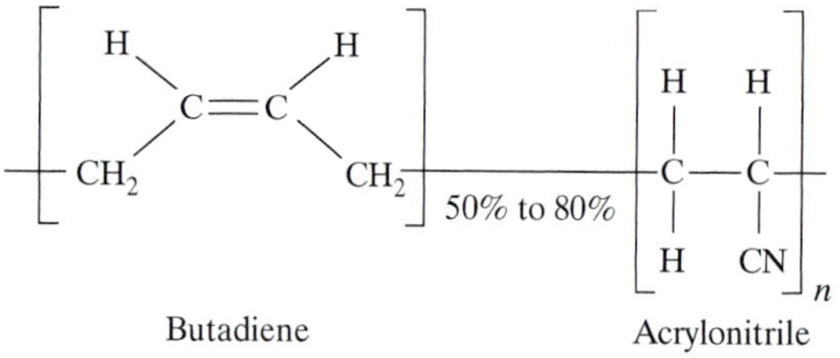

50% to 80%

Butadiene Acrylonitrile

Nitrile rubber (NBR)

Nitrile rubbers are often called buna N rubbers. They are specialty rubbers noted for their ability to resist swelling when immersed in petroleum oils and fuels. They are used for O rings on hydraulic systems; for gasoline hoses, fuel pump diaphragms, transmission gaskets and seals; and even for oil-resistant soles on work shoes.

Butyl Rubber

Butyl rubber is a copolymer of isobutylene and isoprene:

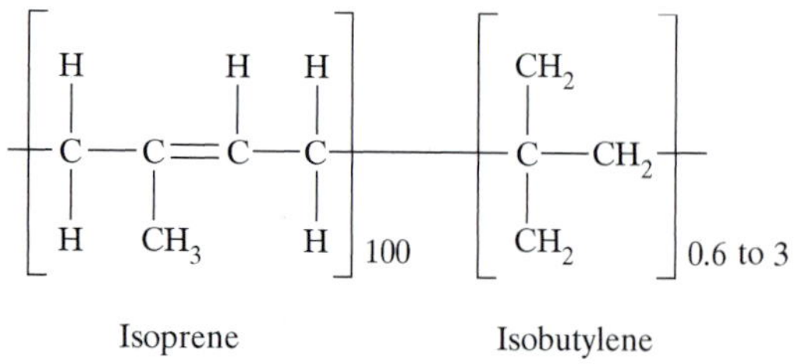

100 0.6 to 3

Isoprene Isobutylene

Butyl rubber (IIR)

Butyl rubber has very low permeability to air and excellent resistance to aging and ozone. These two properties have made it widely used for inner plies in puncture-proof tires, as well as for inner tubes. The addition of chlorine to butyl rubber aids heat resistance and processibility. Chlorinated butyl rubber is called chlorobutyl rubber. Butyl rubbers have poor oil resistance. The chlorine improves oil resistance, but they are still not used for oil immersion applications.

EPM and EPDM Rubber

Ethylene propylene rubber (EPM) is a copolymer of polyethylene and polypropylene:

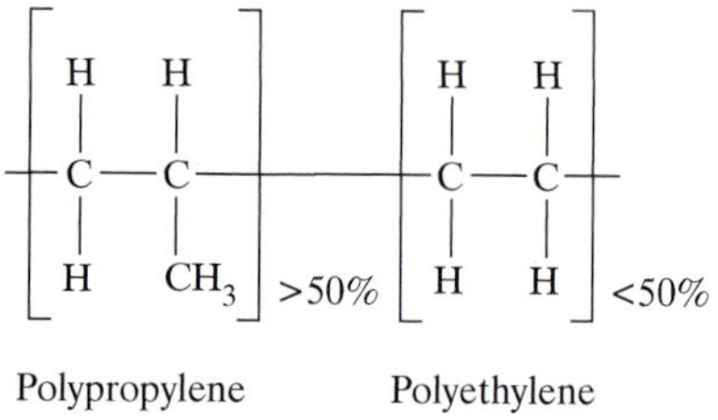

>50% <50%

Polypropylene Polyethylene

EPM rubber

The addition of a diene monomer (polybutadiene) to the copolymer creates EPDM rubber, ethylene propylene diene monomer. Both of these rubber families offer exceptional weathering and aging resistance, excellent electrical properties, and good heat resistance. Both rubbers are widely used for wire insulation, weather stripping, conveyor belts, and other outdoor applications. Like butyl rubber, they are not resistant to petroleum oils, which limits their use in automotive applications.

Chlorosulfonated Polyethylene (CSM)

A trade name for this rubber is Hypalon®. Many of the other rubbers discussed are copolymers or terpolymers. CSM is made by treating polymerized polyethylene with gaseous chlorine and sulfur dioxide to chemically add chlorine and sulfur to the mer:

$$\left[\left(-CH_2-CH_2-CH_2-CHCl-CH_2-CH_2-CH_2- \right)_x -CH(SO_2Cl)- \right]_{x'}$$

CSM rubber (*x* and *x′* are variable)

Fillers such as carbon black and sulfur are not necessary to achieve strength in vulcanization. This rubber has outstanding resistance to weathering, heat, and abrasion. It is used for tank liners and in many chemical plant applications such as pond liners. It is also used for high-voltage electrical insulation. It is considered to be a high-performance elastomer rather than a general-purpose rubber.

Polysulfide Rubber

There are a great number of polysulfide rubber formulations, but their general characteristic is a heterochain of a carbon-based group and sulfur atoms:

$$\left[R-S-S \right]_n$$

The carbon-based functional group can change, as can the number of sulfur atoms. Polysulfides are technical elastomers usually used for special applications. The noteworthy characteristics of polysulfide rubbers are their solvent and oil resistance, impermeability to gases and moisture, and weather and aging resistance. They do not have the tensile strength, resilience, and tear resistance of natural rubber or SBR. They can be made into two-component systems that cross-link after mixing and into systems that cure by oxidation in air or moisture. These latter characteristics make polysulfides suitable for caulking compounds, sealants, puttying agents, and castable shapes.

Silicones

Silicones are essentially inorganic polymers. Instead of the normal carbon atom backbone, this family of materials has a structure characterized by silicon–oxygen bonds. The repeating structure is similar to that of silicate minerals (sand is SiO_2). In fact, it is the silicon-to-oxygen bonding that gives this family of materials its superior thermal stability. The Si—O backbone tends to be linear, and varying properties are achieved by varying the nature of the organic side chains that are bonded to silicon atoms. A silicone structure is illustrated here:

$$\left[\begin{array}{c} -O-Si(R)-O-Si(R)-O-Si(R)-O- \\ | \\ O \\ | \\ -O-Si(R)-O-Si(R)-O-Si(R)-O- \end{array} \right]_n$$

Silicone mer

Silicones are commercially produced as liquids and greases (for lubrication), as elastomers, and as rigid solids. The elastomer grades are available as one-component systems that cure by hydrolysis of functional groups from moisture in the air (these are called RTV resins, for room temperature vulcanizing) or as two-component systems that cure when the components are mixed. Some grades of elastomers are heat cured as in vulcanizing. Polymer composites with silicones as the matrix material can be made from either the elastomer types or the rigid types. Elastomer composites are usually laminates of varying durometer silicone rubbers with woven reinforcements such as cotton cloth. A significant use of these laminates is for printing press blankets and for heat-resistant rubber diaphragms. Silicones are extremely important in the printing industry because in the lower durometers they have a unique ability to transfer ink to other materials; nothing sticks to them.

Polyurethane

The unique feature of polyurethanes is that they can behave as elastomers (rubbers), as rigid hard thermosets, or as injection moldable thermoplastics. A castable elastomer can be designed to suit a particular application by varying the reacting substances. Simple metal molds can be used, and polymerization is accomplished by heating to approximately 500°F (260°C). An elastomer-covered roll, for example, can be made by pouring the urethane resins into a tube around the roll core, baking and then grinding to final diameter. Sheets and blocks are made by pouring the resin into temporary low-cost metal forms. Hardnesses of from 30A (Shore) to 95A can be made. All these hardnesses have elastomeric properties, and they can be used for die springs, forming dies, wear pads, and the like. The injection moldable urethanes are used for ski boots and sports equipment. The abrasion resistance of polyurethane elastomers is a function of its durometer. Low Shore A hardness elastomers have low abrasion resistance as do the rigid polyurethanes with hardnesses on the Shore D scale. A hardness in the range of 90 to 95 Shore A usually produces the best abrasion resistance.

Special-Purpose Elastomers

In addition to the elastomers described in detail, several special-purpose rubbers have characteristics that warrant mention.

Polyacrylic rubber (ABR) can resist hot oils and solvents. It is used for spark plug insulation and transmission seals where these conditions exist.

Fluoroelastomers (FPM) such as Viton®, which is a copolymer of vinylidene fluoride and hexafluoropropylene, have excellent solvent and chemical resistance and can be used at continuous service temperatures as high as 400°F (204°C). FPM is expensive and is not as resilient as most rubbers. Thus it is used only where its temperature or chemical resistance is necessary.

Epichlorohydrin rubber (ECO) has excellent fuel and oil resistance and can be used at very low temperatures. It is suitable for use in snow handling equipment and recreational vehicles. Thermoplastic elastomers are available for injection molding for rubber components. They are predominately olefin copolymers and urethanes by composition. In general, they do not have the rubberlike characteristics of the vulcanized elastomers.

Thermoplastic Elastomers (TPEs)

Conventional rubbers such a EPDM, nitrile, and polybutadiene are cross-linked or vulcanized during molding or fabrication. Such rubbers are often fabricated into components by compression molding. Additionally, some rubbers such as silicones, polyurethanes, polysulfides, and perfluoroelastomers (Viton®) may be cast from a liquid resin and cured via cross-linking. In all instances, once cured, these rubbers cannot be remelted or reprocessed. The fabrication of elastomeric components can be slower and more complicated than other plastic fabrication processes such as injection molding.

Thermoplastic elastomers (TPEs) have both rubber and thermoplastic characteristics. They are flexible like vulcanized rubbers, but are melt processible like a thermoplastic. Many of the thermoplastic elastomers have a *block copolymer* structure. Block copolymers contain rigid thermoplastic and soft thermoset polymer segments within the polymer chain. The rigid segments (thermoplastic) provide melt processibility while the flexible rubber segments (thermoset) provide the elastomeric properties.

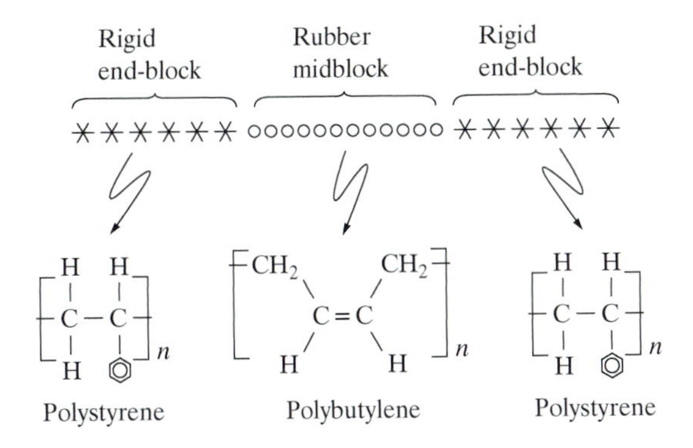

Figure 5–6
Triblock copolymer of polystyrene and polybutylene rubber

Figures 5–6 and 5–7 show the structure of a thermoplastic elastomer based on a copolymer of polystyrene and butadiene rubber. Such a TPE is known as a styrenic block copolymer and is often identified by the acronym SBS. To manufacture SBS, styrene monomer is polymerized with an alkyl–lithium initiator, producing an active polystyrene subgroup or *moiety* onto which a rubber such as butadiene is polymerized. To complete the SBS chain, additional styrene monomer is then added to provide the terminal structure or block. Alternatively, TPEs are also made by blending a rubber such as EPDM with a thermoplastic such as polyethylene or polypropylene. Although these TPEs have soft and hard segments, because they are formed by blending, these TPEs are called *alloy* TPEs.

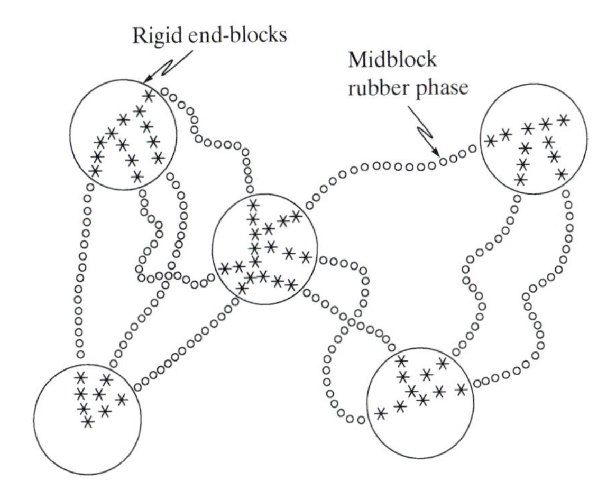

Figure 5–7
Structural morphology of a block copolymer

Most TPEs are available in pellet form in various colors and durometers. They may be processed on conventional injection molding machines. Some advantages of TPEs compared with conventional thermoset elastomers are

1. No compounding/mastication required
2. Simpler processing and faster cycle times
3. No chemical cross-linking
4. Injection moldable, blow moldable, thermoformable, extrudable
5. Secondary processes such as heat welding may also be performed
6. Parts may be designed with living hinges
7. Undercuts and complex shapes are possible with simpler tooling
8. Coextrusion and comolding compatibility
9. Scrap and waste may be recycled.

Some plastic products are molded with integral elastomeric handles and grips. Such products are manufactured by a process called comolding or insert molding. A rigid plastic part is molded and then is transferred to a different mold, where a TPE handle or grip is molded over the plastic body. Personal hygiene products such as razor handles and toothbrush handles are molded in this manner. Many TPEs bond well to the base thermoplastic resin. For example, styrenic block copolymer TPEs bond well when comolded over a polystyrene part. Comolding with a compatible TPE also allows for recyclability. One caveat of TPEs is that due to their melt processibility, they may have lower temperature resistance than most thermoset rubbers. Creep resistance and compression set may also be inferior to thermoset elastomers. More than eight different families of TPEs are commercially available, as shown in Table 5–2.

Table 5–2
Commercially available thermoplastic elastomers

	Elastomer type	Composition		Compression set	Abrasion resistance	Relative cost	Temperature	Chemical resistance	Flexibility/resilience	Strength	Applications
Alloy TPE	Compounded thermoplastic polyolefin	Mechanical blends of polypropylene and ethylene-propylene or ethylene-propylene-diene rubbers		G	F	L	G	G	F	F	Consumer and midperformance: automotive bumper fascia, air dams, valance panels, cable and wire jacketing
Alloy TPE	Melt-processible rubbers	Partially cross-linked polyethylene copolymers compounded with chlorinated polyolefin rubber		F	F	L	F	E	G	F	Midperformance: seals, tubing, hose, handles, pond liners, belting
Alloy TPE	Thermoplastic vulcanizates	Polyethylene or polyproplene mechanically compounded with EPDM rubber that is dynamically vulcanized during blending		G	F	M	G	E	G	G	Midperformance: seals, tubing, hose, handles, wire and cable jacketing
		Rigid end segment	**Flexible middle segment**								
Block Copolymer TPE	Reactor thermoplastic polyolefin	Polypropylene	EPDM rubber	F	F	L	F	G	E	F	Consumer and midperformance: automotive fascia and bumpers, cable and wire jacketing
Block Copolymer TPE	Styrenic block copolymer	Polystyrene	Polybutadiene Polyisoprene Ethylene-butylene rubber Ethylene-propylene rubber	P	G	L	P	P	G	P	Consumer: grips on disposable cameras, razor handles, and medical devices, shoe soles, adhesives
Block Copolymer TPE	Thermoplastic polyurethane	Rigid polyester polyurethane Rigid polyether polyurethane	Long-chain diols	E	E	H	G	G	E	E	High-performance: automotive underhood applications, shoe soles, seals/gaskets, adhesives, conveyor belts, casters/wheels
Block Copolymer TPE	Thermoplastic copolyester	Polybutylene terephthalate	Polytetramethylene ether terephthalate	E	G	H	E	G	E	E	High-performance: seals, belting, and hoses
Block Copolymer TPE	Thermoplastic polyamide	Polyetheresteramide (nylon) Polyetheramide Polyesteramide	Polyether polyol	E	E	H	E	G	G	E	High-performance: seals, cable jacketing, high-toughness athletic equipment

E = excellent, G = good, F = fair, P = poor, H = high, M = medium, L = low.

Many automobiles have color-keyed bumpers and fascia that are molded from thermoplastic polyolefin elastomers (TPOs). Many TPOs are readily paintable, although some require the application of special adhesion promoters. TPOs have excellent flexibility and toughness in temperatures as low as −22°F (−30°C) and resist the higher temperatures experienced in the paint baking ovens. Thermoplastic polyolefin elastomers are typically manufactured by blending a rubber such as EPDM with an olefin such as polypropylene (these rubbers are known as *compounded* TPOs). *Reactor* TPOs are manufactured in a polypropylene reactor through a stepwise copolymerization process. Rubbers produced in this manner have more rubber content and greater flexibility than compounded TPOs.

Thermoplastic polyurethenes (TPUs), are very high-performance TPEs that exhibit excellent abrasion resistence, compression set resistances, and excellent toughness. They are used in applications ranging from seals and bushings to elastic clothing (Spandex® is a TPU), to shoe soles and other wear components. TPUs consist of polyurea rigid segments separated by elastomeric blocks of polyol. Both polyester and polyether TPUs are available. The polyol groups in TPUs can be either aromatic (containing ring structures) or aliphatic (containing linear chains). The aliphatic TPUs are more resistant to UV but are more expensive.

Thermoplastic copolyester elastomers (COPEs) are unique in that the rigid blocks readily crystallize, increasing the strength and temperature resistance of the rubber. Thermoplastic copolyesters consist of tetramethylene terephthalate rigid segments and polyether soft segments. COPEs have high-temperature resistance, strength, chemical resistance, and creep resistance. Aside from the industrial and automotive applications, TPEs are also used as hot-melt and as pressure-sensitive adhesives.

5.7 Selection of Elastomers

In this section we have tried to point out some distinguishing characteristics of the more widely used elastomers. Table 5–3 contains some property information to further aid in selection. In selecting an elastomer for a high-production application, much more detailed information should be obtained from suppliers and handbooks. The principal properties to be scrutinized are physical and mechanical properties and environmental resistance. If a special combination of any of these properties is required, special blends can be obtained if production quantities are sufficient.

If machine design situations require only occasional selection of an elastomer, most needs can be met by weighing the relative merits of only a few rubbers. Most O rings are made from buna N rubber, as are many gaskets for oil sumps and the like. Neoprene is a good general-purpose rubber that can tolerate some oil exposure; it is also flame retardant. EPDM is a similar general-purpose, low-cost rubber, but is has better heat resistance than neoprene. Cast polyurethanes are extremely useful in machine design applications. They have better abrasion and tear resistance than almost all other rubbers, as well as the phenomenal advantage that they can be cast to shape in simple molds with no special tools other than a curing oven. They are extremely useful for roll coverings, testing seals, part fixtures, and similar machine applications. The silicones have similar castability, but they have higher heat resistance and lower tear and abrasion resistance. For extreme corrosion environments, the fluoroelastomer, Viton-type materials can often be justified.

If cost is not a factor, the average designer can be effective with a knowledge of a relatively small number of elastomers. Parts that are to be made in significant quantities will require a more thorough search for the most cost-effective and serviceable elastomer.

Table 5–3
Properties of common rubbers

Property	Natural Rubber (NR)	Styrene Butadiene (SBR)	Chloroprene (Neoprene) (CR)	Nitrile (Buna N) (NBR)	Butyl (IIR)	Ethylene Propylene (EPDM)	Urethane (PUR)	Butadiene (BR)	Fluoro-elastomer (FPM)	Silicone (SI)
Hardness range, Shore A	30–90	40–90	30–95	40–95	40–90	40–90	35–100	40–90	60–90	30–90
Continuous high-temperature limit (°F/°C)	180/62	212/100	302/150	257/125	212/100	302/150	212/100	200/93	437/225	482/250
Continuous low-temperature limit (°F/°C)	−67/−55	−67/−55	−67/−55	−67/−55	−67/−55	−67/−55	−67/−55	−150/−75	−40/−40	−105/−75
Compression set resistance	G	G	G	G/E	F/G	G	F	G	G	E
Abrasion resistance	E	G/E	G/E	E	G	G/E	E	E	G	P
Tensile strength (psi)	4500	3500	4000	4000	3000	2500	5000	3000	3000	1500
Relative cost SBR = 1.00	1.14	1.0	1.25	1.40	1.25	1.00	6.00	1.15	45.00	12.00
Environmental resistance										
Acid, inorganic	F	F/G	G	F	G	G	P	F	E	G
Acid, organic	P	G	G/E	G	G	G	P	F	G	E
Aging (oxygen, ozone, etc.)	P	P	G/E	F/P	G	G/E	G	P	G/E	E
Alcohols	P	G	E	G	G	E	P	G	F	G/E
Aldehydes	F	F/P	G/E	F/P	G	G/E	P	F	P	G
Alkalies	F	F/G	E	F/G	E	E	F/G	G	G	G/E
Amines	F	F	G/E	P	G	G/E	P	F	P	G
Animal oils	F	P	G	E	G	G	G	P	G/E	G
Esters: phosphate	G	P	E	P	E	E	P	P	P	G
Ethers	P	P	P	P	F/P	F	F	P	P	P
Fuels: aliphatic	P	P	P	E	P	P	G	P	E	F
Fuels: aromatic	P	P	P	G	P	P	F/P	P	E	P
Petroleum oils										
High aniline	P	P	G	E	P	P	E	P	E	G
Low aniline	P	P	G	G/E	E	P	G/E	P	E	F
Impermeability to gases	F	F/G	G	G	E	G	F	F/G	G/E	P
Ketones	G	P	E	P	E	E	P	F	P	P
Silicone oils	F	E	E	E	E	E	E	G	E	G
Vegetable oils	P	P	G	E	G	G	F	E	E	E
Water/steam	E	F	E	G	E	E	P	E	P	F

E = excellent, G = good, F = fair, P = poor, H = high, L = low.

Summary

This chapter is intentionally encyclopedic. It should serve as the reference for looking up various plastics when, for example, polysulfone is recommended for a particular part. That section can be reread and the designer can be reacquainted with that plastic or elastomer. The other purpose for stressing chemical formulations is that a user cannot have a good appreciation of differences between various plastics without knowing how their mers differ. Use characteristics do not adequately describe polymers. Effective use of plastics requires a basic understanding of polymerization, whether plastics are thermosetting or thermoplastic, and some idea of the structure and unique characteristics of each plastic. It is not necessary to memorize the structure of each plastic that we have discussed, but it is important to have an understanding of polymer families and basic types. The ethenic polymers, cellulosics, epoxides, amides, acetals, polyesters, silicones, and phenolics have chemical similarities within the family, but the families have fundamental differences in chemical composition and use characteristics.

As will be pointed out in the subsequent chapters, we can effectively use plastics in engineering design if we understand relatively few polymer systems and if we develop a feel for the characteristics of a dozen or so specific polymers. Thus this chapter should serve as a reference on polymer chemistry and as a guide for the selection of polymers and elastomers.

Some of the important concepts covered in this chapter are

- The ethenic polymers are all derivatives of the ethylene molecule.
- The most used plastics of any type are ethenic plastics, PVC, PS, and PE; the reader should remember the acronyms for various plastics.
- PVC, PS, and PE are used for all sorts of consumer items; they are often used for recyclable items, but they have reasonable mechanical properties and they can be used for structural applications if they are used properly.
- The term *engineering plastic* is used to describe plastics that can be used for applications that require high strength and fracture resistance.
- There are no strict criteria for engineering plastics, but some thermoplastic materials that are considered to be engineering plastics are nylons, polycarbonates, acetals, and polyesters. They all compete for the same market.
- The more expensive engineering thermoplastics are materials such as PPS, PI, PAI, and PEEK; these plastics have strengths similar to the other engineering plastics, but they can withstand much higher temperatures. The often have benzene rings as part of their chemical structure.
- The fluorocarbons are part of the ethenic family, and their unique characteristic is chemical and temperature resistance. They usually cannot be used for structural applications because they are relatively low in strength (compared to engineering plastics). They all have the element fluorine as part of their mer structure.
- Important clear thermoplastics are PS, PC, PMMA, PSU, and the cellulosics. They all have different mechanical and chemical characteristics: PS is the cheapest; PMMA has good rigidity and the best optical clarity; PC has the best toughness; PSU has the best temperature resistance.
- The most important thermosetting resins are phenolics, unsaturated polyesters, and ureas. These materials make up over 75% of the tonnage produced in the United States.

- Phenolics are widely used for friction materials and for printed circuit boards; they have good rigidity and strength, and good electrical properties.
- Ureas are mostly used for adhesives and for similiar applications; they are not commonly used for molded shapes.
- Unsaturated polyesters are widely used as the matrix material for fiberglass composites; they are catalyzed to cure and they do not require compression or elevated temperature.
- The phenolics and the ureas have ring structures and they cross-link to form macromolecules.
- An elastomer is a plastic that behaves like a rubber.
- A rubber is a material that has at least 200% elongation at room temperature and it forcibly returns to its original dimension when the force is removed.
- Natural rubber is probably the most "rubbery" rubber, but it is susceptible to degradation in many types of environments; it is fragile.
- Thermoplastic elastomers may be melt processed by injection molding, extrusion, and blow molding. Additionally, they may be reprocessed or recycled.
- Most elastomers have something that they do better than the other elastomers: neoprene has reasonable oil resistance; butyl rubber is used for seals; EPDM can take outdoor exposure; nitrile rubber is good in fuels and oils; silicones have the best temperature resistance; polyurethanes have the best abrasion resistance; SBR has good friction and abrasion resistance, so it can be used for tires; the thermoplastic elastomers are used when injection molding must be used to produce the desired shape.

In addition, it is imperative to remember to question the chemistry of any polymer or elastomer that is considered for an application. Try to understand the material; question the fillers and reinforcements; question the long-term properties; question the toughness, the moldability, the environmental resistance, the flammability. Match your service requirements to polymer and elastomer properties.

Critical Concepts

- Polymers may be categorized based on their application: commodity or engineering. Both categories consist of thermoplastics and thermosets.
- Ethenic polymers have a chain structure based on polyethylene.
- Many polymer families are available. The properties of a polymer are directly related to its molecular structure, additives, and processing.
- Elastomers or rubbers are differentiated from rigid plastics by their high elongation (at least 200% before breaking).
- Thermosetting and thermoplastic elastomer materials are available.
- In selecting polymers, it is important to understand the unique or salient features of each polymer family before a final selection is made.

Terms You Should Remember

ethenic
polyallomer
UHMWPE
olefins
thermoplastic polyester
fluorocarbons
chain stiffening
unsaturated polyester
polysulfone
LCP
polyamide-imide
PEEK
urea
melamine

alkyd
cellulosics
epoxides
phenolics
polyurethanes
polyisoprene
neoprene
silicones
elastomer
vulcanization
polyethylene
polyvinyl chloride
polyvinylidene chloride
polypropylene
polytetrafluoroethylene
polymethyl methacrylate
ABS
SAN
polyacetal
polyamide
polyimide
polycarbonate
thermosetting
thermoplastic
natural resin
polyesters
thermoplastic elastomer
block copolymer
polyphenylene sulfide
polyphenylene oxide
polyetherimide

Case History

Selecting a Wear-Resistant Elastomer for Transport Rollers

An office photocopier uses small rubber rollers to transport paper through the machine. Near the fuser section of the machine (the fuser melts the toner onto the paper to create a permanent photocopy image), the ambient temperature of the rollers can exceed 120°C. The rollers used in the fuser section are designed to rotate while the paper is held stationary (a condition referred to as overdrive). Transport rollers fabricated from polyether polyurethane became sticky and gummy after about one month of use—causing the photocopiers to malfunction.

Analysis and laboratory testing indicate that the high ambient temperatures and frictional heat created during overdrive caused the polymer structure of the polyurethane to revert to a sticky, low-molecular-weight resin. The extended chain structure that makes polyurethane so flexible also renders the chain susceptible to thermal degradation and reversion.

Alternative elastomers were tested under simulated thermal and wear conditions. A millable-gum silicone elastomer displayed a significantly lower abrasive wear rate than the polyurethane while closely matching the polyurethane's kinetic coefficient of friction. The structure of silicone elastomer is very resistant to high temperatures and thermal degradation. Since changing the rollers to silicone, paper transport through the photocopiers is much more reliable.

Questions

Section 5.1

1. Name three commodity thermosets.
2. Explain the difference between commodity and engineering plastics.

Section 5.2

3. Which of the ethenic polymers are homopolymers?
4. What is the substituent for polyvinyl alcohol?
5. Name ten common ethenic polymers.
6. Discuss the effect of density on the properties of polyethylene.
7. Compare and contrast the differences between polypropylene and polyethylene.
8. Explain the influence of the benzene ring on the mechanical and thermal properties of polystyrene.
9. What is the significance of molecular weight in polyethylenes?

10. What is the difference between the soft vinyl used in seat covers and the polyvinyl chloride plastic that is used for plumbing piping?

Section 5.3

11. Compare the differences between acrylonitrile butadiene styrene and polystyrene.
12. Which of the fluorocarbons are injection moldable?
13. What chemical factor is common to all ethenic polymers?

Section 5.4

14. How many types of nylon are there and what are the major grades?
15. Explain the influence of moisture on the mechanical and dimensional characteristics of polyamides.
16. What are the raw materials for cellulosic polymers?
17. What is an engineering plastic? Name three.
18. What is a polyester?

Section 5.5

19. What makes a thermosetting polymer unrecyclable?
20. What are the four most used thermosets and their biggest application (in tons)?
21. What is meant by the term *macromolecule*?

Section 5.6

22. What is the rubber with the best resilience?
23. What is the definition of a rubber?
24. What is a thermoplastic elastomer?
25. Which rubber has the best temperature resistance?
26. What is a block copolymer and why can they be melt processed?

Section 5.7

27. Name a rubber with good oil resistance.
28. Name a rubber with good elevated temperature resistance.
29. What rubber would you use for a punch press die spring? Explain your reasons.

To Dig Deeper

Bhowmick, A. K., and H. L. Stephens. *Handbook of Elastomers*. New York: Marcel Dekker, Inc., 1988.

Brydson, J. A. *Plastics Materials*, 7th ed. Oxford: Butterworth-Heinemann, 1999.

Fontana, M., and N. Green. *Corrosion Engineering*. New York: McGraw-Hill Book Co., 1978.

Goodman, S. H., Ed. *Handbook of Thermoset Plastics*. Park Ridge, NJ: Noyes Publications, 1986.

Harper, C. A. *Handbook of Plastics and Elastomers*. New York: McGraw-Hill Book Co., 1975.

Margolis, J. M., Ed. *Engineering Thermoplastics*. New York: Marcel Dekker, Inc., 1985.

McCrum, N. G., et al. *Principles of Polymer Engineering*. New York: Oxford University Press, 1988.

Mills, N. J. *Plastics Microstructure and Engineering Applications*, 2nd ed. New York: John Wiley & Sons, Inc., 1993.

Morton, M., Ed. *Rubber Technology*. Malabar, FL: Robert E. Krieger Publishing Co., 1981.

Modern Plastics Encyclopedia, an annual. New York: McGraw-Hill Companies, Inc.

Smith, L. *The Language of Rubber*. Oxford: Butterworth-Heinemann, Ltd., 1993.

Szycher, M. *Handbook of Polyurethanes*. Boca Raton FL: CRC Press LLC, 1999.

Ulrich, H. *Introduction to Industrial Polymers*, 2nd ed. New York: Hanser Publishers, 1993.

Van Krevelen, D. W., and W. Hoftyzer. *Properties of Polymers*. Amsterdam: Elsevier, 1976.

Vasile, C., and R. B. Seymour. *Handbook of Polyolefins*. New York: Marcel Dekker, Inc., 1993.

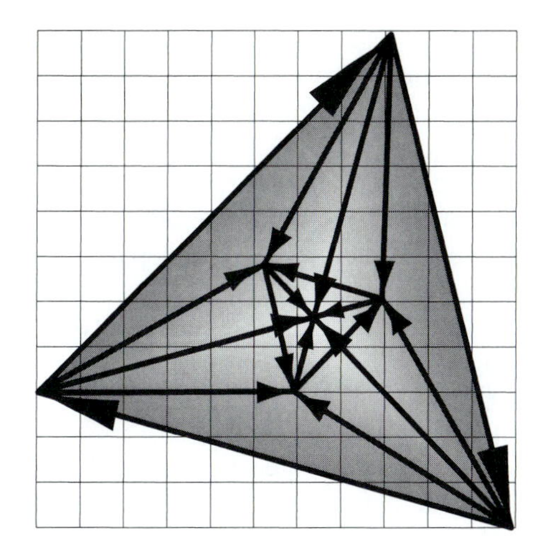

CHAPTER 6

Plastic and Polymer Composite Fabrication Processes

Chapter Goals

1. An understanding of how thermoplastic and thermoset plastics and polymer composites are shaped into parts.
2. Enough knowledge of plastic fabrication processes to allow a designer to select a process for a proposed design.
3. An understanding of how to specify a plastic fabrication process.
4. An understanding of recycling's importance and recycling codes on plastics.

Rationale

Probably the most limiting factor in the application of plastics to a new design is their fabricability. Polytetrafluoroethylene (PTFE; Teflon®) is a great plastic for applications for which chemical resistance is a priority, but not many consumer products are made from this plastic. Why not? It cannot be injection molded, blow molded, cast, or made by any of the processes that prevail in commodity plastics. Parts are only made from PTFE by sintering particles together under significant pressure. This fabricability problem (along with the high cost) limits its application. Continuous fiber-reinforced plastics usually have higher stiffness and strength than unreinforced plastics, but similar to PTFE, the processes that are available for adding continuous fiber reinforcement to plastics are limited in number and none are as low cost as injection molding for making many parts.

If it looks like a particular plastic has properties that will fit your application, you must also consider the fabrication options that exist with this plastic. Is the plastic injection moldable? Does it have a high or a low shrinkage rate in molding? Is it stable in molding? What kind of tooling investment is necessary? What kind of lead time does the fabrication process require? How available is the fabrication capability?

These are the types of questions that need to be answered before finalizing a plastic selection; in effect, what will it take to make the part that you want from the plastic that you selected? An understanding of available plastic fabrication processes and how they apply to different plastics is a necessary requirement for proper selection of plastics for design engineering. What does it take to make a part from polyethylene? What does it take to make the same part from polybenzimidizole? You will benefit from knowing these answers.

6.1 Thermoplastic Fabrication Processes

The previous chapter presented a plethora of plastic options to use in your machine designs. This chapter is intended to give the newcomer to plastics an introduction to the various processes that are used to make thermoplastic plastics, thermosetting plastics, and polymer composites. It will be a review for experienced plastic users. The next chapter will cover how to select one plastic for a particular part under design, but one aspect of selecting a plastic is knowing what form you want the plastic to be in. In metals, the designer has the option of machining the part from a stock shape, casting, extruding, or forging to a near-net shape followed by finish machining, or the part may be shaped to final form by a process such as powder metallurgy or die casting. There are similar options in plastics and composites, but with more options. Shaping into final form is more common for plastics than it is for metals. The reason for this is mostly that plastic fabrication processes often include clever techniques to break plastic flow paths (sprues, runners, risers, and scrim) from parts. Injection molded plastics are often separated from feeders at ejection from the mold. The plastic feeders go off to get recycled, and the parts are conveyed to the next operation. Die-cast metal parts often require a secondary operation to remove feeders. Parts are not easily broken free from the shot. A trim press is used to perform this function.

This section is dedicated to thermoplastic materials, those plastics that can be remelted and reused. Thermoplastics include commodity plastics such as polystyrene and polyethylene as well as engineering plastics such as nylon % and polyphenylene sulfide. Thermoplastic part shapes range from 3-mm-diameter buttons to automobile dashboards. The fabrication processes used (Figure 6–1) range from injection molding, which can produce hundreds of parts per minute, to

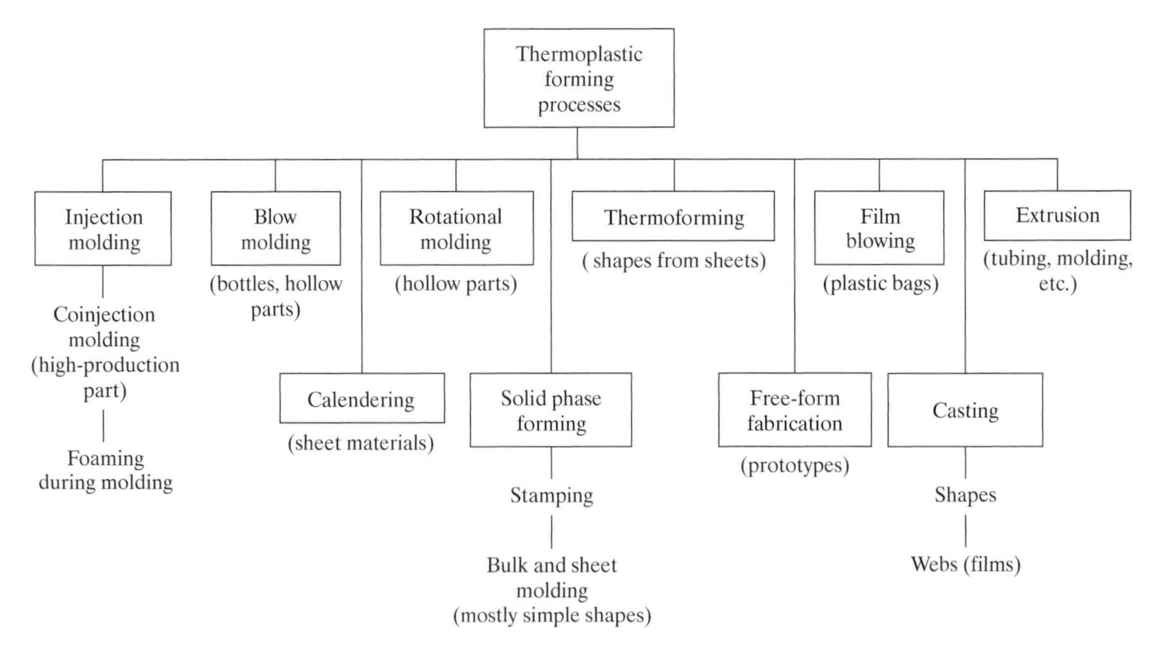

Figure 6–1
Spectrum of forming processes for thermoplastic materials

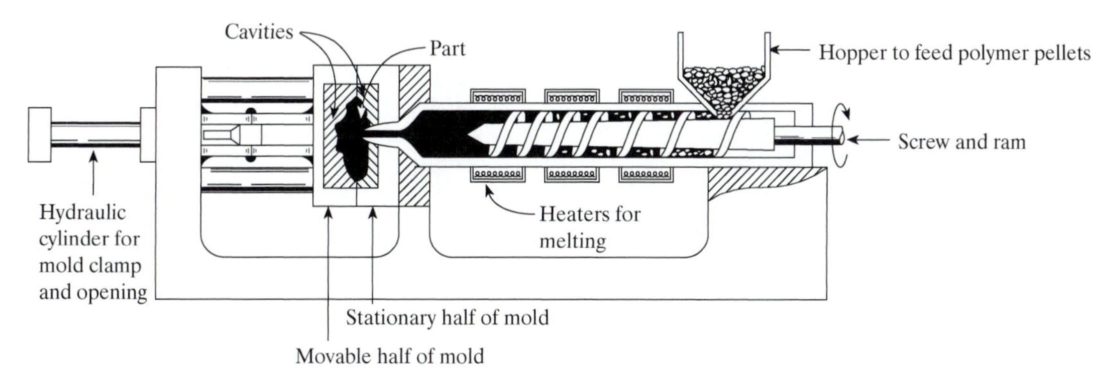

Figure 6–2
Injection molding

casting acrylics, which may take an overnight cure for one part. There are entire books on each process, and we do not have space to go into detail on each of these processes, so we will present a word description (and maybe a sketch) in hopes of gaining some familiarity with the various processes.

Injection molding (Figure 6–2): A process in which granular polymer, usually thermoplastic, is fed from a hopper into a heated barrel where it is melted, after which a screw or ram forces the material into a mold. Pressure is maintained until the part has solidified. The mold is opened, and the part is ejected by some mechanism. This is by far the most important technique for mass production. The major disadvantages of the process are that not all polymers can be processed (some thermosets cannot) and the metal molds are very expensive.

This basic process is also used for *coinjection molding* of two different polymers. There are two extrusion barrels and injection systems. A shot is made with one polymer, and a second shot with a second polymer can be used to surround or surface the part made in the first shot. Coinjection is often done to achieve a cosmetic effect or to alter use properties. For example, a thermoplastic elastomer (TPE) can be injection molded just on the grip area of a plastic drill body after the rigid plastic body is molded.

Another variation of injection molding is structural foam molding. The mold is only partially filled, and the injected plastic expands to fill the mold to produce a part that is lightweight because of the entrapped porosity, but with an integral skin. Foamed polymers have lower weight (and cost) than their nonfoamed counterparts, and the mechanical properties are often comparable. This process is often used on polyphenylene oxide, olefins, vinyls, nylons, and thermoplastic elastomers.

Blow molding (Figure 6–3): A process of forming hollow articles by blowing hot polymer against internal surfaces of a split mold. Usually, a tube *(parison)* of heated polymer is extruded down the center of the closed mold. Air is then injected, and the heated polymer expands in a fairly uniform thickness to form the desired shape. This is the process used to make plastic bottles and containers. It is fast and usually used only on thermoplastic materials.

Calendering (Figure 6–4): The process of forming thermoplastic or thermosetting sheet or film by passing the material through a series of heated rolls. The gap between the last pair of

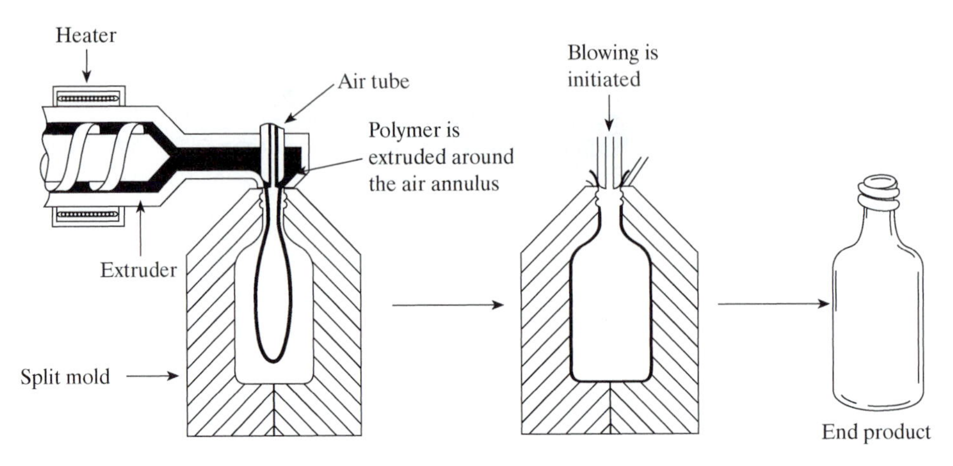

Figure 6–3
Blow molding

heated rolls determines the sheet thickness. The material is usually blended and plasticated on separate equipment. Elastomer sheets, gaskets, and vinyl flooring tiles are often formed with this process.

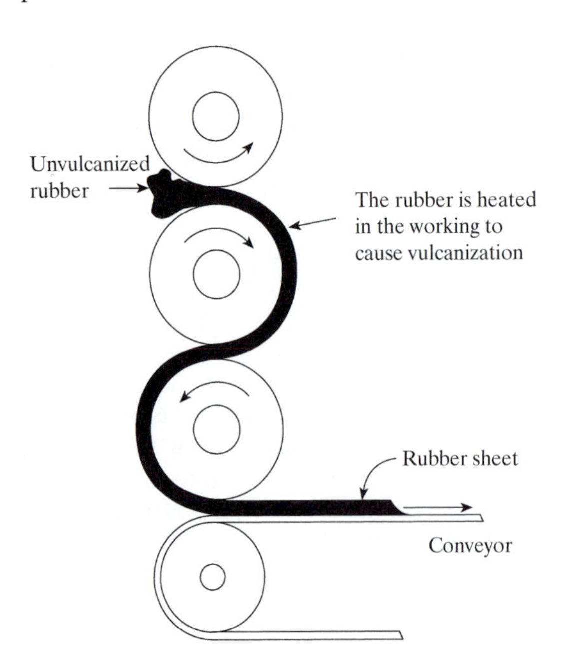

Figure 6–4
Calendering

Rotational molding (Figure 6–5) This process is usually used for making large hollow containers such as fuel tanks, water tanks, floats, and so on. A premeasured quantity of thermoplastic pellets are charged into a closed metal mold. The mold is clamped closed and put into a device that is capable of rotating the mold about two axes. The rotating mold is heated to melt the charge. The molten polymer forms a skin against the mold wall to make the part. Mold heating is stopped, and the mold is air or water cooled to allow part removal. A significant advantage of this process is that it uses relatively low-cost

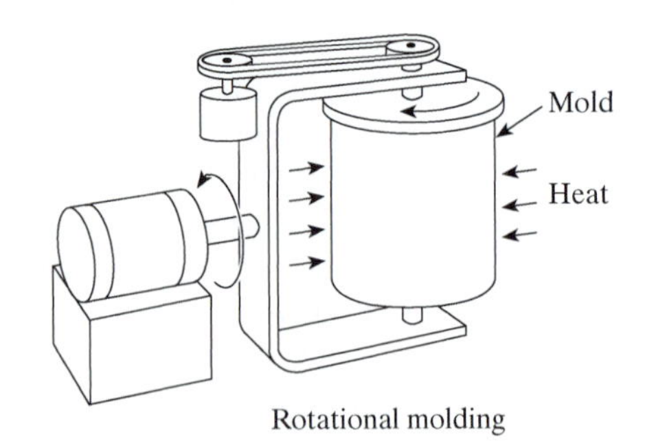

Figure 6–5
Rotational molding

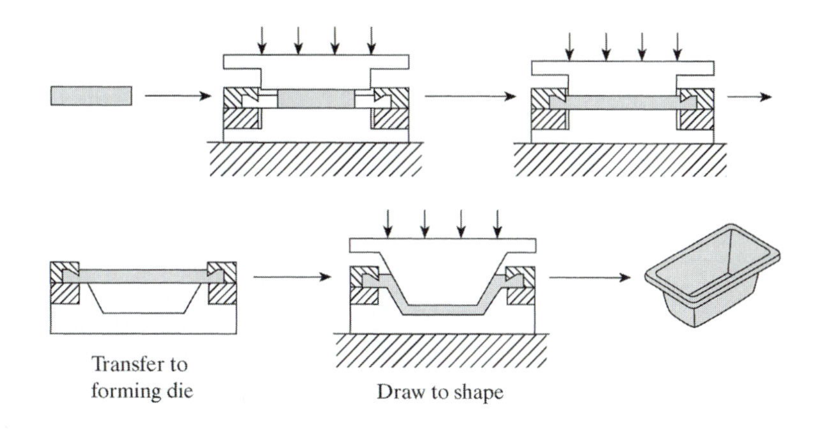
Transfer to forming die
Draw to shape

Figure 6–6
Solid-phase forming of drawn plastic shapes

tooling compared to injection molding and other capital-intensive processes.

Solid-phase forming: Solid-phase forming is a process employed with thermoplastic materials in which a sheet or similar preform is heated to the softening point, but below the melting point, and forged into a drawing ring by a heated die set; the still-warm preform is transferred in the draw ring into a cooled draw die, and a plug draws the part to the finished shape. The steps in this process are illustrated in Figure 6–6. A lubricant is applied to both sides of the preform to assist the draw. Depth-of-draw ratios can be 1.5 or higher, and the forming in the solid phase is said to enhance strength over parts formed by melting techniques.

Simpler shapes can be made by heating sheet and stamping the sheet to make a form such as an arch support for a shoe. Filled or reinforced thermoplastics called sheet or bulk molding compounds can also be formed by press shaping of the preheated material.

Thermoforming (Figure 6–7): A method of forming polymer sheets or films into three-dimensional shapes, in which the sheet is clamped on the edge, heated until it softens and sags, drawn in contact with the mold by vacuum, and cooled while still in contact with the mold.

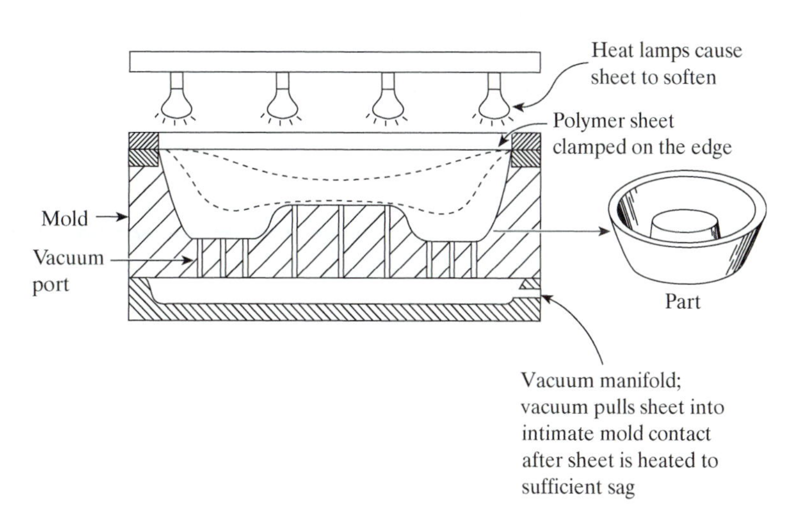
Heat lamps cause sheet to soften
Polymer sheet clamped on the edge
Mold
Vacuum port
Part
Vacuum manifold; vacuum pulls sheet into intimate mold contact after sheet is heated to sufficient sag

Figure 6–7
Thermoforming

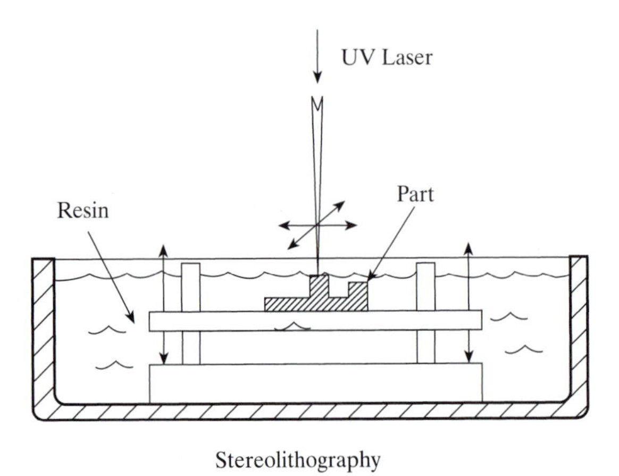

Figure 6–8
Stereolithography

Vacuum is not always necessary. Sometimes the sheet is simply draped over a mandrel. Sometimes matched metal molds are used. Thermoforming is ideal for low-volume production of containers, machine guards, and other parts with suitable shape.

Free-form fabrication: *Stereolithography* is one process in a category of fabrication processes that are termed *free-form fabrication processes*. In this process, illustrated in Figure 6–8, a computer-controlled laser is used to selectively polymerize a liquid resin to form a part. A platen is given a thin coating of liquid resin by dipping or spraying; the laser traces the shape of the part at a vertical slice height of a few microns. This action polymerizes the resin. The part is recoated, and the process is repeated at another slice height. This process is repeated until a three-dimensional part emerges.

Another variation of this process is the laser fusing of powder. A platen is coated with a thin layer of powder, and the laser fuses the powder in the shape of the part. The emerging shape is given another coating of powder, and the process is repeated.

The usual material for stereolithography is a UV-curable polyurethane resin (rigid). The powder process can use nylons, polycarbonate, and a wide variety of thermoplastic materials. Both techniques are used to produce plastic prototypes without hard tools or even drawings. The part can be made from computer data files (computer-aided design, CAD). Functional tests can be made on the part before any capital is invested in tooling. The process is also used to produce short-run tools from thermoplastics. A process variation for rapid prototypes is called three-dimensional (3D) printing. An array of dot print heads deposit wax or low-melt thermoplastics to form a part from CAD drawings. These devices are intended for use in engineering offices.

Extrusion (Figure 6–9): The process of forming continuous shapes by forcing a molten

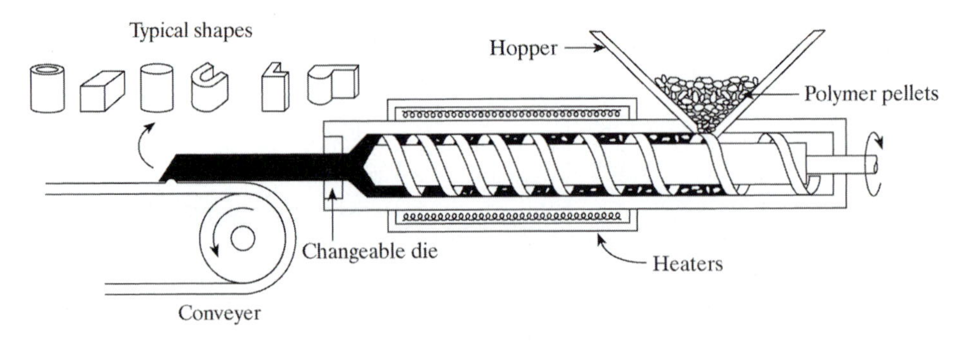

Figure 6–9
Extrusion

polymer through a metal die. *Extrusion* is used to make structural forms such as channels, bars, rounds, angles, tracks, hose, tubing, fibers, films, and countless other forms. It is very fast and usually applied only to thermoplastics. With special techniques, two different polymers or different colors of the same polymer can be coextruded, and plastics can also be foamed during the extrusion process. This process is often used to make specially shaped weather stripping. A variation is used to make plastic bags (blowing).

Summary: Thermoplastic Processes

Many other processes are used to form thermoplastics, but the ones that we have illustrated are the most widely used. Most of these processes require very expensive equipment. An injection molding, blow molding, or film blowing machine can easily cost in excess of $100,000. Extruders, calenders, and stereolithography machines are similar in cost. Thermoforming can be done with very simple equipment (such as wooden forms, a vacuum cleaner, and an oven). Rotational molding and some types of solid-phase forming can also be performed without expensive equipment. The high equipment costs associated with many of these forming processes can be dealt with by using vendors who specialize in particular processes.

6.2 Thermoset Fabrication Processes

Thermoset polymers are typically available as liquid resins or as solid particles, powders, or beads. Many liquid resin systems require two or more components to be mixed together to initiate cross-linking or curing. For example, liquid epoxy often requires a polymer resin (for example, based on bisphenol A and epichlorohydrin) and a hardener such as a polyamide. The resin and the hardener must be mixed together to initiate the cross-linking reaction.

Similar to thermoplastic resins, some thermoset resins are commercially available as solid granules, beads, or pellets containing the necessary additives and fillers for the application at hand. For example, phenolic molding compounds are made by reacting phenol resin, formaldehyde, and an alkaline catalyst in a vessel under temperature-controlled conditions (heat generated from the reaction must be removed). The reaction is allowed to progress until the viscosity of the resin increases. At this point, excess water from the reaction is driven off by vacuum—yielding a viscous resin that is soluble in solvents (known as *A-stage* resin). The A-stage resin is cooled to a solid and ground into a fine powder. Additives such as fillers, colorants, and lubricants are mixed with the powder. The powder is then processed on heated mixing rollers, where some cross-linking begins to occur. When the polymer is nearly insoluble in solvent but still fusible (meltable) with heat and pressure, the reaction is terminated—yielding a *B-stage resin*. The B-stage resin is cooled and rough-ground into granules, or pellets. To mold a part from granulated B-stage resin, the resin is heated until it liquefies and then it is consolidated in a mold by injection or compressing (for example). With time and temperature, the liquefied B-stage cross-links to a rigid solid. The liquefaction and curing process is shown graphically in Figure 6–10(a). Likewise, when a liquid resin is cured under temperature-controlled conditions, the viscosity rises with time as shown in Figure 6–10(b).

Once thermosets are formed they cannot be remelted. This makes molding machine cleanout very difficult, which is the major reason why thermosets are often molded in equipment that is simpler than that used on thermoplastics. Injection molding and extrusion processes are used on thermosets, but the process is complicated by the need to completely remove the polymer from the barrel and runner system at shutdown. If, for example, an extruder is shut down with a thermoset in the barrel that heats and conveys the material, the screw may become

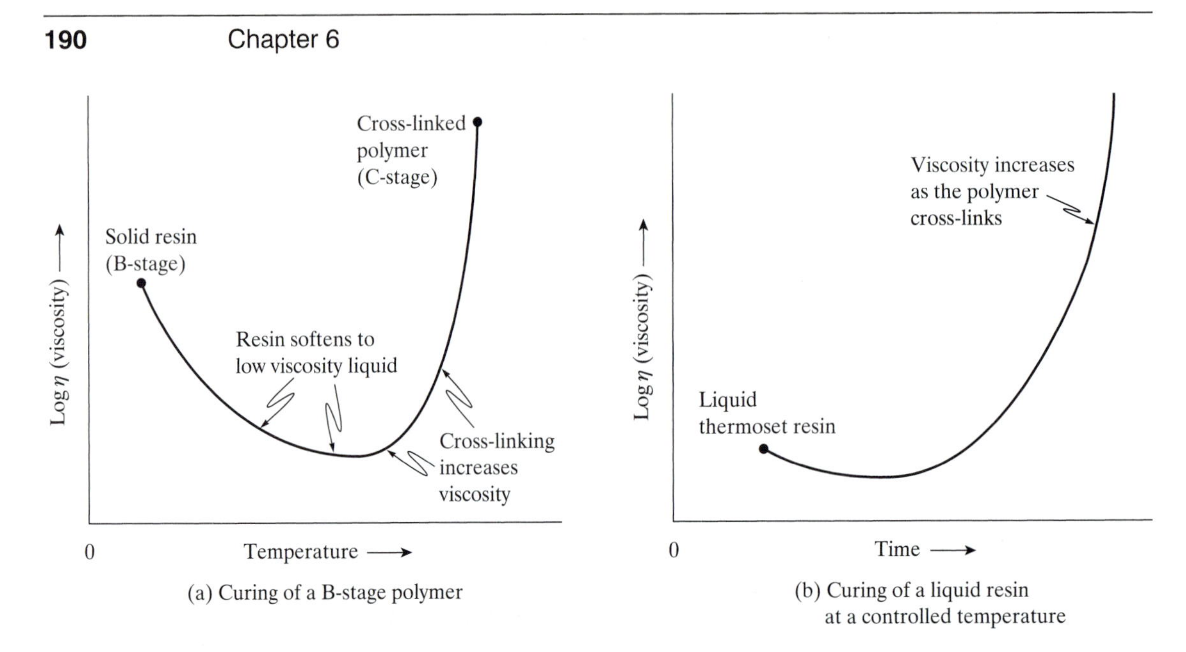

Figure 6–10
Effect of time and temperature on the curing of thermoset resins

an integral part of the barrel. Screws can cost $50,000 and barrels twice that. Thus, thermosets can be formed by some of the thermoplastic processes, but special techniques must be used to deal with the irreversible nature of these materials. Most composites are made with thermosetting resins. These will be dealt with in the next section. This section will cover the processes that are used for thermosets that do not contain continuous reinforcement (Figure 6–11).

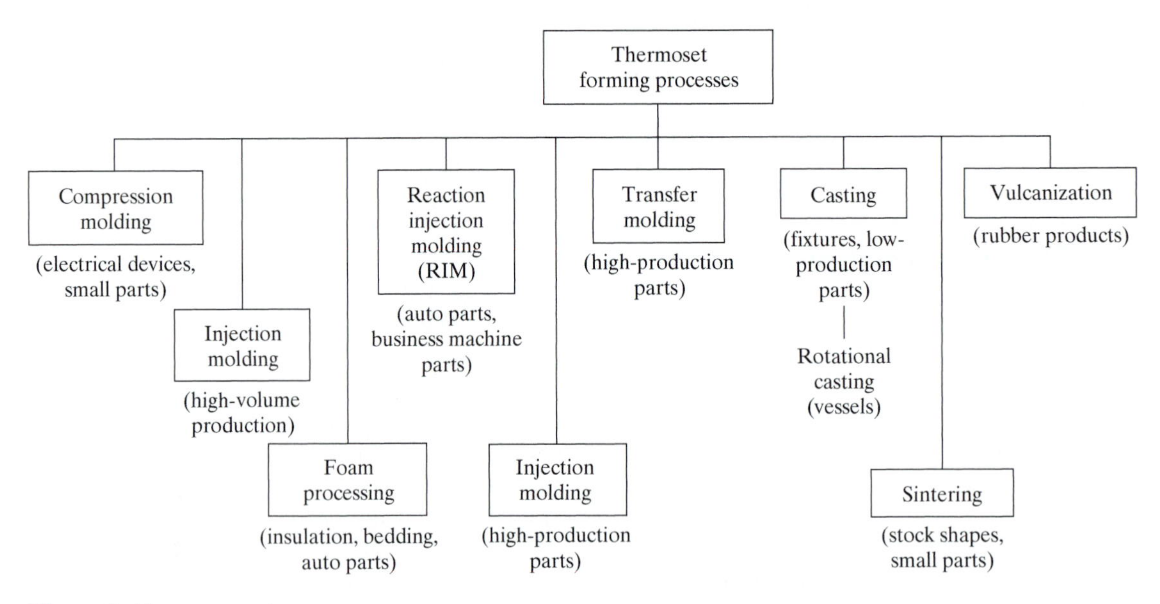

Figure 6–11
Thermoset fabrication processes

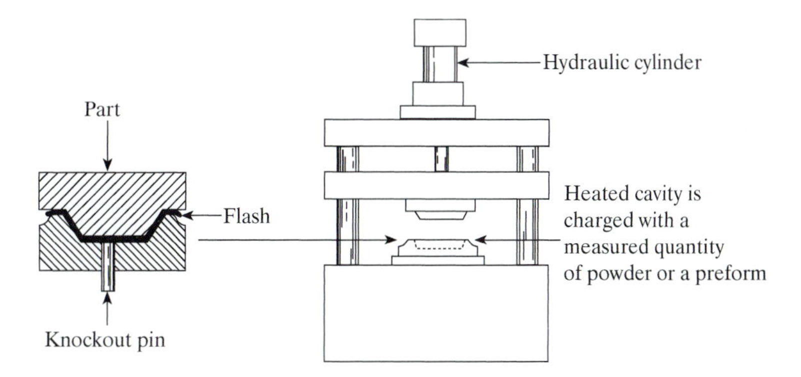

Figure 6–12
Compression molding

Compression molding (Figure 6–12): A method of molding in which the molding material, usually preheated and premeasured, is placed in an open mold cavity, the mold is closed with a cover half or plug, and heat and pressure are applied and maintained until the material has filled the cavity and cures. This is the most widely used process for thermosetting materials. Because curing of thermosetting polymers is a time-dependent reaction, cycle times are usually much longer than those in injection molding.

Injection molding: Thermoset materials may be injection molded in high-production applications. Thermoset injection molding cycles are typically two to three times faster than compression-molding cycles—yielding lower cost parts. Much like thermoplastic materials, a screw or plunger is used to feed the polymer (typically B-stage pellets or beads) though a heated barrel. With heat, the viscosity of a thermoset polymer initially decreases. However, over time the viscosity increases as the polymer cross-links. The intent is to inject the liquid polymer in the mold while it is at its lowest viscosity level. Once in the mold, the polymer is heated long enough to cross-link the resin and form a solid part. The part is then ejected from the mold. It is not uncommon to heat the parts in an oven off-line to further cure the polymer (this in known as *post curing*) to ensure required properties and dimensional stability.

Foam molding: Various techniques are used to produce molded thermoplastic parts with dense skins and high porosity in the core (Figure 6–13). The simplest process involves charging a metal mold with resins that foam and expand during reaction. The expansion of the foaming resin causes the mold to be filled. A variety of similar processes are used for large parts, with the exception that the polymer is injected into the cavity rather than charged into the mold.

Standard injection molding machines can be adapted to produce foamed parts by adding equipment to inject a metered amount of gas or a chemical blowing agent into the polymer near the exit end of the extruder barrel of the injection

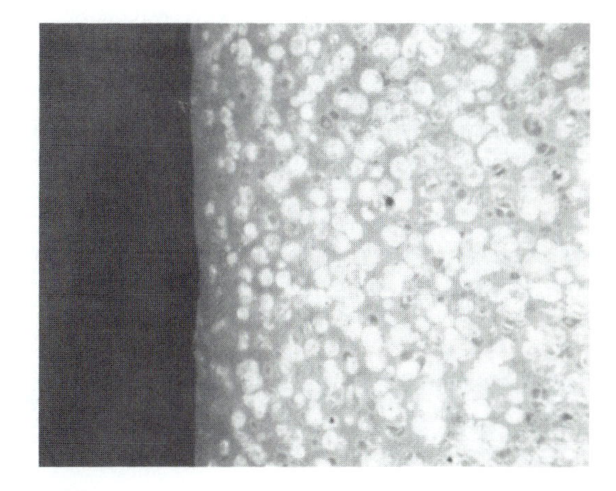

Figure 6–13
Cross section of foam-molded part (×30). Note the dense skin at the surface.

Figure 6–14
Schematic of reaction injection molding

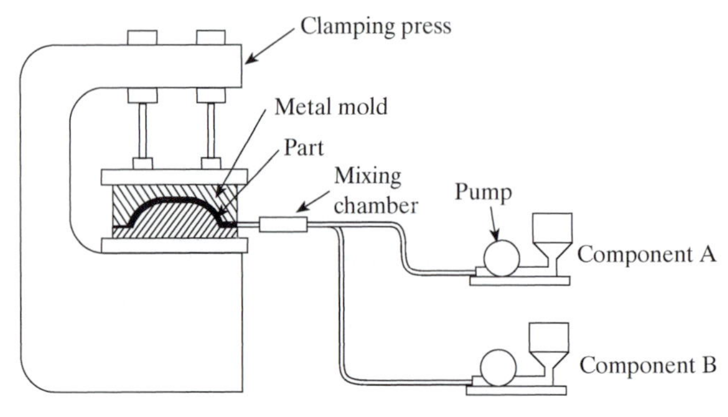

molding machine. The advantage of foam molding is the ability to increase section sizes and part stiffness without using significantly greater amounts of material. Side effects are often less shrinkage and better part tolerances.

Foams such as those used in furniture, bedding, auto seats, and the like are usually made from thermosets that are allowed to foam outside of a mold. The most common foams are polyether or polyester polyurethanes. Components are reacted and applied to a long conveyor belt. The reactants form a huge foam "bun." The bun is sliced to make slabs or cut with a veneer knife to make sheet goods.

Reaction injection molding (RIM): Polymer reactants are pumped under high pressure into a mixing chamber and then flow into a mold at atmospheric pressure. The chemicals expand to fill the mold and to form the polymer. The foaming action of the reactant produces the pressure for replication of mold details [about 30 to 70 psi (207 to 483 kPa)], and the reaction heat speeds the polymer cure (cycle times are usually less than two minutes). A schematic of the RIM process is shown in Figure 6–14. This process is most often applied to large parts (over 1 kg), and polyurethane foams are the most popular molding materials. Many large auto parts, such as bezels, dashboards, and fenders, are made with this process. Fillers can be added to the reactants to improve the mechanical properties of molded parts.

Transfer molding (Figure 6–15): A process, usually applied to thermosetting materials, in which the molding material is put into an open cylinder, heated, and transferred under pressure into the part cavity. It is a modification of compression molding.

Figure 6–15
Transfer molding

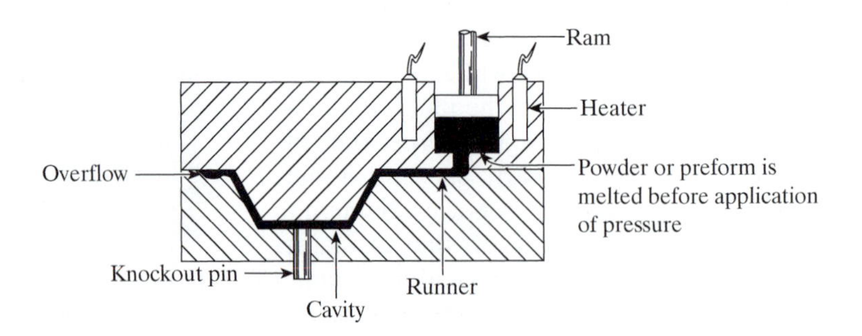

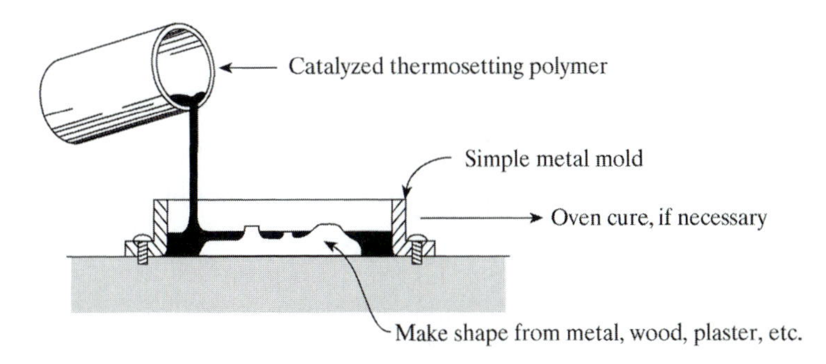

Figure 6–16
Casting

Casting (Figure 6–16): The process of forming solid or hollow shapes from molten polymer or from catalyzed resins by pouring the liquid material into a mold without significant pressure, followed by solidification or curing. The mold is usually open at the top. The *polymer casting* process is widely used on urethane and silicone elastomers to make roll covers, die springs, sheets, and the like. It is also suitable for making jigs and fixtures from filled epoxy or polyester resins.

Rotational casting can be done in tools similar to those used for rotational molding of thermoplastics. Liquid reactants are poured into a closed mold and rotated in two axes until the resins cure. It is not necessary to heat the mold as in rotational molding of thermoplastics.

Sintering: Sintering is the bonding of adjacent surfaces of powder particles. Some of the high-temperature polymers are made into shapes by compressing polymer particles as in compression molding and heating them until the particles join by coalescence. They do not melt and flow as in normal compression and transfer molding. This process is used on some of the fluorocarbons, polyimides, and similar high-temperature plastics.

Vulcanization (Figure 6–17): Most rubbers are vulcanized to form a shape. *Vulcanization* is the application of heat and pressure to cause polymer cross-linking, the process that makes rubber rubbery. Before vulcanization, rubber is sticky. Figure 6–17 illustrates the making of a rubber-covered roll. A tape of partially reacted rubber is wrapped around the roll to the thickness of rubber desired. The taped rubber is shrink-wrapped with a plastic film (to protect the rubber from oxidation) and placed in an autoclave. The steam-heated autoclave produces both the heat and the pressure needed to complete the consolidation of the tape into a solid rubber coating on the roll. The vulcanized rubber coating is then ground to the customer's diameter. This is also used for automobile tires and the other rubber items that we take for granted.

Summary: Thermoset Forming Processes

Our discussion has omitted thermoset forming processes such as making plywood, flake board, and their construction products, which make up

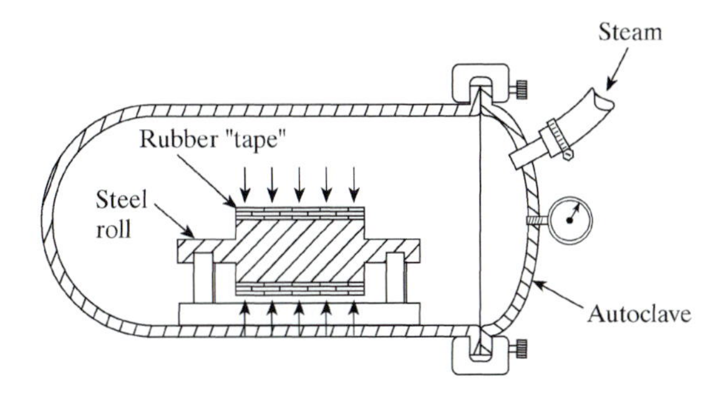

Figure 6–17
Schematic of *vulcanization* of rubber to make a rubber-covered roll

the largest fraction of use of thermosetting polymers. There are two reasons for this: First, they are really composites and thus belong in the next section; second, they are not used by most machine designers, only those in that industry. An important concept to keep in mind with regard to thermoset fabrication processes is that forming/molding cycle times are almost always longer than those for making thermoplastic parts. This is why thermoplastics are used more on a tonnage basis. However, when it comes to heat resistance, the thermosets usually excel. Some thermoplastics have higher use temperatures than thermosets, but they really have little strength at temperatures in excess of 500°F (260°C) and they melt when overheated. Thermosets do not melt. When overheated, they char and are often still usable. So thermosets are somewhat more difficult to form than thermoplastics, but the molding equipment is usually cheaper and these materials simply perform better than thermoplastic materials in many applications.

6.3 Polymer Composites

Reinforcement Types

In Chapter 1, we defined a *composite* as a material composed of two or more different materials, with the properties of the resultant material being superior to the properties of the individual materials that make up the composite. By this definition, an alloy or blend could be a composite, but the accepted use of "polymer" is to mean a material with a continuous *resin matrix* and a controlled distribution of a reinforcing material. Even with this definition some polymer materials, such as liquid crystal polymers, meet the definition, but they are not normally considered to be composites. From the commercial standpoint, composites are made from matrices of epoxy, unsaturated polyester, some other thermosets, and a few thermoplastics. The *reinforcements* are glass, graphite, aramid, thermoplastic fibers, metal, and ceramic. The combination of the two materials makes a composite (Figure 6–18). The

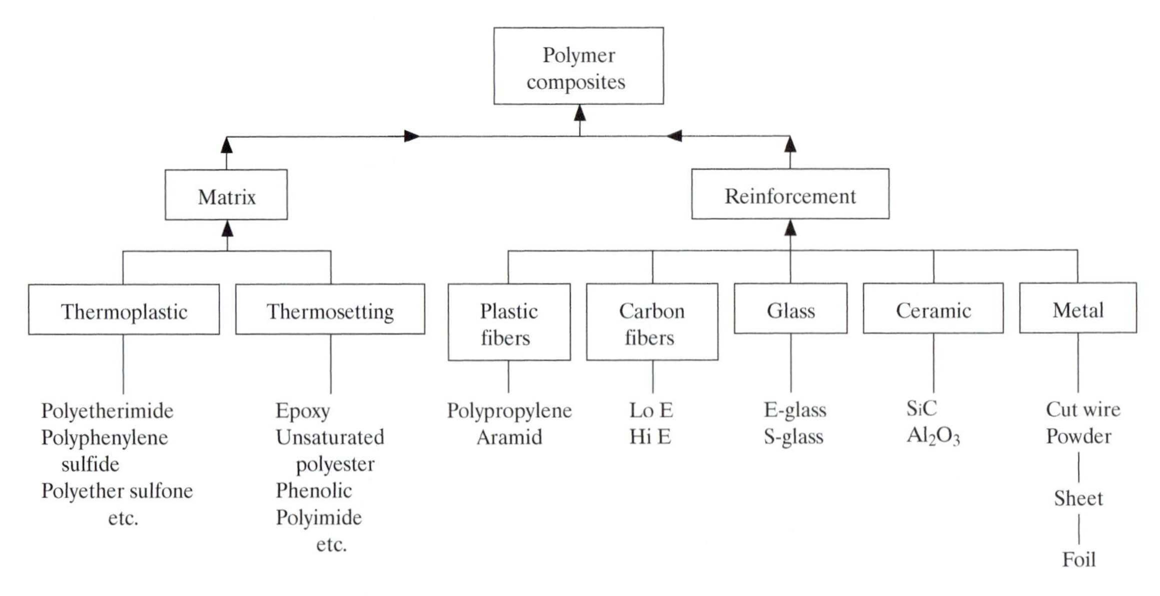

Figure 6–18
Matrix and reinforcement options for polymer composites

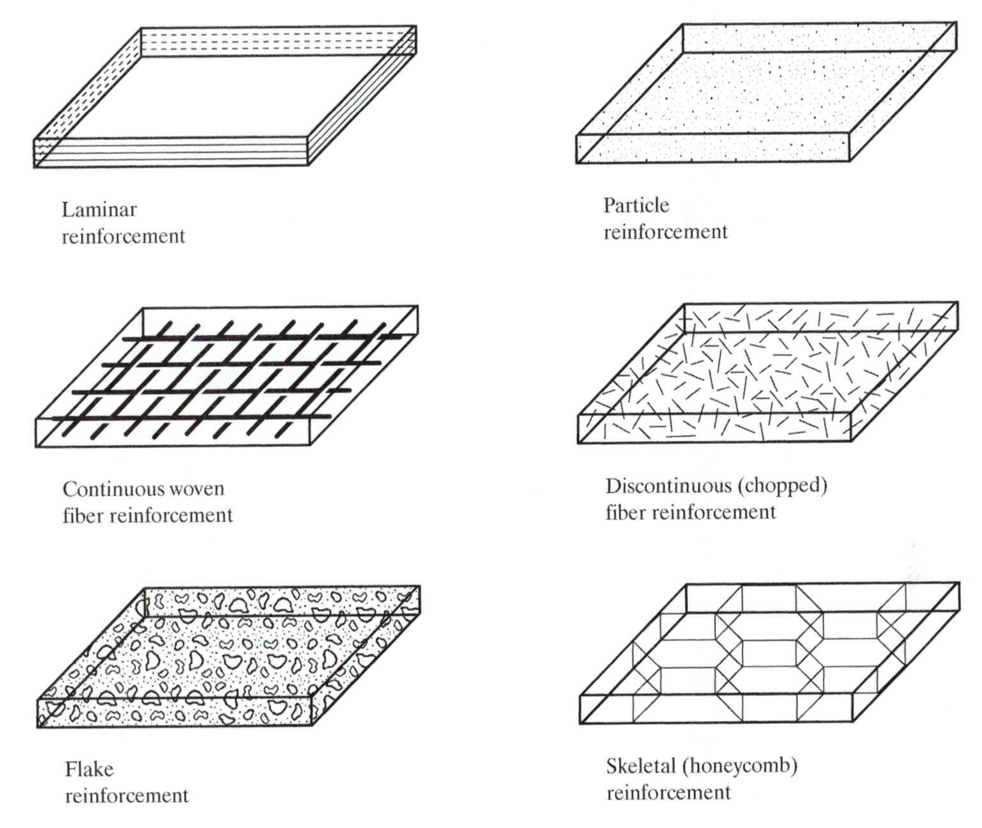

Figure 6–19
Reinforcements used in polymer composites

reinforcement can be continuous, woven, or chopped fiber; conventional composites contain from 20% to 50% by weight of glass or other reinforcement. The percentage of reinforcement in advanced composites can be as high as 70%; these materials usually use epoxy as the matrix material, and graphite is the most common reinforcement. Some reinforcement schemes used in polymer composites are shown in Figure 6–19.

The purpose of adding reinforcements to polymers is usually to enhance mechanical properties. Chopped fibers, flakes, particles, and similar discontinuous reinforcements may enhance short-term mechanical properties, but these types of reinforcements are usually not as effective as continuous reinforcements in increasing creep strength and similar long-term strength characteristics. Continuous reinforcements serve to distribute applied loads and strains throughout the entire structure. These latter types of composites offer the most potential for making polymer-based composites competitive with metals for structural applications.

Cellulose fibers reinforce plants, and they are responsible for the remarkable mechanical properties that are available from wood. Woods are natural composites. The first commercial synthetic polymer composites were phenolic-paper laminates intended for electrical insulation (circa 1915, see Figure 6–20). Probably the most important events in the use of reinforced plastics (RPs) were the development of epoxies,

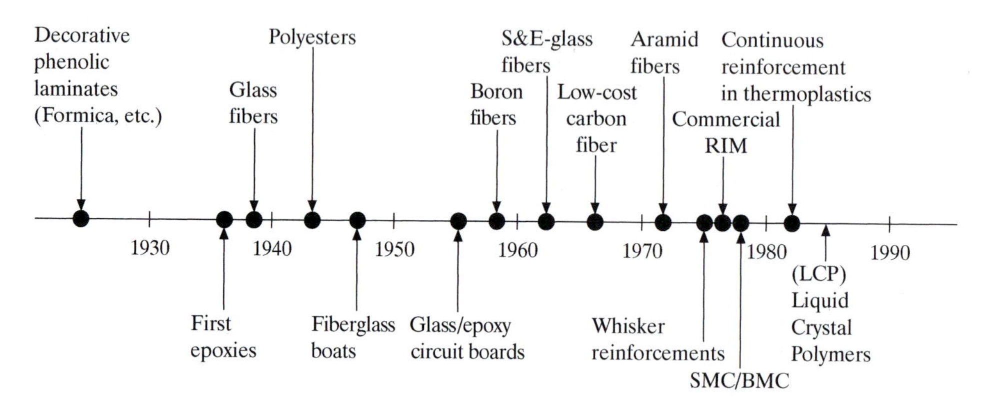

Figure 6–20
Genealogy of polymer composites

polyester resins, and glass fibers. These developments allowed the use of composites for structural items such as boats, piping, and containment vessels. There are many reasons for the use of polymer composites, but most center on their strength and environmental resistance characteristics. As shown in Figure 6–21, the high-performance grades of polymer composites—those with the stronger matrix materials and continuous reinforcements—have specific strengths and specific stiffness ratios that are superior to those of steels and aluminum alloys. *Specific strength* is the tensile strength of a material divided by the density, and *specific stiffness* is the tensile modulus divided by the density. These ratios are used by designers to decide which structural material will provide a desired strength or stiffness with the lowest mass. For example, an I-beam of steel will weigh more than two times as much as a composite I-beam of boron–epoxy for equivalent stiffness. Similarly, an aramid–epoxy lifting device will be four times as strong as a steel device of the same mass. Advanced composites are taking over for aluminum, steel, and titanium for structural components on aircraft for these reasons, and the lower-strength and less costly polymer composites are replacing metals on automobiles because of their lower weight and resistance to atmospheric rusting and road salt corrosion.

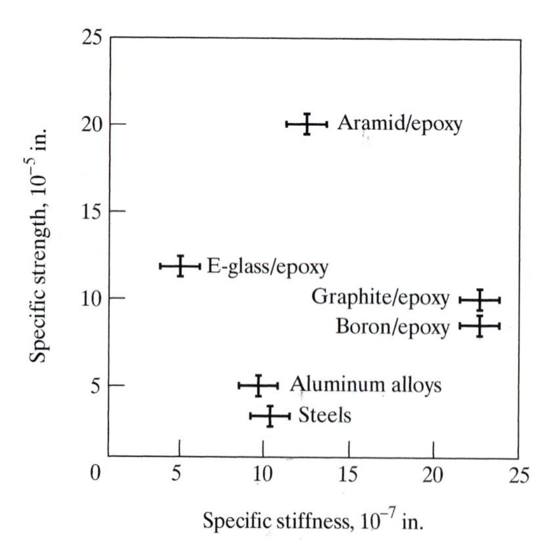

Figure 6–21
Specific strength and stiffness of some cross-ply epoxy matrix composites

The use characteristics of composites depend on the nature of the polymer resin matrix, the nature of the reinforcement, the ratio of resin to reinforcement, and the mode of fabrication. The high-performance composites usually contain over 50% reinforcement. In the remainder of this discussion, we shall describe the common matrix materials, reinforcements, fabrication techniques, and application guidelines.

Matrix Materials

Thermoplastics There are two main types of polymer composite matrix materials, thermoplastic and thermosetting. In 2003, about 90% of the composite market used the thermosetting materials. Up to the late twentieth century, thermoplastic materials were primarily reinforced with chopped glass fibers. The glass reinforcement was usually in the form of short-length fibers (a few millimeters) blended into molding pellets, and parts were injection molded or formed in the conventional techniques. This type of reinforcement does not produce a composite with the types of strengths that can be obtained with the same reinforcements in continuous form. Continuous reinforcement of thermoplastics is a technology that reached commercial importance in the 1990s, but the market is growing as the available reinforcement and matrix products improve.

The technical problem that had to be overcome to make the concept work was to develop the technology to coat continuous reinforcement fibers with a layer of thermoplastic material. The earliest systems drew reinforcement material such as glass through a bath of molten polymer and then wove the polymer-coated filament into cloth or tape for reinforcement of shapes. Tapes of prewoven material could also be dipped in molten polymer for coating, but the wetting problem was more formidable.

If the product to be made is a simple shape such as a cafeteria tray, the thermoplastic-coated reinforcement is placed as a sheet between a matched set of compression molding dies. Heat and pressure are applied, and the part is ejected after it is subjected to a suitable heating and cooling cycle in the mold. Thus the raw material for making a thermoplastic composite is a *prepreg* or previously impregnated cloth that was saturated by the vendor with the matrix material of interest. Making the part from this material involves heating it (with any technique) and then restraining it in the desired shape until it becomes rigid (cools) enough to handle. The term *prepreg* also applies to reinforcement that is impregnated with partially reacted thermosetting matrix resins.

Drawing fibers through molten polymer to coat them is not without problems. Molten plastics have the consistency of honey. Visualize pulling a piece of string out of a jar of honey and trying to get the honey to wet and to coat the string with a thin, uniform coating. Another problem that has had to be reckoned with is the high temperatures needed to get reasonable viscosities out of polymers of interest. For example, some thermoplastics for matrix materials do not develop low enough viscosity for fiber coating until the molten bath is at 650°F (343°C). This high of a temperature produces other problems, such as oxidation of the melt polymer, fume emission, and related problems. In other words, coating fibers from molten baths is not without problems, but it is one of the techniques used to create thermoplastic prepregs for making composites.

A second system that has been developed for coating reinforcements with thermoplastics is to use polymers that can be dissolved in a solvent. The fibers are drawn through a bath of the polymer–solvent solution. The solvent is allowed to volatilize, and the reinforcement coating is reduced to only polymer. This system produces better fiber wetting, but the problem of solvent evaporation must be dealt with. This technique is often used with the amorphous thermoplastics such as polyamide-imide, polysulfone, polyetherimide, and similar materials. One advantage of solution-impregnated prepregs over melt-impregnated prepregs is better conformability and *tack*. The melt-coated reinforcements are normally quite stiff, and this produces problems in making shapes. In our cafeteria tray example, a stiff reinforcement would lie on the open mold like a piece of cardboard; when the mold is closed, the stiff, cardboardlike material may move out of position and an incomplete part may result. Solution-coated materials can have better flexibility as well as some tack that will allow them to conform to a mold and keep their position during the mold closing cycle.

In summary, thermoplastic composites can be made from any of the common reinforcements and from many thermoplastic matrix materials. The most common reinforcements are glass, *carbon fiber*, and *aramid fiber*, and the most common thermoplastics for matrices are polysulfone, polyetherimide, polyamide-imide, polyetherether ketone, polyethersulfone, and polyphenylene sulfide. Composites made from these materials have strength and stiffness approaching that of thermoset composites, and often much better toughness. They can have process economies if used properly. For example, parts can be warm stamped with speeds not possible with thermoset processes that require heating for curing or a long mold cure time. Thermoplastic composites are an emerging technology. Suppliers are still working to make this material easier and more cost effective to work with.

Thermosetting Resins Thermosetting polymer matrices are usually formed from low-viscosity liquids that become cross-linked by combination with a catalyst or by the application of some external form of energy, such as heat or radiation (UV and other types). The earliest composites were made with a phenolic thermosetting matrix. The epoxides followed, then the ureas, the *unsaturated polyesters*, and the silicones, and now there are considerably more. From the usage standpoint, the most important are the first three materials: phenolics, epoxies, and unsaturated polyesters.

Phenolics (PFs) are hard and rigid; they have one of the highest moduli of elasticity of common plastics, and they have good electrical properties. All the normal reinforcements can be used with phenolic resins, but because a major application of these materials is for circuit boards, there are a variety of reinforcements that are specific to needs of the electrical industry. These are called NEMA (National Electrical Manufacturers Association) laminates. There are grades with paper reinforcement, cloth, and glass, and some companies offer grades with aramid fibers as reinforcements. Many automotive brake and clutch pads are molded from phenolics that are reinforced with asbestos, powdered metals, and friction modifiers such as molybdenum disulfide and graphite. Phenolics are extremely useful in machine design in that they are available in standard shapes (rods, plates, strips, and sheets) that can be machined into all sorts of machine components, gears, cams, and structural parts. These laminates have one of the highest compressive strengths of any composite (flatwise). The strength can be in excess of 40 ksi (215 MPa), and they have good stability and machinability.

Phenolic resins are widely used in decorative laminates. A familiar trade name for one of these laminates is Formica®, and they are used for kitchen counters and similar work counters in all sorts of applications. These laminates are layers of paper-type products and decorated paper that are compression laminated into sheet goods. Urea and melamine formaldehydes are similar to phenolic resins in applications and properties, and they are also used for these types of composite laminates. Ureas are the resins used in particle boards that are widely used in furniture and the construction trades.

One useful aspect of phenolic resins in manufacture is that they can be purchased as *B-stage resins*. This means that they will behave as thermoplastics until they are heated to a particular temperature under compression. They then set to final form, and from then on they are thermosetting; they can never be melted again. B-stage resins are only partially catalyzed, and consequently they are only partially cross-linked. The heat and pressure cycle completes the reaction. This means that users can buy phenolic resins in pellet form and mix the pellets with reinforcements of their choosing, and then compression mold the powder–reinforcement blend to make a desired shape.

Epoxies get their name from the epoxy functional group that terminates molecules or that is internal to the structure, cyclically or not. Epoxies are really polyethers, because the monomer units have an ether type of structure with oxygen bonds, R—O—R. The general structure of epoxy types of polymers is shown here:

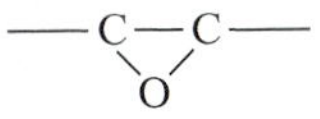

$$\text{—C—C—} \quad (\text{C—C bridged by O})$$

Epoxide functional group

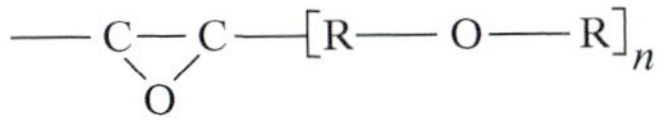

$$\text{—C—C—}[\text{R—O—R}]_n \quad (\text{C—C bridged by O})$$

General formula for an epoxy resin

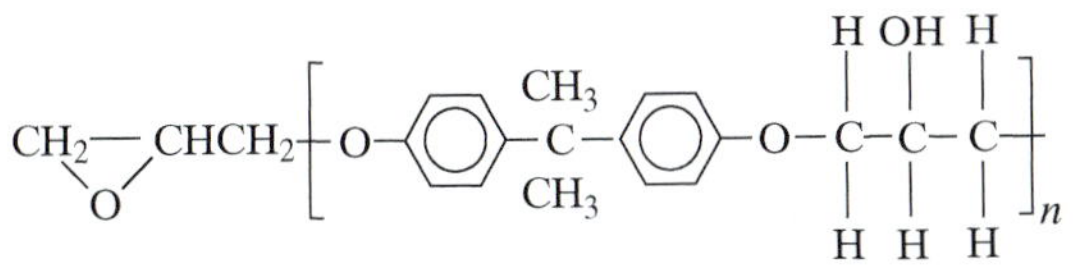

$$CH_2\text{—}CHCH_2\text{—}\left[\text{O—}C_6H_4\text{—}C(CH_3)_2\text{—}C_6H_4\text{—O—}CH_2\text{—}CH(OH)\text{—}CH_2\text{—}\right]_n \quad (CH_2\text{—}CH \text{ bridged by O})$$

Epoxy formed from bisphenol A and epichlorohydrin

Unlike many of the other polymers we have discussed, the chain length of the molecules before cross-linking is relatively short, as short as ten molecules. When cured, these molecules cross-link to form a three-dimensional network, and the catalyst or reacting species becomes part of the structure. This incorporation of the catalyzing agent in the structure is responsible for one of the unique characteristics of epoxies: minimal physical size change on polymerization. Solvents and condensation products are not emitted. The shrink rate can be as low as 0.0001 in./in. (0.01%) of dimension. Low shrink rates make epoxies ideal for adhering to other surfaces and reinforcements. If a material has a strong tendency to shrink on formation, there will be a tendency for its bond to other surfaces to be under a significant shear stress.

Currently, about 25 different grades of epoxy *matrix* resins are commercially available; they differ in molecular structure and the nature of the curing agent. Epoxies used in polymer composites usually have two components, and polymerization commences on mixing.

The actual mechanism of polymerization can be direct linkages of epoxide groups, linkages between epoxide groups and other chain molecules, and epoxy-to-epoxy linkages. Some reactions are caused by catalysis, some are caused by chemical reaction of the mixed species, but the net result is a three-dimensional macromolecule with chemical bonds throughout. The properties of epoxy resins vary with the type of epoxy and the type of curing agent. In fact, it is possible to obtain a fairly wide range of properties with a given resin, depending on the mixture of resin to curing agent and the type of curing agent.

Epoxy resins are probably the most important matrix material for high-performance structural composites. Epoxy is the matrix material that produces the highest strength and stiffness with the stronger reinforcements such as boron and graphite. Its importance in polymer composites is mostly due to its high strength, low viscosity for wetting, and low shrinkage tendencies. There are special grades of epoxy for elevated-temperature service to about 350°F (176°C), but more expensive matrix resins such as polyimides, silicones, and bismaleimides (BMIs) replace epoxies for service temperatures over 350°F (176°C).

Unsaturated Polyesters are styrene–polyester copolymer resins and usually contain inhibitors that allow their storage as liquids for a year or more. Once catalyzed, the resins become rigid in times as short as one minute or as long as several hours. Several other polyester resins are used for reinforced thermosetting plastics (RTPs), RTP composites, including bisphenols, Het acid, and vinyl esters. The last are used for chemical-resistant piping and tanks. They have better corrosion resistance than the general-purpose resin. Bisphenol resins can be modified to impart some resilience to a composite. The Het acid resins are heat stabilized and flame retardant. The vinyl esters are really epoxy based and differ chemically from the other polyesters, but they are usually classified in the polyester resin family. They are used for aggressive chemical environments.

Unsaturated polyester resins are by far the most important materials for general-purpose composite structures and parts. They are the materials used for the familiar fiberglass boat, the Corvette automobile, recreational vehicles, all sorts of storage tanks, pultruded piping, portable toilets, and countless commercial and military applications. These materials are much lower in cost than the epoxies (about $1/lb vs. about $2/lb; $2.2/kg vs. $4.4/kg for the lowest-cost resins). They also have slightly lower strength than the epoxies.

Polyester matrix materials are used with all the reinforcements, but glass is by far the most common. These resins are used for all the manufacturing processes that are used on composites. Big boats are often made by hand lay-up; small boats are made by combining polyester with chopped fiber in a special spray gun, and the hull is formed by simply spraying the mold to the desired hull thickness. Fillers can be added to unsaturated polyesters (UPs) to make a gel coat, which is the smooth, pigmented outer skin on boats, and the like. Tanks and similar structures are made by filament winding or by lay-up techniques for large ones. Pipe and structural shapes are pultruded; fiber reinforcement and resin are coextruded from a die. Parts can be stamped from UP matrix materials when these are supplied as sheet and bulk molding materials. Partially reacted polyester (B stage) comes as a prepreg that is set up by heat when forming is done.

In their simplest form, unsaturated polyesters are supplied as a fairly low-viscosity fluid (similar to cheap maple syrup) that can be clear or pigmented. The resin cures after it is mixed with a catalyst. The catalyst ratio can be as low as 50 milliliters per liter of resin, depending on the degree of mixing and temperature. To make something by hand *lay-up*, the mold surface is coated with a layer of catalyzed resin, the reinforcing cloth is applied to the sticky resin on the mold, the reinforcement is then saturated with more resin, and the part is allowed to cure. The part will set up in anywhere from one hour to overnight (depending on the amount of catalyst added and the temperature). This process description explains why this matrix material is so widely used. It is extremely easy to perform the basic process steps. Low-skill help can be used.

A high degree of technology applies to the design of the reinforcement if you are to optimize a design, but if you are building portable toilets, rigorous calculations on the number of layers and orientation of the reinforcement are not necessary. If a spar for an aircraft is the part under design, calculations will probably be necessary. Whatever the composite application, unsaturated polyesters are almost always the candidates of first choice because they are the lowest cost. If they do not satisfy the design criteria, the more expensive epoxies, polyimides, and so on can be considered.

Silicones are used as a composite matrix resin for special applications. Silicones can withstand service temperatures as high as 600°F (315°C); when used in the form of a low-durometer elastomer, they can have unusual release characteristics. Printing blankets are made from silicones because they will transfer ink completely to other surfaces. Nothing sticks to silicone elastomers. Silicones are also available as rigid thermosets. The high-temperature resistance of these materials is probably the result of the silicon-to-oxygen linkages that make up the backbone of these polymers.

Polyimides, like silicones, are used for special applications, usually high-temperature composites. Polyimide prepregs are available that can be handled and fabricated like other partially polymerized matrix resins. Prepregs can be placed in molds and cured to final form with the application of heat. Service temperatures can be as high as 500°F (260°C). These resins are much more expensive than epoxies and unsaturated polyesters, so their use is usually confined to aerospace and similar applications where there may be more tolerance for the higher cost.

There are no technical limits on the variety of resins that can be used as a matrix for polymer composites. The major criteria for suitability for use are that they have the ability to wet the reinforcement and adhere to the reinforcement. The important matrix materials from the standpoint of commercial availability and desirable properties are unsaturated polyester, epoxies, and phenolics. Other thermosetting matrix materials include polyimides, urea, melamine formaldehydes, furans, and allyls such as *diallyl phthalates* (DAPs). Melamines are widely used with fillers for the familiar unbreakable dinnerware. Ureas are used for *laminating* resins, but the largest tonnage is used for laminating wood products such as plywood. Furans are useful in the chemical process industry for tankage to hold aggressive chemicals. Diallyl phthalate molding resins are often used with glass reinforcing for compression-molded electrical components. They can have better compression properties than the phenolics. Composites for elevated temperature service usually use polyimide, silicone, or BMI matrix resins.

Most of the thermoplastics are available in glass-reinforced form. Glass contents are typically in the range of 10% to 40%. Any thermoplastic can be used, but some of the thermoplastics that are commonly reinforced are polyamides, polycarbonate, polystyrene, acetals, ABS, acrylics, polyethers, polyphenylene oxide, and the fluorocarbons. They are not widely used for matrices with continuous fiber reinforcement because of the problem of reinforcement wetting. If the goal of an application is a high-performance composite, the most suitable matrices will be unsaturated polyester or epoxy.

The common matrices for continuous fiber-reinforced thermoplastic composites include polyether imide (PEI), polyphenylene sulfide (PPS), and thermoplastic polyimides (PIs). If an application involves service temperatures below 200°F (93°C), the most suitable matrices are usually unsaturated polyester or epoxy.

Reinforcements

The spectrum of materials used in polymer composites for reinforcements is shown in Figure 6–22. The first composites were laminates of paper saturated with phenolic resin and compression molded into sheets for electrical applications. Paper reinforcement is still used mostly for phenolic laminates and electrical applications because of favorable heat resistance and electrical insulation characteristics. Rod shapes are made by coiling resin-saturated paper like a roll of wallpaper. Cotton fabrics evolved as the next important reinforcement. Canvas–phenolic composites emerged in the 1930s, and woven cotton fabrics are still widely used as continuous reinforcements in phenolic laminates. A variety of weaves and fiber diameters are used, and these differences translate into different properties in the finished composite. The biggest advantages of paper and cotton reinforcements are low cost and the ease of machining. Hard, inorganic reinforcements such as glass and metal cause excessive tool wear in secondary operations and mating material wear in sliding systems. Paper and cotton are less abrasive reinforcements.

Metals An important technique in the use of metals in polymer composites is the use of metal skeletal structures in the form of honeycomb panels. Beehive pattern honeycombs are often

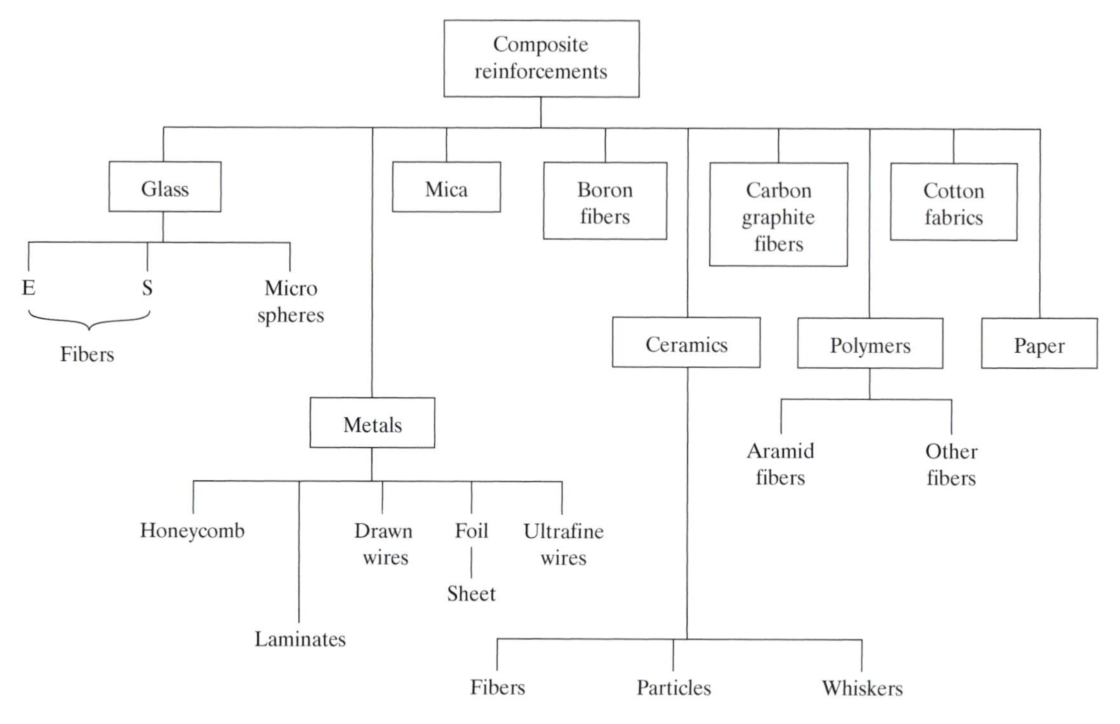

Figure 6–22
Spectrum of reinforcements used in polymer composites; most are available as continuous or discontinuous fibers

made from aluminum in foil thicknesses. These honeycombs form the core of laminates with metals or fabric-reinforced polymers on the facing of the *laminate*. These types of structures have long been important in the aircraft industry, sometimes for structural members of the aircraft, sometimes for interior panels such as doors, seats, headliners, and the like. They can be extremely light for a given section modulus. In fact, the use of metal honeycomb laminate cores may be the most commercially important use of a metal reinforcement in polymer composites. Metal wires or other shapes are not widely used. Metal/plastic/metal laminates have become commercially available. These laminates function much like honeycomb panels, but are easier to manufacture and are less expensive. European automobile manufacturers are developing metal/plastic/metal laminates for autobody panels such as hoods because they may be stamped and deep drawn much like steel sheet.

Asbestos The asbestos that is used as a reinforcement in polymer composites is usually chrysotile asbestos, which is a hydrated magnesium silicate ($3MgO \cdot 2SiO \cdot 2H_2O$). It is a naturally occurring mineral with a fibrous structure that makes it ideal for wetting with low-viscosity resin matrices. Synthetic fibers do not have the leafy structure of these materials. Asbestos can be wet in water and made into fibers that can in turn be woven into fabrics for reinforcements; these can be used either as single-strand filaments or as particles or flakes. They are considered nonflammable and are relatively inert from the standpoint of chemical attack. In the 1970s, asbestos-reinforced phenolics accounted for about 35% of the entire production of polymer composites in the United States (about 3.5 million tons) with much of the usage in automobile brakepads and clutch facings. In the 1980s, a trend started to eliminate asbestos from all products because of a possible link between asbestos and lung cancer. The net result has been a global replacement wherever possible. Glass in chopped fiber form is the usual substitute.

Ceramics Ceramics such as silicon carbide, aluminum oxide, and silicon nitride can be made into small-diameter fibers, *whiskers*, or particles that can be used for polymer composite reinforcements. Whiskers are really single crystals with length-to-diameter ratios up to 10,000. These whiskers can have extremely high tensile strengths, but their length is usually less than 10 mm, which makes them unsuitable for continuous reinforcement. Ceramic reinforcements are not widely used in polymer composites, but they are popular in metal matrix composites.

Polymers In our discussions on polymer crystallinity it was mentioned that currently many research programs are aimed at developing liquid crystal polymers with extremely high strength. A number of these materials, including a number of the olefins, have been used as reinforcements for thermoset and thermoplastic composites. A system of commercial importance is Kevlar®, an aramid fiber with a tensile strength of about 450 ksi (3102 MPa). It is available in continuous fibers, as woven fabrics, and as chopped fiber. These can be used for reinforcement of thermoplastics and thermosets; their main advantages over glass are greater toughness and lighter weight. A 17-ft-long canoe made from a laminate of Kevlar and vinyl ester weighs only 16 lb. A companion polymer reinforcement material is Nomex, a high-temperature nylon. It is not as strong as Kevlar, but it is easier to process and lower in cost.

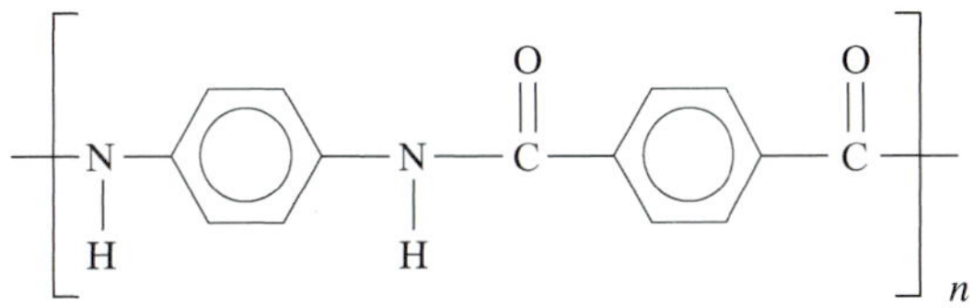

Kevlar

An older form of polymeric reinforcement is polypropylene in the form of continuous fibers for woven reinforcement cloth for unsaturated polyester and epoxy composites. Like Kevlar, the advantage of this type of material over glass is lighter weight and toughness; its strength is comparable to that of glass, but its use as a replacement for glass has not occurred. Polymer reinforcements do not machine like the glass-reinforced composites. The reinforcing fibers tend to melt in sanding operations, and the like. Finally, polymer reinforcements can be porous foams that can be resin impregnated. A common composite technique used in boat building is the use of layers of glass and resin separated by end-grain balsa wood that was saturated in the matrix resin. This system produces a high section modulus and lighter weight than a solid buildup of woven glass layers. End-grain balsa is very expensive, and foamed olefins and other plastics are molded into shapes that are used to replace the end-grain balsa.

Boron Continuous filaments for reinforcement of composites are made by chemical vapor deposition of boron from a boron-rich gas. These *boron fiber* reinforcements have a higher tensile modulus than most other reinforcements, but their high cost has restricted their use to aerospace and military applications.

Carbon–Graphite Amorphous carbon is obtained by heating organic materials, usually in the absence of air. This kind of carbon (carbon black) is used to pigment plastics and to aid vulcanization of rubber. The carbon fibers that are used as reinforcement in polymer composites are obtained by heating precursor fibers of organic materials to very high temperatures in the absence of air and usually under tension. The starting materials are fibers made from rayon, pitch, or polyacrylonitrile (PAN). Pyrolyzing temperatures can range from about 2000°F (1093°C) to as high as 5300°F (2926°C). At the higher temperatures the *fiber* takes on a graphitic structure. Graphite crystals have a hexagonal structure with the basal (base) plane aligned parallel to the fiber axis. When fibers have significant graphitic structure, they can have extremely high strength and modulus. The tensile modulus of elasticity (stiffness) for PAN carbon fibers can be in excess of 100 million psi (758,000 MPa). The highest-stiffness fibers have smoother surfaces than the lower-stiffness fibers, and this means that fibers must be treated to help them bond to the polymer matrix. The lower-stiffness fibers have rougher surfaces and bond better to the matrix. For these reasons it is common practice to use the lower-modulus fibers unless the application absolutely requires the high-modulus fibers.

Carbon fibers (CFs) are grown in diameters smaller than 0.0002 in. (5 μm). They are made into strands for weaving or winding, and they are available as chopped *strand* for use in injection molding resins. The cost of these reinforcements in 2003 ranged from as low as $5/lb ($18/kg) to as high as $200/lb ($440/kg) for the super high-modulus variety. Chopped fibers are the lowest-cost form. Carbon fibers are used where glass reinforcement will not provide the desired stiffness or weight reduction.

Glass The most common reinforcement for polymer composites is glass fibers. The first important structural composites were often improperly termed Fiberglas®, which is a trade name. The acronym *FRP*, for fibrous-glass-reinforced plastic, was established to prevent the misuse of the Fiberglas trade name, and this acronym was replaced by RTP, for *reinforced thermosetting plastic*. The latest acronym to appear in the literature, RP, for *reinforced plastic,* is the most current.

Glass fibers are made by essentially flowing molten glass through tiny holes in dies. Two important types of glass are in wide use for reinforcements: E glass, which is essentially a borosilicate glass named for electrical applications, and S glass (high strength), which is a magnesia/alumina/silica material with higher tensile strength than E glass. Fiber diameters are usually in the range of 0.0002 to 0.001 in. (5 to 25 μm). Both are used for the same types of applications, but the E glass is lower in cost; it can be fabricated at lower temperatures.

Glass reinforcements are available in every imaginable form. The most common forms of glass reinforcement are shown in Figure 6–23. Chopped strand is widely used for reinforcement of thermoplastics and for bulk molding compounds, but the other forms are usually used in large structural composites. Strands usually consist of many individual filaments for use in *filament* winding or in making weaves. Woven glass reinforcements can be obtained in about as many different weaving patterns as are available in clothing. Most are two-dimensional in nature, and new weaving techniques are constantly being investigated. The use of two-dimensional weaves in laminates produces anisotropic strength characteristics. One direction will always be weaker than the others. Making isotropic three-dimensional weaves is a current area of composite research.

Mat cloths are made from randomly intertwined discontinous fibers of moderate length, not unlike a felt. Glass fiber mats require more resin for saturation than do the weaves, but they produce a better surface texture after molding. When bundles of strands are formed into a large, continuous strand, the product is called *roving,* and heavy composites are often made from cloths that are woven from roving.

The bond between the glass reinforcement and the resin matrix is an important part of making a composite that has good mechanical

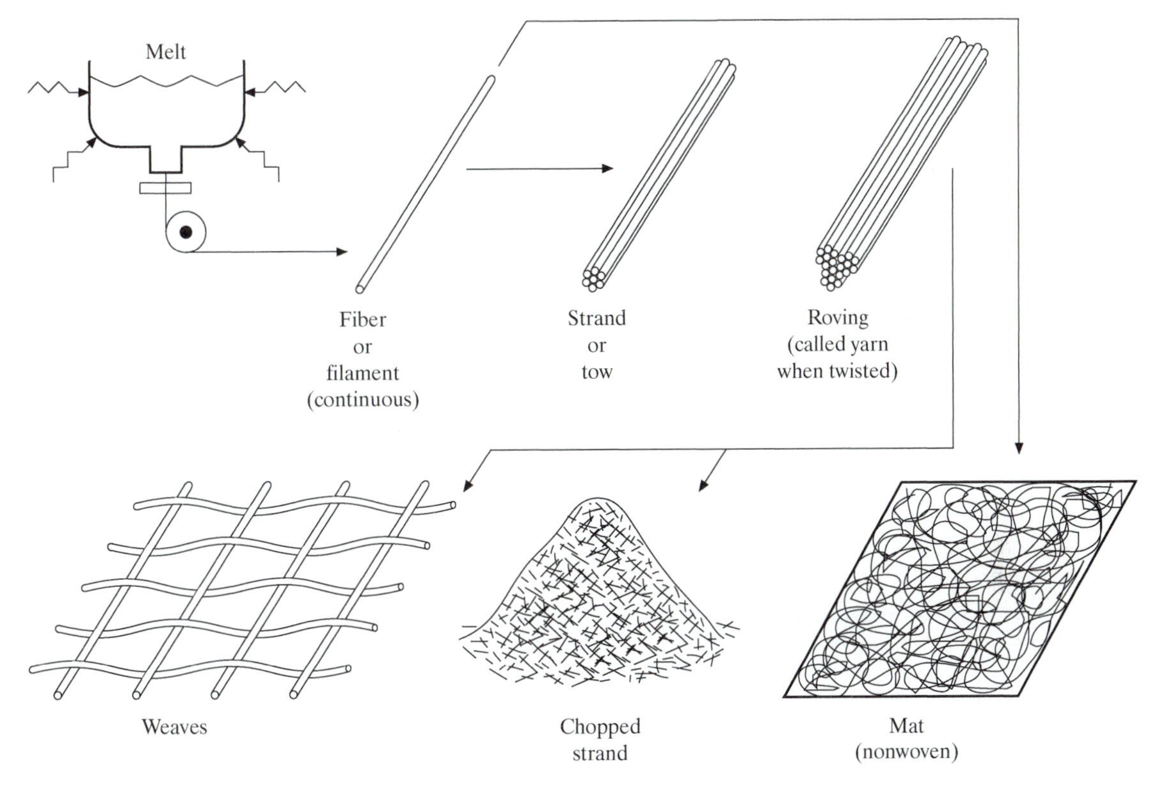

Figure 6–23
Common forms of glass fibers for composite reinforcement

properties. Glasses are often treated by compounds that tend to enhance this adhesion. Silane compounds have the general formula A_3SiB, where A may be a halogen such as chlorine and B is preferably some functional group that will tend to bond to the matrix material. Essentially, the goal in using the silane coupling agents is to have an Si—O type of bond to the glass with the other end of the molecule having an organic molecule that likes to bond to the organic resin matrix. There are other types of coupling agents in commercial use, but from the user's standpoint this can be a design factor if the goal is a high-performance composite. In fact, coupling agents apply to other types of reinforcements as well. The reinforcement cannot distribute operating stresses and strains if it does not bond well with the matrix.

6.4 Composite Fabrication Techniques

The various techniques used to reinforce resins to make a polymer–glass composite are shown in Figures 6–24 and 6–25. Contact molding or *hand lay-up* involves coating a mold or form with a layer of resin; a layer of glass reinforcement is applied, and the reinforcement is thoroughly saturated with resin. The process is repeated until the desired composite thickness is achieved (the maximum thickness is usually

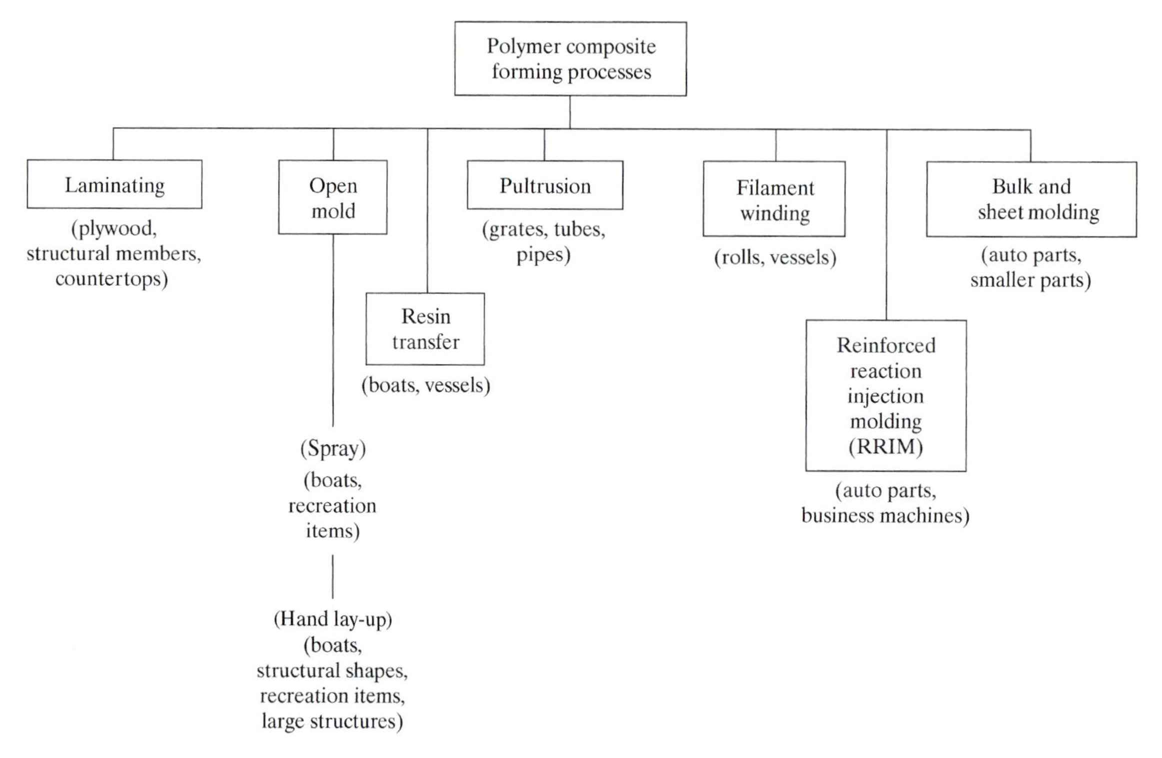

Figure 6–24
Polymer composite forming process

9 mm). The polymer matrix is usually a polyester or epoxide.

Filament winding reinforcing uses special machines to wind the glass reinforcement around a mandrel. The reinforcement is a continuous strand, and the strand is saturated with resin in an in-line bath. The winding pattern can be varied to control the strength characteristics of the composite. This process is widely used for making pipes and tanks to handle chemicals.

Compression molding is similar to the process described previously for unreinforced thermosets, except that special techniques are required to introduce the glass reinforcement into resins that have to be catalyzed and have a limited pot life after catalyzation. In the *sheet molding process*, catalyzed polyester or epoxy resin is kneaded into the glass reinforcement by rollers. Special fillers are added to keep the resin from being tacky, and inhibitors are added to increase the pot life of the catalyzed resin. The finished sheet, called *sheet molding compound* (SMC), consists of resin and reinforcement, and this sheet can be cut to an appropriate size and pressed in a matched mold to make the finished part. The molds are heated to complete the cross-linking of the resin.

A similar product, called *bulk molding compound* (BMC), is produced by adding thickeners to the resin; it is kneaded like dough with chopped fibers to make a compression molding charge that resembles a glob of dough. The heating and pressing are the same as in sheet molding. Both processes can be used for large moldings such as automobile fenders.

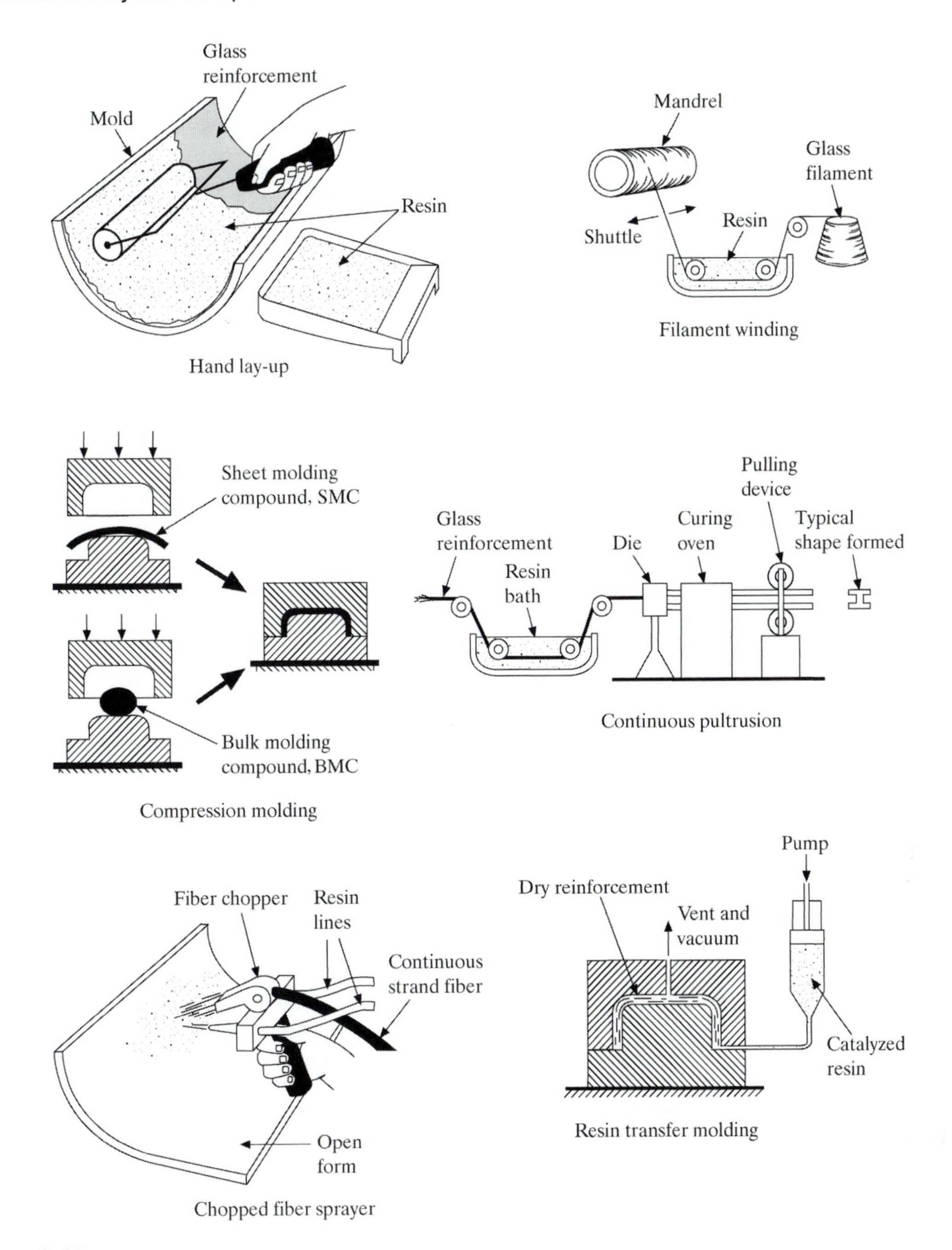

Figure 6–25
Techniques for the fabrication of fiber-reinforced composites

Continuous pultrusion is a process for making glass-reinforced shapes that can be generated by pulling resin-impregnated glass strands through a die. The glass is pulled through a resin bath; it is shaped as it goes through a heated bath, and the resin cross-links in the heated die and combined curing section. Pipes, channels, I-beams, and similar shapes can be generated. Pultrusion structural shapes are frequently used for decking and structural members around corrosive chemical tanks.

Chopped fiber spraying performs the same job as hand lay-up, but it is much faster. Two component resins are mixed in a hand-held gun and sprayed at a mold surface. A chopper is incorporated in the gun. It chops continuous strands of glass into short lengths to act as reinforcement in the composites. This process can be used to make large reinforced composites such as boats, shower stalls, and bathtubs. Chopped fiber reinforcements, however, are not as strong as the hand layups that are reinforced with mat or woven roving.

Resin transfer molding has evolved as a way to speed up contact and to improve the part by having two finished surfaces instead of one. This process requires a close-fitting mold. Glass reinforcement is cut and shaped to the desired thickness in the open mold. The mold is then closed and evacuated, and catalyzed resin is pumped into the bottom of the mold. When the mold is filled, the pump is shut off, the resin line is stopped off, and the part is allowed to cure. This is becoming an important process for the production of large RTP boats. It is replacing hand lay-up.

The use of computers to control filament winding is increasing the complexity of parts that can be wound and the performance characteristics of these parts. For example, computer control can add extra layers of reinforcement or change the pattern in high-stress areas. Advanced systems couple finite-element stress analysis with reinforcement patterns.

Vacuum bag forming (Figure 6–26) is used to shape sheet molding compounds to complex shapes. This process uses atmospheric pressure to do the forming, thus eliminating the high cost of matched metal molds. It is possible to cure the SMC in the vacuum bag rig using temperature-resistant silicone rubbers for the forming bladder, but the more common practice is to use vacuum-bag forming to make a preform and cure the preform in another mold.

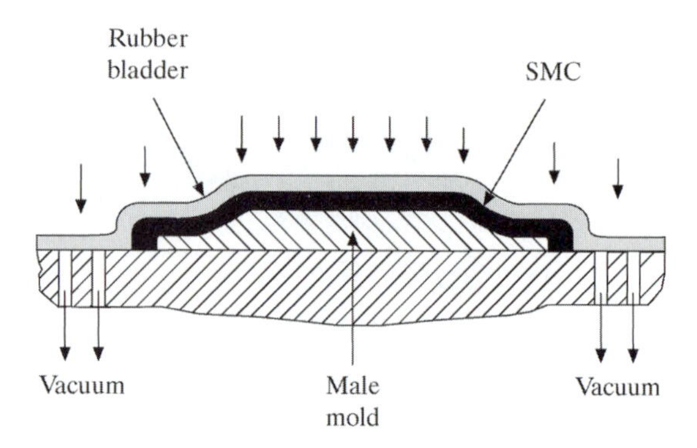

Figure 6–26
Vacuum bag forming

The practice of reinforcing thermoplastic materials means that all the processes that use molding pellets (injection molding and others) can be considered composite fabrication techniques; chopped fibers and particulate reinforcements can be blended into these molding pellets. Filament winding is also done with thermoplastics; a reinforced strip is heated, wrapped on a heated mandrel, and subsequently cooled. Thus just about any thermoplastic or thermosetting plastic fabrication technique can be used to make polymer composites, but the high-performance composites are more likely to be made from thermosetting resins using one of the techniques that were shown in Figure 6–25.

6.5 Application of Polymer Composites

We have discussed what polymer composites are—plastics with some sort of reinforcement to enhance their use characteristics—and how they are formed—by a combination of some matrix and reinforcement—and we have described the various techniques that are used to make usable shapes out of this family of materials. In this section we shall review some of the factors that make polymer composites important engineering materials and present guidelines to help the designer decide if polymer composites should be considered for an application.

Availability

Figure 6–27 is a partial summary of the products that are commercially available in the area of polymer composites. A glance at this illustration will indicate that many thermosetting products are available, but not too many thermoplastic products. Essentially, the important thermoplastic polymer composites are thermoplastic molding resins (usually in pellet form) that contain various volume fractions of reinforcements. These molding resins are available from almost all plastic manufacturers and from many compounders, companies that blend homopolymers and reinforcements. The important reinforcements are chopped glass and graphite fibers. Volume

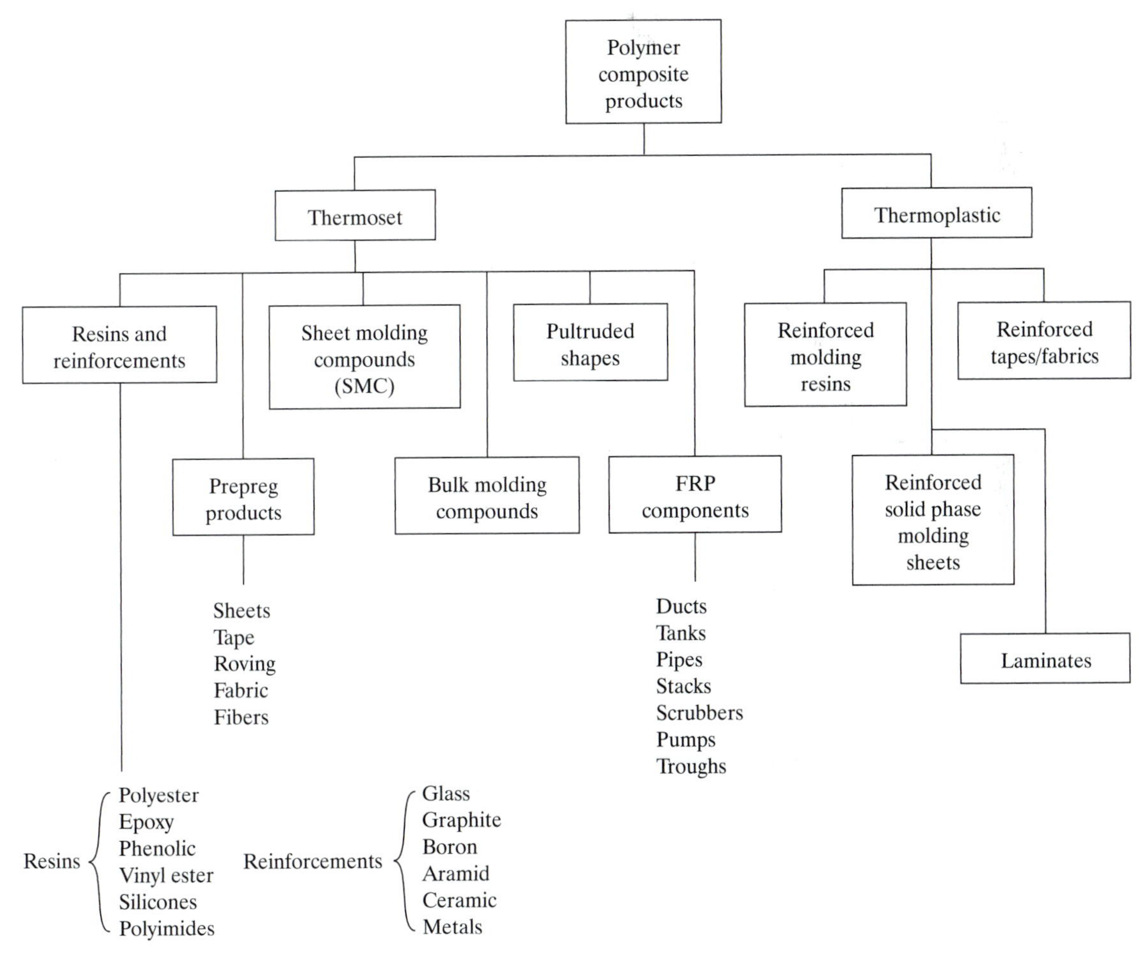

Figure 6–27
Spectrum of commercially available products for forming polymer composites

fractions are usually less than 50%. These materials can simply be ordered from vendor's catalogs, and the vendors will supply processing details.

Thermosetting laminating resins and reinforcements are probably the most important products in the thermosetting category. If a polymer composite is under consideration for some large structure, these are the products that will be used. The user will have to decide on a resin and reinforcement combination, and experienced fabricators can guide the new user through the selection process. Thus from the user's standpoint, there are many possibilities for the use of polymer composites, and there are products available to make just about any type of part or structure that can tolerate the operating limits of polymer resins.

Cost

The decision to use or not to use polymer composites often depends on whether they offer some cost advantage over other candidate materials. Each potential application requires consideration of many factors, but the basic materials used in many polymer composites are low in cost. Glass reinforcements can be low in cost; chopped strand may be only pennies per pound; and woven cloths typically cost a dollar or so per pound. The high-modulus fibers are still expensive, but chopped strand graphite may cost only $10 per pound; aramid fibers can cost $20 per pound, and the ceramic and other very high-modulus fibers may cost $200 per pound. Resins can be as low as $1 per pound (general-purpose polyester) or double or triple digits per pound. The cost of a planned application of polymer composites will depend on the cost of the starting materials and the fabrication costs. The use of polymer composites for a structural member often allows the user to make a structure *monolithic*—one part—and if the same part were made from, for example, metal, the cost comparison should include the cost of the individual parts and their assembly. The corrosion resistance of polymer composites often allows a longer service life. This is also an important cost factor to consider. In general, one of the main reasons for the increasing use of polymer composites is that they are often lower in cost than metals.

Properties

Table 6–1 shows some of the important mechanical properties of thermosetting polymer composites, and Table 6–2 presents similar data on thermoplastic composites. The former table illustrates one of the biggest reasons for using the high-performance polymer composites; they can be stronger and stiffer than high-strength metals, with a weight reduction. The thermoplastic composites do not compete with metals as well, but they fit into the niche between unreinforced thermoplastics and the reinforced thermosets. Sometimes it is desired to replace a metal part with an injection molded plastic, but unreinforced molding resins are simply not strong enough or stiff enough; reinforcing the thermoplastic with chopped glass or carbon fiber may raise the strength enough to allow the replacement to be made and still maintain the economies of injection molding.

The mechanical properties of polymer composites depend on the nature of the matrix and the reinforcement. Boron and graphite fibers produce the stiffest composites. The user must decide whether the additional cost of the high-performance reinforcements is cost effective.

One of the most significant properties of polymer composites is their environmental resistance. It is beyond the scope of this discussion to present details on how each matrix resists the plethora of environments that are likely to be encountered, but corrosion handbooks are available that list this type of information. In general, reinforced thermoplastics will have corrosion characteristics similar to those of the unreinforced matrix material. Continuous reinforced thermosets are somewhat more complicated to deal with. A thermoset polymer composite design for

Table 6–1
Room temperature mechanical properties of polymer composites compared with high-strength lightweight metals (various sources)

Mechanical Property	Boron Epoxy	S-Glass Epoxy	E-Glass Epoxy	E-Glass Polyester
Tensile strength (ksi)	198 (1365)	155 (1068)	70 (482)	50 (344)
Tensile yield strength (ksi)	—	—	—	—
Compressive yield strength (ksi)	255 (1758)	82 (565)	71 (489)	50 (344)
Shear strength (ksi)	9 (62)	—	—	—
Percent elongation	0.7	—	—	—
Tensile modulus (10^6 psi)	31 (214×10^3)	6.4 (44.1×10^3)	4.5 (31.02×10^3)	4.5 (31×10^3)
Density (lb/in.^{3})	0.074 [2.04]	0.066 [1.8]	0.079 [2.2]	0.07 [1.9]
Mechanical Property	**E-Glass Vinyl Ester**	**Carbon Fiber Epoxy (60% CF)**	**Titanium 6A14V**	**Aluminum 7075T6**
Tensile strength (ksi)	55 (379)	80 (303)	145 (1000)	78 (538)
Tensile yield strength (ksi)	—	—	135 (930)	68 (468)
Compressive yield strength (ksi)	—	—	140 (965)	68 (468)
Shear strength (ksi)	—	—	84 (579)	46 (317)
Percent elongation	—	—	6	5
Tensile modulus (10^6 psi)	6.3 (43.4×10^3)	8 (54.9×10^3)	16 (110×10^3)	10.3 (71.1×10^3)
Density (lb/in.^{3})	0.07 [1.9]	0.057 [1.59]	0.168 [4.43]	0.101 [2.8]

Quantities in parentheses are MPa; those in brackets are g/cm^3.

chemical resistance will require design of the laminate. The wetted surfaces are usually resin rich, and special woven fabrics called *veils* are used as the surface for improved appearance. The veil is often backed with a resin-rich mat, and finally the structural reinforcing scheme is determined. A new user may have to rely on fabricator's recommendations in this area. The general-purpose unsaturated polyesters (orthophthalic resins) are not particularly resistant to chemicals, and they are mostly used for structures not intended for chemical contact. Het resin polyesters are

Table 6–2
Room temperature mechanical properties of selected thermoplastics containing 40% chopped glass reinforcement

Property	PA 6/6	PP	PC	PPS	PSF	PES
Tensile strength (ksi) (MPa)	32 (220)	16 (110)	21 (145)	20 (138)	19 (131)	23 (159)
Tensile modulus (10^6 psi) (GPa)	1.9 (13)	1.3 (9)	1.7 (11.7)	2.0 (13.8)	1.7 (11.7)	2.0 (13.8)
Flexural strength (ksi) (MPa)	40 (275)	19 (131)	26 (180)	30 (202)	25 (172)	31 (214)
Flexural modulus (10^6 psi) (GPa)	1.7 (117)	0.9 (6.2)	1.4 (9.6)	1.6 (11)	1.2 (8.2)	1.6 (11)
Compressive strength (ksi) (MPa)	23 (158)	13 (89)	22 (151)	25 (172)	24 (165)	22 (151)
Izod impact (notched) (ft-lb/in.) (J/m)	2.6 (138)	2.0 (106)	2.2 (116)	1.4 (74)	1.6 (85)	1.5 (80)
Heat distortion temperature at 264 psi (°F) (°C)	480 (249)	300 (149)	300 (149)	500 (260)	365 (185)	420 (215)

resistant to many oxidizing media such as acids, but they are not resistant to alkalines and many solvents. They are flame retardant. Vinyl esters are probably the most popular matrix resins for chemical service, but they have a temperature limit of about 250°F (120°C). Furan matrix resins are extremely chemically resistant, but they have poorer fabricability compared to the previously mentioned matrix resins. A factor to consider in the use of unsaturated polyester composites is their susceptibility to surface degradation in sunlight and outdoors and in water. Photo degeneration is caused by UV and other wavelengths of light, which cause chemical reactions in polymer structures. Free radicals are broken, bonds are broken, complex chemical reactions can occur that essentially turn upper layers into a powdering substance that is eroded by weather. The manifestation is usually called "fiber bloom." In glass-reinforced UP composites, the reinforcement starts to show and glass fibers appear in a dull roughened surface. This surface was glossy and smooth when put in service. The damage is usually only 10 μm or so per year, but after 10 years or so of exposure to sunlight, the topsides of the fiberglass boats look like they need a paint job. The damage is usually only cosmetic, but it is still undesirable. Additives are the usual way of slowing this form of degradation, but they do not stop it.

Water attack of UP composites is most prevalent in boats. Water can diffuse through the gel coat and form blisters that are essentially filled with stagnant water that increases in volume when microbiological organisms grow in it. The phenomenon is referred to as osmotic blistering—boat pox to us victims. The mechanism is not known, but the solution is to coat wetted surfaces with a barrier layer of epoxy (usually five to seven layers of epoxy primer). The lack of a solution to osmotic blistering makes unsaturated polyester a potential problem material for boats that will remain in the water for extended periods of time.

As is the case with all material systems, it is the responsibility of the user to decide on the properties that are critical to an application and then match these properties to candidate materials. There are many tabulations of properties of polymer composites, but when it comes to the properties of a specific composite laminate, the user should use data on the specific composite and not generic properties. Each composite structure can be unique; properties depend on the nature of the resin, the cure, and the reinforcement in that composite.

Applications

Where are polymer composites used and where should they be used? About 50% of the total usage of composites in the United States is in the construction and transport industries. The other big users are marine, electrical, military, leisure, and industrial markets. One current big user of FRP is in the marine industry; some 90% of all noncommercial craft are made from unsaturated polyester laminates. Without a doubt this application is the most noteworthy testimonial to the performance of these materials in highly stressed structures. FRP hulls over 25 years old have been cut up and tested for degradation of mechanical properties over this time period. The changes measured were insignificant. The famous testimonial in the transportation industry is the fact that Corvette types of automobiles are still in use and still bring large resale prices after 25 years. Many other testimonials show that polymer composites are suitable for applications requiring long service lives. Many parts on current automobiles are made from reinforced composites, and more parts are converted from metals each year. In the agricultural industry the corrosion resistance and ruggedness of composites are prompting their use for all sorts of structural parts on tractors and similar farm machinery. The appliance industry has used composites for many years for parts that see wet/chemical conditions, air-conditioner parts, dishwasher pumps, laundry tubs, and the like. Countless parts on present-day aircraft are

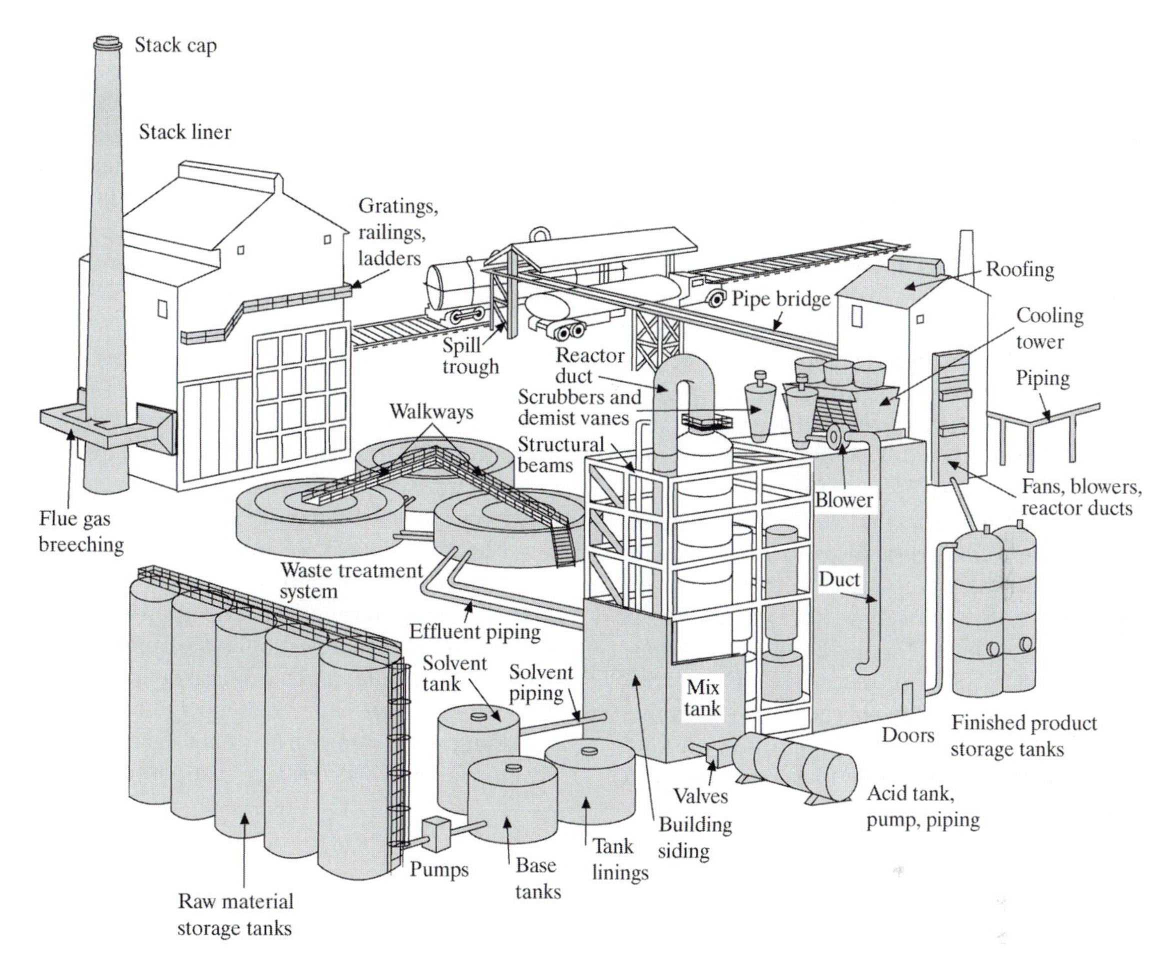

Figure 6–28
Uses of FRP in the chemical process industry
Source: Courtesy of Ashland Chemical Company.

made from composites. Figure 6–28 shows where polymer composites are used in the chemical process industries: piping, roofing, siding, silos, tanks, and all types of structures. Polymer composites are currently a hot item in the construction and housing industry in one-piece showers and tubs, saunas, spas, pools, skylights, and sinks. Finally, the leisure industry is using polymer composites, even the most advanced composites, for an incredible variety of applications. Lightweight kayaks and canoes are being made from aramid composites; tennis rackets, skis, and golf clubs are made from graphite- and even boron-reinforced composites; and off-road vehicles are invariably made from polymer composites, as are jet skis and snowmobiles. We could go on, but it should be clear that these materials are widely used and that they will be used in the future for all types of structures in all sorts of industries.

6.6 Process Specification

We have just described about 30 processes that are used to form and shape plastics and polymer composites. How does a designer pick a process to shape the plastic? The short answer is experience. Many process details must be considered to make an informed decision. We have only introduced newcomers to the more generic plastic-forming processes. We do not have space to discuss these processes in detail. Our recommendation to newcomers in the plastics field is to select a process through discussions with plastic fabricators or use a published standard if one applies (ASTM, MIL, etc.).

We did not list the "universal forming process" in our discussion of forming plastics and composites, but the shaping process that any user can specify is machining from a stock shape. Many thermoplastics, some thermosetting plastics, and some composites are available as stock shapes (Table 6–3). This may be the best forming process for prototypes and for one-of-a-kind parts. An important factor to keep in mind in using stock shapes for prototypes is that the surface properties, and possibly the mechanical properties, will be different if the part is converted to a molding process. This is especially true for structural and wear parts. Molded surfaces of reinforced plastic parts are resin rich; this resin-rich surface and the mold surface finish will not be present on machined parts. This could affect properties. If you want to perform simulated service tests on prototype parts, you should use molded parts if the end part will be molded.

The factors that pertain to selection of forming processes other than machining are illustrated in Figure 6–29. This looks very complicated; it is. It is very difficult to specify a forming process for plastic and polymer composite parts. The first step is to decide on the type of plastic that meets your property needs. The part size, shape, cost, and quantity often determine the specific process or at least narrow process selection to a very few candidates. For example, if you are making one 5000-liter vessel, you immediately limit process selection to the few processes that can be used for big hollow containers (resin transfer of RTP, hand

Table 6–3
Readily available plastics

General-Purpose Plastics	
Polyethylene (HD, LD, UHMW) (PE)	Polymethyl methacrylate (PMMA)
Polypropylene (PP)	Vinylidene fluoride (PVF)
Polyvinyl chloride (PVC)	Polystyrene (PS)
Polyester (PETG)	Acrylonitrile butadiene styrene (ABS)
Engineering Plastics	
Nylon (PA)	Fluorocarbons (PTFE, FEP, PFA, ECTFE)
Acetal (POM)	Polyphenylene oxide (PPO)
Polycarbonate (PC)	Polyetherether ketone (PEEK)
Polyimide (PI)	Polysulfone (PS)
Polyamide-imide (PAI)	
Polyphenylene sulfide (PPS)	
Phenolic (PF)	
Casting Resins	
Polyester (UP)	Silicone (SI)
Epoxy (EP)	Polyurethane (PUR)

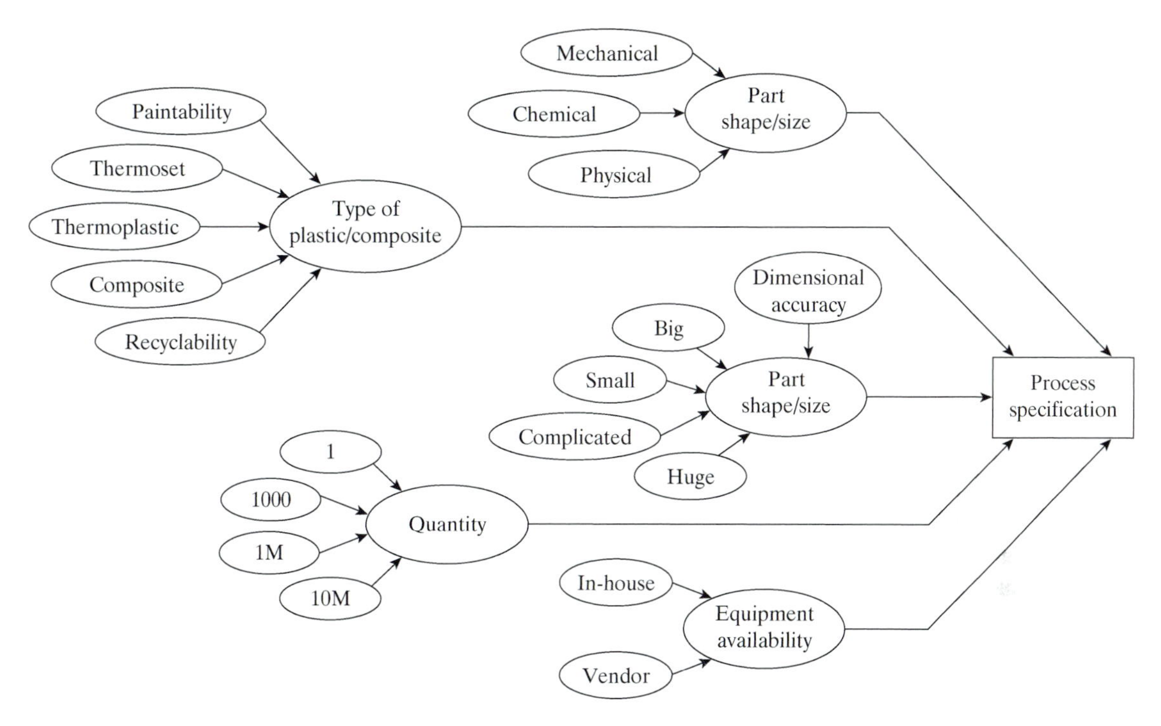

Figure 6–29
Considerations in plastic process specification

lay-up of RTP). If you are designing an automotive part, you may need a high-production process such as injection molding. Here are some process selection considerations that apply to all plastics.

1. Chopped fiber and particulate reinforcements reduce injection moldability.
2. Continuous reinforcement requires the use of thermoset and composite processes.
3. Molded surfaces may have different properties than bulk material.
4. Molded surfaces can contain cosmetic challenges (knit lines, porosity, fiber bloom, etc.).
5. Molded surfaces take on the surface texture of the mold surface (including defects).
6. Properties can be affected by fabrication process; if properties are very important, the fabrication process must be thoroughly specified.
7. Molding resins often require pretreatment such as drying before they are used. These needs add cost and should be considered.
8. Wherever possible, use consensus or published standards for the forming process.
9. Consider the *recyclability* and disposal of the process scrap.

Essentially the only polymer forming process that can be easily specified is machining. All of the others probably need specification of the process details. Most companies in the United States specify the material for plastic parts by generic material and trade name, and then list a process specification that covers process details:

> Material: High-impact polystyrene (Grade A, 300 Ajax Corp., or Sanyu, Vespil Corp.) See process specification X457 for molding details.

The drawing should show approved suppliers if the part is to be made by an outside vendor. Invariably, plastic fabricators will have documented process specifications for every plastic that they work with. Consistently reliable plastic parts require detailed process specifications (molding temperature, pressure, drying, dwell time, mold release, gating, runnering, venting, mold preheat, etc.). Process details affect use properties more than they do in metals and some other material systems.

6.7 Recycling of Plastics

A major concern about the use of plastics for packaging and disposable items is the environmental effect of these materials lasting for centuries in landfills or other places of disposal. Recycling is being promoted worldwide as the way to address plastic items. Many municipalities are establishing recycling programs, and plastic bottles are collected in many U.S. cities. Typically less than 5% of the plastic in municipal waste was recycled, although papers and plastics make up more than one-third of all solid waste. About 20% of paper is recycled, and many metals have very high recycling rates. More than 60% of all the aluminum cans used in the United States in 2003 were recycled, and recycled material amounted to about one-third of the use of aluminum in the United States. Recycling of metals is much more developed than that of plastics. Scrap metal facilities have been in place for decades, and significant amounts of the U.S. usage of metals come from recycled scrap: 40% for copper, 40% for steel, 45% for nickel, 65% for lead, and, of course, the overall winner, gold, which nobody knowingly throws away. It is estimated that more than 95% of all the gold that has been extracted from the earth's crust is still in use.

There are two significant reasons for the poor plastics recycling record: (1) some plastics are thermosetting and cannot be recycled (they cannot be remelted), and (2) there are thousands of grades of thermoplastic materials, and they cannot be mixed when they are remelted. In 1988 the U.S. Society of the Plastics Industry (SPI) established voluntary guidelines to code plastic containers so that they can be easily sorted for recycling. This is the first attempt to solve the second problem mentioned.

There are a great many types of plastics, but almost 90% of the volume of thermoplastics manufactured is made up of only six: polyethylene (high and low density), polystyrene, polyvinyl chloride, polypropylene, and polyethylene terephthalate. For containers, a similar situation exists. About 50% of containers are high-density polyethylene, about 25% are PET (polyethylene terephthalate), and 5% to 10% are from each of the following: polyvinyl chloride, polypropylene, low-density polyethylene, and polystyrene. All other plastics amount to only about 5% to 10% of the total tonnage. In other words, if six plastics could be recycled, it would mean that about 90% of all plastic containers made could be recycled.

Many other factors affect the plastics recycling picture. A limited number of facilities can satisfactorily clean and regrind contaminated materials; some coatings on containers can render them unrecyclable; some containers are composed of several grades of plastics; there are stringent government regulations for levels of contaminants that are allowed when a recycled plastic is used again for a food application, and the like. Notwithstanding these problems, the initiative of the SPI in coding plastics for their recyclability is a major milestone in the goal of universal recycling of thermoplastics. Most of the United States have adopted legislation requiring the use of the SPI code on bottles and containers.

The system is very simple. The SPI is recommending that all 8-oz and larger containers and all 16-oz and larger bottles (up to and including 5/gal) have a symbol molded on the bottom to indicate the type of plastic that they were made from. The symbols established for this purpose are shown in Figure 6–30. A triangular chasing arrow surrounds a number, and there is an

Figure 6–30
SPI recycling symbols. Plastic containers must have one of these symbols on the bottom if they are to be sorted for recycling.

acronym for various plastics under the triangle. The PETE acronym indicates that the material is polyethylene terephthalate; HDPE indicates high-density polyethylene; V indicates polyvinyl chloride; LDPE indicates low-density polyethylene; PP indicates polypropylene; PS indicates polystyrene; and "other" means any plastic that is not one of the above.

In 2003, the SPI system is still preferred in the United States for bottles and rigid containers. Many manufacturers have taken to marking all of their engineering-plastic components for future recycling, waste recovery, reuse, and disposal. The most commonly accepted protocol for marking parts is published in ISO 11469 (International Standards Organization) and ISO 1043 parts 1–4. Generally, parts weighing 50 g or more are marked with a standardized set of symbols bracketed by the punctuation marks ">" and "<." The marking is typically molded into a nonfunctional or nonaesthetic surface of the part. The identification code may also be embossed or indelibly printed on the surface. Although quite complete, the use of these ISO symbols can be complicated to use or decipher. To fully use this labeling system, it is imperative to obtain copies of the ISO specifications mentioned above. Examples of the symbols are shown in Table 6–4. Symbols for the fillers are shown in Table 6–5. Additional symbols for other modifiers are shown in Table 6–6.

There is still significant controversy over plastics recycling. Over 70% of all plastic waste is associated with food packaging. However, recycled plastics are not usually used for food packaging. Ultimately, there is a mismatch in the supply and demand for postconsumer plastic resins—often resulting in excess recycled resin. It is now becoming common practice in Europe to convert recycled plastic into fuel energy via incineration. In the future, there will probably be a balance between recycling and energy recovery.

Table 6–4
Examples of the ISO identification system for plastics

Composition	Example Material	Identification Symbol
Single constituent	Polybutylene terephthalate	>PBT<
Polymer blend or alloy	Polycarbonate and acrylonitrile butadiene styrene where polycarbonate is the main constituent	>PC+ABS<
Fillers or reinforcing additives	Polypropylene containing 30% mineral powder	>PP-MD30<
	Polypropylene containing 15% mineral powder and 25% glass fiber	>PP-(GF25+MD15)<
Plasticizers	Polyvinyl chloride plasticized with dibutyl phthalate	>PVC-P(DBP)<
Flame retardant	Polyamide 66 with flame retardant	>PA66-FR<
Other additives	High-impact polystyrene	>PS-HI<

Table 6–5
Partial list of ISO filler symbols (ISO 1043-2)

Symbol	Material	Symbol	Form
B	Boron	B	Beads, spheres
C	Carbon	C	Chips, cuttings
E	Clay	D	Powder
G	Glass	F	Fiber
K	Calcium carbonate	G	Ground
L	Cellulose	H	Whisker
M	Mineral	K	Knitted fabric
Mx	Metal, where x is chemical symbol for the metal	L	Layer
P	Mica	M	Mat (thick)
Q	Silicon dioxide	N	Nonwoven fabric (thin)
R	Aramid	P	Paper
S	Synthetic organic	R	Roving
T	Talc	S	Scale, flake
W	Wood pulp	T	Cord
Z	Others	W	Woven fabric

Table 6–6
Selected ISO symbols for special characteristics

Symbol	Meaning
B	Block
C	Chlorinated
D	Density
H	High
I	Impact
L	Linear
L	Low
M	Medium
P	Plasticized
U	Unplasticized
X	Cross-linked

Summary

This chapter has intended to show how plastics are fabricated into shapes. We consolidated a former chapter on composites into this chapter because a good portion of the technology of composites is how to make them. Recycling was included because most forming processes produce scrap, and recycling of scrap and discarded parts is an important factor in our world of limited resources and degrading environment. The following are some parting thoughts on the chapter subjects:

- Injection molding requires significant equipment (molding machine and molds) and long lead times to make tools.
- Thermoforming is one of the lowest-cost forming processes from the standpoint of tools and equipment.
- Adding chopped fiber reinforcements to thermoplastics can have a significant effect on moldability.
- Free-form prototyping processes are powerful tools for conceptualizing designs.
- The scrap from thermoset forming processes is not recyclable.
- Foaming is an excellent tool for reducing material costs.
- Molded surfaces may contain cosmetic defects that must be dealt with by painting and other extra-cost techniques.
- Casting is often a low-cost fabrication process for limited numbers of parts.
- Vulcanizing is usually the process of choice for rubbers, but some rubbers (e.g., thermoplastic elastomers, or TPEs) are injection moldable.
- Aramid reinforcement usually produces the highest tensile properties in a composite; boron and graphite produce the stiffest composites.

- Continuous composite reinforcing usually requires design (reinforcement loading, direction, layers, matrix resin, etc.).
- Glass/polyester are the lowest-cost thermoset composites, followed by glass/epoxy. Aramid and carbon reinforced composites are considerably more expensive.
- Reinforced composites can be abrasive to tools used in postforming operations.
- Government building codes may be applicable to large composite structures such as tanks and piping.
- Safety factors on composite structural members are often higher than those for traditional building materials. Safety factors can be as high as ten in tension and five in buckling.
- It is common practice to specify nondestructive testing of large composite structures to ensure integrity.
- Reinforcing of thermoplastics can affect their recyclability.
- Molding processes for recyclable plastics should include accommodations for SPI or other recycling codes.

Overall, forming of plastics into shapes is more "net shape" than for other materials. Hardly any metals can be formed into a finished part with absolutely no secondary operations. This is not the case with moldable plastics. Most are formed to final shape in one of the processes described in this chapter. However, to obtain parts that are usable as formed, it is necessary to be meticulous in process specification and control. We conclude this chapter with an admonition to investigate plastic fabrication processes thoroughly before committing to a process. We have seen costs on injection molded parts go awry when the parts had to be painted to hide flow lines. Make sure that a fabrication process gives you the properties that you need.

Critical Concepts

- Composites are combinations of engineering materials whose properties are intentionally different from the properties of the materials forming the composite.
- Composites usually require more fabrication steps than monolithic materials.
- The most "available" composites use thermoplastic or thermosetting plastics as a key component.
- Injection molding is usually the lowest-cost plastic fabrication process, but tooling can be costly.
- Consider recyclability in selecting a plastic and fabrication process. Thermosets are not recyclable.
- It is the designer's responsibility to specify a fabrication process when selecting a plastic.

Terms You Should Remember

composite
matrix
resin
reinforcements
FRP
laminate
aramid fiber
unsaturated polyesters
specific strength
specific stiffness
filament
filament winding
sheet molding compound
bulk molding compound
B-stage resin
catalyst
pultrusion
lay-up
reaction injection molding
prepreg
silicones
diallyl phthalate
whiskers
boron fiber
carbon fiber
fiber
strand
mat
monolithic
blow molding

calendering
coinjection molding
free-form fabrication
film blowing
extrusion
compression molding
transfer molding
polymer casting
vulcanization
laminating
spray molding
recyclability

Case History

Material for a Reciprocating Slide

A machine design called for a slide capable of carrying a load of 20 lb, reciprocating with a stroke length of 10 mm, and a frequency of 300 cycles per minute. The slide also had to have a wear rate that produces sideways movement in the sliding member of less than 0.1 mm in 10^8 cycles. Not a problem—we thought. Ball bearings never wear. We supported the slide on small diameter ball bearings, and they rolled on hardened steel rails. We built a prototype and life tested it in the laboratory. Severe track wear occured after less than 10^6 cycles; this even occurred with grease lubrication on the rails. Next, we made several slides from self-lubrication plastics, like PTFE-filled acetal, and planned to run them against hardened steel. We machined sliding members from several thermoplastics and none had the dimensional stability that we needed for the application. Next, we tried a composite: Cloth-reinforced phenolic, machined from compression-molded flat stock. The composite fabric was oriented flatwise against the hardened steel counterface, and the system was lubricated with light mineral oil. The system met our wear requirements in laboratory testing. Our final selection was a flat-pad bearing of reinforced phenolic sliding against hardened steel.

Fortunately, the phenolic composite was also suitable for this application from the availability and fabrication standpoint. Only five machines were to be built, so the tooling costs for any molding process could probably not be justified. In addition, the composite that we selected was readily available from suppliers of stock plastic shapes. The material machined easily, and the dimensional stability met design requirements. Ease of fabrication, availability, and freedom from tooling costs were key selection factors.

Questions

Section 6.1

1. List the tool requirements for injection molding plastic spoons.
2. Describe how you would thermoform a container for eggs.
3. List an item in your home that was made by each of the following processes:
   (a) calendering
   (b) blow molding
   (c) rotational molding
   (d) extrustion
   (e) film blowing
   (f) injection molding
4. Detail the process steps used in making the rear bumper on your car.
5. What does the feedstock for injection molding look like?

Section 6.2

6. Why is it that thermosetting plastics cannot be remelted?
7. Name ten thermosetting plastics.
8. Name two rubbers that do not need vulcanization.
9. How is furniture cushioning made?
10. Cite an example in which casting is the preferred forming process. Why?

11. Describe the difference between transfer molding and RIM.

12. Why is the cycle time for compression much longer than for injection molding?

Section 6.3

13. What is the most widely used composite matrix?

14. What is the reinforcer in a liquid crystal polymer?

15. Calculate the specific stiffness of nylon: $E = 1 \times 10^6$ psi, specific gravity = 1.37.

16. Explain the use of specific strength and stiffness in material selection.

17. How is the matrix polymer introduced in thermoplastic composites?

18. What is an advantage of thermoplastic composites over thermosetting composites?

19. Name three common thermoplastic polymer composite matrices.

20. How would you make a pipe from a thermoplastic composite?

Section 6.4

21. How is a hand lay-up made?

22. What is an advantage of thermoset composites over thermoplastic composites?

23. What is a friction material? What kind of composite is it?

24. Why do epoxies adhere better than other resins?

25. What is an advantage of unsaturated polyesters over epoxies as a composite matrix?

26. What are B-stage resins?

27. When would you use a silicone matrix in a composite?

28. What is the advantage of carbon–graphite as a composite reinforcement?

29. Why is glass the most used reinforcement for composites?

30. What is filament winding?

Section 6.5

31. How would you specify a composite pipeline?

32. Compare the costs of glass, carbon, and boron composite reinforcements.

33. Compare the strength characteristics of continuous reinforced UP and chopped strand reinforced UP.

34. Which polymer matrix is most used for chemical resistance?

Section 6.6

35. Write a specification for a gasoline can.

36. Cite ten process variables in the injection molding of a plastic spoon.

Section 6.7

37. What is the recycling code for polyvinyl chloride? For Teflon? For 20% glass-filled nylon 66?

To Dig Deeper

Advanced Composites III, Proceedings of the 3rd Annual Conference on Composites. Metals Park, OH: ASM International, 1987.

Advanced Composites. Published bimonthly.

Agarwal, B. D., and L. J. Broutman. *Analysis and Performance of Fiber Composites*. New York: John Wiley & Sons, 1980.

Bakis, C. E., Ed. *Composite Materials, Volume 14*. West Conshohocken, PA: ASTM International, 2003.

Beland, S. *High Performance Themoplastic Resins and Their Composites*. Park Ridge, NJ: Noyes Corp., 1991.

Brydson, J. *Plastic Materials*, 7th ed. London: Butterworth-Heinemann, 1999.

Delmonte, J. *Technology of Carbon and Graphite Fiber Composites*. New York: Van Nostrand Reinhold Co., 1981.

Engineered Materials Handbook, Volume I, Composites; Volume 2, Plastics. Materials Park, OH: ASM International, 1987.

Goodman, S. H., Ed. *Handbook of Thermoset Plastics*. Park Ridge, NJ: Noyes Publications, 1986.

Jang, B. Z. *Advanced Polymer Composites: Principles and Applications*. Materials Park, OH: ASTM International, 1994.

Lubin, G., Ed. *Handbook of Fiberglass and Advanced Plastics Composites*. New York: Van Nostrand Reinhold Co., 1969.

Modern Plastics Encyclopedia Issue. *Modern Plastics*. Published yearly.

Matthews, F. L., and F. L. Rawlings. *Composite Materials Engineering and Science*. Boca Raton, FL: CRC Press LLC, 2000.

Miracle, D. B., and S. L. Donaldson, Ed. *ASM Handbook, Vol. 21*: *Composites*. Materials Park, OH: ASM International 2001.

Muccio, E. A. *Plastics Processing Technology*. Materials Park, OH: ASM International, 1994.

Slivka, D. C., T. R. Steadman, and V. Bachman. *High Performance Fibers*. Columbus, OH: Battelle, Columbus Div., 1987.

Strong, A. B. *Plastics: Materials and Processing*, 2nd ed. Upper Saddle River, NJ: Prentice Hall, 2000.

The Composite Materials Handbook, MIL 17, West Conshohocken, PA: ASTM International, 2002.

Woishnis, W. A., Ed. *Engineering Plastics and Composites*, 2nd ed. Materials Park, OH: ASM International, 1993.

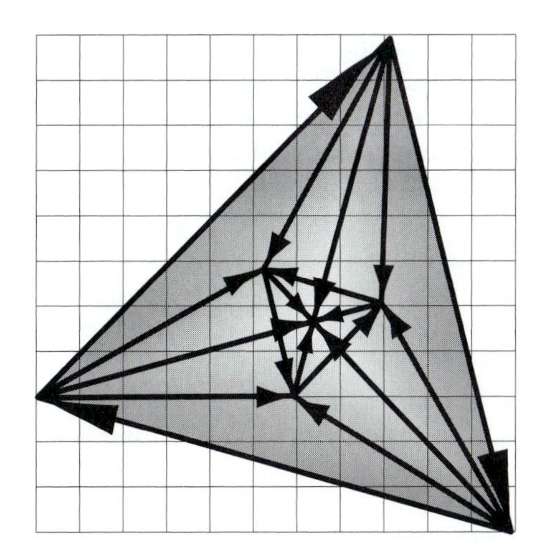

CHAPTER 7

Selection of Plastic/Polymeric Materials

Chapter Goals

1. An understanding of how to select an appropriate plastic for an application.
2. An understanding of adhesives and how to select them.
3. An understanding of paints and how to select them.

Rationale

In previous chapters, our purpose was to give you, the reader, an understanding of the nature, chemistry, properties, and fabrication of plastics and composites. Now it is time to use your understanding of plastics to select a specific plastic for an application; this is the bottom line.

This chapter presents guidelines for using plastic for mechanical components, for applications for which chemical resistance is a consideration, for polymeric coatings, and for adhesives. These kinds of applications are essentially inescapable in engineering. Certainly, as a design engineer, you will be in situations in which you need a plastic to carry a sustained or cyclic load. What are the candidates? What are key selection factors? Similarly, most product design situations require the use of plastics for applications in which they may be subjected to environmental factors that can cause degradation of a plastic, for instance, outdoor exposure. Sunlight is one of the worst enemies of many plastics; you need to know which plastics are candidates for outdoor service.

Of course, designers will encounter protective coatings and adhesives in their career assignments. This chapter presents guidelines on how to pick the right paint and the right adhesive for an application. If you are training to be a software engineer and will never design a machine or build a highway, this selection information will help you to survive home ownership. Home improvement projects invariably require selection of polymer coatings (paints), plastic floor coverings (tile and carpeting), adhesives (for woods, countertops, roofing, etc.), and plastic structural members (in the form of shelves, containers, stools, mailboxes, siding, windows, decking, etc.). Picking the right plastic is indeed important, and its importance increases each year as more and more designs are converted.

7.1 Methodology of Selection

Plastics are used for very different applications, everything from packaging, to artificial arteries, to clothing, to machine parts, to minesweeper ships. Packaging and clothing are huge users of plastics from the tonnage standpoint, but we will not discuss selection for these applications because this text is primarily aimed at a readership of designers and engineers. Most designers concern themselves with machine parts and products—durable goods—made from plastics. How does one address the issue of picking one or two candidate plastics from the 50 or so that we discussed in preceding chapters? We will discuss material selection in more detail at the end of this text, but at this point we will discuss how to select plastics for applications most commonly assigned to plastics: structural parts, bearings and wear parts, products for which environmental resistance is important, and applications in which the electrical or thermal properties of plastics are a major factor.

Essentially, the methodology of selection involves making a list of the properties that you must have for your intended application, then the properties that you would like to have but that are not mandatory. These "must" and "want" properties are then matched with the properties available with the various types of plastics. Figure 2–1 lists the various physical, chemical, mechanical, and dimensional considerations that are reviewed in material selection. This property list can serve as your checklist of properties to consider. Scan the list and check off the properties that you feel are musts and wants. Then go to material data sheets to match your list of desired properties with available properties.

Many property data compilations on plastics are available. The most complete (and expensive) are those on CD-ROM or the Internet that are updated regularly (Figure 7–1). Some databases allow direct uploading of property data to modeling and computer-aided design (CAD) software applications. Many plastic suppliers have their own CD-ROM or Internet databases that can be obtained free on request. They replace the traditional supplier catalogs. We will supply some abbreviated property data in this chapter, but an essential part of the selection process is obtaining reliable property information. The supplier databases and catalogs vary in quality and completeness, but in general they contain reliable and usable information. Their limitation is that they usually only contain information on the plastics that the publisher sells. You will need to check various suppliers to determine which supplier has the best-fit plastic. This problem is usually solved by using generic databases that are purchased from information companies.

Another concern with the use of databases is whether test data generated by different suppliers are measured in the same way. Table 7–1 shows some of the standard (ASTM) tests that are used for measuring plastic properties. Make sure in comparing data on candidate plastics that the property data are generated with the same tests and that the units of measurement are the same. Most plastics databases contain a range of values for most properties. If the range is large, it may be necessary to contact the supplier for more precise data. Most databases in the United States include caveats about not using these data for design. Our litigious society mandates disclaimers and warning labels on just about everything. Most published data are presented in good faith, and precautions are taken to make sure that the data are as accurate as possible. However, we recommend checking the property data with manufacturers before making final selections. Use published data for screening, but check the property data on your final choice with the material manufacturer. Make certain that the data used in the selection process were accurate and up to date. This statement also applies to this text.

Plastics as a class of engineering materials are significantly different from metals and other

CenBASE/Materials Detailed Report

User: CHRIS
Files found = 113 Search for: MOBAY Date: XX-XX-XXXX

ABS/PC Mobay Bayblend FR 1439/I/Unfilled Flame Retardant Good Impact Resistance

Vicat softening point (°F) D 1525	195.00	Compressive STR at yield (kpsi) D 695	—	Brittle temperature (°F) D 746	—
Melting temperature (°F)	520.00	Hardness Rockwell (M) D 785	—	Cost/lb (40,000 lb and up)	—
Mold shrinkage (in./in.) D 955	0.004900	Hardness Rockwell (R) D 785	115.00	Dielectric strength short (V/mil) D 149	760.00
Melt flow rate (gm 10 min) D 1238	24.00	Izod notched $^1/_8$" (ft-lb/in.) D 256	6.00	Dielectric constant at 1 MHz D 150	2.90
Specific gravity/ density D 792	1.170	Izod notched $^1/_4$" (ft-lb/in.) D 256	—	Dielectric constant at 60 Hz D 150	3.00
Elongation at yield (%) D 638	3.50	Water absorption 24 h (%) D 570	0.150	Dissipation factor at 1 MHz D 150	0.007000
Elongation at break (%) D 638	50.00	Deflection temp at 66 psi (F) D 648	195.00	Dissipation factor at 60 Hz D 150	0.005000
Tensile strength at yield (kpsi) D 638	7.700	Deflection temp at 264 psi (F) D 648	180.00	Volume resistivity (ohm-cm) D 257	↑0–E+17 —
Tensile strength at break (kpsi) D 638	5.800	Hardness Shore (D) D 2240	—	Arc resistence (s) D 495	—
Tensile modulus (100 kpsi) D 638	3.800	Hardness Shore (A) D 2240	—	Maximum temp continuous use (°F)	— 60.00
Flexural strength at yield (kpsi) D 790	12.600	Coeff of expansion (in./in./°F) D 696	—	UL temperature index (elec) (°C)	1
Flexural modulus (100 kpsi) D 790	3.600	Conductivity (Btu in./h ft^2 °F) D 177	—	UL94 1 = V-0, 2 = V-1, 3 = V-2, 4 = 5 V, 5 = HB	

Bayblend FR 1439, 1440, 1441 are unreinforced, flame retardant thermoplastic resin blends of MAKROLON polycarbonate and ABS resin. They represent new flame retardant technology that combines easy processibility and excellent property performance. These injection molding grades of Bayblend are naturally opaque and are supplied in pellet form. Offer excellent melt stability and easy processibility. Inherent color stability to fluorescent lighting or filtered sunlight in an office environment. Good impact resistance. UL yellow card recognition. No mold deposit and nonblooming flame retardant system.

Figure 7–1
Typical property data on a plastic
Source: CenBase® (Infodex Inc.)

Table 7–1
Some commonly specified ASTM property tests

Test	Physical Properties
D 792	Test for specific gravity and density by displacement
D 570	Test for water absorption of plastics
D 696	Test for coefficient of linear thermal expansion of plastics
C 177	Test for steady-state thermal transmission properties (thermal conductivity)
D 635	Rate of burning and/or extent and time of burning (horizontal position)
	Mechanical Properties
D 638	Tensile properties of plastics
D 695	Test for the compressive properties of rigid plastics
D 790	Test for the flexural properties of plastics
D 256	Test for impact resistance of plastics
D 785	Rockwell hardness testing of plastics
D 648	Test for deflection temperature of plastics under flexural load
	Electrical Properties
D 150	Test for AC loss characteristics and easily recycled (dielectric constant)
D 149	Test for dielectric breakdown voltage and dielectric strength
D 257	Test for DC resistance or conductance of insulating materials

engineering materials. They have idiosyncrasies that must be dealt with (as do metals, ceramics, and composites). Some of the more noteworthy idiosyncrasies follow.

1. They expand with temperature at a rate that is usually ten times that of metals (coefficient of thermal expansion).
2. Steels are 30 times stiffer than the stiffest unreinforced plastics (modulus of elasticity).
3. Plastics can burn and sometimes give off toxic fumes in the process (flammability, toxicity).
4. Plastics can deteriorate by aging, by exposure to sunlight, and by atmosphere-induced cracking (environmental resistance).
5. Plastics are lighter than metals (density).
6. Plastics are electrical insulators (electrical resistivity, dielectric constant, etc.).
7. Plastics are poor conductors of heat (thermal conductivity).
8. The upper use temperature for 99% of all plastics is less than 500°F (260°C) (heat deflection temperature).
9. Plastics are soft; the hardest plastic is much softer than the softest steel (penetration hardness).
10. Plastics cannot be shaped by the cold-forming processes that apply to most metals (resistance to plastic deformation).
11. Plastics cannot normally be made to the close dimensional tolerances that are common in metals; they are difficult to accurately machine (machinability, stability).
12. Plastics cannot resist impact as well as most metals (impact strength).
13. Plastics may permanently deform in use (creep strength).
14. Plastics may fail under sustained load (stress rupture strength).
15. Plastics are not as strong as most metals (tensile, yield, and shear strength).
16. Plastics may fail under cyclic loading (fatigue strength).
17. Plastics can absorb moisture from their use environment and change; size or properties can be affected (moisture absorption in 24 h).

These idiosyncrasies do not limit applications of plastics. They simply must be recognized and dealt with in design. There is a similar list of idiosyncracies for ceramics and composites. In the event that this list denigrates plastics in your eyes, let us list some of the features of plastics that make them the material of choice for ever-increasing numbers of new products:

1. They are low in cost; the material cost is low, and they can usually be shaped with low-cost processes (injection molding, etc.).
2. They usually do not need to be painted; the color can be "free" and exists throughout the part thickness.
3. They are usually made with processes that yield net shape or near-net shape. Secondary operations such as machining or flash trimming are often unnecessary.
4. They do not rust.
5. They can be easily made cosmetically pleasing with colors and texturing in the forming process.

Needless to say, there are exceptions to every one of these statements (the polyimides are certainly not low in cost), but these statements apply to "mainline" plastics. These are the reasons why plastics are selected for new applications even though they have some frailties compared with metals.

In summary, the selection of plastics is performed in the same way as the selection of other engineering materials. The desired properties are listed, and the properties of candidate plastics are compared to find a match. As a class of engineering materials, plastics are not as strong and stiff as many other engineering materials, and this must be considered in the design. In fact, all of the idiosyncrasies that we listed should be kept in mind in the design process.

7.2 Plastics for Mechanical and Structural Applications

Most plastic components are expected to handle some kind of stress or load. In structural parts such as gears, levers, cams, brackets, and frames, loads are applied to components intentionally. Nonstructural items, such as housings, covers, or other relatively unstressed components, may be incidentally stressed by fasteners, interference fits, and snap fits. In many instances, plastic components are unintentionally stressed. Impact loads, for example, often occur during shipping or dropping. In fact, many consumer components must endure shipping and drop tests. Designers must assess the application and anticipate what types of forces may occur. They must envision worst-case scenarios. For example, if a part may unintentionally be left in a closed automobile in the sun, a polymer that can endure 190°F (87°C) should be used to avoid warping or melting. Commonly used plastics for commodity and engineering applications are listed in Table 6–3.

Because polymers are long hydrocarbon chains held together with covalent bonds, the properties of these materials are much more sensitive to temperature and other environmental conditions than are metals or ceramics. Not only do the applied forces need to be assessed for an application, the time at load, cyclic loading conditions, and anticipated temperatures must be considered. Chemical and environmental exposure also affect a plastic's properties.

The mechanical properties of polymers are largely controlled by the structure of the polymer chain, additives, and processing. Often a preliminary list of potential materials for an application may be made by considering these factors. For example, if a part requires transparency it should be made of an amorphous polymer such as polystyrene, polycarbonate, or polymethyl methacrylate. Components requiring high chemical resistance typically use crystalline plastics such as polyethylene or polytetrafluoroethylene. For injection molded parts requiring close dimensional tolerances, amorphous thermoplastics provide greater dimensional stability.

Aside from polymer structure, polymer properties may be altered with additives and fillers. For example, if a polymer must endure exposure to sunlight and weather, adding certain antioxidants and stabilizers helps ensure resistance to degradation. Once the list of property requirements is developed for an application, a polymer material may be tailored to the application by controlling the structure, by adding additives and fillers, and by copolymerization or alloying with other polymers. Processing methods must also be reviewed to assess the effects of polymer orientation, residual stresses, surface finish, addition of regrind (recycled plastic), weld lines (see Figure 7–2), and other processing factors on the overall properties of the polymer. Table 7–2 compares the relative properties of amorphous and crystalline thermoplastics. The structure, glass transition temperature (T_g), and crystalline melting point (T_m) of various polymers are listed in Table 7–3.

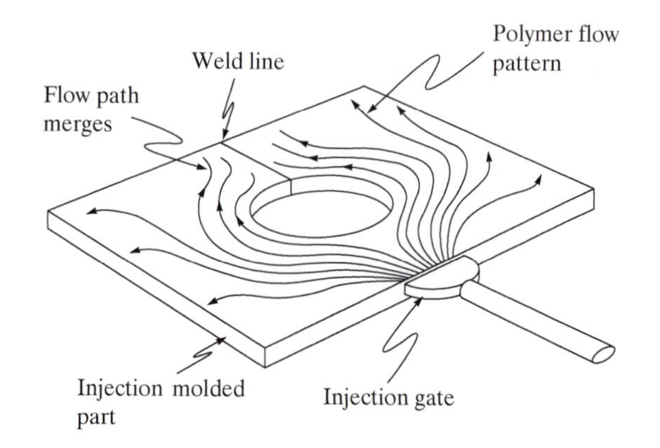

Figure 7–2
Weld lines occur in injection molded parts when the polymer's flow path is interrupted by a feature (such as a hole) and then merges back together. The material strength at the weld line is reduced by 25% or more. Under poor molding conditions the strength of a weld line can approach zero and be visible to customers

Both short-term and long-term (time-dependent) properties of plastics will be reviewed in this section. Short-term mechanical properties for polymers include tensile strength, flexural strength, tensile or flexural modulus, impact toughness, and hardness. Long-term mechanical properties include creep strength, creep modulus, fatigue strength, and stress rupture strength.

Table 7–2
Comparison of the relative properties of amorphous and crystalline thermoplastics

	Relative Properties*	
Characteristic	**Amorphous**	**Crystalline**
Tensile strength	Lower	Higher
Tensile modulus	Lower	Higher
Ductility	Lower	Higher
Toughness	Lower	Higher
Isotropy	Isotropic	Anisotropic
Creep resistance	Higher	Lower
Maximum use temperature	Lower	Higher
Wear resistance	Lower	Higher
Shrinkage	Lower	Higher
Distortion	Lower	Higher
Specific gravity	Lower	Higher
Melting characteristics	Gradual flow above T_g	Sharp melt transition at T_m
Opacity	Transparent	Cloudy to opaque
Chemical resistance	Lower	Higher

*There are some exceptions to these generalizations.

Table 7–3
T_g, T_m, and structure of selected polymers

Polymer	Acronym	T_g (°C)	T_m (°C)	Structure
Low-density polyethylene*	LDPE	—	109	Crystalline
Medium-density polyethylene*	MDPE	—	125	Crystalline
High-density polyethylene*	HDPE	−120	140	Crystalline
Polytetrafluoroethylene	PTFE	−113	327	Crystalline
Polyvinylidene fluoride	PVDF	−45	210	Crystalline
Polypropylene*	PP	−10	170	Crystalline
Polyvinyl chloride*	PVC	80	220	Crystalline
Perfluoroalkoxy	PFA	90	—	Crystalline
Polystyrene*	PS	100	—	Amorphous
Ethylene chlorotrifluoroethylene	ECTFE	100	240	Crystalline
Polymethyl methacrylate	PMMA	105	—	Amorphous
Fluorinated ethylene propylene	FEP	152	290	Crystalline
Polyoxymethylene (Acetal)	POM	−85	170	Crystalline
Polyethylene oxide	PEO	−67	69	Crystalline
Polyamide (Nylon)	PA 11	46	185	Crystalline
	PA 6	50	228	Crystalline
	PA 66	57	265	Crystalline
Polyethylene terephthalate	PET	69	265	Crystalline
Polybutylene terephthalate	PBT	80	232	Crystalline
Polyphenylene sulfide	PPS	90	—	Crystalline
Polyphthalamide	PPA	127	310	Crystalline
Polyetherether ketone	PEEK	143	—	Amorphous
Polycarbonate	PC	145	—	Amorphous
Polysulfone	PSU	190	—	Amorphous
Polyphenylene oxide	PPO	210	—	Amorphous
Polyethersulfone	PES	225	—	Amorphous
Polyamide imide	PAI	272	—	Crystalline
Aliphatic polyketone	PK	5–20	220	Crystalline
Polyimide	PI	330–360	—	Amorphous
Polybenzimidizole	PBI	398	—	Amorphous

*Commodity plastics

Stiffness

In engineering terms, the stiffness of a plastic is normally described as the modulus of elasticity measured in tension. Often handbooks and plastics suppliers report stiffness as flexural modulus. Flexural modulus measures the stiffness of a plastic in bending. Most plastics do not rigorously follow Hooke's law (stress/strain = a constant). Plastics do not exhibit a completely elastic strain region in a stress–strain diagram. It is common to measure the tensile elastic modulus of plastics at low strain levels on a stress–strain curve.

The structure of a polymer as well as additives and fillers control the stiffness. Polymers with aromatic ring structures in the main chain such as polyimide, polyetherimide, and polyphenylene sulfide are much stiffer than straight-chain aliphatic polymers such as polyethylene and polypropylene. Large pendent groups, such as those found in polystyrene, reduce segmental

Table 7–4
Approximate room temperature mechanical properties of various plastics

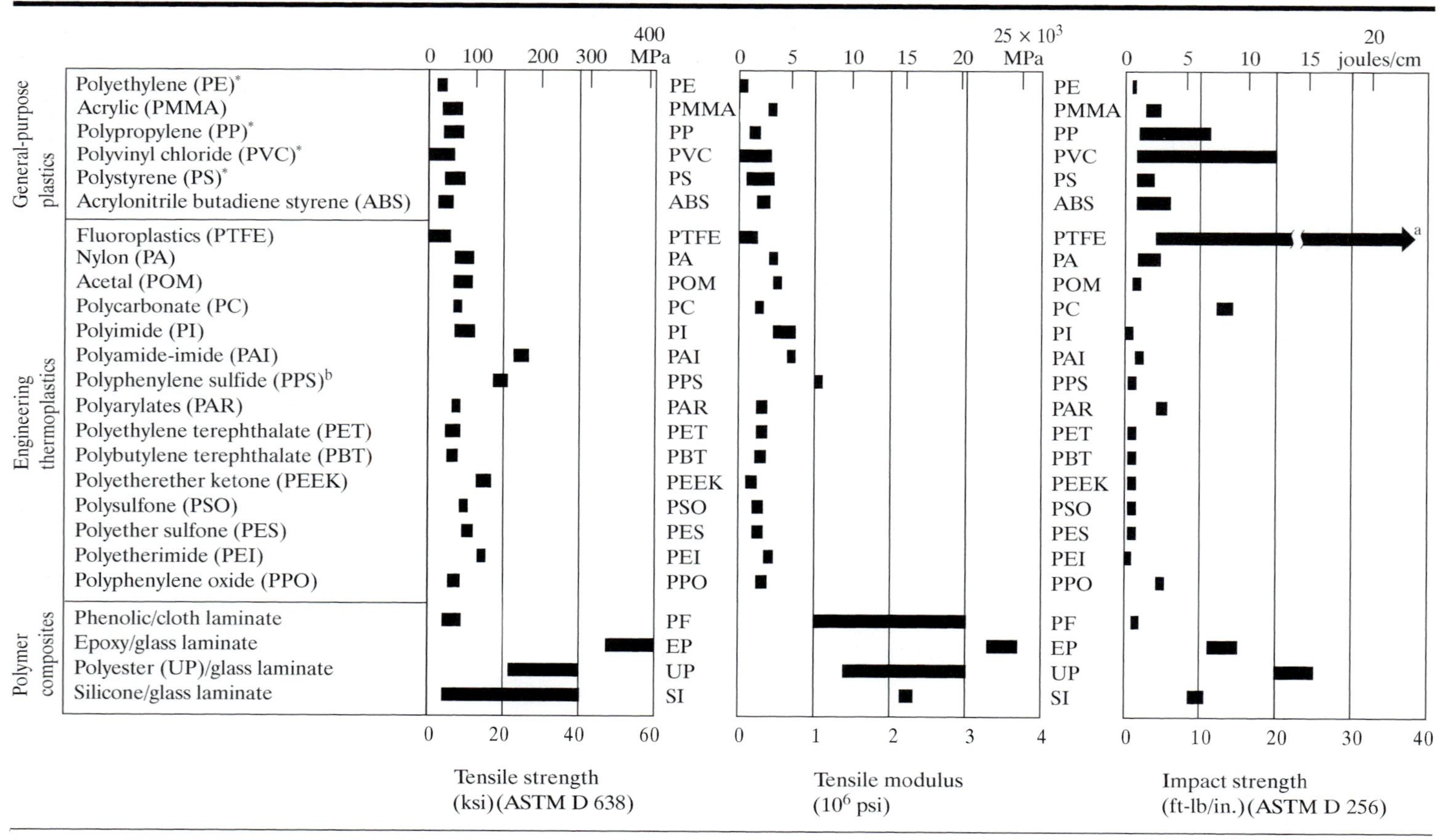

The properties for the general-purpose and engineering plastics are for unfilled and unreinforced grades.
[a]No break
[b]40% glass
*Commodity plastics

motion of the chain and increase stiffness. Fiber fillers such as glass fibers greatly increase stiffness. Particulate fillers such as talc, mica, and milled glass moderately increase the stiffness of a plastic. Additives such as plasticizers and impact modifiers tend to reduce the modulus of a plastic. Higher temperatures and moisture levels in a plastic reduce the stiffness.

Table 7–4 contains a tabulation of typical modulus ranges for various plastics. As can be seen from these data, polyimide with a modulus of 0.6×10^6 psi ($3\text{–}5 \times 10^3$ MPa) is the stiffest unreinforced plastic listed. Aluminum, with a modulus of 10^7 psi (69×10^3 MPa), is on the low end of the range of modulus values for metals. Steels have a modulus of approximately 30×10^6 psi (207×10^3 MPa). From the design standpoint, this means that if you want a plastic part to be as stiff as a steel part it may have to be many times as thick. This usually does not pose a problem. It may appear that because of their lower modulus values, plastics are inferior to metals as structural components, but as pointed out in our discussion in Chapter 6 on polymer composites, the reinforced plastics have specific stiffness comparable to or greater than that of steels and aluminum.

The stiffness of unreinforced plastics is significantly less than that of metals, and the increased deflection must be taken into consideration in design calculations. Parts must usually have thicker sections, but this may not mean a weight gain because of the lower density.

Polymer Creep and Temperature Effects

Polymers, much like metals, will generally behave in an elastic manner when stressed for short periods of time below the elastic limit. If polymers are stressed for extended periods of time or at elevated temperatures, they will creep (flow). The strain produced in a polymer under stress depends on the magnitude of the applied force, the time under load, and the temperature. This change in strain as a function of time is referred to as *creep* or *viscoelastic flow*. As shown in Figure 7–3, when a component is loaded in tension, an initial elastic strain may be measured in the part.

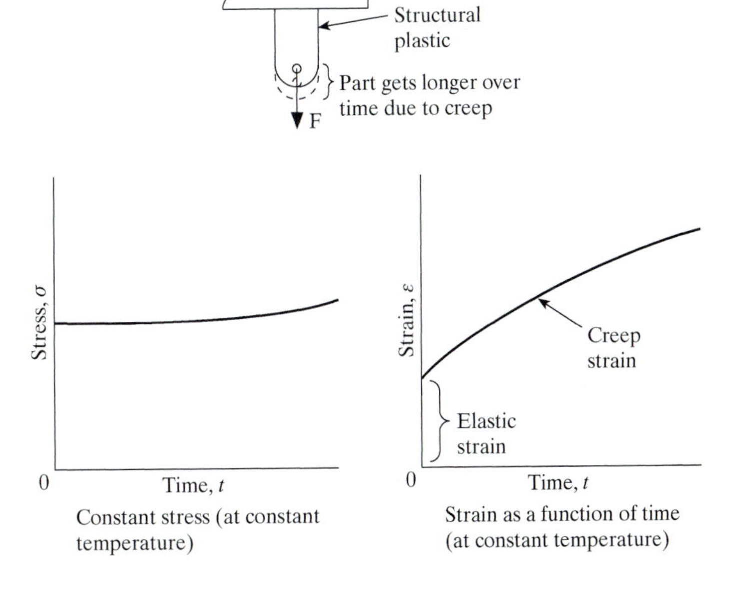

Figure 7–3
Creep of a plastic structural element under a constant stress condition and elevated temperature

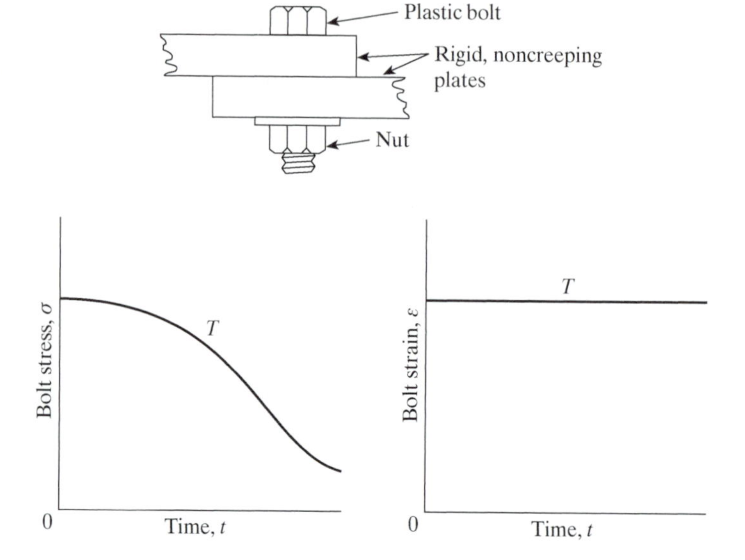

Figure 7–4
Stress relaxation of a plastic bolt. Strain is constant; stress in the bolt decreases (at constant temperature).

However, with time the strain in the part will increase due to creep (the part gets longer). Similarly, creep also occurs in parts loaded in compression or shear (or combinations).

Metals typically creep when exposed to temperatures greater than half of their melting point. Because of their high melting points, most metals do not significantly creep at room temperature, but certain metals, such as lead, tin, and zinc, do. Polymers, though, due to their structure and molecular bonding, require much lower activation energies (temperature) for viscoelastic flow.

It is also possible for applied stresses in a polymer part to relax over time. Figure 7–4 shows a bolted assembly with a plastic bolt and nut. Initially the nut is tightened and a specific preload is developed in the bolt. However, due to creep, the stresses in the bolt relax over time. This phenomenon is known as *stress relaxation*. Stress relaxation generally occurs when a part is loaded to a constant strain value. Stress relaxation may also occur in components stressed in compression or shear. The compressive force on a PTFE gasket, for example, may be reduced over time due to compressive stress relaxation. At some point, the gasket joint may develop a leak due to the relaxed compressive force on the gasket. Aside from time and temperature, creep is also affected by the level of stress, the environment, reinforcements, polymer additives, and the type of polymer (crystalline vs. amorphous) as well as other factors as outlined in Table 7–5. Viscoelastic flow occurs in all engineering plastics, commodity plastics, elastomers, adhesives, and other organic materials.

Viscoelastic creep data for polymers is typically reported as creep modulus or in the form of an isochronous stress–strain curve. Tensile creep, for example, can be measured per ASTM D 2990. A standard tensile bar is fitted with an

Table 7–5
Factors that affect polymer creep or stress relaxation

Polymer structure: crystalline or amorphous
Fillers and reinforcement
Secondary phases or copolymers
Use temperature (relative to T_g, T_m)
Environment (moisture, humidity, chemicals)
Stress level
Glass transition temperature
Heat deflection temperature under load (HDTUL)
Creep modulus

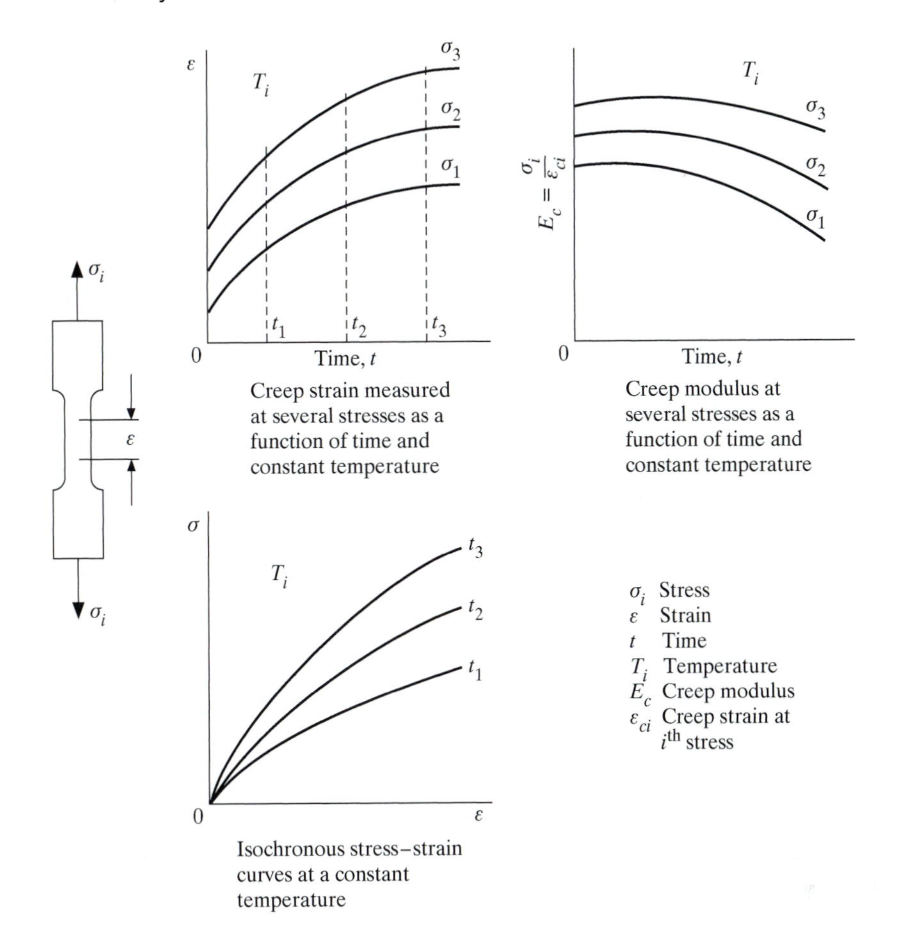

Figure 7–5
Calculation of creep modulus. This type of data is often supplied by plastic compounders.

extensometer or strain gage, and a constant stress is applied to the specimen. The viscoelastic strain due to creep is measured and plotted as a function of time (see Figure 7–5). Creep tests are generally conducted at many stress levels (or at least should be conducted for the anticipated stress on a specific part). Once the creep strain has been measured as a function of time at a given stress and temperature, a creep modulus is calculated as a function of time. The creep modulus at a point in time is calculated by dividing the applied stress by the creep strain at that same point in time. Most plastics suppliers supply creep data in the form of creep modulus curves. If the applied stress in known, the creep strain at a given design life may be calculated by dividing the creep stress by the creep modulus. When selecting a plastic, for creep applications, the creep moduli at a given design life, temperature, and stress may be used as selection indices.

As a general rule of thumb, for elevated temperature applications it is best to stay at least 50°C below the T_g to minimize the effects of creep. Glass reinforcement improves creep resistance, but, with crystalline thermoplastics, glass fiber additions greater than about 10% may not provide any additional increase in creep resistance. Often data sheets on plastics report heat deflection temperatures under load (HDTULs). This value measures the temperature at which a beam, loaded in three-point bending, deflects a specified distance at a predetermined time and stress. Heat deflection temperatures may be used as a first approximation when considering polymers for elevated temperature applications, but they should not be used for design calculations. Figure 7–6 compares the heat deflection temperature and the glass transition temperature of some selected polymers.

A test called the *Relative Temperature Index Test* (RTI) is often used to predict the maximum service temperature a polymer may endure over the life of a part without thermal degradation. A large population of polymer samples (tensile bars and flex bars) is aged in furnaces at different temperatures. Samples are removed from the furnaces at various time intervals and the tensile strength, tensile impact strength, and dielectric strength are measured. The data are plotted as failure time (aging time at a given temperature that causes a 50% reduction in properties) versus inverse temperature (in Kelvin). This plot, known as an *Arrhenius plot*, is then extrapolated to determine the temperature at which there is a 50% reduction in the measured properties after 10 years. This temperature value is known as the Relative Temperature Index (formerly the Continuous Use Rating) and the test is typically conducted per UL 746B. One caveat of the RTI test is that the samples are not stressed during the test. In real product applications, the part will probably be under stress (applied or residual) while exposed to elevated temperature conditions. It is probably prudent to use RTI for ranking materials rather than using the RTI value to predict the actual maximum use temperature of the material. In some instances, the RTI is much higher than the T_g, suggesting that creep could dominate the part life. Clearly the polymer may succumb to creep or stress rupture before the properties degrade per the RTI test. The RTI values for polymers are quite sensitive to additives and manufacturing and process factors. Table 7–6

Figure 7–6
Relation between glass transition temperature and heat deflection temperature for various plastics

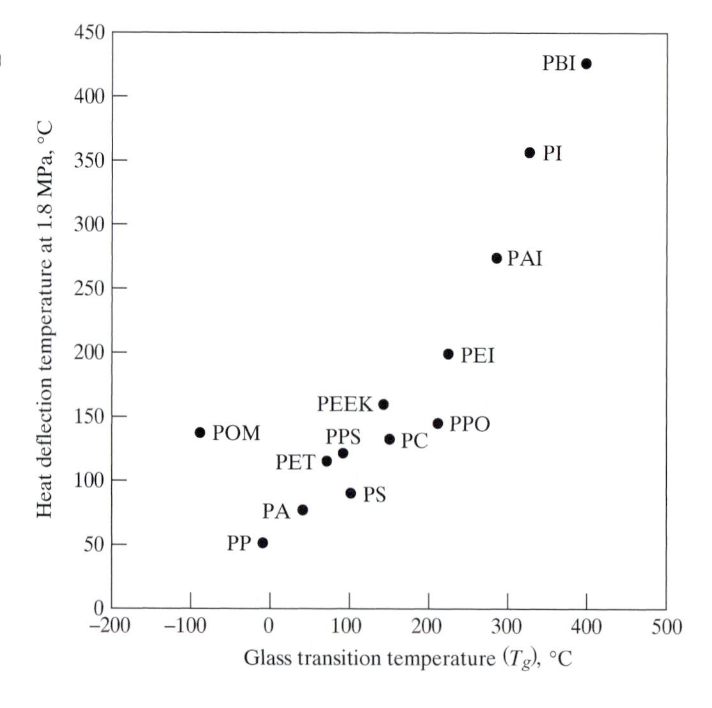

Table 7–6
Physical properties of various plastics

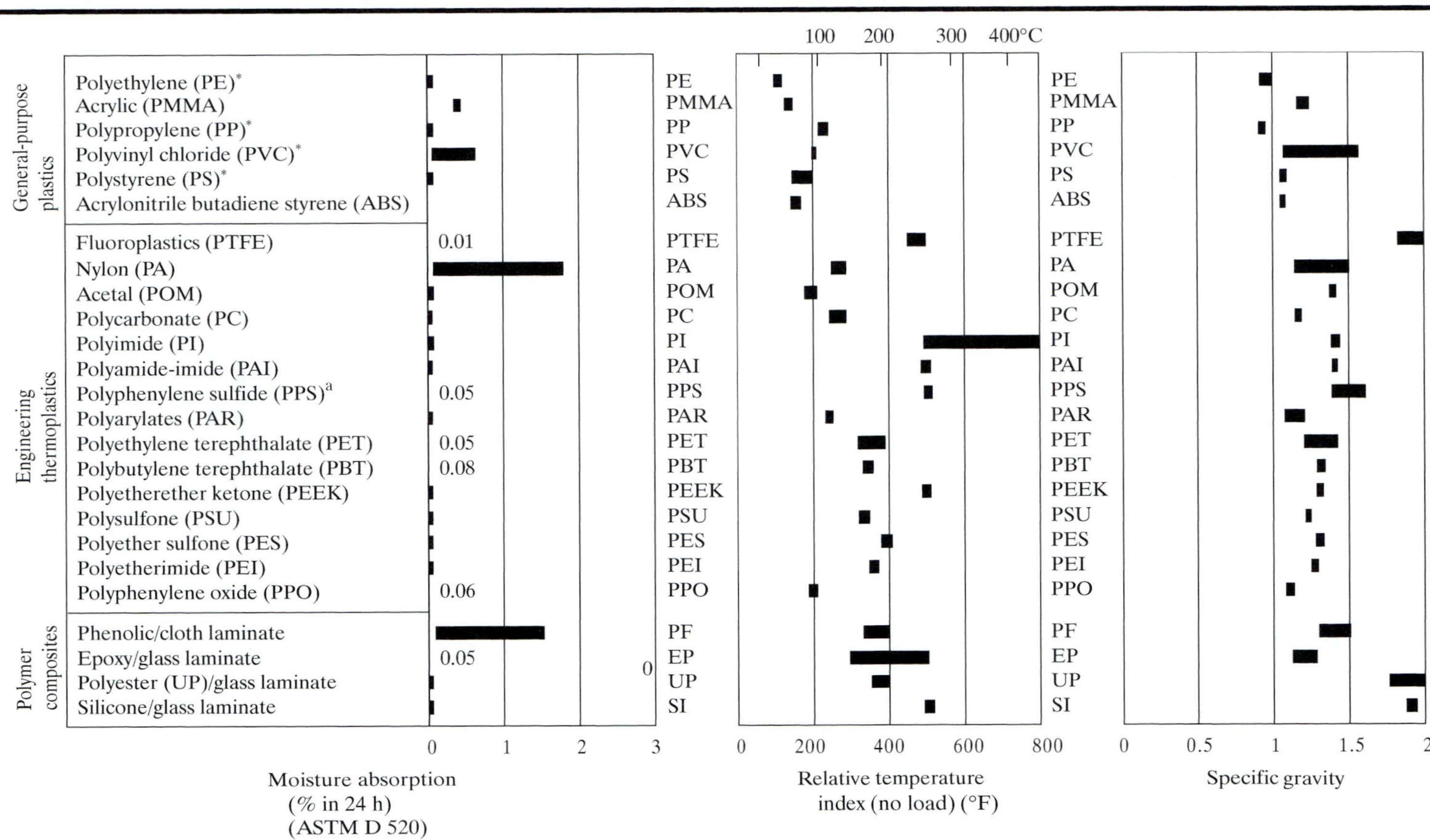

The properties for the general-purpose and engineering plastics are for unfilled and unreinforced grades.
[a]40% glass
*Commodity plastics

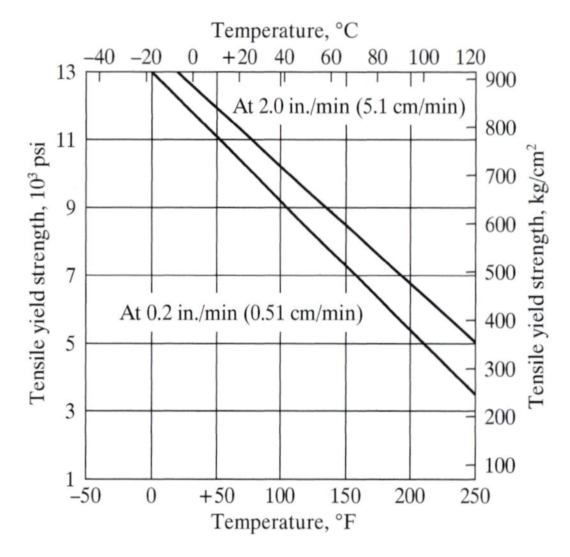

Figure 7–7
Effect of temperature on acetal

Source: *Delrin Design Handbook*, E. I. du Pont de Nemours & Company.

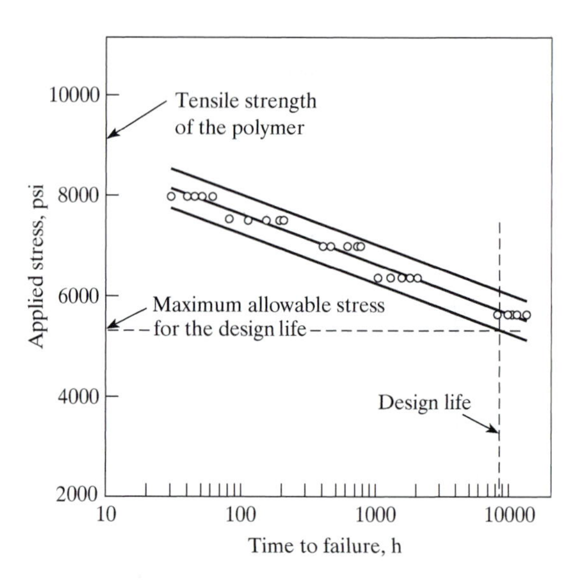

Figure 7–8
Typical stress rupture data for a polymeric material

compares the relative temperature index for various plastics.

Figure 7–7 illustrates the effect of temperature on the short-term tensile strength of acetal resin. Similar effects occur in modulus of elasticity, yield strength, compressive strength, and hardness. In most cases, the impact strength and elongation are improved. The point to be made is that room temperature (20°C) mechanical property data cannot be used in design calculations if the part is going to be used at any other temperature.

Stress Rupture and Environmental Stress Cracking

Stresses and temperatures are low enough in many designs that viscoelastic creep does not have a major effect on component dimensions or functionality. However, when sustained tensile stresses are applied to polymeric materials, stress cracks or crazes may develop over time. These cracks advance and propagate, eventually causing part failure. This failure mode is called stress rupture. Unlike mechanical overload fractures, there is virtually no mechanical deformation associated with stress rupture. Failures of this type usually occur at stresses significantly below the yield or tensile strength of the polymer. All polymeric materials are susceptible to failure by stress rupture. Chemical exposure, temperature, and environment may also influence the resistance of a polymer to stress rupture. Industry surveys indicate that 30% to 40% of all plastic part failures are due to environmental stress cracking and stress rupture.

The time required for a part to fail by stress rupture is directly affected by the level of the stress as shown in Figure 7–8. Temperature and chemical exposure can affect the critical stress to produce stress rupture. The influence of chemicals on stress rupture is often referred to as environmental stress cracking (ESC). Stress cracking agents for polymers are bountiful. Soaps and surfactants cause polyolefins such as polyethylene to stress crack. Numerous chemicals, including some mold-release agents, cause stress cracking in polycarbonates. Even moisture or water can reduce the stress rupture strength of a polymer.

Polymers used in mechanical design when continuous tensile loads are anticipated should use a critical stress rupture strength rather than a yield or tensile strength as the design criterion for mechanical failure. Other factors such as viscoelastic creep should also be considered at higher temperatures. For reliable part design, it is best to ensure that the stresses on the part are lower than the critical stresses for environmental stress cracking or stress rupture (at a given design life).

Aside from applied stresses, sources of stress on a polymeric part include residual stresses from fabrication, assembly stresses from fasteners, interference fits, snap fits, and thermal stresses. Stress rupture and ESC failures often occur at stress concentrations. Plastic parts commonly have sharp corners or molding defects such as knit lines and porosity that act as stress concentrators.

The following factors should be considered to reduce stress rupture and environmental stress cracking tendencies.

1. Lower stresses to values below the critical stress (if material is added to increase section size, be careful of residual stresses from shrinkage).
2. Anneal parts to relieve residual stresses (annealing, however, may affect other properties such as impact toughness).
3. Select an alternative polymer (a different class of polymer or one with a different additive package).
4. Use fiber reinforcement (e.g., glass or carbon).
5. Use metallic components when applicable instead of a polymer (e.g., die castings, precision castings, and powder metallurgical parts are available for near-net-shape parts).
6. Eliminate the stress-cracking agent if possible.
7. Avoid notches and other mechanical stress concentrations.
8. Move stress concentrations such as molding knit lines to areas of lower stress through tool design.
9. The tensile stress in a polymer should not exceed 1/10 to 1/6 of the tensile yield or failure strength of the polymer.
10. Do long-term creep or stress rupture tests whenever it is anticipated that a plastic part will be continuously under stresses that exceed 10% of its yield strength.

Unfortunately, most plastic suppliers have very little published data on stress rupture and environmental stress cracking, so testing is often required. Many of the environmental stress cracking and stress rupture tests were first developed for polyolefin piping. These tests were necessary to provide reliable piping for services such as natural gas and other hazardous chemicals. A number of standard test methods now exist for assessing a polymer's resistance to creep rupture and ESC; however, many of these tests are specifically for polyolefins.

Constant-stress and constant-strain tests are typically used to measure ESC and stress rupture properties of a polymer. Numerous standard test methods exist: ASTM D 1598, D 2837, D 2990, D 1693, D 2552, D 5419, F 1248, ISO 4599, ISO 4600, ISO 6252, and SAE J 2016. For critical applications, tests should be conducted under the anticipated stress state, as the uniaxial test data may not be sufficiently conservative.

When selecting materials for an application in which stress rupture may be a potential mode of failure, it is best to compare candidate materials based on a reduced stress: stress rupture strength divided by static failure strength. The static failure strength, the stress rupture strength, and the reduced stress for a number of polymers

Table 7–7
Stress rupture strength of common polymers*

Polymer Type	Static Fracture Stress, S_x (psi)	Stress Rupture Stress, S_n (psi)	Reduced Stress S_n/S_x
POM homopolymer	10,875	4,350	0.40
POM copolymer	9,570	4,205	0.44
HDPE	4,640	1,160	0.25
PC	9,280	6,815	0.73
PP homopolymer	4,350	1,450	0.33
PP copolymer	3,480	1,450	0.41
PSU	11,745	6,090	0.52
PVC	9,860	5,070	0.51
PMMA (70°F, 50% RH)	12,905	4,930	0.38

*After Teoh, S. H. Predicting the life and design stresses of medical plastics under creep conditions. In H. E. Kambie, and A. I. Yokobori, Eds. Special technical publication 1173: Biomaterials mechanical properties, pp. 77–86. Philadelphia, PA: American Society for Testing Materials, 1994.

are summarized in Table 7–7. Materials with the highest reduced stress will have the best resistance to stress rupture. As a general rule of thumb, a safety factor of six to ten on the failure strength (yield strength for ductile plastics or tensile strength for brittle plastics) of a polymer will generally give a working strength level that lowers the risk of stress rupture failure.

Impact

Engineering components are often exposed to impact conditions. For example, communication devices such as telephones and pagers must endure dropping, automobile components must survive a 5-mph crash, and helmets and guards must not shatter when impacted. Many polymers are inherently brittle compared with metals. Although polymers, as well as most materials, have some measured level of impact toughness, the overall toughness of a component is related to the selection of the materials, the fabrication processes, and the design of the part.

When a plastic is impacted to the point of failure, it fractures in either a brittle or a ductile manner. Brittle fracture occurs with little or no yielding or deformation and may be catastrophic in nature. In ductile fracture, a material will yield significantly before fracturing. Generally, conditions that lead to brittle fracture require less impact energy to initiate fracture than ductile fracture.

Many factors that affect the impact toughness of a plastic component are summarized in Table 7–8. Plastics may undergo a transition from ductile to brittle fracture with decreasing temperature, increasing section thickness, and increasing rate of loading.

Polymer suppliers typically report impact toughness of a material in terms of notched Izod impact strength (ASTM D 256) or Gardner impact strength (ASTM D 3763). As discussed in Chapter 2, the Izod test measures the energy required to fracture a notched beam of a given thickness and temperature. Izod impact tests may be conducted on unnotched samples as well. The impact data are reported in energy per sample thickness. The Gardner impact test measures the energy absorbed when a spherical punch (of given potential energy) is dropped though a 3-in.-diameter (76.2-mm), 0.125-in.–thick (3.2-mm) disk that is clamped around the periphery.

Table 7–8
Factors affecting the impact resistance of plastic components

Category	Factors	Effect
Component design	Part thickness	Thick parts have lower apparent toughness; some polymers such as PC undergo a transition from tough to brittle at a critical thickness.
	Stress concentrations	Sharp corners, notches, scratches, gouges, and tool marks can reduce part impact toughness (*notch sensitivity*). Every corner should have a minimum radius of 0.020 in. (0.5 mm) or 60% of the section thickness.
	Part design	Parts that flex or otherwise absorb impact energy can improve the resistance to impact fracture.
Usage	Rate of loading	Higher loading rates reduce material toughness.
	Temperature	Plastics undergo a relative transition from tough to brittle with decreasing temperature.
	Environment	Parts may have reduced impact strength due to changes in the polymer from the environment (surface crazing, cross-linking, loss of plasticizer).
Processing	Residual stresses	Higher residual stresses reduce impact toughness.
	Orientation	Higher polymer orientation increases impact toughness.
	Molding weld lines	Weld lines act as stress concentrations, lowering apparent material toughness.
	Degree of crystallinity	Higher crystallinity improves impact toughness.
	Annealing	Annealing may reduce residual stresses; however, certain polymers such as PC have reduced toughness when annealed. It is best to check the effects of annealing on toughness.
Material	Material	Certain materials such as PC, PP, PE, PTFE, fluoropolymers, PC, ABS, and HIPS have higher impact toughness.
	Impact modifiers	Impact modifiers such as grafted rubber or plasticizers improve impact toughness.
	Fillers, reinforcement	Mechanical reinforcements such as fibers or particle fillers reduce impact toughness.

When designing a plastic component, one must determine what the requirements must be for impact. If a product may be dropped during use, it must be designed to handle the anticipated abuse. Components and products often experience high loading rates during shipping and handling (possibly under difficult conditions such as low temperature). It is not always possible to predict the impact resistance of a plastic component or product based only on the inherent material toughness as determined by a standard impact test such as the Izod notched beam impact test or the Gardner impact test. This is due in part to the factors outlined in Table 7–8. Often actual parts are impact tested at specific locations on the part (point of impact or areas prone to breakage) using an instrumented impact tester. Measurement of the impact strength of actual parts using instrumented impact tests allows the effects of geometry, molded-in stresses, anisotropy, surface finish, and other factors to be assessed.

Common materials for impact applications include polycarbonate, polycarbonate/ABS blends, nylon, acetal, ABS, high-impact polystyrene, toughened PVC, and polyolefins (PP, PE, PTFE). Impact properties of various polymers are compared in Table 7–4.

Fatigue

If metals are used for cyclic-loaded parts, it is customary to use a fraction of the fatigue strength as the design stress. This same procedure should be used in designing cyclic-loaded plastic parts. However, fatigue data on plastics are scarce. Figure 7–9 presents some data. In metals, if the fatigue data are not available for a particular alloy, the general rule is to use 50% of the tensile strength. No such proportionality constant is readily promoted for use on plastics. In addition to the lack of fatigue data for plastics, because of the lack of perfectly elastic behavior (there is plastic strain at stress below the yield), there is some permanent strain in cyclic loading, even at low stress levels. This strain, especially in complete stress reversals (tension and compression), can result in a hysteresis effect that causes heating of the component. Polymers typically fail in fatigue by crack propagation, by incipient melting due to the temperature rise created by hysteresis heating, or by a combined effect.

To consider fatigue in design, we must try to design to the fatigue strength if the data are available. If they are not, keep design stresses sufficiently low and take into account hysteresis effects and effects of environment, and above all eliminate stress-concentrating notches and corners. Often it is advantageous to conduct fatigue tests on actual parts. Aside from assessing the effects of part surface finish, and residual stresses, testing on actual parts allows the effect of part geometry on hysteresis heating to be evaluated.

Dimensional Stability

Dimensional stability is important in many mechanical designs. Gears, cams, and bearings, for example, must have accurate dimensions that do not change over time or with environment. One disadvantage of plastics is their tendency to absorb moisture from the ambient air and to change size. Moisture-related size changes can preclude the use of plastic for a part when the plastic's other characteristics would make it a

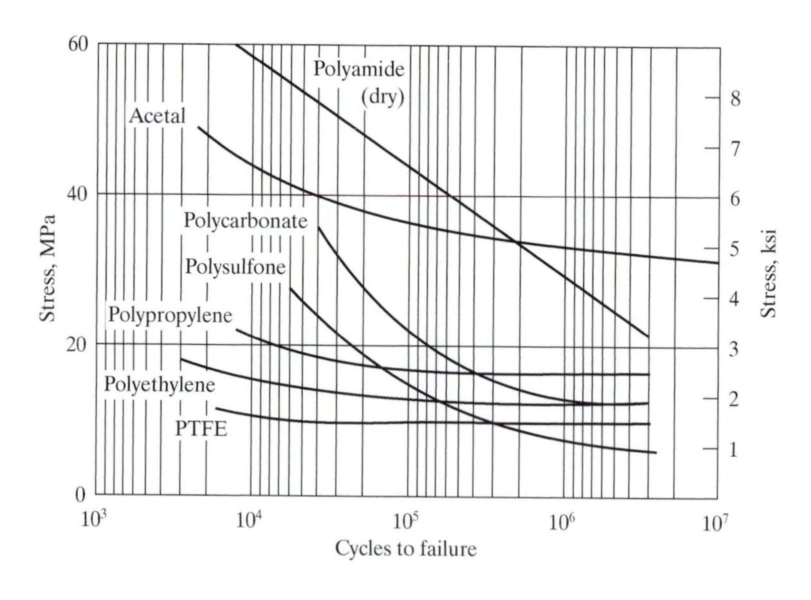

Figure 7–9
Room temperature fatigue characteristics of selected plastics (various sources)

better choice than a metal. Since moisture absorption occurs to a greater or lesser degree in all plastics, the only way to circumvent this problem is to select a material with the lowest absorption rate and then calculate the effect of the volume size change on the critical dimensions. Table 7–6 presents typical moisture absorption ranges for engineering plastics. Moisture absorption in molding resins often degrade the properties of the molded parts. Molding resins need to be kept dry.

The structure of polymers affects dimensional stability. Semicrystalline thermoplastics may change dimension over time. After fabricating a semicrystalline polymer part (for example, by molding or extrusion), both amorphous and crystalline phases are present in the polymer. With time, typically a portion of the amorphous phase changes to a crystalline phase. As this transformation occurs, the volume of the polymer changes. Because the crystalline phase involves chain folding and ordering, the volume of the polymer contracts, causing dimensional shrinkage in the part. The dimensions of a semicrystalline polymer part are often stabilized by postcuring at a specific temperature and time. Polymers that are predominately amorphous are generally more dimensionally stable.

Thermosetting polymers, such as epoxies and phenolics, cure by a cross-linking mechanism. Most parts made from thermosetting resins achieve 95% of their strength in a relatively short amount of curing time, and the parts may be used for their intended application. However, with time, the polymer may continue to cure and cross-link. This extended curing may cause dimensional changes in the part (typically shrinkage). Thermosetting materials are often postcured by heating the parts for an extended period of time (commonly 8 to 24 h) to counteract this phenomenon.

Residual stresses in a polymer part may cause dimensional changes and warpage with time. Annealing processes are typically used to relieve residual stresses and stabilize dimensions. Fixturing may be required to help the part hold its proper shape during the annealing process.

The coefficient of thermal expansion cannot be neglected in designing plastic parts. When bonded or fastened to metal parts, severe distortion may occur on heating. In bearings, this is a factor that can lead to seizure. The coefficient of expansion of metals falls in the range of 5 to approximately 13×10^{-6} in./in. °F (9 to 23×10^{-6} cm/cm °C). The range for polymeric materials is from 5 to 160×10^{-6} in./in. °F (9 to 290×10^{-6} cm/cm °C). Most unfilled plastics expand at a rate of ten times that of steel. Fillers are sometimes added to polymers to help stabilize part dimension. Talc, mica, and milled glass (fine, very short glass fibers) are typically used to improve dimensional stability and control thermal expansion.

Anisotropy

The properties of many reinforced plastics vary with direction of loading. This is *anisotropy*. The strength of fabric-reinforced phenolics is much greater in the direction of the woven filler than the strength measured edgewise. From the design standpoint, this ordinarily does not cause any major problems, because the part can usually be designed to prevent edgewise loading. In unreinforced plastics, anisotropy can also occur, but it commonly causes problems only when, for example, a disk is trepanned out of a thick extruded plate; neither the disk nor the hole will be round after the trepanning because on cooling from the extrusion temperature, uneven cooling may impart residual stresses. When these stresses are relieved in a machining operation, distortion occurs. This condition can usually be overcome by a thermal treatment similar to stress relieving in metals (Table 7–9).

In summary, many plastics have properties that vary with respect to orientation in a standard shape. These can be best dealt with by designing the long axis of a plastic shape as the axis

Table 7–9
Example annealing (stress relieving) treatments for various plastics (verify with plastic manufacturer before using)

Polymer (Unreinforced Grades)	Heating Rate[a]	Annealing Temperature	Hold Time	Cooling Rate	Environment[b]
Acrylonitrile butadiene styrene	30°C/h	The HDT[d] at 1.8 MPa	5 min/mm thickness	25°C/h	Air
Polycarbonate[c]	30°C/h	120°C	As short as possible (flash or infrared anneal). End use testing required to determine optimal time.	25°C/h	Air
Acetal homopolymer	40°C/h	155°C	5 min/mm thickness	25°C/h	Nitrogen
Nylon 6	40°C/h	150°C	5 min/mm thickness	25°C/h	Nitrogen or oil
Nylon 66	40°C/h	175°C	5 min/mm thickness	25°C/h	Nitrogen or oil
PBT	30°C/h	The HDT at 1.8 MPa	5 min/mm thickness	25°C/h	Air
Polyethlyene terephthalate	40°C/h	175°C	5 min/mm thickness	25°C/h	Nitrogen or oil
Polysulfone	40°C/h	165°C	5 min/mm thickness	25°C/h	Air
Polyetherimide	50°C/h	205°C for 2 h	5 min/mm thickness	25°C/h	Air
PPS	50°C/h	205°C for 1 h	5 min/mm thickness	25°C/h	Air
Polyetherether ketone	75°C/h	150°C for 2 h, then heat to 190°C at 95°C/h and hold for 2 h		25°C/h	Air

[a]Multiply by 1.8 and add 32 to convert to °F.
[b]Preferrably, the parts should be immersed in a heat transfer media that will not attack the plastic (oil, glycerin, glycol/water, etc.). Parts should not touch each other. Fixturing may be required to control shape distortion. Shield the parts from radiation heating elements, as melting can occur. (From various sources.)
[c]Polycarbonates can lose impact strength when annealed.
[d]Heat deflection temperature (HDT) at 1.8 MPa

of the principal stresses in the part, and internal stresses can be reduced by stress relieving.

Flammability

The *flammability* of a plastic can be a design consideration if there are any open flames that could possibly come in contact with the part being designed. The thermosetting resins will not melt like the thermoplastics, but they do burn, and some sustain combustion after the source of heat is removed. The term *self-extinguishing* was used to describe a plastic that does not have this characteristic. Over time, flammability tests have become more quantitative. For those plastics that sustain combustion, the flammability can be compared by a burning rate test (ASTM D 635). The propagation of flame on a given-sized specimen is measured in inches per minute. Figure 7–10 illustrates the UL (Underwriters Laboratory) system of quantifying the flammability of plastics. The best rating in these series of tests is 94 V-0: The material does not sustain combustion, and there are no molten drips from the material that can set something else on fire.

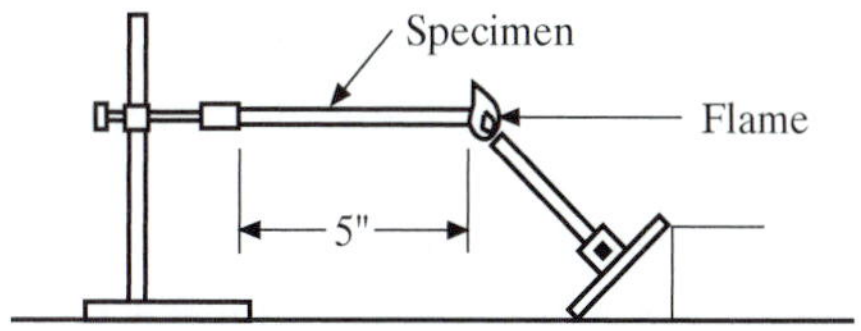

Rating	Maximum burning rate (0.12 to 5-in.-thick samples)	Maximum burning rate (<0.12-in.-thick specimen)	Maximum burn length
94 HB	1.5 in./min/3-in. span	3 in./min/3-in. span	4 in.

	Maximum burning time per flame application (s)	Maximum burning time for ten flame applications (s)	Maximum length of burn (in.)	Flaming drips (that ignite cotton)	Glow time after flame removal (s)
94 V-0	10	50	5	None	30
94 V-1	30	250	5	None	60
94 V-2	30	250	5	Yes	60

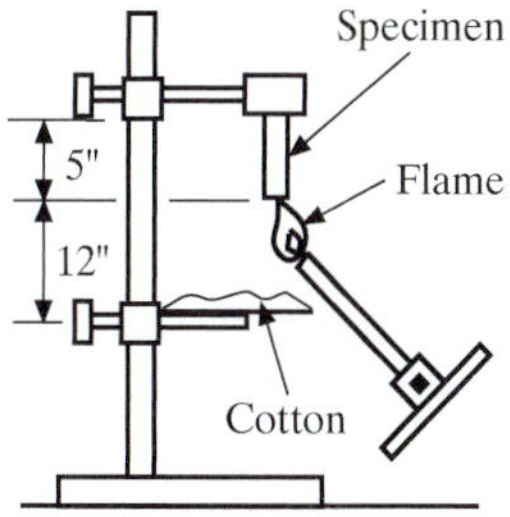

Figure 7–10
ANSI/UL-94 tests for flammability of plastic materials for parts in devices and appliances
Source: Underwriters Laboratories, Inc., Melville, Long Island, NY, 1985.

The limited oxygen index (LOI) test determines the oxygen level required to initiate combustion of a polymer when exposed to a candle flame under specific test conditions. Since air contains approximately 21% oxygen, polymers requiring more than 21% oxygen to combust will probably be resistant to combustion in air (theoretically). As the temperature of a plastic increases, its LOI reduces. As a rule of thumb, an LOI of a least 27% is necessary for fire retardation. Table 7–10 compares the LOI of some commodity and engineering plastics.

Burn rate and combustibility are affected by the composition of the polymer. For example, the higher the hydrogen to carbon ratio in the polymer the greater is the propensity for combustion. Copolymerization sometimes reduces the burn rate. Styrene acrylonitrile, a copolymer of polystyrene and acrylonitrile, has a lower burning rate than the base styrene resin. Some polymers may emit gases that suppress combustion.

Aside from the inherent flammability resistance of a plastic, flame retardants are often

Table 7–10
Limiting oxygen index of selected polymers

Polymer	**Limiting Oxygen Index (%)**
Polyoxymethylene (acetal)	15
Polypropylene	17
Polyethylene	17
Polybutylene terephthalate	18
Polystyrene	18
Polyethylene terephthalate	21
Polyamide (nylon 66)	25
Polyphenylene oxide	32
Acrylonitrile butadiene styrene	32
Polysulfone	30
Fluorinated ethylene propylene	34
Polyether sulfone	36
Polyvinyl chloride	33
Polyamide-imide	46
Polyetherimide	46
Polytetrafluoroethylene	90

Data from various sources.

added to achieve certain flammability ratings. The intent of flame retardants is to increase the plastic's resistance to ignition and to reduce the burning rate. Flame retardants include chemicals such as polybrominated diphenyloxides, brominated phthalate esters, alumina tetrahydrate, and ammonium polyphosphates to name a few. Plastics containing flame retardants such as polybrominated biphenyls and polybrominated diphenyl ethers have come under regulatory scrutiny because of the presence of dioxins and furans. The health aspects of plastic additives should always be a concern.

A common design application in which the melting characteristics or flammability of plastics is important is in tanks and containers used to store flammable liquids. Most government safety codes will not allow plastics to be used. In a fire, these containers will either melt or burn, spill their contents, and add to the fire. Plastics do not have the fire resistance of metals, and this factor must be considered in design.

Mechanical Properties

In general, engineering plastics have ultimate tensile and compressive strengths that are significantly lower than those of metals, especially when compared to tool steels and high-strength steels; but with metals such as aluminum, zinc, copper, and magnesium, the differential in ultimate mechanical strength is much less. Table 7–4 compares the strength ranges of typical plastics.

Tensile strengths in excess of 100 ksi (689 MPa) are common in steels, but if we divide this strength by the density and compare it as a strength-to-weight ratio, we find that the reinforced polyesters and epoxies have higher values than metals.

The hardness of plastics is one mechanical property that is difficult to compare with metals, but the statement can be made that even the softest steel will be harder than the hardest plastic. This statement cannot be made quantitatively in that there is not an accepted hardness tester that can be used on both plastics and metals, but an approximate comparison is illustrated in Figure 7–11. The evidence of this might be best demonstrated by scratch resistance. In spite of the inherent softness of plastics compared to metals, for many applications they perform very satisfactorily in scratching and abrasion resistance. The thermoset laminates (melamines and phenolics) used for tables and countertops resist many forms of abuse, and as a machine workstation surface, they excel.

Although plastics are not as strong or as hard as most metals, if their other characteristics, such as low specific gravity, cost, formability, and machinability, are considered, they may make a better choice. The lower strength can easily be dealt with in design calculations on section size.

Summary

We have discussed a number of plastics that could be successfully used for structural components. Which of the approximately 25 types discussed should be used? For general purposes, those listed in Table 6–3 will probably suffice. For higher loads, engineering plastics should be used. Nylons, acetals, and reinforced phenolics are usually available in standard shapes. This makes them preferred structural plastics for small quantities of machine parts. All three have excellent strength characteristics. The reinforced phenolics have the highest strength, but they are highly anisotropic. Round shapes are usually available only with paper reinforcement. These grades tend to be more prone to internal voids than plate material. Plate material has good properties when loaded on the flat face, but its properties are greatly diminished when loaded edgewise. Thus, they make good high-strength components, but these factors must be considered.

Nylons are second to phenolics in strength characteristics. A readily available grade, 6/6, is

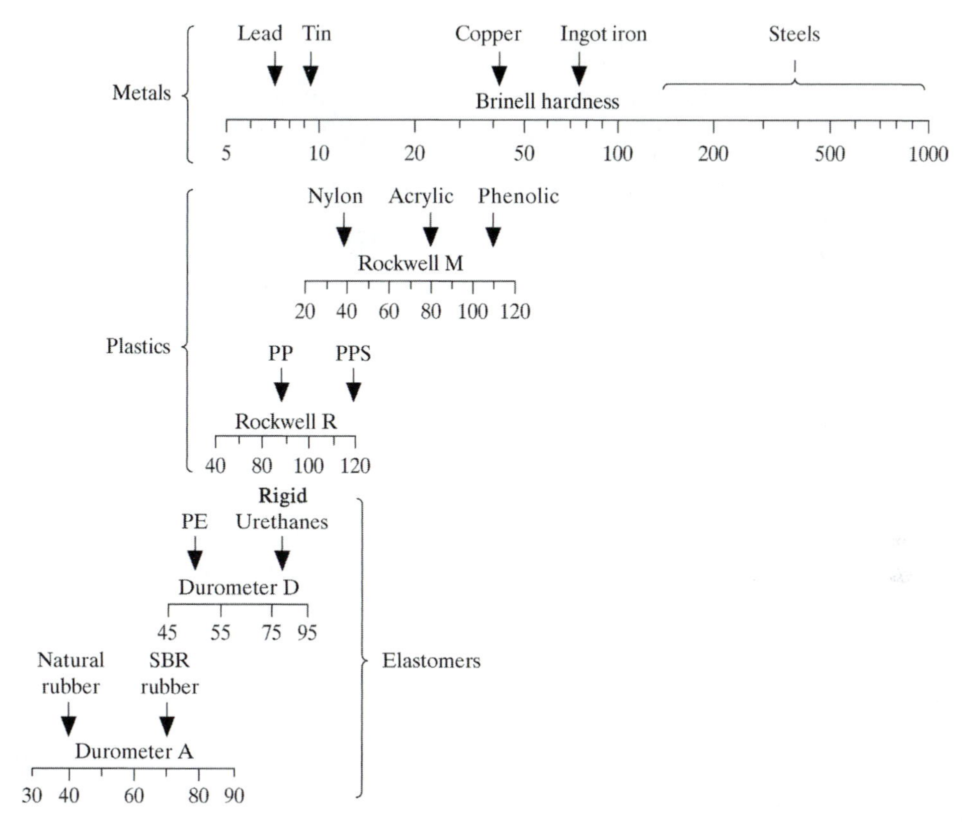

Figure 7–11
Approximate comparison of the hardness scales for plastics, metals, and elastomers

very sensitive to moisture, which makes it difficult to use for parts that require close dimensional tolerances. Acetals are a compromise material for structural components when stability plus strength are design requirements.

The procedure recommended when selecting plastics for structural components involving significant stress levels is to first consider the physical properties that may pertain to your application and then to match strength requirements with the plastics that appear suitable from the property standpoint. As an example, if the part under consideration is a set of mating spur gears, the designer should consider the environmental conditions and establish if chemical attack is a factor, if moisture absorption is a factor, if the coefficient of thermal expansion is a factor, and so on. If none of these factors appears to be important, the strength considerations can be addressed. The beam strength of the gear teeth can be calculated. This value can be compared with the tensile strengths of the candidate plastics. If impact is likely, the impact strength of candidate plastics can be weighed. Once all the physical and mechanical properties have been considered, the designer can then evaluate fabricability and availability of candidate materials.

The selection of plastics for structural applications involves scrutinization of the various properties that we have discussed and the availability of the plastic. Availability depends very much on your quantity requirements. If thousands of parts

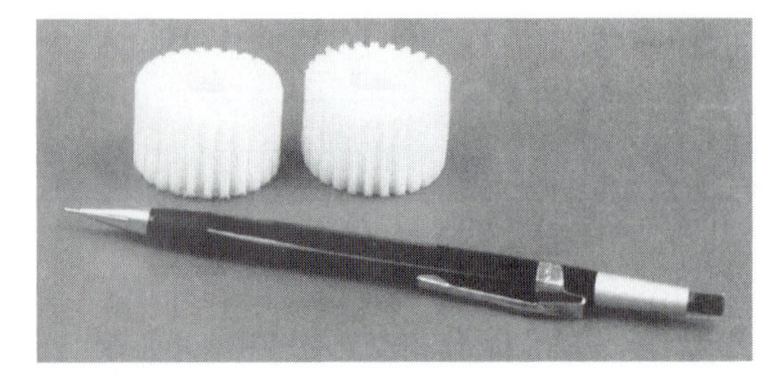

Figure 7–12
Acetal gears

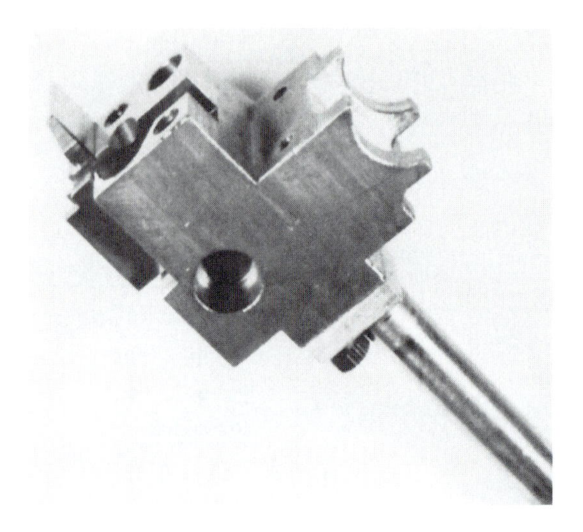

Figure 7–13
Polyimide sliding mechanism

are needed, any plastic may be a candidate. If two spur gears are needed, the designer may need only to weigh the relative properties of the six or so grades of plastic that are available from the local plastic supplier. Some typical examples of structural applications of plastics in machines are shown in Figures 7–12 and 7–13.

7.3 Wear and Friction of Plastics

Wear and friction are systemic properties. Despite many advertisements on materials that make these claims, a surface cannot have wear resistance or low friction. By definition, wear is progressive material removal from a surface produced by relative motion between that surface and contacting surfaces of substances. Similarly, friction is resistance to motion between contacting surfaces when an attempt is made to move one surface with respect to the other. It takes more than just a surface to create wear or friction. It takes a system. Something sliding on a surface.

We describe at least 18 specific modes of wear and erosion in Chapter 3, but at this point it is sufficient to introduce the four major categories of wear that the individual modes fit into:

1. **Adhesive wear.** Wear (W) at least initiated by localized bonding between contacting solid surfaces. Adhesive wear is usually a function of the sliding distance and the applied load:

   $$W \approx k \times (\text{sliding distance}) \times (\text{load})$$
   [the Archard equation]

   where k is a proportionality constant. In metal systems, the denominator contains the hardness of the softer of the two metals. *Specific wear rate* or *wear factor* are terms commonly used for the k term in the wear equation.

2. **Abrasive wear.** Wear produced by hard sharp surfaces or protuberances imposed on a softer surface. This type of wear generally obeys the adhesive wear relationship, but a term related to the sharpness of the abrasive is included.

   $$W \approx k \times (\text{sliding distance}) \times (\text{load})\ 3\ (\tan a)$$

   where a is the included angle of the imposed tip of an abrasive particle.

3. **Erosion.** Wear produced by the mechanical interaction of a fluid (single phase or

multiphase with or without particles or droplets) with a surface. Erosion is usually proportional to the kinetic energy of the impinging fluid/particles/droplets, etc.:

$$\text{Erosion} \approx MV^2/(\text{yield strength of the wearing material})$$

where M = mass of impinging substance (particles, droplets, stream, etc.)

V = velocity of the particles, droplets, etc.

4. **Surface fatigue.** Wear produced by repeated compressive stressing of a surface. This category of wear is usually relegated to rolling element bearings or to systems such as a wheel on a track. Material removal tends to occur in the form of material spalled from one or both of the contacting surfaces. As in most fatigue processes, component life is the usual metric. The fatigue life of a surface is said to be a function of the maximum stress that is encountered in the contact area:

$$W \approx (\text{constant})/(\text{maximum stress})^9$$

Needless to say, these relationships are greatly simplified, but there are no agreed-to models for most wear modes. Existing models are largely empirical. However, people who work in tribology (the science of friction and wear), usually agree to these system "influences."

Abrasion Resistance of Plastics Plastics are not widely used for abrasion resistance, but plastic parts inadvertently see abrasion in countless applications. For example, almost all small power tools used by tradespeople (drills, saws, sanders, routers, etc.) have plastic cases. These tools are dragged over concrete, thrown in the back of trucks, etc. They make contact with hard, sharp particles or surfaces in many ways. They are subject to abrasive wear. Which plastics resist abrasion best? Elastomers are known to have very good resistance to abrasion. They deform rather than cut when a hard, sharp surface is imposed on them. This is why our automobiles have rubber tires. Rubbers resist abrasion quite well. The situation with rigid plastics is not desirable. As shown in Figure 7–14, only a few plastics resist abrasion from sand better than soft stainless steel. In fact, only two materials in this test were more abrasion resistant than the steel: ultra-high-molecular-weight polyethylene (UHMWPE) and a polyurethane with hardness of 90 Shore A. Some very hard and high-strength plastics had very poor abrasion resistance. Glass-reinforced epoxy was poor, as was cloth-reinforced phenolic. These are harder plastics with higher stiffness than most plastics. Even PEEK, a high-performance plastic, did not perform as well as the relatively soft and weak UHMWPE. These tests suggested that abrasion resistance is aided by the ability of a plastic surface to deform when imposed on by an abrasive. Also, if the surface has low friction, abrasive particles tend to slide on the surface rather than plow a furrow. If the surface has high friction with abrasive particles, such as polyurethane, the particles tend to roll through the wear interface rather than plow a furrow.

The two winners in these laboratory tests are the most widely used plastics for abrasion resistance. Many other tests have shown that polyurethanes (in the hardness range of 60 to 90 Shore A) and UHMWPEs excel in abrasion resistance compared to other plastics. Available models do not completely explain why these materials work better than others, but people use them anyway. However, they are not useful at temperatures much above room temperature.

Erosion of Plastics Plastics are often used in systems in which erosion is the predominating mode of wear. Plastics are used for basement sump pumps, which pump sand-laden water. PVC pipes are used for waste plumbing in most houses and commercial buildings. Most dishwashers and washing machines contain plastic pumps that are exposed to particulates in water. Plastics often outperform metals in these applications primarily

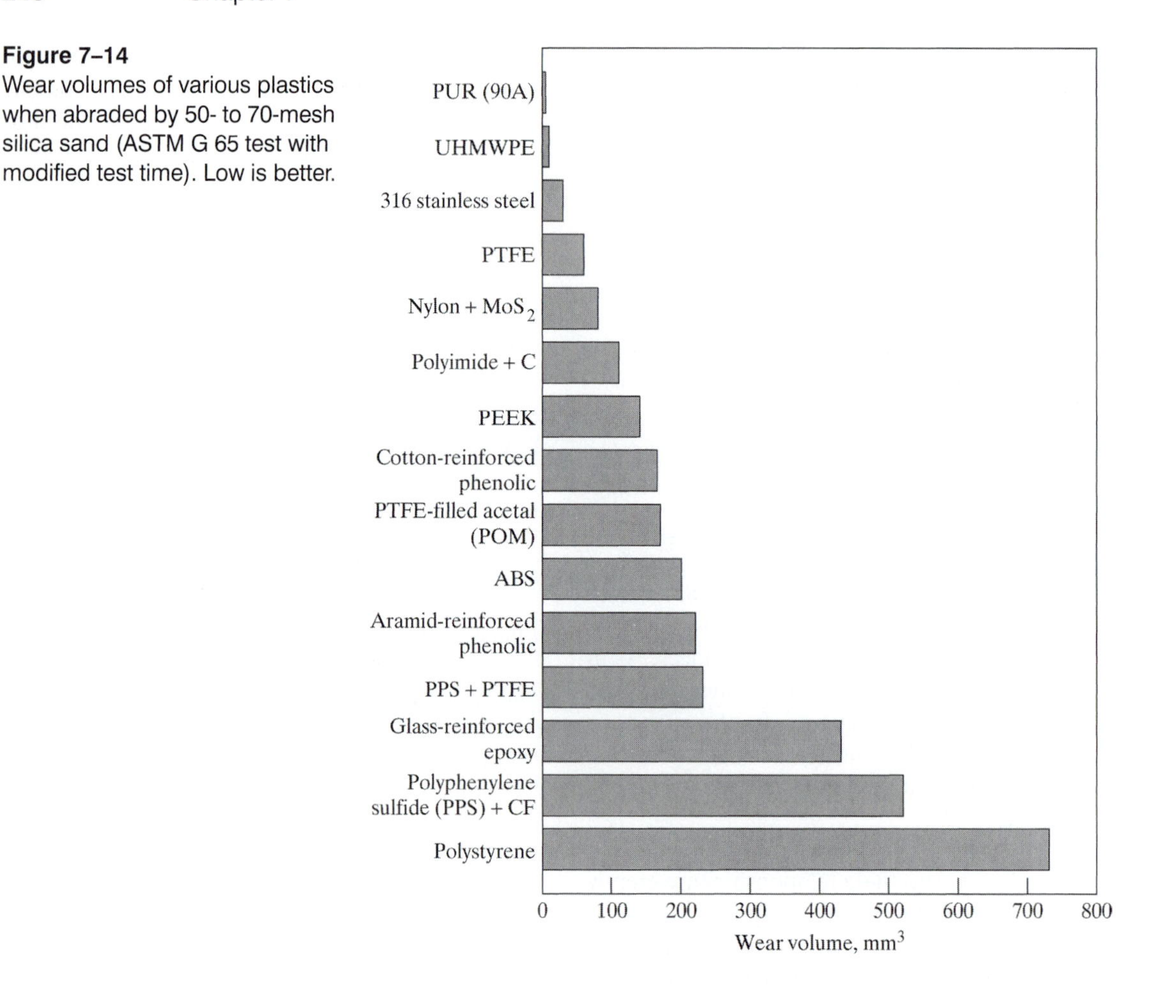

Figure 7–14
Wear volumes of various plastics when abraded by 50- to 70-mesh silica sand (ASTM G 65 test with modified test time). Low is better.

because they are immune to the corrosion component of erosion that usually exists in metal systems. Most metals receive their resistance to corrosion from the presence of passive films on the surface. Impinging fluids or particles can remove protective surface films and corrode the material. This corrosion is often conjoint with material removal due to the action of particles on the surface. In other words, erosion of metal is usually a combination of corrosion and mechanical action from the fluid. This corrosion component is often absent when plastics are used for erosion resistance. Thus plastics are very successful in resisting erosion from water-based slurries and high-velocity water. They are not very resistant to erosion by dry powders and particles (such as sand). In general, plastics excel in handling lightly loaded water slurries, but in heavily loaded slurries and in dry conveying of particulates, plastics usually are less erosion resistant than a soft carbon steel. For example, carbon steel resists erosion in mixing concrete better than most plastics.

Surface Fatigue Plastics are used for balls and races in rolling element bearings and in many types of rollers in which surface fatigue could occur. However, if the tough plastics, such as nylon and acetal, are used, they almost never fail by a mechanism of surface spalling. They usually wear by a mechanism of uniform wear. Ball and roller diameters continually get smaller in use. Cracking and spalling usually do not occur. In other words, the plastics that are normally used for tribological applications do not usually fail by surface fatigue. They fail by adhesive wear rather than by a mechanism that involves fracture of the surface.

Adhesive Wear A more descriptive term for this form of wear is solid-to-solid sliding wear. Plastic-to-metal, plastic-to-plastic, plastic-to-composite, plastic-to-ceramic, plastic-to–any other solid surface are the types or tribosystems included in this wear category. The compatibility of various plastics with various solid surfaces is not very predictable. There are no universally accepted models to predict wear suitability. The general wear equation (Archard) only tells a prospective plastic user that the wear will increase with pressure and total sliding distance. What to do? Plastic suppliers have responded to the need for engineering information on these systems with laboratory compatibility tests. Plastic compounders and neat resin manufacturers have developed a plethora of test data that they use to predict the compatibility of various siding couples. Plastic data sheets often supply information on PV limit and wear factor. We will describe these terms, but the reason compatibility tests are needed is that plastics really do not like to slide against other solid surfaces without the presence of a lubricant. Plastics are poor conductors of heat. Sliding on another solid surface creates heat, which tends to degrade the plastic or make it melt. Additives in the form of particulates, oils, or reinforcing fibers are used to make plastics more resistant to wear degradation. Typical wear and friction additives include:

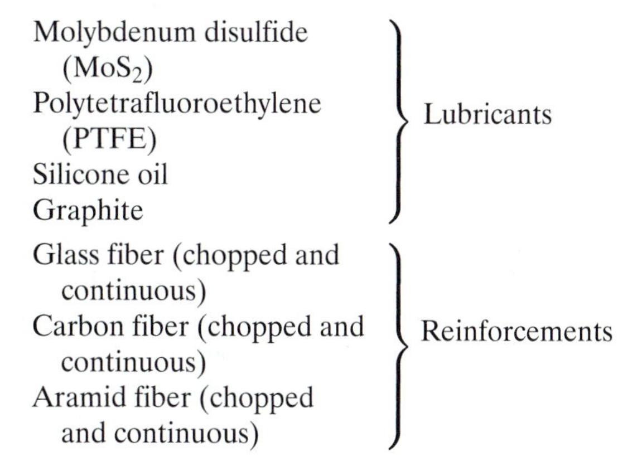

Molybdenum disulfide (MoS_2)
Polytetrafluoroethylene (PTFE)
Silicone oil
Graphite
} Lubricants

Glass fiber (chopped and continuous)
Carbon fiber (chopped and continuous)
Aramid fiber (chopped and continuous)
} Reinforcements

In addition to the above, plastics are blended and alloyed with each other to improve tribological properties. Figure 7–15 illustrates the role of

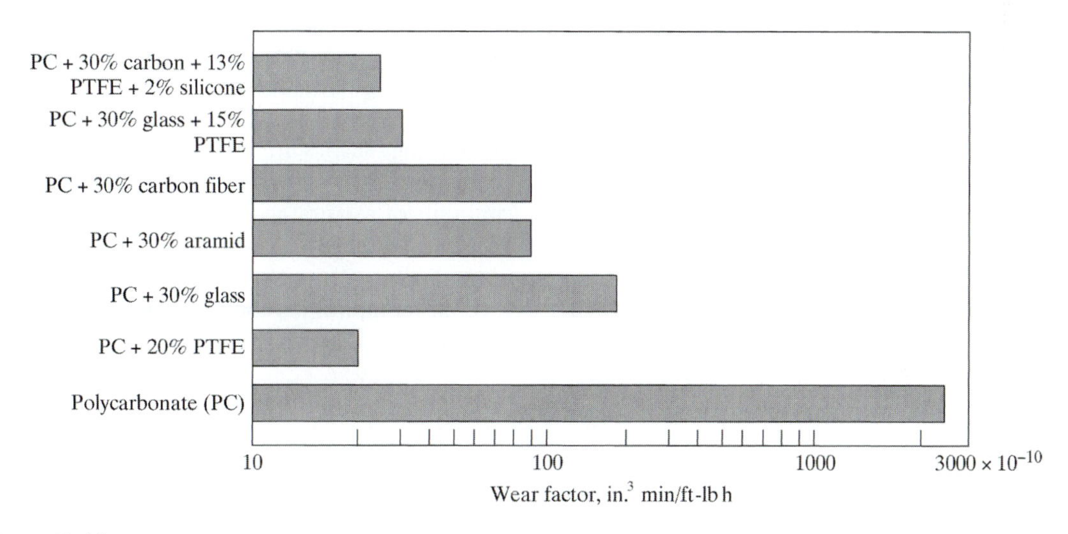

Figure 7–15
Effect of fillers and reinforcements on the wear properties of polycarbonate versus steel in a thrust washer test

additives on polycarbonate (PC). The wear factor, which is a measure of the wear rate in a particular sliding system, goes from around 2500 (in English units $\times$ 10^{10}) for the base resin to as low as 25 with a particular combination of additives.

In summary, plastics will usually produce high wear rates when sliding against themselves or other surfaces unless they are either modified to be self-lubricated or inherently slippery, such as the fluorocarbons or polyethylenes. Plastic suppliers make formulations aimed at various tribosystems. Potential users of plastics for tribosystems must become conversant with plastic wear tests, PV limits, and wear factors. They are your selection tools.

Plastic Wear Tests One of the more popular tests used to measure the wear characteristics of plastics against other solid surfaces is the thrust washer test (ASTM D 3702), which is illustrated in Figure 7–16. The plastic of interest is made into a cup configuration by molding or machining. This cup or thrust washer is rotated against a cylinder of any mating material of interest. The most widely used counterface is 1020 steel with a surface roughness of about 12 μin. (0.3 μm). Tests are conducted at various velocities and apparent contact pressure, and the wear on the rider is measured. Test times are often in the hundreds of hours for the "better" plastics.

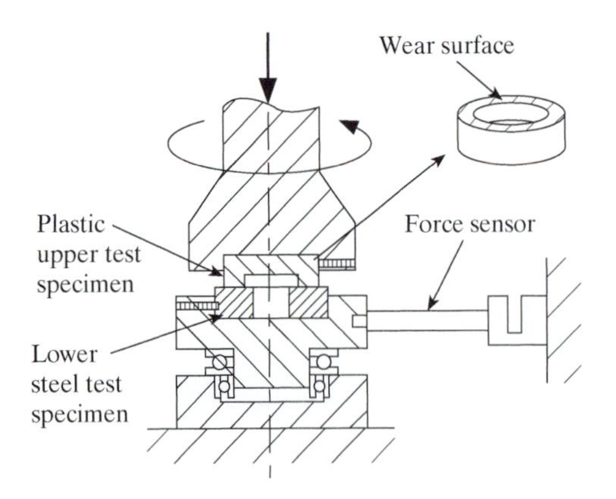

Figure 7–16
Schematic of ASTM thrust washer wear tester for plastics

Candidate plastics for plain bearings are often tested on bushing wear testers. This test uses an actual bushing on a shaft with a known load and sliding velocity. The test metric is wear on the bushing. Test times are frequently more than 500 h.

One of the standard abrasion tests for plastics uses a sandpaper tape to abrade plastic coupons that are affixed to a conveyor that periodically comes in contact with the abrasive tape (ASTM D 1242). Mass loss is the test metric. Another abrasion test uses abrasive-filled rubber wheels rubbing on a flat test sample to abrade material from the surface. The test samples are flat sheets about 100 mm square and a few millimeters thick. The abrading wheels make a groove that is about 300 mm in diameter and 25 mm wide. Mass loss on the flat plastic plate is the metric. This test originated as a means of measuring the durability of floor tiles. There are many other plastic wear tests, but the three mentioned here are the most widely used in the United States to develop empirical wear data on plastics.

Plastic-to-Plastic Wear There are many applications that require a plastic to slide on another plastic. As one might expect, if frictional heating is high between two thermoplastics they will want to melt and friction weld together. If the mating plastics are immiscible, they may not weld but one or both may generate copious amounts of wear debris. In other words, plastic-to-plastic sliding systems are prone to high wear rates. Such systems usually require testing to arrive at compatible couples. Here are some guidelines to use in selecting test candidates:

1. Try to use lubricated grades for both members.
2. If one member is reinforced, the other should contain a lubricant (PTFE, etc.).
3. Avoid rubbing glass-reinforced materials on heat- or carbon fiber–reinforced plastics.

The glass can abrade the carbon fiber–reinforced member or unreinforced counterface.

4. Glass-reinforced plastics may need a glass-reinforced counterface.
5. The potential for fretting damage is high in most plastic couples. Fretting is oscillatory motion of small amplitude as might occur in shipping plastic parts by truck (if the plastic parts touch each other).

Whenever possible, try to avoid mating plastics in tribosystems involving significant sliding and velocity. Plastic credit cards can tolerate many rubs against the plastic guides of the card readers (probably more than we want), but your card would not last very long if this rubbing were continuous. An application in which plastics frequently successfully slide on other plastics is gears. We will discuss this application in a subsequent section. Some couples work well, but many do not. We stand by our admonition to try to avoid plastic-to-plastic systems for continuous sliding.

Plastic-to-Metal Wear Plastic sliding on metal is probably one of the most important applications of plastics because plastics can be self-lubricating. They can eliminate the need to lubricate in service. This not only saves the user from the time required for relubrication, but it also eliminates the need to put in a grease fitting or other features to accommodate a petroleum lubricant. A typical automobile contains several dozen plastic-to-metal bushings and other plastic-to-metal sliding systems. Which plastics slide against which metals? As mentioned previously, the plastic member should be lubricated, so you should select a lubricated grade of plastic. Plastics can wear all counterfaces (including ceramic and cemented carbide), but the harder the counterface, the lower the wear. Most unreinforced plastics or plastics with soft reinforcements such as aramid fibers can run against hardened steel counterfaces (60 HRC) with low system wear, but glass-reinforced plastics should have a counterface that is harder than the glass, which has a hardness of about 700 HV. The only metals that are harder than glass are diffusion-treated steels (borided, TiC treated, etc.). In other words, use very hard counterfaces for glass-reinforced plastics.

In the converse situation, steel shafts can abrade a plastic rubbing on the steel if the steel is too rough. Tests with different roughness counterfaces suggest that there is an optimum counterface roughness for the metal member of plastic-to-metal sliding systems. Wear is lowest when the metal counterface roughness is in the range of 8 to 12 μ in Ra (0.2 to 0.3 μm). Wear is higher with smoother or rougher counterfaces. The surface roughness of the plastic member should ideally be in about this same range. If the surface of a plastic wear part is very rough, it will rapidly wear to create a smooth surface with a roughness in the range that we are recommending. Significant break-in wear produces excessive bushing clearances and may have similar undesirable effects on other plastic-to-metal tribosystems.

Characterization of a plastic for tribological applications is commonly done by determining the plastic's PV limit. PV is an acronym for the product of an apparent pressure (P) and a sliding velocity (V). Units of pounds per square inch (psi) are used for pressure, and velocity has units of surface feet per minute (sfm) in the United States. The metric units for PV are MN/m^2 and m/s. A PV limit is determined by holding either pressure or velocity constant and increasing the other until failure (high wear rates) occurs in one of the laboratory tests. Figure 7–17 shows a PV determination on a reinforced PTFE. It can be seen that wear rates rise sharply when the PV is larger than 10,000. Thus the PV limit is stated to be 10,000. It is common in the United States to leave this term unitless. The higher the PV, the greater the ability of a particular mating couple to resist high sliding velocity and loading. Needless to say, these data must be used with discretion. A PV of 10,000 certainly does not mean that the material in Figure 7–17 can be used at a

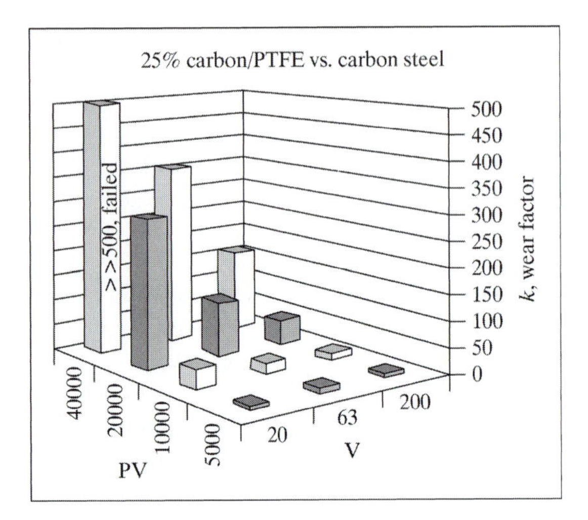

Figure 7–17
Wear test results on 25% carbon fiber–reinforced PTFE sliding on low-carbon steel under varying sliding conditions. The PV limit is 10,000.
Source: Courtesy of LRI Lewis Research, Inc.

sliding velocity of 10,000 feet per minute if the loading is only 1 psi. PV limits only apply to pressures and velocities similar to those used in performing PV limit tests. The proper way to use these data is to aid in comparison of candidate materials. They are not meant to be used to calculate allowable operating conditions.

Another aid available in selecting plastics for wear applications is wear factor. Wear factor is essentially the constant in the wear relationship previously presented:

$$\text{Wear} = k \times (\text{sliding distance}) \times (\text{load})$$

A wear test is run on the thrust washer, bushing, or other tester, and the k is calculated from the measured wear volume. The lower the wear factor, the better the wear resistance. These numbers (U.S.) can range from about 5 to more than 5000. The better plastics have a wear factor of less than 100.

Elevated Temperature Wear One of the advantages of using plastics in tribosystems at elevated temperature is that these materials can sometimes be used at temperatures that are too high for petroleum lubricants. Ordinary mineral oils often turn to tar at temperatures in excess of 400°F (204°C). As shown in Figure 7–7 some plastics can operate above 400°F, but the upper use limit was about 500°F (260°C) in 2004. All plastics lose significant strength at elevated temperatures, but some retain enough to allow their use at high temperatures. The crystalline materials, in general, have better elevated temperature resistance than the amorphous plastics. Carbon fiber is usually the preferred reinforcement for elevated temperature service. Hardened steel (60 HRC) is the usual counterface for plastic bushings used at elevated temperatures. The reinforced fluorocarbons will operate at elevated temperatures with very light loads and moderate sliding speeds. PTFE/glass or PTFE/aramid fabric or similar composite bearings are relatively low-cost, high-temperature bushing materials. PEEK is often better than other thermoplastic bearing materials at the extremes of the maximum use temperature range. Polyimide composites usually have an edge over thermoplastic materials in temperature resistance. Reinforced phenolics work well at temperatures below 400°F (204°C). The usual selection situation exists in picking a plastic for this kind of service. Testing of candidates is ordinarily required if the operating conditions tax the PV/temperature limits of plastics. We recommend screening based on PV and elevated temperature property data. Table 7–11 contains property information on some of the more common plastics for tribological applications.

Plastic Gears The use of plastics for gears may be the second most important engineering application of plastics (behind bushings). Plastic gears can run without external lubrication, are quieter than steel gears, have low inertia, do not rust, and are often much cheaper than metal gears. They can usually be molded to finished shape. Successful use of plastics for gears

Table 7–11
Properties of plastic bearing materials

	Unmodified Polymers					Modified Polymers				
Performance Characteristic or Property	Nylon	Acetal	Fluorocarbon	Polyimide	Phenolic	Nylon, Graphite Filled	Acetal, TFE Fiber Filled	Fluorocarbon, Wide Range of Fillers	Polyimide, Graphite Filled	Phenolic, TFE Filled
Maximum load, projected area (zero speed), psi[a]	4,900	5,200	1,000	10,000	4,000	1,000	1,800	2,000	10,000	4,000
Speed, continuous operation (5-lb load), max fpm[b]	200–400	500	100	1,000	1,000	200–400	800	1,000	1,000	1,000
PV for continuous service, 0.005-in. wear in 1000 h	1,000	1,000	200	300	100	1,000	2,500	2,500 50,000[c]	3,000	5,000
Limiting PV at 100 fpm	4,000	3,000	1,800	100,000	5,000	4,000	5,500	30,000	100,000	40,000
Coefficient of friction	0.20–0.40	0.15–0.30	0.04–0.13	0.1–0.3	0.90–1.1	0.1–0.25	0.05–0.15	0.04–0.25	0.1–0.3	0.05–0.45
Wear factor, $k \times 10^{-10}$ in^3 min/ft-lb h[d]	50	50	2,500	150	250 2,000	50	20	1–20	15	10
Elastic modulus, bending, psi $\times 10^6$[e]	0.3	0.4	0.08	0.45	5	0.4	0.4	0.4	0.63	5
Critical temperature at bearing surface										
°F	400	300	500	600	300–400	400	300	500	600	300–400
°C	204	149	260	316	149–209	204	149	260	316	149–209
Resistance to:										
Humidity	Fair	Good	Excellent	Good	Good	Fair	Good	Excellent	Good	Good
Chemicals	Good	Good	Excellent	Good	Good	Good	Good	Excellent	Good	Good
Density, g/cm^3	1.2	1.43	2.15–2.20	1.42	1.4	1.2	1.54	2.15–2.25	1.49	1.4
Cost index for base material	1.4	1	5	15	—	1.5	6	5	15	—

[a] Multiply by 6.89×10^{-3} to convert to MPa.
[b] Multiply by 0.3048 to convert to m/min.
[c] Exceeds limiting PV.
[d] Multiply by 118.1 to convert to cm^3 min/m-kg h.
[e] Multiply by 6.89×10^3 to convert to MPa.

Source: From R. P. Steijn, "Friction and Wear of Plastics," *Metals Engineering Quarterly*, ASM, May 1967.

requires adherence to PV limits and wear factors, but plastic gears are so widely used that most plastic suppliers have grades of plastics that they know will work well as gears. Some popular plastic gear materials are

Nylon %
Cloth-reinforced phenolic
Acetal homopolymer
PTFE-filled acetal
Acetal copolymer
PPS/40% glass
PET/PBT polyesters
Nylon % + MoS_2

Successful use of plastic gears requires use of calculations to determine if the particular plastic has sufficient strength and stiffness for the application. It is almost always better to mate a plastic gear with metal gear because most plastic-to-metal couples wear less than plastic-to-plastic couples. If glass-filled plastics are needed for the anticipated service conditions, it may be better to use all reinforced plastic gears. A glass-filled gear will abrade all but a very hard metal gear. Another consideration is the effect of moisture on gear dimensions. Some plastics, especially nylons, exhibit significant volume expansion with increases in moisture in the air. Sometimes selection of plastics for gears may be dictated by what is available from plastic gear manufacturers. There are suppliers in the United States that make standard-size plastic gears (usually from nylons or acetal). If large numbers of gears are needed for an application, selection should be based on strength, PV value, wear factors, and manufacturability (moldability).

Plastic Plain Bearings As mentioned previously, the use of plastics for unlubricated plain bearings is one of the most important uses of plastics in design engineering. The wear factors that are published in the literature can be used to calculate service life of plastic bushings. The Archard equation for adhesive wear is rewritten to solve for the time that a bushing will last in service. Wear factors from the literature are substituted for the k in the equation. In Europe the wear factor is commonly referred to as specific wear rate, and it has units of cm^3 min/m-kg h; they use the same equation:

$$t = k\text{PV}T$$

where t = radial wear, in. (cm)
k = wear factor, in^3 min/ft-lb h (mm^3/N m)
P = contact pressure, psi (kN/m^2)
V = sliding velocity, ft/min (m/s)
T = time, h (s)

A frequent cause of plastic bearing failure is seizure due to inadequate operating or *running clearance*. Plastics require greater operating clearances than metal-to-metal sliding systems. Figure 7–18 shows some experimentally determined guidelines. To ensure that these clearances are maintained, it is necessary to consider the reduction in clearance due to interference fitting and thermal expansion. Most plastic bearings are retained by press-fitting them into some type of housing. Since most plastics have a stiffness of less than one-thirtieth that of steel,

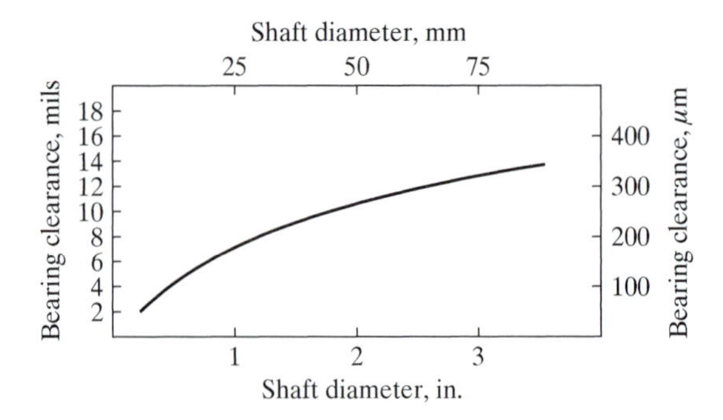

Figure 7–18
Typical running clearances of plastic plain bearings

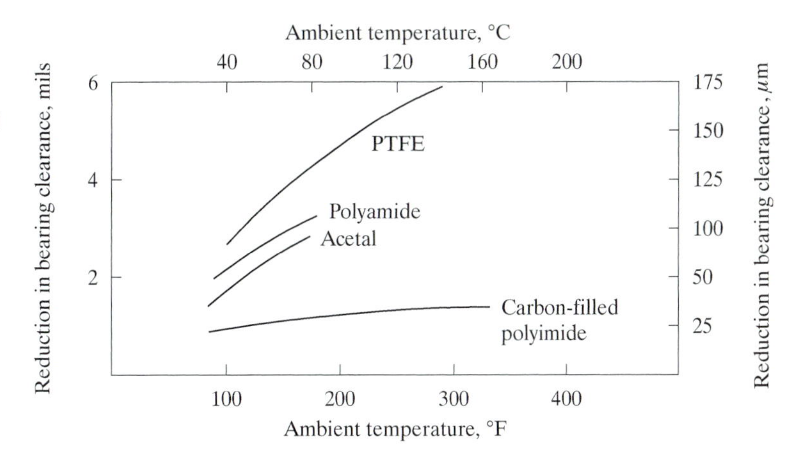

Figure 7–19
Reduction in running clearance on plastic bearings (0.1-in., 2.54-mm wall thickness) retained in steel housing

essentially all the interference goes into reduction in clearance in the bore.

One way to deal with this is to machine or ream the bushing after it is interference fit into its housing. It is important not to overstress the bushing during assembly. There are interference fit equations that can be used to determine proper interferences, but there are guidelines such as 0.1% of the bushing outside diameter interference for a plastic bushing in a metal housing (0.01% for a metal bushing in a metal housing). However, the use of calculations that take into account bushing wall thickness and housing geometry is preferred.

Temperature effects can be taken into consideration by again assuming that any expansion of the bearing due to temperature rises in operation will manifest itself as a reduction in radial clearance. Figure 7–19 shows experimentally determined effects of temperature on some typical plastic bearing materials. A useful technique that is employed to minimize thermal expansion problems in plastic bearings is to keep the wall thickness to a minimum. Shafts under 0.5 in. (12 mm) in diameter can have a wall thickness as small as 0.06 in. (1.0 mm). Bushings as large as 10 in. in diameter (250 mm) can be made with a minimum wall thickness of only 0.31 in. (7.9 mm).

Successful use of plastics for plain bearings requires consideration of factors that control operating clearance. Plastics are very sensitive to this, and seizure will often result if these guidelines are ignored. A variety of plastic plain bearings that has been successfully used in continuous operation in manufacturing machinery is shown in Figure 7–20.

Plastics for Controlled Friction Surfaces

Some engineering designs require low friction between surfaces, some require high friction, and others require a controlled value, be it high or low. The wear resistance of these systems is often not as important as the friction. For example, in a cassette of movie film the film must slide over

Figure 7–20
Plastic bearings: solid polyimide acetal ball bearing, PTFE fabric-lined steel, and nylon-lined steel (left to right)

various surfaces without stick-slip behavior and frictional noise. The film-to-cassette friction must be uniform and controlled, but wear is not a factor because the cassette is discarded after one use. Similar situations occur in sliding parts down a track. Uniform part feeding is often more important than wear characteristics; the track can be made thick to accommodate significant wear. Plastics can be extremely useful in making controlled friction devices, but selecting the right plastic for a device usually requires simulation of service conditions and testing.

The mechanical properties of plastics are significantly affected by temperature. Thus, as sliding velocity and pressure increase, heating occurs and shear strength and hardness decrease. This implies a lowering of friction as conditions become severe. This is exactly what happens with most plastic–metal and plastic–plastic couples. At high loads and velocities, the *coefficient of friction* tends to decrease asymptotically toward a small value. At low loads and velocities, μ is relatively independent of area and the effects of sliding velocity are not severe (Figure 7–21).

In selecting plastics for controlled friction surfaces, it is not possible to rank various plastics because a different ranking will exist for every set of operating conditions. Some selection guidelines based on experience in manufacturing machines are listed next.

1. Avoid self-mating of plastics at high loads and velocities; welding can occur.
2. Plastic–plastic couples usually produce a lower μ but higher wear and debris than a plastic–metal couple.
3. Fluorocarbons and polyethylenes usually have the lowest frictional coefficients against metals and against other plastics (Figure 7–22).
4. At low normal forces (<100 g), surface texture can affect friction.
5. Ambient conditions, especially humidity, can affect plastic friction and wear.
6. Low normal forces promote stick-slip behavior.
7. Wherever possible, use simulative tests to select mating couples (duplicate operating conditions).

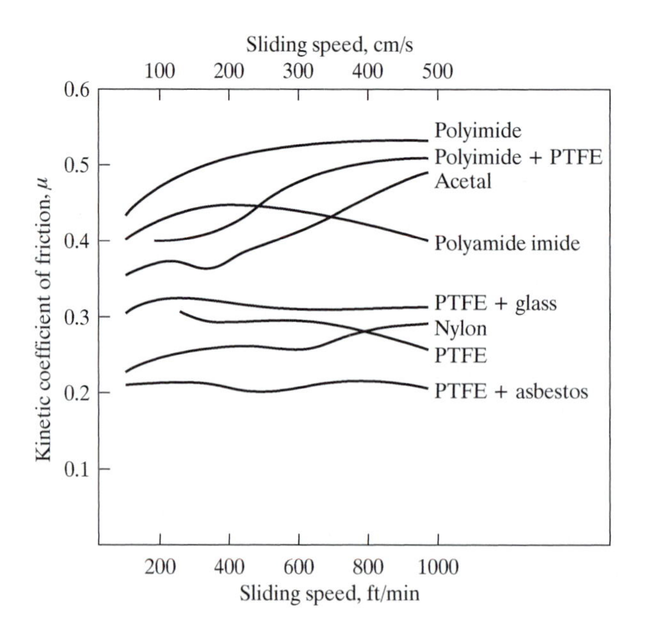

Figure 7–21
Frictional characteristics of various plastics at different sliding speeds; mating material 1020 steel, normal force 50 g

aluminum pits, wood rots, and concrete can pit and stain. Plastics and polymeric materials have a long history of corrosion control. Plastics are not immune to environmental effects, but how they are affected by environments differs from other engineering materials. This discussion briefly reviews some of the most common modes of plastic degradation.

There are many ways in which a polymer may react with its environment. Clearly, if a part must endure exposure to an aggressive chemical solution, then one must verify that the polymer will withstand the effects of the environment. However, other less obvious environments can affect plastics. For example, in an office or home environment, humid air and ultraviolet radiation from fluorescent lights can dramatically affect the service life of some plastics. Sunscreen lotion transferred from a user's hands onto the plastic case of a camera can cause paint coatings to become tacky. Steam or radiation sterilization of polymeric biomedical parts can embrittle the polymer. All environmental factors must be investigated when selecting a polymer for an application.

Permeation: A polymer's interaction with its environment is typically initiated with reactions at the surface. The long molecular chains of polymer macromolecules have a certain amount of kinetic energy that causes movement of the chain segments and main chains. Resulting from this movement is space or *free volume* between the polymer molecules. Liquid chemicals absorb into the surface of the polymer by diffusing into these molecular interstices or free-volume spaces. Often, crystalline polymers are more resistant to chemical permeation because the close-packed arrangement of the molecular chains reduces free volume. Elastomers, on the other hand, have higher free volume than thermoplastics, which makes them more permeable and susceptible to absorbing liquids and chemical species.

In some instances, permeation of a chemical can cause the polymer to swell or grow in dimension. Chemical permeation can also cause a polymer to become soft due to plasticization (see Section 4.3 for additional details). Often food and drink packaging such as plastic beverage bottles use various polymer layers to control unwanted permeation (air and moisture permeation can make food taste stale). In the automobile industry, regulations restrict the amount of gasoline that can permeate through fuel lines and tanks. Materials must be selected to minimize the permeation of refrigerant (sometimes called *fugitive emissions*) in automotive air-conditioning systems. Materials for such applications may be selected based on the permeation rate of the chemical of interest.

Dissolution: In some instances, chemicals can dissolve the polymer chains or portions of the polymer chain by a process called *dissolution*. Additionally, certain constituents such as plasticizers or other additives can be selectively leached from the polymer. Selective leaching of plasticizers can cause a polymer to become brittle whereas selective leaching of an additive such as a stabilizer can lower a polymer's resistance to chain scission, which can also make the polymer weak or brittle. Dimensional change of the polymer component (typically shrinkage) can also accompany dissolution. Applying a solvent such as toluene to remove a price tag from cellulose acetate sunglasses will dissolve the plastic and ruin the sunglasses.

Absorption: By way of permeation, many polymers are prone to absorption of chemical entities. For example, many condensation polymers are hygroscopic and will absorb water from direct exposure to water vapor in air. Hydrolysis can soften or weaken a polymer as well as cause swelling and dimensional changes. Acidic or alkaline conditions can exacerbate hydrolysis. Chemical absorption subsequent to dissolution can cause selective leaching of constituents from the polymer. Swelling may also occur with absorption. For example, natural rubber in contact with kerosene will swell to several times its original volume.

Environmental Stress Cracking (ESC): Another form of chemical degradation is called environmental stess cracking. Certain chemicals cause polymers to crack prematurely (when under sufficient mechanical stress). Environmental stress cracking is similar to stress rupture (discussed in Section 7.2) except that chemical exposure can dramatically reduce the time to failure at a given stress. All polymers are susceptible to ESC; although, thermoplastics in their amorphous state are more susceptible to ESC than are semicrystalline thermoplastics or thermosetting polymers. Polymers with closely packed cyrstals are much more resistant to ESC.

Chemical Attack: Chemicals can attack a polymer by breaking down the molecular bond by a process called *chain scission*. Such reactions can occur directly or by way of catalytic reactions. In some instances, a functional group can be replaced by substituents from the corrodent. For example, sulfate radicals from sulfuric acid exposure can replace hydrogen atoms in the polymer structure—dramatically affecting the polymer's properties. For example, nylon in nitric acid will turn to a gooey mass.

Physical Aging: Besides chemical degradation, polymers also undergo physical aging. In physical aging, there is no change in the polymer's chemical structure or molecular weight; rather attributes such as free volume and crystallinity are affected. For example, some polymers may contine to increase in crystallinity with time, causing dimensional changes (typically shrinkage). This can be a problem for close-fitting components as shrinkage can cause binding or distortion. *Free-volume relaxation* is a process whereby the free volume in the amorphous phase of a polymer is consolidated or densified with time and temperature. Free-volume relaxation can reduce ductility and lower impact toughness. Typically, exposure to elevated temperatures increases free-volume relaxation in polymers. Physical aging can be seen on old car tires. They invariably crack, and surfaces will give off a black smut when rubbed. Rubbers often age poorly as a result of outdoor exposure.

Once a chemical absorbs into the surface there are a variety of consequences. Table 7–12 lists some of the most prevalent chemical effects. Additionally, the effects of chemical exposure on the degradation of polymers are typically exacerbated by the presence of mechanical stress and gas permeation.

Chemical Resistance Testing

It is difficult to predict the effects of long-term chemical exposure of plastics based on short-term testing. This is an industrywide problem that is yet to be resolved. The most practical way of assessing the effect of chemicals and environment on the life and performance of polymers is by using immersion or other exposure tests. Samples, such as tensile bars or coupons, are exposed or immersed in the chemicals or environment of interest for various time intervals and are then tested to determine the effect of the chemical on key properties such as:

Weight	Impact strength
Hardness	Compression set
Thickness or volume	Creep strength
Visual appearance	Stress relaxation
Tensile strength	Fatigue strength
Yield strength	Wear resistance
% Elongation	Dielectric strength
Elastic modulus	Arc tracking resistance
Flexural strength	

Often the data are graphed to determine the effect of time on the degradation of the

Table 7–12
Typical degradation mechanisms for polymers

Degradation Mechanism	Effect on Plastic Material or Part
Absorption	Chemicals are absorbed into the polymer, often accompanied by weight gain and swelling.
Desorption/dissolution	Constituents from the polymer are leached or dissolved into the solvent or chemical. Often dissolution involves the physical removal of polymer chains from the polymer solid.
Embrittlement	Chemical exposure causes the polymer to lose ductility, plasticity, or toughness, typically due to cross-linking, chain scission, or surface crazing.
Environmental stress cracking	In components with sufficient tensile stress, chemical exposure can induce or reduce the time for cracks to initiate and propagate in the polymer.
Permeation	Gases and fluids diffuse into and through the thickness of a polymer.
Change in properties: strength, hardness, elastic modulus, toughness	Some chemicals can plasticize or soften a polymer, whereas others may cause an increase in hardness.
Heat aging	Exposure of polymers to elevated temperatures can cause physical aging which can dramatically affect tensile, impact, dimensional, and electrical properties.
Swelling, change in volume or thickness	Absorption of chemicals, solvents, and water can cause a polymer to swell, resulting in volume or thickness changes.
Crazing	Fine cracks (called *crazes*) develop on the surface or internally. Often the cracks are due to combined effects of stress and chain scission, reduction in molecular weight, or cross-linking. Additionally, the combination of stresses and chemical exposure can cause craze development and subsequent crack propagation.
Oxidation	Oxidation of polymer chains, especially at the surface, can cuase chalking (loose, flaking surface), cracking and crazing, and loss of plasticizer (which can make the part brittle). Additionally, oxidation can cause changes in surface finish, appearance, and color.
Ultraviolet light (UV)	The energy from ultraviolet light can break the chemical bonds in a polymer causing reversion (polymer reverts to a sticky, tacky, or liquid resin). UV exposure can also deplete stabilizers in the polymer. Chain scission, cross-linking, and other degradation mechanisms cause failure by embrittlement, crazing, and other failure modes.
Ozone	Ozone exposure can cause chain scission (and weakening) of polymer chains, particularly double-bonded carbons. In rubbers, sometimes a brittle skin is formed.
Staining	Environmental and chemical exposure can cause polymers to discolor or stain.
Hydrolysis	Water is absorbed into the polymer causing possible loss of strength and stiffness, swelling. Most condensation polymers are hygroscopic; they absorb water from direct exposure to water vapor in air.
Fading/color change/yellowing	Environmental and chemical exposure can cause polymers to fade or change color. Additionally, abusive thermoplastic molding conditions can cause yellowing or color changes due to the loss of heat stabilizers.
Plasticization	Exposure to chemicals can cause a reduction in the elastic properties of a polymer such as elastic modulus, shear modulus, and Poisson's ratio.
Biodegradation	Mold, mildew, and fungus growth on the surface of the polymer.
Reactions to form new compounds	Irreversible alteration of the chemical makeup of a polymer chain causing changes in mass or mechanical properties.
Radiation (e.g., medical sterilization)	Ionizing radiation can cause molecular chain scission (which lowers molecular weight) or increases cross-link density (which increases molecular weight). Substantial changes in mechanical and physical properties are anticipated.
Electrical breakdown due to concomitant effects of electric field and water, ionic contaminants, chemicals	Polymers can absorb water or hydrolyze in the presence of water-based fluids, causing reduced dielectric strength or arc-tacking (electrical tracking) resistance.
Additive degradation	Environmental exposure can leach or otherwise affect certain additives in a polymer, reducing its life or affecting its strength.

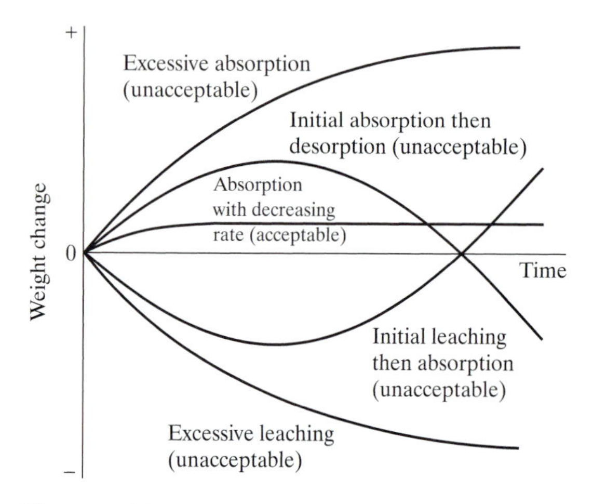

Figure 7–23
The use of weight change to demonstrate the effects of chemical immersion on absorption/leaching characteristics of a polymer

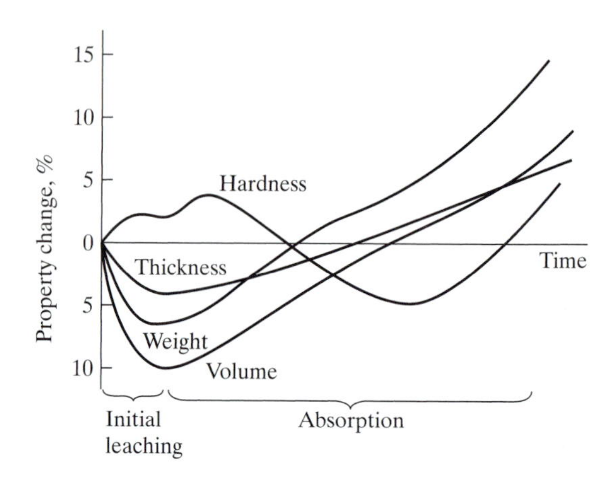

Figure 7–24
Example of property changes in a polymer as a function of immersion time in a chemical environment

polymer and to observe trends. Figure 7–23 shows the effect of chemical exposure on the weight change of a hypothetical polymer. The changes in properties of an elastomer after chemical exposure are shown in Figure 7–24. Table 7–13 provides a guideline for judging the chemical resistance of a polymer to chemical exposure. Typically, polymer chemical resistance tests are conducted in accordance with ASTM standards such as D 1673, D 2990, G 23, D 794, D 1435, D 570, C 619, C 581, and others.

Selection

In selecting a polymer for an application the designer must ask: Will the service environment adversely affect the life of the component? Besides applications of obvious chemical exposure such as a plastic valve used in a chemical plant, one must consider consequential, unintentional, or secondary aspects of exposure. For example, a lubricant applied to a polymer during assembly could cause stress cracking of the polymer. Metal cations from adjoining surfaces can catalyze chemical degradation reactions in the polymer. Adhesives in the uncured state can cause environmental stress cracking of polymers. Mineral oils used in machining can cause subsequent environmental stress cracking. For a given application, each source of chemical exposure must be investigated.

The selection of a polymer for an application involving chemical or environmental exposure

Table 7–13
Guideline for assessing the relative resistance of a polymer to chemical degradation based on a 1000-h chemical exposure test

	Weight Gain	Weight Loss	Thickness Change	Tensile Properties	Hardness	Appearance
Resistant	$< 2\%$	$< 0.5\%$	$< 5\%$	$< 5\%$	< 5 points	No change
Questionable	2% to 5%	0.5% to 2%	5% to 10%	5–20%	5–10 points	Slight change
Not resistant	$>5\%$	$>2\%$	$>10\%$	$>20\%$	>10 points	Significant change

is often based on chemical resistance charts, solubility parameters, and exposure testing. Although solubility parameters are beyond the scope of this text, this method may be used to determine the relative solubility of a polymer in a particular environment. Published chemical resistance tables by plastics suppliers and other sources such as that shown in Table 7–14 often provide a good start toward the selection of a polymer for an application. However, such charts are often based on limited data such as the change in tensile strength after 30 days of exposure. Additionally, small changes in a polymer's formulation can profoundly affect chemical resistance. For example, some vendors may add an ultraviolet stabilizer to make a particular plastic suitable for outdoor exposure when the handbook data suggest that the homopolymer is unsuitable. If pertinent data are unavailable, a corrosion test may be necessary.

7.5 Plastics for Electrical Applications

Everyone knows that plastics are electrical insulators. They are used for countless electrical devices: switches, receptacles, housing for electrical appliances, electrical power tools, and the like. In these roles plastics act as electrical insulators. However, not all plastics provide the same degree of electrical insulation. For each type of electrical application, certain plastics work better than others. Figure 7–25 compares some common plastics in three different electrical properties. *Dielectric strength* is the applied voltage that an insulating material can withstand before insulation breakdown. The units for this property in common use in the United States are volts per mil of thickness; the SI units are volts per millimeter. Thick test specimens tend to produce lower dielectric strengths than thin test specimens of the same material; for this reason it is common practice to specify the test thickness in data handbooks. The designer should use a value that is from tests on thicknesses that are comparable to thicknesses used in the intended service. The larger the dielectric strength, the better the insulator.

Resistivity is a measure of a material's ability to resist the flow of electricity. The resistivity ranges for different families of engineering materials vary over many orders of magnitude, so often different tests are used for metals and for insulators. The term *volume resistivity* is usually used to rate the ability of plastics to resist current flow, and the units of measure are commonly Ω-cm in the United States; the SI unit is the Ω-m. The most common plastic test is performed on a block of plastic with conductors placed on two parallel faces. Surface resistivity tests are also used on plastics and other materials. Some electrical devices such as RF heating units produce current flow on surfaces rather than through the bulk. If this situation exists in an application, this is the property resistivity to evaluate on candidate plastics.

As shown in Figure 7–25, there is a significant difference in the ability of some families of plastics to resist current flow. The range for metals is less than about 100 $\mu\Omega$-cm. Thus most plastics are insulators compared to all metals, but some plastics are much better insulators than others.

The *dielectric constant* is a measure of a material's ability to act as a capacitor. The capacitance of a plastic or other insulator is compared with the capacitance of vacuum. A high number indicates that the material is a good capacitor.

A low number is desired for high-frequency applications; it signifies that the material will have a low rate of electrical heating. The dielectric constant is measured by comparing the capacitance of the insulator of interest with the capacitance of air (assuming the same electrode configuration). Various families of plastics have significantly different dielectric constants.

Additional properties related to the electrical performance of plastics are dissipation factor and arc resistance. *Dissipation factor* is a measure

Table 7–14
Effects of various environments on plastics

Material	Acids		Alkalies		Organic Solvents	Water Absorption (%/24 h)	Continuous Heat Resistance (°F)
	Weak	Strong	Weak	Strong			
ABS (acrylonitrile–butadiene–styrene)	N	AO	N	N	A	0.4	190–230
Acetals (copolymers)	V	A	N	N	R	0.2	185–220
Acrylics							
Methyl methacrylate	R	AO	R	A	A	0.3	140–200
MMA alpha-methylstyrene copolymer	R	AO	R	R	A	0.2	200–230
Acrylic PVC alloy	N	N	N	N	A	0.1	
Allyl resins							
Allyl diglycol carbonate	N	AO	N	SA	R	0.2	212
Diallyl phthalate (mineral or glass filled)	N	SA	SA	SA	N	0.3	300–400
Cellulosic compounds							
Ethyl cellulose	SA	A	N	SA	A	1	115–185
Cellulose acetate, nitrate, propionate, and acetate butyrate	SA	A	SA	A	A	1–7	140–220
Chlorinated polyether	N	AO	N	N	R	0.01	290
Epoxy resins							
Cast	N	V	N	SA	R	0.1	250–550
Glass fiber–filled molding compounds	N	R	N	N	N	0.1	300–500
Fluoroplastics							
Polychlorotrifluoroethylene	N	N	N	N	R	0.0	350–390
Polytetrafluoroethylene	N	N	N	N	N	0.0	550
FEP fluoroplastic	N	N	N	N	N	0.01	400
Polyvinylidene fluoride	V	R	N	N	R	0.04	300
Furan (asbestos filled)	R	AO	N	R	R	0.1	265–330
Melamine–formaldehyde							
Asbestos filler	R	A	R	SA	N	0.1	250–400
Glass-fiber filler	N	A	N	R	N	0.1	300–400
Methyl–pentene polymer	N	AO	N	R	A	0.01	250–320
Nylons							
Type 6, cast	R	A	N	R	R	1.0	180–250
Type 12, unfilled	N	A	N	N	R	0.3	175–260
Phenol–formaldehyde–furfural compounds	R	AO	SA	A	R	0.2	250
Phenylene oxides	N	N	N	N	A	1.07	175–220
Polyallomer	N	AO	N	R	R	0.01	

Polyaryl ether	N	V	N	N	A	0.3	250–270
Polyarylsulfone	N	N	N	N	R	2.0	500
Polybutylene	R	AO	R	R	—	0.01	225
Polycarbonates							
Unfilled	N	SA	V	A	A	0.1	250
ABS–polycarbonate alloy	N	AO	N	SA	A	0.3	220–250
Polyesters							
Linear aromatic	SA	SA	SA	A	R	0.02	600
Alkyd, asbestos filled	N	SA	N	SA	N	0.1	450
Polyethylenes (low, medium, high density)	R	AO	R	R	R	0.01	180–250
Polyimides	R	R	SA	A	R	0.3	500
Polyphenylene sulfides	N	AO	N	N	R	—	400–500
Polypropylenes	N	AO	N	R	R	0.01	190–300
Polystyrenes	N	AO	N	N	A	0.1	150–200
Polysulfone	N	N	N	N	R	0.2	300–345
Silicones	R	V	R	V	V	0.1	500–600
Urea–formaldehyde	A	A	V	A	R	0.5	170
Urethanes (cast)	SA	A	SA	V	V	1.0	190–250
Vinyl polymers and copolymers							
Vinyl chloride and vinyl chloride–acetate	N	R	N	N	V	0.5	150–175
Vinylidene chloride	N	R	R	R	R	0.1	160–200
Chlorinated polyvinyl chloride	N	N	N	N	N	0.1	230

Key: N = no effect; R = resistant, generally; SA = slightly attacked; A = attacked; AO = attacked by oxidizing acids; V = variable behavior, depending on specific media.

Source: From *Corrosion Engineering*. New York: McGraw-Hill Book Co., 1978.

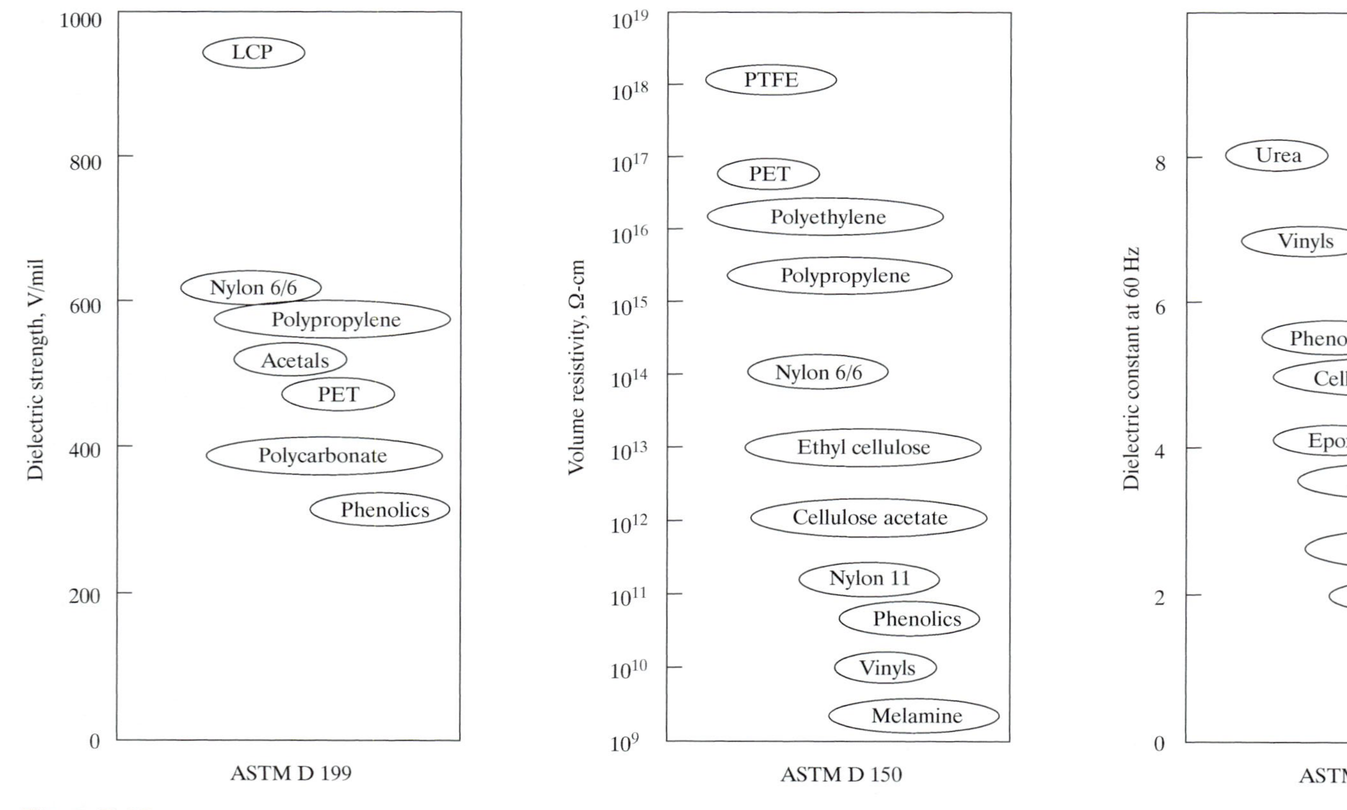

Figure 7–25
Electrical properties of selected plastics

of heating or power loss when the material is used as a capacitor. A low value is usually desirable. The units are often time (s, seconds) per cycle-ohm-farad (per ASTM D 1531). *Arc resistance* is a measure of the ability of a surface to resist breakdown under conditions of electrical stress (an arc). The units of this property are usually the time (s) that a material can withstand the application of an electric arc before it fails.

Other electrical properties can be important for a particular application, but the ones that we have defined are the more commonly used electrical properties. These properties should not be ignored if a design requires a plastic part to interface with some type of electrical device. The more complicated the device is, the more it may be necessary to research the applicability of these properties. Additional properties are tabulated in material handbooks and in plastic manufacturers' literature.

7.6 Polymer Coatings

For centuries, paints have been used for decorative or protective coatings. Early paints were formulated from natural resins, dyes, and pigments. The paints used today are similar, the big difference being the range of polymers that have replaced the natural resins. A *paint* is a suspension of pigment in a liquid that dries or cures to form a solid film. Traditional paints contain a vehicle, a solvent, and pigment. The vehicle forms the film, which traps the pigment. The solvent eases application. Some polymeric coatings are applied by dipping or by spraying parts with a plasticized polymer. Some polymer coating systems (powder coating; etc.) involve heating parts and spraying them with a dry polymer powder that coalesces on the hot part to form a film. Plasma arc processes and the like heat powder particles to the molten state, and they "splat" cool on the substrate to form a coating.

There are also systems where polymer particles are electrostatically charged and sprayed on room temperature parts. These coatings are subsequently fused in a furnace baking operation. Powder coating processes are becoming more important each year because they do not produce environmentally troublesome solvent emissions.

The differences between these various coating systems are the mechanism of film formation and the type of polymer that is being applied. Many important details are involved in surface preparation and in application techniques to arrive at a good polymeric coating. We will only touch on these subjects, but we shall try to present sufficient information on paint systems to allow a designer to select basic coating types for typical design problems.

Film Formation

Some coatings involve chemical reaction, polymerization, or cross-linking; some merely involve coalescence of polymer particles. The various mechanisms involved in the formation of polymer coatings are illustrated in Figure 7–26. These mechanisms must be understood as a first step in the selection of a coating system.

The lacquers and other coatings that are formed by evaporation of a volatile solvent are typified by quick-drying characteristics. Most of the solvents, such as acetone, are highly flammable. Cellulose nitrate is the polymer in some automotive *lacquers*. The paints that form a coating by oxidation of an oil and evaporation of a solvent are best recognized as oil-based paints. Oxygen from the air reacts with the oil vehicles, causing cross-linking to a solid film. The oils used include linseed, tung, tall oil, and others. These paints also contain additives such as drying oils to aid film formation. *Varnishes* are oil paints without pigment. The hardness and gloss of the varnish depend on the oil–resin ratios. When synthetic resins are used instead of natural resins, the paint is often referred to as an alkyd paint. *Alkyds* are essentially polyesters.

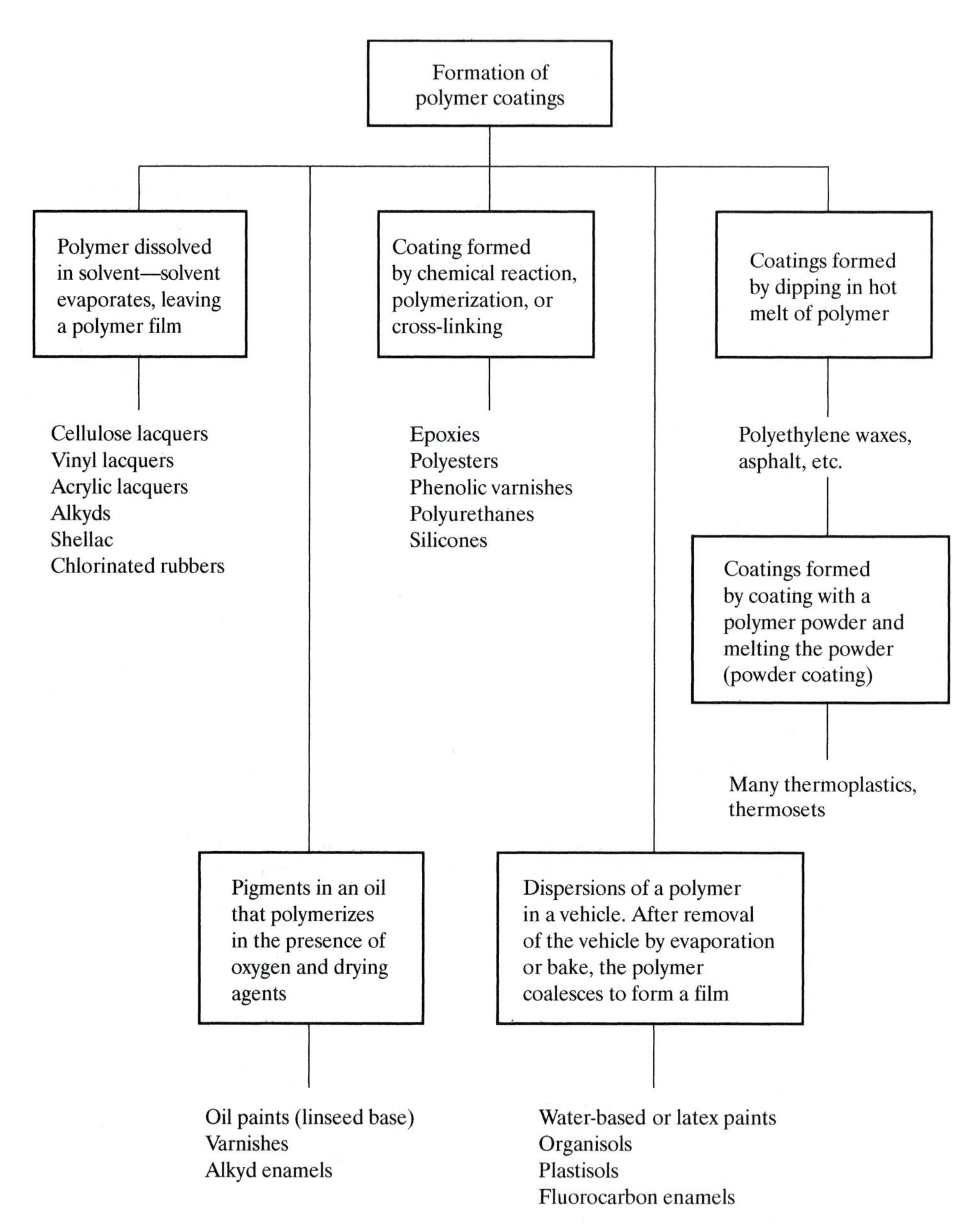

Figure 7–26
Polymer film-forming mechanisms

Many of the coating films that form by chemical reaction, cross-linking, or polymerization are two-component systems. Epoxies and polyester systems frequently cure by action of a catalyst at room temperature. The catalyst can be a significant or insignificant part of the cured film. Included in this class of film formation are coatings that cross-link with curing at an elevated temperature and coatings that cross-link when exposed to radiation.

Radiation curing systems are being promoted as a way of eliminating pollution from paint solvents. Solvents and catalysts are not necessary. Certain polymers will polymerize or crosslink to form a film through the energy supplied by x-rays or gamma radiation, by bombardment with an electron beam, or by exposure to ultraviolet radiation (*UV cure*). The latter are extremely important because they eliminate solvent emissions.

Coatings formed from dispersions of a polymer in water are best recognized as *latex* or water-based paints. They usually contain more solids than other paints. Polymer content can be 60% by weight. The dispersion essentially consists of small polymer particles and pigments. When the water evaporates, the polymer particles coalesce to form a solid film. Water-based coatings eliminate the health problems associated with organic solvents in paints. Some U.S. states are legislating that all paints used in their states be water based.

Organisols and *plastisols* are special types of dispersions. Organisols are polymer particles dispersed in an organic solvent. Plastisols are polymer particles dispersed in a plasticizer.

The last class of film formation, coatings formed directly from polymer—either molten or solid—are more commonly used on heavy coatings (>1 mm). Heavy dip coatings of polyethylene or wax fall into this category. A more refined variation is to preheat metal parts and dip them into a fluidized bed of polymer particles. Powder coating ordinarily entails electrostatic spraying of room temperature parts followed by furnace fusing. The polymer particles melt to form coatings that may be 1 or 2 mils thick (decorative) or as thick as 50 mils (corrosion protection). *Plasma arc spray* (PAS) is used to apply thick coatings (10 mils or more) of thermoplastic materials such as polyethylene or thermoplastic polyester.

In all the preceding systems, the nature of the coating depends on the ratio of pigment to polymer (resin) in the dried film. As mentioned previously, a coating with no pigment is a varnish or *clear coat*. Enamels have a moderate amount of pigment, and high-hiding wall paint has a large amount of pigment. Pigments are commonly fine solids such as carbon black, titanium dioxide, or zinc oxide. Besides adding color and opacity to a paint, pigments often shield materials from UV radiation and improve weathering resistance. Too much pigment can make the paint film weak and chalky. The point to be made from the selection standpoint is that a desired pigment concentration exists for each paint. The average paint user does not know the critical pigment volume concentration for a paint, but if poor service life is observed, this factor could be the cause.

Coating Selection

Each paint and polymeric coating manufacturer has its own formulations. There are no standards for limits on coating components. How then does one go about selecting a coating for a particular application? The first step in coating selection is to list your coating serviceability needs. Is the coating to be only decorative, or does it have to withstand chemicals or some other special environment? Is the coating going to be exposed to outdoor weathering (see Figure 7–27)? Is immersion likely, or will the coating simply be subjected to splash and spill? What temperature range will the coating see? Is cost a factor? Will the coating require flexibility or impact resistance?

The next step in coating selection is to match your service criteria to those of basic types of coatings. It is not possible to select a specific brand of paint or polymeric coating at this point, but if your selection criteria coincide with the properties of an epoxy system, for example, you will then review the relative merits of cost, applicability, performance, and the like for various brands of epoxies and make your final selection. The following coating descriptions include most of the important coating systems that are in commercial use.

Figure 7–27
Outdoor test rack for establishing the weather resistance of polymeric coatings

Vinyls: Paints that employ PVC or a copolymer of PVC and vinyl acetate or vinyl butyral are widely used for general chemical resistance. They are not resistant to many solvents, and their upper use temperature is about 150°F (66°C). Weather resistance is obtained by pigmentation, but for outdoor use, other coatings are superior. When plasticized, vinyl coatings can be extremely flexible, but as is the case with any plasticized coating, the coating can become brittle when the plasticizer migrates out on aging. Vinyls are not particularly hard or abrasion resistant.

The industrial vinyl paints are usually solvent evaporation film formers (lacquers), and they are used on outdoor structures such as tanks, pipes, and steel buildings. They are available in antifouling formulations for boat hull bottoms. Vinyls are also available in latex systems for house paints. They are cheaper than acrylics but do not have the same weather resistance. Vinyls are used for heavy plastisol and organisol coatings for plating fixtures, tool handles, and the like, as well as for decorative powder coatings on appliances and automobile parts. Coil-coated metals for stamped and drawn parts are often coated with thermosetting vinyls. Some countries discourage the use of vinyl paints because they contain chlorine, which is thought by some to have adverse environmental effects.

Epoxies: *Epoxy* paints are the workhorse industrial coatings. They have excellent chemical and solvent resistance, good abrasion resistance, and good adhesion. Epoxies can even be coil coated on metals that will subsequently be blanked and formed into parts. In the chemical process industry, coal tar–epoxy mixtures are used for protection of underground piping and ducts. Epoxy paints can be oil modified to be oxidative film formers, or film formation can be done by cross-linking by catalysis or baking. Because epoxies tend to caulk outdoors, they are not generally applied for outdoor structure protection. They are widely used as machine *enamels*, metal primers, and coatings for equipment in chemical plants.

Urethanes: Polyurethane coatings can cure by catalysis (two component), by absorption and chemical reaction with moisture in the air, and by oxidation in urethane oils and alkyds. The former systems are usually used in industrial-type coatings, and the latter is widely employed in clear wood finishes. Heavy urethane coatings can be used as abrasion-resistant floor toppings. The chemical and abrasion resistance of urethanes makes them popular for machine enamels. Their abrasion resistance even makes them suitable for paints on airplanes, where rain erosion removes normal paints. Clear urethane coatings resist sunlight better than most spar varnishes, but, in general, polyurethanes tend to degrade in sunlight. Urethane/acrylic formulations have been developed for outdoor use. These coatings resist UV damage. Polyurethanes based on aliphatic diisocyanates resist both chalking and discoloration better than other systems. Water-based urethanes (water vehicle), developed in the late

1980s, are widely used for wood floor finishes. The elimination of organic solvents reduces health and emission problems.

Alkyds: Alkyds are polyesters that vary in properties depending on the nature of the organic acid and alcohol from which they are formed. Many hardware store paints (especially the aerosols) are alkyds modified with drying oils. Industrial alkyds usually have smaller percentages of these drying oils. They are good general-purpose paints, but they are not intended for severe service such as resistance to chemicals. Their popularity in decorative paints is due to their gloss and color retention.

Oil-based paints: The natural oils used in oil-based paints include linseed, soybean, tung, castor, and others. These paints are largely oxidative film formers. The oils impart flexibility and gloss to decorative finishes. Oil-based paints have good durability and washability. The dried film consists of natural resins or a combination of natural resins and synthetic resins (alkyds). Most oil-based paints have solvent and chemical resistance characteristics that are inferior to industrial finishes such as vinyls and epoxies. These paints are widely used for exterior wood finishes and interior decorative finishes.

In the 1980s stains became popular in the United States for the final finish on external wood siding. Traditional oil stains were dyes and pigment in linseed oil or the like, but the new "transparent" or "solid" color stains are essentially paints with less than normal pigment. Paints normally contain 50% to 60% solids (binders, pigments, and additives). A solid stain will have only about 35% solids, and a transparent stain may have only 20% solids. Dry-film thicknesses are typically 1.5 and 0.75 mils, respectively. Best outdoor protection of wood is obtained with a 3- to 5-mil finished paint film thickness. This is why stains do not afford the same protection as paints.

Water-based paints: The familiar latex paints are used almost universally for interior and exterior house paints. Film formation is accomplished by particle coalescence. The properties of the film depend on the nature of the polymer that forms the film. The two most widely used polymers are polyvinyl acetate (PVA) and acrylics.

The PVA-based paints are cheaper than the acrylics and have a tendency to breathe. They have moisture permeability, which helps prevent blistering on wood. They are extremely good for interior decorative finishes, but the acrylics have superior weatherability and durability as exterior wood coatings. Water-based paints have the desirable property of no smell or organic solvent emission on drying. Progress is being made on the development of water-based coatings with properties suitable for industrial use. Some water-based polyesters are used for coil-coated steel, but the conversion of industrial coatings to this type of film former will take additional research and development.

Water-based color stains are paints with reduced solids—the same as the solid and transparent oil stains. They do not form as thick of a film as normal latex paints.

Furan: Furan coatings are probably the most difficult to apply (they shrink in curing). They are formed by an acid-induced cross-linking of furfural alcohol. The cross-linked coating is extremely hard and chemical resistant. Because of the difficulty in forming these coatings, they are usually applied to tanks and other chemical process equipment at coating companies that specialize in their use. They are not general-purpose industrial coatings.

Phenolic: Phenolic coatings have many of the properties of the familiar phenolic compression molding resins. They are durable, with good temperature and chemical resistance. As coatings, they are available as oxidation drying varnishes or as industrial coatings that cure by chemical reaction (catalysis or baking). The cross-linked pure phenolic coatings are employed as tank linings and the like. They are brittle and lack caustic resistance, but they are suitable for many types of chemical service.

Special-purpose coatings: In addition to the general coating systems discussed, a number of coatings should be mentioned because they solve special problems. Fluorocarbons can be sprayed or dip-coated to provide extreme chemical resistance or nonstick properties. They are usually applied only to metals, and they are available as pure PTFE, as PTFE-polyimide mixtures, and as conventional enamels that contain PTFE. Vinylidene fluoride coatings are available as nonfade color coatings on exterior building curtain walls. They have a 20-year guaranteed life.

Silicones have the highest temperature resistance of all the coatings; they also can provide nonstick qualities. *Shellac*, a natural resin (secretion from the lac bug) thinned in alcohol, is useful for industrial safety lane paint. It can be removed with alcohol when desired. Plastisol and organisol PVC coatings are useful for high-build metal coatings. There are also a variety of rubber coatings based on various elastomer systems. These coatings essentially have properties similar to those of molded elastomers with the same name.

Inorganic zinc coatings are not polymeric coatings, but they are often used with a polymer topcoat. They are usually silicates with up to 95% zinc filler. They provide excellent outdoor rust protection of steel but are difficult to apply. Surface preparation is extremely critical.

Metallic zinc competes with zinc-containing paints for protection of large structures such as bridges. Welding thermal spray equipment (oxyfuel or arc) can apply a 1- to 10-mil (2.5- to 25 μm) coating of pure zinc on clean steel at the rate of up to 100 lb (45 kg) per hour. In the 1990s this process was still more expensive than paint, but it protects much longer [greater than 30 years for 10 mils (25 μm)].

Flame spraying of plastics such as polyester and polyethylene is increasing in popularity. Thick plastic coatings can be produced without solvent and baking problems.

Summary

We have discussed the general properties of at least a dozen polymeric coatings. To select one of these coatings to do a particular job, the designer should go back to his or her list of coating criteria and determine which coating system provides the best match. Table 7–15 shows a quantitative comparison that assigns a rating number from 1 to 10 (10 is best) on various use properties. This table is only one opinion—there are others. In the chemical process industry, the specific coatings listed in Table 7–16 fulfill most needs.

The final aspect of the use of polymeric coatings is surface preparation and application. Most paints have specific instructions for application. There are no general guidelines other than to follow the manufacturer's recommendations for cleaning, roughening, and priming. Specification of film thickness is usually the designer's responsibility. General industrial environments may require a coating of 5 mils (125 μm) for protection. Structural steel in chemical plants that may be used outdoors and that may see chemical spills may require a dry film of 15 mils (625 μm). Asphalt coatings on buried pipes should be about 90 mils (2250 μm). As a guideline on paint usage, 1 gallon of any liquid applied on a surface to a wet thickness of 1 mil (0.5 mil dry) will cover 1604 ft^2 (150 m^2). Some coatings may have thickness limitations. Mud cracking or peeling could result from a coating that is too heavy. Thus, in coating selection, know your environment, review the properties of the various coating systems, and weigh the selection and application recommendations from reputable paint and coating suppliers.

Table 7–15
Comparative resistance value of typical commercial coating formulations

Condition	Generic Type										
	Neoprene	Vinyl	Saran	Epoxy	Chlorinated Rubber	Styrene Copolymer Blends	Furan	Phenolic	Alkyd	Asphalt	Oil-Based
Sunlight and water	8	10	7	9	7	6	8	9	10	7	10
Stress and impact	10	8	7	3	7	6	1	2	4	5	4
Abrasion	10	7	7	6	7	7	5	5	6	3	4
Heat	10	7	7	9	5	6	9	10	8	4	7
Water	10	10	10	10	10	10	10	10	8	10	7
Salts	10	10	10	10	10	10	10	10	8	10	6
Solvents	4	5	5	8	3	4	10	10	4	2	2
Alkalies	10	10	8	9	10	10	10	2	6	7	1
Acids	10	10	10	10	10	10	10	10	6	10	1
Oxidation	6	10	10	6	9	8	2	7	3	2	1
Total	88	87	81	80	78	77	75	75	63	60	43

Source: From G. S. Weismantel, *Paint Handbook*. New York: McGraw-Hill Book Co., 1981.

Table 7–16
Coatings for the chemical process industry

Application	Coating
On metals for resistance to acids and alkalies and exterior metal surfaces	Polyvinyl chloride (solvent system)
On metals and other surfaces for resistance to solvents	Amine epoxy (two component)
Exterior coating on masonry and decorative interior coatings on wood and metals	Polyvinyl acetate (water based)
Interior semigloss or gloss decorative coatings on wood and metal	Alkyd (oil oxidation)
Clear floor coating and satin finish machine enamel	Moisture curing polyurethane
Oil-resistant gloss enamel for machines, tanks, pipes, etc.	Two-component polyurethane

7.7 Adhesives

Material selection plays an important role when designing an assembly that must be adhesively bonded. Many techniques are used to join two or more components—as summarized in Table 7–17. Mechanical joining methods include screws, fasteners, and snap fits. Although mechanical methods of joining may be fast and can sometimes be removed for repair, often they require extra parts and may add cost and weight to the assembly.

Welding and adhesive bonding processes are alternatives to mechanical joining. Welding heats the interface of the materials to be joined to the melting point and allows fusion to bond the components together. Materials of similar or compatible types are commonly joined by welding.

An adhesive is a substance that functionally bonds two components together through surface attachment. Most adhesives are based on polymer/organic materials. Although soldering and brazing may be considered adhesives, they will not be addressed in this section. Adhesives are useful when vastly different materials are joined (bonding a metal to a plastic) or for materials that are not inherently weldable (thermosetting plastics, foams, wood, paper).

Adhesives perform other functions. For example, adhesives such as silicone and polysulfide rubber are used for sealing and for dampening vibration. The two crankcase halves on a high-performance automobile engine are sealed with a high-temperature silicone sealant. Aircraft manufacturers use adhesives to bond components and to eliminate the weight and expense of rivets and fasteners. Some automakers adhesively bond plastic panels to structural steel space frames with urethane adhesives to eliminate the weight and expense of fasteners while achieving a sealed joint free from vibration noise. Adhesives also allow for uniform stress distribution and minimize stress concentrations caused by screw or rivet holes. One problem with adhesively bonded and welded joints is that

Table 7–17
Joining methods

Mechanical	Fusion	Adhesives
• Snap fits	• Conventional welding	• Bonding
• Fasteners/clips/staples	• Ultrasonic welding	• Potting
• Insert molding or die casting	• Friction/inertia welding	
	• Heat staking	

they are difficult to disassemble. Sometimes this needs to be considered. In this section, basic adhesive systems and their application in engineering design will be discussed.

Mechanisms of Adhesion

Adhesion is not an intrinsic property of adhesives, but basically the response of an adhesively bonded assembly to mechanical separation (a destructive force). Hence, the mechanism of adhesion involves the distribution of stresses within the joint as well as the interaction of the adhesive with the substrate. There is much conjecture over the exact mechanism of *adhesive* bonding; however, there are some prevailing theories.

1. The most basic mechanism is mechanical interlocking between the adhesive and the adherend (substrate). All surfaces, even polished surfaces, are made up of microscopic hills and valleys. As an adhesive wets a surface, the adhesive can become mechanically locked into the microtopography of the surface. The roughness of a substrate may affect the stress distribution at the interface of the adhesive bond as well as energy dissipation during loading. A surface of higher roughness also allows for greater surface contact area. As discussed in a later section, mechanical interlocking may be optimized by using special surface pretreatments. Processes for roughening and pretreating the surface may also remove weak surface layers and contamination that may otherwise compromise the bond's strength.
2. Intimate molecular contact through wetting the adhesive at the interface allows interatomic or intermolecular forces to develop between the adhesive and the adherend. Sometimes primary bonds, such as covalent or metallic bonds, may be formed. Typically, however, secondary electrostatic bonds, such as hydrogen bonds and van der Waals bonds, are most commonly associated with polymer adhesion. It is believed that in any adhesive joint, more than one mechanism is responsible for the intrinsic adhesion of the adhesive.

Surface Preparation

Adhesives bond by surface attachment. The reliability of an adhesive joint is greatly affected by the condition of the surfaces of the *adherends*. At a minimum, the substrates to be bonded must be clean and free from contaminants. Typically, solvent or aqueous cleaning solutions are used to remove the contaminants. With polymers, one must be aware of surface composition changes due to cleaning processes.

The bond strength of many materials may be enhanced with special surface pretreatments. For example, polyethylene is a low-surface-energy plastic. Adhesives generally will not wet the surface of polyethylene or form acceptable bonds. Modifying the surface by exposing it to a corona discharge plasma or by flame treatment alters the surface composition, creating a more reactive surface that is conducive to adhesive bonding.

The reliability of an adhesive bond to aluminum may be enhanced by anodizing the aluminum with phosphoric acid prior to bonding. The anodizing process roughens the surface, and the improved corrosion resistance offered by the anodic layer enhances the bond strength under wet/humid conditions or chemical exposure.

Carbon steel parts are often pretreated by coating the surface with a zinc or iron phosphate conversion coating. Conversion coatings have a microscopically rough surface that allows for excellent mechanical attachment of an adhesive. Conversion coating processes are often performed in bulk, which is more economical than labor-intensive surface treatments such as abrasive blasting or other hand cleaning methods.

These are just a few examples of surface pretreatment techniques for improving adhesive bonds. A brief summary of pretreating processes is outlined in Table 7–18. Pretreating processes should be reviewed with the adhesive supplier

Table 7–18
Surface pretreatments for improved adhesive bonding of materials

Material Category	Material	Typical Surface Preparation Methods*
Metals	Ferrous materials	1. Abrasive blast or abrade 2. Zinc phosphate conversion coating 3. Iron phosphate conversion coating 4. Plate with zinc, iron phosphate 5. Nitric/phosphoric acid etch (ASTM D 2651)
	Aluminum alloys	1. Abrasive blast or abrade and degrease 2. Sulfochromate etch (ASTM D 2654, D 2651) 3. Phosphoric acid anodize (ASTM D 3933, MIL-A-8625) 4. Sulfuric acid anodize (MIL-A-8625)
	Magnesium	1. Abrasive blast or abrade and degrease 2. Chromic acid etching (ASTM D 2651) 3. Anodizing (MIL-M-45202) 4. Chromate conversion coating (MIL-M-3171)
	Copper alloys	1. Abrasive blast or abrade and degrease 2. Nitric acid/ferric chloride etch (ASTM D 2651)
	Zinc alloys and galvanized metals	1. Abrasive blast or abrade and degrease 2. Iron phosphate conversion coating
	Stainless steel alloys	1. Abrasive blast or abrade and degrease 2. Acid etching (ASTM D 2651) 3. Manganese phosphate conversion coating
	Titanium alloys	1. Nitric/hydrofluoric acid etching (ASTM D 2651) 2. Proprietary anodizing processes
Ceramics	Glass, quartz, nonoptical	1. Abrade, clean, dry at 210°F (98°C) for 30 min 2. Combine abrading with chromic acid etch, dry at 210°F (98°C) for 30 min and apply adhesive while part is still hot
	Glass, optical	Degrease/clean only
	Ceramics	Roughen surface by abrading and degrease/clean
	Concrete	1. Abrasive blast (ASTM D 4259) 2. Acid etch and neutralize (ASTM D 4260)
Thermoplastics	CA, CAB, PC, PVC, PS, PMMA, ABS, PA, PET	1. Clean, mechanically abrade, reclean (ASTM D 2093) 2. Plasma or corona treatment 3. Flame treatment 4. Iodine treatment (PA)
	Polyethylene, polypropylene, POM	1. Sulfuric acid/dichromate treatment (ASTM D 2093) 2. Corona discharge treatment 3. Flame treatment
	Fluoropolymers (PTFE, FEP, PFA, ECTFE, and others)	1. Corona discharge treat 2. Sodium naphthalate etch (ASTM D 2093) 3. Proprietary etchants
Thermosetting plastics	Epoxy, phenolic, DAP, polyester, melamine, rigid PUR	Clean, mechanically abrade, reclean
Elastomers	Natural and synthetic rubber, neoprene, chloroprene	1. Clean with methyl alcohol 2. Etch in sulfuric acid and neutralize in caustic

*The surfaces must be appropriately degreased before abrading or blasting (e.g., solvent or other water-based cleaner).

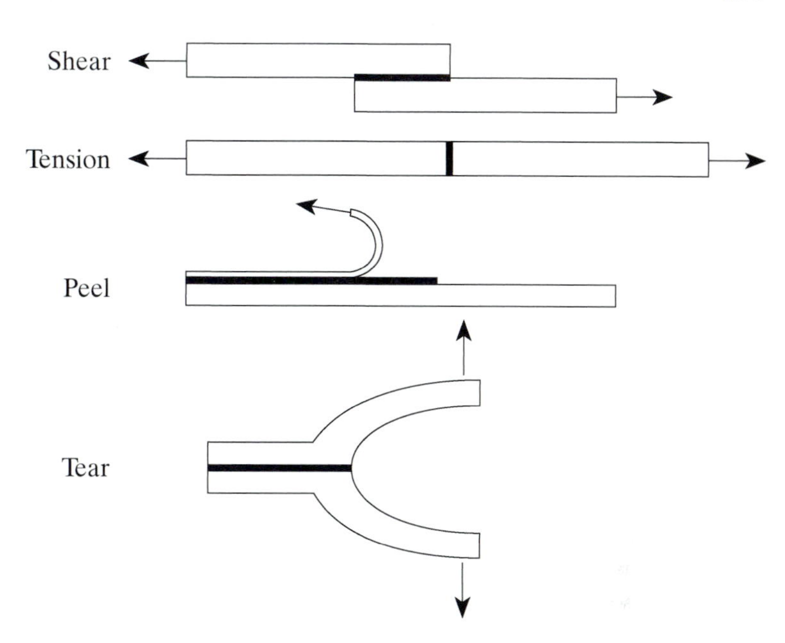

Figure 7–28
Modes of loading adhesive joints

when developing an adhesive system for an application.

Joint Design

Adhesive joint strength is dependent on the distribution of stresses within the joint. Figure 7–28 shows the four modes of loading an adhesive joint. Generally, the highest bond strengths are obtained when a joint is loaded in shear. Adhesive bonds are generally weakest when loaded in peel. As a general rule, the bond area should be maximized while maintaining as much of the applied load in shear or compression (minimizing peel or cleavage forces) as possible.

A number of preferred joint designs are shown in Figure 7–29. Many of these joints translate the applied forces to shear or compressive stresses within the adhesive joint. End conditions may be altered to minimize stress concentrations.

When peel forces must be endured, consider flexible, tough adhesives with low elastic moduli. Even in pure shear situations, flexible, low-modulus adhesives generally provide more uniform stress distribution and often result in higher bond strengths compared with brittle, high-modulus adhesives. In some instances, low-modulus adhesives may not accommodate the applied loads without significant deformation of the joint.

Lap joints are commonly used for adhesive bonding (see Figure 7–29). When designing lap joints, note that the width of a lap joint influences *bond strength* more than the overlap length (for a given bond area, a wider lap joint is stronger than a longer lap joint). The thickness of the adhesive bond line also affects the strength of an adhesive joint. There is usually an optimal bond line thickness for a given adhesive system. Typical bond line thicknesses are in the range of 0.003 to 0.010 in. (75–250 μm). Because adhesives are usually polymers, they are affected by loading rate, time duration under load, and environmental factors such as temperature and exposure to water, moisture, or chemicals.

Joint designs must allow for cost-effective assembly and should not hinder the efficient application of the adhesive to the substrate. The fillet created by expelled adhesive often acts to reduce stress concentrations. Smaller-radius fillets generally increase adhesive joint stresses.

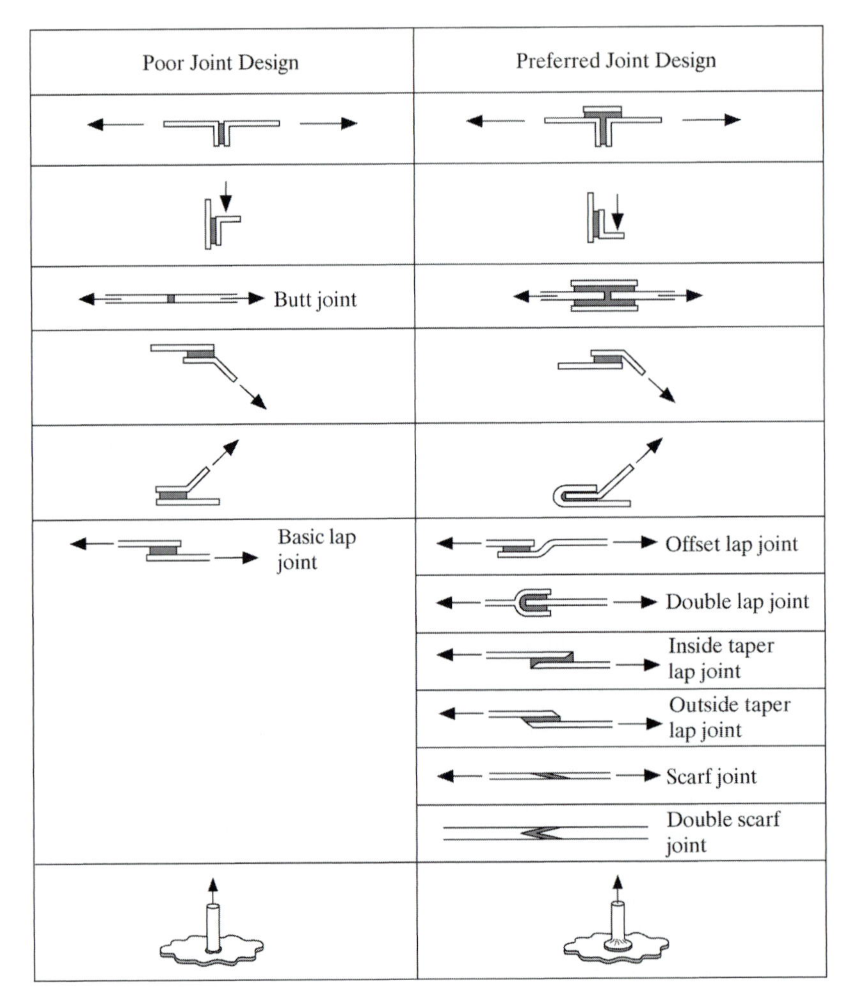

Figure 7–29
Adhesive joint designs

Although beyond the scope of this book, the stresses in an adhesive joint must be evaluated during the design process. Many empirical equations have been developed for various joint designs. Finite-element stress analysis may help optimize adhesive selection and joint geometry.

Adhesive Selection

The process of selecting an adhesive involves reviewing expected service conditions, joint design constraints, manufacturing and assembly processes, cost, toxicity, and other factors. Table 7–19 lists factors to consider when selecting an adhesive. Many adhesive systems are commercially available although it is impossible to review them all in this section. Table 7–20 highlights typical adhesives for engineering applications.

In the area of epoxies, tapes are available that cure with heat application. Hot melt adhesives such as ethyl vinyl acetate can be roll coated for instant bonding. Reactive modified

Table 7–19
Adhesive joint selection factors

Category	Factors of Concern
Materials	Adherends: composition, critical surface energy
	Surface treatments of adherends
	Adhesive system
	Primer (if required)
Joint design	Joint function
	Type of joint (cleavage, peel, shear, tension, compression)
	Joint clearances (bond line thickness)
	Applied loads: creep, stress rupture, ESC, impact, fatigue
	Joint geometry
Environment	Temperature
	Chemicals, environment
	Combined exposure
	Cyclic exposure
Assembly concerns	Application/dispensing method
	Clamping and pressing (logistics and time duration)
	Drying/curing/heating
	Open time/working time of adhesive
	Adherend surface contamination
	Variation in adhesive composition
	Byproducts of the curing reaction can cause outgassing or corrosion
Other aspects	Cost of adhesive, bonding method, and dispensing equipment
	Reliability testing
	Quality control (How do you know the bond is good every time?)

acrylics are suitable for many metal-to-metal and nonmetal couples. They are two-component systems. The adhesive can be precoated to one member and the activator to the other member, and bonding by free radical polymerization occurs when the two adherends are contacted. Two-part urethanes are also used as structural adhesives in some assembly line operations.

The thermosetting adhesives such as the ureas, melamines, resorcinols, and phenolics are used for a wide variety of applications, but, as a class, their advantage over most thermoplastic adhesives is better temperature resistance. Polysulfide rubber is useful for a high-durability adhesive sealant.

Anaerobic adhesives are popular for instant bonding of materials. As the name implies, they cure in the absence of air.

Contact cements as are used for countertops and similar applications are rubber based. They will bond many surfaces, but the bond strength will not approach that of epoxies or other "engineering" adhesives.

Hot melt adhesives, which are thermoplastics that are melted and applied between adherends to make the bond, eliminate the wait required for curing, and these adhesives are becoming more important in high-speed manufacturing operations. A popular hot melt adhesive is ethyl vinyl acetate, but cellulosics and olefins are also widely used. These systems eliminate solvent problems. The use of these materials will increase in the next decade for this reason.

In selecting an adhesive it is desirable to become familiar with the adhesives listed and their idiosyncrasies, properties, and application.

Table 7–20
Summary of the characteristics of common engineering adhesives

Adhesive	Forms/Application Methods	Materials	Advantages	Limitations
Epoxies	Epoxies are available in single- and two-component forms. Single-component forms may be applied as paste, powders, or sheet (exposure to temperature flows the adhesive and cures it). Liquid and paste forms are applied by hand or automated dispensing equipment.	Epoxies are thermosetting polymers based on bisphenol A resin with amine, anhydride, or amide curatives. Polybismaleimide resin is used for high-temperature applications.	High-strength joints Creep resistant May be toughened Temperature resistant Chemical resistant Versatile for mixed adherends Gap tolerant Some are fast curing for high production. Syntactic foams are used for composite applications.	May require temperatures to cure Cure time may be long (2 to 48 h) Not easily disbonded Two-component systems require careful mixing May have poor peel strength May be brittle
Acrylics	Single-component liquids of various viscosity. Fast-setting adhesives, high strength applied by various metering systems.	Acrylics, cyanurates, ultraviolet- or electron-beam-cured systems	Single-component systems Fast curing, high production Optically clear High strength	100% solids One adherent must be transparent for UV/EB curing Uncured adhesive may cause ESC or fogging and contamination
Hot melts	Solid, single-component thermoplastic adhesive that is melted to activate bond. Applied by special hot melt applicators, as powders, or as solvent coatings.	Thermoplastic polyamides, polyesters, polyolefins, polyvinyl acetate, polyurethane	100% solids—no solvents Rapid bond formation Gap tolerant Good for high production Easy disbonding	Limited temperature resistance Poor creep resistance Short open time Low penetration
Pressure sensitive	Transfer tapes may be used to bond adherends together rapidly through the application of pressure.	Rubbers compounded with tackifiers: acrylates, silicones, SBR rubber, natural rubber	100% solids—no solvents Rapid initial bond formation Good for high production—easy application	Limited gap sensitivity Limited heat resistance May creep with static loads

Elastomeric	Single- and two-component systems available.	Based on polyurethane, silicone, butyl, polysulfide, Viton. Contact adhesives based on neoprene or nitrile rubber.	Good sealing Gap tolerant Good impact resistance Good low-temperature resistance Silicones have high-temperature resistance Flexible, tough joints Some are 100% solids.	Longer cure times (2 to 48 h) Two components must be mixed for some systems. Some systems contain solvents.
Water based	Single-component water-based systems. Elastometric caulks, sealants, and adhesives such as wood glue and contact cements.	Resins based on acrylic copolymers, vinyl acetate, latex, vinyl alcohol, various rubber emulsions, and other natural materials. Cures by water removal.	Low order, no organic solvents Low cost and toxicity Some are contact adhesives.	Possible poor moisture resistance Some may not be compatible with plastics Slow drying Poor creep resistance Limited heat resistance Possible shrinkage
Reactive acrylic, (polyester)	Two-component thermosetting systems in paste form. Adhesive is applied to one surface and a curing activator is applied to the other.	Based on polymethyl methacrylate with grafted nitrile rubber. Sometimes mixed with epoxies.	Excellent peel, impact, and lap shear characteristics Good resistance to moisture Excellent bonds with many metals Tolerant to contaminated surfaces Room temperature cure	Poor strength at hot temperatures Possible odor, toxicity, flammability Limited open time Dispensing equipment may be required.
Solvent bonding	Used to bond certain polymers to themselves (e.g., PMMA, PVC, PC ABS, PS). A solvent is applied to both surfaces and they are pressed together.	Various solvents such as methylene chloride, aqueous phenol, and others are used to dissolve the joint interface, allowing adherends to bond.	Possible transparent bonds Relatively rapid initial bond formation	Requires good contact Low gap tolerance High solvent content May cause stress cracking Limited polymers bondable Low open time/limited time for positioning

Table 7–20 can be used as a selection guide. If the adhesive is destined to be subjected to any environment other than room temperature and ambient moisture, then additional investigation of the adhesive is necessary.

Bonding Plastics

Aside from the use of conventional adhesives, many plastics may be fusion bonded or solvent bonded. Options for bonding many common plastics are reviewed in Table 7–21. One of the most efficient means of bonding a plastic is by various welding methods, including ultrasonic welding or staking. If two similar polymers are vibrated together at ultrasonic frequency levels, the interface will melt and create a fusion bond (weld).

Investigate the following aspects when using adhesives on plastics:

1. Any internally or externally applied mold releases that are present on polymer parts may affect joint reliability. Special cleaning may be required to remove mold releases.
2. Additives in the polymer that bloom to the surface may affect adhesion. Plasticizers, solvent residues, stabilizers, and flame retardants can concentrate on the surface of a polymer and adversely affect the reliability of an adhesive bond. When this occurs, consider special cleaning methods or the use of adhesives tolerant to plasticizers.
3. Some adhesives, in the uncured state, may cause stress cracking in polymers. For example, ABS, PC, PMMA, and PS may stress crack in the presence of uncured cyanoacrylates. Minimizing applied and residual stresses, quick and thorough curing of the adhesive, or selecting a non-stress-cracking adhesive will minimize the occurrence of stress cracks.
4. Some adhesives may not wet on low-surface-energy polymers such as PE, PP, fluoropolymers, PA, POM, and others. Special surface treatments (possibly combined with primers) are required to achieve satisfactory bonds with these materials.
5. When polymers are processed by molding or extruding, a weak surface layer may be

Table 7–21
Joining methods for common plastics

Polymer Type	Solvent Bonding	Fusion Welding	Adhesive Bonding
Thermoplastics			
ABS	X	X	X
FEP		X	X[a]
PA	X	X	X
PAI			X[b]
PBT		X	X
PC	X	X	X
PE		X	X[a]
PEEK			X
PEI	X	X	X
PET	X	X	X
PES		X	X
PFA		X	X[a]
PMMA	X		X
POM		X	X[a]
PPS			X
PP		X	X[a]
PPO	X	X	X
PS	X	X	X
PTFE			X[a]
PVC	X	X	X
PVDF		X	X[a]
Thermosets			
Epoxy			X
DAP			X
Melamine			X
Phenolic			X
Polyester			X
Polyurethane			X
Silicone			X[a]
Vinyl ester			X

[a] Low-surface-energy polymer requiring special surface pretreatment to achieve acceptable bond
[b] Special adhesive required

created, affecting bond strength. Sometimes skins on the surface of a polymer must be mechanically removed.

Adhesive Testing and Reliability

One of the greatest challenges in using adhesive bonds in engineering applications is the prediction of long-term performance and joint reliability. Selecting an adhesive system for an application can be a substantial undertaking. Both short- and long-term strength tests must be conducted. Designed experiments should determine the effects of manufacturing variability on joint reliability. These factors must be anticipated when scheduling the development of a new product or component.

Summary

In this chapter we have tried to show how plastics can be applied to specific applications in design engineering. The successful use of plastics in design is predicated on having a basic understanding of polymer chemistry and familiarity with the characteristics of families of plastics. We have tried to show that a few plastics from some of the common plastic families will meet most design needs (Table 7–22). You do not have to learn the names of the 15,000 grades that are commercially available. In addition, it is important to remember some of the idiosyncracies of plastics that make them different from other engineering materials:

- Even the best plastics have a modulus of elasticity that is only 1 million psi (6.9 GPa), compared to 30 million psi (207 GPa) for steel. Expect more deflection from plastic.
- Plastics expand at a rate that is often ten times the rate of metals on heating. This must be taken into consideration in assemblies.
- Plastics cannot be fitted to the tolerance of metals. Sliding parts require running clearances that are preferably at about 10 mils (0.25 mm) per inch of size.
- Plastics cannot be machined to the tolerances customary in metals; they significantly change size with slight changes in environment. Tolerances closer than ±0.5% are often unrealistic.
- Plastics can be flammable to different degrees.
- Plastic structural components should be designed to long-term strength levels. Use creep and fatigue strength rather than short-term tensile properties. Many plastics lose strength or creep with time.
- Almost all plastics have lower toughness than metals. Make a concerted effort to minimize stress concentrations.
- Plastics make great self-lubricating bearings, but to work properly they must be designed with proper running clearances.
- Many plastics rapidly degrade in the outdoor environment because of UV attack. Check UV resistance before using a plastic outdoors.
- Plastics are attacked in many environments; environmental data must be reviewed.
- In selecting polymer coatings, it is important to determine the basic polymer involved in paints and the like. Do not just use a water-based or oil-based paint. Find out if the polymer is a vinyl, acrylic, etc. Then check the durability of the basic polymer system.
- Remember the basic types of adhesives and become familiar with the properties of the few systems that were described in our discussions.

Critical Concepts

- Selection of plastics requires consideration of properties (physical, mechanical, etc.) such as flammability and stress rupture

Table 7–22
Distinguishing characteristics of useful plastics

	Acronym	Maximum Temperature[b]	Cost Factor[a]	Injection Moldable	Distinguishing Characteristics[c]
Commodity Thermoplastics					
Polystyrene	PS	175	38	Yes	Rigid, easy to mold (low shrink)
Polyvinyl chloride	PVC	175	26	Yes	Low cost, can be rigid or flexible, easy to extrude
Polyethylene (LD)*	PE	210	32	Yes	Flexible, good film former, chemical resistant
Polypropylene	PP	250	30	Yes	Flexible, better strength than PE, chemical resistance
Useful Engineering Plastics					
Polyamide (nylon)	PA	300	300	Yes	Heat and oil resistance, good strength, but absorbs moisture
Polyoxymethylene (acetal)	POM	190	210	Yes	Good chemical resistance and mechanical properties (creep)
Polycarbonate	PC	250	400	Yes	Optically clear, high toughness
Polymethyl methacrylate	PMMA	230	200	Yes	Optically clear, lower cost than PC
Polytetrafluoroethylene	PTFE	550	1000	No	Excellent heat and chemical resistance, but low strength
Perfluoroalkoxy	PFA	500	3000	Yes	Similar to PTFE but no porosity and moldable, but higher cost
Acrylonitrile butadiene styrene	ABS	220	80	Yes	Good moldability and impact strength
Polybutylene terephthalate	PBT	350	370	Yes	Good chemical and heat resistance
Polyimide	PI	600	>5000	No[d]	One of the best heat resistances, very expensive
Polyamide-imide	PAI	450	2500	Yes	Higher heat resistance than many thermoplastics
Polyphenylene sulfide	PPS	400	320	Yes	No known solvent at RT, needs reinforcement
Polyetherether ketone	PEEK	600	3000	Yes	Very good heat and wear resistance
Useful Thermosetting Plastics					
Phenol-formaldehyde	PF	400	100	No	Very high compressive strength, heat resistance
Epoxy	EP	400	200	No	Low shrink, good for potting
Unsaturated Polyester	UP	350	150	No	Lowest-cost resin for composites
Polyurethane	PUR	190	110	No[d]	Can be rigid solid, foam, or elastomer.

*LD = Low density
[a] Approximately ¢/lb
[b] No load
[c] From various sources
[d] Some grades are injection moldable.

strength, which may be ignored for elevated temperature applications of metals.

- Plastics can be useful for corrosion applications, but they are not immune to corrosion. Corrosion data must be consulted.
- The tribological properties of plastics are very sensitive to additives and processing. Testing may be necessary to identify the ideal material.
- Paints are simply plastics applied as a film, and their properties correlate with those of their bulk counterparts.
- Adhesives need to be specified by desired strength.

Terms You Should Remember

pultrusion
flammability
self-extinguishing
notch sensitivity
anisotropy
Rockwell R
coefficient of friction
PV
wear factor
bearing pressure
normal force
paint
plastisol
organisol
lacquers
varnishes
shellac
alkyds
enamels
latex
coverage
adhesive
powder coating
UV cure
epoxy
dielectric constant
volume resistivity
dissipation factor
dielectric strength
running clearance
bond strength
adherends
solvent bonding
contact cement
anaerobic
electrostatic
hot melt

Case History

X-Ray Cassette Hinge Fracture

An x-ray cassette designed for medical applications holds a sheet of x-ray film in a light-tight environment during the x-ray process. These cassettes are molded from a carbon-filled, optically opaque polycarbonate resin (carbon filler is used to provide optical opacity because it does not interfere with x-ray transmission). Hospital use conditions are more rugged than anticipated by designers. Although injection molded from a polycarbonate resin chosen for its impact toughness, the cassette's hinge mechanism fractures when the cassette is dropped or mishandled. Significant warrantee returns occurred.

Impact tests were developed that used an instrumented impact tester (which allows fracture-energy measurements) to simulate the hinge failures. The impact tests measure the influence of filler content and morphology, internal and residual stresses, and injection molding conditions. No statistically significant trends appeared. Analysis of the fracture surfaces on failed parts suggested that the polycarbonate was fracturing in a brittle manner. The fracture originated at a relatively sharp corner. It was postulated that the combined effects of the sharp corner and the relative inflexibility of the hinge tab caused high strain rates when the part fell, shifting the polycarbonate's impact properties into a brittle regime. Polycarbonate's impact strength undergoes a ductile-to-brittle transition when the effective material thickness exceeds about 0.10 to 0.13 in. (2.5 to 3.3 mm)

To confirm the hypothesis, sample parts were modified by machining a radius and thinning the cassette's hinge tab. Impact energies increased dramatically and the failure mode changed from brittle fracture to ductile deformation. The mold for the part was modified and impact tests were again conducted to confirm the success of the changes. As a result, hinge-failure warrantee claims dropped to zero.

Questions

Section 7.1

1. How should plastics databases be used in material selection?
2. Specify a tensile test for a plastic.

Section 7.2

3. Explain how weld lines occur and how they influence the strength of a polymer part.
4. Compare the elastic modulus of three thermoplastics to the moduli of aluminum and steel.
5. Explain creep and stress relaxation in polymers. If a supplier provides isochronous stress–strain data for a candidate polymer, how do these data predict creep behavior of a part or component?
6. Calculate the reduction of yield strength from room temperature (the percentage) when an acetal (Delrin) structural member is put into service at 250°F.
7. Explain the significance of the relative temperature index. What candidate polymers would you suggest if the no-load service temperature is 420°F (see Table 7–6).
8. Explain how stress rupture data may be used to design a plastic component that needs to be serviceable for 10 years.
9. List three candidate materials that offer high impact resistance.
10. Explain the effect of stress concentrations on the impact toughness of a plastic component.
11. If a plastic component is cyclically stressed between 0 and 2300 psi and needs to last six million cycles under ambient conditions, what materials in Figure 7–9 would work? If the application required a safety factor of 2 for fatigue failure, what materials would apply?
12. Explain the effect of annealing on the dimensional stability of a thermoplastic.
13. Calculate the change in size that occurs in 1 day due to moisture absorption for a 25-mm-diameter bore in a nylon bushing when it is taken from a 5% relative humidity (RH) environment to an environment with an RH of 100%.
14. Which has better flame resistance, a plastic with 94 V-0 rating or one with a rating of 94 V-2?
15. Name three fire retardant polymers based on limiting oxygen index.
16. What hardness scale is commonly used for elastomers? Which scale is most suitable for a hard polyurethane rubber?

Section 7.3

17. Calculate the radial wear of a 2-in.-diameter polyimide plain bearing after 5000 h of use versus a hardened steel shaft rotating at 100 rpm with a 200-lb load on the shaft.
18. Calculate the proper bore size for an acetal pillow block that will be used on a rotating steel shaft that will have an outside diameter of 1.000 in. at 120°F.
19. Calculate the interference required to hold a 1-in.-outside-diameter, 0.1-in.-wall nylon bushing in a steel plate of infinite size. The required clamping pressure is 100 psi.
20. Estimate the coefficient of friction of a polystyrene part sliding on steel at 100 PV. The PS hardness is in the range of 20 to 40 kg/mm^2; the shear strength of the PS is 5 ksi.
21. Name two plastics that would be suitable for use as plain bearings operating at 300°F.
22. What is the PV limit for PI?
23. Can you use a 1-in. ID PI bearing 2000 rpm and a load of 125 lb?

Section 7.4

24. Explain the effect of permeation in polymers.

25. Explain how environmental stress cracking can affect the serviceability of a plastic component.

26. What is physical aging? Explain how the relative temperature index may be used to rank a polymer's resistance to physical aging.

27. Laboratory testing indicates that the tensile strength of a polymer is reduced by 9% when exposed to automotive motor oil at 80°C for 1000 h. Is this material suitable for this application? Explain.

28. Can nylon 6 be used for service in a strong acid?

29. Is ABS resistant to acetone solvent?

30. Using data from Table 6–1, compare the strength-to-weight ratio of E-glass epoxy and 7075 T6 aluminum.

Section 7.5

31. Name a plastic that is a very good insulator of electricity and one that is a poor insulator.

32. What theoretical thickness of vinyl insulation would be required on an electrical conductor that is designed to operate at a potential of 5000 V?

Section 7.6

33. Describe four different paint systems and their mechanisms of film formation.

34. Recommend a coating system that is resistant to splash and spills of alkalies and solvents.

Section 7.7

35. What is the mechanism of adhesive bonding?

36. To bond PTFE to steel using a suitable epoxy, suggest surface pretreatment processes for the steel and PTFE surfaces.

To Dig Deeper

Bikales, N. M. *Mechanical Properties of Polymers*. New York: Wiley-Interscience, 1971.

Blau, P. J. *Friction Science and Technology*. New York: Marcel Dekker, Inc., 1996.

Ellis, B. *Polymers—A Property Database*. Boca Raton, FL: CRC Press LLC, 2000.

Engineering Plastics, Engineered Materials Handbook, Vol. 2. Metals Park, OH: ASM International, 1988.

Ezrin, M. *Plastics Failure Guide: Cause and Prevention*. New York: Hanser/Gardener, 1996.

Glaeser, W. A. *Materials for Tribology*. London: Elsevier, 1992.

Harper, C. A. *Handbook of Plastics and Elastomers*. New York: McGraw-Hill Book Co., 1975.

Margolis, J. S., Ed. *Engineering Thermoplastics*. New York: Marcel Dekker, Inc., 1985.

Seymour, R. B. *Plastics vs. Corrosives*. New York: John Wiley & Sons, 1982.

Schnabel, W. *Polymer Degradation*. New York: Macmillan, Inc., 1981.

Schweitzer, P. A. *Corrosion Resistance Tables*. New York: Marcel Dekker, Inc., 1991.

Ulrich, H. *Introduction to Industrial Polymers*. New York: Macmillan, Inc., 1982.

Van Krevelin, D. W., and P. J. Hoftyzer. *Properties of Polymers*. Amsterdam: Elsevier, 1976.

Weismantel, G. S. *Paint Handbook*. New York: McGraw-Hill Book Co., 1981.

Yamaguchi, Y. *Tribology of Plastic Materials*. Amsterdam: Elsevier, 1990.

Plastics databases (updated periodically):

Mvision. Santa Ana, CA: MSC Software.

CES4. Cambridge, UK: Granta Design, Ltd.

Engineering Design Database. Pittsfield, MA: General Electric Co.

Corsur II. Houston, TX: National Association of Corrosion Engineers.

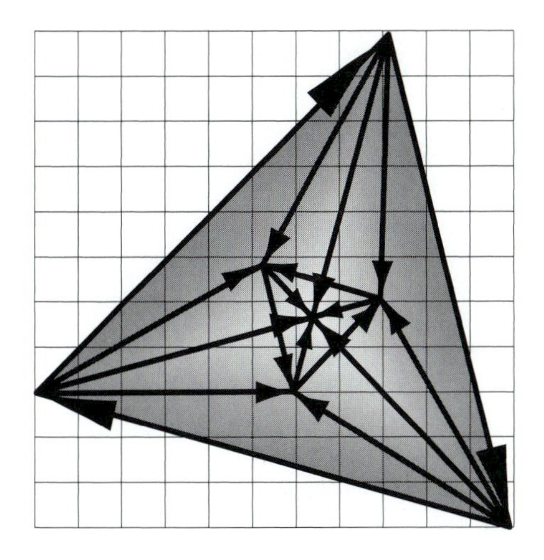

CHAPTER 8

Ceramics, Cermets, Glass, and Carbon Products

Chapter Goals

1. An understanding of the composition and structure of ceramics, glass, carbon products, and cemented carbides.
2. A working knowledge of how to select and use these materials to solve design problems.

Rationale

Ceramics and cermets are becoming the tool materials for the future. Cemented carbides have almost eliminated wear in many woodworking tools (saw blades, router bits, etc.). Coated cemented carbides have essentially displaced high-speed steel for cutting tools, and almost all high-production punch press dies use cemented carbide tooling.

Ceramics are taking over many of the high-temperature machine tasks; for instance, they are substrates for computer chips and used for prosthetic devices. Ceramics are readily available, and designers can consider their use on ordinary design assignments.

Glasses and carbon products have many applications in design engineering, and they can solve many special problems, but designers often do not know where to get them or how to use them.

In summary, this chapter deals with engineering materials that may work in situations in which plastics and metals would fail. They need to be part of every designer's repertoire; sometimes they are able to solve problems that seem hopeless if left to more traditional engineering materials.

8.1 The Nature of Ceramics

To some people, the term *ceramics* means dishes, porcelain figurines, and the like. Most high school continuing education programs have a ceramics course. This course will deal with making ashtrays and figurines. A typical dictionary definition of ceramics is "the art or work of making pottery, tile, porcelain, etc." The spectrum of the things that can be construed to be ceramics is illustrated in Figure 8–1. In Chapter 1 we defined ceramics as solids composed of compounds that contain metallic and nonmetallic elements, and the atoms in the compound or compounds are held together with strong atomic

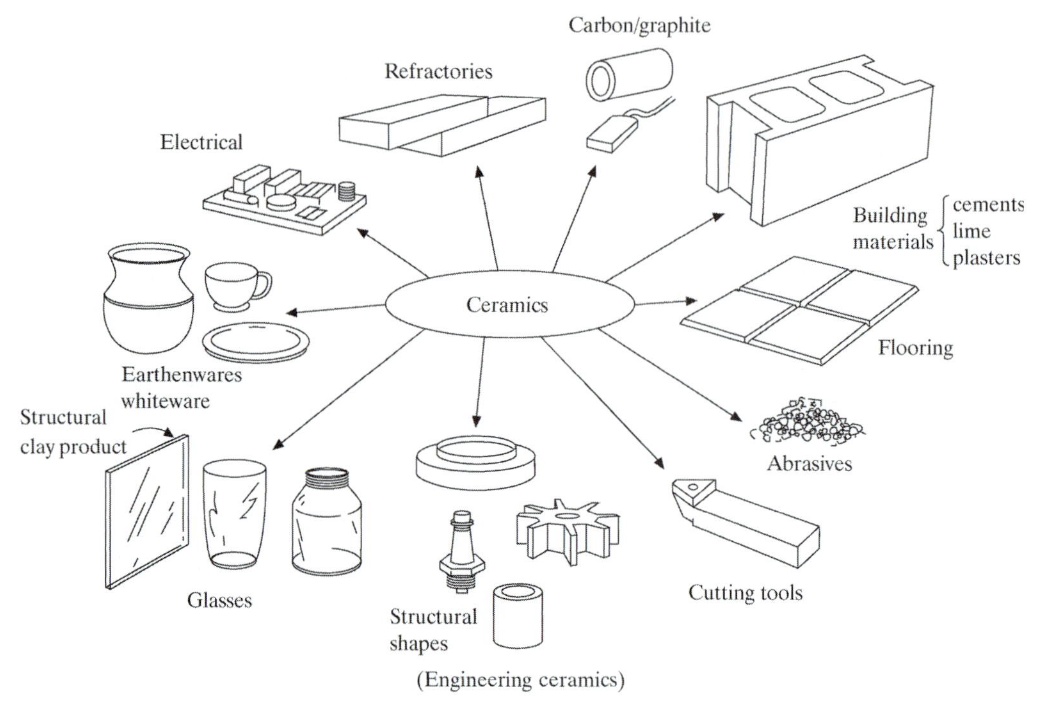

Figure 8–1
Spectrum of ceramics uses

forces (ionic or covalent bonds). This definition does not fit some of the things that are shown in Figure 8–1.

Bricks and clay products are often made from natural clays. Clays are mostly composed of silica and alumina ($Al_2O_3 \cdot 4SiO_2 \cdot 2H_2O$), and the bonding mechanism before firing is usually weak van der Waals forces, an electrostatic attraction between neutral atoms in close proximity. The ceramics with the highest strength, the highest moduli, and the best toughness, such as zirconias, aluminum oxides, silicon carbides, and silicon nitrides, have strong ionic or covalent bonds. It is these materials that meet our original definition of ceramics: compounds composed of metallic and nonmetallic elements with strong ionic or covalent bonds between atoms. In the 1980s these materials were sometimes called structural ceramics, engineering ceramics, or advanced ceramics. In the late 1980s, the Japanese initiated a national effort in this area, and they called ceramics used in engineering applications *fine ceramics*. In 1990, the ASTM Committee on Ceramics (C-28) advocated use of the term *advanced ceramics*. Their definition of this class of materials is "a highly engineered, high performance, predominantly nonmetallic, inorganic, ceramic material having specific functional attributes" (ASTM C 1145). Thus there is no concise definition of the term *ceramic* or variations thereof. We shall use the definition given in Chapter 1, and we shall use the term *engineering ceramics* for the ceramics that are used for tools, machine components, and structural applications.

Now that we have covered the semantics of ceramics, specifically what we will discuss in this chapter is ceramics that have application in ordinary design engineering, cemented carbides, glasses, and carbon–graphite materials. These latter items are not ceramics by our definition of ceramics, but they are used for the types of things that ceramics are used for, and they have some properties that are typical of ceramics. Probably the most significant property that is common to ceramics as well as to cemented carbide, glasses, and carbon–graphite is brittleness. As shown in Figure 8–2, the *fracture toughness* (a measure of crack propagation tendency) of ceramics can be from one to two orders of magnitude lower than that of metals. There probably are no currently commercially available ceramic solids that have the malleability or formability at room temperature that most metals have. A review of the items shown in this illustration (Figure 8–1) will confirm our statement on brittleness. Everything in our sketch is brittle: concrete blocks, ceramic tile, carbide tools, ceramic wear parts, glasses, pottery, computer chips (silicon and the like), bricks, and carbon motor brushes.

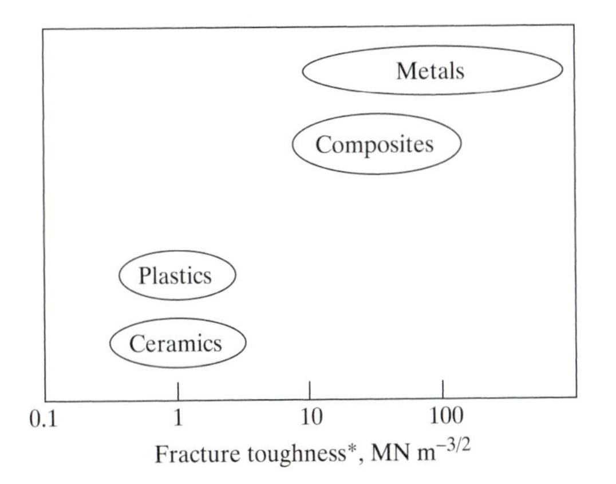

Figure 8–2
Relative fracture toughness of engineering materials

8.2 How Ceramics Are Made

Most of the shapes illustrated in Figure 8–1 require some type of high-temperature firing in their manufacture. Figure 8–3 shows some abbreviated schematics of the manufacturing processes that are used for some of the materials that we discuss in this chapter. The starting material for most ceramics and similar materials is some type of powder that must somehow be "glued" together to make a solid. Some powders can be joined into a useful solid by simply coalescing powder particles together by high-temperature self-diffusion; this is called *sintering*. This operation removes spaces between particles. Other ceramic materials are made by adhering the powders together with glasses; this is called *vitrification*. The first illustration in Figure 8–3 shows concrete products such as sewer pipes and concrete blocks being made from sand and aggregate bonded together with cement. In terms of tonnage, cements (masonry and Portland) are probably the most important ceramic-type materials. Portland cements are combinations of dicalcium silicates and tricalcium aluminates. In curing, these compounds chemically combine with water to become an integral part of the concrete product. Masonry cements are mostly lime, which is calcium oxide. The starting materials for these cements are abundant in nature, and this is the reason for their wide use. Cemented carbides are also essentially ceramic particles glued together, but in their case the glue is the metal cobalt (sometimes nickel). The generic name for this class of materials is *cermets*. They are composites composed of ceramics (cer) and metals (met).

Single crystals of elements such as silicon, boron, and germanium are grown by nucleation and controlled solidification, and these types of materials are extremely important in electronics. They are the substrates for computer chips. Large crystals are grown (sometimes as large as 400 mm in diameter), and they are sectioned

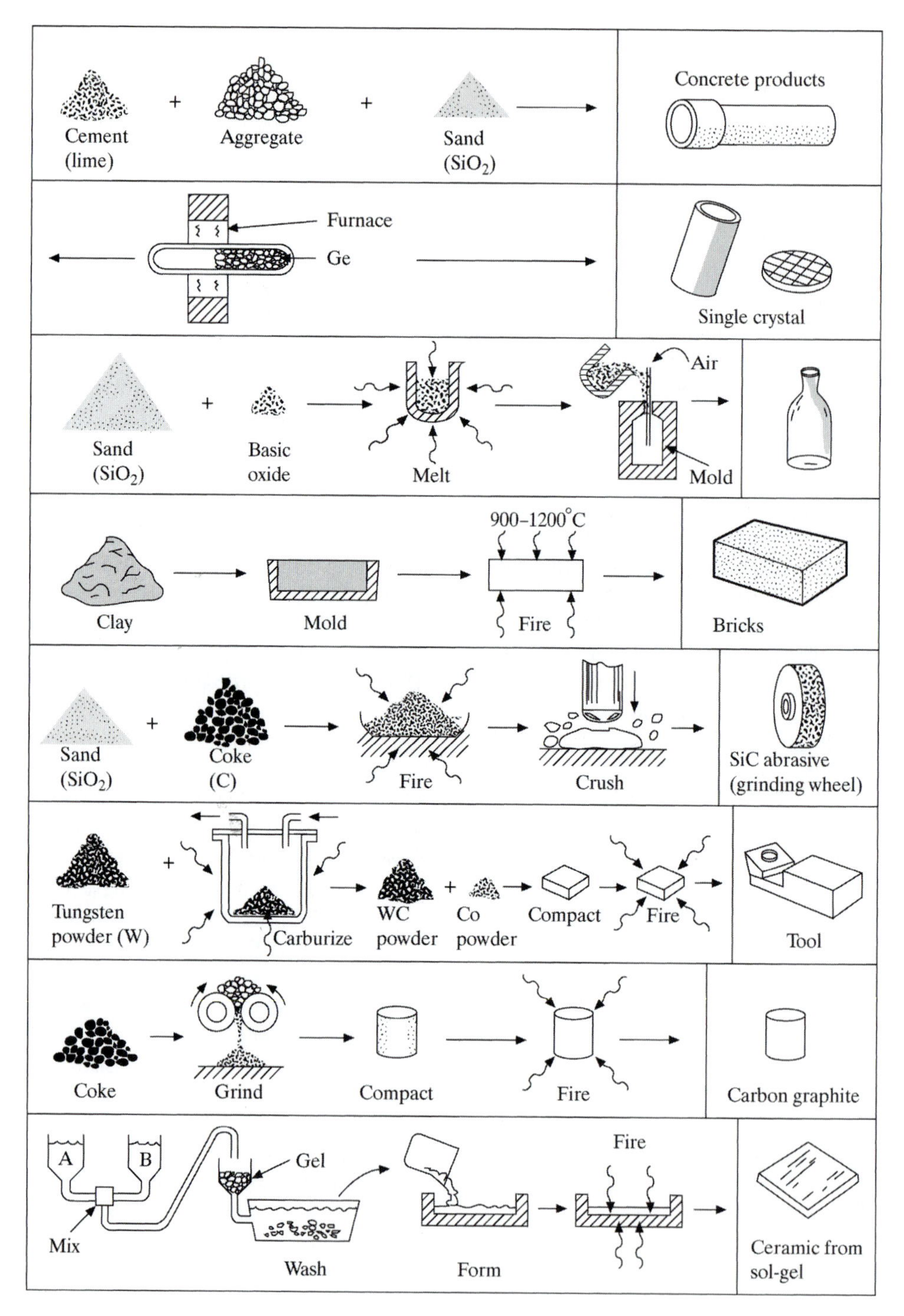

Figure 8–3
Some ceramic fabrication processes

with diamond saws into thin wafers then chips. These materials are more correctly classified as metalloids, materials that can behave like a metal or like an insulator. This is how they work in electronics. A bias voltage can make the material either a conductor or an insulator.

We discuss glasses in more detail in another section, but most glasses are simply amorphous fusion products of inorganic materials. The glasses that are most used are formed by melting sand and basic oxides such as sodium oxide. The molten glass is poured into a shape or blown into a shape.

In terms of tonnage, bricks are extremely important, but there is not much engineering in common bricks. They are simply made by forming clay to a shape and firing the clay to fuse the inorganic material into a solid that can be anywhere from weak and porous to glassy (depending on the firing temperature). Of great industrial importance are refractory bricks, and these can be complicated materials. They are made from clays and mixtures of oxides. Refractory bricks are often the limiting factor in furnaces for melting metals. Metal manufacturers have been searching for ways to improve the life of bricks used for furnace linings for centuries.

Some of the starting materials for ceramics come from the ground in relatively pure form. One such example is titanium dioxide, which is used for sintered ceramic parts as well as for white pigment. This material is simply white beach sand.

Unfortunately, some of the starting materials for ceramics are not so easy to come by. In Figure 8–3 we illustrate one method for making silicon carbide (SiC). Sand, the source of silica, is blended with coke (C), and the entire mass is heated to red heat for a long time (days) to allow interdiffusion to make silicon carbide. The material fuses, and the fused mass is crushed to make particles of different sizes. A finished product from this material may be a grinding wheel. The crushed and sized abrasive is bonded with an organic binder (resin bond) or the particles are vitrified; they are bonded by viscous flow of a glassy substance around the particles. The glass becomes rigid but usually not crystalline on cooling; it becomes an integral part of the grinding wheel.

Tungsten carbide can be made in a way that is similar to the making of silicon carbide—by a diffusion process. Tungsten powder can be carburized. A shape is then made by bonding this powder together with a cobalt or similar metal binder.

Carbon–graphites will be discussed in more detail in a later section, but basically they are made by simply firing an organic material in the absence of air for a long time at very high temperatures.

Finally, in our illustration, the *sol-gel* process is used to form a solid material by gelation of one or more liquids and subsequent hardening of the gel. In our illustration, liquid A is chemically reacted with liquid B to form a *gel*. To most people, the term *gel* means a material that is not a solid or a liquid—something in between. The gels can then be fired to form ceramic shapes. This is an emerging technology. The sol-gel process has the capability of making common ceramics such as aluminum oxide with extreme purity and extremely fine grain, because the starting materials are liquids rather than particles that must be ground to make them fine. Aluminum oxide abrasives for grinding are commercially made by the sol-gel process. This is a major technology breakthrough. Fine abrasive particles can be made without expensive crushing operations. In addition, particle size can be much smaller (<50 nm) than can be obtained by crushing. They can be "nanograin" materials.

Two of the most important ceramic fabrication processes are illustrated in Figure 8–4. Vitrification is most widely used in the making of abrasive wheels and other shapes. It is also used to make a *machinable ceramic*. The machinable ceramics are mostly glass-bonded mica. Mica flakes are bonded together with a glassy binder. The material has considerable porosity, and the

Figure 8–4
Fabrication of ceramic shapes

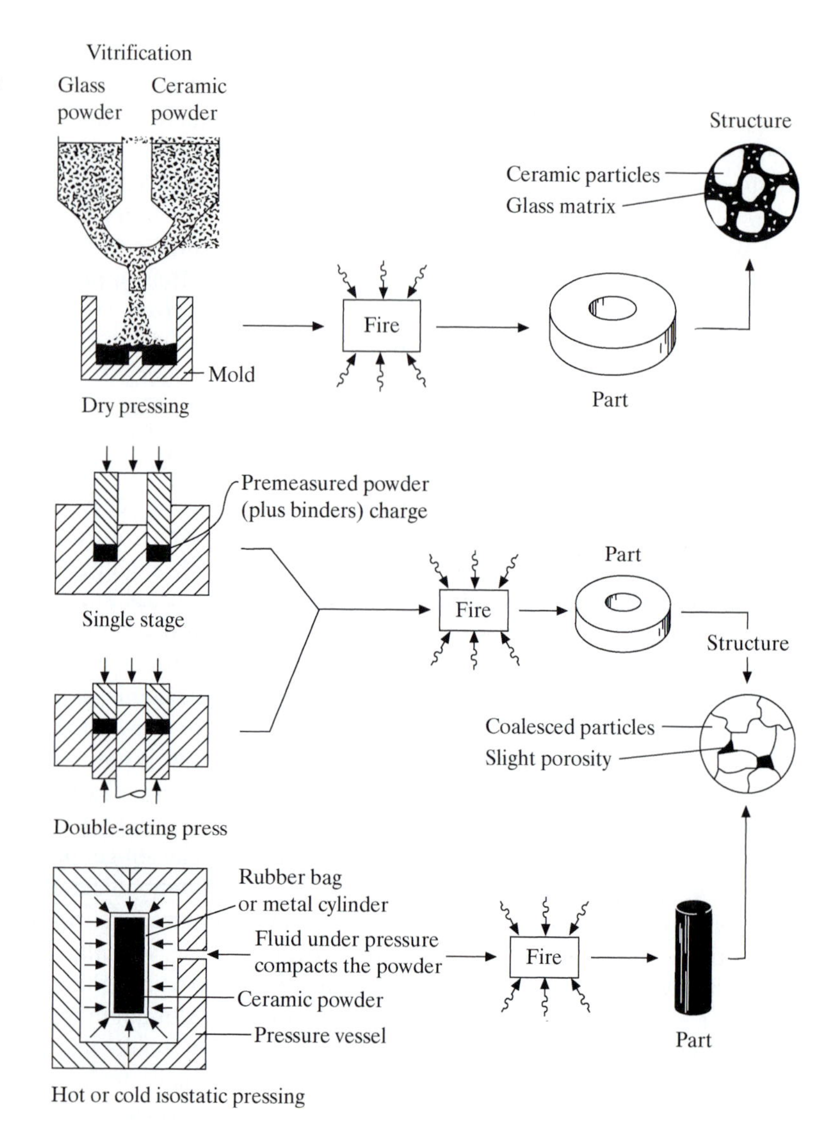

mica fractures easily. These factors produce a friable structure that is machinable with conventional drills, cutters, and the like. The most widely used ceramic fabrication process is *compaction* and sintering (Figure 8–4).

1. The purified ceramic compound is ground or milled until the desired particle size is obtained.
2. The powder may be mixed with binding agents, and it is then densified or compacted by various techniques to the point at which it can be handled but has little strength. This is called the *green state*. Many processes exist to form ceramics into shapes. Tubes, cylinders, and long, simple shapes can be made by extrusion of a ceramic–binder mixture. Small, intricate

shapes can be made by injection molding of ceramic powders mixed with plastic binders.

3. The binders are removed by thermal treatments. Dry pressing of powders by single- or double-acting presses is a low-cost method for making simple shapes. Isostatic pressing is used for large, complicated shapes. Powder and binder are put into a shaped rubber mold or bag, and the rubber mold is subjected to fluid pressure to compact the powder blend into a green shape. A more complicated version of this process is to put the powder into a perishable sheet metal mold and isostatically compact the charge with hot inert gas. This process is called *hot isostatic pressing* (HIP). It can replace the firing operation when used on low-melt glassy substances. For example, HIP is used to make an infrared transmitting lens from lithium fluoride.

   A modification of the hipping process is used to densify reaction-bonded carbide parts. Green parts are surrounded by molten glass in an open-ended cylinder. A piston strikes the glass, turning it into a fluid to produce the hipping action on the part that is suspended in the glass.

   Dishes, whiteware, and pottery-type items are usually made by blending ceramic powders with water and binders and working these with molding presses or spinning devices. Slip casting is used for making vases and other types of hollow ware. The ceramic–water slurry (slip) is cast into a plaster mold. The plaster absorbs the moisture from the slurry, and when a solid skin of dewatered ceramic forms, the mold is turned over and the remaining slip is poured out. These processes are also used for engineering ceramics; if part quantities are adequate it is common to use techniques similar to those used for plastics, such as extrusion and injection molding.

4. After the shaping of parts with one of the foregoing processes, it may be necessary to air dry or furnace heat the parts to drive off binders. This prefiring process densifies the material and increases the green strength.

5. At this point, ceramic parts that call for threaded holes, parallel surfaces, and accurate dimensions are machined. Machining can be done with conventional tooling using special tool bits (carbide, ceramic, or diamond). Special shapes can be created.

6. *Sintering* or firing is the final step in the ceramic densification process. Firing is usually done at a temperature below the melting temperature, but sometimes the process is controlled so that the surface of the ceramic particles starts to melt. The firing process reduces porosity. It also supplies the energy for the processes that cause the ceramic particles to bond into a monolithic solid. Particles experience viscous flow to increase the particle-to-particle contact area. In the contact area, atoms from one particle diffuse into the other contacting particles. The driving force for this interdiffusion is slight chemical dissimilarities between particles and the tendency for a crystal to go to a shape with minimum surface area. Concurrent with the viscous flow and diffusion processes may be vapor reactions and local melting that further fill up porosity in the pressed shape. The interdiffusion between particles is probably the major mechanism of bonding.

There are several important points to remember concerning ceramic fabrication processes:

1. Most ceramics are hard and brittle; thus parts cannot be easily machined from standard shapes. Rough machining must be done in the green or unfired state.
2. Forming ceramics into special shapes by molding usually involves the fabrication of

Table 8–1
Typical ceramics for engineering design

Class	Examples	Uses
Single oxides	Alumina (Al_2O_3)	Electrical insulators (Figure 8–1)
	Chromium oxide (Cr_2O_3)	Wear coatings
	Zirconia (ZrO_2)	Thermal insulation
	Titania (TiO_2)	Pigment
	Magnesium oxide (MgO)	Wear parts
	Silica (SiO_2)	Abrasive, glass, optics
Mixed oxides	Kaolinite ($Al_2O_3 \cdot 2SiO_2 \cdot 2H_2O$)	Clay products
Carbides	Vanadium carbide (VC)	Wear-resistant materials
	Tantalum carbide (TaC)	Wear-resistant materials
	Tungsten carbide (WC)	Cutting tools
	Titanium carbide (TiC)	Wear-resistant materials
	Silicon carbide (SiC)	Abrasives, electronic devices
	Chromium carbide (Cr_3C_2)	Wear coatings
	Boron carbide (B_4C)	Abrasives
Sulfides	Molybdenum disulfide (MoS_2)	Lubricant
	Tungsten disulfide (WS_2)	Lubricant
Nitrides	Boron nitride (BN)	Insulator
	Silicon nitride (Si_3N_4)	Wear parts
	Aluminum nitride (AlN)	Electronic devices
Metalloid elements	Germanium (Ge)	Electronic devices
	Silicon (Si)	Electronic devices
Intermetallics	Nickel aluminide (NiAl)	Wear coatings

tooling for that particular part, possibly making one-of-a-kind parts expensive.

3. The presence of binders (like a glassy phase) in a fired ceramic lowers its strength. If strength is a selection factor, the percentage of theoretical density should be specified and controlled.
4. Sintering of ceramic parts usually involves a size change. They can shrink as much as 30%. For this reason, fired ceramic parts are difficult to make to close dimensional requirements. (Tolerances are typically ±1%, plus allowances for distortion.)

In summary, there are many ways of making ceramic types of materials, but we have described the most common. Engineering ceramics that are made by these processes can be classified as shown in Table 8–1 into oxides, carbide, sulfides, nitrides, metalloids, or intermetallics. *Intermetallic compounds* are compounds that are formed by the combination of two metals. They are not alloys, but sometimes they have metallic properties and sometimes they have properties that are like those of ceramic materials. We shall discuss some, such as nickel aluminides, in a later section.

8.3 Microstructure of Ceramics

We have stressed a number of times that the brittleness of ceramics is due to the strong ionic or covalent bonds that are common in ceramic types of materials. Figure 8–5 uses the Bohr atom to illustrate the concepts of covalent and ionic bonding. Covalent bonds are formed by

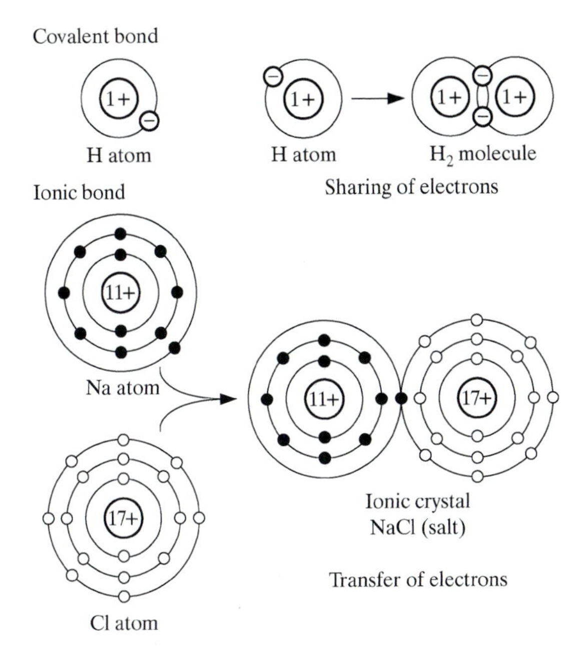

Figure 8–5
Covalent and ionic bonds

sharing of atoms to form molecules. In our example, the valence electrons of two hydrogen atoms come together and the two atoms share the valence electrons. This sharing produces a diatomic hydrogen molecule. This is the natural state for this gas. Most plastics have covalent bonds between carbon atoms to form polymer molecules, and ceramics often have this form of bonding as well. Figure 1–15 illustrates covalent bonding of aluminum and oxygen atoms to form the ceramic *aluminum oxide*.

The formation of an ionic crystal, common salt or sodium chloride, is illustrated in Figure 8–5. The outer valence electron of the sodium atom is donated to the valence orbital of the chlorine atom, and an ionic crystal is formed. These materials have high melting points, and they do not carry current because the electron bonds are so strong. When salt is placed in water, the ionic crystals dissolve, forming chlorine and sodium ions in solution. The ions now can become carriers of current in the water. Water would be an electrical insulator without the presence of dissolved ions. Magnesium oxide is an ionic crystal ceramic of commercial importance.

In crystalline metals, there is a neat geometric configuration of atoms at the corners of a cube or at other definite locations in an imaginary parallelepiped. In polymeric materials, we saw crystallinity develop from an orderly arrangement of long chains of repeating organic molecules. In ceramics, we are dealing primarily with compounds rather than individual elements. Thus for a ceramic to have a crystalline structure, it must maintain the ratio of one element to another dictated by the stoichiometry of the ceramic compound in question. The same types of crystal structures found in metals (for example, face-centered cubic, body-centered cubic, and hexagonal close-packed), as well as some more complex structures, occur in ceramics. Figure 8–6 shows the face-centered cubic structure of magnesium oxide. It illustrates how both crystal structure requirements and

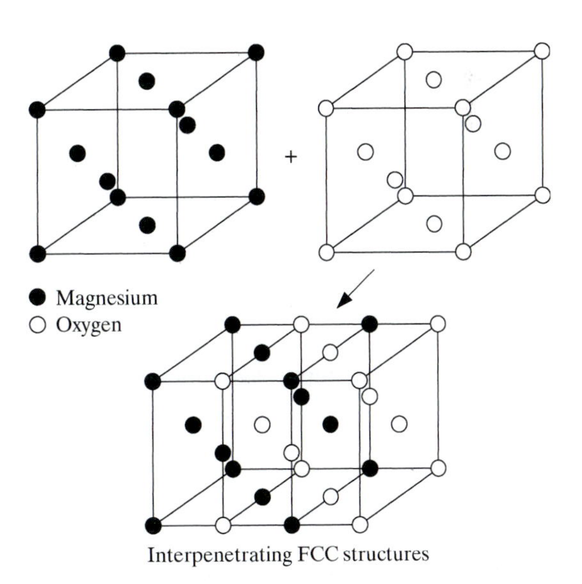

Figure 8–6
Crystal structure of magnesium oxide

chemical stoichiometry can be satisfied in crystalline ceramics. In comparison with a face-centered cubic metal structure such as silver, the magnesium oxide structure must maintain the ratio of one magnesium atom to one oxygen atom. With the exception of impurity atoms, the atomic sites in the metal structure are filled with the same types of atoms.

If a ceramic contains more than one compound, the crystal structure can become even more complex. The important point to remember about the structure of ceramics is that, for the most part, the atoms are bonded by sharing electrons and the atom ratio dictated by the compound or compounds forming the ceramic material must be satisfied by the location of the atoms in the crystal structure. From the practical standpoint, ceramics tend to be brittle and resistant to hostile environments because of the strong bonding forces between atoms. The interesting crystal structures found in various ceramics are of great significance to the ceramist, who can use modifications in crystal structure to alter properties, but it is probably sufficient for the designer to keep in mind that crystal structure can significantly change properties. An example illustrating this point is the ceramic boron nitride. It is a soft, friable insulating material in its hexagonal crystal structure, but in its cubic form it is one of the hardest materials known.

The properties of ceramics can also be affected by composition, phases present, and microstructure. The covalent bonding in ceramics is only between two adjacent atoms. This can lead to directional properties, or anisotropy. Impurity atoms can be in solid solution in ceramics just as they are in metals. Ceramics can also be multiphase—a mixture of, for example, *oxide* types; part of a ceramic may be a single oxide, and a second phase could be a mixed oxide. The relative volume of phases present can be determined by phase equilibrium diagrams similar to those developed for metal alloys.

Probably the most significant microstructure concerns from the user's standpoint are grain size, porosity, and phases present. Most engineering ceramics have a polycrystalline microstructure. The crystallite or grain size can affect properties. Usually, fine-grained polycrystalline ceramics are mechanically stronger than coarse-grained grades. Since most ceramics are made by sintering crystallites together, the microstructures can show voids at the intersection of coalesced grains [Figure 8–7(a)]. The strength of a fired ceramic is almost always lowered by the presence of these sintering voids; the greater the porosity, the lower the strength.

Figure 8–7 presents some typical microstructures that are possible in ceramic types of materials. If we polish and etch the surface of a ceramic and examine it under optical or electronic microscopy at significant magnification ($\times 2000$ to $\times 5000$), these types of microstructures may be present. A single-phase sintered polycrystal ceramic would look like Figure 8–7(a). If more than one phase is present, the structure may appear like Figure 8–7(d). Some ceramics have three, four, or more distinct phases or homogeneous components that make up the whole microstructure. A vitrified structure would show particles held together with some sort of glassy phase [Figure 8–7(b)]. Abrasives are often vitrified. Single crystals and glasses would have no features discernible by ordinary polishing and etching. There are no grain boundaries in single crystals because the crystal is only one grain.

Some microstructures may be single phase, with precipitates of fine particles or needles of a second phase [Figure 8–7(c)]. Some ceramics are reinforced with fibers from other ceramic systems. Cemented particles can show very little glue phase (*binder*) between particles [Figure 8–7(b)], or the glue phase can be as much as a third of the structure. Resin-bonded ceramics would have a microstructure like Figure 8–7(b). Instead of glass the bonding agent could be organic. Ceramic grains in abrasives are often glued together by a phenolic or urea resin. The glue phase is usually kept small because organic binders are much weaker than ceramics of other

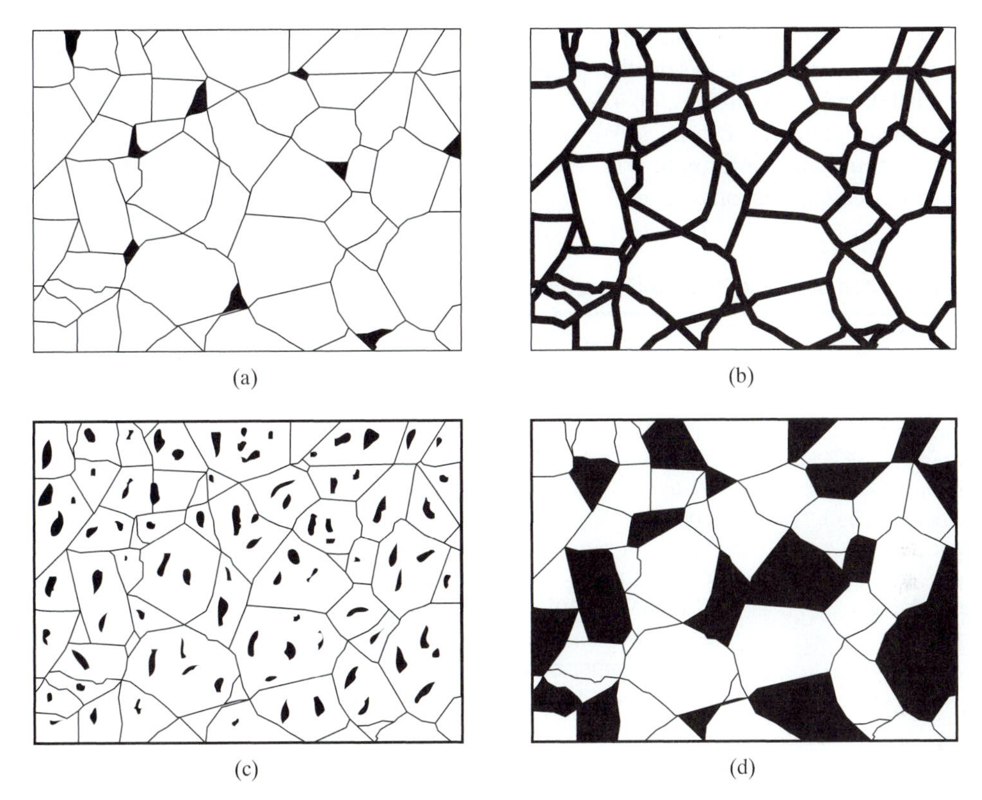

Figure 8–7
Possible microstructures of engineering ceramics

types of bonding agents, such as glass or metal. This type of structure is common in abrasive products, such as grinding wheels.

In addition to the structure options shown in Figure 8–7, ceramics are also available with fiber reinforcement. For example, silicon nitride is available reinforced with silicon carbide whiskers. These are used to strengthen the ceramic. The structure would resemble that of a plastic reinforced with chopped glass or carbon fiber reinforcement.

The user of ceramics usually cannot perform any secondary treatments such as quench hardening to alter the structure of ceramics. Thus it is important to review factors such as porosity, grain size, and composition limits with a ceramic manufacturer. A designer may have a successful application of a particular ceramic on a particular part, and on a repeat order of the same part from the same manufacturer and made from the same ceramic, service failures could occur. The explanation could be due to some of the structure differences just mentioned.

8.4 Properties of Ceramics

Mechanical

We have shown how the atomic structure and microstructure of ceramics differ from metals and plastics; now we shall discuss some of the property differences that set ceramics apart from other engineering materials. We have already mentioned their toughness (impact) limitations; they are brittle. Table 8–2 presents a qualitative comparison of some of the more used selection properties. We shall discuss these to develop a general feeling for the properties of ceramics. We then present more quantitative information on these properties in subsequent sections.

It is extremely difficult to measure the tensile strength of ceramics because the dog-bone-shaped test samples must be made with expensive fabrication processes (diamond grinding, and the like) and because it is difficult to grip tensile coupons in the machine because of their high hardness. It

Table 8–2
Relative property comparison of engineering ceramics to other engineering materials (in bulk shape, not fiber)

Property	Ceramic	Metals	Plastics
Mechanical			
Tensile strength	100 max.*	300 max.*	30 max.*
Yield strength	Does not yield	250 max.*	25 max.*
Elongation	Zero	Up to 50%	Can be 100%
Compressive strength	600 max.*	400 max.*	40 max.*
Impact	Poor	Poor to excellent	Fair
Hardness	Can be 2600*	900 max.*	100 max.*
Tensile modulus	Can be 90×10^6*	Can be 40×10^6*	Can be 1×10^6*
Creep strength	Excellent	Good	Poor
Fatigue strength	OK in compression	Good to excellent	Poor
Physical			
Density	0.1 to 0.6*	0.06 to 0.8*	0.03 to 0.1*
Melting point	>5000°C	Up to 5000°F	<500°C
Conductivity			
Heat	Poor to fair	Good to excellent	Poor
Electricity	Nil to some	Excellent	Nil
Coefficient of expansion (thermal)	Low	High	Very high
Max. use temperature	3000°F	1500°F	500°F
Water absorption	Some	Nil	Some
Electrical insulation	Good to excellent	Nil	Good to excellent
Chemical			
Crystallinity	Usually	Usually	Many are not
Chemical resistance	Excellent	Poor to good	Good to excellent
Composition (raw materials)	Abundant in nature	Some are costly	From oil
Microstructure	From single phase to multicomponent	From single phase to multicomponent	From single phase to multicomponent
Fabricability			
Sheet material	Nil	Excellent	Good
Castings	Nil	Excellent	Fair
Extrusions	Nil	Excellent	Good
Machinability	Nil	Good to excellent	Good
Molding	Fair	Nonferrous only	Excellent

*Arbitrary units

commonly costs from $600 to $800 to make a single ceramic test coupon. For these reasons, performing tensile tests on ceramics is not normally done, and the available data are often based on fewer replicate samples than one would like. There are ceramic fibers with tensile strengths that can be in excess of 500 ksi (3.4 GPa), but in general the tensile strengths of engineering ceramics are less than 100 ksi (0.69 GPa), much less than some metals but better than most plastics. The property that is often substituted for tensile strength is the four-point flexural strength test (MIL-STD-1942). This test uses a cheaper rectangular bar with dimensions of about $3 \times 5 \times 50$ mm, and it is bent as a beam until it breaks. The tensile stress at the outer fibers of the beam at fracture is calculated, and this is used as a measure of tensile strength. It is usually reported as *flexural strength*. If the deflection of the beam is measured in this test, these data can be used to produce a stiffness parameter, flexural modulus. This property is often substituted for the tensile modulus of elasticity.

Ceramics do not have a yield strength. They do not plastically deform before reaching their tensile strength. There is no permanent elongation or reduction in area. This is the property that prevents ceramics from competing with metals and plastics for the market that requires forming of materials by room temperature plastic deformation: deep drawing, bending, forming, and the like. These shaping processes simply cannot be used.

The compressive strength of engineering ceramics can be excellent—better than the compressive strengths of metals, plastics, and composites. Ceramics have the highest hardness of the engineering materials. In fact, diamond (which is considered by most a ceramic) has the highest hardness of any material that we know. The high hardness of ceramics makes some of them suitable for use as tools for working other materials and for use as wear components.

We have mentioned how hard it is to measure the elastic modulus of ceramics, but it is done and ceramics come out as our stiffest engineering materials. Cemented carbides are three times as stiff as steel; silicon carbide, boron carbide, and titanium diboride are more than twice as stiff as steel. This area of ceramics is indeed an asset for engineering materials. When a steel member is deflecting too much in service and the size cannot be changed, the only solution to the problem is to use a material with greater stiffness than the steel. This leaves few choices but ceramics and composites with high modulus reinforcement.

The long-term service characteristics of ceramics—fatigue resistance and creep strength—are good if the parts are made and loaded in modes favorable to ceramics. The tensile fatigue strength of ceramics is poor. This is because the tensile strength of ceramics is in general poor, and a good design would avoid loading a ceramic in tension. The compressive strength of ceramics is very good. The same sort of situation exists in creep and stress rupture; if the ceramic is loaded in compression, the creep and stress rupture characteristics will be excellent. In fact, ceramics can operate at temperatures of 2000°F (1093°C) in furnaces.

In summary, the mechanical properties of ceramics are good in compression, and they can have some special mechanical properties: very high hardness and greater stiffness than other engineering materials. Their greatest weakness is brittleness and *notch sensitivity,* and this means that they should be used in design very much the same way that glass is used. Most people can visualize what this means: no stress concentrations, mount in resilient material, do not bend, and so on. If this is done, the strengths of ceramics can be used without risking failures from their weaknesses.

Physical Properties

The density of some of the engineering ceramics is an important attribute. Some engineering ceramics are lightweight compared to metals (silicon carbide and boron carbide, for example), and for this reason these materials have specific stiffness that is larger than that of steels.

On the other hand, reduced weight can be an indicator of porosity, which lowers the strength of ceramics. Pressureless sintering of ceramics can produce a material with only 95% of theoretical density. There are two points that we are trying to make: ceramics can be lightweight compared to metals, but low density may also be an indicator of too much porosity in a particular grade. The designer should select a ceramic with high theoretical density (99%) if structural strength is a service consideration.

The melting points of ceramic materials are among the highest of all engineering materials. This is one of their strengths, and it is a reason why ceramics can be used for crucibles and refractory bricks.

The thermal conductivity and electrical conductivity of ceramics cover a large spectrum. The superceramic diamond has the highest thermal conductivity, while most ceramics have low thermal conductivity; they are heat insulators. Low thermal conductivity usually produces poor thermal shock resistance. This is a limiting property for some applications. Most ceramics are electrical insulators, but some are semiconducting. In general, ceramics compete with plastics as electrical insulators, but ceramics take the lead when operating temperatures are in excess of 400°F (204°C). Plastic cannot take high temperatures or electric arc damage. Ceramics outperform other materials in high-temperature insulation, but some materials such as silicon carbide are actually fair conductors of electricity at elevated temperatures. The point here is that the electrical and thermal properties of each ceramic should be checked at the temperature of interest to avoid surprises.

With regard to thermal expansion, the properties of ceramics cover a range, but they tend to be lower in expansion than metals and plastics. There are glasses that have a coefficient of thermal expansion of zero.

The maximum use temperatures of ceramics are above those of most other engineering materials. This is indeed one of their strengths.

Water absorption is a property that mostly affects plastics, but it can be a factor with ceramics that contain porosity. Water absorption is not a problem with most ceramics that are near theoretical density. Absorbed moisture can lower electrical insulating properties of porous ceramics.

Chemical Properties

The engineering ceramics are crystalline; some ceramic-type materials such as glasses are amorphous, not crystalline at all. The intermetallic compounds and metalloids are, in general, crystalline. Most ceramics do not respond to the types of heat treatments that are used on metals to change crystalline states; however, alloying, the adding of impurity elements, is often used by ceramists to change crystal structure.

The chemical resistance of ceramics is one strength of this class of materials. The driving force for corrosion is that materials want to return to the state in which they were found in nature. Because ceramics are often oxides, nitrides, or sulfides, compounds found in nature, there is little driving force for corrosion. The engineering ceramics, aluminum oxide, silicon carbide, zirconia, silicon nitride, and others, are very resistant to chemical attack in a wide variety of solutes. The corrosion resistance of ceramics is not always good, however. Each ceramic's corrosion data must be consulted for each chemical under consideration, but in general they are usually resistant. Many ceramics have corrosion characteristics similar to glasses. Glasses are resistant to most acids, bases, and solvents, but a few things like hydrofluoric acid will rapidly attack them. Corrosion data must be consulted.

It is not common practice to state the chemical composition of ceramics as is done for metals. When we purchase a steel alloy, it is common practice to ask for a certificate of analysis that shows the percentages of elements present (C, Cr, Ni, Mo, and so on). This is not normally done with ceramics. A drawing on a ceramic part would probably state something like the following: aluminum oxide, 99% theoretical density min.—Kors grade K33 or equivalent. The specification would not show the percentages of aluminum and oxygen that compose the aluminum oxide ceramic. Some engineering ceramics are made special by processing during manufacture; a manufacturer may make a blend of aluminum oxide and magnesium oxide that has properties that are quite different from pure aluminum oxide. In these instances, there is little recourse but to specify by trade name. The same thing is true with manufacturer processing that involves special treatments of a ceramic. For example, silicon carbide produced by chemical vapor deposition can have different use properties than the same material made by conventional pressing and sintering. Hipping, hot isostatic pressing, of ceramics improves densification, and this type of processing can alter use properties. The microstructure and porosity can vary with the method of manufacture. Thus it is not common to specify chemical composition in specifying ceramics, but it is recommended that density, processing, and additives be part of a specification. Very often ceramics are specified by trade name and manufacturer. In 2003 this was about the only way that you could be assured of getting the same material on each order.

Fabricability

Ceramics would be used much more in machine design and other industrial applications if it were not for their fabrication limitations. If a part is made by sheet metal fabrication processes, these processes will not work at all with ceramics. They cannot be plastically deformed to shapes at room temperatures. Ceramics cannot be melted and cast to shapes like metals, mostly because of their high melting temperatures. Silicon carbide melts at 4700°F (2600°C); aluminum oxide melts at 3659°F (2015°C). Ceramic types of materials are normally used for the molds for casting metals. A new form of material would have to be developed for molds for ceramic materials. There are additional problems in casting ceramics: the furnaces needed to melt a material with a melting point in excess of 3000°F (1648°C). Normal furnaces cannot achieve this type of temperature. Melting would have to be done with arc processes or something like an electron beam. The same situation exists with extrusions. What can be used for tooling to extrude a material at temperatures in excess of 3000°F (1648°C)?

The machining of ceramics is so costly that it is not an option for many machine parts. It was pointed out in our discussion of mechanical properties that tensile test specimens may cost $600 each; the same situation exists for one-of-a-kind parts for machines. It is not uncommon to pay $400 each for a 25-mm-OD, 4-mm-wall, 25-mm-long plain bearing of aluminum oxide. If large quantities of parts are needed and if they are relatively small, it is possible to make these by injection molding or extrusion processes, and the cost can become very low. Unfortunately, most service applications do not require the 100,000-plus annual part requirement that would justify the necessary tooling.

Probably the best example of the successful use of ceramics for a low-cost part is the common spark plug used in automobiles. The insulator is aluminum oxide (because ceramics are electrical insulators that can take the temperatures). On sale, spark plugs can be purchased for less than $1 each. The aluminum oxide insulator on the plug is a fairly complicated shape; how can this ceramic part be produced so cheaply? The answer is that the ceramic insulation is injection molded around the metal components in a molding machine that works similarly to those used for plastics. Aluminum oxide in fine powder form is injected into rubber molds at extremely high pressures. No machining is required; parts are molded and sintered to size. In this way, ceramics can be made at low cost. Many electrical parts are made this way, but it is the norm to have production requirements in excess of one million per year to justify the expense of the tooling required for injection molding. Ceramics often shrink 30% on sintering. To get a part to come out to the right dimensions after sintering, a shrink allowance must be put in the mold; this is the engineering aspect of molding ceramics to size.

In summary, in 2004 most ceramics require expensive processing to bring them to a usable shape. The low-cost processes are injection molding, extrusion, and pressureless sintering. High-performance engine parts and similar structural parts require expensive tooling and processing that are difficult to justify on small quantities of parts. The sol-gel process may someday allow these limitations to be overcome. Parts can be formed in gels that behave like plastics, and sintering converts them to useful ceramic parts. This technology exists in 2004, but to a limited extent—only for small parts such as optical lenses.

8.5 Concrete

Concrete is the most widely used engineering material in our world. Nothing comes close in volume used. It is everywhere: buildings, roads, dams, and so on. It is not officially part of ceramics (per the American Ceramic Society), but from a technical standpoint, it is a masterpiece of ceramic engineering. Concrete is a composite made up of sand, aggregate, and cement. The sand and aggregate are provided by nature; the cement is the amazing part of the system. It bonds the aggregate by forming a ceramic-like structure around each grain of sand and rock fragment. Mortars were discovered by the Greeks and Egyptians thousands of years ago. They used lime that they made by heating chalk ($CaCO_3$) to convert it to CaO, which would harden when mixed with water. The lime (CaO) reacts with water to form CaOH, which in turn reacts with silica (SiO_2) or other particles to form a compound that hardens and can bond other stones, bricks, and the like to make structures. Modern masonry cements still use lime as a key ingredient. Masonry cement is not very strong, so it is usually mixed with sand (SiO_2) and a small percentage of Portland cement for added strength. Up to the nineteenth century, most buildings were masonry bonded by lime-based cements. *Portland cement* was invented at that time, allowing concrete to be made. Bricks and building stones were no longer necessary. Significant structures could be "poured" from concrete. Portland cement is the bonding agent for the sand and aggregate that make up concrete. It bonds by chemical reaction with the water that is added to the dry mix to start the reaction. The chemical reaction starts within minutes and continues for more than 100 days. Most of the strength of a particular mix is achieved after 30 days, and the concrete must be kept wet for this length of time for the best strength.

Chemically, Portland cement is a complex compound of CaO, SiO_2, Al_2O_3, Fe_2O_3, MgO, and amounts of other compounds. It is made by heating a chalk, $CaCO_3$, and a clay at high temperatures. The firing of the clay and chalk or limestone dehydrates these minerals and they are essentially re-formed as "rock" when water is added. Thus concrete is sand and stones embedded in an artificial rock.

Concretes can have an infinite number of compositions, but the following are some typical properties using Portland cement (ASTM C 150):

Density	147 to 150 lb/ft^3 (2435 kg/m^3)
28-Day compressive strength	3000 to 5000 psi (20 to 35 MPa)
Elastic modulus	4 to 22 $\times$ 10^6 psi (30 to 150 GPa)
Approximate tensile strength	580 psi (4 MPa)
Approximate fracture toughness	0.4 ksi $\sqrt{\text{in.}}$ (0.5 MPa m$^{1/2}$)

Concrete is not intended to be used in tension, and when it is used for bridges and floor slabs, steel reinforcement is used to essentially carry the tension loads. Regardless of its brittleness, concrete is probably the most important material in the world for construction of infrastructure and buildings.

8.6 Glasses

Structure

Solids that do not have a three-dimensional, periodic structure are said to be amorphous or glassy (see Figure 8–8). Such materials do not exhibit significant levels of crystallinity. Many classes of materials are capable of forming amorphous or glassy structures under certain conditions, including metal alloys, organic polymers, oxide compounds, and nonoxide compounds. In this discussion we will focus primarily on oxide-based glasses.

A glassy or amorphous inorganic material is basically a super-cooled liquid. From a liquid state, a glass may be formed if the liquid is cooled fast enough to prevent crystallization. For example, some metal alloys are capable of forming metallic glasses if they are quenched rapidly from the liquid state. The quench rate must be extremely high to form a metallic glass because of low melt and crystallization velocity. However, some commercial metallic glasses based on zirconium such as Vitralloy® are now commercially available. To form an oxide *glass*, a mixture of molten oxides is cooled rapidly enough to prevent crystallization. Because many glass-forming oxides have relatively high viscosities even at elevated temperatures, their crystallization rates tend to be fairly low, allowing the formation of glassy structures with relatively slow cooling rates.

Not all oxides can form glasses. The structural relationship between the oxygen and the *cation* of the oxide compound strongly influences

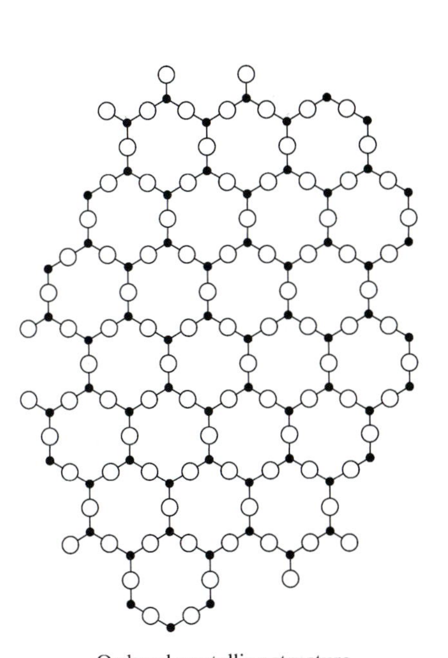

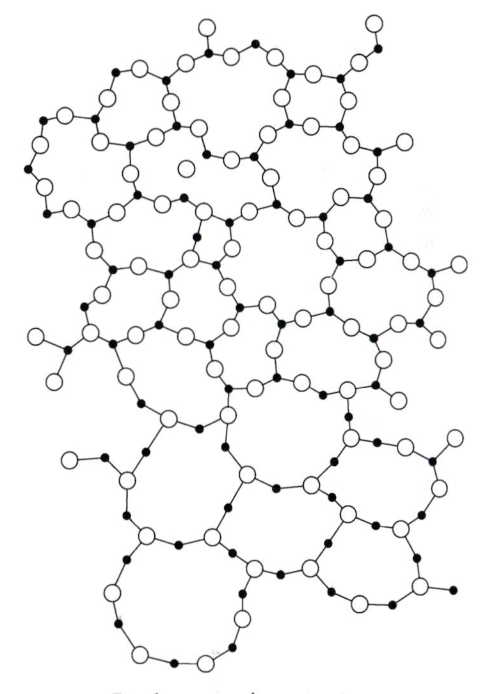

Figure 8–8
Comparison of random and regular structures

Table 8–3
Abbreviated list of oxides commonly used in glass

Main Glass Formers	Conditional Glass Formers	Intermediate Oxides	Network Modifiers
SiO_2	Al_2O_3	TiO_2	MgO
B_2O_3	Bi_2O_3	ZnO	Li_2O
GeO_2	WO_3	PbO	BaO
P_2O_5	MoO_3	Al_2O_3	CaO
	TeO_2	Zr_2O_3	Na_2O
			Y_2O_3
			K_2O

the glass-forming ability of the oxide. As listed in Table 8–3, the *main glass former* oxides, SiO_2, B_2O_3, GeO_2, and P_2O_5, have suitable structures and crystallization rates slow enough to form glasses when cooled from a liquid state with relatively slow cooling rates. *Conditional glass-forming oxides* will form glasses under certain circumstances. *Intermediate oxides* cannot form glasses on their own, but they do form three-dimensional networks when mixed with glass formers. *Network modifier oxides* cannot form a three-dimensional glassy network on their own or when mixed with glass formers. However, they can modify the properties of the glass because they can weaken the glass network by affecting Si—O bonds. These oxides are used to control properties such as the softening point or hardness. Oxides with crystal structures such as A_2O, AO, AO_4, and A_2O_7, where A represents a cation, are so symmetric that they generally crystallize from a melt rather than form glasses.

In general, the glassy state is *metastable*. Because the amorphous state is "quenched in" by cooling a liquid fast enough to prevent crystallization, the glassy state has a thermodynamic driving force to transform to a crystalline state. However, in many glasses the kinetics for the reaction can be so slow that for all practical purposes, the glassy state is stable. Some compositions of glass may be marginally stable over time or with elevated temperature aging, and the glass may transform from an amorphous state to a crystalline state. This phenomenon is known as *devitrification*. Certain classes of ceramics, called glass–ceramics, are based on this devitrification mechanism. With glass–ceramics, an article may be formed in the glassy state by processes such as casting or compression molding and then transformed to a predominately crystalline ceramic via devitrification. Glass–ceramics can have improved mechanical properties over completely amorphous glasses.

Glasses at room temperature are essentially very viscous liquids (with viscosities in excess of 10^{19} poise). For comparison, water has a viscosity of about 0.1 poise [a unit of viscosity; the SI unit is Pascal second (Pa • s), one poise is equal to ten Pascal seconds]. Viscous flow or creep of glass at room temperature is so minute, it occurs on a geologic timescale. However, with increasing temperature the viscosity of glass decreases as shown in Figure 8–9. There are several

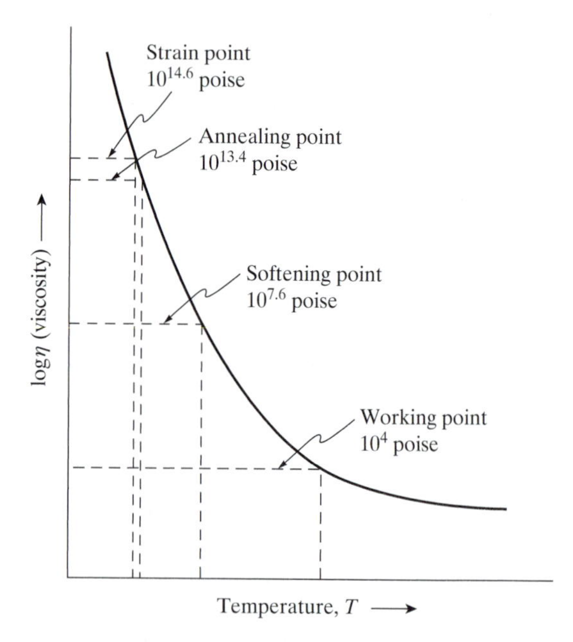

Figure 8–9
Typical viscosity versus temperature curve for glass

Table 8–4
Standard points for glass

Standard Point Temperature	Viscosity (Poise)*	Physical Description
Strain point	$10^{14.6}$	Temperature at which internal stress in the glass is relieved in hours.
Annealing point	$10^{13.4}$	Temperature at which internal stress in the glass is relieved in minutes.
Glass transition temperature	10^{13}	Temperature range at which the glass transitions from a supercooled liquid to a glassy solid.
Softening point	$10^{7.6}$	Temperature at which the viscosity of glass is low enough that the glass slumps under its own weight.
Working point	10^{4}	Temperature at which the viscosity of glass is low enough for forming and sealing.

*1 poise = 10 Pa • S

notable temperatures, called *standard points,* associated with the viscous flow or "melting" of glass. Table 8–4 summarizes the standard points for glass. The *strain point* is about the highest temperature at which a glass may be used for structural applications. Above the strain point, glass will creep appreciably. Most glass-forming operations occur at viscosities between the *softening point* and the *working point*. At the softening point, glass is rather stiff but will yield and flow with relatively little force. Some compression molding processes occur in this temperature range. The viscosity of glass at the working point is much like thick honey or syrup. Glass casting processes may occur at viscosities below 10^2 poise. Once a glass has been formed it is cooled to a temperature slightly above the strain point so that it will retain its shape and resist flow. At this point, the glass article must be *annealed* to relieve any internal stress. Annealing cycles can be very complex and typically include a cool-down regimen to bring the part to room temperature. Glass that is not properly annealed often will fracture or crack upon cooling to room temperature.

When liquid or molten glass cools slowly, the volume (measured as *specific volume*—inverse of density) contracts based on the coefficient of thermal expansion of the material as shown in Figure 8–10. Correspondingly, the viscosity of the glass increases, as shown in Figure 8–9. If the viscosity is still relatively low, structural changes (reorientation) can occur at the same rate as the cooling rate and the glass structure rearranges into a more dense arrangement as the viscosity of the glass increases. At some point the viscosity of the glass increases at a higher rate than the

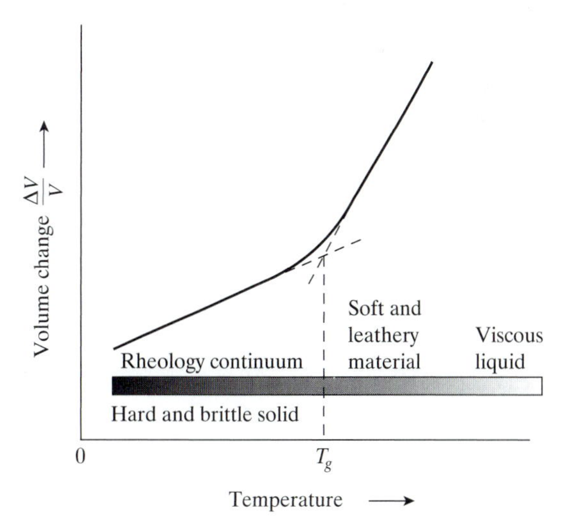

Figure 8–10
Volumetric expansion of glass as a function of temperature. An inflection occurs at the glass transition point, T_g.

rate of structural rearrangement. At this transition temperature, the slope of the specific volume versus temperature curve changes (see Figure 8–10). This change in slope occurs at the *glass transition temperature,* T_g.

Silicon oxide (SiO_2), essentially refined sand, is probably the most common constituent in most commercially important glasses. An excellent glass former, silicon oxide provides a good "base" for many glass compositions. Glasses based on silicon oxide are called *silicate glasses*. Glasses of pure silicon oxide such as silica or fused quartz are quite refractory—their softening temperatures are very high (1200 to 1500°C or higher). Additions of other oxides such as CaO, Na_2O, and K_2O to SiO_2 reduce the working point and viscosity point to much lower levels, enabling a wider array of fabrication processes to be used. Ordinary window glass is usually made from a mixture of sand (SiO_2), limestone ($CaCO_3$), and soda ash (Na_2CO_3). When fused, glass has a composition that is a complex mixture of oxides (Table 8–5), and the structure is similarly very complex. Other glasses are made by varying the

Table 8–5
Composition of commercial glasses

	Major Component (%)								
Type	SiO_2	Al_2O_3	CaO	Na_2O	B_2O_3	MgO	PbO	Other	**Comments**
Fused silica	99								Very low thermal expansion, very high viscosity
96% silica (Vycor®)	96				4				Very low thermal expansion, high viscosity
Pyrex type (borosilicate)	81	2	4	12					Low thermal expansion, low ion exchange
Soda lime									
Containers	74	1	5	15		4			Easy workability, high durability
Plate	73	1	13	13					High durability
Window	72	1	10	14		2			High durability
Lamp bulbs	74	1	5	16		4			Easy workability
Lamp stems	55	1		12			32		High resistivity
Fiber	54	14	16		10	4			Low alkali
Thermometer	73	6		10	10				Dimensional stability
Lead glass tableware	67			6			7	K_2O 10	High index of refraction and dispersion
Optical flint	50			1			19	BaO 13 K_2O 8 ZnO 8	Specific index and dispersion values
Optical crown	70			8	10			BaO 2 K_2O 8	Specific index and dispersion values

Source: From A. K. Lyle, "Glass Composition," *Handbook of Glass*. In F. V. Tooley, Ed., Books for Industry, Inc., New York, 1974.

amount of silica (SiO_2) and by adding other oxides. Colored glasses are made by adding oxides of specific metals: Cobalt oxide yields blue glass, iron oxide (FeO) yields green, and pure selenium yields red. Crystalline glass–ceramics are made from glasses containing lithium oxide, magnesium oxide, and aluminum oxide. In one commercial process, crystallinity is achieved by introducing tiny nuclei of titanium dioxide, a few angstroms in diameter, into the melt. With appropriate thermal cycling, these nuclei initiate crystal growth, and when all the thermal cycling is completed, the resultant material will be about 96% crystalline. Glass–ceramics have mechanical properties and some physical properties that are superior to amorphous glasses.

Even though glasses have a questionable melting point based on physical property measurements, they do melt. To be more correct, on heating, the viscosity decreases to the point at which the glass can be poured into molds and cast to shape. Thus glass parts for machine components are usually shaped in closed metal molds. Tubing is made by extrusion; bottles are made by a technique similar to the blow-molding technique used to make plastic containers. Sheet glass can be made by casting in flat plate-type molds. The less flat window glass is made by drawing a sheet out of a molten pool. Glasses are cast on molten metal and ground to make precision shapes such as lenses.

Properties

The properties of glasses vary with composition, but some general statements can be made.

1. Glasses are harder than many metals (400 to 600 kg/mm^2).
2. Glasses have tensile strength predominately in the range of 4 to 10 ksi (27 to 69 MPa).
3. Glasses have low ductility; they are brittle.
4. Glasses have a low coefficient of thermal expansion compared with many metals and plastics (0 to 5×10^{-6} in./in. °F; 0.54 to 9×10^{-6} cm/cm °C). Steel = 6×10^{-6} in./in. °F.
5. Glasses have low thermal conductivity (3 to 10×10^{-3} cal cm/s °C) compared to metals (steel is 0.1 cal cm/s °C).
6. The amorphous glasses have a modulus of elasticity in the range from 9 to 11×10^6 psi (62 to 76×10^3 MPa).
7. Glasses can be good electrical insulators (electrical resistivity can be higher than 10^{15} Ω-cm).
8. Glasses are resistant to many acids, solvents, and chemicals.
9. Glasses can be used at elevated temperature:

Glass Type	Softening Point	
	°F	°C
Fused quartz (silica)	2870	1580
96% silica (Vycor®)	2732	1500
Crystalline (Pyroceram®)	2462	1350
Soda lime (window)	1274	696
Borosilicate (Pyrex®)	1508	820

10. Glasses are slowly attacked by water and some alkaline solutions.

The optical properties of glasses make them superior to clear plastics for lenses, sight glasses, and windows. The very low coefficients of thermal expansion allow the high-silica glasses to have excellent thermal shock resistance. These glasses can be heated red hot and water quenched without cracking. This makes them a favorite for laboratory glassware and furnace retorts. The crystalline glasses have a modulus of elasticity of approximately 20×10^6 psi (138×10^3 MPa) and good shock resistance. Tempered glass, which fractures in small pieces when broken, is made by heating glass to a point near its melting point

(1000°F; 540°C) and then cooling rapidly. This causes the surface of the glass to be under a compressive stress, and the tendency for breakage as well as the failure mode is altered.

Applications

The applications of glasses in engineering design can have many facets. Besides the window-type applications, glass fibers are used in insulation, for sound deadening, as fillers in plastics, and as reinforcement in plastic laminates. Glasses can be used to shield or to transmit radiation. Photosensitive glasses are chemically machined into intricate shapes for fluidic devices used in machine controls. Glass-lined tanks are used to handle aggressive chemicals. Glasses are widely used in the food industry because of their chemical resistance and their cleanability. Many foods are processed in glass-lined tanks. There is no pickup of taste from the tank walls. Probably the largest use of glass, yet the most ignored, is in building construction. Most new high-rise buildings are predominately glass, which is strong enough to take wind loads and will withstand the outdoor environment indefinitely with the only maintenance being an occasional wash. Chemically strengthened glasses can have tensile strengths as high as 50 ksi (345 MPa). Commercial availability of stronger glasses will further increase the use of glass for structural and architectural applications. Machinable glass–ceramics are also available for use as insulators in machines and similar applications. They have continuous use temperatures as high as 1500°F (815°C). The glass–ceramics are used for stove tops that have heating elements under the glass. The burner regions can be red hot while material of the same piece of glass only a few inches away is at black heat. It is the low coefficient of thermal expansion of these materials that permits this thermal shock without fracture. One of the most important applications of glass in the Internet world is for fiber optic transmission of digital information. Fiber optic cables can carry many times more information than electrical conductors can carry. Glass fiber technology essentially has made possible the information technology era that exists in the new millennium.

In summary, *glasses* are predominately noncrystalline fusion products of inorganic materials, mostly oxides. They have good mechanical properties and a wide range of thermal, electrical, and optical properties. They have many uses in machines, but many more could probably be found with a little more consideration on the part of the designer.

8.7 Carbon Products

Carbon is an unusual element. With a valence of 4, it can behave as a metal or as a nonmetal. In its cubic structure, it is our hardest material, diamond. In its hexagonal form, it is *graphite*, a soft, weak substance that is used as a lubricant. Its lubrication ability is due to the ability of graphite crystals to deform by interplanar shear: The hexagonal crystals slide over each other in their basal plane. Industrial carbon–graphite products are usually combinations of amorphous carbon, graphite, and an impregnant to improve strength.

Making Carbon Shapes

Graphite exists in nature, and it can be mined and used in carbon–graphite products in this form. Industrial carbons are usually made from coke, which is produced by incomplete combustion of coal. Petroleum oils burned in insufficient oxygen will also form carbon in the form of lampblack, but this material will not graphitize. Carbon products are made by mixing carbon or carbon–graphite mixtures with pitch to form a material with a doughy consistency. In this form the material can be extruded into sheets, rods, bars, and the like, or it can be formed into rough part shapes. This material is then sintered at a temperature of about 1800°F (1000°C). The volatiles from the binder and impurities are driven off as gas in this operation, and the material has a considerable amount of porosity. At this point the sintered material may be part graphite, part ash, and part amorphous carbon, depending on the starting material. Most carbon–graphite products are then graphitized by long thermal cycles at temperatures in the range of 4500 to 5000°F (2500 to 2800°C). The sintered and graphitized shapes are then usually impregnated with various materials that will fill up the porosity and improve the mechanical properties.

Hundreds of different grades of carbon graphite are commercially available. They differ in starting material, porosity, degree of graphitization, and type of impregnant or surface treatment. Table 8–6 shows the comparative properties of a variety of grades. The low-permeability grades have different polymer impregnants, and this impregnation makes them less prone to loss of hydrodynamic lubrication in bushings and other similar types of applications. They can contain up to 15% ash, which can make them abrasive to sliding metal counterfaces. Suitable counterfaces are hardened (>50 HRC) steels and chromium plating. The high-thermal-conductivity grades usually have inorganic impregnants that will allow higher use

Table 8–6
Properties of selected grades of carbon graphite

Carbon Grade	**Low Permeability**	**Low Permeability**	**Low Permeability**	**High Thermal Conductivity**	**High Thermal Conductivity**
Impregnant	**Phenolic**	**Furane**	**Polyester**	**Water-Soluble Phosphate**	**Phosphate Glass**
Comparative Properties					
Density (g/cc)	1.80	1.85	1.85	1.91	2.00
Open porosity (vol %)	7	3	1	5	0.3
Permeability (μdarcies)	300	0.5	<0.1	400	<0.1
Temperature limit in air, °F (°C)	500 (260)	500 (260)	450 (232)	1202 (650)	1256 (680)
Hardness (Shore scleroscope)	65	80	75	75	80
Strength, ksi (MN/m^2)					
Compressive	22.5 (155)	25 (172)	27.5 (190)	27.5 (155)	27.5 (190)
Flexural	8 (55)	10 (69)	10.5 (72)	8 (55)	11 (76)
Tensile	5.5 (38)	7.5 (52)	8 (55)	6 (41)	7.5 (52)
Modulus of elasticity 10^4 lb/in.^{2} (10^3 MN/m^2)	21 (3.1)	21 (3.1)	23 (3.3)	14 (2.0)	17 (2.4)
Coefficient of thermal expansion to temp limit, 10^{-6} in./in. °F, 68°F (10^{-6} mm/mm °C, 20°C)	5.2 (2.9)	6.1 (3.4)	5.4 (3.0)	3.8 (2.1)	4.5 (2.5)
Thermal conductivity Btu/h °F ft (W/m °C)	12 (7)	12 (7)	12 (7)	69 (40)	69 (40)
Electrical resistivity 10^{-3} Ω-in. (10^{-3} Ω-cm)	3.8 (1.5)	3.8 (1.5)	3.8 (1.5)	1.5 (0.6)	1.5 (0.6)

Source: From *Materials Engineering,* March 1983. (Additional grades are listed in this reference.)

temperatures than the grades that are impregnated with polymeric resins. The high-strength grades usually contain less graphite than the lower-strength grades. This makes them abrasive in sliding systems. The best counterfaces for these types of carbon–graphites are ceramics or cermets, either solid or applied by thermal spray processes.

Properties

The type of starting material, the processing, and the impregnation allow a wide range of properties in carbon–graphite products. The properties listed in Table 8–6 give the user an idea of the property ranges that are typical, but as a class of materials they have the following characteristics:

1. They sublime and do not melt when overheated.
2. They are electrical conductors.
3. They can be easily machined.
4. They have low hardness.
5. They have low tensile strength.
6. They are brittle.
7. They have distinct temperature limits when used in air.
8. Some grades can retain their strength at temperatures as high as 4000°F (2200°C) in vacuum.
9. All grades are proprietary; there are no standard grades.

An exception to these statements on hardness and machinability are the siliconized grades. These grades have been subject to diffusion of silicon into the surface from silicon monoxide gas at temperatures as high as 3000°F (1650°C). This converts the surface to about 80% to 90% silicon carbide. The remainder of the surface is still graphite. These grades are more difficult to machine, and the surface is hard. The graphite that is still present offers lubrication. These grades are used for sliding systems, but again a very hard counterface may be necessary.

Applications

Industrial applications of carbon products include such things as motor brushes, electrodes, battery cores, heating elements, rocket components, and casting molds. Also of interest is the use of carbon products for sliding applications, such as bushings, seals, and wear inserts. In this latter area, carbon products are particularly useful in machine design. Graphite has been used for many years as a dry-film lubricant. Solids made from graphite materials can have similar self-lubricating characteristics. Wear components fabricated from graphites can have poor friction and wear properties in unfavorable environments. They can be lubricated with petroleum oils, but graphite wear components normally are used only where it is not possible to lubricate or where use temperatures preclude lubrication with petroleum products. Graphite bearings can normally be used at temperatures to 600°F (316°C) with very good wear life and at much higher temperatures with some reduction in service life due to oxidation.

When carbon or graphite slides against a clean metal surface in the absence of water vapor, oxygen, or organic gases, it will wear very rapidly, and the coefficient of friction will be very high (~1.0). Carbon–graphite products, on the other hand, can run against themselves in these environments (for example, in a vacuum or in inert gases) with low friction (~0.3) and favorable wear characteristics. Carbon–graphite products, when sliding against a metal surface, require transfer of graphite to the metal surface for favorable service life. Water vapor and/or other contaminating gases are essential to this transfer film formation. Thus the ground rules for using carbon products for wear applications call for not using them to replace conventionally

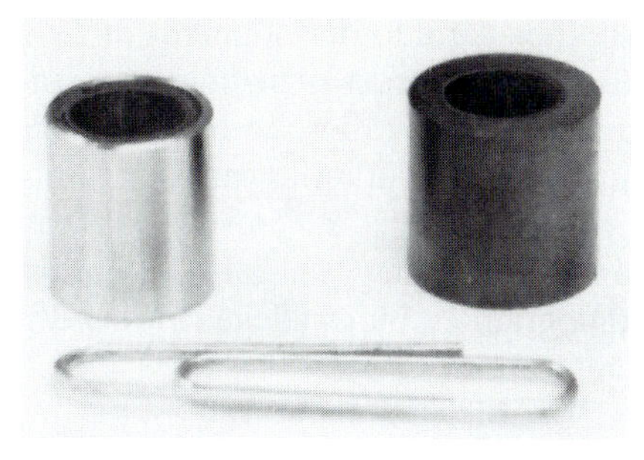

Figure 8–11
Solid graphite and graphite-lined steel bearings

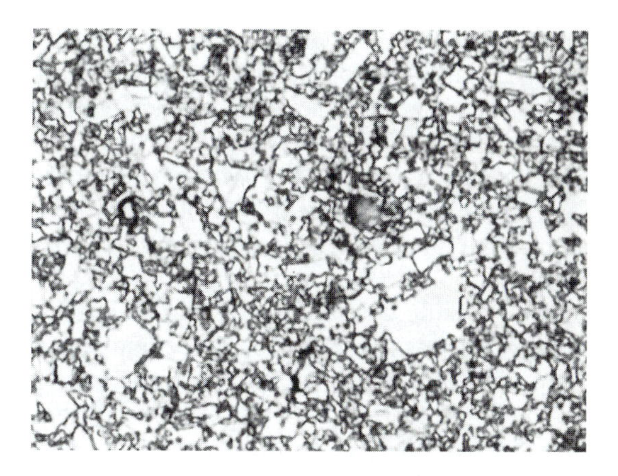

Figure 8–12
Microstructure of cemented carbide (×1500). Light areas are carbide particles; dark areas are cobalt binder.

lubricated bearings unless you have a good reason, and to use them only in atmospheres favorable to graphite-film formation.

We have discussed the limitations of carbon products; let us now discuss their advantages. In the normal unlubricated condition, they have typical PVs on the order of 15,000. Because of their high-temperature resistance, carbon products are probably the only bearing materials that can be used in vacuums and at high temperatures (Figure 8–11). Graphite running against graphite in a vacuum has been used at temperatures up to 3000°F (1649°C). No metal or plastic can do this. Graphite is almost an industry standard for mechanical seals (with a ceramic counterface) on pumps, agitated tanks, and similar applications. Similarly, graphite materials are excellent for sliding electrical contacts. In running carbon products against a metal, it is desirable to make the metal as hard as possible.

In summary, industrial carbon products have unique thermal, electrical, and tribological characteristics that make them useful for solving special problems in machine design.

8.8 Cemented Carbides

Tungsten carbide and some other *carbides* such as titanium carbide are not normally used as solid sintered ceramics, as are aluminum oxide or magnesium oxide; rather, they are used as composites bonded with a metal binder such as nickel, cobalt, chromium, molybdenum, or combinations thereof. Materials that are a composite of a metal and a ceramic are called *cermets*. However, it is common practice in the United States of America to use the term *cermets* for metal-bonded carbides and *cemented carbides* for carbides bonded by cobalt and nickel. Forty-five percent titanium carbide in a steel matrix would be called a cermet. Ninety percent tungsten carbide held together with ten percent cobalt binder would be called a cemented carbide. The cemented carbides most used in machines have tungsten carbide as the primary component, but they may also contain carbides of tantalum, columbium, titanium, and even niobium. The most common composition is tungsten carbide with a cobalt binder (Figure 8–12).

Fabrication Techniques

Cemented carbide cermets are made by compaction techniques. The starting material for the carbide portion of the composite is usually fine tungsten powder. This is converted into tungsten carbide by controlled carburization. The tungsten carbide powder is ball milled with the cobalt or other binder metal to reduce the carbide particle size and to coat the carbide powder with the binder. The coated powder is then classified to the desired particle size range, and it is hydrostatically compressed into billets or press compacted into parts. The compacted material is vacuum or hydrogen sintered into the cemented carbide composite. Hot isostatic pressing is offered as an option to reduce porosity in sintered shapes. However, some manufacturers of cemented carbide densify with a sinter-hip process. Sintering is performed concurrently with hipping. A shape such as a billet (in a steel can) is heated to about 2000°F in a pressure vessel pressurized to about 20,000 psi in argon. Sinter-hip compaction combines steps 3 and 4 in Figure 8–13. Unlike most engineering ceramics, the sintering process produces melting of the binder phase. The mechanism of bonding is similar to brazing of metals. However, a definite alloying between the carbide particles and the binder is produced. Porosity is negligible, and the sintered material has the high hardness, compressive strength, and modulus that make this material unique. The steps involved in the fabrication of most cemented carbides are outlined in Figure 8–13.

Reaction bonded carbides were introduced on a commercial basis in the United States in 1994. These are not cermets, carbide particles bonded with a soft metal, but tungsten carbide particles bonded by a second type of carbide, molybdenum carbide. This particular combination of carbides is produced by a special hipping process that can also be used to make near-net shapes. This type of carbide allegedly has much higher abrasion resistance than WC/Co carbide because the soft metal matrix has been eliminated. Both carbide phases are hard, and the mechanism of bonding is a diffusion and alloying of the two carbide phases. Markets for this product are still developing. A major application is slurry nozzles, in which it allegedly outwears conventional carbides by a factor of at least 10.

Another carbide development in 1994 was the commercialization of nanograin cemented carbide. Powders for these carbides are made by chemical processes and the particles can be less than 50 nm in diameter. Conventional carbides have a lower limit in grain size of about 0.5 μm. The nanograin carbides allegedly outwear conventional carbides.

Properties

Cemented carbides are harder than any tool steel, so they have excellent resistance to low-stress abrasive wear. Ceramics can be just as hard, but care must be exercised when they are used for applications such as cutting tools, dies, and machine parts. They are much more brittle and prone to chipping than cemented carbides. The metal binder allows the impact resistance of cemented carbides to be better than ceramics, but not as good as that of tool steels; the tensile strength can be as high as 200 ksi (1378 MPa).

Cemented carbides can be used as wear parts against hard or soft steels usually with a significant reduction in system wear compared with most metal-to-metal combinations. Some grades can run against themselves; some cannot. Carbide tools are shown in Figures 8–14 and 8–15.

A number of grades of cemented carbides are commercially available. They differ in type and weight percent carbide, type and weight percent binder, and carbide particle size. A description of some of the various grades is provided in Table 8–7. A high binder content is used for impact applications. Because titanium carbide is harder than tungsten carbide, a grade with a significant amount of this carbide will be more abrasion resistant than one with only tungsten carbide and binder. A low binder concentration similarly means greater abrasion resistance, and the grades without metal binders should have abrasion resistance that is better than all WC/Co grades.

The industry grade code shown in Table 8–7 refers to a system developed by the U.S. automotive industry. It was not developed by

Figure 8–13
Making of cemented carbide shapes

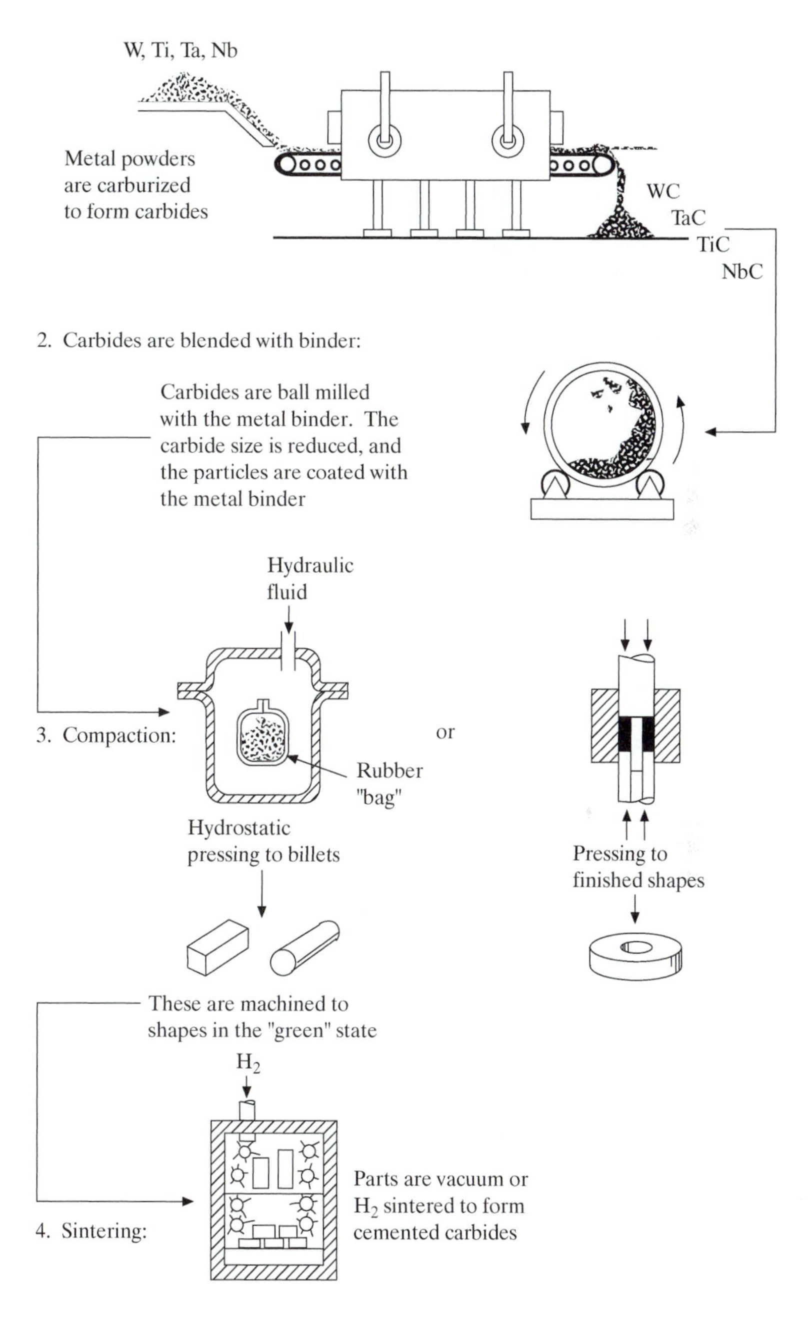
1. Production of carbides:
W, Ti, Ta, Nb
Metal powders are carburized to form carbides
WC
TaC
TiC
NbC
2. Carbides are blended with binder:
Carbides are ball milled with the metal binder. The carbide size is reduced, and the particles are coated with the metal binder
Hydraulic fluid
3. Compaction:
Rubber "bag"
or
Hydrostatic pressing to billets
Pressing to finished shapes
These are machined to shapes in the "green" state
H_2
4. Sintering:
Parts are vacuum or H_2 sintered to form cemented carbides

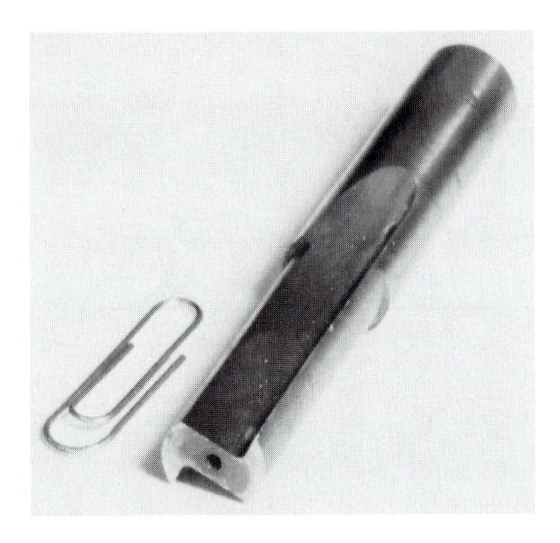

Figure 8–14
Cemented carbide milling cutter

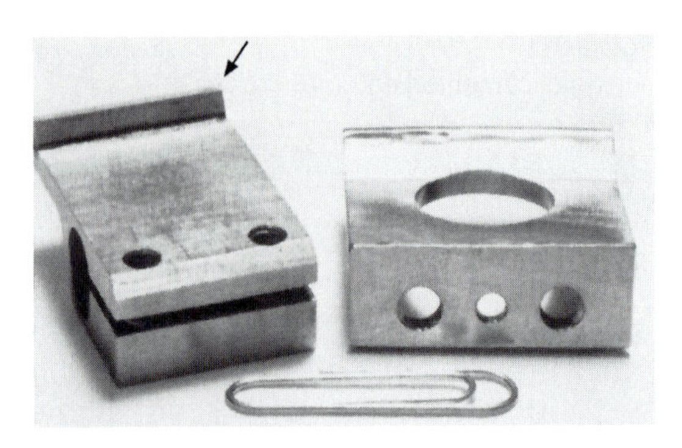

Figure 8–15
Cemented carbide brazed insert used to replace worn steel part (right)

Table 8–7
An example of a carbide grading system

			Example Properties					
			Composition*					
Use Category	Code	Recommended Application	WC	TiC	TaC	Co	Hardness* (RA)	Transverse Rupture trength* (MPa)
Machining of cast iron, nonferrous, and nonmetallic material	C-1	Roughing	94	—	—	6	91	2000
	C-2	General purpose	92	—	2	6	92	1550
	C-3	Finishing	92	—	4	4	92	1520
	C-4	Precision finishing	96	—	—	4	93	1400
Machining of carbon, alloy, and tool steels	C-5	Roughing	75	8	7	10	91	1870
	C-6	General purpose	79	8	4	9	92	1650
	C-7	Finishing	70	12	12	6	92	1750
	C-8	Precision finishing	77	15	3	5	93	1180
Wear applications	C-9	No shock	94	—	—	6	92	1520
	C-10	Light shock	92	—	—	8	91	2000
	C-11	Heavy shock	85	—	—	15	89	2200
Impact applications	C-12	Light	88	—	—	12	88	2500
	C-13	Medium	80	—	—	20	86	2600
	C-14	Heavy	75	—	—	15	85	2750

*Composition and properties are averages from several manufacturers.

carbide manufacturers, but each manufacturer has grades that fit into each of the categories.

Selection Guidelines

As a class of materials, cemented carbides have some unique characteristics:

1. Greater stiffness than steel [E = 60 to 90 × 10^6 psi (413 to 620 GPa)].
2. Higher density than steel.
3. Greater compressive strength than most engineering materials.
4. Hardness greater than any steel or other metal alloy.
5. Tensile strength comparable to alloy steels (200 ksi; 1380 MPa).

The transverse rupture strength values presented in Table 8–7 are a measure of the mechanical strength of the various grades. A high transverse rupture usually means that the material also has a high impact strength (Figure 8–16). The low binder and fine grain grades (<1 μm) usually have the best *abrasion* resistance (Figure 8–17). All grades are resistant to atmospheric corrosion in indoor environments, but there are special grades with nickel–chromium and other binders for severe corrosive environments. Grade C-2 is considered to be the overall general-purpose grade in the United States. If a steel is being replaced with a cemented carbide, this is a suitable starting grade.

There are design limitations. Carbides are usually brazed to other metals for use in tools.

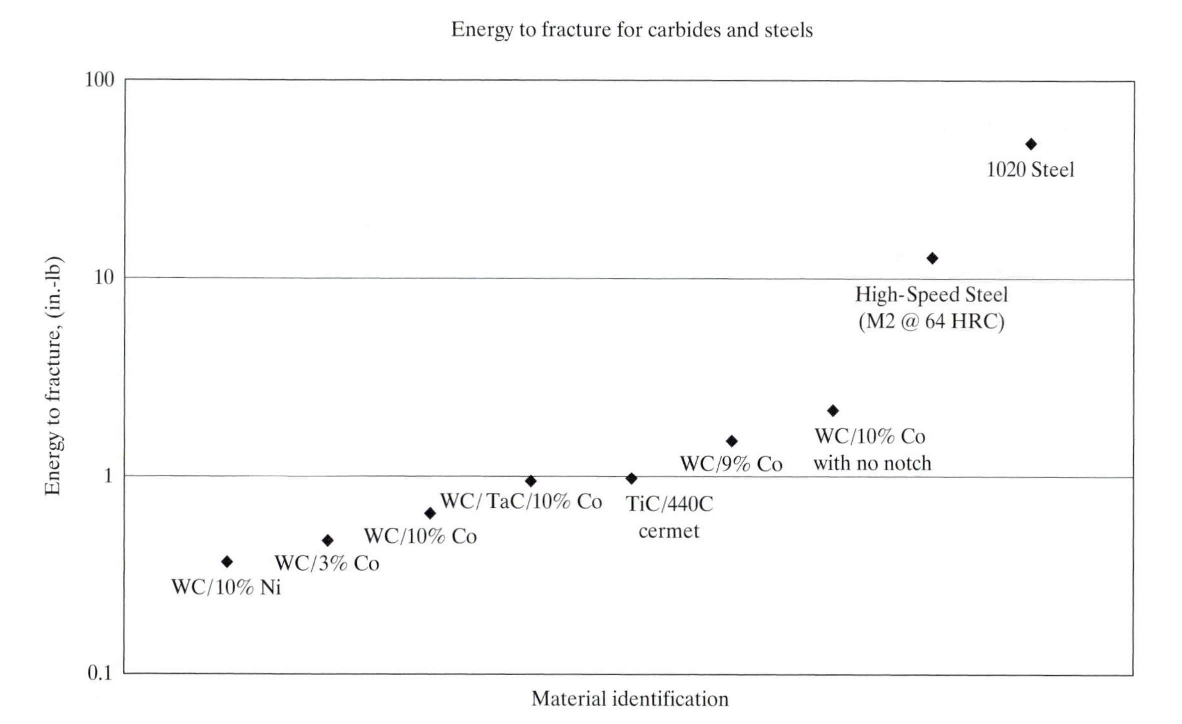

Figure 8–16
C-notch toughness of various types of carbides compared with M2 high-speed steel and 1020 carbon steel for a specific application

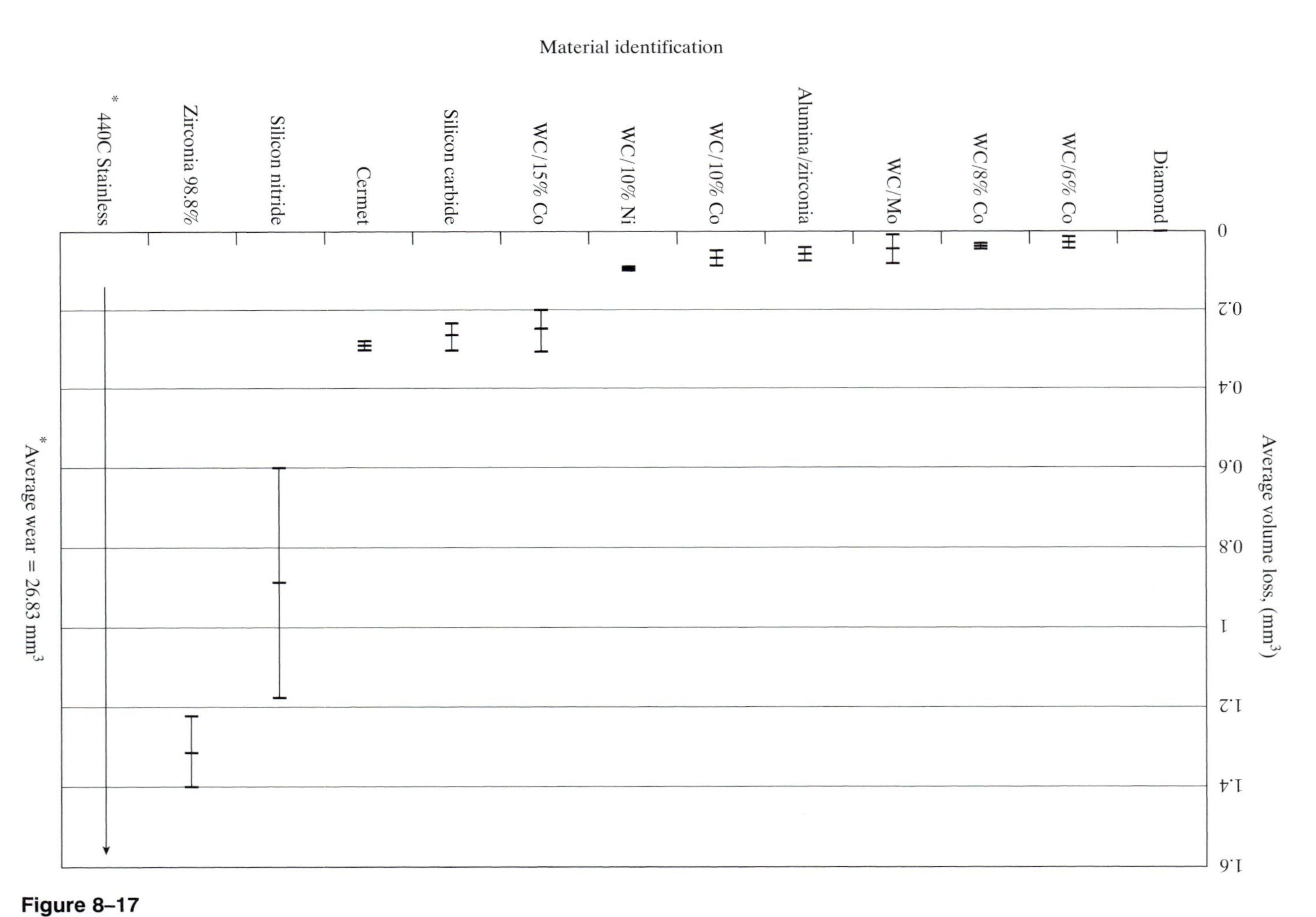

Figure 8–17
Abrasion resistance of various cemented carbides (WC) compared with some ceramics and stainless steel. Average volume loss +/− two standard deviations; tested with 200-g loading mass, 8-h test duration, 30-μm alumina abrasive tape, 7288-m sliding distance.

They have a low coefficient of thermal expansion (less than half that of steel). Care must be taken in brazing and in use to prevent cracking due to expansion stresses. Even though there are impact grades of carbide, the impact that any carbide can withstand is less than for most metals (Figure 8–16).

Probably the biggest design limitation is fabricability. After sintering, carbides cannot be drilled, milled, sawed, or tapped. They can only be ground or electrical-discharge machined. If more than a few parts are required of a given shape, it may be more economical to mold the shape in metal dies. This, of course, involves tooling charges. A second option is to have the part machined in the green state (by the carbide supplier) and sintered to a near-net shape that requires minimal finishing.

The size limit on carbide parts varies with manufacturers, but occasionally parts are made as large as 12 in. (300 mm) in diameter and 48 in. (1299 mm) long. The limit is the compaction and sintering capability of the manufacturer. The dimensional tolerances possible on as-sintered parts or hipped parts are approximately ±0.8% on any dimension. Typical straightness tolerances are 0.003 to 0.004 in./in. of length (0.3% to 0.4%).

The dilemma that the designer faces in using cemented carbides is what grade to use. As previously mentioned, each manufacturer makes a 6%, 10%, 15%, 20%, and 30% Co grade of cemented carbide. They also offer a wide variety of particle mixes (TaC, NbC, MoC, CbC mixed with WC). How do you know which one to use? Can you get the same product from three different suppliers? Unfortunately, the answer to the first question is not easy to accept. The reality of the situation is that one particular grade from one manufacturer will probably give the best performance, but finding that grade and manufacturer is usually done by trial and error. The answer to the second question is probably, No. A 6% cobalt/tungsten carbide material with a 1-μm particle size will probably not perform the same in service as a grade with this same "specification" from another manufacturer. There are differences in processing, and this may show up in performance.

Figure 8–16 shows the fracture toughness of a variety of grades bought to do the same job. All grades but one, the one with the highest fracture toughness, chipped in service. All were recommended by their manufacturers as suitable for the application. The fracture resistance appears to depend on the binder and grain size. A binder of 10% to 15% is usually needed for a keen cutting edge and larger grains seem to promote toughness.

Figure 8–17 compares the abrasion resistance (to aluminum oxide) of some grades of cemented carbide. The high-binder grades have lower abrasion resistance, but the difference between 10% and 6% cobalt is not significant. The ability of different carbide grades to slide against each other is not predictable. Figure 8–18 shows some crossed-cylinder tests on various carbide/steel couples. These data suggest that a suitable counterface for a cemented carbide requires empirical testing.

Our recommendation on grade selection is to use a 6% cobalt grade (C2) for a new application (1- to 2-μm grain size) and fine-tune this grade with more or less binder and different grain sizes and mixes if the C2 grade does not meet expectations.

The final consideration in using cemented carbides in design is economics. Many designers do not even consider cemented carbides because of their reputation as being expensive. On a per pound basis, carbides appear expensive. The tungsten powder that goes into making the carbide may cost over $20 per pound. However, if a carbide part solves an expensive wear problem, even $50 per pound may be an inconsequential basis material cost. A rule of thumb to use in determining if carbides are justified for an application is that carbide costs two to six times as much as a hardened tool steel part. This cost factor can be weighed

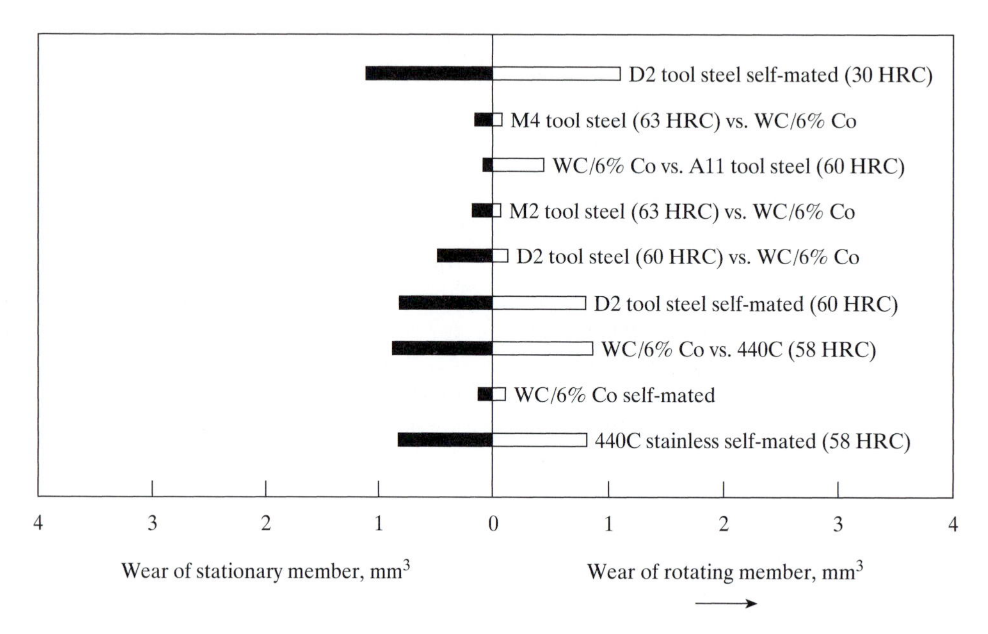

Figure 8–18
Crossed-cylinder wear test results of carbide/steel couples tested per ASTM G 83

against a possible 50-fold increase in service life. Thus cemented carbides should be considered for any machine part that may be subjected to severe low-stress abrasion or for systems in which the other unique properties of carbides offer advantages over other material systems.

8.9 Ceramics for Structural Applications

It is doubtful if aluminum oxide ceramic will ever be used for a wide-flange beam for use in a building, but the most-used materials of any type of structural applications are ceramic types of material, concrete, and glass. Thus, ceramic types of materials are used for structural applications, but what we shall discuss in this section is not any of these items. We shall discuss the use of ceramics for machine or structural parts: pump impellers, turbocharger fans, platens for precision optics, nozzles for extruders, check valve seats, ball valve balls, soldering fixtures, and the like. Ceramics can be good candidates for these types of applications because they can take the harsh environments that these parts often see. However, before ceramics will be used for these kinds of applications, it must be demonstrated to designers that ceramics have the long-term durability needed for these structural applications. In the early 1990s, many materials research laboratories in industry had programs trying to develop applications for which ceramics would be cost effective and provide a service advantage over other materials. Only a few ceramics emerged as contenders for industrial applications: aluminum oxides, silicon carbides, silicon nitrides, zirconias, and a few other ceramic blends. Also, cemented carbides and in some cases the machinable ceramics (glass-bonded micas) are candidates. We shall discuss some of these materials for use as machine parts with the goal of presenting structural use information and guidelines to allow successful use in design engineering.

Aluminum Oxide (Al_2O_3)

Aluminum oxide is commonly made from the mineral bauxite, $Al_2O_3 \cdot 2H_2O$. There are a number of polymorphs of aluminum oxide, but most structural components are made from alpha aluminum oxide, which has a hexagonal structure, or from gamma aluminum oxide, which has a cubic crystal structure. Structural parts are made from aluminum oxide using any of the previously mentioned ceramic fabrication processes, and there are various grades with different impurity concentrations and different levels of porosity. Some refractory applications use sponge aluminum oxide with more than 50% porosity, but most structural grades have porosity that ranges from 10% to less than 0.5%. As we might expect, the strength of these different grades increases with reduced porosity. Typical mechanical and physical properties of a structural grade of aluminum oxide (alumina) are shown in Table 8–8. Some noteworthy properties of aluminum oxide for structural applications are its modulus of elasticity, tensile strength, compressive strength, and strength at elevated temperatures. Its tensile modulus of 53 million psi (3.65×10^{-5} MPa) makes it stiffer than steel. The compressive strength of about 380 ksi (2620 MPa) makes it stronger in compression than many hardened tool steels; its tensile strength at 30 ksi (206 MPa) does not compare too favorably with metals, but this number is not insignificant for ceramic types of materials. At 2000°F (1093°C) alumina retains about 50% of its room temperature strength. Environmental resistance is very good; aluminas are resistant to most oxidizing and reducing media up to their softening points.

As a class of materials, all ceramics should not be loaded in tension. When ceramics are used for structural applications, the loads should be applied in compression and impact should be avoided. Keeping this in mind, aluminum oxide is suitable for any type of load-bearing application that does not violate these design conditions. It is used for rocket nozzles, pump impellers, pump liners, check valves, nozzles subject to erosion, and even for support members in electrical and electronic devices. Shapes are generated by molding to shape or by machining in the green state and then sintering. The size change and distortion that occur in sintering are accommodated by finish machining with diamond grinding wheels. Aluminum oxide is the oldest engineering ceramic, and blends with other ceramics such as zirconia are being investigated to improve on its tensile properties and toughness characteristics, but as a family of ceramics this material will continue to be an important structural ceramic for some time into the future.

Silicon Carbide (SiC)

Silicon carbide is an old ceramic material that is being rediscovered for structural applications. It is somewhat like aluminum oxide in that it has been used for decades for applications other than structural. The most familiar application of silicon carbide is for abrasives for grinding wheels and for bonded abrasive papers. Silicon carbide abrasives are the hardest of the traditional abrasive materials. It has a hardness of about 2400 HK at room temperature, and it tends to fracture to form very sharp particles that produce high removal rates in abrasive machining. Unfortunately, silicon carbide grains tend to be more friable than aluminum oxide; they lose their cutting action quickly as bonded abrasives, and they are not used for general-purpose grinding wheels. They are used for special grinding applications.

Silicon carbide can be made by different processes, but one of the oldest techniques used is to essentially diffuse carbon from coke into sand, SiO_2, by long-time firing at high temperatures.

Table 8–8
Properties of several engineering ceramics

Property	Low-Polytype Sialon (Glassy Phase)	Hot-Pressed Silicon Nitride	Reaction-Bonded Silicon Nitride	Sintered Silicon Carbide	Alumina	Partially Stabilized Zirconia (PSZ)
Room temperature modulus of rupture, MPa (ksi)	945 (137)	896 (130)	241 (35)	483 (70)	380 (55)	610 (88)
Typical Weibull modulus	11	10–15	10–15	10	10	10–20
Room temperature tensile strength, MPa (ksi)	450 (60)	~580 (~80)	145 (21)	299 (42)	210 (30)	466 (67)
Room temperature compressive strength, MPa (ksi)	>3500 (>500)	>3500 (>500)	1000 (~150)	2000 (~300)	2750 (~400)	1850 (~270)
Room temperature Young's modulus, MPa (ksi)	3×10^5 (4.4×10^4)	3.1×10^5 (4.5×10^4)	2.0×10^5 (3×10^4)	4.1×10^5 (6×10^4)	3.6×10^5 (5.3×10^4)	2.0×10^5 (3×10^4)
Room temperature hardness, kg/mm^2 (VPN, 0.5-kg load)	2000	2200	900–1000	2500	1600	1500
Fracture toughness (k_{IC}), MPa $m^{-1/2}$	7.7	5	1.87	3.0	1.75	9.5
Poisson's ratio	0.23	0.27	0.27	0.24	0.27	0.3
Density, kg/m^3 (g/cm^3)	3230–3260 (3.23–3.26)	3200 (3.2)	2500 (2.5)	3100 (3.1)	3980 (3.98)	5780 (5.78)
Thermal expansion coefficient, m/m/K (0 to 1000°C)	3.04×10^{-6}	3.2×10^{-6}	3.2×10^{-6}	4.3×10^{-6}	9.0×10^{-6}	10.6×10^{-6}
Specific heat, J/kg/K (cal/g/K)	620 (0.15)	710 (0.17)	710 (0.17)	1040 (0.25)	1040 (0.25)	543 (0.13)
Room temperature thermal conductivity, W/m/K	21.3	25	8–12	83.6	8.4	2
Thermal shock resistance, ΔTK	~900	500/700	~500	300/400	200	~500

Source: From *Advanced Materials and Processes,* January 1986.

The silicon carbide mass that results is crushed and made mostly into abrasives. There are two principal types of silicon carbide: alpha and beta; the former is primarily hexagonal in structure and the latter is cubic.

Structural parts and monolithic slabs and shapes are typically fabricated from silicon carbide by one of four processes: *pressureless sintering, reaction bonding, hot pressing,* or *chemical vapor deposition* (CVD). All of these processes except CVD use alpha silicon carbide. These processes can either produce parts to near-net shape or produce billets or shapes from which parts may be machined by conventional hard-material machining processes (e.g., diamond cutting and grinding). Components made by pressureless sintering are compacted from SiC powder (plus sintering aids such as boron and carbon) and are heated in an inert gas or vacuum at 2050°C, allowing the particles to bond together.

In reaction bonding, silicon powder is compacted to the desired shape and then reacted with carbonaceous gases at elevated temperatures to form a SiC compound. *Silicon infiltrated* silicon carbide, a variant of the reaction bonding process, is made by making a compact from SiC and carbon powder. The compact is then heated to temperatures above the melting point of silicon and exposed to silicon liquid or vapor. The silicon infiltrates the pores and reacts partially to form SiC. Some excess silicon remains in the pores. Although suitable for applications such as optical mirrors, this two-phase material can have some limitations.

Hot pressing is one of the more common ways to procure SiC. Compacted SiC powder (plus sintering aids such as C and B) is densified by uniaxial pressure during the sintering process. This process usually takes place in vacuum or argon at 2150°C with pressures of 30 MPa. The hot isostatic process is similar except the pressure is applied in all directions.

Chemical vapor deposited silicon carbide is formed by reacting a gaseous vapor with a substrate to form a beta silicon carbide compound. The reactants are coated onto a graphite plate or tool. The layer is then stripped from the graphite tool to yield a monolithic slab of material. Plates as large as 700 mm square by 20 mm thick are routinely manufactured by this method. CVD SiC parts can be made to near-net shape by coating the material onto graphite tooling and then oxidizing the graphite tooling to reveal the silicon carbide component. Many unique components, such as special fixtures for the semiconductor industry, are made in this fashion.

Some noteworthy properties of SiC are the following (see also Table 8–8):

1. Silicon carbide has higher tensile strength, stiffness, and hardness, and lower density than aluminum oxide.
2. Compared with most engineering ceramics (with the exception of various forms of carbon), CVD silicon carbide has the highest thermal conductivity.
3. CVD and silicon-infiltrated SiC are commonly used for precision optical mirrors because of dimensional stability and polishability. CVD SiC can be polished to a surface roughness of 3 to 5 Å Ra—better than almost any engineering material.
4. Silicon carbide is resistant to abrasion and some other forms of wear.
5. Sintered SiC has 3% to 4% porosity. Hot-pressed and reaction-bonded SiC have less than 1% porosity. CVD SiC has "zero" porosity.
6. The coefficient of thermal expansion of SiC is sufficiently low that thermal shock is not usually a problem. With repeated heat cycling from 20 to 800 to 20°C (cooled with inert gas), CVD SiC can develop fine checks or cracks.
7. Silicon carbide is often used at temperatures up to 1500°C in oxidizing or to 2000°C in inert environments. Above 2000°C it undergoes an allotropic phase transformation

from the cubic beta phase to the hexagonal alpha phase. It sublimes at 2700°C.

8. Silicon carbide, especially the CVD grade, is very chemical resistant. For example, surfaces polished to a surface roughness Ra of 5 Å can be exposed repeatedly to immersion in strong hydrofluoric acid solutions without significant roughening or etching of the surface.
9. Probably the most significant drawbacks of silicon carbide are that it is not very tough and the CVD grade can be rather expensive with limited availability of shapes and sizes.

Silicon Nitride (Si_3N_4)

In the late 1980s, silicon nitride received lots of attention and research as a potential material for structural applications. There are a number of manufacturers and many types of this material, but basically there are two main options: reaction-bonded material, which can contain up to 20% porosity, and pressure-sintered materials (hot pressed, hipped, and so on), which can have 100% theoretical density. The reaction-bonded material has the advantage of low size change during firing. The disadvantage of this material is porosity, which lowers strength and other mechanical properties. This grade is often used for furnace parts and the like when high strength is not a critical factor. Cutting tools, wear parts, and structural shapes are usually made from the full dense grades. The properties that distinguish silicon nitride from other engineering ceramics are as follows:

1. No loss of strength in air at temperatures to 1000°C.
2. Greater thermal shock resistance than many other ceramics.
3. Lower density than most other engineering ceramics (one-third the weight of steel).
4. Low thermal expansion.
5. Better toughness than silicon carbide and aluminum oxide.
6. Stiffer than steel (50%).

Silicon nitride is widely used as a cutting tool material; ceramic inserts of this material have been shown to outperform cemented carbides and other ceramics for high-volume machining of cast irons and similar metals used in the automotive industry. It is used for gas turbine parts that must resist thermal cycling, and the use of this material for rolling element bearing applications has been a major success. Hybrid bearings with silicon nitride balls and steel races are widely used for demanding machine applications. The newer fine-grain grades have surface fatigue resistance that made them candidates for balls, rollers, and races for bearings. Other applications include parts for diesel engines, hot extrusion dies, and pump parts for which chemical resistance and wear are required.

Partially Stabilized Zirconia (PSZ)

Partially stabilized zirconia is really zirconium oxide, ZrO_2, that has been blended and sintered with some other oxide such as magnesium oxide, MgO, calcium oxide, CaO, or yttria, Y_2O_3, to control crystal structure transformations. Zirconium oxide or zirconia normally has a monoclinic crystal structure at room temperature and a tetragonal structure at elevated temperatures. The transformation of monoclinic to tetragonal structure results in a volume expansion that can be as high as 4%. If this volume change is allowed to occur (e.g., in hot pressing or sintering), it will probably fracture the part. Stabilization of zirconia with 3% to 20% of another oxide such as MgO will induce the stabilization of a cubic structure that is about halfway between the monoclinic and tetragonal structures in lattice parameters.

One of the most noteworthy properties of PSZ is that it has better fracture toughness than

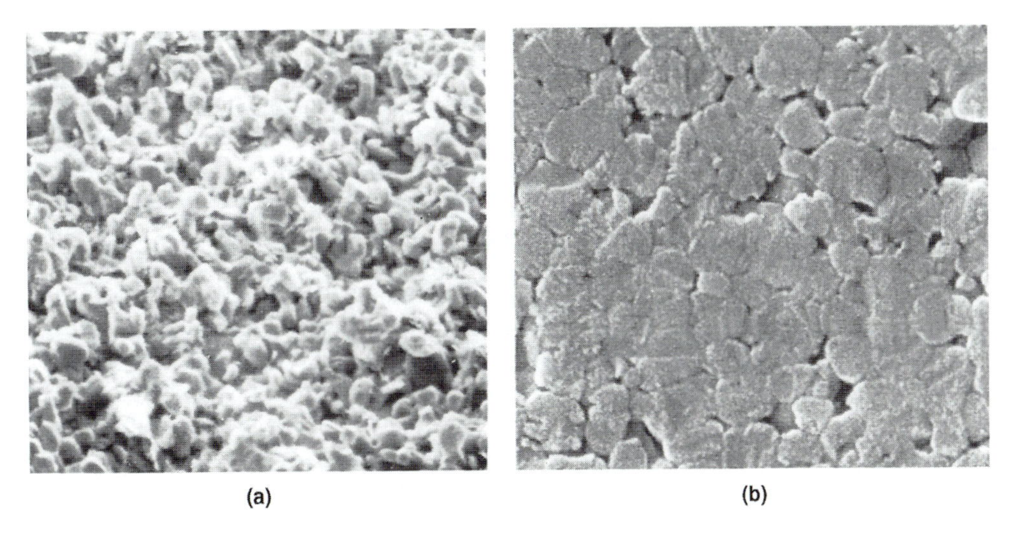

Figure 8–19
Scanning electron micrographs of (a) fracture surface of zirconia (×500) and (b) polished and etched structure of zirconia (×20,000)

the other high-performance ceramics. It is thought that the reason for PSZ's exceptional toughness is that in the partially stabilized condition some percentage of the structure goes through the tetragonal-to-monoclinic transformation when stress is applied. If the volume change that accompanies this transformation occurs at the tip of a crack, it will change the stress field at the tip of the crack and reduce the rate of crack propagation and improve its fracture toughness. PSZ is considered a *transformation-toughened* material.

There are other types of transformation-toughened zirconias: tetragonal polycrystalline zirconia and blends of zirconia and alumina. The tetragonal phase is fully stabilized by additives, and the purpose of the alumina–zirconia blends is to lower the cost (alumina is cheaper than zirconia) and to add some hardness for wear applications. Zirconias are softer than the other ceramics they compete with (SiC, Al_2O_3, and others).

The tensile strength of PSZ is better than that of alumina and some of the other engineering ceramics; its thermal conductivity is among the lowest, making it a good thermal insulator. Its thermal expansion is similar to that of steel, as is its modulus of elasticity. Its toughness and steel-like properties have made PSZ a prime candidate for replacement of metals in internal combustion engines. In other areas it competes with the other ceramics listed in Table 8–8. Figure 8–19 is a scanning electron micrograph of a fine-grain zirconia fracture surface.

Sialon ($Si_3Al_3O_3N_5$)

The low-polytype sialon referred to in Table 8–8 is really a specific ceramic in a relatively new family of ceramics formed by the blending of silicon nitride, silica, alumina, and aluminum nitride. A glassy phase forms in sintering that somewhat limits the elevated temperature applications of this material, but this glassy phase improves the ability of this material to be sintered. It is possible to achieve 100% dense structures with pressureless sintering. This is a significant advantage for structural parts in which high density is needed for toughness and fracture resistance. This material has good mechanical properties, and it is a candidate for many structural applications.

Summary

Ceramics have a very definite place in design engineering. They can offer properties not available in other material systems. Selected properties of ceramics are compared with hardened tool steel and cemented carbide in Figure 8–20.

Carbon–Carbon Composites

Carbon–graphite materials can be used for critical structural applications in the form of carbon–carbon composites. This material is like an "all-carbon fiberglass." The matrix is carbon–graphite and the reinforcement is carbon fibers, the same ones that are used to reinforce polymer matrix composites. These composites are used for applications requiring light weight coupled with high temperature resistance. Carbon remains solid at higher temperatures (>6600°F or 3600°C) than any other engineering material in appropriate atmospheres. Air is not an appropriate atmosphere for carbon–graphite products because they burn in air at temperatures above 1100°F (600°C). Oxidation of carbon–carbon composites is addressed by protective ceramic-type coatings.

Carbon–carbon composites are made from carbon fiber–polymer resin composites such as carbon fiber–phenolic resin. The material is laid up into a part shape in the usual way, and the polymer matrix is converted to carbon by heating in a controlled atmosphere (pyrolysis). There is considerable porosity in the part after this operation, and the porous carbon matrix is densified by intrusion of chemical vapor–deposited carbon or by resin/pitch impregnation. The part is then heated in a suitable atmosphere to pyrolyze the impregnant. The impregnated part may need multiple impregnation/pyrolysis treatments. If the part is to be used for high-temperature service, the surface is protected from oxidation by surface coatings of ceramics produced by chemical vapor deposition or other processes. The end result is a carbon–carbon composite that can have a complex three-dimensional shape and specific strength greater than superalloys at temperatures above 2000°F (1200°C).

Carbon–carbon composites are used for the nose piece and leading edges of the wings on the Space Shuttle and on less exotic applications such as break linings for large aircraft. These materials are very expensive, so they are not used for everyday applications, but they are available for special applications.

Glass–Ceramics

Window glass is amorphous. The silicon dioxide molecules are not ordered, and the mechanical properties are fair but not suitable for many engineering applications. Glass–ceramics are obtained by modifying glasses with nucleating agents that promote formation of an ordered crystalline structure composed of silicon dioxide or other parent glass molecules. Crystallization is induced by a heat treatment process after the desired part is molded into shape. The degree of crystallization can be almost 100% with very little amorphous glass phase remaining. The crystal size is usually less than one micron.

Glass–ceramics can have very low thermal expansion rates. This property allows their use for glass-to-metal seals and other applications that require matching the thermal expansion characteristics of metals. The tensile strength of glass–ceramics can be three times that of the parent glass. Glass–ceramics can be made to accommodate impact and thermal shocks. They are used for cooktops, in which the glass may be red hot in a burner location and cool inches away from the burner. Ordinary glasses

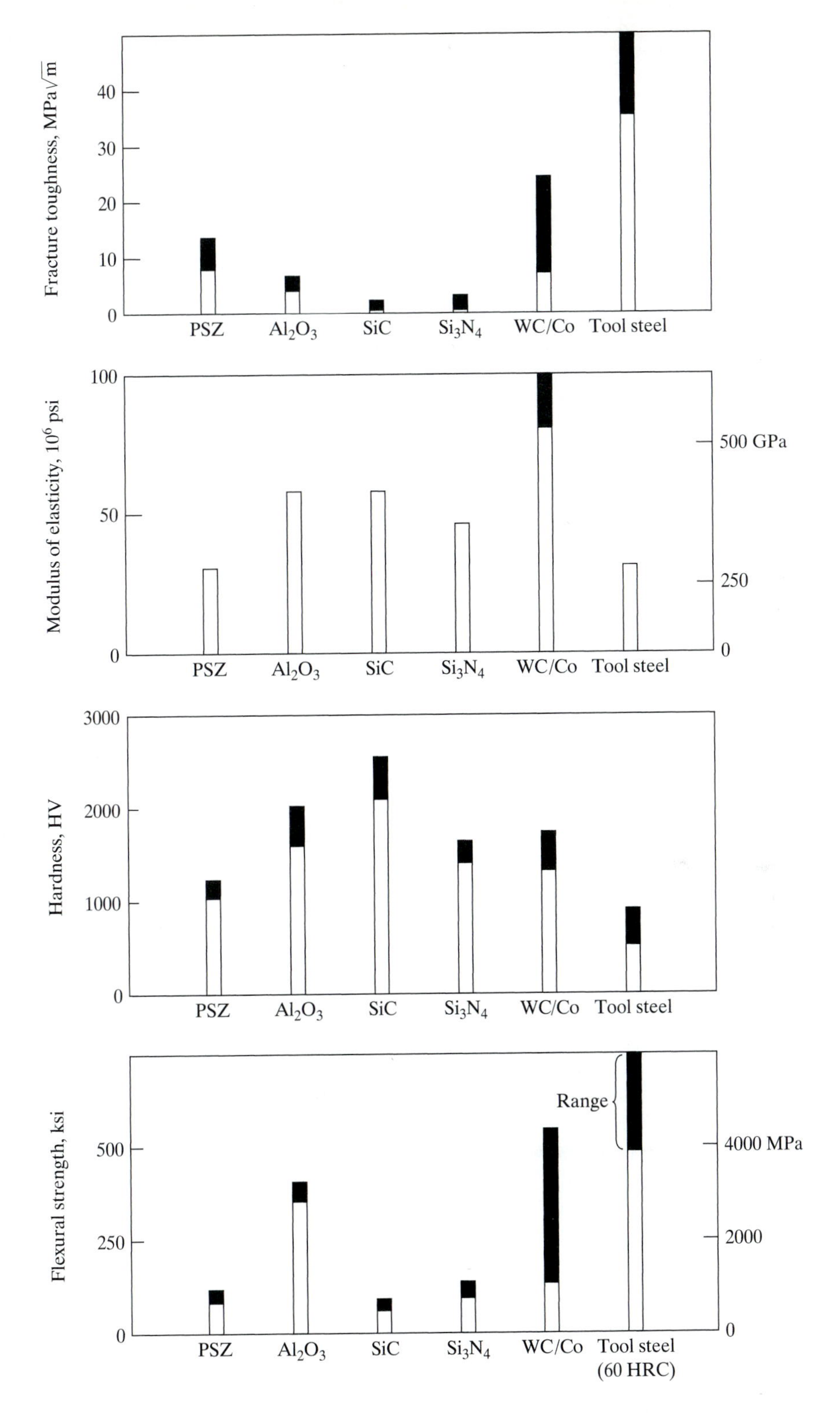

Figure 8–20
Some properties of ceramics compared with cemented carbide and tool steels

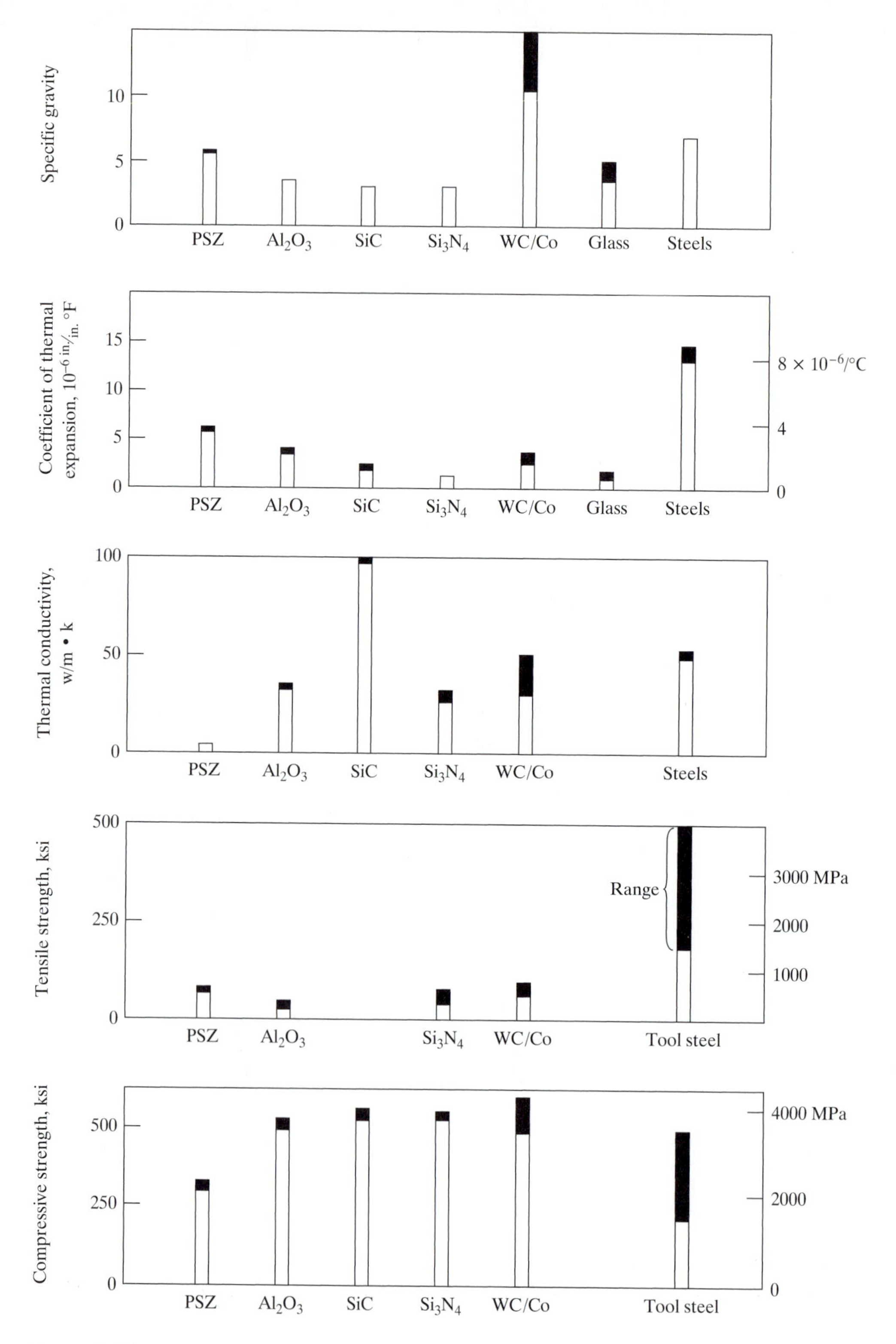

Figure 8–20
continued

could not accommodate this kind of thermal gradient. In summary, glass–ceramics are available with the convenient forming characteristics of glass, but after heat treatment they crystallize and compete with ceramics for structural applications.

Cemented Carbides

All of the technical ceramics that we have mentioned do not compare in mechanical properties with most cemented carbides. The latter invariably have higher tensile strength (118 to 206 ksi; 810 to 1420 MPa); higher stiffness (modulus of elasticity 60 to 100 $\times$ 10^6 psi; 414 to 689 $\times$ 10^5 MPa) and higher compressive strength (400 to 800 ksi; 2750 to 5516 MPa), and their toughness is far superior to that of technical ceramics. Cemented carbides are not normally considered structural materials mainly because their superiority is only at moderate temperatures and their high density is usually counterproductive for structural applications. The trend in structural materials is to have a high specific strength, strength per unit weight, or density. Cemented carbides are as much as five times as heavy as some of the ceramics that they compete with. Cemented carbides are used for structural applications, but these applications are rather limited for the foregoing reasons.

Glass-Bonded Mica

One of the biggest problems in using ceramics and cermets for structural components is that they cannot be machined into parts. Glass-bonded mica has found wide acceptance in most industries for applications requiring better temperature resistance than plastics and electrical and thermal properties similar to those of ceramics. These materials are available in standard shapes (sheets, rods, plates, bars) and parts can be machined from them with techniques comparable to those used for machining aluminum or other easy-to-machine metals. Micas are naturally occurring minerals with different chemical compositions [potash mica is $H_2KAl_3(SiO_4)_3$], and they can be made synthetically. They have excellent dielectric properties. Glass-bonded micas are made by vitrifying tiny mica crystallites with various types of glass. The glass is discontinuous, and this allows the material to be easily machined.

There are different grades of these materials, but some typical properties are as follows:

- Tensile strength from 6 to 8 ksi (41 to 55 MPa).
- Modulus elasticity of 10 to 12 $\times$ 10^6 psi (70 to 83 $\times$ 10^4 MPa).
- Compressive strength of 30 to 50 ksi (206 to 344 MPa).
- Maximum use temperature of 700 to 1300°F (370 to 700°C).

These materials are widely used for structural members that require heat or electrical insulation: supports for heating elements, electrical enclosures, supports for RF devices, and any number of components that require ceramic-like properties without the problems of fabricating a shape from a ceramic (Figure 8–21).

Figure 8–21
Glass-bonded mica part that was turned and milled to shape

In summary, the trend for the start of the new millennium is to use ceramics for structural parts when the application has reached the limits of stiffness, strength, or temperature resistance in metals. There are problems in replacing metals with ceramics because of the inherent brittleness of this class of materials. Some design rules to keep in mind in using ceramics for structural parts are as follows:

1. Try to load in compression; ceramics have relatively low tensile strengths.
2. Ceramics are notch sensitive; radius all reentrant corners.
3. Join to other surfaces with a compliant material to prevent bending.
4. Consider mismatch of coefficients of thermal expansion in joining to other materials.
5. Design parts to minimize machining after sintering.
6. Hold ceramic parts with an interference fit in steel when the part shape allows it.

8.10 Ceramics for Wear Applications

Stone, glass, and clay products have been used for centuries for paving materials, floor tiles, grinding mills, mortars, and similar applications because they resist abrasion. In industrial applications, ceramics have had limited use for wear applications, but the potential for use is outstanding. As abrasives for finishing metals and other materials, ceramics in the form of loose particles or bonded grit have been used for many decades. The infant applications of ceramics for wear are in the field of molded or pressed shapes for mechanical components subjected to wear and in the form of wear-resistant coatings.

Because of their hardness and chemical resistance, ceramics are suitable for systems involving low-stress abrasion and various forms of corrosive wear. Ceramic-to-ceramic and ceramic-to-metal sliding systems can often offer wear advantages over other material systems, but the selection of suitable mating combinations is not intuitive, and laboratory testing may be required to find a combination that works (Figure 8–18). A ceramic face seal is shown in Figure 8–22.

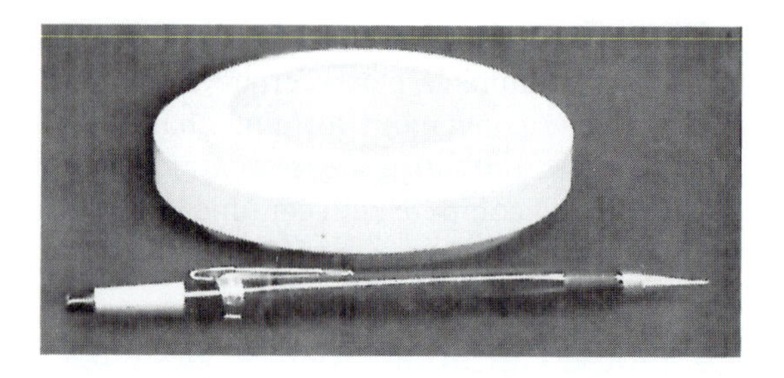

Figure 8–22
Face seal made from aluminum oxide; carbon–graphite is a suitable counterface

Low-stress abrasive wear can usually be reduced by simply increasing the hardness of the surface that is being abraded. If sand is being conveyed in a chute, wear on the chute can be minimized by making the chute from a material harder than sand. This wear solution only holds true, however, for moderate contacting stresses. It is difficult to put a figure on the term *moderate contacting stress*. A rock-crushing operation is considered a high-stress abrasion situation, as is digging rocks with a power shovel. An example of low-stress abrasion is wear produced in plowing a field of sandy soil. The major difference between the two systems is the factor of contact stresses in excess of the yield strength or compressive strength of the abrading substance.

A problem encountered in using hard materials to resist abrasive wear is finding a material that is harder than the abrading substance. As illustrated in Chapter 2 in our discussion of hardness testing, there are far too many hardness scales, and to obtain a material that will withstand wear from ordinary sand, silicon dioxide, it is necessary to relate the hardness of sand to that of steel. Traditionally, the Mohs hardness scale is used to

Table 8–9
Comparison of the hardness of ceramics to metals

	Hardness (kg/mm^2)	Approximate Mohs Number
Polyamide plastic	0.05	
Commercial lead	3	
Industrial carbon	15	
Type 1100 aluminum	20	
Annealed copper	40	
Annealed low-carbon steel	80	2
Magnesium oxide	700	6
Hardened tool steel	740	6
Silicon dioxide (sand)	820	
Chromium plating	900	
Titanium dioxide	1100	7
Zirconium oxide	1150	7
Tantalum carbide	1750	
Tungsten carbide	2000	8
Aluminum oxide	2100	8
Silicon nitride	2200	
Titanium carbide	2500	
Boron carbide	2700	9
Silicon carbide	2800	
Cubic boron nitride	5000	
Diamond	8000	10

measure the hardness of ceramics. It is simply a measure of what scratches what, not a true indicator of the hardness of ceramics compared with other engineering materials. Table 8–9 tabulates the approximate hardness of ceramics (penetration hardness) compared with metals on the same hardness scale.

The most commonly used ceramics for low-stress abrasion are aluminum oxide and chromium oxide. They are available as molded parts or as wear plates that can be adhesive bonded to metal substrates. Cemented carbides can also be applied in these forms. Even more widely used are ceramic coatings applied by plasma arc spray and similar techniques. This technique allows the use of ceramics for wear surfaces on very large components and on mechanical components that are subjected to operating stresses too severe for solid ceramics. Ceramic coatings are often used over solid ceramics simply because the part is too complicated to fabricate by ceramic molding techniques.

Figure 8–23 illustrates some typical machine applications for which ceramics are used to resist wear. These illustrations will serve as a guide for the use of ceramics in everyday design situations. Ceramic coatings that can be applied by thermal spray techniques are shown in Table 8–10. The abrasion resistances of some ceramics are compared in Figure 8–17.

8.11 Ceramics for Environmental Resistance

Glass and glasslike coatings are widely used for environmental resistance at temperatures of 500°F (260°C) and below. Between 500 and 2000°F (260 and 1093°C) some metals have utility, but, in general, ceramics are the only engineering materials that can continuously withstand harsh environments in the visible heat range. Unfortunately, glasses and ceramics are not immune to all types of environments.

Glasses, as coatings or by themselves, have excellent resistance to most mineral acids, excluding hydrofluoric and hot phosphoric acid. The resistance to strong acids exists up to temperatures of 300 to 350°F (149 to 180°C). Most organic acids do not attack borosilicate glasses in any concentration. At room temperature, glass equipment is resistant to most alkaline solutions. This resistance decreases rapidly with temperature, and, in general, glass should not be used for solutions with a pH greater than 10 at 100°C. Organic solutions of almost any type have no effect on glass at room temperature. For these reasons, glass-lined vessels are important in the chemical process industry. They can handle many chemicals as long as the temperature is low.

Figure 8–23
Typical machine applications of ceramics and cermets

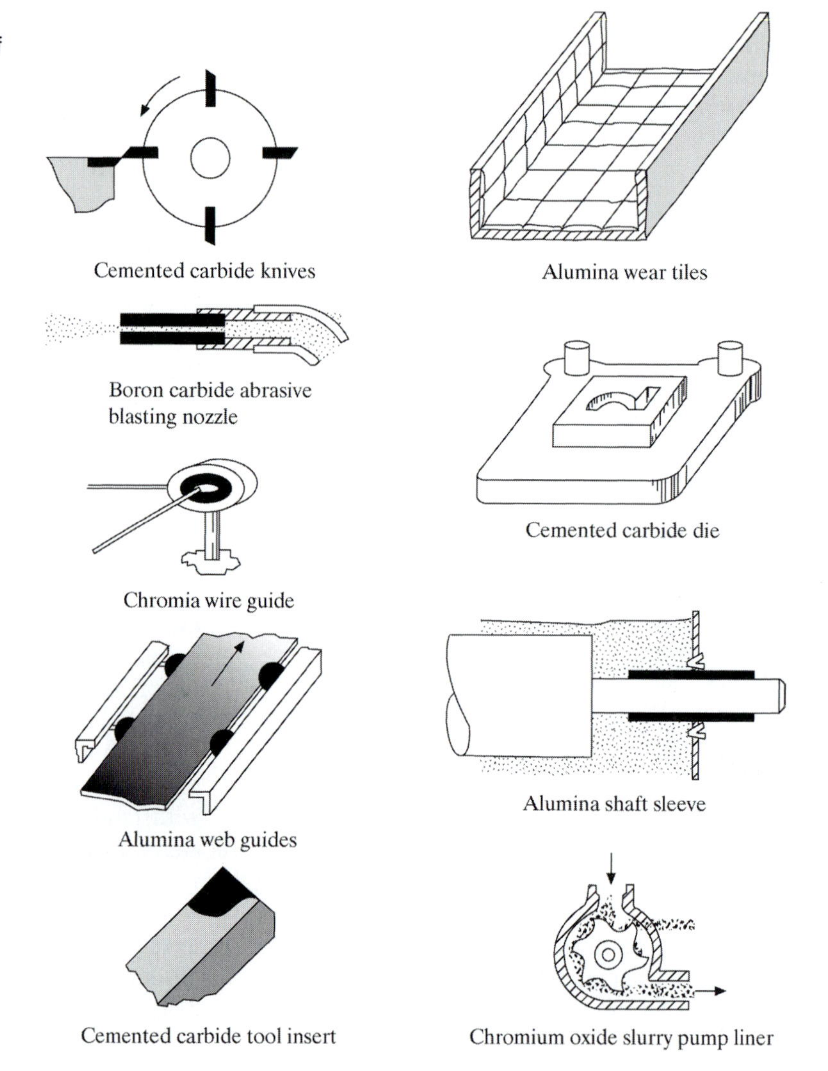

Ceramics in the form of fire bricks and cast shapes have been used for many years for containment of molten metals. This is a very specialized area that pertains only to primary metal producers. With the advent of stringent regulations on chemical disposal, many industries have to develop incinerators and similar systems to combust myriad chemicals. Combustion temperatures can be as high as 2200°F (1204°C), and the gases liberated in combusting complicated organic and inorganic chemicals can be quite corrosive. Ceramics are the only engineering materials that are practical to use at these temperatures. Unfortunately, the corrosion resistance of ceramics at these temperatures is not well documented. In oxidizing atmospheres, the refractory oxides such as Al_2O_3, BeO, CaO, and MgO can withstand continuous use at temperatures in excess of 3000°F (1649°C). Some oxides cannot tolerate reducing atmospheres. When acids and bases are

Table 8–10
Ceramics/cermets that can be applied as coatings on most metal substrates by thermal spray processes

Material	Use
Chromium carbide Chromium oxide Aluminum oxide Titanium oxide Titanium carbide Chromia/silica Tungsten carbide/cobalt	Mostly for wear applications
Zirconium oxide Magnesium oxide Magnesium zirconate	Mostly for physical properties

encountered, the more basic oxides such as MgO, BeO, and $MgAl_2O_4$ are resistant to bases, and the more acidic refractories such as $ZrSiO_4$, SiO_2, SiC, and Si_3N_4 are resistant to acids.

Aluminum oxide and zirconia are fairly resistant to oxidizing acids and bases. Glass and ceramic materials are excellent materials for resistance to harsh environments, but it is essential to determine the specific corrodents that exist. They cannot tolerate all chemicals. The procedure for selecting a glass or ceramic for chemical resistance is to list the composition of a subject environment, estimate service temperatures, and compare corrosion data on candidate materials. If specific data are unavailable, laboratory tests may be necessary.

8.12 Electrical Properties of Ceramics

Ceramics as a class of materials are usually considered to be electrical insulators. Under many conditions they are electrical insulators, but any nonmetallic material can conduct electricity if the conditions are right. When lightning strikes a tree, the electrical charge is readily conducted to ground by normally nonconducting wood. Ceramics are usually poor conductors of electricity because the electrons associated with the atoms that make up the ceramic are shared in strong covalent or ionic bonds. Metals are good conductors because the valence electrons are not tied up; they are free to move throughout the volume of the material. Some ceramic-type materials such as graphite are good electrical conductors and some have high resistivity, but this is used to advantage. For example, silicon carbide is used as heating elements in furnaces. They conduct current, but their high resistivity makes them good resistance heating elements. Some ceramics are semiconductors; they can act as conductors or nonconductors depending on the application of a bias voltage.

When ceramics are to be used as electrical insulators, it is necessary to review several electrical properties to determine just how good an insulator they are. *Volume resistivity* is a measure of the resistance of a known volume of insulating material. It is measured by applying a potential across a shape of known volume and recording the resistance in ohms. The *resistivity* is equal to the area of the shape divided by the thickness times the resistance ($\rho\rho = A/t \cdot R$). The units become ohms per unit length (Ω-cm). Unlike metals, the volume resistivity of ceramics decreases with temperature, largely because when ceramics become conductors they do so by anion or cation movement. Anions and cations are large in size compared to electrons; thus, significant energy is required to produce mobility. Thermal heating can supply this energy. To be a good insulator, a ceramic should have a high volume resistivity at the temperature of intended use.

As in the case of lightning, anything can become a conductor if the driving force for conduction is high enough. The *dielectric strength* of an insulator is a measure of the voltage that will break down an insulator and make it conduct electricity. Ceramics are not as good as plastics in this area. In addition, moisture absorption, surface finish, impurities, and many other factors can lower the dielectric strength and make the ceramic a poor insulator.

The *dielectric constant* (permittivity, capacitivity) is a measure of the capacity of an insulating material to store electrical energy. It is measured by essentially making a capacitor from a material and comparing it to vacuum as the dielectric material in a like capacitor. If you want to use a ceramic as an insulator, this value should be as low as possible. If the ceramic is to be used as a capacitor, this property should be high. Barium titanate has a high dielectric constant.

When a ceramic is used as an insulator in an alternating electrical field it is desirable to have a low *dissipation factor* (loss factor). The dissipation factor is the ratio between the irrecoverable and recoverable parts of electrical energy when an electrical field is established within the material. The irrecoverable energy takes on the form of heat. Thus, if a material has a high dissipation factor, it will heat in use and may not be a good insulator. The dissipation factor in ceramics is strongly influenced by temperature, impurity level, porosity, and frequency. Some insulating materials that maintain a constant dissipation factor at varying frequencies are polyethylene, polystyrene, and fused silica.

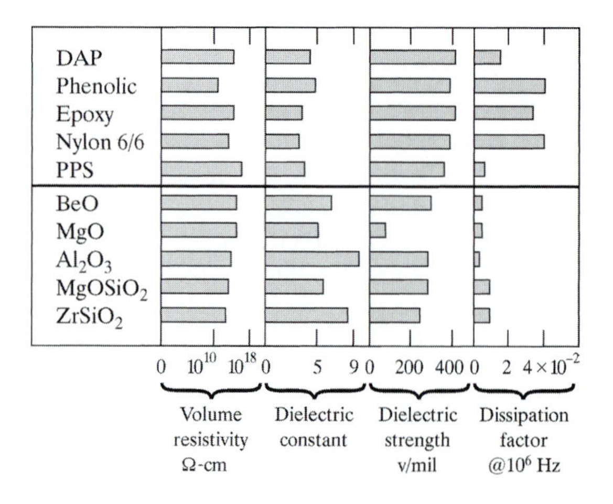

Figure 8–24
Comparison of the electrical properties of ceramics and plastics at 20°C (data from various sources)

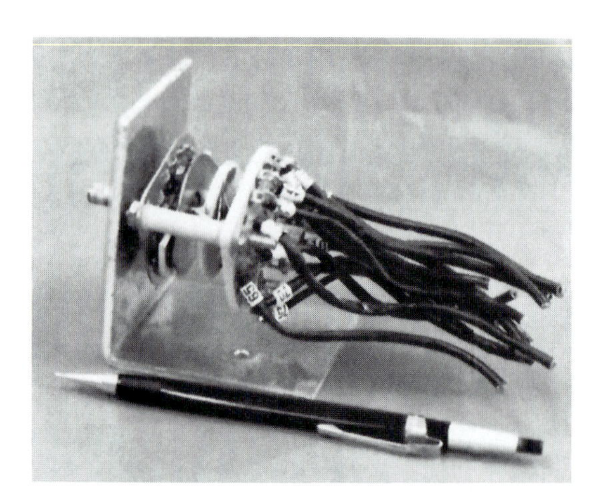

Figure 8–25
Aluminum oxide used as an electrical insulator

In using ceramics for electrical applications it is essential to review these properties for candidate materials at the temperature and electrical conditions of the intended application. In general, ceramics are the best insulators for use at temperatures over 150°C. Plastics—their foremost competitor—are probably better insulators at the lower temperatures. Figure 8–24 compares the electrical properties of some common plastics and some common ceramics. Figure 8–25 shows a typical insulator application of a ceramic (Al_2O_3).

8.13 Magnetic Properties of Ceramics

An important application of ceramic materials in the past few decades has been their use as hard and soft magnetic materials. *Soft magnetic* materials are materials that become magnetized when put into a magnetic field but quickly lose their magnetism when the field is removed. *Hard magnetic materials* are those that can be made into permanent magnets. Prior to the use of ceramics, soft magnetic materials were predominately low-carbon steels, pure iron, or silicon steels. Soft magnetic materials are used for core materials in solenoids and as armature laminates in electric motors. This is a specialized field, but a high-volume application because of the wide use of electric motors and solenoids. Hard magnetic materials are also used in electrical devices, and they have a wide variety of other uses, such as in machine design, magnetic latches, separation devices, and part hold-downs. More than 20 properties of magnetic materials can affect selection. We shall discuss the more important of these properties as they apply to ceramics.

Soft Magnetic Ceramics

The ceramic materials used to replace irons in devices requiring rapid magnetization and demagnetization are usually ferrimagnetic oxides with the general formula ($M^{2+}O \cdot Fe_2^{3+}O_3$), where M^{2+} is a metal ion such as Fe^{2+}, Ni^{2+}, Cu^{2+}, or Mg^{2+}. The generic term for these ceramic compounds is *ferrites*. The ceramics that do not contain metal ions (M^{2+}) are *diamagnetic;* they cannot be made ferromagnetic. The important soft magnetic materials are primarily based on iron oxide, which by itself can have ferromagnetic properties. Ferrites can be made into shapes by the traditional powder pressing and firing technique, or they can be compounded with binders and injection molded into shapes. This latter property as well as weight reduction is making ferrites popular for soft and hard magnetic materials in fractional-horsepower motors for automobiles (windows, heaters, seats).

The magnetic properties that are most important in soft magnetic materials are *permeability* and *hysteresis loss*. Permeability is essentially the ease of magnetization. Hysteresis loss is the energy loss in cyclic magnetization; it is important because materials with a large hysteresis tend to heat up in service. The tests required to measure magnetic properties are very complicated (ASTM A 34, A 341, A 342), but they usually involve measurement of the hysteresis loop of the material. Figure 8–26 shows the shape of a typical hysteresis loop measured in a standard test for magnetic materials. As the magnetizing force H is increased, the induced field is measured. The slope of the initial magnetization curve is the permeability. The greater the slope, the easier the material is to magnetize. The area within the closed loop is a measure of the energy lost during a complete magnetization cycle. This area should be as small as possible if the material is to be free of heat-up during use. The *saturation induction,* B_s should be high if the material is easily magnetized, and the *residual induction* (retentivity) B_r should be low in a soft magnetic material. It is not desirable to have residual magnetism. Also measured from this curve is the *coercive force,* H_c. This is the magnetizing force required to bring the induced magnetism to zero. The value should be low in soft magnetic devices.

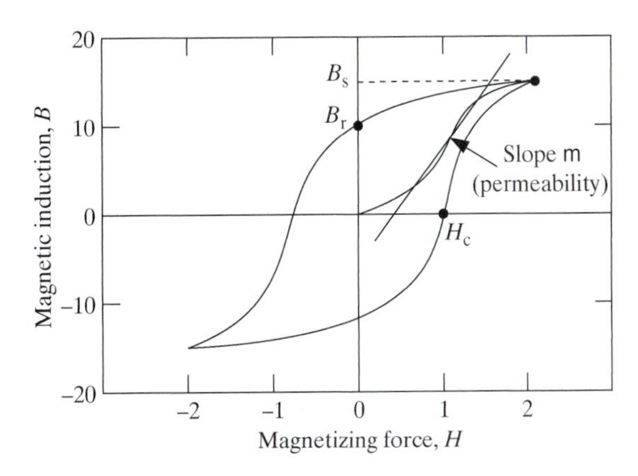

Figure 8–26
Typical magnetization curve for a soft magnetic material

Soft ferrites can have high permeability, low coercive force, and low loss at high frequency. Besides favorable magnetic properties, ferrites offer lower weight than metals and often lower cost, and they do not corrode in many use environments. For these reasons and other more complicated electrical reasons, ferrites have become important as soft magnetic materials.

Hard Magnetic Materials

Hard magnetic materials are used for permanent magnets. The ceramics that make good permanent magnets can also be ferrites ($BaO \cdot 6Fe_2O_3$) or more complex ceramics such as yttrium–aluminum–iron garnet. The most popular are barium and strontium oxides. The desirable properties of hard magnetic materials are the opposite of the soft magnetic materials. A high coercive force and residual induction are desirable. The quality of a permanent magnet can be measured by the *energy product*. This value is the product of the induction B and magnetizing force H in the second quadrant of the hysteresis loop. To be a good permanent magnet, these values should be as high as possible. This value is often compared with density; thus, a material with the highest energy product of a given volume will be the most space saving in use. The units of energy product in the SI system are tesla-ampere/meter (TA/m). A typical Alnico (12Al, 25Ni, 3Cu, Fe remainder) permanent magnet will have an energy product of 1.35 at room temperature compared with a value of 3.5 for some ceramic magnets. Thus, ceramics can be better permanent magnets than some of the traditional metals. Another advantage of hard ceramic materials over metals is that the ceramic powder can be blended with polymers and elastomers, and extruded into shapes such as a combination seal and latch for refrigerator doors.

Both soft and hard ceramic magnetic materials are not without limitations. The fired ceramics are too hard to machine, and they are brittle. Some can become diamagnetic at fairly low temperatures. The Curie temperature of manganese–zinc ferrites is only 300 to 400°F (149 to 204°C). Barium ferrite and some rare earth ceramics have Curie temperatures of about 850°F (450°C). Some of the Alnico magnets have Curie temperatures as high as 2000°F (1093°C).

In general, the limitations of ceramic magnetic materials can be accommodated by design. As a class of materials, they are indeed important in this area.

Summary

In the late 1980s, ceramics became the new steels. Most of the steel companies spent their dollars on restructuring, and new alloys were few and far between. The ceramic companies, spurred on by a very strong Japanese thrust, mounted an offensive on tool materials and high-performance metal alloys. Ceramic companies contacted major industries and tried to demonstrate how ceramics can do things that the metals and plastics cannot, how ceramics can save them money, and how ceramics can improve quality. As of 2003 the jury was still out on how well ceramics will fit into conventional machine design, but without a doubt ceramics are under evaluation in many industries for possible applications. Figure 8–20 compares some of the engineering ceramics with competing materials. It is clear that ceramics have an edge in some properties and at elevated temperatures ceramics are clearly superior to other engineering materials in strength-related properties. The following are some key concepts to keep in mind in considering ceramics for engineering applications:

- Engineering ceramics are not clay products, but mostly oxides, nitrides, and carbides that are sintered to high density.
- Ceramics get their high hardness and brittleness from strong ionic or covalent bonds between atoms.
- Most ceramics are crystalline.

- Ceramics are brittle; strain-to-fracture ratios may be less than 0.1%, compared to 20% for a metal.
- Ceramics are elastic to failure, and they can withstand tensile loads as long as they are in the elastic range.
- Ceramics can have stiffnesses greater than those of steels.
- Ceramics have lower thermal expansion rates than metals and plastics.
- Ceramics, in general, have poor thermal conductivity.
- The critical flaw size to produce failure of a ceramic can be as small as 10 μm; the critical flaw size for metals is typically in excess of 1000 μm.
- Most ceramics cannot be machined (except by grinding) after sintering; consider this in design.
- Ceramics cannot be joined to themselves or to other materials with ordinary welding processes.
- The mechanical properties of ceramics often depend on the grain size and the amount of porosity after firing.
- Glasses have utility in machine design for corrosion resistance and for some of their other properties, such as low thermal expansion.
- Carbon–graphites are excellent plain bearing materials that require no lubrication if used properly.
- Cemented carbides have the highest stiffness and compressive strength of all engineering materials.
- Silicon nitride, aluminum oxide, zirconia, and silicon carbide are the tool materials of the ceramics field.
- The properties of ceramics depend on fabrication techniques. Most ceramics are designated by generic type and approved suppliers:

  Aluminum oxide (Kors 201—grade B)

  The ASTM standard classification system (ASTM C 1286) was not in wide use in 2003.
- Ceramics mated with other ceramics or metals may produce high wear rates; compatibility tests are needed.
- Radius all edges on ceramic parts and minimize stress concentrations.
- Join ceramics with compliant material to other surfaces.
- Design to eliminate machining after firing (if possible).
- Use statistical values for allowable strengths.
- Perform a rigorous stress analysis when using ceramics for structural applications.
- Specify rigorous flaw inspection on critical parts.
- Typical as-fired dimension tolerances are ±1%.
- Specify density for critical applications.

There are many factors to keep in mind when using ceramics, but as a class of materials, ceramics can be invaluable for solving tough service life problems.

Critical Concepts

- Ceramics are crystalline inorganic compounds that tend to be hard and brittle because of strong ionic or covalent bonds between atoms.
- Ceramics cannot be machined with ordinary machining techniques. They are usually molded to a near-net shape and ground.
- Cemented carbides are composed of “ceramic” particles with a metal binder and compete with ceramics for tooling and wear applications.

- Glasses are amorphous solidification products of inorganic materials. They have short-term order between atoms, but not long-term as in crystalline materials.
- Carbon–graphite products are a unique material with useful tribological and electrical properties.

Terms You Should Remember

ceramic
intermetallic compounds
carbides
oxide
vitrification
compaction
sintering
glass
flexural strength
abrasion
absolute hardness
graphite
green state
cemented carbide
binder
carbide grade
fracture toughness
hysteresis loss
hard magnetic material
cermet
transformation toughened
sol-gel
notch sensitivity
tungsten carbide
aluminum oxide

Case History

Zirconia Edge Guides

For decades, chromium-plated 440-C stainless steel was used for the wear plates that guide photographic film through various manufacturing operations, such as coating, slitting, and perforating. These films have chemicals (silver halides) coated on one side, and some have an abrasive, amorphous carbon coat on the backside. The edge of these films rubs on guides that direct their movement, wearing a groove in them. When the groove gets deep, it causes transport problems and the guides need to be removed from service, reground, and replated. The life of edge guides in some operations was as short as a few weeks, prompting the start of a project to find a replacement surface that would last significantly longer. A secondary goal was to eliminate chromium plating because of the environmental concerns with disposal of plating solutions.

Because edge guides come in many different shapes and sizes, it would be impractical to make all of these shapes from solid ceramics that may wear better than chromium. The solution arrived at after many iterations was to "surface" existing guides with an epoxy-bonded strip of 1-mm-thick yttria-stabilized zirconia. All rubbing surfaces were overlayed on existing parts. The zirconia was harder than the chromium (1100 kg/mm^2 vs. 900 kg/mm^2), and the chemical resistance of the zirconia exacerbated the chemical component of the edge erosion. The edge guide wear life went from just weeks to months or years. Zirconia provided a cost-efective solution to a very difficult problem.

Questions

Section 8.1

1. Name two ionic crystals.
2. Name three engineering materials that are joined by covalent bonding.
3. Define the term *ceramic*.

Section 8.2

4. Outline the steps that are involved in making an aluminum oxide plain bearing.
5. What is hipping? When is it used?
6. What is vitrification? When is it used?

Section 8.3

7. What is the effect of crystallinity on ceramics?
8. How many phases are present in alpha aluminum oxide?
9. What is the effect of porosity in ceramics? How can porosity be reduced?

Section 8.4 (see Figure 8–20 also)

10. Compare the fracture toughness of alumina, silicon carbide, and silicon nitride.
11. Calculate the strain to failure of aluminum oxide in tension. The tensile strength is 100 ksi (689 MPa) and $E = 50 \times 10^6$ psi (344 GPa).
12. Compare the thermal conductivity of aluminum oxide to that of steel.

Section 8.6

13. Calculate the strain to failure of borosilicate glass. How does this compare with the value for Al_2O_3?
14. Compare the hardness of glass and hardened steel at 60 HRC.
15. Give an example in which glass is used as a structural material. Explain why it is safe.

Section 8.7

16. Name four industrial items made from carbon–graphite.
17. Compare the weight of a 1-in. (25-mm) cube of carbon–graphite to the same shape made from steel.

Section 8.8

18. Calculate the deflection of a 1-in. square cemented carbide simply supported beam with a span of 30 in. (76 cm) and a center load of 100 lb (45 kg). Compare this deflection with the deflection of the same beam made from 1020 carbon steel ($E = 30 \times 10^6$ psi or 200 GPa).
19. Explain the role of cobalt in cemented carbides.

Section 8.9

20. You are making a punch press die and punch for steel. Which ceramic would you use? Why?
21. Compare the advantages and disadvantages of ceramic and plastic for a frying pan handle. Which would provide the best service life? The lowest cost?
22. What is a carbon–carbon composite?

Section 8.10

23. Explain the concept of hardness of materials as applied to abrasion.

Section 8.11

24. When do ceramics make better electrical insulators than plastics?

Section 8.12

25. What makes a better permanent magnet, a hard magnetic material or a soft magnetic material?
26. What are the three most important rules for designing ceramic parts?

To Dig Deeper

ASM Handbook, Volume 5, Surface Engineering. Materials Park, OH: ASM International, 1994.

Brookes, K. J. A. *Handbook of Hardmetals*, 4th ed. East Barnet Herts, UK: International Carbide Data, 1987.

Chandler, M. *Ceramics in the Modern World*. Garden City, NY: Doubleday & Co., 1968.

Engineered Materials Handbook, Volume 4, Ceramics and Glasses. Materials Park, OH: ASM International, 1991.

Jones, R. W. *Fundamentals of Sol-Gel Technology*. London: Institute of Metals, 1989.

Kingery, W. D. *Ceramic Fabrication Processes*. New York: John Wiley & Sons, 1958.

Kingery, W. D., and others. *Introduction to Ceramics*. New York: John Wiley & Sons, 1976.

Lay, L. A. *The Resistance of Ceramics to Chemical Attack*. U.S. Department of Commerce Report N7926219, 1979.

Musikant, S. *What Every Engineer Should Know About Ceramics*. New York: Marcel Dekker, Inc., 1991.

NIST Structural Ceramics Database, SRB30. Gaithersburg, MD: National Institute of Standards and Technology, 1990.

Richardson, D. W. *Modern Ceramic Engineering*, 2nd ed. New York: Marcel Dekker, Inc., 1992.

Swartz, M. M., Ed. *Engineering Applications of Ceramic Materials*. Metals Park, OH: American Society for Metals, 1985.

Tooley, F. V. *Handbook of Glass*. New York: Books for Industry, 1974.

Van Vlack, L. H. *Physical Ceramics for Engineers*. Reading, MA: Addison-Wesley Publishing Co., 1964.

Yust, C. S., and R. W. Bayer, Eds. *Selection and Use of Wear Tests for Ceramics*, STP 1010. Philadelphia: American Society for Testing and Materials, 1989.

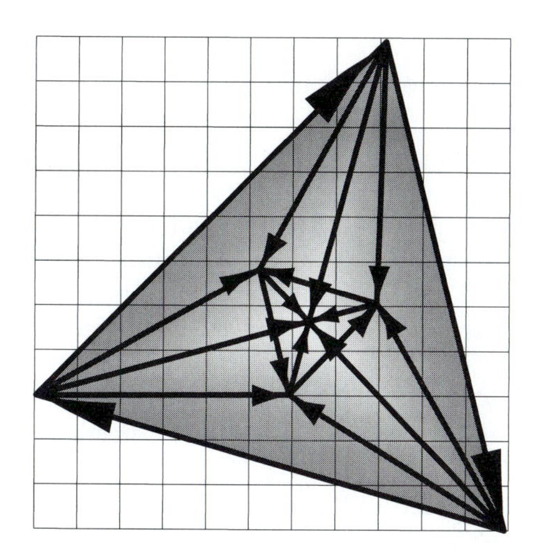

CHAPTER 9

Steel Products

Chapter Goals

1. An understanding of how steels are made.
2. A knowledge of the steel products that are commercially available.
3. An understanding of steel terminology and steel specification procedures.

Rationale

This chapter is our primer on how steels are made and on the various types and forms of steel. Steel helps build our infrastructure, our tools, our means of transportation, our world. Structural shapes are used to construct buildings and bridges; rolled steels in sheet and strip form make the frames and bodies for appliances and vehicles; plate is used for vessels and ships. All of the steel products that are available were developed to meet such design needs. Wire rope was developed to carry loads significantly greater than ropes made from organic materials. Wide flange beams were developed to replace beams that were made by riveting plates into shapes. Tin plate was developed to replace glass and earthen containers.

Designers need to know what steels are available, the language of steels, and the products that are available for use in engineered designs. We will define, among others, important terms (jargon) such as rimming *and* continual casting. *When you have completed this chapter, you should be capable of defining the steel shape needed for an application.*

Material: Carbon steel angle, 3 × 3 × 24 lb*

The term *steel* is a part of everyone's vocabulary, and as a metal it is taken for granted. Our automobiles, tools, and buildings all rely on steel for their manufacture. It seems that this commonplace engineering material has always been with us, but this is really not the case. *Steel* by definition is an alloy of iron and carbon, with the carbon being restricted within certain concentration limits. Surprisingly, this metal so ordinary to

*This is the *U.S. Steel Construction Manual* designation for a hot-rolled carbon steel angle with 3-in.-long legs and weighing 24 lb per foot of length.

us has been in large-scale commercial use for only about 150 years.

Iron as a usable metal dates back at least 6000 years. The earliest iron tools were made by chipping pieces of iron from meteorites containing metallic iron. In the fourteenth century, the technology was developed to melt and cast iron into a useful shape. Prior to this, iron could be retrieved from its ore only by heating it in the presence of charcoal. In this form, it was very high in impurities, and objects were forged from the pasty mass obtained from the rudimentary furnaces. This was known as *wrought iron*; its mechanical properties, because of the slag and nonmetallic inclusions, left something to be desired. Iron in its cast form was used in ever-increasing quantities for several centuries, but it was high in carbon content and not at all malleable and formable as are steels. In the middle of the nineteenth century, a method for reducing the impurity level in iron was developed. It involved blowing air through the impure molten iron. This development ushered in the steel era. The removal of the impurities from the crude iron made iron malleable at room temperature. Forgings could be made. Sheets could be rolled. *Steel* as we know it was finally here. This process, known as the Bessemer process, is no longer used, but it was the forerunner of the basic oxygen process (BOF), which is in wide use today.

It is the purpose of this discussion to familiarize the design engineer with steel production techniques, steel products, designations, and selection factors. Successful use of steels in design requires a working knowledge of the broad range of steel products.

9.1 Iron Ore Benefication

The basic ingredient for steel is one of several forms of naturally occurring oxides of iron. The pure metal unfortunately does not exist in nature. The iron ores native to the United States are concentrated in the Midwest and Southeast. *Hematite* (Fe_2O_3) and *magnetite* (Fe_3O_4) are two principal iron ores, but as our iron reserves have dwindled we rely heavily on taconite, which is a siliceous rock containing fairly low-volume concentrations of iron. The use of these low-grade ores has necessitated concentration of the ore on site to reduce shipping weights to the steel mills. The taconite rocks are *strip* mined and crushed, and the iron-rich portions are concentrated by flotation or other techniques; they are then sintered into small balls about ½ in. in diameter. This process is called *pelletizing*. These balls are shipped to the steel mills. Their iron percentage may be 70% or greater. In the United States most ore is transported to the steel mills by ore boat, which explains the geographic location of many steel mills on waterways. The U.S. steel industry for many years was centered on the Great Lakes. Even today, much of the iron ore used in U.S. steel mills comes from mines near the Great Lakes (the Mesabi range in Minnesota and the Upper Peninsula of Michigan).

At the steel mill the pelletized iron oxide is fed into a *blast furnace*. This marvel of structural engineering is analogous to a chemical reactor. An impure substance goes in the top and comes out pure at the bottom. The furnace is kept at temperature continuously. The charge consists of iron ore, coke, and limestone. Figure 9–1 is a schematic of a typical blast furnace.

The iron ore, being an iron oxide compound, cannot form metallic iron until the oxygen is removed or reduced. The coke in the furnace serves as a source of heat, as well as the source of a reducing gas that will chemically detach oxygen atoms from the iron oxide, yielding the desired iron. Limestone is present in the furnace charge to act as a flux or purifying agent to assist in removing the other impurities from the iron ore. The last and no less essential ingredient in the blast furnace charge is air. Air supplies the oxygen for combustion of the coke, as well as promoting the *chemical reduction* of the iron oxide. The specific chemical reactions that take place in the blast furnace are illustrated in the following:

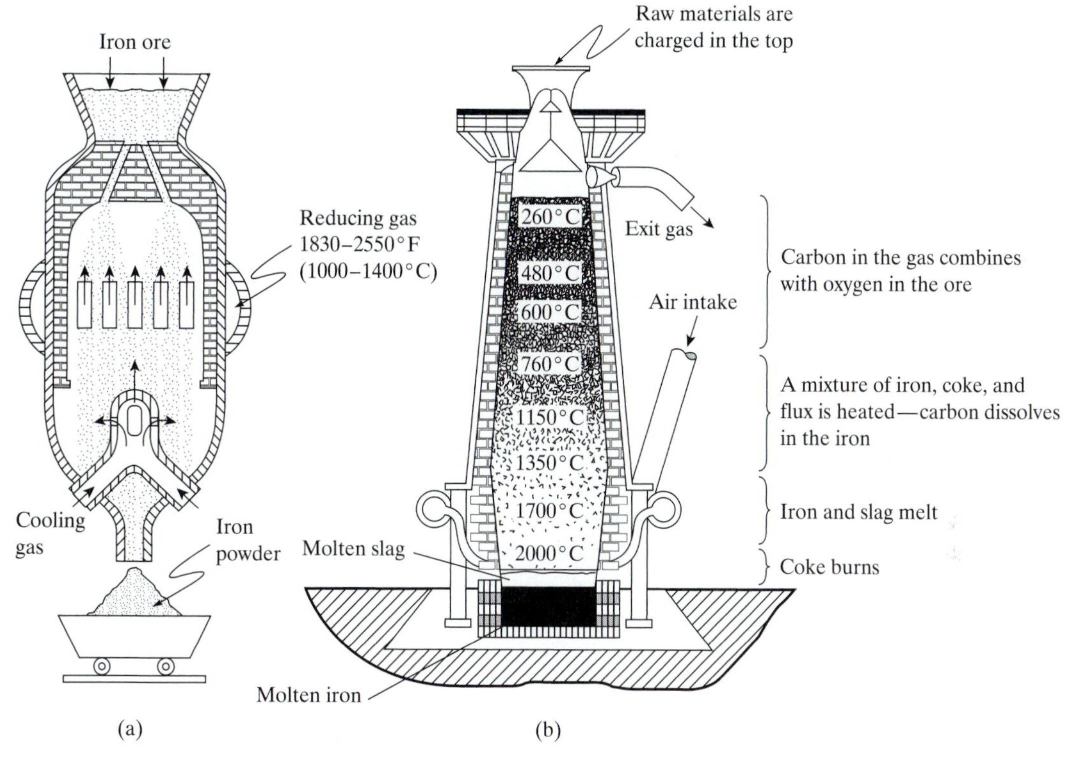

Figure 9–1
Schematic of a direct reduction process (a) and a blast furnace (b)

Production of the reducing gas:

1. C (coke) + O_2 (air) → CO_2 (carbon dioxide)
2. CO_2 + C (excess coke) → 2CO (carbon monoxide)

Reduction of the iron oxide:

1. Fe_2O_3 (ore) + 3C (coke) → 2Fe + $3CO_2\uparrow$
2. Concurrent with the preceding Fe_2O_3 + 3CO → 2Fe + $3CO_2$

Thus the chemical reagents that change the iron oxide to metallic iron are carbon monoxide and carbon, both from the coke.

The metallic iron is tapped at the bottom of the furnace; in this state it is referred to as *pig iron*. It is still impure in that it is high in carbon content and not suitable for making engineering materials, because it would be extremely brittle and weak. The pig iron must be further refined. The slag that accumulates on top of the iron is also tapped, and it is used for such things as earth fill or concrete aggregate. The foregoing description of ore reduction is extremely simplified. Anyone who has toured a steel mill and seen the enormity of a blast furnace and its ancillary equipment will have a better appreciation of the engineering effort involved in this most interesting and important operation.

There are direct reduction processes in use in various parts of the world (Figure 9–1). Eventually these types of processes may replace the blast furnace.

In one direct reduction process, iron ore, coal, and oxygen (from a lance) are reacted in a single melting vessel. Steel rather than iron is the product, and it can go directly to ladles for pouring into ingots or a continuous casting machine. In another process, called the iron carbide process, iron ore in fine particle form is fluidized and reacted with methane and hydrogen at about 600°C. The hydrogen reduces the oxide to metallic iron, which in turn acquires carbon (about 6%) from the methane. The resulting product is the intermetallic compound, iron carbide. This product can be dissolved in molten iron, for example, in a *BOF furnace*, and oxygen is used to reduce the carbon in the entire bath to the desired level for steel. When direct reduction processes transition into widespread commercial use, the costly and environmentally challenging blast furnaces that make most nonrecycled steel will be replaced. The coke ovens that are a necessary part of blast furnace technology are one of the more environmentally difficult processes to control. Partial combustion of coke produces methane and CO, which must be kept contained. This is costly.

In 2004 there were still about 50 blast furnaces in operation in the United States, and some were in the process of being phased out. However, new blast furnaces were being built in other places of the world. The prediction for the future by the U.S. steel industry is that by 2015 about 20% of the iron used in making steel will be produced by direct reduction. The U.S. steel industry already recycles about 80% of all steel, but recycling produces a buildup of impurities in the recycled steel (mostly copper, antimony, and tin) which are not removed in conventional steelmaking processes. Addition of virgin iron through direct reduction processes will help minimills, which use mostly recycled scrap, deal with the problem of impurity accumulation. They are becoming an important part of the steelmaking process.

9.2 Making of Steel

Chemical Composition

Steel by definition is an alloy of iron and carbon, but this statement must be qualified by placing limits on carbon content. When iron–carbon alloys have less than 0.005% carbon present at room temperature, they are considered to be pure iron. Pure iron is soft, ductile, and relatively weak. It is not normally used as an engineering material because of its low strengh, but it is used for special applications, such as magnetic devices and enameling steels (steels that are glass coated such as bathtubs). From the commercial standpoint, steels have a low carbon limit of approximately 0.06% carbon. On the other end of the carbon–content scale, iron–carbon alloys with more than approximately 2% by weight of carbon are considered to be cast irons. Above this carbon level, casting is about the only way that a useful shape can be made from the alloy, because the high carbon makes the iron alloy too brittle for rolling, forming, shearing, or other fabrication techniques. Thus steels are alloys of iron and carbon with carbon limits between approximately 0.06% and 2.0%.

Typically, the carbon content in pig iron may be 4% or 5%, which is too high to use as a steel. In addition to the high carbon content, the pig iron may contain high amounts of silicon, sulfur, phosphorus, and manganese, as well as physical inclusions of nonmetallic materials from the ore. All these substances would be detrimental to a steel's properties if allowed to remain at their uncontrolled high level. Thus to make steel from pig iron, a number of other impurities besides carbon must also be removed. As we shall see, a number of different processes can be used to remove the impurities from pig iron, but they all involve one basic process—oxidation. In the blast furnace, carbon was used in the form of coke to remove oxygen through the mechanism of combination of the oxygen with C and CO. Now oxygen must be used to remove the carbon in the iron. In the molten state the carbon in solution in the iron readily combines with the oxygen that is introduced in the form of air or pure oxygen and forms CO once again. The oxygen, as well as additional fluxing ingredients, reduces the level of the other impurities in the pig iron.

9.3 Steel Refining

Steel refining means making steel from pig iron, scrap, or ore. The process options are illustrated in Figure 9–2. The two primary processes were in use in 2003: the basic oxygen furnace (BOF) and the *electric arc furnace* (EAF). About 60% of the production in the United States is by BOF; the remainder is by *EAF*. Direct reduction processes are still in development mode. It is likely that this process will remain that way until one of the many competing direct reduction processes wins out over the others.

Previous editions of this text showed the open-hearth furnace as the major process for steel refining. The last open-hearth furnace in the United States was decommissioned in 1991. The same situation has occurred worldwide. Open-hearth furnaces were long and broad, but shallow. They could accommodate up to 450 tons of steel (408,240 kg) in a pour. However, they would take up to 10 h to refine a heat of steel. The basic oxygen furnace can do that job in as little as 20 min. Both BOF and EAF furnaces are smaller in size (usually < 350 tons), but they are more economical than the open hearth because of the shorter refining time. This is the reason for the demise of the open hearth.

Other major changes that have occurred in the steel industry in the past 25 years are the change from ingot casting and the trend away from large integrated steel mills to minimills. Refined steel used to be tapped into a ladle, and ingots were poured from the ladle. Continuous casting skips the ingot step. Steel is poured into an intermediate reservoir called a tundish. From there it flows into a water-cooled "continuous mold"; the cast shape that exits the bottom of the mold can be cut and stored for later finishing or it can be continuously finished into product by rolling or other final finishing processes. The process was introduced in the United States in the 1960s; by 1995, over 90% of the steel produced in the United States was continuously cast. In Japan, the percentage was over 95%. The remainder is ingot cast. Large forgings, for example, are still made from cast ingots. In summary, BOF and EAF are the steel-refining processes of the 1990s, and most steel is continuously cast. Integrated steel mills are being slowly replaced by minimills in the United States.

An integrated steel mill is one that has at least one blast furnace, a coke plant, and various steel-refining processes and that makes a number of product lines. These mills usually have annual capacities in excess of 100 million tons. An integrated mill may have a tinmill, a hot-rolling facility, a cold-rolled sheet facility, and possibly even a galvanizing line. These big mills are usually on a waterway to allow transport of ore for the blast furnace by ship. In contrast, a minimill may have only one product line, such as a structural shape (tees, angles, channels, rebar, etc.). Minimills usually convert scrap in a basic oxygen or electric arc furnace (Figure 9–3). They can be located just about anyplace because they do not need a billion-dollar blast furnace and a waterway for ships bringing ore. Three decades ago, steel made from scrap was considered inferior and was relegated to manufacture of barbed wire, highway guard rails, or other noncritical products. This is no longer the case. Steel from scrap is now the norm. Even integrated mills with hot metal facilities use a portion of scrap with most BOF charges. The following sections will describe the BOF and EAF furnaces as well as continuous casting in more detail.

Basic Oxygen Furnace

The basic oxygen furnace is really an adaptation of the first commercial steelmaking process, the Bessemer Converter (circa 1850). The Bessemer process purified steel by blowing air through a vessel filled with molten iron. The oxygen in the air reacted with the carbon in the iron to form

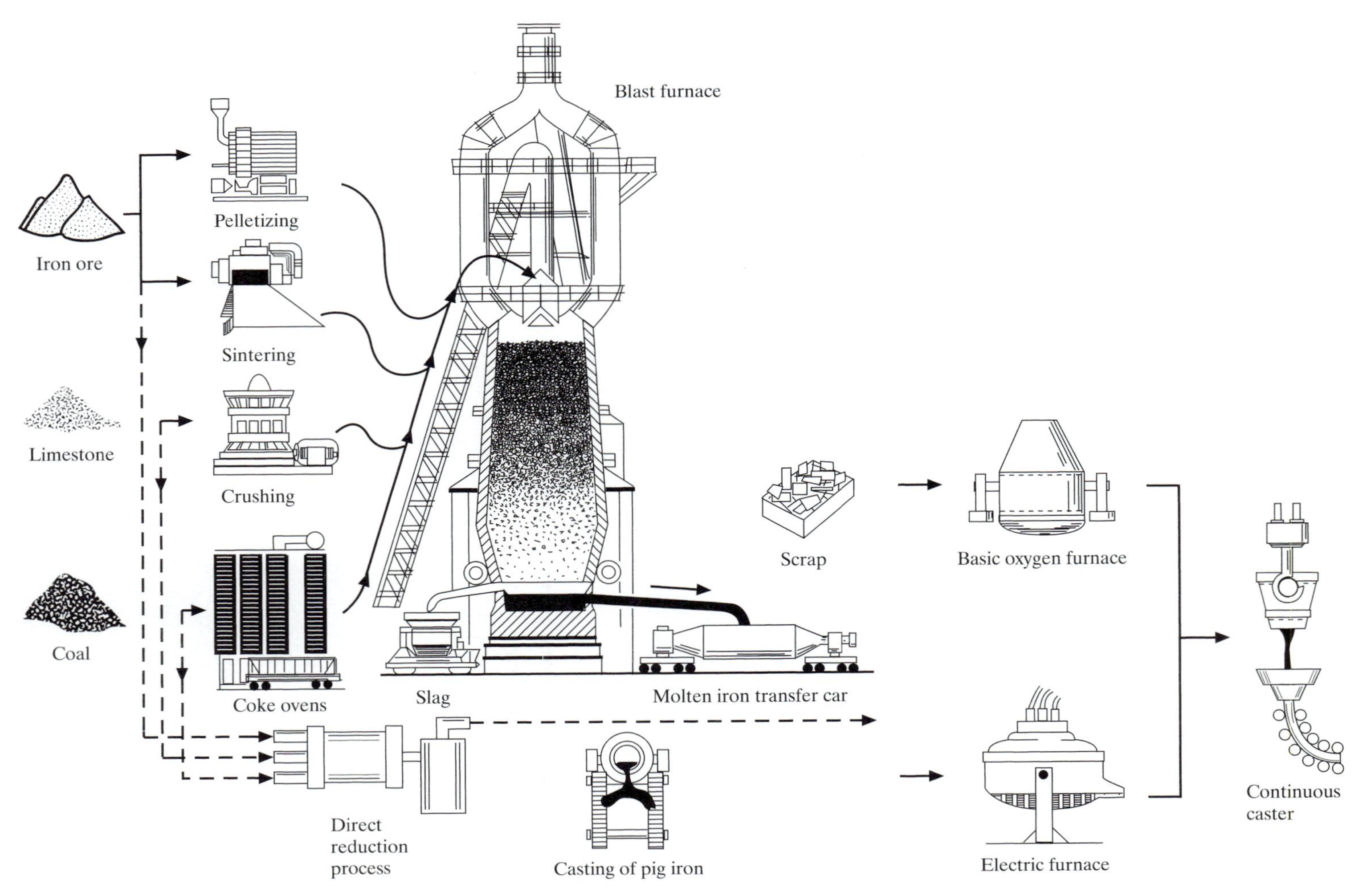

Figure 9–2
Steelmaking process

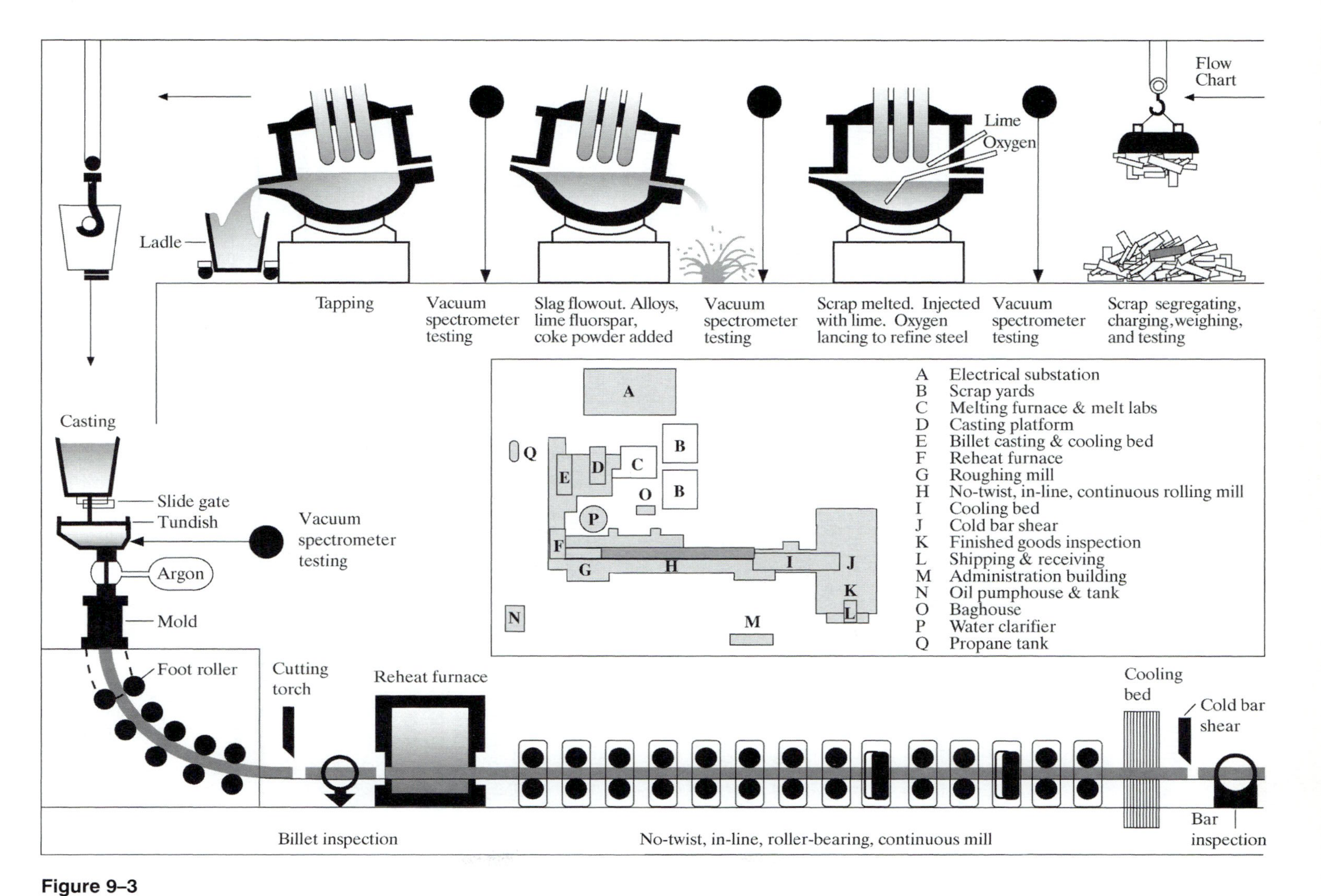

Figure 9–3
Making structural steel shapes from scrap in a minimill
Source: Courtesy of Auburn Steel Co.

Figure 9–4
Open hearth furnace (BOH or AOH)

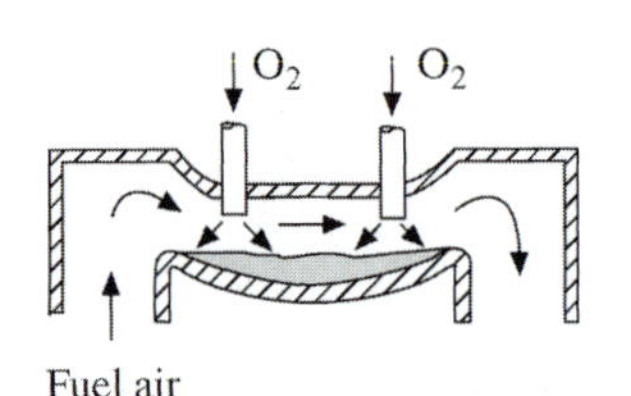

CO and CO_2, thus reducing the carbon in the iron from a few percent to a fraction of a percent, creating steel. The Bessemer furnace was the principal steelmaking process until the start of the twentieth century. The open-hearth process was invented in 1870 and steadily gained in popularity over the Bessemer process. By 1910, the open hearth (Figure 9–4) had become the principal commercial process. It remained that way until it was overtaken by the BOF and EAF (Figure 9–5) after 1960. The open-hearth process is now officially dead in the United States and most other steel-producing countries.

The basic oxygen furnace is just a refractory-lined vessel that can tip to pour. It is charged with scrap, pig iron fluxing agents, and hot metal (molten pig iron) or combinations thereof. When the charge is melted, an oxygen lance (sometimes combined with argon) is brought in or ported from the bottom to reduce the carbon, sulfur, and phosphorus. The argon and fluxes assist in the process. Some furnaces are "blown" from the top, some from the bottom, and some from top and bottom. After the steel is poured, nitrogen is blown into the remaining slag pool to splatter coat the refractory lining with a protective coating of slag. This reduces refractory wear in charging. Alloy additions are often done in the ladle rather than in the BOF. The output of the BOF is sometimes routed to secondary refining operations such as vacuum degassing to further reduce impurities or to modify the chemical composition. In 2003 this was the most important steel-refining process from the tonnage standpoint.

Electric Arc Furnace

The source of heat in the electric arc furnace (Figure 9–5) is an arc that is established between

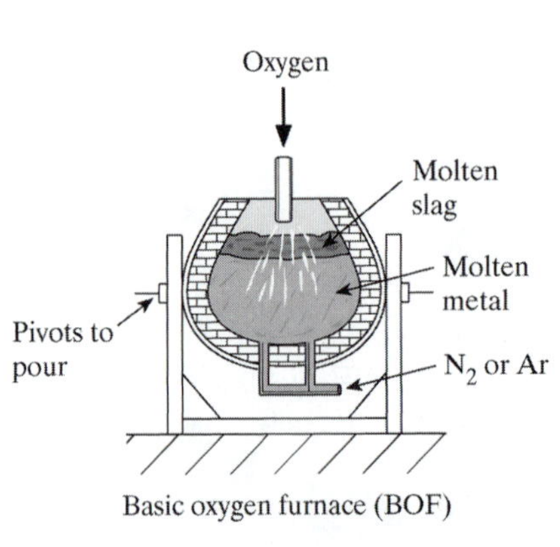

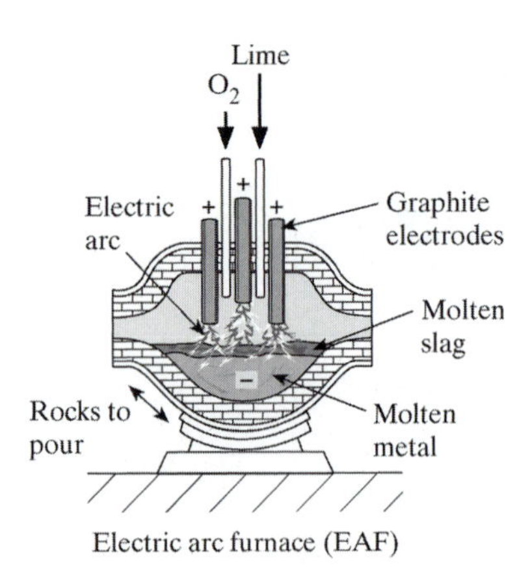

Figure 9–5
Schematic of steel-refining furnaces

the melt and graphite electrodes. The furnace can be charged with scrap or solid pig iron. The arc melts the metal; refining is produced by an oxygen lance introduced into the melt and by the action of fluxing agents. In fact, electric arc furnaces were formerly used to produce alloy and specialty steels. They can be as small as a few tons or as large as 300 tons. The attractiveness of this process for alloy steel is based on the fact that it can be shut down between uses. It does not have to be kept hot to melt the charge. Foundries often melt iron in electric arc furnaces, and they may only run a few hours a day. Electric arc furnaces are tipped to pour the metal and the slag into ladles. The hot metal ladle usually is brought by overhead crane to a continuous caster or to an ingot pouring station. Electric arc furnaces are often used in minimills. They are also currently used for alloy and specialty steels such as tool steels.

Secondary Steelmaking

Commodity steels like those used for structural shapes are usually only refined in basic oxygen furnaces. Oxygen is blown into the melt to reduce carbon, manganese, sulfur, and phosphorus to desired levels. Specialty steels such as high-formability interstitial-free sheet metal, heavy plates for pressure vessels, and steels for critical piping applications may receive secondary steelmaking treatments aimed at altering chemical composition and reducing nonmetallic inclusions. Many strength-of-materials studies have shown, for example, that the fatigue life and toughness of steels are directly proportional to the size and volume fraction of nonmetallic inclusions in the steel.

These inclusions are usually oxides, silicates, sulfides, or aluminas that form during conventional melting and refining (Figure 9–6). The inclusion rating of a piece of steel can be measured by sawing a thin slice from the end of a steel shape and etching it in an acid. A dirty steel will show pits when the inclusions are etched away by the acid. There are also microcleanliness standards for steels that measure inclusion ratings by microscopic examination of a polished sample from a steel shape. These inclusion ratings can be added to a steel purchasing specification so that a designer has the option of specifying a steel of special cleanliness.

Figure 9–7 illustrates some of the many processes that are used in secondary steelmaking. Vacuum degassing is used to remove dissolved gases from steels. The technique illustrated in Figure 9–7 shows molten metal being streamed from an upper vessel to the lower vessel in an evacuated vessel. There are many other ways of vacuum degassing. Ladle stirring is accomplished by bubbling argon from a lance through the melt or from a port in the

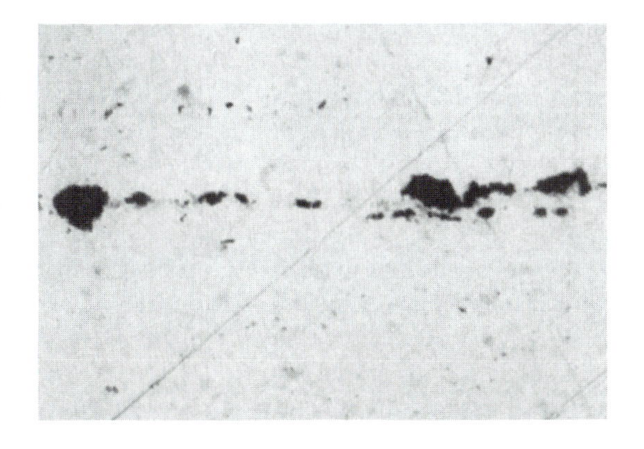

Figure 9–6
Typical nonmetallic inclusions in low-carbon sheet material made from ingot cast steel (×500)

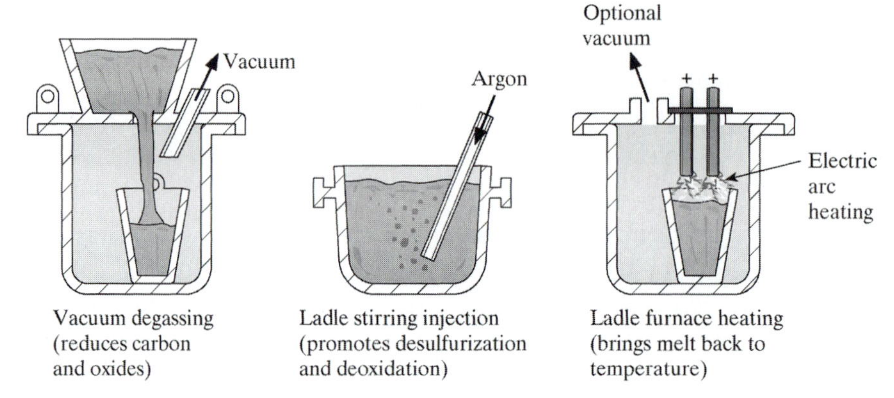

Figure 9–7
Secondary steelmaking processes

bottom of the ladle. Argon and agitation promote sulfur removal and *deoxidation*. Secondary steelmaking processes are usually done in the ladle. The basic refining would have been performed in a basic oxygen or electric arc furnace. Ladle heating is used to restore heat to the metal in the ladle. There are additional secondary steelmaking processes that inject calcium, rare earth elements, gases, and even alloy wire into the ladle. Overall, secondary refining processes play a role in improving steel properties to meet product needs.

Special Refining Processes

Alloy steels, stainless steels, bearing steels, tool steels, and other noncommodity steels are frequently subjected to special refining processes that are intended to modify chemical composition and/or to remove impurities. These processes are illustrated in Figure 9–8. *Vacuum arc remelting* (*VAR*), the most popular technique, involves casting of steel from the BOF or electric furnace into cylindrical ingots. A stub shaft is welded to these ingots, and the ingot is remelted in a vacuum by establishing an arc between the ingot (electrode) and a water-cooled copper mold. The ingot becomes a giant welding electrode. This process is very effective in removal of inclusions because it is very energetic. Every drop of the ingot is exposed to vacuum as the droplets transfer in the arc.

Vacuum induction melting (VIM) is a newer process that is used to melt solid scrap or liquid charges. The charge is placed in a crucible and heated by high-frequency induced currents. This produces convection current mixing of the melt. The entire crucible is in the vacuum, and ingots are also cast in the vacuum.

In *electron beam refining,* molten metal is poured down a tundish (chute) into an ingot mold. The tundish and mold are in a vacuum. As the metal flows down the tundish it is subjected to an electron beam that vaporizes impurities so that they can be removed as vapors in the vacuum.

The processes grouped under chemical reaction are different from the vacuum refining mechanism in that impurities are removed by reaction with some species introduced into the melt; they are not in vacuum. In the *AOD process,* argon and oxygen are introduced into a crucible containing a molten heat from an electric furnace or BOF. The oxygen reduces carbon level (decarburization), sulfides, and other impurities. The argon causes significant stirring to disperse oxides and make them smaller. The argon also promotes removal of dissolved gases.

Electroslag refining is similar to VAR without the vacuum. A VIM or electric furnace melt

Figure 9–8
Special refining processes

Vacuum arc remelting (VAR)
Used for super-alloys and extra clean steels

(+)
Vacuum
Ingot electrode
Electric arc
Water-cooled metal mold
(−)

Vacuum induction melting (VIM)
Used for specialty alloys

Vacuum
Induction heating power supply

Electron beam refining
Used to purify specialty alloys

Ladle
Tundish
Electron beam
Ingot mold

Electroslag melting (ESR)
Used for tool steels and special-purpose steels. The molten slag acts as a cleansing flux

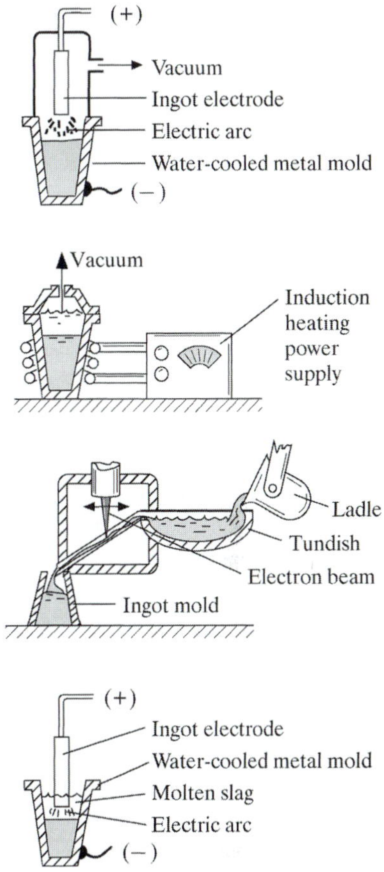

(+)
Ingot electrode
Water-cooled metal mold
Molten slag
Electric arc
(−)

is cast into remelt ingots. These ingots have a stub welded to them, and they are made into electrodes for arc remelting in a water-cooled copper mold. Purification is accomplished when the melting metal from the ingot passes through a molten flux that acts like a welding electrode slag to remove impurities. Shrinkage voids are minimal in *ESR* ingots.

Pouring of Ingots

After a heat of steel is refined, it is poured into a continuous or strand casting machine, or the heat is poured into ingot molds. Less than 10% of the steel in the United States goes into ingot; most is continuous cast. Thus, this phase of the steelmaking process has become less important. However, designers and steel users may encounter a plethora of terms relating to ingot casting.

From the standpoint of the steel user, pouring technique can affect chemical homogeneity and shrinkage voids in the steel. The metallurgical terms for these factors are segregation and pipe. *Segregation* is a variation in chemical composition. It occurs on a microscopic scale in the dendrites that start solidification, and it occurs on a macroscopic scale in the cast shape. One cause for segregation is that the first material to solidify is purer than the last. *Pipe* is a cavity in the top of an *ingot* that is formed by volumetric shrinkage of the metal as it transforms from the liquid to the solid state (Figure 9–9). Both of these ingot defects can find their way into finished product if they are not properly dealt with at the mill.

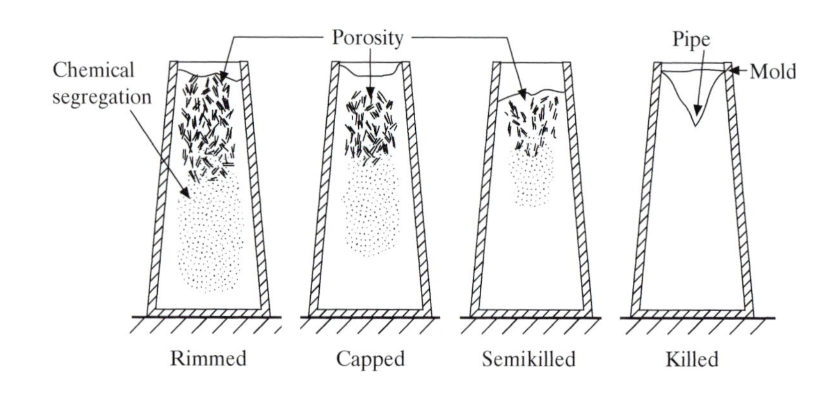

Figure 9–9
Ingot characteristics of various types of steel

When a metal solidifies over a temperature range, part of the metal will be molten and part will be solid. In the case of steels, there is a tendency for the first metal to solidify to be purer than the metal that solidifies last. Ingots that solidify with a skin that is purer than the center produce *rimmed steel.*

Thus, segregation and pipe are a natural part of the solidification process. Fortunately, steel mills have devised ways to minimize their detrimental effects. A portion of the ingot is cropped or cut off to get rid of pipe. Segregation can be minimized by adding elements to the steel, such as aluminum and silicon, that remove dissolved oxygen from the molten metal and alter the ingot solidification characteristics. These are called *killed steels* because the molten metal lies quiet during solidification.

Pipe and porosity are sometimes rolled into sheet or plate, and delamination can occur in forming. Killed ingot steel tends to be less prone to problems with these defects. Continuous cast steels are always deoxidized or killed, and porosity due to pipe is not the concern that it is with ingot cast steel.

Continuous Casting

A schematic of the continuous casting (*concast*) process is shown in Figure 9–10. Steel is refined in the basic oxygen or electric arc furnace, and it is poured into a ladle. The ladle is transported to the caster facility, and it is poured into the caster tundish. The steel flow into the caster is controlled with valves so that the tundish is filled to the top. The tundish in turn feeds the caster. The caster mold determines the semifinished shape (semi) that the caster produces. Figure 9–10 also shows commonly cast shapes. The caster mold is water cooled, and the molten steel stays in the mold long enough to form a solid skin. The skin on the strand becomes like a continuously moving ingot mold. The center of the strand is molten for some distance after the strand exits the mold (from a few meters to as far as 30 m). The strand may require many rollers or a few rollers for support depending on the shape of the strand. A 5-in.-square (32-mm-square) strand would not need as many support rollers as a 4-in.-thick by 30-in.-wide (25- × 193-mm) slab.

The strand is usually cut into designated lengths when the core is fully solidified. This is mainly done with oxy-fuel torches or with hot shears for small strands. Continuous casters do not always look like the one shown in Figure 9–10. They can have different curved paths for the strand, or they can be horizontal or vertical. Vertical strand casters are rare; they are used for casting alloy steels and other metals with limited ductility. Brittle alloys are prone to fracture when bending stresses are high. There could be bending stresses produced in bending the strand from vertical to horizontal, which is the way

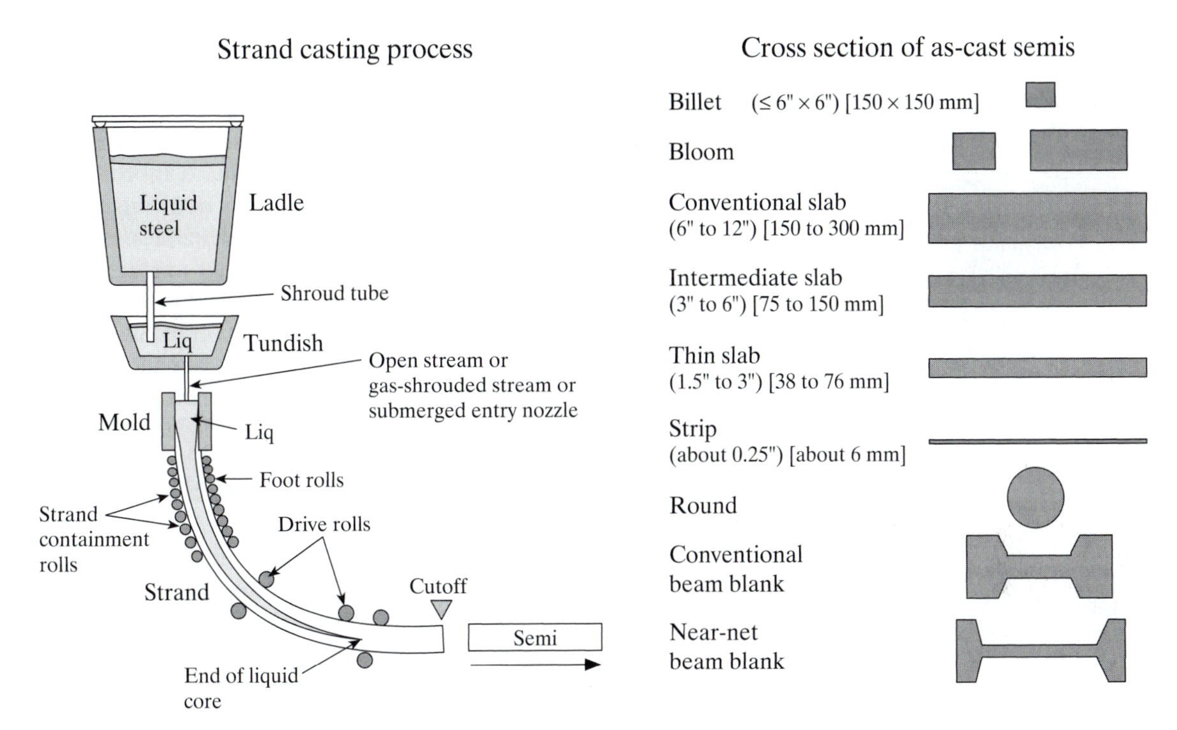

Figure 9–10
Schematic of the strand casting process and shapes produced (semifinished)
Source: Courtesy of E. S. Szekeres of Casting Consultants Inc.

most casters operate. Casters can have multiple strands; four strands is not uncommon. These machines are very expensive, and most mills make it a priority to keep them running continuously. Casters are started by placing a short piece of solidified strand in the mold (called a stool), and molten steel is poured on the closed mold and stool to get a continuous strand. Once the unit is operating it is common to sequence heats of steel so that the machine rarely sees a cold start. Many machines run 90% of the time, and some machines process as many as 3000 heats between shutdowns.

There are steel quality concerns with continuous cast steel just as there are with ingot cast steel. Inclusions, porosity, segregation, and grain size are concerns in the strand. An inherent advantage with strand casting is that the molten core feeds the solidifying metal to prevent solidification shrinkage voids (if the heat transfer in the system is right). There is no pipe to remove as in ingot casting. Strands usually solidify with a grain structure similar to that shown in Figure 9–11. When any metal solidifies against a cold mold, wall dendrites tend to form, and these dendrites grow into grains. A dendrite is a three-dimensional treelike structure. It is nature's way of solidifying. Frost on a window is mostly made up of two-dimensional dendrites. The aspect of dendrites and grains that is incredible, but not visible except with sophisticated tools, is that the unit cells in which the atoms are arranged on the atomic level are all oriented in planes in the same direction. In other words, all of the atoms in a grain have the same spatial orientation. The planes of atoms in

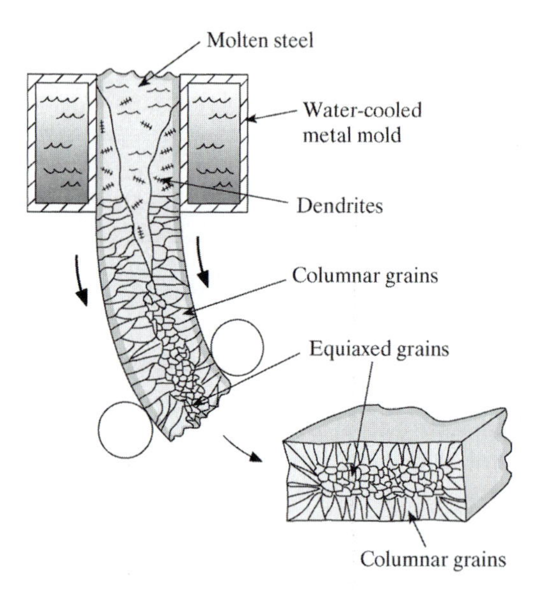

Figure 9–11
Grain size in a strand cast semi

neighboring grains also have common alignments within the grain, but the orientation between grains is different. When grains or crystallites meet, they form a *grain boundary*. The nature of the grains formed in casting solidification and how they are changed in steel finishing processes have a profound effect on steel properties. Grain size is usually measured for quality control purposes in most wrought steels. Grain size relates to strength. Small grain size steels usually have higher strength than a steel with the same composition that has larger grains. The grain size in a wrought steel shape is determined by cutting a cross-sectional piece and then subjecting the cut surface to polishing and chemical etching. Observation under a microscope allows quantification of grain diameter. Steel mills often monitor grain size on the shop floor by macroetching. A cross-sectional slice is cut from the semi and macroetched by grinding the surface and then soaking it in an appropriate acid.

Continuous cast steels usually have columnar grains (like columns) originating at the cold mold walls and the grains become equiaxed (grains with the same shape in the x, y, and z directions) in the center. Continuous casting will continue to be the preferred system for semifinishing steel shapes, and, in the foreseeable future, the process will feed hot- and cold-rolling operations without reheating. In 2003 there are demonstration projects in place to continuously cast strip. The process will go from molten metal to finished coils without interruption. Steel users need to keep in mind how steel is made to troubleshoot steel problems if and when they occur.

9.4 Converting Steel into Shapes

Literally thousands of different steel products are produced by steel mills from a cast shape, either ingots or the shape that exits the continuous casting machine. Ingots start their conversion to a steel product by hot rolling into large shapes called billets, blooms, or slabs, depending on their general *shape*. Very large forgings are made directly from ingots, but such large forgings are not common. Large, unfinished shapes produced in the roughing operations are further worked into a number of basic mill products. Figure 9–12 shows the processing steps involved in converting ingots and continuous cast shapes into finished products. The steel products that are available are listed in Figure 9–13. Each of these product forms has different manufacturing tolerances, different terminology, different surface finishes, different chemical compositions, and, in the case of *cold-finished* products, different tempers. What is available in the United States is shown in manuals published by the Iron and Steel Society under the titles of the products listed in Figure 9–13. There are similar steel product handbooks in other countries. If steel users need to develop exacting specifications on a steel for a given product, these steel product handbooks should be consulted for proper terminology and to determine if their tolerance requirements are within industry standards.

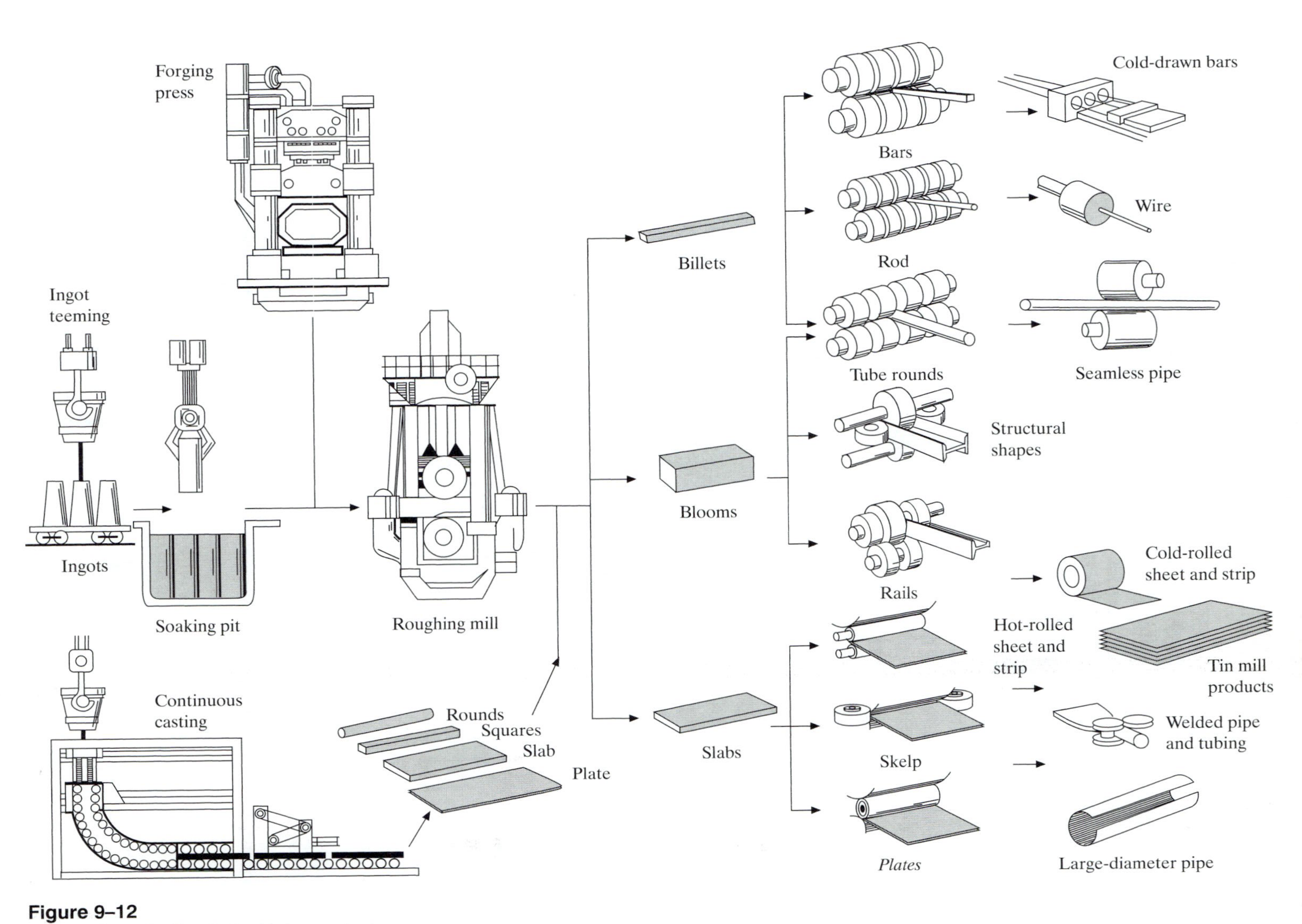

Figure 9–12
Processing of refined steel into products
Source: Courtesy of the American Iron & Steel Institute.

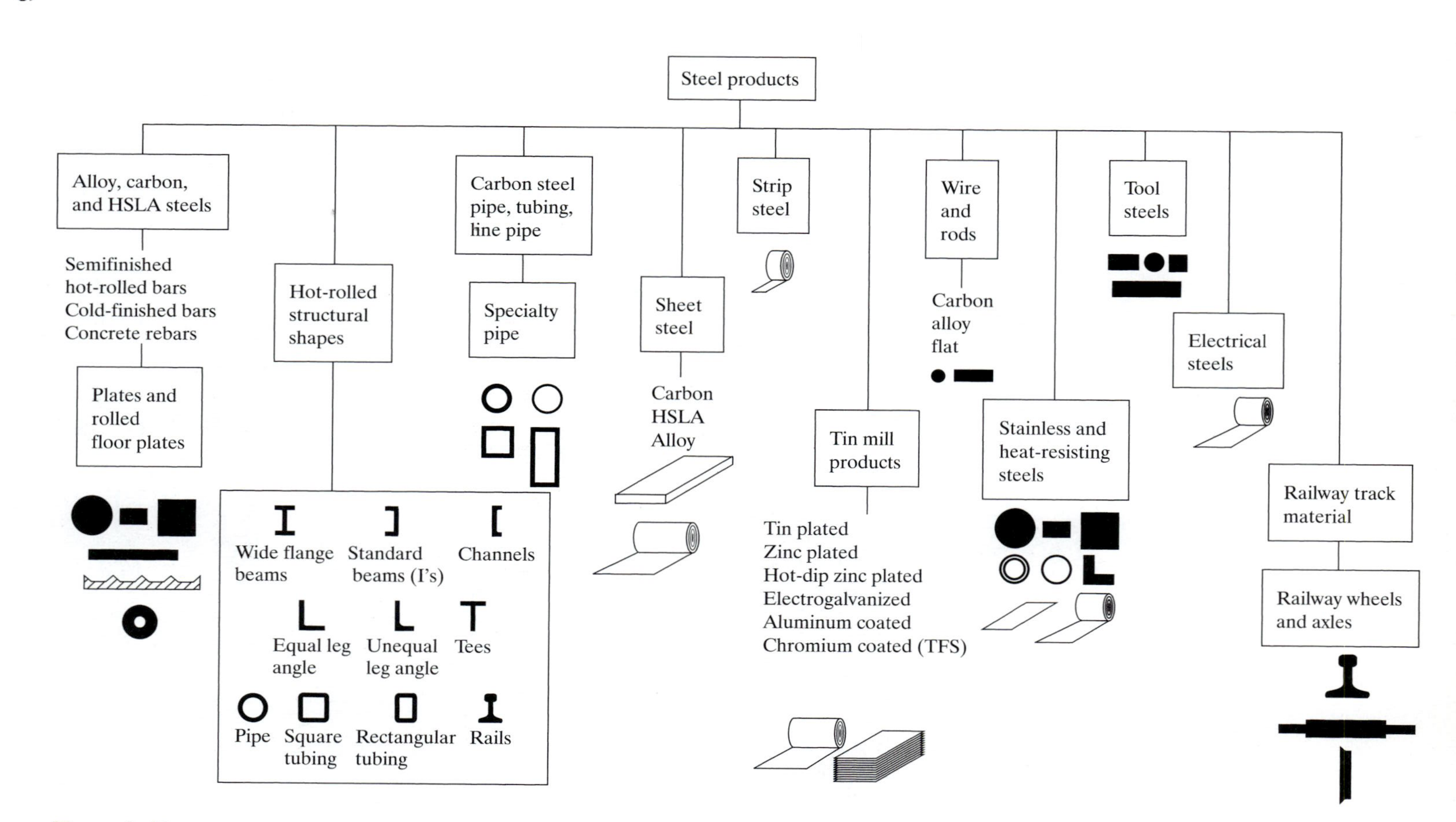

Figure 9–13
Spectrum of steel products

Metallurgy of Mill Finishing

Very important metallurgical principles are involved in the reduction of steels and other metals from ingot form or cast shape to usable product. In Chapter 1 we stated that metals are different from other materials in that the atoms are held together by essentially an electron cloud. This is the metallic bond. The other unique aspect of metals is that they can plastically deform by atomic movements on many atomic planes. Ceramics are brittle because the interatomic bonds are high, and fracture by cleavage can occur when deformation is attempted. The plastic deformation that can occur in metals happens by atomic movement *(slip)* in each crystal or grain that is in the metal shape. The grains or crystallites are in the metal from the solidification process. As shown in Figure 9–11, an ingot or semi solidifies by the growth of dendrites, tree-shaped, three-dimensional structures that grow atom by atom to form a crystal. These dendrites start at nuclei that can be many things, for example, a speck of dust. Dendritic growth of crystals is a significant part of the basic metallurgy. It occurs in all metal systems. When an ingot mold is filled with molten steel, the first thing that happens is the formation of dendrites from nuclei that can be a single atom, an impurity, a protrusion from the mold wall, or the like. This process will continue until there are dendrites throughout the entire volume of liquid. The material will eventually be in a slushy state as the molten metal feeds the growth of these three-dimensional dendrites. When solidification is complete, what were dendrites are now crystals, with the atoms in each crystal having a particular orientation.

If the metal that is solidifying contains some percentage of another metal or element, another phenomenon can happen on solidification. A second *phase* can be formed. A phase is a homogeneous component of a solid, liquid, or gas that is separated from other phases by an interface. If no other phases are present, this is called a single-phase state of matter. If mineral oil is poured in water, it will form a separate phase because the oil is not soluble in the water. The same thing happens in metals. In the ingot mold, if the metal system is conducive to the formation of a second phase, dendrites and grains of each phase will be formed. Figure 9–14 illustrates both the concept of phases and dendrite growth. The photomicrograph shows a slice through a casting of a multiphase cobalt-based alloy. The white areas are cobalt-based dendrite material; it is the cobalt-rich phase, and the dark areas are composed of second- or third-phase components or microconstituents. The dark phases are rich in the alloy elements that were added to the cobalt-based metal.

The role of these basic concepts for converting steel into shapes is this: Two main products come from a steel mill, hot-finished steel and cold-finished steel. The difference between these two product lines is simply what happens to the grains in the steel during mill processing. When a steel ingot is heated red hot and run through a set of rolls or through another tool, the dendritic grains are squashed down. If the steel is red hot, these grains immediately grow back after squashing.

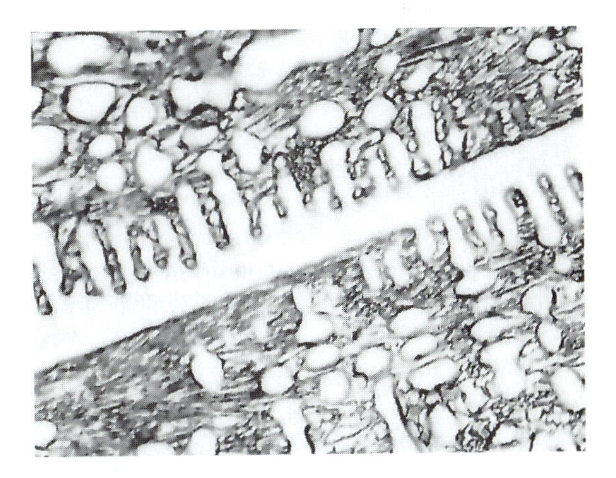

Figure 9–14
Dendrites (comb-shaped areas) in a cast two-phase Co–Cr alloy (×200)

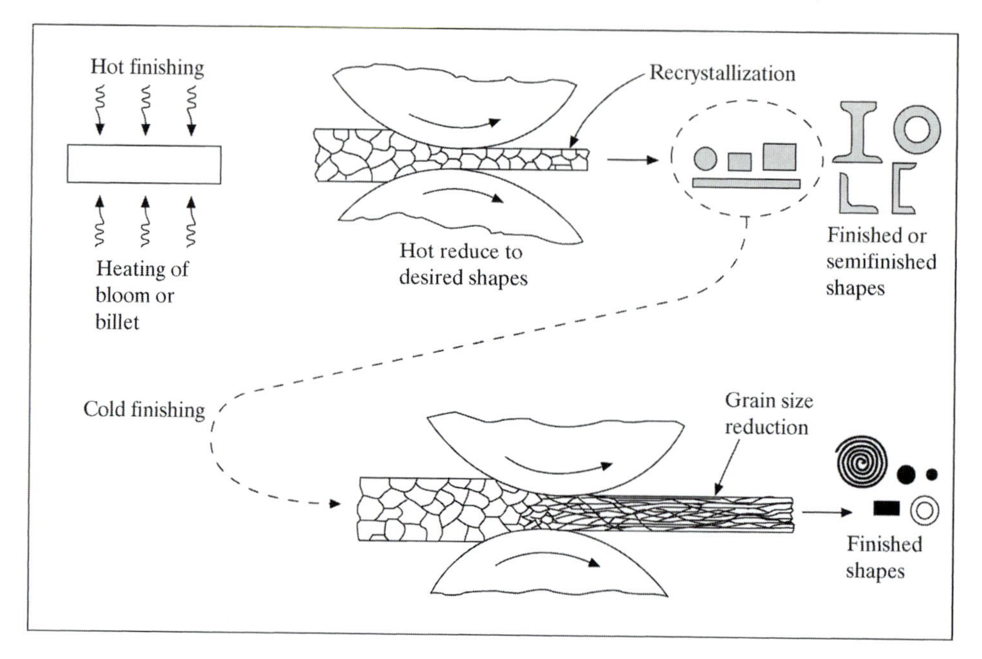

Figure 9–15
Schematic of hot and cold finishing of steels

This is called *recrystallization*. This concept is illustrated in Figure 9–15. If the steel is rolled or worked in some other fashion in the cold condition, this dynamic recrystallization does not occur. The grains get squashed down, and the more the grains are squashed, the harder the steel gets. Eventually, a point will be reached where the steel can no longer be deformed. Fracture will occur if additional reduction or shaping is attempted.

Hot Finishing

In hot finishing, ingots or continuous cast shapes are rolled in the red-hot condition to smaller shapes such as blooms or *billets*. A typical bloom may have a cross section of 8 × 8 in. (20 × 20 cm). Billets are smaller, typically in the range of 2 to 5 in. (5 to 13 cm) on a side. *Slabs* are 2 to 20 in. (5 to 50 cm) thick, with a width that can be 60 in. (150 cm) or more. Slabs are a normal product from continuous casters. When one of these shapes is hot finished into a shape with a smaller cross-sectional area, the grains dynamically recrystallize and the material does not get harder. It stays easy to roll or shape. The crystal defects (dislocations) that reduce a material's ability to deform are annihilated by the energy available from the red-hot material. Hot-finished shapes include sheet, strip, bars, structural shapes, and pipe. These shapes can be the finished product or a starting material for cold finishing. As shown in Figure 9–12, the starting material for a cold-finished sheet steel is a coil of hot-rolled steel (hot band, in steel mill jargon). Hot-finished steel may have a black scale on its surface from its exposure to air in the red-hot condition or a dull, gray surface that is left when scale is removed by spray quenching the red-hot steel.

Cold Finishing

Cold finishing begins with pickling the steel in an acid dip to remove any mill scale that it acquired in the hot-finishing operation. The hot band is then rolled through various types of rolling mills at room temperature. Typically, the hot band may be reduced by 10% to 20% per pass through the rolls. If the goal is to reduce a 0.100-in. (2.5-mm) hot band to a 30-mil (0.75-mm) product, it may have to go through ten rolling operations. There is a limit to the amount of cold reduction that each steel can withstand. If a 100-mil hot band is 1020 steel, it could probably be rolled to a thickness of 0.050 in. (1.25 mm) by going through a cold reduction line that contains five stands of rollers. By that point it would probably be cold worked or *work hardened* to the point where it would fracture if additional cold reduction were attempted. Crystal defects have built up in the squashed-down grains so that further atomic slip is not possible. The material must now be annealed by heating the coil red hot in a batch furnace or in a furnace that is built into the rolling line *(continuous anneal)*. Recrystallization occurs in annealing; the annealing heat supplies the energy for removal of dislocations and the formation and growth of new grains or crystallites. The hot- and cold-finishing processes are the same regardless of the shape produced. Bars can be cold finished only so far before annealing; the same thing is true of wire, rod, and other shapes.

The importance to the user of cold and hot finishing is this: *Hot-finished* steel shapes are soft and the mechanical properties are about as low as we can get from a particular alloy. In addition, the surface finish is dull gray to black and scaled. However, hot-finished steel sheet is available with a pickled and oiled finish that is free from scale and discoloration. Cold-finished steels can be much stronger, and they have a better surface finish. As an example, hot-finished, low-carbon steel may have a tensile strength of 60 ksi (414 MPa), whereas a full, hard, cold-finished shape of the same steel would have a tensile strength of 120 ksi (828 MPa). Because the cold-finished steel is twice as strong, it may be possible for a designer to use only half the weight of steel to get the same strength from the part. This does not always work because of stiffness and section modulus considerations, but, in general, increased use of cold-finished steels has allowed the drastic weight reductions that have occurred in automobiles in the past decade or so. The use of hot- versus cold-finished steel is an important selection consideration for designers.

9.5 Steel Terminology

The selection of steels requires consultation on property information and supplier information on availability. If the designer is to make any sense out of handbook information, he or she should become familiar with the terms used to describe mill processing operations. There are so many terms that it can be very confusing. The following is a tabulation of steel product terms and their definitions:

Carbon steel: Steels with a carbon as the principal hardening agent. All other alloying elements are present in small percentages, with manganese being limited to 1.65% maximum, silicon to 0.60% maximum, copper to 0.60% maximum, and sulfur and phosphorus to 0.05% maximum.

Alloy steel: This term can refer to any steel that has significant additions of any element other than carbon, but in general usage alloy steels are steels with total alloy additions of less than about 5%. They are used primarily for structural applications.

Designation: Many organizations have specifications covering carbon and alloy steels. Many steel specifications were developed by limited interest organizations such as the U.S. military, and their specifications are written with their specific interests in mind. In the United States, the most common generic system for alloy designation used to be a system that had been in place for decades called the *AISI* designation system. The acronym AISI stands for the American Iron and Steel Institute, a division of a technical society. They published steel product manuals that reflected current steel mill alloys and practices.

These publications served as the bible for what alloys are available from U.S. producers, product tolerances, available shapes, available tempers, and so on. In the mid-1990s the AISI stopped publishing their manuals on steel products. This function was turned over to the Iron and Steel Society, an organization consisting of steel manufacturers. These manuals continue to show product forms and dimensions, but they do not consider the information in the manuals to constitute a specification. They defer to other organizations for writing alloy specifications. Table 9–1 lists some of the U.S. specifications applicable to steel products. ASTM International specifications are the most widely used in the United States. ASTM is an international technical society of over 20,000 members who write consensus standards on materials and testing methods. Steel specifications are written by a steel subcommittee; cast irons are addressed by an iron castings subcommittee; and there are subcommittees that write specifications on most other metals. Specifications are published annually in bound volumes, and all 10,000+ specifications are available on CD-ROM. These specifications are available on-line and in most technical libraries around the world.

ASTM steels are designated using the ASTM number and grade (if applicable). UNS (unified numbering system) numbers were developed by ASTM to identify metal alloys by common chemical composition, but this system is not a metal specification. It does not establish requirements for form, condition, property, or quality. Only specifications do this. This text will use the former AISI numbers to identify steels because they are easier to remember than the ASTM specification numbers or the UNS numbers. Another reason for this decision is that ASTM specifications are written for product forms. If you needed a low-carbon steel for a shaft, you would look up the specification for low-carbon steel bars and might arrive at ASTM A 36 as a suitable steel. If later you decide to save weight and use a tube instead of a solid, the same alloy would be specified as ASTM A 512 for seamless tubing and as ASTM A 422 for welded tubing. If you wanted the tube *galvanized*, it would be specified as ASTM A 120. There may be ten ASTM specifications that apply to your 1020 steel shaft, depending on product form and treatments. There are over 1000 ASTM specifications for steel alloys in various product forms. Fortunately, most of the ASTM alloy specifications consider that most people are familiar with the decades-old AISI numbers. Many ASTM specifications have grade designations that use the AISI number. For example, the former AISI 1020 steel can be specified as ASTM A 29 grade 1020. In summary, we will discuss steels using AISI numbers, but we will also list their UNS numbers, which can be related back to ASTM specifications through ASTM DS-56 E, the document describing the UNS numbering system. Engineering drawings should use ASTM designations if those standards are available. If they are not, we suggest using the UNS number with a reference to the previous AISI number: ASTM A 29 grade 1020 steel or UNS G10200 (AISI 1020 steel). Eventually all designers will probably have access to ASTM and other international specifications on the Internet and the most appropriate designation for the country of manufacture can be used. However, this "ideal" system was not yet in place in 2003.

Table 9–1
U.S. specifications applicable to steel products

	Specifications
API	American Petroleum Institute
ASME	American Society of Mechanical Engineers
ASTM	ASTM International (formerly, American Society for Testing and Materials)
MIL	U.S. Department of Defense
FED	U.S. Government
SAE	Society of Automotive Engineers
AMS	Aerospace Materials Specification
AISC	American Institute of Steel Construction

Rimmed: Slightly deoxidized steels that solidify with an outer shell on the ingot that is low in impurities and very sound. These steels can retain a good finish even after severe forming because of the surface cleanliness.

Killed: Strongly deoxidized, usually by chemical additions to the melt.

Galvanized: Zinc-coated steel products. The zinc is applied by hot dipping.

Galvannealed: Zinc-coated and heat treated steel. There are usually paint-adhesion problems with galvanized steels if they are not properly pretreated. The heat treatment given to galvannealed steels creates an oxide layer that allows better paint adhesion.

Sheet: Rolled steel primarily in the thickness range of 0.010 to 0.250 in. (0.25 to 6.4 mm) and with a width of 24 in. (610 mm) or more.

Bar: Hot- or cold-rolled rounds, squares, hexes, rectangles, and small shapes. Round bars can be as small as 0.25 in. (6.4 mm); flats can have a minimum thickness of 0.203 in. (5.0 mm); shapes have a maximum dimension less than 3 in. (76 mm).

Coil: Rolled steel in the thickness range of sheet or strip.

Flat wire: Small hot- or cold-rolled rectangles, often made by cold-reducing rounds to rectangular shape.

Wire: Hot- or cold-drawn coiled rounds in varying diameters, usually not exceeding 0.25 in. (6.4 mm).

Shapes: Hot-rolled I-beams, channels, angles, wide-flange beams, and other structural shapes. At least one dimension of the cross section is greater than 3 in. (76 mm).

Tin plate: Cold-rolled steel with a usual thickness range of 0.005 to 0.014 in. (0.1 to 0.4 mm). It may or may not be tin coated.

Free machining: Steels with additions of sulfur, lead, selenium, or other elements in sufficient quantity that they machine more easily than untreated grades.

Drawing quality: Hot- or cold-rolled steel specially produced or selected to satisfy the elongation requirements of deep drawing operations.

Merchant quality: Steels with an M suffix on the designation intended for nonstructural applications. A low-quality material.

Commercial quality: Steels produced from standard rimmed, capped, concast, or semikilled steel. These steels may have significant segregation and variation in composition, and they are not made to guaranteed mechanical property requirements (most widely used grade).

H steels: Steels identified by an H suffix on the designation and made to a guaranteed ability to harden to a certain depth in heat treatment.

B steels: Steels with small boron additions as a hardening agent. These steels are identified by a B inserted between the first two and last two digits in the four-digit identification number (xx B xx).

Pickling: Use of acids to remove oxides and scale on hot-worked steels.

Temper rolling: Many steels will exhibit objectionable strain lines when drawn or formed. Temper rolling involves a small amount of roll reduction as a final operation on annealed material to eliminate stretcher strains. This process is sometimes used to improve the surface finish on a steel product.

Temper: The amount of cold reduction in rolled sheet and strip.

E steels: Steels with an E prefix on the four-digit designation are melted by electric furnace.

We shall terminate our list of steel terminology at this point, but this list is by no means complete. In addition to the steel products already mentioned, many mills produce alloy steels. These can be hot or cold finished. Other mills produce stainless and heat-resisting steels, tool steels, and specialty products such as electrical steels, railway tracks, high-strength steels, and a variety of coated steels. Each type of steel product has its own characteristics and terminology. If a designer is faced with the situation of coming up with the most economical steel for a high-production part, in the United States it would behoove him or her to make an in-depth study of the possible selections in the *Steel Products Manual* of the Iron and Steel Society. This manual, and its counterparts in other countries, describes the guidelines followed by most manufacturers in the production of each type of steel product. If the design requires steel covered by some other code body, such as ASME, then their materials manuals should be consulted. A thorough understanding of steel products and terminology is essential to proper steel selection.

9.6 Steel Specifications

Some companies elect to make their own standards or identifying system for steels. This is done to apply purchase requirements or quality standards that may be more stringent than industry standards. The Welcom company, for example, may write a document for essentially 1020 steel that contains some special handling or packaging requirement. It identifies this special requirement grade of 1020 steel as W10. This number is used on its drawings as the material of construction and, when purchase orders are let to buy the steel, the company must submit a copy of the W10 standard to steel vendors so that they know what steel is required.

Regardless of the identification system used, essential items need to be addressed in purchasing a steel. They include description, quantity, cross-sectional shape, dimensions, chemical composition, condition, heat treatment, finish, packaging, quality requirements, and any special requirements.

Description: The purchase order must have a description of the steel product written in steel mill terms, found in the *Steel Products Manual*. An example of a proper term would be this: steel, hot-finished, low-carbon, bar, ASTM A 29 grade B.

Dimensions: The purchase order must show the required dimensions and quantity. Again, these should be in steel mill terms. In the United States, standard dimensions, tolerances, and ordering quantity information are also in the *Steel Products Manual*. Sometimes steels are not purchased in familiar units of measure. For example, tin mill products such as beer can stock are sold in units of pounds per base box. A base box is a certain area of cut sheet. These details should be investigated.

Chemical composition: The use of a UNS, ASTM, or other designation number will ensure that the desired steel has a particular chemical composition. If there is a reason to want a special chemical composition, then this must be specified in some type of purchasing document or company standard. Using nonstandard chemical compositions should not be done lightly. This almost always means buying a heat of steel. Sometimes there is a 10,000-lb minimum (or even 40,000 lb), and mills will charge for any extra work involved in filling your order. Try to use standard chemical compositions, and specify that the mill submit a certificate of analysis if you are concerned that the composition may not be what it is supposed to be.

Mechanical properties: If it is important that a particular steel product meet certain tensile, hardness, impact, or other mechanical properties, the allowable ranges and the method of test must be specified.

Dimensional tolerances: Most regulating bodies in the metals industry have commonly agreed to dimensional tolerance limits on their products. If these are adequate for the application, this should be so stated. The tolerance tables appear in the regulating-body handbooks. Such dimensional factors as camber, flatness, edge, and coil set should be specified, as well as thickness, width, and length.

Finish: If you can use a steel product as it comes from the mill without performing your own secondary operations to make the surface acceptable, there can be significant cost savings. Such things as roughness, scratches, and grain must be specified. Most steel mills have actual samples available showing their standard finishes, but a stringent finish specification will require quantitative limits on surface roughness and waviness. A cold-rolled carbon steel could have a surface roughness of 10 to 60 μin. Ra (0.25 to 1.5 μm). If this is important for a desired service, definite finish limits should be stated along with the measuring technique.

Special requirements: In addition to the preceding factors, such things as special heat treatments, coatings, grain direction, packaging, corrosion tests, and forming tests may be called for. It is customary to include limits on allowable defects, such as coil welds, slivers, edge burr, and the like. These special requirements can, however, result in extra charges.

Steel specifications or purchase orders do not have to be lengthy, but they need to address all the factors that we mentioned. Company specifications such as our example of W10 steel usually include all the details about chemical composition, mechanical properties, quality, and the like, making the purchasing function easier. The purchase order may specify only W10 steel and the quantity and size required. Staff reductions in many industries have reduced the manpower available for in-house material standards, and there is a trend to use generic specifications such as ASTM. This means that you, the designer or steel user, must be more precise in writing purchase documents. You may need to address all the factors we mentioned as well as packing and inspection methods for incoming material. You must become familiar with standards for steels such as those published by ASTM and SAE.

In summary, we have described how steels are made, what products are available, steel terminology, and steel specifications. There are two main points to be made from this discussion concerning the use of steels in engineering design:

1. Know what products are available.
2. Specify the desired steel adequately on a drawing and on purchasing specifications.

These two rules, as well as the factors involved in steel specifications, really apply to all metal products and, to some extent, to nonmetal products. In our subsequent discussions on other metals, the following question should always be asked: If I wanted to use this material, how would I specify it on a drawing?

Summary

This chapter may seem misplaced. We have not discussed the basic metallurgy of steels, and we just described how to write a steel specification. The rationale for starting the metallurgy portion of this text with a terminology discussion is that, when we discuss heat treatments of steel, alloy steels, tool steels, and the like, we will be discussing specific steels using the nomenclature that we have introduced in this chapter. When we discuss hardening of carbon steels, we want the reader to know what a carbon steel is, what a bar is, what free machining means, and so on. It is important for metal users to acquire a basic understanding of the products that are available. In the United States and probably elsewhere in the world, if you go to a lumber company and ask for material to build a bookcase, the salesperson will want you to convert your request into specific terms. What wood? What nominal dimensions? What finish? What quality? If you do not know what "2×4, 2×8, $\frac{5}{4} \times 4$, $\frac{3}{4} \times 6$ select" means, you will have a hard time getting the material that you want. The same situation exists with steels. Steel users must learn the language of steel products. Most of our terminology and fundamentals on steel products apply to most other metal systems. The process of making a bar of nickel from ingot is essentially the same as making a bar of steel. The same thing is true of sheet, strip, and the more common metal products. The nonferrous industries have different tolerances than the steel industry on bars, sheets, and so on, but we shall discuss these in chapters on these subjects.

Some of the important fundamentals to remember from this chapter are the following:

- Iron is abundant in nature; about 5% of the earth's crust is iron.
- Steel is made by reducing oxide ores of iron by thermochemical reactions in a blast furnace or in a direct reduction process.
- Pure iron does not have significant industrial use; it is too weak and soft.
- Steel is an alloy of carbon and iron with limits on the amount of carbon <2%.
- Pig iron, which is the product of a blast furnace, is used for making steels, but it is also the starting material for cast irons.
- Currently, most steels are made in the basic oxygen furnace, and a significant amount of the output of these furnaces is continuously cast into semis for mill processing to steel products.
- Steel cleanliness should be a concern when surface finish, high strength, weldability, and structure-related properties are selection considerations.
- Steel melting practice can be specified to control steel cleanliness.
- Ingot cast steel has properties that vary depending on the type of practice used in pouring the ingot.
- Continuous cast steels are deoxidized, and a steel user does not have to be concerned about ingot solidification practice.
- Some steels can have surface imperfections that may show after painting.
- Hot-finished steels have lower mechanical properties than cold-finished steels.
- Hot-finished steels do not work harden in manufacture because the elevated working temperatures produce dynamic recrystallization.
- Grains form in casting solidification, and what happens to these grains in mill processing the metal affects the end properties of the metal.
- Steels should be designated by a number from some standards organization; the ASTM system is the most prevalent in the United States.
- When steel is purchased for a job, the purchaser must question such factors as surface finish, dimensional tolerances, composition limits, mechanical properties, and the like. Neglecting to inquire about one of these may cause the steel to be unsuitable for use.

Essentially, the purpose of this chapter was to make the reader conversant in steel terminology.

Critical Concepts

- Iron ore must be reduced from its oxide form as the first step in steelmaking (usually in a blast furnace).
- Steel is made by refining pig iron from the blast furnace to reduce its carbon content to less than 2%.
- There are "standard" steel products made worldwide (bar, wire, sheet, strip, shapes), but not all steel mills make a full range of these products. Some specialize.
- Steel mill terminology is necessary to properly designate steel on drawings and in specifications.

Terms You Should Remember

reduction
hematite
magnetite
pig iron
wrought iron
steel
segregation
pipe
AISI
EAF
BOF furnace
VAR
blast furnace
electric arc furnace
ESR
killed
rimmed
bar
shape
sheet
strip
galvanized
free machining
pickling
temper
plate
phase
hot-finished
cold-finished
concast
carbon steel
alloy steel
ingot
deoxidation
work hardened
billets
slabs
chemical reduction

Case History

Steel Beam Selection

A machine department wanted to install a large machining center in a small wooden storage building abutting a large shop area. The problem presented to design engineering was how to deal with a steel column supporting the second story of the 24 × 30 ft building. The machine department wanted to eliminate the support column and span the 30-ft centerline with a structural steel shape supported by masonry at the ends. What type of structural shape and size should be used?

The building code required that the upper floor support be capable of supporting 100 lb per square ft. Thus the upper floor must carry a load of 24 × 30 × 100 lb—72,000 lb. We assumed that the load would be supported uniformly by the two side walls and the new center beam. Thus the load on the center beam was 24,000 lb and it was uniformly distributed. Assuming simple support, the maximum bending moment (m) on the new 30-ft center beam would be 2.16×10^6 in.-lb. The building code also specified 13,000 psi as the allowable stress on a structural steel building beam. So inserting this figure into the equation for the stress (σ) in a simply supported beam:

$$\sigma = MC/I$$

where C is the distance from the neutral axis to the outer fibers of the beam and I is the moment of inertia, we end up with a c/I of 166 in.^{3} This is where steel shape selection becomes a factor. In the United States, the American Institute for Steel Construction and others publish tables with c/I values (they actually use a reciprocal term called "s") for commercially available steel shapes. We went to the tables and matched our required c/I with the c/I for various shapes. At least six different shapes would do the job. We selected a 12-in.-high, 12-in.-wide, 133-lb/ft (WF 12 × 12 × 133) wide flange beam. Then we checked with a second formula to see if elastic deflection was within allowable limits. It was. This is the methodology used to select the many structural steel shapes that are available. This job could have been done with channels, an I beam, or possibly with angles. Simple calculations guide the selection process.

Questions

Section 9.1

1. What was the source of iron for tools made 4000 years ago?
2. What is the reducing agent for changing iron oxide to iron?

3. Describe pelletizing and how it is used.
4. Name four materials that enter a blast furnace, and state the function/purpose of each.

Section 9.2

5. Compare the carbon content of steel with the carbon content of pig iron.
6. What is the current steelmaking process for refining pig iron and scrap to carbon steel?
7. What are steel inclusions and where do they come from? What problems can they cause?
8. Name three steel-refining processes for reducing inclusion levels.
9. Is VAR applicable to 1020 steel?
10. What steels are made to a temper?
11. What is a semi? What semi would be used to make steel angles?

Section 9.3

12. Name three steel products that are considered to be shapes.
13. What are grains in steel?
14. What is dynamic recrystallization?
15. Name three ways to work harden steel.
16. What is the difference between hot- and cold-finished steel?
17. What is a metallurgical phase?
18. What are dendrites and what is their importance in steel?

Section 9.4

19. What is the difference between sheet and strip?
20. What is the difference between ingot cast and continuous steel?
21. What is temper-rolled steel?
22. What is a wrought steel?
23. What is a base box of steel?

Section 9.5

24. What factors are required to completely specify a steel on a purchasing specification?
25. Can hot-rolled steel be used for making an automobile hood?
26. What is the difference between a steel designation and a steel specification?

To Dig Deeper

AISI Publications Catalog. Washington, DC (1101 17th Street NW, Suite 1300, ZIP 20036-4700), 1993.

Bowman, N. B. *Handbook of Precision Sheet Strip and Foil*. Metals Park, OH: American Society for Metals, 1980.

Ghosh, A. *Secondary Steel Making*. Boca Raton, FL: CRC Press LLC, 2000.

Lankford, W. I., N. C. Samways, D. F. Craven, and H. E. McGannon. *The Making, Shaping and Treating of Steel*. New York: Association of Iron & Steel Engineers, 1985.

The Metals Black Book, Ferrous Metals, Vol. 1, 3rd ed. Edmonton, Alberta: Casti Publishing, Inc., 1990.

Metals Handbook, 9th ed., *Vol. 1, Properties and Selection*. Iron, Steels and High Performance Alloys. Metals Park, OH: ASM International, 1990.

Steel Products Manuals. Warrendale, PA: Iron and Steel Society, 1975 to 1996. (There are 15 individual publications, one for each product line, e.g., tin mill products, sheet steel, strip steel, etc.)

Thornton, W. A. *Manual of Steel Construction,* 3rd ed. New York: AMA Institute of Steel Construction, 2002.

Viswanathan, R. Ed. *Clean Steels Technology*. Metals Park, OH: ASM International, 1992.

Wegst, C. W. *Key to Steel*, 18th ed. Materials Park, OH: ASM International Verlag, 1998.

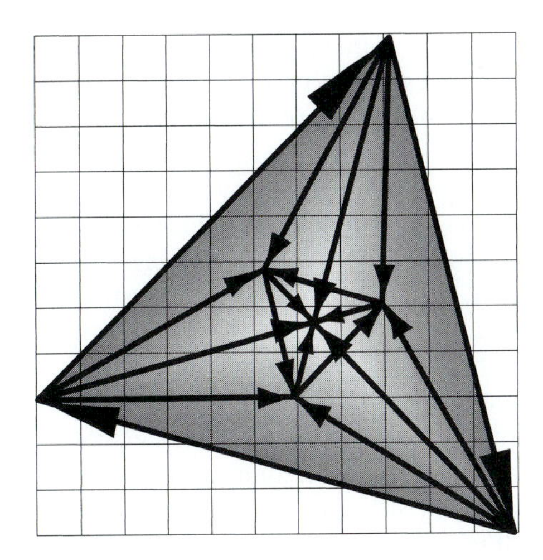

CHAPTER 10

Heat Treatment of Steels

Chapter Goals

1. An understanding of the metallurgical reactions that occur in heat treatment.
2. A knowledge of the common heat treatments used on steels.
3. Guidelines on how to specify heat treatments on engineering drawings.

Rationale

The ability to make steels hard enough to be the tools that shape all other materials is probably the most important gift from nature. How does this happen? Why can't we make all metals hard by heat treatment? How do designers specify hardening of steels? This chapter is intended to address these questions.

As a designer/engineer, you will use many materials in your professional career, but steels will probably be your "most-used" engineering material. Sometimes you will want a steel to be soft and formable; sometimes you will want the hardest steel available; and sometimes you will want high strength and toughness. Heat treatment of steels allows these options.

Artifacts from early civilizations indicate that crude iron was hardened by heat treatment as early as 1000 B.C., but an understanding of the compositional and structural factors controlling the hardening of iron or steel was not achieved until the eighteenth century. It was pointed out in discussing the nature of metals that steels can be hardened through time- and temperature-induced transformation of their crystal structure (allotropic transformation). A number of engineering alloys can be hardened and strengthened by transformation hardening, but steels are unique in the degree of strengthening that can be achieved. Some titanium alloys can be transformation hardened, but their hardness never exceeds 40 HRC. Quench-hardened steels can achieve a hardness of 65 HRC. Strength increases similarly are greater than are possible in other metal systems. This is why steels are unique and why they are probably the most important engineering material.

Steel's ability to become strong and hard in heat treatment is produced by alloying iron with carbon. The changes in solubility of carbon in iron with temperature are used to advantage in quench hardening. It is the purpose of this chapter to describe in detail the role of carbon in hardening through equilibrium phase diagrams, to describe basic heat treating processes, and to show how heat treatments are selected and specified. The concept of equilibrium phase diagrams can be applied to other metal systems that will be covered in subsequent chapters.

10.1 Equilibrium Diagrams

A major contribution to the understanding of metal alloys was the application of equilibrium phase diagrams. Much of the work in this area has been done in the past 100 years. A *phase* is a homogeneous component of a metal alloy. An *equilibrium phase diagram* is a graph showing phase relationships that occur in a metal alloy as it slowly cools from the molten state. *Equilibrium* is said to exist when enough time is allowed for the occurrence of everything that wants to occur. Classical metallurgical texts devote significant attention to the understanding of these diagrams and to their use in calculating theoretical microstructures. In this text, phase diagrams will be given limited coverage, the intent being only to show how they are used to predict response to heat treatments and other thermal cycles.

In previous discussions it was pointed out that metals can be strengthened by alloying elements going into solution in a host metal. Carbon can go into solution in iron just as sugar can dissolve in water. However, in metals this solution can form in the solid state. There are two main types of solid solution: *interstitial* and *substitutional* (Figure 10–1). An orderly structure of crystal cells comprise metals. If the metal is completely pure, there are no other elements present; it is said to be single phase. Under microscopic examination only grain boundaries would be visible. If we wish to make an alloy by adding another element, the added element (*solute*) may go into interstitial or substitutional solid solution, or it may form another phase. If the added solute goes completely into solid solution, the host metal may remain single phase in nature. There is usually a limit to the ability of a host metal to dissolve solute atoms, and this ability varies with temperature. Chicken soup is made by dissolving fats and flavoring ingredients in hot water. When the soup is hot, everything is in solution. When the soup cools to room temperature, the fat will often come out of solution and form a layer of fat-rich phase on top of the liquid. This same sort of thing happens in metals. If a host metal cannot accommodate all of an added solute in solution, a second solute-rich phase will usually form. The phase diagram is a graph with an ordinate of temperature and an abscissa of percentage of alloying element. The graph shows the boundaries of the regions where various phases are present in a *binary* alloy, one with two components, A and B.

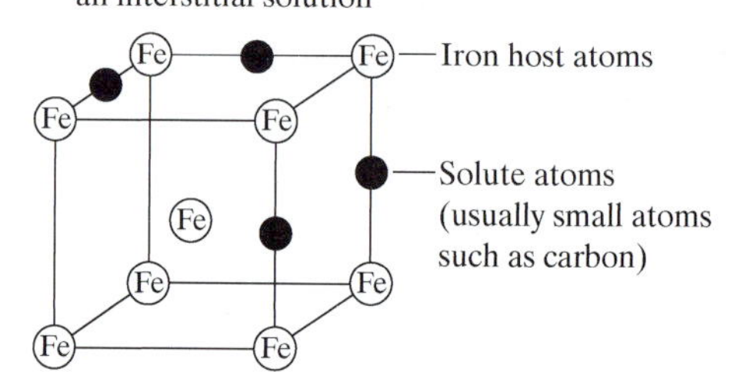

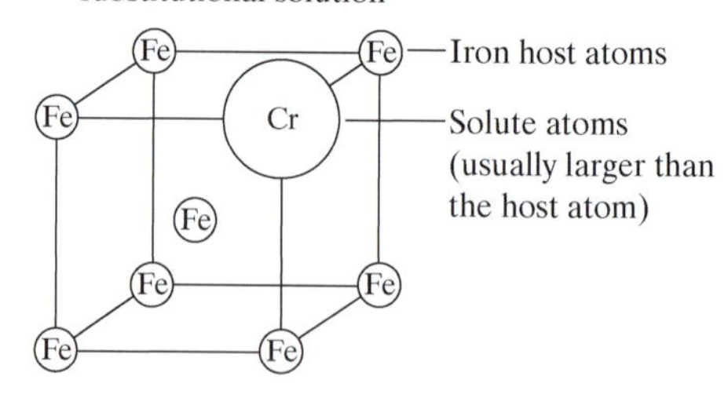

Figure 10–1
Interstitial and substitutional solid solution

If there were such a thing as a pure metal—that is, no foreign elements in it at all—it would have a single melting point. Unfortunately, there are no pure metals because of our limited ability to refine metals to high purity. Thus most metals have other elements in them, and as the concentration of these other elements increases, the resultant alloy may not have a single melting point. Rather, it will have a melting temperature range. In this temperature zone, the metal will be part solid, part liquid—similar to sleet, a mixture of ice and water. A phase diagram of a metal with a single alloying element that is completely soluble through all ranges of composition would appear as in Figure 10–2.

The grain structure of the solid solution of A and B results from the formation of separate crystallites on solidification. The grains form from tiny treelike structures, *dendrites*. If the two metals A and B are not completely soluble through all ranges in composition, second phases will form at grain boundaries. The equilibrium phase diagram for an alloy of the two elements A and B that exhibits partial solubility is shown in Figure 10–3. Solubility of metals and other elements depends on the nature of the host element and the nature of the intended solute. For example, elements with close atomic diameters tend to show solid solubility.

A review of the preceding equilibrium phase diagrams illustrates one additional thought-provoking concept: a composition with a single melting temperature instead of a mush range on solidification. A metal with this composition is called a *eutectic* or easy-melting alloy, and it has a

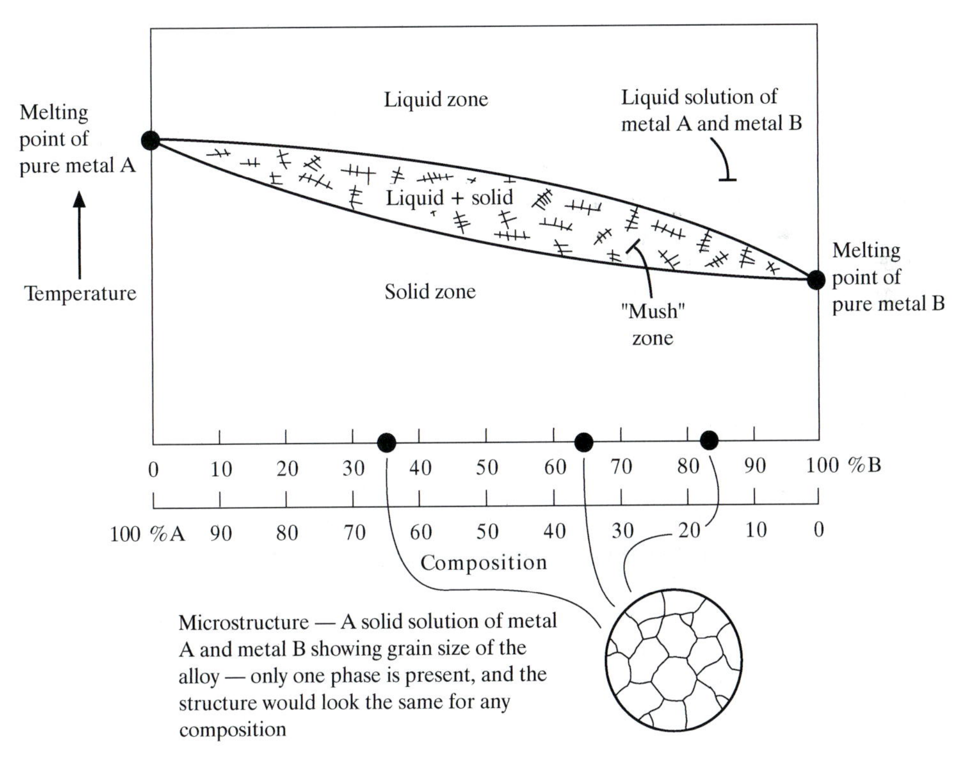

Figure 10–2
Phase diagram of completely soluble metals

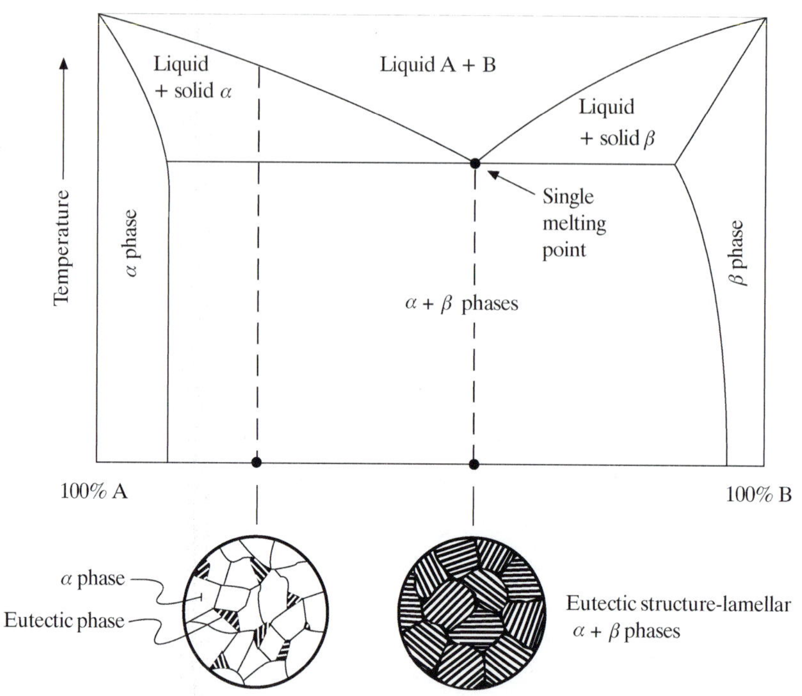

Figure 10–3
Phase diagram of partial solid solubility

single melting point, just as a perfectly pure metal would have. Eutectic alloys typically have a lamellar structure of the two phases that predominate in that alloy. The pearlite component of iron–carbon alloys is similar in concept except that, instead of a transformation from a liquid to a solid at a single temperature, there is a transformation from one type of solid to another type of solid. The composition of iron and carbon that coincides with this point is called the *eutectoid* composition. Pearlite is the resultant structure. One final concept to be mentioned in analyzing phase diagrams is the formation of *intermetallic compounds*. In some metal alloy systems, two metals may chemically combine to form a chemical compound with a definite ratio of one element to the other. A compound that forms in most steels is Fe_3C, *iron carbide*. A phase diagram for an alloy system that forms an intermetallic compound is illustrated in Figure 10–4. The intermetallic compound (A_xB_y) would be a phase, as would the A-rich phase (α) and the B-rich phase (β).

We have defined equilibrium phase diagrams as graphs showing phase relationships in metal alloys. The simple ones illustrated are for alloys formed from two elements (*binary*). There are diagrams for alloys containing three major elements (*ternary*) (Figure 10–5), and even four major elements (*quaternary*). Ceramics systems have similar phase diagrams. How are these diagrams obtained and what does one do with them?

Phase diagrams are obtained by formulating an alloy of pure A and, for example, 1% B, slow cooling it, for example, 100° below its melting temperature, and then quenching it to retain the structure that exists at this temperature. The structure of the solid formed at this point is analyzed metallographically and crystallographically, and the data are plotted. The procedure is repeated until room temperature is reached. A new

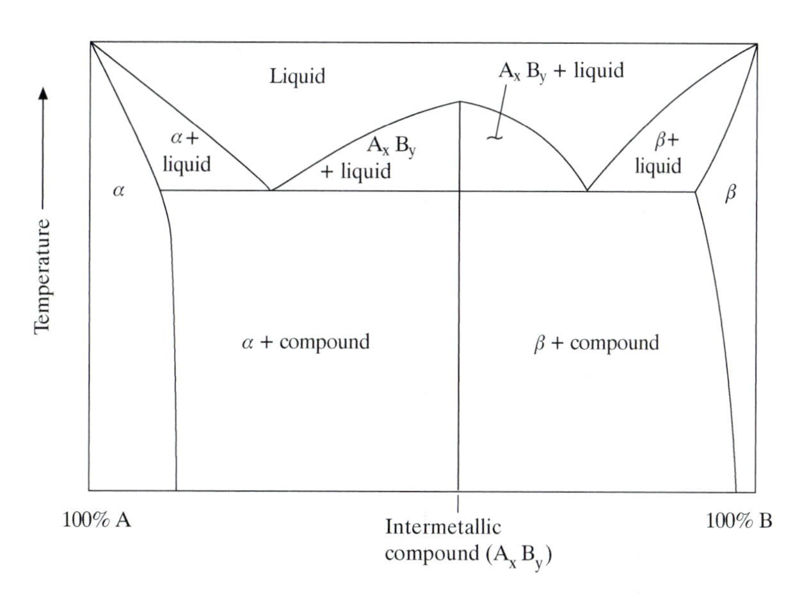

Figure 10–4
Phase diagram with intermetallic compound A_xB_y

alloy is then formulated with, for example, 2% B in pure A, and the procedure is repeated again and again until the diagram shows all the phase relationships existing from pure A to pure B.

Addressing the latter part of our question, phase diagrams can be used to calculate the percentage of various phases present with a given alloy at a given temperature. To illustrate how this works, we put some numerical values on the hypothetical phase diagram in Figure 10–3. This diagram is redrawn as Figure 10–6.

If we want to know the relative fractions of phases that are present in an alloy of 12% B at a temperature of 750°, we start by drawing a *tie line (isotherm)* at the temperature of interest, 750°. This line, *abc*, represents the total alloy at this temperature, and the phase diagram tells us that there are two phases present, α and liquid. To determine the relative percentage of each present, we establish a fulcrum at the composition of interest, 12%, and we ratio the tie line. The fraction of α that is present is

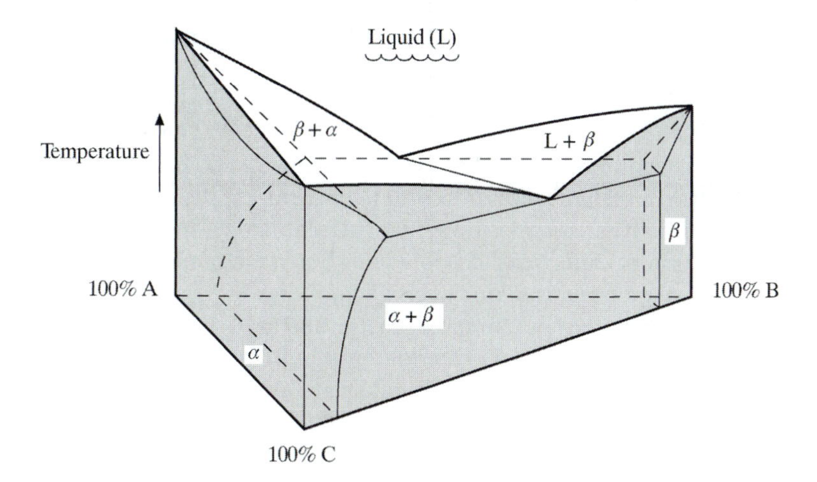

Figure 10–5
Schematic of ternary phase diagram with three major constituents, elements A, B, and C

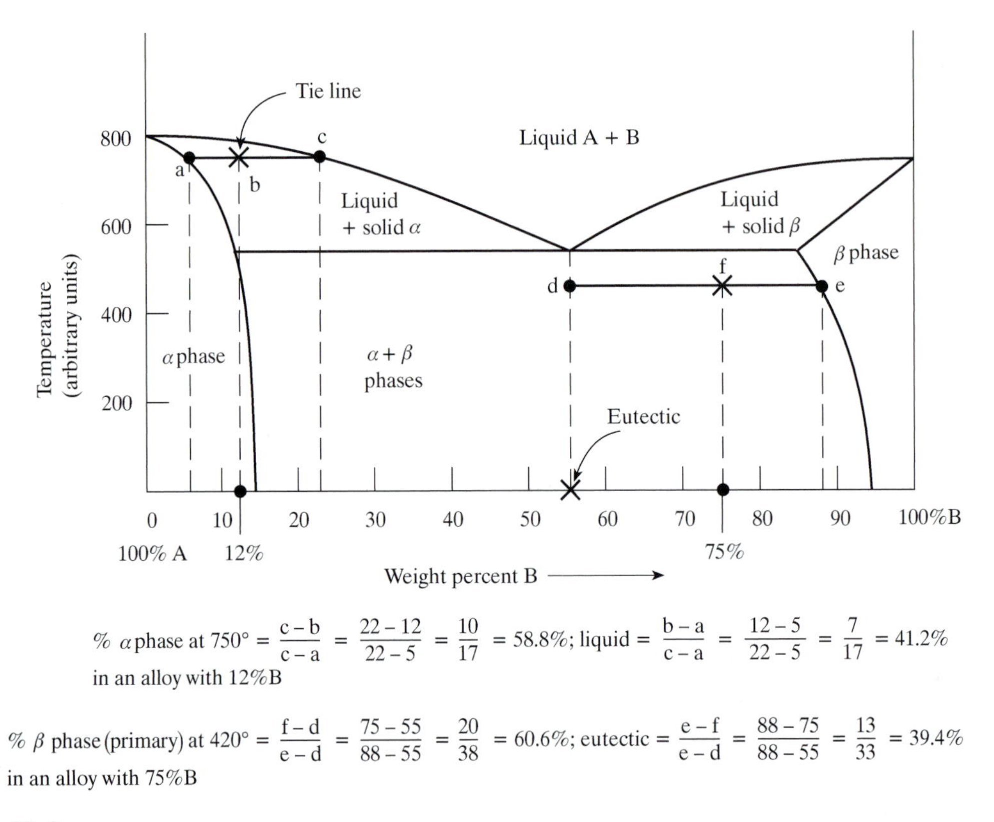

Figure 10–6
Use of tie lines on phase diagrams to calculate relative percentages of phases that are present at a particular temperature

the length of the line to the right of the fulcrum, *cb*, divided by the length of the entire line, *ca*. The percentage of liquid present is the length of the tie line to the left of the fulcrum, *ba*, divided by the length of the entire line, *ca*. In the second example in Figure 10–6, we check the phases present at another temperature and another composition. We calculated the fraction of eutectic and β present, but we could have extended the tie line to the α phase line and calculated the percentage of α and β instead of the fractions of β and eutectic.

These phase relationship calculations are called *lever law* calculations, and this technique applies to all metal systems and alloys that have multiple phases present. These diagrams and lever law calculations are an aid to the metallurgist in predicting *microstructures* that should be present in a given alloy under equilibrium conditions. They are also used to determine if one metal or element is soluble in another. This information is valuable in welding and in formulating alloys. For example, phase diagrams show that very few metals have solid solubility in titanium. This means that titanium cannot be welded to any metal that shows low solubility. Essentially, it can only be welded to itself. Metallurgists use phase diagrams to select materials for new alloys. A material must have significant solubility to be a candidate. Finally, phase diagrams can help predict the long-term stability of an alloy. If a phase diagram shows that a particular composition is not a stable *solid solution*, structure changes can occur with time.

10.2 Morphology of Steel

We previously defined a steel as an alloy of carbon and iron, when carbon is within the approximate limits of 0.06% to 2.0%. The carbon can be in solid solution in iron with a body-centered cubic crystal structure, it can be in solid solution in face-centered cubic iron, or it can be present in the form of a compound with the stoichiometric composition of Fe_3C. This phase is called *cementite*. These are the principal stable phases that can be present in steels; they are summarized in Table 10–1. The relative amounts of these phases present in a particular steel alloy have an effect on most properties. The iron–carbon diagram (Figure 10–7) shows the presence of these phases at different temperatures and at various carbon contents. The microstructure of carbon steels at room temperature is usually a combination of these phases. The phase diagram shows that the structure should consist of a combination of *ferrite* and cementite. However, cementite is largely present in lamellar form, alternate layers of ferrite and cementite. This microstructure is called *pearlite*. Figure 10–8 illustrates the microstructures of steels with varying carbon content. Figure 10–9 shows photomicrographs of microstructures. Thus the microstructures of steels at room temperature can range from almost completely ferrite, to all pearlite, to pearlite plus free cementite. These are the structures that exist under equilibrium conditions. An annealed structure can be considered a structure at equilibrium.

In practice, steels and irons are not usually heat treated under equilibrium conditions. This means that phase diagrams are only used for slow heating and slow cooling, more specifically to determine heating temperatures for hardening and softening.

What happens when a steel is heated or cooled under nonequilibrium conditions? On heating, the phase diagram applies with negligible modification. On cooling, the result is completely different. Nonequilibrium cooling of iron–carbon alloys is used to produce hardening and strengthening. In our discussion in Chapter 1 on the nature of metals, we attributed the ability of steel to harden to its ability to change its crystal structure from body-centered cubic to face-centered cubic at high temperatures (*allotropy*). The face-centered cubic structure can hold more carbon in solution than the body-centered cubic structure; and on rapid cooling, when the iron matrix wants to return to its equilibrium BCC structure, it cannot because of trapped carbon atoms. The net result is a distorted crystal structure called *body-centered tetragonal*. The phase made up of iron–carbon with a body-centered tetragonal structure is called *martensite*. The microstructural changes that occur in hardening a common steel (0.6% carbon) are illustrated in Figure 10–10.

Table 10–1
Principal stable phases of steel

Phase	Crystal Structure	Characteristics
Ferrite	BCC iron with carbon in solid solution	Soft, ductile, magnetic
Austenite	FCC iron with carbon in solid solution	Soft, moderate strength, nonmagnetic
Cementite	Compound of iron and carbon Fe_3C	Hard and brittle

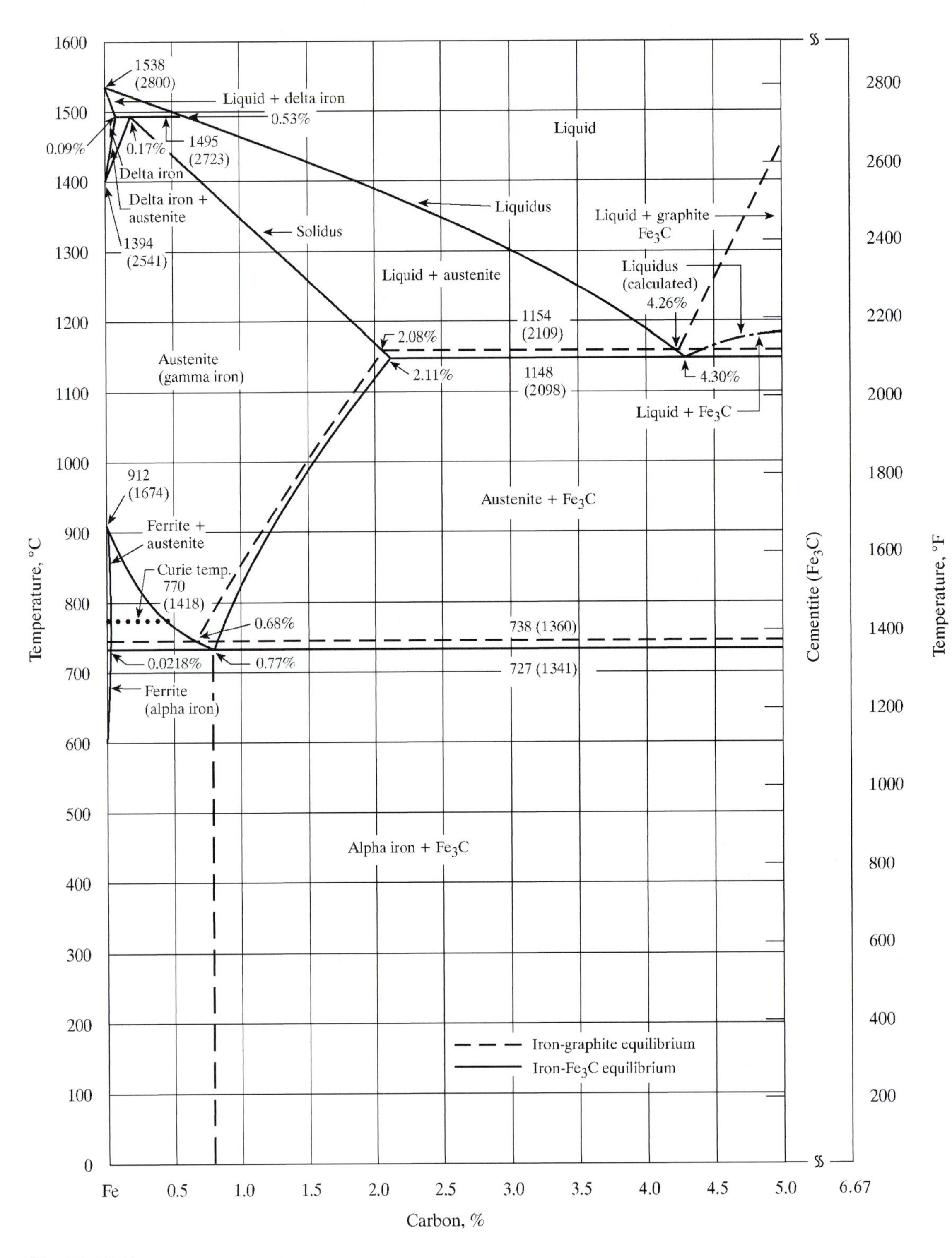

Figure 10–7
Iron–carbon equilibrium phase diagram

Source: *Metal Progress,*© ASM International, 1975

Note: This is only a portion of the iron–carbon equilibrium phase diagram. The far end of the abscissa is graphite (100% carbon). There is a cementite ordinate at 6.67% carbon.

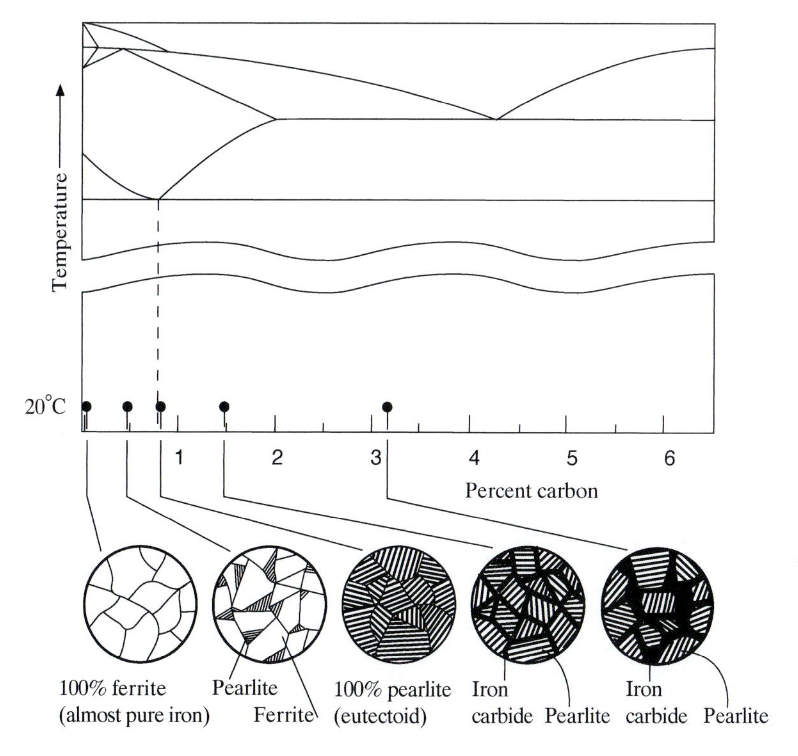

Figure 10–8
Equilibrium microstructure of iron–carbon alloys at room temperature (about ×500)

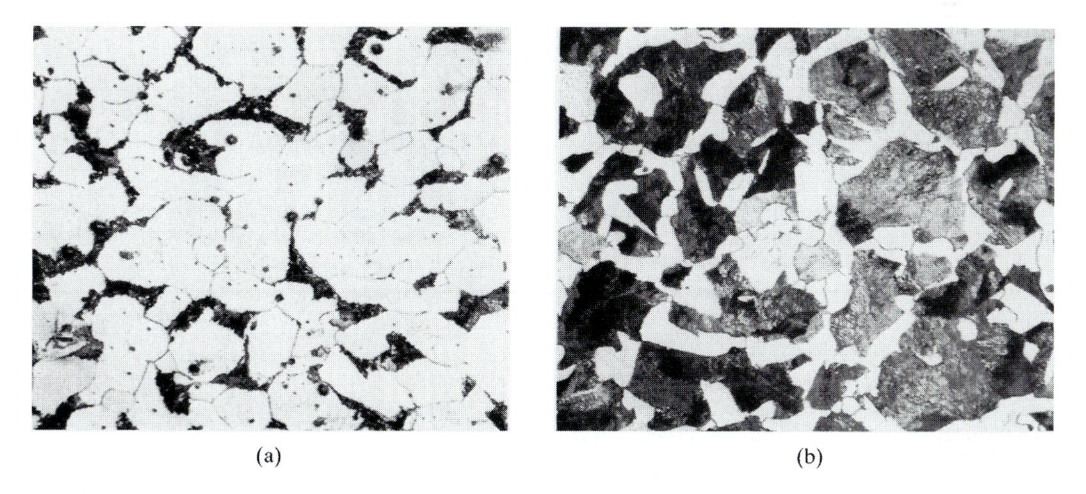

Figure 10–9
Microstructures of annealed carbon steel. (a) About 0.2% C. (b) About 0.6% C. The light areas are ferrite and the dark are pearlite (×400).

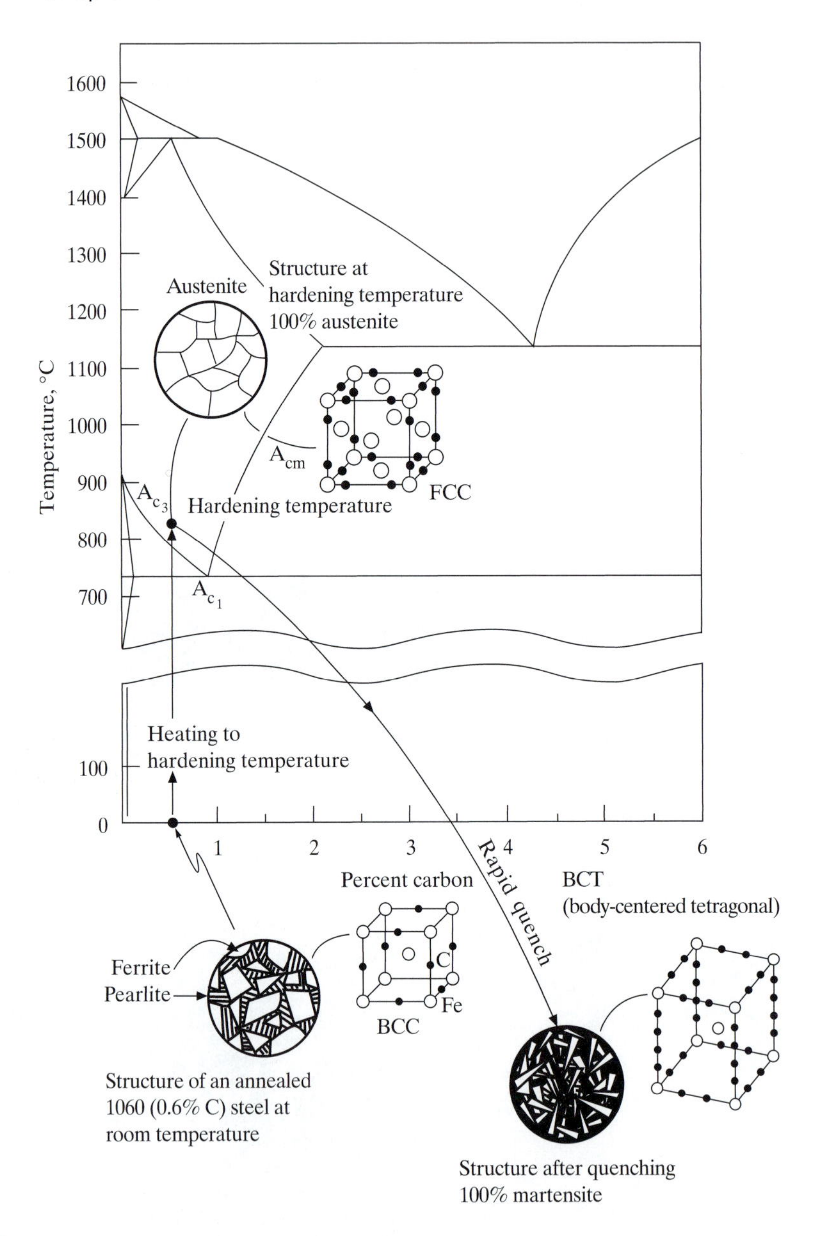

Figure 10–10
Determination of hardening temperature for a 0.6% carbon steel

Now suppose that it is desired to harden a very low-carbon steel, one with 0.01% carbon. Would *quench hardening* be possible? The answer is no. To obtain a fully hardened structure of 100% martensite, in all but thin sections, plain carbon steels with no alloy additions other than carbon require a carbon content of about 0.6% carbon to obtain 60 HRC. Thus we can formulate some basic requirements for hardening carbon steels:

1. Heating to the austenitic temperature range.
2. Sufficient carbon content.
3. A rapid quench to prevent formation of equilibrium products.

Let us summarize our discussion of the morphology of steel by saying that steels are such useful metals because they have the ability to change their crystal structure from something soft and machinable to something very hard and strong. The equilibrium phase diagrams are used to determine hardening temperature and annealed microstructures. Most steels and irons can be hardened if they have sufficient carbon content and are subjected to the proper thermal cycles. Steels that are heat treated in different ways will have different microstructures (Figure 10–11) and properties.

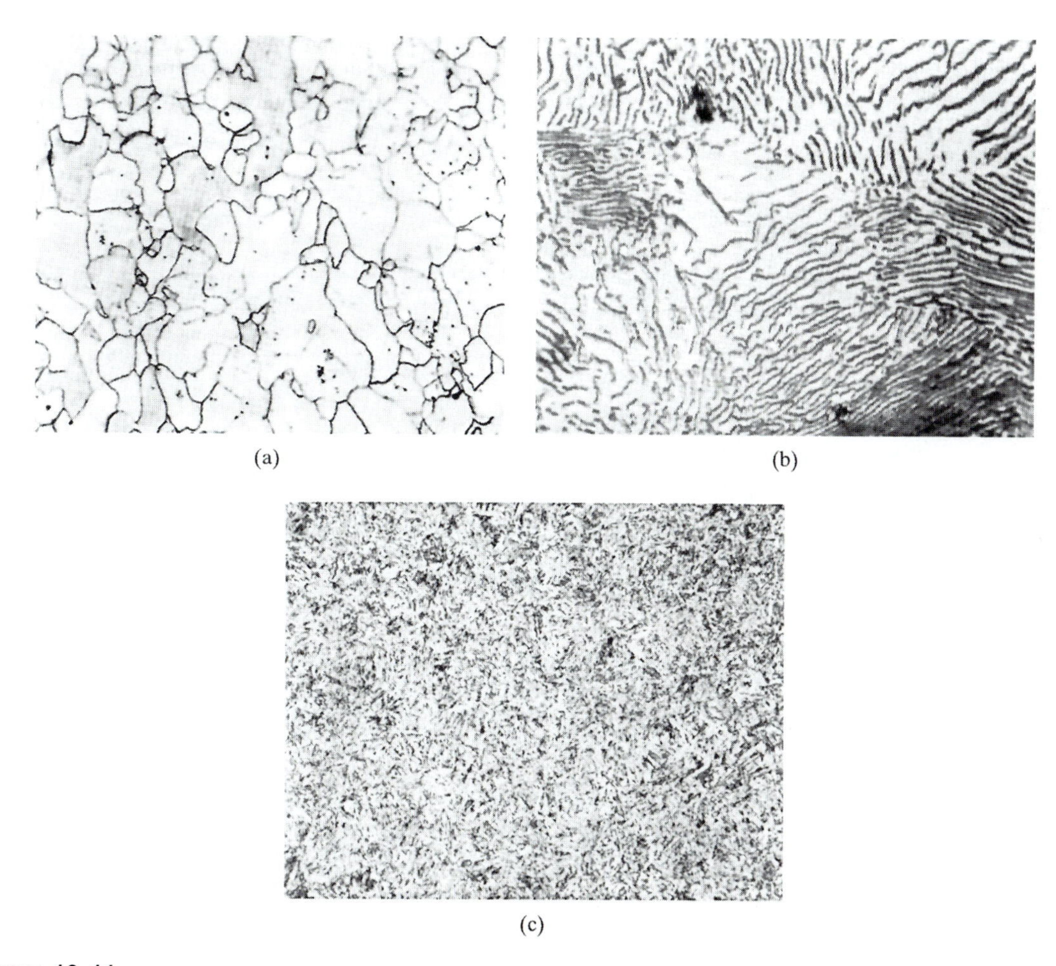

(a) (b) (c)

Figure 10–11
Microstructures of carbon steel. (a) Ferrite: pure iron and low-carbon steel (×100). Austenite has about the same microscopic appearance. (b) Pearlite: an annealed 0.8% C steel is 100% pearlite. The dark lines are Fe_3C; the light areas are ferrite (×1330). (c) Martensite: quench-hardened steels (×500).

10.3 Reasons for Heat Treating

Now that we have established what goes on inside a piece of steel during heat treatment, let us explore the reasons why heat treatments are necessary.

Figure 10–12 arranges the more common steel heat treatments into three major categories. These categories form the basic reasons for heat treating steels. There are a variety of subcategories and many specific heat treating processes. Each process has a feature that sets it apart from the others. All of these processes are commercially available, and many heat treat shops can do most of them. We will discuss the basic reasons for heat treating, and subsequent sections will go into detail on specific processes.

Hardening

Steels with enough carbon and alloy content will direct harden. They can be heated to the austenitizing temperature and quenched to form hard martensite. There is a family of processes that can be used for direct hardening; they will be described later.

If a steel does not have sufficient carbon and alloy content to allow direct hardening, another

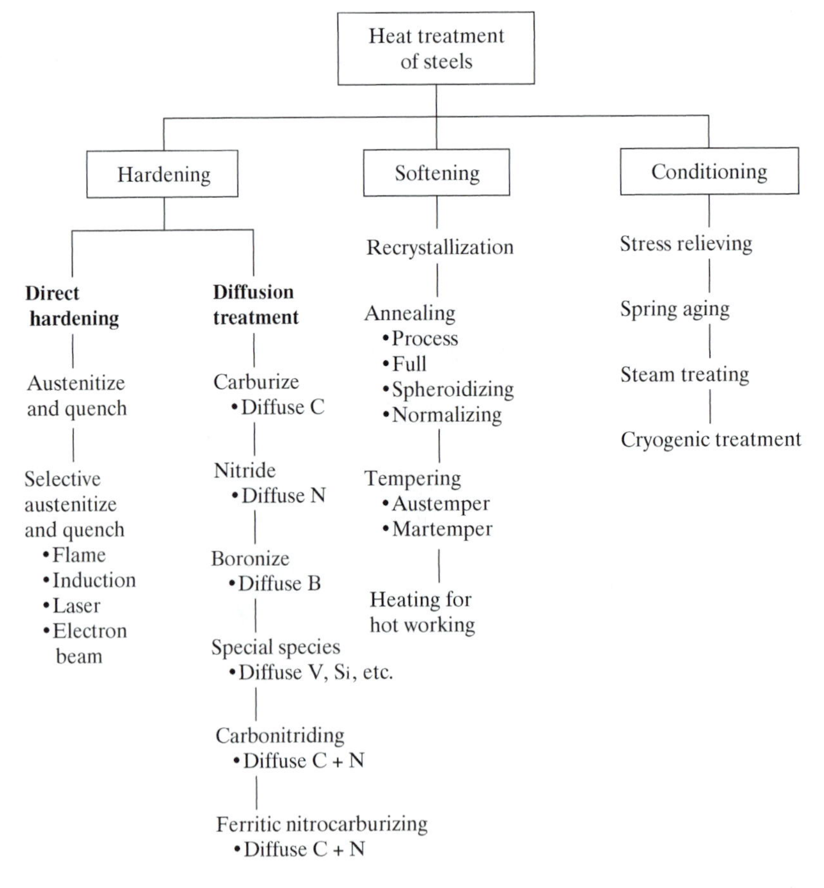

Figure 10–12
Spectrum of heat treatments used on ferrous metals

family of hardening processes applies, diffusion treatments. These processes add chemical species (elements) to the steel so that the surface will harden. They may or may not require a quench to complete the hardening.

Hardening of steels is done for most applications that require the steel to be used as a tool to work other materials (drill, forming tool, saw blade, etc.). Hardening may also be required for an application for which high strength is required—the springs on an automobile, for example, for which high yield strength and toughness are requirements. The stress on a spring cannot exceed the yield strength or the spring will change shape and lower the car. High strength is also needed to prevent creep in this application. Steel springs are usually direct hardened and double tempered to 47 HRC.

Steels are sometimes hardened just to produce dimensional invariance. Gage blocks and other inspection gages need to be very hard (60–65 HRC) if they are to resist wear and scratches that could make dimensions change. Some stainless steels (the martensitic grades) need to be hardened to obtain their maximum corrosion resistance. This will be discussed in more detail in the chapters on stainless steels and corrosion, but quench hardening keeps the chromium in solution in the host iron, and it is this chromium in solid solution that imparts the "stainless" characteristics. Sometimes steels are hardened to improve a physical property such as coefficient of thermal expansion or some magnetic property. Steels used for permanent magnets are usually hardened to make the magnetism last longer. In summary, there are many reasons for hardening steel; examples are everywhere. The important point to remember is that steels are unique in the degree of hardening and strengthening that is possible through heat treatments.

Softening

When steel products are made by cold-rolling or drawing, they work harden and become brittle. If a steel mill is trying to cold-roll 1-mm-thick steel from 10-mm-thick hot band, when they reduce the thickness to 5 mm, the steel may become too hard to roll further without cracking. What to do? The steel is softened by annealing. Annealing is the process of slowly raising the temperature of the steel to the point at which it transforms to austenite (above the Ac_3) and slowly cooling from this temperature (furnace cooling) after a soak to ensure equilibrium at temperature. This is usually done by batch annealing of coils of steel, but some mills have the capability of continuously annealing steel to allow further thickness reductions.

Steels that are as-quench-hardened are often too brittle to use. They are like glass until they are slightly softened by tempering. Tempering is a low-temperature heat treatment that imparts toughness without significant reduction in hardness. Often the hardened part will lose only 1 or 2 Rockwell C hardness points, but the toughness may increase by a factor of 10. Tempering really is an integral part of quench hardening. It is done immediately after the part is cool enough to touch (hand warm) but not stone cold. It softens the hardened structure a slight or significant amount, depending on the desired effect. The softening mechanism differs with tempering temperature, but softening usually starts by diffusion of carbon from the strained BCT structure and formation of iron carbide. A tool that has an as-quenched hardness of 62 HRC may be tempered back to 60 HRC for use. On the other hand, a spring that quenches out at 60 HRC will be tempered (usually double tempered) back to 47 HRC. A structural component such as a high-strength fastener may be tempered back to 36 HRC. Tempering is a variable softening treatment used to customize steel properties (hardness, strength, and toughness) to match an application.

Annealing is performed on hardened steels, but usually only when something goes wrong. If a tool is machined and hardened, and it is discovered that a needed feature has the wrong dimension, it is possible to anneal the part, remachine it, and reharden it. However, this is seldom done because the part will probably change size in the annealing and rehardening process. It is more common to anneal parts to address unexpected heat treating results. If a part has been overtempered, it will not meet the drawing hardness specification. It is possible to anneal it and reharden it. If the part has grind stock on critical dimensions, size change can be dealt with if it occurs.

In summary, steels are softened to improve their malleability for steel mill shaping processes. Steel parts are softened by tempering after quench hardening to improve toughness, and they are softened by annealing to allow rework or welding. There are different tempering and annealing processes. Most of these will be described in more detail in the following sections.

Conditioning

The term *conditioning* is not a standard heat treating term. It is used in this discussion to collect heat treating processes that may have very different purposes but that all involve temperature cycling of steels to alter properties. We will discuss some of these processes in detail because they are important in design engineering. Some of the steel conditioning processes, however, are very specialized, and it is probably adequate to just define them. Designers need to know that they are available. The following are some of the less-used conditioning processes.

Spring aging: When extension or compression springs are wound from high-carbon steel wire, or even from nonferrous wires, they may unwind or otherwise change shape with time. This time-dependent movement is due to anelastic behavior, *recoverable strain*. If the angle changes on the end of a wound spring, the spring may become useless. A simple conditioning treatment for 2 h at 600 to 700°F (315–370°C) will usually remove anelastic behavior. If the spring will change shape with time, this process will make it happen now. New springs can be made to accommodate the winding or unwinding that occurs in the aging treatment. This heat treatment is recommended on any springs and formed parts that need good dimensional accuracy in the formed shape.

Normalizing: The average steel user will probably never specify this process, because it is usually done only at steel mills and foundries. *Normalizing* is the process of heating a steel to the fully austenite region (above the Ac_3 or Acm in Figure 10–10), soaking at this temperature, and air cooling to room temperature. Normalizing is usually performed on hot-worked shapes to make properties and alloy additives uniform in distribution. Castings often locally cool at different rates, and this results in alloy differences or grain size differences in the same part. Normalizing can be used to make both grain size and alloy distribution more uniform. Hot-rolled bars are often normalized. The mechanical properties of normalized steel are better than the properties of annealed shapes, because the air cooling in normalizing is faster than the furnace cooling used in annealing. Faster cooling produces finer microstructural features and higher hardness than annealing. Normalizing may be specified by a designer to make the properties of castings uniform, but this process is not usually specified on steel used to make machine parts. However, when a designer orders, for example, a 100-mm-diameter bar of carbon steel from a steel supplier, it will probably be delivered in the normalized condition rather than the annealed condition. This is not usually a problem, but designers must be aware that normalized steel properties differ from annealed steel properties, and in some design situations this can be important.

Steam treating: The familiar blue-black appearance of guns can be produced by tempering steel in a steam in a temperature range of 650 to 1200°F (340 to 650°C). The exact temperature depends on the alloy and the desired appearance. The blackening is produced by a controlled oxidation of the steel surface. The oxide layer is usually less than 1 μm deep, but it provides a modicum of atmospheric corrosion resistance when saturated with oil. It is widely used on fasteners, drills, and other metal cutting tools. Its role in this capacity is to help prevent chips from welding to the tool surface. There are chemical dips, termed black oxide treatments, that are applied at temperatures below 500°F (260°C) that produce the same blackening effect. Most heat treating shops offer one or both of these processes. They are recommended for carbon and low-alloy steel (chromium less than a few percent). Steam treatment is applicable to more alloys. It is widely used for powder metal parts. The steam penetrates the porosity and reduces it to some degree. Steam tempering should only be used on hardenable steel when the steam-treating temperature is compatible with the required tempering temperature for the part.

Stress relieving: Design engineers are more likely to use this process than annealing or normalizing. Complex machine parts often require stress relieving to achieve dimensional stability. It is usually the designer's responsibility to make a decision about whether stress relieving is needed or not. There are no established rules to base this decision on, but some common examples for which stress relieving may be necessary are as follows:

1. Weldments that require machining of weld deposits.
2. Machining of cold-finished shapes.
3. Castings that require significant machining.
4. Parts with extremely close dimensional tolerances.
5. Long, slender parts machined from heavier shapes.

The stress-relieving process does not involve heating to the critical transformation temperature as required in full annealing and normalizing. Stress-relieving temperatures are usually 100 to 200°F (38 to 93°C) below the transformation temperature. Occasionally, fabrication shops stress relieve parts at temperatures well below 1000°F (538°C). Stress relieving at these temperatures is not usually effective. The mechanism of stress relieving is thermally induced dislocation and defect movement to remove internal strains; in the case of cold-worked materials, recrystallization usually also occurs. Both of these processes require significant energy. As shown in Figure 10–13, the strength of a common construction steel does not change significantly until temperatures reach about 1000°F (538°C). Stress relieving at 600 or 800°F (315 or 426°C) may remove only 10% or 20% of the internal strain in a metal. Heating to the range of 1000 to 1200°F (538 to 648°C) allows removal of up to 90% of all internal strains in carbon steels. Alloy steels may require even higher temperatures.

The cooling rate from the stress-relieving temperature is not usually too critical, but uniform cooling in still air is preferred. The correct soak time at the stress-relieving temperature is a subject of many different opinions, but if the part is at temperature (as measured by a thermocouple on the part) for at least 1 h, this is usually adequate.

Cryogenic treatment: *Deep freezing* is sometimes used to ensure freedom from retained austenite. On quenching a hardenable steel, martensite will start to form at a temperature of around 500°F (260°C). With some steels the transformation may be 100% complete at 200°F (100°C); with other steels the transformation

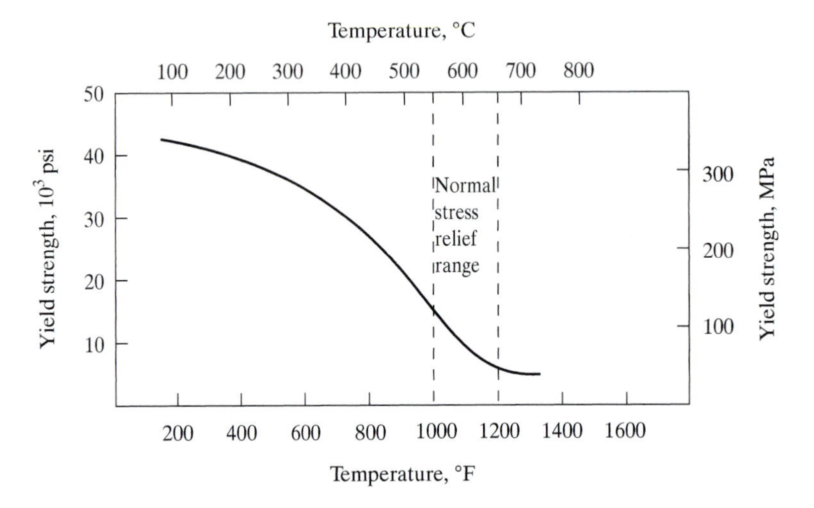

Figure 10–13
Effect of temperature on the yield strength of 1020 steel

may not be complete until a temperature of −50°F (−46°C) is reached. Deep freezing after an oil or air quench is sometimes used to ensure that the structure is 100% martensite. Excessive retained austenite (2% to 10% or more) can reduce the as-quenched hardness as well as affect dimensional stability (the part dimensions can change with time). Certain steels with high alloy or carbon content favor the retention of austenite. Some specialty heat treating firms claim exceptional properties (for a variety of metals) when a deep-freeze treatment of −300°F (−165°C) is used. Most heat treaters use −100°F as a safe temperature to prevent retained austenite. Parts are given a low-temperature [300°F (148°C)] temper before deep freezing to minimize cracking, and they are in the normal manner after deep freezing.

10.4 Direct Hardening

Austenitizing

From the mechanistic standpoint, hardening of a steel requires a change in crystal structure from the body-centered cubic form present at room temperature to face-centered cubic. The iron–carbon equilibrium diagram tells us what temperature we must heat to with a particular carbon content to get the FCC structure. We must heat into the *austenite* or austenite +Fe_3C region. Figure 10–14 is a schematic of the iron–carbon diagram indicating the hardening temperature range for carbon steels. The hardening temperature is close to the transformation temperature to minimize grain coarsening.

Our stated requirements for hardening involve a quenching action to cause carbon to be trapped in the crystal structure. Quenching is normally accomplished by rapidly removing the part from the furnace (after it has soaked for sufficient time to reach the required temperature) and immersing it in agitated oil or water. Some steels can be hardened just by removing the part from the furnace and letting it cool by convection in room temperature air. The ease with which a steel will transform to hardened structure on quenching is called *hardenability*, and it can vary significantly in steels when alloy additions are made. The quenching requirements for a particular steel can be determined by studying diagrams that show phase changes

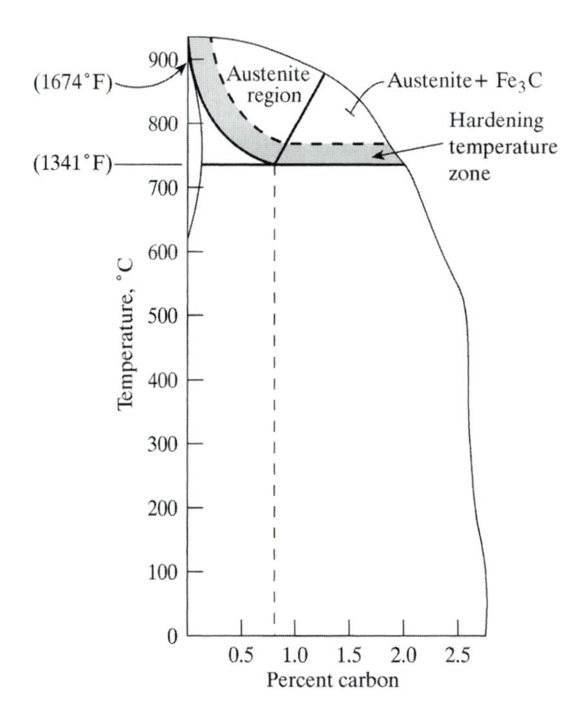

Figure 10–14
Hardening temperature range shown on the iron–carbon diagram

under nonequilibrium conditions. The *isothermal transformation* (IT) *diagram,* which is also called the *time-temperature transformation* (TTT) *diagram,* is a typical tool used by heat treaters to predict quenching reactions in steels. The *IT diagram* is obtained by austenitizing steel specimens of a particular composition and quenching them to various subcritical temperatures, holding them at that point for varying time intervals, then quenching them to room temperature and metallographically analyzing their structures. This procedure is repeated at approximately ten subcritical temperatures until a complete IT diagram is obtained. This procedure may involve 100 structure determinations on samples. It is a tedious process. Fortunately, this work has been done for most of the carbon and alloy steels used by machine designers. Atlases of IT diagrams are readily available. Figure 10–15 is a simplified version of the IT diagram for 0.8% carbon steel.

Another type of diagram used for the same purposes as the IT diagram is called a *continuous cooling transformation* (CT) *diagram.* It is obtained by experimental determination of structure changes by continuous cooling of test samples at various cooling rates to various subcritical temperatures. The end diagrams look similar to the IT diagrams, but continuous cooling usually shifts the beginning of austenite transformation to lower temperatures and to longer times. One advantage of the CT diagrams over IT diagrams is that they often have various cooling rates superimposed on the structure change diagrams. With these data, we can compare a cooling rate on an actual part with cooling rates on the *CT diagram* and thus predict structure changes accurately.

Figure 10–15 was obtained by isothermal transformation experiments, but a continuous cooling curve for a hypothetical part has been superimposed to illustrate how both types of diagrams are used by heat treaters. The dashed line in Figure 10–15 could be obtained by embedding a thermocouple in a part to be heat treated. The part is austenitized at about 1400°F (760°C), and it cools to room temperature in about 2 s. As shown by the IT diagram, this cooling rate was sufficient to form martensite. The part would harden. If the cooling curve of the part was moved to the right of the "nose" of the soft region (austenite plus pearlite), the resultant structure of the part would consist of hard plus soft structure—or entirely soft structure if the curve were shifted far enough to the right.

Heat treaters usually do not monitor the temperature of parts to be hardened with thermocouples, nor do they establish cooling curves superimposed on IT or CT diagrams unless requested to do so by the customer. In most cases a heat treater would look at Figure 10–15 and note the time location of the nose of the

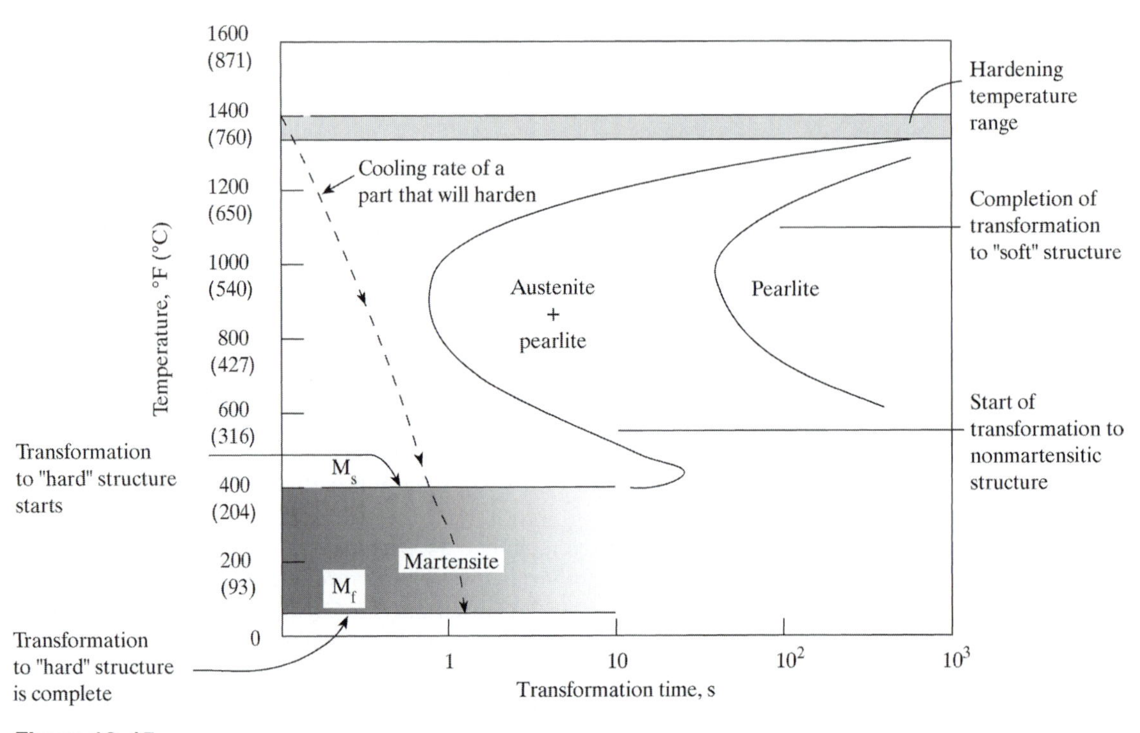

Figure 10–15
Typical time–temperature transformation diagram (IT)

transformation curve. The diagram illustrated tells the heat treater that only about 1 s is allowed to cool the part from 1400°F (760°C) to about 600°F (316°C). He or she knows from experience that the only way to get a cooling rate this fast is to use a water quench. If, for example, we wished to harden a piece of steel that was a cube with a side dimension of 4 in. (100 mm), it would be impossible to get the center of the cube cooled fast enough to meet the 1-s cooling requirement. In this case the surface may harden, but the core of the cube would undoubtedly contain soft transformation products. This is sometimes done intentionally. Heavy parts made from steels with low hardenability are quenched in such a manner that the surface gets hard and the core stays soft. The hard surface provides wear resistance and the soft core provides toughness.

Many other aspects of IT and CT diagrams are used to help heat treaters and metallurgists to harden steels, but designers will probably never have to use these diagrams for anything other than comparing the hardenability of steels. If a designer wants to know if one steel has better hardenability than another steel, he or she has only to compare the time location of the nose on the respective IT or CT curves. If the nose of one is at 10 s and the other is at 30 s, the latter will have better hardenability. It is easier to meet quenching requirements on the latter.

Quenching If a TTT diagram indicates, for example, that a steel must be quenched within 2 s to get it to harden, it may be necessary to use a severe quench. The rate of *quenching* of a part from a heat treating cycle depends on the fluid media used for the quench and the degree of agitation in the media. Water quenches are the most severe, followed by oil, molten salt, and gas quenching. There are many possible quench rates within each of these categories. For example, if 10% sodium hydroxide is mixed in water, it will have a cooling rate that can be twice that of room temperature water. A similar effect is achieved by adding salt to the water (brine quench). Violent agitation improves these quenches even further. Water-soluble oils or polymers can be added to water quenches to make them slower than plain water, or the water can be heated. Oil quenches usually have less than half the quench rate of room temperature water. A molten salt at 500°F (260°C) would be slower than both polymer–water and oil, and a still air or inert gas quench would be even slower than all these. The reason for the large differences in quenching rates that are possible with the liquid quenches is the heat transfer film that forms around the red-hot part on quenching. The addition of salts (brine), polymers, oils, and caustics to water drastically changes the steam cloud that forms around the part. For example, quenching a red-hot part in liquid nitrogen at −350°F (−195°C) will be a slower quench than allowing the part to cool in room temperature air. A vapor cloud will form in the nitrogen, and the part will be insulated.

The slower the quench, the lower the chance of quenching distortion. On the other hand, if the quench is too slow, it will not be possible to get the desired hardness. Part of the skill of the heat treating professional is being correct on the choice of quenching media. Designers usually do not have to concern themselves with quenching media. There is a recommended quenching medium for most alloy and tool steels, but designers should be aware that if problems arise with part distortion or hardness, a possible cause of the problem may be improper quenching media.

Selective Hardening

It was previously pointed out that ferrous materials need sufficient carbon content if they are to respond to quench hardening. In plain carbon steels this level is considered to be about 0.6% for a 100% martensite through moderate sections. If rapid quenching techniques are employed or if the part is thin, this minimum carbon level can be lower, possibly as low as 0.4%. If suitable alloy additions are made, this carbon requirement can be still lower. However, high alloys (some tool steels) may require a soak at the austenitizing temperature to put alloy additions in solution. These alloys are less suitable for the fast *selective hardening* processes such as laser that do not provide a "soak" at the austenitizing temperature. In any case, if a steel has sufficient hardenability, that is, adequate carbon content or carbon plus alloy content, it will respond to selective quench hardening. The transformation mechanism will be the same as in through hardening, and equilibrium diagrams can be applied to determine austenitizing temperatures. The only thing that is different is that the two other requirements for hardening, heating to the austenitizing temperature and quenching, are applied only to selected areas. If only the blade of a knife is heated to the austenitizing temperature, only the blade will harden on quenching.

Flame Hardening *Flame hardening* is the process of selective hardening with a combustible gas flame as the source of heat for austenitizing. As mentioned previously, suitable materials for selective hardening with the flame-hardening process must have sufficient carbon content (usually 0.40%) to allow hardening. Because this process is normally performed on low-alloy or plain carbon steels with low hardenability, quenching after heating to the transformation temperature is usually accomplished with a rapid water quench. Quenching is almost instantaneous. The heating media may be oxygen–acetylene, oxygen–manufactured gas, propane, or any other combination of fuel gases that will allow reasonable heating rates. The hardening temperatures are the same as those required for furnace hardening. Typical flame-hardening systems are illustrated in Figure 10–16. Typical flame-hardened zones on machine components are illustrated in Figure 10–17. The depth of a flame-hardened zone can vary from 1/32 in. (0.8 mm) from the surface to all the way through a section.

The steels that are applicable to flame-hardening techniques include the plain carbon steels with carbon contents ranging from 0.40% to 0.95% and low-alloy steels. Occasionally, high-carbon alloys such as martensitic stainless steels and tool steels are subjected to selective-hardening techniques, but the medium-carbon steels are used more often. Steels with high hardenability have a greater tendency to crack during flame hardening. Cracking tendencies are reduced in alloy steels and tool steels by preheating the parts [to approximately 300°F (149°C)] before hardening and by using an oil- or water-soluble polymer–water quench. The fire hazards of oil quenching, however, preclude wide acceptance of this technique.

Cast steels of compositions similar to those mentioned can be flame hardened. However, it is more common to flame harden cast irons. The graphite structure of cast irons makes them natural candidates for lubricated machine

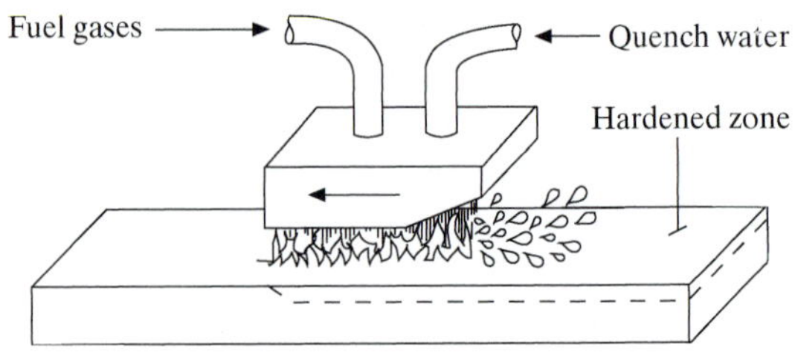

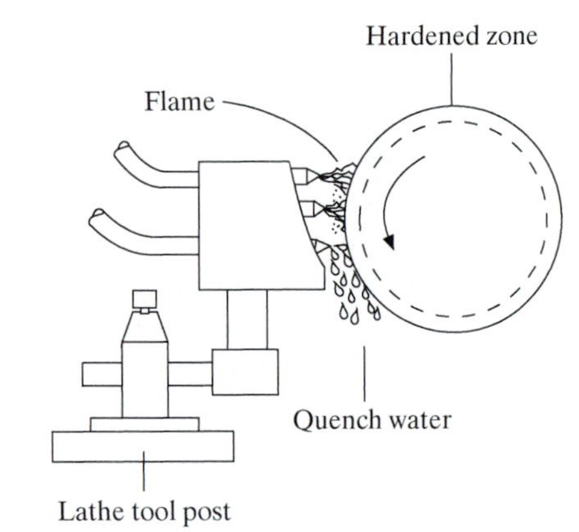

Figure 10–16
Typical flame-hardening systems

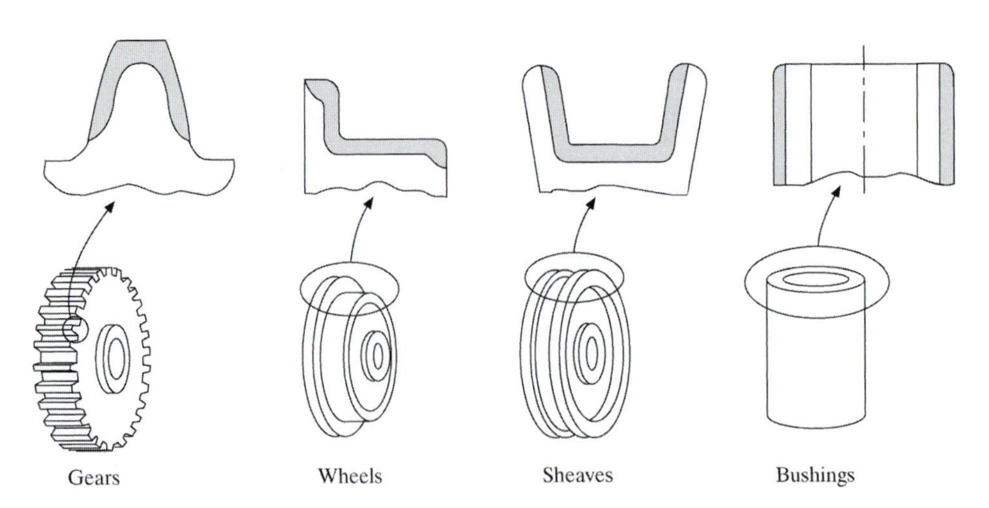

Figure 10–17
Flame-hardening profiles (white areas) for typical mechanical components

components, but in most cases they have relatively low hardnesses. This low hardness can be overcome by flame hardening. The material requirement for hardening cast iron is simply a combined carbon content in the approximate range of 0.40% to 0.80%. Usually, gray irons with moderate alloy contents are used, but the ductile irons are also applicable.

To summarize, flame hardening is a very useful process for selective hardening of plain carbon, alloy steels, and cast irons. A 0.45% carbon steel is probably the most popular flame-hardening steel. If cracking of this steel occurs on hardening, a steel with a lower carbon content can be used. If the as-hardened hardness is too low, a high-carbon-content material can be used. It is desirable to use alloys with the lowest hardenability that will still give the desired hardness. Cracking is by far the most important consideration in flame hardening. Experienced heat treaters have developed techniques to reduce cracking tendencies; do-it-yourself selective hardening with a welder's torch and a pail of water can be unsuccessful. It is best to employ experienced people for this operation. Flame-hardened parts must be tempered after hardening. The tempering temperatures used are determined by the alloy and desired hardness.

Induction Hardening The mechanism and purpose of *induction hardening* are the same as for flame hardening. The primary difference is the source of heat input. In induction hardening an electric current flow is induced in the workpiece to produce a heating action. Every electrical conductor carrying a current has a magnetic field surrounding the conductor. If we were to make a coil of wire and put an electrical conductor inside the coil, an electrical current flow in the coil wire would induce current flow in the core conductor through the magnetic flux lines associated with the outer coil.

Because the core conductor is a dead-end circuit, the induced current cannot flow anyplace; so the net effect is heating of the wire. If we take this system and use an alternating current in the coil, the direction of the current flow will be rapidly changing. The induced current in the core conductor is alternating at frequencies from 60 cycles per second to millions of cycles per second. Low frequencies tend to produce deep heating; high frequencies can produce "skin" heating. The resistance to current flow causes very rapid heating of the core material. Heating occurs from the outside inward. If, instead of a core wire as just cited, a steel round is placed inside the induction coil, induced currents will cause heating of the steel piece to its austenitizing temperature, and hardening can be obtained on quenching. A typical induction-hardening system is illustrated in Figure 10–18.

The system limitation for induction hardening is that the workpiece must be positioned to accept induced *eddy currents* from the induction coil. The coils are usually made of soft copper tubing (filled with running water to prevent

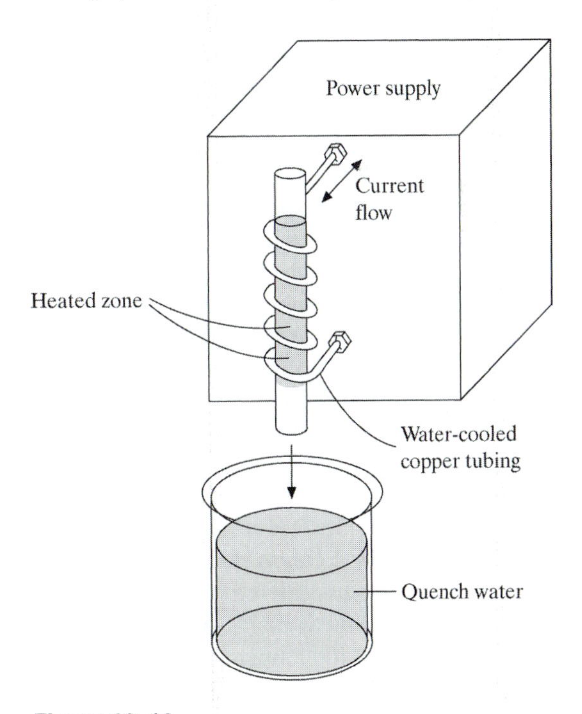

Figure 10–18
Induction hardening

heating), so it is very simple to fabricate coil shapes to fit specific parts. The current-carrying coil should be as close to the part as is practical. Usually, standoff distances are on the order of ⅛ in. (3 mm). The maximum clearance between the workpiece and the coil is a function of the power capacity of the unit, but in no instance is the clearance in the inches category.

A big advantage of this process over flame hardening is heating time. Typically, heating times for substantial part sizes [1 in. (25 mm) in diameter] are on the order of 10 s. On small parts, induction is also more controllable. Flame and quench heads are relatively large, and hardening of an area on, for example, a ¼-in.-diameter (6-mm) shaft may pose a physical manipulation problem. An induction coil could be readily adapted to heating a small shaft.

To summarize, induction hardening is a process of selective hardening using resistance to induced eddy currents as the source of heat. The steels that can be hardened with this process are the same as those used in flame hardening. A steel must be hardenable. Similar to flame heating, this process can also be used for process annealing, brazing, or tempering. The big advantage of this process over flame hardening is its speed and ability to confine heating on small parts. The major disadvantage over flame hardening is the large capital equipment expenditures required. In both types of selective hardening, flame and induction, a useful alloy is a 0.45% carbon steel. Depth of hardening can be as shallow as 0.01 in. (0.25 mm) or through the section. Normal tempering procedures are applied after quenching.

Laser and Electron Beam Hardening Lasers and electron beams (EBs) perform the same function as the flame in flame hardening or the induction coil in induction hardening. These techniques are applicable only to steels that have sufficient carbon and alloy content to allow quench hardening. The laser or electron beam is used to raise the surface temperature of the part to be hardened to above the critical transformation temperature, and the mass of the material provides self-quenching. Because *laser hardening* does not require the vacuum that EB does, auxiliary gas quenching can be used.

Lasers and electron beams are very different forms of energy, but the way that they are used in local hardening is the same. Both have the limitation of covering a relatively small area. Defocused electron beam spot sizes are only about 5 to 10 mm^2. Lasers can be larger, but usually no larger than about 100 mm^2. To harden a large area, the laser or EB has to be computer controlled to raster a pattern on the part. Spots and stripes are the usual patterns. The depth of hardening depends on the mass and composition of the part and the energy density, but hardened depths are usually about 1 to 2 mm (Figure 10–19). Hardnesses are often higher than those obtainable by conventional hardening because of the extremely rapid quench.

Laser and EB hardening have limitations: (1) equipment cost; (2) high alloys may not respond; (3) certain types of lasers are reflected from shiny metal surfaces. This latter problem is overcome by applying black coatings, such as conversion coatings, to parts. EB does not have this problem.

High equipment costs are indigenous to many heat treating processes, but the "high-alloy" limitation means that plain carbon steels and low-alloy steels and irons are the only applicable materials. Tool steels with high-alloy contents usually require soaking at the austenitizing temperature to put alloy elements into solution. Because laser- and EB-hardening scan speeds can be as high as a meter per minute, there is little time for alloy dissolution and the parts do not harden. The current practice is to use these processes only on carbon steels and oil-hardening tool steels. They have, however, the potential to be the heat treating processes of the future. They are extremely fast and energy

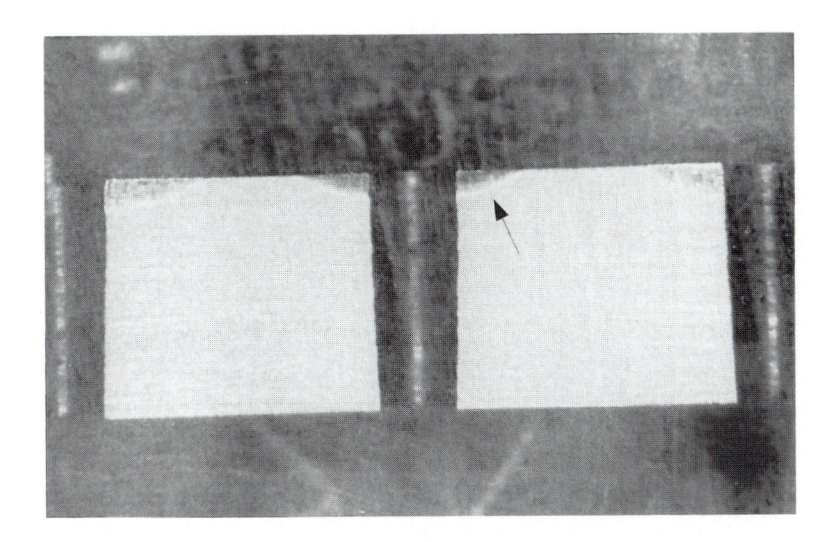

Figure 10–19
Electron beam hardened area around a die hole [0.03 in. (0.75 mm) deep] (×5)

efficient. Both processes are available through commercial laser and EB job shops.

10.5 Diffusion Treatments

If we wish to selectively harden a steel that has insufficient carbon or carbon plus alloy content to allow quench hardening, a diffusion treatment can be applied to add elements to the surface that will make it hard. Quenching may or may not be necessary depending on the nature of the diffusing species. *Diffusion* is the spontaneous movement of atoms or molecules in a substance that tends to make the composition uniform throughout. Basements in most homes are damp because moisture from the soil has a tendency to migrate through the basement walls. If onions are being fried in the kitchen, there is a spontaneous tendency for the smell to permeate the whole house. The driving force for diffusion processes is a concentration gradient. As an illustration, assume that two vessels, one containing salt water, the other pure water, are coupled with a simple tube and valve. When the valve is opened, there will be a mutual migration, tending to make the solution in each vessel equally salty.

As applied to surface hardening of steel, if a steel that is low in carbon content is put in an atmosphere that is rich in carbon, such as carbon monoxide, there is a tendency for the carbon from the gas to diffuse into the steel. The rate at which diffusion occurs in metals depends on the nature of the host metal, the diffusing species, the concentration gradient, and the temperature.

The classical model that applies to diffusion in solids is presented in Figure 10–20(a). Fick's first law states that the diffusion of an element in another material is a function of a *diffusion coefficient* and the concentration gradient. The diffusion constant depends on the material system, essentially the atomic species of the host material. The concentration gradient is simply how much of the solute species you have on one side of the diffusion interface compared to how much of that same species is on the other side of the interface. As applied to carburizing, the concentration gradient would be how much carbon is in the atmosphere on the outside of a piece of steel compared to the carbon content of the steel (how much carbon is already in the steel).

Diffusion coefficients can be calculated from atomic theory. All atoms are vibrating, and

Figure 10–20
Basic concepts for diffusion processes

Diffusion concepts

(a) Model

Fick's law: $J = D\,\frac{dc}{dx}$

where J = flux of atoms (atoms/time/area)
D = diffusion coefficient (area/time)
$\frac{dc}{dx}$ = concentration gradient (c = atoms/volume; x = distance)

Diffusion coefficient: $D = D_0 e^{\Delta H/RT}$
D_0 = a diffusion constant for a material
$\Delta H/RT$ = activation energy for process to occur; ΔH depends on the material system, R is a constant, and T is absolute temperature

(b) How it occurs

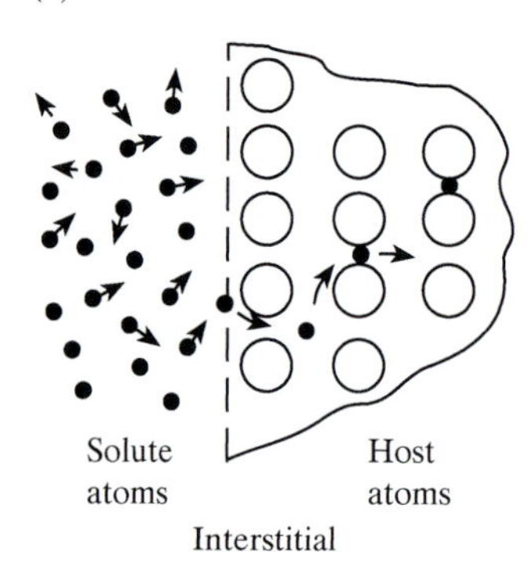

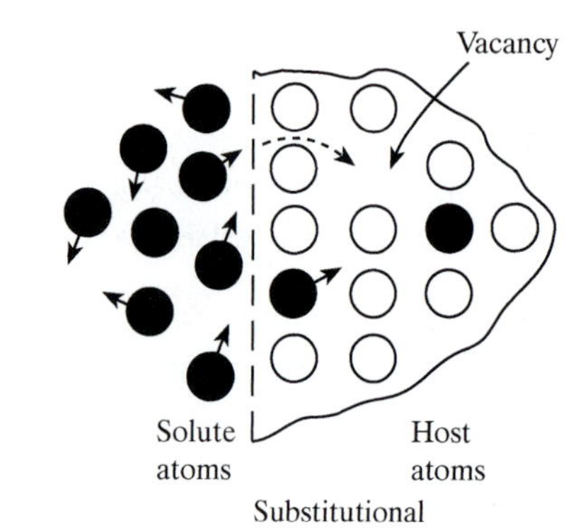

(c) Where it applies

	Solute	Host
Carburizing	C	Low-carbon steels
Nitriding	N	Nitriding steels
Carbonitriding	C + N	Low-carbon steels
Boronizing	B	Low-carbon steels
Chromizing	Cr	Low- and high-carbon steels

they collide with other atoms. These vibrations and collisions can be dealt with mathematically, and theoretical diffusion coefficients can be calculated. On the other hand, these coefficients can be measured by irradiating a metal that is plated on the surface of another metal. The irradiated metal is then diffused into the host material by heating it for a length of time. The diffusion is stopped by removing the material from the furnace, and the depth that the irradiated metal diffused into the host material is measured with devices that measure radiation on a microscale.

There are different ways that a *solute atom* (this is the term used for the diffusing species) can diffuse into a host material. In Figure 10–20(b), we show tiny solute atoms going into the spaces between the host atoms. This is called *interstitial diffusion,* just as we had interstitial solid solution. If we are trying to diffuse big atoms into a host material, they can be too big to occupy interstitial sites. In this instance, *substitutional diffusion* probably occurs. The solute atoms find their way to *vacancy* sites in the host material. *Vacancies* are atomic sites that are supposed to have an atom but do not. They are crystal defects. These vacancies move about with atomic vibrations, so the ones that end up on the surface will receive a solute atom, which in turn migrates inward by movement from one substitutional site to another. There are other ways the solute atoms can diffuse into a host material (for example, down crystal defects such as dislocations and grain boundaries), but the interstitial and substitutional approaches are the most likely.

As shown in Figure 10–20(c), the diffusion processes discussed in this chapter involve limited solute species and host materials, but the basic principles of diffusion apply to all of these processes. Some of the practical manifestations of diffusion theory are as follows:

1. Diffusion processes for hardening steels usually require high temperatures, greater than 900°F (482°C).
2. For diffusion to take place, the host metal must have a low concentration of the diffusing species and there must be a significant concentration of the diffusing species at the surface or elsewhere in the host metal.
3. Diffusion will take place only when there is atomic suitability between the diffusing species and the host metal.

Diffusion of carbon is normally carried out in the temperature range of 1550 to 1750°F (842 to 953°C). Nitrogen diffusion is usually performed at 900 to 1050°F (482 to 565°C). The concentration gradient for diffusion is usually obtained by selecting an atmosphere that readily provides a high solute concentration at the surface. Carbon diffusion treatments may produce a surface that has a carbon content of 1%. If the core material has 0.2% carbon, this provides the necessary concentration gradient for inward diffusion. The third general statement simply means that carbon, nitrogen, boron, and other diffusion-hardening elements do not diffuse with equal facility in various metals. Copper plating is often used as a masking material to prevent diffusion of carbon and nitrogen. There is poor atomic suitability. From the practical standpoint, this means that a designer must question the suitability of candidate materials for intended diffusion-hardening operations.

Carburizing/Processing The *carburizing* process is one of a series of heat treatment processes that involve diffusion of alloying elements into a metal substrate that normally has only a low concentration of that element. Atomic species (atoms) can be diffused into or out of a metal. The diffusion process can also be used to homogenize a metal that has an atomic gradient in alloy within its structure. In carburization, if a piece of low-carbon steel is placed in a carbon-saturated atmosphere at an elevated temperature, the carbon will diffuse or penetrate into the steel, carburizing it. Conversely, if a steel with a relatively high carbon content (~1%) is heated in a furnace that has a carbon-free atmosphere (air), carbon will tend to diffuse out of the steel. This is called *decarburization*.

The purpose of carburization is to provide a hard surface on normally unhardenable steels. There are basically three processes used to provide a suitable carbon gradient to allow inward diffusion: *pack*, gas, and salt. In *pack carburizing* the part to be carburized is packed in a steel container so that it is completely surrounded by granules of charcoal. The charcoal is treated with an activating chemical such as barium carbonate ($BaCO_3$) that promotes the formation of CO_2 gas. This gas in turn reacts with the excess carbon in the charcoal to produce carbon monoxide, CO. The carbon monoxide reacts with the low-carbon steel surface to form atomic carbon, which diffuses into the steel. The carbon monoxide supplies the carbon gradient that is necessary for diffusion. The pack carburizing process is illustrated in Figure 10–21.

The carburizing process per se does not harden the steel. It only increases the carbon content at some predetermined depth below the surface to a sufficient level to allow subsequent quench hardening.

If we have diffused carbon into the surfaces, how are we going to harden the part that is sealed in a steel box? The part may be quenched directly from the pack, but the problems in doing so are obvious. We must break open the pack, pull the part out, and quench it before it slow cools from its austenitizing temperature. It is common to slow cool the entire pack and subsequently harden and temper the part. This technique has a very useful benefit. If it is desired to leave an area soft for machining after final hardening, for example, to locate holes on assembly, the carburized surface can be machined from these areas, and they will remain soft and machinable after final hardening. Masking of parts to prevent carburizing in local areas can also be done by plating the part with copper or nickel or by painting with special paints, but these techniques involve significant hand labor and are usually more expensive than the pack and machine technique.

The disadvantage of reheating for hardening is a slightly softer *case* (55 HRC). Some of the carbon in the case diffuses inward, and the case is lower in carbon than when it came out of the carburizing operation.

There is no technical limit to the depth of hardening with carburizing techniques, but it is not common to carburize to depths in excess of 0.050 in. (1.27 mm) unless more wear than this can be tolerated on the part or unless the contact loads are so great that they may tend to spall a thin hardened case. The time required for various case depths depends on the carburizing media, diffusion temperature, and alloy. High diffusion temperatures tend to speed up the carburizing process, but grain coarsening may occur that will reduce the mechanical properties of the core. Typical carburizing times are illustrated in Figure 10–22.

Gas carburizing can be done with any carbonaceous gas, but natural gas, propane, or generated gas atmospheres are most frequently used. The source of the diffusing species is carbon from CO that is produced from the starting gas of CH_4, C_3H_2, and others. Most carburizing gases are flammable if not explosive, and controls are needed to keep carburizing gas at 1700°F

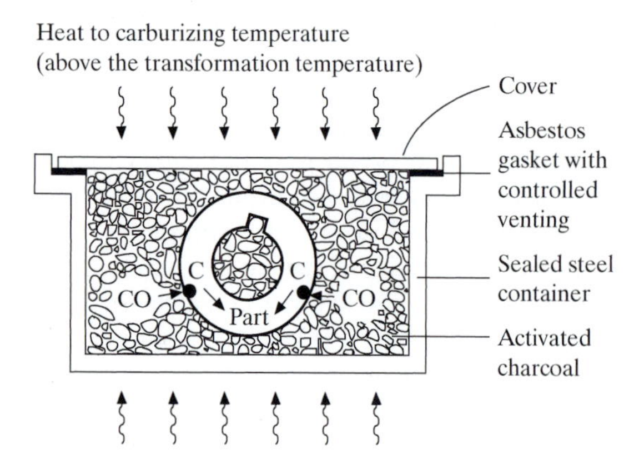

Figure 10–21
Pack carburizing

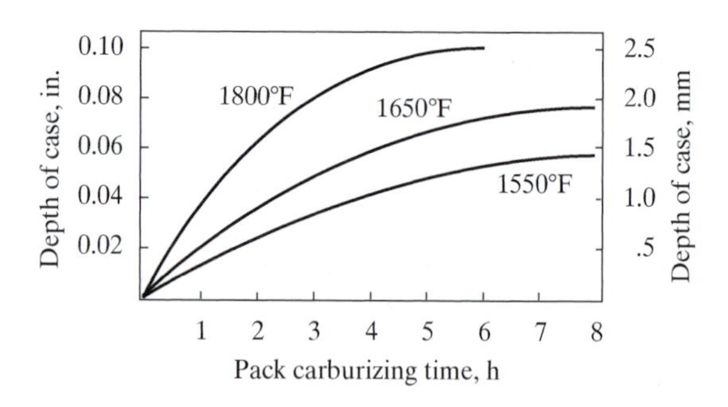

Figure 10–22
Effect of carburizing temperature on case depth

Source: G. M. Enos and W. E. Fontaine. *Elements of Heat Treatment*. New York: John Wiley & Sons, Inc., 1963.

(937°C) from contacting air (oxygen). Production gas carburizing is performed in batch-type retort furnaces, in vacuum furnaces, or in conveyorized continuous furnaces. The former provide quenching problems; the latter are more commonly used. In vacuum carburizing, parts are heated to the carburizing temperature in vacuum or inert gas. Propane or other hydrocarbon gas is then introduced as the source of carbon. Treatment times can be shorter than with gas carburizing. Gas carburizing is really a form of pack carburizing except that the carburizing gas comes from hydrocarbon gases instead of from activated charcoal. The carburizing times are similar to those of pack carburizing. The advantage of this process over pack carburizing is an improved ability to quench from the carburizing temperature. *Conveyor hearth* furnaces make quenching in a controlled atmosphere possible.

Salt or *liquid carburizing* is performed in internally or externally heated molten salt pots. The carburizing salt usually contains cyanide compounds such as sodium cyanide, NaCN. The carbon from the cyanide provides the diffusing species. Because heating by liquid convection is faster than heating by gas convection, the cycle times for liquid carburizing are shorter than the times for gas or pack carburizing. This is the main advantage of liquid carburizing over the gas and pack processes. A disadvantage of the process is that salt pots usually require batch processing. Conveyor systems for treating large volumes of parts are not readily adaptable. The more significant disadvantage of liquid carburizing is used-salt disposal. Eventually, the salt bath efficiency will diminish and the liquid carburizing medium must be replaced. Environmentally safe disposal of cyanide compounds is troublesome and expensive. Suppliers of heat treating equipment are working on environmentally clean salts that do not contain cyanide. These developments may solve disposal problems. In general, liquid carburizing is a favorite carburizing process with job shop-type heat treaters. Most of their work involves batch processing; equipment requirements are less than for gas carburizing, and fast cycle times provide quick part turnaround.

All the carburizing processes (pack, gas, vacuum, and salt) require quenching from the carburizing temperature or a lower temperature, or reheating and quenching. Parts are then tempered to the desired hardness. The case depths obtainable with the various processes are comparable, and the selection of a particular process (gas, salt, or pack) really depends more on economic and nonmetallurgical considerations.

Carburizing/Case Depth One question of concern in carburizing—or, for that matter, in diffusion treatments in general—is how we determine the *case depth* that has been achieved on a carburized part. A very simple tool used to measure case depth is a small sample of steel being carburized with a filed notch in it. This is carburized with the part to be hardened, and after carburizing and hardening the sample is fractured at the notch and the visible indication of hardened zone on the fracture surface is measured with a machinist's scale or caliper. The more accurate way of measuring depth of surface hardening is to conduct a *microhardness* survey across a polished section of the part or a test sample. Hardness measurements are made at increments of 0.001 in. (25 μm) or so from the surface. The point at which the hardness falls below 50 HRC is the *effective case depth* of hardened case. The *total case depth* is the depth from the surface that the case hardness is above the core hardness of the steel. This concept is illustrated in Figure 10–23.

Carburizing/Material Requirements So far we have discussed what carburizing is, how it is done, masking, and measurement of case depth. Let us now review the material requirements for carburizing. We pointed out in our discussion of alloy steels that certain steels carburize better than others. All carbon steels can be carburized. Obviously, there is no reason to carburize the

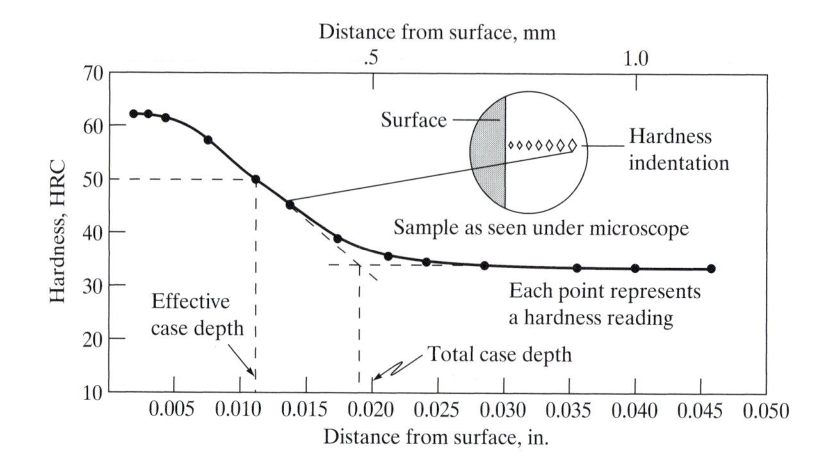

Figure 10–23
Determination of carburized or carbonitrided case depth by microhardness survey

high-carbon grades, so most carburizing is performed on the 0.1% to 0.3% carbon grades of plain carbon steels. The carbon–manganese and free-machining steels can also be readily carburized. The mass effect on quenching, however, makes the use of alloy carburizing steels desirable on section thickness greater than ½ in. (13 mm). Carburizing is not normally used on stainless steels, nor is it commonly done on cast irons, so its application is limited to low-carbon and low-alloy steels.

Nitriding The *nitriding* process is one of the most interesting and useful surface-hardening techniques. In this process, monatomic nitrogen is diffused into the surface of the steel being treated. The reaction of the nitrogen with the steel causes the formation of very hard iron and alloy nitrogen compounds. The resulting nitride case can be harder than even the hardest tool steels or carburized steels. The outstanding advantage of this process over all other hardening processes previously mentioned is that subcritical temperatures are used and hardness is achieved without the oil, water, or air quenching required of other heat treating processes. As an additional advantage, hardening is accomplished in a nitrogen atmosphere that prevents scaling and discoloration.

The source of the nitrogen for the diffusion process is most commonly ammonia. Parts to be nitrided are placed in a *retort* (racked to provide good gas circulation), and the retort is heated to the nitriding temperature, which is usually between 925 and 1050°F (500 and 570°C). Either nitrogen or ammonia is flowing during the heat-up cycle. At the nitriding temperature the ammonia dissociates by the following reaction:

$$2NH_3 \rightarrow 2N + 3H_2\uparrow$$

The nitrogen diffuses into the steel, and the hydrogen is exhausted. A basic nitriding setup is illustrated in Figure 10–24. In our example, the ammonia dissociates when it comes in contact with the hot part. Some heat treaters use an external dissociator, a device that breaks down ammonia into nascent nitrogen and hydrogen. External *dissociation* provides an extra measure of process control. Ion nitriding is the newest version of the process.

Because nitriding does not involve a quench, parts may be slowly cooled in the retort. The depth of hardened case obtained in nitriding is a function of the steel, the nitrogen potential of the atmosphere, and the time at the nitriding temperature.

As mentioned in our discussion of carburizing, case depths are best measured by a microhardness traverse. Nitrided surfaces have somewhat different case characteristics than most

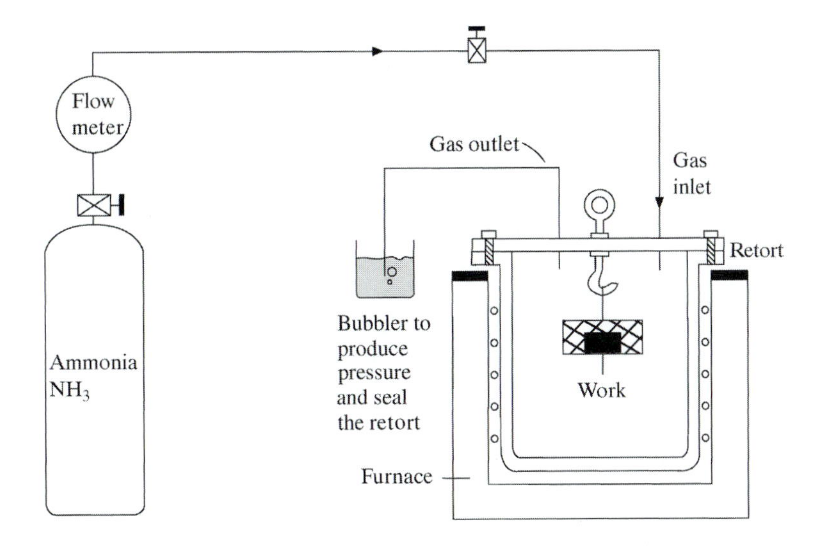

Figure 10–24
Schematic of a gas nitriding system

other surface-hardened steels. The case can include a *white layer*, which is brittle and prone to spalling, a hard nitride layer, and a *diffusion zone* of decreasing hardness. A typical nitrided surface is illustrated in Figure 10–25.

The white layer has a detrimental effect on the fatigue life of nitrided parts, and it is normally prevented or removed from parts subjected to severe service. In cases where it is present, the thickness can be 0.0003 to 0.002 in. (7.5 to 50 μm), depending on the heat treater and the nitriding time. There are ways of controlling or eliminating the white layer. For example, a two-stage gas-nitriding process, proprietary salt-nitriding process, and some plasma processes virtually eliminate the white layer. Masking in nitriding can be accomplished with the same types of techniques used for masking in carburizing. Copper plating is the most common technique.

The selection of steel for nitriding is of utmost importance. Nitrogen diffusion and the formation of hard nitrides are enhanced by the presence of such elements as aluminum,

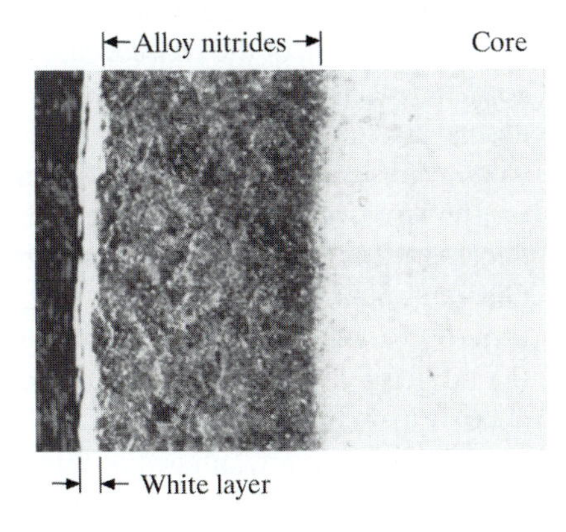

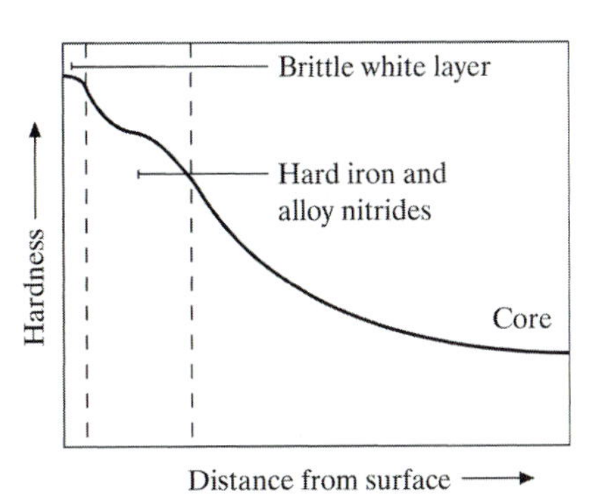

Figure 10–25
Hardness of a nitrided case

chromium, molybdenum, vanadium, and tungsten. Steels such as plain carbon steels that do not have significant amounts of these elements are not well suited to gas nitriding. The most commonly nitrided steels are chromium–molybdenum alloy steels and a class of alloys called Nitralloys in the United States. These latter alloys contain approximately 1% aluminum, which greatly enhances the formation of hard nitrides. Proprietary techniques are available for nitriding stainless steels, but corrosion resistance may be reduced. The maximum hardness obtainable in the alloy steels (SAE 4140 and 4340) is approximately 55 HRC, with an attainable case depth (over 50 HRC) in the range of 0.005 in. (0.125 mm). The Nitralloys can have a surface hardness of 70 HRC, with case depths as great as 0.020 in. (0.5 mm) (over 60 HRC). The nitriding characteristics of these steels are compared in Figure 10–26. Because only the first few thousandths of an inch of nitrided steels are very hard, it is common not to perform any grinding operations after nitriding. If the white layer must be removed, it should be done by lapping. Prevention is the preferred option. One very important factor to be considered in nitriding steels is prehardening of steels prior to nitriding. The high nitride hardnesses just mentioned cannot be obtained on annealed material. The low-alloy steels and Nitralloys must be prehardened before nitriding. A tempering temperature at least 50°F (10°C) higher than the nitriding temperature must be used to prevent tempering size changes during nitriding. Often, tool steels with high tempering temperatures (hot work, high speed, and so on) are nitrided for additional surface hardness. In addition to very high hardnesses, most people feel that nitriding provides additional corrosion resistance. This is questionable, but nitrided steel will have lower rusting tendencies in ambient air than unnitrided steel.

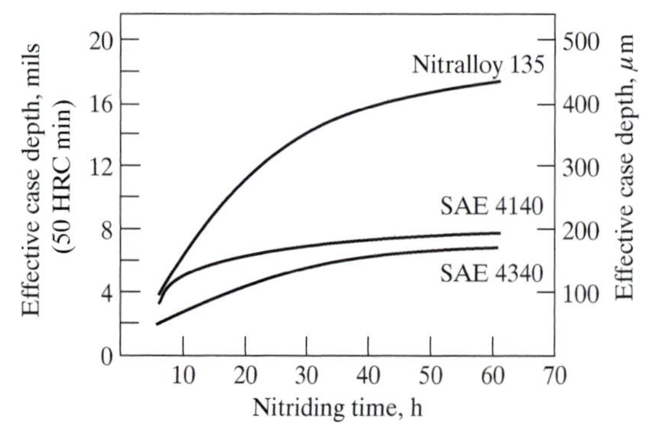

Figure 10–26
Typical gas nitriding times (for system in Figure 10–23)

In the late 1970s, *ion nitriding* was introduced in the United States. Ion nitriding is performed in what is essentially a glow discharge vapor deposition unit. Parts are racked on a fixture that is electrically connected to a high-voltage dc power supply. The fixture is placed into a vacuum furnace (internally or externally heated). The parts are made the cathode, and the walls of the vacuum furnace are made the anode or positive member of the electrical circuit. The furnace is sealed and pumped down, and the charge is heated to the nitriding temperature, 660 to 1015°F (350 to 550°C). When the parts are at temperature, the heating elements are shut down, a mixture of hydrogen and nitrogen (ammonia) is introduced, and the dc power supply is energized to produce a plasma field around the work pieces (glow discharge). The plasma supplies the energy to maintain part temperatures. The dc potential causes ionization of the process gases. The plasma contains electrons, nitrogen ions, monatomic gas atoms, and some gas molecules. The action of nitrogen ions and monatomic nitrogen impinging on the work-pieces supplies the nitrogen that is necessary for nitrogen diffusion into the work. Gas pressures are usually in the range of 0.2 to 3 torr (30 to 400 Pa), glow discharge voltage is in the range of 400 to 1000 V, and current density is about 1 mA/cm^2.

The alleged advantages of this process over the ammonia retort processes are shorter times (60% to 70% of the time for conventional nitriding), elimination of the white layer, a somewhat tougher case, and better process control. The major limitations of this process are the need for good fixturing to allow electrical contact and gas flow and the high cost of the equipment. Eventually ion nitriding or a similar variation of the process may completely replace ammonia nitriding. Carbonaceous gases can even be combined with nitrogen to produce nitrocarburization. Whatever the evolution of this process, the concept of surface hardening by diffusion of nitrogen into steels will undoubtedly continue for many years. It is a low-distortion process for producing cases as hard as 70 HRC on a variety of steels. The wear resistance of these alloy nitride and iron nitride cases can be superior to the wear resistance of most other steels.

A consideration to keep in mind when using nitrided surfaces is that these processes involve a growth of dimension on treated surfaces. The formation of hard nitride compounds in the steel causes a size change; the magnitude of this change is a function of the amount of nitrogen that diffuses into the steel, but for normal cases of 0.005 to 0.020 in. (125 to 500 μm), the growth will usually be in the range of 0.0005 to 0.001 in. (12.5 to 25 μm) per surface. Bores may be reduced in size by 0.002 in. (50 μm), and outside diameters may grow by that amount. The normal way to deal with this growth is grinding of 0.001 to 0.003 in. (25 to 75 μm) from nitrided surfaces after treatment. This grind will also remove any white layer that may have formed. Size change can be minimized by keeping cases thin [0.001 to 0.003 in. (25 to 75 μm)]. Thin cases also have less white layer, if any.

Carbonitriding In the carburizing process, the diffusing hardening element is carbon. In nitriding, the diffusion involves nitrogen. As the name implies, *carbonitriding* is a surface-hardening process that involves the diffusion of both nitrogen and carbon into the steel surface. The process is performed in a gas-atmosphere furnace using a carburizing gas such as propane or methane mixed with several percent (by volume) of ammonia. The organic gas serves as the source of carbon; the ammonia serves as the source of nitrogen. Why mix these two basic processes? Carbonitriding is performed at temperatures above the transformation temperature of the steels being treated (1400 to 1600°F; 760 to 871°C); however, the nitrogen allows lower temperatures than are required for gas carburizing as well as faster cycles and a less drastic quench. Often, a gas quench is used, and seldom is it necessary to use a quench medium as drastic as water. The use of a less drastic quench significantly reduces distortion. A typical carbonitriding system is illustrated in Figure 10–27.

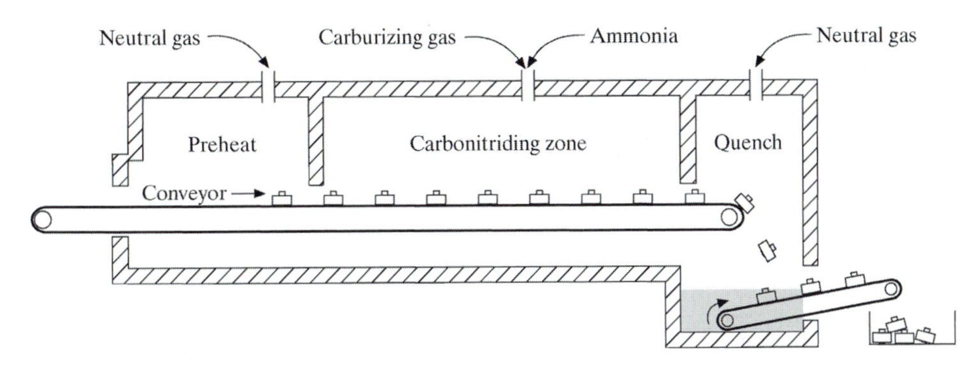

Figure 10–27
Conveyor hearth carbonitriding

The case hardnesses attainable with carbonitriding are similar to those possible with carburizing, approximately 60 to 65 HRC at the surface. They are not as hard as nitrided nitriding steels. The depth of case attainable depends on the steel and the processing parameters, but most cases fall within a thickness range of 0.003 to 0.030 in. (0.07 to 0.75 mm). The case consists of hard nitrogen compounds as well as hard martensite. The steels that are commonly carbonitrided are the low-carbon and low-carbon alloy steels. The dimensional changes occurring in this process are comparable to those that might occur in carburizing. In general, it is not suitable for parts of extreme accuracy. The fact that this process involves heating above the transformation temperature followed by quenching raises the possibility of quenching distortion. Carbonitrided parts are usually tempered to reduce their brittleness. In this respect, they are treated like carburized steels, but a higher tempering temperature may be required because of the presence of nitrides in the case microstructure.

In summary, carbonitriding is a gaseous diffusion-hardening process involving the diffusion of both carbon and nitrogen into the steel. It has advantages over carburizing in that lower heat treating temperatures are required and a less drastic quench is needed. Its advantage over nitriding is that it can be used to get significant case depths (~0.010 in., 0.25 mm) on plain carbon steels. Its disadvantage compared to nitriding is that it produces more distortion because it involves heating through the transformation range and quenching. This process is adaptable to low-carbon steels, powdered-iron compacts, and medium-carbon, low-alloy steels. From the design standpoint, it might best be applied to hardening large volumes of parts in a production operation.

Cyaniding Similar to carbonitriding, *cyaniding case hardening* involves the diffusion of both carbon and nitrogen into the surface of the steel to be treated. The source of the diffusing element in this instance is a molten *cyanide salt* such as sodium cyanide. It is a supercritical treatment involving temperatures in the range from 1400 to 1600°F (760 to 870°C). An oil or water quench is required to achieve high surface hardness. Normally, only low-carbon steels are cyanided. Case depths are normally less than 0.010 in. (0.25 mm), but depths as great as 0.030 in. (0.75 mm) may be achieved. Diffusion times are usually on the order of 1 h, considerably less than that required for carbonitriding. Processing times may be as short as 15 min. This is one of the reasons for the popularity of this case-hardening process on general machine components intended for moderate wear and service loads. Because a water or oil quench is required, as well as heating above the transformation temperature, there is the possibility of noticeable distortion in hardening.

Cyaniding is done in a molten salt heated by immersed electrodes or externally heated. In the latter case, the cyanide salt in contained in a simple steel pot. Cyanide salts in powder form are also available. It is possible to obtain a shallow case on low-carbon steel by simply heating the part to red heat and sprinkling the salt or immersing it in the granular powder followed by a quench. This technique allows surface hardening of low-carbon steels with no equipment expenditures other than an oxy-fuel torch or hot-air furnace to heat the part.

This hardening process is not currently in wide use. The case is usually very thin and the distortion noticeable. In engineering design it would be best to avoid the use of this process unless the production requirements make it economically attractive. Nitriding provides lower distortion, and carburizing better case depths. Table 10–2 compares case characteristics on the various surface-hardening processes.

Special-Purpose Surface Treatments Low-carbon steels normally do not respond to low-temperature ammonia gas nitriding cycles. In the late 1970s, a number of furnace manufacturers in the United States developed furnace and gas mixtures that allowed the formation of a

Table 10–2
Comparison of diffusion surface-hardening treatments

Process	Diffusion Elements	Process Temperature °F(°C)	Case Depth Range, in.	Case Hardness, HRC	Applicable Basis Metal	Characteristics
Pack carburizing	Carbon	1500–2000 (816–1093)	0.010–0.120 (0.25–3 mm)	45–65	Low-carbon steels and carburizing alloy steels	Use for cases over 0.03 in. (762 μm) deep
Salt carburizing	Carbon	1450–1650 (788–899)	0.003–0.12 (0.07–3 mm)	45–65	Low-carbon steels and carburizing alloy steels	Accurate control of case depths
Gas carburizing	Carbon	1500–1800 (816–982)	0.003–0.12 (0.07–3 mm)	45–65	Low-carbon steels and carburizing alloy steels	Best suited to high-production jobs
Carbonitriding	Carbon + nitrogen	1300–1650 (704–899)	0.001–0.03 (25–762 μm)	50–60	Low-carbon and low-alloy steels	Suitable for thin cases on large-quantity jobs
Cyaniding	Carbon + nitrogen	1400–1600 (760–871)	0.001–0.03 (25–762 μm)	50–60	Low-carbon and low-alloy steels	Thin cases; water quench required and parts may distort
Gas nitriding	Nitrogen	925–1050 (495–566)	0.003–0.03 (75–762 μm)	50–70	Alloy steels, tool steels, and some stainless steels	Lowest-distortion diffusion process; nitriding steels are preferred
Chromizing	Chromium	1500–2050 (816–1121)	0.003–0.006 (75–150 μm)	40–70	Carbon steels	Produces hard case on high-carbon steels, corrosion-resistant surface on low-carbon steels
Ferritic nitrocarburizing	Carbon + nitrogen	950–1250 (510–675)	0.0001–.001 (2.5–25 μm)	40–60	Carbon steels, alloy steels, and some stainless steels	Produces thin, hard surface for scuffing resistance
Salt nitriding	Nitrogen	950–1050 (510–566)	0.0001–0.005 (2.5–127 μm)	50–70	Tool steels, alloy steels, carbon steels, stainless steels	Commonly used on hard tool steels; useful on stainless steels but lowers corrosion resistance

thin, hard layer on low-carbon steels and some stainless steels. The hardened layer is usually less than 1 mil (25 μm) thick and hardness depends on the alloy, but hardnesses in the range of 45 to 65 HRC have been recorded. This process is identified by a number of trade names, but the generic name for the process is *ferritic nitrocarburizing* (FNC). This process is done at temperatures below 1250°F (675°C), usually at 1060°F (570°C); thus, the structures of ferrous substrates to be treated are ferritic. The process is called nitrocarburizing because it involves the diffusion of both carbon and nitrogen from gas mixtures consisting of ammonia and a hydrocarbon gas. Processing times are usually short (about 3 h), and batch-type furnaces with a controlled atmosphere cooling zone are usually used. This process is also performed in a fluidized bed reactor. Heating is accomplished by contact with heated fluidized alumina particles. The process gases (ammonia + carbonaceous gas) also do the fluidizing. This type of treatment can be faster than normal gas atmosphere processing. This process is best suited for high-production "skin" hardening of low-carbon steel parts. It is not a replacement for the heavier cases obtainable with conventional processes.

The term *cementation* is used in heat treating to mean the introduction of one or more elements into the surface of a metal by high-temperature diffusion. Most of the diffusion processes just discussed would be covered by this definition, but it is common practice to use the term to cover the diffusion processes that involve elements other than carbon and nitrogen. One of the earliest commercial processes for surface hardening by diffusion of another species was *chromizing*: diffusion of chromium into the surface of steels. This process can be done by packing parts in a solid compound (*pack cementation*) that gives off a chromium-rich gas at the diffusion temperature [up to 2000°F (1093°C)]. A retort and gas can also be used. If chromizing is performed on low-carbon steels, the surface of the steel can acquire the corrosion-resistant properties of a ferritic stainless steel. If it is performed on high-carbon steel, the chromium combines with carbon in the steel to form a hard, wear-resistant case from 50 to 150 μm deep.

Boronizing is a similar process that is receiving more attention than other processes of this type. It is a pack cementation process that uses compounds that produce boron-rich gases that allow diffusion of boron into the surface. It can be done on low-carbon and high-carbon and alloy steels, but for the most part its use is on low-carbon steels. It is an attempt to obtain extremely wear-resistant surfaces on low-cost steels. Although surface hardnesses can be over 70 HRC, this process has not replaced conventional carburizing and nitriding processes. Typical treatment temperatures are in the austenitizing range for steels, and cooling from these high temperature often produces significant part distortion.

Pack cementation processes can also be used to diffuse aluminum, silicon, and beryllium for special applications such as high-temperature oxidation. The most unusual process in this category is *Metalliding*®. This process is performed by electroplating elements on a metal in a molten salt bath. The bath temperature and plating rate are maintained such that the plated species diffuses in at the same rate as it is plated. Thus, there is usually no coating but a surface that is high in concentration of the diffusing species. This process can produce surfaces rich in almost any element: Be, B, Al, Si, Ti, W, V. This process is extremely difficult to control from the standpoint of salt and atmosphere purity. It is used only where other, simpler diffusion processes do not work.

The TD (Toyota Diffusion) process was developed in 1968 in Japan, and in the past 36 years its use has been expanded to the point at which it is readily available from treatment companies or it is available with licensing. It produces a thin, hard case of carbide compounds by diffusion of refractory metals into substrates containing

Figure 10–28
Abrasive wear of diffusion surface treatments compared with unhardened 1020 steel using the ASTM G 65 dry-sand rubber wheel abrasion test, procedure C

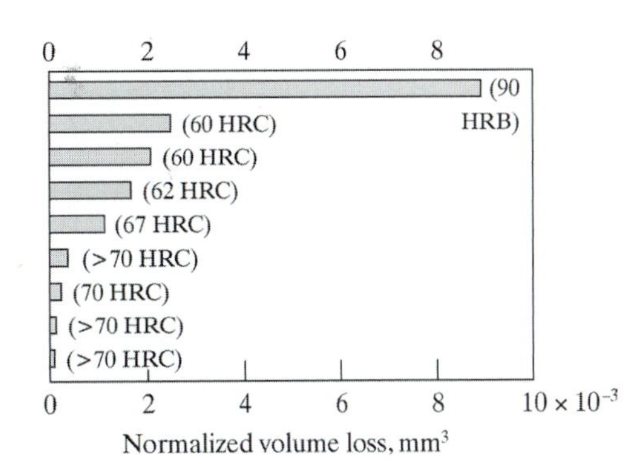

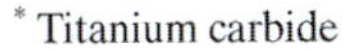

carbon. The usual species is vanadium, and the case is about a 0.001-in.-thick (25-μm) layer composed of vanadium carbide. The diffusing species are added to a borax ($Na_2B_4O_7$) bath maintained at temperatures in the range of 1500 to 1800°F (800 to 1000°C). Treatment times are from less than an hour to as much as 10 h, and the usual substrates are tool steels in the form of dies, cutting devices, and other tools that may benefit from a very hard skin on a hardened tool material substrate.

TD process coatings are competitive with titanium carbide (TiC) coatings, which have been available in Europe and the United States since the 1970s. Titanium carbide coatings are applied by pack cementation or chemical vapor deposition, but all these treatments are done at temperatures at or above normal hardening temperatures for most tool materials. The high temperature of application produces a very strong possibility of distortion in cooling from the process temperature. Because the coating is very thin [about 0.0001 in. (2.5 μm)], there is usually not sufficient case depth to allow grinding after treatment to eliminate process distortion. Essentially, this treatment is done on annealed substrates, and the treatment temperature is also the substrate-hardening temperature. Parts are quenched from the process temperature using the normal quench for the steel in question, and the parts are subsequently tempered. These treatments provide an extra measure of abrasion resistance on tools (Figure 10–28), but their use is usually confined to tools that can tolerate, for example, 0.005 in. (125 μm) or more of distortion without effect on performance. These high-temperature diffusion processes are difficult to apply successfully to tools made to high degrees of precision.

Vacuum carburizing is an evolution of gas carburizing. Parts are heated to carburizing temperature in a retort-type vacuum furnace (Figure 10–29). When the parts are at temperature, a controlled amount of purified natural gas is introduced to provide the source of the carbon for diffusion into the work. The major advantage of vacuum carburizing is scale-free parts and carburizing cycle times that may be only a fraction of the time required for conventional gas carburizing. The disadvantage is high equipment costs.

Salt nitriding is usually performed in proprietary salts, and the processes are given special names such as Tufftriding®, Melonite®, and Nu-Tride®. The former process will produce a nitride case on most ferrous metals that is free of the white layer common in gas nitriding. In addition, alloys that are difficult to gas nitride often respond to this process. Tufftriding and similar processes are useful for thin cases on tool steels for improved scuffing resistance.

Melonite is a process aimed at eliminating the environmental problems that exist in using cyanide salts as the source of nitrogen for salt nitriding. In this process, parts are preheated to the

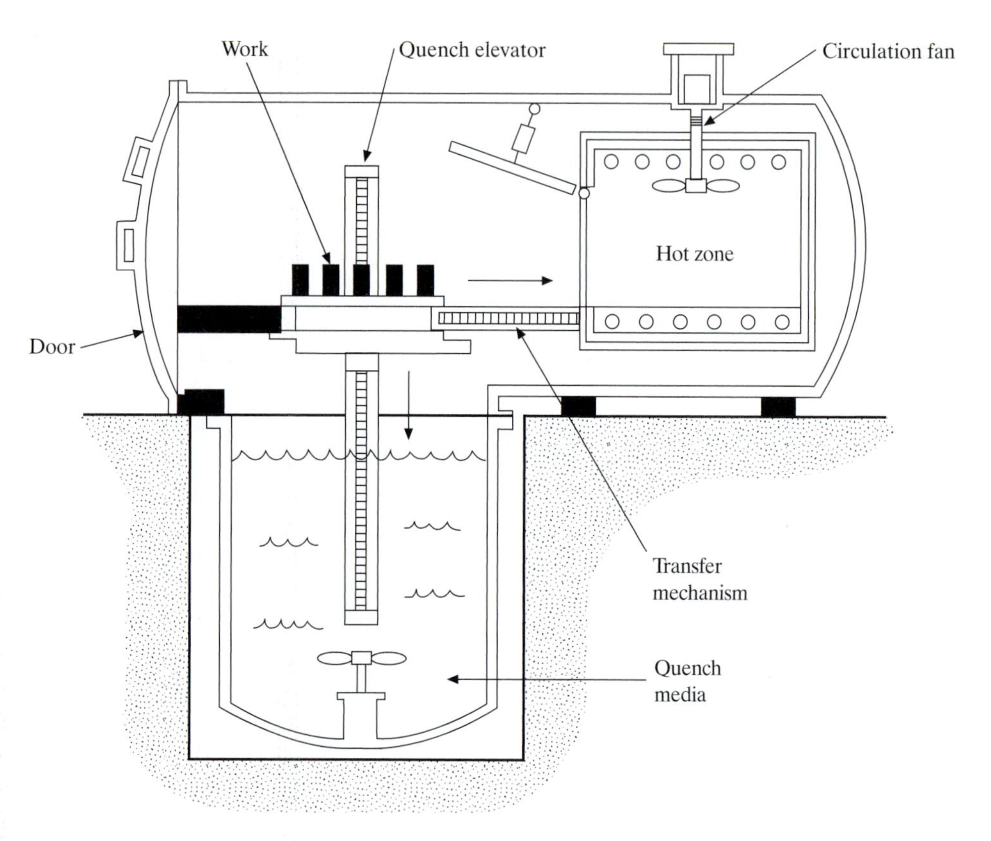

Figure 10–29
Vacuum furnace with integral quench tank

nitriding temperature of about 1075°F (580°C) in a separate neutral atmosphere furnace. They are then transferred to the nitriding salt bath, where the soak is reduced because of the preheat. The final steps are a salt quench at about 700°F (371°C) and an oil or air quench to room temperature. The salt quench reportedly removes any cyanide that may have remained on parts after removal from the nitriding salt. The uses of this process are similar to those of Tufftriding; the reported advantage over Tufftriding is a reduction of cyanide pollutants that must be washed from Tufftrided parts. This process produces a black surface that has a degree of rust resistance. There are other proprietory case-hardening processes, but all involve the diffusion of an elemental species into the surface to increase hardness or improve other surface properties.

10.6 Softening

Recrystallization

As was pointed out in the previous chapter, if a metal is cold worked, these grains deform, become elongated, and in doing so harden and strengthen a metal. There is a limit to the amount of cold work that a particular metal can be subjected to. In rolling steel into thin sheets, you can reduce the cross-sectional area only so much before it gets too hard to roll. At this point it would be desirable to return the grains' size to their original shape. Heat treatment can accomplish this. The transformation of cold-worked grains to an undistorted equiaxed shape is called *recrystallization*. Very large, coarse grains can also be refined by recrystallization. In the former case, the driving force for the formation of new grains is the stored energy of cold work; in the latter case, the change in crystal structure from BCC to FCC can cause the formation of new, refined grains or crystallites. The goal in either case is to achieve a microstructure and properties that are more favorable for future processing of the metal. Recrystallization heat treatments are essential if a steel is to be subjected to severe cold working in rolling, drawing, and the like.

Annealing

The annealing heat treating process consists of heating a steel to its austenitizing temperature and then cooling it at a slow enough rate to prevent the formation of a hardened structure. A rule of thumb on the required cooling rate for annealing is 100°F/h, but some steels require a rate as low as 30°F per hour. This means part temperature, not furnace temperature. In most cases, the required slow cool is achieved simply by turning the furnace off after soaking at the annealing temperature and letting the closed furnace cool down to the ambient temperature. Sometimes the parts are removed from the furnace and allowed to cool slowly, packed in sand or lime. The annealing temperatures for plain carbon steels are the same as the hardening temperatures. The structure transforms to austenite or austenite plus cementite at the annealing temperature. A slow cool will give approximate equilibrium conditions, so the room temperature transformation products will be those shown on the iron–carbon diagram: ferrite and pearlite, or pearlite and cementite. Alloy steels behave similarly, but the phase diagram and temperatures will be different.

The softening that occurs in annealing takes place through a number of mechanisms. If a part is being annealed to change the structure from hard martensite to a machinable structure such as ferrite and pearlite, the softening is accomplished by diffusion of carbon from the metastable martensite and re-solution of the carbon in austenite at the annealing temperature. *Diffusion* is the spontaneous movement of atoms in the crystal structure of a metal. Martensite is hard because there is an overabundance of carbon atoms trapped by quenching in a crystal structure that wants to be BCC. Diffusion of carbon in steels is controlled by temperature; by the time a martensitic steel reaches the annealing temperature, most of the carbon that was trapped in martensite has diffused out. At the annealing temperature the structure transforms to austenite, and all the carbon goes into free cementite or into solution in austenite. An annealing operation that involves heating to the austenite region is called a *full anneal*. Heat treaters sometimes use a *process anneal* that involves heating to a point just below the austenite transition temperature. It does not soften the steel as effectively as a full anneal, but since a change in crystal structure does not take place, it usually produces less distortion than a full anneal. The lower process time also lowers cost.

If a part is being annealed to remove internal strains from cold work, welding, or some fabrication process, the softening mechanism predominately involves removal of internal strains by thermally assisted dislocation movement. If we were to hammer one surface on a piece of steel, the grains near that surface would become distorted and the hammered surface would plastically deform; bowing would probably occur. If we wish to restore the piece and its microstructure to an unstrained state, new grains must be formed and some portions of the piece must plastically deform.

As a part is heated to its annealing temperature, it becomes weaker. Atomic mobility increases, dislocations accumulated in the strained metal *lattice* start to disappear, and on a macroscale the distorted part adjusts itself to a neutral, unstressed condition. By the time the annealing temperature is reached, most of the internal strains have been eliminated and the distorted grains have probably recrystallized; that is, new, undistorted grains are formed.

When the critical transformation temperature is reached, the crystal structure transforms to austenite or austenite plus cementite (for high-carbon steels). At this temperature, all distorted grains and internal strains are removed. Soaking at the annealing temperature will also allow homogenization of any chemical gradients in the metal. Diffusion is easy, and if the metal surface has a lower or higher carbon content than the core, the carbon gradient serves as a driving force for homogenization by diffusion.

Tempering

Anyone who has witnessed a blacksmith harden a piece of steel will recall that the steel was reheated after quenching. It has been known for many centuries that quench-hardened steels are useless. They are so brittle that the slightest impact or bending load will cause fracture. *Tempering* is a subcritical heat treating process used to improve the toughness of quench-hardened steels.

The proper tempering temperature for a particular steel depends on the composition of the steel and the desired properties. As can be seen in Figure 10–30, tempering has an effect on tensile strength, ductility, hardness, and impact strength. Most steels destined to be used as tools are tempered at a temperature that will reduce the as-quenched hardness by only 2 to 4 points on the Rockwell C scale. The effect of such a tempering operation on strength is usually slight. Springs and parts subject to impact are usually tempered into the mid-40s on the Rockwell C scale for maximum toughness.

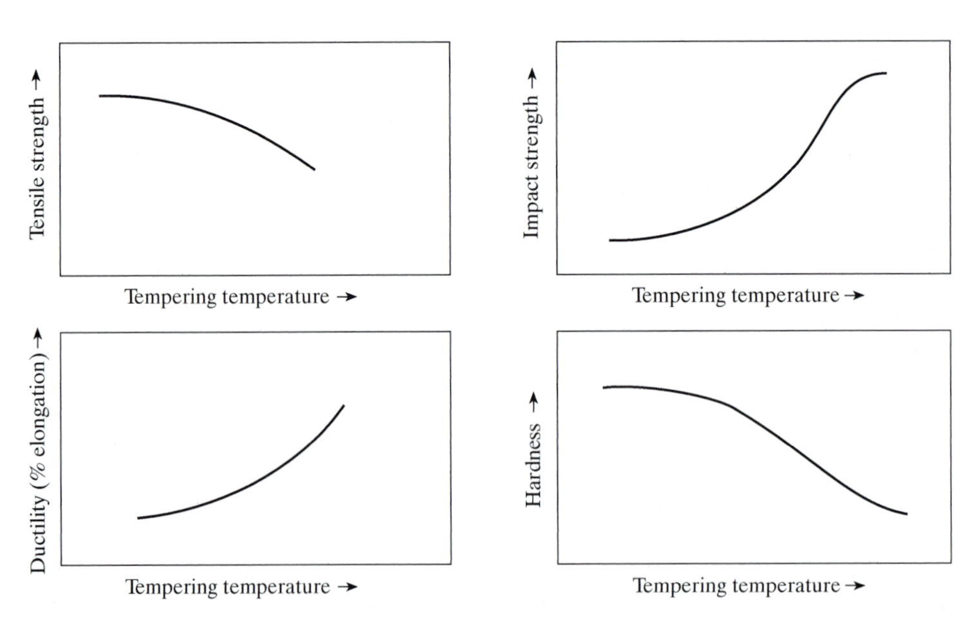

Figure 10–30
Effect of tempering on the properties of a quench-hardened carbon steel (1040–1060)

The normal procedure for tempering is to quench the part; while it is still warm, it is put in a furnace at the desired tempering temperature. When temperature sensors indicate that the part is at temperature, it is soaked for about 2 h and then air cooled. The time at the tempering temperature is not overly critical. After 2 h, continued heating has little effect.

The exact mechanism of tempering depends to some extent on the nature of the steel. In the plain carbon steels, tempering of martensite has been shown to involve diffusion of carbon atoms from martensite and the formation of carbide precipitates and concurrent formation of ferrite. At low tempering temperatures, a fine transition carbide (not the equilibrium Fe_3C carbide) is distributed parallel to martensite platelets. Most quenched structures contain some *retained austenite*. Tempering also causes some retained austenite to transfer to cementite and ferrite. At higher tempering temperatures, the transition carbide transforms to the equilibrium Fe_3C carbide, and prolonged tempering times will cause agglomeration of fine carbides into discrete Fe_3C spheroids. In addition to austenite and carbide reactions, tempering can include reduction in the strain of the distorted martensite by thermally induced dislocation motion processes.

Of more practical concern, tempering causes a physical size change in a part that was quench hardened. This size change is usually a contraction, and it is undoubtedly the result of the structure change that occurred.

To the designer the exact mechanism and procedure for tempering may not be important, but it is important that all quench-hardened steels be tempered; the designer should specify working hardnesses on parts that are within the normal tempering range for the alloy selected. When a designer specifies a hardness on a drawing, that person is essentially telling the heat treater how he or she wants the part tempered. Occasionally, designers call for hardnesses so high that the only way that they can be obtained is by using the steel as quenched. If a designer is not sure of the normal tempered hardness for a steel, he or she should ask a heat treater or refer to a steel handbook. Do not confuse heat treat tempering with temper rolling of metals; they are not the same (rolling tempers are measures of cold work).

Other Softening Heat Treatments

The processes discussed to this point are the most important, but there are additional processes that the designer should be aware of. *Spheroidizing* is usually a subcritical heat treatment. It is used to agglomerate iron carbide in the structure of carbon steels. A metallographic section of a spheroidized steel would show second phases present as spherical particles. It is infrequently used by most heat treaters, but the process is widely used in tool steel mills to improve machinability by agglomerating alloy carbides that are present. Tool steels often are sold in the spheroidized condition. This process is similar to a process anneal except that soak times are usually very long.

Austempering is really a quenching technique. It involves quenching a hardenable steel from the austenitizing temperature to a temperature slightly above the temperature at which martensite starts to form (*M_S temperature*), possibly 600°F (315°C), and below the temperature at which an annealed type of structure would result (Ac_3). The part is held at this temperature until transformation is complete, then the part is quenched or air cooled. The resulting structure is usually *bainite*, which looks something like martensite under a microscope, but is coarser. Crystallographically, it is ferrite containing cementite. Austempering is usually used to reduce quenching distortion and to create a tough and strong steel (Figure 10–31). Springs, retainers, automotive seat belt components, and lawnmower blades are commonly austempered. Because the hardness range achievable with austempering ranges from 30 to 50 HRC, austempered steels are not as wear resistant as a

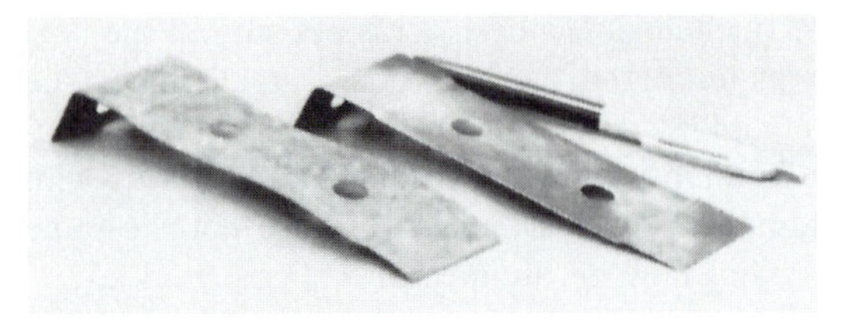

Figure 10–31
Part on the left was water quenched; the other was austempered

fully hardened (60 HRC) martensitic steel. Only certain carbon and alloy steels are amenable to austempering and there are limitations on the thickness of the steel (typically not more that about 6 to 9 mm).

Martempering or marquenching is similar to austempering, only the part is quenched from the austenitizing temperature to a temperature a few degrees above the martensite start temperature (M_S); it is held until the temperature is uniform in the piece, and then it is oil or air quenched to room temperature. This process is used to minimize quenching distortion. The thermal shock of quenching from, for example, 1600°F (871°C) to 400°F (204°C) is less than the shock of a quench from 1600°F to room temperature oil. The heat treating media for martempering and austempering are usually molten salts.

Hot-Working Operations

Most metal shapes produced by steel mills are at least rough shaped at elevated temperatures. Heat treating is required to bring the rough metal shapes to the proper temperature for hot-forming operations. Forging, hot heading, hot rolling, roll welding, and the like are all performed at temperatures of sufficient magnitude to prevent the formation of distorted grains that will harden the metals. Hot-working operations require dynamic recrystallization. Working at the proper hot-working temperature provides this.

This discussion of steel heat treating processes can be summarized by reviewing the temperature ranges for heat treating superimposed on the iron–carbon diagram (Figure 10–32) and by showing a schematic of the thermal cycles for common heat treating processes (Figure 10–33).

10.7 Atmosphere Control

Many types of equipment are used for heat treating operations, but most production heat treating is done in electric or gas-fired furnaces or in electric-heated molten salt baths. If a steel part is heated in a furnace in which it will be in contact with air, it will oxidize. If the temperature is over about 1000°F (538°C), a significant scale will build up. This will happen at normal austenitizing temperatures, and surface damage may be as deep as 0.020 in (0.5 mm). If the part has finished machined surfaces, it would be ruined, so most high-temperature [over 1000°F (538°C)] heat treating operations are performed in a *controlled atmosphere*. Electric furnaces can be evacuated or purged with an inert or nonoxidizing gas to control oxidation. The combustion products in gas-fired furnaces can be controlled to minimize oxidation, and salt baths are chemically controlled so that they are neutral, neither oxidizing nor reducing. Additional special techniques are employed, such as packing parts in cast-iron chips, packing in stainless-steel bags or foil, and heating in special carbon block containers. Whatever the technique used, something should be done to minimize surface degradation during high-temperature heat treating operations.

When should the design engineer be concerned with the heat treating atmosphere? If a part is to be hardened and it is apparent that oxidation during heat treating can affect serviceability, the designer must assume responsibility for noting atmosphere control on the drawing. Most heat treaters will assume responsibility for sufficient atmosphere control to prevent loose flaky

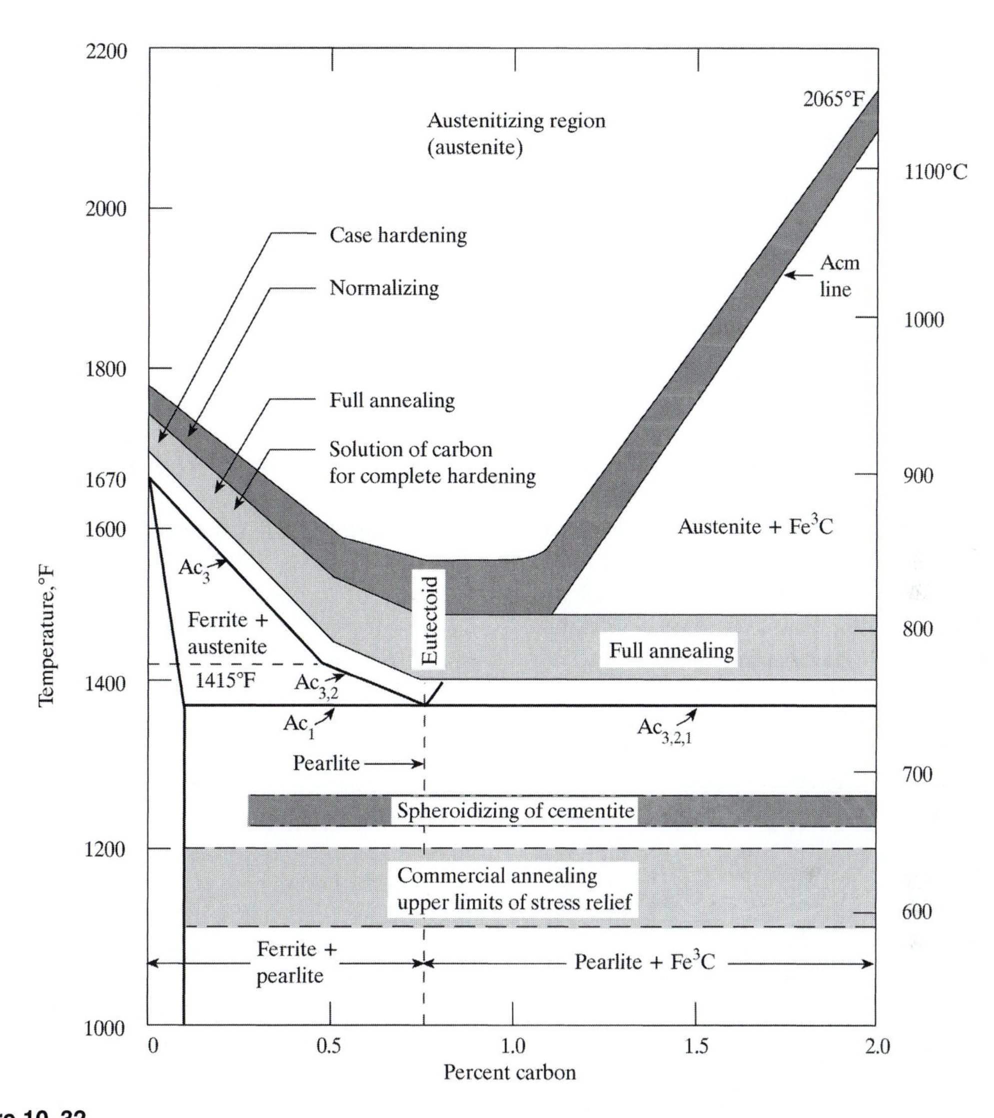

Figure 10–32
Temperature ranges for various heat treatments
Source: G. M. Enos and W. E. Fontaine, *Elements of Heat Treatment*, New York: John Wiley & Sons, Inc., 1963.

scale and softening due to surface decarburization, but the heat treat customer will normally get a part with a thin oxide and some discoloration in high-temperature heat treatments (or the scale will be removed by abrasive blasting, or the like). If a part is to be finish machined and then hardened, it may be necessary to call for *bright hardening*. This is especially true if there are areas or a part that will not be finished after heat treat but are still critical to function or appearance. In most heat treating shops, the term *bright harden* or *bright anneal*, *temper*, or *stress relieve* means that the operation must be performed in a vacuum or hydrogen atmosphere furnace. The use of bright heat treatments will not eliminate any distortion that may occur in the thermal processing

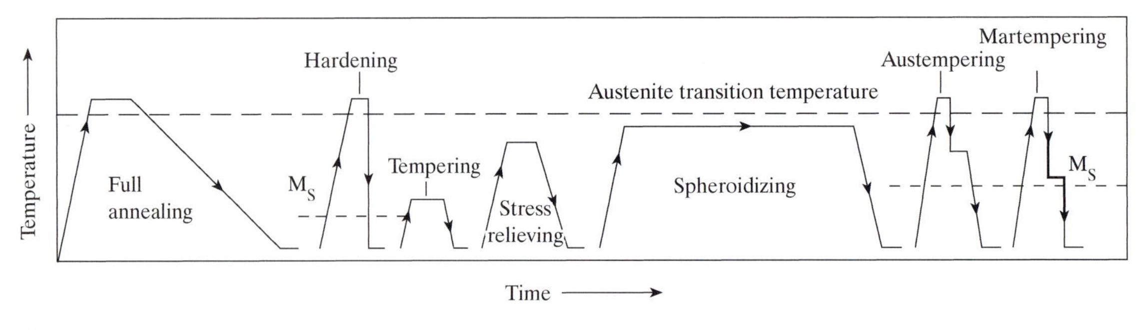

Figure 10–33
Heat treating thermal cycles

or size changes due to formation of martensite, but it will eliminate discolorations and size change due to oxidation (Figure 10–34). Thus bright heat treating should be specified on a drawing if it is warranted by the nature of the part. If there are functional regions on a part that will not be ground after hardening, that part may be a candidate for bright hardening.

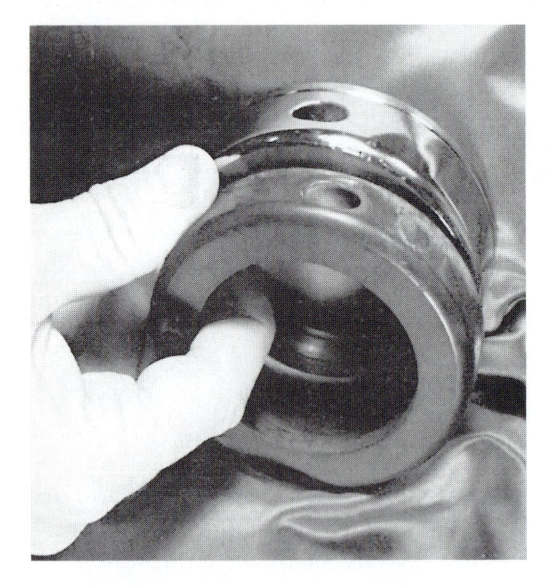

Figure 10–34
The held part was annealed in a carbon block. The black oxide is tight and adherent but still must be removed for this application. The other part was bright hardened in hydrogen.

10.8 Cost of Heat Treating

Many metalworking factories have in-house (captive) heat treating operations, but those that do not, use commercial heat treaters. Most industrial cities have a number of these. Some shops are divisions of large, nationwide heat treating corporations; some are locally owned operations. Some heat treaters specialize in certain types of heat treating such as vacuum or salt bath heat treatment, but most commercial heat treaters in the United States offer quench hardening, annealing, tempering, stress relieving, age hardening of nonferrous metals, carburizing, carbonitriding, flame hardening, vacuum heat treating, and ferritic nitrocarburizing. The facilities in captive heat treating departments in large factories depend on the product needs of the firm.

The cost of heat treating in a captive shop is determined by the cost of labor plus overhead, divided by the number of parts processed. Usually a particular heat treatment will have a labor charge such as 0.03 h/part. The plant hourly rate then determines the per-part cost. Some commercial heat treaters do production hardening of parts by the thousands for nearby metalworking industries. In such instances, the commercial heat treater will often price the heat treating operation per part. For example, parts small enough so that several fit in your hand can be hardened for a cost between \$0.03 and \$0.04 per piece. Special processes, of course, may have significantly different costs.

Commercial heat treaters also do heat treating of single parts and small quantities of parts. These heat treatments are often priced per unit mass, for example, $/lb. The following are examples of some heat treating prices that prevailed in 2003 in one area of the United States:

Hardening*	
Alloy steels	$0.90/lb
Tool steels	$1.60/lb
High-speed tool steel	$1.90/lb
Annealing*	$0.90/lb
Stress relieving*	$0.70/lb
Carburizing	$0.85/lb
Nitriding	$1.00/lb
Age hardening (nonferrous metals)	$0.90/lb

*Add 15% for vacuum processing

This type of pricing is most often employed on work received from tool and die companies where the normal fare is to harden small quantities of tool components. Vacuum heat treatment adds about 20% to the preceding prices. Most commercial heat treaters also employ minimum charges and service extras—for example, a minimum charge of $75 and a $35 surcharge for overnight service. The minimum charge is in place to accommodate shops that want to harden, say, only one dowel that weighs 2 ounces (0.06 kg). The per-pound rate structure would mean it would cost only $0.08 to harden the part. No commercial heat treater that wants to stay in business can harden a dowel for $0.08; minimum charges encourage customers to accumulate work before sending it for heat treating. If you want a special heat treatment that precludes inclusion of other customer parts in a furnace this is called buying the furnace. Costs can be in the range of $60 to $200/h.

In summary, heat treating costs depend on the situation; sometimes they can be insignificant and sometimes substantial. It usually costs only pennies (per piece) to harden small steel parts, but it can be expensive to harden a tool-steel injection molding cavity that weighs 700 lb (317 kg). Heat treating costs should be considered in designs that require substantial amounts of heat treated parts.

10.9 Selection and Process Specification

We have pointed out that under normal circumstances the designer is typically not responsible for specifying heat treating temperatures (except for special cases) but that the designer is responsible for the process and process limitations, if any. What is the proper drawing notation? To answer this question, let us assume a material of construction to be a hardenable carbon steel with about 0.8% carbon, and let us cite the proper notation for various processes using hypothetical hardnesses:

Purpose of Heat Treatment	Drawing Notation
Soften hardened part	Full anneal to 95 HRB maximum in an atmosphere to prevent carburization or decarburization per AMS 2759.
Remove welding stress	Stress relief anneal in an inert atmosphere at 645 to 655°C for 2 h plus 1 additional hour for each 25 mm of material thickness per AMS 2759. The maximum heating and cooling rate shall be 150°C/h.
Harden a part for wear resistance	Harden and temper to 58 to 60 HRC per AMS 2759.

Often a nationally recognized specification such as the Aerospace Materials Specifications (AMS) is referenced in the heat treating note to ensure that the parts are heat treated to a well-described process.

The other heat treatments, such as austempering, spheroidizing, and martempering, are usually used only for special applications, and additional drawing details may be required. These treatments should be used only after consulting a heat treater or metallurgist. Frequently, it is helpful to the fabricator to include the heat treating specification in an overall fabrication scheme:

Fabrication Sequence Example

1. Rough machine
2. Stress relieve
3. Semifinish machine (+grind allowance)
4. Harden and temper to 58 to 60 HRC
5. Finish grind

A notation like this helps to convey how you want the part made without insulting the fabricator's intelligence by pointing out what size drill or lathe bit to use. Always supply the heat treater with a satisfactory drawing designation on the type of steel. Do not use trade names. It is absolutely impossible to get a satisfactory heat treatment if the heat treater does not know the composition of the steel. See Table 10–3 for examples of common heat treating specifications.

All the diffusion processes discussed plus some additional proprietary processes are summarized in Table 10–2. The abrasion resistance of surfaces treated with some of these processes is compared in Figure 10–28. There are pros and cons to each, but the average designer can meet most diffusion treatment needs with carburizing by any technique and conventional gas, or ion nitriding. The other processes can be employed for special applications. Carburizing is usually the best process to use for surface hardening of low-carbon steels. If heavy sections are required, a more suitable substrate is a low-carbon alloy steel. Nitriding is a lower-distortion process than carburizing, but it requires the use of chromium–molybdenum alloy steels or preferably Nitralloy-type steel substrates. The TD and TiC diffusion processes have the best abrasion resistance.

Flame hardening is the preferred process for heavy cases or selective hardening of large machine components. The shallowest case that can be obtained with flame hardening is about 0.06 in. (1.5 mm); induction cases can be as thin as 0.01 in. (0.25 mm). Induction hardening works best on parts small enough and suitable in shape to be compatible with the induction coil. It is more suited to production-lot applications. Flame- and induction-hardened surfaces are usually employed over the diffusion processes when heavy cases are required. Heavy cases are desirable for surface fatigue applications such as wear from rolling elements.

The drawing specifications for diffusion treatments should include the following:

1. Basis metal
2. Process (and special sequences)
3. Case depth
4. Case hardness (if process involves tempering)
5. Applicable heat treatments
6. Masking (if desired)

Examples of surface-hardening treatments are shown in Table 10–2. Selective hardening requires the same types of notations with the possible addition of inspection techniques (see Table 10–2).

Every designer needs to be familiar with surface and selective hardening. Most machines require these processes on some parts. The basic advantage of all these techniques over through hardening is that only the surfaces subjected to wear are hard and brittle, and the remainder of the part can be tough to resist impact loading. Other advantages are that a part can often be made from a lower-cost material and distortion in hardening is lower.

Table 10–3
Summary of typical ferrous hardening processes

Heat Treatment		Example Steel Material	Example of a Typical Drawing Specification (hardnesses, case depths, and other parameters must be adjusted for specific applications)
Softening processes	Annealing	Type 1018 (UNS G10180)	Full anneal to 95 HRB maximum in an atmosphere to prevent carburization or decarburization per AMS[a] 2759 (add fixturing note as required).
	Stress relief annealing	Type 1018 (UNS G10180)	Stress relief anneal in an inert atmosphere at 645 to 655°C for 2 h plus an additional hour for each 25 mm of material thickness per AMS 2759. The maximum heating and cooling rate shall be 150°C/h. (Stress relieving is typically performed between the rough and semifinish machining processes; add fixturing notes as required.)
Through-hardening processes	Neutral hardening	Type 4140 (UNS G41400)	Harden and temper 35 to 40 HRC per AMS 2759 (commonly for alloy and carbon steels).
		Type D2 (UNS T30402)	Harden and double temper (2 h each) to 58 to 60 HRC. Cool to −75°C for 1 h between the first and second temper (common for some tool steels).
		Type M4 (UNS T11304)	Harden and triple temper (2 h each) to 60 to 62 HRC. Cool to −75°C for 1 h between the first and second temper (common for high-speed tool steels).
	Austemper	Type 1075 (UNS G10700)	Austemper to 40–45 HRC (for certain carbon and alloy steels).
	Martemper	Type E52100 (UNS G51320)	Martemper to 58–60 HRC.
	Bright hardening	Type 420 (UNS S42000)	Harden and double temper (2 h each) to 50 to 52 HRC per AMS 2759. Cool to −75°C for 1 h between the first and second temper. The part surface shall be protected from oxidation such that the surfaces shall be as bright as the as-received condition (used in special cases for tool steels or hardenable stainless steels).

Table 10–3
continued

Heat Treatment		Example Steel Material	Example of a Typical Drawing Specification (hardnesses, case depths, and other parameters must be adjusted for specific applications)
Surface-hardening processes[b]	Carburize	Type 8620 (UNS G86200)	Carburize 0.75–1.00 mm deep per AMS 2759. Temper at 175°C. Surface hardness shall be 60 HRC (equivalent) minimum. Effective case depth shall be 50 HRC (equivalent) at 0.75–1.00 mm deep.
	Carbonitride	Type 1018 (UNS G10180)	Carbonitride 0.13–0.25 mm deep per AMS 2759. Temper at 175°C. Surface hardness shall be 60 HRC (equivalent) minimum. Effective case depth shall be 50 HRC (equivalent) at 0.13–0.25 mm deep.
	Nitride	Nitralloy® 135M (ASTM A355 Class A)	1. Rough machine 2. Harden and temper to 24–32 HRC (or use prehardened Nitralloy 135M) 3. Finish machine 4. Nitride 0.25–0.40 mm deep per AMS 2759 finish grind to remove white layer (0.025–0.050 mm).
	Ferritic nitrocarburize	Type 1018 (UNS G10180)	Salt-bath ferritic nitrocarburize, compound zone shall be 0.004–0.008 mm deep per AMS 2753B (typically performed on finished machined parts).
	Gaseous ferritic nitrocarburize	Type 1018 (UNS G10180)	Ferritic nitrocarburize and oil quench, compound zone shall be 0.004–0.008 mm deep per AMS 2757B (different quench media may be used depending on the part, typically performed on finished machined parts).
Selective-hardening processes	Flame hardening	Type 1095 (UNS G10950)	Flame harden and temper areas shown 1.00–1.50 mm deep. Surface hardness shall be 58 HRC minimum. Inspect for cracks.
	Induction hardening	Type 1095 (UNS 10950)	Induction harden and temper areas shown 1.00–1.50 mm deep. Surface hardness shall be 58–62 HRC.

[a] AMS is the acronym for a U.S. government military specification that presents details for cleaning, heating rate, temperature measurement, certification, etc.
[b] Most diffusion processes can be selective through the use of masks.

Summary

This chapter contains some hard-core metallurgy that applies to most steels. It is extremely important that material users have a feel for what heat treatments do and how they affect part performance. It is also important to know how heat treatments are done so that they can be properly specified and sequenced into fabrication schemes. The following are some summary thoughts on this subject:

- Equilibrium diagrams provide profiles of alloy systems: the phases present, heat treat temperatures, compositions to avoid, temperatures to avoid, and so on.
- The concept of solid solubility must be understood; many heat treating operations are based on the solubility characteristics of metals (quench hardening, precipitation hardening, and the like).
- Many metals of industrial importance are multiphase, and the relative amounts of various phases present determine the properties of the alloy.
- The stable phases in soft steel at room temperature are ferrite and cementite; martensite is the hard phase.
- The iron–carbon diagram is probably the most important reference on the metallurgy of carbon steels.
- The requirements for hardening a steel are (1) heating to the proper temperature, (2) sufficient carbon content, and (3) adequate quench. All three must be met.
- Quench-hardened steel always requires tempering to prevent brittleness.
- Stress relieving is a subcritical process, but adequate temperature must be used for it to be effective [1200°F (650°C) for most carbon steels].
- Each hardenable steel has quenching requirements that must be met; IT diagrams are used to predict quenching requirements.
- Flame and induction hardening require the use of hardenable steels.
- Diffusion processes usually apply only to steels with insufficient carbon content for through hardening.
- Some metallic coatings such as electroplated copper provide a barrier to the diffusion of carbon and nitrogen. They are used for masking areas to be left soft. There are also special paints that perform this function.
- Carburizing does not harden a steel; quenching after carburizing is necessary to get hardening.
- Nitriding is the lowest-temperature diffusion-hardening process, and it does not require a quench.
- Induction hardening is usually most efficient on small parts.
- Diffusion takes place in all metals at elevated temperatures if there is a concentration gradient in the metal or in the atmosphere surrounding the metal.
- Carbon and nitrogen diffusion processes are available in most heat treating facilities; the diffusion processes involving other species, such as boron, chromium, and vanadium, are less widely available and the processes may be proprietary.
- Diffusion treatments and selective hardening are most commonly used when a through-hardened part would be too brittle for the intended service.
- All heat treatments over 1000°F (538°C) must be done in protective atmospheres if a part's surface or dimensions are important. Oxidation will occur.
- Stress relieving should be considered on most parts with close dimensional tolerances.
- Heat treating drawing notes should show the type of steel, the desired process, the desired hardness, and any special processing, such as deep freeze or double temper.

Most of the decisions in heat treating are made by the heat treater (temperatures, quenching media, and so on), and designers must be familiar enough with the available processes to let the heat treater know what is wanted in unambiguous terms. If the drawing specification is clear, the parts will usually come out right.

Critical Concepts

- Quench hardening of steel involves atomic interactions—movement of solute atoms and crystal structure changes.
- The iron–carbon diagram is the source of the temperature information needed to heat treat carbon steels.
- Some steels can be direct hardened. Others require diffusion processes to harden the surface (low carbon, <0.3%).
- Carbon content is the primary factor that determines a steel's ability to quench harden.

Terms You Should Remember

iron carbide
phase
intermetallic compound
solid solution
lattice
equilibrium diagram
ferrite
austenite
cementite
pearlite
microstructures
dendrites
eutectic
eutectoid
quench hardening
recrystallization
stress relief
tempering
diffusion
IT diagram
M_S temperature
deep freezing
tie line
austempering
martempering
CT diagram
quenching
binary
interstitial
substitutional
diffusion
case hardening
selective hardening
flame hardening
induction hardening
eddy currents
carburizing
pack
nitriding
white layer
diffusion zone
carbonitriding
cyaniding
normalizing
spheroidizing
process anneal
hardenability
controlled atmosphere
annealing
solute
pack cementation
ferritic nitrocarburizing
ion nitriding
microhardness
case depth
cyanide salt
bright hardening
case
Metalliding
chromizing
boronizing
liquid carburizing
laser hardening
conveyor hearth
retort
decarburization
dissociation
vacancy
diffusion coefficient
allotropy

Case History

Emergency Heat Treatment

Many U.S. manufacturers outsource heat treating to companies who specialize in this area. However, subcontracting means that it may take a week under ideal conditions to get parts hardened. Everyone in the machine building department was well aware that heat treating took at least a week and planned their work accordingly. However, on a debugging run of a new machine, a drive spindle showed significant wear produced by belt rubbing on the face of the spindle. The machine was scheduled for shipping to the customer the day after the debugging run. A simple hardened flange could solve the wear problem, but the delivery schedule did not allow time to harden the remedial flange at the normal vendor.

The shop had a furnace capable of reaching hardening temperatures, but without atmosphere control. Heating an A2 tool-steel part to 1800°F in air would oxidize the part to uselessness. The protective atmosphere problem was solved by packing the part in cast-iron chips in a small stainless-steel cup and wiring the stainless-steel lid on the cup. The container was austenitized, quenched (water), and tempered. The emergency part came out only slightly discolored, and it was the required hardness of 60 HRC.

It is possible to harden steels with any furnace that will get hot enough, but if the hardening atmosphere is not controlled, the parts will be ruined. In this emergency job, we used the CO given off by reaction of the graphite in the cast iron with oxygen to produce a protective atmosphere—an old heat treater's trick. The heat treating atmosphere is a significant part of the process.

Questions

Section 10.1

1. Using Figure 10–6, calculate the percentage of α phase present at room temperature in a 50% B alloy.
2. What phases are present in a 55% A alloy at 600°? (Refer to Figure 10–6.)
3. What is the eutectoid composition in carbon steels? (Refer to Figure 10–7.)
4. How many microconstituents are there in a ternary alloy?

Section 10.2

5. Name the phases that are present under equilibrium conditions in a 0.4% carbon steel at room temperature.
6. What phases are present in a 0.8% carbon steel at 800°C?
7. Can a steel with 0.06% carbon content be quench hardened? Explain your answer.
8. How does the lattice of martensite (BCT structure) vary from the lattice of body-centered cubic iron?
9. What are the microconstituents of a 1.00% carbon steel that has been austenitized and furnace cooled to room temperature? Calculate the fraction of each named microconstituent.

Section 10.3

10. What heat treatments would you use to make a ball bearing race that is 6 in. (15 cm) in outside diameter? A section through the ring would have approximate dimensions of 0.5 × 0.5 in. (1.3 × 1.3 cm).
11. What is the driving force for inward diffusion of species in diffusion hardening processes?

Section 10.4

12. What is the hardening temperature for a 1% carbon steel?
13. What is the stress-relieving temperature for a 1% carbon steel?
14. What will be the quenched structure of a 1-in.-diameter bar of 1040 steel when a thermocouple in the center of the bar indicates a quench rate of 100°F (38°C) per second? Assume that Figure 10–15 applies.
15. Compare the hardness of a half-hard temper cold-finished steel bar (0.25- × 1-in. section, or 12.5 × 50 mm) to the hardness of a 0.8% carbon steel that has been oil quenched from 1550°F (843°C) and tempered at 300°F (149°C).
16. A steel part is hot forged to shape and allowed to air cool from the forging temperature. What kind of structure will the metal have?
17. You want to make a part from a hardenable tool steel. What type of heat treatment (if

any) should you ask for on the bar as you receive it from the steel supplier?

18. You want to make a flat spring from 1% carbon steel strip. You are troubled with excessive distortion. What heat treatment might solve the problem?
19. Compare the quenching capability of oil with that of 10% sodium hydroxide in water.
20. What is the purpose of deep freezing in the heat treatment of steels?
21. You want to harden a complicated part that cannot be ground after hardening. What hardening process should you specify?
22. You are going to make one million screwdriver blades. Write the fabrication and heat treating specification that you would use on the part drawing.

Section 10.5

23. What is the minimum carbon content required to produce a hardness of 60 HRC in a carbon steel?
24. Can a steel with 0.4% carbon be carburized?
25. The activation energy for the diffusion of carbon in steel at room temperature is 20; the diffusion constant is 1.6×10^{-4} m^2/s. Calculate the diffusion coefficient at 1000°C and compare this number with the value at room temperature.
26. Name two elements that will diffuse in steel as interstitials.
27. What is the minimum depth of case that can be specified with flame hardening?
28. How hot does a 1040 steel part get when it is flame hardened?
29. What is the role of frequency in induction hardening?
30. What is the shallowest case depth possible with induction hardening? With carburizing?
31. Where does the carbon come from in pack carburizing?
32. How long would it take to get a 1-mm carburized case depth at about 800°C, compared with about 1000°C treatment temperature?
33. How is effective case depth determined on a surface-hardened part?
34. What is the diffusion temperature for nitriding?
35. What is the maximum case depth possible with nitriding?
36. What would be the approximate treatment time to get a 0.5-mm nitride case on Nitralloy 135?
37. Where does the nitrogen come from in ion nitriding?
38. What is the process temperature and what are the diffusion species in ferritic nitrocarburizing?
39. What diffusion process would you use to silicide the surface of a piece of steel?
40. What is the best process for hardening the teeth on an 8-in. pitch diameter, 14-pitch gear with a face width of 1 in.?
41. What are three suitable substrates for nitriding?
42. Write a fabrication sequence for a long, slender nitrided 4140 steel shaft.

Section 10.6

43. What is the recrystallization temperature for 1020 steel?
44. What is the difference between annealing and normalizing?
45. What is the mechanism of annealing?
46. How do you determine the correct tempering temperature for a hardenable steel?
47. What is the purpose of martempering? Austempering?

Section 10.7

48. What is bright hardening?

49. How do you quench a steel in vacuum hardening?

Section 10.8

50. How do you specify bright hardening?

51. Write a heat treat/fabrication sequence for a tool-steel blanking die for a flat washer.

To Dig Deeper

Alloy Phase Diagrams. Materials Park, OH: ASM International, 1992.

ASM Handbook, Volume 4, Heat Treating. Metals Park, OH: ASM International, 1991.

Ashby, M. F., and D. R. Jones. *Engineering Materials*. Elmsford, NY: Pergamon Press, 1980.

Atkins, M. *Atlas of Continuous Cooling Transformation Diagrams for Engineering Steels*. Materials Park, OH: American Society for Metals, 1980.

Boxer, H. S. *Practical Heat Treating*. Materials Park, OH: ASM International, 1986.

Brooks, Charlie R. *Heat Treatment of Ferrous Alloys*. New York: Hemisphere Publishing Co., 1979.

Brooks, C. R. *Principles of the Heat Treatment of Plain Carbon and Low Alloy Steels*. Materials Park, OH: ASM International, 1996.

Carburizing and Carbonitriding. Materials Park, OH: American Society for Metals, 1977.

Enos, George M., and William E. Fontaine. *Elements of Heat Treatment*. New York: John Wiley & Sons, 1966.

Heat Treater's Guide: Practices and Procedures for Irons and Steels, 2nd ed. Materials Park, OH: ASM International, 1995.

Krauss, George. *Steels: Heat Treatment and Processing Principles*. Materials Park, OH: ASM International, 1990.

Metals Handbook, Volume 4, Heat Treating. Materials Park, OH: ASM International, 1991.

Shrivastava, S., and F. Specht, Eds. *Heat Treating*. Materials Park, OH: ASM International, 2002.

Source Book on Nitriding. Materials Park, OH: American Society for Metals, 1977.

Vander Voort, G., Ed. *Atlas of Time-Temperature Diagrams for Irons and Steels*. Metals Park, OH: ASM International, 1991.

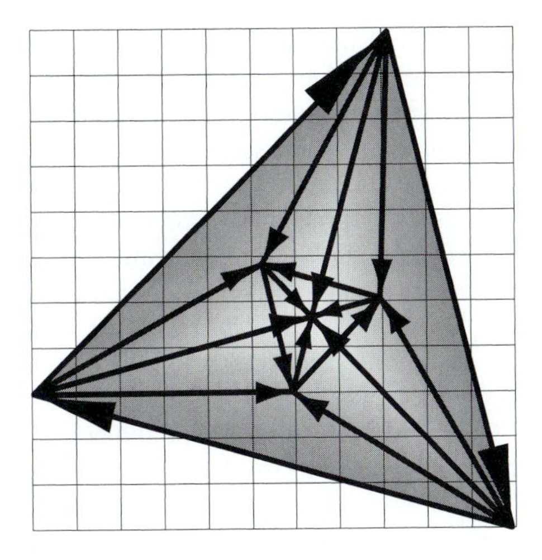

CHAPTER 11

Carbon and Alloy Steels

Chapter Goals

1. An understanding of how cold finishing and alloy additions alter steel properties.
2. A knowledge of identification systems used on carbon and alloy steels.
3. An understanding of the differences among various carbon and alloy steels and development of a repertoire of steels for general use in design.

Rationale

The previous chapter, on heat treatment of steels, hopefully made the case for the importance of steels as engineering materials. No material is more important. Steels have allowed society to advance to where we are in 2004. We take automobiles, long bridges, and high-rise buildings for granted, but they would not exist without steel. Engineers need to know some of the details about steels to continue using them in ways that make life easier and society better. Designers need to know what cold-finished steels are, the varieties that are produced in steel mills, differences between them, and how to specify them. Designers should know what happens when steels are "improved" with alloy additions. Knowing how alloying works helps in understanding why there are so many types of steels. Finally, none of this knowledge can be implemented without understanding steel designation systems. How do you specify a particular steel on an engineering drawing so that the part can be made anywhere in the world? This chapter is intended to make our readers proficient in selecting and specifying steels.

Carbon steels are simply alloys of iron and carbon, with carbon as the major strengthening agent. The handbook of the U.S. steel industry, the *Steel Products Manual,* published by the Iron and Steel Society, more rigorously describes *carbon steels* as steels with up to 2% carbon and only residual amounts of other elements except those added for deoxidation (for example, aluminum), with silicon limited to 0.6%, copper to 0.6%, and manganese to 1.65%. Other terms applied to this class of steels are *plain carbon steels, mild steels, low-carbon steels,* and *straight-carbon steels*. These steels make up the largest fraction of steel production. They are available in almost all product forms: sheet, strip, bar, plates, tube, pipe, wire. They are used for high-production items such as automobiles and appliances, but they also play a major role in machine design for base plates, housings, chutes, structural members, and countless machine components.

Alloy steel is not a precise term. It could mean any steel other than carbon steels, but accepted application of the term *alloy steel* is for a group of steels with varying carbon contents up to about 1% and with total alloy content below 5%. The *Steel Products Manual* defines alloy steels as steels that exceed one or more of the following limits: manganese, 1.65%; silicon, 0.60%; copper, 0.60%. A steel is also an alloy steel if a definite concentration of various other elements is specified: aluminum, chromium (to 3.99%), cobalt, molybdenum, nickel, titanium, and others. These steels are widely used for structural components that are heat treated for wear, strength, and toughness. They are the types of steels used for axle shafts, gears, and hand tools such as hammers and chisels.

A family of steels related to but different from alloy steels is *high-strength, low-alloy (HSLA) steels.* This term is used to describe a specific group of steels that have chemical compositions balanced to produce a desired range of mechanical properties. These steels are also called *microalloyed* steels. Some of these steels also have alloy additions to improve their atmospheric corrosion resistance. They are available in various products, but usage centers about sheet, bar, plate, and structural shapes. The yield strength is usually in the range of 42 to 70 ksi (289 to 482 MPa), with the tensile strength 60 to 90 ksi (414 to 621 MPa). The primary purpose of these steels is weight reduction through increased strength. Smaller section sizes are possible.

In addition to carbon, alloy, and high-strength, low-alloy steels, there are *tool steels,* steels for special applications such as pressure vessels and boilers, *mill-heat-treated* (quenched and tempered or normalized) *steels,* and *ultra-high-strength steels*. All these steels can be useful in engineering design, but the most important are undoubtedly the carbon and low-alloy steels and tool steels. Tool steels are important enough to warrant a chapter of their own (Chapter 12). This chapter will be primarily concerned with the selection of carbon and alloy steels. Limited coverage will be given to some of the other groups mentioned. More specifically, we shall discuss alloy designation systems, the metallurgy of carbon steels, alloying effects, and selection considerations. The overall purpose of this chapter is to provide an understanding of the spectrum of steel alloys and to aid the designer in the selection of steel. Carbon and alloy steels will be the most commonly used, but when these steels do not meet design requirements, some of the more special steels should be employed. All the steels discussed in this chapter are considered to be in the wrought form. Cast steels will be covered in Chapter 16.

11.1 Alloy Designation

In the 1970s and 1980s the steel industry was very active in the area of alloy development. The high-strength, low-alloy, quenched and tempered, and some of the ultra-high-strength steels were developed in this period. They meet industry needs for weight reduction, higher performance, and, in many cases, lower costs. The disadvantage from the designer's standpoint is that it became difficult to categorize steels in an orderly fashion to aid selection. The common denominator for the steel systems defined in our introductory remarks is use. They are the types of steels that would be used for structural components. Figure 11–1 is an attempt at categorization. Even with the abundance of special-purpose steels, the workhorses are and will continue to be for some time the wrought ASTM and AISI/SAE carbon and alloy steels. Fortunately, these steels have an understandable and orderly designation system. We shall describe this system in detail and make some general comments on the other steel systems shown in Figure 11–1.

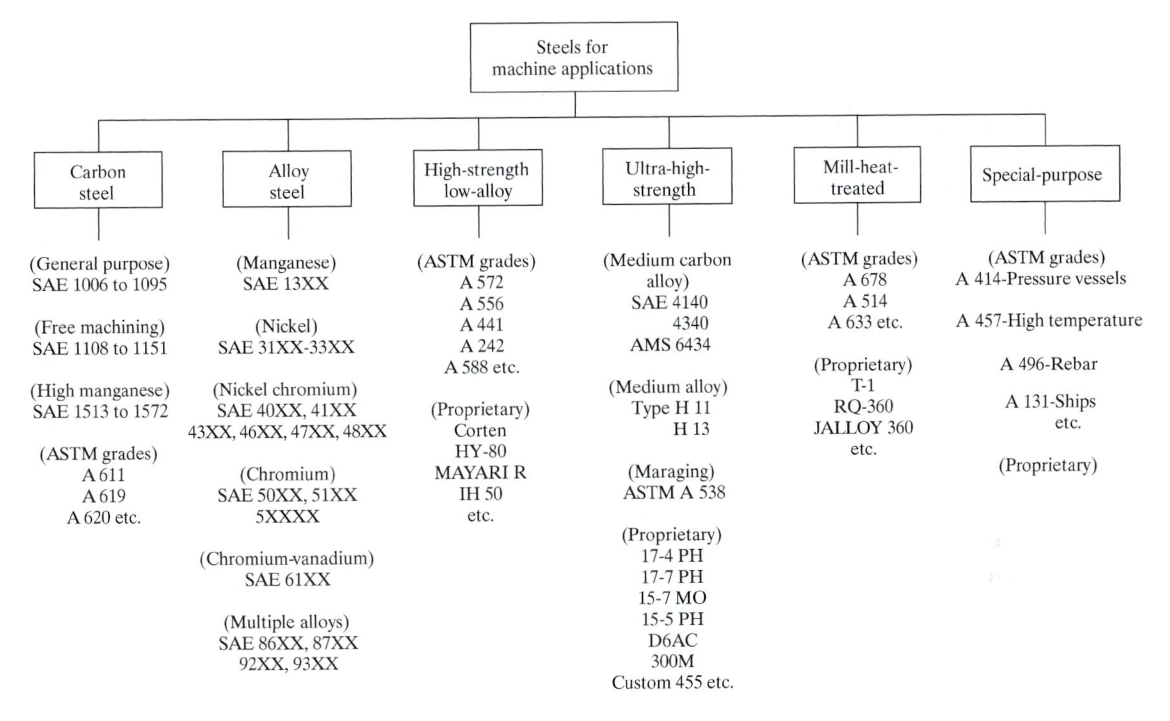

Figure 11–1
Steels for machine design (excluding tool steels)

Carbon and Alloy Steels

The most widely used identification system for carbon and alloy steels in the United States is the system adopted by the American Iron and Steel Institute (AISI) and the Society of Automotive Engineers (SAE). This system usually uses only four digits. The first digit indicates the grouping by major alloying elements. For example, a first digit of 1 indicates that carbon is the major alloying element. The second digit in some instances suggests the relative percentage of a primary alloying element in a given series. The 2xxx series of steels has nickel as the primary alloying element. A 23xx steel has approximately 3% nickel; a 25xx steel has approximately 5% nickel. The last two digits (sometimes the last three) indicate median carbon content in hundredths of a percent. A 1040 steel will have a normal carbon concentration of 0.40%. The classes of steels in this system are shown in Table 11–1.

In addition to the four digits, various letter prefixes and suffixes provide additional information on a particular steel (see Table 11–2).

There were AISI/SAE systems for identifying other classes of steels, such as stainless steels and tool steels, but we shall cover these in later discussions.

The ASTM International has published voluntary industry consensus specifications on many types of ferrous and nonferrous metals. Each specification number covers a particular alloy or family of alloys. For example, ASTM A 36 covers carbon steels for structural applications. There are several grades in the specification, and to specify the use of one of these alloys, the grade designation would be included (for example, ASTM A 36 grade B). The last two digits of an ASTM number are the year of the latest

Table 11–1
Major groups in the SAE steel designation system

Class	SAE Series	Major Constituents
Carbon steels	10xx	Carbon steel
	11xx	Resulfurized carbon steel
Alloy steels		
Manganese	13xx	Manganese 1.75%
	15xx	Manganese 1.00%
Nickel	23xx	Nickel 3.50%
	25xx	Nickel 5.00%
Nickel–chromium	31xx	Nickel 1.25%, chromium 0.65 or 0.80%
	33xx	Nickel 3.50%, chromium 1.55%
Molybdenum	40xx	Molybdenum 0.25%
	41xx	Chromium 0.5 to 0.95%, molybdenum 0.12 to 0.20%
	43xx	Nickel 1.80%, chromium 0.50 or 0.80%, molybdenum 0.25%
	46xx	Nickel 1.80%, molybdenum 0.25%
	48xx	Nickel 3.50%, molybdenum 0.25%
Chromium	51xx	Chromium 0.80, 0.88, 0.93, 0.95, or 1.00%
	52xxx	Chromium 1.45%
Chromium–vanadium	61xx	Chromium 0.80 or 0.95%, vanadium 0.10 or 0.15% min.
Multiple alloy	86xx	Nickel 0.55%, chromium 0.50%, molybdenum 0.20%
	87xx	Nickel 0.55%, chromium 0.50%, molybdenum 0.25%
	92xx	Silicon 2.00%, or 1.40% and chromium 0.7%
	93xx	Nickel 3.25%, chromium 1.20%, molybdenum 0.12%
	94xx	Manganese 1.00%, nickel 0.45%, chromium 0.40%, molybdenum 0.12%
	94Bxx*	Nickel 0.45%, chromium 0.4%, molybdenum 0.12%

*Denotes boron.

revision. Most steels are covered by ASTM specifications: sheet, strip, wire, and steels used for structural applications such as building beams, columns, piping, tanks, and pressure vessels. ASTM specifications are available in bound volumes in all technical libraries and in most engineering offices. The ASTM designations for carbon and alloy steels use the AISI designations as grades. For example, AISI/SAE 1020 steel is ASTM A 29 grade 1020.

In an attempt to come up with a designation system that would apply to all metal alloys, steels included, the ASTM and the SAE have developed a Unified Numbering System (UNS). The UNS does not include specifications for compositions and properties of alloys, it merely provides a number to identify an alloy that is covered by specifications written by AISI or some other society or trade association. Where possible, the five-digit UNS number is preceded by a letter that indicates what metal system is involved: Axxxxx is used for aluminum alloys, Pxxxxx is used for precious metals. The five digits often coincide in some fashion with the identification system that they are replacing. For example, type 1020 steel is covered by UNS G10200. The first four digits come from the AISI/SAE system, and a 0 is added as the last digit. An outline of the entire Unified Numbering System is shown in Table 11–3.

The UNS is described in detail in the ASTM DS-56E specification. The system is discussed in this text because some property handbooks have adopted it in identifying alloys, and at least one

Table 11–2
Letter prefixes and suffixes in steel identification

Prefix	Meaning
E	Made in an electric furnace
X	Composition varies from normal limits
Suffix	**Meaning**
H	Steel will meet certain hardenability requirements
Other Letters	**Meaning**
xxBxx	Steel with boron as an alloying element
xxLxx	Steel with lead additions to aid machinability

Table 11–3
Outline of the Unified Numbering System for metals

UNS Number	Alloy System
Axxxxx	Aluminum and aluminum alloys
Cxxxxx	Copper and copper alloys
Exxxxx	Rare earth and rare element–like metals and alloys
Fxxxxx	Cast irons
Gxxxxx	Former AISI and SAE carbon and alloy steels
Hxxxxx	Former AISI and SAE H-steels
Jxxxxx	Cast steels (except tool steels)
Kxxxxx	Miscellaneous steels and ferrous alloys
Lxxxxx	Low-melting metals and alloys
Mxxxxx	Miscellaneous nonferrous metals and alloys
Nxxxxx	Nickel and nickel alloys
Pxxxxx	Precious metals and alloys
Rxxxxx	Reactive and refractory metals and alloys
Sxxxxx	Heat and corrosion resistant (stainless steels)
Txxxxx	Tool steels, wrought and cast
Zxxxxx	Zinc and zinc alloys

trade organization the U.S. Copper Development Association, has adopted UNS numbers as the official identification system for all copper alloys. We have opted not to show UNS numbers in our discussions of all metals other than copper alloys. Thus this is the extent of our discussion of the UNS metal identification system.

As mentioned in Chapter 9, in the mid-1980s, the AISI deferred steel designation and specification to other organizations. The AISI/SAE designation system for carbon, alloy, and tool steels is now only the *"SAE identification"* system. We will use SAE designations for steel wherever possible because the system is much easier to understand and remember than ASTM alloy designations and UNS alloy identification numbers. Many countries have their own designation systems but all steel producers have products that fit into the steel categories that we will discuss. ASTM specifications and designations are recognized worldwide.

High-Strength, Low-Alloy Steel

There is a significant amount of commercial competition in this family of steels, and trade names are often used to designate high-strength, low-alloy steels, but it is wise to avoid this practice. The most accepted practice is to separate these alloys by minimum tensile properties for a given section thickness. Chemical compositions are published, but the ranges are wide and alloy additions are balanced to meet mechanical properties rather than composition limits. The preferred system to use in specifying one of these alloys on an engineering drawing is to use an ASTM designation number followed by the strength grade desired. ASTM specifications A 242, A 440, A 441, A 588, and A 572 cover some of these steels in structural shapes. ASTM specifications A 606, A 607, A 715, A 568, A 656, A 633, A 714, and A 749 cover some of these steels in sheet and strip form. A typical drawing notation for one of these steels is

Material: Steel, ASTM A 242, grade 70

Ultra-High-Strength Steel

The discriminating characteristic of this class of steels is high strength. Yield strengths usually exceed 175 ksi (1206 MPa). Certain of the alloy steels are considered to be ultra-high-strength steels (for example, 4140, 4340), as are some of the former AISI tool steels, H11 and H13. In these cases, proper designation is achieved by using the SAE system. Some steels are strictly proprietary (for example, D6AC, 300M, 17-4PH). In this instance there is no recourse but to use the trade name on the drawing along with the name and address of the manufacturer. A very useful class of ultra-high-strength structural steels is the 18% nickel *maraging steels*. These are covered by ASTM specification A 538, and this specification can be used for alloy designation.

Mill-Heat-Treated Steel

Proprietary and ASTM specifications cover these alloys. The ASTM specifications again are the preferred method of designation. For example, ASTM specification A 678 covers quenched and tempered carbon steel plates for structural applications. ASTM A 663 covers "Normalized High-Strength Low-Alloy Structural Steel," and A 514 covers "High Yield Strength, Quenched and Tempered Alloy Steel Plate Suitable for Welding." There are abrasion-resistant grades with hardnesses between about 300 and 400 HB, but trade names often are used to specify these grades (for example, Jalloy 360).

Special-Purpose Steel

Many steels have been developed for special applications. There are steels intended for high-temperature and low-temperature service, for springs, for pressure vessels, for boilers, for use in concrete, for railroads, and for all sorts of service conditions. In addition to steels for specific types of service, there are steels with various coatings (for example, tin plate, terne, and aluminized). These can be applied to any type of service, and specifications center on base metal and coating characteristics. The *Steel Products Manual* has product specifications on coated steels, rail steels, and steels for various building applications. There is no numbering system on these steels and product information is best obtained in the United States from the *Steel Products Manual*. The ASTM has specifications on many coated products, rail products, and steels intended for particular types of service. The ASTM specification number can be used for specifying these steels. There is an ASTM index that supplies information on the availability of specifications on these types of steels. Special-purpose steels and many steels from the other categories mentioned are covered by competitive alloy-designation systems developed by various government regulatory bodies. These alloy-designation systems are usually mandated for government work, but in private industry the predominating alloy-designation systems for steels used in machines are the SAE and ASTM systems.

11.2 Carbon Steels

We previously defined a carbon steel as an alloy of iron with carbon as the major strengthening element. The carbon strengthens by solid solution strengthening, and if the carbon level is high enough, the alloy can be quench hardened. As discussed in Chapter 1, all metals, in addition to strengthening by alloying, can be strengthened by mechanical working, or cold finishing, as it is more appropriately termed. Carbon steels are available in all the mill forms (bar, strip, sheet shapes), and an important selection factor pertaining to carbon steels is whether to use a cold- or hot-finished product. Similarly, the designer must make a decision about whether to use a hardenable or nonhardenable alloy. Thus, in this section we shall discuss carbon steels with regard to selection criteria for hot- and cold-finished products and hardening of carbon steels.

Cold Finished

In Chapter 9 we described how cold finishing caused work hardening by grain reduction and buildup of dislocation density. This phenomenon occurs in most ductile metals; it is one of the marvels of metal. The strength of a metal can be increased by as much as a factor of 100 by simply reducing the cross section of a shape by rolling, drawing, swaging, or some related process. The mechanism by which this strengthening occurs is not as simple as we have implied. It is not making big grains small that strengthens. It is the complex atomic interactions that occur in the crystalline structure of metals.

Figure 11–2 shows some idealized stress–strain diagrams for a metal and a ceramic. In Chapter 2 we showed a typical stress–strain diagram as one would generate in tensile testing a piece of low-carbon steel. It would look something like *abg* in Figure 11–2. There is an anomaly in traditional stress–strain curves that may not be apparent. The stress at failure (point *g*) is lower than the stress reached when the sample is stretching. How can this be? The explanation is that stress is calculated on the original cross-sectional area of the test sample, but the sample is really necking down, so its area is smaller than when the test started. People who perform tensile tests of metals have learned to accept this, and because everybody tests this way there is no problem. However, if we were able to measure the instantaneous area of a tensile sample in testing, the curve generated would be the true stress–strain diagram. The part fails at the highest stress. The area used in the calculation of true stress is the instantaneous area, and the true strain is

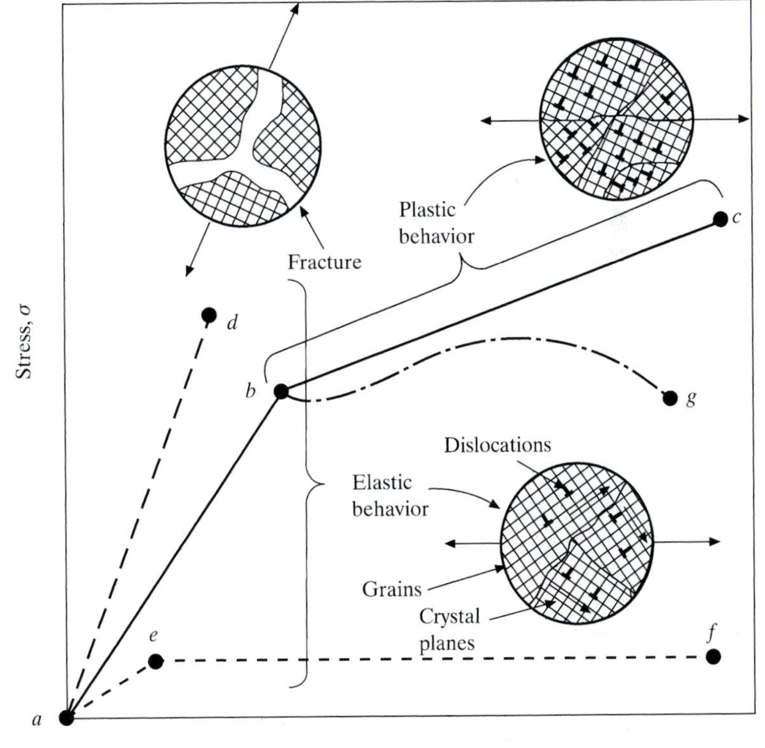

Figure 11–2
Curve *abg* is the room temperature stress–strain curve for a ductile steel. Curve *abc* is the true stress–strain curve for the same steel. Curve *ad* is the true stress–strain curve for a ceramic. Curve *aef* is the true stress–strain curve for a steel at hot-working temperatures.

calculated as the natural log (ln) of the instantaneous length divided by the original sample length. These curves are now frequently generated by newer tensile test machines that use computers to process the test data.

How does all this relate to cold finishing? When a material is cold finished by whatever process, it is undergoing essentially a tensile (or compression) test. If we were to measure a material's response to cold-finishing operations, we would essentially be operating in the *bc* region of the true stress–strain curve, *abc*. In the elastic part of the curve (*ab*) the deformation is recoverable. If we stop loading the material, the strain that was induced is recoverable. On the atomic scale some dislocations may have been produced, but little happened crystallographically to change the structure that was present before loading. In the plastic region of the stress–strain curve, *bc*, many things happen on the atomic scale. Grains are elongated; large numbers of dislocations are produced; they intersect each other; they collide with alloy elements; they spontaneously multiply; they change slip planes. They are said to be turbulent in motion. The net effect of all these dislocation interactions is a strengthening of the material. Figure 11–3 shows the strengthening that can occur in carbon steel in cold finishing. Different metals behave differently under plastic flow conditions. The slope of the plastic flow curve (on a true stress–strain curve), *bc,* is called the work-hardening exponent. The length of the curve is also a measure of the ability of a material to plastically deform. To have good malleability in the cold condition, the metal must be capable of a lot of stretch (strain) between the yield point and the tensile strength. Face-centered cubic metals and some of the body-centered cubic metals have good plastic flow characteristics. The hexagonal structure metals such as magnesium and titanium can have dislocation motion only on the planes that are parallel to their base planes (basal). They have very limited formability.

Ceramic materials have stress–strain diagrams like *ad* in Figure 11–2. They are perfectly elastic up to the point of fracture. There is no ability to plastically deform. The crystallites in the microstructure cleave, and there is none of the dislocation strengthening that occurs in the ductile metals. A stress–strain curve for a steel when it is red hot may look something like *aef*. The stress required for plastic flow is very low, and there is no *work hardening*. The material is completely plastic. It is obvious that plastic flow is much easier at elevated temperatures, and this is the reason that much of the shaping of steel products is done when the steel is red hot; this is hot working. Dynamic recrystallization occurs, and dislocations and other defects are annihilated.

Other dislocation phenomena happen in cold work, but it is important to remember that the ability to work harden is a very special property that belongs only to certain metals, and

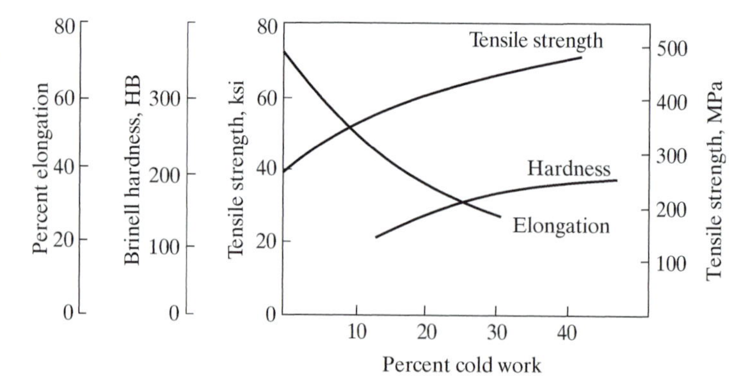

Figure 11–3
Effect of mechanical working on the properties of iron

Source: L. H. Van Vlack, *Elements of Material Science,* Reading, MA: Addison-Wesley Publishing Co., Inc., 1959

carbon steels are so useful from the materials standpoint because they have favorable plastic flow characteristics.

Steel sheet, bar, strip, and special shapes may be purchased from mills in various degrees of cold-work strengthening. This is accomplished by rolling sheet cold through large mills or by drawing through dies that reduce their cross-sectional area. The thickness or diameter goes down, the sheet or strip gets longer, and the strength goes up, depending on how much reduction is produced in rolling. Most carbon steel products are available as *cold-rolled* product. Typically, cold-finished products do not receive reductions greater than about 10%, but the strengthening effect can yield as much as a 20% increase in tensile strength over the annealed condition. The yield strength (at 10% reduction) may be increased as much as 50%. If a one-of-a-kind part is being designed with cold-finished material and it is overdesigned to the point that strength is not critical, it is probably adequate to specify cold-rolled xxxx steel. If strength is critical, it may be necessary to specify a desired degree of cold finishing. In bars and sheet, the temper designation system is one-quarter, one-half, three-quarters, full-hard, and *skin-rolled* temper. In strip and tin mill products a number system is used to designate temper or degree of cold work. Table 11–4 illustrates the difference in mechanical properties for these various strip tempers. The numbering system for tin mill products is different from the strip system; T1 temper is the softest temper. It is uncommon to get sheet materials in the harder tempers, because the rolling of large widths would require extraordinary rolling mills.

From the standpoint of fabricability, the harder the temper, the less the ability to cold form into the desired shape (*formability*). In general, full-hard materials cannot be bent even 90° without a generous bend radius. Skin rolling, besides providing a good finish, eliminates *stretcher strains*. Stretcher strains are road map–type lines on a severely deep drawn sheet. They come from

Table 11–4
Mechanical properties of carbon steel strip with maximum 0.25% carbon

Temper	Rockwell Hardness	Nominal Tensile Strength [psi (MPa)]
No. 1 hard temper	B-85 minimum	90,000 (620)
No. 2 *half-hard* temper	B-70 to 85	65,000 (450)
No. 3 quarter-hard temper	B-60 to 75	55,000 (380)
No. 4 skin-rolled temper	B-65 maximum	50,000 (350)
No. 5 dead-soft temper	B-55 maximum	50,000 (350)

nonuniform atomic slip in the area of the yield point. By yielding the surface of a sheet before forming, this objectionable defect is eliminated. Figure 11–4 illustrates the effect of temper rolling on formability.

All steels can be cold finished, but it is not common to severely cold work high-carbon or alloy steels because it is difficult to do; the material has poor formability, the strengthening effect is not as pronounced, and it will be lost if the steel is subsequently heat treated. The most commonly used cold-finished carbon steels are 1006 to 1050, 1112, 1117, and other free-machining steels. The benefits are closer size tolerances, better mechanical properties, and improved surface finish (Figure 11–5). The disadvantages are lower ductility, greater instability during machining operations, and possibly higher price. Weighing these factors, it is usually desirable to make general machine parts such as brackets, chutes, guards, and mounting plates from cold-finished steels unless the part requires significant machining of surfaces.

The different types of cold-finished sheet steels that are commercially available are listed

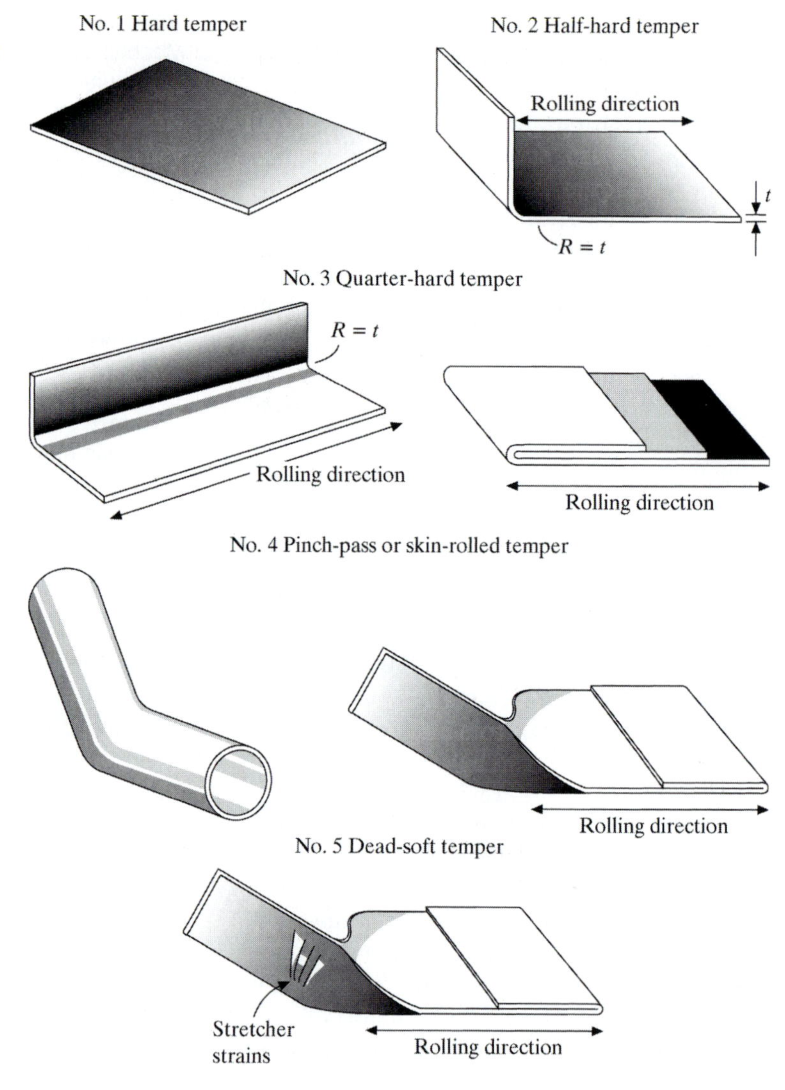

Figure 11–4
Formability of various tempers of 0.25 C max steel strip
Source: *Steel Products Manual—Carbon Steel Strip,* © American Iron and Steel Institute, New York, 1971

in Table 11–5. Many sheet steels are used for items such as appliance housings and automobile panels. In these types of applications a prime material requirement is the ability to be formed or drawn into shapes. Thus cold-rolled sheet steels are available in four basic types that vary in forming characteristics and strength characteristics. Only the structural quality (SQ) has definite mechanical property requirements. The other grades have varying degrees of formability, and it has been common practice to use the lowest (and lowest-cost) grade that will do the job. High-strength sheet steels are available for applications in which weight reduction is a concern. These steels are discussed in Section 11.5. ASTM specifications can be used for drawing designation:

Material: ASTM A 366 commercial quality sheet steel

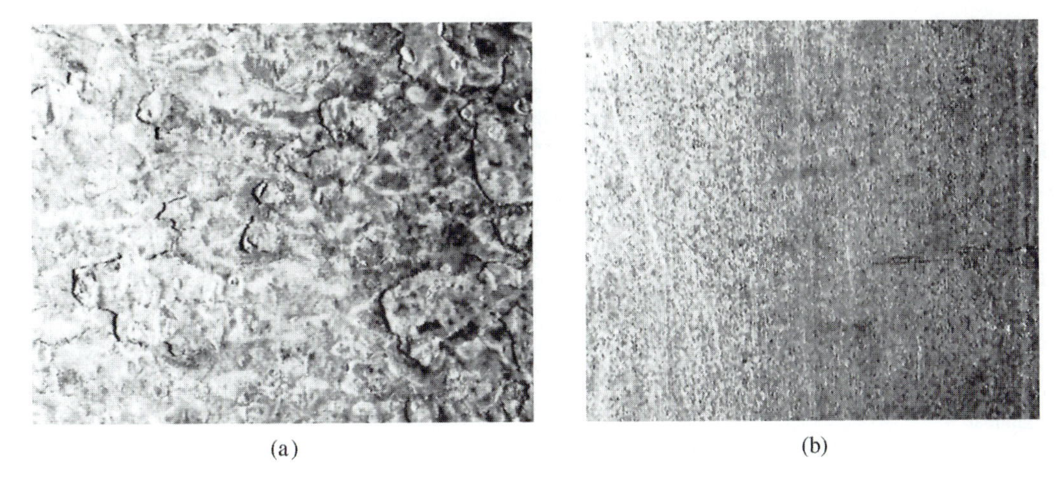

(a) (b)

Figure 11–5
Surface finish of (a) hot-rolled and (b) cold-rolled 1020 steel plate (×4)

If severe drawing is a design concern, there are sheet steels available with very high elongations (up to 50%) that may do the job. Interstitial-free (IF) steels are sheet steels developed in the 1990s to address higher-formability sheet steels. Carbon is the major interstitial alloy addition that affects steel strength. IF steels are not really free of interstitial carbon, but they are called interstitial free because the carbon content is very low, less than 10 parts per million. Microalloying with titanium or niobium is used to keep yield strengths in the range of 30 to 40 ksi (200 to 300 MPa). Higher-strength grades are available. Other specialty sheet steels include bake-hardening steels, which get stronger in paint curing, dual-phase, high-strength steels (martensite in a soft ferrite matrix) and ultra-high-strength steels that can be precipitation hardened to tensile strengths as high as 166 ksi (980 MPa). The goals of these steels is weight reduction in automobiles.

Table 11–5
Different qualities of cold-rolled steel sheet (AISI)

AISI Type	ASTM Specification	Approximate Hardness (HRB)	Characteristics
Commercial quality (CQ)	A 366	60 max	For moderate bending, drawing, and welding
Drawing quality (DQ)	A 619 A 619M	50	For drawing and severe forming
Drawing quality special killed (DQSK)	A 620	50	For severe drawing and forming nonaging (killed)
	A 620M	[20 ksi (138 MPa) min yield strength (YS)]	
Structural quality (SQ)	A 611	6 grades, YS 25 to 80 ksi (173 to 552 MPa)	Minimum tensile and yield strengths

Table 11–6
Tempers available on tin mill steels

Temper	Aim Hardness, HR 30T	Approximate Yield Strength* ksi (MPa)	Description
T–1	46–52	25–42 (172–289)	Soft for drawing
T–2	50–56	34–46 (234–317)	Moderate drawing
T–3	54–60	40–52 (275–358)	Shallow drawing
T–4 CA	58–64	48–60 (330–414)	General purpose, some stiffness
T–5 CA	62–68	56–68 (386–469)	Increased stiffness
DR–8	73–mean	75–85 (517–586)	Limited ductility
DR–9 CA	76–mean	85–105 (586–724)	Highest strength, very low ductility

*Longitudinal.
CA, Continuously annealed; DR, double reduced.

Tin mill products are low-carbon steels, generally in the thickness range of 0.006 to 0.015 in. (0.15 to 0.37 mm), that are cold finished to various tempers and may or may not be coated. The following are some of the common coatings available:

Tin plating: one side or both sides, the same thickness on both sides or differentially coated (different thickness on each side).

Chromium plated: This product is also called TFS or tin-free steel, but there is no tin involved in the process; it is a very thin chromium coat [0.3 μin. (7.5 nm)] on both sides.

Dichromate coated: a thin chemical conversion coating for paint adhesion.

Tin mill steel without a coating is called *black plate*. The plated coatings are applied by continuous web electrodeposition. These steels are available in large coils (up to 15,000 lb) or in cut sheet form. The largest volume use of these steels is for beverage cans, and for this application cut sheets are usually supplied. The unit of measure for steel quantity is the base box. A base box can be various weights, but it will consist of cut sheets with a total surface area of 217.78 ft^2. Different thicknesses are specified by different weights per base box (10-mil-thick material has a weight of 90 lb per base box; 15-mil material is specified as 135 lb per base box material). The various tempers available are listed in Table 11–6. Different finishes are also available. These are described in the *Steel Products Manual-Tin Mill Products*.

Hot Finished

Hot finishing simply means that a steel product was produced by a shaping process that is done above the recrystallization temperature of the alloy. With carbon steels, hot finishing is usually done in the temperature range from 1500 to 2300°F (815 to 1260°C). The purpose of hot finishing is obvious. It is simply easier to deform steel into a particular shape when it is red hot than when it is at room temperature. Hot-rolled steel products thus are cheaper than cold finished. There are fewer steps involved in the manufacturing process. Cold rolling requires pickling of hot-rolled material and rerolling. Structural shapes such as I-beams, channels, angles, wide-flange beams, and heavy plates are almost exclusively produced by hot-rolling operations. Any steel alloy can be *hot finished*, but the bulk of the steels produced are carbon steels with carbon contents of less than 0.25%. Hot-finished products usually have lower strength than cold-finished products. The grain deformation that strengthens cold-finished products does not occur when a steel is red hot. When grains are deformed, they immediately recrystallize and return to their undistorted shape.

The biggest disadvantage of hot finishing over cold finishing is the poor surface finish and looser dimensional control. The oxide scale on some hot-finished shapes almost precludes their use as received in making machine components that require any kind of accuracy. The same limitation prevails in dimensional considerations. Some advantages of hot-finished products other than low cost are better weldability and stability in machining. In fact, these are two important reasons for using these steels in machine design. Welding on cold-finished products causes local annealing, which negates the value of the cold working in strengthening. Machining of cold-finished products unbalances the cold-working stresses, and the part is apt to distort badly. Because there are low residual stresses from hot working, hot-finished products are preferred over cold-finished for parts requiring stability during and after machining operations.

Typical carbon steel grades used for machine bases, frames, and structural components are SAE 1020, 1025, and 1030. Normal (merchant) grades of hot-finished steel do not have rigorous control on composition or properties. If the mechanical properties of the steel must be guaranteed, it is best to specify a hot-finished shape to ASTM specifications. ASTM specification A 283 covers several strength grades of structural-quality steel. A 284 covers machine steels, A 36 covers bridge and building steel, and A 285 covers steels for flanges and fireboxes. The proper drawing specification would show the ASTM specification number and the strength grade. The mechanical properties of some construction steels are shown in Figure 11–6. The available types of hot-rolled steel in sheet form are listed in Table 11–7.

Hardening

In previous discussions on hardening it was pointed out that a requirement for hardening is sufficient carbon content. With carbon steels, to get 100% martensite in moderate sections, a carbon content of about 0.6% is desirable. This does not mean, however, that any size part can

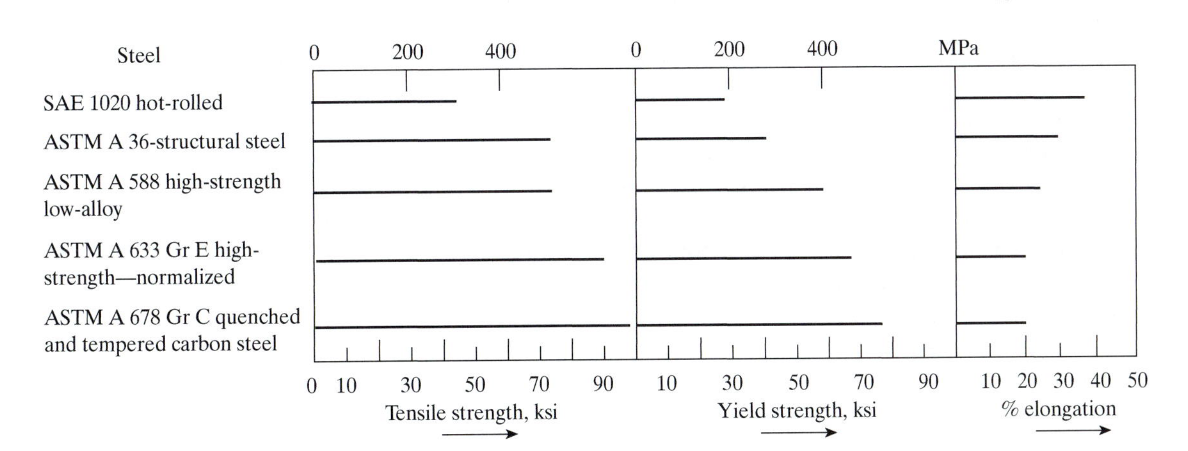

Figure 11–6
Mechanical properties of construction steels

Table 11–7
Different qualities of hot-rolled steel sheet

AISI Type	ASTM Specifications	Approximate Yield Strength (MPa)	Characteristics
Commercial quality (CQ)	A 569 (0.15C max) A 569 (0.15–0.25C)	38 ksi (262)	May contain surface imperfections
Drawing quality (DQ)	A 621	27 ksi (186)	Better drawing properties than CQ; some surface defects
Drawing quality special killed (DQSK)	A 622	24 ksi (165)	Best drawability; some surface defects
Structural quality (SQ)	A 570	6 grades, 30 to 50 ksi (207 to 345)	Same surface as CQ but controlled mechanical properties

be made from 1060 steel and that it will harden to 60 HRC. Carbon steels have poor hardenability, and it is difficult to meet quenching requirements. In our discussion of time–temperature transformation diagrams, it was pointed out that each steel has certain time requirements on quenching if hardened structure is to be obtained. These diagrams, for example, tell us that a 1080 steel must be quenched from its hardening temperature of about 1600°F (870°C) to a temperature below about 1000°F (530°C) in less than 1 s. This rapid cooling rate can be achieved in water with thin section, sheet metal, and bars up to 1 in. in diameter (25 mm), but on heavier sections only the surface may harden. As bar diameter or section size increases, even the surface will not harden. As shown in Figure 11–7, hardenability increases with carbon content. Maximum hardenability is achieved at about 0.8% carbon. Hardenability decreases somewhat as carbon content is increased over 1%, because carbon tends to promote the formation of ferrite.

The hardening situation with carbon steels is even worse than that implied by Figure 11–7. If hardness readings were taken in the center of the bar, these data would show that even the 1095 steel reached a hardness of only about 40 HRC. Thus, plain carbon steels have low hardenability, and large parts will harden only on the surface, if at all. Rapid heating and quenching techniques, such as flame or induction, work very well on these steels and overcome hardenability limitations. Thin sections, such as flat springs and wire springs, are well suited for

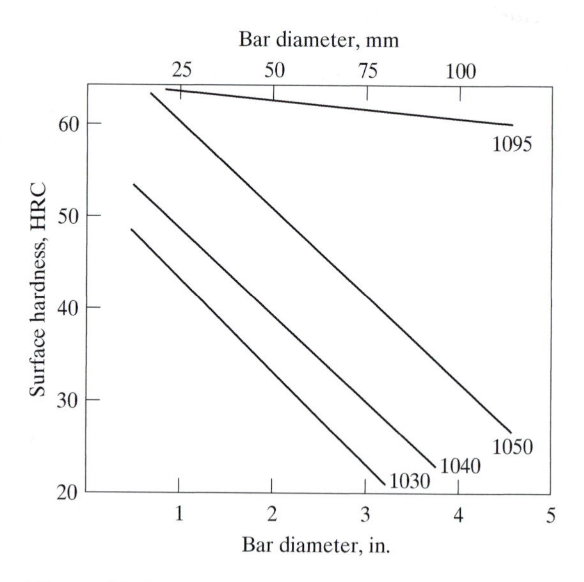

Figure 11–7
Approximate maximum surface hardness of carbon steels of varying bar diameters (water quenched)

manufacture from carbon steels. SAE 1080 to 1090 steels are commonly used for dowel pins, springs, knife blades, doctor blades, and the like. Large parts made from 1040 to 1060 or their free-machining counterparts 1140 to 1151 are commonly flame or induction hardened. All the low-carbon (<0.3% carbon) grades can be carburized and quench hardened to obtain hardened surfaces.

Weldability

All metals can be welded to themselves by at least one of the many processes that are available. Similarly, any metal can be welded to another metal, but this does not mean that the weld will have usable properties. It is a design rule that titanium cannot be welded to other metals (except for several exotic metals). Titanium can be fusion welded to steel, but the weld will be like glass. Tool steels can be welded, but unless special precautions are taken, nine times out of ten the weld will crack. The weldability of metal refers to these kinds of welding effects or after-effects. The *weldability* of a particular metal combination is the ease with which the weld can be made and the soundness of the weld after it is made.

The primary factors that control the weldability of metals are chemical *composition,* the base metal composition, and the composition of the filler metal, if any. A fusion weld cannot be made between dissimilar metals unless they have metallurgical solubility. Titanium cannot be welded to most other metals because it has limited solubility and tends to form brittle compounds. Phase diagrams can be consulted to determine if dissimilar metals can be welded. If the phase diagrams show low solid solubility, the weld probably cannot be made. The combination will have poor weldability.

With metals that have adequate solubility to be fusion welded, poor weldability will be manifested by such things as cracking of the weld, porosity, cracking of the heat-affected zone, and weld embrittlement. Many things can cause these weldability problems. Metals with high sulfur contents (*free machining*) crack in the weld because the sulfur causes low strength in the solidifying metal (hot shortness). Welding of rusty or dirty metals can cause weld porosity; welding coppers with high oxygen content causes embrittlement; and welding of hardenable steels can lead to cracking in the heat-affected zone.

Of all the potential weldability problems that can occur, two stand out as the most common and most troublesome:

1. Arc welding resulfurized steels: 11xx carbon steels.
2. Arc welding hardenable steels: 1030 to 1090 carbon steels, alloy steels with a *carbon equivalent* (CE) >0.3%.

High-sulfur steels (>0.1%) are widely used for parts that require significant machining. They save money; they are a real aid in lowering machining costs. However, there is no way to avoid weld-cracking problems with these steels other than to establish a hard and fast rule never to arc weld them. Brazing and soldering are acceptable, but not arc welding to themselves or to other metals.

Welding hardenable steels causes a weldability problem because there is a high risk of cracking in the weld or adjacent to the weld. When a weld is made, it and the heat-affected zone go through a thermal cycle not unlike a quench-hardening cycle. It makes no difference whether the parts are hardened or soft before welding. Melting of the steel requires a temperature of about 3000°F (1650°C). The hardening (*austenitizing*) temperature range for most steels is 1500 to 2000°F (815 to 1093°C). Obviously, the weld is going to be cycled in this temperature range, as is some of the metal adjacent to the weld. The mass of the metal being welded serves as a quenching medium, and hard martensite will form either in the weld, or in the heat-affected zone, or both. This structure will be brittle, and if the weld is under restraint, the

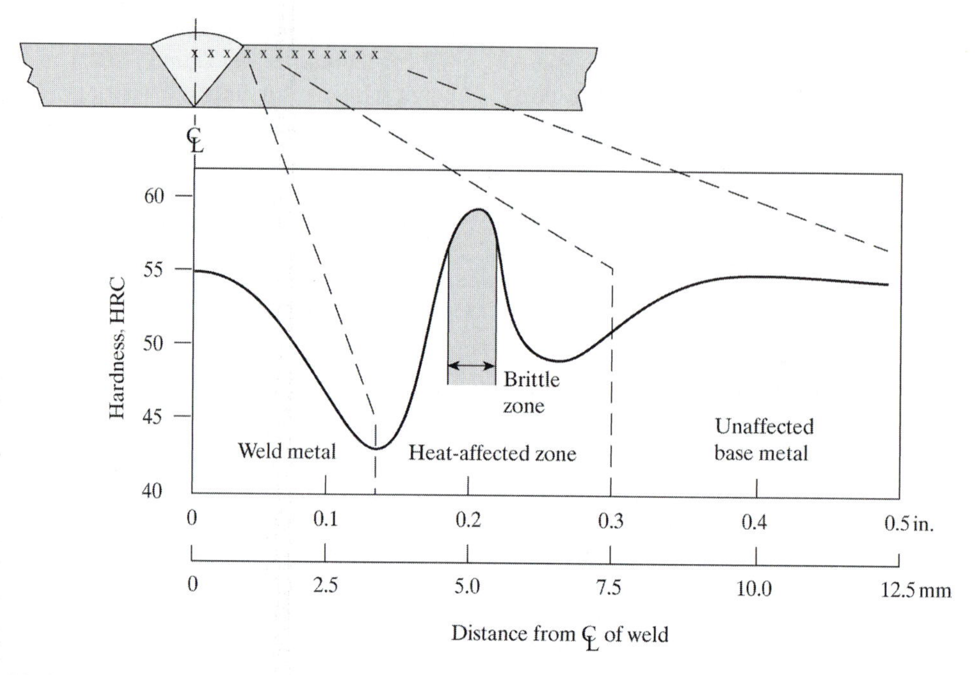

Figure 11–8
Hardness profile through an autogenous weld on A2 tool steel

brittle structure will crack. It cannot deform. Figure 11–8 illustrates with a *hardness traverse* how brittle martensite can form in the heat-affected zone of a hardened steel.

An obvious way to prevent this cracking is to prevent the quench that leads to martensite formation. Preheating the parts helps to accomplish this. If the part survives the welding operation without cracking, it may be prone to cracking in service if any brittle martensite formed. The solution to this is to postheat or temper the weld. An 1100 to 1200°F (593 to 648°C) stress relieve is even more desirable.

Thus there are ways to prevent cracking in welding hardenable steels, but the risk is so high that the designer is wise to prevent fusion welds on hardenable steels if at all possible. The steels that have a high risk of cracking in welding are alloy steels, tool steels, and plain carbon steels with carbon contents greater than 0.3%. Because alloy steels have higher hardenability than carbon steels, weldability can be a problem when the carbon equivalent is greater than about 0.3%.

$$\text{Carbon equivalent (CE)} = \%\text{C} + \frac{\%\text{Mn}}{6} + \frac{\%\text{Ni}}{15} + \frac{\%\text{Cr}}{5} + \frac{\%\text{Mo}}{4} + \frac{\%\text{V}}{4} + \frac{\%\text{Cu}}{15}$$

Note: This empirical relationship predates wide use of boron steels and nitrogenated steels. B and N terms possibly could be added.

Physical Properties

We have concentrated on the strength characteristics because they are usually a primary selection factor in using carbon and alloy steels. However, some physical properties are worthy of note. One of the most useful physical properties of carbon and most alloy steels is ferromagnetism;

Table 11–8
Most frequently used carbon steels

Nonhardenable	Free Machining	Hardenable
SAE 1006/1010	SAE 1112	SAE 1040
1020 (ASTM A 29)	1140 (hardenable)	1045
ASTM structural grades	1150 (hardenable)	1050
A 36, A 611, etc.		1060
ASTM sheet & strip		1080
A 651, A 794, A 621, etc.		1095

they will be attracted to a magnet. Thus they can be used in magnetic devices such as motors and solenoids, and they can be held in machining with magnetic chucks. The physical property that makes steel the most useful metal is the modulus of elasticity. Carbon and alloy steels have a value of 30 million psi (207 GPa), making them the stiffest common engineering metals. The only engineering metals that have higher stiffness are some nickel alloys, beryllium alloys, tungsten, and molybdenum.

The thermal conductivity of carbon and alloy steels is about 27 Btu ft/h ft^2°F (47 W/m K). Electrical conductivity is about 15% of that of pure copper; they are not good conductors of heat or electricity, but they are better conductors than stainless steels and some of the high-alloy steels. The coefficient of thermal expansion is about 7×10^{-6} in./in./°F (12.6×10^{-6} m/m/K). All these physical properties are about the same for high-strength, low-alloy steels, mill-heat-treated steels, and low-alloy ultra-high-strength steels. Because physical properties are often of less importance in selection of steels than mechanical properties, we shall not discuss them further. If they are important for a particular application, they are well covered in handbooks.

Summary

The mechanical properties of carbon steels vary significantly. Cold finishing can cause significant strengthening, and with the hardenable grades the properties are dependent on the tempered hardness. From the selection standpoint, a designer will probably most often use the alloys shown in Table 11–8. Cold-finished products should be used where you can take advantage of the higher strength and closer dimensional tolerances. Hot-finished shapes are preferred when freedom from residual stress is desired and where welding is contemplated. Additional considerations that will aid selection are as follows:

1. Use 1006 to 1020 steels for parts requiring severe forming (temper rolled or annealed).
2. AISI 1030 to 1060 steels are sometimes used to get higher strength without heat treatment. This should be done only if parts do not require welding (there are cracking tendencies). High-strength, low-alloy steels may be better choices.
3. Parts made from cold-rolled steels should be ordered to a temper.
4. Use free-machining steels for parts requiring significant machining but no welding (they crack due to hot shortness).
5. Use 1040 to 1060 steels for flame or induction hardening.
6. Use 1080 to 1095 steels for small parts requiring through hardening. They cannot be welded.

7. **7.** Use ASTM grades of carbon steels for pipelines, buildings, boilers, pressure vessels, and the like. They have guaranteed mechanical properties.
8. **8.** Do not weld any steel with a carbon or CE content over 0.3% without special welding techniques.
9. **9.** Do not weld steels with a sulfur content in excess of 0.06% or with a phosphorus content over 0.04%.
10. **10.** Cold-finished products have the best finish of the available carbon steel products, as well as better machinability than hot-finished products. Their higher hardness prevents the "gummy" machining behavior of hot-finished steels.

11.3 Alloy Steels

We have defined alloy steels as a group of mill products that meet certain composition limits. They are the 13xx, 4xxx, 5xxx, 6xxx, 8xxx, and 9xxx series in the SAE designation system. The same alloys are specified in ASTM A 29 and A 29M (M specs are metric). Before discussing the selection of specific alloys in this system, it would be well to discuss some of the basic metallurgical concepts involved in alloying steels. How do alloy additions strengthen? Which properties are altered? How is the effect of alloying measured? Why do we need alloy steels, since carbon steels can be hardened?

Attacking the first question, there are several things that can happen when alloy atoms are added to a pure metal. The atoms can go into a random solid solution, either substitutionally [Figure 11–9(a)] or interstitially [Figure 11–9(b)], or the atoms may want to pair with the host atoms in some definite proportion [Figure 11–9(c)] or stay by themselves [Figure 11–9(d)]. We have already discussed the strengthening effects of alloying elements going into solution. Their presence impedes dislocation motion. Ordered arrangements of alloy atoms with some stoichiometric relationship of host atoms to alloy atoms leads to

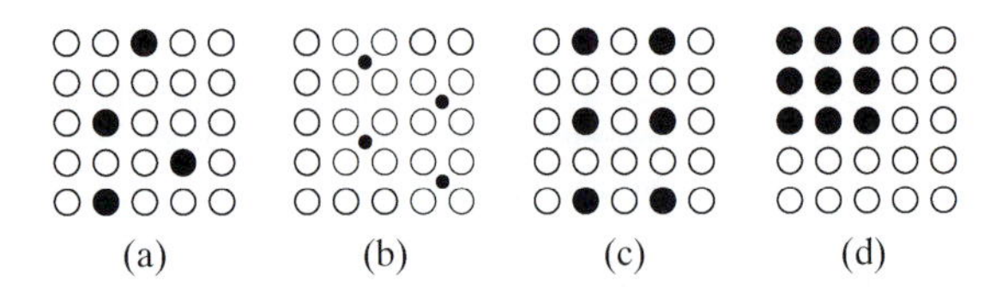

Figure 11–9
Distribution of alloy atoms (black) in a pure metal atomic lattice. (a) Substitutional solution (random). (b) Interstitial solution (random). (c) Ordered substitutional solution. (d) Clustering of like alloy atoms.

the formation of compounds with ionic or covalent bonds rather than metallic electron bonding. The formation of compounds in metals is common, and their occurrence can have a strengthening effect, as does a pearlite structure in steel (pearlite is a compound of Fe_3C layered with ferrite), or it can sometimes cause brittleness. Some intermetallic compounds cause alloys to become useless as structural materials. Finally, as in Figure 11–9(d), the solute atoms may give rise to the formation of separate phases, rich in solute. The phases can have a good or bad effect on properties, depending on the nature of the phase.

Which of these things happens when alloy is added depends on a number of thermodynamic factors, such as the effect on free energy. Other factors controlling the effect of alloying are the relative size of the solute atoms compared with the host atoms, the electrochemical nature of solute atoms, and even the valence of the alloy atoms. Alloy steels are really quite complex in that commercial alloys invariably contain more than one major alloying element. The exact atomistic effects of adding, for example, manganese, molybdenum, and chromium to make a 4000 series alloy steel are not fully understood, but the net effect of these additions on phase equilibrium, transition temperatures, and properties is known. Isothermal transformation diagrams have been established on most alloy steels, and these can be used to determine the net effect of alloying on the quenching requirements of a steel. Figure 11–10 shows TTT diagrams for

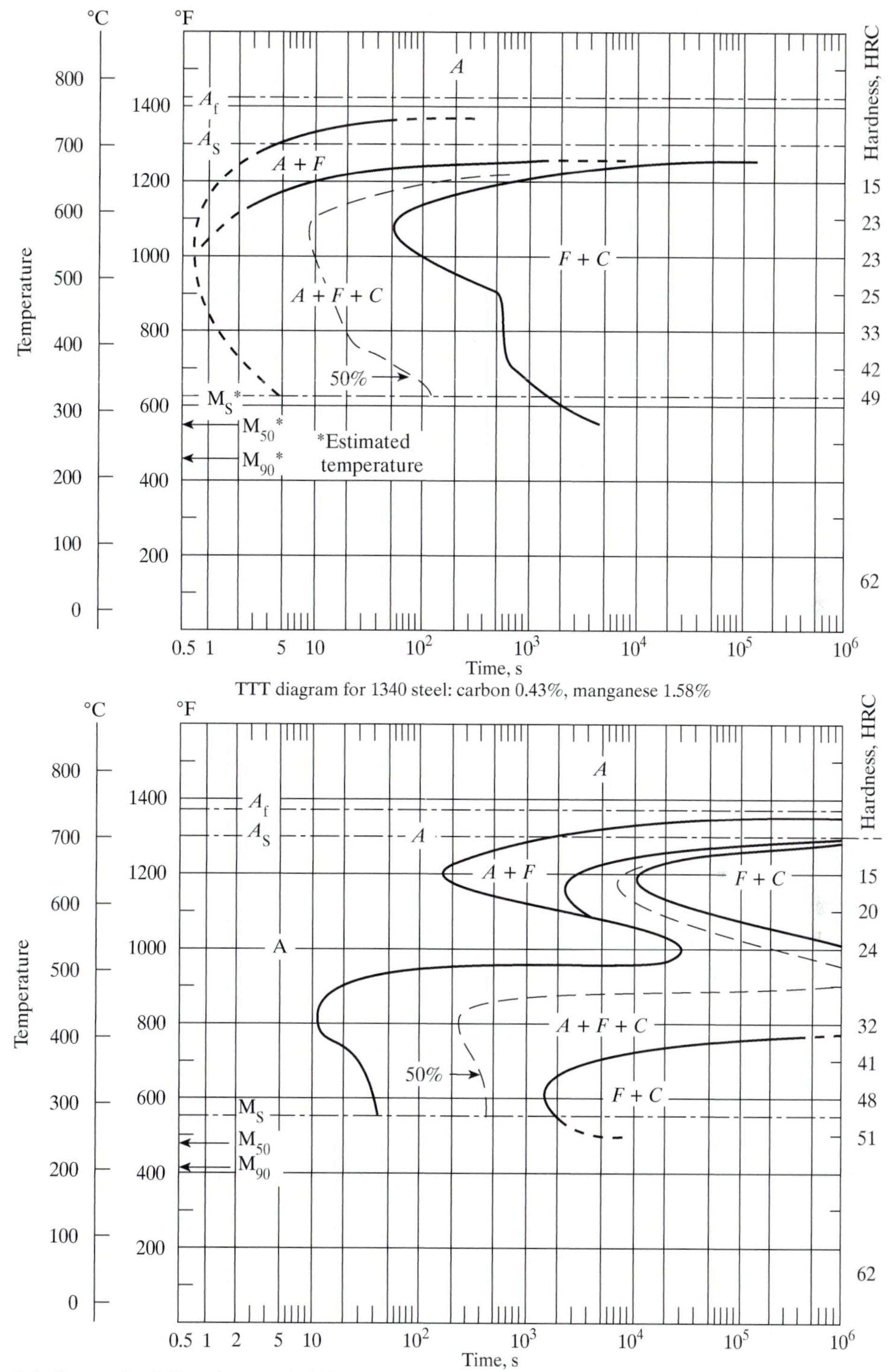

Figure 11–10
TTT diagrams for two steels with the same carbon content but different alloy content
Source: Copyright © United States Steel Corp., 1963

two steels with the same carbon content but differing contents of major alloying elements. The diagram for 1340 steel indicates that the heat treater has to quench the steel very rapidly to get it to harden (<1 s). The effect of the nickel, chromium, and molybdenum combination is to lessen the quenching requirements. An oil quench can probably meet the 10-s requirement indicated on the TTT curve for the 4340 steel.

The phase diagrams would similarly be changed by alloy addition. Iron–carbon alloys (carbon steels) are binary alloys. When an alloy steel contains four major alloy additions, a quaternary phase diagram is needed to determine phase relationships. The austenite transition temperature changes, as do the phases that can be present. The equilibrium diagrams for alloy steels will probably never be used by design engineers, but it is important to realize that heat treating temperatures for alloy steels will differ from those for carbon steels.

Probably the most important effect of alloying steels is to alter hardenability. The TTT diagrams are an indicator of the hardenability of a steel, but an equally useful tool is the *Jominy* end quench test. In this test a specified-sized bar is heated to a prescribed austenitizing temperature and water quenched on one end in a special fixture under prescribed conditions. The quenched end cools rapidly, and the rate of cooling is progressively less as cooling moves toward the other end of the bar. After the bar cools, a small flat is ground on it and the hardness is measured at various distances from the quenched end. These data are plotted on a curve of hardness versus distance (Figure 11–11), and the curve serves as a measure of hardenability. It can also be used to calculate the section size that will through harden for a given alloy. The factors that control the hardenability are the alloy content, the cooling rate, and the *grain size* of the steel. Thus if a steel has only 0.20% carbon but also contains 1% molybdenum, the carbon equivalent will be 0.45 and the material will have the hardenability of a steel with 0.45% carbon rather than 0.2% carbon. This equation is normally used in welding to determine if a particular steel requires preheating, but it shows which elements are more important in improving hardenability.

There are other reasons for adding alloying elements to steels besides improving hardenability. Sometimes they improve corrosion characteristics, physical properties, or machinability. Copper improves corrosion resistance. Sulfur and phosphorus additions improve machinability by producing small inclusions in the structure that aid chip formation. Sulfur contents over about 0.06%, on the other hand, make steels unweldable due to hot shortness. Thus there are good and bad effects of adding alloying elements. Table 11–9 summarizes the effects of common alloying elements in alloy steels.

As can be seen, most of the common alloy elements used in alloy steels tend to increase hardenability. They do so by altering transformation temperatures and by reducing the severity of the quench required to get transformation to hardened structure. Why the emphasis on hardenability? First, if a steel has very low hardenability, it is possible that it will not harden at all in heavy sections. Second, a severe quenching rate leads to distortion and cracking tendencies.

11.4 Selection of Alloy Steels

As shown in Table 11–10, numerous alloy steels are commercially available. In fact, added to this list are H steels (controlled hardenability steels), boron steels, nitrogenated steels, and free-machining grades. How does one go about selecting one of these for a machine application? One step that can be taken in categorizing these steels is to separate them into grades intended for through hardening and grades intended for carburizing. All the grades with carbon contents of about 0.2% or less are intended for carburizing. The remainder of the alloys through harden,

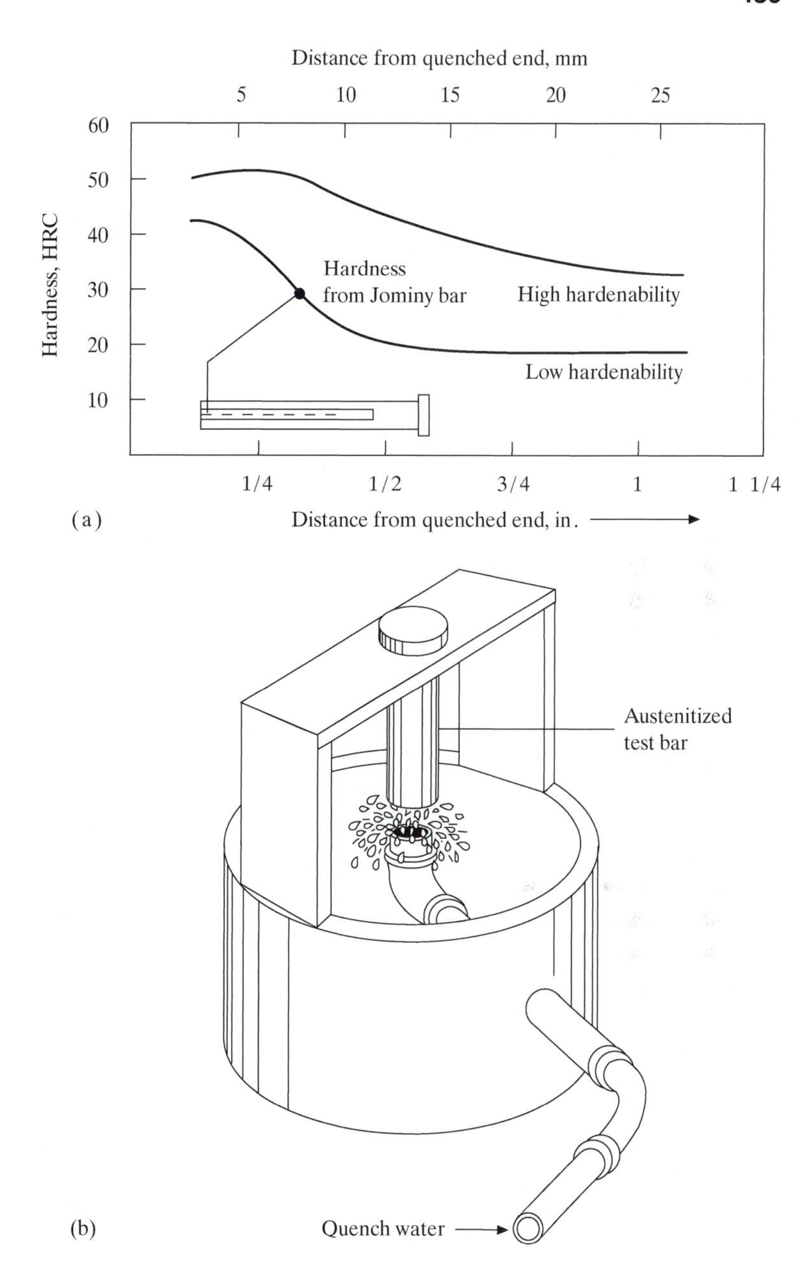

Figure 11–11
(a) Use of Jominy data to measure hardenability.
(b) Jominy end quench fixture (ASTM A 255 and A 304 describe test details).

but they do so to different degrees; they have differing hardenabilities. The carburizing grades also have different hardenabilities and differing core strengths after carburizing. There are also some special grades, such as the aluminum-bearing grades, that are intended for nitriding application. These are the Nitralloy-type steels discussed in Chapter 10.

The physical properties of most grades do not vary significantly from those of carbon steels;

Table 11–9
Effect of alloying elements

	Typical Ranges in Alloy Steels (%)	Principal Effects
Aluminum	<2	Aids nitriding Restricts grain growth Removes oxygen in steel melting
Sulfur and phosphorus	<0.5	Adds machinability Reduces weldability, ductility, and toughness
Chromium	0.3–4	Increases resistance to corrosion and oxidation Increases hardenability (significant effect) Increases high-temperature strength Can combine with carbon to form hard, wear-resistant microconstituents
Nickel	0.3–5	Promotes an austenitic structure Increases hardenability (mild effect) Increases toughness
Copper	0.2–0.5	Promotes tenacious oxide film to aid atmospheric corrosion resistance
Manganese	0.3–2	Increases hardenability, lowers hardening temperature Promotes an austenitic structure Combines with sulfur to reduce its adverse effects
Silicon	0.2–2.5	Removes oxygen in steelmaking Improves toughness Increases hardenability
Molybdenum	0.1–0.5	Promotes grain refinement Increases hardenability Improves high-temperature strength
Vanadium	0.1–0.3	Promotes grain refinement Increases hardenability Will combine with carbon to form wear-resistant microconstituents
Boron	0.0005–0.003	Added in small amounts to increase hardenability
Lead	<0.3	Added only to aid machinability
Nitrogen	<0.1	Acts like carbon in strengthening

Elements that promote austenite formation: Mn, Ni, N, Co, Cu, C
Elements that promote ferrite formation: Cr, Mo, V, W, Ti, Zn, Cb, Ta, Si

Table 11–10
Chemical compositions of ASTM A 29/SAE steels

ASTM A 29 Grade or SAE Number	UNS Number	C	Mn	Ni	Cr	Mo	Other Elements
1330	G13300	0.28/0.33	1.60/1.90	—	—	—	—
1335	G13350	0.33/0.38	1.60/1.90	—	—	—	—
1340	G13400	0.38/0.43	1.60/1.90	—	—	—	—
4023	G40230	0.20/0.25	0.70/0.90	—	—	0.20/0.30	—
4027	G40270	0.25/0.30	0.70/0.90	—	—	0.20/0.30	—
4028	G40280	0.25/0.30	0.70/0.90	—	—	0.20/0.30	S 0.035/0.050
4037	G40370	0.35/0.40	0.70/0.90	—	—	0.20/0.30	—
4047	G40470	0.45/0.50	0.70/0.90	—	—	0.20/0.30	—
4118	G41180	0.18/0.23	0.70/0.90	—	0.40/0.60	0.08/1.15	—
4130	G41300	0.28/0.33	0.40/0.60	—	0.80/1.10	0.15/0.25	—
4137	G41370	0.35/0.40	0.70/0.90	—	0.80/1.10	0.15/0.25	—
4140	G41400	0.38/0.43	0.75/1.00	—	0.80/1.10	0.15/0.25	—
4142	G41420	0.40/0.45	0.75/1.00	—	0.80/1.10	0.15/0.25	—
4145	G41450	0.43/0.48	0.75/1.00	—	0.80/1.10	0.15/0.25	—
4147	G41470	0.45/0.50	0.75/1.00	—	0.80/1.10	0.15/0.25	—
4150	G41500	0.48/0.53	0.75/1.00	—	0.80/1.10	0.15/0.25	—
4320	G43200	0.17/0.22	0.45/0.65	1.65/2.00	0.40/0.60	0.20/0.30	—
4340	G43400	0.38/0.43	0.60/0.80	1.65/2.00	0.70/0.90	0.20/0.30	—
E4340		0.38/0.43	0.65/0.85	1.65/2.00	0.70/0.90	0.20/0.30	—
4617	G46170	0.15/0.20	0.45/0.65	1.65/2.00	—	0.20/0.30	—
4620	G46200	0.17/0.22	0.45/0.65	1.65/2.00	—	0.20/0.30	—
4720	G47200	0.17/0.22	0.50/0.70	0.90/1.20	0.35/0.55	0.15/0.25	—
4815	G48150	0.13/0.18	0.40/0.60	3.25/3.75	—	0.20/0.30	—
4820	G48200	0.18/0.23	0.50/0.70	3.25/3.75	—	0.20/0.30	—
5120	G51200	0.17/0.22	0.70/0.90	—	0.70/0.90	—	—
5130	G51300	0.28/0.33	0.70/0.90	—	0.80/1.10	—	—
5132	G51320	0.30/0.35	0.60/0.80	—	0.75/1.00	—	—
5140	G51400	0.38/0.43	0.70/0.90	—	0.70/0.90	—	—
5150	G51400	0.48/0.53	0.70/0.90	—	0.70/0.90	—	—
5160	G51600	0.56/0.64	0.75/1.00	—	0.70/0.90	—	—
E51100		0.98/1.10	0.25/0.45	—	0.90/1.15	—	—
E52100		0.98/1.10	0.25/0.45	—	1.30/1.60	—	—
6150	G61500	0.48/0.53	0.70/0.90	—	0.80/1.10	—	V 0.15/min
8615	G86150	0.13/0.18	0.70/0.90	0.40/0.70	0.40/0.60	0.15/0.25	—
8617	G86170	0.15/0.20	0.70/0.90	0.40/0.70	0.40/0.60	0.15/0.25	—
8620	G86200	0.18/0.23	0.70/0.90	0.40/0.70	0.40/0.60	0.15/0.25	—
8622	G86220	0.20/0.25	0.70/0.90	0.40/0.70	0.40/0.60	0.15/0.25	—
8630	G86300	0.28/0.33	0.70/0.90	0.40/0.70	0.40/0.60	0.15/0.25	—
8637	G86370	0.35/0.40	0.75/1.00	0.40/0.70	0.40/0.60	0.15/0.25	—
8640	G86400	0.38/0.43	0.75/1.00	0.40/0.70	0.40/0.60	0.15/0.25	—
8642	G86420	0.40/0.45	0.75/1.00	0.40/0.70	0.40/0.60	0.15/0.25	—
8645	G86450	0.43/0.48	0.75/1.00	0.40/0.70	0.40/0.60	0.15/0.25	—
8720	G87200	0.18/0.23	0.70/0.90	0.40/0.70	0.40/0.60	0.20/0.30	—
8822	G88220	0.20/0.25	0.75/1.00	0.40/0.70	0.40/0.60	0.30/0.40	—
9260	G92600	0.56/0.64	0.75/1.00	—	—	—	Si 1.80/1.20
9310		0.08/0.13	0.45/0.65	3.00/3.50	1.00/1.40	0.08/0.15	—

thus the selection procedure can be reduced to weighing mechanical properties and hardening characteristics. If an alloy steel is to be used for a load-bearing member, the properties of prime importance may be yield strength and toughness. If the part is intended for a wear application, an alloy might be desired that can harden to 60 HRC on the surface with a particular section thickness. Keeping these things in mind, let us categorize alloy steels into through-hardening and carburizing grades and discuss the relative merits of some of the more common grades.

Through Hardening

At least 40 SAE alloy steels are considered to be through hardening. They are those with a carbon content greater than about 0.3% (xx30). The differences in hardenability on some of the more popular alloys can be seen in Figure 11–12. From the use standpoint, if large section parts are required, an alloy such as 4340 should be used because it will harden through in heavy sections. All the alloys shown would develop comparable hardness and strength in small sections. It should also be noted that a carbon steel with the same carbon content as the alloy steels will not through harden with an oil quench even in thin sections. All the hardness–versus–bar diameter curves in Figure 11–12 would shift upward with a water quench and alloys with only 0.3% carbon may develop respectable hardnesses, but the risk of cracking and distortion is such that water quenching is not advisable unless economy dictates the use of a lean alloy and water quench. When serviceability rather than cost is the prime selection consideration, the designer can probably get along nicely using only one or two alloys: SAE 4140 and 4340. The mechanical properties of all the alloy steels depend on the condition of heat treat (the tempered hardness). As can be seen in Figure 11–13, the mechanical properties vary slightly with alloy at a given hardness, but these two alloys have excellent strength and toughness at most hardness levels, and at about 30 HRC the toughness is comparable to that of a low-carbon steel but the yield strength is still four or five times as high as that of mild steel. Thus these two alloys are widely used for shafting, gears, and other power transmission components hardened and tempered to 30 to 32 HRC. They can be finish machined after hardening, and this provides another advantage. The distortion in heat treatment is eliminated. Why two alloys? SAE 4340 has better hardenability than 4140. Why use 4140 at all? It is true that 4340 would suffice for all applications, but it is common warehouse practice to stock 4140 in small sizes where high hardenability is not essential and to stock 4340 in larger-section products. These two alloys plus a 4150 or 8740 simply give the designer a little latitude in availability.

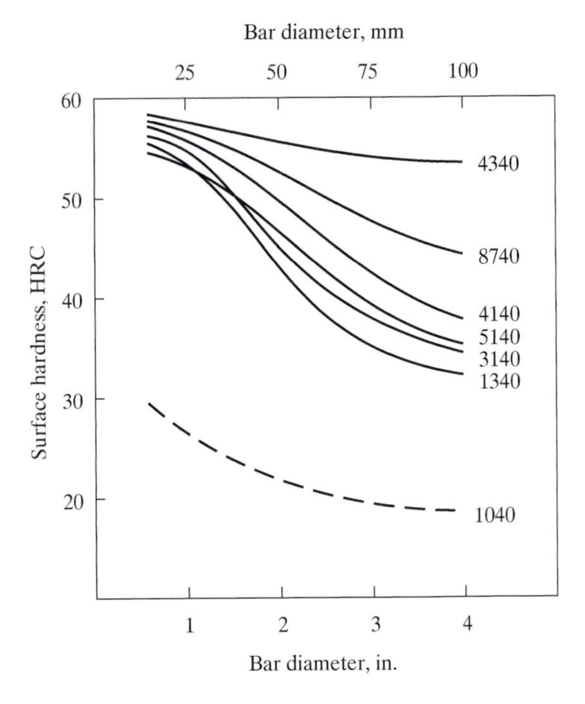

Figure 11–12
Hardening characteristics of some alloy steels, as oil quenched

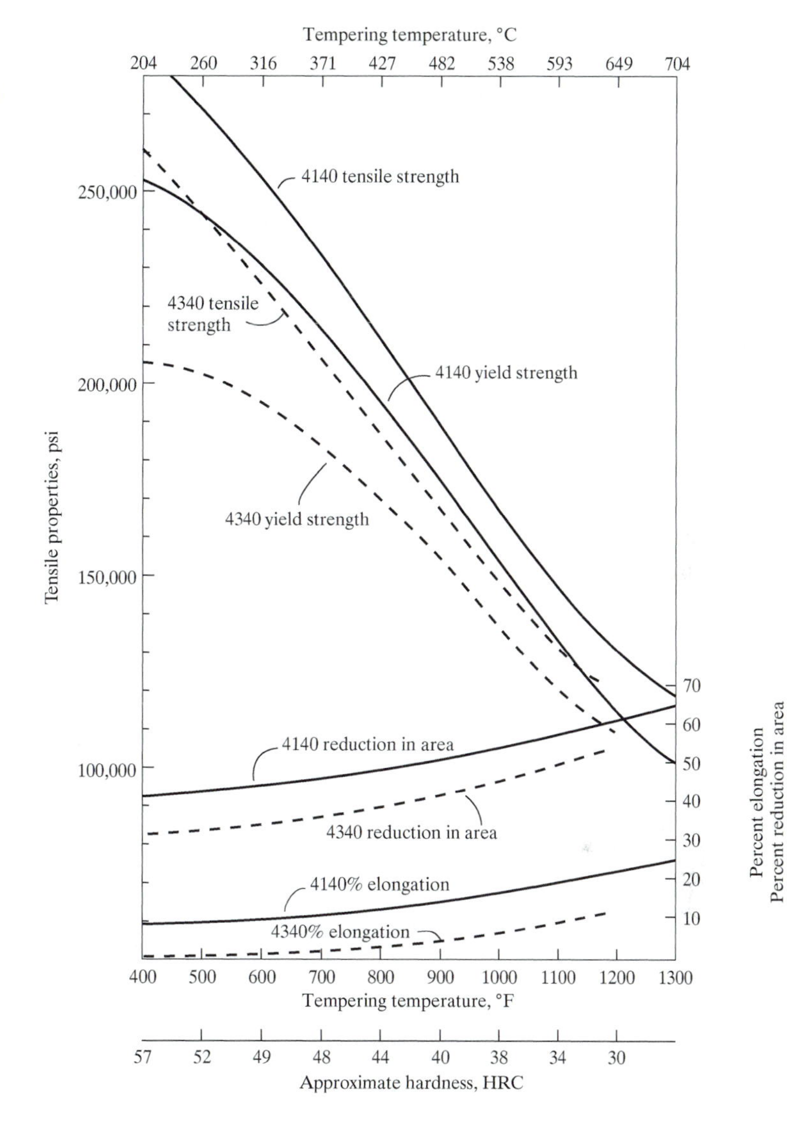

Figure 11–13
Effect of tempering on the mechanical properties of 4140 and 4340 steels (0.505 bar, oil quenched)

Source: *Modern Steels,* Bethlehem, PA: Bethlelem Steel Corp., 1980

Carburizing Grades

Similar to ASTM/SAE through-hardening grades, about 20 grades of alloy steels are intended for carburizing. They all will do about the same thing. They have better hardenability than low-carbon steels, so it is easier to get high-hardness deep cases. In addition, they offer better mechanical *core properties* than are possible with low-carbon steels. Figure 11–14 compares the case hardness characteristics for two carburizing alloy steels with 1020 steel. These data show why alloy steels are needed on heavy sections. Even with a water quench, a 5-in. (125-mm) bar of 1020 steel will not develop a hard case. Figure 11–15 compares the *core strength* on alloy steels with 1020 steel; again there is a significant advantage.

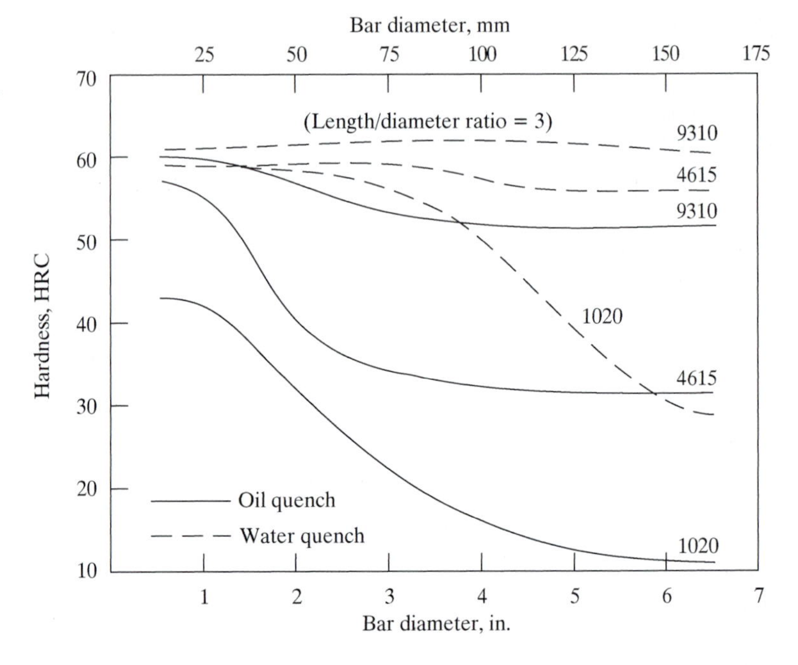

Figure 11–14
Surface hardness of steels carburized 0.04 in. (1 mm) deep, oil quenched and water quenched

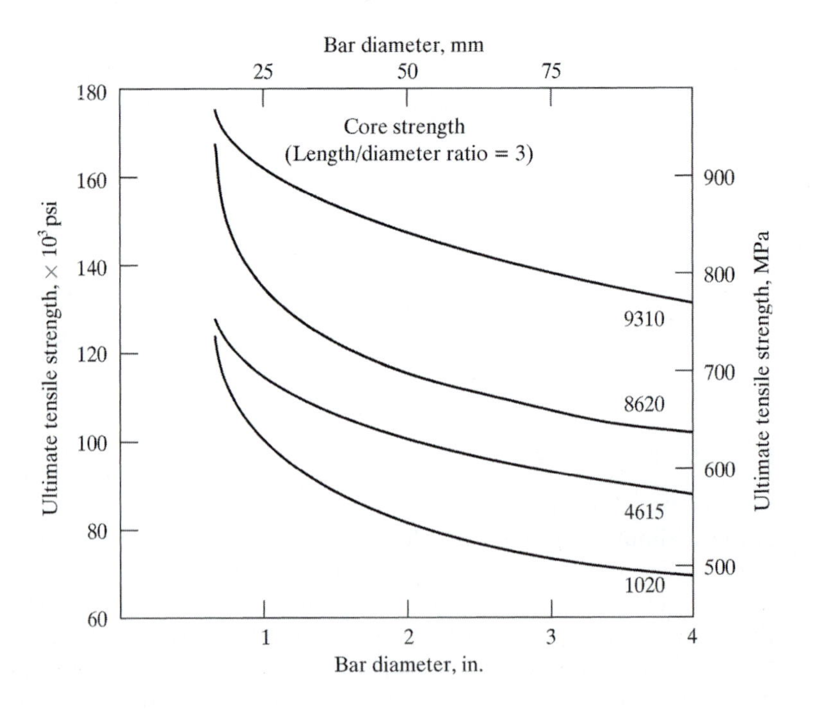

Figure 11–15
Core strength of carburized steels (oil quenched)

Of the 40 or so *case hardening alloys* that are available, the two alloys listed are sufficient for most applications. The 8620 is more readily available than the 9310, but the latter has higher core strength if it is needed. If an 8620 is not readily available, suitable substitutes are 4617, 4620, 8615, 8612, and 8622. The 9310 alloy is somewhat set apart from the others, and there are no really close substitutes. Thus designers can meet most carburizing needs with three alloys or their equivalents: 1020, 8620, and 9310. The 1020 alloy should be used only on small sections, less than about 1 in. (25 mm) in diameter in thickness. On heavier parts, use the 8620 alloy; if a hard case and the highest core strength are needed, use a 9310. If free-machining characteristics are required in a carburizing steel, the 11xx low-carbon alloys can be carburized, and there are even free-machining grades of alloy steel (41L40 leaded). Thus most alloy steel needs in engineering design can be met with very few alloys.

H Steels

Most of the power transmission components in automobiles are made from alloy steels. These steels are moderate in cost and can be hardened to provide the strength and wear resistance needed for shafts, gears, and similar components. Because a part manufacturer may make thousands or even millions of the same gear or cam, it is essential that each batch of steel harden the same as the previous batch. To provide steels with the same hardenability in each and every heat, the U.S. steel industry established *H steels,* alloy steels with guaranteed hardening characteristics. H steels are available in most of the standard grades, such as 4140, 4340, and 1340, but the composition ranges are adjusted to provide the desired hardenability. The H steels are identified by an H suffix on the standard designation: 4140H, for example. The UNS designation has an H prefix: 4140H = UNS H41400. These steels should be used over the standard alloy steel grades when repeatable hardening characteristics are essential on high-production parts.

B Steels

Boron and nitrogen are next to carbon in the periodic table and, as we might expect, they behave like carbon as hardening agents in steel. There is a family of about ten SAE steels that contain small amounts of boron (0.0005% to 0.003%) as an alloying element. They are identified by the standard SAE designation, but with a B in the center of the alloy number: 50B44 is an AISI 5044 steel with boron added. The UNS designation for boron steels has a G prefix and a 1 after the SAE number. SAE 50B60 = UNS G50601.

The incentive for producing boron steels is that with low- to medium-carbon contents an extremely small amount of boron can conceivably provide the same hardenability improvement as adding, for example, 1% of an expensive alloying element such as molybdenum or chromium. This can provide a significant cost reduction on high-production parts. There are many metallurgical considerations associated with the use of boron. Boron lowers hardenability when added to high-carbon steels: Too much boron can cause formation of Fe_2B compounds, which lower toughness; nitrogen must be kept low or it will combine with boron to form boron nitride compounds. Thus boron strengthening is not widely used because of these factors, but for suitable applications it can provide a hardenable steel with good formability, machinability, and heat treating characteristics at a significantly lower material cost compared to standard grades.

11.5 High-Strength Sheet Steels

Since about 1975 there has been a trend toward weight reduction in all forms of vehicles for fuel savings. Because a significant amount of the weight of automobiles and trucks is in sheet steel components (about 45%), various steel producers started commercial production of sheet steels with minimum yield strengths higher than the 30 ksi (206 MPa) that is typical on low-carbon sheet steel. Many of these steels are proprietary in nature, but higher strengths are usually achieved by microalloying, melting practice, heat treatment, or a combination of these. ASTM specifications cover a number of these steels: ASTM A 568, A 572, A 606, A 607, A 715, and A 816. There are also many proprietary grades, and ultra-high-strength sheet steels with tensile strength as high as 140 ksi (980 MPa) are under development.

The structural-quality, low-alloy, and *weathering grades* of these steels have solid solution strengthening by alloy additions as a part of their strengthening mechanism. *Dual-phase* steels usually contain some alloy additions such as manganese or silicon, but the strengthening effect employed in these steels is the formation of martensite or bainite in the ferrite matrix. After cold rolling, dual-phase sheet steels are continuously heated to the temperature region at which the structure is part austenite and part ferrite (see Figure 10–7). A temperature in the range from 1360 to 1400°F (730 to 760°C) is typical. The steel is then cooled at a rate sufficient to cause the austenite to transform to martensite or lower bainite. This results in a soft, ductile ferrite matrix containing islands of quench-hardened martensite. The strengthening effect is almost directly proportional to the volume percentage of hard structure. Yield strength can be increased to as high as 100 ksi (700 MPa) while still maintaining formability. In addition, these steels can be further strengthened by a postforming age hardening.

Bake-hardening sheet steel is steel that increases in strength when heated for short times (as short as 30 s) at elevated temperatures [in the range of 340 to 1110°F (170 to 600°C)]. These steels have very low carbon content (interstitial free) and copper contents up to 1.5%. These steels are continuously annealed and subsequently age hardened to get microscopic copper precipitates (3 nm) that strengthen the steel. Tensile strength can be increased from 60 ksi (420 MPa) to as high as 95 ksi (660 MPa). Most ingot-cast sheet steels age harden during paint baking due to reversion of carbon atoms to dislocations, but the strengthening effect was not as pronounced as in these copper-bearing steels.

Deoxidation practices are used to enhance the properties of some grades of high-strength, low-alloy (HSLA) steels. As mentioned in our discussion of melting practice, killed steels have finer grain size and less chemical segregation than nonkilled steel. The sulfide inclusion control referred to means the use of processes such as ladle inoculation to reduce the size and volume fraction of sulfur inclusions. Calcium argon degassing (CAD) is one such process. Electric furnace–melted steels are transferred to a degassing vessel. Calcium is added to the melt to scavenge sulfur and make it float up to the slag. Argon is then blown into the charge to mix, agitate, and further complete the sulfide control. Rare earth elements can also be used in this type of process, but the net effect in all cases is a greatly reduced volume fraction of sulfides, and the ones that remain are globular rather than *stringers*. The objectives of these inclusion control processes are elimination of anisotropy and improved toughness.

The area of application of all high-strength sheet steels is usually on structural components for which weight and material cost savings can be realized by reducing part thickness without lowering strength. This can be done with the higher strengths of these steels. If weight reduction is a design priority, then the use of these stronger steels should be investigated through steel mills.

11.6 High-Strength, Low-Alloy Steels

These steels are defined in the U.S. *Steel Products Manual* as a "group of steels with chemical composition specially developed to impart higher mechanical properties and in certain of these steels materially greater resistance to atmospheric corrosion than is obtainable from conventional carbon steels." Unfortunately, how higher mechanical properties are produced varies significantly with the manufacturer. Chemical composition and processing are the usual factors that are varied to increase mechanical properties, but these alloys are made primarily to a designated strength level rather than to rigorous chemical composition specification.

In general, HSLA steels are not hardened by heat treatments. They are supplied and used in the hot-finished condition; they have low carbon contents (<0.2%); they have a microstructure that consists of ferrite and pearlite; they all contain about 1% manganese to produce solid solution strengthening of the ferrite, and they usually contain small concentrations (<0.5%) of other elements for additional strengthening or other effects. This is called *microalloying*.

HSLA steels are primarily intended for structural-type applications in which weldability is a prime selection requirement. This is why the carbon and alloy content of these steels is low. They do not have the hardenability of alloy steels, and even though they contain significant amounts of alloy (Mn), they are not considered to be alloy steels. Table 11–11 is a tabulation of some HSLA steels showing some of the alloying elements that are used for strengthening.

In the United States, many bridges and similar structures are made from medium-strength HSLA steels with yield strengths less than 50 ksi (395 MPa). ASTM A 36 steel is the lowest-strength construction steel. It has a carbon content of about 0.3% and a yield strength of about 36 ksi (250 MPa). With microalloying techniques (adding small percentages of vanadium and niobium) this steel becomes ASTM A 572, Grade 50 with a minimum yield strength of 50 ksi (345 MPa). Higher strengths [up to 100 ksi (690 MPa)] are attainable by combinations of microalloying and process changes such as rapid cooling from hot working temperatures. These steels provide significant weight reductions, but they bring risks of welding problems. Weldability and toughness become more of a concern as strength increases.

Table 11–11
ASTM high-strength, low-alloy steels

ASTM Grade	Minimum Yield Strength ksi	(MPa)	Major Alloying Elements	Typical Forms
A 242 (2 grades)	42–50	(290–345)	Mn, Cu, Cr, Ni	Structural bars, plates, shapes
A 440	42–50	(290–345)	Mn, Cu, Si	Structural bars, plates, shapes
A 441	40–50	(275–345)	Mn, V, Cu, Si	Structural bars, plates, shapes (heavy)
A 572 (6 grades)	42–65	(290–450)	Mn, Nb, V, N	Structural bars, plates, shapes
A 588 (10 grades)	42–50	(290–345)	Mn, Nb, Cu, Cr, Si, Ti	Structural bars, plates, shapes
A 606 (4 grades)	45–50	(240–345)	Mn	Sheet and strip
A 607 (6 grades)	45–70	(290–485)	Mn, Nb, V, Ni, Cu	Sheet and strip
A 618 (3 grades)	50	(345)	Mn, Nb, V, Si	Structural tubing
A 633 (5 grades)	46–60	(320–410)	Mn, V, Cr, N, Cu	Structural shapes for low-temperature service
A 656 (2 grades)	80	(550)	Mn, V, Al, N, Ti	Plates for vehicles
A 715 (4 grades)	50–80	(345–550)	Mn, V, Cr, Nb, N	Sheet and strip

Copper is one of the most important alloying elements in HSLA steels. In small concentrations (<0.5%), it produces solid solution strengthening of ferrite, as well as a side benefit that is even more important—the atmospheric corrosion resistance of the steel can be improved by a factor of 4. Copper-bearing HSLA steels are called *weathering steels* (ASTM A 242), and they achieve their improved corrosion resistance by the formation of a tenacious oxide film. This film takes on the appearance of copper, but it is really rust that does not crack and flake like conventional rust. Chromium, nickel, and phosphorus are usually used in these alloys to assist in the formation of the protective oxide film. If the copper content is over about 0.75%, these alloys can be precipitation hardened by heating in the range from 1000 to 1100°F (540 to 600°C).

These steels have been widely used for structural applications (bridges, towers, railings, and stairs), and they are left unpainted. It was thought that the adherent rust would limit the corrosion to just this layer. After about two decades of use, using these steels in the unprotected condition is losing popularity. In places where water is allowed to accumulate and in salt atmospheres, the rust does not stop at the thin surface layer; it can be more substantial. It is current practice to protect these steels from corrosion at least in areas that are conducive to the accumulation of corrodents.

Niobium, titanium, vanadium, and nitrogen are used as microalloying additions in a significant number of HSLA steels (A 575, A 607). Their role is to form precipitates that restrict grain growth during hot rolling. Concentrations of these elements are usually less than about 0.2%. Thus the use of these elements for strengthening does not significantly increase manufacturing costs.

Rare earth elements such as cerium, lanthanum, and praseodymium are used in HSLA steels to obtain shape control of sulfide and oxide inclusions. These elements tend to combine with sulfides and oxides and essentially reduce their plasticity so that they do not elongate to form stringers during hot rolling. Elongated sulfides and oxides lead to lamellar tearing in high-restraint welds. The globular sulfides and oxides that are produced in shape-controlled steels are less detrimental.

Aluminum and silicon are used in HSLA steels for the usual reason—deoxidation. They are used to kill steels and reduce chemical segregation. Silicon also contributes to solid solution strengthening of ferrite. Calcium additions such as in the CAD process are also used for control of sulfides and oxides.

In summary, the hot-rolled HSLA steels have minimum yield strengths in the 40- to 80-ksi range (275 to 550 MPa) and tensile strengths in the range from 55 to 90 ksi (379 to 620 MPa) with good formability and acceptable weldability. These strength levels are achieved by major alloy additions of manganese, microalloying, inclusion shape control, grain refinement, deoxidation, and controlled rolling. The mechanisms of strengthening vary from simple solid solution strengthening of ferrite to complex reactions between microalloying additions and dislocations. As a class of steels they should be used wherever their high strengths or their corrosion resistance is an advantage. Many proprietary alloys are considered to be in the HSLA category, but wherever possible these alloys should be specified on engineering drawings by ASTM number and strength grade.

11.7 Special Steels

Austenitic Manganese Steels

This family of steels is sometimes referred to as Hadfield steels after their inventor. They were introduced in the United States in 1892, and they continue to be used for special applications. In our discussions of phase diagrams it was pointed out that alloy additions can cause substantial changes in phase equilibrium. Austenitic manganese steels contain from 1% to 1.4% carbon and from 10% to 14% manganese. The high manganese content causes the austenite phase that is normally stable in steels at temperatures in excess of about 1400°F (760°C) to be the stable phase at room temperature.

Austenitic manganese steels are available as castings, forgings, and hot-rolled shapes. Their yield and tensile strengths are comparable to the HSLA steels, but their unique use characteristic is their ability to work harden when impacted. They are supplied with an annealed hardness of about 200 HB. With impacting, the surface hardness for a depth of 1 or 2 mm can increase to 500 HB.

They have application in design for wear plates on earthmoving equipment, rock crushers, ball mills, and the like where the predominating wear mode is gouging, abrasion, and battering. They are weldable, which allows easy application in high-wear areas. There are a number of grades of these steels (ASTM A 128), but their use characteristics are all similar.

Mill-Heat-Treated Steels

Some mill-heat-treated steels are approximately high strength, low alloy in composition, but they are supplied from the mill in other than hot-rolled or hot-rolled and annealed condition. They are either normalized or quenched and tempered. Yield strengths can be as high as 110 ksi (758 MPa). Some of these steels are alloyed with manganese and silicon; some are plain carbon in composition. They are available in most forms suitable for structural use. They are usually proprietary, but there are some ASTM specifications that can be used for designation (ASTM A 321, A 514, A 577, A 633, and A 678).

These steels have good toughness and formability, but welding heavy sections requires special welding precautions. A particularly useful group of these alloys is the abrasion-resistant grades. They have a hardness in the range of 250 to 400 HB. This hardness is not high enough to resist scratching from hard particles, but the high yield strength [about 100 ksi (689 MPa)] minimizes denting and gouging in handling rocks, coal, and similarly aggressive media. These steels are widely used in truck bodies and ore-handling equipment. In machine design they find application where the high strengths can be used for weight reductions and for making hoppers, chutes, and loading platforms where the abrasion-resistant grades will provide resistance to abuse.

Three mill-heat-treated steels that are widely used for very high-strength structural fabrications are designated in the United States as HY-80, HY-100, and HY-130. The numbers stand for the minimum yield strength of these alloys in ksi. The nominal compositions of these alloys are as follows:

%	C	Mn	Si	P	Cr	Ni	Mo	Others
HY-80	0.18% max	0.25%	0.25%	0.25% max	1.4%	2.6%	0.4%	
HY-100	0.20 max	0.25	0.25	0.25 max	1.4	2.9	0.4	
HY-130	0.12 max	0.75	0.30	0.01 max	0.65	5.0	0.5	0.015 V%

As in the HSLA steels, the carbon is kept low to achieve weldability. They have a predominately martensitic structure, and special welding techniques are required. They are used for high-strength, high-toughness plates for pressure vessels, nuclear reactors, and submersible vessels.

Ultra-High-Strength Steels

When all the steels that we have just discussed still do not meet desired strength requirements, the ultra-high-strength steels can be employed. This is a catchall category in that it includes alloy steels, tool steels, steels with a high concentration of alloying elements, and stainless steels. The through-hardening alloy steels that we recommended for general use, SAE 4140 and 4340, when heat treated to high-strength levels, are

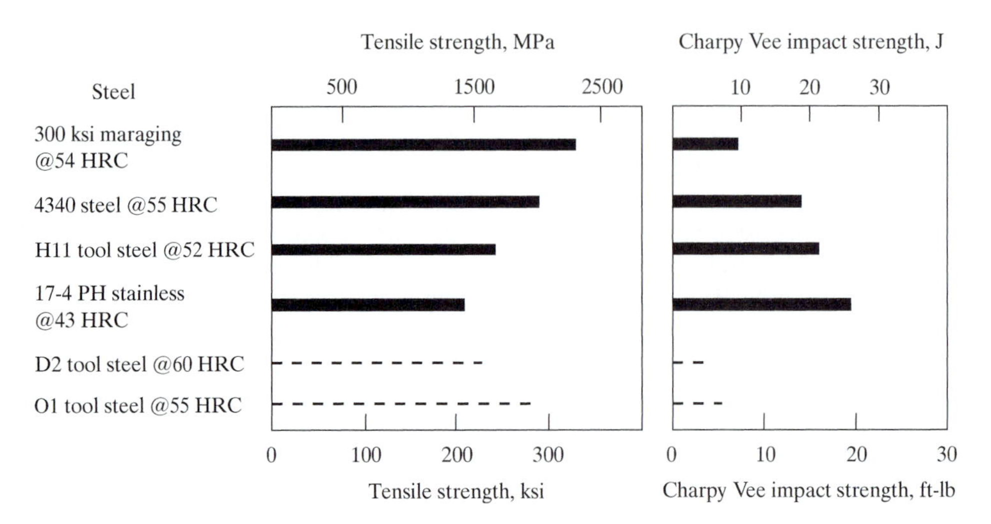

Figure 11–16
Tensile strength and impact strength of ultra-high-strength steels compared with O1 and D2 tool steels

considered to be ultra-high-strength steels. Some modified versions of 4340 are also included: AMS 6434, D6AC, and 300M are the usual designations. AMS is a U.S. government specification number; the others are trade names. Two hot-work tool steels, SAE H11 and H13, have sufficient strength and toughness to be used for structural applications. These steels will be discussed further in Chapter 12. The stainless steels will similarly be discussed in subsequent chapters. In the high-alloy category, a particularly useful group of steels in design is the *18% nickel maraging steels*.

All of these steels are available in some of the standard wrought forms, but the bulk of these steels are supplied as forged shapes. Many of these alloys are also melted using vacuum, VAR, and other special techniques aimed at improved cleanliness. The common denominator for ultra-high-strength steels is high strength with measurable toughness. Most tool steels have high strengths, but they are so brittle that they are unusable except where bending is not expected. Their elongation in a tensile test is almost immeasurable. The strength and toughness characteristics of some ultra-high-strength steels are shown in Figure 11–16.

The yield strength of these steels even in the annealed condition is usually several times that of carbon steels; thus formability is limited. Welding is risky on some alloys, but it is quite good on the maraging nickel and stainless steels. These latter steels are usually supplied from mills in the martensitic or semiaustenitic condition, and hardening to maximum strength is accomplished on most grades by a low-temperature aging process [900 to 1150°F (428 to 620°C)]. As shown in Table 11–12, the maraging steels come in different strength grades. It is common to designate these alloys by tensile strength; a 300-ksi maraging steel is an 18% nickel steel with a minimum tensile strength of 300,000 psi. The more appropriate designation would be the ASTM specifications and grade: ASTM A 538, Grade A, B, or C.

Because we are talking about diverse steel systems, it is difficult to put selection into an orderly pattern. Experience in design has shown that three alloys, 4340 and the 250- and 300-ksi grades of the 18% nickel maraging steels, will meet most needs. The 250-ksi maraging steel

Table 11–12
Composition and properties of three maraging steels (ASTM A 538 specification)

Chemical Requirements (%)	Grade A (200)	Grade B (250)	Grade C (300)
Carbon, maximum	0.03	0.03	0.03
Nickel	17.0–19.0	17.0–19.0	18.0–19.0
Cobalt	7.0–8.5	7.0–8.5	8.0–9.5
Molybdenum	4.0–4.5	4.6–5.1	4.6–5.2
Titanium	0.10–0.25	0.30–0.50	0.55–0.80
Silicon, maximum	0.10	0.10	0.10
Manganese, maximum	0.10	0.10	0.10
Sulfur, maximum	0.010	0.010	0.010
Phosphorus, maximum	0.010	0.010	0.010
Aluminum	0.05–0.15	0.05–0.15	0.05–0.15
Boron (added)	0.003	0.003	0.003
Zirconium (added)	0.02	0.05	0.05
Calcium (added)	0.05	0.05	0.05
Tensile requirements			
Tensile strength, ksi, minimum	210 (1448 MPa)	240 (1655 MPa)	280 (1930 MPa)
Yield strength, 0.2% offset, ksi	200–235 (1379–1619 MPa)	230–260 (1586–1793 MPa)	275–305 (1891–2100 MPa)
Elongation in 2 in., minimum	8	6	6
Reduction in area for round specimens, % minimum	40	35	30

Source: Reprinted with permission of ASTM International.

has solved innumerable breakage problems in injection molding cavities. It also provides an extra measure of toughness for high-loaded shafts with unavoidable stress concentrations. It has solved problems on components with operating stresses over 100 ksi (689 MPa). The alloy steel 4340 is more available and cheaper to use when the service is not quite as severe.

11.8 Selection and Specification

One of the big ten steel producers ran an ad in a trade magazine with a lead line that read, "Choose from one-quarter million basic combinations of steel bar chemistry and processing." This is undoubtedly a true statement, and the number of combinations would be even greater if the ultra-high-strength steels were included. How is the designer to cope with such an awesome multitude of steels? You need only one for the part on your Computer-aided design terminal.

We have discussed a wide range of steels, but familiarization with a few specific steels from these various categories will suffice. If a primary design criterion is strength, as shown in Figure 11–17, the strength ranges overlap. If a design calls for a yield strength of, for example, 100 ksi (68.9 MPa), the job could be done with a hardenable carbon steel. If lowest cost is a second criterion, this may be the proper choice. If the design is a large structure such as a vessel or a tank, a mill-heat-treated steel will eliminate heat treatment difficulties. If the design is for a few power transmission shafts, alloy steels would be a better choice. They cost more than carbon

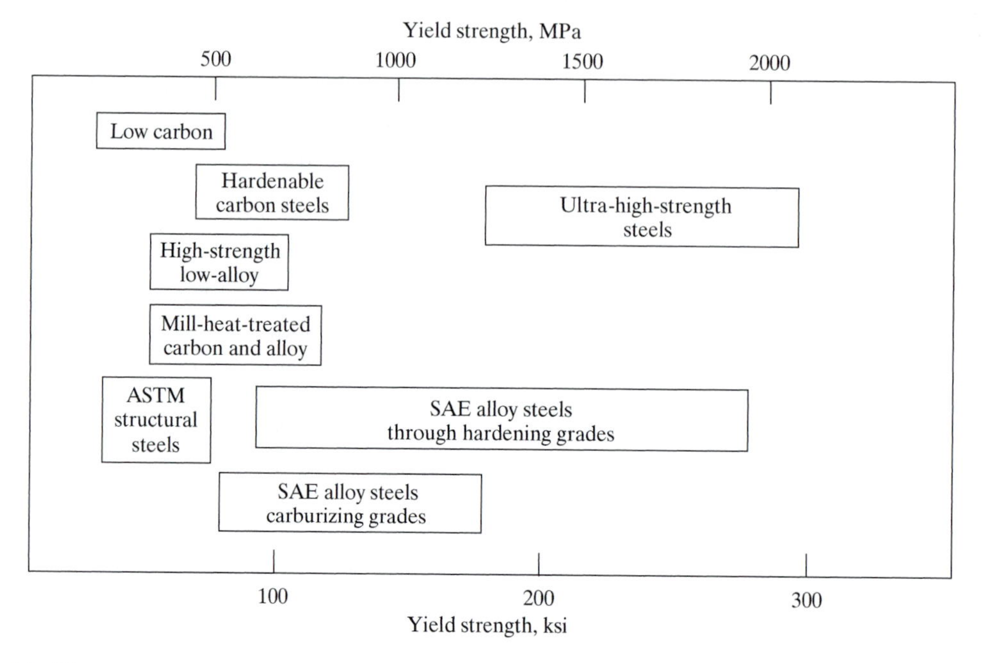

Figure 11–17
Yield strengths possible with various steels

steels, but they have better safety in hardening because of the oil quench. This weighing of relative merits must be done to arrive at a single material choice.

If the application under design calls for a wear-resistant surface, the carbon and alloy steel systems must be analyzed for hardening characteristics (Figure 11–18). If an application requires a surface that a cam follower or rollers run against, a surface hardness in the range of 60 to 62 HRC will be required. Carbon steels can be hardened to this level only in thin sections. The more likely choice, however, would be a carburized low-carbon or carburizing grade of alloy steel. Carburized surfaces with a case depth of 0.030 to 0.040 in. (0.75 to 1 mm) are usually adequate for this type of application. An alternative, if the part is applicable, would be a flame- or induction-hardened 4140 steel. Ultra-high-strength steels, even though they have hardnesses in the low 50s Rockwell C, are not very wear resistant because of their low carbon levels. They should be avoided for wear applications. Mill-heat-treated abrasion-resistant steels with hardnesses in the range of 30 to 40 HRC are suitable for the pounding, high-stress type of abrasion that might be needed in a solids conveying device or part chute. They would not be good choices for metal-to-metal wear applications. The other steels with hardnesses below about 20 HRC (~200 HB) are not suitable for wear applications.

If cost is a primary selection factor, the relative alloy content is a fair indication of cost. Plain carbon steels are the lowest cost, then high-strength, low-alloy and mill-heat-treated steels, then alloy steels. The most expensive steels are the high-alloy, ultra-high-strength steels. These can cost as much as $15/lb, compared with $2/lb for alloy steel.

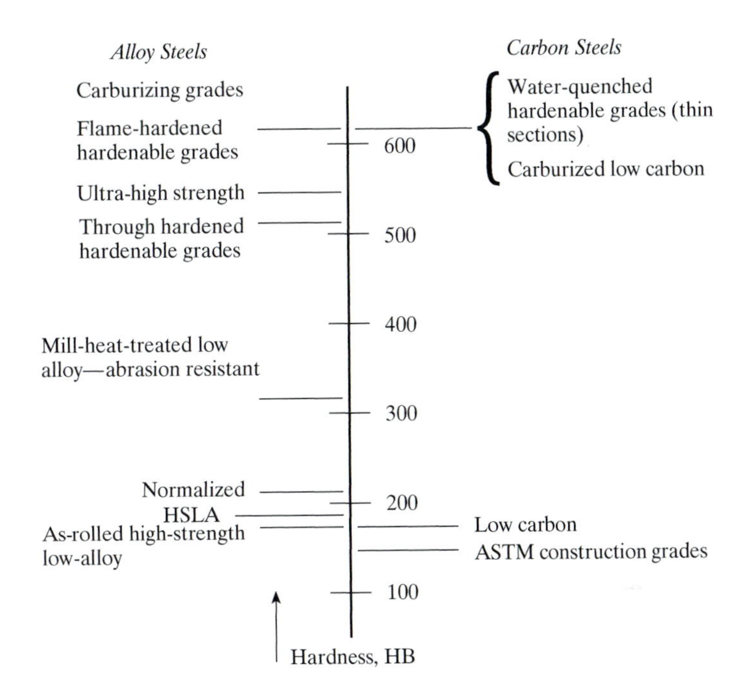

Figure 11–18
Hardness obtainable in various steel systems

Thus, in selecting carbon and alloy steels, there is a spectrum of steels with varying properties, cost, and utility. The designer has eight categories and thousands of steels to choose from within these categories. The selection situation is not as bad as it appears. Most designers get along with a few carbon steels, a few alloy steels, and an occasional high-strength steel. The following are the most frequently used carbon and alloy steels in the United States.

1. SAE 1010: formed sheet metal parts.
2. SAE 1020: general machine applications.
3. SAE 1040: flame- or induction-hardened parts.
4. ASTM A 36: structural steel.
5. SAE 4140: high-strength machine parts.
6. SAE 4340: high-strength machine parts.
7. SAE 8620: carburized wear parts.

This is an abbreviated list, but familiarity with these steels is a good starting point in coping with the selection of carbon and alloy steels.

Proper specification of a carbon or alloy steel on engineering drawings can be done by statement of the SAE, ASTM, or other designating body name and the condition of heat treatment or grade. Some basic elements for ordering steels to an ASTM specification are the following:

1. ASTM specification number.
2. Name of material (high-strength, low-alloy sheet, for example).
3. Grade or temper (if applicable).
4. Condition (oiled or dry, as required).
5. Type of edge (mill, cut, slit, rolled, and so on).
6. Finish (cold rolled, hot rolled, matte, bright, for example).

7. Dimensions (thickness, width, length, cut sheet size, and so on).
8. Coil size (inside diameter, outside diameter, and maximum weight).
9. Application (include drawing).
10. Packaging (coil on pallet, bars in boxes, cut sheet on skids, for example).

There are other necessary elements for different steel products, but the ASTM specifications provide a good guide on the essential elements for a particular type of product. Proper ordering information is the final part of a material selection, and it should be done with due consideration. ASTM designations should be used for international designs that may be manufactured anywhere in the world.

Summary

This chapter contains information on the steels that will be most used by machine designers. They are the steels for frames, base plates, angle iron, wide-flange beams, shafting, gears, pipes, tubes, extension springs, flat springs, and countless other devices that make up a machine design or building design or the design of most structures. Carbon and alloy steels make up probably 90% of the tonnage of all steels.

We have pointed out the differences among various grades of carbon and alloy steels, and demonstrated how to specify them. Some important concepts to keep in mind are the following:

- SAE steels that have designations starting with 1 are carbon steels; if the designation starts with another integer, it is an alloy steel.
- The last two digits of the SAE designation represent the carbon content in hundredths of a percent; 20 = 0.20%.
- The SAE steels are used mostly in machine design; structural steel and steel for specific applications are covered by ASTM specifications.
- ASTM A 29 specification can be used to designate carbon and alloy steels, but this specification merely references the SAE designations (e.g., SAE 1020 steel is ASTM A 29 grade 1020). For these steels, the SAE system is the preferred designation system in the United States. The ASTM system is preferred for international applications.
- The UNS numbers for most carbon and alloy steels are the SAE designations with a G prefix and a 0 suffix; SAE 4140 steel is UNS G41400.
- Cold-finished steels can be ordered to different degrees of cold work (tempers). The harder tempers can often save costs (beverage cans seem to get thinner each year, but they still work; the steel is getting stronger).
- Hot-finished steel is more stable in machining, and it is lower in cost than cold-finished steel; its disadvantage is poor surface finish.
- There is a difference between hardenability and hardness. Hardenability is the ease with which a material can be hardened. Hardness is simply the degree of hardening.
- Section thickness has a profound effect on the ability of a piece of steel to harden. When a section is thicker than 0.5 in. (12 mm), we should start to consider a material's hardenability.

In the United States the most common carbon and alloy steel designation systems are the SAE and ASTM systems; there are equivalents to most SAE and ASTM steels in other countries.

Critical Concepts

- Alloy additions to steels can be done for various reasons, but most increase hardenability and strength.

- There are at least ten classes of alloy steels corresponding to different intended uses.
- Phase diagrams show microstructure changes that occur with alloy additions.
- Carbon and alloy steels are usually lower in cost than most other steels that can be quench hardened (if cost is a selection factor).

Terms You Should Remember

grain size
half-hard
skin rolled
formability
dual phase
austenitizing
Jominy
weldability
core properties
hardness traverse
SAE identification
free machining
composition
carbon steel
alloy steel
ultra-high-strength steel
case hardening
maraging steel
weathering steel
high-strength, low-alloy (HSLA) steels
mill-heat-treated steels
work hardening
hot finished
cold rolled
carbon equivalent
core strength

Case History

Selection of a Steel for a Shaft

One of the duties of a metallurgist in a manufacturing operation is to investigate and resolve metal failures. One failure analysis concerned a 1-in.-diameter type 304 stainless steel shaft about 3 ft long with a ¼ in. wide × ¼ in. deep keyway 3 in. long on both ends. The shaft broke in half at the shaft side of one of the keyways. This shaft was driving a conveyor, and one end was coupled to the conveyor drive gear. The other end went into a flexible coupling, then the motor. We were asked to tell the production department why the shaft failed and what to do to pervent recurrence.

Visual inspection of the fracture surface indicated that cyclic bending fatigue was the cause of the failure. A machine inspection uncovered significant misalignment between the motor and conveyor drive—too much for the flex coupling to accommodate. Our recommendation was to reduce the misalignment from millimeters to less than 0.1 mm and to replace the 304 stainless steel shaft with 4140 steel hardened to 30 HRC.

Austenitic stainless steels are notoriously poor in bending fatigue situations, while 4140 and 4340 steels at 30 HRC have exceptional fatigue resistance as well as significantly higher strength than 304 stainless. Hardened alloy steels should be used for all power transmission applications if long service life is a design factor.

Questions

Section 11.1

1. What is the difference between 4140 steel and 4340 steel?
2. What is the SAE equivalent to a UNS G11310 steel?
3. What is the carbon content in 52100 steel?
4. List three carbon steels that can be hardened to 60 HRC.
5. What is the difference between an ASTM steel specification and an SAE designation?

Section 11.2

6. What increase in strength will be produced in pure iron by roll reducing it from 10 to 7 mm in thickness?

7. What is the maximum hardness possible by cold finishing a 1010 steel?
8. What is the minimum bend radius permissible on No. 4 temper steel strip?
9. Name three microscopic effects in steels that produce strengthening.
10. In what condition of heat treatment are hot-finished steels when they are sold to users?
11. An aluminum scaffold deflects 3 in. under the weight of two painters. What would be the deflection if the scaffold were made of 1020 steel? Hardened 1080 steel?
12. Why does hot-finished steel machine more easily than cold-finished steel?

Section 11.3

13. Can an alloy steel with 0.18% Cr, 0.2% Ni, 0.1% C, and 0.02% Mn be arc welded?
14. A Jominy test bar has a hardness of 45 HRC 20 mm from its quenched end; does this steel have high hardenability?
15. What is the tensile strength of 4140 steel that has been hardened and tempered at 1000°F?
16. A design requires a steel with a tensile strength of 325,000 psi (2240 MPa). What steel would meet this requirement?
17. In dual-phase steels, what two phases are present in the microstructure?

Section 11.4

Based on Figure 11–12, what is the approximate maximum surface hardness attainable for 4340, 8740, and 4140 steels? Using Table 11–1 and Table 11–10, compare the chemical compositions of these three steels. Why do all three of these steels have nearly the same surface hardness?

Section 11.5

18. What is the range of Rockwell hardnesses available in mill-heat-treated steels?
19. How fast (seconds) must you quench 4340 steel to prevent formation of soft phases in hardening?

Section 11.6

20. What are the differences between a high-strength, low-alloy steel and a carbon steel?
21. You want to case harden a 2-in.-diameter shaft for wear resistance. What steel would you select for the shaft?
22. Write a purchasing specification for the steel to be used for blanking of 10^7 flat washers. The finished washers will be 2 mm thick and 10 mm in outside diameter. They must have a good appearance.
23. What is the role of boron in steel? Vanadium? Copper?
24. Specify a steel for building a cellular telecommunications tower.

Section 11.7

25. Where would you use an austenitic manganese steel?
26. What are the distinguishing characteristics of nickel maraging steels?

To Dig Deeper

Bringas, Wayman M. *The Metals Black Book,* 4th ed. Edmonton, Alberta: Caso Publications, 2000.

Davis, J. R., Ed. *ASM Specialty Handbook: Carbon and Alloy Steels.* Materials Park, OH: ASM International, 1995.

Frick, J., Ed. *Woldman's Engineering Alloys,* 9th ed. Materials Park, OH: ASM International, 2000.

Davis, J. R., Ed. *Alloying: Understanding the Basics.* Materials Park, OH: ASM International, 2001.

Guy, Albert G. *Introduction to Materials Science.* Marietta, OH: R.A.N. Publishers, 1972.

Leslie, W. C. *The Physical Metallurgy of Steels.* New York: Hemisphere Publishing Co., 1981.

Metals and Alloys in the Unified Numbering System, SAE HS 1068, ASTM DS-56E. Warrendale, PA: Society of Automotive Engineers, and Philadelphia: American Society for Testing and Materials, 1993.

Metals Handbook, Volume 1, Properties and Selection: Irons, Steels and High Performance Alloys. Metals Park, OH: ASM International, 1990.

Modern Steels and Their Properties. Bethlehem, PA: Bethlehem Steel Corporation, 1980.

Steel Products Manual, Hot Rolled Structural Shapes, H-Piles and Sheet Piling. Book No. MN 41/211. Warrendale, PA: Iron and Steel Society, 1991.

Steel Products Manual, Plates: Rolled Floor Plates: Carbon High Strength Low Alloy, and Alloy Steel. Book No. MN 41/212. Warrendale, PA: Iron and Steel Society, 1991.

Steel Products Manual, Sheet Steel: Carbon, High Strength Low Alloy, Alloy Uncoated, Metallic Coated, Coil Coated, Coils, Cut Lengths, Corrugated Products. Book No. MN 41/214. Warrendale, PA: Iron and Steel Society, 1996.

Steel Products Manual, Strip Steel: Carbon, High Strength Low Alloy and Alloy. Book No. MN 41/215. Warrendale, PA: Iron and Steel Society, 1995.

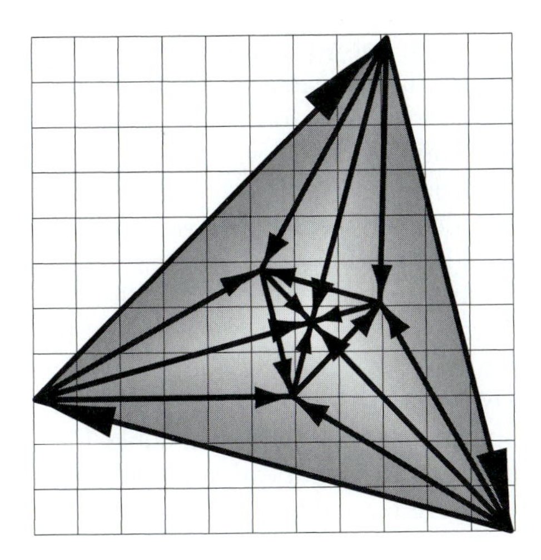

CHAPTER 12

Tool Steels

Chapter Goals

1. An understanding of the metallurgy of tool steels.
2. A knowledge of the properties of the various types of tool steels to aid in selection.
3. An understanding of how to specify tool steels and their heat treatment.

Rationale

We have pointed out how carbon and alloy steels make up the bulk of steel production and how they are used in machines and products. Now we move on to the steels that have evolved over the past century or so as the "premium steels," the steels with hardness and toughness capabilities not found in carbon and alloy steels. These steels are used to shape, form, and machine the other steels.

Any designer who is asked to develop a tool for perforating a metal, molding a plastic, machining a metal, bending a metal—just about any shaping task—will need to use a tool steel. For this reason, we dedicate an entire chapter to this class of steels. We need to show how tool steels differ from alloy steels, that there are various categories for different applications, and how to select them. Tool steels are usually too expensive to use for making consumer products and structures; they are for tools—punches, dies, knives, molds, reamers, saws, end mills, lathe bits—whatever it takes to shape a raw material into a form that can be sold or a part that goes into an assembly that is sold. These are important steels for designers to know.

Tool steels are an extension of the spectrum of steels discussed in the preceding chapter. They are defined by U.S. steel producers as "carbon or alloy steels capable of being hardened and tempered." Many alloy steels would fit this loose definition. *Tool steels* usually contain significantly more alloying elements than alloy steels. However, the real factor that discriminates tool steel from carbon or alloy steel is the manufacturing practice. Tool steels are melted by electric furnace techniques, and some are even remelted by VAR, VIM, or ESR techniques. These melting practices impart cleanliness and alloy control not obtainable with the melting processes that are used on most carbon and alloy steels. In addition to melting practice, tool steels are usually made in small heats and are subjected to quality controls and inspections not used on merchant-quality and construction steels. In general, they are available as bar, rod, and large forged shapes. Bars are sometimes cut from plate. ASTM A 597 lists specifications on nine cast tool steels, but availability is limited.

It is the purpose of this chapter to give the designer the necessary information for selecting the right tool steel for any design situation that requires this kind of steel. To achieve this goal, we shall discuss tool steel identification, metallurgy, properties, selection, and specification.

12.1 Identification and Classification

Approximately 70 types of tool steels are listed in the U.S. steel products manuals, and other steel-producing countries also make a variety of grades. Why so many? Tool steels evolved from alloy development aimed at hardening heavy sections and using less severe quenching media. One reason for so many types of tool steels is evolutionary development over a period of 90 years. The second reason is the wide range of needs that they serve. Tool steels have properties that permit their use as tools for cutting and shaping metals and other materials both hot and cold. If one grade did not work, another was developed. We shall try to focus our discussion on the grades that are most widely used and most available. Probably fewer than 20 tool steels are readily available in the United States. We continue to list grades that are no longer promoted by producers to help designers identify tool steels that may be specified on old drawings. We will point out in our discussion of various grades which ones have the best availability.

The U.S. tool steel classification system is based on use characteristics. Tool steels intended for use in hot-work applications such as forging are put in one category; tools intended for use in cold-work dies, forming tools, and the like are in another category. There are four basic categories, and one category contains grades intended for special purposes. A prefix letter is used in the alloy identification system to show use category, and the specific alloy in a particular category is identified by one or two digits. The tool steel use categories are illustrated in Figure 12–1, and the identification letters for specific alloys are listed in Table 12–1. The complete alloy identification number consists of the use letter and the alloy number:

S1	a shock-resistant tool steel
D2	a cold-work tool steel
H11	a hot-work tool steel
M42	a high-speed steel

There are tool steels equivalent to the U.S. tool steels in most steel-producing countries, but it is difficult to find cross-reference information. If a part under design in the United States may potentially be manufactured in another country, it may be prudent to use ASTM or UNS designations on the part drawing as opposed to the traditional U.S. alphanumeric designation. ASTM specifications are international in scope. The ASTM A 681 specification covers all of the AISI tool steels with the exception of the W, M, and T series; the W series is covered by ASTM A 686. The M and T series are covered by ASTM A 600. Nine grades of cast tool steel (A2, D2, D5, M2, S5, S7, H12, H13, and O1) are covered by ASTM A 597. The ASTM specifications essentially use the former AISI tool steel designations. For example, AISI D2 tool steel can be designated as ASTM A 681, type D2 or ASTM A 681, type T30402; W1 tool steel is ASTM A 686, type W1 (Grade A or C) or ASTM A 686, type T72301. The latter is the UNS number. These numbers have been included in Table 12–2. We have elected to use the more manageable former AISI designation in our subsequent discussions, but

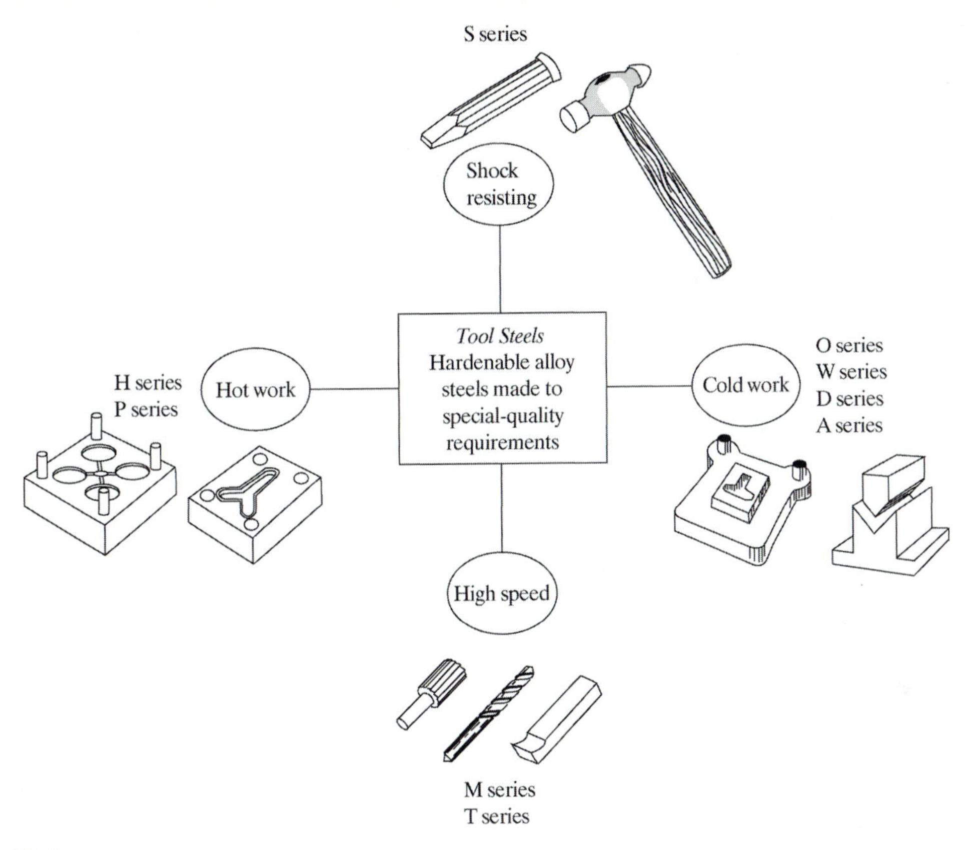

Figure 12–1
Tool steel categories

Table 12–1
Tool steels

	Prefix	Specific Types
Cold work	W (water hardening)	W1, W2, W5
	O (oil hardening)	**O1**, O2, **O6**, O7
	A (medium alloy air hardening)	**A2**, A4, **A6**, A7, A8, A9, A10, A11
	D (high C, high Cr).	**D2**, D3, D4, D5, D7
Shock resisting	S	S1, S2, S4, S5, S6, **S7**
Hot work	H	H10–H19 chromium types (**H11 and H13**)
		H20–H39 tungsten types
		H40–H59 molybdenum types
High speed	M	Molybdenum types: **M1**, **M2**, M3-1, M3-2, M4, M6, **M7**, **M10**, M33, M34, M36, M41, **M42**, M46, M50
	T	Tungsten types: T1, T4, T5, T6, T8, **T15**
Mold	P	Mold steels: P6, **P20**, P21
Special purpose	L	Special-purpose steels: L2, L6

Note: The most available grades are shown in bold in this and other tables in this chapter.

Table 12–2
UNS identification numbers for U.S. tool steels

Tool Steel Category	Type	UNS Number
High-speed steels	M series	
	M1	T11301
	M2	T11302
	M3/1	T11313
	M3/2	T11323
	M4	T11304
	M6	T11306
	M7	T11307
	M10	T11310
	M33	T11333
	M34	T11334
ASTM A 600	M36	T11336
	M41	T11341
	M42	T11342
	M46	T11346
	M50	K88165
	T series	
	T1	T12001
	T4	T12004
	T6	T12006
	T8	T12008
	T15	T12015
	Chromium type	
Hot-work steels	H10	T20810
	H11	T20811
	H12	T20812
	H13	T20813
	H19	T20819
ASTM A 681	Tungsten type	
	H21	T20821
	H22	T20822
	H23	T20823
	H24	T20824
	H26	T20826
	Molybdenum type	
	H42	T20842
Cold-work steels	Water hardening	
	W1	T72301
ASTM A 686	W2	T72302
	W5	T72305
	Oil hardening	
	O1	T31501
	O2	T31502
	O6	T31506
	O7	T31507
	Air hardening	
	A2	T30102
	A4	T30104
	A6	T30106
	A7	T30107
	A8	T30108
	A9	T30109
	A10	T30110
	High-carbon, high-chromium	
ASTM A 681	**D2**	T30402
	D3	T30403
	D4	T30404
	D5	T30405
	D7	T30407
Shock resisting	S1	T41901
	S2	T41902
	S4	T41904
	S5	T41905
	S6	T41906
	S7	T41907
Mold steels	P6	T51606
	P20	T51620
	P21	T52621
Special purpose	L2	T61202
	L6	T61206

Most widely available grades are shown in bold.

keep in mind that an ASTM designation for a tool steel is also valid and preferred for international trade.

12.2 Tool Steel Metallurgy

Comparison with Carbon and Alloy Steels

In the 2000s the tool materials used by industries facing international competition are alloy steels, tool steels, *cermets*, and ceramics. Each of these materials has different characteristics, advantages, and disadvantages. Progressive companies use all these materials for machine components, forming tools, cutting devices, and the like, that is, for implements that form and shape other materials. We have already discussed the characteristics of ceramic, cermets, carbon, and alloy steels. We shall start our discussion of tool steel metallurgy by comparing tool steels with these materials.

One problem that exists in discussing the metallurgy of tool steels is that, because there are six major categories of tool steels (hot-work, cold-work, high-speed, mold, special-purpose, and shock-resistant steels), it is difficult to make statements that apply to all of these steels. The composition and properties of the steels in these categories vary significantly. One common thread is the quality factor, which we have previously mentioned. These steels are made in smaller lots than carbon and alloy steels. Billets are often hand ground before size reduction to remove defects. Some receive special heat treatments to spheroidize carbides, and in general they are treated differently. Some additional factors that set them apart from carbon and alloy steels are that they cost more than alloy steels, they have better hardenability, some have better heat resistance, most are easier to heat treat, and most are more difficult to machine than carbon and alloy steels. There is probably a tool steel that could be cited as an exception to each of these statements, but this is about all that we can say about tool steels when such diverse steels are put in the same grouping.

Figure 12–2 is an attempt to illustrate some of the metallurgical differences between carbon steel, alloy steels, and tool steels. We have mentioned how the manufacturing procedures are different with regard to quality requirements; as is shown in Figure 12–2, tool steels are not generally available in all the forms that carbon and alloy steels are. Most tool steels are sold as hot-finished shapes such as rounds and bars. Flat ground stock is available in a number of alloys. Cold-finished sheets are not available because it is too difficult to cold roll or cold finish these materials. Most commonly used shapes such as rounds and bars are also available in machined finishes, ground plates, rounds, and bars. Alloy steels are most widely supplied in the hot-finished form. Some alloys are available in plate form (mostly abrasion-resistant alloys and high-strength, low-alloy grades). Alloy plates are used for items such as off-road truck bodies and similar applications where battering is a service condition. Carbon steels, on the other hand, are available in cold-finished sheet and strip and as a wide variety of shapes (essentially all mill shapes).

The chemical compositions of most tool steels are significantly different from alloy and carbon steels. Some tool steels have compositions that fit into the composition ranges that we defined for carbon and alloy steels, but most tool steels have alloy concentrations that are significantly higher than those found in carbon and alloy steels. The range of alloying elements is also larger. Some tool steels can contain elements such as tungsten, cobalt, and vanadium that are usually not present in alloy steel. Carbon steels essentially do not contain any significant amounts of alloy elements except those added for deoxidation or machinability purposes. One important factor that should be kept in mind about the alloy content of tool steels is that these alloy additions usually do not impart

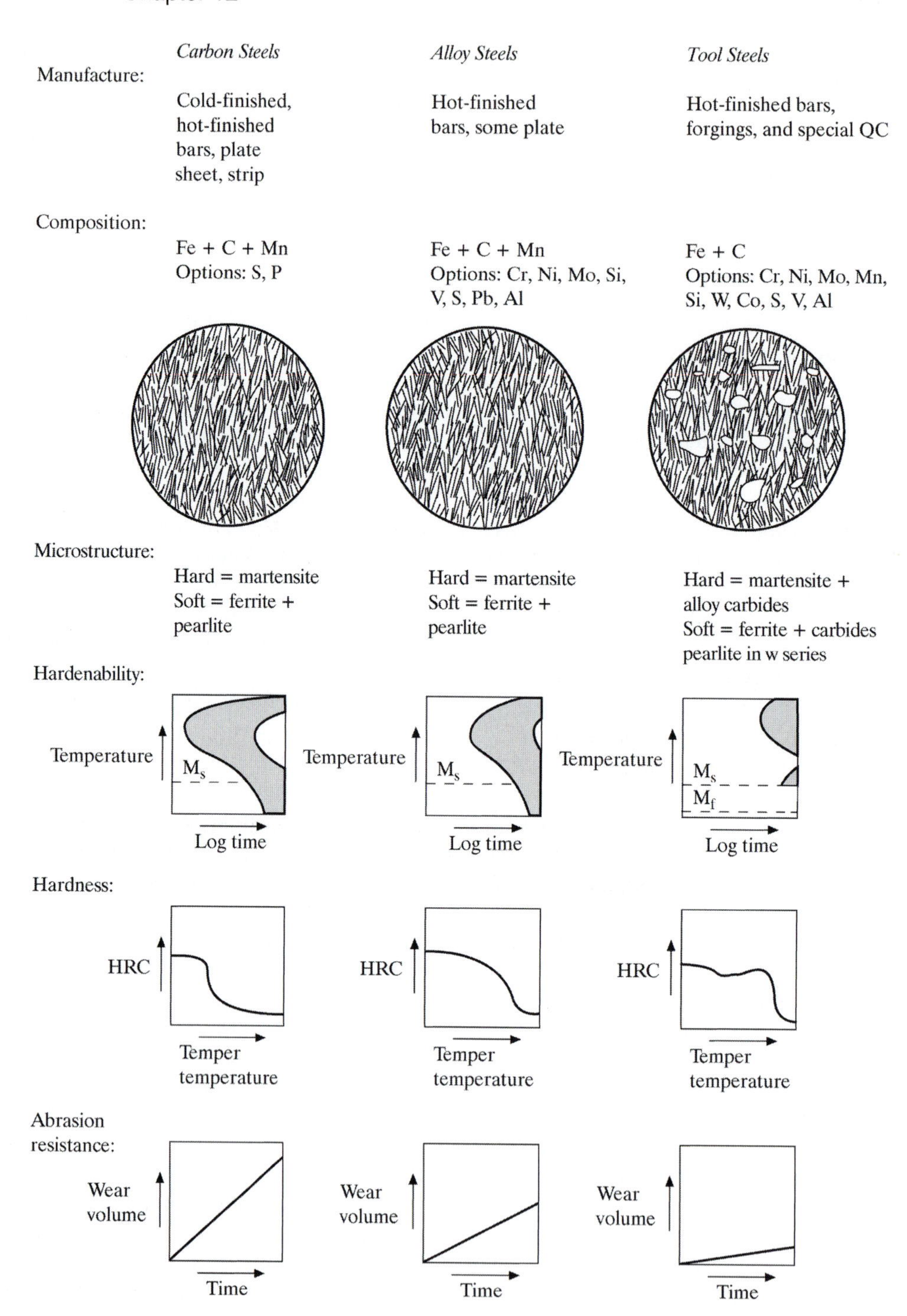

Figure 12–2
Comparison of some of the metallurgical characteristics of tool steels with those of carbon and alloy steels

corrosion resistance to these steels. None of them should be considered to be similar to stainless steels, even though some grades can have as much chromium in them as stainless steels. The reason for this is that the alloy elements are usually combined with carbon to form carbides. The alloy is not in solid solution to make a corrosion-resistant (passive) surface. Type D2 tool steel is a special case in which the chromium does impart a modicam of atmospheric rust resistance. In the annealed condition, D2 has most of its chromium present in the form of M_{23} C_6 carbides. After hardening most of the smaller carbides are dissolved in the *matrix* to assist in resisting rusting in air.

The most significant metallurgical difference between tool steels and the other steels is their microstructure. Figure 12–3 contains a schematic of *microconstituent* that would be present in these different steels when examining a cross section with optical microscopy at about 400 diameters magnification. Figure 12–4 shows actual microstructures of important tool steels. A fully hardened carbon steel or alloy steel would have only martensite as the predominating matrix. Some tool steels have this same structure, but the wear-resistant grades of tool steel (high speed and cold work) have a hardened structure of martensite with some volume fraction of alloy *carbides*. The volume fraction may be as low as 1% and as high as 30%. We shall discuss this in more detail later, but essentially the presence of alloy carbide in the microstructure of many grades of tool steel is one of the most significant factors that makes these materials stand out from other steels.

Figure 12–2 shows schematically how the TTT diagrams of tool steels differ from those of alloy and carbon tool steels. Essentially, the higher alloy content gives these steels significantly better hardenability, and this is why we previously made the statement that these steels are easier to heat treat. The quenching requirements are less severe for obtaining a fully hardened structure. This means less distortion and less chance of getting heat treating cracks from a severe quench. Tool steels in the hardened condition can have the same hardness as fully hardened carbon and alloy steels, but the tempering characteristic of the hardened structures may vary significantly (Figure 12–2). Tool steels can be much more resistant to softening in use from heat. Some tool steels experience secondary hardening. When heated to 800°F (426°C), some tool steels actually get harder. Alloy and carbon steels have little resistance to softening when heated. Their tempering characteristics are such that they readily lose hardness when used at temperatures over about 400°F (204°C).

Finally, in Figure 12–2 we show significant differences in carbide morphology. These differences are responsible for the superior abrasion resistance of tool steels compared to other steels. Not all grades of tool steels contain massive carbides, but the cold-work and high-speed tool steels have them, and they have significantly better abrasion resistance than carbon and alloy steels. These carbide phases can be vanadium, chromium, molybdenum, or tungsten carbides. All these carbides are harder than the martensite matrix. The more carbides are present, the greater the abrasion resistance; the bigger the carbides are, the better the abrasion resistance. In Figure 12–3, we have sketched the typical microstructures that are possible in tool steels and in cermets, a tool material that competes with tool steels. The tool steels that have compositions similar to alloy steels have carbides in their martensitic matrix that are so fine they are not normally resolved when viewed with optical microscopy at about ×400 (type I). As alloy and carbon content increase, the volume fraction of carbides will increase, but the shapes of the carbide phases can also vary. Some groups of tool steel tend to have the spheroidal carbides (types II and III) and some have blocky carbides (types IV, V, and VI). These blocky carbides are usually elongated in the direction of rolling, and they tend to form in

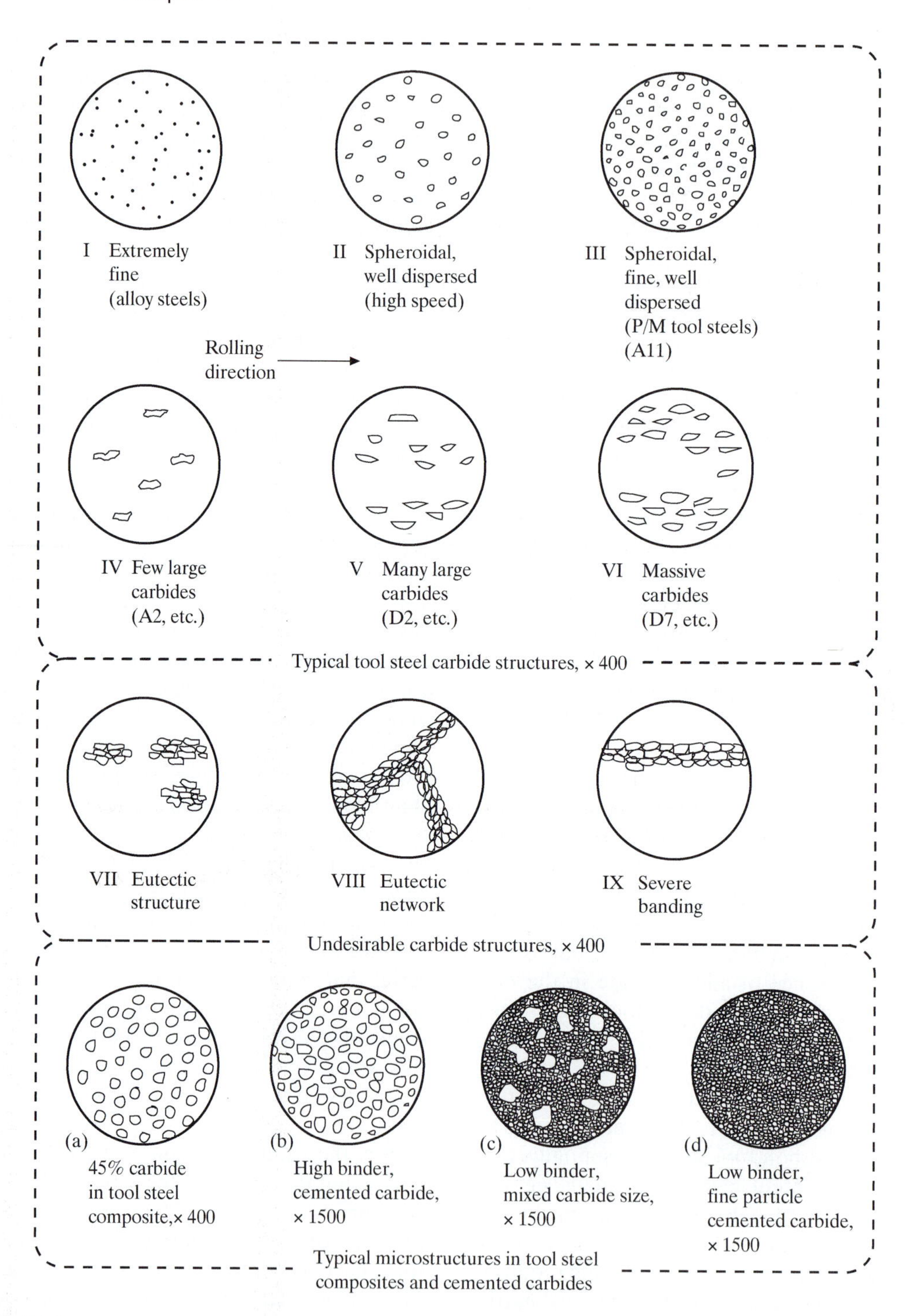

Figure 12–3
Carbide morphologies common in various tool materials: tools steels (type I to IX) and cermets (a) to (d)

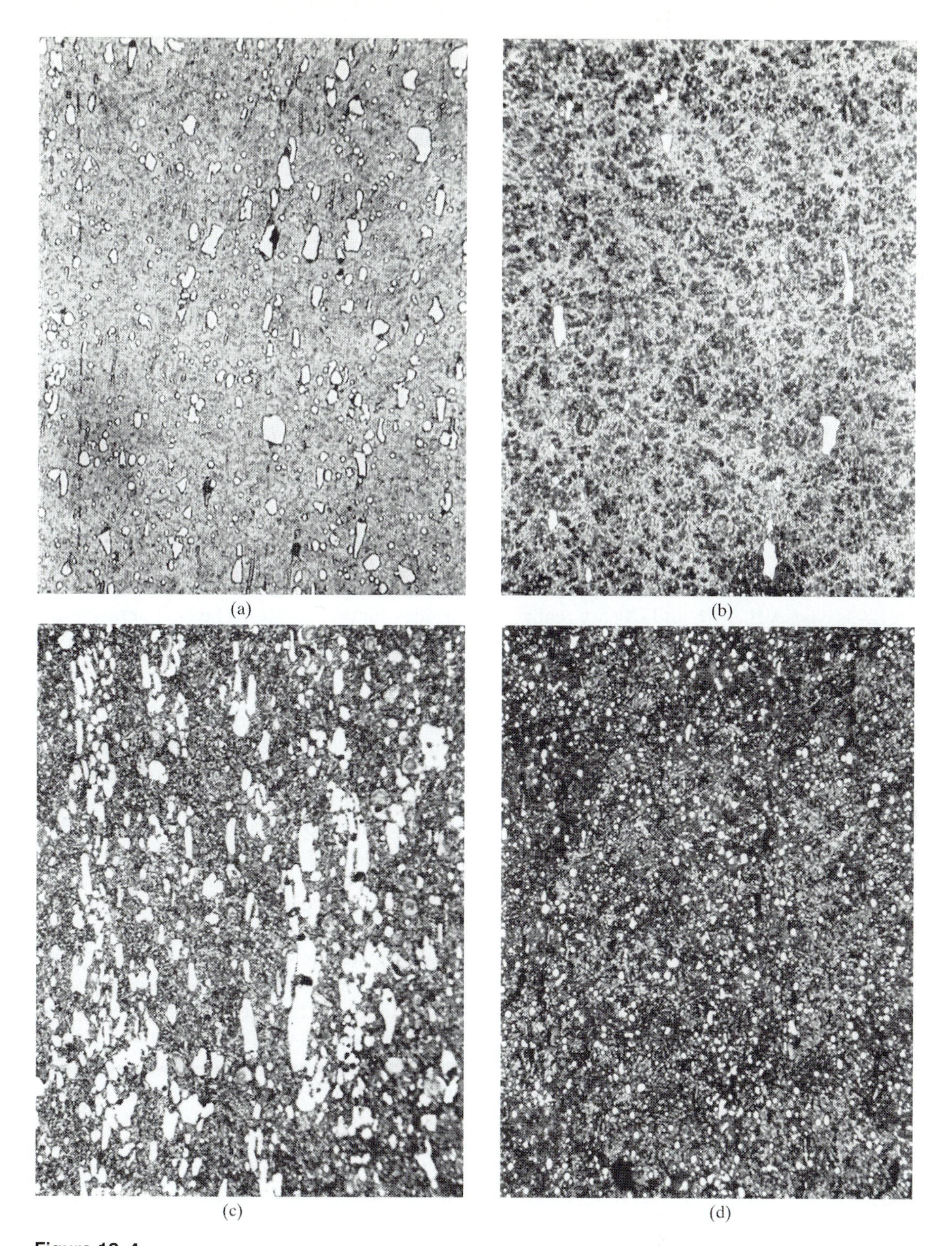

Figure 12–4
Carbide particle distribution in several tool steels (×400). (a) Type D2, (b) type A2, (c) type A7, and (d) type M3 (the white particles are carbides).

bands that are also aligned with the rolling direction. High-speed steels tend to have type II carbides. The tool steels that are made by powder metal techniques tend to have even finer spheroidal carbides (type III). The cold-work tool steels have structures of types IV, V, and VI. The more carbon and alloy present, the more carbides. Some *air-hardening tool steels* such as A7 and D7 have massive carbides in their structure like type VI (unless they are made by powder metal (P/M) techniques). In the United States, the alloys like A7 and D7 with massive carbide have been largely replaced by powder metal alloy like A11. Ingot cast shapes from A7 and D7 were very difficult to grind and had poor toughness. Figure 12–4 shows photomicrographs of some tool steel carbide patterns.

The other carbide possibilities illustrated in Figure 12–3 (types VII, VIII, and IX) are really tool steel defects in wrought forms. Eutectic carbides are often formed on casting. This eutectic structure looks somewhat similar to the pearlite structure that is common in carbon steels, layers of matrix and carbide. Hot working (wrought) is supposed to break up cast structures and produce a uniform distribution of carbide phase (particles), but bar size also affects carbide distribution. For example, 6-in.-diameter bars (150 mm) tend to have "hooked" carbides, 2-in.-diameter bars (50 mm) tend to have carbide "streaks," and 1-in.-diameter bars (25 mm) may have well-dispersed carbides. Sometimes there are islands of this type of structure (type VII); sometimes the eutectic follows the grain boundaries to give what is termed *network carbide*. Finally, we show severe *banding* of blocky carbides. All these are undesirable structures caused by lack of hot work (large bars) or problems in processing or alloying. Massive carbides and many of them produce the best abrasion resistance. The well-dispersed fine carbides like type II produce favorable *metal-to-metal wear* resistance against many counterfaces.

The structures of some of the competitors of tool steels are presented in the last four sketches in Figure 12–3. Sketch Figure 12–3(a) is the structure of a carbide–tool steel cermet. These are usually proprietary materials, and they are tool steels containing up to 50% volume fraction of carbides that are added in manufacture (Figure 12–5). These are often titanium carbide, and they tend to be spheroidal. Shock-resisting grades of cemented carbides can have about 20% cobalt binder and 80% carbide (mostly WC). Their structure is like sketch Figure 12–3(b). The low-binder grades of cemented carbide, those with less than 10% cobalt binder, usually have a structure that looks like sketch Figure 12–3(c) or (d). The carbides can be uniform in size or they can be heterogeneous, that is, some large, blocky carbides in a matrix that is mostly fine carbides Figure 12–3(c). The best abrasion resistance is usually obtained with the last structure: fine, uniformly sized carbides and low binder. The other structures, however, may have certain advantages over the fine structure, such as better toughness.

To summarize the subject of tool steel metallurgy compared to carbon and alloy steels, tool steels are easier to heat treat (better hardenability), they are superior in abrasion resistance (presence of carbides), they harden deeper, and many grades are less prone to the softening effect of heat (they have high tempering temperatures). The wear-resistant grades rely heavily on carbide-phase morphology for their improved service life over carbon and alloy steels. The grades that do not have large volume fractions of carbides, such as the mold steels, are formulated to produce other property advantages, such as deep hardening, heat resistance, fatigue resistance, and the like. In the following section we shall discuss tool steels as separate groups and the differences in specific grades, but the foregoing are the significant metallurgical differences that separate tool steels from carbon steels, alloy steels, and cermets.

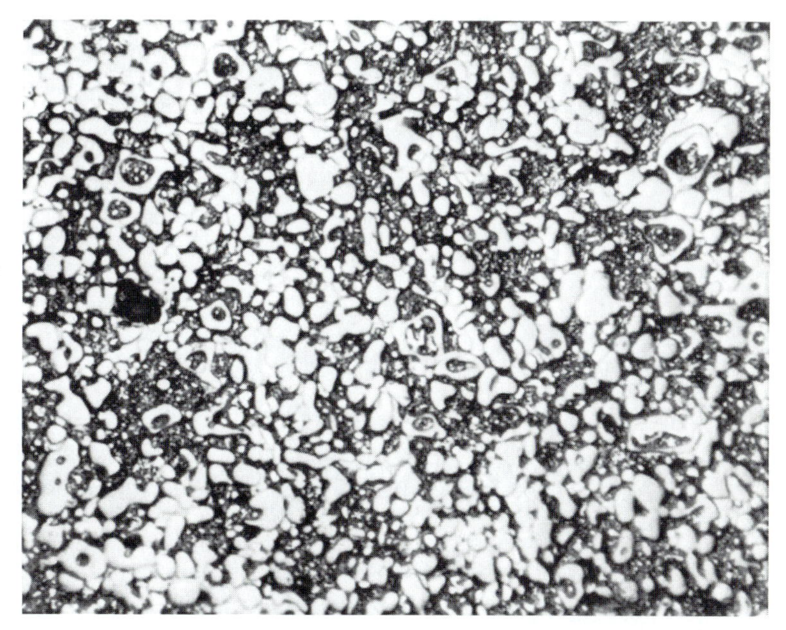

Figure 12–5
Microstructure of a steel-bonded carbide (cermet) at ×800. The matrix is a martensitic, stainless steel (black). There is about 45 volume % titanium carbide (TiC) phase present for wear resistance. The TiC is about twice as hard as the stainless steel matrix. These cermets are made with powder metal (P/M) techniques

12.3 Chemical Composition of Tool Steels

There are so many tool steels that it may seem awesome to learn something about all of them. We have shown how tool steels differ in carbide structure; we came up with six types of carbide structures that these steels can have. Thus, from the microstructure standpoint, they are not that complicated. The same type of simplification can be done with chemical composition. The various tool steel alloys fit into the five major use categories (cold-work, hot-work, shock-resisting, high-speed, and mold steels). We are leaving out the special-purpose grades (L) because they are not in wide use.

The five groups of tool steels have compositional characteristics that allow us to simplify the understanding of the compositional differences of tool steels. As shown in Figure 12–6, each subgroup in the five tool steel categories presents compositional similarities. For example, five D-series die steels are commonly available. Basically, all five have the same type of microstructure (massive carbides, but different amounts), and they differ by only minor changes in chemical composition. They all have between 1% and 2% carbon and 12% chromium. So why are five alloys available? Why not just one? The various alloys evolved from special requests from users. The number of grades is becoming fewer because of the economics of the tool steel business. Air-hardening tool steels are replacing oil and water-hardening grades, and D2 is the only D-type available from some tool steel producers. One hundred ten tool steel types were listed in U.S. handbooks in 1980. In 2003 probably less than 20 types are commercially available in the United States (even though more types are shown in handbooks).

Figure 12–6 lists the compositional similarities that exist in each of the major categories of tool steels. This should simplify readers' understanding of the chemical composition. The hot-work steels have medium

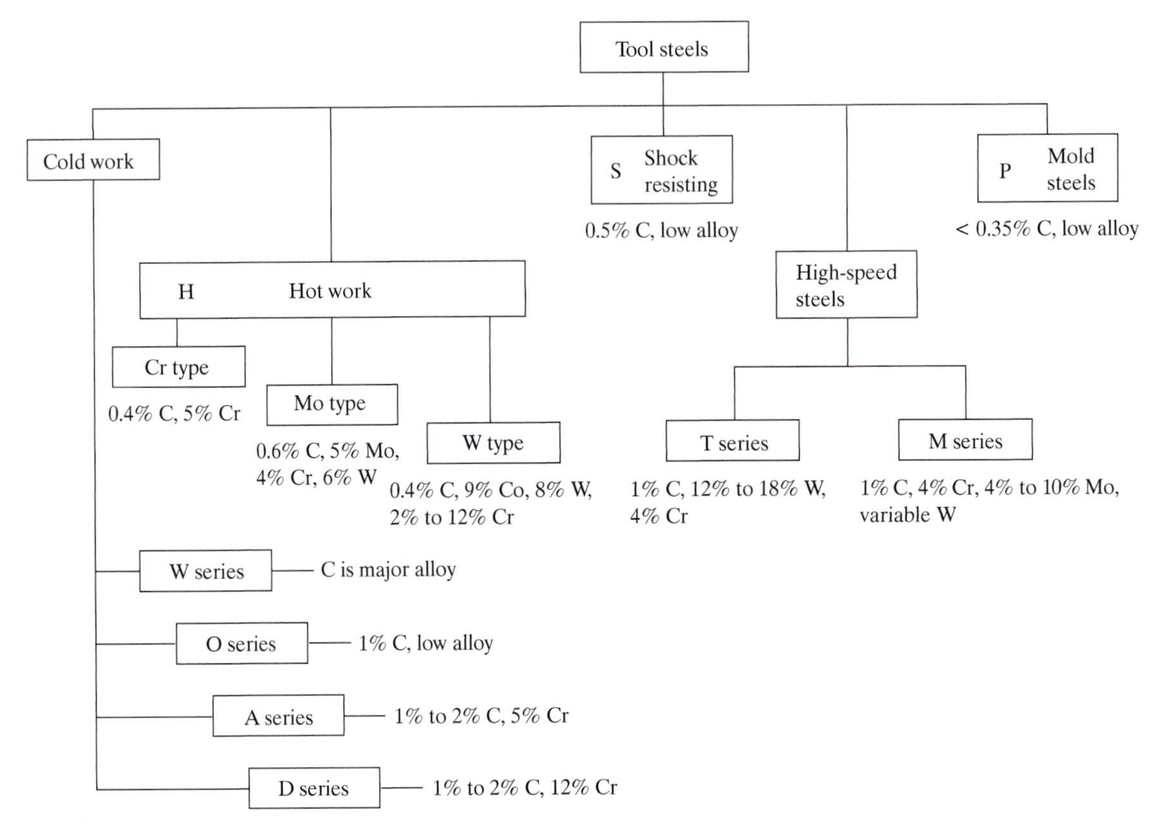

Figure 12–6
Comparative nominal chemical compositions of tool steels

carbon contents and various amounts of refractory metals such as chromium, tungsten, and molybdenum. The shock-resisting steels have about 0.5% carbon and a few percent of other alloying elements. The high-speed steels have about 1% carbon, and there are types that have tungsten as the major alloying element (T series) and types that have molybdenum as the major alloying element (M series). The mold steels have very low carbon and only a small amount of other alloying elements. The point is that you do not have to memorize the compositions of 70 alloys to understand the chemistry of tool steels. Remembering the similarities pointed out in Figure 12–6 will probably suffice. When A-series tool steels are discussed, think about tool steels that have from 1% to 2% carbon and 5% chromium; when T-series steels are discussed, remember that tungsten is the major alloying element. When the hot-work steels are discussed, remember that they have a relatively low carbon content but contain refractory metals in their composition for heat resistance.

In the next sections we will state some discriminating facts about all the commonly used tool steel alloys. Keep in mind the compositional similarities.

Cold Work: Water-Hardening Grades

Three grades of *water-hardening tool steels* (W) are in common use in the United States:

Nominal Composition (Wt. %)*			
Type	**C**	**Cr**	**V**
W1	0.6–1.4	—	—
W2	0.6–1.4	—	0.25
W5	1.1	0.5	—

*Iron is the remainder in this and subsequent lists.

These steels are essentially carbon steels. They will not contain alloy carbides in their hardened structure. The W2 alloy contains vanadium for grain refinement and for reduction of quench cracking tendencies. W5 contains chromium to improve hardenability. All W tool steels are shallow hardening and require a water quench.

Because of their low alloy content, W tool steels are usually the lowest-cost tool steels. They are very prone to cracking because of the water quench and are only used where their shallow hardening characteristics are desired. Some types of tools perform best with a soft core for toughness and a hard skin for wear resistance. Use of W-series steels is declining in many industries.

Cold Work: Oil-Hardening Grades

The *oil-hardening tool steels* that are commercially available are in the United States are the following:

Nominal Composition (Wt. %)						
Type	**C**	**Mn**	**Si**	**W**	**Mo**	**Cr**
O1	0.90	1.00	—	0.50	—	0.50
O2	0.90	1.60	—	—	—	—
O6	1.45	0.80	1.00	—	0.25	—
O7	1.20	—	—	1.75	—	0.75

These alloys have sufficient hardenability to allow an oil-quenching medium in hardening. Types O1 and O2 are similar in all use properties. Type O2 has a slightly lower austenitizing temperature than O1. A low hardening temperature reduces distortion in quenching.

Type O6 tool steel has sufficient carbon content to cause about 0.3% free graphite to form in the martensitic matrix of the hardened structure. It is called a graphitic tool steel. The graphite is intended to help lubricate when this steel is used in metal-to-metal sliding systems, but more important, the graphite makes this steel one of the easiest to machine. A tool made from this tool steel may be lower in cost than one from any other tool steel.

Type O7 contains a significant concentration of tungsten to promote formation of tungsten carbides, which improve abrasion resistance and edge retention in cutting devices.

Of the group, type O1 is the most widely used alloy, and typically it is used for short-run punch-press tooling.

Cold Work: Medium-Alloy Air-Hardening Grades

Most A-series air-hardening tool steels have between 5% and 10% total alloy, and they have sufficient hardenability to air harden in section thicknesses up to 6 in. (150 mm). Eight alloys are commercially available in the United States:

Nominal Composition (Wt. %)								
Type	**C**	**Mn**	**Si**	**W**	**Mo**	**Cr**	**V**	**N**
A2	1.00	—	—	—	1.00	5.00	—	—
A4	1.00	2.00	—	—	1.00	1.00	—	—
A6	0.70	2.00	—	—	1.25	1.00	—	—
A7	2.25	—	—	—	1.00	5.25	4.75	—
A8	0.55	—	—	1.25	1.25	5.00	—	—
A9	0.50	—	—	—	1.40	5.00	1.00	1.50
A10	1.35	1.80	1.25	—	1.50	—	—	1.80
A11	2.45	0.50	0.90	—	1.30	5.25	9.75	—

Type A2 is the most popular and most widely available steel. The chromium and molybdenum provide hardenability and promote the formation of alloy carbides in the microstructure.

Types A4 and A6 have manganese as the principal hardenability agent. They have lower austenitizing temperatures than the 5% chromium grades (lower distortion). Type A6 is readily available.

Type A7 has an excess of carbon and high vanadium to promote the formation of massive alloy carbides for wear resistance. The trade-off for improved wear resistance is poor *machinability*. This alloy has limited availability.

Types A8 and A9 contain a reduced carbon content compared to the other grades for improved toughness. The A9 alloy contains an additional 1% V to enhance wear life.

Type A10, like type O6, contains excess carbon to allow free graphite in the hardened structure. However, it has poor availability.

Type A11 is a grade of tool steel introduced in the late 1970s. It is made by powder metallurgy practice (P/M), and it contains 10% vanadium to promote the formation of wear-resistant vanadium carbides (VCs) in the microstructure.

Cold Work: High Carbon, High Chromium

The D-series tool steels have mixed hardening characteristics. Some are air hardening; some are oil hardening. All contain about 12% chromium and over 1.5% carbon:

	Nominal Composition (Wt. %)				
Type	C	Mo	Cr	V	Co
D2	1.50	1.00	12.00	1.00	—
D3	2.25	—	12.00	—	—
D4	2.25	1.00	12.00	—	—
D5	1.50	1.00	12.00	—	3.00
D7	2.35	1.00	12.00	4.00	—

Type D2 is the most widely used steel in the United States for cold-work tooling. It air hardens with low distortion in section thicknesses as large as 10 in. (250 mm). It contains very large chromium carbides in its hardened structure, which promotes abrasion resistance. Type D3 has higher carbon to produce even more alloy carbides and wear resistance. This alloy requires an oil quench, which is a disadvantage from the standpoint of distortion. D4 is a high-carbon version of D2 for greater abrasion resistance. D5 has a cobalt addition to promote resistance to temper softening. D7 is a high-chromium version of A7. It contains massive alloy carbides. Some of the carbides are of the VC type, which are harder than the chromium carbides in D2. It has excellent abrasion resistance, but poor machinability.

Of these steels, the D2 alloy is preferred for most applications, and it is the most widely available. It has excellent wear characteristics and an outstanding record for freedom from cracking and minimal size change in hardening.

Shock-Resisting Tool Steels

The S series of tool steels was originally developed for chisel-type applications, but the number of alloys in this category has evolved to include steels with a broad range of tool applications. There are six alloys listed in U.S. handbooks.

	Nominal Composition (Wt. %)					
Type	C	Mn	Si	W	Mo	Cr
S1	0.50	—	—	2.50	—	1.50
S2	0.50	—	1.00	—	0.50	—
S4	0.55	0.80	2.00	—	—	—
S5	0.55	0.80	2.00	—	0.40	—
S6	0.45	1.40	2.25	—	0.40	1.50
S7	0.50	0.60	—	—	1.40	3.25

This family of tool steels contains low carbon content compared with the cold-work steels. They never achieve the hardness levels of cold-work steels (58 HRC maximum), but they have better toughness. Tungsten is added to the S1 alloy to provide better wear resistance. It is the original chisel steel. Types S2, S4, S5, and S6 have silicon as a major hardenability agent. They have similar use characteristics, but subtle differences in hardening characteristics.

Type S7 is significantly different from the other S steels. The high chromium and molybdenum content makes this steel air hardening in substantial section thicknesses [4 in. (100 mm)]. The toughness and air-hardening characteristics have made this steel extremely popular for injection molding cavities. It is the most important and available S-series tool steel.

As a class, the S-series steels have low abrasion resistance because they do not have the hardness and alloy carbides of cold-work steels. These steels are used where toughness is a more important selection factor than wear resistance.

Hot-Work Tool Steels

There are about 12 hot-work tool steels listed in U.S. references. They are categorized by major alloying element into three subgroups. The molybdenum types and tungsten types are aimed at a specific industry—primary metalworking. They are used for working red-hot ingots, blooms, and the like. We shall discuss only the chromium types because the W and Mo series are infrequently used.

The addition of refractory metals such as W, Mo, Cr, and V tends to form carbides on tempering that inhibit dislocation motion and reduce the softening effects of heat. These metals also produce a secondary hardening or hardness increase on tempering. The chromium hot-work tool steels resist softening at temperatures up to 800°F (430°C), and the tungsten and molybdenum grades resist softening at temperatures up

	Nominal Composition (Wt. %)						
	C	Si	W	Mo	Cr	V	Co
Chromium types							
H10	0.40	1.0	—	2.50	3.25	0.40	—
H11	0.35	1.0	—	1.50	5.00	0.40	—
H12	0.35	1.0	1.50	1.50	5.00	0.40	—
H13	0.35	1.0	—	1.50	5.00	1.00	—
H14	0.40	1.0	5.00	—	5.00	—	—
H19	0.40	—	4.25	—	4.25	2.00	4.25
Tungsten types							
H21	0.35	—	9.00	—	3.50	0.50	—
H22	0.35	—	11.00	—	2.00	0.40	—
H23	0.30	—	12.00	—	12.00	1.00	—
H24	0.45	—	15.00	—	3.00	0.50	—
H26	0.50	—	18.00	—	4.00	1.00	—
Molybdenum types							
H42	0.60	—	6.00	5.00	4.00	2.00	—

to 1150°F (620°C). The compositions of H10 to H13 are very similar, and they have similar hardening and use characteristics as well. H14 and H19 contain high tungsten for improved hot hardness and hot *erosion* resistance, and they are forging die steels. H11 and H13 are the most widely used hot-work tool steels. H11 is also used for structural applications, and H13 is used throughout industry for general hot-work applications, plastic and nonferrous injection molding cavities, extrusion dies, and the like. H13 is the most popular and available hot-work tool steel, and it will meet most hot-work needs. In the homogenized condition it has the toughness of a "super-duty" steel.

High-Speed Tool Steels

High-speed tool steels derive their name from their original intended application: to make tools that can machine other metals at high cutting speeds. They contain the highest percentage of alloying elements of any of the tool steels. There are more than 30 different alloys in the United States, some with molybdenum as a major alloying element (M series) and some with tungsten as the major alloying element (T series). Some high-speed steels are presented in the following list:

Nominal Composition (Wt. %)						
Type	C	W	Mo	Cr	V	Co
M1	0.85	1.50	8.50	4.00	1.00	—
M2	0.85[a]	6.00	5.00	4.00	2.00	—
M3-1	1.05	6.00	5.00	4.00	2.40	—
M3-2	1.20	6.00	5.00	4.00	3.00	—
M4	1.30	5.50	4.50	4.00	4.00	—
M7	1.00	1.75	8.75	4.00	2.00	—
M42[b]	1.10	1.50	9.50	3.75	1.15	8.00
M50	0.85	—	4.00	4.00	1.00	—
T15	1.50	12.00	—	4.00	5.00	5.00

[a]Other C contents available.
[b]Available with 0.3% or 0.55% silicon.

The carbon and the chromium in these alloys provide hardenability. Vanadium up to about 1% is used for grain refinement. Higher concentrations promote the formation of vanadium carbides that provide wear resistance. Tungsten is used to produce resistance to softening at elevated temperatures. Even the M series contains a significant amount of tungsten. Cobalt is present in M42 and T15; its role is to provide resistance to the softening effect of heat. It does so primarily through matrix strengthening rather than through carbide formation.

Type M1 is the original molybdenum high-speed steel. M2 has more carbon and vanadium than M1 for improved wear resistance. Type M3-1 was developed to have better grindability than M2. Type M3-2 was developed to have better wear resistance than M3-1 (more carbon and vanadium), and type M4 is supposed to have the highest wear resistance of the M grades listed (highest carbon and vanadium).

All the high-speed steels can be hardened to higher working hardnesses than other tool steels (62 to 67 HRC), and they contain fine, well-dispersed carbide particles to enhance wear resistance. In addition, they all maintain their hardness in use temperatures up to 1000°F (540°C). They are the industry standards for cutting tools. Types M1, M2, M7, and M42 are probably the most popular and most used for twist drills. Type M50 is a low-alloy, high-speed steel more often used for rolling element bearings than for tools (available VIM-VAR). A number of M-type grades and T15 are usually made by P/M techniques. M3 type 1 and T2 are in essence obsolete.

Mold Steels

Many types of molds are used in industry, but the molds referred to in the title of this group of steels are plastic injection molding cavities, holding blocks, and related tooling. The commercially significant P-series steels are as follows:

Nominal Composition (Wt. %)					
Type	C	Mo	Cr	Ni	Al
P6	0.10	—	1.50	3.50	—
P20	0.35	0.40	1.70	—	—
P21	0.20	—	—	4.00	1.20

Type P6 mold steel has very low carbon content. It cannot be quench hardened. The carbon and alloy content is low to allow hubbing of mold details. The desired mold shape is pressed into the steel with a hub that is usually made from a hardened high-speed steel. Thus mold cavities can be made without machining. Hubbed cavities are then carburized to make a production injection molding cavity.

Type P20 is the most widely used grade of mold steel. It has sufficient carbon and alloy content to through harden to about 30 HRC in heavy sections. In this condition it is widely used for massive holding blocks for die casting and injection molding cavities.

Type P21 can be quench hardened and solution heated to about 25 HRC and then hardened to about 40 HRC by precipitation of aluminum intermetallic compounds in the hardened steel matrix. It is a special-purpose alloy.

Special-Purpose Tool Steels

The L-series tool steels are most often used for structural applications rather than for tool applications. Two grades are commercially available in the United States.

	Nominal Composition (Wt. %)				
Type	**C**	**Mo**	**Cr**	**V**	**Ni**
L2	0.50–1.10[a]	—	1.00	0.20	—
L6	0.70	0.25[b]	0.75	—	1.50

[a]Various carbon contents are available.
[b]Optional.

Type L2 is available in a range of carbon contents for different degrees of hardenability. Both L2 and L6 are usually oil hardened to hardnesses in the range of 30 to 45 HRC, and they are used for applications such as press brake dies, wrenches, and riveting tools.

To summarize our discussion of tool steel metallurgy, the chemical composition differences of the various groups are aimed primarily at producing different hardenabilities and different microconstituents. To a large extent, the use characteristics of tool steels are determined by their hardness characteristics and by the nature of the carbides that are present.

12.4 Steel Properties

Before discussing the details of tool steel selection, we must establish a set of property terms for rating steels. If a designer wants to design a punch-press die, he or she must weigh many service factors in establishing the design. These service factors also apply to the materials of construction. If you have the assignment to make a tool that can make 50 holes per minute in a piece of steel, you must not only design a suitable mechanism, but also you must select a tool steel that can withstand the abuse of this service. To determine which tool steel will work, you compare the properties of each tool steel. A system has been established in the United States that many steel users follow; it rates all tool steels against each other on a dozen or so properties that pertain to hardening and use characteristics.

In hardening characteristics, tool steels are rated on ease of hardening, cracking tendencies, distortion, and resistance to *decarburization*.

In use characteristics, such things as wear resistance, machinability, softening, and toughness are compared. In the remainder of this section we shall explain the significance of each of these tool steel properties.

Hardening Characteristics

Safety in Hardening This property rates tool steels on the probability or relative risk of cracking during the hardening operation. A number of tool steel defects, such as grinding burn and decarburization, can lead to cracking in hardening; but some steels, especially the ones requiring a water quench, are prone to cracking even without defects. The prime cause of heat treat cracking is differential volume change on going from a soft to a hard structure. The austenite to martensite reaction in hardening results in a volume expansion of several percent. If a thin section on a part experiences transformation and its attendant volume expansion before a heavy section on the same part, severe local stresses can be developed, which in turn lead to cracking. Something as simple as a hole in a shaft could lead to cracking on hardening.

As you might expect, this property is closely related to the quenching medium required on a particular steel. Water-hardening steels are very susceptible, oil-hardening steels are less susceptible, and air-hardening steels are the least susceptible to cracking in hardening.

Depth of Hardening This property is a ranking of hardenability. From the practical standpoint, it tells the user how deep a part will harden from the surface. The tool steels with large amounts of alloy usually have good depth of hardening, whereas the low-alloy tool steels are shallow hardening. Alloy additions make steels deep hardening by a combination of mechanisms. Some alloy additions refine grain size; some raise the austenitizing temperature; some tend to promote the formation of alloy carbides. The desired net effect is to make transformation to soft structures sluggish, thus making it easier to quench heavy sections without the core being soft.

A deep-hardening steel is desirable on severe service applications such as punch-press dies, die-casting cavities, and similar tools. Shallow-hardening steels are often preferred for shafts, dowel pins, gears, and similar parts that may see impact loading in service. The soft core imparts a measure of toughness not available in the through-hardening steels.

In general, if a part is to be subjected to high operating stresses, it should be through hardened. Figure 12–7 shows the relative hardenability of some common tool steels.

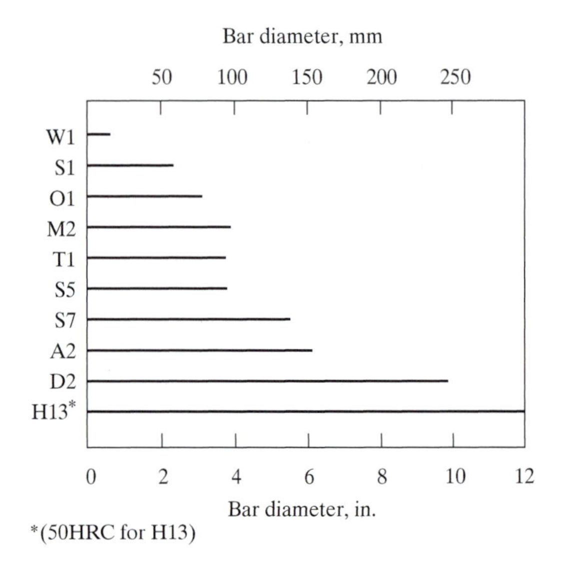

Figure 12–7
Estimated bar diameter that will through harden to 55 HRC

Size Change in Hardening This property refers to the net size change that will occur in a part after hardening and tempering. It is one of the most important of all tool steel properties but probably the most difficult for which to get quantitative information. We can show that the theoretical size change on going from soft structure to 100% martensite is on the order of 4% expansion, but in the actual cases some steels even shrink. The net size change varies with tempering temperature, geometry, and anisotropy. Tool steel literature often includes size change data on a standard shape. Figure 12–8 illustrates a typical size change curve. The problem and danger of using these data is that they hold true only for that particular geometry. Any other shape will exhibit a different size change.

What then is the value of this property? Even though it is not quantitative for different shapes, it reflects the experience of many service case histories. The size change versus tempering temperature curves are useful for selecting a tempering temperature that gives the lowest size change.

Resistance to Decarburization In our discussion of heat treating, it was pointed out that heating a steel to its hardening temperature in air can cause severe problems in the form of oxidation of the surface coupled with removal of carbon or decarburization of the surface. This tool steel property rates steels on their ability to resist loss of carbon at the surface during hardening. Why is this harmful? Very simply, a decarburized tool steel will have surface properties

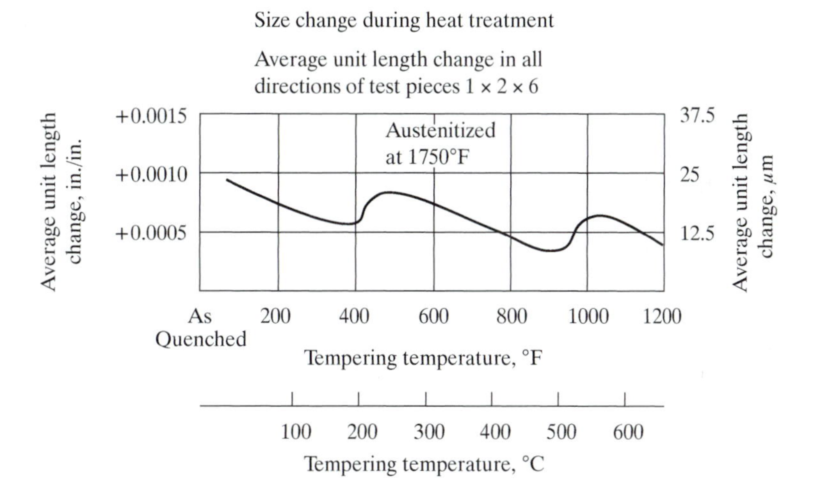

Figure 12–8
Size change during hardening for type A2 tool steel
Source: Courtesy of Latrobe Steel Company

not much better than a piece of ordinary 1020 steel. It will be soft. An example of a decarburized surface is shown in Figure 12–9.

Besides being soft, decarburization can lead to cracking on quenching. Because the surface has a different chemical composition than the core, it will go through its phase transformation and related size change before or after the core, thus setting up high internal stresses and possible cracking.

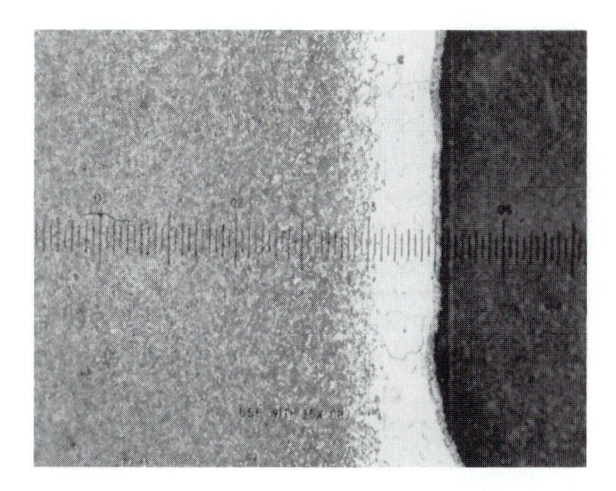

Figure 12–9
Decarburization [0.005 in. (0.127 mm) deep] on 4140 steel; the white area is soft (×100)

Some may say that it is the heat treater's job to use a protective atmosphere and prevent decarburization. This may be true, but the fact of the matter is that most heat treating facilities are job-shop operations, and the customer could not afford sophisticated atmosphere controls on every piece of tool steel treated. This selection factor is not as important as some of the others to be considered, but it is not one to be ignored. Heat treaters need all the help they can get. If two tool steels look alike, they will both do the job; but if one has better resistance to decarburization than the other, then use it over the other.

Use Characteristics

Resistance to Heat Softening Just as tempering is used to soften a tool steel slightly to reduce brittleness, the heat generated in use may cause excessive softening. Anyone who has ever taken a deep cut on a piece of steel in a lathe will have an appreciation of the temperatures that can develop in working metals with tool steels. It is not uncommon in the example cited to have red-hot chips peel off a bar in the lathe. If the chips are red hot, what is the temperature at the surface of the tool steel bit that produced the chips?

This is a severe type of service, but even in punch-press-type operations, a tool steel can get quite hot at the surface. If the surface temperatures exceed the original tempering temperature of the steel, it will soften and lose mechanical properties and wear resistance. Essentially, this property reflects the tempering characteristics of tool steels. Figure 12–10 shows the effect of temperature on the hardness of several tool steels.

The important factor from the standpoint of selection is the hardness that a steel will maintain at some high temperature. If this information is not readily available, the tempering characteristics will serve as an indication of hardness at elevated temperatures, or *red hardness*. High-speed steels have exceptional red hardness compared with other tool steels. The low-alloy W series is very poor in this regard. Red hardness is important as a selection factor in hot-working operations, machining, and many high-production metal-working operations.

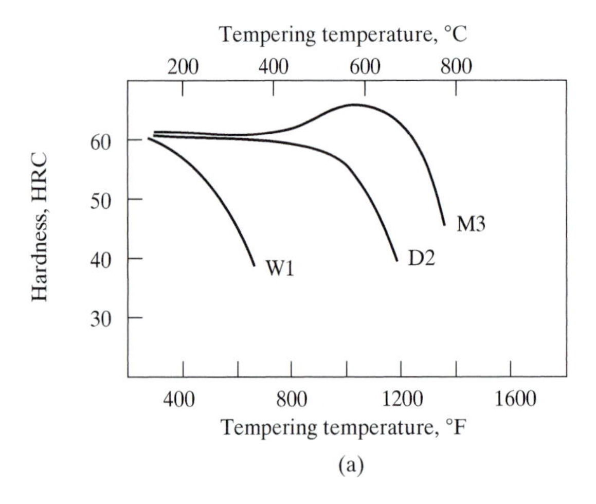

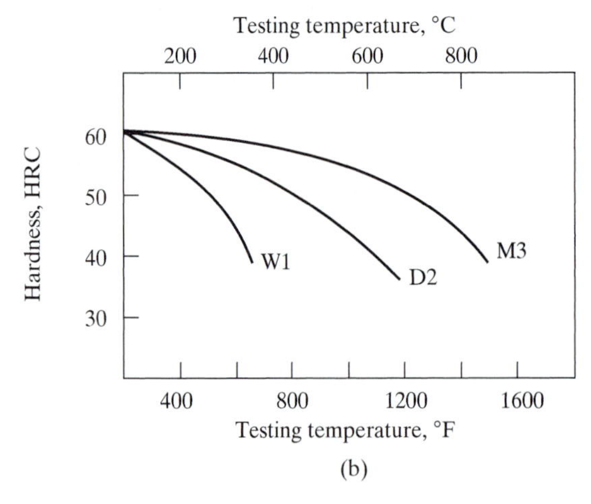

Figure 12–10
(a) Effect of tempering on hardness. (b) Hardness at elevated temperatures.

Wear Resistance This tool steel property is self-explanatory in theory. In practice, it often means someone's subjective opinion on how a tool will last in service. There is no standard wear test for tool steels, and there are probably no known service applications in which all the 70 or so tool steels were rated on their relative resistance to wear in service. This property indeed must be taken with a grain of salt. Vendor catalogs on tool steels invariably show a relative wear resistance rating, but tool steels do not just "wear." As explained in Chapter 3, many types of wear can occur.

It is not really possible to say that one tool steel is more resistant to wear than another tool steel without qualifying this statement to define what type or types of wear are to be encountered. When selecting a tool steel for a wear application, the first step is to ask how this tool might wear. Define the type of wear that is likely to occur, and then compare how various tool steels will resist this type of wear. We will present some laboratory wear data on tool steels in the section on selection.

Toughness Toughness is another self-explanatory property—on the surface, anyway. In our previous discussions on measuring impact resistance, we mentioned that a number of tests are used to measure toughness:

1. The Charpy V-notch impact test.
2. The notched or unnotched Izod impact test.
3. The dynamic tear test.

Table 12–3
Comparison of Charpy V-notch and Izod tests on tool steels

	Charpy V		Unnotched Izod	
Steel	(ft-lb)	(J)	(ft-lb)	(J)
D2 at 60 HRC	<2	(2.7)	30	(40.7)
M3 at 64 HRC	~5	(6.8)	50	(67.8)
H13 at 50 HRC	10	(13.6)	150	(210)
S7 at 56 HRC	15	(20.3)	250	(350)

Most tool steels are brittle. Their impact strengths in a notched condition are almost immeasurable. Without the notch, at least a measurable impact value can be obtained, but these data can be misleading. Table 12–3 gives a comparison of Charpy V and unnotched Izod tests on some tool steels.

Previously, we defined a brittle steel as any steel with an impact strength lower than 15 ft-lb (20.3 J) at room temperature. Thus we say that all tool steels are brittle unless proven otherwise in a Charpy V test. The relative toughness comparison of tool steels usually reflects unnotched Izod values, if any tests at all. These tests have value in selecting the best tempering temperature and for rating tool steels among themselves, but just because a tool steel comparison rates a steel as having the "best toughness," do not misinterpret this to mean that it can be battered like a piece of carbon steel. The shock-resisting steels and some mold and hot-work tool steels have measurable Charpy V toughness, but the other tool steels are not to be abused. They cannot tolerate sharp corners and notches when used in situations involving significant dynamic loads. Figure 8–16 compares the notched toughness for a variety of tool materials. Cemented carbides were about ten times more brittle than high-speed tool steel, and high-speed steel was about ten times more brittle than unhardened carbon steel.

Machinability Our final tool steel property, like wear resistance and toughness, is intuitive in meaning, but this property is just as nebulous as wear and toughness. There is no industry-accepted test for rating the machinability of metals, or of any material for that matter. Again, ratings on machinability are qualitative, but they do reflect users' experience; from that standpoint, they can be used with fair confidence.

Comparison of Tool Steel Properties We have discussed the nature of tool steels and we have shown that there are about 70 AISI types of tool steels, each with different characteristics; some differences are subtle, some are sizable. Let us at this point compare these steels using the properties covered in the foregoing discussion. Table 12–4 presents the U.S. steel industry property ratings of most of the commonly used tool steels. These data compare tool steels to each other and not to other steels, such as carbon or alloy steels.

We may draw several conclusions concerning the relative properties of tool steels:

1. Steels within a particular group (for example, W or A) have essentially the same properties.
2. The air-hardening tool steels have the best safety in hardening and the lowest *distortion*. Water-hardening steels are poor in these respects, and oil-hardening steels fall in between these extremes.
3. High-speed steels as a group have the poorest toughness and machinability.
4. The A, D, H, and high-speed steels are *deep hardening*.
5. Resistance to decarburization varies even within groups of steels, and it mostly affects mill processing. Most heat treaters know how to prevent decarburization in hardening.

Table 12–4
Tool steel properties*

Tool Steel Type	Quenching Media	Safety in Hardening	Depth of Hardening	Resistance to Decarburization	Distortion in Hardening	Resistance to the Softening Effect of Heat	Wear Resistance	Toughness	Machinability
Cold work									
W1	Water	Poorest	Shallow	Highest	High	Low	Fair	High	Best
O1	Oil	Good	Medium	High	Low	Low	Medium	Medium	Good
A2	Air	Best	Deep	Medium	Lowest	High	Good	Medium	Fair
D2	Air	Best	Deep	Medium	Lowest	High	Very good	Low	Poor
Shock resisting									
S1	Oil	Good	Medium	Medium	Medium	Medium	Fair	Very good	Fair
S7	Air	Best	Deep	Medium	Lowest	High	Fair	Very good	Fair
High speed									
M1	Oil, air, salt	Fair	Deep	Low	Air or salt, low; oil, medium	Very high	Very good	Low	Fair
M2	Oil, air, salt	Fair	Deep	Medium	Air or salt, low; oil, medium	Very high	Very good	Low	Fair
M3	Oil, air, salt	Fair	Deep	Medium	Air or salt, low; oil, medium	Very high	Very good	Low	Fair
M4	Oil, air, salt	Fair	Deep	Medium	Air or salt, low; oil, medium	Very high	Best	Low	Poor to fair
Hot work									
H11	Air	Best	Deep	Medium	Very low	High	Fair	Very good	Fair to good
H12	Air	Best	Deep	Medium	Very low	High	Fair	Very good	Fair to good
H13	Air, salt	Best	Deep	Medium	Very low	High	Fair	Very good	Fair to good
Mold steels									
P20	Oil	Good	Medium	High	Low	Low	Fair	Good	Fair

*The complete table is presented in the *Steel Products Manual—Tool Steels*, The Iron and Steel Society, Warrendale, PA, 1988.

6. The H, D, and high-speed steels are the best for resistance to softening.
7. The high-speed and D series all have very good *abrasion* resistance.

Other conclusions may be drawn, but the preceding are general statements that everyone should know about the various classes of tool steels.

12.5 Tool Steel Selection

The information in Table 12–4 is intended to be used as a guide for the selection of tool steels. This list contains only the more popular tool steel alloys. The complete table, which is published in the United States in the Tool Steels section of the Steel Products Manual by the Iron and Steel Society, contains 70 or so tool steels. The recommended procedure for the selection of a tool steel for a particular application is to list the service properties required and then to compare the properties of the tool steels to your property list and determine which alloy has the best match. Table 12–4 lists five selection factors that pertain to heat treating characteristics and four that pertain to service conditions. For example, a design problem may be to select a steel for a punch-press punch and die that will be used for a short-run blanking operation. The production requirement is only 3000 special 1010 steel flat washers with a thickness of 0.04 in. (2 mm). To address this problem you should list the properties required of the tooling and then find the steel with the best match. The property list may look like the following:

1. Low cost
2. Hard enough to get a clean-cut edge
3. Sufficient strength and toughness that it does not break

This is a meager list, but it is appropriate. You are not asking for much with regard to service requirements. The property comparison in Table 12–4 does not list cost, so the table does not help here; with regard to hardness, the designer will want a hardness of 60 HRC. Again the table has a deficiency; it does not show the recommended *working hardness* of these steels. Some cannot be hardened to 60 HRC. The table does not quantify toughness, but any of the steels listed will have adequate toughness and strength to blank such thin steel. Thus Table 12–4 has some limitations with regard to selection, but it does compare the properties of the various tool steels to each other. It will help to show the difference between, for example, an M2 and an M4 tool steel.

Our recommendation on selection is to list the properties required and then start to narrow the candidate field down by asking questions like these:

1. Do I need corrosion resistance?
2. Do I need weldability?
3. Is material cost a significant factor?
4. Is machining cost significant?
5. Is the part going to be subject to battering?
6. Will temperature resistance be needed?
7. Is wear resistance important?
8. If it is, what is the wear mode?
9. Is there a potential for size change problems in hardening?
10. What hardness do I need?
11. What is the anticipated service life?
12. What is the severity of service?
13. Can the part be ground after hardening?
14. Is high strength a factor?
15. How many of these parts are required?
16. Size—do I need to through harden a heavy section?

Other questions could be asked, but the point is that questions like these let you develop a list of service requirements.

No designer, or for that matter metallurgist, is going to be familiar enough with all 70 or so tool steels available in the United States. Most designers and metallurgists develop a list or repertoire of tool materials that they become familiar with, and they make their selection from this reduced list. The following is a proposed repertoire that works well in the chemical process industry. A list for another industry may look different.

A Designer's Repertoire of Tool Materials

Tool Steels

1. O1: medium abrasion resistance, oil hardening.
2. A2: medium abrasion resistance, air hardening.
3. D2: good abrasion resistance, air hardening.
4. A11: excellent abrasion resistance, salt or pressure quench (from vacuum) hardening
5. S7: air hardening, shock resisting.
6. H13: hot-work mold steel.
7. M2: general-purpose high-speed steel.

Nontool Steels

8. 4140 Alloy steel: structural steel.
9. 4340 Alloy steel: structural steel.
10. 420 Stainless steel: mold steel.
11. 440C Stainless steel: corrosion-resisting tool material.
12. C2 Cemented carbide: excellent abrasion resistance.

Of this list, only seven are tool steels. The others are tool materials from other classes of material that compete with tool steels. If a designer asked several metallurgists, "If you owned a tool and machine company, what steels would you stock to meet 90% of your needs?" each metallurgist would come up with a different list, but the list would contain far fewer than 70 tool steels. The list given is one such list. It will not work for all industries, but it will work for most. Some materials on the list may not be necessary. For example, in many automotive parts plants, corrosion is not a concern because parts and tools are usually covered with oil from machining and forming lubricants. There is no need to use stainless steels for tooling. On the other hand, the auto plant is likely to have a plastic molding operation, and the 420 stainless steel would be a strong candidate for water-cooled plastic injection molding cavities. Whatever the industry, a repertoire of tool materials can be developed. Your list of service requirements should be compared with the properties of the tool materials in your tool material repertoire.

An oil-hardening steel was included in our repertoire of tool materials because it machines more easily than the other cold-work steels and its wear characteristics are adequate for short-run tooling or for normal service machine parts. The major disadvantage is more distortion in hardening than air-hardening steels, but the size change difference is inconsequential on parts with heavy, uniform section size.

One of the best punch-press die steels is D2 steel. Because it air hardens, it rarely distorts or cracks during hardening. It has sufficient wear resistance for severe service components. It is hard to machine, so it should not be used unless warranted by service requirements. Type A2 tool steel has much better machinability than D2, and it is also air hardening (low distortion). Thus this steel is a compromise between the O series and the D series.

The A11 steel is more abrasion resistant than the other cold-work steels because of the presence of a significant volume fraction of VC-type carbides that are obtained by adding about 10% vanadium to what is essentially an air-hardening tool steel. This steel is between the other cold-work steels and cemented carbides or other cermets in abrasion resistance (Figure 12–11). It is used when it is known that steels like D2 do

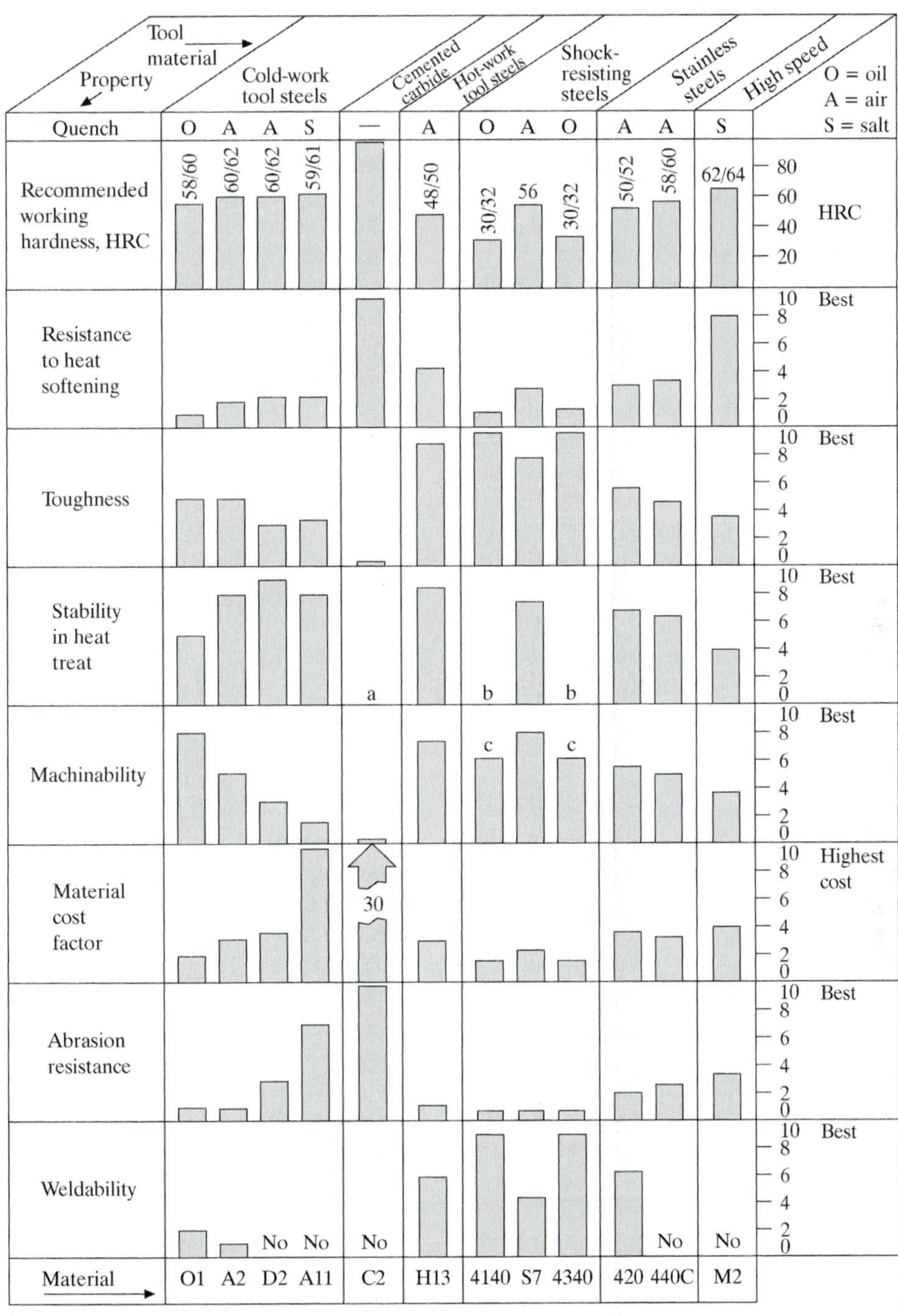

[a] Does not require hardening
[b] Prehardened to 30 HRC
[c] At 30 HRC

Figure 12–11
Comparison of use properties of various tool materials. Ratings apply to material heat treated to the recommended working hardness.

not provide adequate service life and we do not want to get into the fabrication problems of using carbides.

Cemented carbide or steel–carbide composites are generally accepted as the material to use if you are working a highly abrasive substance or if you want tooling to make 50 million parts or so without sharpening. They are expensive to fabricate, but justifiable on some jobs. Carbide cannot be machined and requires diamond grinding or electrical discharge machining. Steel–carbide composites can be machined before hardening, but they are difficult to machine.

An S-series tool steel is desirable to have on hand to use on machine components that will be subject to battering. This is a chisel steel. S7 is normally used at a hardness of 56–58 HRC, but it is still not very wear resistant. All the cold-work and high-speed tool steels are brittle. This material is not as brittle.

Hot-working operations are a specialty field. Forging, hot piercing, hot extrusion, and the like require very special tooling. However, many plants have plastic, zinc, or aluminum molding operations that involve thermal cycling of tooling. An H-series steel, such as H13, is a good general-purpose material to use for these molds. It is not very wear resistant, but it has good toughness and fair machinability.

Every designer should have at least one high-speed steel in his or her material repertoire. Most of the high-speed steels used as cutting tools are from the M series, in particular M1, M2, and M3. We have selected M2 over the other types merely because use experience has shown it to be a good general-purpose steel for special cutting tools, knives, and choppers. It has somewhat greater wear resistance than M1 and M10, but slightly lower machinability. As shown in the property tables, the high-speed steels are superior in red hardness. Tempering temperatures for heat treating high-speed steels are mostly in the range of 1000 to 1100°F (540 to 600°C). Thus these steels can be used at temperatures approaching this range without significant softening. They are not normally used for hot-working operations, however, because of their very poor machinability compared to the H-series steels and because of their poor toughness.

Probably the main reason for the seeming overabundance of high-speed steels (>30) is that users who have special, repetitive machining jobs are always looking for longer-lasting tools. Even subtle differences are sought; thus the mills modify and change to help a particular user and a new type is born. Troublesome machining jobs would probably warrant experimentation with the various available high-speed steels, but the designer who only occasionally needs a high-speed steel could subsist on a single type.

The nontool steels on our list, 4140 and 4340, are essential materials in design. At hardnesses in the range of 30 HRC, they excel as a shafting material. Type 4140 is used for diameters or thicknesses less than 1½ in. (40 mm), and 4340 for greater diameters or thicknesses. They have toughness comparable to that of low-carbon steels and sufficient hardness and compressive strength to hold accurate dimensions and to resist the abuse of set screws and pressed-on bearings. They are available in plates for backup materials for knives, punches, dies, and so on. They are also available in free-machining types.

As mentioned previously, stainless steels could be omitted from a designer's repertoire if corrosion were never a factor, but the 420 steel hardened to 50 to 52 HRC is very popular for plastic-injection molds. The rust resistance prevents scaling in cooling passages and allows shorter molding cycles. Type 440C is a cutlery stainless steel. Some of the knives in your kitchen may be made from this material. It has wear properties comparable to those of D2, with the added advantage of rust resistance. This has proved to be a big advantage on devices with acute angle edges that could be significantly attacked by rusting in high-humidity air.

Figure 12–11 shows a semiquantitative comparison of our selected tool steels. The number of heat treat factors was reduced, and several property categories have been added compared to the Steel Products Manual table (Table 12–4).

1. Abrasion resistance
2. Machinability
3. Weldability
4. Cost factor
5. Recommended working hardness

The heat treat factors were reduced in number mostly because of space limitations on the chart, but modern heat treat methods have greatly reduced concerns about heat treating cracking and decarburization. The properties that we added are fairly self-explanatory, but the reasons for adding these could stand some qualification.

Wear Properties

Figure 12–12 shows laboratory test data on the metal-to-metal wear characteristics of various tool steels sliding against a hardened stainless steel counterface. These test results would be different if the counterface were another material. If the counterface were 1020 steel, both the counterface and tool steel wear rates would be higher; both would be lower if the counterface were a cemented carbide. The data show that all the high-hardness tool steels behave in a similar way. This is the case with high-hardness counterfaces. In general, the tool steel combinations that work well in metal-to-metal sliding systems have fine alloy carbides and high hardness (over 60 HRC). Counterface wear will be high if a tool steel with massive carbides is sliding in contact with a steel with fine carbides. If the A7 or D7 types with massive carbides are to be used, they should be self-mated. ASTM G 83 crossed cylinder wear tests indicate that the low system wear can be obtained with a mating couple of M2 or M4 tool steel at 62–64 HRC mated with a cemented carbide (6% Co). Type D2 tool steel at 60 HRC is similarly a suitable counterface for cemented carbide. In general, if hardened tool steels must

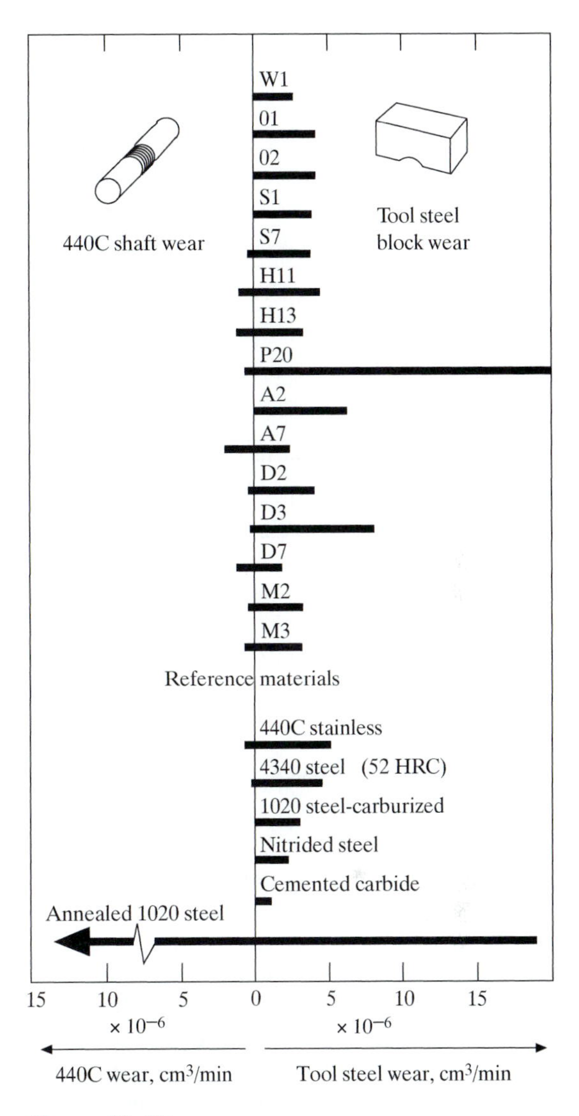

Figure 12–12
Metal-to-metal wear characteristics of various tool steels (at maximum working hardness) versus 440C steel at 58 HRC

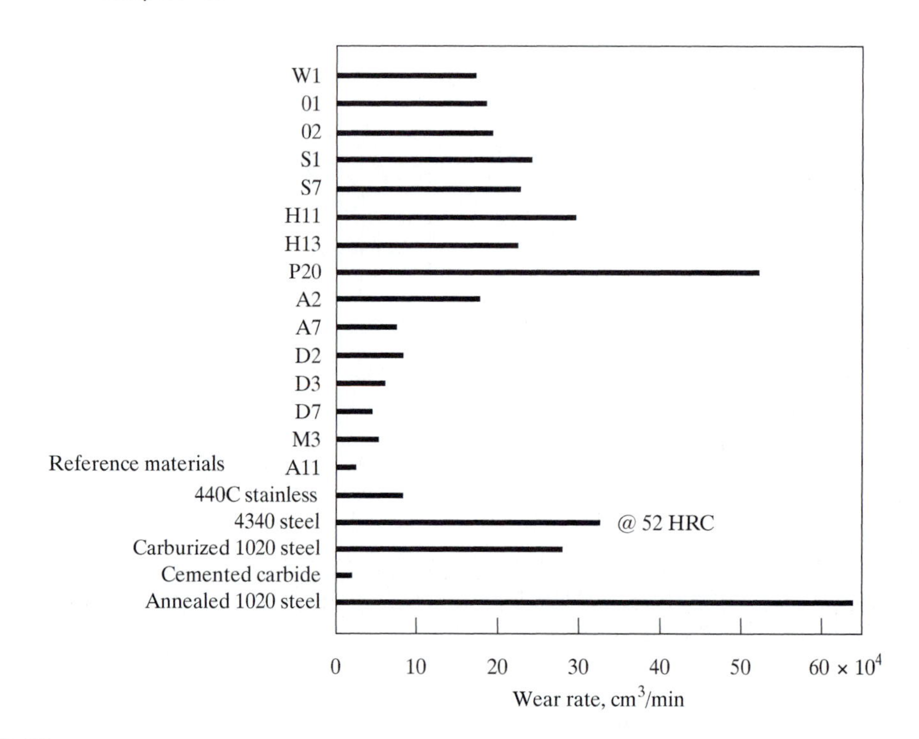

Figure 12–13
Abrasion rates of tool steels (at maximum working hardness) versus 60-mesh silica (ASTM G 65)

slide against one another, material cemented carbide is often a low-wear counterface.

In most metal working tooling, the predominating mode of wear is abrasion and *surface fatigue*. In blanking of steel in a punch-press die, abrasion occurs from dirt particles on the oily steel and from hard inclusions in the steel (Fe_3C, oxides, and so on). Conjoint with the abrasion, the repetitive compressive stressing of the die edges produces a tendency for microscopic surface fatigue of the edge. Figure 12–13 shows some laboratory test data on the abrasion resistance of some tool steels as measured by a dry-sand abrasion test. These tests show that the abrasion resistance is proportional to the steel hardness, the hardness of the carbides in the steel structure, the size of the carbide particles, and the volume concentration of carbides. If two steels have the same hardness and one contains a significant concentration of carbides and the other has none, the former will have greater abrasion resistance. Similarly, if one steel contains predominately chromium carbides with a hardness of 1800 HK_{500}, it will not have the abrasion resistance of a tool steel with the same volume fraction of carbides that are of the VC type with a hardness of 2400 HK_{500}. The current trend in tool steels is to design the carbide structure. Some cold-work (A11) and high-speed tool steels are currently being made by powder metallurgy techniques that provide significant control over carbide type, size, volume fraction, and distribution (Figure 12–14).

As we have shown in Figure 12–4, conventional tool steels contain varying volume fractions of carbides; the H, P, L, and S tool steels have hardened structures that are essentially

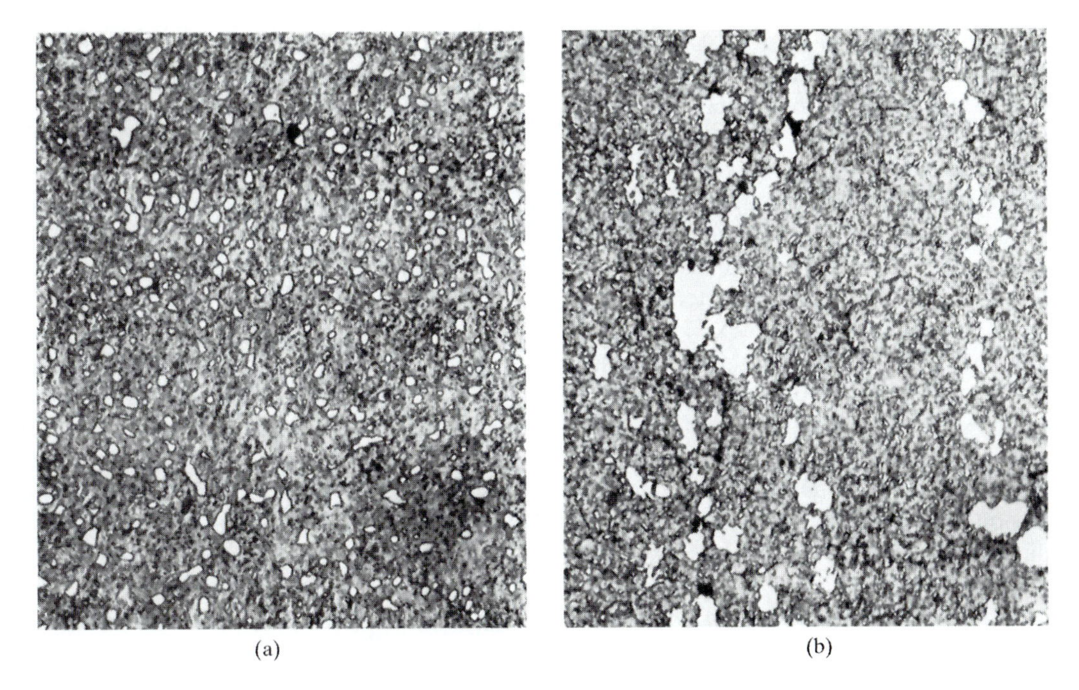

Figure 12–14
Microstructures of type 440C stainless steel (×800) made by (a) powdered metal process and (b) conventional melting and rolling

tempered martensite with alloying elements in solid solution. There are no large carbide particles in the structure to enhance abrasion resistance. The W and O steels contain a relatively small volume fraction, maybe only 2% or 3%, of very fine carbides, and the carbides present may be of the Fe_3C type with a hardness of less than 1000 kg/mm^2. The A series have bigger carbides and more of them, and they may be alloy carbides such as chromium carbide; the D-series steels contain over 10% massive chromium carbides with a hardness as high as 2000 kg/mm^2. The high-speed steels contain about the same volume fraction of carbide as the D-series tool steels, but they may be vanadium or tungsten carbides. There is over 10% VC carbide in A11 tool steel. Abrasion resistance is increased as the volume fraction of carbide increases and as the hardness of the carbide increases.

Conventional wrought tool steels typically contain less than 10% hard carbides in their structure, and these carbides are usually segregated in stringers aligned with the rolling direction. If alloy additions are increased to produce a higher carbide fraction, the mechanical properties of the steel may tend to deteriorate by, for example, the formation of network carbides. Segregation can be eliminated only by techniques such as cross rolling, which is too expensive to be used for commercial production of tool steels. When tool steels are made by powder metallurgy (P/M) techniques, the carbide volume fraction can be as high as 50%; the types of carbides present can be more easily controlled, and the carbide distribution tends to be uniform rather than segregated. A number of specific processes are used to make these tool steels; some are made by bonding carbide particles with tool steel and other steel matrices (steel–carbide composites or cermets). They are sintered and hot isostatic pressed (hipped) into a full-density material. Another technique is to atomize a tool steel composition into small particles, sealing the particles into a metal or glass container and hipping or pressurizing the atmosphere around the container to form a billet. The billet is then forged and worked into product forms as in conventional tool steel manufacture. The atomization process makes the carbide particles uniform in size, and the rapid cooling that occurs in atomization allows compositions to be made that would be troublesome in conventional pouring of ingots. The net result of powder metallurgy tool steel techniques is tool materials with fine, well-dispersed carbides, and the carbides that are present may be of types not possible with conventional tool steel melting practice.

From the user's standpoint, tool steels can be obtained that have greater than normal volumes of carbide and a uniform distribution of these carbides. In addition, if the carbides are harder than the carbides in conventional tool steels, the abrasion resistance can be further enhanced. As shown in Figure 12–15, the bulk hardness of the hardest steels is about 900 on the absolute hardness scale. If a design situation requires, for example, resistance to abrasion from titanium dioxide (a common white pigment used in plastics), it is desirable to make tool materials harder than the TiO_2, which has a hardness of about 1100 kg/mm^2, but steels cannot have a bulk hardness greater than 900. What to do? If a tool steel is designed to have a significant volume fraction of tungsten carbides, vanadium carbides, or chromium carbides, the steel will start to develop significant resistance to the abrasive titanium dioxide. This is exactly what is being done with powder metallurgy tool steel and with steel-bonded carbides. Structures are being designed with carbide phases that enhance wear resistance. All "(A eleven)" tool steel contains 10% vanadium, which forms vanadium carbides to enhance wear resistance. There are other tool steels, high-speed steels, and stainless steels made with similar practices, and the steel-bonded carbide tool materials can have a tool steel matrix with up to 50% titanium carbide in the structure. These tool materials are an essential part of modern tool materials, and they are used when the conventional high-hardness D-series and high-speed steels do not provide desired abrasion resistance. The trade-off is higher cost; they can cost $10/lb or more and are more difficult to grind. These factors must be considered in the selection process.

Machinability

The machinability rating in Figure 12–11 was obtained by combining machining and grinding data from catalogs from tool steel manufacturers. The tool materials are rated among themselves. The best rating was given to O1 tool steel. The cost of machining this steel would still be about twice the cost of machining the steel standard for easy machining, type 1112 sulfurized carbon steel. In other words, if it cost $1 to machine a part from 1112 steel, it would cost $2 to produce that part in O1 tool steel. The cost of machining the part from A11 would be about $8, and the cost of making the part from cemented carbide would be even more.

There is no standard test for machinability, but the designer should keep in mind that the carbide phases that we claim are necessary in tool steels for abrasion resistance also diminish machinability. These carbides are still present in tool steels in the annealed condition; they dull tools and, in general, make tool steels more expensive to machine than steels that do not have a significant volume fraction of carbides.

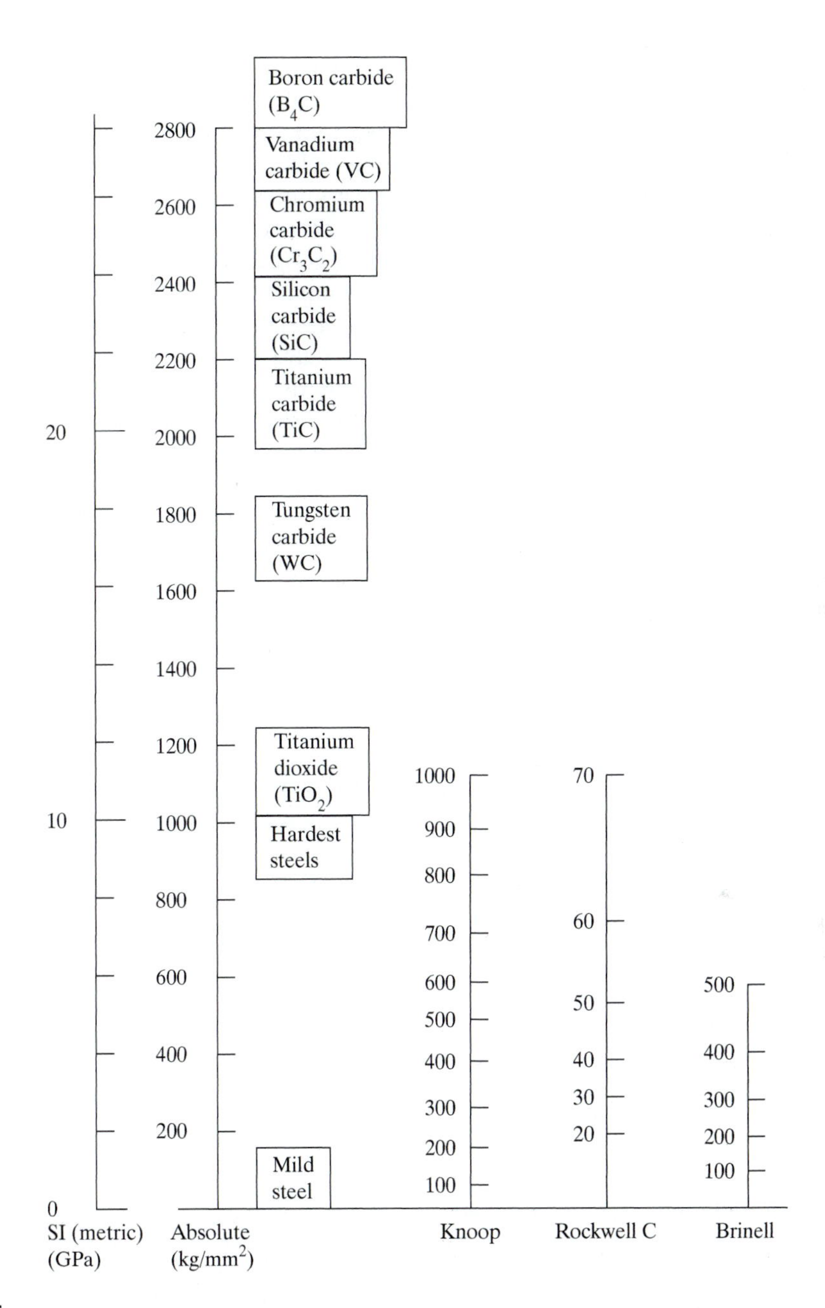

Figure 12–15
Common hardness scales showing their relationship to absolute hardness and the hardness of various wear-resisting materials. The Knoop and Vickers scales are different tests but the numbers vary by less than 10% (300 to 1004 VHN = 311 to 946 KHN).

Weldability

Figure 12–11 shows "no" as the response of some tool materials to weldability. Cemented carbide is simply not weldable by conventional methods; it is composed of ceramic particles held together with a metal binder. It is extremely brittle, and it is incapable of withstanding thermal shocks of the type that occur in welding processes. The tool steels that are labeled unweldable in Figure 12–11 were so labeled because they have an extremely high risk of cracking on cooling from the welding temperature. The others have a risk of cracking that is proportional to the rating assigned in Figure 12–11. Cooling from the welding temperature is the same as a quench from a hardening temperature. Some materials, as we have seen, are very prone to cracking with drastic quenches. Mass effect cooling can produce the equivalent of a liquid quench. The brittle structure that results from the rapid quench is strained by the shrinkage stresses from the weld solidification, and this produces a combination of effects that tend to crack tool steels (and any hardenable steel) when they are welded.

The best approach to the welding of tool steels is never to do it. *Never* do it by design. If welding must be done to save a mismachined part, or the like, it should be done only with the knowledge that there is considerable risk of cracking, and then special procedures should be used. It is best to weld hardenable steels in the annealed condition; slow cool from the welding temperature, and re-anneal after welding. If the parts are already hardened, they must be preheated to their original tempering temperature before welding. They must be maintained at this temperature during welding. After welding, the parts are slowly cooled to hand warm and immediately tempered again at the original tempering temperature. Wherever possible, use a nonhardenable filler metal such as type 310 stainless steel. If the properties of the weld must match the steel, use a matching composition filler metal. Always dye-penetrant inspect for cracks after welding. Never weld free-machining tool steels; they crack from hot shortness; the weld is weak when cooling from the weld temperature.

Two of the tool materials in our list actually have good weldability if the precautions that we mentioned are used. They are H13 tool steel and type 420 stainless steel. Their favorable welding characteristic is a significant reason for their popularity as mold steels. Molds are frequently altered because of design changes, and the like. The use of tools with acceptable weldability permits these changes to be made. Thus weldability can sometimes be an important selection factor, and this is why it was added to our selection information.

Cost

The costs of tool steels vary by the day, so it cannot be definitely stated that A2 costs $2/lb and H13 $2.25/lb. However, cost ratios among the various tool steels may remain fairly constant, barring any unforeseen alloy scarcity. If plain carbon 1020 steel is selling for $0.50/lb, a medium-carbon steel may cost $1/lb, a standard alloy steel, such as 4140, $1.50/lb, an oil-hardening tool steel, $2/lb, and high-alloy air-hardening steels, $3/lb. Carbides are usually special ordered and may require die charges for shaping of the blank. A comparative price per pound might be $30/lb, assuming no die charges. The cost of a tool steel should be neglected unless you are using it for tooling that weighs thousands of pounds. If you are building a die that weighs 30 lb, it makes very little difference if your material costs are $60 or $100 because you probably have several thousands of dollars of labor into machining the die. It is much more economical to use a steel with good service life and good heat treating characteristics than it is to use a lower-grade tool steel and take your chances on getting a successful tool.

Working Hardness

The last point to be made concerning the tool steel property table is that tool steels should be used at a particular working hardness. Often designers will call for hardnesses not possible in a particular steel. Sometimes designers want toughness, and they ask for a lower hardness than is normally used on a particular steel. This is fine except that some tool steels become more brittle at 50 HRC than they are at 60 HRC.

When two tool steels are used in sliding contact, there is a misconception that they should have dissimilar hardnesses. In general, wear is lowest if both are hardened to their maximum working hardness (and lubricated). Sometimes mating parts are made with a hardness differential to make the less expensive part persishable with respect to the other member in the sliding system. Sliding a 50 HRC member on a 60 HRC member usually results in more wear on both members than if they were both 60 HRC. At 50 HRC, the lower-hardness member is more prone to adhesive wear, which damages both members. If the goal is to make one member of a sliding system wear before the other member, use the maximum working hardness on both members, but mix alloy types. For example, make the more expensive member from M3-2 high-speed steel at 63 HRC and the perishable member from O1 tool steel at 60 HRC. These hard steels will usually wear by a polishing mode of wear as opposed to an adhesive/galling mode, as would occur if one member were as soft as 50 HRC. The high-speed member usually will still wear less than the O1 member because it has more hard carbides in its microstructure than the O1 member. The more foolproof way to protect an expensive wear member is to make the expensive member from a hard tool material (>60 HRC) and the mating member from a good grade of bronze or cast iron (with proper lubrication) or a self-lubricating plastic.

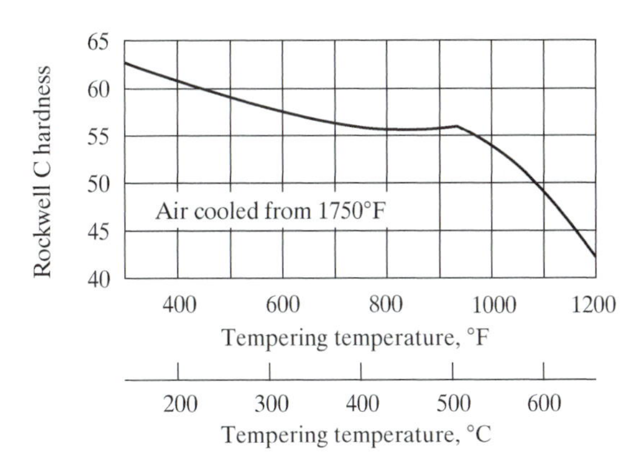

Figure 12–16
Tempering characteristics of A2 tool steel
Soure: Courtesy Latrobe Steel Company.

Another problem with indiscriminate lowering of hardness is that some hardness ranges are almost impossible to obtain from the heat treating standpoint. Figure 12–16 shows the *tempering curve* for A2 tool steel. This steel is normally tempered at 400 to 500°F (200 to 260°C), yielding a hardness range of from 58 to 62 HRC. If a designer specifies a hardness of 50 HRC, it means that the heat treater will have to go to a tempering temperature of 1100°F (600°C). If the treater overshoots this temperature by 50°F (18°C), you may end up with a hardness of only 45 HRC. Furnace temperatures cannot be that accurately controlled. Thus it is unwise to experiment with working hardnesses without referring to toughness versus tempering temperature and hardness versus tempering temperature data for the particular steel being considered.

Sample Selection Problems After establishing a repertoire of tool steels (or adopting the one recommended), the designer is faced with the question of selecting one from this group of 11 or so materials for a particular machine or tool component. The following are examples of selection for typical design problems.

Problem:

Material for a punch-press die: 0.030-in.-thick (0.76-mm) 1010 steel, 5 million parts required.

Solution: Referring to our modified selection table, you would immediately look at the cold-work materials. Why not use O1? Oil hardening gives more distortion than air hardening, and it has lower wear resistance than A2, D2, and carbides. How about A2? Maybe; but D2 still has better distortion characteristics and wear resistance (abrasive). How about carbides? Not enough production to justify the cost. The best selection would be D2.

Problem:

Shaft for a gear train: must carry high loads and stop fast.

Solution: The "must stop fast" requirement puts this into the category of a shock-resistant material. How about S1? Do you need a hardness of 45 HRC? Probably not. The 4140 and 4340 steels at 30 HRC have far greater toughness than S1, and they have the added advantage of being able to be machined after hardening. This negates the effect of distortion during hardening. Which of these two steels is better? If the shaft is less than 1½ in. (38 mm) in diameter, use 4140. Use 4340 if the shaft is over 1½ in. (38 mm) in diameter. Thus our selection would be 4140 or 4340 steel.

Problem:

A material for a plastic-injection molding cavity for unfilled polystyrene.

Solution: Intuitively, you may be drawn to the hot-work category, but molds for plastics do not often fail by a mode of thermal fatigue, so cold-work steels with a high tempering temperature may be used. However, H13 would be our recommendation, not so much from the standpoint of heat resistance, but because it has much better toughness than the air-hardening tool steels. The strains encountered in clamping and the strains from molding pressures may lead to cracking if a brittle tool steel is used. If a filled plastic with known abrasive characteristics is involved, the more wear-resistant A2 steel may have been our choice. If the die is for a low-accuracy, low-cost item such as a throwaway display, a P20 prehardened steel at 30 HRC would make the cheapest mold. Thus our selection is H13; but if the material and molding conditions were different, we might have selected another material.

Problem:

A tool material to deburr a drilled hole in steel plate.

Solution: Almost any of the cold-work tool steels can hold an edge and cut steel, but the superior steels for cutting operations are the high-speed steels. They hold an edge better and longer. Thus we would select M2. If there were 50 million holes to consider and the burr had an uncomplicated shape, carbide could have been justified.

12.6 Specification of Tool Steels

Hardness

After completing the ritual of filling in the "Material" box on a drawing, what additional information should be included on the drawing? Hardness absolutely; and not a single value, but a range of at least two points HRC. The specified hardness should be the recommended working hardness for that material and no other range unless you have a good reason for asking for it. Check the tempering data to make sure that your special hardness request is reasonable from the heat treating standpoint.

Stock Size

If one of your design duties is to specify the stock size of the tool steel for a desired part, an important consideration is stock allowance for scale and decarburization (*decarburization allowance*). In the rolling operations at the steel mill, the steel is red hot and in an unprotected atmosphere. A heavy oxide scale builds up, and the surface of the steel is depleted of carbon. Even cold-finished materials may have some decarburized surface. Thus if you are going to harden any tool steel, you must machine the surface. The only exception would be if you specified one of the commercially available decarburization-free stocks. These are machined by the warehousing company. How much stock must be removed to get rid of decarburization? The U.S. steel producers have established recommended machining allowances and maximum allowable depths of decarburization. Table 12–5 shows the maximum depth of decarburization to be expected on various-sized bars. A quick review of these data will illustrate that indeed you cannot use stock sizes on hot-worked steels. The decarburization limits are not as great on cold-finished bars, but Table 12–5 will serve as a conservative guide on how much stock must be removed before hardening a tool steel.

Heat Treating

The minimum note on a drawing specifying a tool steel is "Harden and temper to xx HRC." If the part involved has a length to diameter or cross section greater than 10 (1 in. diameter, 10 in. long), there will be a strong tendency for distortion in hardening. The distortion tendency can be greatly reduced by using an air-hardening steel, but it may be well to stress relieve the part after rough machining. The drawing would then read as follows (using Type A2 tool/steel as an example):

Table 12–5
Minimum allowance for machining hot-rolled square and flat bars; minimum allowance per side for machining prior to heat treatment

		Specified Width [in. (mm)]				
Specified Thickness, in. (mm)		**0 to ½ (0 to 12.7), incl.**	**Over ½ to 1 (12.7 to 25.4), incl.**	**Over 1 to 2 (25.4 to 50.8), incl.**	**Over 2 to 3 (50.8 to 76.2), incl.**	**Over 2 to 3 (76.2 to 101.6), incl.**
0 to ½ (0 top 12.7),	A	0.025 (0.64)	0.025 (0.64)	0.030 (0.76)	0.035 (0.89)	0.040 (1.02)
incl.	B	0.025 (0.64)	0.035 (0.89)	0.040 (1.02)	0.050 (1.27)	0.065 (1.65)
Over ½ to 1 (12.7 to 25.4),	A	—	0.045 (1.14)	0.045 (1.14)	0.050 (1.27)	0.055 (1.40)
incl.	B	—	0.045 (1.14)	0.050 (1.27)	0.060 (1.52)	0.075 (1.90)
Over 1 to 2 (25.4 to 50.8),	A	—	—	0.065 (1.65)	0.065 (1.65)	0.070 (1.78)
incl.	B	—	—	0.065 (1.65)	0.070 (1.78)	0.085 (2.16)
Over 2 to 3 (50.8 to 76.2),	A	—	—	—	0.085 (2.16)	0.085 (2.16)
incl.	B	—	—	—	0.085 (2.16)	0.100 (2.54)
Over 3 to 4 (76.2 to 101.6),	A	—	—	—	—	0.115 (2.92)
incl.	B	—	—	—	—	0.115 (2.92)

Source: From *Steel Products Manual–Tool Steels*, © Iron and Steel Society, Warrendale, PA, 1988.

Sequence

1. Rough machine
2. Stress relief anneal in an inert atmosphere at 645 to 655°C for 2 h plus 1 additional h for each 25 mm of material thickness per AMS (Aerospace Material Specification—United States) 2759. The maximum heating and cooling rate shall be 150°C/h.
3. Finish machine leaving *grind allowance*
4. Harden and double temper (2 h each) 58 to 60 HRC. Cool to −75°C for 1 h between the first and second temper
5. Finish grind

Some typical shapes that might warrant stress relieving are illustrated in Figure 12–17. Heat treating will cause scaling and size change; the question that arises is, How much stock should be left on for finish grinding? This is usually the prerogative of the machinist, but Table 12–6 shows some typical values for oil-hardening steels. These could probably be reduced to half the listed value for air-hardening steels.

In quenching tool steels from the hardening temperature, the steel transforms from austenite to martensite. However, the high carbon and alloy content of some tool steels can cause the martensitic reaction to be sluggish. As a result, some residual austenite may be retained in the martensitic microstructure (called *retained austenite*). Retained austenite depresses the hardness of the as-quenched martensite to some extent. Double tempers (2 h each) with a deep freeze in between the first and second temper are often performed on cold-working tool steels to ensure low levels of retained austenite, improved dimensional stability, and enhanced wear resistance. After the first temper and deep freeze, some of the retained austenite transforms to martensite. The second temper serves to temper the additional martensite formed from the retained austenite. If measures, such as the double-temper process,

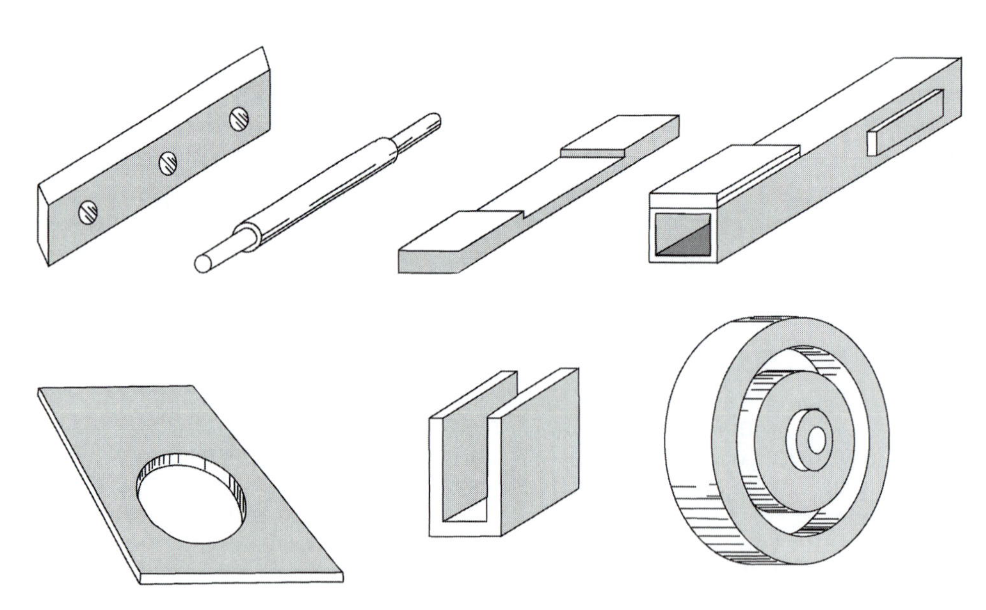

Figure 12–17
Typical parts that may require stress relieving

Table 12–6

Grind allowances for distortion in hardening for oil-hardening tool steels

External Surfaces

Grind Diameter (in.)	Part Length (in.) 3	6	9	12	18	24	30	36	48	60
	Overall Stock Allowance for Parts to be Hardened (in.)									
Up to 1/2	0.014	0.018	0.020	0.022	0.024	0.026	0.028			
3/4	0.014	0.016	0.018	0.020	0.022	0.024	0.026			
1	0.014	0.016	0.018	0.020	0.022	0.024	0.026	0.036		
1 1/2	0.014	0.016	0.018	0.020	0.022	0.024	0.026	0.036		
2	0.014	0.014	0.018	0.018	0.020	0.020	0.024	0.026	0.030	
3	0.014	0.014	0.016	0.016	0.018	0.018	0.020	0.024	0.030	
4	0.014	0.014	0.016	0.016	0.018	0.018	0.020	0.024	0.030	0.036
	Overall Stock Allowance for Parts not Hardened (in.)									
All	0.010	0.010	0.016	0.016	0.016	0.025	0.025	0.025	0.025	0.025

Internal Surfaces

Hold Tolerance ±0.001 in. on Holes to 1-in. Diameter; ±0.002 in. on Holes Over 1-in. Diameter

Hole Diameter (in.)	Hole Length (in.) 1	2	3	4	5	6	7
	Overall Stock Allowance for Parts to be Hardened (in.)						
Up to 1/2	0.008	0.010					
1	0.010	0.012	0.014				
3	0.014	0.016	0.018	0.020	0.024		
4	0.018	0.020	0.022	0.022	0.026		
4–8	0.022	0.024	0.026	0.028	0.030	0.032	
8–12	0.030	0.032	0.034	0.036	0.038	0.042	
12–16	0.038	0.040	0.042	0.044	0.046	0.050	0.055
17–20	0.038	0.042	0.044	0.046	0.048	0.052	0.058
	Overall Stock Allowance for Parts Not Hardened (in.)						
Up to 3	0.008	0.010	0.014	0.016	0.016	0.016	
4–8	0.016	0.016	0.016	0.016	0.016	0.016	
8–20	0.020	0.020	0.020	0.020	0.020	0.020	0.020

Flat Surfaces

Material Thickness (in.)	Part Length (in.) 1 to 4	4 to 9	9 to 14	14 to 19	19 to 24	24 to 30
	Overall Stock Allowance for Parts to Be Hardened (in.)					
To 3/16	0.012	0.015				
3/16–5/16	0.012	0.015	0.018			
5/16–1/2	0.012	0.015	0.018	0.022		
1/2–3/4	0.012	0.015	0.018	0.020	0.024	
3/4–1 1/4	0.012	0.015	0.016	0.018	0.020	
1 1/4–2	0.012	0.015	0.016	0.018	0.020	0.024
	Overall Stock Allowance for Parts Not Hardened (in.)					
All	0.012	0.015	0.018	0.020	0.022	0.024

SI Conversions: 1 in. = 25.4 mm; 0.001 in. = 25.4 μm.

are not taken to reduce the retained austenite level in a tool steel component, the austenite can transform during use, causing dimensional changes in the part. This occurs because the austenite has different crystallographic dimensions than tempered martensite. High-speed tool steels often are tempered three times (for 2 h each) to ensure minimum levels of retained austenite and maximum hardness. Again, a deep freeze is usually performed between the first and second tempers.

If you are designing a complex shape such as an injection molding cavity with many milled contours, you may not have the option of grinding after hardening. If you must harden to size, *bright hardening,* that is, hardening in a vacuum or protective gas atmosphere, will at least remove the possibility of size change due to scale formation. All steels experience a volume change on hardening, and bright hardening will not eliminate this, but the quenching in vacuum or hydrogen furnaces is usually less severe than in any other type of hardening, and thus it minimizes quenching strains. To specify the use of bright hardening will certainly help in hardening to size, but the only true way to harden to size is to make one part, harden in a protective atmosphere, measure the size change, and adjust your machining dimensions accordingly on subsequent parts.

One final point regarding the heat treatment of tool steels: Certain shapes of parts can be very prone to *quench cracking*. The chief rule in designing with tool steels is to avoid sharp internal corners and surface stress concentrations such as stamp marks and the like. They are sources of cracking in hardening as well as stress concentrations in service. Figure 12–18 illustrates some typical crack-sensitive part shapes.

In most of the examples cited, the cracking tendency is caused by sharp corners plus heavy sections joined by thin sections. The thin section transforms before the heavy, and the volume-change strains cause cracking in the areas of stress concentrations—the sharp corners.

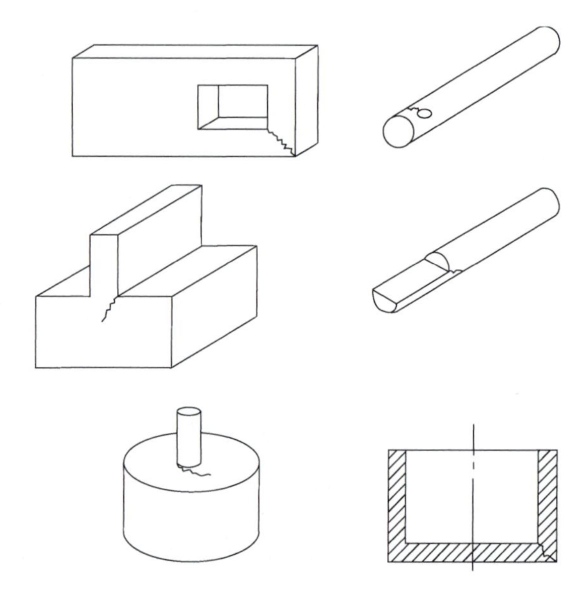

Figure 12–18
Typical crack-sensitive shapes

12.7 Tool Steel Defects

Occasionally, a tool steel part has some detrimental defects, but internal defects are actually rare. The following are some possible manufacturing defects:

1. Seams and laps
2. Forging bursts
3. Inclusions
4. Laminations
5. Excessive decarburization

Of these, decarburization is the most common, and all the defects will probably exhibit themselves (if they are present) through the mode of cracking during hardening. The only way to protect against integrity defects is to use nondestructive testing techniques such as radiography and ultrasonic inspection. Decarburization requires metallographic examination or removal of material from the surface [about 0.005 in. (0.1 mm)] and a hardness recheck. The most insidious causes of problems in the successful use of tool steels, however, are user-caused defects:

1. Grinding burn
2. *Electrical discharge machining damage*
3. Hydrogen embrittlement

Almost every tool designer will encounter the failure of a tool steel during his or her career. So it is wise at least to mention how these defects can be detected and prevented.

Grinding burn comes from rehardening of the steel during excessive grinding. Most die steels can only tolerate less than 0.001 in. (25 μm) of thickness removal per pass of the wheel. If operators try to take a 0.005-in. (125-μm) stock removal, they may see the surface of the part turn black. This is no problem for them—they can take a light cut and make it look good again. However, the damage may have already been done. The high heat from excessive grinding can cause the surface to rise to a temperature in the range of its hardening temperature. On cooling, the surface reforms martensite, but this is untempered martensite and is extremely brittle. Usually, fine cracks (perpendicular to the grinding direction) undetectable to the naked eye form in the brittle, rehardened surface. Figure 12–19 shows a typical "burnt" surface.

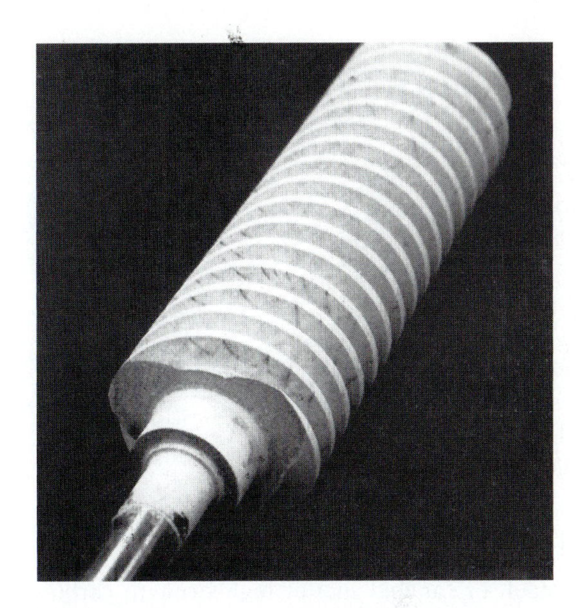

Figure 12–19
Grinding burn cracks (dark lines) on a tool steel worm

In service, grinding cracks usually propagate until you have several parts where once you had only one. The cure for this user-induced defect is intuitive: cautious grinding. If you suspect grinding damage (you may see telltale coloration), you can use a dye penetrant to inspect for cracks or apply an etchant that will bring out rehardening of the surface.

Electrical discharge machining (EDM) damage is similar to grinding burn in what it does to a tool steel surface. This process is widely used to shape hard-to-work tool steels. It involves a spark discharge between the electrode and the workpiece. Material removal is accomplished by actually melting a thin layer of the surface and washing it away with the flow of dielectric from the electrode. A *recast layer* (Figure 12–20) is formed on the surface, which is essentially rehardened, untempered martensite. It is porous, rough, and cracked and forms a stress concentration.

To prevent service failures from electrical discharge machining damage, it is best to remove several thousandths of an inch (0.1 mm) from the surface by grinding. If you have a massive part with no sharp corners that will not be subjected to high operating stresses or fatigue, you can forget about this surface damage. Most tool makers lap or polish the surface to try and remove the damaged layer. An etching process can be used with proper controls. The main point to be made is that this damage can cause failure of highly stressed parts, and it is best if it can be removed.

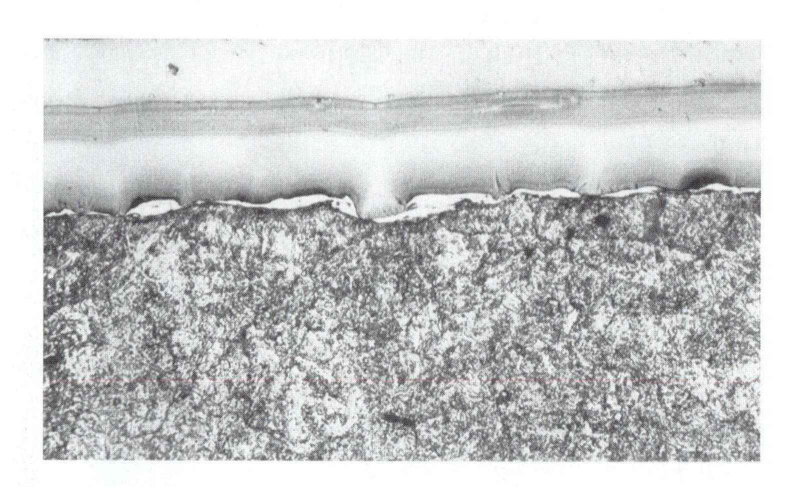

Figure 12–20
EDM damage (white) on 420 stainless steel (×400)

The final user-caused defect to be discussed is *hydrogen embrittlement*. If any hardened tool steel, or for that matter high-strength carbon steel, is to be subjected to an electroplating operation, it can and probably will pick up hydrogen from the plating reaction. This causes embrittlement. It can be easily reduced by a 4-h bake at 300 to 400°F (150 to 200°C). It is recommended that any hardened steels that are electroplated be baked to prevent this embrittlement.

Summary

In this discussion, as well as throughout this book, we have tried to emphasize that the effective use of materials in design is predicated on a basic understanding of material systems and familiarity or intimacy with a few materials in each system. Thus the answer to effective use of tool steels is an understanding of what they are (a group of alloy steels), what the differences are between each class (major alloying elements), and why they perform better than other steels as tools (hard microconstituents and in some cases just a tough microstructure).

Some additional concepts to keep in mind are the following:

- The abrasion resistance of tool steels depends on the amount of carbide phase present, as well as the carbide size and distribution.
- It is the nature of the carbide phase that discriminates tool steels from each other and from other steels.
- The cold-work and high-speed steel properties depend heavily on their carbide morphology.
- The P, L, H, and S series of tool steels do not contain the large carbide particles present in the cold-work and high-speed tool steels.
- Hot-work tool steels have resistance to softening and toughness as their main strengths.
- P-series tool steels are low carbon, and they are either used at about 30 HRC or case hardened.
- Shock-resisting steels have about 0.5% carbon and a few percent alloy; they cannot normally be hardened above 56 to 58 HRC.
- There are many cold-work tool steels, but they all are intended for punch and die types of applications.
- The high-speed steels are harder than all other tool steels (65 to 68 HRC); they are intended for metal cutting tools.

- Try to use the most available grades (it usually lowers cost).
- Most tool steels have their optimum properties at the recommended working hardness.
- Always try to pinpoint the wear mode that will occur in a tool; pick a steel that is good for this specific wear mode.
- Never weld tool steels by design; avoid it where possible.
- Cemented carbides and similar cermets will usually provide longer service life than tool steel, but fabrication costs are higher.
- The use of P/M tool steels usually provides abrasion resistance intermediate between carbides and other tool steels.
- Grind allowances and stock allowances for decarburization are often the designer's responsibility.
- Avoid designs that may be cracking-prone in hardening.

Terms You Should Remember

tool steel
microconstituents
water-hardening tool steels
air-hardening tool steels
oil-hardening tool steels
high-speed tool steels
matrix
carbides
depth of hardening
distortion
decarburization
red hardness
machinability
abrasion
metal-to-metal wear
erosion
deep hardening
electrical discharge machining damage
recast layer
tempering curve
working hardness
decarburization allowance
quench cracking
grind allowance
grinding burn
cermet
banding
surface fatigue

Critical Concepts

- Tool steels are made to different quality standards than carbon and alloy steels.
- Tool steels are categorized by use characteristics (cold work, hot work, shock resisting, etc.).
- Tool steels, cemented carbides, cermets, and, to a lesser degree, ceramics are the standard materials of construction for tools that shape, machine, and work other materials. Of these materials, tool steels are the lowest in cost.
- Tool steel selection requires the identification of anticipated wear modes and toughness needs for the application under consideration.

Case History

Tool Steel for Punch and Die

A new testing device produced a continuous loop of abrasive finishing tape. The number of passes that the loop made was recorded by passing a light through a hole in the tape to a receptor and a computer that recorded the number of "sightings" of light through a 4-mm-diameter hole in the tape. The testing department requested a recommendation for a punch die to perforate the test tape.

The tape had a thickness of 0.2 mm; 30 μm aluminum oxide was coated on one side of the tape and the nonabrasive side was polyester

(PET). The punch was to go through the backside of the tape, so the D2 tool steel at 60 HRC seemed an adequate punch material. Type A11 tool steel was recommended for the die. It is more abrasion resistant than D2, so it should resist abrasion by the abrasive coating. Six months after implementing our tooling recommendations, we learned that the tool couple had worked as planned and there was no measurable wear on the tools after hundreds of uses.

Questions

Section 12.1

1. What is the significance of the letters in the former AISI tool steel identification system?
2. State the ASTM designation for a D2 tool steel.

Section 12.2

3. What is the range of carbon content in tool steels?
4. What is the difference in hardenability of tool steels and alloy steels? Carbon steels?
5. Compare the volume fraction of carbide in type A7 tool steel to the amount in M3 (Figure 12–4).
6. What is secondary hardening? Which steels have it?
7. Name three tool steels that have a type II carbide structure; name two with a type VI structure.
8. What is the difference between a tool steel and a cermet? Name two types of commonly used cermets.
9. Write the drawing designation for a water-hardening drill rod.

Section 12.3

10. What are the primary elements in high-speed steels that provide red hardness?
11. What is the role of vanadium in A11 steel?
12. What is a steel-bonded carbide?

Section 12.4

13. Calculate the size change that would occur in a 12-in. (600-mm) opening in an A2 tool steel die when it is tempered at 500°F (260°C).
14. Compare the effect on hardness of using W1 and M3 tool steels at 800°F (426°C).
15. A hardened tool steel is used as a die to blank 1010 carbon steel. What type of wear will be likely in this type of service?
16. How can decarburization affect tool steels?
17. What steel would you use if you wanted an 8-in. section to through harden?
18. If a cutting tool made from M3 was quench hardened and tempered at 1050–1100°F, what would be the resulting hardness? If the same tool was used at 1200°F in service what would be the resulting hardness?

Section 12.5

19. Compare the cost of making a part from M2 steel to the cost of making the same part from O1 tool steel.
20. You want to bright harden a punch and you cannot grind after hardening. What is the best tool steel to use?
21. Which tool steel would have the best abrasion resistance to silica?
22. When would resistance to decarburization be an important selection factor?
23. What is the role of chromium carbides in D2 tool steel?
24. What preheat temperature should be used when welding a D2 tool steel?

Section 12.6

25. You want to make a 3-in.-diameter punch from O1 tool steel. What size bar should you order from the warehouse?

26. You are hardening an A2 tool steel plate with dimensions of 0.75 × 4 × 16 in. (20 × 100 × 400 mm). What grind allowance would you specify?
27. What is the best way to eliminate quench cracks in tool steels?
28. What is grinding burn, and why it is undesirable?
29. What is the recommended working hardness for D2? for M2? for H13? for S7?
30. What is retained austenite, and how is it detected? Controlled?
31. Explain the role of sharp reentrant corners in quench hardening.

To Dig Deeper

ASM Handbook, 10th ed., Volume 7, Powder Metal Technologies and Applications. Materials Park, OH: ASM International, 1998.

Brookes, J. A. *Hardmetals and Other Hard Materials*, 2nd ed. Hertfordshire, UK: International Carbide Data, 1992.

Davis, J. R., Ed. *Tool Materials*. Materials Park, OH: ASM International, 1995.

Edwards, R. *Cutting Tools*. London: The Institute of Materials, 1993.

Hoyle, G. *High Speed Steels*. London: Butterworths & Co., Ltd., 1988.

Metals Handbook, 10th ed., Volume 1, Properties and Selection: Irons, Steels and High Performance Alloys. Metals Park, OH: ASM International, 1990.

Metals Handbook, 10th ed., Volume 4, Heat Treating. Metals Park, OH: ASM International, 1991.

Pocketbook of Standard Steels. Warrendale, PA: Iron and Steel Society, 1996.

Roberts, G., and R. Cary. *Tool Steels*, 4th ed. Metals, Park, OH: American Society for Metals, 1980.

Roberts, G., G. Krauss, and R. Kennedy. *Tool Steels*, 5th ed. Materials Park, OH: ASM International, 1998.

Steel Products Manual—Tool Steels. Warrendale, PA: Iron and Steel Society, 1988.

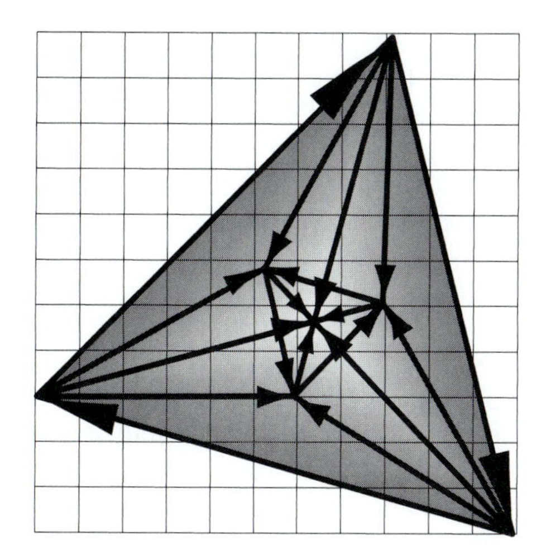

CHAPTER 13

Corrosion

Chapter Goals

1. An insight into the nature of corrosion and its various manifestations.
2. A knowledge of corrosion-measuring techniques.
3. Design guidelines for corrosion control.

Rationale

The effect of an environment on the life of an engineering material cannot be ignored. If you buy a $30,000 automobile with a black elastomer bumper and after 18 months or so it turns dull and discolored, you may become unhappy with your purchase. Early plastic bumpers commonly did this. In the medical, food, and chemical industries, all materials involved in processing feedstocks are subject to attack by their environments, and a long-term material survival strategy must be developed. Environmental degradation (corrosion, and like effects) is a design consideration for every part designed on a computer-aided design (CAD) terminal. You, the designer or engineer, must decide: Do I need to take some action to protect this part or in some way ensure survival for its intended service life in a particular environment? You are responsible if the part corrodes in three weeks, if the part fails by environmentally induced cracking after two years, or if it turns to powder in 10 years.

This chapter is intended to raise your awareness of environmental effects on engineering materials and to prevent failures caused by the action of an environment on your material selection.

What is *corrosion*? To many people corrosion means rust. Rust is a byproduct of corrosion, but a simple definition of corrosion is deterioration of a material or its properties because of reaction with its environment. Sometimes the deterioration is a weight gain; sometimes it is a weight reduction; sometimes the mechanical properties are affected. Most corrosion processes are electrochemical in nature; some are not.

Why is corrosion prevention important? If a designer is concerned with, for example, an assembly machine that sees no chemicals and spends its life in a controlled indoor environment, he or she may ignore all corrosion considerations in a design. This can be unwise. Almost all environments can cause corrosion. Normal room air can rust most steels. Sometimes even a very small amount of rust will affect serviceability.

If a machine will be dripping with oil in service, chances are corrosion may not occur; but if parts in a machine or device are to be unlubricated in service, it may be wise to use corrosion-prevention measures on every part where surface damage or property deterioration could affect serviceability.

According to a 1998 study funded by the Federal Highway Administration, the annual cost of corrosion in the United States exceeds $276 billion. Sectors with the highest corrosion costs include drinking water/sewage systems, automotive, defense, and highway bridges. Probably the lion's share of corrosion costs for automotive, highway-bridge, and water/sewage systems are caused by the corrosive effects of deicing salt and saltwater environments. Other consequences of corrosion, such as the environmental impact of scrapping and replacing rusty cars and the economic and environmental implications of construction delays due to road and bridge replacement, exacerbate the above cost estimate. An aging and degrading infrastructure decreases the nation's ability to compete in a global economy, especially against younger societies who observe modern corrosion-control practices.

13.1 The Nature of Corrosion

Why Corrosion Occurs

The question of why corrosion occurs can be answered by calculations involving free energy, thermodynamics, and reaction kinetics. Corrosion in simple terms occurs in metals because most metals are not in their natural state until they return to the ore form in which we found them. Almost all metals are found in nature in the form of an oxide, sulfide, or some other metal compound. Only a few metals such as gold occur in nature in the metallic form. Thus all metals that are derived from ores want to return to that form, and corrosion is the reaction process that occurs, allowing this return to nature.

In the case of plastics and ceramics, the return to nature is not so apparent. Corrosion occurs in plastics because the bonds between the organic molecules making up the plastic can be affected by various environments. Plastics do not rust. Deterioration is often in the form of a loss of properties. A common example of corrosion of plastic-type materials is the aging of rubber. Invariably, automobile tires craze and crack when they get to be more than 3 or 4 years old. This is usually due to attack by ozone in the air.

Ceramics are relatively inert to most environments because they are often oxides, sulfides, and other compounds that occur in nature. Thus there is little driving force to make these materials corrode. Another reason that plastics and ceramics behave differently from metals is that they are usually poor conductors of electricity. Many of the corrosion processes that occur in metals are electrochemical in nature; they require the flow of electrons. Ceramics, being poor conductors, are not as susceptible to electrochemical corrosion as metals. They do corrode, but not as readily. A common example of corrosion of a ceramic-like material is the glass-lined household water heater. You are fortunate if the glass lining lasts 10 years. There is nothing in the tank but water, but that is all that is needed; glass is slowly attacked by common, ordinary water. Thus the cause of corrosion in metals is a tendency to return to their original state; in plastics and ceramics, corrosion occurs simply because some environments chemically react with the molecules making up the plastics, ceramics, or glasses.

Electrochemical Reactions

The corrosion of metals is an *electrochemical* process. Electrochemical processes involve the interaction of ions (charged atoms) and electrons.

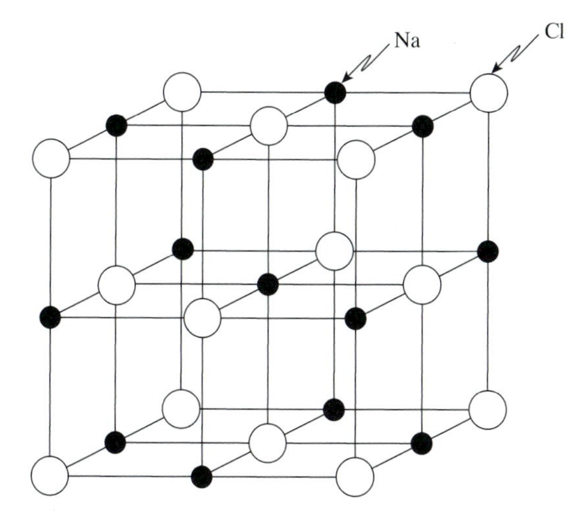

Figure 13–1
Crystal structure of sodium chloride, NaCl. The chlorine anions are arranged in the cubic close-packed position. The sodium cations fill the octahedral interstitial positions.

Before discussing electrochemistry, we need to introduce a few concepts.

Sodium chloride (or table salt) is a white, water-soluble solid (usually in granular form). This solid is bonded by ionic bonds and crystallizes in the cubic rock salt structure as shown in Figure 13–1. If sodium chloride granules dissolve in water, the sodium and the chlorine atoms separate (or *dissociate*) to form the ion species: Na^+ and Cl^-. An ion with a positive charge (such as Na^+) is known as a *cation,* and one with a negative charge (such as Cl^-) is known as an *anion*. Hence, when solids such as metals, salts, acids, oxides, or bases dissolve in liquid or aqueous solutions, the dissociated species exist as ions. Solutions containing dissolved ions are called *electrolytes*. Unlike reacted compounds such as sodium chloride, metals exist in an elemental state. When a metal corrodes it does not just dissolve in a liquid like an ice cube in water. For the metal to dissolve in the liquid it must transform from a solid atom to an ion, but in the process, it liberates electrons causing the metal to become polarized or charged.

If a sample of iron is placed in dilute hydrochloric acid, a corrosion reaction occurs. The iron goes into solution as an ion and forms an iron chloride complex with the chlorine in the acid. Correspondingly, hydrogen gas is formed at the iron's surface while bubbles of hydrogen gas grow sufficiently large to float to the surface. An example of this corrosion reaction is shown in Figure 13–2. The chemical equation for this reaction can be summarized as

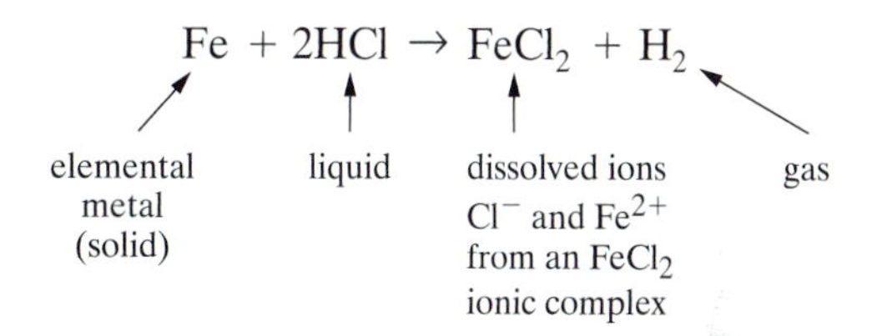

Because the choride ion does not really enter into the chemical reaction (it does not change oxidation state in this reaction), a simplified equation can be written:

$$Fe + 2H^+ \rightarrow Fe^{2+} + H_2$$

This equation simply states that elemental iron reacts with the hydrogen ions in the acid to produce dissolved iron ions plus hydrogen gas. This chemical equation consists of two *partial processes,* an *anodic* process and a *cathodic* process, shown respectively as

$$Fe \rightarrow Fe^{2+} + 2e^- \quad \text{(anodic partial process)}$$
$$2H^+ + 2e^- \rightarrow H_2 \quad \text{(cathodic partial process)}$$

For an atom of iron to transform into an ion, it must liberate two electrons (denoted as e^-). When iron is converted from its elemental form into Fe^{2+} it is said to be *oxidized* and the corresponding reaction is called the *anodic* partial process.

The hydrogen ion in the acid must accept an electron to become elemental hydrogen. However, because hydrogen is most stable as a *diatomic* molecule, it immediately combines

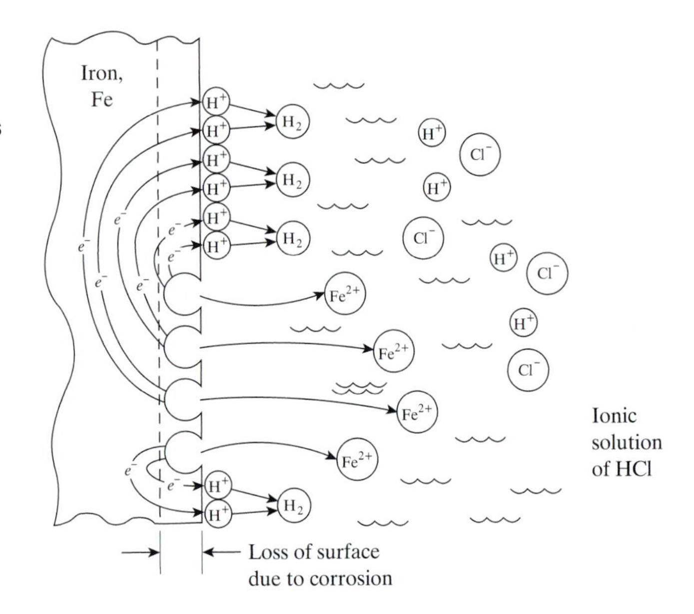

Figure 13–2
The corrosion of iron in hydrochloric acid. As iron dissolves in the solution, excess electrons become available to react with the hydrogen ions in the acid (*corrodent*)

with a similar elemental hydrogen atom to form H_2. In this reaction, the hydrogen is said to be *reduced,* and the corresponding reaction is called the *cathodic* partial process.

In both of these partial processes, a reaction takes place between an ion and an electron. Chemical reactions between ions and electrons, known as *electrochemistry,* are the foundation of corrosion science. Electrochemical metallic corrosion processes require the simultaneous occurrence of both *oxidation* and *reduction* partial processes. Stop either half of the oxidation/reduction process and the corrosion reaction halts. Accelerate either half of the process and the corrosion rate increases. For metallic corrosion to occur, it requires a system or environment in which the electrons liberated during the metal oxidation process can be accepted by another process or system.

At the interface between the iron and the acid solution (see Figure 13–2), an iron cation, Fe^{2+}, is liberated from the surface of the elemental iron into the liquid. Correspondingly, two electrons are generated in the metallic iron. These electrons, which are conducted back to the body of the iron, are available to reduce the hydrogen ions in the acid solution to elemental hydrogen. Once reduced, the elemental hydrogen recombines to form a diatomic hydrogen gas molecule, H_2. Generally, when a metal corrodes, the surface is covered with many microscopic anode (oxidation) and cathode (reduction) sites. Grain boundaries, inclusions, and other surface defects or features can be incipient anode or cathode sites.

The rusting of steel in water produces a similar reaction. For example, when an article made from steel is left outside to rust, the water from rain and condensation contains some dissolved oxygen (from the air). The anodic corrosion reaction is the oxidation of iron:

$$Fe \rightarrow Fe^{2+} + 2e^- \quad \text{(anodic partial process)}$$

And the cathodic or reduction reaction is the reduction of oxygen:

$$O_2 + 2H_2O + 4e^- \rightarrow 4OH^- \quad \text{(cathodic partial process)}$$

In this instance, the oxygen dissolved in the water is reduced to an ionic state, where it forms a hydroxyl complex with the hydrogen in the water. The overall corrosion reaction may be obtained by combining the anodic and cathodic partial processes:

$$2Fe + 2H_2O + O_2 \rightarrow \underbrace{2Fe^{2+} + 4OH^-}$$

$$\downarrow$$

$$2Fe(OH)_2 \downarrow \text{ ferric hydroxide precipitate}$$

On the right side of the equation the iron ion and the hydroxide (OH^-) react to form a ferric hydroxide precipitate (a familiar orange-colored compound). If the surface of the corroded iron is allowed to dry, the ferric hydroxide dehydrates and forms a ferrous oxide, Fe_2O_3 (red rust):

$$2Fe(OH)_2 \rightarrow Fe_2O_3 \cdot 3H_2O$$

The point to be made here is that ions other than hydrogen may be reduced as part of the corrosion process.

Corrosion Thermodynamics and Redox Potentials

As discussed earlier, a metal is compelled to return to its native or natural state through corrosion. However, to understand and predict when corrosion can occur, we need to review the chemical thermodynamics of corrosion. For corrosion to occur, both oxidation and reduction chemical reactions must occur simultaneously. They do not, however, have to occur on the same metal surface. An excellent example of this concept is an electrical battery. A simple battery may be created by placing bars of zinc and graphite in an electrolyte such as a sodium chloride solution (see Figure 13–3). If a voltmeter

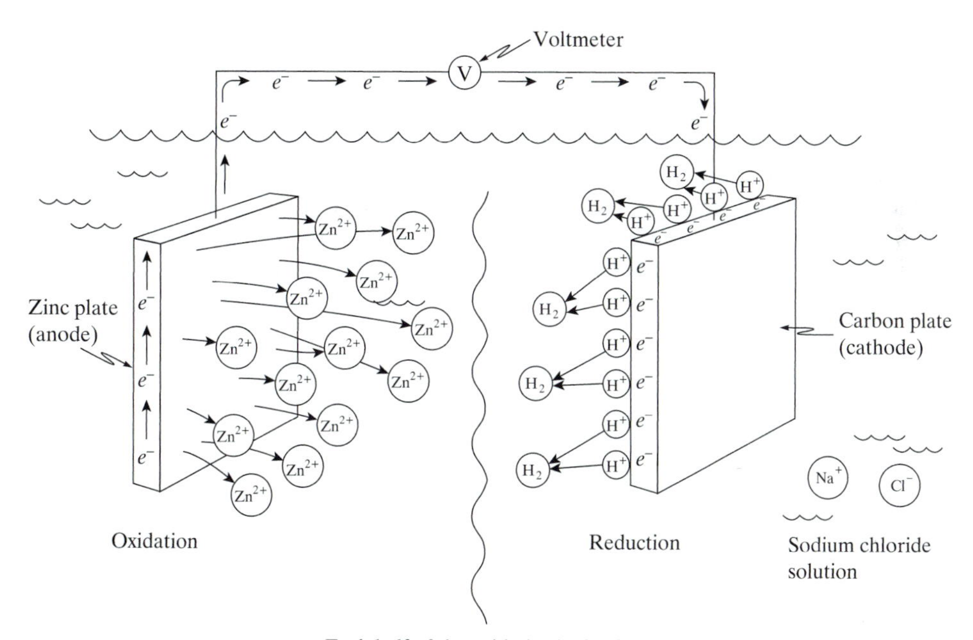

Figure 13–3
Model of an electrochemical battery. When zinc is electrically coupled to carbon, the electrons liberated during the zinc oxidation process flow to the carbon cathode.

were connected to the cell, we would find that there is a voltage or potential difference between the zinc and the carbon. Replace the voltmeter with a light bulb, and it would light up. In this example, the anodic reaction,

$$Zn \rightarrow Zn^{2+} + 2e^-$$

occurs on the zinc surface and the cathodic reaction

$$2H^+ + 2e^- \rightarrow H_2$$

occurs on the cathode. Hence each half of the oxidation/reduction reaction has a voltage or potential associated with it. In practice, batteries are much more complicated. However, fundamentally this is how they work.

A technique, called *standard half-cell potentials* or *oxidation/reduction* (*redox*) *potentials*, is used to standardize the potentials or voltages of the half-cell reactions. For example, in Figure 13–4 a cell is constructed where a zinc electrode is placed in a solution with a standard concentration of zinc ions. An adjacent cell contains a platinum electrode in a solution with a controlled hydrogen ion concentration. This platinum electrode is used as a reference and is known as a *hydrogen reference electrode*. A special high-impedance voltmeter measures the voltage difference between the two electrodes. The potential of the cell, E_{cell}, is calculated as the potential difference between the reduction half-cell potential ($E_{reduction}$) and the oxidation half-cell potential ($E_{oxidation}$):

$$E_{cell} = E_{reduction} - E_{oxidation}$$

The potential of the hydrogen reduction process that occurs at the hydrogen reference electrode provides a reference or datum for all other redox processes. By convention, the hydrogen reduction process has a reference voltage, E_{H_2}, of 0 volts. In the case of the cell in Figure 13–4, the voltage of this cell, E_{cell}, is about 0.763 volts. Because $E_{H_2} = 0.000$ volts, the redox potential of the zinc oxidation process (E_{Zn})

$$Zn \rightarrow Zn^{2+} + 2e^-$$

is −0.763 volts relative to the hydrogen reduction reaction.

This technique is used to measure the redox potential of the oxidation and reduction process of many metals, and metal ions, as well as other species such as oxygen. Charts listing the redox potentials of metals and chemical species are

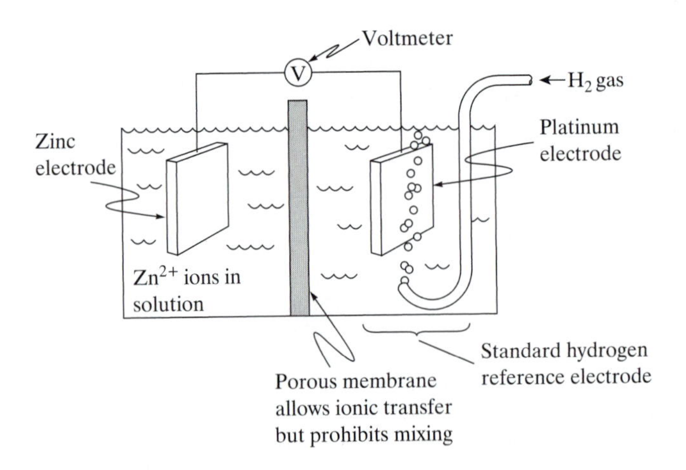

Figure 13–4
Sketch of the arrangement for measuring the half-cell potential of zinc

published in various handbooks. Some typical redox potentials are

Equilibrium Reaction	Potential, V^*_{SHE}
$Au^{3+} + 3e^- = Au$	+1.49
$O_2 + 4H^+ + 4e^- = 2H_2O$	+1.23
$Pt^{2+} + 2e^- = Pt$	+1.20
$Fe^{3+} + e^- = Fe^{2+}$	+0.77
$O_2 + 2H_2O + 4e^- = 4OH^-$	+0.40
$Cu^{2+} + 2e^- = Cu$	+0.34
$2H^+ + 2e^- = H_2$	0.00
$Ni^{2+} + 2e^- = Ni$	−0.25
$Fe^{2+} + 2e^- = Fe$	−0.44
$Zn^{2+} + 2e^- = Zn$	−0.76
$Al^{3+} + 3e^- = Al$	−1.66

$^*V_{SHE}$ = voltage relative to a standard hydrogen electrode.

The redox potential voltages help determine if a corrosion reaction has the thermodynamic driving force to spontaneously occur. For any two oxidation/reduction reactions, the most negative half-cell will be oxidized and the most positive will be reduced. For example, from this list of redox potentials we can conclude that the metals Fe and Zn, with redox potentials more negative than hydrogen, will corrode in deaerated acid solutions (free of dissolved oxygen). Copper, alternatively, has a redox potential positive to hydrogen—therefore it does not corrode in deaerated sulfuric acid. However, if the sulfuric acid has dissolved oxygen in it, the copper will spontaneously corrode because its half-cell potential is more negative than the oxygen reduction reaction.

In thermodynamic terms, the tendency for a metal to corrode in a given solution is determined by the change in *free energy*, ΔG. Free energy is the difference in energy between the initial and the final state or a process. If the ΔG for a given system is negative, the process occurs spontaneously. If ΔG is positive, the process will not occur spontaneously without additional energy. For example, if we have the corrosion reaction for iron in a dilute HCl acid solution:

$$Fe + 2H^+ \leftrightarrow Fe^{2+} + H_2,$$

does this reaction proceed to the left or to the right? The direction that gives a lower or negative free energy state (ΔG) is the spontaneous direction of the process. As an analogy, if a ball is placed at the top of a ramp or inclined plane, it will roll down the ramp and come to rest at the bottom. At the top of the ramp, the ball is in a higher energy state than the ball at the bottom of the ramp. Hence the free energy for this system is negative or spontaneous for the ball rolling down the ramp. When resting at the bottom of the ramp, the ball requires energy to move up the ramp. Pushing the ball up the ramp is not spontaneous and has a positive free energy. Although advanced treatments are beyond the scope of this book, the free energy for electrochemical corrosion can be calculated from the redox potential using the following equation:

$$\Delta G = -nFE_{cell}$$

where ΔG = change in free energy

n = number of moles of electrons involved in the reaction

F = *Faraday*'s constant, the charge on one mole of electrons (96,487 coulombs).

E_{cell} = the potential difference of the reduction and oxidation half-cell potentials: $E_{cell} = E_{reduction} - E_{oxidation}$

The following examples show the propensity for copper corrosion in a sulfuric acid solution:

1. Copper immersed in deaerated H_2SO_4:

$$Cu + H_2SO_4 \leftrightarrow CuSO_4 + H_2$$

Half-cell reactions:

$Cu^{2+} + 2e^- \leftrightarrow Cu$ $\quad E_{reduction} = 0.34\ V_{SHE}$

$2H^+ + 2e^- \leftrightarrow H_2$ $\quad E_{oxidation} = 0\ V_{SHE}$

$$E_{cell} = E_{reduction} - E_{oxidation} = (0\ V_{SHE}) - (0.34\ V_{SHE}) = -0.34\ V_{SHE}$$

$$\Delta G = -(2\,\text{mol}\,e^-)(96{,}487\,\text{C/mol}\,e^-) \times (-0.34\,\text{V}_{\text{SHE}}) = 65611\,\text{J}$$

Corrosion of copper is not spontaneous in deaerated H_2SO_4, no corrosion occurs.

2. Copper immersion in H_2SO_4 with dissolved oxygen

$$2\,\text{Cu} + 2\text{H}_2\text{SO}_4 + \text{O}_2 \leftrightarrow 2\text{CuSO}_4 + 2\text{H}_2\text{O}$$

Half-cell reactions:

$$\text{O}_2 + 4\text{H}^+ + 4e^- \leftrightarrow 2\text{H}_2\text{O} \qquad E_{\text{oxidation}} = 1.23\,\text{V}_{\text{SHE}}$$

$$2\text{Cu}^{+2} + 4e^- \leftrightarrow 2\text{Cu} \qquad E_{\text{reduction}} = 0.34\,\text{V}_{\text{SHE}}$$

$$E_{\text{cell}} = E_{\text{reduction}} - E_{\text{oxidation}} = (1.23\,\text{V}) - (0.34\,\text{V}) = 0.89\,\text{V}_{\text{SHE}}$$

$$\Delta G = -(4\,\text{mol}\,e^-)(96{,}487\,\text{C/mol}\,e^-) \times (0.89\,\text{V}_{\text{SHE}}) = -343494\,\text{J}$$

Corrosion of copper is spontaneous in H_2SO_4 with dissolved oxygen.

Free energy can be valuable in calculating the thermodynamic tendency for corrosion. However, there are cases where corrosion does not occur even though ΔG is negative. The passivation of some metals, for example, can reduce the corrosion rate to very low levels even though ΔG is negative. The thermodynamic free energy value does not provide any indication of the kinetic aspects of the corrosion process (such as corrosion rate). In many instances, metals do corrode in their environment, but the corrosion rate is low enough that the part will last for its intended design life.

Electrolytes

An essential requirement of electrochemical corrosion is the presence of an electrolyte. An *electrolyte* is simply a fluid that conducts electricity. Fluids conduct by migration of ions. Water is a good conductor because it usually contains mineral ions, hydrogen ions, and hydroxyl ions (OH^-). If these ions were removed, the solution would not carry electricity and electrochemical corrosion could not occur.

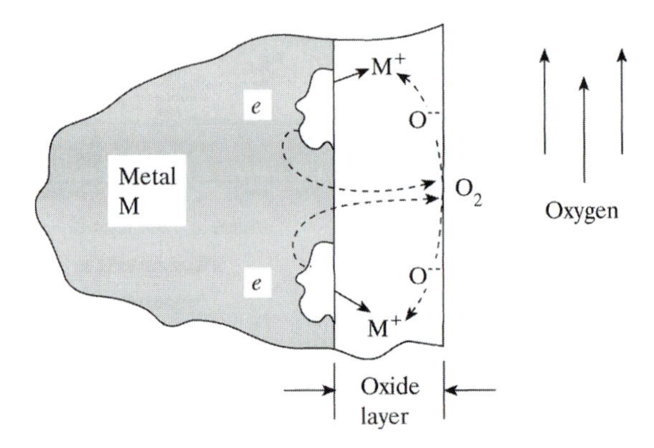

Figure 13–5
Oxide film as an electrolyte

In atmospheric corrosion the electrolyte comes from the humidity in the air. Rusting in desert-type atmospheres is low because the humidity is low, and dry air is a poor electrolyte.

Most metals that are used at high temperatures, such as in furnaces, oxidize and deteriorate with time. Even this type of corrosion is believed to be electrochemical in nature, the electrolyte being the metal oxide or coating (Figure 13–5). The oxide layer conducts metal ions from the metal surface. The metal ions combine with oxygen cations produced at the oxide layer–oxygen interface to yield a metal oxide. This is why oxide layers in furnace atmospheres do not protect. They keep getting thicker or spall, and the parts get thinner.

Summary

The following are the basic requirements for electrochemical corrosion of metals:

1. The metal is oxidized at the anode of an electrolytic cell.
2. Ions are reduced at the cathode.
3. Oxidation and reduction processes occur simultaneously and at the same rate.

4. There is a potential between the anode and cathode.
5. An electrolyte must be present.
6. The electrical path must be completed (somehow the anode must be coupled to the cathode).

If any of these conditions is not met, electrochemical corrosion will not occur. The implication is that the designer can have control over some of these factors and, hence, control over electrochemical corrosion.

13.2 Factors Affecting Corrosion

Many factors can control the tendency for a material to corrode; however, material properties and the nature of the environment are probably the most important.

Material Properties

Corrosion is not an intrinsic material property. It is the result of a material's interaction with a particular environment; take away the environment and there is no corrosion. Among the material properties that affect corrosion, chemical composition has an obvious effect. Everyone knows that different materials, such as gold, chromium, polyethylene, and iron, respond differently to exposure to water. Iron corrodes; the others do not. The reason that iron rusts when the others do not is that there is a favorable thermodynamic driving force for the reaction. All chemical and electrochemical processes are controlled by thermodynamics. It requires heat to melt ice, to heat water, and to vaporize water. The same processes in reverse give off heat. A certain amount of energy is needed for each of these changes of state. The same is true of corrosion processes. Gold will not corrode in water because the energy considerations are not favorable for corrosion to occur. Thus materials do not have intrinsic corrosion characteristics, but the chemical makeup of materials does determine the energy considerations necessary for corrosion to occur.

Redox Potentials Redox potentials, as discussed in Section 13.1, provide a means of determining if a corrosion process of a particular metal is favored from the thermodynamic standpoint. The potential measured is a function of the metal, and thus there is some relationship between the nature of a material and the possibility of its corroding.

Metallurgical Factors The concept of redox potentials is somewhat abstract, but there are other material properties that have a more obvious effect on corrosion. It was previously stated that electrochemical corrosion requires an anode and a cathode. In pure metals, anode areas are usually grain boundaries or even smaller microheterogeneities. In alloys and common metals with impurities, many metallurgical factors can affect corrosion:

1. Chemical segregation
2. Presence of multiple phases
3. Inclusions
4. Cold work
5. Nonuniform stresses

There are others, but the reason that these metallurgical factors affect corrosion is the same: they create anodic areas. If one area of a casting chills much faster than the bulk of the casting, it may be purer (chemical segregation). When immersed in a corrodent, the pure area coupled with the impure area provides a cell action. The same thing is true with multiphase alloys. If a metal has a large-volume fraction of a less-corrosion-resistant phase, it will become the anode in a corrosion cell. This is common in austenitic stainless steel castings. A second phase (ferrite) usually occurs in what is supposed to be a single-phase alloy (austenite), and corrosion tendencies are increased. If inclusions in metals are metallic, they can lead to galvanic reactions; if they are inert, they can sometimes inhibit corrosion. Wrought iron contains about 3% slag (silicates), which does not corrode in many liquids. It is believed that this is why wrought irons do not rust in water as much as steels.

Local differences in grain size in metals can promote a corrosion cell action. The local grain reduction that can result from a peening operation has been known to produce corrosion problems. Even without reducing grain size, stresses in a metal can cause corrosion effects that would not occur if the stress were absent. Local corrosion has been caused by such seemingly harmless things as having a bolted assembly immersed in a corrodent. The area under the bolt head is under a compressive stress; the stressed region of the metal reacts differently to the corrodent than the unstressed area, and cell corrosion can occur.

Thus many metallurgical factors can affect corrosion; these are illustrated in Figure 13–6.

Passivity Many important engineering metals derive their corrosion resistance from a surface condition that inhibits electrochemical action between the metal and its environment. This surface condition is known as *passivity*. In its simplest form, passivity exists when a metal is unattacked in an environment that is known to be capable of causing attack of that particular metal. The classic example of this condition is steel in nitric acid. Dilute solutions will cause rapid attack, whereas concentrated solutions have little visible effect. The explanation of this occurrence is that concentrated nitric acid causes a surface film to form that imparts passivity. There are more sophisticated definitions of passivity derived from electrode kinetic studies of corrosion (anodic polarization), but at present, let us assume that it is the formation of a protective film.

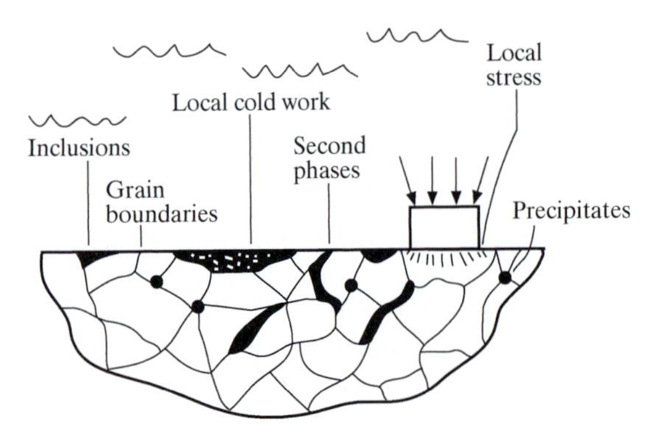

Figure 13–6
Metallurgical factors that can affect corrosion

There are various theories on how passive films are formed. One theory is that a simple insoluble oxide film is formed on the metal surface with oxygen from the electrolyte. Another theory is that the film is simply absorbed gas that forms a barrier to diffusion of metal ions from the substrate. A more sophisticated theory is called the electron-configuration theory. Proponents of this mechanism claim that passivity is due to electron vacancies in the *d* shell of the metal atoms. Some metals that have a tendency to form passive surfaces are chromium, nickel, cobalt, iron, and molybdenum. All of these metals have unfilled *d* shells. When the metal surface is in an electrolyte that can supply oxygen, the oxygen atoms will combine with metal atoms on the surface to form a thin oxide film. This is not a bulk film as in the oxide or adsorbed gas theories, but a monomolecular film possibly less than 10 angstroms (Å) thick. A thicker film may form (100 Å), but passivity comes from the thin initial film. The real mechanism is not very important to machine designers; the important point to remember is that some metals form a passive film and this film imparts corrosion resistance by acting as a barrier between the metal surface and the corrodent.

What is the significance of passivity to the practicing designer? Any operational conditions such as surface abrasion, impingement, or high-velocity solution contact that tend to remove the passive film by mechanical action will lead to corrosion. For many years it was common practice to passivate stainless steels to form a passive film. This was usually done by soaking the parts in dilute nitric acid. It has been found that a passive film will spontaneously form in air in a matter of minutes on freshly abraded stainless steel. Thus passivation is not needed for film formation but rather to remove metallic contaminants that may interfere with film formation.

Environment

Chemical Nature Metals, plastics, and ceramics of different types react differently to various environments. It is possible to categorize environments somewhat. The following is a list of some of the major classes of environments that materials may be subjected to:

1. Acids (oxidizing and reducing)
2. Bases
3. Salts (acid, neutral, alkaline)
4. Gases
5. Solvents

Some materials like oxidizing environments, some like reducing, some like alkaline, and so on down the list. Figure 13–7 is an illustration of how various metals react to oxidizing and reducing environments. The important point to remember is that the designer must always try to find out as much as possible about an intended environment. Some gases are oxidizing; some are reducing. The same thing can be said about acids. The *pH* (positive hydrogen ion) concentration of an environment will serve as a measure of the strength of acids or bases. A solution with a pH of 7 is neutral, pH values from 7 to 14 are alkaline in increasing strength, and solutions with pH values less than 7 are acid, with 0 being the strongest. The presence of halogen ions (chlorine, iodine, and others) in solutions can lead to corrosion in solutions that would not normally cause corrosion. It is important to investigate the nature of an environment. It is a necessary first step in corrosion control.

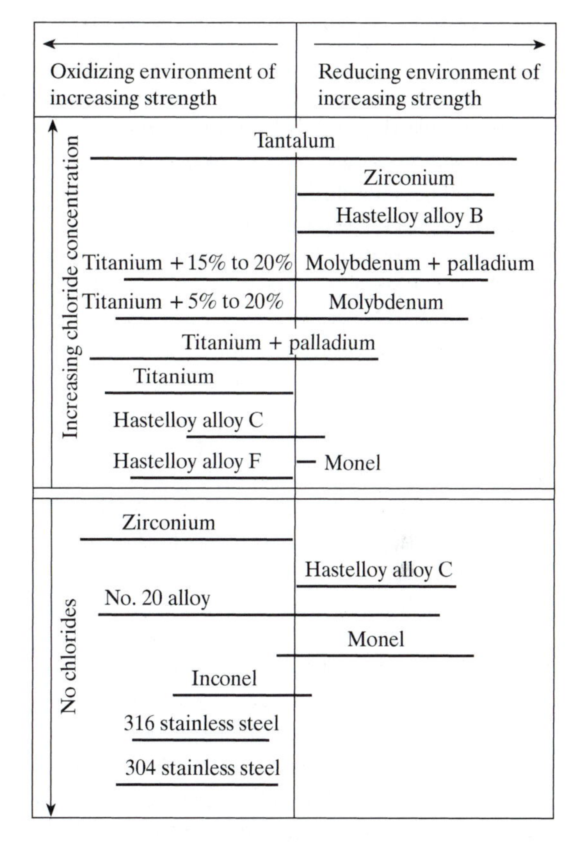

Figure 13–7
Satisfactory operating ranges for various metals in oxidizing and reducing environments
Source: Courtesy of Carpenter Technology Corp.

Operating Conditions Some of the more essential operating conditions to consider in selecting a material for corrosion resistance are the following:

1. Intended service life
2. Temperature
3. Velocity
4. Concentration
5. Impurities
6. Aeration

Increasing the temperature of most chemical reactions increases the rate of the reaction. The same thing is true for most corrosion systems. The velocity of a corrodent over a metal surface similarly affects corrosion rate. This is especially true for the metals that derive their corrosion resistance from a passive film on the surface. The mechanical action of high corrodent velocity can remove the passive layer. Solution concentration has a strong effect, but not always in the direction perceived. As mentioned in our discussion of passivity of iron in nitric acid, dilute solutions sometimes cause high corrosion rates, and concentrated solutions, low corrosion rates. The effect of concentration is not intuitively predictable. Corrosion data must be consulted.

The role of impurities in a corrodent similarly is not always predictable. If the impurity is abrasive particles, it may accelerate corrosion by removal of passive films. If the impurity is a trace chemical such as water in an organic solvent, the result may be pitting attack. Many times corrosion failures are due to impurities present in some solution. Periodic solution analysis is the best way to determine if potentially dangerous impurities are present.

The effect of *aeration* on corrosion can be none, increased corrosion, or decreased corrosion. Metals that do not depend on passive surface films for their corrosion resistance usually show increased corrosion rates in aerated solutions. Copper is an example of such a metal. If a metal such as stainless steel, which relies on a passive film for corrosion resistance, is corroding (active), aeration will help reestablish the passive film and corrosion will be reduced. If the metal is not corroding, the passive film is intact and solution aeration will have no effect. In any case, aeration is a factor to consider in the handling of corrodents.

Polarization When corrosion reactions are studied under controlled conditions in a special apparatus, it is possible to monitor the voltage and current between a corroding metal and a reference electrode. The voltage measured in such a setup is called *corrosion potential*; the current measured is the *corrosion current*. The change in potential with increasing or decreasing corrosion current flow is called *polarization*. As we shall see later, polarization can be used to investigate corrosion characteristics, but it should be included in this discussion because polarization affects corrosion. Several types of polarization are normally seen in laboratory investigations of corrosion: (1) activation polarization, (2) concentration polarization, and (3) *IR* drop.

Activation polarization occurs when an electrode reaction does not occur as fast as it should. Some additional time or energy is needed for activation. This type of polarization is often seen at the cathode in a corrosion cell where hydrogen is being reduced. The reduction reaction is

$$2H^+ + 2e \rightarrow H_2\uparrow$$

Hydrogen ions are adsorbed on the metal surface; they pick up electrons from the metal, combine into diatomic hydrogen, and go off as gas bubbles. This reaction can sometimes be sluggish, and thus polarization is said to occur. The gas bubbles at the cathode keep other hydrogen ions from reaching the surface, and the corrosion rate decreases.

Concentration polarization occurs when there are not enough of the reacting species at an active electrode. Using the example of hydrogen reduction at a cathode, if there are not enough hydrogen ions available for reaction at the cathode, the corrosion potential will decrease. This concept is illustrated in Figure 13–8.

The last form of polarization, *IR drop,* simply means a voltage or potential drop in the electrochemical cell due to resistance in the electrolyte. The buildup of corrosion products in an etching operation is an example of *IR* drop polarization. The flow of the corrodent to the metal surface is impeded by the surface film, and corrosion decreases. Thus polarization can affect corrosion processes by altering the reaction

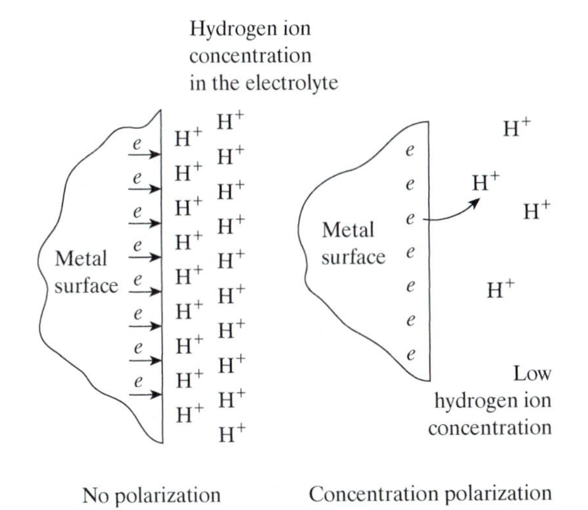

Figure 13–8
Concentration polarization

rate at the anode or cathode in a corroding system. Sometimes this effect is desirable (slows corrosion), sometimes not (interferes with electroplating operations).

Summary

We have chosen to concentrate on a few of the more important factors that affect corrosion. Intuitively, we must expect material properties and the environment to affect corrosion. As shown in Figure 13–9, physical conditions such as design and applied stress affect corrosion, and these factors will be covered in subsequent discussions. The primary recommendations that can be made regarding factors that affect corrosion are (1) know your environment, and (2) know your material.

13.3 Types of Corrosion

The prevention and analysis of corrosion failures require recognition of the type of corrosion that can occur in a system. Most workers in the field try to limit the number of basic types of corrosion to eight:

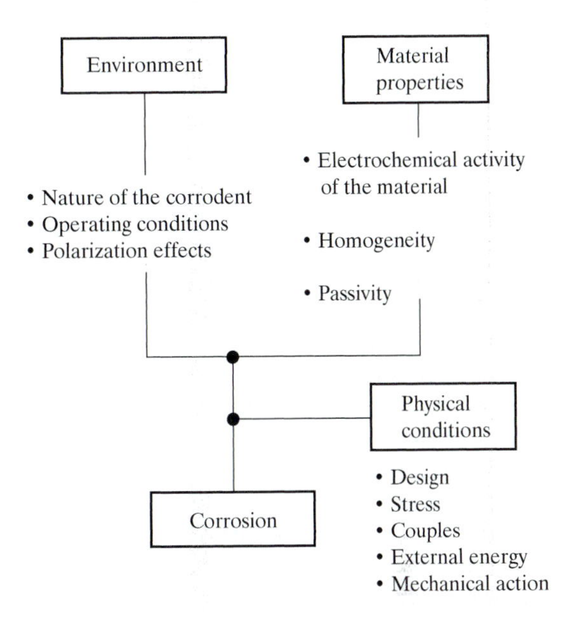

Figure 13–9
Factors affecting corrosion

1. Uniform
2. Pitting
3. Crevice
4. Galvanic
5. Stress corrosion cracking
6. Intergranular attack
7. Dealloying
8. Erosion

In this section we shall define each of these types of corrosion and say something about the mechanisms responsible.

Uniform

The simplest form of corrosion is *uniform attack* of all surfaces exposed to a corrodent (Figure 13–10). It can be electrochemical in nature or simply *direct attack*.

Figure 13–10
Uniform corrosion

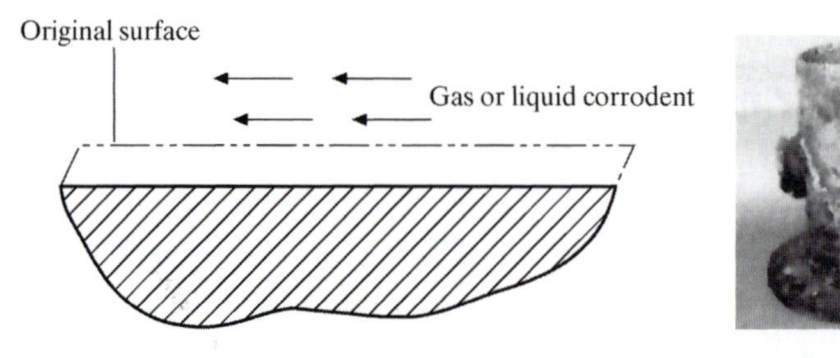

Most plastics and ceramics, since they are poor conductors of electrons, corrode by direct attack. Electrochemical cell action does not play a role. Sometimes the attack is not complete dissolution. As an example, nylon in strong oxidizing acids turns soft and gooey. When metals undergo direct attack, there is no oxide or similar surface film that can promote electrochemical cell action. An example of film-free direct attack of a metal is molybdenum at elevated temperatures in air [over 1400°F (760°C)]. No apparent scale of film is formed; the metal slowly disappears (sublimes). Similar situations can occur with metals used to handle liquid metals, with metals in molten salts, and with metals in certain corrosive gases. Chlorine gas causes direct attack of a number of metals.

The more typical situation of uniform corrosion involves electrochemical reactions. Ordinary atmospheric rusting can be uniform in nature. The cell action is between metal grains, grain boundaries, and the like. A metal pickling operation usually results in uniform attack; at least, this is the type of attack desired. Uniform corrosion can be prevented by removing the electrolyte or by using a material that is unaffected by the corrodent. The removal of the electrolyte in atmospheric rusting means lowering the relative humidity (RH) to some low value, possibly less than 30%. Choosing a material that is not affected by a specific corrodent is not always as easy as it sounds; consultation of detailed corrosion data is the best place to start.

Pitting

By definition, *pitting* is simply local corrosion damage characterized by surface cavities (Figure 13–11). This is a particularly insidious form of corrosion because the usefulness of a part may be lost by removal of a relatively small amount of metal. If a single pit perforates the side of a tank of pipe, serviceability is lost until the pit is repaired.

Pitting in many cases is simply caused by the chemical nature of environments. Halogen-containing solutions, brackish water, salt water, chloride bleaches, and reducing inorganic acids are known to be solutions that tend to produce pitting. Similarly, certain metals such as stainless steels are particularly prone to pitting attack.

There is no universally accepted theory on the mechanism of pitting. Laboratory tests have confirmed that it is electrochemical in

Figure 13–11
Pitting of copper in stagnant water

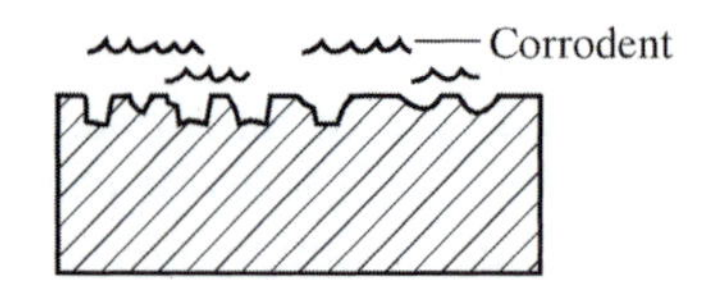

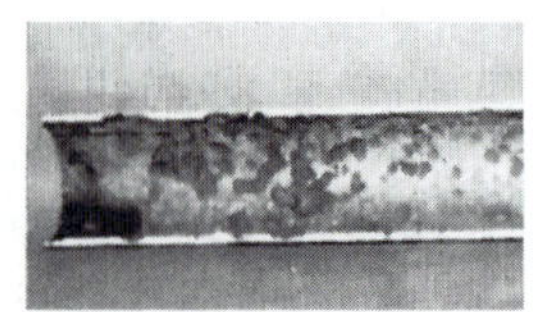

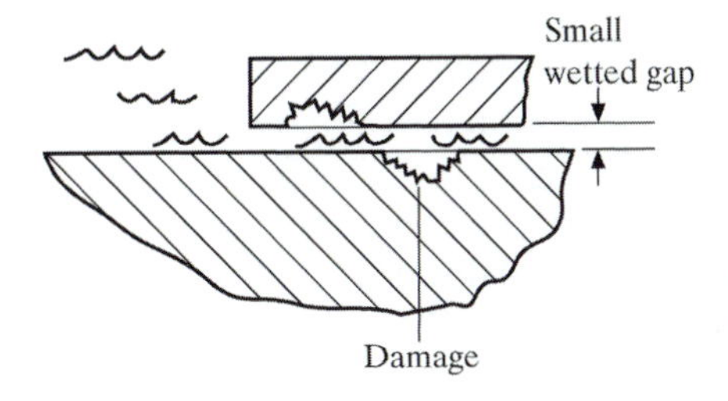

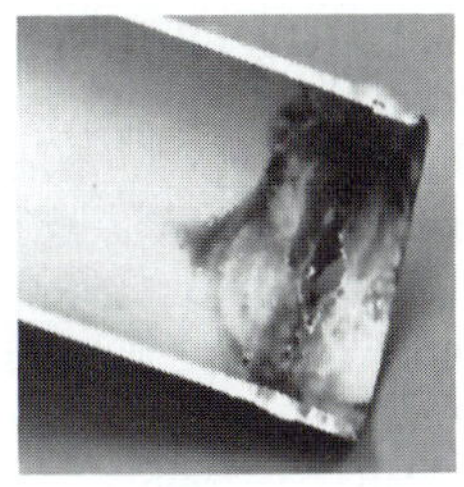

Figure 13–12
Crevice corrosion in an incomplete penetration pipe weld

nature, the areas around pits show little attack, it occurs more in stagnant solutions than in moving solutions, and pits seem to be self-propagating. The observations imply that pitting starts at some nuclei, such as small breaks in a passive film, inclusions, or scratches. The active metal will begin to corrode. The solution at the incipient pit becomes concentrated with metal ions. Additional ions from the corrodent migrate to the pit area and stimulate further metal dissolution. It is like a hyperactive corrosion cell in one local area.

The accepted practice in dealing with corrodents that tend to cause pitting is to avoid the use of metals that are prone to pitting—not a very creative solution, but it works. As an example, austenitic stainless steels pit in salt water, so most designers tend to use copper alloys, bronzes, brasses, Monels, and other materials that have lower pitting tendencies. Some use carbon steels in salt water. The overall corrosion rate is much higher than with stainless steel, but the mode of attack is more uniform than pitting, and perforation is not as likely. There is no good solution to pitting, but proper material selection can minimize the effect.

Crevice Corrosion

Crevice corrosion (Figure 13–12) is a local attack in a crevice between metal-to-metal surfaces or between metal-to-nonmetal surfaces. One side of the crevice must be exposed to the corrodent, and the corrodent must be in the crevice. Crevice corrosion commonly occurs in poorly gasketed pipe flanges and under bolt heads and attachments immersed in liquids. It is generally believed that the chemistry of the corrodent changes in the stagnant area of the crevice (Figure 13–13). Intuitively, we would expect the oxygen content of the corrodent to be different in the stagnant area of the crevice than out in the moving bulk solution. If the solution chemistry differs between the crevice and the bulk solution, a solution concentration cell can exist, with the area within the crevice being the anodic site. There is evidence that this may not be the predominating mechanism. Some investigators believe that the solution chemistry changes in the crevice are such that the solution per se is more aggressive. A decrease in pH has been measured in crevices of metals immersed in seawater. The neutral seawater became acidic in the crevice. The exact chemical reactions occurring in

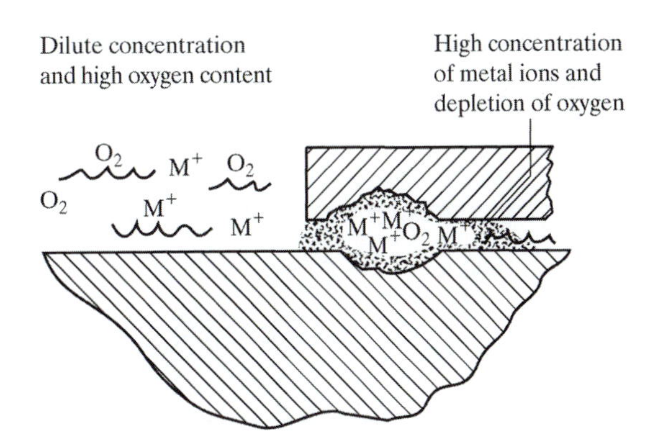

Figure 13–13
Crevice corrosion

various metal-solution systems probably vary, but it can be agreed that crevice corrosion is caused by changes in corrodent chemistry or electrochemical activity in the crevice area.

Like pitting, crevice corrosion can be very destructive because the damage is so localized. Stainless steels and aluminums are noticeably susceptible to crevices and stagnant areas. Good gasketing works, but the term *good* should be emphasized. If any solution at all wicks into a joint, crevice corrosion will occur. Do not use gaskets that can absorb the corrodent. Avoid ungasketed faying surfaces and surface deposits.

Galvanic Corrosion

If two dissimilar metals are electrically connected in an electrolyte, they meet the requirements of an electrochemical cell; if they are significantly dissimilar, one metal will become anodic and corrode by galvanic cell action (Figure 13–14). This form of corrosion also occurs if external energy is applied in two immersed metals or if similar metals are electrically coupled in solutions with dissimilar chemistry.

Using techniques mentioned in our discussion of the electrochemical nature of corrosion, a *galvanic series* of metals in an electrolyte can be established. Table 13–1 is a galvanic series of metals in seawater. The farther apart two metals are in this listing, the greater the potential for corrosion when they are coupled in an electrolyte. Magnesium to steel is a very bad combination, with magnesium being the metal that will take all the attack. Monel to stainless will show negligible activity.

The other factor that controls galvanic corrosion is the relative size of the anode and cathode. If the anode is small with respect to the cathode, the attack on the anode will be great. If the situation is reversed, the attack will be low even though the difference in the galvanic series is great. As an example, an aluminum bolt in a steel plate is very susceptible to galvanic attack because there is a small anode (the aluminum bolt) and a large cathode (the plate). The reverse situation, a steel bolt in an aluminum plate, would be less susceptible to noticeable attack, because the anode is so large compared with the cathode.

Galvanic corrosion is easily controlled by not mixing metals that are immersed in an electrolyte. If a metal mix is unavoidable, the two metals should be electrically insulated so that there is no path for current flow between the two (Figure 13–15).

Stress Corrosion Cracking

Stress corrosion cracking is different from other corrosion processes in that the material deterioration is due not to significant material removal

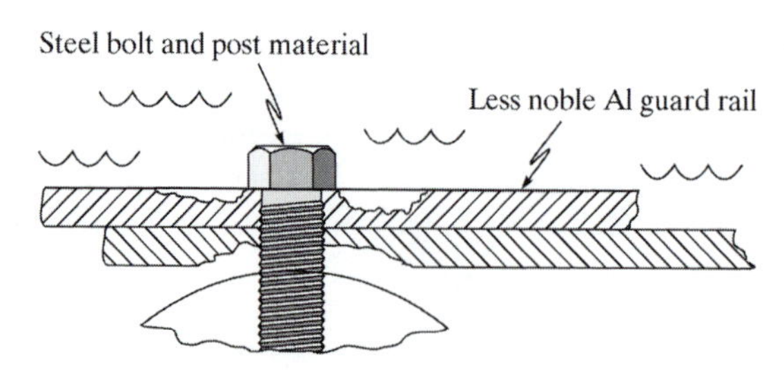

Figure 13–14
Galvanic attack of aluminum guard rail bolted to a steel post

Table 13–1
Galvanic series of some metals in seawater

↑	Magnesium and magnesium alloys
	CB75 aluminum anode alloy
	Zinc
	B605 aluminum anode alloy
	Galvanized steel or galvanized wrought iron
	Aluminum 7072 (cladding alloy)
	Aluminum 5456, 5086, 5052
	Aluminum 3003, 1100, 6061, 356
	Cadmium
Anodic	2117 aluminum rivet alloy
or least	Mild steel
noble	Wrought iron
(active)	Cast iron
	Ni-Resist
	13% chromium stainless steel, type 410 (active)
	50–50 lead tin solder
	18-8 stainless steel, type 304 (active)
	18-8 3% Mo stainless steel, type 316 (active)
	Lead
	Tin
	Muntz metal
	Manganese bronze
	Naval brass (60% copper, 39% zinc)
	Nickel (active)
	78% Ni, 13.5% Cr, 6% Fe (Inconel) (active)
	Yellow brass (65% copper, 35% zinc)
	Admiralty brass
	Aluminum bronze
	Red brass (85% copper, 15% zinc)
	Copper
	Silicon bronze
Cathodic	5% Zn, 20% Ni, 75% Cu
or most	90% Cu, 10% Ni
noble	70% Cu, 30% Ni
(passive)	88% Cu, 2% Zn, 10% Sn (composition G-bronze)
	88% Cu, 3% Zn, 6.5% Sn, 1.5% Pb (composition M-bronze)
	Nickel (passive)
	78% Ni, 13.5% Cr, 6% Fe (Inconel) (passive)
	70% Ni, 30% Cu
	18-8 stainless steel type 304 (passive)
	18-8 3% Mo stainless steel, type 316 (passive)
	Hastelloy C
	Titanium
	Graphite
	Gold
↓	Platinum

Source: F. L. LeQue, *Introduction to Corrosion,* National Association of Corrosion Engineers, Houston, TX, 1970, p. 27.

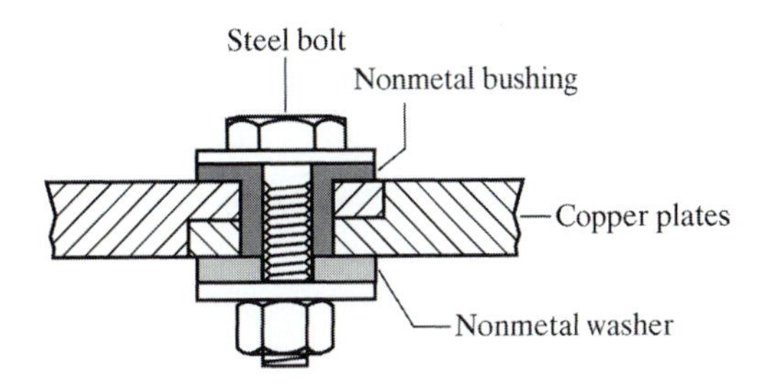

Figure 13–15
Galvanic insulation of a bolt

but to cracking (Figure 13–16). It is defined as spontaneous corrosion-induced cracking of a material under static stress, either applied or residual. It is a form of *environmentally assisted cracking* (EAC), which is cracking of a material due to the combined effect of the environment and tensile stresses. *Hydrogen embrittlement, caustic embrittlement*, and *liquid metal corrosion* are also forms of EAC.

Unfortunately, the mechanism of this phenomenon is not well understood. It occurs in many plastics, aluminum alloys, copper alloys, magnesium alloys, carbon steels, stainless steels, titanium, and other metal alloys. Stress corrosion cracking is strongly influenced by the condition of heat treatment and stress in the part. High applied or residual stresses increase tendencies for stress corrosion cracking. The corrodents and environmental conditions that produce cracking in these metal systems are different. Stainless steels are prone to cracking at elevated temperatures [122 to 140°F (50 to 60°C)] in chloride-containing solutions; steels crack in caustic solutions; aluminums crack in chloride solutions; acrylics crack in chlorinated solvents; copper alloys crack in ammonia atmospheres and sometimes even in neutral water. This corrosion is time dependent, sometimes taking months to occur.

Fortunately, this form of corrosion happens only when surfaces are under tensile stresses. Thus a common remedy is to eliminate the stress

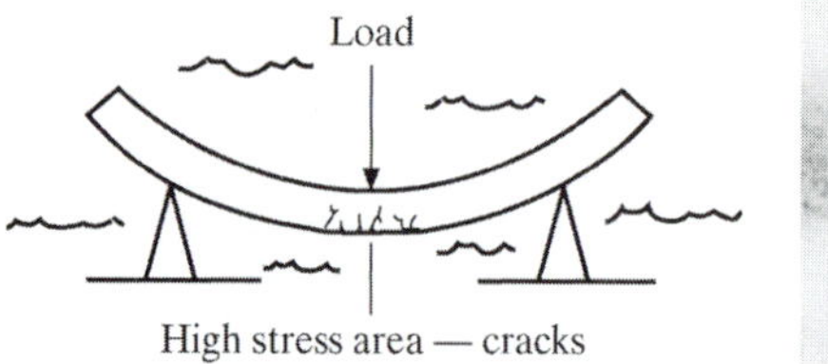

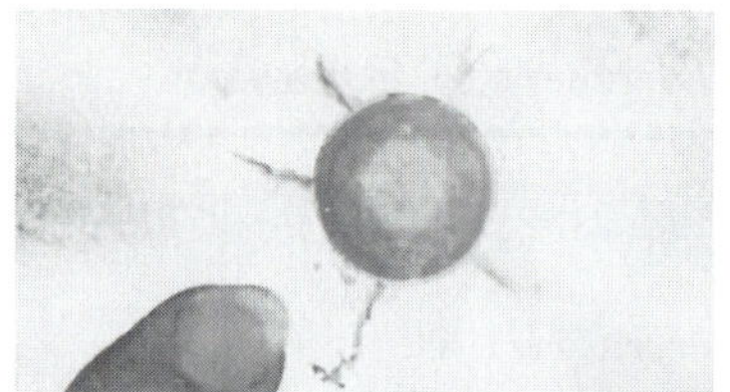

Figure 13–16
Stress corrosion cracking of stainless steel

factor. Stress relieving heat treatments are a tool to lower the stress level. Pure metals tend to be immune. The environments that cause stress corrosion cracking seem to be environments that produce low overall corrosion of the material. The most-used prevention technique for stress corrosion cracking is to consult available corrosion data and avoid the environment–material combinations that tend to cause stress corrosion cracking. Some environments that cause stress corrosion cracking of stainless steels are shown in Table 13–2.

Intergranular Attack

Intergranular corrosion occurs preferentially at grain boundaries in metals (Figure 13–17). The cause of this type of corrosions is usually alloy segregation at grain boundaries. Grain boundaries are somewhat higher-energy areas than the grains (because of atomic disarray), and in uniform corrosion they may be anodic to the grains, but not strongly so. If something is done to an alloy to make grain boundaries chemically dissimilar to the grains, intergranular attack can occur.

The most commonly encountered example of intergranular attack is *sensitization* of stainless steels. Any time that austenitic stainless steels are heated in the temperature range from about 750 to 1550°F (400 to 850°C), chromium carbides tend to form in the grain boundaries. The carbides are formed from carbon and chromium that were in solid solution in the alloy. When chromium diffuses from the matrix to form these carbides, it depletes the matrix material adjacent to the grain boundary so far that it becomes essentially a low-alloy steel. Many environments will then readily attack the chromium-depleted grain boundary regions.

The most common cause of sensitization in stainless steels is welding. When a fusion weld is made, the molten metal reaches temperatures in the area of 3000°F (1649°C) and metal adjacent to the weld deposit cycles through the sensitizing temperature range. If a stainless steel is to be subjected to media that cause intergranular attack, precautions should be taken to prevent soaking in the forbidden temperature range or modified grades should be used. Low-carbon grades are available that resist sensitization by virtue of the fact that there is not enough carbon present to form chromium carbides. Another approach is to use alloy *stabilization;* scavenger elements, such as titanium and niobium, are added to tie up carbon atoms and prevent chromium depletion.

Intergranular attack also occurs in some high-strength aluminum alloys and in some copper alloys. It is a particularly debilitating form of corrosion in that progressive cases can result in grain boundaries being corroded out through thick sections, leaving a weak, spongy metal.

Dealloying

This is a corrosion process whereby one constituent of metal alloy is preferentially removed

Table 13–2
Aqueous environments that can cause stress corrosion cracking in stainless steels

Material	Class	Cl^-, acid	Cl^-, neutral	Cl^-, oxidizing	Br^-	I^-	OH^-	F^-	S^{2-}	S^{2-}/Cl^-	$S_4O_6^{2-}$	SO_3^{2-}	SO_4^{2-}	CrO_4^{2-}	NO_3^-	NH_3	Ultrapure $H_2O + O_2$	Seawater
405	Ferritic		1				1		4					1	5	1	1	x
18-2		4	1	4			3		1	1						1		x
26-1		4	1	4			4		2				5	1	1	1		x
26-1S		4	1	4			4		2				5	1	1	1		x
29-4		1	1	1			4		2					1	1	1		4
430			1						1	1				1	1	1		x
434			1							3								
431	Martensitic, quenched and tempered or precipitation hardened	5	5	5			4	2	5				2	1	1	1		x
410		5	2		5	5	2	5	5	5				1	2	1	2	x
Ca-6NM			5						5	2								x
440 A, B, C		5	5	5					5									x
PH15-7Mo		5	4					1	5				5	1	1	1		x
17-4PH		5	4					1	5	5			5	1	1	1		x
PH13-8Mo		5	4						5									x
17-7PH		5	5					1	5									x
Custom 450		4	1	4			4	2	2	2			2	1	1	1		x
Custom 455		4	1	4			4	2	2				2	1	1	1		x
202	Austenitic	5	5	5		1	4	1	2		3	3	1	1	1	1		x
216		5	5	5		1	4	1	2		3	3	1	1	1	1		x
216L		5	5	5		1	4	1	2		3, 4	3, 4	1	1	1	1		x
Nitronic 50		5	4	5		1	4	1	2		3	3	4	1	1	1	1	4
Nitronic 60		5	5	5		1	4	1	2		3	3	1	1	1	1		x
304		5	5	5		1	4		2	5	3	3	4	1	1	1	3	x
304L		5	5	5		1	4	1	2	5	3, 4	3, 4	4	1	1	1	3, 4	x
309S		5	5	5		1	4	1	2		3	3	1	1	1	1	3	x
310S		5	5	5		1	4	1	2		3	3	1	1	1	1	3	x
316		5	5	5	4	1	4	1	2	5	3	3	1	1	1	1	3	x
316L		5	5	5			4		2		3, 4	3, 4	1	1	1	1	3, 4	x
317		5	5	5		1	4	1	2		3	3	1	1	1	1	3	x
317L		5	5	5		1	4	1	2		3, 4	3, 4	1	1	1	1	3	x
321		5	5	5		1	4	1	2	5	1	1	1	1	1	1	1	x
347		5	5	5		1	4	1	2	5	1	1	4	1	1	1	1	x
329	Duplex	4	5	5		2	4	2					1	1	1	1		x
3RE60		4	1	5			3		4	4			1	1	1	1	3	4
18-18-2		4	4	5		1	4	1					1	1	1	1		x

Code: 1 Resistant
2 Resistant unless cold-worked or hardened
3 Resistant unless sensitized
4 Resistant except at high temperatures and concentrations
5 Nonresistant
x Not recommended for this environment

Source: *Chemical Engineering,* reprinted with permission.

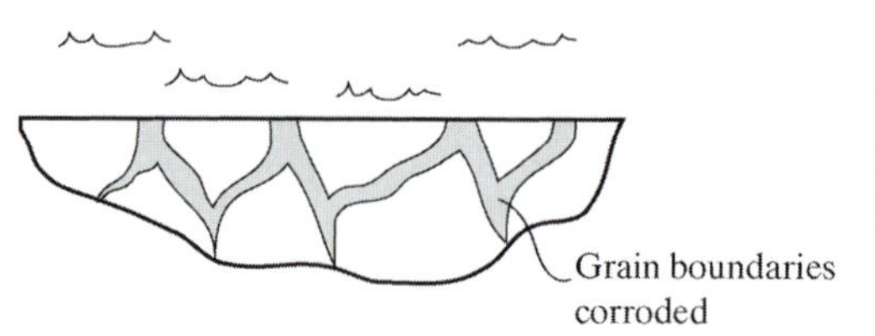

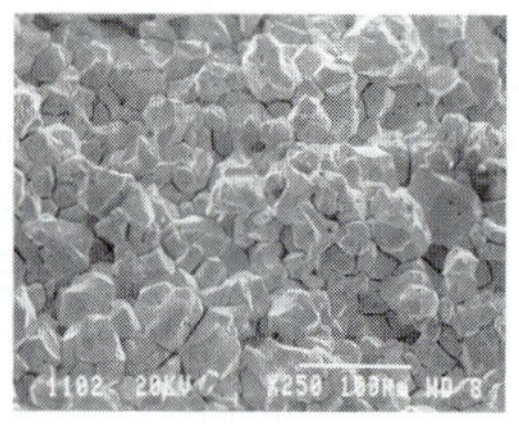

Figure 13–17
Scanning electron photomicrograph of intergranular corrosion on the surface of sensitized austenitic stainless steel. Improper annealing caused the sensitization. Note the gaps between the grain boundaries. The corrodent dissolved the material adjacent to the grain boundaries allowing the grains to fall out of the surface. ASTM A 262 describes susceptibility tests (×250 original magnification).

from the alloy, leaving an altered residual microstructure (Figure 13–18). A number of alloy systems are susceptible to this process, but the most commonly encountered is yellow brass. The removal of zinc from brasses is called *dezincification*. The net damage is usually a mechanical failure, because the metal remaining after *dealloying* is usually weak and spongy, an effect similar to intergranular attack.

One theory as to the mechanism of dezincification is that the brass dissolves and the copper in the brass is spontaneously replated back into the surface without the zinc. Zinc metal ions go into solution and are carried away.

Other examples of dealloying are selective leaching of aluminum from aluminum bronzes, selective leaching of nickel from cupronickels, and *graphitization*, the dissolution of iron from gray cast irons leaving only the graphite. The aluminum bronze problem is solved by adding about 4% nickel to the alloy; nickel dealloying can be controlled by control of environmental conditions and solution velocity, but graphitization is extremely difficult to control. It is probably the most costly form of dealloying. Most large-diameter underground water and fire mains are made from gray iron, and the intended service life is from 30 to 100 years. When one of these pipes fails under a highway, repair costs are seldom less than $10,000. Most corrosion failures of buried cast-iron pipes are caused by graphitization. Graphitization occurs in corrosive soils, cinder fill, and soils contaminated with industrial wastes. The material solution to

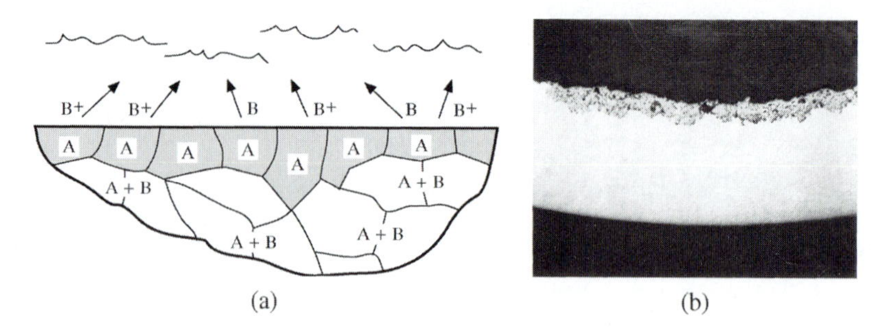

Figure 13–18
(a) Dealloying in the form of dezincification of a brass part; zinc has been removed from the surface, which leaves weak copper, (b) (×40)

graphitization is to use austenitic nickel alloy cast-iron pipes, but this solution is usually too expensive for the typical municipal water department. It is current practice still to use gray-cast-iron pipes for water and fire mains; but in new installations and in failure areas, the corrosivity of the soil is measured, and if found to be potentially corrosive, the pipes are wrapped with polyethylene film before installation.

The general solution to dealloying is to avoid the environments that cause this form of attack or to avoid susceptible alloys. Dezincification can be eliminated by not using yellow brasses in hot-water systems. Copper-based alloys with at least 85% copper will be immune. Poly wrapping of cast-iron pipes has been practiced for at least 20 years, and to date it appears effective.

Erosion

In Chapter 3 we defined the major types of wear. Most of the types of wear that we listed under the category of erosion are also considered forms of corrosion:

1. Liquid impact
2. Liquid erosion
3. Slurry erosion
4. Cavitation erosion

There has been an argument about jurisdiction over these types of material deterioration for some time. The tribologists claim that they are wear processes; the corrosion engineers claim that they are corrosion processes. We shall repeat their description in this chapter to placate both groups.

Liquid Impingement This is material removal due to the action of an impinging stream or droplets of fluid. It does not necessarily have to be a liquid; steam or similar "gases" can cause this effect. The mechanism of attack in some metals is removal of protective films (passive films), which leads to accelerated corrosion. The effect can be combined with material removal by the mechanical action of the impinging stream per se. This type of attack is prevented by designing baffles or similar devices to prevent the *impingement* action.

Liquid Erosion This form of corrosion is similar to impingement except that the fluid flow is parallel to the surface. The mechanism in metals is removal of films or metal by mechanical action plus corrosion of the active metal. Liquid erosion is often characterized by grooves or waves in the metal surface. The best remedy for liquid erosion is lowering of the fluid velocity. Most metals have a fluid velocity tolerance level, so it is possible to minimize liquid erosion by using metals that can tolerate high fluid velocities, such as 316 stainless steel, titanium, and some aluminum bronzes.

Slurry Erosion This is material removal due to the combined action of corrosion and wear. The source of the wear and an accelerator of the corrosion is abrasive particles dispersed in a slurry.

The mechanism of this type of erosion is the same as in impingement and liquid erosion, with the exception that the presence of abrasive particles makes the removal of metal and protective films even more energetic. There are very few metals with protective films strong enough to resist the erosive action of slurries of silica, alumina, and similar hard compounds.

The most prevalent systems for handling abrasive slurries are ceramics and elastomers. Ceramics are often immune to chemical attack, and they can be hard enough to resist the action of abrasive compounds. Elastomers deform to resist the cutting action of sharp abrasive particles and thus they usually have good abrasion resistance. Ceramic- or elastomer-lined pumps and piping are commercially available. Plastic piping may work for some slurries. Corrosion is not a factor but the plastics may erode from the mechanical action of particles.

Cavitation *Cavitation* is material removal by the action of imploding bubbles.

Materials that resist cavitation have tenacious passive films, high hardness, and high strength. Unfortunately, not many metals have these characteristics. Titanium and corrosion-resistant cobalt-based alloys appear to have the best cavitation resistance in a wide range of environments.

Summary

We have devoted a considerable amount of discussion to describing a number of types of corrosion, and there are still some types that have not been touched. The ones covered, however, are by far the most prevalent forms. For brevity, we shall only define those that were not covered before:

Caustic embrittlement: Embrittlement of a metal by an alkaline environment (a form of stress corrosion cracking).

Corrosion fatigue: Reduction of fatigue strength by a corrosive environment.

Underdeposit attack: Corrosion under or around a localized deposit on a metal surface (a form of crevice corrosion).

Exfoliation: Scaling off of a surface in flakes or layers as the result of corrosion.

Filiform corrosion: Corrosion that occurs under organic coatings on metals as fine, wavy hairlines (Figure 13–19).

Stray current corrosion: A form of attack caused by electrical currents going through unintentional paths.

Microbiological corrosion: Corrosion promoted by the presence and growth of living organisms such as algae, fungi, and bacteria. This form of corrosion is most common in soils and in cooling water systems where there is likelihood of stagnant water. Biocides are the usual remedy.

We emphasize the types of corrosion because the designer must be aware of the types of situations that can lead to corrosion. He or she must recognize that crevices can cause an abnormal corrosion rate in feed water, that slight concentrations of chlorine can cause stress corrosion cracking in stainless steel, and that coupling widely dissimilar metals can cause galvanic attack. If the designer knows about these things, he or she can often design around them. Second, when corrosion failures occur in service, a knowledge of the corrosion type and mechanism is essential for a decision on remedial action.

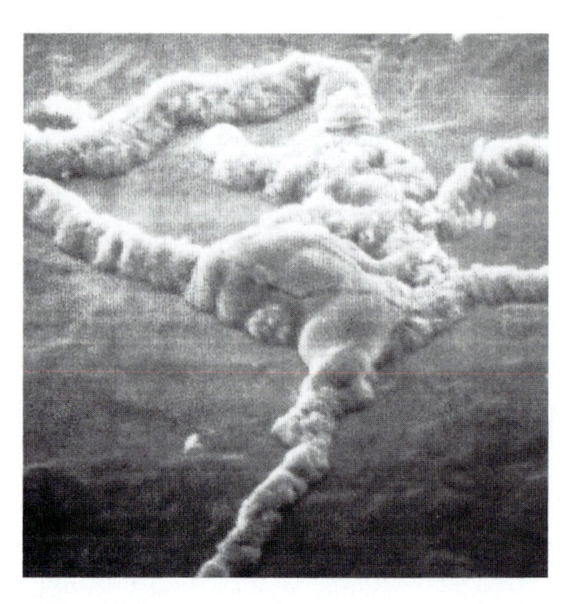

Figure 13–19
Filiform corrosion, a form of crevice corrosion, on chromium-coated steel. The attack is under the coating, and it progresses as filaments (scanning electron micrograph, ×1000).

13.4 Determination of Corrosion Characteristics

It is not necessary to run a corrosion test every time a new environment or construction material is encountered. Extensive data exist on the corrosion characteristics of all kinds of materials in all kinds of environments. A typical tabulation of corrosion data is shown in Figure 13–20. Similar data are now available from

Figure 13–20
Typical corrosion data

Source: Courtesy of the National Association of Corrosion Engineers

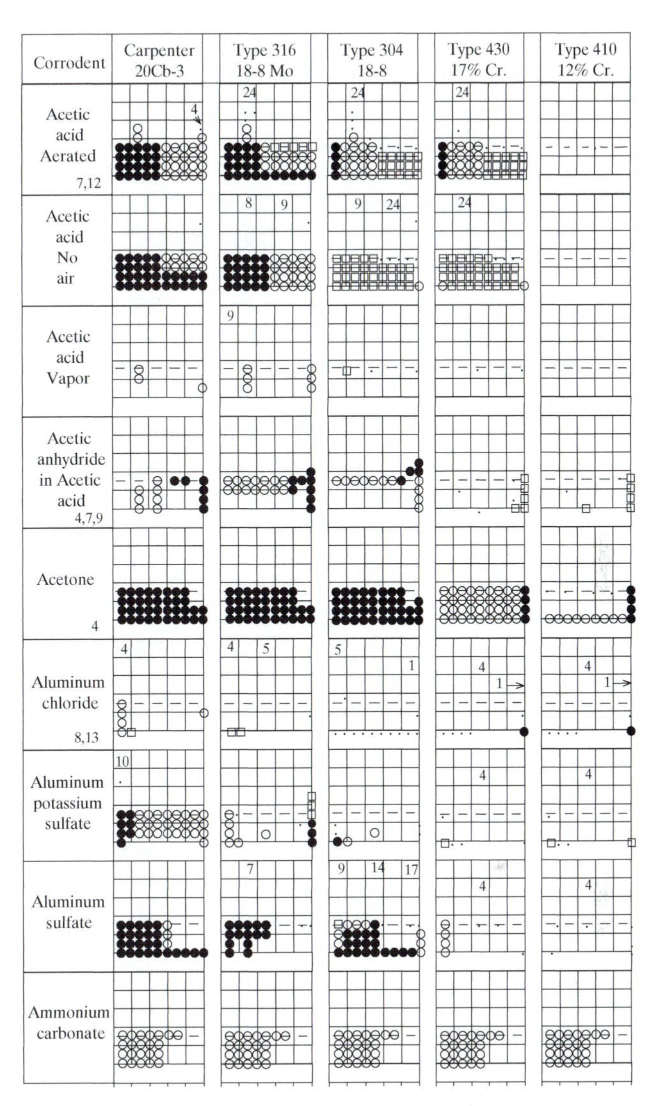

CODE
● Corrosion Rate less than 0.002" per year
○ Corrosion Rate less than 0.020" per year
□ Corrosion Rate from 0.020" to 0.050" per year
× Corrosion Rate greater than 0.050" per year

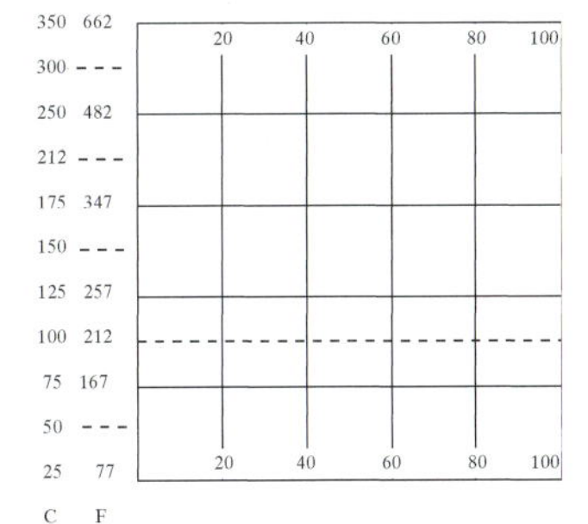

computer databases. To use these data, the designer establishes the chemical nature of the corrodent and its concentration, temperature, and other pertinent information, such as aeration and state of stress. The reaction of various metals to this specific environment can then be obtained from the corrosion data survey. Take the example of acetic acid; the dark circles show low corrosion rates. We can see that the metals listed have only low corrosion rates under limited operating conditions. It is typical to present corrosion rates in units of surface removal or penetration per unit time. The units can be inches per year (*ipy*), mils per year (mpy), millimeters per year (mm/yr), micrometers per year (μm/yr), or milligrams per square decimeter per day (mdd). The units of penetration per unit time can be particularly useful in that they can be used to determine the useful service life of a piece of equipment. If it is determined that a vessel will have a corrosion rate of 1 mm/yr, the wall thickness is 10 mm, and the minimum wall thickness required by operating stresses may be 5 mm, this means the life of the vessel will be 5 yr. It is common practice to measure vessel-wall thickness each year with ultrasonic thickness gauges, and thus the corrosion rate prediction can be double-checked.

If existing corrosion data do not apply to a particular material and environment, laboratory corrosion tests can be justified. Much of the existing data were obtained by relatively simple immersion tests, as shown in Figure 13–21 and Figure 13–22. The simple immersion in a beaker test is the most widely used test for uniform and pitting corrosion. Susceptibility to galvanic attack can be measured by electronic devices that measure corrosion potential and current. An easy test for susceptibility to crevice corrosion is to put a rubber band around a sample and immerse it. The rubber band simulates a crevice. There are many tests for stress corrosion; any technique for imposing a tensile stress in the sample can be used. The final test shown in Figure 13–21 is a technique for determining the effect of fluid velocity on corrosion. Abrasive particles may be introduced to the fluid to measure slurry erosion behavior.

In addition to these laboratory tests, there are many techniques and devices for performing in situ corrosion tests. Clever devices exist to electronically measure corrosion rates in corrosive production environments. One such device is a wire probe that is inserted into a vessel or pipeline. As the wire corrodes, its resistance increases. Periodic checks with a resistance-checking device yield a corrosion rate. The wire is made from the material of interest. Electronic devices measure pitting and wall thinning. Thus a host of techniques and devices can be used to monitor corrosion characteristics. A limitation of many of these devices and laboratory tests is that it may take weeks, months, or even years to get the desired corrosion rate data. This is especially true of stainless steels and other corrosion-resistant materials. A popular technique currently used by corrosion engineers is polarization studies. This technique usually yields corrosion data in a matter of hours.

Because most corrosion is electrochemical in nature, electrical techniques can be used to monitor corrosion. Test samples can be made to be the anode in a special electrochemical cell, and a corrosion "fingerprint" of a particular metal in a particular environment can be obtained by electronically monitoring corrosion current as the cell potential is varied. These tests are called *polarization studies*. Figure 13–23 is an illustration of a polarization setup and the types of data that are generated in anodic polarization. These techniques are described in detail in ASTM G 5 and G 61 standard test methods, but essentially they involve making a small pin from candidate materials. The pin is placed in a special holder and put in a glass cell containing the

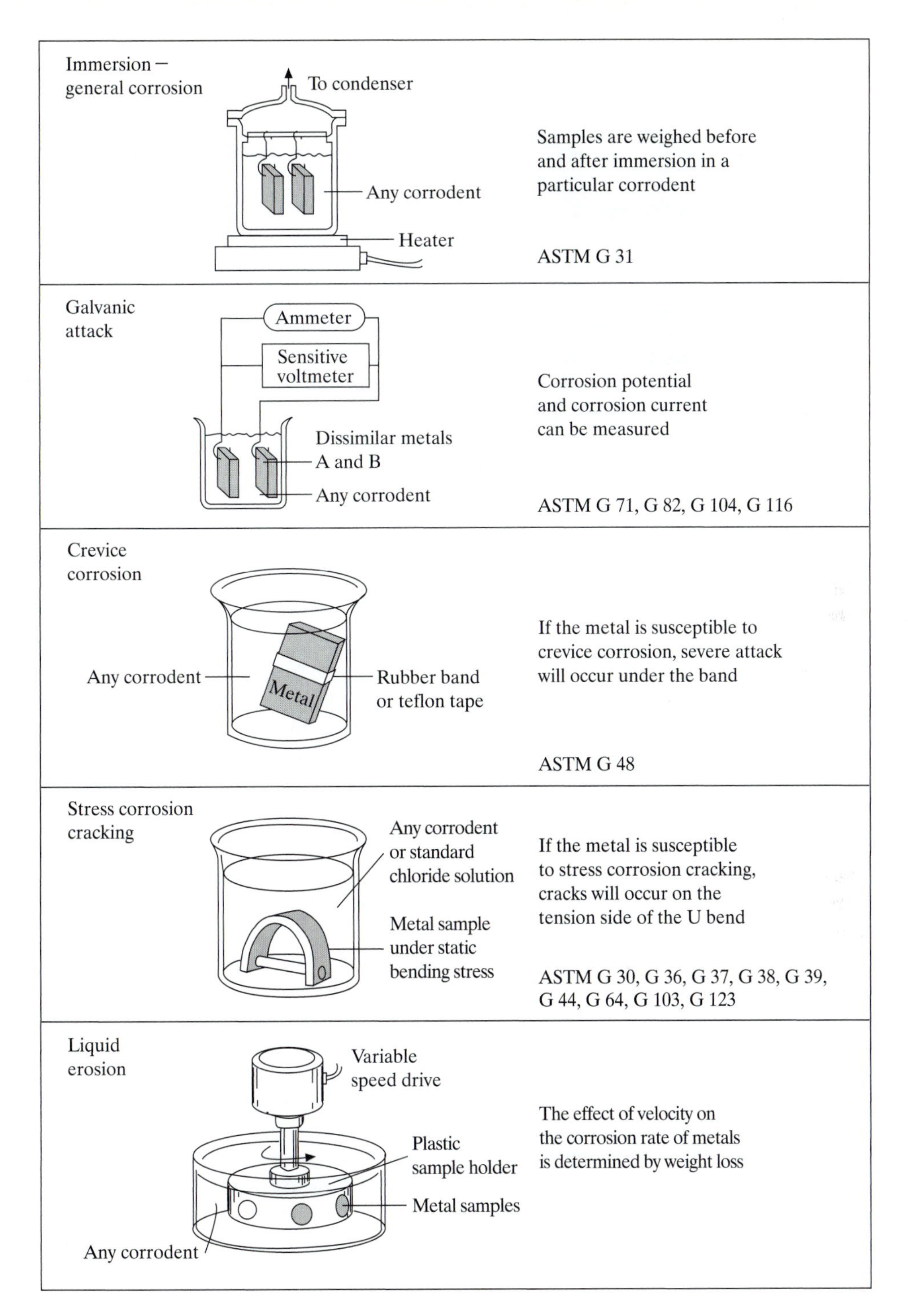

Figure 13–21
Laboratory corrosion-testing techniques

Figure 13–22
Immersion corrosion tests

solution of interest. Two other electrodes are inserted into the cell, and the electrodes are connected to the necessary power supplies and instrumentation. Potentiostats are commercially available that are computer controlled, and the corrosion experiment can be programmed into the computer. It will perform the experiment in about an hour, plot the data, and make calculations on such things as corrosion rate and corrosion potential. Determination of corrosion rates by simple immersion tests typically takes 30 days at minimum.

Polarization essentially drives the oxidation or reduction reaction of the metal specimen through an applied potential relative to a reference electrode. Two common reference electrodes are saturated calomel (SCE) and platinum–hydrogen gas. The specimen is said to be polarized if the potential of the specimen is moved from the corrosion potential through the application of current. If the potential is electrode positive to the corrosion potential, the specimen is being anodically polarized and the oxidation reaction is occurring on the specimen surface. Application of a current that makes the potential electrode negative to the corrosion potential cathodically polarizes the surface of the specimen, thereby driving the reduction reaction; the specimen is protected. The linear portions of these curves may be extrapolated to determine the corrosion current, which can in turn be converted to a corrosion rate [Figure 13–23(i)].

Two main types of techniques are used to study corrosion characteristics: (1) The corrosion current is varied over several decades and the potential of the sample with respect to a reference electrode is continuously recorded; this is *potentiostatic polarization*. (2) The potential with respect to the reference electrode is varied over a predetermined range and the corrosion current is continuously recorded; this is *potentiodynamic polarization*. Both techniques produce curves that can be interpreted to deduce the types of corrosion factors listed in Figure 13–23.

The important point of this discussion is that the material user should use published data on corrosion characteristics to aid material selection; but if appropriate data are not available, which is often the case, sophisticated laboratory techniques can be used to screen candidate materials and to develop pertinent corrosion rate information. Electrochemical polarization techniques can be used to determine corrosion rates of a metal in a particular solution, the effects of temperature on corrosion processes, the kinetics or mechanism of a corrosion process, passivity of a metal in a solution, and even the electrical parameters that must be used to cathodically protect a metal from corrosion in an environment. Polarization is one of the state-of-the-art techniques for dealing with corrosion processes.

13.5 Corrosion Control

The use of existing corrosion data is the first step in solving corrosion problems or preventing potential corrosion problems. To simplify the subject of corrosion control, there are three factors that have about equal weight in importance: material selection, environment control, and

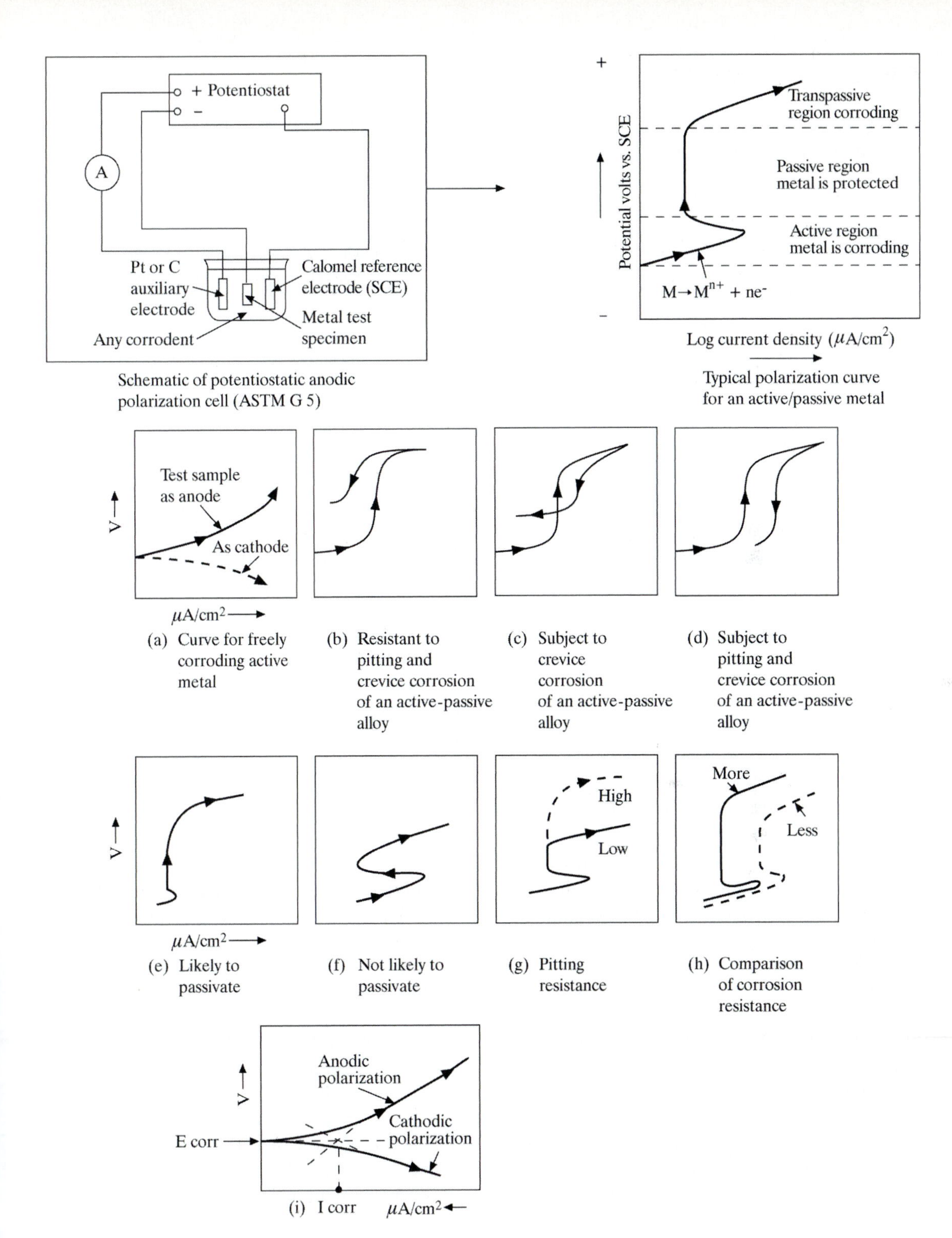

Figure 13–23
Schematic of a potentiostatic polarization setup (ASTM G 5). Figures (a) to (i) are examples of the shapes of curves that are obtained and their interpretation.

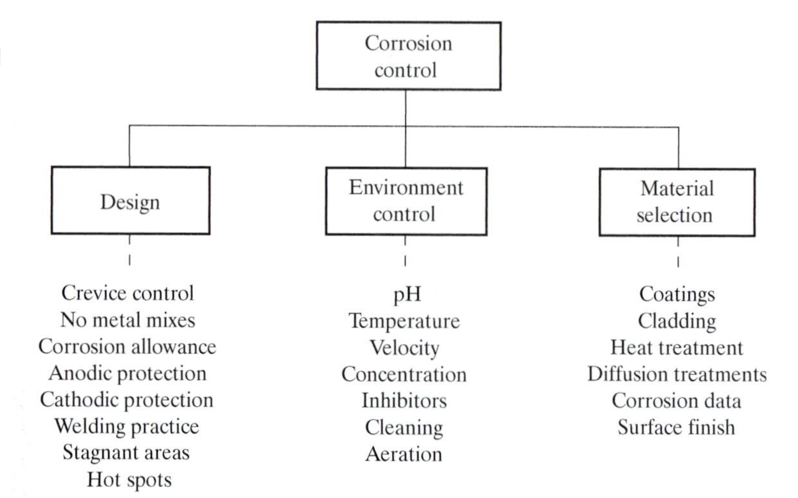

Figure 13–24
Techniques for corrosion control

design. Some specific corrosion factors under these headings are shown in Figure 13–24.

In this section we shall discuss corrosion control principles applicable to many metal systems. Some corrosion information on specific metals is given in the chapter covering specific materials.

Material Selection

If after consulting corrosion data it has been determined that no economically justifiable metal will work, surface treatments can be considered. The simplest coating for corrosion control is a conversion coating. These are simple immersion treatments, and they most often apply only to atmospheric corrosion. They are not effective in chemicals. The same thing can be said of paints, even though paint manufacturers would dispute this statement. Effective corrosion control with polymer coatings requires that they be thick ($\approx$1 mm), pinhole free, impermeable, and inert to the desired environment. This is also true of nonmetal, nonplastic coatings such as ceramics and glasses. Platings are used for corrosion control, but this is very risky. Metal platings are never crack free or pinhole free. They usually are adequate for atmospheric corrosion, but caution is required in their use in chemicals. Metal coatings that are anodic to the substrate, such as zinc galvanizing, are very effective for control of atmospheric corrosion because they are not sensitive to pinholes.

Sometimes diffusion treatments can be used to provide corrosion resistance. It is well known that nitriding will help prevent atmospheric rusting of steels. Chromizing is effective in improving the corrosion resistance of steels. On the other hand, nitriding of stainless steels lowers their chemical resistance.

There are instances when heat treatments can help corrosion resistance. When stainless steels are sensitized by welding or similar operations, they can be returned to their unsensitized condition by a solution anneal and water quench. The heterogeneity of castings can also have an effect on corrosion resistance. Some cast alloys should be annealed or normalized before using them in corrosive environments.

One final item concerning materials: Surface finish should be controlled. There is little documentation on the effect of roughness on corrosion rates, but experience has shown that the money spent in grinding tank walls to a roughness level of 0.5 to 1 μm Ra may be well worth it. A low-roughness surface greatly aids

Figure 13–25
continued

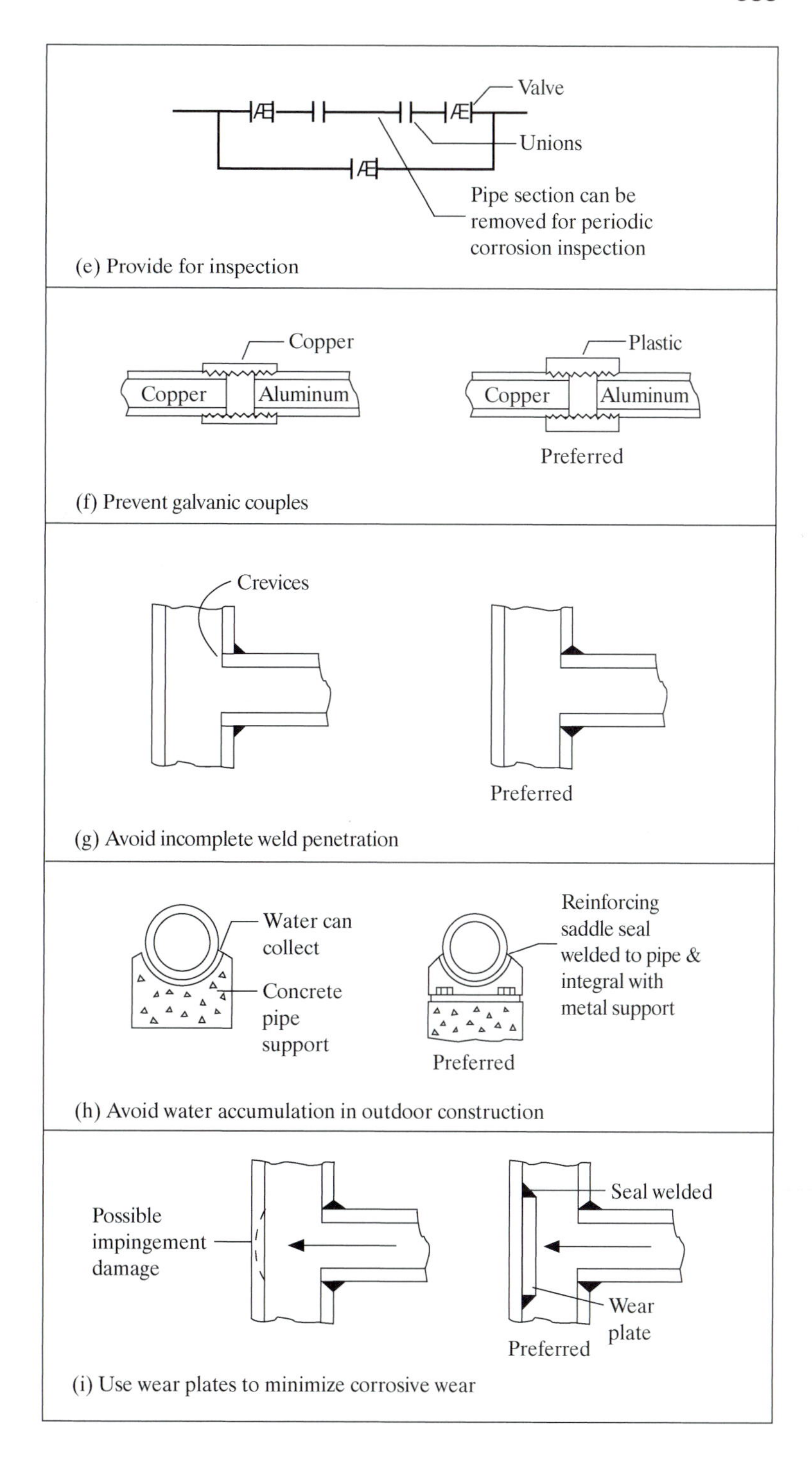

is a solution. If a gasket sticks out from a mechanical joint, the gasket itself can form a crevice.

Figure 13–25(c) shows some design pitfalls that affect cleanability of tanks. If a tank cannot completely empty, the residue corrodent can concentrate and produce abnormal corrosion effects.

Figure 13–25(d) shows another way that corrodents can concentrate. If fill tubes are above the liquid level, the corrodent can dribble down the side of the vessel, evaporate, and form residues or precipitate salts. The solution is to put the fill tube into the liquid. Similarly, splashing on the hot walls of steam-heating coils can cause corrodent evaporation and concentration. The solution is to fully immerse the coil.

A useful technique for inspecting for corrosion and microbiological activity is shown in Figure 13–25(e). Without shutting down a process, a short length of pipe can be removed by turning two valves and disconnecting two unions. Corrosion monitoring is a part of corrosion control.

Galvanic couples can be prevented by using insulating materials, as shown in Figure 13–25(f).

Incomplete welds, as shown in Figure 13–25(g) can cause crevice corrosion, concentration of electrolytes, and differential aeration cells; they are generally poor practice in systems involving corrosive environments. Weldments should also be designed so that it is easy to finish grind welds. Welding backup rings that do not completely fuse in should be avoided in pipe welds.

Care should be taken in pipe and vessel supports to prevent recesses where solutions can collect. Procedures similar to those shown in Figure 13–25(h) should also be used on vessel supports.

Figure 13–25(i) shows that design procedures can be used to prevent impingement effects from fluids. Similar techniques can be used on slurries. The best design, however, would provide a fluid path that does not involve any impingement.

The examples cited are but a few of the design techniques that can be used to control corrosion. Good design practices may involve some extra construction costs, but compared to vessel or pipeline replacement costs, these monies are well spent.

One final method of corrosion control is making electrochemical cell action work in your favor. We have shown that coupling an active metal with a more noble metal will result in current flow and corrosion of the active metal. The principle of galvanic corrosion can be used to provide *cathodic protection*. As shown in Figure 13–26, if a buried steel pipe is connected with an insulated copper conductor with buried anodes (large ingots) of magnesium or another active metal, the anode metal will corrode in deference to the pipe. An impressed direct current between an inert electrode and the pipe can be used as an alternative to sacrificial anodes.

The same technique is used in vessels such as home water heaters. There are even special magnesium drain plugs for automobile engine crankcases that provide cathodic protection.

A modification of this technique is to fasten an active metal to the component to be protected [Figure 13–26(c)]. Zinc anode plates are available for steel boat hulls, and zinc collars are available that can be put on boat propeller shafts to take the bulk of corrosion and prevent galvanic attack between the more noble metal mixes (the bronze shaft and steel hull).

A newer and more sophisticated technique is *anodic protection* [Figure 13–26(d)]. In anodic protection, the potential of the object to be protected is controlled so that the metal stays passive. These systems are relatively complicated and require laboratory tests to establish the active–passive behavior of the system of interest, but these systems are used on vessels and chemical process equipment. There is significant potential for future use of this technique in chemical process industries.

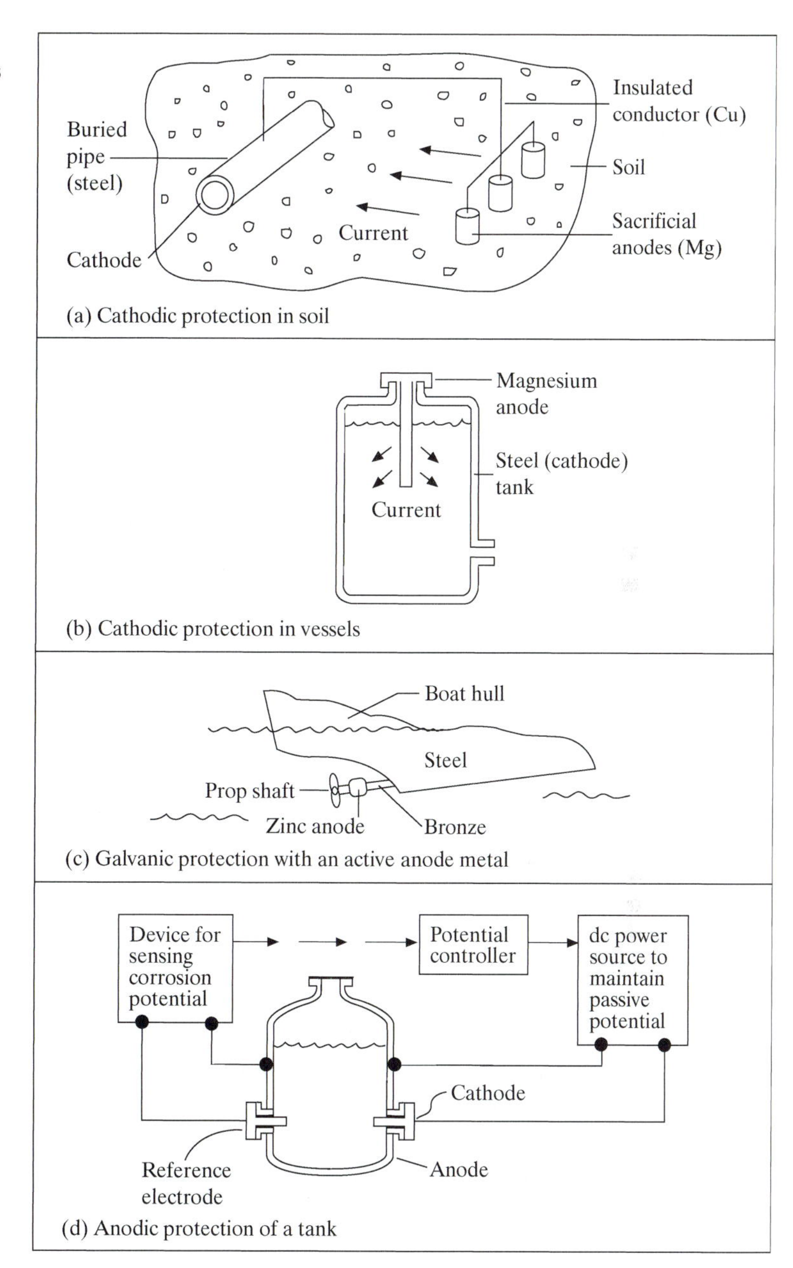

(a) Cathodic protection in soil

(b) Cathodic protection in vessels

(c) Galvanic protection with an active anode metal

(d) Anodic protection of a tank

Figure 13–26
Corrosion-protection techniques

Summary

The bulk of this chapter has been concerned with corrosion theory and the things that affect corrosion. This is intentional: It is not possible to discuss in a single chapter the response of various materials to various environments. However, if the fundamentals of corrosion are understood, existing corrosion data can be used to solve many potential corrosion problems. We have drawn attention to the importance of knowing your environment and material characteristics. Too often, corrosion failures occur because someone neglected to calculate fluid velocity or used the wrong heat treatment. Finally, our discussion of designing for corrosion was intended to point out some of the improper design practices that have caused innumerable corrosion failures in industry. There are far too many repeats. Thoughtful design is essential to corrosion control. Some important corrosion concepts to keep in mind follow:

- Corrosion of metals is usually electrochemical in nature; it requires an anode, a cathode, and an electrolyte.
- Plastics and ceramics are immune to many forms of corrosion because they are not good conductors of electricity (electrochemical effects do not apply).
- The corroded member in a corrosion cell is the anode.
- The severity of a galvanic metal couple depends on the anode-to-cathode size ratio and the relative activity of the coupled metals.
- Passivity is a prerequisite for the corrosion protection of many metals.
- Contaminants must be removed (by pickling, grinding, and the like) if passive films are to be continuous.
- Some materials are good in oxidizing media; some are good in reducing media. Know your environment.
- Probably the most common type of corrosion is crevice corrosion; it is preventable by eliminating stagnant areas.
- Corrosion of a material in an environment is not intuitive; corrosion data must be consulted.
- When significant data cannot be found on a material–environment couple, corrosion testing should be done.
- The corrosion control technique with the biggest payback is usually proper design.
- Many plastics are attacked by sunlight.
- Galvanic couples are not likely to cause corrosion damage in 50% relative humidity (RH) room air. The problem occurs when there is a good electrolyte present (immersed).
- Noble metal and plastic coatings are ineffective for corrosion control unless they are pinhole free.
- Metal coatings that are anodic to the substrate (zinc or steel) can protect even if there are breaks in the coating.

Critical Concepts

- Corrosion of metals is usually an electrochemical reaction.
- Environmental degradation of plastics is usually produced by molecular reactions (e.g., chain scission) produced by contact with a particular environment (e.g., sunlight).
- Solving corrosion problems starts by identification of the type of corrosion that has occurred or might occur.
- Corrosion tests must simulate process conditions and solutions to be valid.

Terms You Should Remember

corrosion
free energy
electrochemical
corrosion
  potential
redox
electrolyte
Faraday
equilibrium
passivity
polarization
oxidation
reduction
corrodent
aeration
direct attack
uniform attack
pitting
crevice corrosion
galvanic corrosion

stress corrosion cracking	underdeposit attack
intergranular corrosion	exfoliation
dealloying	ipy
cavitation	sensitization
impingement	dezincification
corrosion fatigue	inhibitors
	cathodic protection
	anodic protection

Case History

Pitting Corrosion of Stainless Steel Springs

Small garter springs are used in x-ray developing machines to clamp pairs of rollers together. The rollers are used to feed x-ray film through the processing machine. In a newly designed machine, the garter springs were fabricated from spring-temper Type 316 stainless steel. A short time after customers started using the new machines, garter springs in the washer/dryer section failed by pitting corrosion after only 1 week (springs would break due to deep pits). Apparently, chlorides from the potable water used in the washer section dried on the springs. The combination of concentrated chloride environment, crevices created by the spring coils, and warm temperatures (60° C) caused pitting corrosion in the springs. The failure mode was confirmed by scanning electron microscopy and energy dispersive spectroscopy.

Corrosion tests suggested that an alloy with a PREN of at least 35 was required for the application (see Section 13.6). UNS 31600 stainless steel, which is sensitive to chloride pitting, has a PREN of 22. After review of alloys with suitable pitting corrosion resistance and high tensile strength (springs require high tensile strength) a cobalt–chromium–nickel alloy called Elgiloy® (UNS R30003) was selected. Since switching to Elgiloy, there have been no pitting corrosion failures of garter springs. The final specification:

> Elgiloy UNS R30003 per AMS 5833B, after winding stress relieved 1 h at 385 ±5°C and pickle per ASTM A 380 Code D.

Questions

Section 13.1

1. Write the chemical equation for dissolution of zinc in sulfuric acid.
2. What is the electrolyte in high-temperature oxidation?
3. Calculate the free energy for zinc in hydrochloric acid, where $Zn + 2H^+ \leftrightarrow Zn^{2+} + H_2$. Does the zinc corrode?
4. If iron is freely corroding in a hydrochloric acid solution, what effect would the addition of a salt that inhibits the cathodic partial process have?
5. Calculate the free energy change when zinc is immersed in ferric chloride. Is the reaction spontaneous?
6. Why do bubbles form on the surface when steel is immersed in 80% hydrochloric acid?

Section 13.2

7. What is the significance of an aqueous environment with a pH of 4 on carbon steel?
8. When should passivation be specified?
9. Name a reducing environment and a material that can resist attack in such an environment.
10. Name an oxidizing environment and a material that is resistant to that environment.

Section 13.3

11. What is the galvanic effect of fastening aluminum siding to a house using steel nails?
12. What causes stress corrosion cracking in stainless steels?
13. What is the most common form of dealloying in gray cast iron?
14. How is sensitization prevented in stainless steels?
15. What type of corrosion is dezincification?
16. How can stress corrosion cracking be prevented in copper alloys?
17. What is the critical amplitude for fretting corrosion?
18. What is a common cause of intergranular corrosion?

Section 13.4

19. What is the corrosion rate of 304 stainless steel in 30% aerated acetic acid at 350°F (176°C)?
20. What is the significance of corrosion current?
21. How does soil resistivity affect corrosion of buried pipes?

Section 13.5

22. Steel corrodes in brackish water at the rate of 10 mpy. What should be the corrosion allowance to get 10 years' service from a steel vessel in this environment?
23. How can corrosion of carbon steel be mitigated in a closed water cooling system?
24. What is the minimum thickness of plating required to make steel resistant to continuous immersion in freshwater?
25. What is the best anode material for cathodic protection of a buried cast-iron pipe?
26. How can you monitor corrosion in a piping system?
27. How can the rusting of carbon steel be prevented in overseas shipments?

To Dig Deeper

ASTM Standard G15, Terminology Relating to Corrosion and Corrosion Testing. West Conshohocken, PA: American Society for Testing and Materials, 1997.

Babelon, R., Ed. *Corrosion Tests and Standards: Application and Interpretation*. Philadelphia: American Society for Testing and Materials, 1995.

Chemical Compatibility, Vol. 1, Book A; Vol. 2, Book B. New York: Plastics Design Library, 1990.

*Cor*Sur Users Guide,* Vol. 1, Ver. 1. Houston, TX: National Association of Corrosion Engineers, 1986 (computer database).

Craig, B. D., Ed. *Handbook of Corrosion Data*. Metals Park, OH: ASM International, 1989.

Evans, Ulick R. *An Introduction to Metallic Corrosion*. Metals Park, OH: American Society for Metals, 1981.

Fontana, Mars C., and Norbert D. Green. *Corrosion Engineering*. New York: McGraw-Hill Book Co., 1978.

Hammer, Norman E. *Corrosion Data Survey*. Houston, TX: National Association of Corrosion Engineers, 1981.

Hayes, G. S., and R. Babelon, Eds. *Laboratory Corrosion Tests and Standards,* STP 866. Philadelphia: American Society for Testing and Materials, 1985.

Kane, R. D. *Environmentally Assisted Cracking: Predictive Methods for Risk Assessment,* STP 1401. West Conshohocken, PA: ASTM International, 2000.

Lai, G. Y. *High Temperature Corrosion of Engineering Alloys*. Metals Park, OH: ASM International, 1990.

Metals Handbook, 9th ed., Volume 13, Corrosion. Metals Park, OH: ASM International, 1988.

McIntyre, D. *Forms of Corrosion: Recognition and Prevention*. Houston, TX: NACE International 1997.

Piron, D. L. *The Electrochemistry of Corrosion*. Houston, TX: National Association of Corrosion Engineers, 1991.

Renzo, D. J. *Corrosion Resistant Materials Handbook,* 4th ed. Park Ridge, NJ: Noyes Data Corp., 1985.

Schweitzer, P. A. *Corrosion Engineering Handbook*. New York: Marcel Dekker, Inc., 1996.

Seymour, R. B. *Plastics vs. Corrosives*. New York: John Wiley & Sons, 1982.

Talbot, D., and J. Talbot. *Corrosion Science and Technology*. Boca Raton, FL: CRC Press LLC, 1999.

Townsend, H. E. *Outdoor Atmospheric Corrosion,* STP 1421. West Conshohocken, PA: ASTM International, 2002.

Uhlig, Herbert H. *Corrosion and Corrosion Control,* 3rd ed. New York: John Wiley & Sons, 1985.

Von Baeckmann, R., et al., Eds. *Handbook of Cathodic Corrosion Protection: Theory and Practice of Electrochemical Processes*. Houston, TX: Gulf Publishing Co., 1997.

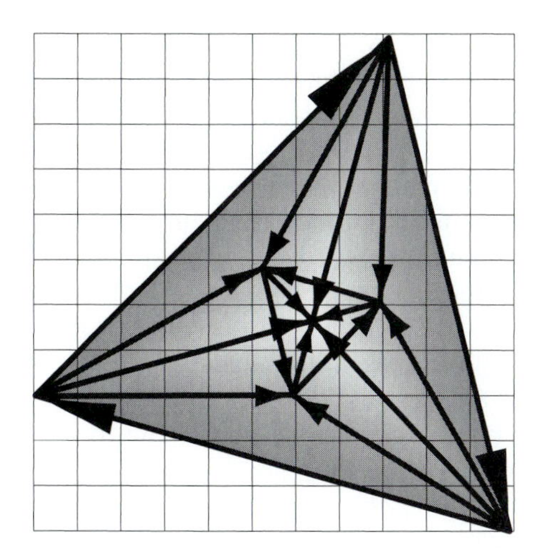

CHAPTER 14

Stainless Steels

Chapter Goals

1. An understanding of stainless steels: their metallurgy, basic types, and differences.
2. Selection guidelines for using stainless steels in design.

Rationale

Behind the scenes stainless steels make possible the many instruments needed for sterile conditions in the food and medical industries; they support the offshore oil rigs that bring us petroleum; they contain the fissionable materials in nuclear reactors that bring us electricity. They work faithfully and mostly silently to allow our civilization to be what it is today.

Stainless steels have sufficient corrosion resistance to resist rusting in countless applications; they can resist high temperatures better than most other materials; and they have fabrication characteristics that allow their use in cold-formed shapes, in weldments, in machined shapes, and in cast shapes. Stainless steels are extremely useful materials in design engineering whenever corrosion or even rust alone is a concern. This chapter will provide the designer with guidelines on which ones to use for various applications.

Stainless steels are alloys of iron, chromium, and other elements that resist corrosion from many environments. A steel cannot qualify for the stainless prefix until it has at least 10% chromium. If the corrosion of steels with differing chromium contents is measured in oxidizing environments, there is a tendency for the corrosion rate to decrease as the chromium concentration increases. The effect of chromium on corrosion of steel in water is shown in Figure 14–1. Another qualifier for a definition of stainless steel is that the steel must exhibit passivity. Some tool steels have chromium contents of 12%, but they are not considered to be stainless steels because the high carbon content and other alloy additions prevent passivity. Thus *stainless steels* are steels with at least 10% chromium that exhibit passivity in oxidizing environments.

The history of stainless steels is rather interesting. Chromium was first reduced from its ore in the late seventeenth century, and many

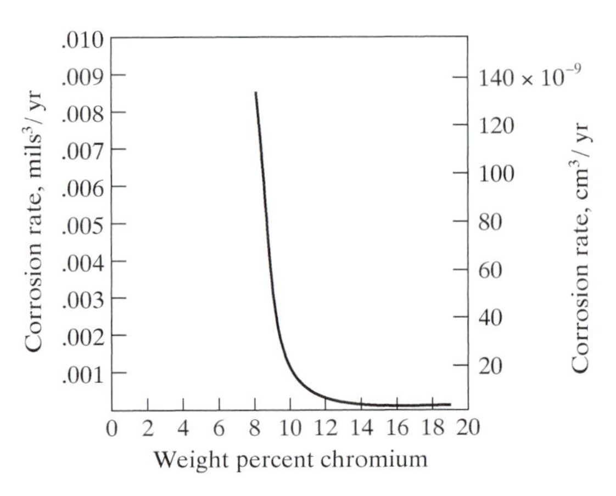

Figure 14–1
Corrosion rate of iron–chromium alloys in intermittent water spray at room temperature
Source: W. Whitman and E. Chappel, *Industrial and Engineering Chemistry Fundamentals*, vol. 13 (1926) p. 533.

subsequent experimenters alloyed chromium metal with iron in the laboratory; but it was not until the early part of the twentieth century that anyone noticed the commercial potential for iron–chromium alloys in reducing corrosion. One reason for this was that sulfuric acid was thought to be a good solution to use for measuring the corrosion resistance of metals. If a metal did not react with sulfuric acid, it was corrosion resistant. Unfortunately, iron/chromium stainless steels are not well suited for use in *reducing media* such as sulfuric acid. Thus the erroneous conclusion was drawn that chromium does not improve the corrosion resistance of iron. This myth was laid to rest by the work of a number of investigators in the early 1900s, and in 1915 stainless steels were born as a commercial reality. The first stainless steels were straight iron–chromium alloys and were used for cutlery.

By 1930 most of the types of stainless steel in use today (martensitic, austenitic, ferritic) were in commercial production. There were many technological problems to be overcome in manufacturing stainless steels. Chromium has a great affinity to combine with carbon and oxygen. If the carbon gets high, the structure of stainless steels is affected, and without special techniques, chromium in steel melting can be lost by oxidation. The technique arrived at to successfully manufacture stainless steels is to melt low-carbon steel scrap (sometimes with iron-ore additions) in an electric furnace and add chromium in the form of low-carbon ferrochromium. Ferrochromium is a hard, brittle compound containing about 80% chromium and 20% iron. This technique is basically still in use. Current technology includes this process in addition to argon–oxygen decarburization (AOD) melting and ladle treatments to control carbon and nitrogen.

It is the purpose of this chapter to explain the nature of stainless steels, describe the different types, and show how to select stainless steels for engineering applications. We shall briefly describe the metallurgy of stainless steels to illustrate that the differences in stainless alloys are primarily in their microstructure. The more important properties will be discussed, but our discussion will be minimal because the previous chapter discussed many of the corrosion characteristics of stainless steels.

14.1 Metallurgy of Stainless Steels

As pointed out in our discussions of the iron–carbon equilibrium diagram, the major microstructural components possible in steels are the following:

1. *Ferrite:* body-centered cubic (BCC) iron
2. *Cementite:* an iron–carbon compound, Fe_3C
3. *Pearlite:* a lamellar composite of ferrite and cementite
4. *Austenite:* face-centered cubic (FCC) iron

In addition to the foregoing, hardened steel can have a martensitic structure [body-centered tetragonal (BCT) iron]. When chromium is added to iron, the equilibrium diagram changes, as do the possible structures. The iron–chromium phase diagram is shown in Figure 14–2. At room

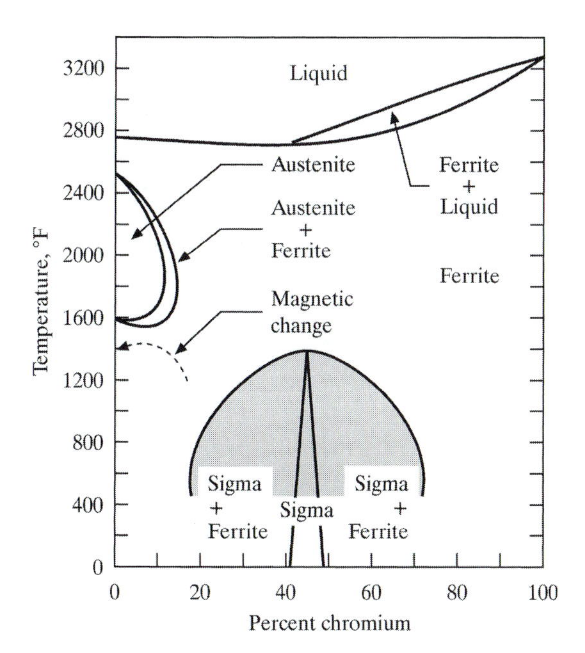

Figure 14–2
Iron–chromium diagram

Source: A. W. Grosvenor, *Basic Metallurgy*, © American Society for Metals, 1962. Used with permission.

temperature, iron–chromium alloys will have a ferritic structure up to about 20% chromium. At this point a phase called sigma starts to appear. *Sigma phase* is a hard, brittle compound of 50 atomic % iron and 50 atomic % chromium. This is an undesirable phase; it reduces corrosion resistance and lowers mechanical properties. When nickel is added to an iron–chromium alloy, austenite can be the equilibrium structure at room temperature. Thus some stainless steels can have a ferritic structure, some can be austenitic, and some can be hardened to a martensitic structure. Stainless steels are categorized by their microstructure, and within each group there are a number of alloys with slightly different compositions. The U.S. stainless steel producers use a three-digit designation number that applies to 66 stainless steels. Thirty-nine are austenitic, the remainder are ferritic or martensitic. In addition, there are 4 precipitation-hardening alloys and more than 30 other alloys that are identified by UNS numbers. The use characteristics differ for each of these groups. In addition to the "standard" types, there are at least 100 proprietary alloys. Some of these proprietary alloys have duplex structures, part ferrite, part austenite; some have even more complex structures. They are often designed for a particular type of service application.

Ferritic

Ferritic stainless steels have a body-centered cubic ferrite microstructure, low carbon contents (usually $<0.2\%$), and chromium contents predominately in the range of 16% to 20%. Because these steels contain carbon, a true description of the equilibrium phases that can be present in this group of stainless steels should be shown on a *ternary* diagram, a phase diagram for varying amounts of three elements (C–Fe–Cr). Carbon and nitrogen have the effect of expanding the *gamma field*, the high-temperature range showing the presence of austenite. Figure 14–3 shows the gamma loop expanded because of an approximate 0.2% carbon addition. It can be seen that the equilibrium structure of low C + N ferritic stainless steels is ferrite at room temperature as well as at all temperatures up to the melting point. Thus these steels will not go through a crystal structure change, and they cannot be quench hardened. Some analyses violate this, but, in general, these steels are considered to be nonhardenable. Ferritic alloys can have up to 20% chromium, and as seen on the equilibrium diagram, this gets into the sigma-phase area. Careful analysis and processing controls are needed to minimize the tendency for sigma-phase formation.

Ferritic stainless steels historically have not seen wide application for piping, tanks, and structural components because of their poor weldability and notch sensitivity. They could not

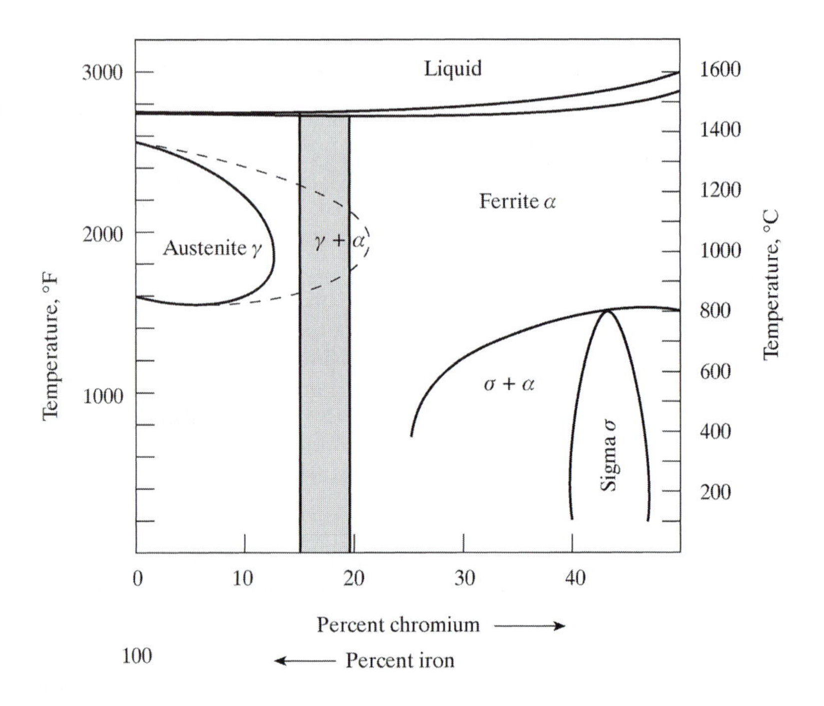

Figure 14–3
Iron–chromium phase diagram for alloys with a carbon content of about 0.2%; the shaded area shows the composition for ferritic stainless steels

be welded and formed in section thicknesses greater than sheet metal. The poor weldability results from the formation of embrittling phases and carbide precipitation on cooling from welding temperatures. It has long been known that ferritics are subject to *475°C embrittlement,* a phenomenon involving decomposition of ferrite to two separate BCC phases that lower ductility and impact strength when the alloys are heated in the temperature range from 650 to 950°F (343 to 510°C). Some ferritics are also subject to the formation of brittle sigma phase when they are heated in the temperature range from 1000 to 1600°F (540 to 870°C). For this reason, the use temperature of ferritics for pressure vessel construction has always been limited to 650°F (343°C). Some grades will also form martensite in the weld-heat-affected zone or weld, which also produces brittleness. The second aspect of the ferritic welding problem is reduced corrosion resistance, caused by welding-induced carbide precipitation.

The justification for using ferritic stainless steels over the easier-to-fabricate austenitic stainless steels is that they are not as susceptible to stress corrosion cracking. In the 1970s, steelmaking practices were developed that minimize some of the weldability limitations of the ferritics. If carbon and nitrogen levels are reduced below 100 ppm, welds show improved ductility and minimal sensitization due to carbide precipitation. There are about six proprietary alloys with low carbon contents, and they are called the *special ferritics*. They contain between 18% and 30% chromium and from 1% to 4% molybdenum. Their carbon and nitrogen levels are kept low by AOD processing, vacuum induction melting, electron beam melting, and related processes. These ferritics have excellent corrosion resistance in certain environments such as seawater, and they are relatively immune to stress corrosion tendencies. There are still welding and fabrication limits with these alloys; thus they are used only for special applications.

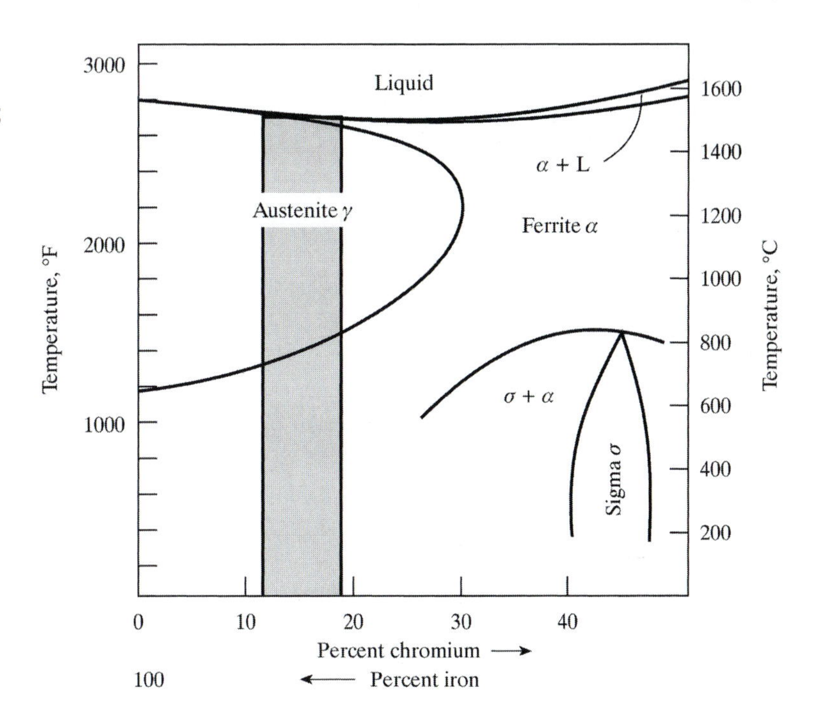

Figure 14–4
Iron–chromium phase diagram for alloys with about 1% carbon; the shaded area shows the composition for martensitic stainless steels

Generally, they are available only in relatively thin gages.

Martensitic

Martensitic stainless steels have a chromium range of approximately 12% to 18%, and carbon content can be as high as 1.2%. The role of the higher carbon content is to further expand the gamma loop such that, on heating, these steels will transform to austenite, making quench hardening possible (Figure 14–4). As mentioned in our discussion of carbon steels, a carbon content of at least 0.6% is required for a fully martensitic structure. The large alloy additions of chromium in the martensitic stainless steels improve the hardenability to such a degree that 100% martensitic structures are possible with carbon contents as low as 0.07%. The martensitic stainless steels with approximately 1% carbon have enough carbon to form 100% martensite, as well as a large volume percentage of hard *chromium carbides* in the structure. There is an overabundance of carbon.

Austenitic

Austenitic stainless steels are significantly more complex in nature than the ferritics and martensitics in that they have at least four major alloying elements: iron, chromium, carbon, and nickel. The carbon content is as low as it is commercially feasible to obtain; chromium can range from 16% to 26% and nickel content is usually at least 8%, but can be as high as 24%. The effect of the nickel is to promote a completely austenitic structure *('austenitizer')*. Figure 14–5 shows the approximate phase relationships with an 18% chromium content and varying amounts of nickel. In the 8% to 10% nickel range, the phase diagram shows that austenite and ferrite are the equilibrium structure. In practice, most austenitic stainless steels have a completely austenitic structure.

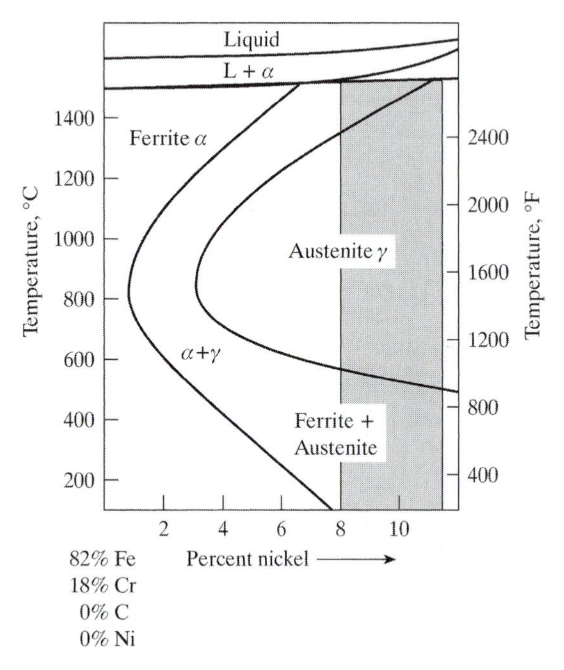

Figure 14–5
Section of the Fe–Cr–Ni ternary diagram for an alloy of 18% chromium; the shaded area shows the composition range for austenitic stainless steel

Source: C. A. Zapffe, *Stainless Steels*, © ASM International, 1949, p. 136. Used with permission.

The difference between the equilibrium diagram and reality is due to the fact that all commercial alloys are used in the annealed condition; annealing requires a water quench, so the austenitic structure is quenched in. It is really *metastable*, but at normal use temperatures the structure remains austenitic. Austenitic stainless steels have a strong tendency to *cold work* or *work harden*. The energy of deformation promotes the transformation of the metastable austenite to martensite. Austenitic stainless steels are incontrovertibly austenitic in most commercial forms. The austenite structure is brought about by nickel additions or in some cases by nickel and manganese additions. Manganese is also an austenitizer.

PH Alloys

The term *PH* is a common acronym for *precipitation hardening*. *PH alloys* are a fairly recent development (within the past 50 years), and there are several types: martensitic, semiaustenitic, and austenitic. The martensitic PH stainless steels are low-carbon, chromium–nickel alloys with chromium-to-nickel ratios like 13 to 8, 15 to 5, and 17 to 4. The composition is controlled so that some of these alloys have a predominately martensitic structure in the annealed condition (usually air cooled from the annealing temperature). This is a low-carbon martensite (carbon is usually <0.1%), and the annealed hardness is relatively low (as low as 25 HRC). These alloys can be fabricated at this hardness and then aged to higher hardness and strength by a low-temperature aging process at around 1000°F (540°C). Age hardening, or, more correctly, precipitation hardening, is accomplished by adding elements such as titanium, aluminum, or copper in small percentages (several percent). The precipitating elements are in solution in the austenite at the annealing temperature, and they want to come out of solution in the low-temperature phase, martensite. The aging heat treatment causes the precipitation to occur. The precipitates further strain the BCT martensite lattice, and hardening and strengthening occur.

The semiaustenitic PH alloys have typical chromium–nickel ratios of 17 to 7 and 15 to 7. These alloys have a much more complicated heat treatment than the martensitic PH alloys. They are essentially austenitic in the annealed condition. A conditioning heat treatment and a deep freeze make them martensitic, and then they are precipitation hardened like the martensitic PH steels. They are capable of higher strengths and hardness than the martensitic PH alloys.

The last class of PH stainless steels is the precipitation-hardening austenitics. They simply have an austenitic structure, and an aging treatment produces a sometimes metallographically visible coherent precipitate formed from aluminum, titanium, or some other minor alloy addition. Strengthening is minimal, so these alloys are not widely used.

Duplex Alloys

We have stressed the importance of chromium and nickel in determining the phases present in stainless steels, but many other elements also have a profound effect on the microstructure of stainless steels. Elements such as chromium, silicon, molybdenum, vanadium, aluminum, niobium, titanium, and tungsten tend to promote the formation of ferrite. Nickel, cobalt, manganese, copper, carbon, and nitrogen tend to promote the formation of austenite. As we might expect, composition adjustments of ferritizers and austenitizers in stainless steels allow the formation of structures that are part ferrite and part austenite. For many years the composition of cast alloys and welding filler metals has been formulated to provide some ferrite (as much as 50%) in the structure of austenitic alloys. This was done to reduce hot cracking.

In the late 1970s a dozen or so proprietary wrought ferrite–austenite alloys were introduced. They are called *duplex alloys*. The introduction of ferrite in an austenitic alloy can increase the yield strength to as much as twice that of the all-austenitic counterpart. The other advantages are improved resistance to stress corrosion cracking and improved weldability. Most of the duplex or *dual phase* alloys contain very low carbon content ($\leq$0.03%), between 20% and 30% chromium, about 5% nickel, and other ferritizers and austenitizers to obtain the desired structure. Modern duplex alloys also contain about 0.12% nitrogen, which aids in leveling corrosion differences between the austenite and ferrite phases. Chemical compositions are usually controlled to produce alloys that have ferrite and austenite in the range from 40% to 60%.

These alloys are not immune from stress corrosion cracking, and the presence of ferrite can lead to accelerated attack in some environments. Thus duplex alloys are not without disadvantages, but as more is learned about their corrosion characteristics they are starting to play an important role in the stainless steel family. At present, like the special ferritics, duplex alloys are largely used for special applications. Some alloys are designed for specific applications such as offshore drilling rigs.

Proprietary Alloys

The trend in the 1970s and 1980s to use new melting practices such as the argon-oxygen decarburization (AOD) process to produce stainless steels has allowed heretofore impractical compositions to be obtained. The carbon content, for example, can be as low as 0.02%; this low carbon content has helped to reduce some of the welding problems in ferritics, and it has allowed austenitics that are less susceptible to sensitization in welding, *stress relieving,* and elevated temperature service.

Lower carbon contents reduced some problems and created some new ones; the low-carbon alloys tended to have tensile and yield strengths that are noticeably lower than those of conventional grades of stainless. This problem was solved by the same equipment that is used in the AOD process to reduce the carbon. After deoxidation, ladle innoculation of nitrogen was developed. Nitrogen additions strengthen both the duplex alloys and the austenitics, and also help to promote the desired ratios of ferrite-to-austenite in the duplex alloys. Nitrogen contents as high as 0.3% have been used in the proprietary duplexes and in the *super austenitics*. Conversely, nitrogen must be kept low in most ferritics because of adverse effects on ductility. Some of the proprietary compositions that are currently commercially available are tabulated in Table 14–1.

Table 14–1
Nominal compositions of some special/proprietary stainless steels developed for improved serviceability over conventional grades

UNS Designation	Type	Composition (Percent by Weight Maximum Unless Otherwise Specified)*									Application
		C	Mn	P	S	Si	Cr	Ni	Mo	Others	
Austenitics											
N08020	20 Cb-3	0.06	2.00	0.035	0.035	1.00	19/21	32.5/35	2/3	3/4 Cu	Superior SCC resistance
N08024	20 Mo-4	0.03	1.00	0.035	0.035	0.50	22.5/25	35/40	3.5/5	0.15/0.35 Cb	Better pitting resist than 20 Cb-3
N08026	20 Mo-6	0.03	1.00	0.03	0.03	0.50	22/26	33/37	5/6.7	2/4 Cu, 0.1/0.16 N	Resists hot chlorides, Low pH
N08028	Alloy 28	0.03	2.50	0.03	0.03	1.00	26/28	30/34	3/4	0.6/1.4 Cu	
N08031	Alloy 31	0.015	2.00	0.02	0.01	0.30	26/28	30/32	6/7	1.0/1.4 Cu, 0.15/0.25 N	
N08366	AL-6X	0.03	2.00	0.04	0.04	1.00	20/22	23.5/25.5	6/7		Resists chloride pitting
N08367	AL-6XN	0.03	2.00	0.04	0.03	1.00	20/22	23/25.5	6/7	0.18/0.25 N, 0.75 Cu	Resists chloride pitting
N08700	JS700	0.04	2.00	0.04	0.03	1.00	19/23	24/26	4.3/5	8X C/.4 Cb	Resists SCC
N08904	904L	0.02	2.00	0.045	0.035	1.00	19/23	23/28	4/5	1/2 Cu, 0.1 N	Resists reducing acids
N08925		0.02	1.00	0.045	0.03	0.50	19/21	24/26	6/7	0.8/1.5 Cu, 0.1/0.2 N	
S20910	22 Cr, 13 Ni, 5 Mn	0.06	4.00/6.00	0.04	0.03	1.00	20.5/23.5	11.5/13.5	1.5/3	0.2/0.4 N, 0.10/0.3 V 0.1/0.3 Cb	High strength, good corr. resist.
S21000	SCF-19	0.03	5.00	0.025	0.003	0.40	20	18	5	0.85N	High strength, SCC resistance
S21300	15-15 CC	0.25	15.0/18.0	0.025	0.003	0.40	16/21	1.10	1.10	1.4 N, 0.56 Cu	Improved SCC resistance
S21904	21 Cr, 6 Ni, 9 Mn	0.03	8.0/10.0	0.04	0.03	1.00	19/215	1.10	—	0.15/0.40 N	High-temperature oxidation resistance
S24100	18 Cr, 2 Mo, 12 Mn	0.15	11.00/14.00	0.06	0.03	1.00	16.5/19	5.5/7.5	—	0.45 N	Higher strength than 304
S28200	18-18 Plus	0.15	17.00/19.00	0.04	0.03	1.00	17/19	0.5/2.5	0.75/1.25	0.4/0.6 N	2× strength of 304

S31254	254 SMO	0.02	1.00	0.03	0.01	0.75	19.50/20.50	17.50/18.50	6/6.5	0.5/1.0 Cu, 0.18/0.22 N	
S32654	654 SMO	0.02	2/4	0.03	0.005	0.50	24/25	21/23	7/8	0.3/0.6 Cu, 0.45/0.55 N	
S34565	45655	0.03	5/7	0.03	0.01	1.00	23/25	16/18	4/5	0.1 Cb, 0.4/0.6 N	
Ferritics											
S44627	E-Brite	0.1	0.4	0.02	0.02	0.4	25/27	0.5	0.75/1.5	0.15 N, 0.05/.2 Cb, 0.2 Cu 0.5 Ni + Cu	Resists SCC
S44735	29-4C	0.3	1.0	0.04	0.03	1.0	28/30	1.0	3.6/4.2	0.045, Ti + Cb = 6 × (C + N) min.	Resists chlorides
S44660	SC-1	0.03	1.0	0.04	0.03	1.0	25/28	1/3.5	3/4	0.04 N, 0.2/1.0 Ti + Cb, 6 × (C + N) Ti + Cb	Resists seawater
S44800	29-4-2	0.01	0.3	0.25	0.02	0.2	28/30	2/2.5	3.5/4.2	0.02 N, 0.15 Cu, 0.025 C + N	Resists seawater
Hardenable											
S35500	Alloy 355	0.1/0.15	0.5/1.25	0.04	0.03	0.5	15/16	4/5	2.5/3.25	0.07/0.13 N	Can quench or precipitation harden
S45000	Custom 450	0.05	1.0	0.03	0.03	1.0	14/16	5/7	0.5/1	1.25/1.75 Cu	High strength, like 304 resistance
S45500	Custom 455	0.05	0.5	0.04	0.03	0.5	11/12.5	7.5/9.5	0.5	1.5/2.5 Cu, 0.1/0.5 Cb, 0.8/1.4 Ti	Higher strength than 450
S35000	Alloy 350	0.07/0.11	0.5/1.25	0.04	0.03	0.5	16/17	4/5	2.5/3.25	0.07/0.13 N	Can quench or precipitation harden
Duplex											
S31803	AL 2205	0.03	2.0	0.03	0.02	1.0	21/23	4.5/6.5	2.5/3.5	0.08/0.2 N	High strength, austenite + ferrite
S32550	Alloy 2855	0.04	1.5	0.04	0.03	1.0	27/28	4.5/6.5	2.9/3.9	0.1/0.25 N, 1.5/2.5 Cu	Good in reducing acids
S32950	7-Mo Plus	0.03	2.0	0.035	0.10	0.6	26/29	3.5/5.2	1.0/2.5	0.15/0.35 N	40% austenite in ferrite, for SCC

*Balance is iron.

As a natural part of the investigation of the use of the AOD process to produce cleaner stainless steels, it was determined that the nitrogen additions had a beneficial effect in resistance to some forms of corrosion, and the use of nitrogen also allowed molybdenum concentrations to be raised to as high as 6% (super austenitics). It is well known that molybdenum reduces pitting tendencies in stainless steel. Thus by reducing the carbon content and adding nitrogen and molybdenum, families of stainless steels have been created with better resistance to stress corrosion cracking and pitting. It has been estimated that about half of the corrosion failures of stainless steels are caused by stress corrosion cracking (SCC) or pitting. Thus there is economic justification to develop alloys that have improved resistance to these forms of corrosion. By the mid-1980s there were more than 50 commercially available proprietary alloys for severe service conditions. Most of these alloys have been assigned a UNS number and have been incorporated into ASTM specifications, and this is the correct way to specify these alloys on drawings and purchase specifications. These alloys are still new enough that their resistance to particular environments should be investigated on a case-by-case basis; it is also recommended that potential users thoroughly investigate the availability of required product forms (*Steel Products Manual* in the United States) before committing one of these alloys to be used for a large installation. For example, you may build a tank from one of these materials and find that piping and valving for the tank are not available from that alloy. These new alloys should be used with due investigation of these factors.

Summary

Stainless steels can be very complicated from the metallurgical standpoint because they can contain significant concentrations of eight or more different elements. It is hard to visualize an eight-sided phase diagram. Users of stainless steel should remember that there are several classes of stainless steel, categorized by their microstructures (see Figures 14–6 & 14–7).

1. *Ferritic:* iron–chromium + low carbon
2. *Martensitic:* iron–chromium + higher carbon
3. *Austenitic:* iron–chromium–nickel + low carbon
4. *PH:* iron–chromium–nickel + low carbon + precipitating element
5. *Duplex:* part austenite, part ferrite

It should also be remembered that, because of these different microstructures, some stainless steels are quench or age hardenable and some are not.

14.2 Alloy Identification

Wrought stainless steels are commonly identified in the United States by the three-digit system that originated with the American Iron and Steel Institute (AISI). The first digit indicates the classification by composition type. The 200 series has chromium, nickel, and manganese as the primary alloying elements. The 300 series is chromium–nickel alloys; the 400 series is straight chromium alloys.

The 500 series is used for low-chromium alloys, and they do not have enough chromium to be considered as stainless steels. They are intended to be heat-resisting steels. Thus they are not included in lists of stainless alloys. The PH stainless steels were originally identified by 600 series numbers, but in the late 1970s this identification was dropped in favor of UNS numbers. There are UNS numbers on all grades, but the 200, 300, and 400 identification numbers are still widely used by steel suppliers and their customers for most alloys. The UNS numbers for stainless steels usually consist of an S prefix, the AISI type number, and two zeros; the UNS

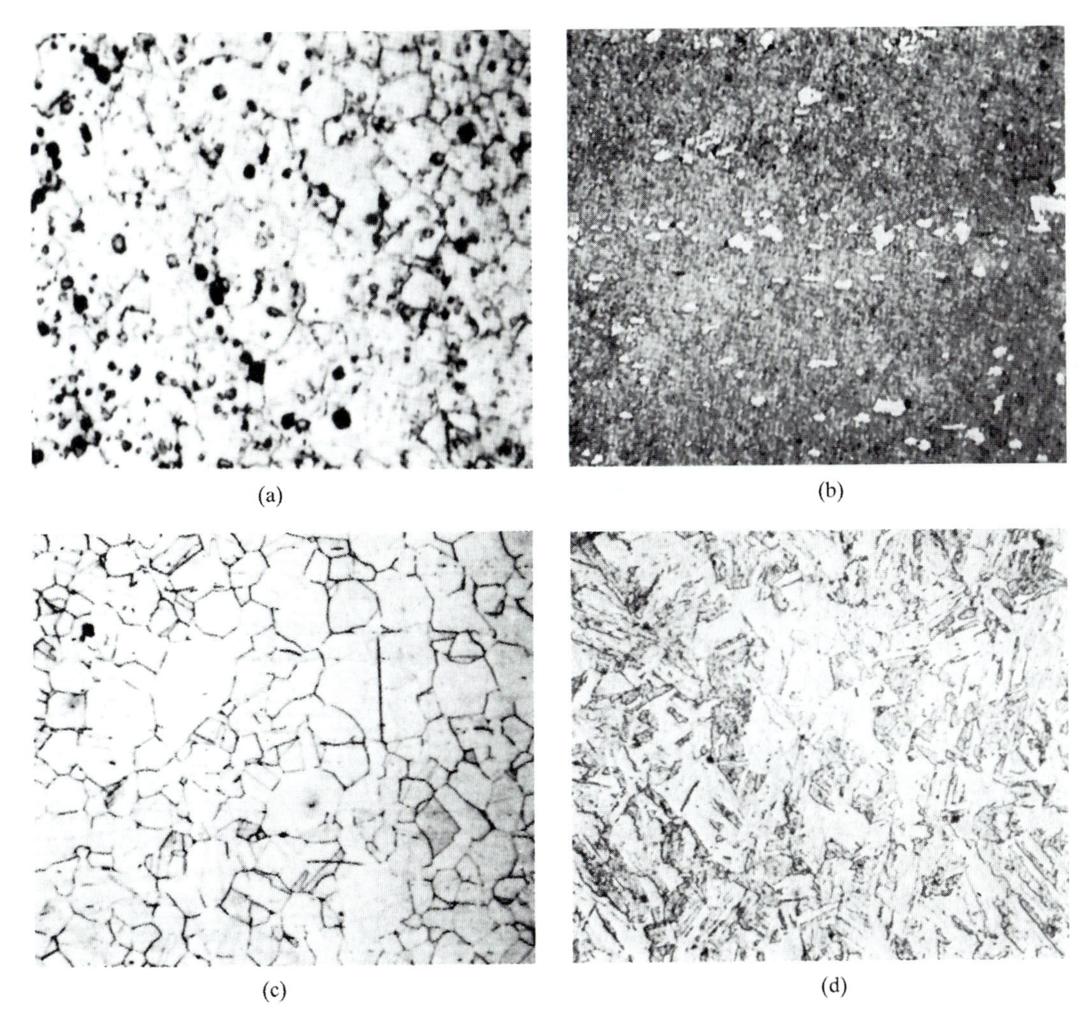

Figure 14–6
Microstructure of stainless steel: (a) Type 430 ferritic (ferrite + carbides; ×1200). (b) Type 440 martensitic (martensite + carbides; ×500). (c) Type 316 austenitic (austenite grains; ×150). (d) Type 15-5PH (low-carbon martensite; ×250).

number for 316 stainless steel is S31600. There are also ASTM designations for stainless steel alloys, but some standards are written for specific types of service. ASTM A 480 covers cold-rolled sheet and strip; ASTM A 313 covers wire; ASTM A 268 covers ferritic stainless tubing. The ASTM specifications are most often used when a fabrication is performed under jurisdiction of a U.S. code body. The three-digit designations are easier to use. They are appropriate in the United States unless there is a reason for another designation system. For international trade, ASTM specifications and UNS numbers are appropriate.

The most commonly used wrought stainless steels are listed in Table 14–2. One factor that may lead to confusion is that the ferritic and martensitic alloys are both 400 types. The

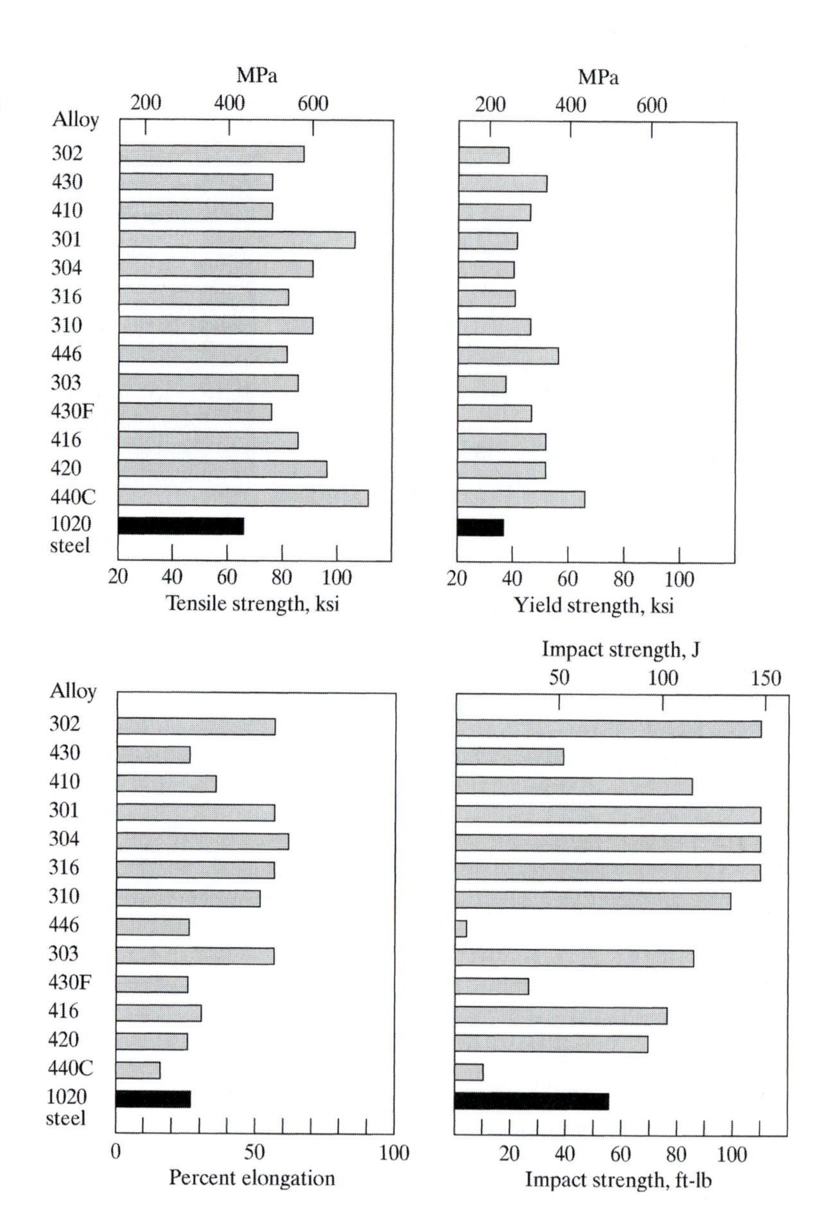

Figure 14–7
Mechanical properties of some stainless steels; typical values for annealed bar and plate

use characteristics are very different. Caution should be used in order to not select a ferritic when a hardenable alloy is desired. The confusion factor in the austenitic types is the 200 series. These alloys are not in wide use, but they are also austenitic.

Table 14–2 can be useful in figuring out the differences among the different alloys. As shown in the table, there is a basic ferritic, a basic martensitic, and a basic austenitic composition. These compositions are in a sense doctored to obtain most of the other alloys. The last two digits in the three-digit system have no significance to the stainless steel user.

The most commonly used identification system in the United States for cast stainless

Table 14–2
Stainless steels categorized by structure and chemical composition

Group	General Properties	Hardenability	Type	Analysis Built Up from Basic Type
Chromium–iron	Martensitics: Nonrusting tools and structural parts	Hardenable by heat treatment	403	Cr 12% adjusted for special physicals
			410	**Basic type, Cr 12%**
			414	Ni added to increase corrosion resistance and physicals
			416	S added for easier machining
			416Se	Se added for easier machining
			420	C higher for cutting purposes
			420F	S added for easier machining
			422	Mo, V, and W added for strength to 1200°F
			431	Cr higher and Ni added for better resistance and properties
			440A	C higher for cutting applications
			440B	C higher for cutting applications
			440C	C still higher for wear resistance
	Ferritic: Used for elevated temperature and nonrusting architectural parts	Nonhardenable	405	Al added to Cr 12% to prevent hardening
			409	Low Cr for auto exhaust
			429	Less Cr for better welding than 430
			430	**Basic type, Cr 17%**
			430F	S added for easier machining
			430F Se	Se added for easier machining
			434	Mo added for pitting resistance
			436	Mo, Cb, and Ta added for heat resistance
			439	Cr added for improved corrosion resistance
			442	Cr higher to increase scaling resistance
			444	Stabilized for welding
			446	Cr much higher for improved scaling resistance
Chromium–nickel	Austenitic: Used for chemical resistance	Hardenable by cold work	301	Cr and Ni lower for more work hardening
			302	**Basic type, Cr 18%, Ni 8%**
			302B	Si higher for more scaling resistance
			303	S added for easier machining
			303Se	Se added for easier machining
			304	C lower to avoid carbide precipitation
			304L	C lower for welding application
			304N	N added to increase strength
			304LN	Low C, N added
			305	Ni higher for less work hardening

(continued)

Table 14–2
continued

Group	General Properties	Hardenability	Type	Analysis Built Up from Basic Type
			308	Cr and Ni higher with C low more corrosion and scaling resistance
			309	Cr and Ni still higher for more corrosion and scaling resistance
			309S	Lower C than 309
			310	Cr and Ni highest to increase scaling resistance
			310S	Lower C than 310
			314	Si higher to increase scaling resistance
			316	Mo added for more corrosion resistance
			316F	0.1% S for improved machining
			316L	C lower for welding applications
			316N	N added to improve strength
			316LN	Low carbon, nitrogen added
			317	Mo higher for more corrosion resistance and strength at heat
			317L	C low for welding applications
			318	Cb, Ta added to avoid carbide precipitation
			321	Ti added to avoid carbide precipitation
			332	Ni added to resist carburization
			347	Cb, Ta added to avoid carbide precipitation
			348	Similar to 347, but low Ta content (0.10%)
			384	High Ni for easier cold heading
Chromium–nickel–manganese			201	N and Mn partially replace Ni
			202	**Basic type, Cr 18%, Ni 5%, Mn 8%**
			205	N and Mn partially replace Ni
Precipitation hardening (PH)	Martensitic and semiaustenitic; combination of chemical resistance and high strength	Hardenable by precipitation heat treatment	S17400	17% Cr, 4% Ni (17-4), high-strength alloy
			S17700	17% Cr, 7% Ni (17-7), higher strength than 17-4
			S15500	Lower Ni than 17-4 to reduce ferrite
			S13800	Lower Cr, higher Ni than 17-4 for reduced *anisotropy*
Duplex	Austenite plus ferrite	Not normally hardenable by heat treatment	329	**Basic type, Cr 25%,** Ni 4%, like 316 but less SCC
			S32550	3% Mo for pitting resistance
			S32950	Higher strength than 316

Source: Adapted from *Stainless Steel Handbook,* Allegheny Ludlum Steel Co.

steels is the system of the Alloy Casting Institute (*ACI*). Stainless steel alloys that are predominately used for corrosion applications have C as the first letter in an alphanumeric alloy-designation system. Alloys primarily for heat or oxidation resistance are identified by a two-letter system with the first letter being H. ASTM International has detailed specifications covering the analysis and properties of cast stainless steel: ASTM A 296, "Corrosion-Resistant Iron–Chromium and Iron–Chromium–Nickel Alloy Castings for General Applications," and ASTM A 297, "Heat-Resistant Iron–Chromium and Iron–Chromium–Nickel Alloy Castings for General Applications."

There may or may not be a wrought equivalent for a cast alloy. As shown in Table 14–3, most alloys do have wrought equivalents, but the cast-alloy designation or UNS number should always be specified on a drawing or material purchase order. The ASTM or three-digit type is the correct designation to use for drawings calling for wrought alloys. Chemical compositions of the wrought alloys are shown in Table 14–4.

Table 14–3
Wrought equivalents of cast stainless steels (C) and heat-resistant steels (H)

Cast-Alloy Designation ASTM A 743 Grade	Wrought Alloy Type	Cast-Alloy Designation ASTM A 297 Grade	Wrought Alloy Type
CA-15	410	HA	—
CA-40	420	HC	446
CB-30	431, 442	HD	327
CB-7Cu	17–4PH	HE	—
CC-50	446	HF	302B
CD-4MCu	—	HH	309
CE-30	312	HI	—
CF-3	304L	HK	310
CF-8	304	HL	—
CF-20	302	HN	—
CF-3M	316L	HT	330
CF-8M	316	HU	—
CF-8C	347		
CF-16F	303		
CG-8M	317		
CH-20	309		
CK-20	—		

Source: Alloy Casting Institute, Steel Founders Society of America.

14.3 Physical Properties

The physical properties of the various stainless steel alloys vary from alloy to alloy; a number of physical properties are important to note because they can affect many design applications.

Density

Because these alloys are iron based, the density is about the same as carbon or low-alloy steels.

Structure

We have already made the point that stainless steels have ferritic, martensitic, austenitic, or duplex structures. To the designer concerned with applications, the structure is only important as it affects mechanical properties. However, the fact that some stainless alloys have an austenitic structure means that these alloys are not ferromagnetic. They cannot be used anywhere that a material must respond to a magnetic field. Austenitic stainless steels are troublesome to grind on machines with magnetic chucks. The ferritic, martensitic, and PH stainless steels are ferromagnetic.

Conductivity

Stainless steels are fairly poor conductors of both heat and electricity compared with carbon steels. Thermal conductivity is usually less than half that of carbon steels. The electrical resistivity of stainless steels can be as much as six times

Table 14–4
Chemical composition of wrought stainless steels

UNS Number	TYPE Number	C	Mn	P	S	Si	Cr	Ni	Mo	Other Elements
Austenitic Stainless Steels (Chemical Composition, Percent Maximum unless Otherwise Shown)										
(S20100)	201	0.15	5.50/7.50	0.060	0.030	0.75	16.00/18.00	3.50/5.50		N 0.25
(S20200)	202	0.15	7.50/10.00	0.060	0.030	0.75	17.00/19.00	4.00/6.00		N 0.25
(S20300)	203	0.08	5.00/6.50	0.040	0.18/0.35	1.00	16.00/18.00	5.00/6.50	0.50	Cu 1.75/2.25
(S20500)	205	0.12/0.25	14.00/15.50	0.060	0.030	0.75	16.50/18.00	1.00/1.75		N 0.32/0.40
(S30100)	301	0.15	2.00	0.045	0.030	0.75	16.00/18.00	6.00/8.00	—	N 0.10
(S30200)	302	0.15	2.00	0.045	0.030	0.75	17.00/19.00	8.00/10.00	—	N 0.10
(S30215)	302B	0.15	2.00	0.045	0.030	2.00/3.00	17.00/19.00	8.00/10.00	—	
(S30300)	303	0.15	2.00	0.20	0.15 (min)	1.00	17.00/19.00	8.00/10.00	—	
(S30323)	303Se	0.15	2.00	0.20	0.060	1.00	17.00/19.00	8.00/10.00	—	Se 0.15 min
(S30400)	304	0.08	2.00	0.045	0.030	0.75	18.00/20.00	8.00/10.50	—	N 0.10
(S30403)	304L	0.030	2.00	0.045	0.030	0.75	18.00/20.00	8.00/12.00	—	N 0.10
(S30409)	304H	0.04/0.10	2.00	0.045	0.030	0.75	18.00/20.00	8.00/10.50	—	N 0.10
(S30430)	304Cu	0.08	2.00	0.045	0.030	0.75	17.00/19.00	8.00/10.00	—	Cu 3.00/4.00
(S30451)	304N	0.08	2.00	0.045	0.030	0.75	18.00/20.00	8.00/10.50	—	N 0.10/0.16
(S30453)	304LN	0.030	2.00	0.045	0.030	0.75	18.00/20.00	8.00/12.00	—	
(S30500)	305	0.12	2.00	0.045	0.030	0.75	17.00/19.00	10.50/13.00	—	
(S30800)	308	0.08	2.00	0.045	0.030	1.00	19.00/21.00	10.00/12.00	—	
(S30900)	309	0.20	2.00	0.045	0.030	1.00	22.00/24.00	12.00/15.00	—	
(S30908)	309S	0.08	2.00	0.045	0.030	1.00	22.00/24.00	12.00/15.00	—	
(S31000)	310	0.25	2.00	0.045	0.030	1.50	24.00/26.00	19.00/22.00	—	
(S31008)	310S	0.08	2.00	0.045	0.030	1.50	24.00/26.00	19.00/22.00	—	
(S31600)	316	0.08	2.00	0.045	0.030	0.75	16.00/18.00	10.00/14.00	2.00/3.00	N 0.10
(S31603)	316L	0.030	2.00	0.045	0.030	0.75	16.00/18.00	10.00/14.00	2.00/3.00	N 0.10
(S31609)	316H	0.04/0.10	2.00	0.045	0.030	0.75	16.00/18.00	10.00/14.00	2.00/3.00	N 0.10
(S31620)	316F	0.08	2.00	0.20	0.10 (min)	1.00	16.00/18.00	10.00/14.00	1.75/2.50	N 0.10

(S31651)	316N	0.08	2.00	0.045	0.030	0.75	16.00/18.00	10.00/14.00	2.00/3.00	N 0.10/0.16
(S31653)	316LN	0.030	2.00	0.045	0.030	0.75	16.00/18.00	10.00/14.00	2.00/3.00	N 0.10/0.16
(S31700)	317	0.08	2.00	0.045	0.030	0.75	18.00/20.00	11.00/15.00	3.00/4.00	N 0.10
(S31703)	317L	0.030	2.00	0.045	0.030	0.75	18.00/20.00	11.00/15.00	3.00/4.00	N 0.10
(S31725)		0.03	2.00	0.045	0.030	0.75	18.0/20.0	13.50/17.50	4.0/5.0	N 0.10 Cu 0.75
(S32100)	321	0.08	2.00	0.045	0.030	0.75	17.00/19.00	9.00/12.00	—	Ti 5(C + N) min 0.70 max
(S32109)	321H	0.04/0.10	2.00	0.045	0.030	0.75	17.00/19.00	9.00/12.00	—	Ti 4(C + N) min 0.70 max
(S32900)	329	0.08	2.00	0.040	0.030	0.75	23.00/28.00	2.50/5.00	1.00/2.00	
	332	0.08	2.00	0.040	0.030	0.75	19.00/23.00	30.00/34.00	—	Ti 0.60 Al 0.60
	334	0.08	1.00	0.040	0.030	0.75	18.00/22.00	18.00/22.00	—	Ti 0.60 Al 0.60
(S34700)	347	0.08	2.00	0.045	0.030	0.75	17.00/19.00	9.00/13.00	—	Cb 10 × C min 1.00 max
(S34709)	347H	0.04/0.10	2.00	0.045	0.030	0.75	17.00/19.00	9.00/13.00	—	Cb 8 × C min 1.00 max
(S34800)	348	0.08	2.00	0.045	0.030	0.75	17.00/19.00	9.00/13.00	—	Cb + Ta 10 × C min 1.00 max
										Ta 0.10 max Co 0.20 max
(S34809)	348H	0.04/0.10	2.00	0.045	0.030	0.75	17.00/19.00	9.00/13.00	—	Cb + Ta × C min 1.00 max
(S38400)	384	0.08	2.00	0.045	0.030	1.00	15.00/17.00	17.00/19.00	—	

Ferritic Stainless Steels
(Chemical Analysis, Percent Maximum unless Noted Otherwise)

(S40500)	405	0.08	1.00	0.040	0.030	1.00	11.50/14.50	0.6		0.10/0.30 Al
(S40900)	409	0.08	1.00	0.045	0.045	1.00	10.50/11.75	0.5		(6 × C) Ti, min 0.75 max
(S42900)	429	0.12	1.00	0.040	0.030	1.00	14.00/16.00	0.75		
(S43000)	430	0.12	1.00	0.040	0.030	1.00	16.00/18.00			
(S43020)	430F	0.12	1.25	0.060	0.15(min)	1.00	16.00/18.00			
(S43400)	434	0.12	1.00	0.040	0.030	1.00	16.00/18.00		0.75/1.25	
(S43600)	436	0.12	1.00	0.040	0.030	1.00	16.00/18.00		0.75/1.25	(5 × C) Cb min, 0.7 max
(S44200)	442	0.20	1.00	0.040	0.030	1.00	18.00/23.00			
(S44600)	446	0.20	1.50	0.040	0.030	1.00	23.00/27.00			0.25 N

(continued)

Table 14–4
continued

					Martensitic Stainless Steels (Chemical Analysis, Percent Maximum unless Noted Otherwise)					
UNS Number	**TYPE Number**	**C**	**Mn**	**P**	**S**	**Si**	**Cr**	**Ni**	**Mo**	**Other Elements**
(S40300)	403	0.15	1.00	0.040	0.030	0.50	11.50/13.00	0.75		
(S41000)	410	0.15	1.00	0.04	0.030	1.00	11.50/13.00	1.25/2.50		
(S41008)	410H	0.08	1.00	0.040	0.030	1.00	11.50/13.50	0.6		
(S41400)	414	0.15	1.00	0.040	0.030	1.00	11.50/13.50			
(S41500)	415	0.05	0.5/1.00	0.30	0.30	0.60	11.50/14.00	3.5/5.5	0.5/1.00	
(S41600)	416	0.15	1.25	0.060	0.15 (min)	1.00/3.00	12.00/14.00			
(S42000)	420	0.15 (min)	1.00	0.040	0.030	1.00	12.00/14.00			
(S42010)	420	0.15/0.30	1.00	0.040	0.030	1.00	13.00/15.00	0.25/1.00	0.4/1.00	
(S42020)	420F	0.15 (min)	1.25	0.060	0.15 (min)	1.00	12.00/14.00		0.60	
(S42200)	422	0.20/0.25	0.5/1.00	0.025	0.025	0.50	11.00/12.50	0.50/1.00	0.9/1.25	0.20/0.30 V
(S43100)	431	0.20	1.00	0.040	0.030	1.00	15.00/17.00	1.25/2.50		0.75/1.25 W
(S44002)	440A	0.60/0.75	1.00	0.040	0.030	1.00	16.00/18.00		0.75	
(S44003)	440B	0.75/0.95	1.00	0.040	0.030	1.00	16.00/18.00		0.75	
(S44009)	440C	0.95/1.20	1.00	0.040	0.030	1.00	16.00/18.00		0.75	
					Precipitation Hardening Stainless Steels (Chemical Analysis, Percent Maximum unless Noted Otherwise)					
(S13800)	13-8	0.05	0.10	0.010	0.008	0.10	12.25/13.25	7.50/8.50	2.00/2.50	0.90/135 Al 0.010 N
(S15500)	15-5	0.07	1.00	0.04	0.03	1.00	14.00/15.50	3.50/5.50		2.50/4.50 Cu 0.15/0.45 Cb
(S15700)	15-7	0.09	1.00	0.04	0.03	1.00	14.00/16.00	6.50/7.75	2.00/3.00	0.75/1.5 Al
(S17400)	17-4	0.07	1.00	0.040	0.030	1.00	15.50/17.50	3.00/5.00		0.15/0.45 Cb, 3.00/5.00 Cu
(S17700)	17-7	0.09	1.00	0.040	0.040	1.00	16.00/18.00	6.50/7.75		0.75/1.50 Al

See *Steel Products Manual, Stainless and Heat Resisting Steels* for complete list.

that of carbon steel. This is a consideration in applications involving heat transfer or electrical current flow.

Expansion

This property has caused a number of application problems. The austenitic alloys have a coefficient of thermal expansion that can be 50% greater than that of carbon steel. It is not uncommon to hear a designer complain that a stainless part is unstable; it is distorting in service. If the part is mechanically fastened to a carbon steel part and the part experiences a little heat in service, bowing results. This is a bimetal assembly. The ferritics, martensitics, and PH stainless steels have expansion characteristics similar to those of carbon steels.

Modulus of Elasticity

Stainless steels have tensile moduli slightly lower than those of carbon and alloy steels, 28 to 29 compared with 30×10^6 psi (207 GPa). Thus for comparable section sizes stainless steels will have slightly more elastic deflection than most steels. Austenitic alloys that are severely cold worked as in drawn wire may have a modulus as low as 22×10^6 psi (151 GPa)

This lower stiffness often causes problems with coiled or flat springs. A designer might experience rusting problems with a steel spring and try to solve the problem by using a stainless spring. It does not perform the same. The lower modulus is the reason. Caution must be used in substituting stainless steels for steels where elastic deflection is a consideration.

14.4 Mechanical Properties

The mechanical properties of stainless steels are very important in that the largest tonnage goes into the chemical process industries, where they are used to contain corrosive materials or for corrosion-resistant structures. Chemical plants use stainless steels for tanks, piping, pressure vessels, stills, valves, pumps, ductwork, and the like. These types of applications usually require good strength, toughness, and formability. The nuclear industry is a large user of stainless steel; stainless is used in piping and heat exchangers operating at high-stress levels and often at high temperature.

The most popular stainless steels used in these types of industries are the 300-series austenitics. Types 304 and 316 are the biggest sellers on a tonnage basis. As can be seen from the mechanical-property comparison in Figure 14–7, they are much better than mild steel in tensile properties, ductility, and toughness. The ductility (percentage of elongation) of some ferritics is less than in the 300 series, and the toughness is mediocre at best. Two of the martensitics shown, types 420 and 440C, have lower ductility and toughness than the austenitics, but these alloys are used more as tool steels than structural alloys, and they are usually heat treated to high hardness. Low toughness is expected in these types of alloys.

After hardening, the tensile strength of type 440C is over 250 ksi (1724 MPa). The hardness can be 57 to 60 HRC. The 440 family and 420 steels are used for their strength and wear resistance in cutting devices, punches, and dies, and even as injection molding cavities. Mechanical properties in the hardened condition are shown in Table 14–5. The other martensitic alloys, types 403, 410, 414, 416, and 431, do not achieve the strength and hardness ranges of 420 and 440. Maximum hardness is usually about 42 HRC (Figure 14–8). They are used for structural applications rather than wear applications.

The austenitic alloys can be hardened only by cold work, but as shown in Table 14–5, the strengthening effect can be substantial. A full-hard type 301 will have a hardness of about 40 HRC and a tensile strength of 185 ksi (1250 MPa). Cold-rolled sheet and strip are commonly used for springs and miscellaneous fasteners and retainers. As a matter of interest, the highest

Table 14–5
Mechanical properties of cold-worked austenitic and heat-treated martensitic stainless steels

Hardened and Tempered Mechanical Properties (Bar)				
AISI Type Number	**410**	**416**	**420**	**440-C**
Yield strength, (lb/in.^{2})[a]	70,000–150,000	70,000–150,000	70,000–215,000	90,000–250,000
Ultimate tensile strength, (lb/in.^{2})[a]	95,000–200,000	95,000–200,000	110,000–245,000	130,000–265,000
Elongation in 2 in.	30–10%	25–10%	25–7%	12–2%
Hardness, Brinell	200–425	200–425	250–550	275–600
Impact strength, Izod ft-lb[b]	110–20	60–20	60–5	10–3
Cold-Worked Mechanical Properties (Sheet and Strip), Types 201 and 301 (Minimum Values)				
Temper	**Quarter-hard**	**Half-Hard**	**Three-Quarters Hard**	**Full Hard**
Yield strength, (lb/in.^{2})[a]	75,000	110,000	135,000	140,000
Ultimate tensile strength, (lb/in.^{2})[a]	125,000	150,000	175,000	185,000
Elongation in 2 in.	25%	15%	12%	8%
Hardness, Rockwell C	25	32	37	41

[a] Multiply by 6.8948 to convert to kPa.
[b] Multiply by 1.355 to convert to joules.
Source: Courtesy of Republic Steel Corporation

tensile strengths available on any steel have been recorded on small-diameter cold-drawn austenitic stainless wire. Tensile strength can be as high as 400 ksi (3760 MPa). The ability to work harden is most prominent in the 301 alloy.

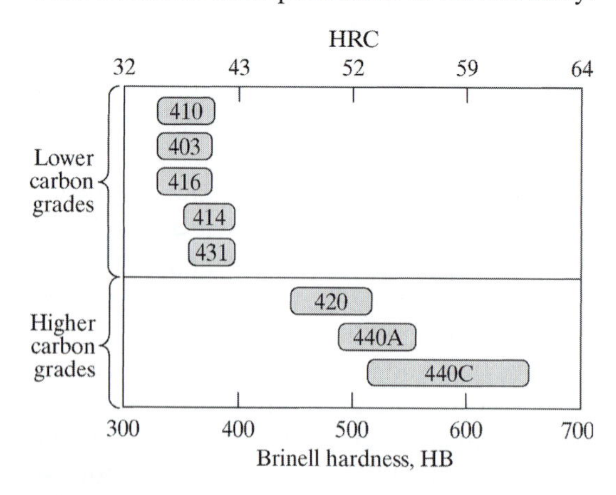

Figure 14–8
Hardness ranges for various stainless steels; typical values for hardened bar and plate
Source: Courtesy of Republic Steel Corporation

The ability to work harden decreases as the nickel content increases over about 9%. Type 305 alloy with 12% nickel is considered to be free spinning (a low work-hardening rate). The tensile strengths obtainable in the various families of stainless steel are shown in Figure 14–9.

The PH stainless steels have become the leading alloys for high-strength applications. These alloys can have tensile strengths around 200 ksi (1880 MPa) and still have good toughness and resistance to crack propagation. They are widely used in the aircraft and aerospace industry for structural components. On machines, they are very useful for base plates, springs, and highly stressed structural members. They often replace the 410, 416, and similar types of alloys. The advantage over these alloys is the simple heat treatment. Quench hardening is not required. PH stainless steels can be heat treated to hardnesses as high as 48 HRC, but because of their low carbon content they are not as wear resistant as the higher-carbon martensitics (even at the same hardness).

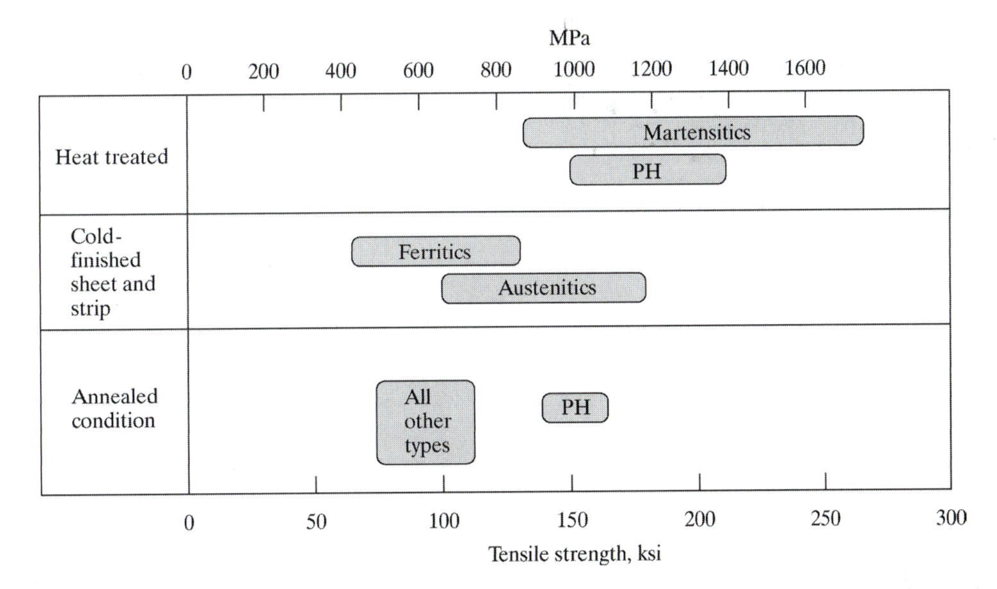

Figure 14–9
Tensile strength ranges for families of stainless steels

One final comment on mechanical properties: Because stainless steels have good oxidation resistance, they are often used in furnaces where they must carry loads at elevated temperatures. As can be seen from Figure 14–10, the creep characteristics of the austenitics are far superior to those of carbon steels or the ferritic grades; as a matter of fact, they compete with some of the "witch's brew" specialty high-temperature alloys.

The conclusion can be made that the mechanical properties of stainless steels are some of their best points. They have the strength, toughness, and formability to meet a wide range of structural applications.

14.5 Fabrication

Wrought stainless steels are available in most of the forms and shapes that are available in carbon steels: bars, plates, strips, tubing, pipe, and the like. They are fabricated into parts and structures by a wide variety of processes. Most fabricators charge extra in addition to the material cost to fabricate something from stainless steel as compared with carbon steel, because some stainless steels (the ones that readily work harden) require heavy-duty tooling and special techniques. We shall mention a few of the idiosyncrasies of stainless steels in the areas of forming, machining, welding, and heat treatment to help the designer assess fabrication costs.

Forming

As shown in Figure 14–7, most austenitic stainless alloys have high elongations in tensile tests, which means that they can stretch a lot without fracturing. The ferritic stainless grades, as a group, are not as formable as drawing quality carbon steels. Most of the alloys, with the exception of the quench hardenable 400 series, can be bent flat in thin gages. One problem exists, however: Most stainless alloys have higher yield strengths than carbon steels, which means greater spring-back after forming. To compensate for the extra spring-back, higher forming forces (possibly 50%) are needed compared with carbon steels. For drawing and stamping,

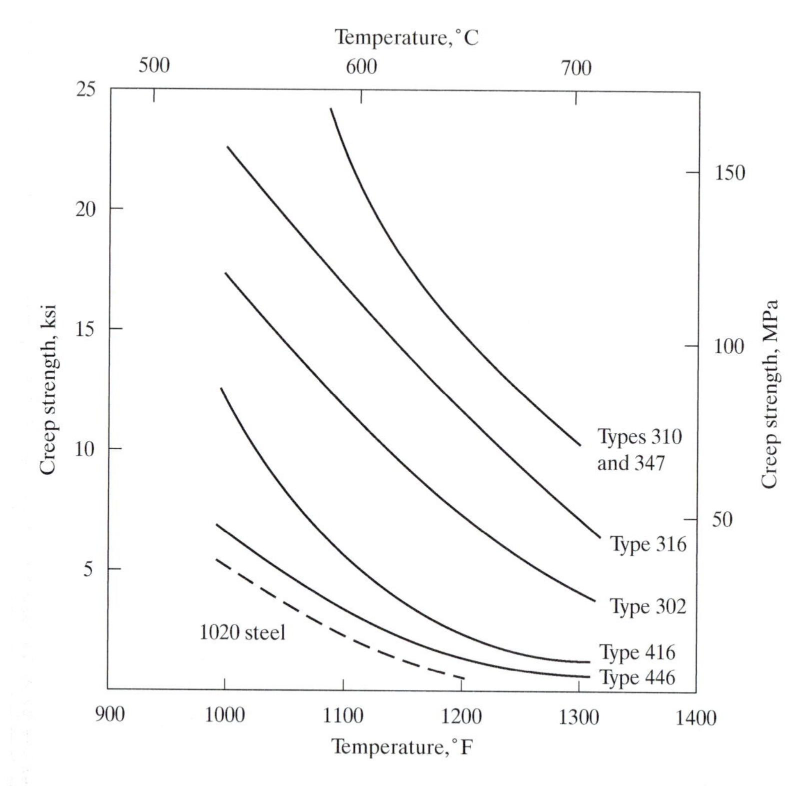

Figure 14–10
Creep strength of stainless steels (the stress that produces 1% elongation at 10,000 h of loading)
Source: Courtesy of Republic Steel Corporation

possibly 150% to 200% extra press capacity is needed compared with carbon steel. The higher forces also result in more tool wear than would be encountered with carbon steel. Types 304 and 316 have good drawing properties. Deep vessels are frequently made from these alloys. Stainless tends to gall or pick up on draw dies. This can be minimized by cladding draw dies with aluminum bronze.

The PH alloys have limited formability. Drawing is not possible, and sharp 90° bends are questionable in some of the cold-worked tempers (condition C).

In spite of these limitations, cold forming of stainless steels is very prevalent and the problems are few with adequate equipment. One factor that is helpful in forming stainless sheet and strip to shape is the availability of special mill finishes. If the goal is to make a large sheet metal container such as an institutional cooking kettle and the inside must have a good finish for cleanability, a *2B finish* can be ordered on the sheet to be used. It will come protected in paper and be bright, with a surface roughness as low as 4 μin. (0.1 μm). Table 14–6 shows some of the finishes that are available to assist in fabricating parts with good finishes.

Machining

The stainless alloys that are not modified for improved machinability have less than 50% of the machinability of AISI B1112 steel, the traditional steel screw machine stock. The *free-machining* 400 series alloys 416 and 430F allegedly have 90% of the B1112 machinability. Type 303, which is the free-machining austenitic, has about 60% of the machinability of B1112.

Most published machinability information has been obtained on automatic screw machines and on machines for which all operations are carefully controlled and monitored. In

Table 14–6
Surface finishes available on stainless steel sheet, strip, and plate

Mill-Rolled Finishes
Cold-Rolled Strip
No. 1 finish, cold rolled, annealed, and pickled
No. 2 finish, bright cold rolled
No. 2 finish, bright annealed
Sheets
No. 1 finish, hot rolled, annealed, and pickled
No. 2B finish, bright cold rolled
No. 2D finish, dull cold rolled
No. 3 to 8 different degrees of polish (8 is most reflective)
Hot-Rolled Plates
Hot rolled
Hot rolled and annealed
Hot rolled, annealed, and pickled
Mill-Polished Finishes (On One or Both Sides)
Sheets and Plates
No. 3 finish, intermediate polish (low grit)
No. 4 finish, standard polish
No. 6 finish, dull satin
No. 7 finish, high luster polish
No. 8 finish, mirror polish (highest luster)

job-shop types of machining operations, stainless steels are absolutely no problem if handled properly. Slow speeds, sharp tools, and positive feeds are the secrets to machining stainless steels. The ferritics are gummy, and rigid tool setups are needed. The austenitics tend to cold work, and with these alloys, continuous feed is absolutely essential. Many times novice machinists will hesitate on the downfeed while drilling. If this is done, you are "finished." The drill point cold-works the dwell area to hardnesses as high as 50 HRC, and further drilling is impossible. This does not happen if the operator does not hesitate in the downfeed. In milling operations, positive chip loads are essential; "kissing" the surface will cause work hardening and distortion.

The PH alloys, because of their high hardness in the as-received condition (17-7 condition C can have a hardness of 47 HRC), require carbide tooling and careful control on speeds and feeds. It is very difficult to drill and tap holes smaller than no. 6 thread (4 mm). This usually poses no problem; most design applications can work around the necessity of small screws.

The PH alloys, martensitics, and ferritics grind without problems, but the nonferromagnetic nature of the austenitics requires special hold down practices in surface-grinding operations.

The free-machining grades of stainless steel should be considered (types 430F, 416, and 303) whenever substantial machining is required. The sulfur additions are really helpful. The biggest disadvantage of these grades is that some environments cause more corrosion on these grades than would have occurred on their nonsulfurized counterparts.

Pickling and Passivation

Stainless steels are basically iron–chromium alloys that contain a minimum of 10% chromium. Alloy additions of nickel, molybdenum, and other elements are also incorporated for enhanced properties. Stainless steel alloys achieve their corrosion resistance primarily from the formation of a very thin chromium oxide surface layer referred to as a *passive film*. Thermodynamically, chromium oxide is a very stable oxide that forms spontaneously on stainless steels upon exposure to air and moisture. However, for maximum corrosion resistance, the chromium oxide (or passive) film must be uniform, continuous, and free from defects. Defects such as iron contamination or welding scale disrupt the passive film, creating sites for corrosion. Free iron can be scrubbed onto the surface on stainless steel by handling with steel tools, such as drilling with a steel bit or forming in a steel die. Although the free iron is invisible, once exposed to water or other solutions the stainless

Table 14–7
Surface defects that may reduce the corrosion resistance of stainless steels

Surface Defect	Sources	Typical Cleaning Method
Free iron	Machining with steel tools Press forming or blanking with steel tools Handling with steel equipment Grit blasting/glass-bead blasting with contaminated media Handling with stainless steels tools that have previously handled steel Cleaning with steel items such as wire brushes or steel wool Rolling mills at the steel producer Tumbling, deburring, and lapping	Passivate
Heat treating scale	Stress relieving, hardening, annealing, mill scale	Pickle
Welding/cutting scale	Heat scale (discoloration), spatter, arc strikes, undercuts, flux, laser cutting, torch cutting, spot welding	Pickle
Mechanical defects	Scratches, grinding marks, metal chips and burrs	Mechanical polish or electropolish
Rust	Rust on stainless steel parts will continue to rust until removed	Pickle
Casting defects	Carbon pick-up from casting molds reduces the corrosion resistance on the surface	Pickle
Other contaminates	Greases, oils, crayon, or grease pen marks	Alkaline clean

steel readily rusts. Discolored scale from heat treating or welding also rusts if left on the surface of stainless steels. Table 14–7 lists common surface defects that may reduce the corrosion resistance of stainless steels.

Depending on the level of surface contamination, *pickling* or *passivating* may be required. In some instances, mechanical and electrochemical cleaning may also be used. Pickling or acid descaling is typically performed to remove heavy, tightly adhering oxide films such as those produced by heat treating, welding, hot-forming, and other high-temperature processes. It is also used to remove rust deposits. Generally, solutions containing sulfuric acid or nitric-hydrofluoric acid are used for pickling. Aside from removing the oxide contaminants, pickling will remove some of the base metal and may affect parts with close tolerances (± 0.01 mm or less). Pickling may also slightly roughen the surface of smooth stainless steel. In instances where the surface finish of austenitic stainless steels is critical, special solutions may be used for pickling. For example, a solution of phosphoric acid and a surfactant will remove rust and scale without adversely affecting surface finish.

For larger items or in instances where aqueous pickling solutions cannot be used, pickling pastes and sprays are commercially available. The areas of concern are locally coated with the paste and allowed to react with the surface. After sufficient time, a neutralizing paste is applied. Once the neutralizing reaction has mitigated, the paste may be cleaned off with water.

Passivating is most commonly used to remove iron contamination from stainless steels. Solutions of nitric acid, phosphoric acid, citric acid, and other organic acids are routinely used to passivate stainless steels. In most instances, passivating will not dissolve the base metal or roughen the surface. Passivating solutions will not remove oxide scales such as heat discoloration or rust.

Another means of cleaning stainless steel surfaces is electropolishing. Using special chemicals and electrical current, the surface of the part is uniformly corroded away, leaving a smooth, polished finish free from contamination. Electropolishing removes free iron contamination and creates a smooth, polished surface. The highly polished surfaces created by electropolishing are more resistant to corrosion than rough surfaces. Scale such as heat discoloration and rust must be removed mechanically or by pickling prior to electropolishing.

It is possible to remove oxide scales from heat treating, welding, and other high-temperature processes by mechanical means such as grit blasting, bead blasting, wire brushing, and grinding. Rust deposits may also be removed by these means. Some cleaning processes, such as grit blasting, have a tendency to roughen the surface. Roughened surfaces may reduce the corrosion resistance of the stainless steel. Abusive grinding may also lower the corrosion resistance. It is good practice to pickle or passivate parts after mechanical cleaning to remove any traces of iron contamination or scale.

Good corrosion resistance on stainless steel can also be achieved by preventing the formation of scale during fabrication. In welded stainless pipe and duct systems, for example, an inert gas such as nitrogen or argon may be used to prevent scale formation on the inside of the pipe while the pipe is TIG welded along outside diameter. Shielding with an inert gas is called *inert gas backing*. In heat treating operations, vacuum or inert gas atmospheres can be used to prevent scale formation.

Pickling and passivating processes use strong acid solutions that require special safety precautions. Disposing of spent solutions containing chromium, iron, nickel, and molybdenum must also be considered. Hence, serious consideration should be given to the application of chemical passivating treatments to stainless steels. For example, if a stainless steel is purchased from the steel mill in the passivated condition and dedicated machining equipment uses carbide cutting tools and grinding wheels to make the part, it may not be necessary to passivate the part. In such instances, a good alkaline cleaner may be sufficient to clean the part.

Different stainless steels require different pickling and passivating solutions to clean and passivate the surface. For example, 200 and 300 series austenitic stainless steel may be passivated in a nitric acid solution at 55°C. However, free-machining versions of these alloys should be passivated in a nitric acid solution with sodium chromate for corrosion inhibition. Without the chromate addition, the parts would develop a black/gray appearance upon immersion, ruining the passive surface film. A detailed guideline to pickling and passivation solution compositions is published in ASTM A 380 and ASTM A 97.

It is best to utilize a published standard such as ASTM: "Passivate per ASTM A 380 Code F" or "Pickle per ASTM A 380 Code E" to specify a cleaning procedure for stainless steel. Refer to ASTM A 380 or A 97 for the proper code for the stainless steel alloy and cleaning method of interest. It is also important to indicate the level of cleanliness required in the specification, for example, "After finishing, parts shall be clean, bright and free from, iron contamination, rust, heat scale, weld spatter/flux, dirt, debris, oils/organic contamination, or other injurious defects." Tests to evaluate surface cleanliness may also be specified (see Table 14–8). It is important to keep in mind that parts may be recontaminated after passivating or pickling. Purchasing specifications should ensure that appropriate precautions be taken to prevent recontamination.

Welding

Assorted welding problems exist with the different grades. Fusion welding should be avoided if possible on the martensitics. Modern ferritics are fusion welded. Thousands of tons are resistance or arc welded annually for automobile exhaust

Table 14–8
Tests for evaluating the cleanliness of stainless steels

Test Method	Brief Procedure	Response
Water-wetting/drying test	The part is exposed to wet–dry cycles for 24 h.	Any free iron will begin to rust.
High-humidity test	The part is exposed to 95%–100% relative humidity (RH) at 100–115°F for 24–26 h.	Any free iron will begin to rust.
Copper sulfate test	The part is immersed in or coated with a copper sulfate solution.	Free iron or iron oxide scale will turn copper colored. Best for PH, ferritic and austenitic stainless steels. Martensitic or lower Cr ferritics (<16% Cr) do not respond well.
Ferroxyl test	The part is immersed or coated with a solution of potassium ferricyanide and nitric acid.	Free iron, rust, and oxide scales will turn blue. This test is very sensitive.

systems, but mostly the ferritics are prone to grain growth at welding temperatures. This makes the metal weak, and cracking may result under solidification stresses. Second, the impact strength of coarsened, heat-affected basis metal may be very poor. Third, straight chromium stainless steels are subject to 475°C *embrittlement* and sigma phase formation. Certain of the special ferritics can be readily welded, but weldability of each alloy should be investigated with the alloy supplier.

The martensitics are not prone to *475 brittleness* or sigma-phase formation like the ferritics. Their problem is formation of hardened martensite during the quenching action that can follow welding. Cracking may result. The 440 family is absolutely unweldable by any process known (this statement reflects many unfortunate personal experiences). Besides all these ills, the free-machining grades of straight chromium stainless, 430F and 416, crack in the weld owing to hot shortness.

All of these problems can be overcome, and ferritics and martensitics are successfully welded, but there are potential problems, and welding of these grades should not be attempted without manufacturer-approved welding procedures.

The austenitics have excellent weldability by almost all welding processes. The free-machining grades (type 303 and others) should not be welded for the same reason mentioned previously. The only remaining problem with welding the austenitics is *sensitization*. As discussed in the previous chapter, heating stainless steels in the temperature range from approximately 750 to 1550°F (400 to 850°C) causes local chromium depletions the formation of chromium carbides, and a corresponding loss of corrosion protection.

There is a tendency to sensitize during welding and heat treating, and all stainless alloys are susceptible. The degree of sensitization is a function of the alloy type (ferritic, martensitic, and so on), the carbon content, the time that the material was exposed to the sensitizing range, and whether the alloy is *stabilized* with small amounts of Ti, Nb, or some other element. Figure 14–11 illustrates the approximate sensitizing region for two alloys, 304 and 304L. These data show how the use of L-grade steels helps to prevent sensitization. If a straight 304 stainless steel is welded, the heat-affected zone of the weld will cycle through the sensitizing range. If the L grade were used, it is very likely that the welding would not allow the part to stay in the

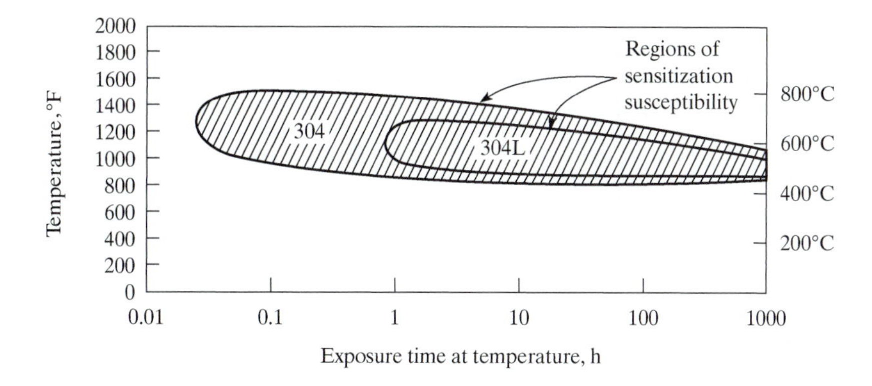

Figure 14–11
Exposure times at elevated temperatures that will produce sensitization of 304 and 304L stainless steel. These regions should be avoided in welding and heat treating.

sensitizing range for an hour, and thus it would not be sensitized. The L grade could be sensitized, however, if, for example, the weldment was stress relieved after welding.

It is probably apparent that it is very easy to sensitize a stainless steel through welding or through heat treatment. Sensitization, however, does not mean that the material loses all corrosion resistance. If a part only has to resist atmospheric rusting, most of the grades of stainless will do this when sensitized. The same is true about rusting in an outdoor environment. If a sensitized stainless is to be used in a chemical environment, it should be tested in that environment for attack in the sensitized condition. It may or may not be resistant. In general, it is good practice to use L grades or stabilized grades (321 and 347) of stainless for parts that are going to be welded or stress relieved and then put into chemical service.

The PH stainless steels weld well with few problems. Matching composition rods are preferred, and postweld heat treatments are required if the weld must be as strong as the basis metal. The normal practice is to weld in the age-hardened condition and re-age after welding. This provides about 75% joint efficiency.

Resistance welding is possible with all grades, but all the precautions mentioned still apply. It may also be necessary to abrade the faying surfaces to break down the oxide films.

Flame cutting stainless steels is very difficult because of the refractory oxide that forms near the melting range. It is best not to consider the use of normal flame cutting. Plasma cutting units, which have a much higher flame temperature, work effectively. If torch cutting is required, plasma is the process that should be used. In any case, it is good practice to grind away cut surfaces that had been molten to avoid possible contamination. Laser and electron beam cutting are possible in thin sections. Water jet (with abrasive) also works well.

Heat Treating

Because ferritic and duplex stainlesses do not quench harden, the only useful heat treatment is annealing. This is done to remove stresses. Precautions must be taken to prevent sigma and 475 brittleness. Usually, parts are slow cooled to ~600°C and then quenched.

The martensitic stainless steels use the same types of heat treatments that are used on alloy steels and tool steels. Types 403, 410, 414, and 416 are oil hardening; the others can usually be air hardened. Metallurgists prefer to call the low-temperature postquenching operation

Table 14–9
Heat treatment for martensitic stainless

Treatment	Purpose	Process
Full anneal	Maximum softening	1400 to 1600°F (800 to 900°C) slow cool
Process anneal	Soften hardened parts	1200 to 1400°F (650 to 800°C) air cool
Hardening	Harden and strengthen	1700 to 1950°F (920 to 1050°C) air or oil quench
Stress relieve	Increase toughness of hardened parts	300 to 750°F (120 to 400°C) air cool
Temper	Convert retained austenite	1000 to 1200°F (520 to 650°C) air cool

stress relieving rather than tempering. Alloy steels are tempered to reduce the carbon level of the martensite. In stainless steels it is argued that, because the carbon is already low in the martensite, the toughening mechanism is simply removal of thermally induced stresses. Tempering in martensitic stainless steels means heating at relatively high temperatures to transform retained austenite. Some typical heat treatments for martensitic stainless steels are shown in Table 14–9. The most used heat treatments are hardening followed by stress relieving. All other processes tend to sensitize the stainless and lower the corrosion resistance.

Normally, only two heat treating processes are used on austenitic stainless steels: annealing and stress relieving. Annealing is performed at temperatures in the range from 1800 to 2000°F (1000 to 1100°C), followed by a water quench. The quench is imperative if sensitization is to be prevented. Most stainless steel shapes are purchased in the annealed condition. Thus most people avoid this heat treatment. Water quenching machined parts causes severe distortion. On complex machined shapes or weldments, stress relieving may be required for stability. Some low-temperature processes [650 to 850°F (340 to 450°C)] are used without fear of sensitization, but they are really ineffective. Effective stress relieving requires temperatures above 1600°F (870°C), followed by as rapid a cooling rate as the part will permit. Sensitization will occur, but L-grade or stabilized alloys can be used to minimize this effect. The designer should question the necessity of stress relieving when the part is still on the boards. If it seems that stress relieving may be necessary, a low-carbon or stabilized grade of stainless should be specified.

PH stainless steels have very definite recipes that must be rigidly adhered to. Types 17-4, 15-5, and 13-8 are solution treated, quenched, and age hardened. The age-hardening temperature may be adjusted to achieve different strengths and toughnesses. Types 17-7 and 15-7 require solution treatment, conditioning, deep freezing, and aging. Again, the aging can be varied. There are also variations in the conditioning treatment. Typical PH stainless steel heat treatments are illustrated in Figure 14–12. There are two important points to be made concerning heat treatment of these alloys: (1) there are noticeable size changes, and (2) the high-temperature conditioning treatments on 17-7 and 15-7 cause scaling if performed in an air furnace. The problems are easily overcome if recognized. The size changes are simply allowed for in machining allowances. The scaling is usually overcome by doing the conditioning on rough machined parts. The aging heat treatment does not cause significant discoloration even in air furnaces. The best approach to handling PH stainless steels is to harden completely and then finish machine. The low-carbon martensite structure can usually be handled with cemented carbide tooling.

Summary

Many fabrication factors have been brought to light in this discussion; the intent was not to scare

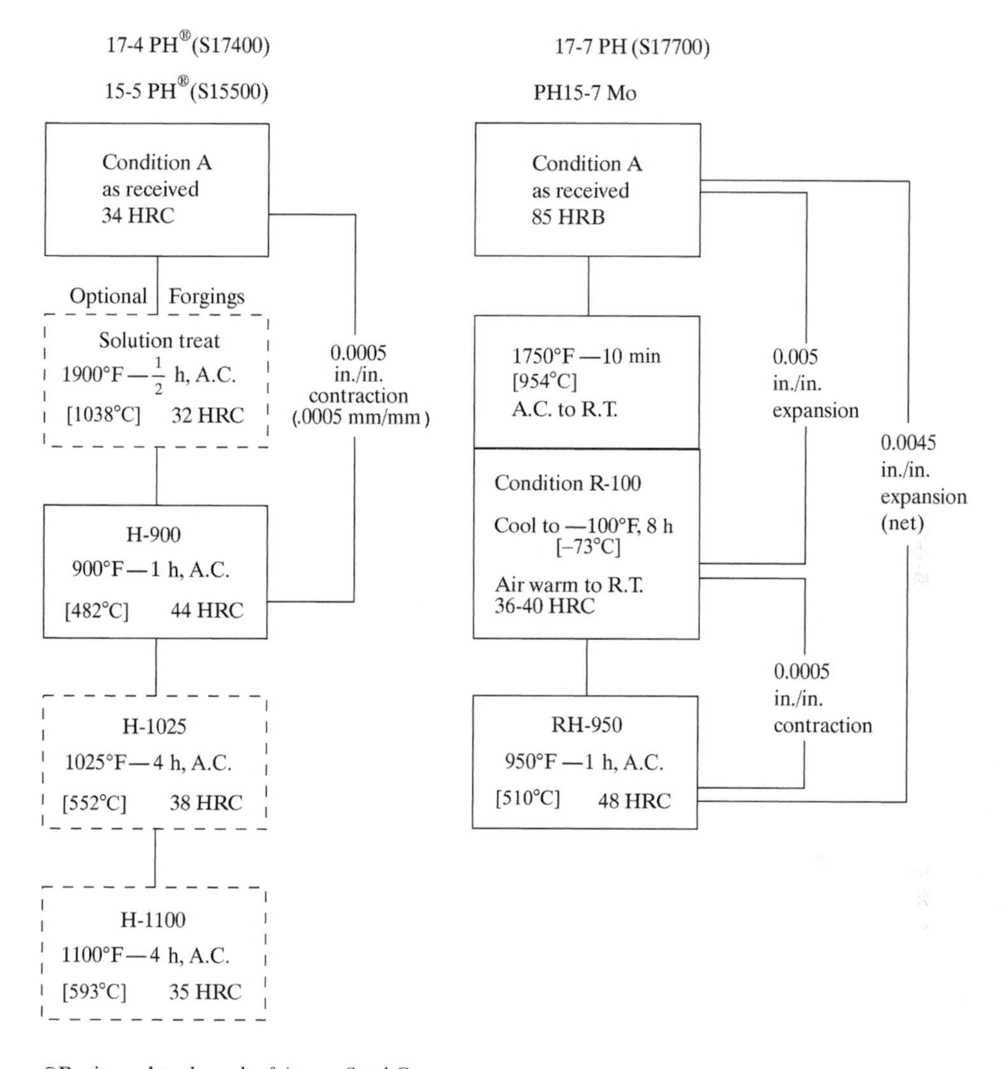

Figure 14–12
Typical heat treatment for PH stainless steels. See ASTM A 564 for additional heat treatment details.
Source: Courtesy of Armco Steel Corporation

the potential user but to educate. Fabrication of stainless steels is really easy as long as you take the time to review recommended precautions and practices. The chemical process industries have for many years used stainless steels for everything from I-beams to Faraday cups. They can be fabricated.

14.6 Corrosion Characteristics

Stainless steels owe their corrosion resistance to the chromium that is in solid solution in the various alloys. Chromium has an affinity for oxygen, and in solution in stainless steels it assists in the formation of a passive surface film that inhibits corrosion. The four major types of stainless steels—ferritic, martensitic, austenitic, and PH—are obtained by adding elements to a basic iron–carbon–chromium alloy or by varying the amounts of these elements. These composition changes also affect corrosion. The high-carbon alloys have lower corrosion resistance than the low-carbon grades. Nickel additions change the structure and allow higher chromium contents without getting into strength problems. Additions of sulfur or selenium for easier machining lower corrosion resistance. Additions of columbium, tantalum, and titanium prevent sensitization. Molybdenum additions reduce pitting tendencies. Thus the varying chemical compositions in stainless steels produce widely different corrosion characteristics. It is not possible to make blanket statements such as "Stainless steels are good in nitric acid." Some alloys are good in some concentrations at some temperatures. Correct corrosion-resistance data can be obtained only from corrosion-data surveys such as those referred to in the previous chapter. The best that we might do in this discussion is to make some general statements on the corrosion limitations of stainless steels and then comment on their suitability in some common environments.

Limitations

In Chapter 13, eight forms of corrosion were described. A number of the examples used for describing these processes were stainless steels. These metals are the most widely used for corrosion applications, yet they are fraught with frailties. This may seem incongruous, but if other metals were used in many types of corrosive environments, the list of frailties might be much longer. The major limitations or weaknesses of stainless steels are the following:

1. Stainless steels are prone to pitting in some environments.
2. They perform best in oxidizing environments.
3. They are susceptible to crevice corrosion.
4. They are prone to attack in chloride solutions and reducing acids.
5. Some are prone to stress corrosion cracking in halides, principally chlorides, and certain other atmospheres.
6. They are very susceptible to intergranular attack when sensitized.
7. They are susceptible to galvanic corrosion on a microscale when their structure consists of two phases (for example, ferrite in austenite).

Fortunately, we have learned to recognize the environments and situations that lead to these problems. They can be designed around.

Environments

Stainless steels surpass all other material systems in usefulness in corrosive environments. Table 14–10 presents some guidelines on the general resistance of specific alloys to types of environments. In the following we shall comment on how stainless steels react to some of the common environments that might be encountered in machine design.

Atmospheric Corrosion All the classes of stainless steels have excellent resistance to outdoor exposure, barring the presence of corrosive gases such as chlorine or sulfur. Test samples have been exposed to rural atmospheres for as long as 50 years with no measurable deterioration. Seaside atmospheres can cause attack because of chlorides, but the effect is usually minor. Alloys are available that will resist attack.

High-Temperature Oxidation As shown in Figure 14–13, some grades of stainless can be used in continuous service in temperatures up

Table 14–10
Relative corrosion resistance of AISI stainless steels*

AISI Type	UNS Number	Mild-Atmospheric and Fresh Water	Atmospheric		Salt water	Chemical		
			Industrial	Marine		Mild	Oxidizing	Reducing
201	S20100	×	×	×	—	×	×	—
202	S20200	×	×	×	—	×	×	—
205	S20500	×	×	×	—	×	×	—
301	S30100	×	×	×	—	×	×	—
302	S30200	×	×	×	—	×	×	—
302B	S30215	×	×	×	—	×	×	—
303	S30300	×	×	—	—	×	—	—
303 Se	S30323	×	×	—	—	×	—	—
304	S30400	×	×	×	—	×	×	—
304L	S30403	×	×	×	—	×	×	—
304Cu	S30430	×	×	×	—	×	×	—
304N	S30451	×	×	×	—	×	×	—
305	S30500	×	×	×	—	×	×	—
308	S30800	×	×	×	—	×	×	—
309	S30900	×	×	×	—	×	×	—
309S	S30908	×	×	×	—	×	×	—
310	S31000	×	×	×	—	×	×	—
310S	S31008	×	×	×	—	×	×	—
314	S31400	×	×	×	—	×	×	—
316	S31600	×	×	×	×	×	×	×
316F	S31620	×	×	×	×	×	×	×
316L	S31603	×	×	×	×	×	×	×
316N	S31651	×	×	×	×	×	×	×
317	S31700	×	×	×	×	×	×	×
317L	S31703	×	×	×	×	×	×	—
321	S32100	×	×	×	—	×	×	—
329	S32900	×	×	×	×	×	×	×
330	N08330	×	×	×	×	×	×	×
332	S33200	×	×	×	—	×	×	—
334	S33400	×	×	×	—	×	×	—
347	S34700	×	×	×	—	×	×	—
348	S34800	×	×	×	—	×	×	—
384	S38400	×	×	×	—	×	×	—
403	S40300	×	—	—	—	×	—	—
405	S40500	×	—	—	—	×	—	—
409	S40900	×	—	—	—	×	—	—
410	S41000	×	—	—	—	×	—	—
414	S41400	×	—	—	—	×	—	—
416	S41600	×	—	—	—	—	—	—
420	S42000	×	—	—	—	—	—	—
420F	S42020	×	—	—	—	—	—	—
422	S42200	×	—	—	—	—	—	—

Table 14–10
continued

AISI Type	UNS Number	Mild-Atmospheric and Fresh Water	Atmospheric Industrial	Atmospheric Marine	Saltwater	Chemical Mild	Chemical Oxidizing	Chemical Reducing
429	S42900	×	×	—	—	×	×	—
430	S43000	×	×	—	—	×	×	—
430F	S43020	×	×	—	—	×	—	—
431	S43100	×	×	×	—	×	—	—
434	S43400	×	×	×	—	×	×	—
436	S43600	×	×	×	—	×	×	—
440A	S44002	×	—	—	—	×	—	—
440B	S44003	×	—	—	—	—	—	—
440C	S44004	×	—	—	—	—	—	—
442	S44200	×	×	—	—	×	×	—
444	S44400	×	×	×	—	×	×	—
446	S44600	×	×	×	—	×	×	—
—	S13800	×	×	—	—	×	×	—
—	S15500	×	×	×	—	×	×	—
—	S17400	×	×	×	—	×	×	—
—	S17700	×	×	×	—	×	×	—

*An × indicates resistance to an environment. This is a guide and not a substitute for quantitative corrosion data.
Source: *Steel Products Manual—Stainless and Heat-Resisting Steels,* The Iron and Steel Society, November 1990

to 2000°F (1100°C) in oxidizing atmospheres, such as in hot-air furnaces. Temperatures for intermittent use are somewhat lower. Safe temperatures in reducing fuel gases are approximately the same.

Sulfuric Acid At room temperature, types 316 and 317 show low corrosion rates at low concentrations (<10%) and at very high concentrations (>95%). At higher temperatures, they are resistant only to very dilute solutions (<1%).

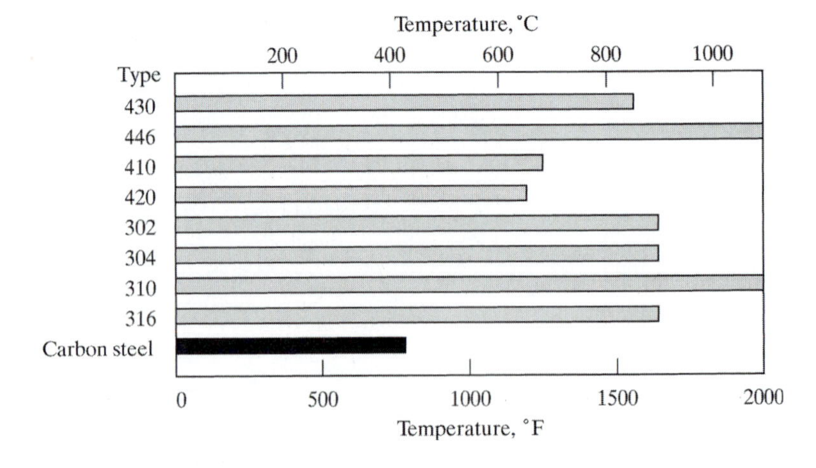

Figure 14–13
Maximum continuous-use temperature for stainless steels without excessive scaling in an oxidizing environment
Source: Courtesy of Republic Steel Corporation

Some of the proprietary alloys offer significantly improved resistance.

Nitric Acid Type 430 can be used at room temperature in concentrations up to 80%. Types 321 and 347 can be used at room temperature in all concentrations. Type 316 behaves about the same. Temperatures up to the boiling point can be used if the concentration is kept below 70%. Proprietary alloys are available for use with a wide variety of nitric acid conditions.

Phosphoric Acid Most grades are resistant to dilute solutions at room temperatures. Proprietary grades are available for more severe conditions.

Organic Solvents All classes are resistant as long as the solvents are uncontaminated.

Gasoline Types 410, 416, and 430 have fair resistance. Types 302, 304, and 316 have excellent resistance if the gasoline is uncontaminated with water, or the like.

Chloride Service Many service environments such as seawater, brackish water, potable water, and bleach solutions contain *chlorides*. Chlorides and chlorine-containing ions can cause pitting and crevice corrosion to stainless steel alloys. If the chloride-containing environment is above about 120°F (48°C), stress corrosion cracking is also a concern (if tensile stresses are present). Type 316 gives fair service in chloride-containing environments, although it is susceptible to pitting. Type 304 is not suitable when the chloride concentration is above about 50 ppm. It is totally unsuited for seawater service. Ferritic and PH alloys are generally not recommended for chloride service, particularly for stagnant conditions. Some of the proprietary super austenitic and super ferritic stainless steels provide exceptional resistance to pitting, crevice corrosion, and stress corrosion cracking in chloride-containing environments (see Table 14–1). An index for measuring the relative pitting resistance of stainless steel alloys is the *Pitting Resistance Equivalent Number* (PREN). Though empirical in nature, the PREN calculates the pitting resistance of austenitic and duplex stainless steel alloys based on the amount of chromium, molybdenum, and nitrogen in the alloy (these constituents have the strongest influence on pitting resistance).

$$\text{PREN} = \%\text{Cr} + 3.3(\%\text{Mo}) + 16(\%\text{N})$$

Although several forms of this equation exist (the nitrogen coefficient can range from 11 to 30) for specific environments, this equation is most widely used. Figure 14–14 ranks the PREN of various high-performance stainless steels. As an example, some seawater applications require the stainless steel alloy to have a PREN greater than 40.

Neutral Water Most grades show no attack (304, 316, 410, 430). Distilled water is similar. Velocities up to 100 ft/s can be tolerated in austenitic piping and pumps. A minimum velocity of 5 ft/s is recommended to prevent fouling, which can lead to underdeposit pitting. Chloride contents as low as 30 ppm can lead to SCC in austenitics.

Hydrochloric Acid Most concentrations cause rapid attack of most types.

Hydrofluoric Acid Stainless steels are not recommended for use in solutions of any concentration.

Acetic Acid Type 316 has good resistance at temperatures below the boiling point. Concentrations can be as high as 98%.

Food Products The austenitic grades have good corrosion resistance to fruit and vegetable

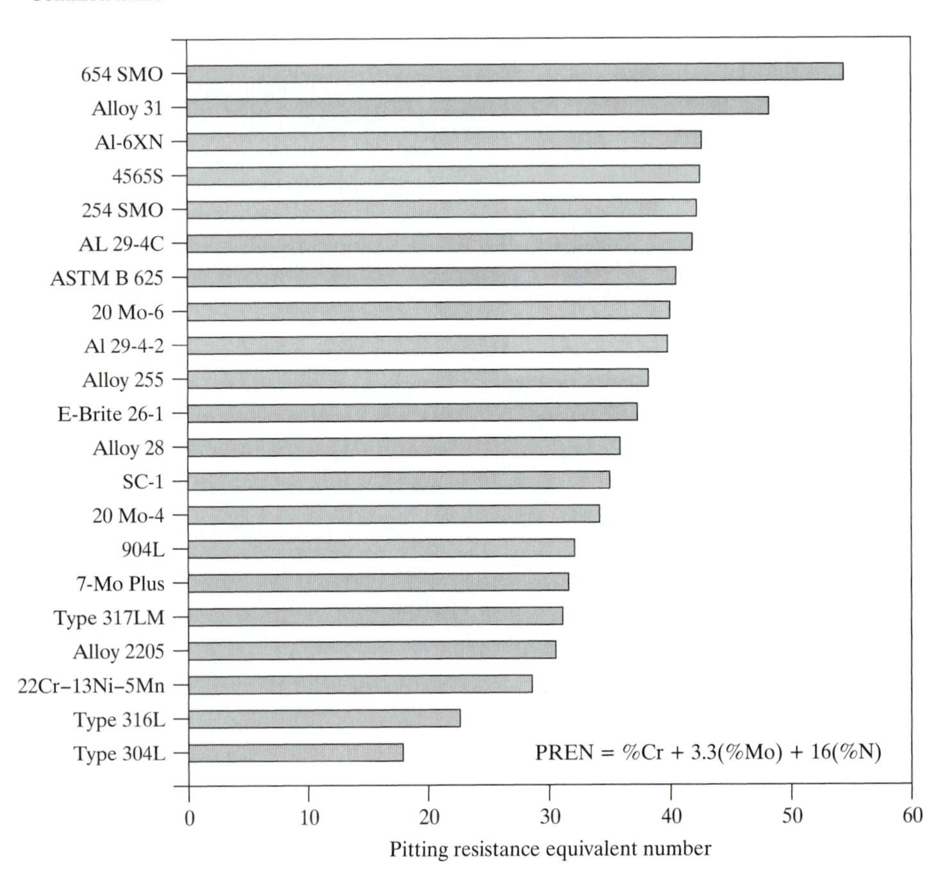

Figure 14–14
Relative pitting resistance of high-performance stainless steel alloys. Higher PREN values indicate greater resistance to pitting corrosion.

juices, and most types resist attack by dairy products. The presence of chlorides may cause pitting.

Bleaches All types and grades can be attacked. Their use is not recommended. Some of the super ferritics and super austenitics are acceptable for some solutions.

Alkalies All classes and types are resistant at room temperature.

The high-nickel grades (austenitics) are preferred at elevated temperatures.

We could go on indefinitely mentioning specific environments, but the ones mentioned will indicate that stainless steels can satisfactorily handle many environments in spite of the various forms of corrosion that can occur. It is still recommended before making a final choice of a specific stainless steel for a specific environment that corrosion data be consulted (NACE Corrosion Data Survey and others).

14.7 Alloy Selection

General Characteristics

It should be apparent at this point that stainless steels are very useful materials in engineering design. A wide range of mechanical properties exists, and the corrosion and oxidation resistance of these alloys is better than most other material systems. How do you select the right alloy for a particular application? Each class of stainless steels is intended for certain applications, and the types within each class have particular traits that set them apart from others in that class. Figure 14–15 shows the spectrum of stainless steel alloys and their primary areas of application. The ferritic alloys are somewhat lower in cost than the other alloys (Figure 14–16), and thus they find wide application for nonstructural applications. Type 430 is widely used for such things as household appliances and automobile trim. Type 430F

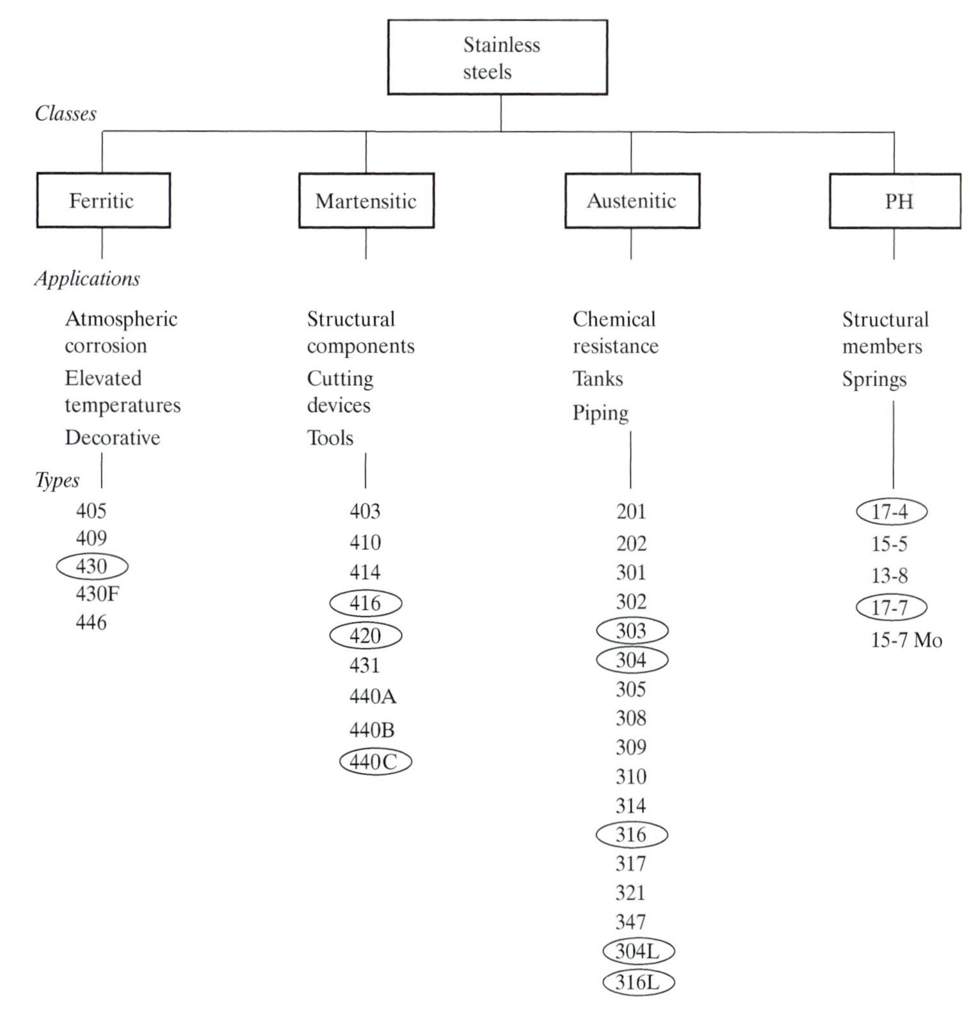

Figure 14–15
Widely used types of stainless steel

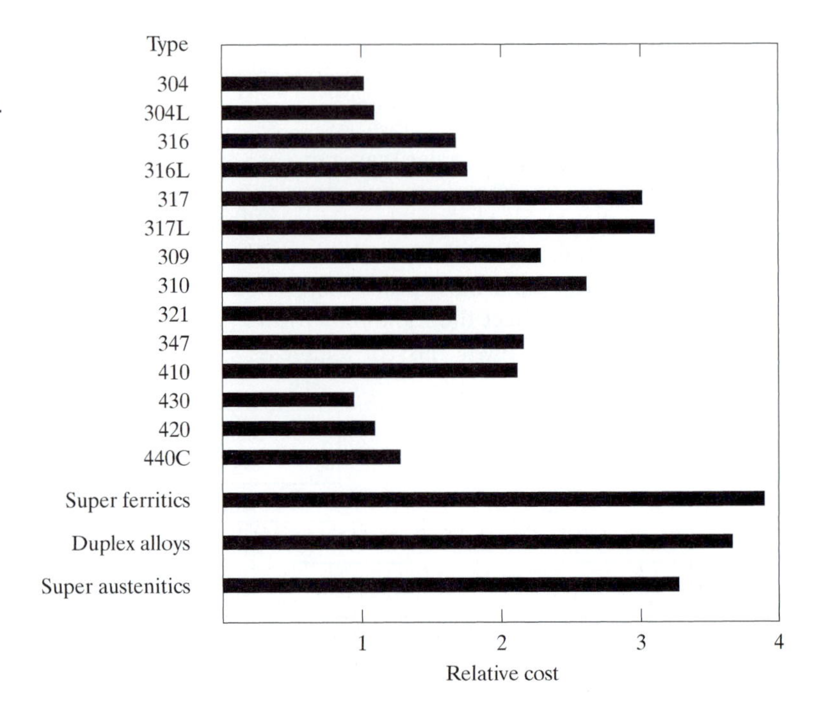

Figure 14–16
Relative cost factors for various stainless steel (hot finished bar).
Source: Courtesy of Republic Steel Corporation

is free machining, so it is used for fasteners and other parts that require screw-machine types of operation. Type 446 has very good *oxidation resistance*, so it sees use in furnace parts.

In the martensitic class, type 403 is used for elevated-temperature turbine parts. Types 410, 414, and 420 are used in cutlery, surgical instruments, and some tools. Type 416 is free machining, so it is used for fasteners and the like. Type 431 has the best corrosion resistance of the group; it is used where strength and corrosion resistance are required. The 440 family is used for tools.

Only two stainless steels have sufficient hardness and carbon content to be very useful for wear applications, types 440C and 420. The abrasion resistance of some stainless steels is compared with carbon steel and a tool steel in Figure 14–17. Even though the PH stainless steels have hardnesses above 40 HRC, they have poor wear characteristics for most forms of abrasion and sliding wear. The 440C alloy is capable of the highest hardness and abrasion resistance of any conventional stainless steel at 58 to 60 HRC. Types 420 and 440C are available in proprietary grades that are modified by P/M processing and chemistry changes for improved abrasion resistance. For example, 420 stainless steel containing about 6% vanadium provides better abrasion resistance than even some tool steels. These alloys are widely used for cutting tools and knives.

There are many of austenitic alloys, but most are a modification of the original 18-8, chromium–nickel composition. Types 201 and 301 readily work harden; they are often used in cold-rolled sheet and strip for flat springs and in cold-drawn wire form for corrosion-resisting wire springs. The 201 is like the 301 alloy in properties, but manganese replaces nickel as an austenitizing element. From time to time there are nickel shortages, and the manganese 200 series alloys provide an alternative composition.

Types 302 and 202 are the basic 18-8 alloys, which, with type 304, are what might be

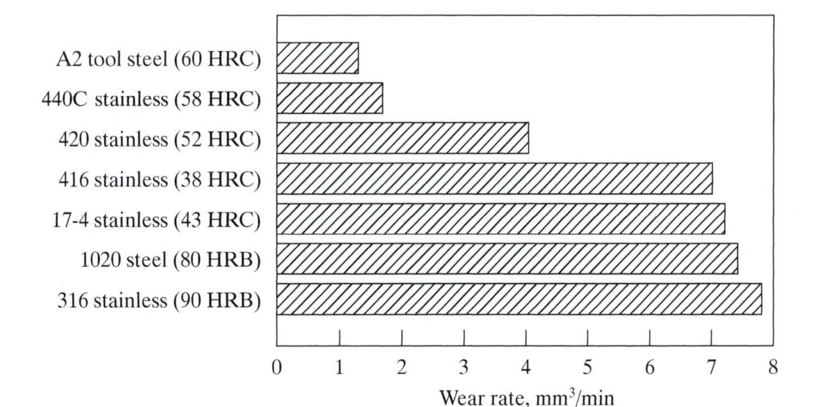

Figure 14–17
Relative wear rates of stainless steels in a dry-sand–rubber wheel abrasive test (modified ASTM G 65 procedure)

considered general-purpose alloys. They are used for tanks, architectural items, sanitary piping and wares, and any number of applications. They do not work harden as much as types 301 and 201, and thus they can be used for deep-drawn parts. Type 304 has a lower carbon content than 301 or 201. It is less susceptible to sensitization during welding than 301 and 201. It is the most widely used stainless steel based on tonnage. Only type 409, a special grade developed for automotive exhaust systems (catalytic convertors, etc.), competes in tonnage.

Type 303 has sulfur additions to make it free machining. It cannot be welded or readily pickled, but the free-machining characteristics justify its use on bolts, shafts, nuts, and mechanical components requiring excessive machining or large production quantities. The chemical resistance is lower than that of the nonsulfurized austenitics. Corrosion rates should be carefully checked before use.

Type 305 has a high nickel content that minimizes work hardening. It is considered a free-spinning alloy. That is, it can be used for spun shapes without fear of excessive work hardening. It is also used for cold-headed fasteners.

Type 308 contains relatively high percentages of chromium and nickel, which imparts an extra degree of corrosion resistance over 302- and 304-type alloys. It is frequently used for welding filler metal on 302 and 304 stainless steels.

Types 309 and 310 contain even more chromium and nickel than type 308 does. They are used for special chemical-process equipment. They have better corrosion resistance than most of the austenitics, but they also cost more (Figure 14–16). Type 310 is frequently used as the "universal" welding filler metal. Because it always stays austenitic and resists cracking, it is often used to weld carbon steels to stainless steels and for other dissimilar metal combinations. These alloys are also used for oxidation resistance.

Type 314 is basically a type 310 with about 2% silicon added to improve resistance to sulfuric acid and similar environments.

Type 316 is the most commonly used alloy for chemical service. It has about 2% molybdenum in it to improve corrosion characteristics in reducing media and to improve pitting resistance. It is used for everything imaginable in the chemical-process industries.

Type 317 is a "souped-up" 316 with higher percentages of chromium, nickel, and molybdenum. It has better pitting resistance than 316, but it costs more and is not as readily available in all forms.

Types 321 and 347 are essentially 18-8 stainless (like type 302), only stabilizing elements have been added to reduce sensitization in welding.

Types 304L and 316L are becoming *the* stainless steels to use for chemical service when sensitization may be a problem. If a structure will need a stress relief heat treatment, these grades are the best to use. The same is true if a given structure will involve high welding heat inputs. Shielded metal arc provides the lowest heat input in moderate-section-thickness weldments.

The precipitation-hardening alloys vary somewhat in chemical resistance, but they have better corrosion resistance than the martensitics and ferritics and are almost as good as the austenitics.

Type 17-4 was one of the early PH alloys; it often has a two-phase structure that causes anisotropy. Type 15-5 has replaced 17-4 in that respect, and type 13-8 is even better. All these alloys are particularly useful for structural components, such as shafting and agitators, that are immersed in chemicals. Their strength and stability also make them useful for general machine applications.

Type 17-7 is more commonly used for corrosion-resistant structural components that must be made from sheet, strip, and wire. If purchased in the cold-worked condition (condition C), springs and similar parts can be hardened by a simple 1-h age at 900°F (482°C). Distortion and discoloration are negligible.

We have discussed briefly most of the wrought stainless alloys. The same things can be said about the cast alloys. If there is a wrought equivalent to a cast alloy, the foregoing statements on usage will apply.

A Design Repertoire

Let us at this point address the problem of reducing our inventory of stainless steels. We have described about 40 alloys. The average designer will not become intimate with such a large number. Which alloys can be used to meet 90% of a designer's needs? Referring to Figure 14–15, it can be seen that some alloy types are circled. This is a proposed repertoire of specific alloys. These alloys are listed in Table 14–11 with a comment on when each might be preferred over the others. This list could probably be reduced further, but most of the alloys are usually on the shelf in any stainless steel warehouse. Thus, a list of 11 alloys should not be difficult to cope with.

When specifying an alloy on an engineering drawing, the three-digit alloy type, UNS, ASTM, or ACI designation should be used. The complete specification should also include applicable heat treatments and surface finish requirements.

Table 14–11
A repertoire of stainless steels

Type	Uses
430	For rust resistance on decorative and nonfunctional parts (bezels, nameplates, etc.)
416	Harden to 30 HRC and use for jigs, fixtures, and base plates
420	Harden to 50 to 52 HRC and use for tools that do not require a lot of wear resistance (injection molding cavities, nozzles, holding blocks, etc.)
440C	Harden to 57 to 60 HRC and use for cutting devices, punches, and dies
303	For fasteners and shafts where only rust or splash and spill resistance are needed
304 and 304L	For all types of chemical immersion (L grades for welding)
316 and 316L	For all types of chemical immersion where 304 is not adequate (L grades for welding)
17-4 PH	For highly stressed fasteners, shafting, agitators, and machine supports; age harden
17-7 PH	Harden to condition CH900 and use for chemical-resistant springs

Summary

Some designers may have occasion to use stainless steels every day; others might see the need only once a year. Most of the stainless steel produced (60%) goes into consumer products; the remainder goes into the chemical process industry. Whatever the case, every designer should know these alloys, what they can do, and their limitations. Successful use requires consultation of applicable corrosion data and a careful check of the environment and installation to make sure that there are no conditions that would lead to stress corrosion cracking, pitting, crevice corrosion, or any of the other insidious forms of corrosion that can occur. There is nothing mysterious about stainless steels, but they do have idiosyncrasies that need to be taken into consideration:

- Type 304 is the most widely used type.
- Stainless steels are often susceptible to local corrosion in the form of pitting, SCC, intergranular corrosion, and so on.
- The austenitics provide the best combination of corrosion resistance and fabricability for most applications.
- Stainless steels are usually not resistant to halide environments, reducing environments, and bleaches.
- Stainless steels are usually resistant to oxidizing environments.
- The proprietary grades of stainless steels are usually employed when standard grades have been tried and did not meet expectations.
- Situations that potentially cause sensitization should be avoided or dealt with by alloy selection.
- Most stainless steels have mechanical properties that are better than those of carbon steels: higher strength, toughness, and so on. However, austenitics are not normally recommended for fatigue applications.
- Stainless steels have a slightly lower modulus of elasticity than carbon and alloy steels.
- Annealed austenitic grades are not ferromagnetic; ferritics and hardenable and cold-worked austenitics are ferromagnetic.
- Stainless steels usually cost at least twice as much as carbon steel; the ferritics are the lowest-cost grade.
- All grades are resistant to atmospheric corrosion.
- Corrosion data should be consulted or testing should be done to pick grades for immersion in chemicals. The service environment must be duplicated.

Critical Concepts

- An iron–carbon–chromium alloy must be capable of forming a passive surface to be considered a stainless steel.
- There are four major categories of stainless steel (ferritic, martensitic, austenitic, and PH), and each grouping has different application characteristics.
- Each stainless steel has different corrosion characteristics, and selection should be based on corrosion data for the alloy of interest in the application environment.
- Most stainless steels have fabrication characteristics that need to be considered in process specifications.
- Stainless steels cost significantly more than carbon steels.

Terms You Should Remember

stainless steels	PH
ferritic	sigma phase
austenitic	ternary
martensitic	gamma field

metastable	oxidation resistance
austenitizer	chromium carbides
precipitation hardening	pickling
free machining	anisotropy
2B finish	ACI
cold work	sensitization
passivating	dual phase
stress relieving	embrittlement
stabilized	chlorides
reducing media	work harden
	475 brittleness

Case History

Stainless Steel Sailboat Rigging

A former co-worker brought a rust-spotted steel plate (with dimensions of 3 × 30 × 400 mm) to the materials engineering laboratory and asked a metallurgist to perform a failure analysis. The stainless steel part was a stern chain plate from a 20-ft sailboat. The chain plate broke causing the boat to lose its mast. This part fastens the stainless steel wire ropes that support the mast to the hull.

The chain plate was analyzed and found to be made from type 304 stainless steel. A metallographic examination of the fracture indicated that the part had failed due to SCC around the hole where a turnbuckle attached the rigging to the chain plate. The rust spots in the chain plate coincided with corrosion pits.

Apparently, this sailboat was used for years in a northern U.S. freshwater lake. When the owner retired, he moved to Florida and started sailing on the ocean (Tampa Bay). The failure occurred in its first summer in Florida. Why did SCC occur?

All austenitic stainless steels are susceptible to stress corrosion cracking in salt water if a critical temperature is reached—usually about 130°F (38°C)—and if the stress level is high enough. The chain plate was certainly under high stress; most sailors tension the fore and aft stays to the point of almost driving the mast through the bottom of the boat. They do this to improve sailing performance. Thus the required stress was present. The salt spray supplied the necessary corrodent (chlorine ions), and the sun apparently supplied the necessary temperature. Stainless steels are great materials, but they do not perform well in certain environments.

Questions

Section 14.1

1. Name the possible microstructures in all grades of stainless steel.
2. What makes a stainless steel "stainless"?
3. What type of structure would be present in an iron–chromium alloy with 45% chromium?
4. Referring to Figure 14–3, what chromium content would be necessary to make a ferritic steel that was quench hardenable?
5. Type D2 tool steel contains 12% chromium. Is it a stainless steel? Explain.
6. What are duplex alloys? How are they achieved?

Section 14.2

7. Name the basic ferritic, austenitic, and martensitic stainless grades and state their composition.
8. How would you designate the use of wrought 17-4 stainless steel on an engineering drawing?
9. What is the correct designation for a type 304 stainless steel casting?
10. Explain the UNS numbering system for stainless steels.

Section 14.3

11. Calculate the diameter change that a 20-mm-diameter type 316 stainless steel shaft would experience when it is put in service at 500°F (260°C). Compare this to the size change that carbon steel would experience under the same conditions.
12. Calculate the difference in deflection of stainless and carbon steel cantilever beams 3 m long (I = 0.004 mm^4) carrying a concentrated end load of 10 kg.
13. State the types of stainless steel that will be attracted to a magnet (ferromagnetic).

Section 14.4

14. Compare the yield strength of type 304 stainless to that of 1020 steel. Compare 17-4 to 1020.
15. How many stainless steels can be hardened to hardnesses greater than 50 HRC? Name the alloys that are applicable.
16. How hard can stainless steels get by cold working?

Section 14.5

17. Name two free-machining stainless steel alloys.
18. What solution would you use to passivate a 416 stainless steel?
19. Would a 304 stainless steel be sensitized if it was laser welded? Explain.
20. Name the stainless steels that are unweldable.
21. How does the machinability of stainless steel compare with that of carbon steel (1020)?
22. What is the hardest stainless steel?

Section 14.6

23. Will a sensitized stainless steel rust in 60% RH room air?
24. Name three environments that will cause rapid corrosion of any stainless steel.
25. You want to make a blade for a mechanized meat slicer. What stainless steel would you use? Why?
26. You want to make a deep-drawn vessel similar to a sink. What grade of stainless would you use? Why?
27. How do you prevent stress corrosion cracking of stainless steel?
28. Write a passivation specification for 304 stainless steel fasteners.

To Dig Deeper

ASM Handbook, Volume 1, Properties and Selection: Irons, Steels, and High-Performance Alloys. Materials Park, OH: ASM International, 1990.

Beddoes, J., and J. G. Parr. *Introduction to Stainless Steels*, 3rd ed. Materials Park, OH: ASM International, 1999.

Chawla, S. L., and R. K. Gupta. *Material Selection for Corrosion Control*. Materials Park, OH: ASM International, 1994.

Davis, J. R., Ed. *Stainless Steels*. Materials Park, OH: ASM International, 1994.

Davis, J. R., Ed. *Heat Resistant Materials*. Materials Park, OH: ASM International, 1997.

Davis, J. R., Ed. *Alloy Digest Sourcebook: Stainless Steel*. Materials Park, OH: ASM International, 2000.

Parr, J. G., A. Hanson, and R. A. Cula. *Stainless Steel*. Materials Park, OH: ASM International, 1985.

Peckner, D., and I. M. Bernstein. *Handbook of Stainless Steel*. New York: McGraw-Hill Book Co., 1977.

Sedriks, A. John. *Corrosion of Stainless Steels*. New York: John Wiley & Sons, 1996.

Steel Products Manual, Stainless and Heat Resisting Steels. Warrendale, PA: Iron and Steel Society, 1990.

Zapffe, Carl A. *Stainless Steels*. Cleveland, OH: American Society for Metals, 1949.

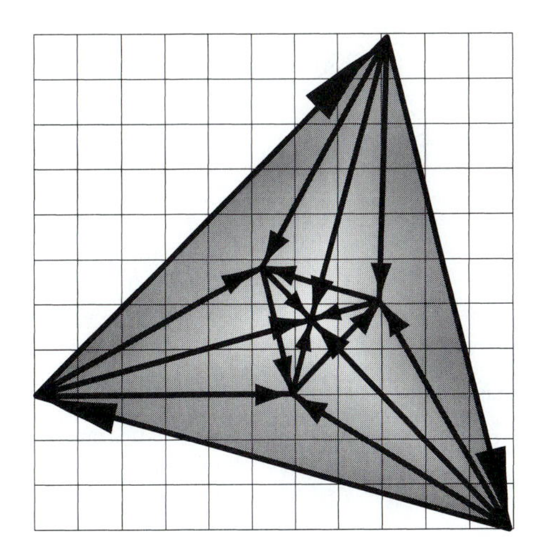

CHAPTER 15

Cast Iron, Cast Steel, and Powder Metallurgy Materials

Chapter Goals

1. A knowledge of how castings are made.
2. Guidelines for casting design.
3. An understanding of the various types of cast irons and steels.
4. A knowledge of powder metals.
5. Selection information on ferrous castings and powder metal parts.

Rationale

All engineers will in some way have to deal with cast irons, cast steels, and metal parts made with powder metallurgy techniques. There was not enough space to make separate chapters for each of these, so the somewhat diverse subjects of metal casting (foundry processes) with powder metal processes (P/M) have been combined because they both are used to shape metals for final salable shape or to near-net shape.

Of course, most metals can be cast to shape, but, from the tonnage standpoint, cast irons and cast steels are the most important. Cast irons are still key to the automobile (for engine blocks) and countless piping and pump applications. Cast steels are used worldwide in rail transportation (wheels, switch gears, etc.). Compacting metal powders in molds and sintering them to make a shape is used as a manufacturing process for countless small parts in devices that are used in our daily lives, for instance, door locks, hidden parts in automobiles such as plain bearings, window lift mechanisms, hinges, etc. These parts are everywhere.

Thus this chapter is necessary because designers need to know what cast irons and steels are and when to use them, and they need to know the same things about powder metals. In many cases, cast irons and steels compete with powder metals. If a designed metal part looks as if it could be made by casting, also consider a powder metal. When several parts can fit in your hand, Metal injection molding (MIM) and P/M become competitive processes.

In summary, cast irons, cast steels, and powder metals are required elements of engineering materials that designers and engineers need to know about.

To use a casting or not is often a matter of personal judgment. Some designers like to use them; some do not. Similarly, some industries use large numbers of castings; some seldom use any. Present-day welding technology allows this choice. Complicated parts can be made as weldments or castings. Economics should be the factor that determines the choice between a casting or a fabrication. Sometimes a quantity of one is sufficient volume to justify the hardware required to go the casting route. Thus every designer should consider the use of castings every time it looks like a part may require extensive shaping by machining or when a cast alloy may offer some property advantages over wrought alloys.

It is the purpose of this chapter to familiarize the designer with the most commonly used casting processes and ferrous casting alloys. The objective is to provide enough information so that the designer can make an intelligent decision on casting technique and alloy if he or she decides to choose casting. The format will be to discuss casting processes and design techniques, and then to focus on the most important casting systems: cast irons and steels. Obviously, nonferrous castings such as aluminum and copper are also important, but these will be discussed in subsequent chapters. The casting process and design information will still apply. This chapter will conclude with a discussion of powder metallurgy, which competes with casting as a fabrication process for small parts.

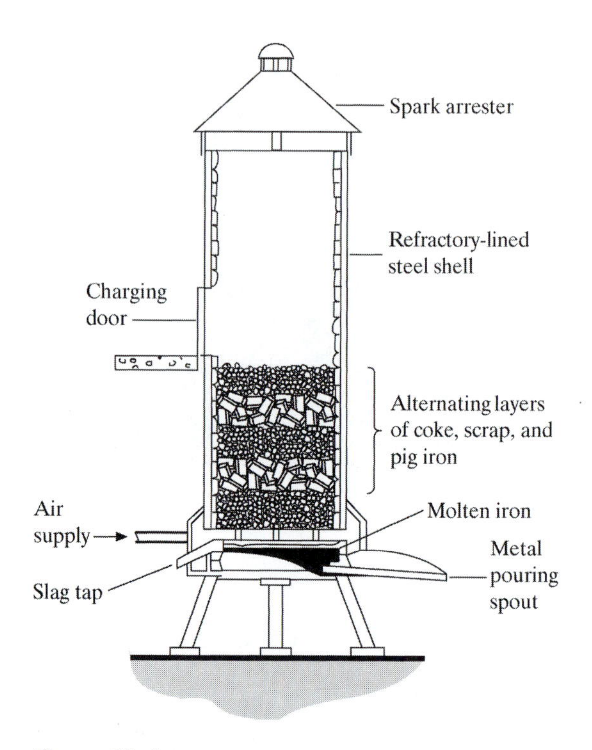

Figure 15–1
Cupola used to melt cast iron

15.1 Casting Processes

A *casting* is simply a shape achieved by allowing a liquid to solidify in a mold. The casting takes the shape of the mold. Many techniques are used to melt metals to make castings. Cast irons are often melted in a *cupola*, as shown in Figure 15–1. Other melting techniques include electric-arc, induction, open-hearth, and reverberatory furnaces; almost any technique may be used to melt a metal for casting. Castings can be made in sand molds, metal molds, *graphite* molds, and plaster molds, by pouring, low-pressure injection, high-pressure injection, spinning, and so on. The casting process can affect a part design, the alloy, the soundness, and even the surface finish. This is why the designer should have a basic knowledge of how each casting process works. Without going into unnecessary detail, the most commonly used casting processes are shown in Figure 15–2.

It is usually the responsibility of the design engineer to select a casting process. A complete casting drawing will show the process and alloy designation, as well as pertinent postcasting treatments. Each process has advantages and disadvantages. Some factors to consider in evaluating a casting process are

1. Tolerance
2. Surface finish
3. Size limitations

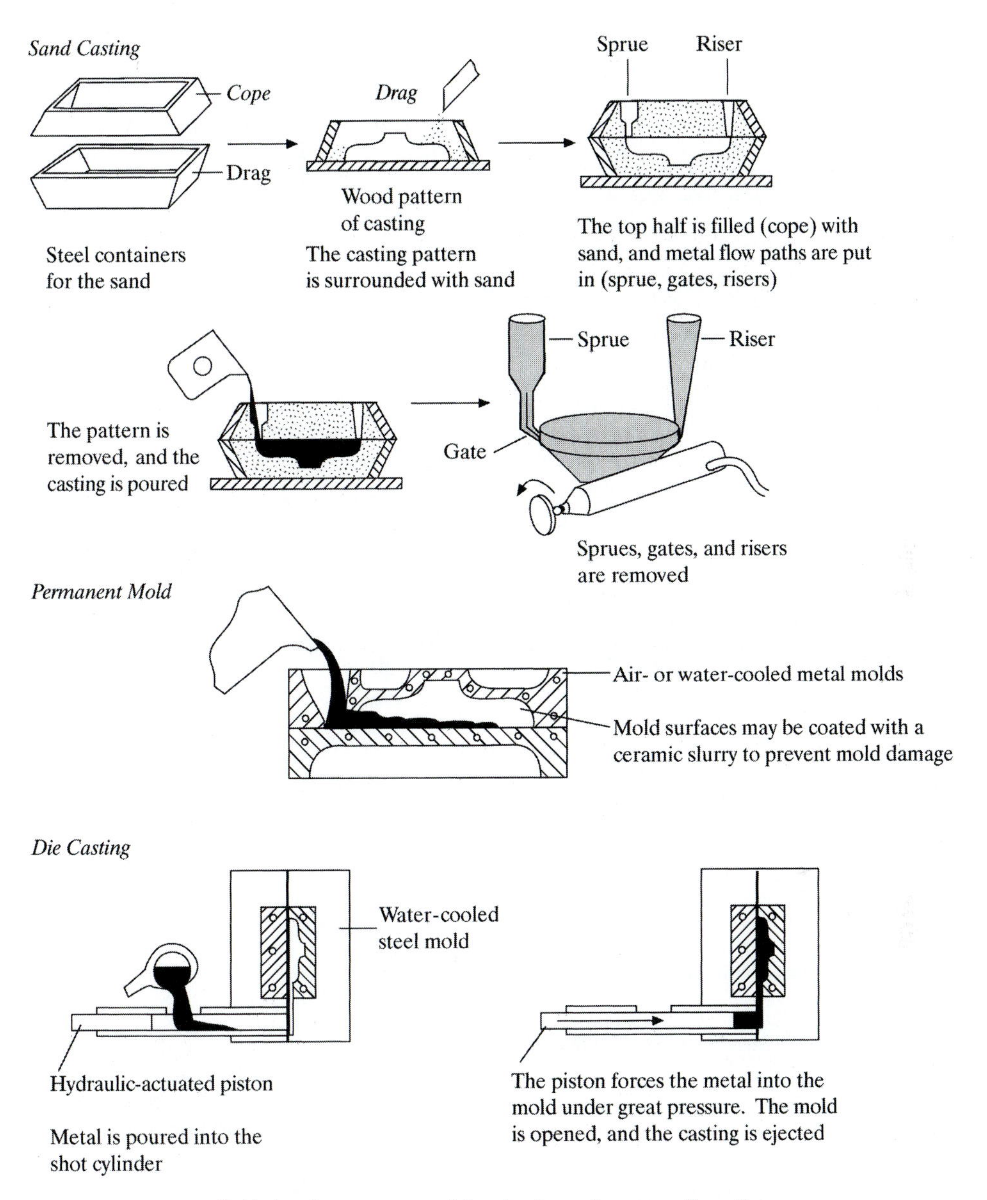
Sand Casting
Cope
Drag
Steel containers
for the sand
Drag
Wood pattern
of casting
The casting pattern
is surrounded with sand
Sprue
Riser
The top half is filled (cope) with
sand, and metal flow paths are put
in (sprue, gates, risers)
The pattern is
removed, and the
casting is poured
Sprue
Riser
Gate
Sprues, gates, and risers
are removed
Permanent Mold
Air- or water-cooled metal molds
Mold surfaces may be coated with a
ceramic slurry to prevent mold damage
Die Casting
Water-cooled
steel mold
Hydraulic-actuated piston
Metal is poured into the
shot cylinder
The piston forces the metal into the
mold under great pressure. The mold
is opened, and the casting is ejected
Cold chamber process used for aluminum & copper alloys. See
Figure 18–3 for submerged plunger machine used for zinc.

Figure 15–2
Common casting processes

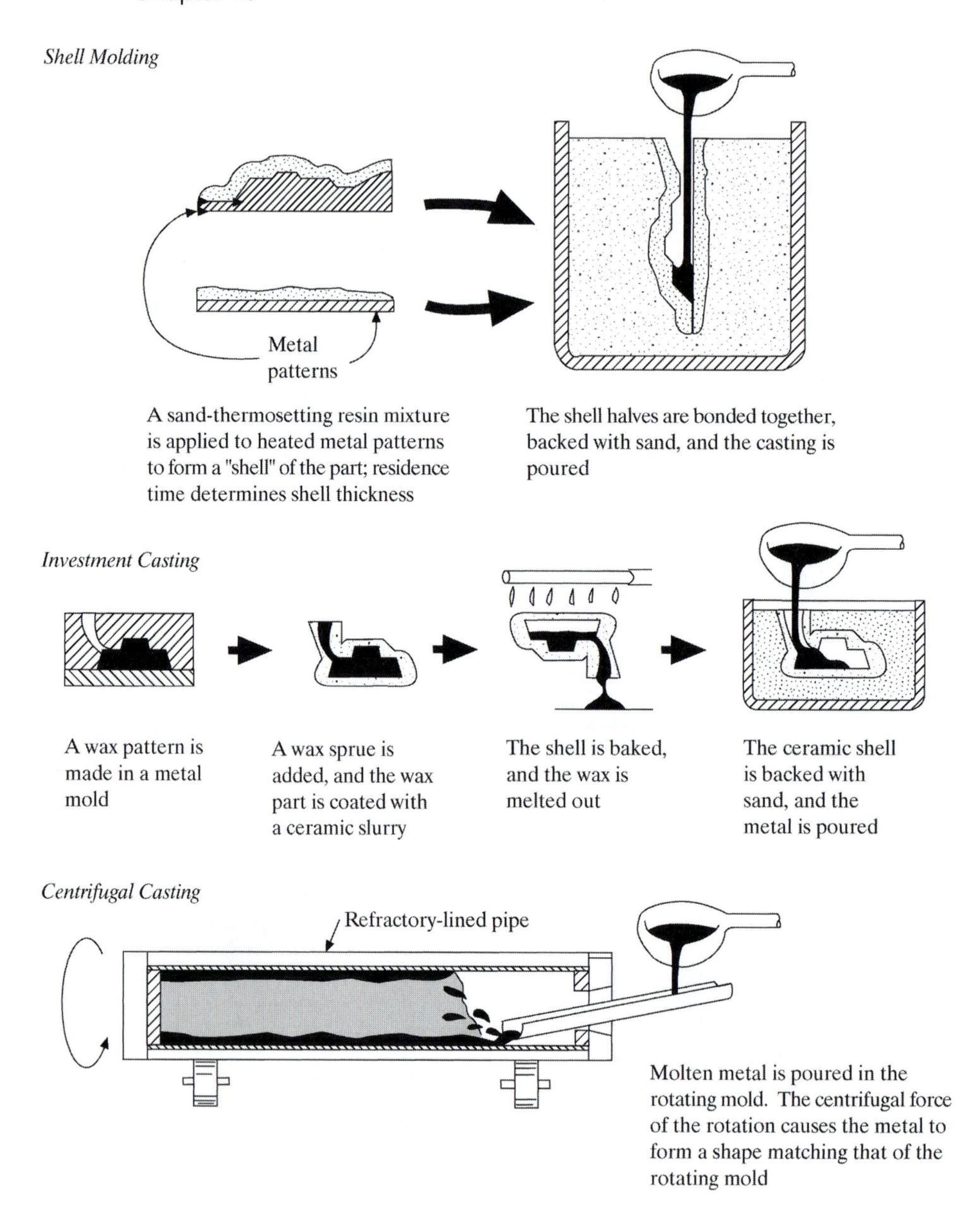

Shell Molding
Metal
patterns
A sand-thermosetting resin mixture
is applied to heated metal patterns
to form a "shell" of the part; residence
time determines shell thickness
The shell halves are bonded together,
backed with sand, and the casting is
poured
Investment Casting
A wax pattern is
made in a metal
mold
A wax sprue is
added, and the wax
part is coated with
a ceramic slurry
The shell is baked,
and the wax is
melted out
The ceramic shell
is backed with
sand, and the
metal is poured
Centrifugal Casting
Refractory-lined pipe
Molten metal is poured in the
rotating mold. The centrifugal force
of the rotation causes the metal to
form a shape matching that of the
rotating mold

Figure 15–2
continued

4. Materials that can be cast
5. Casting cost
6. Availability
7. Anticipated quantities
8. Wall thickness

Each casting process described has different characteristics in these areas. Process selection should be based on a comparison of these factors. We shall briefly discuss common casting processes with these factors in mind.

Tolerances

The commercial tolerances for some casting processes are shown in Table 15–1. The possible tolerances depend on the nature of the process and in many cases the nature of the metal being cast (melting point, shrinkage). The data in Table 15–1 show different tolerances on sand castings and die castings for different metals. The reason for the dependence of tolerance on the nature of the alloy is that some metals have high-volume shrinkage rates on solidification; others have low *solidification, shrinkage* rates. It is a law of nature that metals shrink or contract on going from the liquid to the solid state and on cooling from high temperatures. Nothing can be done about it but to apply the shrink rate to the pattern dimensions. Typical *shrinkage allowances* for sand-cast metals are shown in Table 15–2. The shrinkage rate can be as low as 1% or as high as about 5%. This factor is taken into consideration in arriving at tolerances for the various casting processes.

The lowest tolerances of any of the casting processes are on die casting. This process, unfortunately, is limited on a commercial basis to nonferrous metals. Investment castings have the lowest tolerances of the processes that apply to ferrous as well as nonferrous metals. *Sand casting* has the largest tolerance.

Surface Finish

Surface finish should be of concern to the designer if some casting surfaces are to be functional as cast or if the casting has to have low roughness for such purposes as cleanability or appearance. Any of the casting processes that have coarse sand as the mold surface will have high roughness. Table 15–3 shows typical surface roughness ranges for some of the casting processes. Small die castings and investment castings often have finishes lower than the ranges cited. It is not uncommon to achieve surfaces good enough to allow assembly of as-cast parts. Very seldom are sand-cast surfaces good enough to allow attachment of other parts with a respectable fit. Figure 15–3 shows the surface finish possible on a typical *investment casting*.

Size Limitations

The casting processes with the best tolerances and surface finish do not apply to large parts. An extremely large die casting may weigh 40 lb (18 kg). Investment castings seldom exceed 30 lb (13 kg), except in aircraft and aerospace applications. If a part under consideration weighs more than these limits, chances are another process should be considered.

Slenderness is also a consideration. Investment castings are seldom longer than 18 in. (450 mm). Die castings have been made 4 or 5 ft long (~1.5 m), but this is an unusually expensive casting. The molding machines used on die-cast auto dashboards can cost $2 million, and the mold cost is not far from this. There is no real limit to the size of sand castings. Shell-molded parts are not usually large, because large parts would cause handling problems with the resin-bonded shells. A typical shell-molded part is a camshaft for an automobile.

Most cast-iron pipes are centrifugally cast. Typical lengths are 20 ft (6 m), and diameters as large as 24 in. (600 mm) are not uncommon. The

Table 15–1
Tolerances for several casting processes (small parts)

Sand Castings (Various Metals)		Sand Castings (Gray Iron)	
Alloy	**Tolerance, in. (mm)**	**Dimension, in. (mm)**	**Tolerance, in. (mm)**
Aluminum	±0.031 (0.79)	up to 8 (203)	±1/16 (1.6)
Beryllium copper	±0.062 (1.57)	up to 12 (356)	±3/32 (2.4)
		up to 24 (610)	±5/32 (4.0)
		up to 36 (914)	±3/16 (4.8)
Copper alloys	±0.093 (2.32)	over 36 (914)	±1/4 (6.4)
Steels	±0.062 (1.57)	across parting line add ±0.015 to 0.065 (.37 to 1.62)	
Magnesium	±0.031 (0.78)		
Malleable iron	±0.031 (0.78)		

Resin Shell (Most Alloys)		Investment Castings (Most Alloys)	
Dimension, in. (mm)	**Tolerance, in. (mm)**	**Dimension, in. (mm)**	**Tolerance, in. (μm)**
		0.000 to 0.500 (0 to 12.7)	±0.003 (75)
0.000 to 1.00 (0 to 25.4)	±0.010 (0.25)	0.500 to 1.000 (12.7 to 25.4)	±0.005 (125)
1.00 to 3.0 (25.4 to 76.2)	±0.016 (0.41)	1.00 to 1.50 (25.4 to 38.1)	±0.007 (175)
3.00 to 6.0 (76.2 to 152.4)	±0.030 (0.76)	2.00 and over (>50.8)	±0.005 in/in. (0.005 mm/mm)
6.0 to 12 (152.4 to 304.8)	±0.062 (1.57)	Straightness	±0.005 in./in. (0.005 mm/mm)
12 and over (>304.8)	±0.093 (2.36)	Hole positioning	±0.005 in./in. (0.005 mm/mm)

Die Castings

Alloys	Commercial Tolerance up to 1″, (μm)	Additional Tolerance, in.(μm) Over 1″ to 12″, (25 to 300 mm)	Over 12″ (300 mm)
Zinc	±0.003 (75)	±0.001 (25)	±0.001 (25)
Aluminum	±0.004 (100)	±0.0015 (37)	±0.001 (25)
Copper	±0.007 (175)	±0.002 (50)	
Magnesium	±0.004 (100)	±0.0015 (37)	±0.001 (25)

Additional tolerance must be added for dimensions across parting lines and for large areas. (See casting associations for detailed tolerance tables.)

Source: Reprinted from *Machine Design*, August 1949. © 1949 by Penton-IPC, Inc., Cleveland, OH.

Table 15–2
Shrinkage allowances for sand castings

Material	Shrinkage Allowance, in./ft*
Gray iron	0.125
Bronze	0.187
Nickel	0.250
Aluminum	0.156
Copper	0.312
High-alloy steel	up to 0.312
Magnesium	0.156
Nickel	0.250
Carbon steel	0.250
Stainless steel	0.250
Titanium	0.312
Zinc	0.012

*Multiply by 0.846 to convert to cm/m.

size limitation on centrifugal castings is the rolling mechanism for the molds.

Materials

Some metals, such as tungsten and molybdenum, cause casting problems because of their high melting temperature (Figure 15–4). They cannot be melted by conventional techniques. As mentioned previously, *die casting* is seldom used for ferrous metals. Steel cavities cannot withstand the high molten metal temperatures, and expensive refractory metal molds are needed.

Table 15–3
Surface roughness ranges for some casting processes

Process	As-Cast Roughness (Ra), μ in. (μm)
Die casting	20–100 (0.5–2.5)
Investment casting	50–100 (1.25–2.5)
Shell molding	250–700 (6.25–17)
Sand casting	250–1000 (6.25–25)

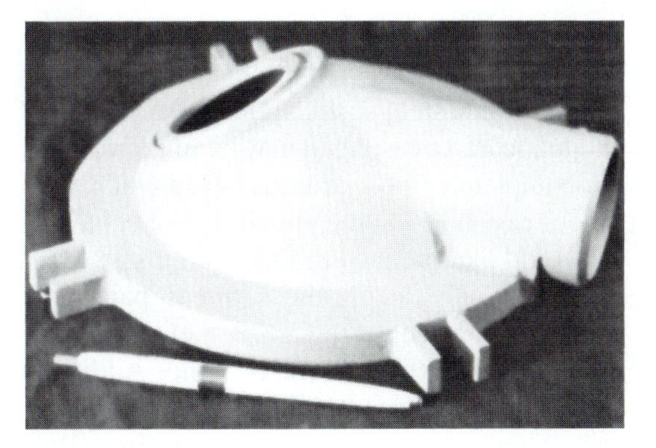

Figure 15–3
Stainless steel investment casting

Permanent mold and *centrifugal casting* can be done on ferrous metals because a refractory mold wash is used to prevent deterioration of the steel or cast-iron mold surfaces. *Shell molding*, investment casting, sand casting, centrifugal casting, and permanent mold casting can be applied to all common metal systems.

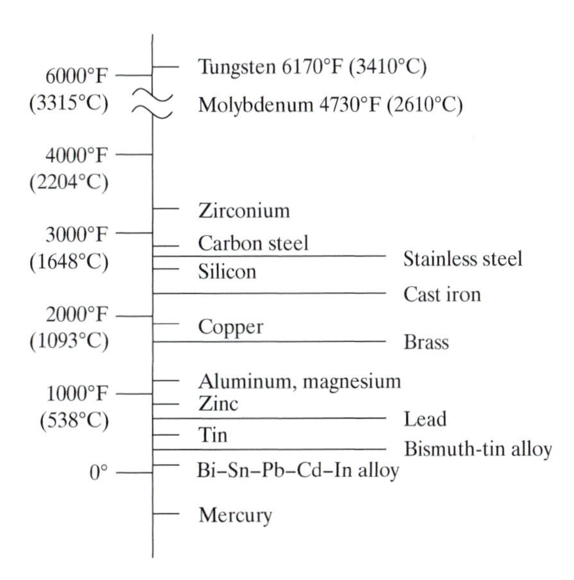

Figure 15–4
Approximate melting points of some common metals
Source: Courtesy of Talbot Associates, Inc.

In addition to differing melting characteristics, each metal has a different castability. If a wrought alloy steel such as 4130 steel is melted and made into a casting, it may be full of voids or not completely filled, or it may even crack. This is poor castability. Some metals do not behave as we would like them to in the molten state. It is best to use only casting alloys for cast parts. The chemical compositions of many metals are doctored to make them more castable. Additions are made to increase fluidity and reduce shrinkage. There is a casting alloy approximately equivalent to many wrought alloys. These are the alloys that should be specified by the designer; do not specify wrought alloys for castings unless there is no cast equivalent.

Table 15–4
Hardware requirements for casting processes

Casting Process	Hardware Required
Sand	Wooden pattern of the part
Investment	Metal mold (matching halves) for casting of the wax pattern
Centrifugal	Metal mold (usually split) of the shape to be rotated
Permanent mold	Coated/or water-cooled metal mold and mechanisms for activating cores and opening the mold
Die casting	Water-cooled metal cavities, machined metal holding blocks, and an ejection mechanism
Shell molding	Metal master is often required for making the shells

Casting Cost

The cost of casting versus a machined part versus a weldment is a question frequently faced by design engineers. There is no easy answer, and the best answer is simply a quotation from various foundries on each method of manufacture. However, some statements can be made concerning what is involved in making a casting with various techniques. Sand casting requires a wooden pattern, whereas die casting requires intricate, matched metal cavities. Table 15–4 lists the hardware requirements for casting processes.

All these processes, with the exception of sand casting, require metal molds. This usually means high hardware costs. In order of productivity, however, sand casting probably ranks last. A 20-lb (9-kg) sand casting may require several hours to cool to the point at which it can be shaken out. Cycle times for centrifugal castings and permanent mold castings can be shorter. Investment castings are often ganged so that a few or even hundreds can be made in a single pour. Die casting is by far the fastest casting process. Cycle times may be as short as 10 s, and multiple-cavity molds can make any number of parts in a cycle.

Thus, if only a few castings are required, sand casting may be the cheapest process. The long cycle time can be tolerated. The tooling costs for investment casting can often be easily amortized on as few as 50 castings. Centrifugal casting similarly fits the needs of low volumes. The other casting processes usually require quantities of thousands to override the high hardware costs.

Availability

The hardware requirements of some of the casting processes discussed often lead to problems of delivery. It takes considerable time to build the hardware for a die casting. Sand casting, the simplest process, uses a wooden pattern for prototypes, steel for production. Stereolithography and related processes can be used to produce a plastic pattern for rapid prototyping. Thus, in the consideration of castings versus other manufacturing techniques, it should be kept in mind that a sand casting may take 1 month to get, investment castings may take 3 months, and die castings and permanent mold castings may require 6 months' lead time before the first part is produced.

Summary

We have described most of the commonly used casting processes and some of their idiosyncrasies. There is no clear-cut method for selecting a particular casting process. It is recommended, however, that the designer consider each of the factors that we discussed (tolerances, surface finish, size) and question himself or herself on the requirements in each of these areas. The answers to these questions should point the way to the appropriate casting process for a part.

15.2 Casting Design

Most metal forms start as castings. Wrought steels evolved from cast ingots. The same thing is true for most other wrought metals. What goes on in a mold during the process of solidification can have an effect on properties. The mechanical and thermal processing steps involved in making wrought metal shapes can obliterate the effect of the casting process, but in using cast shapes the effects of the casting process are still there to be reckoned with.

Mechanism of Solidification

Metals in the molten state are characterized by atoms in a state of randomness or disarray. In the solid state, metals normally have a neat, orderly arrangement of atoms. They have a crystal structure. On a more macroscopic scale, metals are made up of grains, and the atoms in each grain have the same crystal orientation. We have seen how cold work can cause these grains to deform and strengthen a metal. Let us address the question of how these grains form. Grains are formed on solidification by atoms attaching themselves to nuclei of one form or another and then usually growing in a treelike form called a *dendrite*. A dendrite becomes a grain when solidification is complete. It is not really understood why dendrites are nature's way of crystal growth, but the fact that they occur needs little documentation. Anyone who has watched frost form on a window in the winter sees a dendritic, treelike crystal growth. Three major things happen when a casting solidifies: (1) Grains are formed, (2) melt chemistry changes during freezing, and (3) shrinkage occurs. The formation of dendrites and grains is important because they affect properties. Dendrites may start to form on a cold surface of a mold. They may start to grow from impurities, or they may nucleate at random. If dendrites grow on the mold wall, a *columnar* casting structure may be obtained (Figure 15–5). If the dendrites form randomly, the casting will have random grains, as shown in Figure 15–6. If the casting solidifies very slowly, coarse grains will occur. Usually, the coarse grains and columnar grains are undesirable

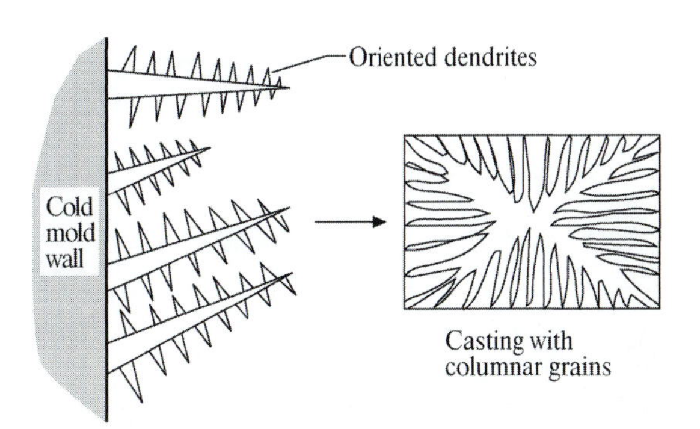

Figure 15–5
Oriented dendrites

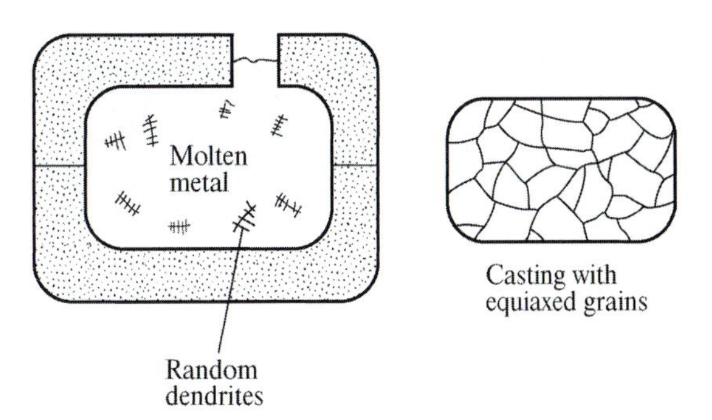

Figure 15–6
Random dendrites

Figure 15–7
Fractured casting showing columnar grain structure

because they make mechanical properties directional. Sometimes an oriented structure is desired. For example, turbine blades are directionally solidified to get columnar grains in the long direction. In most situations this is undesirable (Figure 15–7).

The second major occurrence in solidification is a change in melt chemistry. When metals solidify, the first solid to solidify is purer than the last. Foreign elements are essentially ejected as solid forms. In pure metals this is of no consequence, but engineering metals are not pure. They all contain other elements. The net result of this phenomenon is *segregation* (variation of chemical composition on a microscopic scale). Segregation is important in that it alters physical and mechanical properties. A casting that solidifies throughout at the same time will have fine dendrites (grains) and little segregation. The practical problem is that this type of solidification is difficult to achieve. The casting user is often at the mercy of the foundry concerning grain size and segregation, but the designer should recognize these factors because they can be the cause of service problems.

The final occurrence in the solidification process, shrinkage, is of crucial importance for a number of reasons, including: It affects the dimensional changes of the casting and increases the tendency for solidification shrinkage defects.

We have shown that shrinkage allowances on some metals are much greater than on others. High shrinkage rates make it harder to predict size changes in casting and also increase the possibility of complex internal stresses in the casting after cooling from the casting temperature.

Solidification shrinkage and porosity are the major sources of casting defects. Porosity can come from gas bubbles entrapped by the molten metal when it is poured into the mold. Microporosity is also possible when the treelike arms of dendrites intertwine and seal off feed paths for molten metal. Gross solidification shrinkage occurs when the major metal feed paths solidify before the remainder of the casting. The sources of porosity and shrinkage voids are shown schematically in Figure 15–8.

Figure 15–8
Sources of casting defects

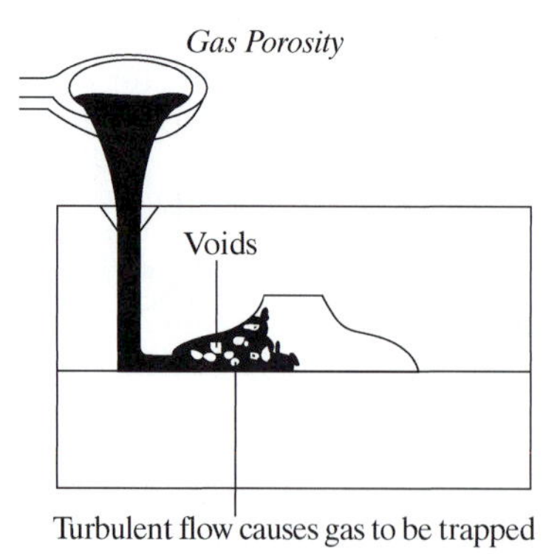
Gas Porosity
Voids
Turbulent flow causes gas to be trapped

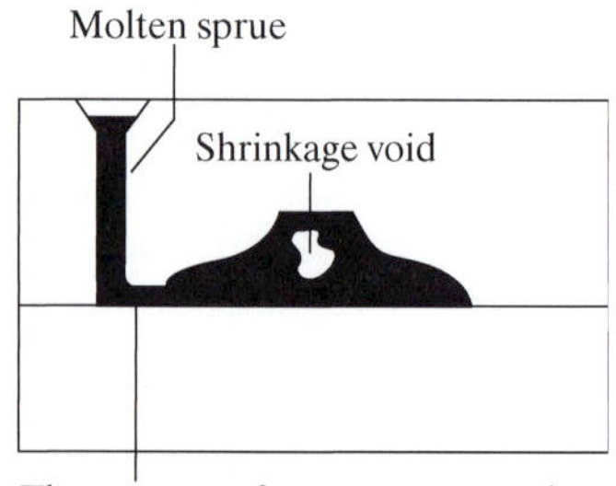
Solidification Shrinkage
Molten sprue
Shrinkage void
The gate area freezes prematurely, preventing the flow of molten metal to the heavy center section of the casting

What can be done about solidification shrinkage voids? The best solution is to design the casting so that gates and *risers* are the last areas to solidify. This is easier said than done, but progressive foundries have gone so far as to use electrical analogs and computer heat-transfer models to ensure that castings solidify in this manner.

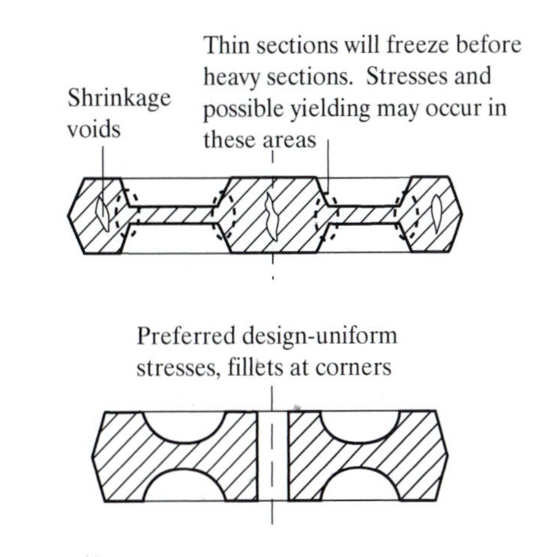

Figure 15–9
Shrink in heavy sections

Property Considerations

The property requirements of the part should be listed. A casting alloy is selected with these requirements as a primary basis. As we shall see when we discuss specific casting alloys, the mechanical and physical properties of cast metals often differ from those of their wrought equivalents. However, contrary to the opinions of some designers, castings are not brittle. A fine-grain casting with a postcasting normalizing treatment to reduce chemical segregation will have mechanical properties equivalent to any wrought counterpart of that alloy. The main point to be made is that castings should be engineered with all the considerations and design calculations used on any other component.

Casting Design Hints

Experience has shown that many casting problems are related to design, and the design errors seen by foundrymen are usually repetitive. The designer should try to predict how the metal may feed into the mold to make the desired shape. Feed paths should be as uniform in cross-sectional area as possible. Thick sections connected to heavy sections lead to casting stresses and solidification voids (Figure 15–9). Some methods for improving casting design are shown in Figure 15–10.

15.3 Gray Iron

Cast iron as an engineering material dates back to at least the fourteenth century. Before that time, most ferrous objects were made from wrought iron obtained by crude reduction of iron ore. *Wrought iron* is really pure iron with several percent of slag. Steels as we know them occurred in the nineteenth century, along with improved techniques for making cast irons. There are five major types of cast iron (Figure 15–11). Gray iron is the oldest type, but it is still widely used.

Today the term *cast iron* refers to a family of iron, carbon, and silicon alloys with the carbon content greater than can be accommodated in solid solution; a graphite or iron carbide phase is usually present. The spectrum of present-day cast irons is shown in Figure 15–12. *Gray iron* is a high-carbon, iron–carbon–silicon alloy. *Malleable iron* is similar in chemical composition, but its structure has been altered by thermal treatments to give it measurable ductility. *White iron* has a hard as-cast structure. *Ductile iron* is similar to malleable iron in ductility, but this ductility is achieved by ladle additions. *Alloy irons* usually are gray or white irons altered by alloy additions to make them hard or corrosion resistant. Thus, *cast iron* is a term that defines a family of iron–carbon alloys and not a specific material.

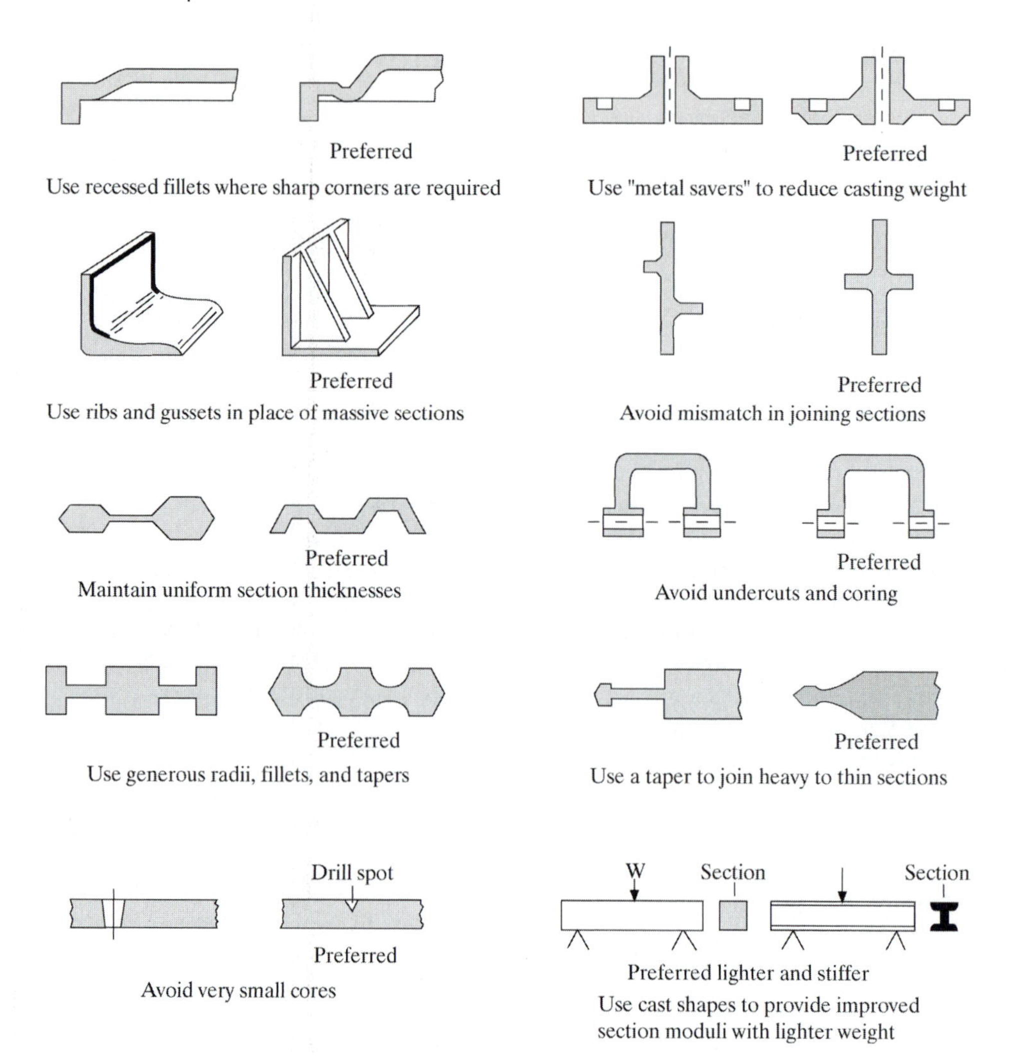
Preferred
Use recessed fillets where sharp corners are required
Preferred
Use "metal savers" to reduce casting weight
Preferred
Use ribs and gussets in place of massive sections
Preferred
Avoid mismatch in joining sections
Preferred
Maintain uniform section thicknesses
Preferred
Avoid undercuts and coring
Preferred
Use generous radii, fillets, and tapers
Preferred
Use a taper to join heavy to thin sections
Drill spot
Preferred
Avoid very small cores
W
Section
Section
Preferred lighter and stiffer
Use cast shapes to provide improved section moduli with lighter weight

Figure 15–10
Methods for improving casting design

Metallurgy of Gray Iron

A similarity among all cast irons is that they have carbon contents higher than about 2%. Iron–carbon alloys with lower carbon contents are considered steels. The carbon in most steels is either in solution or combined in carbide form. This is not the case with cast irons. The solubility limit of carbon in austenite is 1.7%. If an iron–carbon alloy has a carbon content of 3% or 4%, up to 1.7% of this carbon can go into solution, and the extra carbon is usually present in the room temperature structure in the form of graphite. Thus the carbon content of gray iron and other cast irons can be broken down into the component that is in solution in iron (combined carbon) and the part that is in the form of graphite. The accepted practice is to state *total carbon content* and *combined carbon content*. Gray iron can be considered to be a steel with graphite in it.

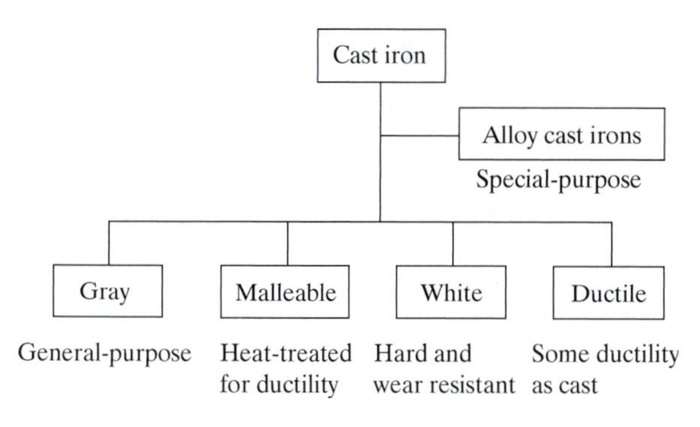

Figure 15–11
Spectrum of cast irons

The chemical composition of gray iron ranges from 2% to 4% total carbon with at least 1% silicon (Figure 15–12). The steel matrix of gray iron usually will have a microstructure of ferrite, pearlite, or martensite. The graphite is really present in the form of three-dimensional, rose-like structures, but metallographic sections show the graphite as flakes because metallographic studies require cross-sectioning of the specimen.

The shape, size, and distribution of the graphite flakes can have an effect on properties. Thus it is something closely scrutinized by foundry staff. An illustration of graphite distribution in gray iron is shown in Figure 15–13.

The iron–carbon equilibrium diagram, which was used to described phase relationships in steel, shows that cementite should be a predominating phase in high-carbon (>2%) iron–carbon alloys. If this diagram was projected to 100% carbon, it could be called the iron–graphite diagram. Graphite is really the thermodynamically stable form of carbon in all iron–carbon alloys.

The silicon in gray iron aids graphite formation. The chemistry and cooling rate are controlled in most commercial cast irons to prevent the formation of cementite-rich phases (ledeburite). The important gray irons have a steel matrix containing graphite flakes of varying size and volume distribution. The carbon content of the matrix is usually less than 1%. A ferrite matrix will yield a low-strength gray iron; a pearlitic matrix, higher strength; and quench hardening of a high-carbon matrix iron yields a

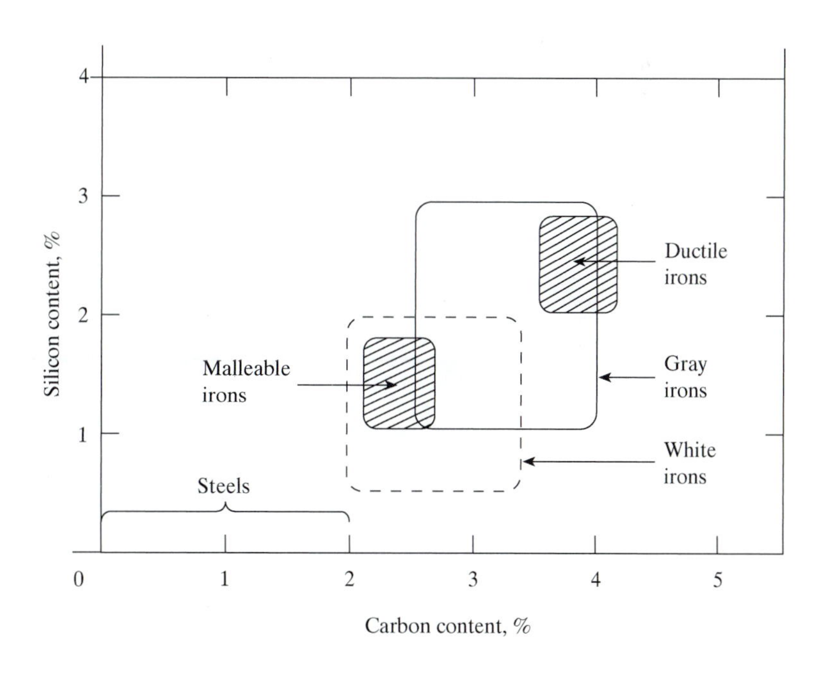

Figure 15–12
Approximate ranges of carbon and silicon in various types of cast irons

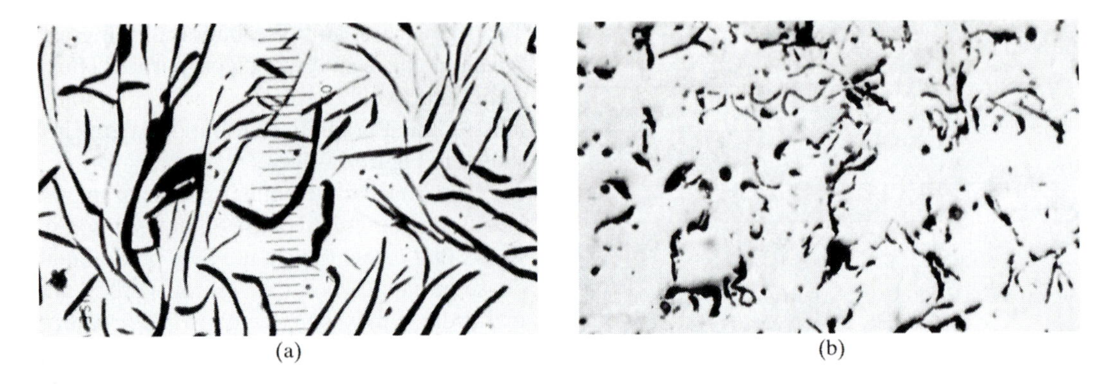

Figure 15–13
(a) Coarse and (b) fine graphite in gray iron (×100). ASTM specification A 247 explains interpretation of graphite morphology

high-strength martensitic matrix. Figure 15–14 shows the structure of a pearlitic gray iron before and after quench hardening. The pearlitic structure produced a hardness of 180 HB. The martensitic structure had a hardness of 600 HB.

Alloy Designation

It is not common practice to identify various gray irons by their chemical composition, because their microstructures are equally important. A common industry-accepted system for identification of gray-iron alloys is a class designation (ASTM A 48). There is a class 20, a class 30, and so on. The 20 in the designation stands for the minimum tensile strength in kips (1000 lb) per square inch. The common grades are shown in Table 15–5.

The complete alloy designation should have an A, B, C, or S suffix to indicate the size of the tensile test specimen used in measuring the tensile strength. Right or wrong, it is common practice to

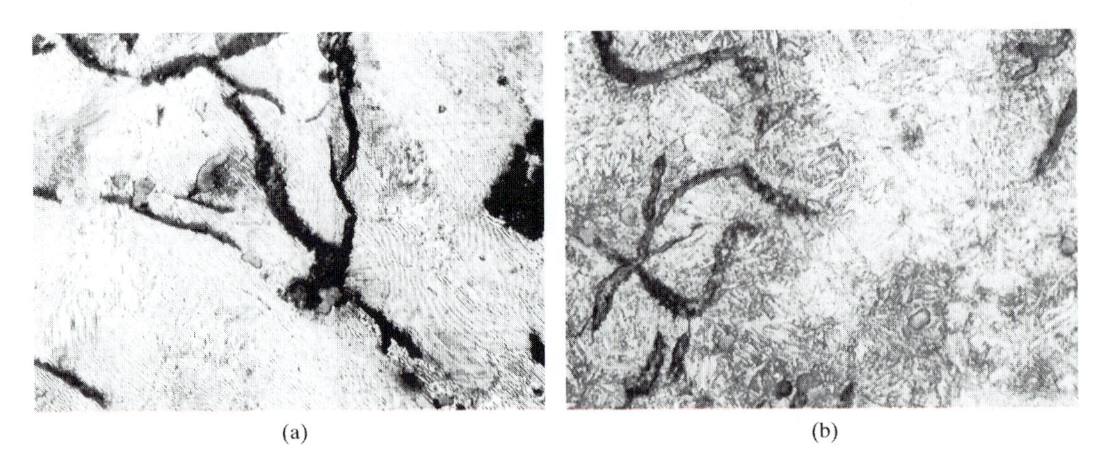

Figure 15–14
Microstructure of pearlite gray iron (a) before (b) after quench hardening (×500)

Table 15–5
Grades of gray-iron alloys

Class No.	Minimum Tensile Strength (psi)	MPa
20	20,000	138
25	25,000	172
30	30,000	207
35	35,000	241
40	40,000	276
45	45,000	310
50	50,000	344
55	55,000	380
60	60,000	414

neglect this suffix, at least in discussing these alloys. A drawing designation for a gray-iron casting might read as follows:

> *Material:* Gray iron, ASTM A 48 class 20 (state casting process: sand, investment, and so on)

Physical Properties

Because gray irons are essentially steels with graphite in them, some of their physical properties are similar to those of plain carbon steels. The thermal conductivity and coefficient of thermal expansion values are close to those of steels. The presence of graphite in the structure, however, does affect other physical properties such as damping capacity, electrical resistivity, magnetic characteristics, and even corrosion resistance.

The term *damping capacity* is often mentioned in discussing cast irons. It refers to the ability of a material to suppress elastic deflections or vibrations. There is very little quantitative information on this property, but because gray irons contain up to 10% by volume of graphite, it is said to have high damping capacity. The graphite essentially absorbs vibrations. Because of this property, cast iron has always been a favorite for machine bases and the like.

Another effect of the graphite in the microstructure of gray irons is to cause size change with time: pearlite may transform to graphite and ferrite, causing a growth. At elevated temperatures, internal oxidation of graphite can cause growth. As long as operating temperatures are kept below about 750°F (400°C), this phenomenon can be neglected.

Electrical resistivity is affected by the nature of the graphite in gray irons. A coarse graphite structure will have higher resistivity than a fine structure. All the graphite-containing cast irons have higher electrical resistivities than carbon steels.

All the common gray irons are ferromagnetic, but there are such things as austenitic cast irons, and, of course, these are not ferromagnetic. The other magnetic properties, such as permeability, coercive force, and hysteresis loss, are affected by the type of graphite structure in the particular type of gray iron.

The corrosion resistance of gray cast iron in most environments is better than that of carbon steels. When corrosion starts, some of the steel matrix dissolves, leaving the graphite flakes standing proud from the surface. They are more noble than the matrix. This relief effect aids the formation of a tenacious surface film, which in turn reduces the rate of overall attack. This is particularly true if the film is insoluble in the corrodent.

Gray irons are widely used for piping systems. They perform well in alkalies, seawater, and some acids. They are almost a standard material of construction for pipes for potable water and fire mains. Fifty years of service is not uncommon, and some old cities have cast-iron systems that have been in continuous service for hundreds of years. Most older homes in the United States have cast-iron piping on sewer connections. Cast-iron piping is being rediscovered in the United States for sanitary piping drains in new construction. It is quieted than plastic piping in flushing and draining because of its damping characteristics.

Mechanical Properties

We have already discussed the range of tensile strength available in gray iron. The class designations coincide with the minimum tensile strength. Noticeably absent in the discussion of mechanical properties of gray irons are data on yield strength, the reason being that gray irons are brittle. There is negligible plastic deformation in a tensile test, and thus the yield strength and tensile strength are usually one and the same. Figure 15–15 illustrates a typical tensile test stress–strain diagram on mild steel and gray iron. Important points to note are that there is no significant strain before failure in gray iron and the stress–strain curve has no linear portion. Gray irons do not obey Hooke's law very well. The graphite in the structure causes microslip at all stress levels. It does not have true elastic behavior. The modulus of elasticity of gray iron (slope of the curve) is lower than that of steel. Because the stress–strain is not linear, it is not even possible to determine the modulus of elasticity using the technique used on steels.

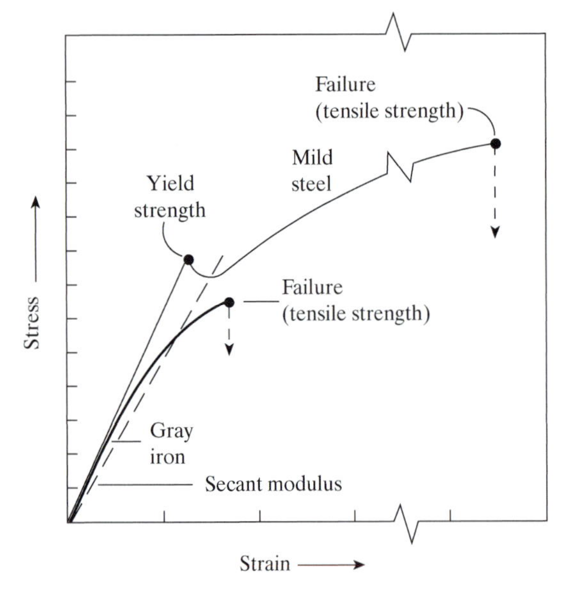

Figure 15–15
Comparison of the stress-strain curves for steel and gray cast iron

The modulus of elasticity (*secant modulus*) derived from a secant line of the strain curve can vary from 12 to 20×10^6 psi (83 to 138×10^3 MPa). Gray irons should not be used for tensile loading situations, if avoidable.

On the other hand, the compressive strength of gray iron is one of its better mechanical properties. Typically, the compressive strength can be three to five times the tensile strength. The shear strength is about equal to the tensile strength (from 1 to ~1.5 times).

Annealed ferritic gray iron typically has a hardness in the range of 110 to 140 HB; the pearlitic class 30 and class 40 irons have a hardness range of about 140 to 190 HB. Class 60 irons can be as hard as 350 HB, and the quench-hardened white irons can have hardnesses as high as 600 HB.

In addition to low ductility, gray irons have very poor toughness. The toughness is too low to measure with a typical notched bar in a swinging pendulum machine. Unnotched impact tests are run, and data are available on the various grades, but it is best to design with gray-cast-iron parts in such a manner that they never receive shock loads. Stress concentrations should also be avoided; there should be no sharp reentrant corners.

Gray cast irons have fatigue characteristics not unlike those of carbon steels. Typical fatigue strengths are about 40% of the tensile strength.

One of the most valuable properties of gray cast iron from the standpoint of design is wear resistance. Gray iron is not any better than medium carbon steel in resistance to abrasion, *fretting*, and some forms of corrosive wear, but the graphite in the structure provides assistance in resisting metal-to-metal wear. Gray iron is very resistant to seizure when used for threads, sliding devices, worms, and the like. Mated against a hardened steel and lubricated, gray iron provides a very low-wear sliding counterface. A testimonial to the metal-to-metal wear resistance of gray iron is its continued use for engine blocks in automobiles.

Another typical wear application for cast iron is in gears. Very large gray-iron gears are often used in combination with hardened steel pinions. Cast irons are similarly used in the quench-hardened condition. Gear teeth are sometimes flame hardened.

The graphite in the structure of gray iron is responsible for its metal-to-metal wear resistance. The graphite itself is a lubricant, and when it is leached from the surface by wear, the hole that remains serves as a reservoir for lubricant. No metal-to-metal sliding combination works well without lubrication, and cast iron is no exception. However, the graphite does improve performance when lubrication is inadequate.

Heat Treatments

All the heat treatments that apply to carbon steels also apply to gray irons. The bases for the various heat treating cycles are equilibrium diagrams and TTT curves. However, because gray irons contain silicon plus other alloy elements, the iron–carbon diagram is altered. Gray irons are really at least ternary alloys. The graphite present in the microstructure has some effects that warrant a few comments on some of the heat treating cycles.

Normalizing The main purpose of normalizing is to increase strength and hardness. It involves heating above the critical temperature for the particular alloy and air cooling. Normalizing is usually used to produce grain refinement. Because gray irons do not usually rely on grain size for properties, the effects of normalizing are rather to remove internal stresses and to increase strength by reducing segregation in the matrix.

Annealing Annealing is performed on gray-iron castings to remove internal stresses and to provide optimum machinability. Gray iron with

Table 15–6
Typical annealing temperatures

Type of Iron	Annealing Temperature, °F (°C)	Purpose
Plain or low alloy	1300–1400 (700–760)	Change pearlite to ferrite + graphite
Alloy irons	1450–1650 (790–900)	Change pearlite to ferrite + graphite
Mottled or chilled iron	1650–1750 (900–955)	Maximum machinability

a structure of ferrite and graphite has better machinability than any other ferrous metal. It is said to be 20% more machinable than B1112 free-machining steel. Annealing will convert *free cementite* and combined carbon to graphite. Thus a casting with a pearlite structure or even a structure containing free cementite can be converted to ferrite and graphite if held for sufficient time at the annealing temperature.

Annealing lowers tensile strength, possibly by 25%, but a grade of iron can be chosen such that this lower strength meets design requirements. Annealing may be essential if machining difficulties occur owing to the presence of *chill* or free cementite. Some typical annealing cycles are shown in Table 15–6.

Stress Relieving The most common causes of internal stresses in castings are the following:

1. Nonuniform cooling of different-sized sections.
2. Nonuniform cooling due to chill in the mold or in heat treating operations.
3. Castings shrinking onto rigid *cores*.

A stress relief is the minimum treatment for a casting that will receive significant machining or severe service. Stress relieving is done at many temperatures.

Up to about 1960, many automobile engine manufacturers practiced seasoning of castings, letting engine blocks and similar large castings sit outdoors in the weather for 6 months to a year. The thermal cycling of the outdoor environment was thought to produce a stress relief. Modern just-in-time manufacturing practices would not allow money in the form of casting inventory to be tied up like this, so this practice has been discontinued. There is no agreement on whether this practice was effective, but the fact that gray cast iron is never completely elastic lends credence to the possibility that this technique could have merit. The point that applies to present-day application of gray cast iron is that these materials can distort with time at room temperature if casting stresses are significant. If dimensional stability is a service requirement, it would be desirable to thermally stress relieve castings before machining. The normal stress relief range is 750 to 1150°F (400 to 619°C). Approximately 75% of the residual stresses should be removed at these temperatures. All residual stresses will be removed at 1200°F (650°C), but there is risk of dissolving pearlite to form graphite. If strength is an important design consideration, a stress-relief temperature of 1000 to 1050°F (538 to 565°C) is usually recommended.

Quench Hardening Because cast irons by definition have high carbon contents (greater than 2%), they can all be quench hardened. However, for optimum hardening characteristics it is best to have an iron with combined carbon in the range of 0.5% to 0.7%. A 100% pearlite matrix with fine graphite is the optimum structure for hardening. If a structure is ferrite plus graphite, as is the case with the weak classes of gray iron, a long soak and high austenitizing temperature may be needed. Because there is not enough carbon in ferrite to allow a fully martensitic structure (~0.6%), the carbon must come from dissolved graphite. Such alloys cannot be flame hardened or induction hardened because there is insufficient soak time at the austenitizing temperatures to allow the graphite to dissolve. Thus, high-strength or alloy cast irons with a pearlite matrix should be used for hardened gray-iron castings.

The advantages of quench hardening are improved strength, hardness, and wear characteristics. Care should be used, however, to ensure proper tempering after hardening. Unlike carbon steels, the as-hardened strength of cast gray irons decreases. Increased strength is obtained only by tempering. The mechanical property effects of quench hardening cast iron are illustrated in Figure 15–16.

The correct tempering temperature for a particular gray iron depends on the class, but temperatures in the range from 700 to 800°F (370 to 430°C) are common. A common heat treating practice used for strength and wear resistance combined with fabricability is to harden and temper to 300 to 350 HB. This practice can

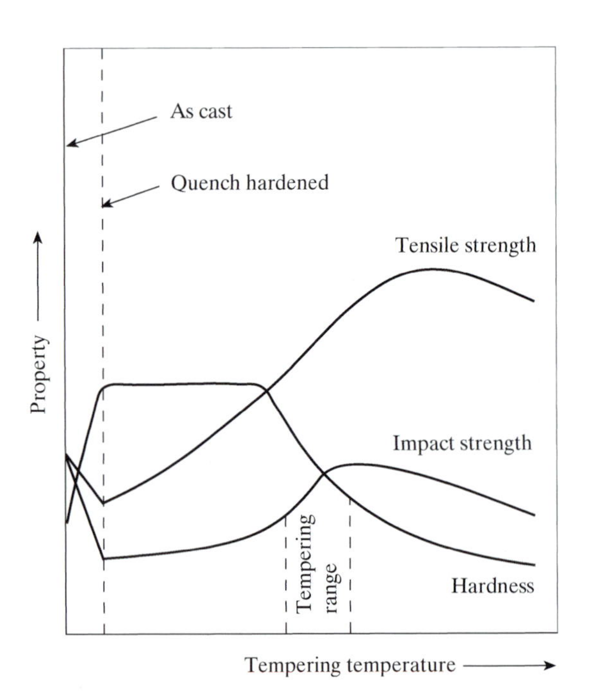

Figure 15–16
Effect of tempering temperature on the properties of quench-hardened gray iron

yield a 50% increase in strength and toughness over as-cast conditions, yet the part can still be machined. Heat treating distortion is eliminated in machining. On gray irons, the size change during quench hardening can be 0.5% expansion.

Special heat treating techniques such as austempering and martempering can be applied to cast irons, but the use of these and other special quench-hardening techniques should be discussed with the foundry before they are used. Each cast-iron alloy has different heat treating characteristics.

Gray cast iron is an excellent material to use for cams, gears, and machine ways; for these applications it is advisable to at least stress relieve the casting before final machining. If good wear characteristics are desired, use the castings in the quenched and tempered to 300 to 350 HB condition. If maximum wear resistance is required, flame or induction hardening of wear areas should be considered. Annealing is used when machinability is a bigger concern than mechanical properties.

Welding Cast irons are prone to cracking during welding for the same reasons that we mentioned in our discussions of tool and alloy steels. Cast irons should not be welded by design. If a casting fractures, repair welding is frequently attempted. The risk of getting additional cracking is high, but welders that are experienced in cast iron welding can have a good repair success rate. Two main processes are used, arc welding with high-nickel filler metal and torch brazing with copper-based filler metal. The procedure for arc welding involves preheating to about 700°F (370°C), welding with minimum weld size, and peening of the deposit, followed by a slow cool to room temperature and a temper at about 600°F (315°C). The preheat and postheat are not needed with torch brazing because, by the time the part is hot enough to melt the brazing alloy, it will be adequately preheated. The secret to success with this process is to slowly heat the entire casting with the process is to slowly heat the entire casting with the torch up to at least 700°F (370°C) before concentrating the torch heat at the weld joint. It is also important to apply a suitable brazing flux. After brazing, the part should be slow cooled, but a temper is usually not necessary because with slow torch heating there is little tendency to get a quench that would cause local hardening. In general, welding on all cast iron should be avoided, but when it is unavoidable these types of procedures should be employed.

15.4 Malleable Iron

By definition, *malleable iron* is a cast iron that has been thermally treated so that it has significant ductility. In the United States, malleable cast iron is made by converting a white iron to either a ferrite or a pearlite matrix, with the excess carbon present in the form of small, flowerlike nodules called *temper carbon*. The structure of a ferritic malleable iron is shown in Figure 15–17.

Metallurgy

Malleable irons are iron–carbon–silicon alloys, with carbon usually in the range of 2% to 3%

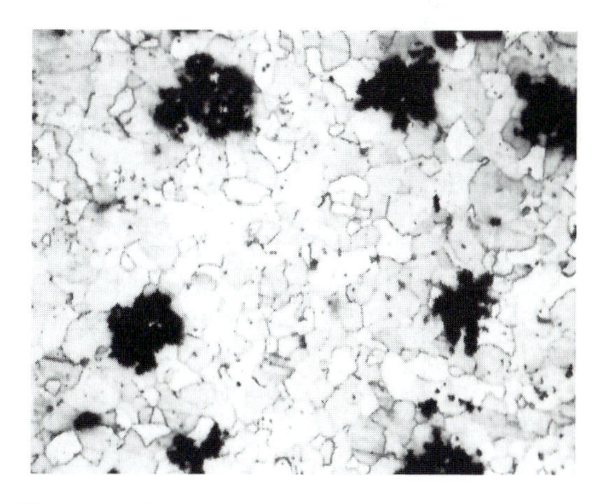

Figure 15–17
Ferritic malleable iron: Black spots are “temper” carbon (graphite); matrix is ferrite (×150)

and silicon in the range of 1% to 1.8%. The raw materials for malleable iron are white-iron scrap (foundry returns), scrap iron and steel, and pig iron. Melting can be done in a cupola or any suitable furnace. It is important to control the chemistry of the melt so that white-iron castings are obtained as poured. White iron consists of pearlite and free cementite; it is very hard and brittle. Because these castings are so brittle, it is possible to fracture gates, *sprues*, and risers from castings before malleablizing. The heat treatment process of converting white iron to malleable iron is illustrated in Figure 15–18. Castings are slowly heated to a temperature well above the critical transformation temperature, possibly 1600 to 1800°F (870 to 980°C). This is the *first stage of graphitization*. Free cementite and pearlite dissolve in austenite. Eventually, graphite nodules or temper carbon starts to form. Because continued heating would cause flake graphite to form, the castings are then cooled to a temperature in the critical temperature range or slightly below. This starts the *second-stage graphitization* and temper carbon and ferrite form. The castings are then either air cooled or oil quenched, depending on the desired strength level. By alloying or by different heating and cooling rates, the structure after malleablizing can be pearlite and temper carbon instead of ferrite and temper carbon. This type of structure is more favorable for subsequent flame-hardening or quench-hardening operations.

One factor limiting the use of malleable iron for one or two castings is that the heat treating requires exacting controls on the thermal cycles in addition to long heat treating times. The primary graphitization may require 40 h or more. Secondary graphitization may require an additional 30 h. Thus the costs may be high for a few castings. The entire malleablizing cycle may take over 100 h.

Alloy Designation

A number of regulatory bodies have alloy designation systems for malleable iron, but the most common is probably the specifications of the ASTM International. ASTM A 47 specification covers "Malleable Iron Castings." The alloy designation system in this specification is a five-digit

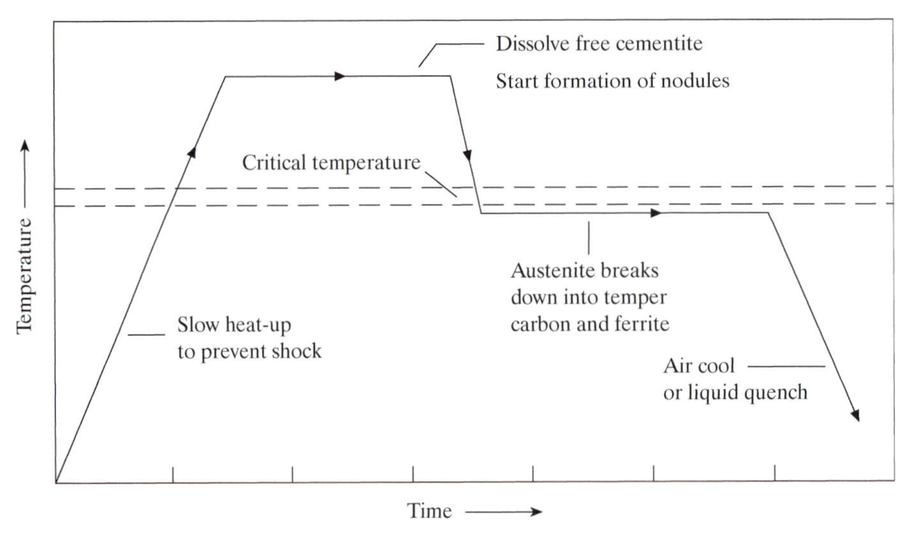

Figure 15–18
Typical heat treating cycle for malleablizing white iron

number, with the numbers corresponding to various mechanical properties. A grade 32510 malleable iron has a minimum yield strength of 32.5 ksi and 10% elongation. Grade 35018 has a minimum yield strength of 35 ksi and 18% elongation.

There are additional ASTM specifications on pearlitic malleable iron (A 220), pipe fittings (A 338), cupola malleable iron (A 197), and automotive malleable iron castings (A 602).

Properties

Most of the physical properties of malleable iron are similar to those of gray iron. There are, however, important differences between gray iron and malleable iron in mechanical properties. The ductility of malleable iron is undoubtedly the most important of these differences. Up to 20% elongation means that malleable iron can bend without breaking. It can be loaded in tension, and it can withstand impact loads. The tensile strength of ferritic malleable iron can be 53 ksi (365 MPa). The pearlitic grades can have tensile strength as high as 100 ksi (689 MPa). Impact strengths of ferritic malleable iron can be as high as 15 ft-lb (20 J) Charpy V-notch. Pearlitic malleable iron may have an impact strength of 5 to 10 ft-lb (7 to 13 J).

Another important difference is the modulus of elasticity. A value of 25×10^6 psi (172×10^3 MPa) is typical for ferritic grades, and it can be as high as 27×10^6 psi (186×10^3 MPa) for pearlite malleable iron. This puts the stiffness of malleable iron almost in the range of steels, an important design consideration.

The machinability of malleable iron is excellent below 300 HB. It is comparable to that of B1112 steel. It is also extremely stable because of the long annealing heat treatments.

The wear characteristics are still good because of the temper carbon in the microstructure. The pearlitic malleable grades can be quench hardened to over 500 HB for metal-to-metal wear and abrasion resistance.

The fatigue strength can be as much as 60% of the tensile strength, and the compressive strength as cast can be four times the tensile strength. Heat-treated, compressive strengths can be as high as 400 ksi (2758 MPa).

Heat Treatment

Because malleable iron involves extensive annealing in its manufacture, the only heat treatment commonly used is quench hardening. Malleable irons can be flame hardened, induction hardened, and quench hardened like steels. The pearlitic grades are used for quench hardening. Tempering after hardening is necessary, and procedures are similar to those used on gray iron.

Stress relieving can be performed as in gray irons, but there is usually no need for such treatments because most residual stresses are removed in the malleablizing treatment.

15.5 Ductile Iron

Ductile iron is cast iron with nodular or spheroidal graphite. The nodules are about the same as those in malleable iron (temper carbon), except they are more perfect spheres. Ductile iron is a relatively recent addition to the cast-iron family (about 1948); as the name implies, its function is to avoid the brittleness of gray and white irons and the excessively long heat treatments required to make malleable iron. Ductile iron has good ductility as cast.

Metallurgy

Chemically, ductile iron is a carbon–iron–silicon alloy with carbon in the range of 3% to 4% and silicon in the range of 2% to 3%. Many ductile irons also contain significant nickel additions. Ductile iron can be melted in a cupola or most of the other furnaces used for gray irons. The charge is usually scrap, pig iron, and ductile iron returns. Because the composition is similar to that of gray iron, the molten iron during solidification would normally form white iron or gray iron with flake graphite. Formation of nodular iron is accomplished by adding magnesium or cerium to the melt in the ratio of a few pounds per ton. The temperature of the melt is much higher than even the boiling temperature of magnesium, and the reaction is energetic.

Other materials such as ferrosilicon are often added to promote graphitization and to control nodule size. This is commonly done by ladle *inoculation* (thrown into the ladle, not the furnace). The net result of these additions is the formation of spherical graphite in a matrix that can be ferritic, pearlitic, or even martensitic, depending on the melt chemistry and process controls. Both the chemical composition and the melting practice must be carefully controlled if the proper structure is to be achieved.

Alloy Designation

As is the case with the other types of cast irons, many control bodies have designation systems for ductile irons. The most commonly used system is a series of hyphenated numbers that corresponds to minimum mechanical properties. ASTM specification A 536 covers five ductile iron grades:

1. Grade 5 (60-40-18)
2. Grade 4 (65-45-12)
3. Grade 3 (80-55-06)
4. Grade 2 (100-70-03)
5. Grade 1 (120-90-02)

The first two or three digits stand for the minimum tensile strength in kips per square inch (ksi), the middle digits for the minimum yield strength in kips per square inch, and the last two for percentage of elongation in a tensile test. There are additional ASTM specifications on austenitic ductile irons:

A 395	Ferritic ductile iron pressure-retaining castings (high temperature)
A 439	Austenitic ductile iron
A 476	Ductile iron castings for paper mill dryer rolls
A 571	Austenitic ductile iron for pressure-retaining parts (low temperature)
A 716	Ductile iron culvert pipe

Properties

The most notable differences in physical properties between gray iron and ductile iron are in magnetic and electrical properties. For example, the electrical resistivity of ductile iron is much lower. A review of the designation system presents a good picture of the important differences between ductile iron and the other cast irons in mechanical properties. There is a wide range of tensile and yield strengths, and the ductility in the ferritic grade, 60-40-18, is almost as good as that of mild steel. The hardness can range from 140 to 350 HB. The modulus of elasticity is in the range of 23 to 25 $\times$ 10^6 psi (158 to 172 $\times$ 10^3 MPa). The compressive strength is about equal to twice the tensile strength. The impact strength of the lower-strength grades is almost as good as that of mild steel, and all grades have better impact strength than gray irons. The fatigue strength is in the range of 40% to 50% of the tensile strength.

Corrosion resistance of ductile irons is similar to that of gray iron: The graphite in the structure lends some assistance in resisting corrosion in many media. Many alloyed varieties of ductile iron have good corrosion resistance in a variety of media.

Machinability is dependent on hardness, but the ferritic grades machine almost as well as B1112 screw machine steel.

Heat Treatment

All the heat treatments that apply to gray irons apply to ductile irons as well. They can be annealed, stress relieved, normalized, and quench hardened. The quench-hardening characteristics depend on the combined carbon content. As is the case with all cast irons, a pearlitic matrix is desirable for best results. Hardnesses of up to 600 HB are possible by quench hardening. The 80-55-06 and 100-70-03 grades are preferred for hardening.

Heat treatment is sometimes a part of the manufacture of a particular grade of ductile iron. Grade 60-40-18 is annealed, 100-70-03 is in the normalized condition, 120-90-02 is quenched and tempered, and 65-45-12 and 80-55-06 are used as cast.

An innovation in ductile iron that reached commercial reality in the early 1980s is *austempered ductile iron* (ADI). ADI is really an entire system for making castings with the strength of hardened and tempered steels while obtaining ductility and toughnesses that can even be superior to that of steels. Austempering of steels, as previously discussed, involves austenitization of a hardenable steel and quenching to a temperature above the temperature at which martensite is formed (M_s) and holding it there until the structure transforms to bainite, a fine mixture of ferrite and cementite. Austempered steels are noted for their excellent toughness. Austempered ductile iron is produced by a similar heat treatment, but the structure that is produced is different and the ductile iron may require alloy additions such as nickel and molybdenum to achieve the desired ADI structure.

There are basically five types of ADI in the ASTM A 897 specification produced by austenitizing ductile iron castings. The five ASTM grades of ADI are designated by their nominal mechanical properties. Grade 1 (125-80-10) has a tensile strength of 125 ksi (861 MPa), a yield strength of 80 ksi (551 MPa), and 10% (min.) elongation.

A typical heat treatment to produce ADI would be to austenitize for 1.5 to 2 h at 1500 to 1515°F (816 to 824°C), followed by a quench to 675 to 725°F (357 to 385°C) with a 30- to 45-min soak at this temperature and an air cool to room temperature. A coarse acicular or bainitic structure will be formed. Higher-strength grades use lower quench temperatures and different austenitizing temperatures. They have finer bainite or martensitic structures. Martensite is not desirable, and techniques are used to suppress its formation.

The mechanical properties of ADI are compared to conventional ductile iron and some steels in Table 15–7. A review of this property

Table 15–7
Comparison of the mechanical properties of austempered ductile iron (ADI) with cast irons and cast steels

		Cast Iron			Steel	
	Austempered Ductile Iron	Gray (A48)	Malleable (A47)	Ductile (80-55-06)	Cast (A27)	Forged (AISI 8620)
Hardness, Brinell	269–555	140–290	110–156	196–255	131	341–388
Ultimate strength (10^3 psi)*	125–230	20–55	50–53	80–100	63	167–188
Yield strength (10^3 psi)*	95–185	—	32.5–35	55–75	35	120–149
Elongation (%)	10–<1	—	10–18	—	30	12–14

*Multiply by 6.89 to convert to MPa.

Source: From *Materials Engineering*, reprinted with permission.

comparison will make the advantages of this heat treating process readily apparent. ADI castings can have tensile strengths as high as 230 ksi (1585 MPa). Grade 125-80-10 castings can have a hardness in the range of 269 to 321 HB; Grade 200-155-1 castings have a hardness in the range of 388 to 471 HB. Grade 200-155-1 castings are used for gears and power transmission parts. They reportedly have wear characteristics that are superior to 600 HB quenched and tempered steel gears. Grade 125-180-10 castings are used for applications that require higher toughness and resistance to fatigue. As a class of materials, austempered ductile iron castings have the potential for being cost-effective replacements for expensive hardened and tempered forged steel components. ADI processing improvements are still being developed to make the heat treatments simpler and more foolproof, but it has been a viable, commercially available process since the early 1980s. These materials are specified by ASTM number and grade:

> *Material:* Austempered ductile iron sand casting, ASTM A 897 Grade 125-80-10

15.6 White Alloy Irons

White irons get their name from their fracture appearance: they look white, as gray irons look gray. Unalloyed white irons contain carbon in the range from 2% to 4%, silicon from 0.5% to 2%, and about 0.5% manganese. These irons solidify with a graphite-free structure in thicknesses less than about 4 in. The structure is predominately free cementite and pearlite, possibly with some austenite (Figure 15–19). Hardness can be as high as 600 HB. Gray iron will form a white-iron structure when rapidly cooled from the casting temperature. In this instance, it is usually called chilled iron. This is an important material for engineering design. Rolls, wear plates, pump linings, balls, and the like can be made with extremely wear-resistant surfaces in the as-cast condition. Chilled iron can be made by casting in metal or graphite molds or by using metal inserts in sand molds (Figure 15–20).

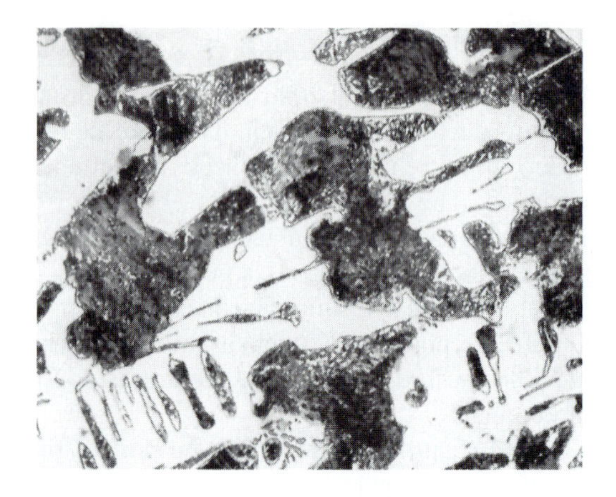

Figure 15–19
Microstructure of white iron; fine pearlite (dark) and free cementite (light); hardness 62 HRC (×750)

When alloy elements such as nickel, molybdenum, and chromium are added to various irons, it is possible to get chilled-iron-type structures in heavy sections. There is a large family of welding hard-surfacing filler metals that are essentially white irons with 10% to 30% chromium in their composition. These alloys are very wear

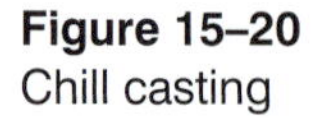

Figure 15–20
Chill casting

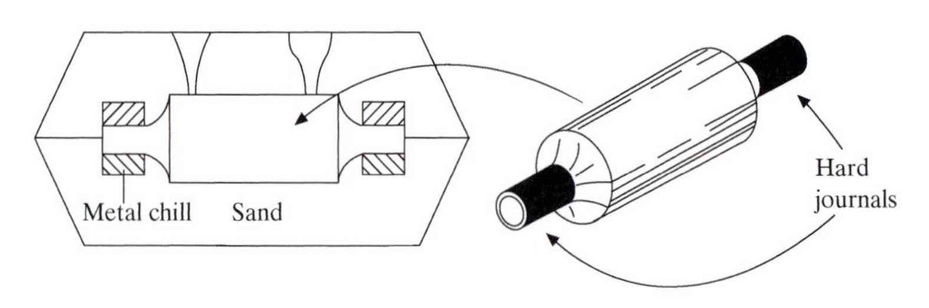

resistant, and they are in wide use in mining and agriculture for abrasion applications. It is not possible to describe all the alloy cast irons available; there are too many. Some are designed for corrosion resistance, some for oxidation resistance, some for creep strength, and so on. One of the more commonly used alloy irons is Ni-Hard. It has a hardness in the range from 525 to 600 HB and is used for wear plates, grinding balls, rolls, and the like. ASTM specification A 532 covers eight grades of abrasion-resisting cast irons. These alloys are designated by the ASTM grade:

> *Material:* Abrasion-resisting cast iron, ASTM A 532 Type A Ni–Cr–HiC.

In summary of our discussion of cast irons, there is an iron that can be used for almost any type of application. A wide range of properties is available, and these property variations are obtained by process controls, composition controls, and thermal treatments of a basic alloy of iron–carbon and silicon.

15.7 Steel Castings

Some designers are reluctant to use cast irons because they think that all cast irons are brittle. We have seen that ductile irons and malleable irons overcome the brittleness of gray and white irons, but there are some valid limitations to the use of cast irons. They all have lower stiffness than steels, and the toughness and ductility of steels can be much better than that of cast irons. Steel castings can be used where the range of properties available in cast irons will not meet design requirements. The modulus of elasticity of steel castings is compared with cast irons in Figure 15–21; the toughness is compared in Figure 15–22.

It is not possible to discuss the metallurgy of steel castings in a few paragraphs because all steels can be cast; casting is the first step in the production of all wrought steel products. Steel castings are not used as often as cast irons, however, because they are harder to cast. Cast irons melt at lower temperatures: They have lower shrink rates, and the carbon and silicon contents

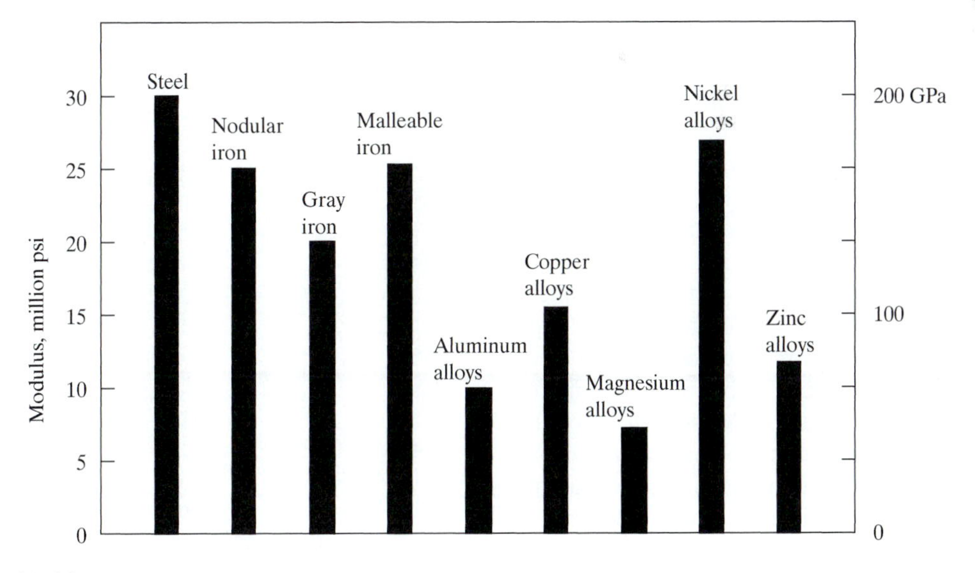

Figure 15–21
Modulus of elasticity of various cast metals
Source: C. W. Briggs, *Steel Castings Handbook*, Steel Founders' Society of America

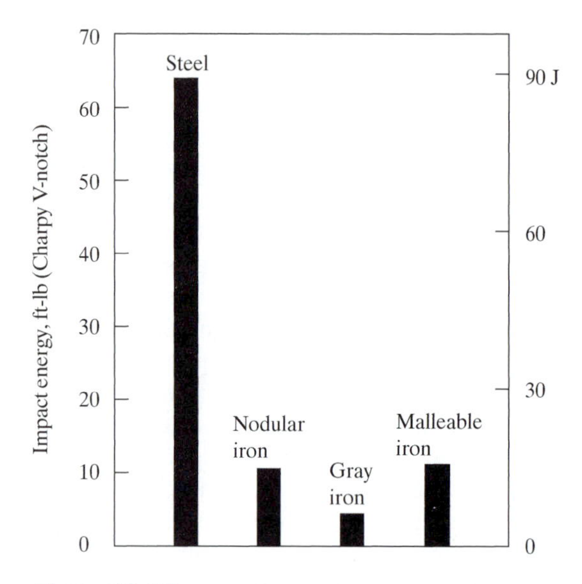

Figure 15–22
Charpy V impact strength of ferrous castings at 70°F

Source: C. W. Briggs, *Steel Castings Handbook*, Steel Founders' Society of America

impart fluidity. The likelihood of cold shuts, incomplete filling, and shrinkage voids is lower in cast irons than in cast steels.

In the previous chapter it was pointed out that stainless and heat-resistant alloys are available in cast form. Castings are also made from carbon steels, alloy steels, tool steels, and special steel alloys.

There are casting alloys that coincide with many of the common U.S. carbon and alloy steels (former AISI steels), but their chemical composition is slightly altered to improve casting characteristics. In most cases, the cast steel grades will contain 0.30% to 0.65% silicon and 0.5% to 1.00% manganese. Most of the wrought steel grades do not contain this much silicon, and the manganese content may be different. ASTM A 148 specification covers alloy steels similar to the wrought alloy steels such as SAE types 1030, 1330, 4135, and 4335. This specification covers 15 grades of hardenable alloy steel with different strengths. The different grades are designated by numbers that reflect their tensile and yield strengths. ASTM A 148 grade 105-85 is a cast alloy steel with a tensile strength of 105 ksi (223 MPa) and a yield strength of 85 ksi (585 MPa). The casting supplier varies the chemical composition and heat treatment to produce the specified casting grade. The ASTM specification number and grade is the appropriate designation system for cast alloy steels.

There are at least 20 ASTM specifications on cast steels, and many are written for various types of casting service requirements. ASTM A 757 covers castings for low-temperature service; A 56, steam turbine castings; A 217, castings for high-temperature service; A 486, castings for highway bridges; and A 487, castings for pressure service.

There are many more like these. The ASTM specifications that cover most of the alloys that are similar to the wrought AISI alloys are listed in Table 15–8.

Table 15–8
Wrought alloy equivalents for steel castings

ASTM Specifications	Wrought Alloy Equivalents
A 426	Low-carbon steels: 1010, 1020
A 352, A 356	Medium-carbon steels: 1030, 1040, 1050
A 217	
A 352-LC3; A 426- CP2, 5, 5b, 11, 12, 15, 21	Low-alloy, low-carbon steels: 1320, 4110, 8020, 2315, 4120, 8620, 2320, 4320
A 27, A 356; A 148, A 389; A 216, A 486; A 217, A 487	Low-alloy, medium-carbon steels: 1330, 1340, 2325, 2330, 4125, 4130, 4140, 4330, 4340, 8030, 8040, 8630, 8640, 9530, 9535
A 352; A 297, A 608; A 743, A 744	Stainless steels and high-chormium alloys
A 890	

Table 15–9
Properties of ASTM A 27 steel castings

Chemical Composition (Maximum Percent unless Range Is Given)										
Specification	**Grade**	**C**	**Mn**	**P**	**S**	**Si**	**Ni**	**Cr**	**Mo**	**Other Elements**
ASTM A 27	N-1	0.25	0.75	0.05	0.06	0.80	0.50	0.40	0.20	Cu 0.30
	N-2	0.35	0.60	0.05	0.06	0.80	0.50	0.40	0.20	Cu 0.30
	U-60-30	0.25	0.75	0.05	0.06	0.80	0.50	0.40	0.20	Cu 0.30
	60-30	0.30	0.60	0.05	0.06	0.80	0.50	0.40	0.20	Cu 0.30
	65-35	0.30	0.70	0.05	0.06	0.80	0.50	0.40	0.20	Cu 0.30
	70-36	0.35	0.70	0.05	0.06	0.80	0.50	0.40	0.20	Cu 0.30
	70-40	0.25	1.20	0.05	0.06	0.80	0.50	0.40	0.20	Cu 0.30

Specification and Heat Treatment			Mechanical Properties (Minimum unless Range Is Given)			
Specification	**Grade**	**Heat Treatment**	**Tensile Strength [ksi (MPa)]**	**Yield Strength [ksi (MPa)]**	**Elong.[a] in 2 in. (%)**	**Red.[b] of Area (%)**
ASTM A 27	N-1	—	—	—	—	—
	N-2	A, N, NT, or QT	—	—	—	—
	U-60-30	—	60 (414)	30 (207)	22	30
	60-30	A, N, NT, or QT	60 (414)	30 (207)	24	35
	65-35	A, N, NT, or QT	65 (448)	35 (241)	24	35
	70-36	A, N, NT, or QT	70 (483)	36 (248)	22	30
	70-40	A, N, NT, or QT	70 (483)	40 (276)	22	30

Source: Courtesy Alloy Casting Institute, Steel Founders' Society of America.
[a] Elong. = elongation
[b] Red. = reduction

Table 15–9 lists the designation number and properties of some general-purpose steel castings (ASTM A 27). The designation numbers for most alloys coincide with the tensile strength and yield strength. ASTM A 27 grade 60-30 has a minimum tensile strength of 60 ksi and a yield strength of 30 ksi.

The properties and heat treatments for steel castings depend on the particular alloy. In general, these are the same as those used on the wrought alloys. One additional property advantage of steel castings over cast irons is weldability. Cast low-carbon steels, stainless steels, and low-alloy steels can be readily welded. Cast irons of any type cannot be welded without high risk of cracking because of their high carbon content.

Cast tool steels, alloy steels, and HSLA steels are becoming very popular in design. Most investment casting houses are pouring tool steels, and it is often economical to make small wear-resistant parts by this process. Cast stainless steels have always been widely used for pipe fittings, valves, pumps, and the like. All the common casting processes can be used. These materials should be designated with their respective ASTM type and grade. Copies of these standards are available from ASTM, and complete sets of ASTM standards are contained in most engineering libraries and are on-line.

Table 15–10
Comparative properties of ferrous castings

Material	Tensile[a] Strength, ksi	Yield Strength, ksi	Compressive[a] Strength, ksi	Percent Elongation %	Modulus[b] of Elasticity, 10^6 psi	HB[c]	Minimum[d] Casting Section, in. (mm)	Characteristics
Gray iron—ASTM A 48								
Class 20	20	—	95	Nil	12	175	0.125 (3.1)	Good for thin sections and low-strength castings
Class 35	35	—	135	Nil	16	200	0.187 (4.7)	For structural parts, pumps, pipes, bases, etc.
Class 40	40	—	135	Nil	18	210	0.250 (6.3)	For medium-strength castings; can be hardened
Ductile iron—ASTM A 536								
Grade 5								
60-45-18	60	45	120	18	25	175	0.187 (4.7)	Excellent mechanical properties; can tolerate bending
Grade 3								
80-55-06	80	60	160	6	25	235	0.187 (4.7)	High strength, some ductility; hardenable
Malleable iron								
ASTM A 47-77 Grade 32510	50	32	208	10	26	125	0.125 (3.1)	Ferritic, free from casting strains
ASTM A 220-76 Grade 60004	80	60	240	3	26	225	0.125 (3.1)	Pearlitic, free from casting strains; hardenable
White iron								
ASTM A 532								
Grade A, Nitr–HiC	45	—	—	Nil	25	550	0.250 (6.3)	Extreme abrasion resistance; grinding balls, slurry pumps, etc.
Steel castings								
ASTM A 27								
Grade 60-30	60	30	50	24	30	156	0.250 (6.3)	For high stiffness, shock loading, and for welded fabrications

[a]1 ksi = 6.89 MPa. [b]1 psi = 6.89 kPa. [c]Brinell hardness. [d]Increases for large castings.

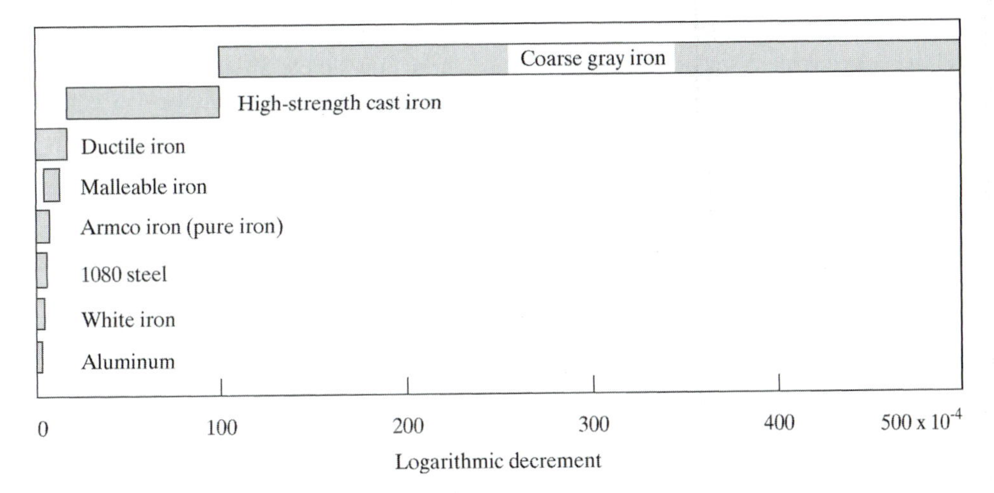

Figure 15–23
Damping capacity of cast irons as measured by damping of successive vibrations
Source: Talbot Associates, Inc.

15.8 Casting Selection

We have shown that there are various types of cast irons and cast steels, gray irons, white irons, ductile irons, malleable irons, carbon steels, alloy steels, and so on. Within each of these categories there are also many specific alloys. How is the design engineer to go about selecting one specific alloy from the hundreds that are available? A repertoire of ferrous castings should include an alloy or two from each system. Table 15–10 is a list of nine casting alloys that will meet most design applications.

Gray irons should be considered for machine parts that require good machinability, stability, *damping capacity* (Figure 15–23), and good availability, but that do not involve shock loads, bending, or high stresses. The outstanding property of gray iron is its self-lubricating metal-to-metal wear characteristic. Gray irons can also be hardened for additional wear resistance.

Ductile irons are extremely useful for machine frames, mounting plates, gears, cams, and other mechanical components that may get an occasional shock load. They also provide as-cast strengths higher than those of gray irons and many mild steels. Machinability and wear characteristics are good, but not as good as those of the gray irons. An example of a ductile iron machine component is shown in Figure 15–24. They are well suited for large, heavy-wall parts.

Malleable iron is not as available as gray iron or ductile iron. It is a special casting material best suited to small, thin-wall parts. If many units are to be made, malleable iron provides

Figure 15–24
Shaft (4-in.-diameter) housing made from high-strength ductile iron

the best combination of machinability and stability of all the cast ferrous alloys.

White iron is a specialty item. It is used for tough wear problems, especially abrasion resistance. It is widely used in machinery for hard rolls.

Steel castings are used after a designer is certain that a cast iron will not do the job. Steel castings are used for the tough material problems.

15.9 Powder Metals

Powder metallurgy is the field of engineering materials that deals with making parts and products with metal powder as the starting material. This is a very old field in concept; precious metal shapes have been made for centuries by compacting metal powders into shapes and heating them to get the metal particles to fuse together without melting of the actual powder—a solid-state bond between the particles—or by a controlled amount of melting at particle contacts. Powder metallurgy as it is used in modern industry started in the United States in the 1920s with the development of powder metallurgy (P/M) bronze plain bearings. These bearings were made by compaction and sintering of bronze powders in such a manner that the compacted bushing had up to 30% porosity. This porosity was then vacuum impregnated with oil and the bushing would essentially be lubricated for life. World War II brought on scarcities of many engineering materials, and it was learned that sintered powder iron could be used to make all sorts of mechanical components. Further, the sintered P/M irons had properties that rivaled parts machined from wrought steels. Today ferrous powders comprise about 60% of the tonnage of powders used for P/M parts. The remainder of the tonnage is made up of at least a dozen other metals and alloys. This is a growing industry, and powder metallurgy will continue to compete with casting, machining, stamping, and fine blanking as a way of producing parts to near-net shape or to finished size.

In the remainder of this chapter we will describe how shapes are produced by powder metallurgy, available materials, design considerations, and typical applications.

P/M Processes

The basic principles of powder metallurgy are the same as the principles previously described in our discussions of ceramics and cermets. Ceramic shapes are made by mixing ceramic powders with binders such as waxes, cold compacting them, and then sintering the powders to form a shape. P/M shapes are made in a similar fashion, only binders are not normally used and compaction is done by special presses, usually mechanically actuated. The steps in the process are illustrated in Figure 15–25. Annealed metal powders, usually in the size range from 10 to 80 μm (0.0004 to 0.003 in.), are blended to achieve the desired powder mix and to add compaction lubricants. Compaction is performed in hardened steel or cemented carbide tooling machined to produce the desired part shape. The powder is automatically fed into the open die with a hopper device, and excess powder is removed by a mechanically actuated arm. The upper and lower compaction tools then simultaneously compact the powder with pressures that can be in the range from 10 to 60 ksi (69 to 414 MPa). The upper compaction punch retracts, and the lower punch rises up to eject the compacted shape. This shape is said to be in the green state. It is relatively weak but strong enough to withstand sliding down chutes and falling into part hoppers. The green parts are then sintered in a reducing atmosphere at some temperature that is well under the melting point of the powder.

The mechanism of bonding is similar to that described in sintering of ceramics. The compaction pressure causes plastic deformation of the powders to maximize bonding surface area. The powders fuse at contact points by mutual diffusion or in some cases by incipient melting of

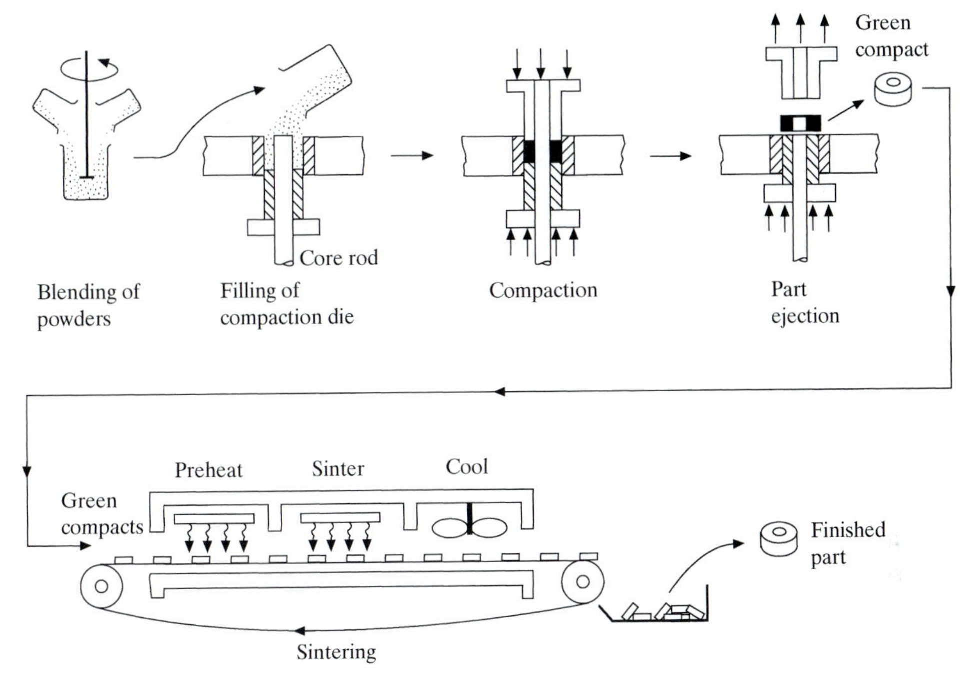

Figure 15–25
Basic steps in making parts by powder metallurgy

the powder or a component of a powder mixture (see Figure 15–26). The compaction lubricant is added to increase the density by facilitating compaction and to prevent adhesion of powders to the tooling. If the lubricant is a stearate (which is often the case), it is driven off in the sintering process. If graphite is used as the lubricant, it may be retained in the finished compact, and in some cases it can be used to add carbon content to ferrous powders to enhance their strength. The sintering atmosphere is usually a generated gas atmosphere. For ferrous powders the sintering furnace atmosphere can be hydrogen, dissociated ammonia, nitrogen + hydrogen, an endo gas, or vacuum. Whatever the atmosphere, it must be capable of preventing oxidation and capable of reducing oxides that may be present on the powder surfaces. The finished P/M shape can have interconnected porosity as high as 30%, and the porosity is controlled by

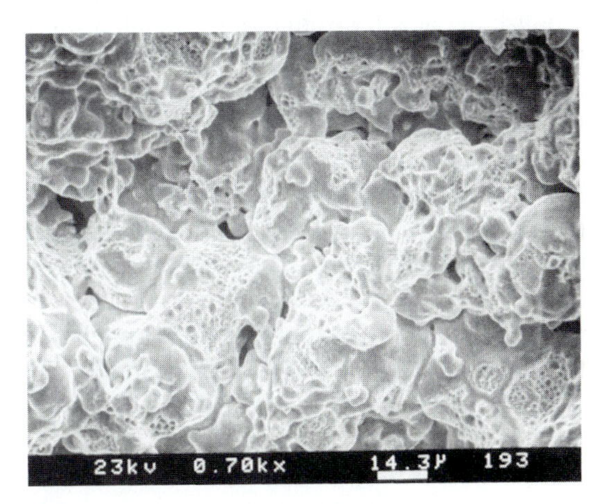

Figure 15–26
Scanning electron photomicrograph (×700 original magnification) of the fracture surface of a powder metallurgical steel part. Note the grains and the porous structure.

the nature of the powder, the degree of lubrication, and the compaction pressure. It is the user's responsibility to specify the desired degree of porosity. This is done by requesting a particular density in grams per cubic centimeter and the percent interconnected porosity. For example, a 100% dense iron compact will have a density of 7.8 g/cc, essentially the density of iron. A compact with a density of 6.2 will weigh 6.2 g/cc and contain 20% total porosity. The interconnected porosity could be less than 20%.

P/M Materials

About two-thirds of the powder metals used in the United States are ferrous. Between 5% and 10% are copper-based alloys or pure copper, about 10% are aluminum powders, and the remainder of the powders produced are stainless steels, nickel-based metals, cobalt-based materials, and refractory metal mixtures. Obviously, from these figures, ferrous powders are the most important from the standpoint of use in industry.

Powders for P/M parts are made by many techniques. One of the simplest processes is to atomize the powder from molten metal. Molten metal is poured from a ladle into an orifice type of device at the top of a shot tower, a chimney-like cylinder with the orifice at the top. Gas jets around the orifice assist atomization. The powders are quenched to solid condition by water jets or by convection cooling in falling from the top of the shot tower. The resultant particle shapes are mostly round. Chemical reduction of iron ore to metal is another widely practiced technique. Iron ore, usually Fe_3O_4, is mined and crushed to fine particle shape, and treatment of the ore by a reducing medium such as CO at high temperatures can reduce the fine particles of ore to essentially pure iron. This is similar to the process that takes place in a blast furnace. Ferrous particles made with this technique tend to be more spongelike and irregular in shape than atomized powders, and this type of shape is preferred when good green strength is critical. Powders can also be made by milling (brittle materials are crushed by running through rolls), by electrodeposition (thin deposits are made on electrodes and scraped off in particle form), and by many other techniques.

The most widely used ferrous powder materials are pure atomized irons with a carbon content of less than 0.03%, low-carbon steels with a carbon content of about 0.3%, moderate-carbon-content steels, 0.5% to 0.8% carbon steel and two types of stainless steel, type 316 and type 410. The low-carbon irons can be used for magnetic applications where the low carbon content is necessary for favorable magnetic properties. The low-carbon steels are used for general-purpose compacts, and the higher-carbon powders are used for applications in which higher mechanical strengths are required. Tensile strengths in excess of 200 ksi (1034 MPa) can be obtained after heat treatment. The stainless steels are used when corrosion resistance is a factor, and aluminum powders are sometimes used for their light weight and resistance to atmospheric rusting. The pure coppers are used for motor brushes and for similar applications in which high electrical conductivity is required. The bronzes are used for the familiar oil-impregnated plain bearings.

Many powder blends are proprietary in nature and are designated on drawings by trade names. There are two generic systems in the United States for specification of powders: MPIF or ASTM numbers. MPIF is the acronym for the Metal Powder Industries Federation, which is an umbrella organization for powder metallurgy trade organizations and technical societies. ASTM is an international consensus standards organization. Generic specifications can use either the MPIF or the ASTM number, and both will be recognized by powder suppliers and P/M job-shops. Table 15–11 is a tabulation of the ASTM powder specifications.

The ASTM specifications cover the mechanical properties that can be expected from a

Table 15–11
ASTM specifications for sintered P/M materials

ASTM Designation	Description
B 255	Sintered bronze structural parts
B 282	Sintered brass structural parts
B 438	Sintered bronze bearings (oil impregnated)
B 439	Iron-based sintered bearings (oil impregnated)
B 458	Sintered nickel silver structural parts
B 612	Iron–bronze sintered bearings (oil impregnated)
B 715	Sintered copper structural parts for electrical conductivity
B 782	Iron–graphite sintered bearings (oil impregnated)
B 783	Materials for ferrous powder metal structural parts
B 817	Powder metallurgy titanium alloy structural components
B 823	Materials for nonferrous powder metallurgy structural parts

particular powder compacted and sintered to a particular density. Heat treatments can be applied to alter mechanical properties, and powder particle size may be specified if the service requirements call for this amount of detail. A complete specification requires designation of the powder type, the compacted density, the percent interconnected porosity, and any postsintering treatments. The MPIF standard 35 provides a designation system for metal powders that relates to chemical composition and minimum strength:

F-000-10 is a pure iron P/M with a minimum yield strength of 10 ksi (69 MPa).

FC-0205-30 is an iron–carbon–copper P/M with a minimum yield strength of 30 ksi.

This system is used in the United States for ferrous P/M structural parts, and it is included in some of the ASTM P/M standards. Nonferrous P/M parts can be specified using an ASTM designation (Table 15–11).

P/M Heat Treatments

The carbon contents of ferrous P/M compacts are usually reported as combined carbon; this includes the carbon in solution in the powder particles and the carbon in the form of graphite that may have been added to the compact as a compaction lubricant. If the carbon content or carbon and alloy content in the particles of a P/M part meets the requirements for quench hardening, the sintered parts can be quenched and tempered in the same manner as wrought steels. The porosity that is usually present in the P/M parts precludes the use of salt bath heat treating techniques because the salt will become entrapped in the pores. Similar problems can occur with oil quenching. For these reasons, quenching may be performed by cooling atmospheres in furnaces or by using water quenching. Oil quenching, however, can impart corrosion protection and lubricity. Diffusion-hardening processes are widely used for P/M parts that require hardening. Applicable processes include carburizing, nitriding, and carbonitriding.

Because of their porosity, the effectiveness of a hardening heat treatment is difficult to assess with conventional penetration hardness testing techniques. The particles may be 60 HRC, but a Rockwell C test on the compact may show a bulk hardness of only 40 HRC. The most effective check on a hardening technique is to make a metallographic section and measure particle hardness with microhardness indentation techniques; bulk hardness and particle hardness are measured.

Impregnation with lubricant is a very common postsintering process. This is usually done by pressure impregnation. Parts are batch loaded into a basket in an impregnation vessel. The vessel is evacuated and, while still under vacuum, the parts are lowered into the oil and the vessel is pressurized. The vacuum and pressure cycles force the lubricant into the interconnected porosity. Porosity in P/M parts can also interfere with plating operations. The plating chemicals can leach out with time and attack or lift the plating. P/M parts requiring plating are usually impregnated with polymeric materials that wick into the porosity and then cure without forming a surface skin that could interfere with plating.

P/M Applications

Most powder metal parts are relatively small in size. Typically, parts are less than 2 in. in diameter (50 mm) and weigh only grams. Most are made with some degree of porosity. These are not limits; tool steel ingots made by powder metal techniques can weigh several tons, and hipped P/M parts can be 100% dense. The largest growth in the P/M market is in large structural parts such as gears and pressure plates for automotive applications. Most P/M parts are made in mechanical presses as illustrated in Figure 15–25. A typical part made on one of these presses is illustrated in Figure 15–27. A metallurgical cross section of a ferrous P/M part is shown in Figure 15–28. Normally, P/M parts have a projected surface area (the area seen by the end of the compaction punch) of less than 20 in.^{2} (129 cm^2). The factor that limits part size in mechanical compaction presses is the compaction pressures that are required to achieve a moderate green strength. With a compaction pressure of 30 ksi (207 MPa), a part with a projected area of 20 in.^{2} (129 cm^2) would require a press force of 600,000 lb (4.16 Pa); a very large press would be required. Thus most P/M parts are small in size and have geometries that have no undercuts. Conventional presses can core holes in the direction of compaction, but molding in undercuts would prevent parts from being ejected from the die. Typical compaction presses do not have facilities for pulling cores that would be required for undercuts. Some typical P/M parts are illustrated in Figure 15–29. Note that these parts do not have any undercuts that would prevent them from being ejected from a compaction die.

Figure 15–27
Typical P/M part, a small gear for a camera mechanism

Figure 15–28
Metallurgical cross-section of a powder metallurgical steel part. The dark areas are porosity and the light areas are ferrite with pearlite (unetched, × 100 original magnification).

A growing trend in the powder metal industry is to use metal injection molding (MIM) to

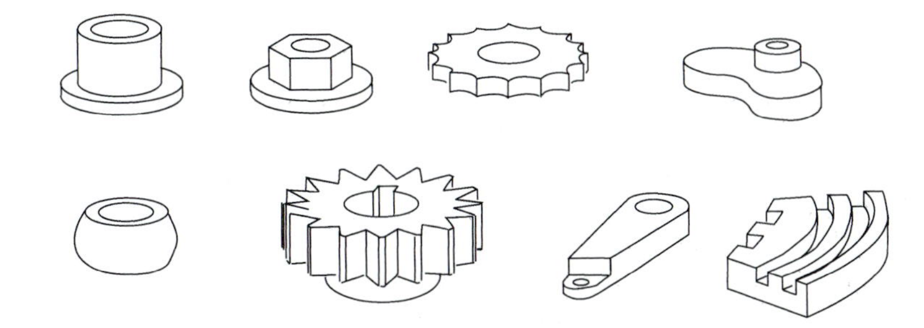

Figure 15–29
Some typical part shapes that can be made by P/M press compaction

make P/M parts with undercuts and complex shapes similar to those produced with injection molding of plastics. In this process, the metal powder is blended with a special plastic or wax and injected as if it were a plastic. The plastic or wax binder is removed by baking or chemical processes, and the part is sintered to its final density (94% to 99% of theoretical). MIM powders are usually finer in particle size than those used in conventional P/M. Figure 15–30 shows a typical injection molded iron powder part (with undercuts and details not possible with conventional P/M powder compaction techniques). This process was first commercialized in the 1980s and has been rapidly expanding ever since. The alloys that can be readily injection molded are similar to those used in conventional P/M (17-4 stainless steel, 4140 steel, carbon steel, tool steels, etc.), but powders like ceramics that are difficult to press can easily be injection molded. Most MIM parts are small in size [<150 g and 5 in. (125 mm)] have complex shapes and close dimensional requirements (0.3/0.5%), and require a good surface finish, Economic quantities are similar to plastic injection molding. Enough volume is needed to amortize The tooling costs. Usually several thousand parts are needed to amortize tools. Tool costs are similar to those for injection molding of plastics and the process is best suited to complex geometries for which machining or other secondary operations are eliminated.

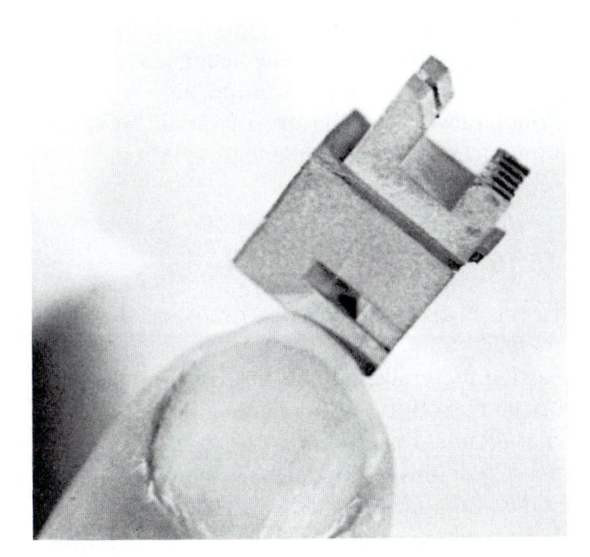

Figure 15–30
Injection molded (MIM) P/M iron part

Very large P/M parts are made by hot or cold isostatic pressing techniques. Ten-inch-diameter (250 mm) by 100-in. (2.54-m)–long ingots can be molded from powders by hot isostatic pressing: Powder is poured into a steel can, the can is welded shut and placed in a pressure vessel, the pressure in the vessel is raised to about 20,000 psi (138 MPa), and the interior of the vessel is heated to about 2000°F (1093°C). The pressurizing medium is usually an inert gas such as argon. This is one method of *hipping* or *hot isostatic pressing*. Parts can be made to 90% to 100% density with this type of technique. Cold isostatic pressing can be used to compact large P/M shapes. These shapes are subsequently sintered in the normal fashion. Powders can be heated and extruded to form tubing in a manner that is not unlike the techniques used to make plastic piping. Powders can be rolled into sheets that are sintered and subsequently cold rolled. Another way to obtain full-density parts is to compact parts and sinter them with conventional techniques, and then hip them as a batch process. Full-density parts can also be made by hot or cold forging after conventional compaction and sintering. Since the 1980s there has been great interest in making full-density P/M parts for automotive applications. Full density is often required for fatigue resistance. The porosity in conventional P/M parts produces stress concentrations that have an adverse effect on fatigue and fracture toughness. However, whatever the application, the properties available with the various P/M techniques are well documented, and these available properties will meet a wide variety of service needs.

P/M versus Other Processes

Powder metal fabrication processes fit into the spectrum of manufacturing processes in two basic ways: press-manufactured P/M parts are cost-effective for small parts (parts small enough that several will easily fit into your hand), and the processes suited for large parts (hipped, isostatic molded, and so on) can be used to produce near-net shapes that are competitive in cost and properties with forgings, castings, and parts machined from solid. The major competitor with small P/M parts is investment casting. Investment casting has the advantage of full density, but the disadvantage that it is a slower manufacturing process with less tolerance control and parts usually require some secondary finishing operations. Most P/M parts are used as pressed and sintered; bronze P/M parts may require a sizing operation after sintering. The production rate on a typical P/M compaction press is at least 50 parts per minute (800-ton presses), and it may be 100 or more parts per minute (100-ton presses). It is a process suitable for the production of parts where millions are required. Sintering in conveyor hearth furnaces makes P/M a continuous process that is well suited to large production volumes. The equipment costs for P/M pressing of small parts are mating compaction dies made from hardened steels. These dies are really inserts in general press tooling, and for many parts both die halves can be made by EDM machining the desired shape into 4-in. (10-cm)–diameter, 4-in. (10-cm)–long standard die inserts. In other words, tooling is not complicated, and the tooling costs on a new part may be only $1000 to $10,000.

Large-hipped P/M parts are usually handled as single pieces, and tooling and processing costs may be high. These techniques are most suitable for applications for which castings or forgings do not offer the desired use properties. For example, full-density hipped turbine buckets can have mechanical properties that are superior to the same part made from wrought or cast metals. They do not have the anisotropy that is indigenous to the latter processes. P/M manufacture of small parts will probably compete with investment casting for at least the next decade. In the same time frame, P/M techniques will see increased application for full-density large parts such as automobile cams and gears. Hipping will be increasingly applied to castings to improve their properties. Without a doubt, powder metallurgy will remain an important part of materials engineering into the foreseeable future.

15.10 Process Selection

In this chapter we described most of the commercial casting processes and discussed ferrous casting materials, notably cast irons and steels, We discussed the powder metallurgy processes that compete with castings as a manufacturing process. In this concluding section we will review some of the factors that are important in the selection of these processes.

Castings can have tolerances as large as ±0.1 in. (±2.5 mm) and as small as ±0.003 in.

(±0.075 mm); the latter tolerances are produced by investment casting and the former would be typical of very large cast-iron sand castings. If the parts are small, P/M will produce the lowest tolerances. The surface finish of castings follows an almost identical pattern; investment castings and P/M parts have the best surface finish, and sand castings have the roughest surface finish. The size capabilities of castings range from many tons for sand castings to grams for small investment castings. Investment castings are usually never larger than 50 lb (23 kg), and P/M parts are seldom larger than 35 lb (16 kg) (except P/M ingots and other intermediate shapes). The materials available in cast form are most metals, but cast irons have the highest usage on a tonnage basis. About a dozen metals are commonly made into shapes by P/M techniques, but ferrous metal powders are the most frequently used. Cast irons have better castability than most other ferrous metals, and this is the reason for their wide use. Ferrous powders are the most used powders for P/M parts because they have higher strength than the nonferrous metals and their cost is lower than that of the other powder materials.

The cost of making parts by casting or P/M depends on the process and the number of parts required. Each process has certain hardware requirements that must be amortized on the number of parts. If few parts are required, these hardware costs become a significant selection factor. If millions of parts are required, the processes that produce parts to final shape may be the lowest-cost manufacturing technique. P/M parts can be ready for use after sintering. In this respect this process has an edge over all the casting processes; they all require at least removal of sprues and risers. Injection molding allows complex P/M shapes.

Table 15–12 is a chart comparing some selection factors for a variety of casting processes, and these casting processes are compared with P/M processing of parts. Review of these data can serve as a guide in the selection of casting processes. All these processes are commercially available. The processes with the simplest hardware requirements can be obtained with the smallest lead time; the processes like P/M and investment casting that require machined metal tooling and tooling design may take longer. Die casting dies require the most complex tooling; they may require a lead time of 12 to 22 weeks. On the other end of the spectrum, sand casting that requires only a wood pattern may require a lead time of only 2 weeks. There are foundries in all parts of the country that specialize in cast irons. Nonferrous foundries, die-casting job shops, and P/M suppliers may not be located in every city, but they are available in all regions of the United States and in many other countries, and there are no availability restrictions. In summary, casting and P/M are essential processes in materials engineering, and designers and material users should develop a feeling for the capabilities of the various processes. Their use should be considered for almost every design. Sometimes casting is the most cost-effective process to use even when only one part is required.

Summary

Some of the important concepts to keep in mind regarding castings are as follows:

- Each casting process has size and dimensional tolerance limitations; this is probably the most significant difference among processes.
- Each casting process involves costs for tooling, patterns, and related hardware; these cost differences must be considered in the selection of a casting process.
- A key concept of casting design is to try to make all wall thickness uniform; avoid heavy sections connected to light sections. In ideal casting design, the gate is the last part of the casting to solidify.

Table 15–12
Comparison of various casting processes

	Metals	Size Range	Tolerances[a]	Tooling Costs[b]	Part Price[b]	Design Freedom[b]	Surface Roughness	Normal Minimum Section Thickness[c]	Ordering Quantities
Investment (lost wax)	Most castable metals	Fraction of an ounce to 150 lb	±0.003 to ¼ in. ±0.004 to ½ in. ±0.005 per in. to 3 in. ±0.003 in. for each additional inch	3	1	1	63–125 RMS[d]	0.030 in. (small areas) 0.060 in. (larger areas)	Aluminum: usually under 1000 Other material: all quantities
Powder metal	Iron and iron alloys, copper, bronzes, brass, stainless steel, aluminum	Limitation on surface area (up to 20 in.^{2}) ferrous; larger for nonferrous	Sized: ±0.005 in./in. Unsized: ±0.004 in./in. Direction of pressing, ±0.005 in.	3	4–5	5	Equivalent to 16–90 RMS in solid metals	0.030–0.090 in. (depending on density and length ratio)	Usually 10,000 and up; larger parts sometimes as low as 1000
Plaster mold	Aluminum, brass, bronze, zinc, beryllium copper	Normally up to 500 in.^{2} area; some foundries capable of going much larger	One side of parting line, ±0.005 in.^{2} up to 2 in. Over 2 in., add ±0.002 in./in. Across parting line, add ±0.010 in.	4	1	2	63–125 RMS	0.070 in.	Usually low; often used to prototype for die castings. Average: 50 to 250 pieces
Ceramic mold	Most castable metals	5 to 350 lb	One side of parting line, ±0.005 in. up to 2 in. Over 2 in., add ±0.004 in./in. Across parting line, add ±0.010 in. (assumes epoxy or metal patterns)	5, Wood 4, Epoxy or metal	1	3	80–125 RMS	⅛ in.	Low to medium; often used for tooling for other casting methods, forging dies, plastic injection molds, special machine parts
Die casting	Aluminum, zinc, magnesium, and limited brass	Not normally over 2 ft^2, some foundries capable of larger sizes	Al and Mg ±0.002 in./in. Zinc, ±0.0015 in./in. Brass, ±0.005 in./in. Add ±0.001 to ±0.015 in. across parting line depending on size	1	5	4	32–63 RMS	Al: 0.03 in. small parts, 0.06 in. medium parts Mg: 0.03 in. small parts, 0.045 in. medium parts Zn: 0.025 in. small parts, 0.040 in. medium parts	Usually 2500 and up

Permanent mold	Aluminum zinc, some brass, bronze, lead, and gray iron	Limitation mainly foundry capabilities, aluminum and copper base; ounces to 100 lb; ferrous: 60 lb max.	Aluminum: basic ±0.015 in. Add ±0.002 in./in. If across parting line add ±0.010 to ±0.030 in. depending on size Copper base: similar to investment; iron: ±0.03 in. basic	2	Nonferrous 3–4 Ferrous 4	Nonferrous 3 Ferrous 5	Aluminum: 150–250 RMS Copper base: 150–200 RMS Ferrous: 200–350 RMS	Aluminum: 0.100 in. for small areas, up to 3/16 in. or more for large areas Copper base: 0.060 in. Ferrous: 3/16 in. for small areas, 1/4 in. normal	Minimum one day's run (100–1000 depending on size)
Metal injection molding	Mostly ferrous and copper base	0.0005 to 0.22 lb	±0.005 in./in.	2	1	1	45 RMS	0.015 (small areas)	Usually 10,000 and up
Resin shell mold	Most castable metals	Normal maximum 550 in.^{2} usable mold area; depends on equipment at each foundry	Nonferrous: ±0.008 in./in. decreasing with size Ferrous: ±0.010 in./in. Add ±0.005 to ±0.10 in. across parting lines	2	2	2	Nonferrous: 125–250 RMS Ferrous: 200–350 RMS	Nonferrous: 3/32 in. Ferrous: 1/8 in.	Nonferrous: usually 100 and up Ferrous: usuallay 1000 and up
Sand casting	Most castable metals	Limitation mainly foundry capabilities; ounces to many tons	Nonferrous: ±1/32 to 6 in. Add ±0.003 for each additional inch Ferrous: ±1/32 to 3 in. ±3/64 in. from 3 to 6 in. Across parting line, add ±0.020 to ±0.090 in. depending on size (assumes metal patterns)	Wood 5 Aluminum 4 Iron 2	3–4	2	Nonferrous: 150–350 RMS Ferrous: 300–700 RMS	Nonferrous: 1/8–1/4 in. Ferrous: 1/4–3/8 in.	All quantities

[a]These are for critical dimensions and should not be specified unless necessary.
[b]1 most, 5 least.
[c] Influenced by size and surface area.
[d] RMS = root mean square roughness in microinches.

Source: Reprinted with permission of Talbot Associates, Inc., P.O. Box 416, Springfield, NJ 07081.

- Cast irons have better casting characteristics than most other metals (even nonferrous); they should be the cast materials of choice when ferrous castings are required.
- Cast irons differ significantly in properties because different microstructures are produced by varying silicon and carbon contents and by differing thermal treatments.
- Each grade of cast iron has an application niche; each does something better than the others.
- All cast irons are quench hardenable, but the most desirable microstructure for hardening is a fine pearlitic matrix.
- Welding of all cast irons involves the risk of cracking, and it should be avoided if possible.
- Steel casting complement cast irons; they are used when the stiffness or toughness of cast irons is inadequate. They can have the stiffness, toughness, and weldability of wrought steels.
- Castings should be specified on engineering drawings by their ASTM or similar designation number.
- Powder metals compete with castings on small parts for near-net shape or net shape.
- Powder metals are normally made from iron, bronze, nickel, or aluminum alloy powders.
- Most powder metal processes do not produce 100% density, and this usually reduces toughness.

Castings are the oldest way of shaping metals, and castings should be candidates for any part that will require substantial or difficult machining to fabricate. P/M competes with these same types of applications.

Critical Concepts

- Cast irons are a family of iron/carbon/silicon alloys with widely differing properties depending on structure.
- Cast steels have higher stiffness than cast irons.
- Casting are usually used to produce near-net shapes; powdered metal parts are most often made to final shape.
- The decision on whether to use castings or powdered metal parts is often determined by the quantity required.

Terms You Should Remember

casting	damping capacity
cupola	investment casting
cope	centrifugal casting
drag	die casting
sand casting	shell molding
sprues	shrinkage allowances
risers	solidification shrinkage
permanent mold	dendrite
columnar	graphite
cores	temper carbon
gray iron	damping capacity
wrought iron	secant modulus
malleable iron	graphitization
free cementite	inoculation
white iron	chill
ductile iron	casting

Case History

Investment Castings for Filter Clamps

A chemical manufacturing department requested that a materials engineering laboratory investigate ways to reduce the cost of filter clamps. The department uses hundreds of these clamps, which are very expensive. These clamps hold filter material in a pack, and many of these packs are inserted into a large chamber. The clamps consist of three cast and machined parts. Two parts of the clamp are 316 stainless steel and resemble large flat washers (about 6 in. in diameter and 0.15 in. thick). The third part is made from cast silicon nickel, a very expensive alloy that allegedly resists galling when threaded into the 316 stainless steel parts. All three parts are cast to near-net shape and machined. One of the flat washers contains female threads, and the silicon nickel part has the male threads needed for clamping the filter material between the two washers.

With minor part redesign, the three cast and machined parts were replaced with two investment cast and precipitation hardened 17-4 stainless steel parts at greatly reduced cost. Male threads are cast on one member, and female threads on the other. Machining is no longer necessary on either part. In fact, the cast-to-shape clamps work better than the expensive machined parts, because the hardened threads are less prone to handling damage in cleaning. Many times, castings can save money and work better than other options.

Questions

Section 15.1

1. Compare the surface roughnesses of sand castings and investment castings.
2. What casting process would you use to cast an inside diameter (ID) of 3 in. in a disk with an outside diameter of 8 in. and a thickness of 1 in.? The tolerance on the ID is ± 0.005 in. The material is aluminum.
3. What is the largest casting that can be made with each of the following processes: die casting, investment casting, and permanent mold casting?
4. What will be the shrinkage on a 100-mm-diameter cylinder cast from steel? Cast from cast iron?

Section 15.2

5. How is grain refinement achieved in castings?
6. What is the cause of solidification shrinkage?

Section 15.3

7. How do you weld cast iron?
8. What is the chemical composition of gray cast iron?
9. How does the strength of a class 30 gray iron compare with a common construction steel such as SAE 1020?
10. What is the approximate fatigue strength of a class 60 gray iron?
11. What is the maximum surface hardness that can be obtained in a class 45 cast iron?
12. What are the matrix options with gray cast irons?
13. Why is gray cast iron weaker in tension than compression?

Section 15.4

14. What happens when a white iron casting is malleablized?
15. Compare the stiffness of gray iron, malleable iron, and carbon steel.
16. When would you use malleable iron in preference to other cast irons?

Section 15.5

17. What is the difference between ductile iron and malleable iron?
18. Explain the designation system for ductile irons.
19. What is ADI?

Section 15.6

20. Where are white irons used?
21. How hard are white irons?

Section 15.7

22. How would you specify a cast steel with yield strength of at least 30 ksi?
23. What are three advantages of cast steel over gray cast iron?

Section 15.8

24. Select a casting alloy for a water pump on an automobile. Explain your choice.
25. You want to cast a wheel for a railroad car. What process and alloy would you use? Why?
26. When would you use investment casting?

Section 15.9

27. Can a powdered metal process be used to make an 8-in.-pitch diameter gear with a thickness of ½ in. and a bore of 1 in.?
28. What is the size limit for investment casting? For sand casting? For MIM?
29. What is the difference between conventional P/M molding and injection molding?

Section 15.10

30. You want to make 1 million key-chain ornaments. What casting process would you use? Why?
31. You want to make 500 large, 1-in.-diameter, 10-in.-long eyebolts. What process would you use? Why?
32. What density would you use for a powder transmission gear made with the P/M process?
33. How do you control porosity in P/M parts?

To Dig Deeper

Angus, H. T. *Cast Iron: Physical and Engineering Properties*. London: Butterworths, 1976.

Davis, J. R., Ed. *ASM Specialty Handbook, Cast Irons*. Materials Park, OH: ASM International, 1996.

Material Standards for P/M Structural Parts, MPIF Standard 35. Princeton, NJ: Metal Powder Industries Federation, 1990–91 edition.

Metals Handbook, 9th ed. Volume 7, Powder Metallurgy. Materials Park, OH: ASM International, 1980.

Metals Handbook, Vol. 15 Casting, Materials Park, OH; ASM International, 1988.

Metals Handbook, 10th ed. Volume 1, Properties and Selection: Irons, Steels, and High Performance Alloys. Materials Park, OH: ASM International, 1990.

Steel Castings Handbook, 6th ed. Des Plaines, IL: Steel Founders' Society of America, 1995.

Walton, Charles F. *The Gray Iron Castings Handbook*. Cleveland, OH: Gray Iron Founders' Society, 1958.

Steel Castings Handbook, Supplement 2. Des Plaines, IL: Steel Founders Society of America, 1991.

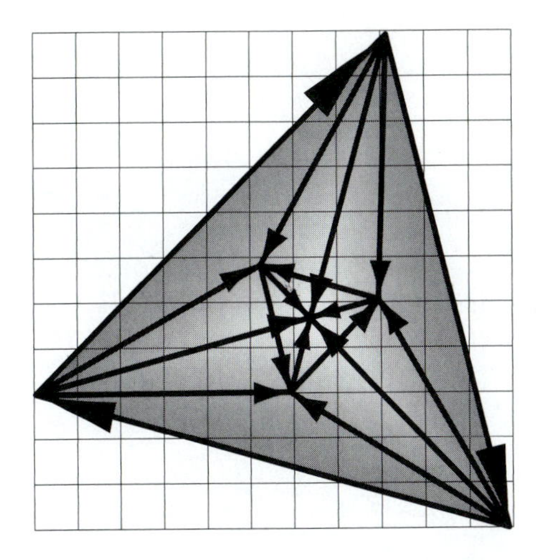

CHAPTER 16

Copper and Its Alloys

Chapter Goals

1. An understanding of the nature of copper and copper alloys, their metallurgy, and the products available.
2. A knowledge of copper and copper alloy designation and properties.
3. Guidelines to aid in selection of copper and copper alloys for engineering design.

Rationale

Why do we dedicate a whole chapter to copper? Sentimentality may play a role—it is our oldest tool material. Ancients only had precious metals and a few soft metals at their disposal. When strength for hardware and tools was needed, copper alloys could be used (bronzes); when malleability to form cooking vessels, and the like, was needed, pure copper could be used. In addition, copper alloys do not rust or readily deteriorate from environmental effects.

Copper alloys are still very important for these reasons, and that is why a whole chapter is devoted to them. However, in the twenty-first century, the property that makes copper one of the most important metals in our lives is its high electrical conductivity. It would not be possible to have the multitude of electrical devices that we take for granted without copper conductors. No other material has yet become serious competition. Aluminum has been tried and failed; precious metals cost too much; and superconductors are still at the laboratory stage. In our current society, other indispensible uses of copper are for plain bearings (large, small, P/M) and for copper piping for potable water. Copper alloys are mostly buried deeply in equipment or structures, but they are there performing their job, mostly as unseen heroes. Designers need to be familiar with the properties of copper alloys and know where to use them to meet all of our current demands.

Historically, copper was one of the first engineering metals because it, unlike most other metals, can occur in nature in the metallic form as well as in an ore. Mass copper can be obtained by simply crushing rocks and physically removing chunks of solid copper. Relatively little of this form of copper is left today, and most of today's supply of copper is obtained from

Figure 16–1
An open-pit copper mine in Utah, in the United States. The black spots in the center of the photo are trucks, each carrying 300 tons of ore.

copper ores that at best may contain only 1.5% copper by weight (Figure 16–1). This is probably the root cause of the high cost of copper compared to "commodity" metals like steel and cast iron. The price of copper and copper alloy varies. It is usually about five times the cost of low-carbon steel.

In this chapter the most commonly used copper alloys will be discussed together with their metallurgy and properties. Then two of the most important application areas for copper alloys, wear and corrosion resistance, will be discussed.

16.1 Extraction of Copper from Ore

In previous discussions it was pointed out that there is no such thing as a pure metal. We shall use the term *pure copper*; it is not really pure, but rather copper with 99.9% by weight copper. Copper is made from low-grade sulfide ores by several steps:

1. Concentration of crushed and milled ores by flotation and other physical separation techniques.
2. Roasting of the ore in special furnaces to remove volatiles and to change the ore into a form more suited to further concentration. The ore does not melt in this process. It merely changes in chemical nature.
3. *Smelting* or melting of roasted ore to produce a *matte*, which is around 30% copper. The smelting operation is simply melting in a controlled atmosphere. This further purifies the ore by, for example, converting iron in the ore to a sulfide that can be removed by subsequent oxidation.
4. Converting the molten copper matte (Cu_2S) to metallic copper is usually accomplished by blowing air or oxygen through the matte in furnaces, not unlike those used in steelmaking. The purification mechanism is *oxidation* of copper sulfides to form metallic copper. The conversion produces *blister* copper with some dissolved oxygen and other impurities. Typically, the product contains 98% to 99% copper.
5. Electrolytic refining is used to further reduce impurities. It essentially involves making anodes from blister copper and plating it out as cathode that is 99.9% (including Ag)

pure copper. When the copper cathodes are melted and cast as cake billet or wire bar, the product is known as *electrolytic tough pitch (ETP)* copper (out UNS C11000).

Low-grade oxide ores can be converted to copper by a *solvent extraction* process wherein acids leach copper from the ore and after some concentrating steps the copper-bearing liquid is used as the electrolyte in *electro winning* or plating cells. Almost pure (99.5%) copper is plated onto cathodes of copper, stainless steel, or titanium.

16.2 Alloy Designation System

The system commonly used in the United States to designate copper alloys is the use of UNS numbers or ASTM specifications, which also reference the UNS numbers. The use of UNS numbers started in 1968. Prior to that, copper alloys in the United States were designated by a numbering system developed by the Copper Development Association (*CDA*), an organization of U.S. copper and brass producers. CDA 101 to 799 were numbers assigned to wrought alloys, and CDA 800 to 999 were assigned to cast alloys. The alloy designation numbers currently have no connection with respect to properties or composition. Each simply specifies a unique alloy with a composition that is different from all other alloys.

The UNS numbers for copper alloys are essentially the CDA numbers with a C prefix and a suffix of two zeros. CDA alloy 122, for instance, is UNS C12200. The UNS number is the appropriate system to use in the United States for designating a particular copper alloy. ASTM specification, which reference the CDA numbers, should be used for international alloy designation. There are about 180 ASTM specifications on copper alloys, which apply to specific product forms. For example, ASTM B 36 covers nine copper–zinc brasses in plate, sheet, strip, and rolled bar form. ASTM B 148 covers aluminum bronze sand castings; ASTM B 176 covers copper alloy die castings, and so on. The same UNS alloy may be covered by another ASTM spec in another product form such as wire. For example, ASTM B 36 covers UNS C12200 in bar form, and another ASTM standard may cover UNS C12200 in wire form.

There are currently about 200 wrought alloys and 130 cast alloys in commercial production in the United States. The complete listing of these alloys is available in the United States from the CDA, with most being covered by ASTM product specifications. A proper drawing specification would be:

> *Material:* ASTM B 36, UNS C22000, H 01
> bar (¼ hard), 10 mm × 10 mm × 3 m

The H 01 suffix on the lower UNS number references the temper designation obtained from ASTM specification B 601 (see Table 16–1). This specification presents an alphanumeric system for specifying 145 different tempers or mill treatments. The use of these temper suffixes provides the most accurate material designation. They should be used when exactness in alloy designation is necessary—for example, if you are ordering stock for millions of stampings. The ASTM B 249 specification must be referenced in detailed purchasing documents for wrought bar rod and shaped copper alloys because it contains industry-accepted tolerances for wrought copper alloy shapes. In the United States the UNS could be used on a drawing, but this number does not constitute a specification, it only references a chemical composition. The ASTM specification lists properties and tolerences that completely specify a material.

Wrought Copper Alloys

Because copper is such an old engineering material, over the years a very confusing vocabulary has developed in alloy identification. What is the

Table 16–1
A selection from ASTM B 601 temper designations for copper and copper alloys

Designation	Description
Cold-Worked Tempers	
H 01	$\frac{1}{4}$ hard
H 02	$\frac{1}{2}$ hard
H 03	$\frac{3}{4}$ hard
H 04	Hard
Cold-Worked and Stress Relieved	
HR 01	H 01 + stress relief
HR 02	H 01 + stress relief
HR 03	H 01 + stress relief
HR 04	H 01 + stress relief
As-Manufactured Tempers	
M 01	As sand cast
M 06	As investment cast
M 07	As continuous cast
M 10	As hot forged and air cooled
M 11	As hot forged and quenched
M 20	As hot rolled
Annealed Tempers	
010	Cast and annealed
025	Hot rolled and annealed
030	Hot extruded and annealed
061	Annealed
070	Dead soft annealed
080	Annealed to $\frac{1}{8}$ hard
081	Annealed to $\frac{1}{4}$ hard
082	Annealed to $\frac{1}{2}$ hard
Solution-Treated Temper	
TB 00	
Precipitation-Hardened Temper	
TF 00	TB 00 + precipitation hardened
Quench-hardened tempers	
TQ 00	Quench hardened
TQ 50	Quench hardened and temper annealed

difference between a brass and a bronze? What are cupronickels, red brasses, or nickel silvers?

A *brass*, by age-old definition, is a copper alloy containing zinc as the principal alloying element. The term *leaded brass* is used to describe copper–zinc–lead alloys, and the term *tin brass* is used to describe copper–zinc–tin alloys.

The term *bronze* usually means a copper alloy with tin as the major alloying element. However, this family of copper alloys has evolved into a group that may have elements other than tin as the major alloying element. Some bronzes have aluminum as the major alloying element; manganese bronzes have zinc.

The alloys of copper and nickel, formerly called *cupronickels*, are now called copper–nickel alloys.

The *nickel silvers* are copper alloys with nickel and zinc as the principal alloying elements. The new terminology for nickel silvers is copper–nickel–zinc alloys. The spectrum of wrought copper alloys and the chemical compositions of some of the commercially available wrought alloys and major alloy categories are shown in Table 16–2.

Cast Copper Alloys

The definitions for brass, bronze, copper–nickel, copper–nickel–zinc, and *high-copper* alloys generally hold true for cast copper alloys. As shown in Table 16–3, the categories of alloys are almost the same, but the cast alloys often have significantly different chemical compositions from their wrought counterparts. For example a common red brass has a composition of 85% Cu, 15% Zn. The popular cast red brass has a composition of 85% Cu, 5% Pb, 5% Sn, 5% Zn. Tables 16–2 and 16–3 show that many copper alloys contain lead (Pb). Users should be aware that in the United States there is a trend to avoid using lead-bearing alloys in pumps, piping, and valves that handle potable water. There are claims that lead leaches out of these alloys and creates a health risk to potable water users. The use of lead-bearing solders for assembly of copper

Table 16–2
Nominal composition of some wrought copper alloys

Coppers (99.3% Cu Min.)		Chemical composition (%)	
UNS No.	Description	Cu Min.	Others
C11000	Electrolytic tough pitch (ETP)	99.9	
C12200	Phosphorus deoxidized (DHP)	99.9	0.015–0.04 P
C15000	Zirconium copper	99.8	0.10–0.2 Zr

High-Copper Alloys (96% Cu Min.)		Cu	Others
C16200	Cadmium copper	Rem	0.5–1.2 Cd
C17000	Beryllium copper	Rem	0.2 Si, 1.6–1.79 Be, 0.2 Al
C17200	Beryllium copper	Rem	0.2 Si, 1.8–2 Be, 0.2 Al
C17500	Beryllium copper	Rem	0.1 Fe, 2.4–2.7 Co, 0.2 Si, 0.4–0.7 Be, 0.2 Al
C18000	High-strength copper	96.4	2.5 Ni, 0.7 Si, 0.4 Cr
C18200	Chromium copper	99.8 min.	0.6–1.0 Cr, 0.1 Fe, 0.1 Si, 0.05 Pb

Copper–Zinc Alloys (Brasses)		Cu	Pb	Fe	Zn	P	Others
C23000	Red brass (85%)	84–86	0.05	0.05	Rem[a]		
C26000	Cartridge brass 70%	68.5–71.5	0.07	0.05	Rem	0.02–0.05	
C27000	Yellow brass 65%	63–68.5	0.1	0.07	Rem		
C28000	Muntz metal	59–63	0.3	0.07	Rem		
C32000	Leaded red brass	83.5–86.5	1.5–2.2	0.1	Rem		0.25 Ni
C36000	Free cutting brass	60–63	2.5–3.7	0.35	Rem		
C37000	Free cutting muntz metal	59–62	0.8–1.5	0.15	Rem		
C38500	Architectural bronze	55–59	2.5–3.5	0.35	Rem		
C44300	Admiralty, arsenical	70–73	0.07	0.06	Rem		0.8–1.2 Sn, 0.02–0.06 As
C46500	Naval brass, arsenical	59–62	0.2	0.1	Rem		0.5–1.0 Sn, 0.02–0.06 As
C48500	Naval brass, leaded	59–62	1.3–2.2	0.1	Rem		0.5–1 Sn
C66700	Manganese brass	68.5–71.5	0.07	0.1	Rem		0.5–1.5 Mn
C69400	Silicon red brass	80–83	0.3	0.2	Rem		3.5–4.5 Si

Bronzes		Cu	Pb	Fe	Sn	Zn	P	Others
C50500	Phosphor bronze 1.25%	Rem	0.05	0.05	1–1.7	0.3	0.03–0.35	
C51000	Phosphor bronze 5%	Rem	0.05	0.10	4.2–5.8	0.3	0.03–0.35	
C52400	Phosphor bronze 10%	Rem	0.05	0.10	9–11	0.2	0.03–0.35	
C54400	Leaded phosphor bronze	Rem	3.5–4.5	0.10	3.4–4.5	1.5–4.5	0.01–50	
C60800	Aluminum bronze 5%	Rem	0.1	0.10				5–6.5 Al, 0.02–0.35 As
C61400	Aluminum bronze D	Rem	0.01	1.5–3.5		0.2	0.015	6–8 Al, 1 Mn
C65100	Low-silicon bronze B	Rem	0.05	0.1	1.2–1.6	0.2		0.1 Al, 0.8/2.0 Si
C65500	High-silicon bronze A	Rem	0.05	0.8		1.5		0.5/1.3 Mn, 2.8/3.8 Si

Copper–Nickel Alloys		Cu	Pb	Fe	Zn	Ni	Mn	Others
C70600	Copper nickel 10%	Rem	0.05	1–1.8	1.0	9–11	1.0	
C71000	Copper nickel 20%	Rem	0.05	1.0	1.0	19–23	1.0	
C71600	Copper nickel 30%	Rem	0.05	0.5	0.05	29–33	1.0	

Copper–Nickel–Zinc Alloys		Cu	Pb	Fe	Zn	Ni	Mn	Others
C74500	Nickel silver 65–10	63.5–66.5	0.1	0.25	Rem	9–11	0.5	
C75200	Nickel silver 65–18	63.5–66.5	0.05	0.25	Rem	16.5–19.5	0.5	
C77000	Nickel silver 55–18	53.5–56.5	0.05	0.25	Rem	16.5–19.5	0.05–0.35	

[a]Rem = remainder

Table 16–3
Nominal composition of some cast copper alloys

Coppers (99.3% Cu Min.)		Chemical Composition (%)	
UNS No.	**Description**	**Cu Min.**	**Others**
C80100	Pure copper	99.95	

High-Copper Alloys (94% Cu Min.)		Cu Min.	Ag	Be	Co	Si	Ni	Fe	Al	Sn	Pb	Zn	Cr
C81500	Beryllium copper	95		0.45–0.8	2.4–2.7	0.15	0.2	0.1	0.1	0.1	0.02	0.1	0.1
C82000	Chromium copper	98				0.15		0.1	0.1	0.1	0.02	0.1	0.4/1.5
C82500	Beryllium copper	95.5		1.9–2.15	0.35–0.7	0.2/0.35	0.2	0.25	0.15	0.1	0.02	0.1	0.1

Copper–Zinc Alloys (Brasses)		Cu	Sn	Pb	Zn	Fe	Ni	P	Al	Si	Other
C83300	Low-zinc brass	92–94	1–2	1–2	2–6						
C83600	Red brass 5-5-5	84–86	4–6	4–6	4–6	0.3	1	0.05	0.005	0.005	0.25 Sb, 0.08 S
C84500	Semi-red brass	77–79	2–4	6–7.5	10–14	0.4	1	0.02	0.005	0.005	0.25 Sb, 0.08 S
C85200	Leaded yellow brass	70–74	0.7–2	1.5–3.8	20–27	0.6	1	0.02	0.005	0.005	0.2 Sb, 0.05 S
C85500	Yellow brass	59–63	0.2	0.2	Rem[a]	0.2	0.2			0.05	0.2 Mm
C85700	Leaded yellow brass	58–64	0.5–1.5	0.8–1.5	32–40	0.7	1		0.8		
C86300	Manganese brass	60–66	0.2	0.2	22–28	2–4	1		5–7.5		2.5–5 Mn
C87500	Silicon brass	79 min.		0.5	12–16				0.5	3–5	

Bronzes		Cu	Sn	Pb	Zn	Fe	Sb	Ni	S	P	Al	Si	Mn
C86100	Manganese bronze	66–68	0.2	0.2	Rem	2–4					4.5–5.5		1.5–5
C86700	Leaded manganese bronze	55–60	1.5	0.5–1.5	30–38	1–3		1			1–3		1–3.5
C87300	Silicon bronze	94 min.		0.2	0.25	0.2						3.5–4.5	0.8–1.5
C90200	Tin bronze	91–94	6–8	0.3	0.5	0.2	0.2	0.5	0.05	0.05	0.005	0.005	
C90300	Tin–zinc bronze	86–89	7.5–9	0.3	3–5	0.2	0.2	1.0	0.05	0.05	0.005	0.005	
C92410	Leaded tin bronze	Rem	6–8	2.5–3.5	1.5–3	0.2	0.25	2.0			0.005	0.005	
C94300	High-lead tin bronze	68.5–73.5	4.5–6	22.25	0.8	0.15	0.8	1.0	1.0	0.08	0.005	0.005	
C94800	Nickel–tin bronze	84–89	4.5–6	0.3–1	1–2.5	0.25	0.15	4.5–6	0.05	0.05	0.005		0.2
C95200	Aluminum bronze	86 min.				2.5–4					8.5–9.5		
C95500	Nickle–aluminum bronze	78 min.				3–5		3–5.5			10–11.5		3.5

Copper–Nickel Alloys		Cu	Pb	Fe	Ni	Mn	Si	Nb	C
C96200	Cupronickel 10%	84.5–87	0.03	1–1.8	9–11	1.5	3	1	0.1
C96300	Cupronickel 20%	Rem	0.03	0.4–1	18–22	1.0	0.7	1	
C96400	Cupronickel 30%	65–69	0.03	2.5–1.5	28–32	1.5	0.5	0.5–1.5	0.15

Leaded Coppers		Cu	Sn	Pb	Fe	Ag
C98200	Leaded copper	73–79	0.5	21–27	0.35	
C98600	Leaded copper	60–70	0.5	30–40	0.35	1.5

Copper–Nickel–Zinc Alloys		Cu	Sn	Pb	Zn	Fe	Sb	Ni	S	P	Al	Mn	Si
C97300	Nickel silver (high lead)	53–58	1.5–3	8–11	17–25	1.5	0.35	11–14	0.08	0.05	0.005	0.5	0.15
C97400	Nickel silver	58–61	2.5–3.5	4.5–5.5	Rem	1.5		15.5–17				0.5	
C97800	Nickel silver (low lead)	64–67	4.5–5.5	1–2.5	1–4	1.5	0.2	24–27	0.08	0.05	0.005	1	0.15

[a]Rem = remainder

plumbing have been banned in the United States since 1986 for the same reason. This should be considered in the alloy selection process.

16.3 Copper Products

Almost 80% of all copper produced is used in the form of pure copper in shapes used in plumbing, electrical wiring, and architectural applications. The remainder is used in the many alloy forms. Pure copper and copper alloys are available in the wrought form of sheet, strip, bar, shapes, rod, wire, tubing, and pipe. Sheet and strip products can be obtained in varying degrees of cold work (quarter-, half-, three-quarters, and full-hard tempers) as well as in spring tempers. It is the user's responsibility to select the necessary temper or mill heat treatment and specify it on purchasing documents and drawings (ASTM B 601 lists the options). Temper requirements are mostly determined by required mechanical properties. If a part to be made from sheet requries significant forming, the user will probably want to use an annealed temper (O). If a part will be flat blanked, a cold-worked temper (H) may be appropriate.

Cold-finished sheet and strip have clean, scale-free surfaces with almost a polished surface in some of the thinner gage products. The only finish specification commonly used is "cold finished." The heavier products such as plate and pipe still have scale-free surfaces suitable for many uses in the as-received condition.

The tempers and finishes available in wrought products obviously do not apply to cast products, but clean, scale-free castings are the normal product. The surface appearance is a function of the casting process used, and all the common casting processes, including die casting, are used for copper alloys. Plumbing fittings are often sand cast, and they have a relatively rough surface. On the other hand, investment-cast copper has a surface suitable for decorative applications and even jewelry.

16.4 Metallurgy

Pure Copper

Copper as an element has a face-centered cubic (FCC) crystal structure. It is an excellent conductor; it is malleable and easily cast and shaped. As mentioned previously, copper is significantly affected by very small amounts of impurity elements. This effect is primarily in physical properties, but the mechanical properties of some alloys can be increased by relatively small alloy additions. For example, adding 0.8% chromium can double the strength of pure copper.

Tough pitch copper contains about 0.04% oxygen (from the refining operations). The oxygen is relatively insoluble in the copper, and it is present in the form of cuprous oxide. Large amounts of oxygen, such as the 1% present in blister copper, can make copper brittle. Even trace amounts of oxygen, such as the 0.04% present in tough pitch copper, cause *embrittlement* by internal oxidation or reduction of sulfides when the alloy is heated to temperatures over about 700°F (370°C) in reducing environments. This might occur in furnace parts or in welding or brazing. Hydrogen atmospheres can cause blistering of the oxygen-bearing copper. The hydrogen diffuses into the copper and combines with the oxygen to form water vapor. High pressures develop and blisters appear on the metal surface.

Copper producers go to great lengths to keep oxygen low in copper products that may experience high temperatures. About 0.02% phosphorus can be added to remove oxygen. These are the *phosphorus deoxidized coppers* (*DHP*). *Phosphor bronze*, as the name implies, also contains phosphorus. The solubility of phosphorus in copper is only about 0.5%; thus the phosphorus level is usually kept very low to prevent the formation of phosphorus-rich microstructural components. Small additions of phosphorus slightly increase tensile strength, but they drastically reduce electrical conductivity.

Pure copper is alloyed with many other elements to produce minor changes in properties. Sulfur, tellurium, selenium, and lead are added to improve machinability. All are soluble in small amounts, and machinability is improved by the action of dispersed sulfides, tellurides, and lead inclusions. Chromium is added for strengthening, as are other elements such as zirconium, cadmium, and tin. Some U.S. copper ores contain silver as a natural impurity. Silver has excellent solubility with copper, and the silver remains with the copper during refining to produce silver-bearing pure copper, which has improved heat resistance over other coppers. Total alloy additions in pure copper are usually less than 1%.

Special *oxygen-free coppers* are available that were not deoxidized by phosphorus additions. These coppers have the highest electrical conductivity. Because the oxygen content is very low (<10 ppm), these alloys are not subject to embrittlement at elevated temperatures and can be readily welded or brazed.

Pure copper usually is a single-phase alloy. Its outstanding properties are its thermal and electrical conductivities. The atomic reason for the high conductivity is the metallic bonds between atoms. In early discussions on the nature of materials, we defined a metal as an element that has the atoms held together with an electron cloud; the electrons from one atom are free to move to the next, and so on, throughout the entire crystal and throughout the entire shape. Pure copper is the epitome of the metallic bond. There is a very high degree of electron mobility. Many alloy elements affect the conductivity of pure copper even though they are present in very low concentrations (Figure 16–2). The commercial definition of pure copper is copper with at least 99.88% copper, with silver counted as copper. When a few percent of alloy is added to copper for strengthening reasons, these types of alloys are called high-copper alloys, and the principal alloying elements for high-copper alloys are beryllium, chromium, zirconium, and tellurium copper. All of these high-copper alloys remain single-phase FCC copper, and the alloys elements are usually in substitutional solid solution, except when they are made to precipitate in age hardening. If high concentrations of zinc or aluminum are added to copper, a second phase can form. In copper–zinc alloys this second phase is called beta, and it has a bodycentered cubic (BCC) structure (Figure 16–3). The most widely used copper alloys are single-phase alloys, and the copper in this phase has a face-centered cubic structure, with most of the alloy in solid solution in the FCC structure.

Beryllium Copper

The addition of small amounts of beryllium to copper creates a family of high-copper alloys with strengths as high as alloy steel. The high strength is obtained by the *precipitation hardening* (*PH*) mechanism, similar to the PH stainless steels. In binary alloy systems, precipitation hardening can occur if the alloying element has high solubility at high temperatures and low solubility at low temperatures. Precipitation hardening consists of heating the alloy to some elevated temperature, liquid quenching, and aging at some moderate temperature. At least 20 elements theoretically could be used to make a precipitation-hardening copper alloy, but the only systems in wide commercial use are copper–chromium and copper–beryllium. Figure 16–3 contains a portion of the copper-beryllium phase diagram. It can be seen that the solubility of beryllium in copper is about 2.7% at 870°C (1600°F). At room temperature, the solubility is only about 0.5%. If an alloy of about 2.0% beryllium is water quenched from its annealing temperature of about 800°C (~1470°F), an excess amount of beryllium will be trapped in solution. Equilibrium conditions dictate that only 0.5% can be retained in solid solution at room temperature. The solid solution is metastable, and beryllium can be made to precipitate from solid solution by a simple aging heat treatment at

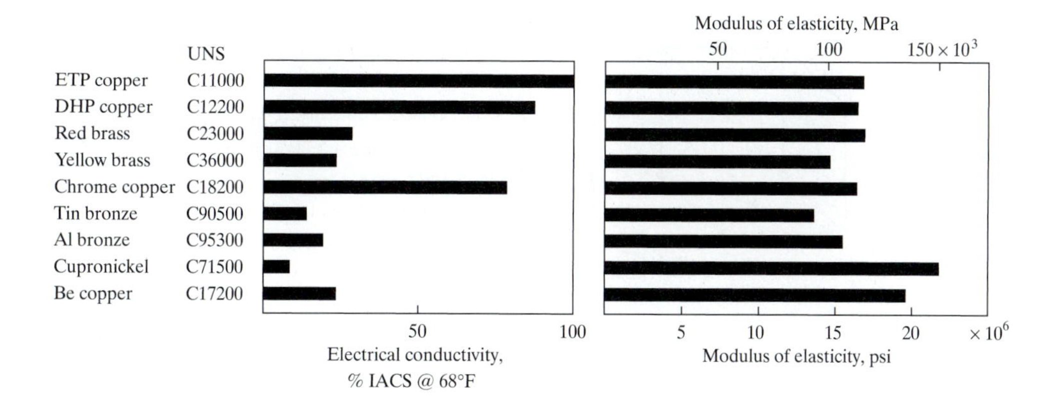

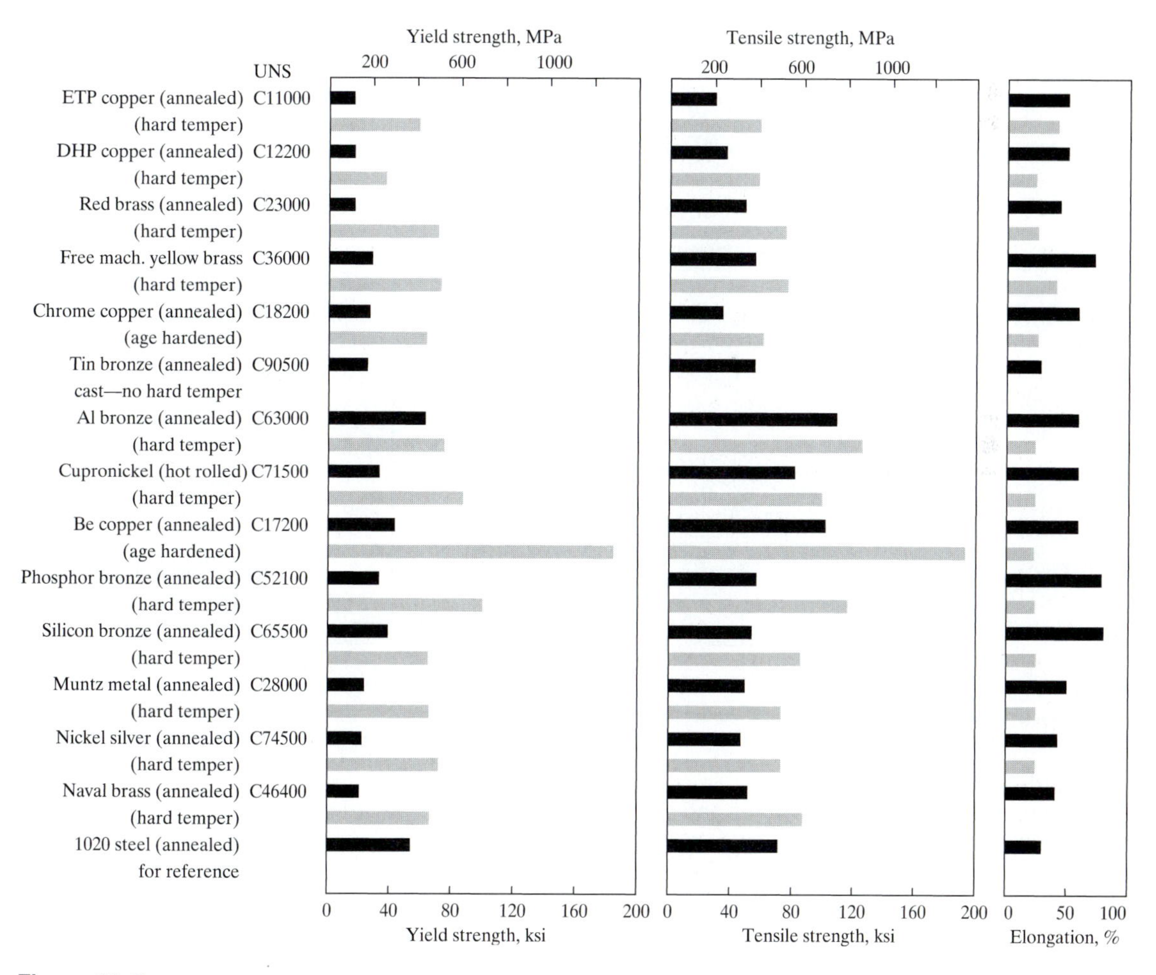

Figure 16–2
Properties of some copper alloys

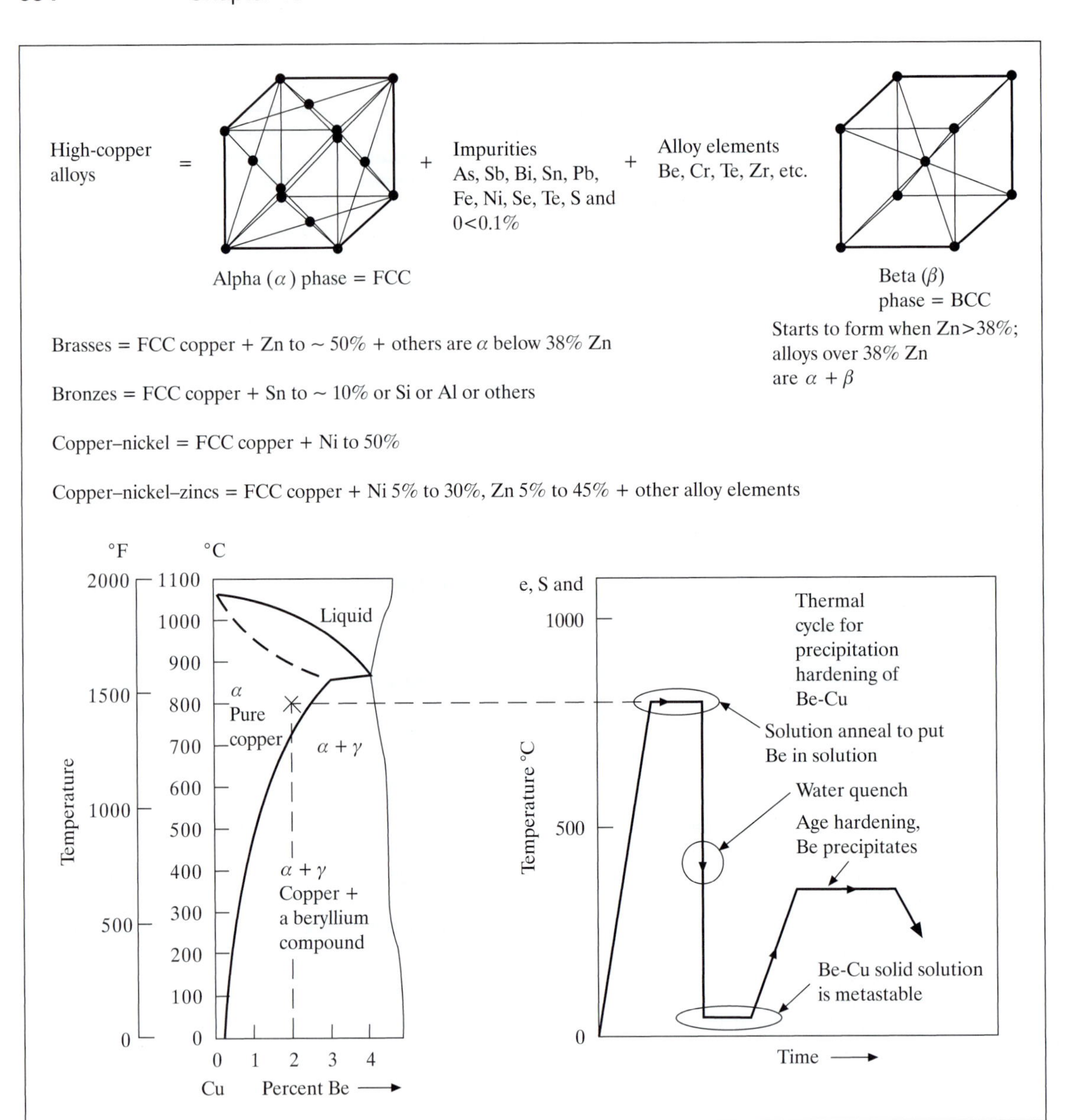

Figure 16–3
Metallurgy of copper alloys—structures and compositions; precipitation hardening of BeCu alloys

Source: Part of BeCu phase diagram from *Metals Handbook*, Volume 8,© ASM International, 1973. Used with permission.

about 315°C (600°F). The beryllium atoms strain the crystal lattice as they precipitate, and hardening and strengthening occur. In the case of *beryllium copper*, the electrical conductivity also improves.

Thus precipitation hardening is an important strengthening heat treatment for copper alloys. Many copper alloys can show some precipitation-hardening effects, but the most dramatic improvement in mechanical properties is obtained with alloys of copper and from about 1.5% to 2.00% beryllium. Some alloys use cobalt as a beryllium substitute, but the strengthening mechanism is the same.

Brass

Plain brasses, binary alloys of copper and zinc, can have two different types of crystal structure. An *alpha brass* is a solid solution of copper and zinc with a FCC crystal structure. Copper–zinc systems have this structure up to about 38% zinc. With higher zinc contents, *beta* structure, which is BCC, starts to form. Brasses with zinc contents near 40% can have a structure of alpha and beta, but at around 50% zinc the structure is entirely beta. The properties of the alpha and beta structures differ considerably. The alpha brasses with high zinc contents (~30%) are referred to as *yellow brasses*. The most common *beta brass*, with a composition of 60% copper–40% zinc, is called *Muntz metal*.

Yellow brasses are frequently alloyed with several percent of lead to improve machinability (leaded yellow brasses). The lead is relatively insoluble and plays no major role in altering phase relationships.

Red brass contains about 15% zinc. With such a low zinc content, the alloy is more copper colored than yellow, thus the term *red brass*. In the cast form, red brass usually is an alloy of 85% copper, 5% zinc, 5% tin, and 5% lead.

Many brasses contain other alloying elements, such as aluminum, tin, nickel, and arsenic, for special purposes.

Bronze

As previously stated, present-day bronzes can be almost any copper alloy, but we shall adhere to the older definition: copper–tin alloys. These alloys are often referred to as phosphorus bronzes because they frequently contain a noticeable amount of phosphorus (~0.3%) from the refining operation. Tin is relatively insoluble in copper. At room temperature only about 1% tin can be dissolved in copper in a primary solid solution. When the tin content gets to about 10%, delta phase, the hard, brittle copper compound appears as a discrete phase. This phase is really present in tin concentrations down to about 1%, but it usually cannot be resolved by optical metallographic techniques. The effect of delta phase is to strengthen and harden by lattice distortion.

Silicon bronze is an important alloy for marine applications and high-strength fasteners. Silicon is soluble in copper up to about 4%. Most of the commercial silicon bronzes contain less than this amount of silicon, plus small amounts of other elements that promote the formation of more complex structures. In general, the effect of silicon is to strengthen, harden, and improve corrosion resistance.

Aluminum bronzes contain no tin, and today they are more correctly termed copper–aluminum–iron alloys. Aluminum is soluble in copper up to about 8%, and a single-phase solid solution will be formed in this range. Two-phase aluminum bronzes usually contain 8% to 12% aluminum. Iron is added to aluminum bronzes to increase the strength and hardness. The ternary alloys of copper–aluminum–iron can be quench hardened and temper annealed (TQ50 in the ASTM B 601 temper designation specification).

Cupronickels

Nickel has complete solid solubility in copper. A single-phase solid solution will form with any combination of the two elements, a rare metallurgical happening (see Figure 10–2). These alloys are ductile, and they are hardened and strengthened by cold work. The most important cupronickel alloys between 10% and 30% nickel and small amounts of iron. They have better corrosion resistance than many other copper alloys in seawater.

Nickel Silvers

Copper–nickel–zinc alloys, as they are more correctly termed, have concentrations in the approximate ranges of 45% to 75% copper, 5% to 30% nickel, and 5% to 45% zinc. Ternary alloys of copper–nickel–zinc generally have structures similar to brasses. They are ductile, soft, single-phase alloys. The two-phase alloys are less ductile and harder. Nickel silvers have moderate strength in the cold-worked condition, so they find application as springs and mechanical components, but one of the most important uses of these alloys is for aesthetic purposes. With the right combination of nickel and zinc, alloys can be obtained with the appearance of silver, thus the name nickel silver. These alloys are used for nameplates, bezels, eyeglass frame, silver-plated cutlery, and silverware. If the silver plate wears off it is not too noticeable because nickel silver can match the color of silver.

16.5 Properties

Physical

It is estimated that about 50% of all copper used in the United States goes into electrical conductors as wire and cable. In this application, electrical conductivity is usually the most important selection factor. Rather than use resistivity values, it is common practice to rate the conductivity by *% IACS*, which means percent of the *International Annealed Copper Standard*. As can be seen from the property tabulations in Figure 16–2, tough pitch copper (ETP) rates 100%. DHP copper, which has only about 0.02% phosphorus, has a conductivity of only about 80% IACS. Thus ETP or pure coppers are usually used for conductors. Chromium copper, which has better elevated temperature strength than ETP copper, is usually used for resistance welding electrodes because it still has about 80% of the conductivity of pure copper. Yellow brass has relatively poor conductivity; beryllium copper, chromium copper, or chromium/zirconium coppers are commonly used for springs and contacts that must carry electricity. About 60% of BeCu production is in strip form for these types of applications.

The modulus of elasticity of most copper alloys is about 17×10^6 psi (117×10^3 MPa). Cupronickels have the highest modulus at about 22×10^6 psi (151×10^3 MPa). This positions copper alloys between aluminum and steels in stiffness.

Some other physical properties of importance are color, spark resistance, nonmagnetism, and melting point. The pleasing color of copper alloys is a prime reason for using these alloys for decorative and architectural applications. The color possibilities are wide, and alloying elements dictate the color. Age-hardened beryllium copper alloys and nickel/aluminum bronze (UNS C63000) are used for hand tools in chemical plants where nonsparking properties are required for safety.

The relatively low melting temperatures of copper alloys simplify fabrication by casting, hot forming, or fusion welding. The melting range is generally between the temperatures of 1500 to 2000°F (815 to 1100°C). Metal mold die casting and extrusion are possible. Low forging temperatures contribute to long die service life. Copper alloys are widely used for brazing dissimilar metals, and copper alloys themselves can readily be fusion welded by all the conventional processes. The exceptions are the alloys with high lead, zinc, or oxygen content (for example, ETP). The deoxidized oxygen-free grades should be used if full-strength ductile welds are required. Welding should be avoided on high-lead alloys.

Mechanical

Pure copper has the lowest tensile strength of all the alloys shown in Figure 16–2. The role of the various alloying elements used in copper systems is usually to increase strength and hardness. All the ductile copper alloys can be cold worked to improve strength, but beryllium copper is the best when it comes to strength. In the age-hardened condition, the tensile strength can be as high as 200 ksi (1379 MPa).

The ductility of wrought copper alloys depends on the degree of cold work. In the annealed condition, they all have high elongations and good formability. Wrought beryllium copper after heat treatment has poor ductility, but forming can usually be done in the annealed or cold-worked condition. The ductility of some of the cast bronzes can be as low as 5% elongation.

Low temperatures have little effect on the ductility of copper alloys; they do not get brittle. Most copper alloys soften when used at temperatures over 400°F (200°C), and the oxygenbearing coppers embrittle when exposed to reducing atmospheres. Use at elevated temperatures requires careful consideration of elevated-temperature property data. In addition to lowering strength, copper and copper alloys oxidize significantly in air at temperatures above 500°F (260°C). They should not be used for elevated-temperature applications without consideration of this factor. As shown in Figure 16–4, the ability to cold work the wrought alloys produces wide strength ranges with each major alloy system.

Copper alloys never get very hard, as shown in the lower graph in Figure 16–4. Beryllium coppers can achieve a hardness of 40 HRC. One of the most severe mechanical applications of a copper alloy is the use of beryllium copper for cavities in plastic injection molds. The cavity must withstand high compressive loads and injection pressures that are often as high as 20 ksi (138 MPa), but the hardness is not as important in this application as compressive and tensile strength.

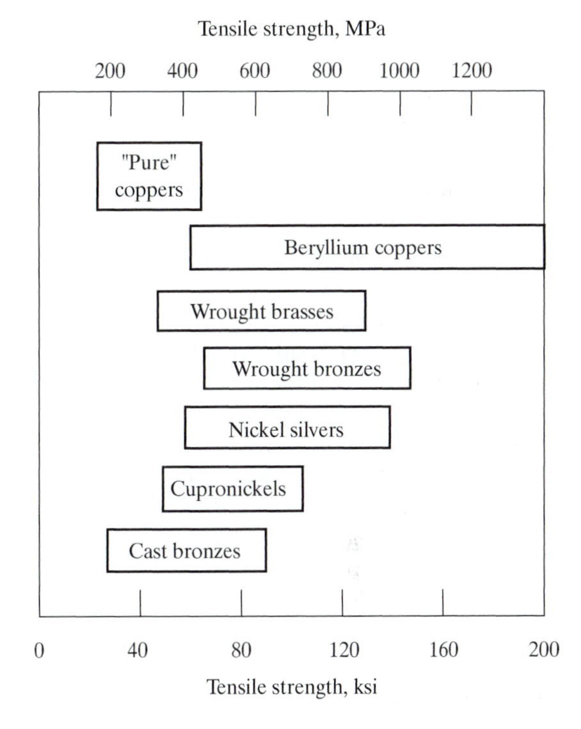

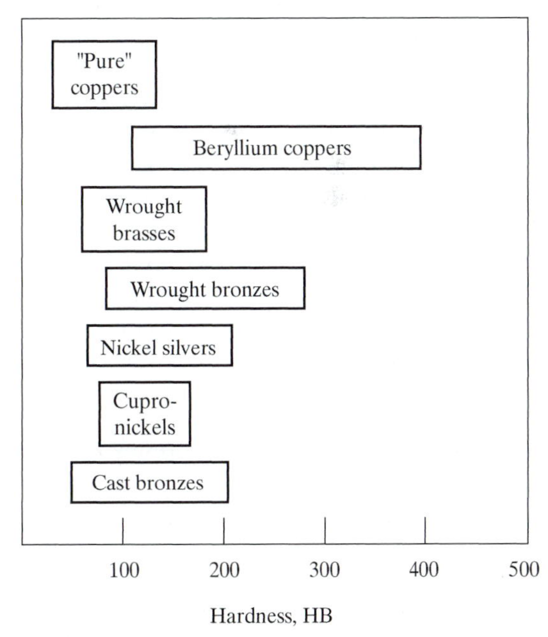

Figure 16–4
Tensile strength and hardness ranges for various copper alloy systems at room temperature

16.6 Heat Treatment

Pure copper and copper alloys are usually used in the as-processed condition, but occasionally heat treatments are needed to modify properties. Annealing is used to soften wrought alloys when formability problems arise. Annealing temperatures are usually in the range from 800 to 1200°F (430 to 650°C).

Stress relieving is used to provide dimensional stability on parts subject to significant machining or to lower residual stress levels on parts that may be subject to atmospheres that cause *stress corrosion cracking*. Stress relieving is usually performed for about 1 h at 400 to 500°F (204 to 260°C). On cold-worked alloys, careful control is needed to prevent significant lowering of strength.

The age-hardening heat treatments for beryllium copper are summarized in Table 16–4. Other alloys can also be solution treated and age hardened: chromium copper and zirconium copper. Solution-treating temperatures are as high as 1800°F (981°C), and aging is performed in the temperature range from 600 to 900°F (315 to 482°C). The strengthening effect of age hardening on these other alloys, however, is not as significant as with the beryllium coppers. All the age-hardenable alloys, with the exceptions of C82000 and C82200, achieve a hardness of about 40 HRC+ when age hardened to maximum strength. Aluminum bronzes are quench hardened to form a "copper" martensite. They can be used as quenched or they can be tempered after quench hardening.

16.7 Fabrication

A very large percentage of the tonnage of copper in industry is used in the form of wire and tube (plumbing and air conditioning) where machining and other fabrication processes are of little concern; wire and tube are often just cut to length when they are used. All the wrought copper alloys have good formability sheet and strip. Even in full-hard tempers, most brasses and copper alloys, including beryllium copper, can be readily formed into shapes in thin gages. Bend allowances must be increased when thicknesses get above about 0.06 in. (1.5 mm), but, in general, wrought coppers, brasses, copper–nickels, phosphor bronzes, and nickel silvers are very malleable. Annealed tempers of copper alloys have the best formability. The cast alloys are not normally formed, but if the occasion arises, the formability of the specific alloy should be researched in the copper literature. Some of the cast alloys can have low ductility.

All the copper alloys are machined at one time or another, and the various alloys have different machining characteristics. Figure 16–5 illustrates the relative machinability of some copper alloys rated against leaded free-machining carbon steel. The best of the copper alloys, free-cutting brass (alloy C36000), is allegedly five times easier to machine than the best carbon steel. These data are based on screw machine turning and facing operations, and they may not apply to other machining operations such as tapping or milling. These data suggest that many copper alloys are easy to machine. From the standpoint of personal experience, leaded yellow brasses are probably the best metals of any type to machine. The high-copper alloys, on the other hand, are among the worst to machine. They tend to "grab" tools, and machine tool setups must be extremely rigid. Diamond tools are frequently used to turn large copper cylinders. They resist the tool pickup and welding common with the high-copper alloys. Unleaded bronzes are also not user friendly. Overall, all copper alloys can be machined with normal machine shop practices, but the unleaded alloys are much more difficult to machine than the leaded alloys. The leaded brasses are excellent to machine. Copper researchers have demonstrated the suitability of bismuth and selenium for improving the machinability of lead-free copper alloys. These alloys became commercial in 1996.

Table 16–4
Standard solution treating and aging practice for beryllium copper alloys

UNS No.	Nominal Composition	Solution Treating* Temperature		Aging Temperature		Time (h)	Expected Tensile Strength	
		°F	°C	°F	°C		MPa	ksi
Wrought alloys								
C17000	Cu–1.7Be–0.25Co	1425–1475	775–800	575–650	300–345	2–3	1240–1380	180–200
C17200	Cu–1.9Be–0.25Co	1425–1475	775–800	575–650	300–345	2–3	1310–1490	190–215
C17300	Cu–1.9Be–0.4Pb–0.25Co	1425–1475	775–800	575–650	300–345	2–3	1310–1490	190–215
C17500	Cu–2.5Co–0.5Be	1650–1750	900–955	850–900	455–480	2–3	830–1040	120–150
C17510	Cu–1.8Ni–0.5Be	1650–1750	900–955	850–900	455–480	2–3	830–1040	120–150
Casting alloys								
C82400	Cu–1.7Be–0.25Co	1425–1475	775–800	650	345	3	1070 min.	155 min.
C82500	Cu–2Be–0.5Co	1425–1475	775–800	650	345	3	1070 min.	155 min.
C82510	Cu–2Be–1Co	1425–1475	775–800	650	345	3	1070 min.	155 min.
C82800	Cu–2.75Be–0.5Co	1425–1475	775–800	650	345	3	1040 min.	150 min.
C82000	Cu–2.5Co–0.5Be	1650–1750	900–955	850–900	455–480	3	660 min.	96 min.
C82200	Cu–1.8Ni–0.4Be	1650–1750	900–955	850–900	455–480	3	660 min.	96 min.

*Time for solution treating thin strip and wire products is held to 10 to 20 min at temperature in order to restrict grain growth. Time for solution annealing heavier wrought and cast products is 1 h at temperature per inch or fraction of inch of section thickness. All alloys are water quenched immediately from the solution treating temperature.

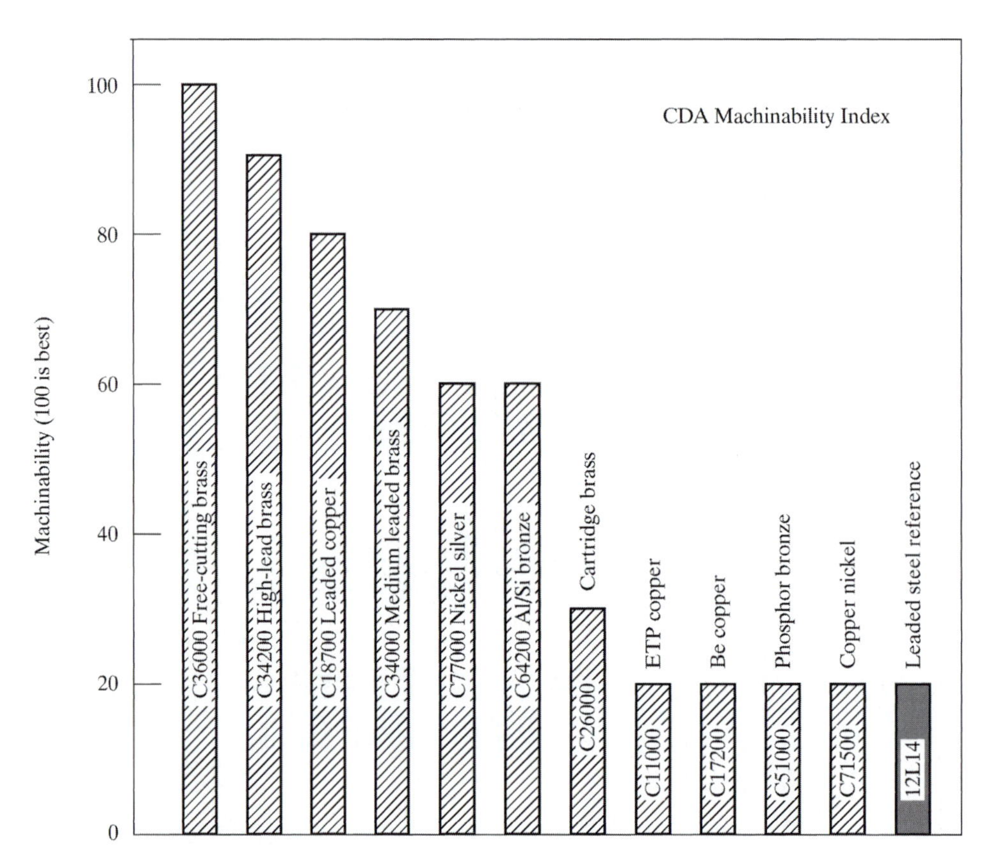

Figure 16–5
Relative machinability of selective copper alloys

The beryllium coppers may require occupational health concern. When beryllium is handled in a workplace, the U.S. occupational health organization, OSHA, requires monitoring the particulate level in ambient air, and the amount in the air must be controlled. Beryllium allegedly produces respiratory problems if inhaled. Beryllium coppers contain less than 2.5% beryllium, but some government regulatory bodies apply the same rules to beryllium copper as to pure beryllium. Essentially, it is recommended that dry grinding and similar operations that could produce airborne particulate be avoided. If parts are machined with flood coolant, chances are likely that airborne particulates would never be produced, but some industries still monitor the air as a precaution. Users of these alloys should determine if there are any health codes that apply to fabrication processes and take appropriate precautions.

Copper alloys can be welded but they weld together with different degrees of probable success. Table 16–5 gives the weldability of copper and other metals and shows how the weldability differs with the different processes. The data in Table 16–5 are for an alloy welded to itself. Copper alloys cannot be welded to steels, but they can be welded to nickel-based alloys. Essentially, copper has limited solubility in steel and many other metals, and its weldability limitation

Table 16–5
Weldability of some metals

	Welding Processes								
Base Metals Welded	**Shielded Metal Arc**	**Gas Tungsten Arc**	**Plasma Arc**	**Submerged Arc**	**Gas Metal Arc**	**Flux Cored Arc**	**Electroslag**	**Braze**	**Gas**
Aluminums	C	A	A	No	A	No	Exp	B	B
Copper-based alloys									
Brasses	No	C	C	No	C	No	No	A	A
Bronzes	A	A	B	No	A	No	No	A	B
Copper	C	A	A	No	A	No	No	A	A
Copper nickel	B	A	A	No	A	No	No	A	A
Irons									
Cast, malleable nodular	C	No	No	No	C	No	No	A	A
Wrought iron	A	B	B	A	A	A	No	A	A
Lead	No	B	B	No	No	No	No	No	A
Magnesium	No	A	B	No	A	No	No	No	No
Nickel-based alloys									
Inconel	A	A	A	No	A	No	No	A	B
Nickel	A	A	A	C	A	No	No	A	A
Nickel silver	No	C	C	No	C	No	No	A	B
Monel	A	A	A	C	A	No	No	A	A
Precious metals	No	A	A	No	No	No	No	A	B
Steels									
Low-carbon steel	A	A	A	A	A	A	A	A	A
Low-alloy steel	B	B	B	B	B	B	A	A	A
High and medium carbon	C	C	C	B	C	C	A	A	A
Alloy steel	C	C	C	C	C	C	A	A	A
Stainless steel (austenitic)	A	A	A	A	A	B	A	A	C
Tool steels	No	C	C	No	No	No	No	A	A
Titanium	No	A	A	No	A	No	No	No	No
Tungsten	No	B	A	No	No	No	No	No	No
Zinc	No	C	C	No	No	No	No	No	C

Metal or process rating: A, recommended or easily weldable; B, acceptable but not best selection or weldable with precautions; C, possibly usable but not popular or restricted use or difficult to weld; No, not recommended or not weldable; Exp, experimental.

Source: After H. B. Cary, *Modern Welding Technology*. Upper Saddle River, NJ: Prentice Hall, 1979.

reflect its solid solubility characteristics. Copper can be brazed to steels, and copper alloys can be brazed to each other with minimal problems. Two types of brazing alloys are used on copper alloys: silver-based alloys that melt in the temperature range from about 1000 to 1400°F (590 to 760°C) and the copper–phosphorus and copper–silicon alloys that melt in the range of about 1400 to 1650°F (760 to 900°C). Both consumables produce excellent results, and the normal application process is an oxy-fuel torch.

Copper tube is often joined by soldering. Tin/lead solders have been used for many years, but lead-bearing solders are no longer permitted in many parts of the United States because of the concern for a health hazard from the lead. Tin/antimony or other tin-based alloys are currently used as replacements for tin/lead solders.

Various surface finishes can be obtained on wrought copper alloys at the mill, and many additional cosmetic finishes apply. Copper is an excellent substrate for electrodeposited coatings such as chromium and nickel. There are no diffusion treatments for hardening the surface of copper alloys other than some less common processes such as metalliding.

16.8 Wear Resistance

Copper alloys have been used for wear applications for centuries. Today this area remains one of the most important uses of copper alloys. Plain bearings of copper alloys are widely used in fractional-horsepower motors, in marine applications, and in very large, heavily loaded journals such as paper mill rolls, railroad train bearings, and steel mill roll bearings. Wrought copper alloys are used for stamped gears, cams, and escapements in clocks, timing mechanisms, switch gears, and cameras. Brasses, bronzes, and beryllium coppers are the most widely used copper alloys for wear components. As emphasized earlier, there are different modes of wear, and the design engineer should always start a design by assuming a predominating mode of wear. Copper alloys are often subjected to cavitation and other forms of corrosive wear, but in machine applications the most important modes of wear to be concerned with are abrasion and metal-to-metal wear. In this discussion, we shall concentrate on these areas.

Abrasion/Erosion

Copper alloys can be subjected to *erosion* by hard particles when they are used as pipe for carrying slurries, for valves or impellers in pumps carrying sand-filled water, and for similar applications. Many times copper alloys are used for plain bearings in dirty machinery for which embeddability is important. If grit gets into a bearing, it will not cause *seizure* if the grit embeds itself into the soft bearing materials.

Traditional theories of wear propose that the abrasion resistance of metals is inversely proportional to penetration hardness. As shown in Figure 16–4, the hardness range possible in copper alloys is from about 30 to 400 HB. Experimental data show that hardness is not a valid indicator of abrasion resistance of copper alloys. Annealed 1020 steel (with a hardness of 180 HB) has better abrasion resistance than even the hardest copper alloys. The conclusion of laboratory tests is that none of the copper alloys has good resistance to hard abrasives, and in this type of abrasion situation there is little difference between the various copper alloys.

To the designer concerned with applications, it is best to avoid the use of copper alloys if the service involves exposure to abrasive substances. Copper alloys are not very abrasion or erosion resistant.

Metal-to-Metal Wear

Bronzes are the most commonly used copper alloys for metal-to-metal wear systems. Beryllium coppers and high-strength bronzes in the

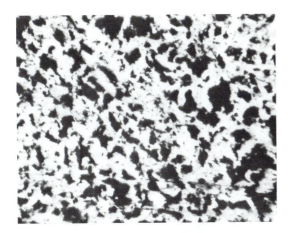

Figure 16–6
Porosity in a P/M bronze (dark areas) (×20)

hardened condition are used for plain bearings in aircraft because they withstand the high bearing pressures. Yellow brasses are used for a wide variety of applications in consumer goods, but they are seldom used on high-performance machines. The most popular bronze alloys are tin bronze, high-leaded and *leaded tin bronze*, aluminum bronze, and tin bronze in sintered powdered metal (P/M) form. The P/M alloys are made with the techniques described in the previous chapter. Sintered *P/M bronze* has significant porosity (Figure 16–6), which allows impregnation of lubricants. The bronze powder can be any alloy, but tin bronze is preferred.

There is no simple answer to the question of which bronze has the best metal-to-metal wear characteristics. Some bearing manufacturers claim that high-lead tin bronzes provide the best friction and wear characteristics. Others claim that high lead contents (>10%) have no beneficial effect. The lead is insoluble, and it allegedly smears over the surface and lubricates. Conflicting theories also exist on the role of tin concentration and zinc. Some studies show that tin contents over 4% do not reduce wear. Other studies show that metal-to-metal wear resistance increases with increasing tin content up to 15%. Some tin-bearing bronzes also contain several percent of zinc. It has not been clearly established if these zinc additions have any beneficial effect. About the only thing agreed to by bearing manufacturers and researchers is that the metal-to-metal wear characteristics of copper alloys are highly dependent on chemical composition and the nature of the microstructure.

Essentially, bronzes and similar copper alloys are used as plain bearing materials because under conditions of shaft–bushing contact they have less tendency to gall to a steel shaft than a steel–steel couple. They need lubrication for satisfactory operation, and the wear that occurs from the contact at startup and shutdown is a function of the alloy couple.

Figure 16–7 shows experimental data on the metal-to-metal wear properties of a variety of copper alloys sliding against hardened steel. As might be expected, the oil-impregnated P/M bronze showed the lowest wear on itself as well as on the mating material. These results were repeated in lubricated tests in which the oil impregnation provided no advantage over the other test combinations. Approximately the same ranking was obtained when the tests were run with a soft 1020 steel counterface. The system wear rate, however, was much higher, and the tin bronzes produced noticeable counterface wear. This was undoubtedly due to the presence of hard delta phase.

Tests conducted to measure the frictional characteristics versus hardened steel showed a trend of decreasing coefficient of friction with increasing sliding velocity and normal force. Aluminum bronze and straight tin bronze, however, showed significant increases in friction with increasing velocity and normal force.

The conclusions drawn from these laboratory tests, as well as from service experience, were used to establish the following guidelines for the selection of copper alloys for metal-to-metal wear applications:

1. Copper alloys should be mated with fully hardened steel (60 HRC).

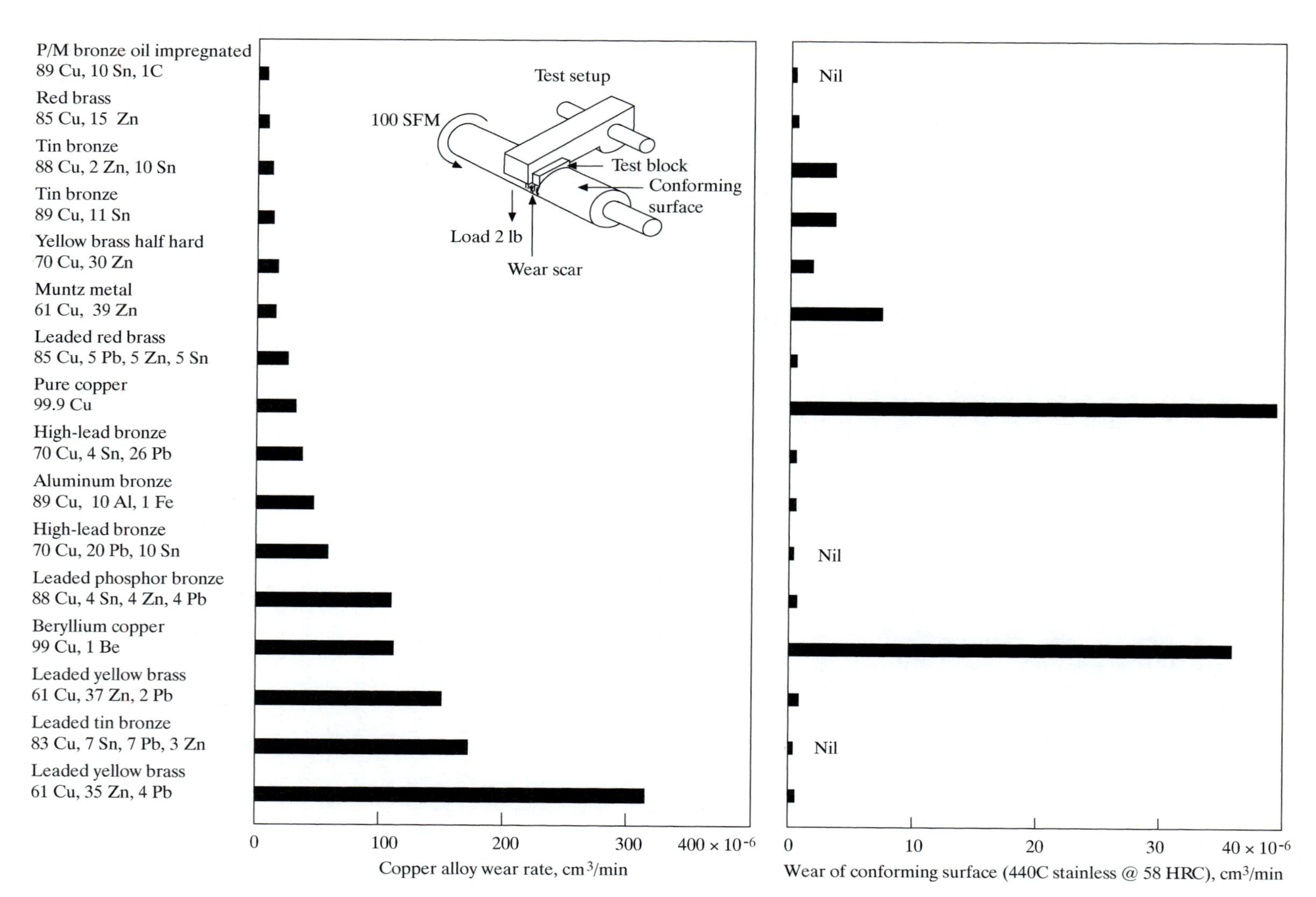

Figure 16–7
Wear of copper alloys sliding against 440C stainless steel at 58 HRC

2. If sliding against soft steel is unavoidable, high-lead tin bronzes should be used.
3. Use adequate lubrication on all copper-alloy-to-steel sliding systems.
4. Use unleaded tin bronzes (P/M, cast, or wrought) or aluminum bronze in preference to other copper alloys when the counterface is hardened steel.
5. Use high-lead tin bronzes if a low coefficient of friction against steel is required.
6. Avoid pure copper and yellow brasses for metal-to-metal or abrasive wear systems.

Some specific alloys that can be used for wear applications are listed in Table 16–6. The table implies that P/M tin bronzes would fulfill all metal-to-metal wear requirements. These bearings are available from many manufactuers in many sizes, but unfortunately the P/M process is limited in the size of part that can be economically manufactured. Plate material can be no larger than about 12 in. (30 cm) in any dimension, and sintered bearings are seldom larger than about 3 in. (76 mm) in inside diameter. Cast or wrought alloys such as the foregoing are needed for large parts or for parts that require significant machining. Wear surfaces cannot be machined on P/M parts without risk of smearing over the porosity that holds the lubricant.

The guidelines just outlined will not solve all wear applications of copper alloys, but they will serve as a basis for narrowing the selection field from several hundred available copper alloys to a manageable few.

16.9 Corrosion

About one-third of all the copper alloys produced are used for tube and pipe that carry corrodents of one form or another. Copper alloys are also used in electrical devices and in architectural applications where they are subjected to atmospheric corrosion. Copper alloys are widely used in marine applications (propellers, bushings, hardware, and heat exchangers) where they must resist attack by seawater. In the chemical process industry, copper alloys are subjected to a wide range of acids, bases, and organic solutions. In all these applications the corrosion resistance of copper alloys is an important selection consideration.

Copper alloys are susceptible to a number of types of corrosion: uniform, pitting, intergranular, dealloying, cavitation, and stress corrosion cracking. As was emphasized a number of times, the corrosion characteristics of a specific alloy in a specific environment require consultation of detailed corrosion data, but some general statements can be made concerning the major alloy systems. We shall discuss general corrosion characteristics for the more important alloy systems: pure copper, brasses, bronzes, and copper–nickels.

Table 16–6
Copper alloys for wear applications

Alloy (%)	Application
C90700 (89 Cu, 11 Sn)	Metal-to-metal wear versus hardened steel, well lubricated
C94300 (70 Cu, 25 Pb, 5 Sn)	Metal-to-metal wear versus soft steel and for poorly lubricated systems
P/M tin bronzes, oil impregnated	Metal-to-metal wear versus hard or soft steel with no external lubrication [maximum allowable bearing pressure, 1500 psi (10.3 MPa); maximum PV = 50,000 (psi-ft/min) for continuous speed, 20,000 for variable speed, 10,000 as a thrust bearing]

Pure Copper

Most copper plumbing systems are made from joinable DHP copper (alloy C12200). The corrosion characteristics of the various coppers are about the same. They are very resistant to corrosion by neutral waters with moderate hardness. Corrosion rates are typically less than 1 mil (25 μm) per year. Excessive water velocity will cause erosion, but if the water velocity is kept below about 4–5 ft/s (1.2–1.5 m/s), erosion will not occur. Copper is resistant to seawater. The corrosion rate is about 1 to 2 mil (25 μm) yr and slightly higher in moving seawater.

Pure copper has been used for centuries for roofing, gutters, and other architectural applications. The only attack caused by most atmospheres is the formation of a green *patina*. This coloration is primarily due to sulfur compounds in the air. The patina is a combination of copper compounds, mostly sulfates. General material removal is usually less than a few mils per century, but pitting (from airborne chemicals) or solder joint failures can make these roofs require repair after 100 years.

Pure coppers are not very susceptible to stress corrosion cracking. However, they readily oxidize at elevated temperatures in oxidizing environments. Heavy scales can form, and internal oxidation can occur. At temperatures over 1500°F (815°C), the scale becomes nonadherent and gross material attrition occurs. Oxide growth is insignificant at temperatures lower than about 500°F (260°C).

Copper is resistant to some low concentrations of nonoxidizing deaerated acids such as HCl and H_2SO_4. It is not resistant to oxidizing acids or hot or aerated reducing acids.

The resistance to alkaline solutions is good with the exception of ammonium hydroxide. Resistance to most organic solutions and solvents is usually acceptable. Gasoline causes negligible attack. In the galvanic series, copper ranks close to stainless steels. High-copper alloys are relatively noble, but they should not be coupled with carbon steels, aluminum, or other active metals. Bronze propellers on steel ships require special treatments to prevent galvanic attack of the steel near the prop shaft.

One of the most useful properties of high-copper alloys in aqueous environments is their resistance to biological growths. The ship *U.S.S. Constitution* has a copper-sheathed bottom. After 200 years it is still afloat, and it never has to be drydocked for bottom cleaning. The copper hull prevents fouling by marine growths. A copper surface immersed in water slowly corrodes and liberates metal ions (Cu^{2+}). These metal ions in turn poison the marine creatures that tend to grow on any object immersed in freshwater as well as salt water. Antifouling properties are best on the pure coppers, but most high-copper alloys are superior to all other metals in this regard. The antifouling characteristics of copper alloys make them suitable for water storage vessels, cladding of marine equipment, and water pumps that see intermittent service.

Brasses

There are a wide range of brass compositions, and the alloys with less than about 15% zinc have corrosion characteristics similar to those of pure copper. As the zinc content increases over 15%, the susceptibility to *dezincification* and stress corrosion cracking increases. Dezincification is one of the most insidious forms of corrosion in brasses; it occurs in seawater, in neutral water at elevated temperatures, or when stagnant water conditions occur.

Brasses with less than 15% zinc do not dezincify. Alloys with greater than 15% zinc are susceptible to dezincification (Figure 13–18), and beta or alpha–beta brasses with over 37% zinc are very prone to dezincification, especially in seawater. Bronzes (except those that contain over 15% zinc) are not susceptible to dezincification. Among the most commonly used cast alloys for plumbing valves, pump casings, and the like, are red brasses with a composition of 85% copper, 5% lead, 5% tin, and 5% zinc (C83600). This material is not considered to be susceptible to dezincification. Dezincification can also be prevented by using inhibited alloys (C44300, C44400, C46500, C46600). In addition to dezincification potential, the high-zinc brasses are prone to stress corrosion cracking in a variety of environments. Ammonia environments are known to cause stress corrosion cracking in yellow brasses, and shapes with high degrees of residual stress are particularly prone. Deep-drawn shapes may crack when they get old (season cracking) without apparent exposure to environments other than air. Essentially, yellow brasses are not suitable materials for long-term use in areas where there is potential for dezincification or stress corrosion cracking.

There is a corrosion advantage in using high-zinc brasses. The tendency for *impingement* attack is lower. Muntz metal, admiralty metal, naval brass, and other copper alloys with about 40% zinc are used for impingement resistant tubing in heat exchangers. High zinc content, however, increases tendencies for stress corrosion cracking. This is a selection consideration.

Bronzes

Bronzes with more than 5% tin are especially resistant to impingement attack compared to pure copper. Their resistance to stress corrosion cracking is also better than that of brasses. Tin bronzes have good resistance to seawater as well as to neutral water.

Aluminum bronzes are more oxidation resistant at elevated temperatures than most other copper alloys. Resistance to impingement attack is also better than that of brasses.

Silicon bronzes have most of the corrosion characteristics of pure coppers, but better mechanical properties and fair weldability (there are some cracking concerns). This makes these alloys candidates to use in welded tanks and vessels.

Cupronickels

There are various cupronikel alloys, but probably the most commonly used alloys are 70% Cu–30% Ni (C71500) and 90% Cu–10% Ni (C70600). They have the best general resistance to aqueous corrosion of all the common copper alloys. They are very resistant to impingement and stress corrosion cracking, and thus they are widely used for heat exchanger tubing.

Because there are hundreds of commercial copper alloys and an infinite number of potential corrodents, it is not possible to be more specific in statements on corrosion resistance. Specific data abound in the corrosion literature, and they should be consulted on all important selection problems. Experience in the chemical process industry has shown that the most common corrosion failures of copper alloys are caused by erosion and dezincification in neutral water systems. Both of these types of failure can be easily prevented. Erosion can be prevented by keeping water velocities low (<5 ft/s) or by using alloys such as cupronickels. Dezincification most commonly occurs in yellow brass valves, stems, and seats in hot-water systems. Red brasses, high-copper alloys, or inhibited brasses will solve this type of problem. It is recommended that designers keep these corrosion examples in mind when using copper alloys for corrosion resistance.

16.10 Alloy Selection

A significant factor limiting the use of copper alloys in design is that there are too many of them. The copper industry is reducing the availability of alloys but over 300 are still listed as commercially available in the United States. The average designer could solve most application problems with about ten alloys:

1. ETP copper, C11000 (99.9% Cu)
2. DHP copper, C12200 (99.9% Cu, 0.027% P)

3. Beryllium copper, C17200 (98% Cu, 2% Be)
4. Red brass, C23000 (85% Cu, 15% Zn)
5. Yellow brass, C26000 (70% Cu, 30% Zn)
6. Free-cutting yellow brass, C36000 (62% Cu, 35% Zn, 3% Pb)
7. Tin bronze, C90700 (89% Cu, 11% Sn)
8. High-lead tin bronze, C94300 (70% Cu, 5% Sn, 25% Pb)
9. One or two alloys for specific design situations

When cost considerations are important or where an unusual environment is encountered, it may be necessary to scan the complete list of copper alloys to arrive at the best selection. To illustrate how the occasional user of copper alloys might use this abbreviated repertoire, let us look at some common applications.

Tanks and Vessels Red brass (C23000) is a good material with low tendency for dezincification. It can be easily soldered, welded, and brazed. Silicon bronze (C65500) may be necessary if high strength is required in a welded fabrication. Both alloys have good resistance to biofouling.

Tubing, Piping, and Heat Exchangers Pure copper DHP (C12200) is the most widely available and should be used unless there is some reason not to. If water velocities are high, Muntz metal (C28000), naval brass, or cupronickels are preferred.

High-Strength, High-Conductivity Mechanical Components Chromium copper (C18200) has about 80% of the conductivity of pure copper with higher strength. Beryllium copper (C17200) has the highest strength, but only about 30% of the conductivity of pure copper.

Springs and Electrical Contacts Cold-worked ETP (C11000) copper will give the best electrical conductivity. Beryllium copper (C17200) is the very best material to use from the strength and fatigue standpoint for coiled and flat springs. For lower-cost springs, where the expense of heat treating is undesirable, phosphor bronze (C51000), silicon bronze (C65500), nickel silvers, and yellow brass (C26000) can be used in the cold-worked condition.

Wear Components In sliding systems, the tin bronzes (C90300–C93700) run best against a mating surface of a hardened steel. Tin bronzes in powdered-metal form make excellent oil-impregnated plain bearings. Aluminum bronzes are used for galling resistance on worms and stainless steel draw dies. They are also widely used as welded surfacings to rebuild worn cams, dies, tracks, and so on. Pure coppers, high coppers, and yellow brasses should not be used for wear applications. Beryllium coppers can be used in lubricated sliding systems subjected to very high loads. They are even frequently used for plastic injection molding cavities.

None of the copper alloys has resistance to abrasive wear. They are best suited to metal-to-metal wear systems.

Valves, Fittings, Heavy Castings Leaded red brass (87% Cu, 5% Pb, 5% Zn, 5% Sn) is a frequently used material for cast copper piping valves and fittings. It also has good bearing characteristics mated with steel and when lubricated. Varify applicable regulations before specifying alloys containing lead.

Screw Machine Parts Free-cutting brass (C36000) is an excellent material to use for parts that require significant amounts of machining. If certain areas of these parts require wear resistance, they should be inserted with hardened steel or more wear-resistant alloys. Consider dezincification and stress corrosion cracking when using these materials. Varify applicable regulations before specifying alloys containing lead.

Electrical Conductors ETP (C11000) copper is widely used for bus bars, wiring terminals, and the like. There are pure coppers with even higher conductivity than ETP if the application warrants the extra cost.

In summary, copper and copper alloys are indispensable engineering materials. They have unique corrosion characteristics, strengths that can be greater than those of steels, and electrical and thermal conductivity exceeding most other engineering materials.

Summary

Proper specification of copper alloys should include the ASTM/UNS designation number, the product form (sheet, strip, and so on), nominal dimensions, dimensional tolerances, finish temper, and any special requirements, such as deburred edge. Some metallurgical and property factors to keep in mind are the following:

- Copper is the oldest engineering metal, and it still has properties that make it indispensable (high conductivity, good malleability, and good corrosion resistance).
- Copper alloys are not well suited to elevated temperature service [$>$500°F (260°C)].
- The major categories of alloys are high-copper alloys, brasses, bronzes, copper–nickel alloys, and copper–nickel–zinc alloys.
- Beryllium coppers are the highest-strength alloys because of their ability to be precipitation hardened. Other precipitation-hardening alloys are chromium and zirconium copper.
- Aluminum bronzes can be quench hardened by martensite transformation. They are next to beryllium coppers in strength.
- Most copper alloys have a FCC crystal structure, and most are single phase. High-zinc (beta) brasses and some bronzes can be two phase.
- The electrical and heat conductivity of copper is significantly reduced by the presence of impurities and alloying elements.
- Copper alloys have good machinability and fabricability, but the best machinability is obtained with the leaded alloys or the new bismuth-containing free-cutting alloys.
- Copper alloys are not very abrasion resistant, but some (bronzes) have good metal-to-metal wear characteristics in lubricated sliding systems.
- Copper alloys do not rust, and some of these alloys have exceptional resistance to seawater corrosion.
- The largest use for copper alloys is in electrical conductors and plumbing tubing. In these applications they have few competitors from the standpoint of suitable properties and reasonable cost.
- Consider environmental/health regulations in using lead-containing alloys for potable water.
- Copper has excellent resistance to biological growth and fouling.

Critical Concepts

- Copper has unique conductivity characteristics.
- Beryllium copper can have strength comparable to steels and be nonferromagnetic and nonrusting.
- Copper alloys have unique resistance to biological growth.
- Brasses can have better machinability than most other engineering metals.

Terms You Should Remember

DHP
ETP
CDA
brass
bronze
nickel silver
copper nickel
high-copper
aluminum bronze
silicon bronze
red brass
yellow brass
phosphor bronze
embrittlement
beryllium copper
precipitation hardening
alpha brass
beta brass
% IACS
leaded tin bronze
impingement
erosion
P/M bronze
seizure
dezincification
stress corrosion cracking
oxidation
patina
smelting

Case History

Beryllium Copper for Latch Springs

Cassettes used to hold x-ray film (to protect the film from light exposure) were made with a small plastic door in one corner of the side of the cassette that faces the x-ray source. The door can be opened and closed by the x-ray technician allowing for patient identification on the piece of film in the cassette. The sliding door is latched in the open or closed position by a flat spring made from beryllium copper (CDA C17200) and precipitation hardened to 40 HRC after forming. These springs must not break, corrode, or elastically deform in use or handling, and they must work every time for many years (probably more than ten).

Beryllium copper was selected because it meets this design requirement. Beryllium coppers are unique because they have good formability compared to other spring materials and they can be hardened at a temperature low enough [600°F (315°C)] to minimize distortion or size change in hardening. Beryllium coppers are therefore extremely useful in design engineering for springs and flexures.

Questions

Section 16.2

1. What is the difference between a brass and a bronze?
2. What is the difference between a CDA number and a UNS number for copper alloys?
3. How can a cast alloy be recognized by its UNS number?
4. What is the difference between an ASTM specification for a copper alloy and a UNS designation?

Section 16.4

5. What is the solution treating temperature for a beryllium copper with a beryllium concentration of 1% (see Figure 16–2)?
6. What causes strengthening in precipitation hardening?
7. What is the crystal structure of a beta brass?
8. Can a copper–nickel alloy be precipitation hardened?
9. Name a quench hardenable copper alloy.

Section 16.5

10. Is copper heavier than steel (the specific gravity is 8.96)?
11. Twelve-gage copper wire has a diameter of 0.8 in. What diameter wire of yellow brass would be necessary to have equivalent resistivity to the copper wire?

12. Compare the stiffness of copper to that of aluminum and steel.

13. What is the hardest copper alloy? How did it get that hard?

14. How does ETP copper compare to 1180 aluminum in thermal conductivity?

15. Can you weld copper? How?

Section 16.8

16. How would you make copper abrasion resistant?

17. What degree of porosity should be specified in a P/M bronze plain bearing?

18. What is the proper hardness for a counterface for a lubricated phosphor bronze plain bearing?

19. Compare the metal-to-metal wear characteristics of red brass and leaded yellow brass.

20. Compare the machinability of C11000 copper with C36000 brass.

21. When would you use a high-lead bronze?

Section 16.9

22. What is the mechanism of dezincification?

23. Which brasses are resistant to dezincification? Name one.

24. Which copper alloys are suitable for service in seawater at room temperature?

Section 16.10

25. What copper alloy would you use for a flat spring that had to be a good electrical conductor and required significant cold forming?

26. What is the best copper alloy to use for a welded tank?

27. What is a satisfactory copper alloy for plumbing tubing?

28. What is the biggest application for copper alloys?

To Dig Deeper

Bearing Design Program for Hydrodynamic and Boundary Lubricated Bearings, Version 2.0 (computer selection of copper bearings). New York: Copper Development Association, 1988.

Joseph, G., and K. Kundi, Eds. *Copper: Its Trade, Manufacture, Use, and Environmental Status*. Materials Park, OH: ASM International, 1999.

Leidheiser, H., Jr. *The Corrosion of Copper, Tin and Their Alloys*. New York: John Wiley & Sons, 1971.

Mendenhall, J. Howard. *Understanding Copper Alloys*. E. Alton, IL: Olin Corporation, 1977.

Metals Handbook, 10th ed, Volume 2, Properties and Selection: Nonferrous Alloys and Special-Purpose Materials. Materials Park, OH: ASM International, 1990.

Standards Handbook—Cast Copper and Copper Alloy Products—Part 7—Alloy Data. Greenwich, CT: Copper Development Association, 1996.

Standards Handbook, Part 2—Alloy Data, Wrought Copper and Copper Alloy Mill Products, 8th ed. Greenwich, CT: Copper Development Association, 1997.

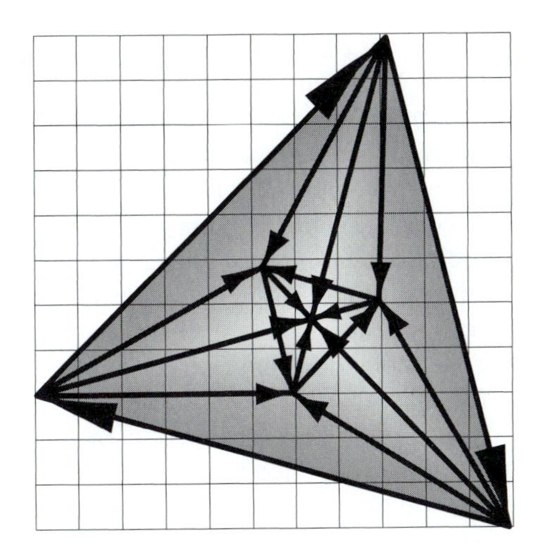

CHAPTER 17

Aluminum and Its Alloys

Chapter Goals

1. An understanding of aluminum alloy systems and alloy designation.
2. A working knowledge of heat treatment and coating of aluminum.
3. Guidelines for alloy selection.

Rationale

Aluminum is probably second only to steel in importance in our modern world. Steel makes large structures and tools possible. Aluminum makes possible structures that would be too heavy to be useful if made from steel—aircraft, 50-knot fast ferries, missiles, satellites—and architectural components that do not rust, a problem of steels—automobile trim, building panels, windows, siding, doors.

Every home center in the United States sells aluminum and steel shapes (for homeowner projects). This is not true of other metals or even plastics. Aluminum can replace steel in many instances, and in addition it is lighter and more corrosion resistant and a better conductor of heat and electricity.

Aluminum alloys are essential engineering materials, and designers need to know some of their idiosyncrasies, how to designate them on drawings, and how to use them so that they perform as expected. The curious aspect of this metal is that it is relatively new to our world. It has only been available as a "commodity metal" for about 50 years, not a long time in evolutionary terms. It is a remarkable metal and it needs to be a part of every designer's repertoire of engineering materials.

Aluminum was first produced in the laboratory in 1825 by reducing aluminum chloride. However, wide acceptance of aluminum as an engineering material did not occur until World War II. Aluminum is the most abundant metal in nature. Some 8% by weight of the earth's crust is aluminum. Many rocks and minerals contain a significant amount of aluminum. Unfortunately, aluminum does not occur in nature in the metallic form. In rocks, aluminum is present in the form of silicates and other complex compounds. The ore from which most aluminum is currently extracted, *bauxite*, is a hydrated aluminum oxide. Referring back to our discussion of ceramics, you may recall that aluminum oxide as a ceramic material is extremely hard and chemically inert. Because of the inertness of aluminum compounds, it took almost 60 years of research to find an economically acceptable way of making aluminum from ore.

From 1825, when aluminum was discovered, to about 1890, aluminum was produced on a small scale by complex and expensive chemical reductions of aluminum compounds. In 1850 the cost of aluminum was about \$500/lb. Improvements in processing techniques from 1850 to 1860 lowered the price to around \$25/lb, but these prices still made aluminum almost a precious metal. In 1886 a patent was issued to Martin Hall for a process of reducing aluminum oxide ore to aluminum by electrolysis in a molten salt. In this process, cryolite, a sodium–aluminum–fluorine compound (Na_3AlF_6), is melted, and aluminum oxide (Al_2O_3) is dissolved in the molten salt. The salt now contains aluminum ions (Al^{3+}). Carbon electrodes are put in the bath and current flow is established between the electrodes. Aluminum ions are reduced to metallic aluminum at the cathode, and oxygen is produced at the *anode*. Because the bath is at a high temperature [about 1750°F (953°C)], aluminum formed at the cathode melts [the melting point of pure aluminum is 1220°F (659°C)] and collects at the bottom of the cell. It goes to the bottom because aluminum is denser than the molten salt. A schematic of the Hall cell is shown in Figure 17–1.

This type of cell allowed the cost of aluminum to be greatly reduced and commercial exploitation to be possible. The aluminum industry started in the United States in about 1890 with the founding of the Pittsburgh Reduction Company. This company became the Aluminum Company of America in 1907. It was the sole U.S. producer of aluminum until World War II, and usage of aluminum was still relatively small. The needs of the war spurred production of aluminum. Other producers came on

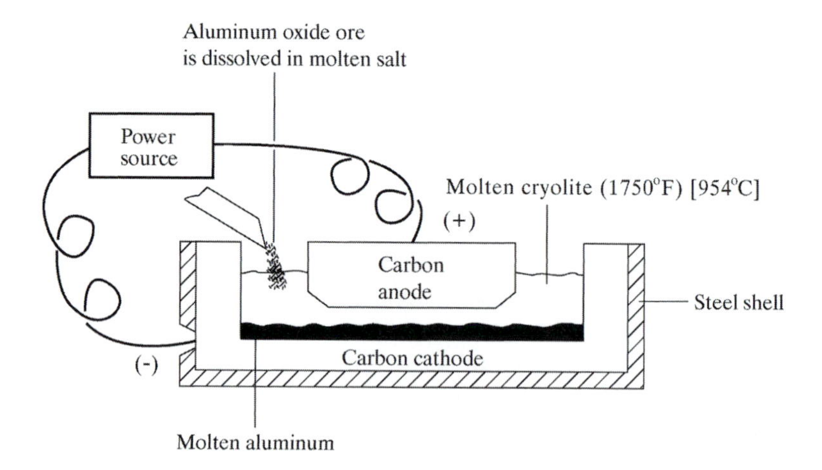

Figure 17–1
Cell reduction of aluminum oxide to aluminum

Process reactions:

1. Alumina is dissolved
   $Al_2O_3 \rightarrow 2\ Al^{3+}$(aluminum ions) + $3\ O^{2-}$(oxygen cations)
2. At the anode
   $2Al^{3+} + 6e \rightarrow 2\ Al$
3. At the cathode
   $3\ O \rightarrow 1^1/_2\ O_2 + 6e$

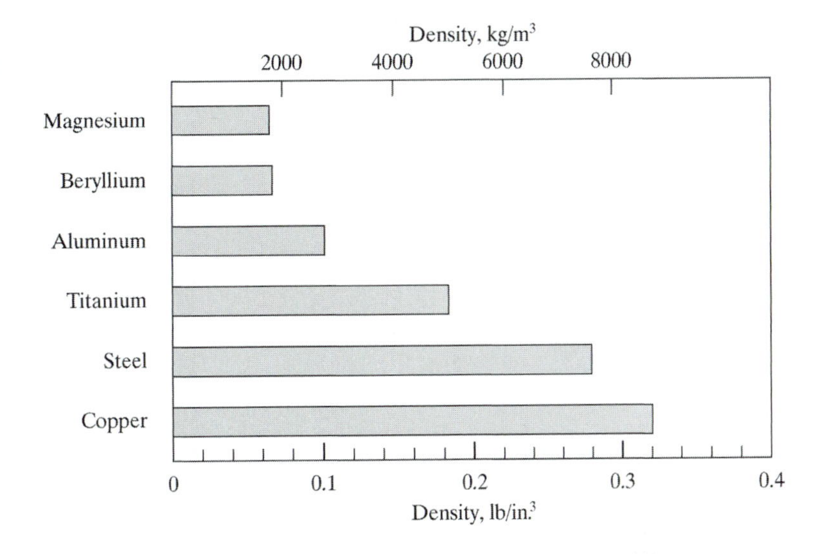

Figure 17–2
Density of aluminum compared to other metals

the scene and usage has steadily increased each year.

The price of aluminum has gone from the \$500/lb in the 1850s to as low as \$0.15/lb. The price in the United States in 2003 was as low as \$0.65/lb for pure aluminum ingot. Thus improvements in refining techniques have made aluminum a reasonably priced engineering material, and its properties allow it to be used in machines, construction, domestic products, aircraft, automobiles, and other design applications.

17.1 General Characteristics

Aluminum is a good electrical conductor; it is ductile and can be readily cast and machined. It has a face-centered cubic structure as do other "metallic" metals, such as copper, silver, nickel, and gold.

Several properties set aluminum apart from other metals. First, it is lighter than all other engineering metals except magnesium and beryllium. It has a density of about 0.1 lb/in.^{3} (2990 kg/m^3). A comparison with other metals is shown in Figure 17–2.

A second important property of aluminum is its thermal and electrical conductivity. It has about 60% of the conductivity of pure copper (IACS). Because of its lower density, aluminum has a higher conductivity than copper per unit mass. For example, a 10-mm-diameter aluminum wire will have the same resistivity as a 6-mm-diameter copper wire and still be about 13% lighter than the copper wire. This is an important consideration in long power-transmission cables.

The third property that is responsible for the wide use of aluminum alloys is their corrosion resistance. Aluminum is not widely used for chemical resistance, but for applications involving atmospheric corrosion resistance it is probably the most widely used metallic material. Architectural applications of aluminum are everywhere—railings, windows, frames, doors, flashing, and so on.

The mechanical properties of aluminum alloys are important, and we shall discuss these properties and aluminum's corrosion resistance in more detail in subsequent sections. The following are some of the noteworthy advantages of using aluminum.

1. One-third the weight of steel
2. Good thermal and electrical conductivity

3. High strength-to-weight ratio
4. Can be given a hard surface by anodizing and hard coating
5. Most alloys are weldable
6. Will not rust
7. High reflectivity
8. Can be die cast
9. Easily machined
10. Good formability
11. Nonmagnetic
12. Nontoxic
13. One-third the stiffness of steel

About 25% of the aluminum produced in the United States is used for containers and packaging when freedom from toxicity, strength, lightness, and corrosion resistance are important. About 20% of aluminum production is used for architectural applications such as windows and siding. Again, corrosion characteristics are a principal reason for its use in this application. Ten percent of aluminum production is used for electrical conductors, and the remainder is used for durable goods in industry, in consumer products, in vehicles, and in aircraft.

17.2 Alloy Designation

Wrought

The most commonly used alloy designation system in the United States is that of the Aluminum Association. For wrought alloys, it is based on four digits corresponding to the principal alloying elements:

Commercially pure aluminum (99% min.)	1000
Copper (major alloying element)	2000
Manganese	3000
Silicon	4000
Magnesium	5000
Magnesium and silicon	6000
Zinc	7000
Other elements	8000
Unused series	9000

The second digit in this system designates mill control or lack of same on specific elements. The last two digits have no significance, except that in the 1xxx series they coincide with aluminum content above 99% in hundredths. A 1040 alloy has 99.4% aluminum. The complete specification of wrought aluminum alloys involves a suffix that indicates the degree of cold work or thermal treatments. Table 17–1 lists some of the most commonly specified conditions. A complete listing of temper designations in the United States can be obtained from the Aluminum association. These numbers can have letters and numbers after them to indicate additional treatments. The H letter is followed by one, two, or three digits to indicate degrees of cold working (*strain hardening*). The T is followed by one, two, or three digits to indicate various thermal treatments:

xxxx-H1	Strain hardened only
xxxx-H2	Strain hardened and partially annealed
xxxx-H3	Strain hardened and stabilized by low-temperature thermal treatments
xxxx-H4	Strain hardened and lacquered or painted

The digit following H1, H2, or H3 indicates the degree of strain hardening. A 1 indicates the smallest amount of cold work, and an 8 indicates maximum cold work or full-hard condition:

Table 17–1
Temper designations

xxxx-F	As fabricated, no special controls
xxxx-W	Solution heat treated (used only on alloys that naturally *age harden*)
xxxx-O	Annealed (wrought alloys only)
xxxx-H	Strain hardened (cold worked to increase strength), wrought alloys only
xxxx-T	Thermally treated to produce effects other than F, O, or H

xxxx-H_2	Quarter-hard
xxxx-H_4	Half-hard
xxxx-H_6	Three-quarters hard
xxxx-H_8	Full-hard

A third digit can be used to indicate a variation of the two-digit temper designation in which properties are slightly different from those of the two-digit temper.

The meanings of the numbers following the T temper designations are as follows:

xxxx-T1	Cooled from a hot working temperature and *naturally aged*
xxxx-T2	Cooled from an elevated temperature, cold worked, and naturally aged (means annealed for cast products)
xxxx-T3	Furnace solution heat treated,* quenched, and cold worked
xxxx-T4	Furnace solution heat treated,* quenched, and naturally aged
xxxx-T5	Quenched from a hot-work temperature and furnace aged
xxxx-T6	Furnace solution heat treated* quenched and furnace aged
xxxx-T7	Furnace solution heat treated* and stabilized
xxxx-T8	Furnace solution heat treated,* quenched, cold worked, and furnace aged
xxxx-T9	Furnance solution heat treated,* quenched, furnace aged, and cold worked
xxxx-T10	Quenched from an elevated temperature shaping process, cold worked, and furnace aged

Additional digits can be added to T1 through T10 temper designations to indicate significant variations.

*Solution heat treatment is achieved by heating cast or wrought products to a specified temperature, holding at that temperature long enough to allow constituents to enter solid solution, and cooling rapidly enough (quenching) to hold the constituents in solution.

xxxx-T51	Stress relieved by stretching
xxxx-T510	Stress relieved by stretching with no further processing
xxxx-T511	Stress relieved by stretching and minor straightening
xxxx-T52	Stress relieved by compression
xxxx-T54	Stress relieved by stretching and compression

A complete alloy specification includes the alloy and its treatments:

1. 3003-H38 (3003 alloy cold finished to full hard temper and stress relieved by a low-temperature treatment, H38)
2. 6061-T6 (6061 alloy, solution heat treated* and furnace age hardened, T6)

Cast Alloys

In the United States cast alloys have been identified by a four-digit identification number with the last digit separated by a decimal. A letter prefix is occasionally used to signify alloy or impurity limits. The first digit indicates the alloy group. The second and third digits identify an alloy within a group, and the last digit indicates product form. A last digit of 0 indicates a casting; a digit of 1 indicates ingot form. The designations for groups of cast alloys are shown in Table 17–2.

Table 17–2
Cast aluminum alloy designations

Case Alloy Designation	Major Alloying Elements
1–99 (old system)	Aluminum + silicon
1xx.x	99.5 min. aluminum
2xx.x	Copper
3xx.x	Silicon + copper or magnesium
4xx.x	Silicon
5xx.x	Magnesium
6xx.x	Unused series
7xx.x	Zinc
8xx.x	Tin
9xx.x	Other element

The thermal treatment suffix used on wrought alloys is also used on cast alloys. A complete alloy specification must include this suffix, e.g., 355.0-T6. The preferred practice for cast alloy designations is to use UNS numbers for alloy designations, and the Aluminum Association/ASTM temper designations (For example, ASTM B 108 alloy A03550-T5). ASTM B 108 lists permanent mold alloys; ASTM B 85 lists die casting alloys; ASTM B 26 lists sand casting alloys; and ASTM B 686 covers high-strength alloys.

17.3 Aluminum Products

There are hundreds of commercially available aluminum alloys, but the more readily available alloys are listed in Figure 17–3. It can be seen that aluminum alloys can be cast by all the common casting techniques. Investment castings are made from alloys in the permanent mold category. Castings are used for everything from engine blocks to camera parts.

Wrought aluminum products include foil, sheet, plate, bar, rod, wire, tubing, powder metals, and structural shapes such as I's, channels, and angles. Open and closed die forgings are used for many aerospace and aircraft applications. *Extrusions* account for between 10% and 20% of all aluminum products. They are widely used for special shapes for everything from pencils to sailboat masts. They can be extremely useful in machine design for chutes, moldings, and part nests.

Wrought aluminum products are commercially available with a wide range of special finishes. These include mechanical finishes, chemical finishes, and coatings. Mechanical finishes include cold finished, buffed, and textured. Chemical finishes include such things as etched, bright dipped, and chemical conversion coatings. Anodizing, painting, plating, and chemical conversion coatings are included in the coating category. There are number designation systems for these surface finishes, but there are so many of these finishes that the best way to learn about them is to request a finish manual from an

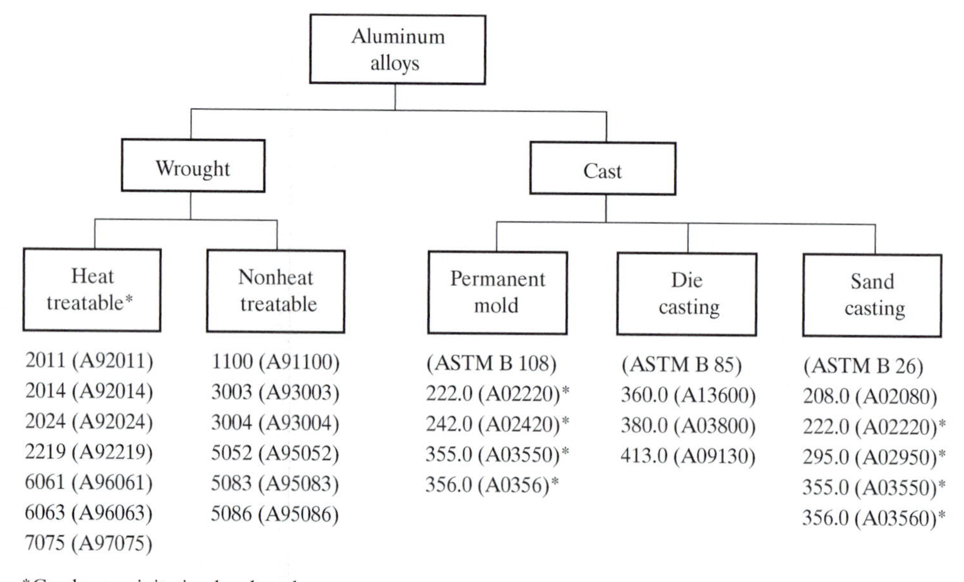

Figure 17–3
Useful cast and wrought aluminum alloys—Aluminum Association designation and UNS number

aluminum supplier. These manuals are usually free and contain actual samples of aluminum with these finishes. If a proposed application could benefit from a pretreated surface, mill finishes should certainly be investigated.

17.4 Metallurgical Characteristics

Aluminum is miscible in the liquid state with many metals, but solid solubility of alloy elements is typically only a few percent. Intermetallic compounds form and become a phase in the structure of the aluminum alloy. No element is completely soluble in aluminum in the solid state. More often than not, the intermetallic compounds that form with large alloy additions are hard and brittle. They often have a deleterious effect on mechanical properties. The percentage of alloy elements in useful aluminum alloys does not exceed about 15%. The most important alloying elements in aluminum alloy systems are copper (2xxx), manganese (3xxx), silicon (4xxx), magnesium (5xxx), and zinc (7xxx).

The *binary* phase diagrams for these elements all have similarities. They show good solubility of the alloying element at elevated temperatures but low solubility at room temperature. The aluminum–copper phase diagram is shown in Figure 17–4. The shaded portion, labeled α, is the region where the alloying element is completely soluble in the solid state with aluminum. In the example of Al–Cu, copper is soluble up to 5.65% at the eutectic temperature of 548°C. At room temperature, the solubility is less than 0.02%. The importance of this limited solubility at room temperature is that precipitation hardening (PH) is possible. These systems are similar in makeup to the beryllium copper and PH stainless steels previously discussed. An aluminum–copper alloy with, for example, 4% copper can be precipitation hardened by heating to a temperature of high solubility, say 500°C (930°F); the copper will go into solid solution.

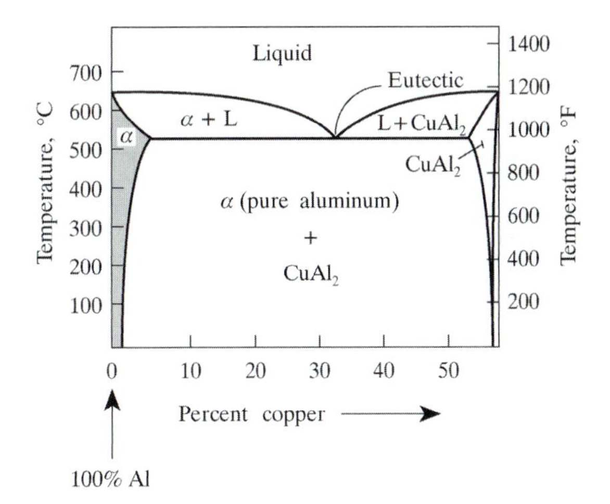

Figure 17–4
Portion of the aluminum–copper equilibrium diagram

The alloy is then quenched in water. The copper wants to come out of solution, but the water quench prevents it. Age hardening is accomplished by holding at some low temperature like 200°C (400°F) for a prescribed time interval such as 20 h. The copper will *precipitate* from solution and form a copper/aluminum compound ($CuAl_2$). This strains the metal lattice, and strengthening occurs.

The other important aluminum alloying elements—silicon, magnesium, manganese, and zinc—strengthen in a similar manner. When they are added in combinations, as in the 6xxx series alloys, ternary diagrams must be consulted for the actual solubility conditions and temperatures; but it is safe to say that the precipitation phenomenon is still due to the solubility conditions just described. The specific solubility limits of the important aluminum alloying elements are listed in Table 17–3.

Besides their role in precipitation hardening, alloying elements in aluminum produce such effects as solid solution strengthening, improved machinability, and corrosion resistance. Table 17–4 lists some of the effects, good and bad, of some alloying elements in aluminum.

Table 17–3
Solubility limits of the important aluminum alloying elements

Alloying Element	Solubility in Pure Aluminum at Room Temperature (%)	Solubility at the Solidification Temperature, °C (°F)
Cu	0.02 (wt)	5.65% at 548 (1018)
Mg	2.5	14.9% at 450 (842)
Mn	0.3	1.8% at 659 (1217)
Si	0.1	1.65% at 577 (1071)
Zn	2	82% at 382 (716)

Table 17–4
Effects of alloy elements

Alloying Element	Effects
Iron	Naturally occurs as an impurity in aluminum ores; small percentages increase the strength and hardness of some alloys and reduce hot-cracking tendencies in castings; reduces pickup in die-casting cavities.
Manganese	Used in combination with iron to improve castability; alters the nature of the intermetallic compounds and reduces shrinkage; the effect on mechanical properties is improved ductility and impact strength.
Silicon	Increases fluidity in casting and welding alloys and reduces solidification and hot-cracking tendencies; additions in excess of 13% make the alloy extremely difficult to machine; improves corrosion resistance.
Copper	Increases strength up to about 12%, higher concentrations cause brittleness; improves elevated temperature properties and machinability; concentrations over 5% reduce ability to hard coat.
Magnesium	Improves strength by solid solution strengthening, and alloys with over about 3% (0.5% when 0.5 % silicon is added) will precipitation harden; aluminum–magnesium alloys are difficult to cast because the molten alloy tends to "skin-over" (dross) in contact with air.
Zinc	Lowers castability; high-zinc alloys are prone to hot cracking and high shrinkage; percentages over 10% produce tendencies for stress corrosion cracking; in combination with other elements, zinc promotes very high strength; low concentrations in binary alloys (<3%) produce no useful effects.
Chromium	Improves conductivity in some alloys, and in small concentrations (<0.35%) it acts as a grain refiner.
Titanium	Naturally occurs as an impurity in aluminum ores, but it is intentionally added to some alloys as a grain refiner.
Lead/Bismuth	Added to some alloys to improve machinability; 2011 and 6062 are screw machine alloys containing Pb and Bi.
Zirconium	Used as a grain refiner in some aerospace alloys.
Lithium	Added to some aerospace alloys (Space Shuttle fuel tanks) to reduce weight. These alloys need a protective atmosphere when being cast.

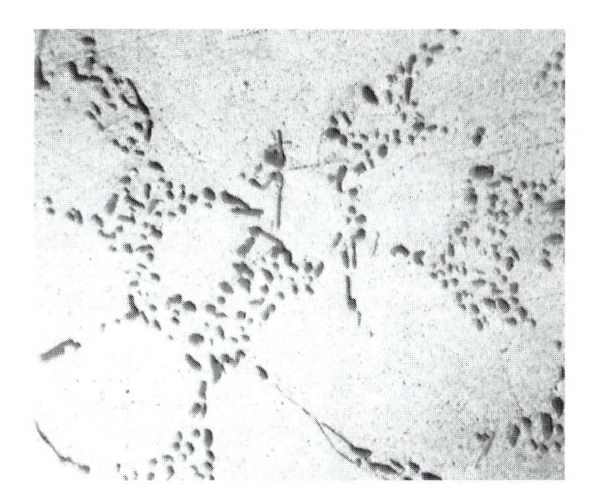

Figure 17–5
Unetched microstructure of an aluminum casting alloy (355) showing *intermetallic* compounds between grains. These second phases are formed because of the low solubility of many alloying elements in aluminum.

A number of other elements are added for special purposes, but the foregoing are the most important because of their roles in fabricability and strengthening. The appearance of a typical microstructure of an aluminum alloy is shown in Figure 17–5. The most useful aluminum alloys are pure aluminum with small concentrations of copper, silicon, magnesium, manganese, or zinc.

17.5 Heat Treatment

We have briefly mentioned age hardening, and this is probably the most important heat treatment process, but stress relieving and annealing are also important in using aluminum alloys. Typical thermal cycles for aluminum heat treatments are illustrated in Figure 17–6.

Annealing is performed on cold-worked alloys to aid forming. Sheet metal parts subject to deep drawing operations may require anneals between drawing steps. Castings with varying cross sections may solidify with different structures because of differences in cooling rates. Annealing will homogenize the structure so that mechanical properties are uniform. Heat treating temperatures vary for each alloy, and heat treatment cycles should conform closely with handbook recipes (see *Aluminum Standards and Data in the U.S.*). Figure 17–6 shows a wide temperature range for many alloys, and it is intended as an illustration, not as a heat treatment recipe reference. The high thermal conductivity of aluminum allows abrupt heat-up and cooling. An anneal to soften heat-treated alloys is the only heat treatment that requires a very slow cool. Usually, parts are cooled to about 500°F (260°C) at a rate of 50°F/h (28°C/h). Where applicable, Aluminum Association suffixes should be used to specify thermal treatments (temper designations). An anneal on a wrought alloy is specified with an O suffix. A T2 suffix indicates annealing on cast alloys.

Solution treatment is the first step in the age-hardening process. Its purpose is to put the precipitating elements in solid solution. The water quench traps these elements in solution. The water quench required in solution treatment may cause distortion. Thus it is wise to purchase wrought alloys in the age-hardened condition from the mill, where they have equipment to perform straightening. When it is necessary to solution treat after machining, generous stock allowances should be provided for final machining. A W suffix is used to designate a solution-treated condition. This is an unstable condition, and parts are not normally used in this condition because properties change with time. An exception would be an alloy that naturally age hardens at room temperature. In the aircraft industry, rivets are sometimes *solution treated*, refrigerated to prevent aging, and then used cold. They can be easily peened in the W temper, and they naturally age to maximum hardness after installation (T3).

Stress relieving is a very important tool for the designer who wants close dimensional control on aluminum parts. Temperatures are low,

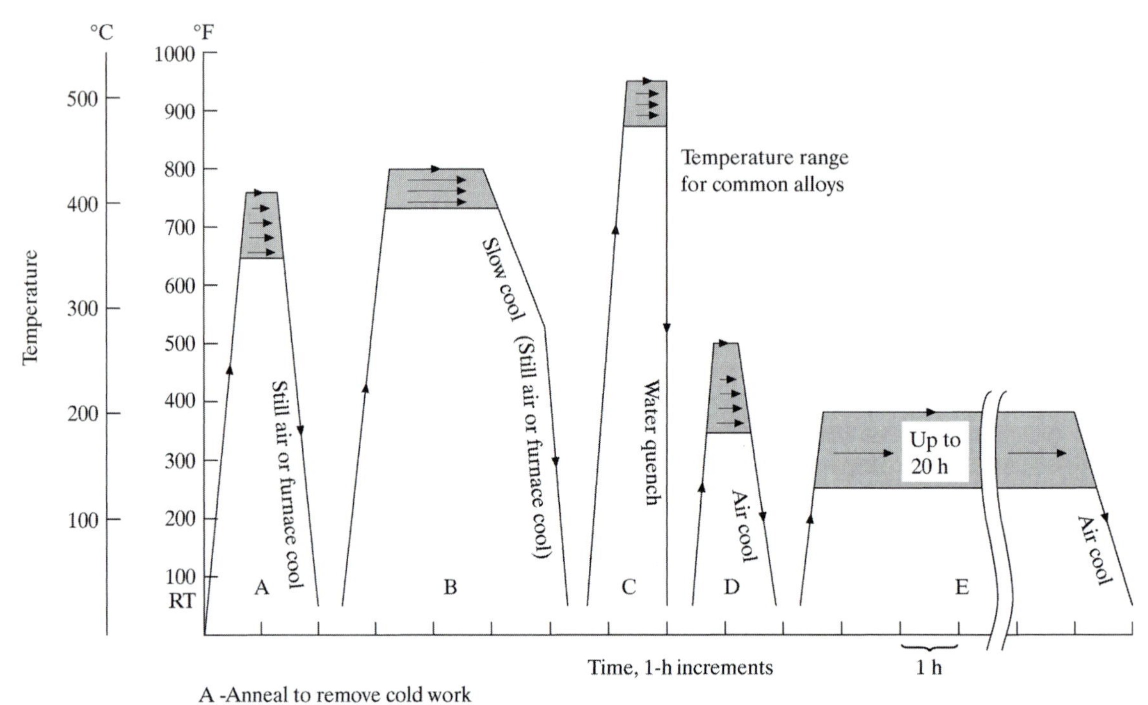

Figure 17–6
Typical heat treating cycles for aluminum alloys

and soak times can be as short as 15 min. Long, slender parts, weldments, large flats, rings, and complex shapes can be rough machined, stress relieved, and finished. Stress relieving removes residual stresses remaining from mill operations and stresses imparted in machining. One word of caution: Stress relieving can reduce the mechanical properties of age-hardened alloys, so if tensile properties are critical it may be well to avoid stress relieving or measure the strength reduction produced by a particular stress-relieving operation.

The last heat treatment cycle in Figure 17–6, age hardening (*thermal aging*), is the lowest-temperature operation, but times can be 20 h or even longer. Time recommendations for specific alloys are shown in aluminum handbooks and should be adhered to. ASTM B 597 presents details for the heat treatment of most commercial alloys. The tensile properties usually peak at a certain aging time. Too short an aging cycle or too long a cycle cause lower tensile properties. As we might expect, elevated-temperature use of age-hardened aluminum alloys causes *overaging* and should be avoided. Short excursions up to the aging temperature range are usually not too detrimental because precipitation hardening is time dependent. Aerospace alloys 7075, 7050, and others are sometimes intentionally overaged (T751). Although this overage decreases the tensile properties, it significantly increases resistance to cracking (fracture toughness) and corrosion.

The suffix for solution treating and age hardening in a furnace (artificial) is T6. This temper designation applies to both cast and wrought alloys.

Designers will not usually be involved in actual heat treatments of aluminum parts, but they are responsible for specification of thermal treatments and temper designation. If possible, purchase heat treatable aluminum alloys in the heat-treated condition (T6, etc.) and limit in-house treatments to simple stress reliefs.

17.6 Surface Treatments

Aluminum without some surface treatment is like good wood without varnish. The wood is strong and may make a good structural member, but it does not look as good as it could, and it is susceptible to wear and weather. This is also the situation with aluminum. Aluminum can be protected with the conventional platings and organic finishes that can be applied to most metals: paints, vitreous enamels, organisols, and the like. However, unlike many other metals, aluminum has the ability to develop a tenacious and relatively thick oxide coating by making the aluminum the anode in a plating-type electrochemical cell. The generic name for such coatings is *anodizing*. This is by far the most commonly applied surface treatment for aluminum.

A typical anodizing setup is illustrated in Figure 17–7. The mechanism of anodizing is an electrochemical conversion of the surface of the aluminum from metallic aluminum to aluminum *oxide*, Al_2O_3. In converting the surface to essentially a ceramic compound, anodizing causes a small amount of surface dissolution such that on thin anodized coatings there is a penetration of two increments of depth to one increment of surface buildup. In thick anodic coatings (*hardcoating*) there is equal penetration and growth. This is a very important point for the design engineer to consider, because it affects dimensional tolerances (Figure 17–8).

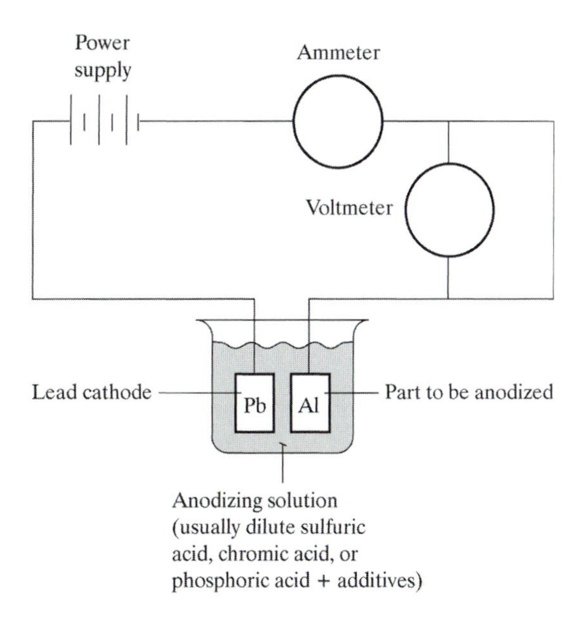

Figure 17–7
Basic anodizing system

In our discussion of ceramics it was pointed out that aluminum oxide is extremely hard ($\sim$2000 kg/mm^2). The anodically produced aluminum oxide is porous, so it is difficult to get accurate measurements on its hardness, but in almost all instances it is harder than the hardest steel ($>$1000 kg/mm^2). This means that it is very hard on cutting tools if this coating has to be machined after anodizing. Thus it is common practice to account for the coating thickness in part dimensions and machining after anodizing is avoided. Threaded holes can be particularly troublesome with the thick (2 mil; 0.05 mm) anodized coatings; the coating buildup may make it difficult to insert the threaded member. The solution to this problem is to mask holes during anodizing.

Not all aluminum alloys anodize well. Alloys containing over 3% copper (2011, 2017, 2024) and alloys containing over 5% silicon (319, 333, most die casting alloys, 328) do not readily accept thick anodized coatings (hardcoating). They discolor and may form a dark smut. The deleterious effect of silicon means that many cast alloys do not hardcoat well.

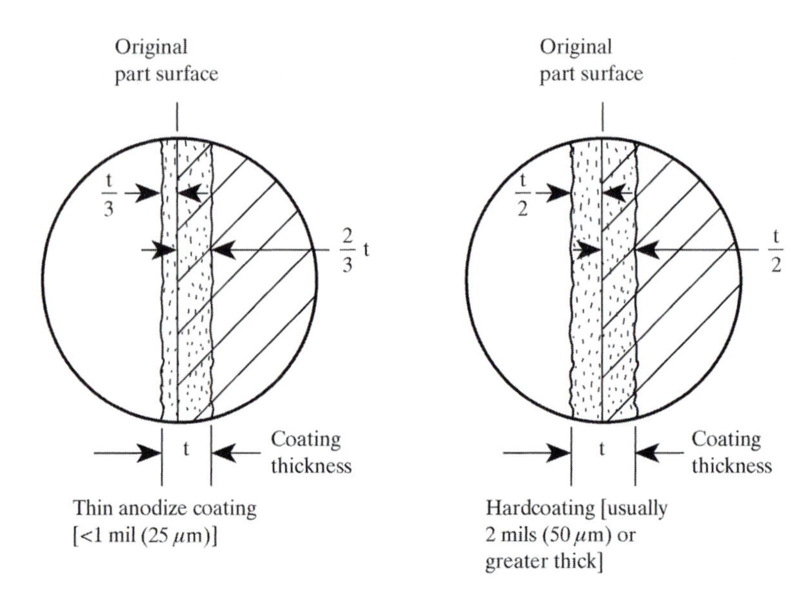

Figure 17–8
Dimensional changes in anodizing and hardcoating

Almost all aluminum alloys will accept thin anodization (<1 mil; 25 μm), but their suitability for hardcoating varies (Table 17–5). Hardcoating is anodizing, but a special type of anodizing. It has been termed *hardcoating* to distinguish it from conventional anodizing. Hardcoating baths are usually aqueous solutions of sulfuric and oxalic acid, and they are chilled to about 40°F (4°C). The coating can be 2 mils thick (50 μm) or thicker, and it is black to gray-black in appearance. Conventional anodizing is done in a sulfuric acid bath at room temperature and at current densities that are lower than used in hardcoating. Some of the typical coating thicknesses used in clear anodizing and hardcoating are shown in Table 17–6. If a design calls for the most wear-resistant, thickest coating on aluminum, the drawing should specify "2-mil (50-μm) thick hardcoat on areas shown" (or all over).

Anodized coatings tend to grow perpendicular to the surface, and coatings are thin on sharp corners. For the best service life, all outside corners should be radiused. The thicker the desired coating, the more generous the radius should be. A typical radius is 0.030 in. (0.76 mm) for a 1-mil (25-μm) coating. A 2-mil (50-μm)-thick coating would require a 0.060-in. (1.5-mm) radius.

Anodized coatings have varying degrees of porosity (Figure 17–9). For the best corrosion resistance, it is desirable to reduce the porosity

Table 17–5
Hardcoating compatibility

Suitable for Hardcoating	Less Suitable
5052	2024
6061	355
1100	7075
3003	

Table 17–6
Thickness guide and suitable alloys

	Anodic Coating Thickness	
Purpose	in.	μm
Decorative	0.0001 to 0.0003	2.5 to 7.6
Light industrial	0.0003 to 0.0005	7.6 to 12.7
Heavy industrial	0.0005 to 0.0007	12.7 to 17.7
Wear resistance	0.0015 to 0.0020 (hardcoating)	38.1 to 51

Figure 17–9
Cross section of a hardcoated aluminum part (black area) (×400)

by a sealing operation. Immersion of the anodized coating in hot water converts the oxide to a hydrated form that swells and reduces the pore size *sealing*. If a *dye* is put in the sealing water *dying*, the color will permeate the coating. Architectural shapes and automotive trim are often colored by this process. These coatings do not chip or peel as paints do because they are an integral part of the metal surface.

Hardcoating unfortunately increases surface roughness. A surface with a 5-μin. Ra (0.12-μm) roughness would increase to possibly 20-μin. (0.5-μm) Ra on thicker [0.002 in. (0.05 mm)] hard anodized coatings. On thin coatings [<0.001 in. (0.025 mm)], this roughening can be inconsequential. For structural and architectural applications, the roughening coincident with thick anodized coatings can be ignored, but if a hardcoating is intended as a wear surface in a machine, the rough aluminum oxide surface can cause mating metal surfaces to wear rapidly. For this reason, hardcoating should not be used for bushings for rotating shafts. It will wear the mating rotating shaft. Hardcoating can be successfully used for bushings and gibs for linear motion because the wear area is usually larger. Anodized surfaces are not normally used against rolling elements, balls, and rollers, because there is a tendency to deform the aluminum substrate, and cracking of the coating will result.

In summary of our discussion of aluminum coatings, many coatings are commercially used. These include organic coatings, chemical conversion coatings, and platings, but the most important are anodizing and hardcoating. These processes are performed like plating processes, but the nature of the coating is an electrochemically induced aluminum oxide. This coating provides a good corrosion barrier and wear resistance. It should be considered each time aluminum is used on a machine component or a structural member.

17.7 Corrosion

Anodizing, hardcoating, and conversion coatings are very effective in preventing atmospheric corrosion, but there are many applications where surface treatments just cannot be applied: large structures, parts that have to be machined or formed on assembly, and parts on which coatings would affect electrical conductivity (an anodized coating is an insulator). There are three general types of environments that aluminum alloys might be subjected to: indoor and outdoor atmosphere, water, and chemicals. Discussion will be limited to a few comments on each of these areas. Detailed corrosion data must be consulted for resistance to specific chemicals and unsual environments.

Atmospheric Corrosion

Indoors in a typical manufacturing environment, any aluminum alloy should show no deterioration for indefinite periods of time. Every home-owner probably has an example to attest to this statement. Many household appliances have aluminum on them in the form of trim, handles, housings, or nameplates. Look at the oldest one you can find. Chances are it will look as it always has.

Outdoors, the environmental resistance of aluminum depends on the chemicals in the air and the moisture conditions. All aluminum alloys are attacked by immersion in salt water or salt sprays, as automibiles in northern states experience from salted highways. However, corrosion test panels on beach sites around the world show that the corrosion rates are extremely low when the environment is a salty air. After 20 years of exposure, total corrosion penetration on most wrought alloys is less than about 5 mils (125 μm). This is about the same as in industrial atmospheres. Outdoor corrosion rates are high for the first 2 years or so until the surface is unformly covered with an oxide corrosion product or *patina*. After this coating formation period, the corrosion rate is very low.

Cast alloys behave similarly; however, the corrosion rates are somewhat higher. The material attrition on seaside test racks can be as high as 20 mils (500 μm) in 20 years. As is the case with wrought alloys, there is little difference between different alloys. The one factor that comes to light in almost all corrosion applications of aluminum is that pure aluminum has the best atmospheric corrosion resistance, as well as resistance to almost all other corrosive environments.

Most of the elements used to strengthen and harden aluminum alloys form second phases in the microstructure. These microconstituents can have significant dissimilarity with the matrix in the galvanic series, and they can serve as anodes or cathodes in a miniature galvanic cell. Because pure aluminum has none of these microconstituents, it is not prone to this microcell action. The aluminum–copper alloys have the worst tendency for cell action between microconstitutents and the matrix; the alloys with manganese, magnesium, and silicon have the lowest tendency (solution potentials are very close to aluminum).

All the alloys that contain significant amounts of copper, silicon, zinc, or magnesium are susceptible to stress corrosion cracking, but pure aluminum and the alloys with low-alloy concentration are considered to be immune. The alloys of main concern are the 2xxx and 7xxx series. Stress corrosion cracking can occur in moist environments and in halide environments. Seawater causes pitting as well as stress corrosion cracking. In moist environments, stress corrosion is often controlled by minimizing the residual stress in components. Using aluminum in the stress-relieved condition (Tx5x) is usually sufficient protection. Extrusions are often supplied in this condition. Mill processing and extrusion often result in a warped shape. The extrusions are stretched several percent in straightening, and this axial yielding equalizes and lowers the state of stress throughout the extrusion. There are ratings on the corrosion resistance of many alloys, and the best precaution is to consult these data. Alloys that are extremely resistant are the 1xxx, 3xxx, and 6xxx (6061 is particularly resistant).

Anodizing improves the *atmospheric corrosion* resistance of aluminum significantly. The patina or gray oxide that forms on weathered aluminum will not occur if the anodized film is intact. Thus aluminum and its alloys are very much contenders for application for components where atmospheric corrosion is important. This explains their wide use in architectural shapes.

Water Corrosion

Aluminum is not widely used in plumbing systems because water containing heavy metal ions such as copper, lead, tin, and nickel will lead to pitting. This is true even with neutral water. As a general rule, aluminum is resistant to water solutions in the pH range of 4.5 to 8.5. Aluminum is resistant to distilled and deionized water because the heavy metal ions have been removed in distillation. All aluminum alloys and even anodized surfaces are attacked by waters with a high pH. Parts have been destroyed in detergent cleaning operations of short duration. Most detergents, even hand soaps, can cause pH levels to be high enough to produce noticeable attack. Coatings such as anodizing reduce, but do not eliminate, damage.

The 5xxx series are acceptable for use in seawater, but aluminum alloys are prone to pitting in seawater. Many aluminum boats and outboard motors are used in seawater, but complete corrosion protection usually requires cathodic protection or coatings. Painting is the usual method of protection. Freshwater rinses after use are recommended by some marine engine manufacturers.

Referring to previous discussions of the galvanic series, aluminum is anodic to most metals. Thus aluminum and its alloys should not be coupled with other metals immersed in water or any electrolyte. It is closest in the galvanic series with magnesium and zinc, and these are the safest couples. Stainless steel fasteners and attachments in aluminum fabrications do not usually cause a problem if the anode-to-cathode ratio is favorable [a large area of aluminum (anode), and a small stainless steel cathode (screws)].

Problems with Aerospace Alloys

Stress corrosion cracking is a problem found in 2xxx and 7xxx alloys generally used for aerospace applications. This condition occurs over long periods of time in metal that is being stressed while in a mildly corrosive environment. The test for this condition is described in ASTM G 47 and in effect loads an aluminum part while the part is immersed in an aqueous salt solution.

Exfoliation corrosion is another problem for aerospace alloys 2xxx and 7xxx. The test for exfoliation corrosion is found in ASTM G 34 and is accomplished by simply submerging samples in a harsh salt solution. The test is intended to simulate years of seashore environment contact.

Both of these tests are only used periodically because the industry has substituted the T751 and T7511 overaged tempers for the stronger but more corrosion-susceptible T651 and T6511 tempers. Much simpler and more reliable electrical conductivity testing is now used to control the amount of overage and thus the resistance to stress corrosion cracking and exfoliation corrosion.

Chemical Corrosion

Strong caustic solutions are probably the worst enemy of aluminum alloys, but they are also attacked by mineral acids, both oxidizing and reducing. One exception is nitric acid. At room temperature, pure aluminum is resistant to concentrated solutions (over 80%). Apparently, the oxidizing nature of this acid causes the protective oxide film to be an effective barrier to corrosion. Pure aluminum is also resistant to a range of ammonia solutions, many organic acids, sulfur atmospheres, and refrigerant-type fluorocarbon gases. At room temperature most aluminum alloys are resistant to a wide range of petroleum products and solvents (alcohols, acetone, benzene, ethylene glycol, formaldehyde, methyl ethyl ketone, gasoline, and toluene). However, there is a danger of exothermic reaction in the use of aluminum in handling chlorinated solvents such as methylene chloride and carbon tetrachloride. Dry halogen gases have little effect on aluminum, but if moist, these same gases cause attack.

Aluminum is widely used in handling food products. It resists attack by many types of food, but, more important, it is nontoxic. Everyone is familiar with aluminum pots and pans and aluminum foil.

The subject of chemical resistance cannot be adequately covered by general statements, but an important fact for the designer to keep in mind is that pure aluminum has better corrosion resistance than any of the aluminum alloys. Alloy microconstituents impair the protective oxide film. If the microconstituents are made to concentrate at grain boundaries, stress corrosion cracking may result. This cracking can occur in environments that would normally be considered only mildly corrosive.

17.7 Alloy Selection

There are specific reasons for the existence of the 200 or so alloys that are commercially available. Each does some job better than other alloys, but it is quite a task for the designer to become familiar with this many alloys. In previous discussions it was pointed out that aluminum alloys can be categorized as heat treatable or non–heat treatable and by casting technique. In addition, there are alloy systems in each of these categories. The average designer can probably meet most design needs with a repertoire of a half-dozen or so wrought alloys, one sand and permanent mold casting alloy, and one die casting alloy (an asterisk identifies age-hardenable alloys)

1. *Wrought:*
   1100 (pure Al)
   2024* (4.4% Cu, 0.6% Mn, 1.5% Mg)
   3003 (1.2% Mn, 0.12% Cu)
   5052 (2.5% Mg, 0.25% Cr)
   6061* (0.6% Si, 0.27% Cu, 1.0% Mg, 0.20% Cr)
   6063* (0.4% Si, 0.7% Mg)
   7075* (2.5% Mg. 1.6% Cu, 0.3% Cr, 5.6% Zn)
2. *Sand cast:*
   355.0* (5.0% Si, 1.2% Cu, 0.5% Mg, 0.25% Cr)
3. *Die cast:*
   380.0 (8.5% Si, 2.0% Fe, 3.5% Cu, 3.0% Zn)

The pure aluminum alloy will have the best corrosion resistance, and corrosion resistance will decrease with alloy content. The 2024, 3003, 6061, and 7075 alloys are available with high-purity aluminum cladding, at least on one side, to improve corrosion performance. However, these alloys (Alclad) are used predominately in the aircraft and aerospace industries.

For design applications, certain mechanical and physical properties are usually required. We shall discuss selection with these properties in mind.

Mechanical Properties

Probably the most important factor to consider in selection of aluminum alloys for structural applications is the modulus of elasticity. Aluminum alloys, as shown in Figure 17–10, are not as stiff as steels. It is imperative to take this into consideration in replacing steel parts with aluminum. Elastic deflections will be three times as great as with steel of equal section size. Copper and titanium have better stiffness than aluminum, but there is no plastic (excluding carbon and boron filament reinforced) that can approach the stiffness of aluminum. Occasionally

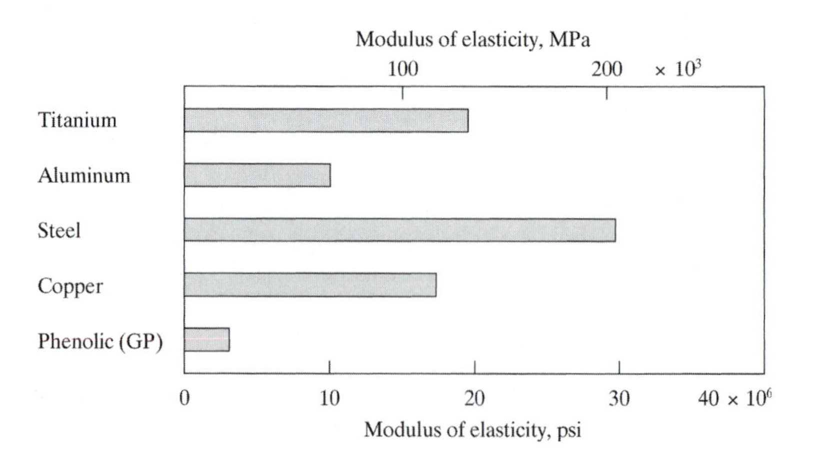

Figure 17–10
Stiffness of aluminum compared with other engineering materials

designers want to lighten a component, so they use aluminum. If section sizes can be made smaller, a steel part can be made as light as aluminum because of the higher modulus and available strength. The specific stiffness (modulus/density) of aluminum is about the same as that of steel. The modulus does not vary significantly with alloy or form.

The hardness of aluminum alloys varies from about 20 HB to about 120 HB, but this usually has no significance in selection for mechanical properties. The hardest aluminum alloy is softer than a soft steel. Hardness is not an indication of wear resistance. It is even difficult to use the hardness of aluminum to check conditions of heat treatment or temper. The most significant use of hardness is as an indicator of machinability. If an alloy has high hardness, it is likely to have better machinability than an alloy with low hardness. Anodizing and hardcoating provide a surface hardness comparable to that of the hardest steels. This surface hardness provides a measure of wear resistance, but it cannot be machined. A hardcoated surface sliding against a hardened tool steel or another hardcoated surface can be an acceptable wear combination. Because anodizing also roughens a surface, it is not recommended for sliding against a soft metal. It will cause the soft metal to wear. Similarly, anodized surfaces should not be used for bushing-type applications.

The tensile properties of aluminum alloys vary from a tensile strength of 13 ksi (90 MPa) for pure aluminum (1100) to as high as 98 ksi (676 MPa) for a zinc–copper–magnesium alloy in the age-hardened condition (7001-T6). The pure aluminum alloys, the 1350 and 1100 grades, are used for applications for which strength is not overly important. The 1350 (electrical conductor) grades are used for wires, bus bars, and the like. Type 1100 is used for reflections, nameplates, flashing, roofing, and applications for which good corrosion resistance is important.

The other non-heat-treatable alloys, the 3xxx and 5xxx series, have tensile strengths in the range from 16 ksi (110 MPa) to around 50 ksi (345 MPa). The 3003 alloy is a general-purpose alloy that has good formability, weldability, and corrosion resistance. It is widely used for sheet metal parts on machines (covers, chutes, guards, wireways, switch boxes, and the like).

If these same types of properties are desired but a higher strength is needed, the 5xxx series is a good choice. Type 5052 is widely used for tanks, structural members, and parts requiring corrosion resistance.

In the heat treatable grades, the 6xxx series is considered to be medium strength; the 2xxx and 7xxx series are high-strength alloys. The trade-off for the higher strength is fabricability. Alloys such as 6061 and 6063 have good weldability and moderate formability, while high-strength alloys like 2024 and 7075 have lower weldability and poor formability. In design, the former alloys are used for structural components unless the strength of the 2xxx or 7xxx series alloys is absolutely necessary. The specific parts made from the age-hardenable alloys range from machine frames to ultrasonic transducers.

The strength of aluminum casting alloys depends on the casting technique. The metal mold processes that provide fast cooling rates usually give higher strength from an alloy compared with a sand casting. Casting alloy systems include alloys of aluminum and Si, Cu–Si, Mg, Si–Mg, Zn–Mg, and Si–Ni–Mg–Cu. Some alloys are age hardenable (333.0, 355.0, 356.0, 520.0), but the tensile strength even in the age-hardened condition is still less than in many wrought alloys. In addition, the ductility is usually less than for wrought standards (<10% elongation). A designer who only occasionally needs an aluminum casting could get along with one alloy, 355.0, an age-hardenable general-purpose sand and permanent mold alloy. It has good weldability, machinability, and reasonable strength. It is an alloy of 5% silicon, 1.3% copper, and 0.5% magnesium.

For similar general-purpose applications where a die casting is required, alloy 380.0 will usually be satisfactory. It has good castability and an as-cast tensile strength as high as 48 ksi (331 MPa).

In the 1970s and 1980s a considerable amount of research was expended in improving the mechanical properties of aluminum alloys. There were many avenues for this research, but two that were brought to commercial reality in the mid-1980s are lithium–aluminum alloys and metal matrix composites with aluminum as the matrix.

Lithium is the lightest metal, with a density of 0.019 lb/in.^{3} (0.53 g/cm^3) compared with aluminum's density of 0.1 lb/in.^{3} (2.8 g/cm^3). Adding 1% lithium to an aluminum alloy can lower the density by 6%. In the aerospace industry there is a constant quest to reduce the weight of airframes. Aluminum–lithium alloys with 2% to 3% lithium allow a weight reduction in many airframe components of 10%. Lithium is being added to standard cast and wrought alloys such as 2024. Lithium increases the modulus of aluminum alloys when added in small amounts, but it can have a deleterious effect on toughness and ductility. Much of the research that is being done on the aluminum–lithium alloy systems is aimed at solving the toughness problems. In the late 1980s the 2% to 3% lithium aluminum alloys had about the same strength as their nonlithium counterparts and a modulus of about 11.6 million psi (79,980 MPa) compared to 10 (68,950 MPa) for the conventional aluminum alloys. Fracture toughnesses and elongations were slightly lower than those of their nonlithium counterparts. In 2003, several aluminum–lithium alloys were commercially available, and they were being used for weight reduction in aircraft and space craft. The lithium content was most commonly between 1% and 2% (UNS A98090, A20970, etc.).

Since the 1960s metals have been strengthened and stiffened by the incorporation of reinforcements in the form of continuous fibers, particles, or whiskers. Adding particles to metals was called *dispersion strengthening*; this family of engineering materials is now called *metal matrix composites (MMCs)*. There was a pervasive trend in the 1970s and 1980s to reduce the weight of automobiles, and the use of aluminum in automobiles increased many times in this same time period. If an aluminum could be made to have the strength and toughness of steel and the ferrous metals, most car parts would be made from such a material. This is one of the goals of aluminum metal matrix composites. Aluminum oxide, silicon carbide, and carbon and other fibers are added to aluminum to perform the same function that glass fibers perform in FRP structures. The reinforcements are much stronger than the matrix, and the function of the matrix is to transfer the strains to the stronger and stiffer reinforcements. Polymer composites can be made very strong and stiff, but because they have a polymer matrix they cannot take the elevated temperatures that are frequently encountered in many types of service, particularly automobile engine components. This area of application is being addressed by metal matrix composites.

There are a number of ways of reinforcing aluminum composites. One of the simplest approaches is to fill a casting mold with carbon or similar fibers in mat form and to pressure cast the aluminum such that it intrudes the reinforcement. Reinforcements in whisker or particle form can be added to aluminum by P/M techniques, and the resultant parts can be used as sintered or hipped to full density. This technique can also be used to make ingots that are subsequently processed in the normal rolling and hot-working techniques. Continuous fibers can be added to aluminum as a reinforcement with a technique that is similar to pultrusion. Carbon or boron filaments are pulled through a molten aluminum bath and the resulting wires can be hot formed into plates or other usable shapes.

Tensile strengths as high as 250 ksi (1723 MPa) have been achieved with 35% volume fraction of SiC fibers. The tensile modulus of this particular composite is 31 million psi (214 GPa), over three times the stiffness of

Table 17–7
Approximate properties of commonly used wrought and cast aluminum alloys[a]

Aluminum Alloy	Nominal Chemical Composition %	Typical[b] Tensile Strength ksi (MPa)	Typical[b] Yield Strength ksi (MPa)	Typical %[b] Elong.	Typical Tensile[b] Modulus 10^6 ksi (GPa)	Typical Uses
1060 UNS A91060	99.60 min. Al	0–10 (70) H18–19(130)	4 (30) 18 (125)	43 6	10 (69)	Welded tanks, chemical equipment
1100 UNS A91100	99.0 min. Al, 0.12 Cu	0–13 (90) H18–24 (165)	5 (35) 22 (150)	35 5	10 (69)	Sheet metal parts, (ducts, guards, etc.)
1145 UNS A91145	99.45 min. Al	0–11 (75) H19–29 (165)	5 (35) 21 (145)		10 (69)	Foil, heat exchanger fins
2014 UNS A92014	4.4 Cu, 0.8 Si, 0.8 Mn, 0.50 Mg, Bal Al	0–27 (185) T6–70 (485)	14 (95) 60 (415)	18 13	10.6 (73)	Vehicle and aircraft frames
2024 UNS A92024	4.4 Cu, 1.5 Mg, 0.6 Mn, Bal Al	0–27 (185) T361–72 (495)	11 (75) 57 (395)	22 13	10.6 (73)	Fasteners, aircraft parts
2219 UNS A92219	6.3 Cu, 0.3 Mg, 0.06 Ti, Bal Al	0–25 (170) T87–69 (475)	11 (75) 51 (395)	18 10	10.6 (73)	High-strength weldments—use up to 316°C (600°F)
3003 UNS A93003	1.2 Mn, 0.12 Cu, Bal Al	0–16 (110) H18–29 (200)	6 (40) 27 (185)	35 7	10 (69)	Sheet metal parts, (ducts, guards, hoods)
3004 UNS A93004	1.2 Mn, 1.0 Mg, Bal Al	0–26 (180) H32–31 (215)	10 (70) 25 (170)	22 16	10 (69)	Sheet metal parts, tanks
4043 UNS A94043	5.2 Si, Bal Al					Welding filler metal
5052 UNS A95052	2.5 Mg, 0.25 Cr, Bal Al	0–28 (195) H34–38 (260)	13 (90) 31 (215)	27 12	10.2 (70)	Sheet metal parts, tubing
5083 UNS A95083	4.4 Mg, 0.7 Mn, 0.15 Cr, Bal Al	0–42 (290) H321–46 (315)	21 (145) 33 (230)	22 16	10.3 (71)	Pressure vessels, boats, aircraft parts
5086 UNS A95086	4.0 Mg, 0.45 Mn, 0.15 Cr, Bal Al	0–38 (260) H32–42 (290)	17 (115) 30 (205)	22 12	10.3 (71)	Pressure vessels, boats, aircraft parts
6061 UNS A96061	1.0 Mg, 0.6 Si, 0.28 Cu, 0.20 Cr, Bal Al	0–18 (125) T6–45 (310)	8 (55) 40 (275)	27 13	10.0 (69)	Fixtures, frames, rolls, shafting
6063 UNS A96063	0.7 Mg, 0.4 Si	T1–22 (150) T6–35 (240)	13 (90) 31 (215)	 12	10 (69)	Extruded shapes—architectural, pipe, tube, channels, etc.
355.0 (cast) UNS A03550	0.25 Cr, 1.2 Cu, 0.6 Fe Max, 0.5 Mn, 5 Si, Bal Al	T6–20 (140) T51–18 (125)	32 (220) 25 (170)	2	—	General-purpose sand/perm. mold castings
3380 (die cast) UNS A03800	3.5 Cu, 2 Fe, 8.5 Si, 0.5 Mn Max, Bal Al	23 (160)	46 (320)	2.5	—	General-purpose die casting alloy
201 (cast) UNS A02010	0.1 Si, 4.6 Cu, 0.45 Mn, 0.35 Mg, 0.25 Ti, Bal Al	T7–50 (345)	60 (415)	3	—	High-strength sand casting—see ASTM B 686

[a]At various tempers, O, H18, T6, etc.
[b]From various sources—not for design.

conventional aluminum. The density of most aluminum metal matrix composites is about the same as unreinforced aluminum. Thus the increased strength and stiffness of these materials will produce higher specific strengths and stiffnesses than are possible with conventional unreinforced metals. Dispersion-strengthened (15 volume % Al_2O_3) pure aluminum for power transmission conductors can have the strength and creep strength of copper but be lighter for the same current-carrying capacity.

The problems of adding reinforcements to aluminum are the same as those that exist in polymer composites: adhesion of the matrix to the fibers, the cost of high-strength fibers, and directionality of properties. PVD coatings of reinforcements are widely used to increase the reinforcement–matrix bond; free market conditions control the reinforcement costs, but since the early 1980s there has been a steady decline in the cost of carbon fibers and some of the other reinforcements. The problem of directionality must be dealt with by design—design of the reinforcement or design of the loading of parts that have directional properties. In any case, in 1995 a number of automobile manufacturers employed SiC-reinforced aluminum connecting rods in their production engines. Silicon carbide whisker-reinforced billets are commercially available from aluminum mills for processing into shapes. Thus this process is a commercial reality, and reinforced aluminum should be considered when it is apparent that the desired strength or stiffness properties cannot be obtained with conventional aluminum alloys. Aluminum metal matrix composites cost as much as $50 per pound in 2004, so this is a selection factor.

To summarize the selection of aluminum alloys for mechanical properties, we have tried to show that a very wide range of properties is available, but that a repertoire of about six alloys will meet most design needs. The range of mechanical properties available on popular aluminum alloys is shown in Table 17–7. The relative machinability of some alloys is shown in

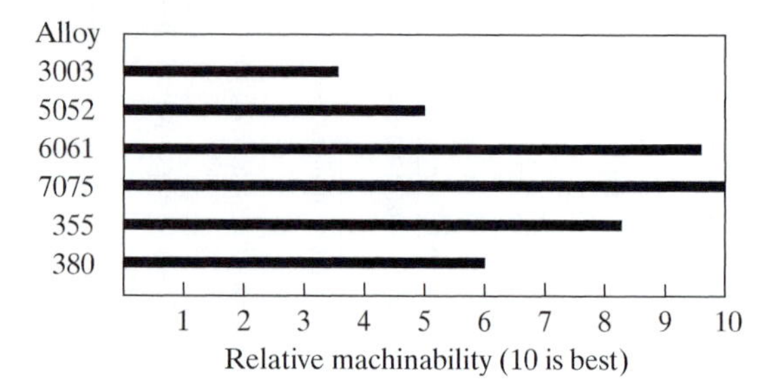

Figure 17–11
Machinability of aluminum alloys in the age-hardened condition or full-hardened condition

Figure 17–11. However, if screw machine types of operations are required, the free-machining aluminum alloy 2013 might be added to this repertoire. The weldability of these alloys, with the exception of 2024, 7075, and 380 alloys, is acceptable. The filler metal most commonly used to weld aluminum alloys is alloy 4043.

If the property data just presented are insufficient, there is a wealth of mechanical property data available in aluminum handbooks. These should be consulted.

Physical Properties

There is no significant variation in most physical properties from one alloy to another, with the exception of electrical and thermal conductivity. The specific gravity of wrought alloys varies from about 2.65 to 2.80; for cast alloys the range is about 2.50 to 2.95 (2.5 or less for Li–Al alloys). The modulus of elasticity varies from 10 to 10.4×10^6 psi (70 to 72 GPa).

Coefficients of expansion for wrought alloys vary from about 12.4 to 13.3; for cast alloys the range is about 10.5 to 14.0 [in./in. °F $\times 10^{-6}$ (18.7 to 25 m/m K)].

For applications involving heat or electrical conduction, it may be necessary to carefully scrutinize conductivity data (Figure 17–12). As might be expected, the electrical conductor EC grade has the best properties.

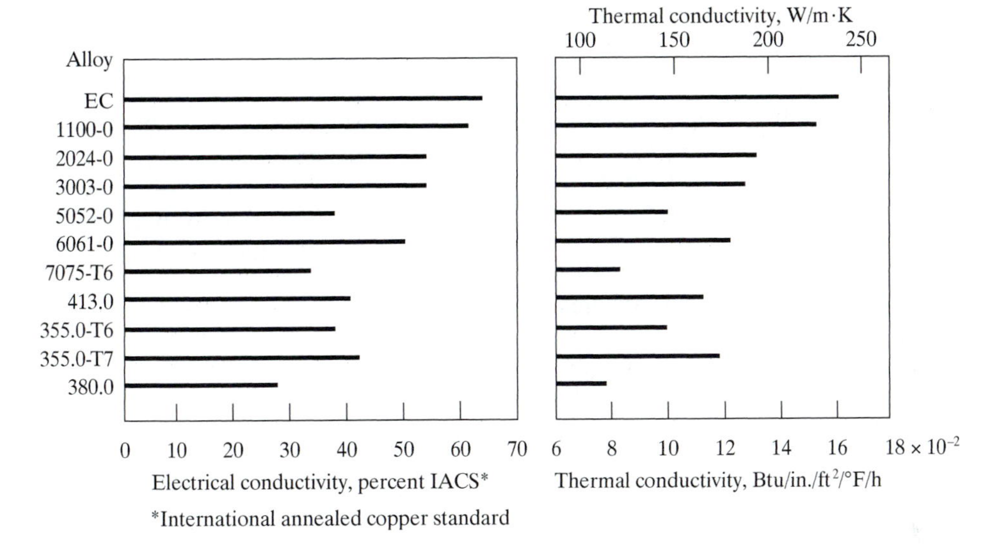

Figure 17–12
Electrical and thermal properties of aluminum alloys

Summary

Many aluminum alloys were not even mentioned in our discussions, and this is intentional. Aluminum is somewhat like copper in that there are more alloys than designers can become familiar with (more than 600 UNS alloys). Our repertoire of 17 or so will meet most design needs, but this could be carried a step further. If faced with the situation of taking care of all aluminum design needs with only one alloy, type 6061 in the T6 or T651 condition would be a good choice. It has good machinability, weldability, and moderate strength, and it accepts hardcoating. Thus the occasional user of aluminum alloys could get along with one alloy (if formability is not a significant factor). Engineering design is not usually simple enough to permit a repertoire of one alloy, but aluminum alloys are essential and the designer should be on intimate terms with at least a few alloys.

As an engineering material, aluminum has become almost indispensable. It is widely used in machine design for all sorts of structural parts (Figure 17–13). It is extremely useful in consumer items because of its fabricability and pleasing cosmetic appearance. In commercial architectural applications, aluminum has almost cornered the market on window and door frames. In electrical wiring, it has all but replaced more expensive and heavier copper for high-current distribution lines. Weight-reduction mandates have greatly increased the use of aluminum in automobiles. In 1978, 141 lb (64 kg) of aluminum were used in the average U.S. automobile. In 1999, the amount had

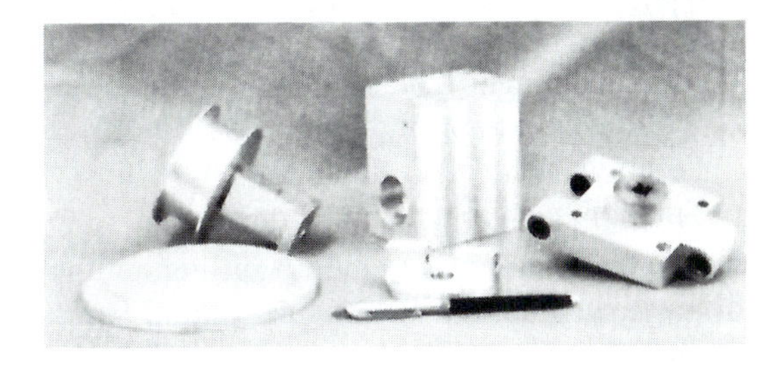

Figure 17–13
Typical aluminum machine components

increased to 250 lb (73 kg), an increase of more than 75%. The useful properties and reasonable cost of aluminum have resulted in such increased use that in 2003 aluminum was the second most widely used metal (second only to steel).

Some important concepts to keep in mind are the following:

- Aluminum alloys have higher conductivities (electrical and thermal) than most metals, and they are usually cheaper than the alloys that are better conductors (copper, silver, gold, and so on).
- Aluminum alloys must be specified with a designation system that specifies temper or heat treatment (the suffixes are important).
- Aluminum extrudes very easily; most bars and structural shapes are extruded and straightened. This gives them good dimensional characteristics.
- Aluminum is easy to die cast; die casting should be considered for intricate parts with high production quantities.
- One feature of aluminum alloys that makes them attractive as engineering materials is the ability of many alloys to be age hardened. They have excellent mechanical properties.
- Avoid continued heating of age-hardened alloys over 100°C if high strength is a service factor. They soften at relatively low temperatures.
- Anodizing and hardcoating are desirable ways to protect aluminum from tarnishing.
- Hardcoating is the preferred coating for aluminum when abrasion is a service consideration. Remember that some alloys are difficult to hardcoat.
- Aluminum is susceptible to pitting and accelerated corrosion in seawater and halide environments. These environments should be avoided or corrosion suitability should be established.
- The age-hardened aluminum alloys usually machine better than the non-heat–treated aluminum alloys.
- ASTM specifications can be used for international alloy designation:

  Material: ASTM B 221 alloy A96063 extruded tube

Critical Concepts

- Aluminum has limited solubility with other metals. It cannot be welded to them with most fusion processes.
- Some aluminum alloys can be age hardened, some cannot.
- Aluminum is lighter than most metals, but it also has lower stiffness.
- Aluminum competes with magnesium and beryllium as a lightweight metal.
- Anodizing and hardcoating greatly improve the serviceability of aluminum alloys, but some alloys accept these treatments better than others.

Terms You Should Remember

strain hardening
solution treated
thermal aging
naturally aged
anodizing
hardcoating
extrusions
age harden
anode
oxide
dying
sealing
bauxite
precipitate
intermetallic
binary
overaging
atmospheric corrosion
MMCs
dispersion strengthening
patina

Case History

Abrasion Tester Made from Aluminum

Development work for either product or manufacturing processes often requires laboratory testing of concepts and tests to develop data for solving problems. Invariably, research and development requires constructing equipment for laboratory testing. In this case, a significant wear problem required development and construction of a tester for screening candidate cemented carbides for use as perforating tooling. A device was designed for screening, called a *loop abrasion tester*. The test machine required more than 30 parts and aluminum, 6061-T6, was used as the material for all machine parts except those used in power transmission. The drive shaft was made from steel and it rotated in steel rolling element bearings. The machined aluminum parts were fastened with hardened steel cap screws.

Why did we use aluminum? Why did we opt for 6061-T6? One of a kind or prototype machines can usually be made fastest and at the lowest cost by machining parts from an easy-to-machine material like aluminum. It is also important to select a readily available material. Type 6061 in the T6 condition is stocked in almost every metal warehouse in the United States. It is a "favorite" metal of many machinists because of its favorable machining characteristics in the T6 condition. In addition, it does not rust, and it can be anodized or hardcoated if needed. Aluminum is often the most cost-effective material for machines and fixtures needed for test purposes or for necessary research and development fabrications.

Questions

Section 17.1

1. What is the function of the molten cryolite in the Hall cell?
2. A steel beam with a cross section of 1 in. × 4 in., 60 in. long is replaced by aluminum. What aluminum cross section is needed for stiffness equal to the steel?

Section 17.2

3. What does the 38 mean at the end of the alloy specification 3003-H38?
4. What does T651 mean at the end of 6061-T651?
5. Is 355-T6 a cast or wrought alloy?
6. Write the UNS designation for 355 alloy in the T5 temper.

Section 17.3

7. What is the significance of aluminum's ability to be die cast?
8. What manufacturing process is used to produce an aluminum sailboat mast?

Section 17.4

9. What is the precipitate that forms when a 2014 aluminum alloy is solution treated and aged at 350°F (176°C)?
10. What type of crystal structure does aluminum have?
11. What are the two primary methods for strengthening aluminum?
12. What is the purpose of solution treating?
13. What phases are present in the microstructure of type 1100 aluminum?

Section 17.5

14. How do you specify age hardening of 6061 aluminum on an engineering drawing?
15. What is the stress relief temperature for 6061 alloy?

Section 17.6

16. What is the mechanism of anodizing?
17. What is the difference between clear anodizing and hardcoating?

18. What precautions must be taken in using hardcoating on precision parts?
19. Can aluminum be electroplated?

Section 17.7

20. Would aluminum be a suitable material for a roof on a house (from the standpoint of atmospheric corrosion)? Explain your answer.
21. Is aluminum suitable for continuous immersion in water? Explain your answer.
22. Name two metals that could be used to provide cathodic protection of aluminum.
23. Would aluminum be a suitable material for a tank to contain a strong detergent?

Section 17.8

24. Compare the thermal conductivity of aluminum to that of carbon steel.
25. Using data from Figure 11–6 and Table 17–6 compare the specific strength (tensile strength/density) of A36 steel and 6061-T6.
26. An aluminum tube 40 in. long (100 cm) is clamped into a steel frame and the assembly is put into service at 200°F (93°C). What stress will develop in the aluminum member? [The area is 0.4 in^2 (2.5 cm^2).]
27. Aluminum has high electrical conductivity. Why isn't it widely used to wire electrical devices?
28. Compare the specific cost ($/$in.^3$) of aluminum at $0.60/lb, steel at $0.50/lb and ABS plastic at $1.25/lb.
29. List three reasons why aluminum is rarely used for auto bodies and frames.

To Dig Deeper

Aluminum Publication and Audiovisual Guide. Washington, DC: The Aluminum Association, 1994.

Aluminum Design Manual. Washington, DC: The Aluminum Association, 2000.

Aluminum Standards and Data, 1997. Washington, DC: Aluminum Association, 1997.

Crepeau, P., Ed. *Light Metals 2003*. Warrendale, PA: TMS Publications, 2003.

Dalta, J. *Key to Aluminum Alloys,* 5th ed., Dusseldorf: Aluminium Verlag, 1998.

Dalta, J., Ed. *Key to Aluminum Alloys*, 6th ed. Dusseldorf: Aluminum-Verlag, 2002.

Davis, J. R., Ed. *ASM Specialty Handbook. Aluminum and Aluminum Alloys*. Materials Park, OH: ASM International, 1993.

Hatch, John E., Ed. *Aluminum: Properties and Physical Metallurgy*. Materials Park, OH: ASM International, 1984.

Kammer, C., Ed, *Aluminum Handbook*, 15 ed. (1st English). Dusseldorf: Aluminum Verlag, 1999.

Kaufman, J. G. *Introduction to Aluminum Alloys and Tempers*. Materials Park, OH: ASM International, 2000.

Metals Handbook, 9th ed. Volume 2, Properties and Selections: Non Ferrous Alloys and Special-Purpose Materials. Materials Park, OH: ASM International, 1991.

Metals Handbook, 9th ed. Volume 13, Corrosion. Materials Park, OH: ASM International, 1987.

Polmear, I. J. *Light Alloys, Metallurgy of Light Metals*. Materials Park, OH: ASM International, 1981.

Standards for Aluminum Permanent Mold Castings. Publication #18. Washington, DC: The Aluminum Association, 1992.

VanHorn, Kent R. *Aluminum, Volume 1, Properties, Physical Metallurgy and Phase Diagrams*. Materials Park, OH: ASM International, 1967.

Wernick, S., and others. *The Surface Treatment and Finishing of Aluminum and Its Alloys*, 5th ed. Materials Park, OH: ASM International, 1987.

Zolensos, D. L., Ed. *Aluminum Casting Technology*, 2nd ed. Washington, DC: The Aluminum Association and American Foundrymans Society, 1992.

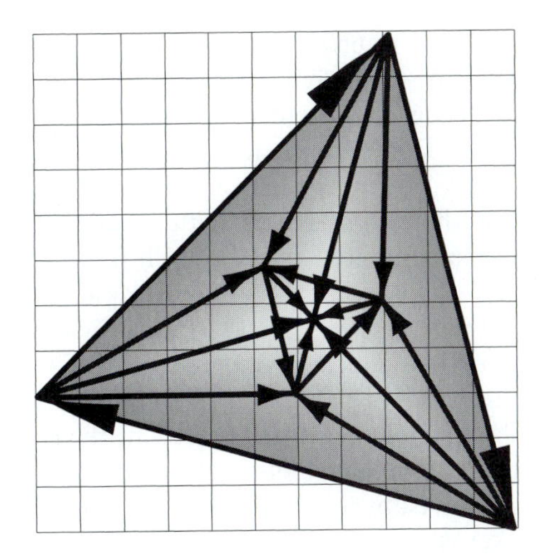

CHAPTER 18

Nickel, Zinc, Titanium, Magnesium, and Special Use Metals

Chapter Goals

1. An understanding of the metallurgy and properties of some less widely used metals.
2. Selection information of these metals and how they can be used to solve design problems.

Rationale

This chapter is our metal "attic." Designers and engineers can meet most of their engineering material requirements with the materials discussed up to this point, which include, a wide range of plastics, some ceramics and cermets, all of the common steels and cast irons, and important aluminum–copper alloys. Probably 90% of the materials used in your day-to-day design activities will come from these materials—thus we dedicated whole chapters to these material systems. However, there are occasions for which these traditional engineering materials will not meet service requirements. There are other metals commercially available that can be used to meet special design requirements.

For instance, we would not have jet engines for airplanes without nickel-based superalloys for engine parts; highway guard rails without the low-cost corrosion protection of steel made possible by zinc coatings; satellites without titanium for structural components; miles of underground piping without protective magnesium anodes to protect the system from corrosion; incandescent light bulbs without one of the refractory metals—tungsten; many orthopedic prosthetics made from cobalt-based alloys; and the many forms of satellite communication without lightweight beryllium reflectors. Of course, countless things would be different in the world and in our daily lives without precious metals like gold and silver.

Thus this chapter is about the metals that you may not need for everyday design jobs but may need for special designs. This chapter will cover most of the important metals that have unique properties.

In the metals category, we have covered only a half-dozen or so alloy systems. However, there are about 80 elements that are considered to be metals. In this chapter, we shall briefly discuss some of the metals that have industrial significance but that are not used on an everyday basis by most designers. They are the metals that are invoked for special problems, for design situations for which more common metals have inadequate properties. The specific systems to be covered are nickel, zinc, magnesium, titanium, and refractory metals, cobalt, beryllium, gold, and silver.

18.1 Nickel

Nickel as an element was discovered in 1750 but had limited uses until the twentieth century, when it became an essential alloying element in stainless steels and high alloys. Much of the world supply of nickel is obtained from deposits in Ontario and other provinces in Canada. Nickel occurs in nature in the form of sulfides or oxides. The Canadian ore is primarily a sulfide with a nickel content of less than 3%. The ore is usually deep mined, and the normal flotation and concentration processes are used to arrive at the starting material for making metallic nickel. The *beneficiation* process is somewhat like that used for copper. The sulfide ore is roasted to convert the sulfide to an oxide. The oxide is then smelted to produce a nickel–iron matte. The matte is processed with blast-furnace techniques to remove the iron and further concentrate the nickel. The final step is to electrolytically purify the nickel by essentially plating it onto cathode plates. The refined nickel is then cast into ingots. A good portion of the total production of nickel in this country leaves the mill in ingot form. About 57% of all nickel is used in austenitic stainless steels. About 10% is used in electroplating, and the remainder is used in nickel alloys, high-nickel specialty metals, copper–nickel alloys, and ferrous alloys that have nickel additives.

Physical Properties

Nickel and most high-nickel alloys have the same physical appearance as steels. *Pure nickel* has a melting point of 2650°F (1455°C) and a modulus of elasticity of 30×10^6 psi (207 GPa). Both of these qualities are similar to those of steel. It is ferromagnetic up to 680°F (360°C), and its density is 8.9 g/cm^3 compared with 7.85 g/cm^3 for most steels. The coefficient of thermal expansion is not far from that of steel: 7.4 compared with 6.5×10^{-6} in./in. °F (15.1×10^{-6} m/m K). Thermal conductivity is slightly better than that of steel—49.6 compared with 27 Btu/h ft^2 °F/ft (47 W/m K)—and electrical conductivity is about 16% that of pure copper. Carbon steels have only about 8% of the conductivity of pure copper. The physical properties of pure nickel are for the most part similar to those of wrought steels, with the exception of electrical and thermal conductivity. Because of its relatively good electrical conductivity, nickel is often used for high-temperature electrical conductors and wire terminals.

Nickel alloys are often used because they have unique physical property characteristics. Some high nickel–iron alloys (80–20, 50–50) have magnetic permeability characteristics that make them superior for magnetic shielding of electronic devices. Some alloys excel as magnetic materials for tape recording heads. Alloys of iron and nickel are used for parts requiring controlled thermal expansion. *Invar*—iron with 36% nickel—has one of the lowest coefficients of thermal expansion of all metals (1×10^{-6} in./in. °F) in a limited temperature range. Other iron–nickel alloys have expansion rates that are altered to be the same as that of glass for glass-to-metal sealing applications.

Nickel alloys with chromium as the major alloying element (80% Ni, 20% Cr; 75% Ni, 20% Cr + others; 70% Ni, 30% Cr) are widely used for resistance heating elements. This family of alloys is often recognized by the trade name "Nichrome," but they are manufactured by a number of suppliers. They have high resistivity and good oxidation resistance.

Metallurgy

There are at least eight nickel alloy systems of major commercial importance:

1. Pure Nickel: Nickel 200, 201, 211, 270, 210, 213, 305, and Duranickel 301.

2. Nickel–copper alloys: Monel 400, Monel K-500, Ni–Cu alloys 410, 505, and 506.
3. Nickel–chromium alloys: Inconel 600, 601, 702, X750, 690, 671, Incoloy 804, Nichrome.
4. Nickel–iron alloys: Hy-mu 80, Permalloy, Ni span C, Invar, Elinvar, NiResist type 3, 5.
5. Nickel–molybdenum alloys: Hastelloy B, B-2, N, Chlorimet 2.
6. Nickel–chromium–molybdenum alloys: Hastelloy C, 276, C-4, C-22, X, F, Inconel 625, Chlorimet 3.
7. Nickel–chromium–molybdenum–iron–copper alloys: Hastelloy G, G30, Inconel 825.
8. Nickel–based superalloys: Inconel 600, 601, 617, 718, X750, Udimet 500, 700, Waspaloy, Hastelloy S, Mar-M-200, PS Mar-M-200.

There are about 220 UNS numbers and more than 100 ASTM specifications for nickel alloys, but because there are few U.S. manufacturers and because many alloys are proprietary, it has been the practice in the United States to identify alloys by trade name. The preferred designation is the ASTM specification and UNS number.

The various types of pure nickels exist because properties can vary slightly by impurity level. All the "pure" nickel alloys contain at least 95% nickel; all have a single-phase face-centered cubic (FCC) structure; all can be cold worked to increase strength; and some alloys (Duranickel) can be strengthened by age hardening. None of the nickel alloys responds to allotropic transformation quench hardening.

Nickel is completely soluble in copper, and vice versa. The nickel–copper alloys are also single phase, and they are called *Monels* in the United States. Some Monels (K-500) can be precipitation hardened.

Nickel exhibits good solid solubility with iron, chromium, manganese, molybdenum, tungsten, and zinc. The nickel–chromium, nickel–iron, and other alloy groups are usually single-phase solid solutions of combinations of these elements. Some are age hardenable, and the plethora of different alloys evolved to meet varying strength and corrosion characteristics. The nickel-based superalloys (*Inconels* and *Incoloys*) can contain up to a dozen different alloying elements, and there are no metallurgically common characteristics other than that they are nickel based. Most of these alloys were developed for a particular, severe service condition such as high strength at elevated temperature. Inconel 718, for example, has a tensile strength of 120,000 psi (827 MPa) at 1400°F (760°C).

Mill Products

Standard mill forms include sheet, strip, rods, bars, and plate tubing; some shapes are available in all the major alloy systems. Nickel alloys are also available as forgings, castings, and as powders for P/M parts.

Many nickel alloys are covered by code body specifications (ASTM, AMS, SAE, MIL). As an example, ASTM specification A 494 covers 14 grades of nickel alloy castings. A drawing specification for a Monel casting might read:

Material: ASTM A 494; grade M35-1

Many times the cast equivalent of a wrought alloy has different properties and a composition that is different from that of the wrought alloy. The Alloy Casting Institute (ACI), a U.S. casting industry organization, designations for some of the more important nickel-based alloys are presented in Table 18–1. The ACI designations or ASTM designations are preferred for drawings and purchasing specifications for cast alloys.

Corrosion Resistance

Nickel in stainless steels does not really contribute to corrosion resistance. It is present mainly to produce an austenitic structure. Pure nickel, however, has some unique corrosion characteristics. It is extremely resistant to caustics. This includes all concentrations and temperatures up to several hundred degrees Fahrenheit

Table 18–1
ACI casting designations for some nickel-based alloys (See also ASTM A 494)

Trade Name	UNS Number	ACI Cast Equivalent
Hastelloy C-4	N08020	CW2M
Inconel 625	N06625	CW6MC
Inconel 600	N06600	CY40
Nickel 200	N02200	CZ100
Monel 400	N04400	M35-1
Hastelloy B2	N10665	N7M
Hastelloy C22	N06022	—
Carpenter 20Cb3	N08020	CN7M

(150°C+). Nickel does not rust, and it is resistant to neutral waters and seawater. Its resistance to acids is mixed, and detailed corrosion data should be consulted.

Nickel–chromium alloys of the Inconel and Hastelloy types are noted for their resistance to corrosion at elevated temperatures. These alloys are nickel based, but the chromium concentration can be as high as 40%. The corrosion resistance in general increases with chromium content. Inconel 600 resists oxidation in furnace-type atmospheres up to 2150°F (1170°C). It is particularly resistant to stress corrosion cracking in chloride atmospheres. Inconel 625 has good corrosion resistance to a wide variety of chemicals, with the noteworthy characteristic that it is not susceptible to carbide precipitation during welding and subsequent intergranular attack.

Incoloy 825, a nickel–iron–chromium alloy, is resistant to a variety of reducing acids and oxidizing chemicals. It is exceptionally resistant to sulfuric and phosphoric acids and to seawater.

Monel 400, a nickel–copper alloy, was previously mentioned for its resistance to seawater, but it is also used for handling sulfuric and hydrochloric acid, and it has unusual resistance to deaerated hydrofluoric acid.

A number of nickel-based alloys of the Hastelloy type are widely used for particularly severe corrosion problems. Hastelloy B-2, a nickel–molybdenum alloy, is resistant to hydrochloric acid at all concentrations and at temperatures up to the boiling point. Hastelloy C-22, a nickel–chromium–molybdenum alloy, has the best overall corrosion resistance of the nickel–chromium–molybdenum alloys (C-276). It has exceptional resistance to pitting and stress corrosion cracking; in addition, it resists both oxidizing and reducing environments.

Welding grades and consumables are available in these alloys that have been modified to prevent sensitization in welding.

Most of the alloys mentioned can be fabricated with the welding, machining, and forming techniques used for austenitic stainless steels. Thus they find wide application in tanks, heat exchangers, furnace parts, flue stacks, and general chemical process equipment. Good corrosion data are available on these alloys, and it is not difficult to determine how they behave in very specific environments.

Alloy Selection

An important area of application for nickel-based alloys is corrosion control. However, the mechanical and physical properties of nickel alloys are often the determining factors in alloy selection. *Nichromes*, nickel–chromium alloys, have high electrical resistance and good oxidation resistance, and they are widely used for resistance heating elements. Inconel X750 is used for its high-temperature mechanical properties (Figure 18–1). It resists creep at operating temperatures up to 1500°F (815°C). These alloys are not used enough by the average designer to warrant a detailed discussion of each nickel-based alloy. For selection purposes, we shall simply list the more commonly used alloys and state where they might be used (Table 18–2). As can be seen from this tabulation, important uses of nickel-based alloys are for corrosion resistance and electrical-magnetic properties.

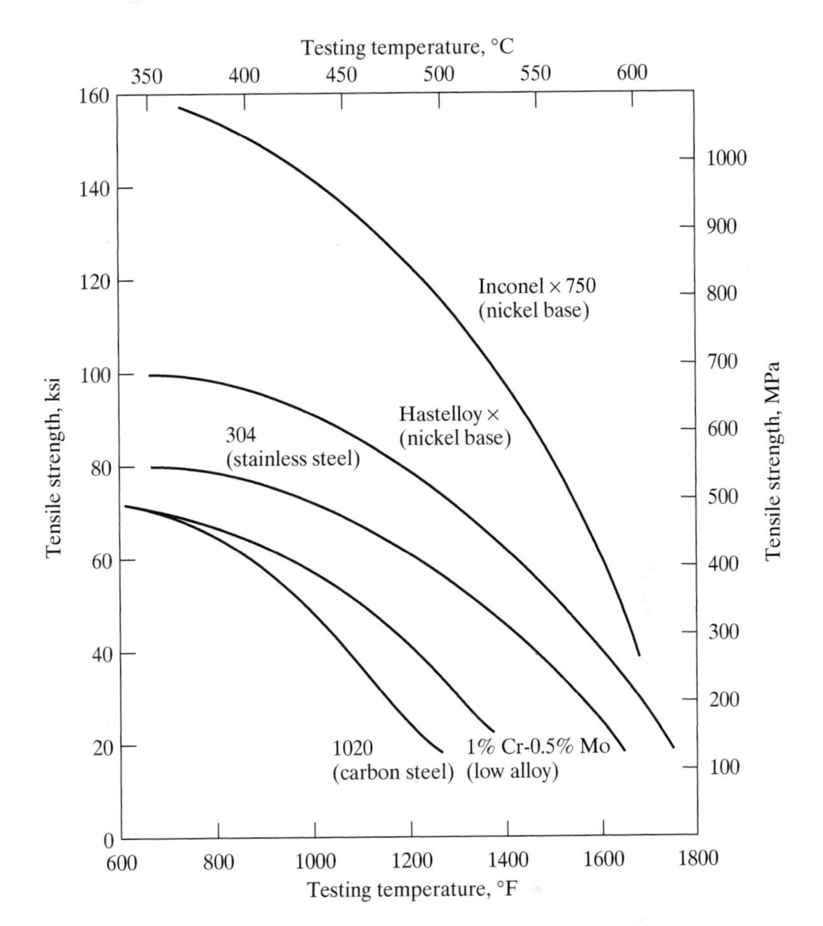

Figure 18–1
Short-term tensile strength of various alloys

Some nickel alloys have tensile properties comparable to those of alloy steels in the annealed condition. The highest-strength nickel alloys are precipitation hardened. The heat treatment for age hardening is similar to those used on copper and aluminum: solution treat, quench, and age. Aging temperatures are fairly high, 1150 to 1600°F (620 to 871°C). Most nickel-based alloys can be cold worked to higher strength, and their response to cold work is somewhere between that of carbon steel and austenitic stainless steels.

Other applicable heat treatments besides age hardening are annealing to produce maximum softening, and stress relieving or stress equalizing at low temperatures [500 to 900°F (260 to 482°C)] to partially remove cold-work stresses.

Most nickel-based alloys never achieve high hardness. Even age-hardened Inconel and Monel never achieve hardnesses much over 300 HB. The exceptions are nickel-based hard-surfacing welding filler metals. These alloys contain chromium and boron (*Ni Cr B* alloys), and hardnesses as high as 600 HB are possible. They have excellent abrasion and metal-to-metal wear resistance. Pure nickels, Inconels, Incoloys, and *Hastelloys* are not normally used for wear applications. Some of the softer nickel-based alloys have achieved a reputation for good resistance to *galling*. Cast silicon nickel is used for valve parts and adjusting devices when corrosion and

Table 18–2
Properties of some important nickel alloys

Alloy Family	UNS Number	Major Elements (%) Nominal Composition	Comments
Pure nickel	N02200	99.0 Ni min. (Nickel 200)[a]	Resistant to strong caustics ASTM B 160 (bar and rod), B 161 (pipe), B 162 (sheet)
Nickel–copper	N04400	63 Ni, 31 Cu, 2.5 Fe max., 2 Mn max. (Monel 400)[a]	Resistant to neutral waters, seawater, and some acids ASTM B 127 (plate), B 165 (tubes), B 164 (bar)
	N05500	66 Ni, 1 Fe, 29.5 Cu, 2.7 Al (K Monel)[a]	Age-hardening grade of Monel 400 ASTM F 467, F 468 (medical)
Nickel–chromium	N06600	72 Ni, 15.5 Cr, 8 Fe (Inconel 600)[a]	Resistant to oxidizing and reducing atmospheres at high temperature ASTM B 166 (bar), B 168 (plate), B 167 (tube)
	N06625	61 Ni, 21.5 Cr, 2.5 Fe, 9 Mo + 3.6 Cb and Ta (Inconel 625)[a]	High strength and toughness to 1800°F (980°C) ASTM B 443 (sheet), B 444 (pipe), B 446 (bar)
	N07718	52.5 Ni, 19 Cr, 18.5 Fe, 3 Mo, 5 Cb and Ta (Inconel 718)[a]	Age hardenable, oxidation resistance to 1800°F (980°C) ASTM B 670 (plate)
	N07750	73 Ni, 15.5 Cr, 7 Fe, 2.5 Ti (Inconel X750)[a]	Age hardenable with good corrosion and oxidation resistance ASTM B 637 (bars, forgings)
	N06004	60 Ni, 16 Cr, 1 Mn, Remainder Fe (Nichrome)	High oxidation resistance and resistance for electric heating elements ASTM B 344
	N07718	60 Ni, 20 Cr, 5 Fe, 1 Al, 3 Ti (Rene 41)[b]	Good resistance to oxidation and sulfidation; turbine/jet alloy
Nickel–chromium–iron	N08800	32.5 Ni, 46 Fe, 21 Cr, 0.4 Cu (Incoloy 800)[a]	Strong, resistant to oxidation, carburization, and sulfidation ASTM B 407 (pipe), B 408 (bar), B 409 (plate)
	N08825	42 Ni, 21.5 Cr, 30 Fe, 3 Mo, 2.2 Cu (Incoloy 825)[a]	Resistant to a wide variety of corrosive chemicals (reducing and oxidizing) ASTM B 423 (pipe), B 424 (sheet), B 425 (bar)
Nickel–molybdenum	N10665	65 Ni, 28 Mo, 2 Fe (Hastelloy B-2)[b]	Resistant to hydrochloric acid and other reducing acids ASTM B 333 (plate), B 335 (bar)
Nickel–chromium–molybdenum	N06455	62 Ni, 16 Cr, 3 Fe, 16 Mo (Hastelloy C-4)[b]	Resistant to oxidizing and reducing acids, high-temperature stability ASTM B 574 (bar), B 575 (plate), B 619 (pipe), B 622 (tube)
	N10276	48 Ni, 16 Cr, 16 Mo, 6 Fe, 4 W (Hastelloy C-276)[b]	Resistant to chlorine-contaminated hydrocarbons, oxidizing, and reducing acids ASTM B 366 (fittings), B 564 (forgings), B 575 (plate)

[a] Registered trade name of Huntington alloys.
[b] Registered trade name of Haynes International.

galling resistance are important. Figure 18–2 illustrates the use of 3% silicon nickel for a large nut that must resist galling versus soft stainless steel. Illium®, a nickel–chromium–molybdenum–copper alloy and Waukesha® alloys are also used for these types of applications.

In the latter half of the 1980s, nickel aluminide *intermetallics* have come under significant study and have entered commercialization. Many metal alloys form intermetallic compounds in various temperature and composition ranges. Nickel forms an *intermetallic compound* with aluminum, and this material was normally considered to be useless as an engineering material because of its brittleness. It was determined that this brittleness problem could be overcome by the addition of small percentages of boron (about 200 ppm). This discovery has allowed nickel aluminides to be produced that have useful engineering properties; in fact, they have exceptional high-temperature properties.

One intermetallic that has been commercialized is Ni_3Al (76% Ni, 24% Al). This intermetallic has a FCC structure like nickel, and the aluminum atoms take up sites on the cube corners. Having specific sites for an alloy additive produced what is called an *ordered structure*. The unique feature of this alloy is that its yield strength doubles (to 130 ksi; 900 MPa) as the temperature is raised to about 1500°F (800°C). Most metal alloys get weaker as the temperature is raised (Figure 18–1). Nickel aluminide does not have to be heat treated to develop high strength. It will take some time for this material to become commonplace, but since 1990 it has been under evaluation by the aircraft and aerospace industries. Nickel aluminide castings are commercially available.

As mentioned, the largest use of nickel is as an alloying element in stainless steel, steel, coppers, and other metals. The use of nickel-based alloys is limited to special applications, with corrosion and elevated temperature resistance as primary application areas. The data that we have supplied may be inadequate for application to specific parts. Our intention is to let the designer know that there are nickel-based alloys available for corrosion problems when stainless steels and other alloys fail. Similarly, there are nickel-based alloys that have special properties useful in sophisticated magnetic and electrical devices.

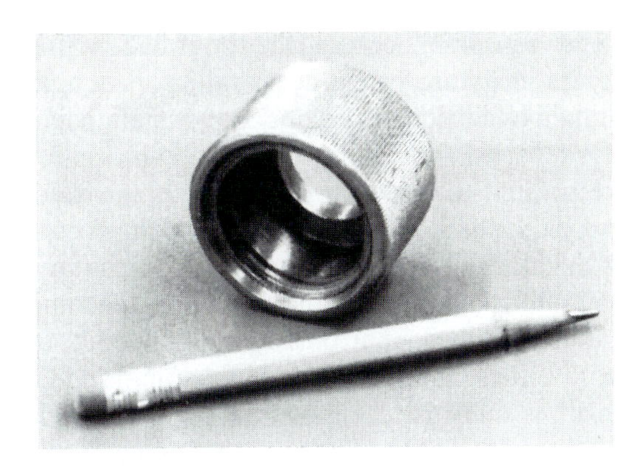

Figure 18–2
Silicon–nickel nut

The use of nickel alloys for a particular application may require more study of property information than we have presented here; this is especially true for corrosion and elevated-temperature applications. The point that we have tried to make with the limited information that we presented is that nickel-based alloys are the normal materials used for tough corrosion problems and for superalloy applications such as rotor vanes in jet engines. If you are faced with a severe service application, it may be that there is a nickel-based alloy that can fit the need. These alloys are specified with ASTM standards and identified with UNS numbers. Nickel alloy manufacturers can supply a complete listing of available alloys along with property information.

18.2 Zinc

Zinc was used for centuries before it was officially "discovered." Zinc ores were combined with copper ores and refined to produce the alloy that we know today as brass. As a pure metal, zinc was identified in about 1750. Zinc occurs in nature primarily in sulfide ores that contain a few percent of zinc. Beneficiation can be similar to that for copper: The sulfide ores are roasted to convert the sulfide to an oxide, and the oxide is reduced with carbon from coal to yield the metal and CO or CO_2. Because zinc ores often contain other metals such as lead and cadmium, subsequent processing is needed to remove these metals. Fractional distillation in a retort is one process that is used, but this is being replaced with techniques involving chemical leaching of the zinc from the ores with acids. The dissolved zinc is then plated out on electrodes.

Zinc has a hexagonal close-packed structure (*HCP*). It has limited solid solubility with other elements, and only "pure" zincs have single-phase structures. Melting temperatures for zinc alloys range from about 700 to 900°F (370 to 482°C). Pure zinc has a specific gravity of 7.14, compared with 7.85 for steel. It is a dense material. All zinc alloys are soft (<150 HB), and hardening processes, both thermal and mechanical, are usually not applied. Alloys are usually used as cast or wrought.

Pure zinc and high-zinc alloys (over 98% Zn) can be cold rolled, drawn into wire, and extruded. The pure zinc alloys usually contain a fraction of a percent of copper or titanium. Wrought zincs (ASTM B 69 lists seven alloys) are used for jar covers, engraving plates, architectural panels, and miscellaneous applications usually involving corrosion resistance. A number of superplastic wrought zinc alloys contain from 20% to 24% aluminum. Strips and sheets of these alloys can be "warm" formed by vacuum forming and similar plastic forming operations. They can have extremely high elongations (100%+).

Because of their low melting point, zinc alloys are adaptable to casting in a wide variety of molds: plaster, metal, graphite, and even silicone rubber. *Die casting* is by far the most important casting process, but alloys are available for gravity casting. Gravity-cast alloys contain from 8% to 28% aluminum and up to 2.5% copper. They have tensile strengths in the range of 30 to 60 ksi (206 to 414 MPa) with fair ductility and impact strength (ASTM B 791 and B 792).

An important aspect of the use of zinc alloys in engineering design is that unlike most other metals they never behave elastically under stress. They creep at all stress levels and temperatures, and there is no definite modulus of elasticity (based on classical definitions). The pseudo modulus value used in calculations is about 3×10^6 psi (20×10^3 MPa). Extensive creep data are available on zinc alloys, and the lack of elastic behavior is dealt with by using creep data to determine deflections with time. Creep of zinc die-cast alloy has been addressed by alloy development. In the 1980s an alloy designated as A Cu Zinc was developed by automotive researchers that has as much as ten times the creep resistance of conventional AG 40A alloy. A Cu Zinc is an aluminum (3%), copper (5.5%), zinc alloy that has improved mechanical properties because of a copper-rich phase in the zinc matrix. The tensile strength can be as high as 59 ksi (407 MPa), and the yield strength is 49 ksi (338 MPa) making it stronger than most zinc alloys.

Commercial zinc alloys are commonly designated by UNS numbers and ASTM specifications. The average designer will most likely use zinc in only two forms: as die-cast parts or as coatings for corrosion control. We shall address these applications in the remainder of this section.

Die Castings

Zinc is alloyed with aluminum, copper, lead, and cadmium, but the solubility of these metals in zinc is extremely low. Their function is to produce strengthening or to improve fabricability. There are a number of commercial die-casting alloys, but the most commonly used alloys are ASTM B 86 grades AG 40A (UNS Z33520) and AG 41A (UNS Z33531) (Table 18–3). Seventy

Table 18–3
Commonly used die-casting alloys

Casting Alloys						
Alloys						
Alloy Group			**Zinc**		**Zinc–Aluminum**	**Aluminum**
		ASTM B 86			ASTM B 791	ASTM B 85
Alloy designation	Gr. AG 40A	Gr. AC 41A	Gr. 40B	Gr. AC 43A	Gr. ZA 8	Gr. 380
UNS number	Z 33520	Z 35531	Z 35523	Z 35541	Z 35635	A 03800
Mechanical Properties						
Tensile strength, psi	41,000	47,600	n.a.	52,100	54,200	47,000
Yield strength, psi (0.2% offset)	—	—	—	—	42,000	23,000
Compressive strength, psi	60,000	87,000	60,000	93,000	>87,000	—
Shear strength, psi	31,000	38,000	31,000	46,000	40,000	27,000
Elongation % inch per inch	10	7	13	8	8	3.5
Hardness, BHN	82	91	80	100	103	80–85
Impact strength, ft-lb (Charpy)	43	48	43	n.a.	31	3
Physical Properties						
Density lb/in.^{3}	0.24	0.24	0.24	0.24	0.227	0.098
Melting range °F	718 728	717 727	718 728	715 734	707 759	1000 1100
Coefficient of thermal expansion μin./in./°F	15.2	15.2	15.2	15.4	12.9	12.1
Thermal conductivity, Btu/h ft^2 °F/ft	65.3	62.9	65.3	60.5	66.3	55.6
Electrical conductivity, % IACS	27	26	27	25	27.7	27

Source: Adapted from *Design News*, June 20, 1988. Reprinted with permission.

percent of all zinc die castings in the United States are made from the AG 40A alloy.

There are maximum limits on elements such as iron, lead, cadmium, and tin (0.1%, 0.005%, 0.004%, and 0.003%, respectively). Concentrations larger than these will have an adverse effect on corrosion resistance. Intergranular attack can occur in many environments, causing the die casting to be extremely weak.

The big advantage of die-cast zinc over the other metals that can be die cast, such as magnesium, aluminum, and copper alloys, is that the casting operation is done at low temperatures (720°F; 382°C). This allows the use of *submerged plunger* machines (Figure 18–3) and gives long *cavity* life. The shot system and cavities in aluminum, magnesium, or copper alloy die-casting systems may need replacement after only

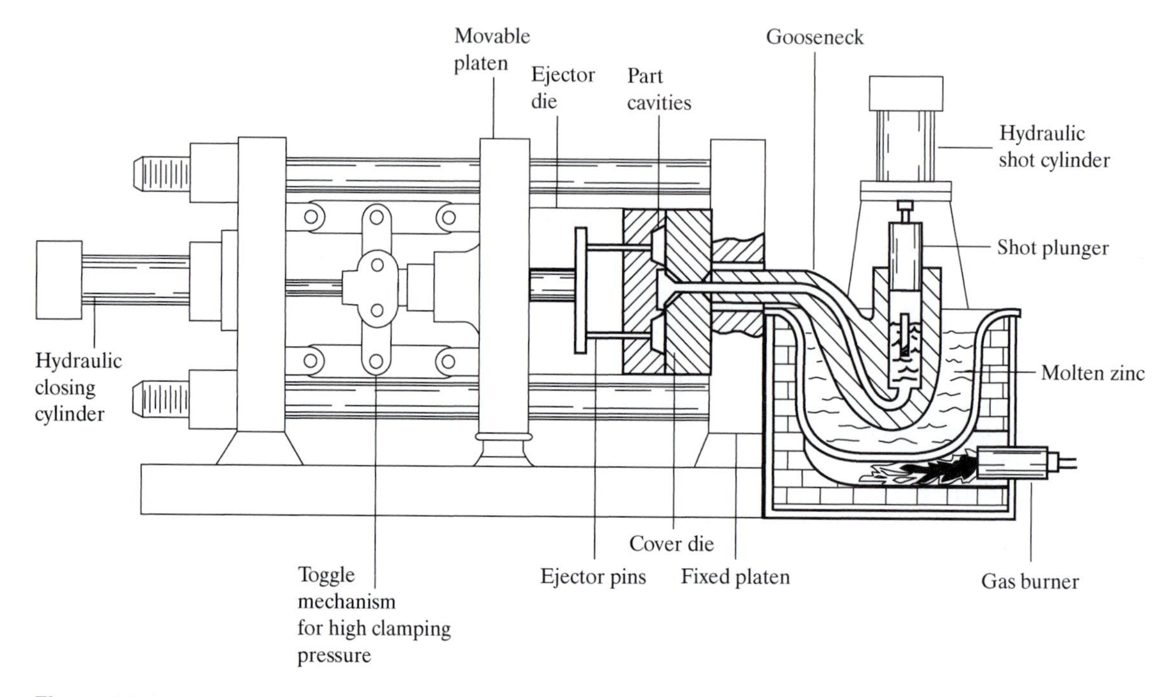

Figure 18–3
Zinc die-casting machine; the shot plunger injects the molten metal, and ejector pins push the part from the mold when it opens

100,000 or so shots due to *thermal fatigue*. Zinc cavities have been used continually for 25 years, making millions of parts per year with no signs of degradation due to thermal fatigue or liquid metal erosion. Zinc die castings can be very inexpensive.

How can zinc die castings be used in design? Die castings are not economical for small production quantities. Tooling can be expensive, but modular cavities are used by many job-shop die casters, often making it reasonable in cost to use this process for part quantities as low as 500 units. Thus the designer should investigate tooling costs before specifying this process. Even without designing for zinc die castings, most machines use a sizable number of zinc die castings in the form of knobs, electrical devices, clamping devices, safety devices, and even hinges.

In summary, zinc die castings are frequently used in design. They are inexpensive and have mechanical properties in many respects superior to those of plastics. They can be used for all types of machine components and as housings, chassis, and the like (Figure 18–4) in consumer goods. The two die-casting alloys mentioned should suffice as a design repertoire. The ASTM B 86 AG 40A alloy is characterized by excellent retention of impact strength and good long-term dimensional stability. The AG 41A alloy has higher strength and better castability than the former alloy. If creep is a design concern, the A Cu Zinc alloy can be considered.

Zinc Coatings

Pure zinc has a corrosion rate of less than 0.0001 in./yr (2.5 μm/yr) in normal urban atmospheres. Uncoated steel will rust at a rate as high as 30 times this value. For this reason, zinc coatings have been used for many years to protect steel

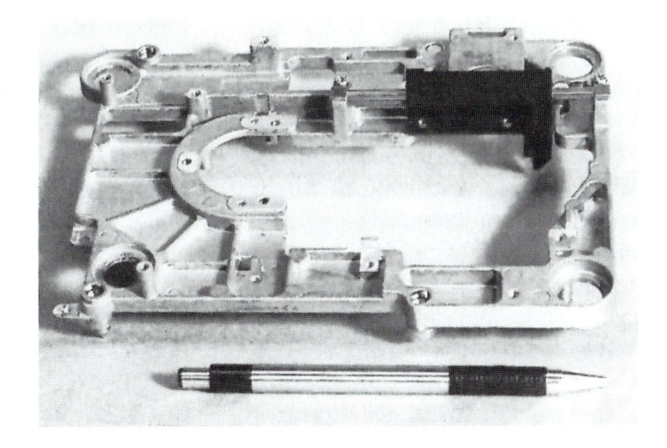

Figure 18–4
Zinc die-cast mechanism chassis. Note exceptional detail, ribbing, cored holes, bosses, and thin wall design for light weight. This is a good casting design.

against atmospheric corrosion. Zinc is put on steel surfaces by hot-dip *galvanizing*, electrodeposition, diffusion coating, metal spraying, and zinc-rich paints. Because zinc has a low corrosion rate, a coating of 0.003 in. (75 μm) may provide protection for as long as 30 years in nonindustrial, noncoastal environments. There is an even more important aspect of zinc as a protective coating. In our discussions of the galvanic series of metals, it was pointed out that zinc is anodic to most metals. When a zinc-coated part has a coating defect, such as a scratch or machined area, the exposed steel will be protected by the nature of the zinc *anodic coating*. It will be corroded in preference to the exposed steel. This is illustrated in Figure 18–5. More *noble coatings* or platings such as copper or nickel will cause the reverse effect. Pitting will

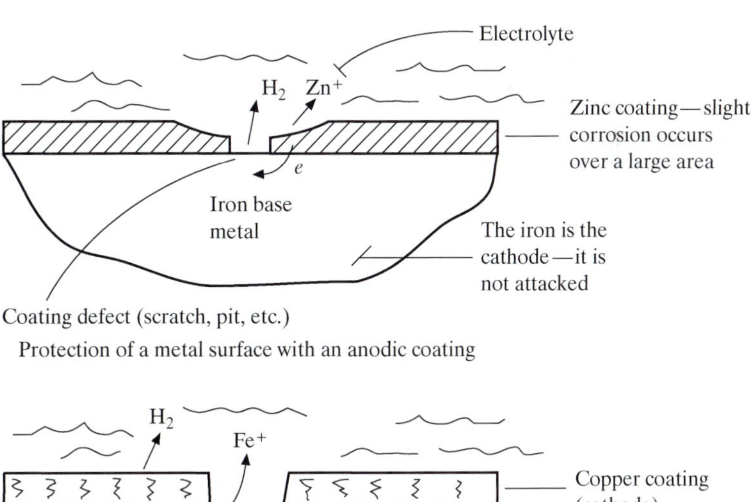

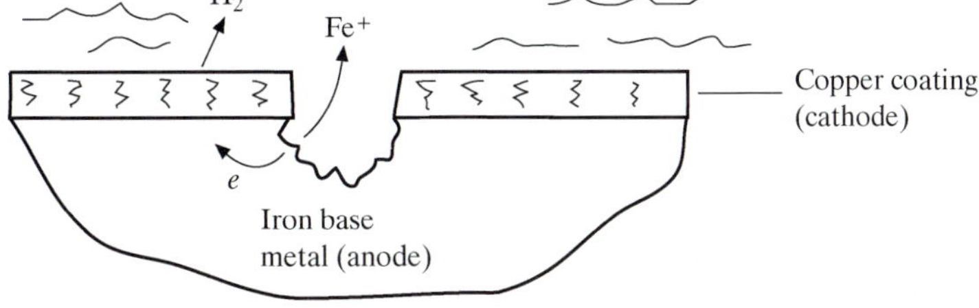

Noble coatings cause pitting at coating defects

Figure 18–5
Corrosion protection with zinc coatings

take place at defects. When there is a large anode area of zinc surrounding a scratch or defect, the corrosion cell action on the zinc will be small. There is a favorable anode–cathode ratio. The sacrificial nature of zinc coatings makes them well suited for applications where scratches and defects are probable.

The chemical resistance of zinc is not particularly good. It is attacked by acids and strong alkalies. The recommended pH range for aqueous solutions is about 7 to 12. It is not affected by most solvents and organic fuels. Immersed in neutral water, zinc has fair resistance to corrosion. In room temperature hard water, the rate of corrosion may be as low as about 0.001 in./yr (25 μm/yr). At one time, galvanized pipe was the standard for potable water pipes in homes.

In designing for zinc coatings, it is common to specify a coating thickness of about 5 mils (125 μm) for the hot-dip process. Galvanized steel mill products are made with varying coating thicknesses. The current practice for specifying zinc thickness on sheet materials is to specify the zinc coating weight per unit area (ounces per square foot). These specifications can get confusing if the coating is on one side, or in dealing with tin mill products. Some guidelines to follow are (1) mill-galvanized sheet usually has about a 1-mil coating (25 μm), and (2) a coating of 1 ounce of zinc per square foot is equivalent to a thickness of 1.7 mils (43 μm).

Hot-dip galvanizing is practical on mill products and on relatively small machine parts. Large structures can be zinc coated with metal spray techniques or zinc-filled paint systems. The paints do not provide the same degree of protection as galvanizing. Many paints do not adhere well to fresh zinc coatings. If zinc is to be used under a paint system, many mills offer heat treated *galvannealed* products. Hot-dip sheet is annealed at a temperature high enough to cause the zinc to convert to a protective zinc oxide. The characteristic galvanize spangle disappears in this process.

Zinc coatings have been with us for a long time, and they continue to be an effective means for control of atmospheric corrosion. Many automobile body panels and structural members are zinc coated prior to painting. Since the 1990s, over 500 lb (227 kg) of zinc-coated steel was used in each vehicle made in the United States. Eighty percent was hot-dip coated; 20% was plated (electrogalvanized). Galvanizing a 55% aluminum–1.5% silicon–zinc alloy (ASTM A 792) allegedly offers better corrosion protection than conventional zinc coating for metal buildings and steel roofing materials. Galfan (ASTM A 875), which is a 95% zinc–5% aluminum alloy, is said to have better corrosion resistance and formability than conventional galvanize and it is intended for parts that require severe forming.

Zinc coatings are traditionally the largest single use of zinc in the United States. More than 50% of the world's production was used in coatings. This trend will probably continue for the foreseeable future. Zinc-coated steel is replacing wood for building framing, and metal roofs are experiencing a resurgence. Zinc coatings have a definite place in engineering materials and design.

18.3 Titanium

Titanium is probably the newest engineering metal. As an element it was discovered in 1791, but it was not produced in metallic form until 1910. It remained a laboratory curiosity until commercial processes were developed for its manufacture in the 1940s.

Titanium is abundant in nature; about 1% of the earth's crust is titanium. Almost all common rocks contain percentages of titanium dioxide, TiO_2. Deposits of pure TiO_2, rutile, are fairly common, and this mineral is one of the principal titanium ores. The technological delay in making titanium metal is due to the fact that this metal is extremely reactive. It readily combines with many other elements to form compounds that are unusable as engineering materials. The Kroll process was developed during World War II to the point at which commercial production of titanium was possible. This process involves converting TiO_2 to titanium tetrachloride, $TiCl_4$, by chemical techniques, and then reducing $TiCl_4$ gas with sodium or molten magnesium in an inert gas retort to metallic titanium. The titanium produced in this step is relatively impure and is further refined by chemical leaching or by vacuum melting techniques. Production processes have been improved significantly, and metal costs are low enough to allow the use of titanium for many special design situations. The price of titanium was as low as $10/lb by the 1980s. In 2003, it was in the range from $8 to $15/lb. New processes for beneficiation developed in the late 1980s continue to promise further reductions in cost.

Physical Properties

As a pure metal, titanium has a melting point higher than that of steel, 3040°F (1671°C); a specific gravity of 4.5; a coefficient of thermal expansion of 5×10^{-6} in./in. °F (10.8×10^{-6} m/m K); a thermal conductivity of 11.5 Btu/h ft^2 °F/ft (20 W/m K); and a tensile modulus of elasticity as high as 18×10^6 psi (12.7×10^4 MPa).

The physical properties of prime importance are its density and modulus. Titanium weighs only about half as much as steel, 0.16 versus 0.28 lb/in.^{3} (4.5 vs. 7.87 g/cm^3). Its mechanical properties can be better than those of many alloy steels, and thus it has a very high specific strength. The same thing is true about stiffness. It has a much higher modulus than the other light metals, magnesium and aluminum.

Titanium is considered by some metallurgists to be a *refractory metal* because it is situated in the periodic table with other metals with high melting points (Figure 18–6). We prefer not to call titanium a refractory metal because it is not usually used for high-temperature applications such as some of the high-melting-point metals (e.g., tungsten and molybdenum). It is considered to be a light metal, and it often competes with low-melting-point metals such as aluminum and magnesium for structural applications in aircraft.

Metallurgy

At room temperature, pure titanium has a hexagonal close-packed crystal structure. This structure is called *alpha-phase titanium*, and it is stable up to about 1620°F (882°C), at which point the structure changes to body-centered cubic (BCC). The BCC structure is called *beta-phase titanium*. Only a few metals have this ability to change crystal structure on heating (allotropic transformation). Recalling our discussions of steels, if steel

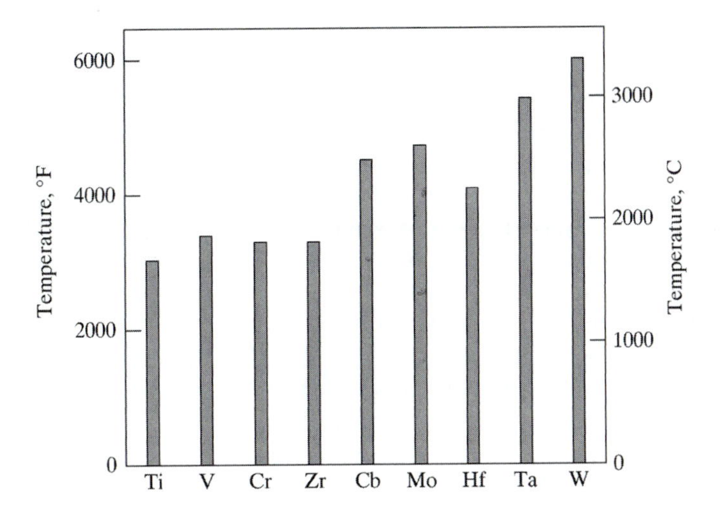

Figure 18–6
Melting points of refractory metals

is heated to above its transformation temperature and quenched, high-strength martensite is formed. This can also be done with titanium, but the strengthening effect is not as pronounced as in steels.

Commercially pure titanium (CP) will not respond to quench hardening. There are several commercially available grades of pure titanium, and they differ in physical and mechanical properties depending on their chemical composition. Elements such as carbon, oxygen, nitrogen, and hydrogen can go into interstitial solid solution in titanium and cause strengthening. The different grades of pure titanium have increasing concentrations of these interstitials, and their strength increases accordingly. The highest strength grade of pure titanium contains about one weight percent of these interstitials.

Titanium has limited solid solubility with other metals, but it has a strong tendency to combine with other metals and to form brittle intermetallic compounds. Titanium will weld nicely with steel, but the fused zone will be as brittle as glass and unusable. This is because the weld becomes an iron–titanium intermetallic compound. Thus titanium cannot be fusion welded to other metals, and alloy additions to pure titanium have to be controlled such that intermetallics are not formed.

As a side note, there are useful applications of titanium intermetallic compounds. P/M techniques have been developed to make titanium aluminides that are similar in properties to the nickel aluminides previously discussed. Two intermetallics that are used for high-temperature applications are TiAl (Ti, 40% Al) and Ti_3Al (Ti, 16% Al). Cast and wrought second- and third-generation titanium aluminides appear to have potential in high-temperature turbine and automotive engine applications.

Alpha grades of titanium are pure titanium that is solid solution strengthened by small additions of elements such as aluminum, tin, nickel, and copper. These alloys are not quench hardenable, but they have higher strengths than CP titanium.

Alpha–beta alloys, as the name implies, are titanium alloys that have a structure that is partially alpha and partially beta. Elements such as molybdenum, vanadium, columbium, and tantalum, when added to pure titanium, tend to promote the room temperature presence of beta phase. An alloy such as Ti–6Al–4V, which contains 6% aluminum and 4% vanadium, has a two-phase structure, about one-half alpha and one-half beta phase. When this alloy is heated to a temperature of about 1725°F (955°C), it transforms completely to beta structure. When it is water quenched to room temperature, part of the beta phase is *metastable*; it wants to transform to alpha, but this is prevented by the water quench. This is called *solution treating*, and the alloy has high strength in this condition, but hardness and strength can be further increased by aging for 4 h at about 1000°F (539°C). In the aging operation, discrete zones of alpha phase precipitate from the metastable beta phase. Thus alpha–beta alloys are considered to be precipitation-hardening alloys. There are various alpha-beta alloys with different strength levels and even different precipitation-hardening mechanisms, but the 6Al–4V alloy is the most important of the group. A lower alloy version (3% Al, 2.5V) of this alloy provides better formability but lower strength (80% of Ti–6Al–4V).

Beta alloys are produced by adding larger amounts of beta stabilizing elements such as Mo and V, to make beta the stable phase at room temperature. Beta alloys have good ductility and formability when they are not heat treated. Some beta alloys can be age hardened to cause precipitation of alpha phase or intermetallic compounds. This can produce very high strengths, but ductility and toughness are reduced. A common beta alloy has a composition of 13% vanadium, 11% chromium, and 3% aluminum.

As a family, titanium alloys can be essentially pure titanium, solid solution strengthened alpha phase, alpha–beta dual phase, beta phase, or transition structures called near alpha and near beta. The CP and alpha alloys cannot be age hardened; many of the other alloys can. About 70% of all titanium used is the 6Al–4V alpha–beta alloy. Most of the remaining 30% of titanium production is made up of the four grades of commercially pure titanium (CP). There are more than 40 different grades of titanium in commercial use, but the CP and 6Al–4V grades are the most available and the most useful.

Mill Products

Titanium and its alloys are manufactured in bar, wire, plate, sheet, strip, tubing, extrusions, and in cast or forged shapes. Availability of all these forms depends on the alloy, but U.S. warehouses usually handle only a few alloys, and these are predominately in sheet, plate, or bar form.

Four grades of pure titanium and about 40 alloys are available. The industry-accepted practice for specifying titanium is to use the elemental symbols for the major alloying elements and digits to indicate their concentration in the alloy:

Ti-6Al–4V = 6% Al, 4% V, Ti remainder

Ti-8Al–1Mo–1V = 8% Al, 1% Mo, 1% V, Ti remainder

This system is acceptable, but it does not ensure accurate control to these nominal analyses and does not cover the various grades of pure titanium. It is preferred to use designation numbers of government and ASTM specifications. ASTM specification B 265 covers sheet, strip, and plate forms of the more common titanium alloys (see Table 18–4).

These ASTM designations are recommended on engineering drawings unless government specifications (MIL, AMS, and others) apply to the parts under design:

Material: ASTM B 265 Grade 5 (Ti-6Al-4V)

Table 18–4
Titanium alloy specifications

Alloy	ASTM Designation
Commercially pure	ASTM B 265* Grade 1 B 265 Grade 2 B 265 Grade 3 B 265 Grade 4
Ti–6Al–4V	ASTM B 265 Grade 5
Ti–5Al–2.5Sn	ASTM B 265 Grade 6
Ti–0.15 Pd	ASTM B 265 Grade 7
Ti–0.15 Pd	ASTM B 265 Grade 11
Ti–0.3Mo–0.8Ni	ASTM B 265 Grade 12
Other Specifications	
ASTM B 299	Titanium sponge
ASTM B 337	Seamless and welded pipe
ASTM B 338	Seamless and welded tube
ASTM B 348	Bars and billets
ASTM B 363	Welding fittings
ASTM B 367	Titanium castings
ASTM B 381	Titanium forgings
ASTM B 600	Cleaning and descaling
ASTM F 67	Surgical implants (pure Ti)
ASTM F 136	Surgical implants (6Al–4V ELI)

*ASTM B 265 covers 16 grades of titanium alloy sheet, strip, and plate.

Corrosion Resistance

Titanium and its alloys have excellent corrosion resistance to seawater and aqueous chloride solutions over a wide range of temperature and concentration. Most alloys are resistant to a wide range of oxidizing media such as nitric acid. Reducing acids such as hydrochloric and sulfuric will corrode most alloys, but alloying pure titanium with noble metals such as palladium will extend its resistance to mildly reducing acids.

Titanium is not toxic, and it is often used for corrosion problems in handling foods.

Because titanium is noble to most other metals in the galvanic series, rapid attack of the other metal will occur in metal-to-metal couples immersed in an electrolyte. Coupling titanium with other metals should be avoided if immersion in an electrolyte is a service condition.

Titanium is corroded by some environments that austenitic stainless steels are resistant to, but in many cases titanium is more corrosion resistant. Its corrosion characteristics are the reason why it is seeing wide application in the chemical process industry.

Fabrication

The lowest–strength grades of titanium have fair formability, but the high-strength grades and alloys require the use of bend radii as great as five times the thickness. Many times, forming must be done in the temperature range from 400 to 600°F (204 to 315°C). The low-strength grades of pure titanium have annealed hardnesses in the range of soft steels, but the alloy grades have hardnesses typically greater than 30 HRC, and machining can be troublesome. Turning all hardnesses is easy, but milling and drilling operations are more difficult. It is difficult to drill and tap small holes in alloy grades. It is best to limit threads to 65% full thread. Operations that produce titanium fines must be controlled. They can be *incendiary*. Flood cooling is recommended for grinding and abrasive cutoff operations.

Titanium and its alloys cannot be welded in air, and they cannot be welded to other metals. Heating in air causes formation of a brittle titanium–oxygen compound, and mechanical properties are affected. Welding must be done in a vacuum or inert gas. Welding to other metals produces brittle compounds, and cracking results. Pure titanium welds satisfactorily using shielding procedures. The 6% aluminum, 4% vanadium alloy has fair weldability, and other alloys range from excellent weldability to unweldable. Gas tungsten arc welding in an inert gas chamber is the usual technique applied.

The heat treatments that apply to titanium are annealing, stress relieving, solution treating, and aging. These treatments are performed for the usual reasons. The annealing temperature range is about 1300 to 1500°F (704 to 815°C); stress relieving is done at 1000 to 1200°F (528° to 648°C); and solution treating is done at about 1750°F (953°C) followed by a rapid quench (6 s maximum). Age hardening usually involves a soak of a few hours at about 1000°F (538°C). Scale from heat treat operations is somewhat difficult to remove because nitric–hydrofluoric pickle baths will attack titanium. Molten salts are the most effective descalers.

Alloy Selection

Solution-treated and aged titanium, 1% aluminum, 8% vanadium, 5% iron, can have a tensile strength of 210 ksi (1448 MPa). Dividing this value by the density will produce a *specific strength* greater than all other metal alloys. Titanium alloys are usually used for only two reasons: (1) their specific strength and (2) their corrosion resistance. We have already discussed their corrosion resistance. Alloy selection for corrosion applications should be done with the assistance of detailed corrosion data.

Alloy selection for mechanical properties can be confusing because of the number of alloys, but this number can be reduced because many alloys are not available unless purchased in large mill quantities. The average designer can subsist on five alloys (see Table 18–5). Grades 1 and 2 are the best to use when formability is an important consideration. The higher-strength grades are used when formability is not important. The pure grades are the best to use for corrosion applications unless the environment calls for an alloy grade. The higher-strength grades of pure titanium and the 6% aluminum, 4% vanadium alloy (grade 5) should be used where strength is needed.

Table 18–5
Properties of titanium alloys

		Typical Properties[a]			
Alloy (ASTM B 265)	UNS	Tensile Strength, ksi (MPa)	Yield Strength, ksi (MPa)	Percent Elongation %	Hardness
Grade 1	R50250	40 (276)	30 (207)	25	70 HRB
Grade 2	R50400	50 (345)	40 (276)	22	80 HRB
Grade 3	R50550	65 (448)	55 (379)	20	90 HRB
Grade 4	R50700	80 (552)	70 (483)	15	100 HRB
Grade 5	R56400	130 (896)	120 (827)	10	35 HRC
Grade 5[b]	R56400	160 (1103)	150 (1034)	8	37 HRC

[a] At room temperature.
[b] Solution treated and age hardened.

Titanium has a good high-temperature strength compared with the other light metals, aluminum and magnesium. The latter should not be used over about 250°F (121°C), but titanium can be used up to about 800°F (426°C) in air (Figure 18–7). Higher temperatures lead to oxygen embrittlement.

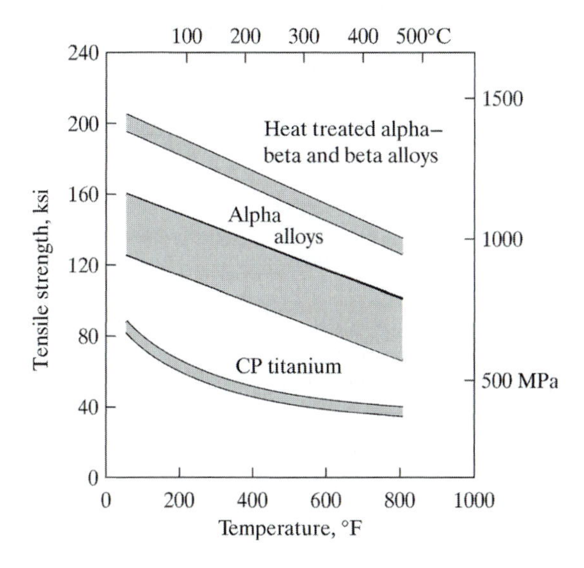

Figure 18–7
Tensile strength of titanium alloys at elevated temperatures

All titanium alloys have poor wear characteristics. This statement includes abrasion, metal-to-metal wear, and solid-particle erosion. They are very resistant to liquid erosion and cavitation. In fact, titanium is frequently used for ultrasonic devices that operate in corrosive liquids.

Applications for titanium abound. In the area of corrosion, it can be used for tanks, piping systems, and ductwork. Even in the cost-conscious automotive industry, titanium is used for automatic soldering machines and for plating racks and fixtures. Titanium and its alloys are widely used in aircraft for their strength and fatigue resistance, and on industrial machines they can be used wherever mass effects must be reduced and high strength is needed: frames for flying shears, high-speed rolls, quick-acting latches, pump shafting, clutches, high-temperature springs, torsion bars, and the like. Titanium is no longer an exotic metal. It can be used by the average designer.

18.4 Magnesium

Some 2% of the earth's crust is made up of the element magnesium. It is present in significant quantities in seawater and in many common rocks, such as limestone and shale. It was identified as an element in 1755, but it was not commercially produced in the United States until about 1920. The ore for producing magnesium can be magnesium chloride from seawater or salt mines, or it can be rocks containing complex magnesium compounds. A number of commercial processes are in use for making magnesium. The electrolysis process essentially involves plating of magnesium out of a molten bath of magnesium chloride. Pure magnesium forms at the cathode, and chlorine gas is evolved at the anode. The thermal processes heat ore (Dolomite) in retorts and magnesium is recovered as a condensed gas.

Physical Properties

Magnesium has the distinction of being the lightest engineering metal (excluding exotic metals such as Be and Li), with a specific gravity of 1.738. It also has one of the highest coefficients of thermal expansion, 14×10^{-6} in./in. °F (25×10^{-6} m/m K), and fairly high electrical and thermal conductivity [~31% IACS and 79 Btu/h ft^2 °F/ft (137 W/m K)]. The modulus of elasticity of most alloys is about 6.5×10^6 psi (4.5×10^4 MPa), and it is said to have high damping capacity. The melting point is 1202°F (650°C). It has a white appearance.

Metallurgy

As a pure metal, magnesium is very weak; the tensile strength can be as low as 10 ksi (69.7 MPa). For this reason, engineering uses of magnesium usually involve an alloy. The primary alloying elements are aluminum, manganese, and zinc. Less common elements such as thorium, zirconium, and rare earths are also used. These elements all have relatively low solubility. Aluminum, manganese, and zinc all have decreasing solubilities with decreasing temperature. Thus age hardening can be produced as in aluminum, copper, and other nonferrous systems. Aluminum is probably the most important alloying element. It has a solubility of 12.7% at the eutectic temperature, and only about 2% at room temperature. Small concentrations yield solid solution strengthening, and high concentrations lead to the ability to age harden.

Magnesium has a close-packed hexagonal (HCP) crystal structure. The significance of this from the user's standpoint is that in this type of structure, dislocations can move on fewer planes than they can in, for example, a metal like copper, which has a face-centered cubic (FCC) structure. This in turn means less ability to plastically deform by dislocation movement. The manifestation is poor ductility. This material is difficult to roll and to form. In fact, cold finishing is extremely limited. If a formed shape is desired in sheet, it will have to be hot formed. A simple 90° bend with even a bend radius of two times the material thickness is usually not possible; it cracks. Poor formability is a limitation of magnesium for some applications.

The addition of zinc in small percentages (<1%) improves castability and corrosion resistance. Manganese aids corrosion resistance. The rare earths and other elements are used in small concentrations for solid solution strengthening and to improve elevated-temperature properties.

Magnesium Products

More than half the magnesium produced in this country is used in powder and ingot form for alloying with aluminum. About 30% is used for chemical processes and other nonstructural applications. The largest part of the remainder is used in the aircraft, automotive, and durable goods industries in all the common forms: castings of all types, wrought sheet, plate, strip, bar, wire, extrusions, and forgings.

The most widely used alloy designation system in the United States is that of ASTM International. There are specifications covering sheet and plate (B 90), forgings (B 91), ingot (B 92), and castings (B 93, B 94, B 199, B 403). The alloy designation system uses numbers and letters. The first two letters indicate the major alloying elements; the second part of the designation indicates the nominal amounts of the principal alloying elements; the third part is a letter from A to Z (except I and O) to indicate a different alloy with the same nominal composition. Suffixes like those used in the aluminum system follow the alloy designation to indicate condition and heat treatments:

AZ	63
Aluminum and zinc are principal alloying elements.	6% Al 3% Zn
A	**T6**
First alloy with this composition	Solution treated and age hardened

Some additional code letters are the following:

B	Bi
C	Cu
D	Cd
E	rare earth
F	Fe
H	Th
K	Zr
L	Li
M	M
N	Ni
P	Pb
Q	Ag
R	Cr
S	Si
T	Sn

A second alloy with a composition similar to the example cited would have a B after the number: AZ63B. Typical suffixes are the following:

F	As fabricated
0	Annealed
H1X	(12 = ½ hard, 14 = ¼ hard, 16 = ¾ hard, 18 = full hard)
H2X	(22, 24, 26, and 28, strain hardened and partially annealed)
T4	Solution heat treated
T5	Artificially aged only
T6	Solution heat treated and artificially aged

Selection

Table 18–6 lists some of the more commonly used magnesium alloys and their properties. As can be seen from these data, magnesium alloys have respectable mechanical properties. Tensile strengths are comparable to many aluminum

Table 18–6
Magnesium alloys

Alloy	UNS Number	Form	Nominal Composition (%)	Typical Properties: Tensile Strength, ksi (MPa)	Typical Properties: Yield Strength, ksi (MPa)	Typical Properties: Percent Elongation %
AZ 31B-H24	M11311	Sheet and plate	3 Al, 1 Zn	39 (290)	20 (220)	7
AZ 91A-F	M11911	Die castings	9 Al, 0.7 Zn	39 (227)	24 (152)	5
AZ 91C-T6	M11914	Sand castings	9 Al, 0.7 Zn	34 (276)	10 (159)	3
AZ 31B-F	M11311	Extrusions	3 Al, 1 Zn	35 (262)	22 (200)	8

alloys, but the ductility is lower. In sheet forming, the material must be heated to 400 to 600°F (204 to 315°C) to get adequate formability.

Another factor that limits the application of magnesium is that aluminum has a higher specific strength than magnesium. If the tensile strength of the strongest magnesium alloy, 50 ksi, is divided by its density of 0.065, a specific strength ratio of 770 is obtained. Compare this with 7075-T6 aluminum, which has the ratio 830. Doing this same thing with the modulus of elasticity, the ratios are 124 for magnesium and 118 for aluminum. Thus it does have a slightly higher *specific stiffness* than most other metals including steel, but usually this does not justify the cost of magnesium compared with aluminum. It costs about twice as much per unit weight.

Magnesium does show an advantage over other metal systems in weight reduction when section size is limited. A high-strength aluminum can have a higher specific strength than magnesium, but the high-strength properties cannot be put to use on long, slender parts. Minimum sections are required to prevent buckling.

Another redeeming property of magnesium in addition to its light weight is its machinability. It has better machinability than any metal, reportedly 5 to 20 times better than the machinability of B1112, a free-machining steel.

The atmospheric corrosion resistance of magnesium is not as good as that of aluminum, but it is better than that of steel, and outdoor exposure is quite common. Seacoast atmospheres or contact with salts, however, can cause rapid pitting. Magnesium automobile rims can be destroyed in one northern U.S. winter if salt is used on the roads. The use of high-purity AZ91D has reduced the propensity for salt corrosion, but coatings are still sometimes necessary to prevent road salt corrosion. Magnesium is attacked by most acids except hydrofluoric. It is fairly resistant to caustics and many solvents and fuels. Chemical conversion coatings can be used to improve corrosion resistance.

Magnesium is the most active engineering metal in the galvanic series, so it should not be coupled with other metals unless the intent is to use the magnesium as a sacrificial anode, as is done in cathodic protection of buried pipes and water tanks.

The elevated-temperature properties of some magnesium alloys are better than those for aluminum. Use temperatures as high as 700°F (371°C) have been applied. Precautions are necessary in heating magnesium to very high temperatures. It is incendiary. Magnesium filings, powder, or chips, if ignited, will burn with extreme heat, and the fire cannot be put out with water. Special techniques are required. Flood coolant must be used in machining and grinding operations to prevent ignition.

In spite of some of the less favorable aspects of magnesium, it is a material that should be considered for use where low *inertia* or weight reductions are required. The four alloys previously listed will suffice as a repertoire for the infrequent user. They are the most frequently used alloys in each product category. Magnesium's machinability is excellent, its formability somewhat limited, but it can be readily welded with the processes that use protective gases.

The popularity of hand-held electronic devices in the 1990s has spurred interest in magnesium die casting. Laptop computers, pagers, cell phones, cameras, and similar devices are often made from magnesium castings, mostly die castings. Magnesium usually has much better mechanical properties than plastics: It is lightweight and it provides effective electromagnetic radiation shielding. A die casting improvement that is contributing to the expanded use of magnesium die casting is the use of thixotropic molding. Instead of injection molten metal, machines have been developed to inject the metal in the "slush" state, typically 60% solid, 40% liquid. The injection properties of the slush are more like those of a plastic, and the molding machines resemble plastic injection molding machines. The slush flows are less turbulent than molten metal, and castings have lower porosity. It is also claimed that wall thicknesses as thin as 1 mm are possible, Most of the normal die-casting alloys are applicable. In summary, magnesium is a useful metal to have in your design repertoire. It can provide weight savings in many applications, and often it can do a better job than plastics for housings for electronic devices.

18.5 Refractory Metals

The term *refractory* means resistant to heat. *Refractory metals* are a group of heat-resistant metals that have melting points over 3000°F (1650°C) (see Figure 18–6).

Some of the precious metals such as platinum, rhodium, rhenium, and osmium have melting points over 3000°F (1650°C), but they are considered to be precious metals first and are not from appropriate periods in the periodic table. From the commercial standpoint, the most widely used refractory or heat-resisting metals are zirconium, molybdenum, tungsten, and tantalum. Tungsten and molybdenum are often used for their high continuous-use temperatures, 5070°F (2800°C) and 3450°F (1900°C), respectively. A substantial amount of niobium is used as an alloy addition to other metals: steels, copper alloys, molybdenum, titanium, and aluminum. As a "pure" metal it is used for its chemical resistance and its neutron transparency in reactors. ASTM specifications are used to designate these alloys: ASTM B 392 (bar), B 393 (sheet), and B 394 (tubes). Hafnium has limited industrial use. It is an impurity in zirconium. It does not have neutron transparency, and it must be removed from niobium or zirconium if they are to be used in nuclear reactors. There are two ASTM specifications on hafnium alloys: B 737 (rod and wire), and B 776 (sheet, strip, plate).

Zirconium

Zirconium was discovered in the late eighteenth century and was isolated during the 1820s. Van Arkel & DeBoer produced purified zirconium metal in the 1920s using the iodide decomposition process. In the 1940s magnesium reduction of zirconium tetrachloride was perfected as a means of producing highly purified zirconium.

Production of zirconium begins with zircon-containing beach sand, which is separated by standard ore-dressing techniques and chlorinated to produce zirconium tetrachloride, which in turn is reduced to zirconium metal with magnesium. Because the source material for zirconium also contains hafnium, special hafnium-removal processing is required for zirconium that is to be used in nuclear applications.

The greatest use of zirconium today is in water-cooled nuclear reactors. The next largest use is for chemical-processing equipment, followed by incendiary ordnance and gettering applications. Zirconium is available in mill products covered by a number of ASTM specifications: B 495 (ingots), B 523 (tubes), B 550 (bar, wire), B 551 (flat rolled products), B 653 (fittings), B 658 (pipe), and B 752 (castings).

Properties Zirconium is a ductile, fabricable metal. It can be machined, formed and welded. Density is 0.234 lb/in.^{3} (approximately 17% less than steel). Thermal conductivity of zirconium is approximately one-third that of steel. Zirconium has a low coefficient of thermal expansion (~½ that of carbon steel), and a low modulus of elasticity (also ~½ that of carbon steel).

Like titanium, zirconium has a hexagonal close-packed crystal structure (called *alpha*) at room temperature, which undergoes a reversible allotropic transformation to body-centered cubic structure (called *beta*) at ~1600°F (~870°C). Small amounts of impurities, notably oxygen, strongly affect transformation temperature. Alpha-stabilizing elements (e.g., Al, Sn, Hf, N, and O) raise the alpha-to-beta transformation temperature. Beta-stabilizing elements (e.g., Fe, Cr, Ni, Mo, and Nb) depress the alpha-to-beta transformation temperature.

The most common zirconium alloys—Zircalloy-2 and Zircalloy-4—are used in nuclear reactor applications, taking advantage of their excellent corrosion resistance to high-temperature steam and superheated water and their ability to transmit neutrons. The Zr–2.5 Nb alloy—as well as Hf-free, unalloyed Zr—also are used in nuclear applications.

For industrial applications, unalloyed zirconium is the preferred composition. It has fairly high strength: yield strength = 30 to 50 ksi (207 to 345 MPa); tensile strength = 55 to 80 ksi (380 to 550 MPa). Zirconium, like titanium, is a reactive metal, thus requiring welding to be done in a dry box or under localized inert gas shielding.

Zirconium has excellent corrosion resistance to acids, alkalies, organic compounds, and salt solutions. Specifically, it has excellent resistance to HCl [≤40%, ≤212°F (100°C)], HNO_3 [≤95%, ≤400°F (204°C)], and H_2SO_4 (≤70%, ≤boiling point). Welds and weld heat–affected zones are somewhat less resistant to attack in H_2SO_4 than is the parent metal. Zirconium has excellent resistance to natural seawater, chloride salt solutions, and brackish waters (provided no ferric or cupric ions are present).

Zirconium is not resistant to hydrofluoric acid at any concentration. Stress corrosion cracking of zirconium occurs in methanolic HCl, mixtures of methanol and iodine, solutions containing ferric or cupric ions, concentrated HNO_3, and liquid mercury.

Zirconium process equipment (columns, reactors, heat exchangers, scrubbers, mixers, and the like) commonly are produced in both solid and clad construction. Zirconium is approved for use in construction under the ASME Boiler & Pressure Vessel Code.

Molybdenum

The physical properties that make molybdenum useful in design are a low coefficient of thermal expansion, 2.8×10^{-6} in./in. °F (5.1×10^{-6} cm/cm °C), about half that of steel alloys; good thermal and electrical conductivity, 30% that of copper (IACS); stiffness greater than steel, $E = 46 \times 10^6$ psi (317×10^3 MPa); and a high melting point that provides strength at elevated temperatures. Molybdenum is heavier than steels. Its specific gravity is 10.2 compared with 7.9 for steel, but this property is usually not of particular value in most applications. About 90% of the production of molybdenum in the United States is used in steel mills, and the like, for alloy additions. Five percent is used in chemical manufacture, and metal mill products make up the remainder.

Molybdenum is available in most of the common wrought shapes: wire, tube, bar, sheet, forgings, castings, and P/M forms. Pure molybdenum has good strength and ductility. The tensile strength can be as high as 100 ksi (689 MPa) at room temperature, and it can be formed and machined with conventional tools. The hardness is in the range from 150 to 185 HB. There are a few molybdenum–tungsten alloys, but the only alloy of commercial importance in the United States is *TZM*, which contains 0.5% titanium, 0.07% zirconium, and 0.03% carbon. The strength of TZM is as much as 5% greater than pure molybdenum at most temperatures.

The high-temperature strength of molybdenum is responsible for most of its industrial applications. As shown in Figure 18–8, TZM retains good tensile properties at temperatures

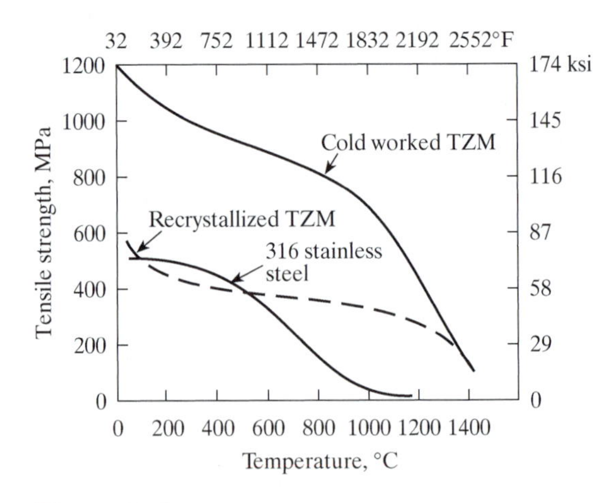

Figure 18–8
Short-time tensile strength of TZM compared with type 316 stainless steel

in excess of 2000°F (1093°C). No steel or nickel-based superalloy can compare with molybdenum. The heretofore unmentioned problem with this material is that it cannot be used at these temperatures in air. If molybdenum is used at temperatures over 1400°F (760°C) in air, it rapidly oxidizes and the oxide sublimes; it suffers direct attack. Thus molybdenum and TZM are used for high-temperature applications, but only in vacuum or inert atmospheres, or when protected by oxidation-resistant coatings.

In industry, molybdenum is used for all sorts of lamp parts and electrical tube components. It is used for vacuum evaporation boats, thermocouple sheaths, and radiation shields. It is extremely important in vacuum heat treating furnaces. It is used for shields, grates, and structural components in the furnaces.

In large shapes, TZM is used for zinc, aluminum, and brass die-casting cavities. It resists metal erosion better than steels do. It has wide application for molds in glass manufacture. It is also used for extrusion and forging dies. It is even used as a wire spray welding buildup material. Thus the applications of molybdenum are widespread, and the average machine designer may have occasion to use it. ASTM B 386 and B 387 specifications can be used to designate molybdenum alloys on drawings. These specifications cover six grades of "pure" molybdenum as well as the TZM alloy.

Tungsten

As a pure metal, tungsten (also known as *wolfram*) has several unique physical properties. It is the heaviest engineering material (Figure 18–9), it has the highest melting point of all metals, and it has the highest modulus of elasticity of all metals [59×10^6 psi (40.6×10^4 MPa)]. In the area of mechanical properties, it has the distinction of being the hardest pure metal. Annealed, the hardness can be as high as 400 HB. Tungsten has high temperature strength characteristics that make it suitable for furnace applications, but because it is expensive and has poor fabricability, molybdenum is a preferred material.

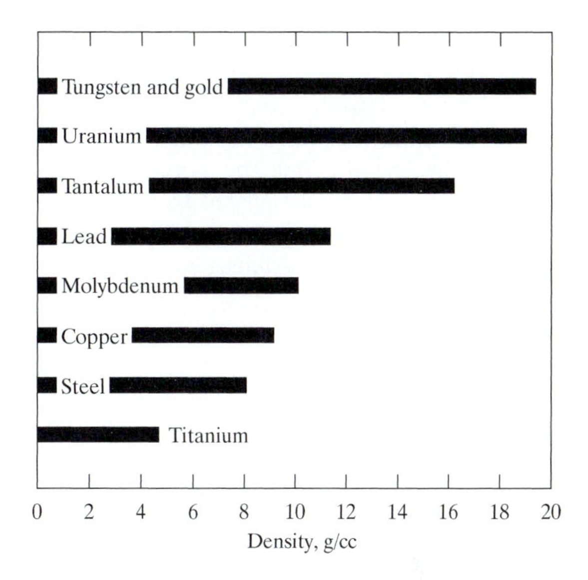

Figure 18–9
Densities of various metals

About two-thirds of the U.S. production of tungsten is used to make tungsten carbide for use in cemented carbide manufacture. About 15% is used for alloy additions, 15% is used for tungsten mill products, and the remaining fraction is used for miscellaneous applications.

Everyone knows of the importance of tungsten as a filament material in light bulbs, but in industrial applications tungsten is important as inert gas welding electrodes and as an electrode material for electrical discharge machining. An increasingly important application of tungsten is its use for high-inertia devices and for balancing masses. Because of its high density, a flywheel or balance weight made from tungsten will take up only one-third the space required for a steel component. This often solves size limitation problems. The use of tungsten for inertial devices has been made easier by the development of tungsten–nickel–copper alloys. These alloys are tungsten with up to 10% copper, nickel, or combinations thereof. They are much more machinable than pure tungsten, and the specific gravity is still about 17. Thus there are some applications of tungsten for relatively ordinary machine design situations. Tungsten plate, sheet, and foil can be specified by ASTM B 760. The more machinable grades (four) can be designated by referring to ASTM B 777; class 1 alloy contains 90% tungsten; class 4 alloy contains 97% tungsten.

Tantalum

Tantalum has a high density and melting point like tungsten, but its major use is in the electronics industry for capacitors (foil and wire). There is a significant industrial use based on the excellent corrosion resistance of tantalum and its alloys. Emerging military applications involve high-density tantalum and tantalum alloys in ballistic armor penetration. Tantalum mill products are covered by ASTM specifications B 364 (ingots), B 365 (rod and wire), B 521 (tube), and B 708 (flat rolled products).

High-purity electron beam or vacuum arc-melted tantalum is highly malleable, but it has relatively low strength [yield strength ~ 20 ksi (138 MPa); tensile strength ~ 30 ksi (207 MPa)]. Strength can be increased by cold working or, more usually, by alloying with tungsten (2.5 weight % or 10 weight %) or with niobium (40 weight %).

Tantalum and its alloys have excellent corrosion resistance, relatively good thermal conductivity, and relatively easy fabricability. This combination makes these materials practical for construction of special chemical vessels, piping, and heat exchangers. As with other reactive metals, welding of tantalum must be done in the absence of air and moisture. This normally is accomplished either in dry boxes or by local shielding with inert gas. Tantalum and tantalum alloys seldom are used for massive metal parts because of their high cost (~$150/lb) and high density. However, use of tube or sheet forms often is cost effective, as is cladding on heavy steel substrates.

Another way to circumvent the high cost of tantalum is to use it as a coating. There are companies in the United States that use the metalliding process (Chapter 13) for plating the surface of parts with a continuous layer of 5 to 15 mils (125 to 375 μm) of tantalum (Metalating®).

Tantalum is inert to almost all organic and inorganic compounds at room temperature. The exceptions are strong alkalies, fuming sulfuric acid, HF, hot oxalic acid, and acids containing fluorides. Tantalum is less chemically resistant at elevated temperatures, but its chemical resistance still is better than that of most stainless steels, titanium, zirconium, and superalloys. The use of a metal for a chemical-resistance application requires consultation of specific corrosion data. This also is true with tantalum. If there is an application that requires better corrosion resistance than stainless steel, tantalum or tantalum alloys should be investigated as possible candidates. Tantalum has almost become standard in the chemical-process industry for patching glass-lined vessels. It behaves similarly to glass in its inertness to many media.

Niobium

Niobium (often called *columbium* in the United States) is a Group V element, as is tantalum. As such, niobium and tantalum have similar thermal expansions and thermal conductivities. Both have a BCC structure and are similarly ductile and malleable. Note that niobium has a significantly lower melting point than tantalum and approximately half the density of tantalum. ASTM specifications B 391 (ingots), B 392 (bar, rod, wire), B 393 (flat rolled products), and B 394 (tubes) cover the available mill products forms.

Niobium is used as an alloying element in specialty steels and in nickel- and cobalt-based alloys. It also has applications in the aerospace and nuclear programs. Niobium, niobium–titanium, and niobium–tin alloys are used as superconductors. Because of its excellent corrosion resistance to liquid metals, niobium is used in sodium vapor lamps for highway lighting.

Niobium provides good corrosion resistance to a wide variety of process environments.

It generally parallels the corrosion resistance of tantalum, although limited to somewhat lower temperatures and concentrations than tantalum. Because of its lower price and density, however, niobium can be a cost-effective substitute for tantalum. Application, fabrication, and welding techniques for niobium are the same as those used for tantalum.

18.6 Cobalt

Cobalt is an essential metal in industry because it plays a key role in cemented carbides, wear-resistant alloys, chemical-resistant alloys, and other important areas like prosthetic devices. It is between iron and nickel in the periodic table, so it has properties that are similar. Cobalt is obtained from a variety of ores in a variety of ways, but it is often a byproduct of other mining operations such as copper, zinc, and nickel. Cobalt has the look and feel of steel. The specific gravity is 8.71 compared to 7.87 for iron. It has about 16% of the conductivity of copper. It is ferromagnetic, and its expansion characteristics, tensile modulus, thermal conductivity, and melting point are similar to iron.

Cobalt is not widely used as a pure metal for structural parts because it is relatively expensive (sometimes >$20.00/lb) and sometimes it is scarce on the worldwide market. It has a hexagonal crystal structure (HCP) below 788°F (417°C) and a cubic structure up to its melting temperature of 2723°F (1495°C). A use that is very important in manufacturing is as a binder in cemented carbides. Cemented carbides are the tool materials of choice for most metalworking operations. They can contain from 3% to 30% cobalt as the binder for tungsten carbide or other carbide particles. Nothing else works quite as well. For many years a cobalt-based alloy called *Vitalium* was used for many types of medical prostheses. It is also widely used in dental appliances and medical tools. It has the necessary in vitro corrosion resistance, durability, and stiffness. The mechanical properties of pure cobalt are similar to steel. It has good ductility and tensile and yield strengths resembling unhardened steel.

Cobalt-based wear- and chemical-resistant alloys are useful in machine design and general use. Stellites® are a family of cobalt–chromium–tungsten alloys that are resistant to a variety of wear modes: abrasion, metal-to-metal, galling, liquid erosion, and cavitation. In addition, they are very chemical resistant—usually better than austenitic stainless steel. Pure cobalt has poor corrosion resistance, but combined with chromium, it becomes very good. There are maybe a dozen or so Stellites® commercially available. Only two, alloys 6B and 6K, are available in wrought form. The others are available as castings and powders for P/M parts. The 6B alloy has a hardness of 43 HRC and the 6K is about 48 HRC as rolled. Heat treating is not necessary, and these alloys do not get soft if heated for brazing and the like. They are metallurgically quite complicated because of the number of alloy additions, but basically, they can have hardnesses from about 30 HRC to 65 HRC depending on alloy and carbon content.

		Normal Chemical Composition (Wt %)									
Stellite® Alloy	UNS	Cr	Ni	Fe	Mo	W	Si	Mn	C	Co	Typical Cast Hardness
6 B/6	R30006	30	2.5	3	1.5	4	0.7	1.4	1	Rem[a]	39–43 HRC
12	R30012	30	3	3	1	8.3	2	1	1.4	Rem	47–51 HRC
1	R30001	31	3	3	1	12.5	2	1	2.4	Rem	50–55 HRC
3		31	3	3	1	12.5	1	1	2.4	Rem[b]	51–58 HRC

[a]Rem = remainder
[b]+0.1 B

These alloys are also available as hardfacing rods for welding deposition. Some important corrosion-resistant cobalt-based alloys are:

Alloy	UNS	Normal Chemical Composition (Wt %)								
		Cr	Ni	Fe	Mo	W	Si	Mn	C	Co
Haynes 25	R30605	20	10	3		15	0.4	1.5	0.1	Rem
Ultimet	R21233	26	9	3	5	2	0.3	0.8	0.06	Rem + (0.08N)
Haynes 188	R30188	22	22	3		14	0.4	1.25	0.1	Rem + (0.03La)

Detailed corrosion data should be consulted for application information (supplier's Web site—Haynes International, Deloro Stellite, Elgiloy, etc.). The low-carbon alloys are used mostly for chemical resistance. The Stellites are used for wear applications. The lower hardness alloys have the better corrosion/erosion resistance, and the harder alloys the best abrasion resistance. In general, cobalt-based alloys are used where normal stainless steels do not work. They are "superalloys." There are other cobalt-containing metals such as the 18% nickel maraging steels that are of significant industrial importance. Thus cobalt is an essential member of the special metals family and cobalt-based superalloys should be considered for special applications. They can be specified by UNS number or industry alloy designation.

18.7 Beryllium

In Chapter 16, it was pointed out that alloying copper with beryllium produces a metal with very useful spring properties. However, beryllium has some unusual properties that make it an important metal in the pure form. The heat shield on the spacecraft that John Glenn used to make his orbital flight was made from beryllium. The properties that set it apart from other metals are

1. It is the second lightest structural metal (specific gravity of 1.85 compared with 1.74 for magnesium and 2.7 for aluminum).
2. It has the highest stiffness of any light metal [42×10^6 psi (290 GPa)].
3. It has the ability to easily transmit x-rays and other useful "nuclear" properties that make it useful for military applications.

All of these properties make it a very important material for aerospace applications. It has the highest specific stiffness and strength comparable to high-strength aluminum. If a steel part is still too heavy for an application, beryllium may be the only metal to do the job. The downside of beryllium is that it is expensive. Sometimes its price in the United States is as high as $260/lb ($118/kg). Also, in particulate form, it can be a perceived health threat, so working with beryllium, in some countries, requires monitoring of the air where fabrication processes occur. In spite of these limitations, beryllium's role as a light, stiff metal means that it will almost always have industrial importance, albeit cost-limited in sphere.

18.8 Gold

Gold has been a most valued metal as far back as records go. Gold articles have been discovered in Egyptian tombs 3000 or more years old. The reasons why this metal became valued are probably its beauty and permanence. It does not degrade or change with time. It does not oxidize, and it is not attacked by acids or alkalis (only aqua regia—a mixture of nitric and hydrochloric acid). Its color and brilliance are very pleasing to most eyes. A more practical reason for its popularity in early civilizations is that it was available in metallic form in nature. It was learned very early that the nuggets that were picked from stream beds and other places could be melted into larger shapes and easily worked into decorative objects. It is alleged that King Croesus (of "rich as Croesus" fame) developed small gold disks of uniform weight as the basis for barter, paying wages, and accumulating wealth. He invented coinage around 550 B.C. The gold that Croesus accumulated from a nearby river also contained silver, so he also instituted silver coins, probably at lower value. Archeological digs in the region of Turkey that was Croesus' kingdom indicate that he had a gold refinery. He purified gold by heating it in salt to remove silver and copper. These "tramp" metals were also recovered from condensed gases from the salt treatment. Essentially, this same process was used for another 2000 years. When King Croesus' country was conquered by Persian armies, they adopted his coinage idea and spread the concept widely through the other lands that they conquered.

Machine designers will probably not get good management reviews by specifying gold for motor brackets, but there are places where gold is an essential engineering material. It can be very cost-effective when used as an electrodeposit, foil, or inlay. It has unique properties that set it apart from other metals.

1. It does not oxidize in air.
2. It is one of the best electrical conductors (75% IACS).
3. It is extremely malleable. (Gold leaf can be hammered to a thickness of <500 Å.)
4. It is easily electrodeposited.
5. It reflects 90% or more of IR radiation.

All of these properties are useful in the electrical/computer devices that are part of the new millennium. The things that are plugged into computers (inside and out) usually do not function well without low-resistance electrical contact. Often, gold is necessary to ensure this low resistance.

Gold is a much better conductor than many copper alloys. In addition, copper and copper alloys oxidize in air as do most other metals. The malleability of gold is due to its FCC crystal structure and the nature of its electronic configuration (the 6*s* electron has a high probability of being in a completed shell). A 10-g rod of gold can be drawn into more than 2 km of wire. Gold goes a long way. In spite of its big price (which varies daily), it can be used where unique properties are needed. Gold is an essential material in space exploration and satellites. Gold contacts resist the harsh environment of space, and gold reflectors are used to protect devices from the sun's radiant energy as well as to use this energy to generate electrical energy. Gold can be alloyed with many other elements to make soldering and brazing filler metals with melting temperatures from about 700 to 1600°F (371 to 871°C). These alloys can be very useful for joining metals that have use temperature restrictions. The term *carat* applied to gold refers to the fractional percentage of gold in twenty-fourths. A 10-carat gold alloy would contain 10/24 or 41% gold. Pure gold is 24 carat. The remainder in gold alloys could be any metal, but copper and silver are the usual additions.

In summary, gold is an essential part of engineering materials, and it can be specified (at least in plating form) for many "ordinary" engineering challenges. It is probably more valuable to mankind for its technical usefulness than for its beauty. However, jewelry dealers (and our spouses) may take issue with this statement.

18.9 Silver

Silver is similar to gold in that it can occur naturally in metallic form, but most silver is obtained as a byproduct of beneficiation of other ores such as copper, lead, and zinc. Silver has the following distinguishing properties:

1. The highest thermal conductivity of any metal.
2. The highest electrical conductivity of any metal.
3. It is the best reflector of visible light.
4. Silver halides are light-sensitive.

Silver has many properties similar to gold:

	Silver	Gold
Specific gravity	10.5	19.32
Melting point	1863°F (962°C)	1945°F (1063°C)
Thermal conductivity	242 Btu ft/h ft^2 °F (4.29 W/cm °C)	181/Btu ft/h ft^2 °F (3.19 W/cm °C)
Coefficient of thermal expansion	10.9 in./in. °F (18.6×10^{-6} cm/cm °C)	7.9 in./in. °F (14.2×10^{-6} cm/cm °C)
Tensile modulus	11×10^6 psi (61 GPa)	12×10^6 psi (83 GPa)
Tensile strength (annealed)	22 ksi (152 MPa)	19 ksi (131 MPa)

Its malleability is similar to that of gold. Silver, like gold, has been used since ancient times for jewelry, for coinage, and as a measure of wealth. Probably the reason why even today silver is less valuable than gold is that it tarnishes in air. It reacts with sulfur in the air to form silver sulfide, which is black. Thus it is not as "noble" as gold.

From the industrial standpoint, silver is widely used as a brazing alloy for joining metals. Its thermal and electrical conductivity make it useful for electrical applications. The use of silver to make silver halides as the light-sensitive ingredient has been significant since photography was invented in 1839. Silver halides react with light to form a silver precipitate. The degree of reaction is proportional to the amount of light to which the silver halide layer was exposed. The value of silver halide photography in industry and consumer use is well known.

There are many silver alloys, especially for jewelry. Sterling silver in the United States means that 92.5% of the metal is silver. The remainder can be any other metal, but copper is the most common. Silver is widely used for plating of decorative items, and its use in dental amalgams is of great importance. Machine designers should remember that silver can be used when electrical or thermal conductivity become limiting factors. It is also useful as an electroplate to prevent galling of threaded fasteners and other places for which petroleum lubricants cannot be used. Finally, silver brazing is the joining process of choice for many industrial applications such as attaching cemented carbide tips to steel cutting tools. Even though it is a precious metal, it is needed to keep machinery running and for making many commercial products.

Summary

In this chapter we have discussed only another 13 of the 80 or so metal systems. Those discussed are used by designers on a regular basis, but probably not frequently. Nickel alloys have many different applications: high temperature resistance, corrosion resistance, and special physical properties. Titanium is useful for corrosion problems and structural applications; magnesium is useful for weight reduction; and zinc is important for die castings and as a coating for atmospheric corrosion resistance. The refractory metals are used for even more specialized applications. No matter how infrequently a designer uses these metals, they should be part of his or her material repertoire. They can solve problems for which the traditional metals may fail.

The following are attributes to remember:

- Nickel is a better conductor of heat and electricity than steel. Silver is the best.

- Nickel-based alloys can often solve corrosion problems that defeat stainless steels.
- Nickel-based alloys are used as superalloys for extreme service at elevated temperatures.
- Zinc is the easiest metal to die cast.
- Zinc coatings are anodic to carbon steels and can provide better atmospheric rust protection than most other coatings.
- Titanium is a light metal that can have a tensile strength comparable to that of a hardened alloy steel.
- Gold does not oxidize.
- Beryllium is the lightest structural metal.
- Cobalt alloys offer unique wear and corrosion characteristics.
- Titanium has good resistance to seawater.
- Titanium is stiffer than Al and Mg.
- Magnesium is the lightest common metal (lithium and beryllium are lighter, but they are too exotic and expensive for normal use).
- Magnesium has poor formability and is best used in cast form or for machined parts.
- Molybdenum has useful high-temperature properties if heating is done in the absence of air.
- Tungsten is one of the heaviest metals, and it has utility in machines for high-inertia components.
- Tantalum is extremely corrosion resistant and can be considered for severe corrosion applications.

Critical Concepts

- Material problems may require the use of metals other than the "commodity" metals.
- Nickel alloys can have corrosion and high-temperature properties that are better than most other materials.
- Magnesium can often "beat" aluminum in a weight-reduction application.
- Zinc is a faithful rust protection on steel, and zinc die castings can beat plastics as the most cost-effective material for some applications.
- Titanium has very useful strength and corrosion characteristics.
- The refractory metals can be useful for high-temperature and some corrosion applications.
- The use of precious metals is sometimes justified in machine design by their unique properties.
- Cobalt-based metals can have useful corrosion or wear properties

Terms You Should Remember

beneficiation	Ni Cr B
pure nickel	Invar
Monels	galvanizing
Inconels	anodic coating
Incoloys	galvannealed
Hastelloys	noble coatings
HCP	alpha-phase titanium
TZM	beta-phase titanium
wolfram	specific strength
Nichromes	specific stiffness
galling	metastable
die casting	incendiary
submerged plunger	solution treating
cavity	refractory metal
intermetallic compound	thermal fatigue
	inertia

Case History

Tungsten Die-Casting Cores

A severe erosion problem was occurring with H13 tool steel core pins that were subjected to impingement from molten aluminum in die-casting cavities used in making parts for an automotive fuel system. The pins were in about a dozen molds, and they were situated directly in the path of and only a few millimeters away from the gate to the mold. Neither the gate nor the core pin could be relocated because of design constraints. The core pins were about 10 mm in diameter, and after about 30,000 shots (parts) the pins would develop erosion notches as deep as a few millimeters. This notching from metal impingement eventually prevented part ejection. The molds would then have to be removed and the pin replaced—a very costly operation. The normal cavity service life for dies without erosion problems is about 100,000 shots.

This problem was solved by making the problem core pins from solid tungsten. They then machined more difficultly than steel pins, but the cost was comparable to the hardened steel because heat treating was not necessary. The tungsten pins were made from sintered (P/M) rounds, and they had a hardness of about 40 HRC. Tungsten is brittle, but these pins are not impacted in service. They formed mounting boltholes for the part and only molten aluminum touches them in service. Tungsten core pins were a cost-effective solution to a very serious production problem. Erosion was never again a concern in these molds.

Questions

Section 18.1

1. Can nickel alloys be hardened? Explain.
2. Specify a nickel alloy for use in seawater.
3. Is nickel aluminide a metal? Explain.
4. Is nickel ferromagnetic?
5. Compare the density, thermal conductivity, and elastic modulus of pure nickel (200) and 1020 steel.

Section 18.2

6. Compare the elastic modulus of zinc with those of the other metals that can be die cast (there are four that are commonly die cast).
7. Why is zinc the easiest metal to die cast?
8. What will be the galvanic reaction of zinc plating on copper? On carbon steel?
9. Where would you use wrought zinc? why?

Section 18.3

10. Compare the specific strength of titanium to that of carbon steel, aluminum, and magnesium.
11. What metals can titanium be welded to?
12. Describe the quench-hardening procedure for titanium–6Al–4V.
13. What grade of titanium should be used for a formed part?
14. How does titanium compare to steel for a shaft on a golf club?
15. Titanium has good resistance to seawater. Why not build ships from it?

Section 18.4

16. In magnesium AZ31B, what does the 31 mean?
17. You have a 6061-T6 aluminum structural member 1 in. (25 mm) in diameter and 30 in. (750 mm) long. You want to reduce its weight, but you want its stiffness to remain the same. Will magnesium meet this requirement?
18. What physical property of magnesium presents a safety concern?

19. Why does magnesium form so poorly?

20. Cite three applications for which magnesium has been successfully used in automobiles.

21. What engineering materials are lighter than magnesium?

Section 18.5

22. Name two molybdenum alloys and explain where they can be used.

23. Compare the specific stiffness of molybdenum to that of steel.

24. Can refractory metals be welded?

25. You want a counterweight for an application. It must weigh 40 lb and have a thickness of ½ in. What diameter should the weight be if made from tungsten? From steel? From lead?

26. Cite two applications where tungsten's high-temperature properties are utilized.

27. You would like to make a 500-gal. (2000-l) tank to store an extremely corrosive liquid. Would you specify tantalum for the tank? Explain.

28. Cite an important application of zirconium.

To Dig Deeper

Baker, H. and M. Avedesian, Eds. *ASM Specialty Handbook: Magnesium and Magnesium Alloys*. Materials Park, OH: ASM International, 1999.

Davis, J. R., Ed. *ASM Specialty Handbook: Nickel, Cobalt and Their Alloys*. Materials Park, OH: ASM International, 2000.

Donachie, M. J., Jr. *Titanium: A Technical Guide*, 2nd ed. Materials Park, OH: ASM International, 2000.

Engineering Properties of Zinc Alloys. New York: International Lead Zinc Research Organization, 1981.

Everhart, John L. *Titanium and Titanium Alloys*. New York: Van Nostrand Reinhold Co., 1954.

Friend, W. Z. *Corrosion of Nickel and Nickel-Base Alloys*. New York: John Wiley & Sons, 1980.

Huntington Alloys: Handbook, 5th ed. Huntington, WV: International Nickel Co., 1970.

Kaplan, H. *Magnesium Technology 2003*. Warrendale, PA: TMS Publications, 2003.

Kim, Y.-W., H. Clemens, and R. Rosenberger, Eds. *Gamma Titanium Aluminides*. Warrendale, PA: TMS Publications, 2003.

Kleefisch, E. W. *Industrial Applications of Titanium and Zirconium STP 728*. Philadelphia: ASTM, 1981.

Metals Handbook, 9th ed, Volume 2: Properties and Selection: Nonferrous Alloys and Special-Purpose Materials. Metals Park, OH: ASM International, 1991.

Polmear, I. J. *Light Alloys: Metallurgy of Light Metals*. Metals Park, OH: ASM International, 1981.

Porter, Frank C. *Corrosion Resistance of Zinc and Zinc Alloys*. New York: Marcel Dekker, Inc., 1994.

Roberts, C. Sheldon. *Magnesium and Its Alloys*. New York: John Wiley & Sons, 1960.

Zelikman, A. N., and others. *Metallurgy of Rare Metals*. Washington, DC: Natural Science Foundation (NASA TTF-359), 1966.

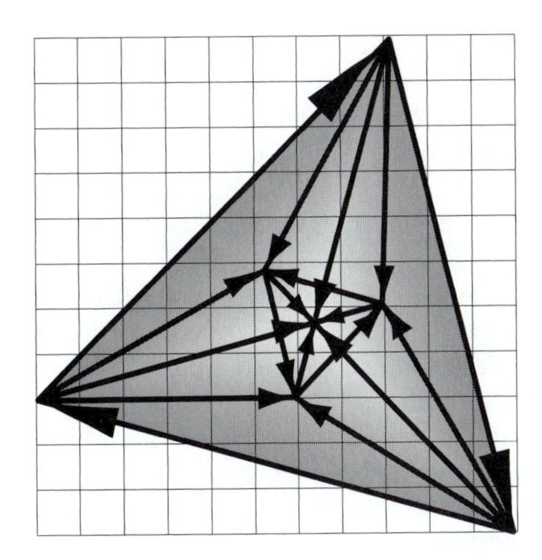

CHAPTER 19

Surface Engineering

Chapter Goals

1. A knowledge of surface engineering processes.
2. An understanding of how coatings, surface treatments, and surface modifications interrelate.
3. An understanding of when to use various coatings and surface treatments and how to specify them.

Rationale

Surface engineering *is a term that came into use in the United States in the 1970s. It was coined to include the many processes and materials that people used to alter the functional surfaces on solids used in machine design. There are surface considerations involved in every design. As the designer, you must decide the texture that you want on a molded ceramic or plastic part, the paint or varnish on a wood part, the machining finish on a machined part, the plating or corrosion protection on a steel part—all manufactured parts need something done to improve their surface, even if it is just cleaning.*

Surface engineering can include things such as sterilization for medical reasons, or painting and coloring for cosmetics. Our discussion will address only the aspects of surface engineering that apply to machine design—how to make surfaces harder, more corrosion resistant, more durable, or how to alter a physical property. Designers need to know what is available for improving working surfaces of materials, because many times a design budget cannot afford to make the part from a fancy (expensive) material but can afford the lesser cost of altering its surface. Sometimes bulk materials are not available for the job. For example, titanium nitride coatings are very widely used to improve the wear characteristics of a part but titanium nitride is currently not available in bulk form. Designers can diamond coat a hole punch to reduce wear, but it would not be possible to make the punch from solid diamond. Surface engineering allows designers material options not affordable or practical from bulk materials. It is a field of great importance, and its importance increases each year. It often allows companies to do more with less.

Surface engineering is a multidiscipline engineering activity aimed at tailoring the properties of engineering materials to improve their function or service life. The most common reasons for altering the surface of a material are listed in Figure 19–1.

As applied to metals, surface engineering includes processes such as diffusion treatments, selective hardening, plating, conversion coatings, thin film coatings, hardfacing, thermal spray coatings, high-energy treatments such as laser processing, and organic coatings such as paints and plastic laminates. Essentially all of the processes illustrated in Figure 19–2 apply to metals. Platings, solid-lubricant coatings, paints, and laminates apply to plastics. In fact, these coatings apply to any engineering material—metals, plastics, ceramics, cermets, and ceramic composites.

1. Alter appearance
   - Color
   - Surface texture/gloss
   - Lay
   - Clean/remove oxide, rust, etc.
2. Alter dimensions
   - Rebuild worn surface
   - Add a protective layer
   - Fix mismachining
3. Alter properties
   - Improve wear resistance
   - Improve heat resistance
   - Improve fatigue resistance
   - Improve corrosion resistance
   - Friction characteristics
   - Biological condition
   - Optical properties
4. Cost
   - Lower cost
   - Add value to raise price

Figure 19–1
Reasons for surface engineering

The surface engineering processes that apply to ceramics include paints, dry-film lubricants, plating, PVD wear coatings, and high-energy treatments such as ion implantation.

There are surface engineering processes that apply to any engineering material, and they are used to enhance properties on surfaces on which things are happening. This chapter will describe some of the surface engineering processes listed in Figure 19–2 that have not been described in previous chapters. Surface treatments such as nitriding and carburizing that were discussed earlier will be included among selection considerations. The overall goal of this chapter is to provide the reader with a repertoire of surface engineering processes that can be called on to solve design problems. When there is evidence that a bulk material will not meet an application requirement, a surface engineering process might be called on to make the system work. For example, regardless of what automobile salespeople tell you, steel bumpers are stronger and stiffer than the plastic bumpers promoted since the early 1990s. But steel rusts, and rusty steel does not promote automobile sales. Steel makes a fine bumper when it is chromium plated or protected with a substantial powder coating. Unfortunately, in the United States in 2004, the only vehicles with "real" coated steel bumpers were trucks and sport utility vehicles. However, steel will return to passenger cars in triumph when a surface engineering process is developed to make it a more cost-effective and aesthetically pleasing surface than molded plastic.

Figure 19–2 lists more processes than can be fully described in a single chapter, so we will start by discussing both the most important coating processes and the most important processes that improve a surface without application of a coating; these are termed *surface treatments.* Some applications, such as gears, simply cannot tolerate a surface buildup. They are often carburized or flame hardened to improve durability.

Figure 19–3 shows our estimate of the relative usage of surface engineering processes. Though the steel case hardening processes that

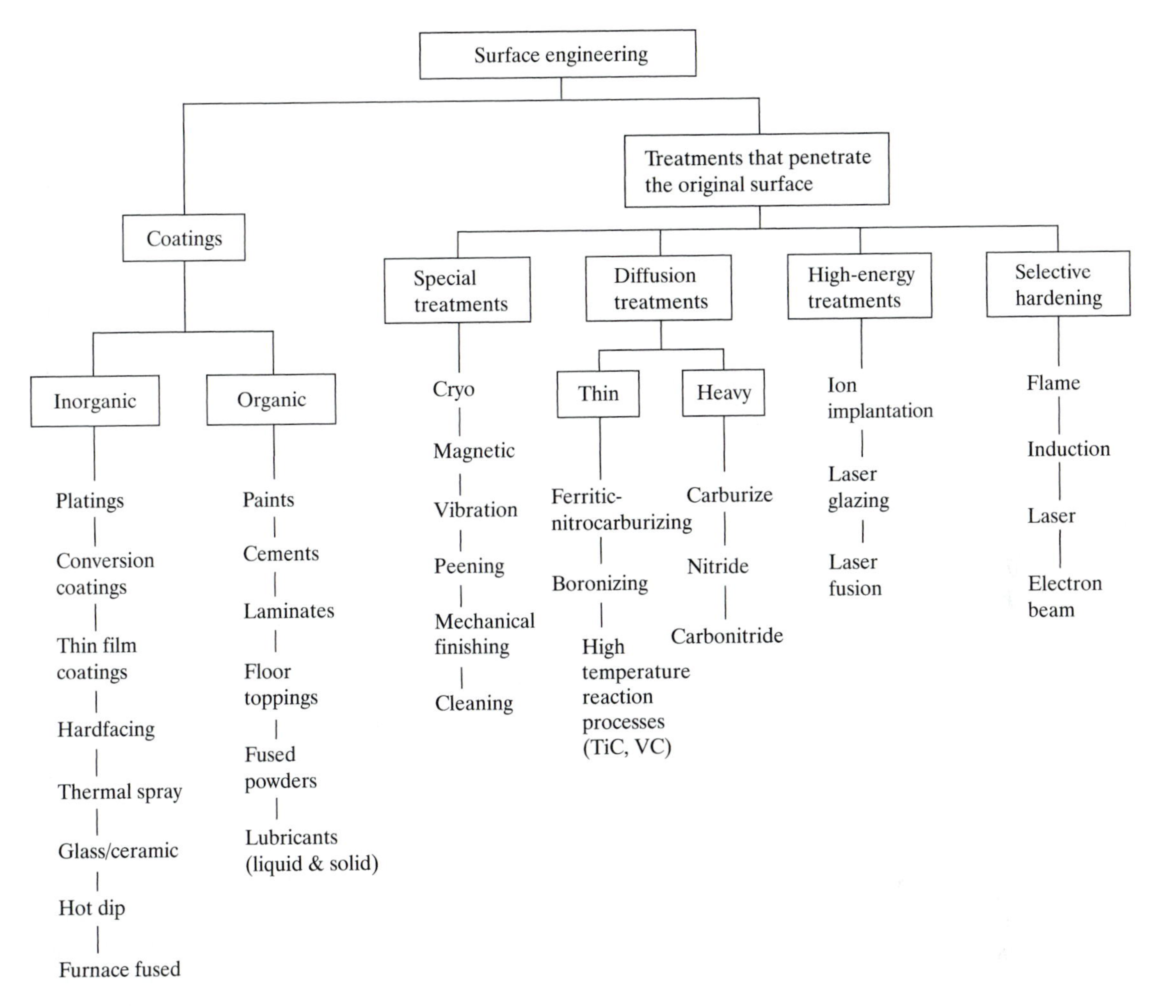

Figure 19–2
Surface engineering processes

were covered in Chapter 9 on heat treating will not be discussed, they will be included in selection discussions because all of these processes compete.

Our coverage of surface engineering processes will reflect their relative importance in design engineering, and thus we will only discuss surface engineering processes that apply to machine building and product engineering. Medical applications, such as lasers used for many types of surface engineering of skin and body parts, are beyond the scope of this book. Our coverage starts with a discussion of cleaning and mechanical finishing because these apply to most surfaces. Parts need to be cleaned before being coated with anything, and most surface treatments as well require that surfaces be treated to be free of contaminants. Similarly, mechanical finishing is performed on most surfaces that are used as tools to form and shape parts: injection molding cavities, casting patterns, embossing rolls, work rolls, and the like. Our discussion will

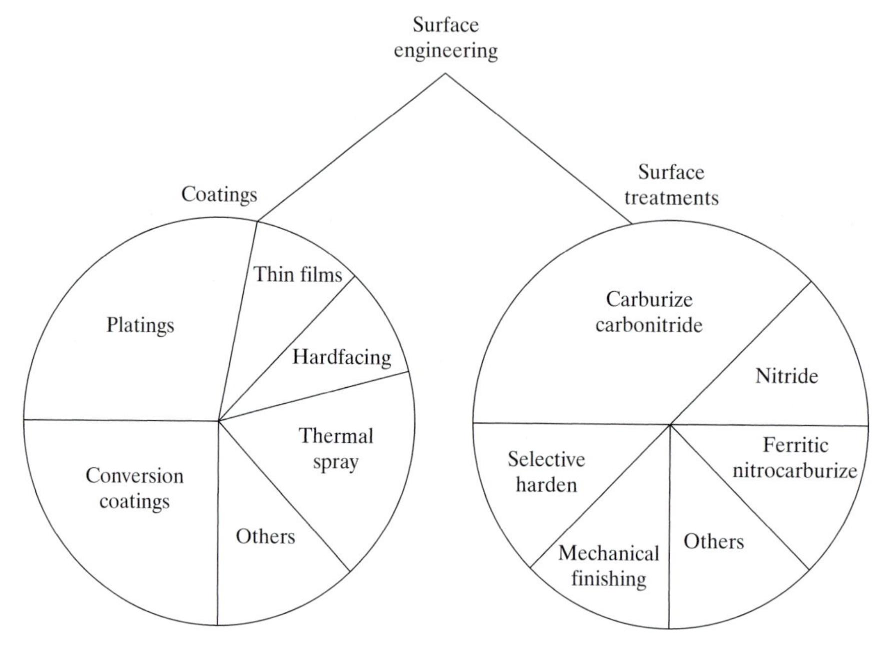

Figure 19–3
Estimated usage of surface engineering processes

then proceed to important coating processes and on to surface treatments. The remainder of the chapter will be dedicated to the selection of surface engineering processes and materials and the way to specify them on engineering drawings.

19.1 Cleaning

Cleaning of engineering materials is becoming much more technical each year as environmental regulations ban solvents and other chemicals that have been used in the past. Metal parts that are machined must be cleaned of machining fluids before painting. Plastic and ceramic parts may need cleaning for adhesive bonding (like PVC plumbing fittings). How do you clean these materials? How do you know if they are clean?

There are, fundamentally, only a few ways to remove material from a solid: to dissolve it, to fracture it off, to burn it off, or to vaporize it off. Solvents are used to dissolve contaminants from a surface. They can be organic, aqueous, or even liquefied gases. Fracture, as a cleaning process, is practiced with abrasive blasting. Painting of structural members such as bridges usually requires sand blasting to "white metal." The abrasive media fracture contaminants from the surface. There are processes that clean by blasting a surface with dry ice (CO_2) particles. This solves emission and residual blasting debris problems. The particles become gas after impact. Burning contaminants from surfaces is performed frequently in self-cleaning ovens, which are widely used in U.S. households. In industry, it is common to oblate plastic from extruder screws and filters that have been removed from service. Cleaning by vacuum works by a mechanism of converting contaminants to a gas that is pumped away.

There are pros and cons to each of these cleaning techniques, and each probably has applications for which it is the most cost-effective technique. For many years in the United States, the cleaning method of choice for machined metal parts was vapor degreasing with a chlorinated solvent such as trichloroethylene. Parts were placed in wire baskets suspended above the boiling solvent. The solvent forms a vapor, leaving contaminants behind in the liquid. The vapor condenses on the parts and drips back into the solvent tank. Only clean vapor reaches the work. Contaminants remain in the liquid phase. There are perceived environmental and health risks with chlorinated organic solvents, and their use is declining in the United States and other countries. In fact, the use of certain chlorinated organic solvents is against the law in some countries. They are being replaced with water-based cleaning agents and less contentious solvents. Some industries have replaced solvent cleaning with the more equipment-intensive techniques previously mentioned, cryogenic liquids, and vacuum systems.

The effectiveness of a particular cleaning technique can be assessed with analytical surface analysis techniques. We will discuss these in more detail in a subsequent section of this chapter, but one technique that was developed specifically to measure film presence, not its composition, is photo-electron emission. A small probe is brought in proximity to the surface. The surface is bombarded by concentrated ultraviolet radiation, which in turn stimulates emitted radiation from electron shell movements. The probe contains a collector to quantify this emitted radiation. Contaminating films suppress surface emission, and this signal quantifies the degree of contamination. Figure 19–4 compares the films left by various organic solvents on aluminum foil that has never been exposed to any environment after rolling. These data suggest that many cleaning solutions will contaminate surfaces by leaving

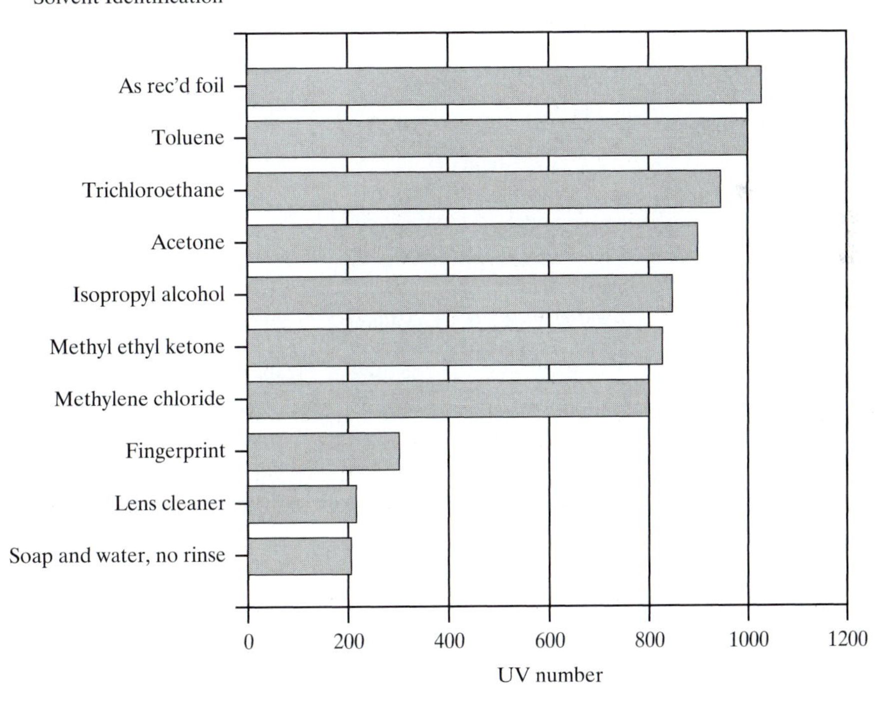

Figure 19–4
The tendency of various solvents to leave a contaminating film. The lower the UV number, the thicker the residual film after cleaning. Toluene left the thinnest film.

a residue after evaporation. Most cleaners do this. If a surface must be atomically clean for a particular application, it may be necessary to use analytical techniques such as photo-electron emission that can detect the presence of very thin contaminating films (<50 nm).

In summary, effective cleaning of surfaces to remove dirt, oils, fingerprints, and even oxides can be required for some coating processes such as vapor-deposited thin films. Cleaning technique and cleanliness verification must be made a part of the process if good adhesion is a service concern.

19.2 Mechanical Finishing of Surfaces

The bulk of this chapter is about ways that surfaces can be hardened, coated, or otherwise altered to improve their functionality. However, almost any surface has a desired surface texture that is a requirement over and above whatever else is done to that surface. For example, an extremely hard surface may be needed on a part, but it may also have a requirement that it be dull and nonreflecting. A dull surface could be created by chemical etching, by deposition of a matte plating, by oxidation, and so on. There are a lot of ways to produce a nonspecular surface. Mechanical finishing is one way that it could be produced. Our definition of mechanical finishing is the intentional alteration of the surface texture or properties of a surface by mechanical action of another contacting solid or solids. Figure 19–5 illustrates many of the processes that can be considered to be types of mechanical finishing. We will briefly describe some of the more important ones because machine designers need to specify these processes on drawings or specifications if they are needed for function or appearance.

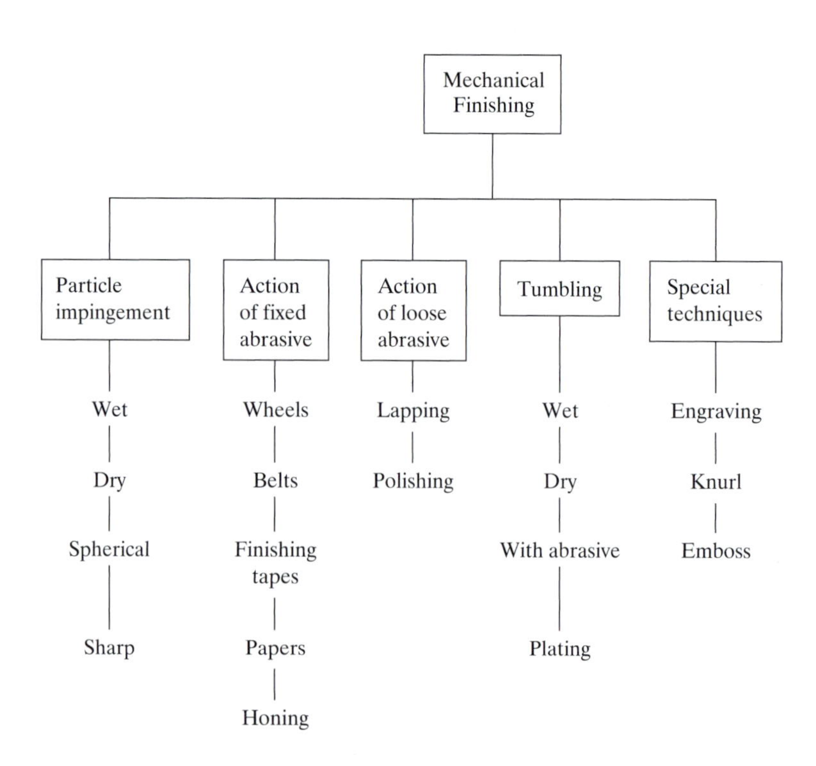

Figure 19–5
Mechanical finishing processes

Particle Impingement This category of mechanical finishing includes processes that alter a surface by the mechanical action of particles that forcibly strike a surface after being propelled at the surface by the action of a fluid. The fluid can be a gas or liquid. Sandblasting is the simplest example of this mechanical finishing process. It can produce surfaces with a random roughness in the range of 0.1 μm to 8 μm Ra. These surfaces are usually dull. Fine abrasives (<10 μm in diameter) do not alter most metal surfaces when propelled by a gas jet. Air resistance slows them too much. Putting fine particles into a water/air stream makes them energetic enough to dull surfaces but keep the roughness low, possibly less than 0.1 μm. On the far end of the "blasting" processes is peening, in which spheres are impacted on the surfaces. Sometimes large balls are used and the surface becomes cold worked as well as textured.

Fixed Abrasive Texturing Grinding with wheels fits into this category, as does rubbing a surface with abrasive belts or papers. Belt sanding with automatic machines is commonly used to produce different scratched patterns on one or both sides of sheet materials. Once-through abrasive finishing tapes can be imposed on a cylindrical surface with a tape finishing head that attaches to a lathe tool post. Finishing tape can be fine enough to produce mirror finishes (specular) or as rough as about 4 μm Ra. The surface is altered by a mechanism of particle scratching.

Honing is the word used to describe the use of fixed abrasive stones to produce a final surface on a metal part. Cylinder walls are often honed with spring-loaded stones to change size or produce a particular texture. For example, high-performance automobile engines often have a surface that is honed to produce a cross-scratch pattern that allegedly helps to hold oil and thus reduce ring wear. The last operation in sharpening a knife is often called honing. A fine stone is carefully rubbed on the cutting edge to lower edge roughness and remove any burrs that may fracture in service and produce nicks.

Fixed abrasive finishing of surfaces is really the most widely used mechanical finishing technique. The important factor for designers to remember in specifying this process is to only ask for what you need in roughness, and if it is important to function you need to detail the orientation (lay) of the surface texture.

Three-Body Abrasion Three-body abrasion textures by the action of hard particles rolling and sliding on a surface. The particles producing surface deformation are imposed on the surface to be treated by another body—either compliant or hard. Abrasive particles imposed with a hard lap (metal) often result in a dull surface—pitting occurs. When the abrasive becomes fixed as in cast iron laps, the texturing mechanism becomes more like a fixed abrasive. Polishing similarly can involve a loose or fixed abrasive. Many shops polish with rouge on a sheepskin or cloth wheel. This kind of polishing produces wavy surfaces and errors of form. Polishing is preferably done with fixed abrasives of decreasing size. For example, to get a mirror finish on a part, a sequence of fixed abrasives like the following will work on most metals:

240 grit silicon carbide
320 grit silicon carbide
400 grit silicon carbide
600 grit silicon carbide
3 μm aluminum oxide
>1 μm aluminum oxide

The abrasive sequence for plastics and ceramics may be different, but the procedure is the same—abrade the surface with each until it is uniformly covered with the scratches from that size abrasive. Repeat this with the next finer size, and the next. The submicron finishing abrasive could be applied by a compliant medium such as cloth, but this abrasive is too small to produce dimension changes, so it will not produce surface waviness and the errors of form that are characteristic of buffing.

Lapping is typically performed to produce flat, smooth surfaces. If a part is to be made to an accuracy of plus or minus several micrometers, it will need low-roughness flat surfaces are needed just for measurement accuracy. The peak-to-valley distance of asperities on a ground surface (0.5 μm) may be as high as 2 μm, so measurements can be off by 4 μm if, for example, you are measuring thickness and sometimes the measuring device contacts the surface peaks and sometimes the valleys. Cast iron charged with abrasive is the traditional lap material. Abrasive tends to lodge in the graphite flakes. Lapping can produce a matte pitted surface or a polished surface depending on the media and procedure used. If flatness is desired, special techniques need to be employed to prevent nonuniform wear of the lap.

In summary, loose abrasive mechanical finishing is a necessary part of surface engineering, but one of the more difficult to control.

Tumbling A common production finishing technique is to put parts in a rotating barrel or similar device and let them hit each other jumbled inside the barrel. This is called tumbling. It removes burrs and rounds over edges, and flat surfaces are characterized by small dig marks and scratches from impact by part edges. Tumbling can be done wet or dry, and stones or other media can be put into the "charge" to produce different appearances. Tumbling is seldom used on small quantities of parts because it takes many parts just to charge the tumbling unit. Similarly, it is seldom used on large parts or plastic parts. Injection molds can be altered to produce spiral surface textures and radiused part edges. A novel use of tumbling is to plate parts with soft metals. The process of tumbling parts jumbled in a barrel to which soft metal powder is added is call mechanical plating. If powders such as aluminum, cadmium, zinc, and so on are put in a barrel tumbling machine, the part impacts adhered powder on the part surface, and all surfaces that can be uniformly impacted will become plated. The process is called *mechanical plating*, and it can be used on hardened steel parts that may embrittle in plating chemicals (hydrogen embrittlement).

As a mechanical finishing process, tumbling should be reserved for applications involving many small parts. It can alter appearance and remove sharp edges very effectively.

Special Processes Engraving is mechanically producing a pattern or texture on a surface by pressing a tool into the surface and deforming it. Rolls are frequently engraved to produce textures on sheet goods or web products. When these patterned rolls are used to produce a pattern on a work piece, it is called embossing. Many composition building materials are embossed with patterns such as wood grain. Knurling is a related process that is applied to parts that can be rotated. A knurl is a cylindrical tool made by machining shallow grooves, usually at angles, to the axes of rotation.

There are countless variations of the mechanical finishing processes that we have described. In most cases, the mechanism of surface alteration is simply plastic deformation with negligible material removal. The surface deformation can be in the form of indents or grooves. These treatments can be applied to surfaces to improve function (friction, reflection, adhesion, etc.) or simply to alter appearance. The depth of those treatments can be only a few micrometers or a fraction of a millimeter. The machine designer should be aware that they are available to be used to engineer surfaces.

19.3 Organic Coatings

Chapter 7 described the types of paints and related organic coatings that are available to protect surfaces. There are also cements that can be applied like hardfacing, and they are often epoxies filled with hard particles. There are polyurethanes that can be poured to form thick abrasion- or

Table 19–1
Organic coatings with significant utility in surface engineering

Coating	Used For
Fluorocarbons	
Polytetrafluoroethylene (PTFE)	Release, corrosion protection to 500°F (320°C)
Teflons (PTFE/PI blend)	Wear/friction reduction—more durable than PTFE, lower use temperature
PTFE/mica (mica filler)	Release, more durable than PTFE
PTFE filled enamels	Wear/friction reduction, lower cure temperature than PTFE
Silicones	Release
Solid-film lubricants	
MoS_2/organic binder	Lubrication of sliding systems
Graphite/organic binder	Lubrication of sliding systems
Conductive paints	Static dissipation, electrical conduction

corrosion-resistant floor toppings. Vinyl laminates are available to protect metal and other surfaces from scratching and to improve surface appearance or to provide other functions.

Powder coatings of thermoplastic materials can be applied to a thickness of a few mils (75 μm) to improve the appearance and corrosion resistance of materials. They are sprayed with electrostatic equipment; the parts are then baked to produce a finished coating. The use of plasma spray processes to apply heavy coatings [10 mils (250 μm)] of selected thermoplastic materials (nylons, polyesters, etc.) for corrosion applications is increasing each year. *Plasma* sprayed coatings do not require baking, and thus they are probably more environmentally friendly than other organic coatings. Heated particles fuse and coalesce on impact with the substrate to form the coating.

Application of solid-film lubricants is performed like the application of paint. Most have organic binders that either air dry or are baked to cure. Solid-film lubricants are usually proprietary in nature, and they often contain intercalative lubricants (crystalline materials that have atomic planes of easy glide or deformation) such as graphite, molybdenum disulfide, or tungsten disulfide. The lubricating crystals are bonded to the surface with various binders such as epoxies, phenolics, or cellulosic lacquers. They are usually applied in thicknesses of less than 0.002 in. (50 μm), and for some applications they allow metal-to-metal (or other solid-to-solid) sliding systems to operate with no lubricant other than the bonded dry-film lubricant. In fact, many of these coatings are removed by application of liquid lubricants. Because these coatings are proprietary, selection usually requires consultation with manufacturers or companies that specialize in application of these materials. Table 19–1 is a list of some solid-film lubricants and release coatings that have utility in machine design. These coatings and the other organic coatings that we discussed are a useful, permanent part of a designer's repertoire of surface engineering tools.

19.4 Electroplating

In our discussions of corrosion phenomena it was shown that, when a metal corrodes, the metal that is dissolved goes into solution in the form of metal ions. An ion is an atom that has lost its electrical neutrality. It cannot be seen; it is part of the solution. Normal drinking water contains trace amounts of many metal ions. If metal ions are reduced by chemical or electrochemical techniques, they will revert to metallic form. Three important plating techniques—*electroplating*, *electroless plating*, and *immersion plating*—all involve converting metal ions in solution to metal atoms at a surface. The difference between these systems is in how the metal-ion reduction is accomplished and the properties of the coatings. Some metals, like nickel, can be coated by all three plating processes. The properties and applicability of nickel produced by these different systems varies. Nickel electroplates have properties that are different from electroless nickel platings and immersion nickel platings. For this reason we shall discuss these plating systems separately.

Theory of Electroplating

Because many objects that we use in our everyday lives are coated by electroplating techniques, we tend to take the process for granted. However, electroplating is not very old as a commercial process. It came into use about the same time as electrification, in the late nineteenth century. No single person is credited with invention of the process, but Michael Faraday in the early nineteenth century is responsible for developing the basic laws that govern electrodeposition. Basic electroplating employs two electrodes immersed in an electrolyte and connected to a power supply (Figure 19–6). If the electrolyte and metal combination are favorable,

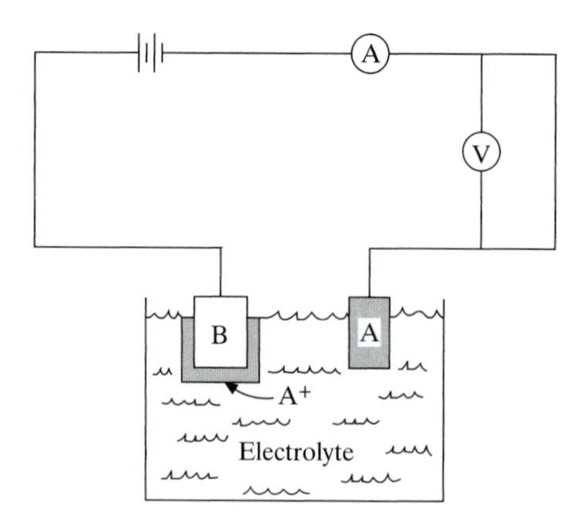

Figure 19–6
Schematic of plating cell

metal A will dissolve, forming metal ions A^+. With direct current from a battery or other suitable source as the driving force for the reaction, ions of A will migrate to B. The A^+ ions are reduced (their valence state is reduced) by gaining an electron from metal B. Metal A is the anode. Metal B is the cathode, and it is the member that gets coated.

Faraday established a law that states that the amount of metal deposited in this reaction is a function of the amount of current flowing in the system (measured by the ammeter A) and the electrochemical equivalent of the metal that is being plated:

$$\text{weight of metal reacting} = kIt$$

where k = the electrochemical equivalent of the anode material
I = the current in amperes
t = time

The electrochemical equivalent of an element is equal to the atomic weight of that element divided by its apparent valence (the valence in the reaction involved). Cadmium has an atomic weight of 112 and a valence of 2. Thus, the electrochemical equivalent is $^{112}/_{2}$, or 56. The electrochemical equivalent of nickel is $^{59}/_{2}$, or 29.5. Thus, according to *Faraday's law*, for a given quantity of current in ampere-hours, almost twice as much cadmium as nickel will be deposited. This is assuming the same process efficiency. The point to be made is that electroplating rates depend on the nature of the metal being plated. Using Faraday's law, the theoretical time (*deposition rate*) needed to produce a 1-mil plating (25 μm) of different metals would be approximately as follows:

1. Silver, 6 min.
2. Cadmium, 10 min.
3. Copper, 18 min.
4. Nickel, 19 min.
5. Chromium, 26 min (trivalent).
6. Chromium, 1 h (hexavalent).

This is assuming the same areas and current density (the amount of current per unit area) and 100% efficiency. Because in practice plating baths do not operate at high efficiencies, all these metals except chromium plate at about the same rate, about 1 mil/h (25 μm/h). Chromium baths typically plate at a rate of about 0.3 mil/h (7.5 μm/h).

Electrolytes

The electrolytes for electroplating are usually water solutions of salts of the metal to be electroplated. Copper sulfate ($CuSO_4$) dissolved in water is used as an electrolyte for copper plating. The copper sulfate is formed by the reaction of copper and sulfuric acid ($Cu + 2H_2SO_4 \rightarrow CuSO_4 + 2H_2O + SO_2\uparrow$). Copper sulfate is a blue, crystalline substance that looks like rock salt. When it is dissolved in water, it forms copper and sulfate ions, Cu^{2+} and SO_4^{2-}. Both of these ions migrate in an electrolytic cell when the cell is connected to a current source. Solutions must contain ions if electrical conduction is to occur. Thus an electrolyte for plating is usually a water solution containing ions of the metal to be plated. The concentration of salt needed for a particular plating operation is obtained from recipes tabulated in handbooks. Most plating electrolytes were developed by trial and error. Some plating electrolytes have sufficient metal-ion concentration that it is not even necessary to have an anode of the material that is to be electrodeposited. An inert electrode such as carbon could be used. The requirements of a good plating electrolyte are the following:

1. The metal to be plated must dissolve in it.
2. It must be a good conductor of electricity.
3. It must be free of impurities that could be codeposited with the metal to be plated.

The source of current for plating can be anything from a battery to a motor generator. Direct current is usually used, but alternating current is used for special applications. For a plating bath to work properly, a certain current density must be maintained. Current density is simply the amount of current flowing, divided by the surface area of the object to be plated. If a part to be plated has an area of 1 ft^2 and 100 A is flowing in the plating cell, the current density will be 100 A/ft^2. It is important to have a uniform current density to get a uniform plating thickness. Current density tends to be high on corners and low in holes. This is why platings tend to build up on corners and often do not "throw" into holes. Some plating electrolytes tend to throw better than others, and as we shall see later, this can be a selection factor in choosing a type of electroplate.

Metals

There are 80 or so metals in the periodic table of elements, and probably techniques are available to electrodeposit any of them. The Metalliding process that was discussed in Chapter 11 has the capability of depositing strange elements such as yttrium and tellurium. However, only about 30 metals are electrodeposited from aqueous solutions, and of those 30 only about 10 are of wide industrial importance: copper, chromium, gold, iron, silver, tin, zinc, nickel, cadmium, and platinum. Some alloys of these metals can be plated; but if an alloy is to be electrodeposited, the electrochemical equivalents of the component alloy elements must be about the same. Brasses and bronzes can be plated as alloys because the alloy elements fulfill this requirement (Cu $^{64}/_{2}$, Zn $^{65}/_{2}$, Sn $^{119}/_{4}$).

Substrates

All the aforementioned metals can be electrodeposited on any other metal, but the bond strength depends on the chemical nature of the *substrate*. A good electroplate will have a bond strength in shear in excess of 20,000 psi (138 MPa). Plating adhesion is a difficult parameter to measure, but a simple test is to grind the edge of a deposit with a coarse wheel on a pedestal grinder. The grinding forces will peel off a poorly bonded deposit. The mechanism of plating adhesion is not well documented, but plated deposits never have the bond that could be obtained by a fusion deposit. If iron is plated on iron, it will not be bonded as well as a layer of iron fusion welded on. Plating adhesion is probably part mechanical and part atomic bonding. If two atoms are brought in very close proximity, there is a net attraction between their nuclei. To get this atomic interaction between the plating and the substrate, the substrate must be atomically clean. There can be no adsorbed oxide layers or films of any sort. This is where the platability of metals becomes a factor. Atomically clean surfaces are approached in electroplating operations by acid pickling or electrochemical cleaning immediately prior to plating. Very corrosion-resistant metals such as titanium, stainless steels, tantalum, and even aluminum often have passive surfaces composed of oxide films. It is difficult to remove these, and thus these metals are difficult to plate. Titanium and aluminum are unplatable without special techniques. Plastics can be electroplated if they are first made conductive by immersion or autocatalytic metal deposition. The most suitable plastics for electroplating are ABS, polypropylene, polysulfone, polycarbonate, polyester, and nylon.

The physical nature of a surface (roughness, waviness, and the like) determines what the electrodeposited surface will look like. If a shiny chromium plating is necessary, the metal must be polished before the plating operation. Most electroplatings do not level surface roughness or bridge large inclusions or defects. Hardened steels (>30 HRC) are susceptible to *hydrogen embrittlement* from plating. Hydrogen diffuses into the steel and makes it brittle (like glass). When hard steels must be electroplated they should have a hydogen bake at about 500°F (260°C) after plating. This is not 100% effective so it is best to avoid plating hardened steels.

Plating Properties

Some electroplated deposits are porous, some are *microcracked*, some contain impurities, some contain pits. The occurrence of these is controlled by plating conditions, and in general they are undesirable. However, some additional aspects of plating properties are important from the selection standpoint. When some metals are electrodeposited, they become hard. Chromium as a pure metal has a hardness of only 150 kg/mm^2. As an electrodeposit it can have a hardness as high as 900 kg/mm^2, harder than any other metal. Nickel behaves the same way, and so does iron. The origin of this hardening is internal microstrains that develop in the plating operation. The exact mechanism of this hardening is not known, but electroplaters do know how to control it. Nickel and iron electrodeposits can be specified with a hardness anywhere from 100 kg/mm^2 to as high as 600 kg/mm^2. Chromium can be obtained with a range of about 500 to 900 kg/mm^2. The artificial sweetener saccharin is added to nickel baths to control hardness. High bath temperatures will produce the softer chromiums. Thus electroplaters can give the user different plating hardnesses (within limits).

Another effect of plating is the formation of macroscopic stresses in the deposit. If a thin metal strip is masked on one side and plated on the other, it could do nothing, bend concave, or bend convex. When electrodeposits form, they can cause distortion of parts. Similarly, if a deposit forms with a high tensile stress in it, it could crack in service. As an example, a nickel-plated aluminum roll was being washed in hot water. The plating split and popped off. Apparently the deposit was under large tensile stresses, and the additional stress caused by the expansion of the aluminum substrate caused spalling. The point of importance to the designer is that if macroscopic stresses in an electrodeposit are unwanted, it may be possible to alter the state of stress with process controls. Stress-free deposits can be achieved with a number of metals.

Other physical and mechanical properties can also be altered by electrodeposition, but the two that we have discussed are the most important. With regard to such things as electrical properties and corrosion characteristics, the designer can assume that these properties will be the same as for the pure metal before electrodeposition.

Design Considerations

One of the most important factors limiting the use of electroplating is *corner buildup*. As a rule of thumb, the electroplate thickness at the corners or end of a part is twice the thickness of the coating in the center (Figure 19–7). Thus heavy deposits will require a machining or grinding operation if dimensions are critical. Another rule of thumb is to limit plating thickness to 0.001 in. (25 μm) if dimensions are important. Edge buildup still occurs, but if the 0.001 in. (25 μm) is measured at the ends, the center will be only 0.0005 in. (12.5 μm) thinner. This is usually not troublesome. Edge buildup is illustrated in Figure 19–8. In addition to keeping plating thickness low, radiused edges prevent high current densities that cause corner buildup (Figure 19–9).

Plating into recesses requires widening of the recess opening or using an electroless process, in which the *throwing power* of electricity is not a factor. Masking of parts to prevent plating can be made easier for the electroplater if the designer follows some simple guidelines:

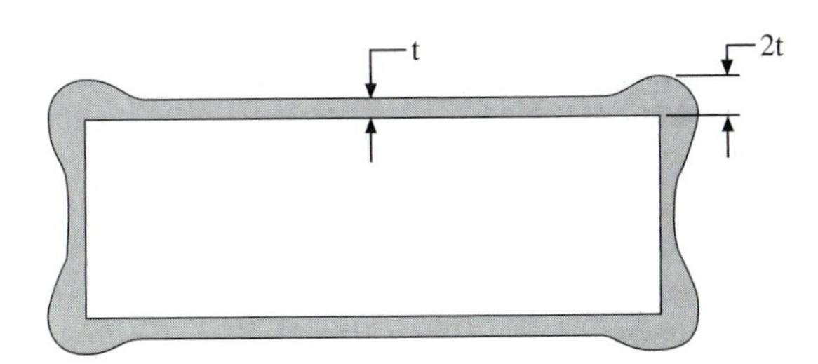

Figure 19–7
Corner buildup

Figure 19–8
Edge buildup on chromium plate 3 mils (75 μm) thick (×4)

1. Avoid masking small screw holes. Large holes can be masked cheaply and easily with rubber stoppers. Small holes require more hand labor.
2. Avoid plating of assemblies of dissimilar metals. Occasionally, galvanic corrosion will occur, but, more important, the surface activation treatments required for plating may differ.

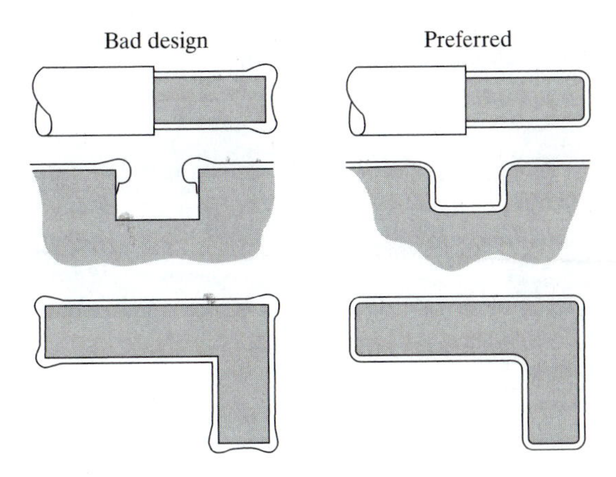

Figure 19–9
How to reduce corner buildup

3. Avoid plating specifications that require hand painting of plating stop-off. This is expensive and inaccurate. Try to design the part so that masking can be done by wrapping the part in tape.
4. Provide the plater an area for electrical connection for plating. If a part must be covered 100%, it can only be done by using needle-like electrical connections. Arcing can occur at the needle connections; a tab or eyebolt hole will make the job easier. As shown in Figure 19–10, hanging on hooks is a common way to hold parts to be plated. Designers should consider how a part will be held if it is to be plated.
5. Use *selective plating* or *local plating* if a part is too large or complicated to fit in a plating tank. A pump circulates electrolyte to a carbon anode brought into proximity with the area to be plated (Figure 19–11). The electrolyte is collected in a drip pan and recirculated. This process is useful for rebuilding worn shafts, and the like.

A last factor that deserves design consideration is *rinsability* after plating. Many plating solutions can be corrosive. If they are trapped in a plated assembly, they can leach out with time and discolor or ruin a part. Drain holes can be used on tack-welded assemblies; on weldments, seal welding is necessary if the crevice is wider than 1 in. (25 mm). If the crevice can be rinsed from only one side, the crevice should not exceed ½ in. (12.5 mm).

Figure 19–10
Racking of parts in a plating tank

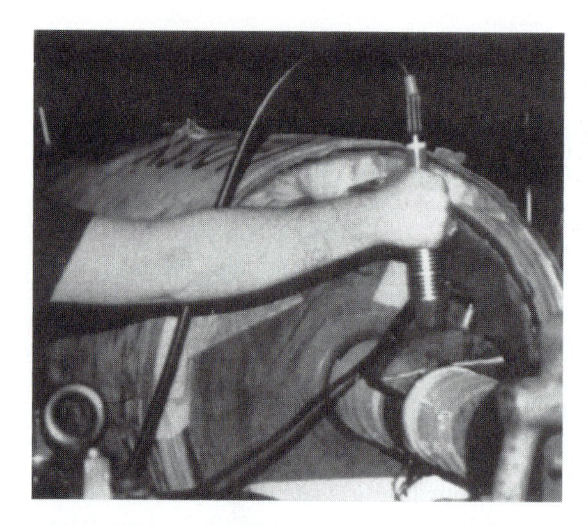

Figure 19–11
Selective electroplating of nickel on the worn journal on a paper-making roller

19.5 Other Metallic Platings

Immersion Plating

Immersion plating is mentioned only because it is part of the spectrum of metallic coatings. It is not industrially important because most immersion platings are too thin to be useful or have poor properties and adhesion. Immersion plating is the simplest plating process. The object to be plated is immersed in the plating bath and is plated without current or external energy. The classic example is plating of steel with copper in a copper sulfate bath. The mechanism is that iron is dissolved by the copper sulfate solution. The iron ions formed from corrosion replace copper ions in the solution; copper ions in turn are reduced to metallic copper by gaining electrons from the corroding iron. Usually, the plating wipes off. Thus it is not a useful process for machine use, but it is sometimes industrially used as a first step in plating hard-to-plate metals, such as aluminum or magnesium, and plastics.

Nickel can be plated from a nickel chloride bath using this same principle. Like the copper immersion plating, it is not very adherent or useful by itself, but it is often the first step in conventional electroplating, a pretreatment.

Autocatalytic (Electroless) Plating

Electroless plating started with a patented process developed for nickel in about 1945. It does not involve any external electrical energy. The part is simply immersed in the plating bath and an adherent coating develops. This process is different from immersion plating in that the mechanism of plating is a chemically induced reduction of metal ions. Electroless nickel is deposited from an aqueous solution containing nickel salts, a reducing agent, and chemicals to control pH and reaction rates. When a suitable substrate is put in this bath, it acts as a catalyst or aid in causing the nickel ions in solution to be reduced by the reducing agent. The ions are not picking up electrons from the cathode, as is the case in electroplating. The reducing agent is causing the metal ion reduction, and the metal to be coated serves as the catalyst. This electroless plating technique is also called *autocatalytic plating*, and this is now the preferred term. The reaction with a nickel sulfate bath and a sodium hypophosphite reducing agent is

$$NiSO_4 + 2NaH_2PO_2 + H_2O \xrightarrow[\text{catalyst}]{\text{heat}} \text{Ni plating} + 2NaH_2PO_3 + H_2SO_4$$

The best catalysts for this process are iron, nickel, cobalt, and palladium, but most metals will respond to electroless nickel plating because an immersion deposit of nickel forms and then the nickel surface becomes the catalyst for the reduction. Thus the term *autocatalytic*. The plating does not stop when it is completely covered with nickel. It continues to build. The rate is normally from 0.0005 to 0.001 in./h (12.5 to 25 μm). The heat that is shown to be part of the reaction comes from the solution. It is near the boiling point of water [212°F (100°C)].

Electroless nickel deposits are dense and relatively hard (43 to 55 HRC). They are not pure nickel, but nickel–phosphorus alloys containing from 4% to 12% phosphorus. They respond to heat treatment. An age hardening for about 2 h at 650°F (343°C) will increase the hardness to about 65 HRC. The coating is uniform in thickness, and all wetted surfaces including blind holes get plated.

Other electroless systems are now used besides nickel, but the nickel system is extremely useful in engineering design. For dimensional repairs, it gets around the edge buildup problem. It is more corrosion resistant than cadmium and zinc, in addition to being more durable because of its hardness. It is widely used to provide atmospheric rust resistance for machine frames, base plates, fixtures, and any number of machine parts. Plating thicknesses are usually kept below 0.002 in. (50 μm) because of the slow deposition rate. A coating of about 0.001 in. (25 μm) is usually suitable for normal industrial uses.

Electroless nickel plating can be filled with various particles to form a composite coating. Electroless nickel plating filled with from 15% to 40% PTFE is widely used for metal-to-metal wear applications. If two metals rub together and conventional oils and greases cannot be used, electroless nickel/PTFE plating can be applied to one or both members to mitigate wear. Electroless nickel can also be filled with hard particles such as silicon carbide, boron carbide, aluminum oxide, and diamond. These coatings are intended for abrasive wear situations. The normal particle loading is about 15%. These coatings are available through plating vendors in the United States. The usual thickness is in the range of 10 to 25 μm.

19.6 Electropolishing

Electropolishing is an electrochemical process, analogous to electroplating, used to polish the surfaces of metallic parts and components. The parts are placed in racks and immersed in special electrolytes. However, contrary to electroplating, the parts are made the anode and stainless steel (or another metal such as lead) is used as the cathode. Hence the current from the rectifier is used to remove metal from the workpiece surfaces in a controlled electrodissolution process.

The surface roughness of the part is smoothed and leveled in this process. Simultaneously, the part achieves a bright, mirror surface finish. Control of the electrical current and voltage is used for accurate metal removal. Typically 0.0001 to 0.0025 in. of material is removed from a surface.

Aluminum, stainless steel, copper alloys, steel, and nickel alloys are commonly electropolished. Often, electropolishing is used to polish stainless steel components for industries for which corrosion resistance and cleanability are paramount (the food, dairy, medical, pharmaceutical, and photographic industries). Tanks, vessels, and even the insides of pipes and tubes are routinely electropolished. Highly polished stainless steel surfaces are generally more corrosion resistant than rough surfaces, particularly for localized corrosion such as pitting.

Electropolishing is sometimes used to debur special or delicate parts. One of the differentiating features of electropolishing is that it may be used on irregularly shaped or complicated parts (see Figure 19–12). It is typically not as labor intensive as mechanical polishing and is amenable to high-production volume applications. Often the cost of racking and fixturing of the parts is a significant portion of the electropolishing cost. Where feasible, electropolishing can be performed on bulk parts loaded in rotating baskets.

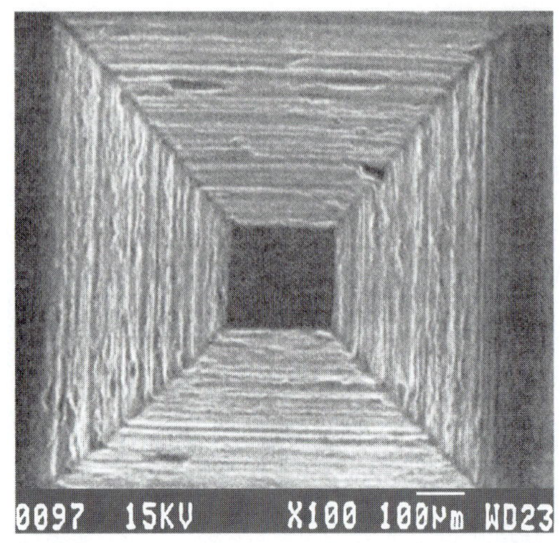

Knurl tooth before electropolishing (note the machining marks)

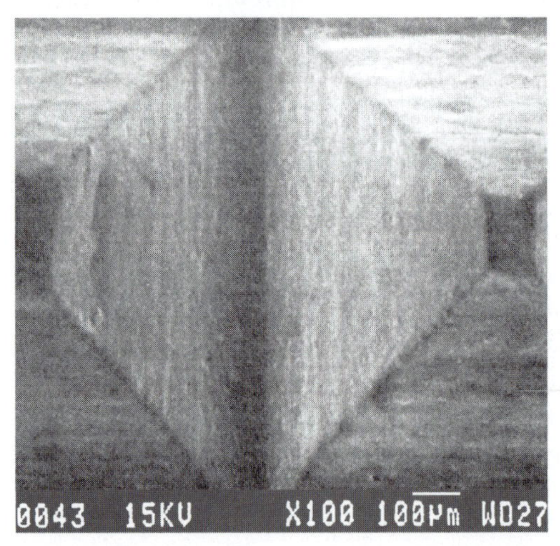

Knurl teeth after electropolishing (machining marks are smooth)

Figure 19–12
Electropolishing to remove tool marks

19.7 Photoetching

Photoetching is a technique used to chemically etch very fine and accurate features into a surface. Essentially, the part surface is masked with a photosensitive paint (photoresist), tape, or polymer film. A high-contrast or binary image is exposed on the photoresist, and the resist is developed much like a photographic negative. Depending on the type of photoresist, various forms of energy are used to expose the resist, including lasers, UV light, infrared light, white light, and x-rays. Typically, high-contrast negatives, metallic master masks, or lasers are used to expose the photoresist. Once developed, the mask remains in locations where no etching is to take place and the resist is dissolved off other surfaces where chemical etching is desired. The part with the developed resist is then placed in a suitable etching solution such as a 140°F ferric chloride solution. When exposed surfaces are etched to the desired depth, the part is removed and rinsed to stop the etching process. Then remaining photoresist is removed from the part surface.

Very fine and accurate features and structures may be created by this technique. It is not uncommon to create features as small as 10 μm. Figure 19–13 shows an example of a photoetched part. In some instances, the photoetching process does roughen the surface of the metal.

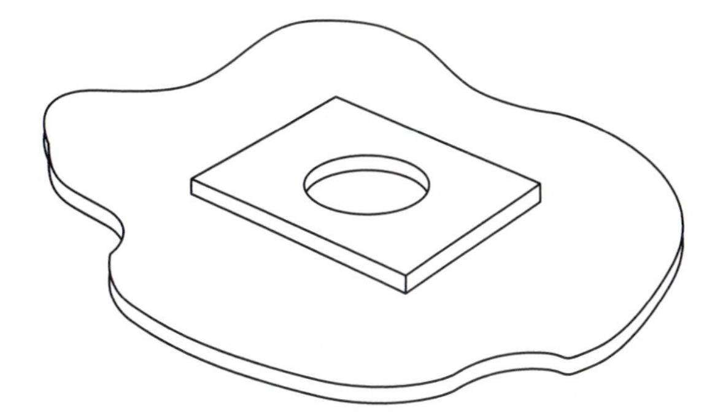

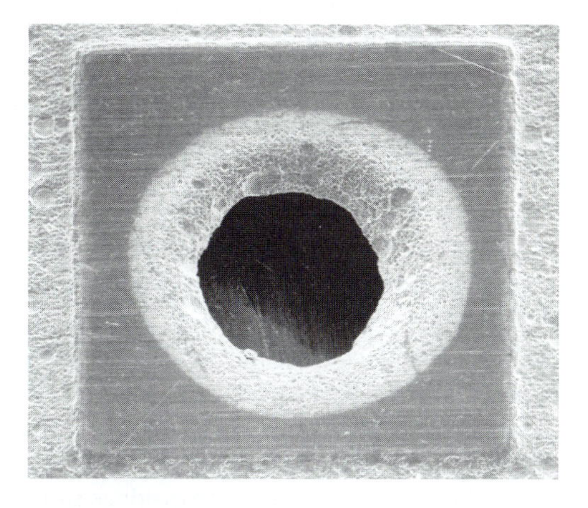

Figure 19–13
Very small features (0.2-mm hole) photoetched in Type 304 stainless steel

Photoetching is a key manufacturing process in the electronics industry. To create the copper circuit traces on printed circuit boards (PCBs), the surface of the PCB substrate (typically an epoxy/fiberglass composite) is completely plated with copper and a photoresist is applied to the surface. The photoresist is exposed and developed to reveal the desired circuit paths. The board is then placed in an acid solution to dissolve the unwanted copper. Even traces for microcircuits and integrated circuits use photoetching methods.

Other applications for photoetching include printing plates (for newspapers and such), toy models and hobbyist applications, flexible printed circuit boards, trophies, plaques, and signs. Very large parts may be processed in this manner. For example, large aircraft structures are sometimes photoetched to remove material and create certain geometric features that would otherwise be difficult or costly to machine. Photoetching is relatively fast and is amenable to high-production applications. It is often used in conjunction with other electroplating processes in the making of computer chips.

19.8 Conversion Coatings

Oxide Coatings

One of the oldest and simplest oxide coatings for metals is gun-bluing-type oxidation by heating steel parts at about 700°F (370°C) in a steam atmosphere. The reaction between the steam and the steel is the formation of a thin, tightly adherent oxide with no appreciable thickness. This type of coating has little useful corrosion resistance or wear resistance, but it does create microporosity on the surface, which absorbs oils; an oiled gun bluing will provide some measure of atmospheric corrosion resistance.

Chemical baths are available that produce this same type of coating by simple immersion techniques. A black oxide coating can be produced on copper, steel, and most stainless steels. The oxide coating is really a corrosion product. Silverware turns black from exposure to room air; the corrosion product is black. These chemical baths cause a very slight amount of corrosion of the surface, and the corrosion product is a thin [<0.0001 in. (2.5 μm)] oxide with good adhesion. On carbon steels some measure of indoor rust resistance is produced, and on stainless steels and copper alloys these coatings are usually used for a esthetic purposes or to reduce light reflectance.

Black oxide treatments are usually done with proprietary chemicals. Pastes are available that can be rubbed on steel surfaces to produce similar coatings. These coatings are not to be confused with black anodize coatings on aluminum. An anodize is an oxide produced by electrochemical conversion, and it is much thicker, more corrosion resistant, and more wear resistant than chemical conversion oxide coatings.

Phosphate Coatings

Phosphate coatings are thin adherent coatings produced by the chemical conversion of a metal surface to a phosphate compound. These coatings are produced by spraying or immersion in a heated (but not boiling) solution of dilute phosphoric acid and additives, usually of a proprietary nature. The phosphoric acid is usually less than 1% in concentration. It attacks the metal surface, a slight amount of the metal goes into solution, and this neutralizes a surface layer of the phosphating solution, precipitating metal–phosphate crystals on the metal surface. Because the coating is formed from the surface, it is very adherent, with porosity that is a function of the nature of the crystals formed.

Chemically, the phosphate crystals formed on the part can be iron, zinc, or manganese phosphates. Different baths produce different types of phosphate coatings. Zinc phosphate coatings are usually crystalline in nature and heavier than iron phosphate coatings. Coating thickness is usually measured by weight gain. The normal units are milligrams per square foot (0.09 m^2). Coatings can be as thick as 3000 mg/ft^2 [~2 mils (50 μm)]. When impregnated with oils or waxes, these coatings provide some measure of atmospheric corrosion resistance.

Iron phosphate coatings are usually thin compared with zinc and manganese phosphate coatings [30 to 90 mg/ft^2 (323 mg/m^2 to 968 mg/m^2)], and they are primarily used to assist paint adhesion. The natural porosity of these coatings provides some tooth to hold paints to metal surfaces.

Manganese phosphate coatings are the heaviest of the group [1000 to 4000 mg/ft^2 (11,111 to 44,444 mg/m^2)], and their major application is as a break-in wear coating. Rubbing machine parts such as gears, cams, sprockets, and slides never mate perfectly when they are made. There will always be some high spots, some coarse machine marks, some defects. All phosphate coatings are porous. When a machine part is phosphated, the surface layer is weak and spongy. It will absorb oil to aid lubrication, and its spongy nature causes a healing of surface scratches and waviness. Surface pits are filled in by the compressed phosphate coating, and these conforming effects combined with the oil impregnation prevent metal-to-metal contact. After running, the phosphate coatings will usually disappear from the wear surfaces, but the conditioning effect of the coating will often cause the overall wear life of a phosphated part to be as much as twice that of a part that was put in service with naked steel.

All the phosphate coatings will produce this effect, but manganese phosphate is more widely used for wear applications. This coating has a pleasing black appearance, and with oil impregnation it is often the only rust protection given to machine parts.

Phosphate coatings can be applied to cast iron, carbon steel, low-alloy steels, zinc, cadmium, aluminum, and even tin plate, but they are usually applied only to steels. The higher the alloy content, the harder it is to get a good coating. Stainless steels, tool steels, and other high alloys do not coat at all; they are immune to the phosphoric acid. There are many factors to be controlled to maintain good coatings (cleaning, pickling, solution chemistry, rinsing), but the designer need not be concerned with these.

Phosphate coatings should be used on all wear parts that could benefit from assistance during break-in. If dimensional loss cannot be tolerated, the designer should make this known. The part tolerances will dictate coating thickness. If a part has tolerances to ±0.0005 in. (12.5 μm), the coating can be kept thin enough so that size change is not detectable. Of all the nonmetallic coatings, phosphating is one of the most useful (Figure 19–14).

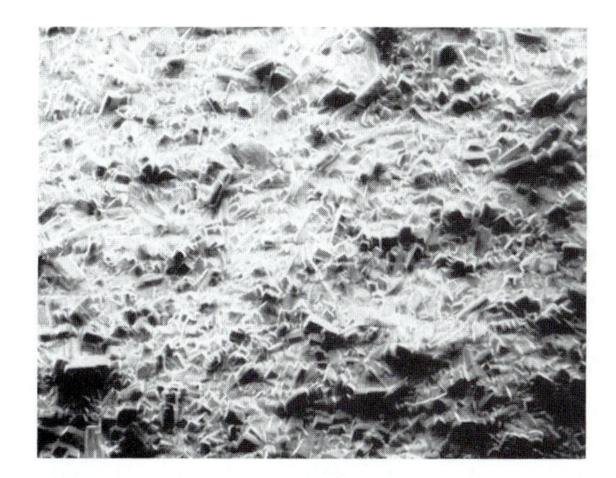

Figure 19–14
Scanning electron photomicrograph of zinc phosphate conversion coating (×1500)

Chromate Coatings

Anyone who has purchased screws, hinges, or electrical hardware has seen a typical application of a chromate conversion coating. The yellow-brown appearance on many hardware items is a *chromate coating*. It helps protect metal from atmospheric corrosion. Chromate conversion coatings are similar in concept to phosphate conversion coatings, except the protective film is formed by the reaction of water solutions of chromic acid or chromium salts such as sodium or potassium dichromate. The coating is formed when the solution attacks the metal to be coated; metal ions go into solution, giving off electrons. The electrons allow reduction of chromium ions to an amorphous film on the metal surface. The film has some solubility and would go back into solution if left too long in the chromating bath. Immersion times are normally in seconds, and the coating is fragile until it is dry. Coating thickness can be as much as 0.5 mil (12.5 μm).

This coating is applicable to zinc, cadmium, magnesium, and aluminum, and it is even used as a post treatment for phosphated surfaces. It is widely applied to zinc, magnesium, and aluminum castings, but other frequent uses are on zinc and cadmium electroplates and on galvanized steel. The corrosion protection afforded by these coatings depends on the thickness, but atmospheric corrosion protection is excellent as long as the part is not continuously immersed. Chromate coatings are somewhat water soluble.

In engineering design, this coating would be used if atmospheric corrosion would affect the serviceability of any of the metal surfaces mentioned. Because it forms fast and can be applied by immersion, brush, or swab, it is usually an inexpensive coating. This factor makes it suitable for large quantities of parts. As a class of coatings, conversion coatings range from chromate coatings less than a micrometer thick for corrosion protection to hardcoated aluminum (an electrochemical conversion coating) that may produce a wear coating 150 μm thick. They form from the reaction of a metal surface and a chemical bath. These coatings are widely used. Chromates, phosphates, and black oxide are most often used to provide a modicum of atmospheric corrosion resistance. Anodizing, hardcoating, and manganese phosphate coatings are often used for wear applications.

In the United States, chromium (and many other heavy metals) emissions to air and water are controlled by regulatory agencies. Fume scubbers are required to reduce fugitive emissions from plating and conversion coating baths, and there are strict limits on chromium concentrations in industrial wastewater. These environmental restrictions are reducing the use of chromium plating and chromate conversion coatings. A recent development in chromate coatings is the use of "dried in place" coatings. In this process, chromate solution is roller coated on to the work (for example, a metal sheet or strip) and the coating is dried on the surface like a paint coating. The immersion bath with its disposal problems is eliminated. Chromate coatings are used to enhance paint adhesion on nonferrous and ferrous metals. The dried in place coatings are not as thick as the immersion chromate coatings, but for many applications they are adequate.

19.9 Thin-Film Coatings

Physical Vapor Deposition Coating

The acronym for *physical vapor deposition* is PVD. There are a variety of PVD processes, and as a class they all involve atom-by-atom, molecule-by-molecule, or ion deposition of various materials on a solid substrate. The PVD processes of most industrial importance are thermal evaporation, sputtering, and ion plating. Thermal evaporation is the simplest and oldest of the PVD processes. It is illustrated schematically in Figure 19–15. All these processes are performed in vacuum systems.

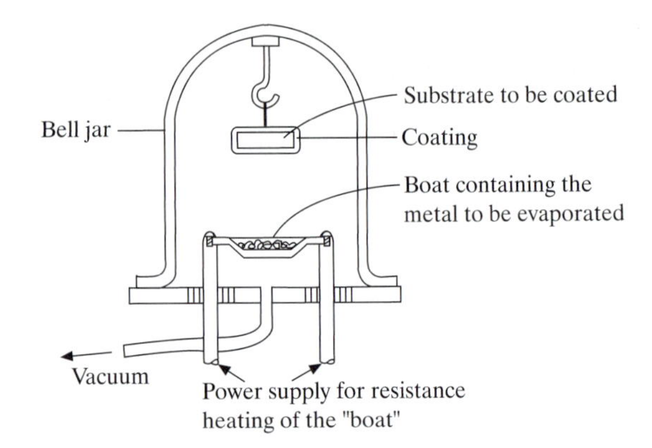

Figure 19–15
Schematic of thermal evaporation deposition

In *thermal evaporation* the metal to be coated is placed in a refractory metal boat or a crucible. The boat is heated to melt the evaporant, or an electron beam is directed into the crucible to melt the evaporant. The evaporant forms an atomic cloud (shaped like an ice cream cone, with the tip of the cone at the source) that coats all surfaces in the line of sight of the boat or crucible. This process is widely used to produce decorative coatings on plastic parts that are supposed to resemble shiny metal. For example, many automobile trim parts are plastic with a PVD coating of aluminum. A lacquer coating is applied over the evaporated coating to provide a modicum of corrosion protection.

At the other extreme, this process is used to apply relatively thick [0.04-in. (1-mm)] coatings of heat-resistant materials—MCrAlY's, alloys of a metal (M), chromium (Cr), aluminum (Al), and yttrium (Y)—on jet engine parts. These coatings are not normally used by average machine designers. The thin [0.00004 in. (0.5 μm)], decorative coatings are too fragile for wear coatings, and the heavy thermal evaporation coatings are specialty coatings not for general use. If a design situation requires a decorative coating or a coating on optics, for example, thermal evaporation may be a candidate surface engineering process.

High-technology coatings such as ceramics, metal alloys, and organic and inorganic compounds are applied by *sputtering*. In dc sputtering the workpiece and the substance to be coated are connected to a high-voltage dc power supply. When the vacuum chamber has been pumped down, a controlled amount of argon or another gas is introduced to establish a pressure of about 10^{-2} to 10^{-3} torr. The current supply is energized, and a plasma is established between the work and the material to be coated. The gas atoms (usually argon) are ionized, and they bombard the material to be coated (target). The energy of the impinging ions causes atoms of the target material to be sputtered off, and they are transported through the plasma to form a coating on the work. Direct-current sputtering is used when the evaporant (target) is electrically conductive. Radio-frequency (RF) sputtering, which uses an RF power supply rather than a dc power supply, is used when the evaporant (target) is a nonconductor such as a polymer.

A third variation of the PVD process is *ion plating*. In this process, metal is evaporated as in thermal deposition, and a plasma is also established to ionize the evaporating species. Evaporant ions bombard the substrate with such energy that they physically implant into the substrate to produce an extremely strong coating bond strength.

Sputtered and ion-plated coatings are used in design for very thin [<3 μm (0.12 mils)] coatings for electrical, optical, and wear-resistant applications. Hard thin-film coatings (<3 μm) are widely used to enhance the wear properties of tools.

If a tool such as a punch press die is wearing faster than anticipated, the face of the tool can be TiN coated to add abrasion resistance. TiN is harder than the hardest metal.

Diamond and diamond-like coatings by PVD or *chemical vapor deposition* (CVD) are commercially available, but in 2003 they were not as widely used as TiN and TiCN. They may require experimentation to determine if they will adhere to a particular substrate.

The PVD coatings most widely used in machine design are titanium nitride (TiN), titanium carbonitride (TiCN), and diamond-like carbon (DLC). The latter coatings are also called amorphous hydrogenated carbon (AHC) coatings because they often do not have the same atomic bonding as diamond and they can contain up to 30% hydrogen (from the hydrocarbon gas used to produce these coatings). Many vendors in the United States and in most industrialized countries apply these coatings. They are applied to punches and dies, cutting tools, forming tools, injection molding tools, and countless other tools. Most are applied to add additional hardness to a substrate that is already hard, but some versions of these coatings are intended to be lubricious to help sliding wear systems.

Titanium nitride is gold in color, and the DLC (or AHC) coatings are often black. The colored coatings are often preferred for many tool applications because users can visually see when they are removed in use. The hardness of some of these coatings are compared here:

Coating	Hardness (HV)	Color
Titanium nitride	2900	Gold
Zirconium nitride	2800	Gold
Titanium aluminum nitride	2600	Brown
Titanium carbonitride	4000	Silver
Chromium nitride	2500	Silver
Amorphous DLC	1000 to 5000	Black

These coatings as are usually applied with a thickness in the range of 1 to 2 μm. They are often applied over other coatings so that there is not a thin coating on a much softer substrate, like a thin layer of ice on top of 6 in. of snow. For example, titanium nitride coated on H13 tool steel at 53 HRC may be more prone to surface fatigue failure than the same coating applied over a 25-μm-thick nitride layer on the H13. Graded coatings reduce the effects of property mismatch. There are models that help select

Figure 19–16
Nodules of TiN on PVD-coated TiN. These can be abrasive when softer materials slide against them.

coatings and thicknesses using relative hardness and elastic modulus data as the model input.

Many sputtered coating processes, especially those that involve arc melting of the evaporant, produce microscopic nodules (Figure 19–16) on the coated surface. These nodules can have diameters as large as several micrometers, and they make coated surfaces abrasive to conforming surfaces. These nodules are commonly called *macros* in the United States, and the presence of these surface features should be a selection factor in choosing a coating vendor. Macros are usually undesirable for metal-to-metal sliding systems. On the other hand, they are usually beneficial when the coating is applied to cutting tools. The nodules reduce tendencies for chips to weld to the coated surface.

Coating temperature is another selection factor. Some coating processes are performed at 300°F (148°C), some require 800°F (426°C). The higher temperature processes usually produce optimum coating properties, but an 800°F (426°C) coating temperature will soften many tool steel substrates. Most cemented carbides can tolerate this coating temperature with no ill effects. If softening of the substrate is a concern, the coating temperature should not exceed the original tempering temperature of the part being coated.

These coatings can be specified on drawings with the type of coating, the thickness, and the allowable process temperature:

> Coat area shown with 1 to 2 μm of titanium nitride, maximum process temperature 300°F. (Approved suppliers are Braden Corp. and Gulden Coatings Inc.).

The DLC/AHC coatings may not be as commercially established as the sputtered TiN and TiCN coatings. The coatings that contain substantial hydrogen are affected by humidity. Some coatings contain additional species such as silicon to reduce the humidity sensitivity. There are no standard DLC or AHC coatings in the United States. Each supplier produces a coating that is different in properties, structure, and adhesion. There are scores of commercially available coatings; users need to decide which of these will meet performance expectations.

Chemical Vapor Deposition Coating

This process is known by the acronym CVD. It has an advantage over other coating systems in that heavy coatings can be produced on metals as well as nonmetals such as glass and some plastics. The metal coatings can be dense and ductile with good adhesion. The basic requirement of the process is the availability of a metal compound that will volatilize at a fairly low temperature and decompose to a metal when it comes in contact with a substrate at some higher temperature. A schematic of the equipment required for this process is shown in Figure 19–17.

An example of chemical vapor deposition systems is nickel deposited from nickel carbonyl ($NiCO_4$). Coatings as thick as 0.1 in. (2.5 mm) can be applied at a rate of up to 0.01 in. (0.25 mm) per hour. One application of this coating that may find use in machine design is a heavy

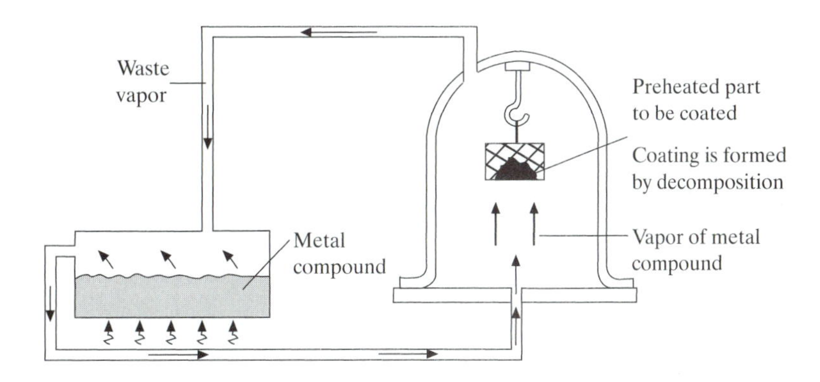

Figure 19–17
A CVD system

nickel coating (~0.1 in.) on glass containers to make them explosion or shatter resistant.

In the 1990s, diamond CVD coatings came into vogue on cutting tools. The CVD process has an advantage over the PVD process in that it is not line of sight. All surfaces in the reaction chamber get coated. There are disadvantages: A separate process and reaction must be developed for each coating. Some of the gases used are extremely toxic and dangerous. Probably the biggest disadvantage is that the reactions involved in applying the coatings may be done at temperatures in excess of 1300°F (700°C). This temperature will soften most tool steels. Thus the most suitable substrates for CVD coatings of this type are cemented carbides, which will not soften at this temperature.

A newer CVD process involves plasma assist, and this process is called *plasma-assisted chemical vapor deposition*. This process is used to apply diamond and diamond like carbon coatings. Some processes are very low temperature; for example, there are processes for applying silicon carbide barrier coatings on plastic films and semiconductors. The hard wear coatings are usually applied in the thickness range of 1 to 4 μm (0.04 to 0.16 mils). On the other extreme, CVD is used to produce bulk shapes of high-purity silicon carbide. Reactants are deposited on a chamber wall to a thickness of tens of millimeters. The coating is stripped and cut into shapes for use as ceramic parts. Even diamond films can be created.

Thin-film coatings show much promise as a surface engineering tool. They are already key to the manufacture of many electronic devices; they are used to apply dopants, sealants, and the like, to chips and other microelectronic parts. In product design, thin PVD coatings of pure metals can improve the aesthetics of plastic parts. Thermal evaporation is a low-cost process, but all these processes are normally batch processes because of the vacuum chamber requirement. These processes are not well suited to a continuous operation.

19.10 Surface Analysis

Part of surface engineering is analyzing the chemical or atomic nature of a surface, or even analyzing its state of stress (residual stress). Surface cleanliness is important in all of the coating processes that were reviewed. If coating adhesion becomes a problem in the use of a surface coating, there are a plethora of surface analytical processes that can be used to research these problems. Figure 19–18 lists some of the more common surface analysis processes used on metals. In addition to these, Fourier transform infrared analysis (*FTIR*) is useful in identifying organic contaminants on a surface and x-ray diffraction (*XRD*) is the process used to analyze the state of residual stress at a surface or to identify an unknown chemical substance (surface or bulk).

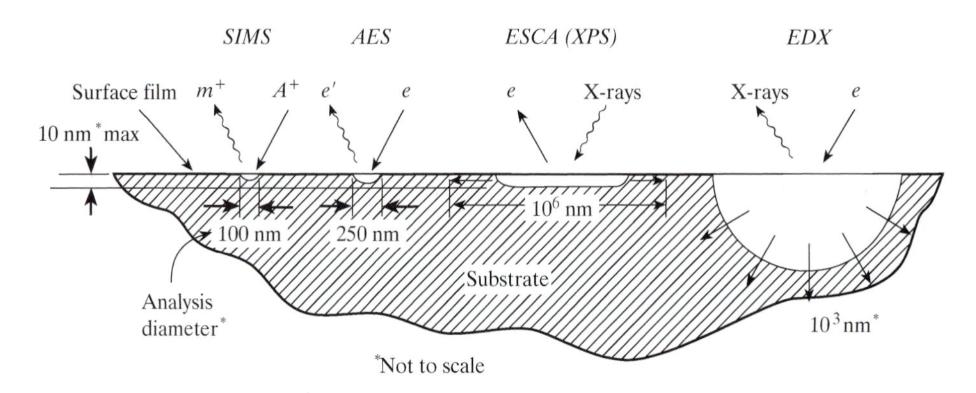

Figure 19–18
Comparative analysis volume of different surface analysis techniques. The analysis volume is the depth of penetration times the area over which the analysis is taken. The analysis diameters are not to scale.

Raman spectroscopy is used to examine diamond coatings to see if they really have the atomic bonding that distinguishes diamond from graphite and other forms of carbon. As mentioned previously, photoelectron emission is a tool to determine surface cleanliness.

Essentially, surface analytical techniques bombard surfaces with electrons, photons, or ions or forms of electromagnetic radiation (x-ray, etc.) and detectors measure the material response by detection of the same things: electrons, photons, ions, x-rays, etc. The differences between the various surface analytical processes depend on the impinging medium and what is detected. Each process has some advantage over the others. Most surface analysis labs have several processes, and they use whatever process is best for the suspected contaminant or surface component. Users of surface engineering processes may have occasion to investigate coating or other surface problems. The following are brief descriptions of some of the more common processes to aid in their selection and use.

Scanning electron microscopy (*SEM*) is used like optical microscopy to simply show the topological features of surfaces. It can be used for magnifications that are outside the realm of optical microscopy ($>\times 3000$), and it also gives a depth of field that shows features that cannot be resolved by conventional optical techniques. Many SEM units are now equipped with an *energy dispersive x-ray* capability, *EDX*, that will produce quantitative information on the chemical nature of surfaces. The surface to be analyzed is bombarded by an electron beam, and the x-rays that are produced are used to identify the elemental nature of the surface at the point of impingement. As shown in Figure 19–18, the area of the surface that is analyzed is small compared to some other processes, but the depth of the analysis is so great, about 1 μm, that this technique really produces analysis of the substrate. This process is most effective on metallic substrates, and it is most commonly used for analyzing particles or phases that come to the surface.

Electron spectroscopy for chemical analysis (*ESCA*), also known as x-ray photoelectron spectroscopy (XPS), involves bombarding the surface with low-energy x-rays to emit photoelectrons that are collected and analyzed for information on the elemental nature of the surface. The analysis depth is a few nanometers, and the area of analysis is quite large, about a square millimeter. It can be used to analyze all types of films on most types of substrates. This process is used where the large area of analysis is beneficial.

Secondary ion mass spectroscopy (*SIMS*) can be used to remove layers of the surface and profile the chemical changes as the layers are sputtered off. The surface is bombarded with a beam of ions of some gas such as argon or neon. The action of the beam sputters atoms from the surface in the form of secondary ions, and these are detected and analyzed to produce information on the elemental nature of the surface. The depth of the analysis is usually less than a nanometer, making this process the most suitable for analyzing extremely thin films. The area of analysis can be as large as 0.1 μm. This process is often used to profile the composition through the thickness of films.

Auger electron spectroscopy (*AES*) is also called SAM (for scanning Auger microprobe) in devices that allow movement of the area of analysis. It produces the smallest analysis spot size of the four techniques illustrated by Figure 19–18. The surface is bombarded with an electron beam, and the action of this beam produces electron changes in the target atoms; the net result is the ejection of Auger electrons, which are characteristic of the element from which they were emitted. Because of the small depth of analysis and the small spot size, this process is most often used to chemically analyze microscopic surface features.

The chemical nature of the surface of a material is often part of surface engineering and these four processes are the most popular ones used to produce this information.

19.11 Hardfacing

Hardfacing is applying with welding techniques a material with properties superior to those of the material to which it is applied (substrate). The term was coined in the United States in the 1970s by a subcommittee of the Welding Research Council of the United States, and it was intended to include both fusion and nonfusion welding processes. Fusion processes require melting of the hardfacing and the substrate, whereas the latter require melting only of the material to be deposited (as it is sprayed onto the substrate). *Thermal spray coatings* is the term used to describe coatings that are melted and sprayed or cold sprayed with welding techniques. Fusion hardfacing processes include the following:

	Process	***Acronym***
Arc processes	Shielded metal arc welding	SMAW
	Flux core arc welding	FCAW
	Gas metal arc welding	GMAW
	Gas tungsten arc welding	GTAW
	Submerged arc welding	SAW
	Plasma arc welding	PAW
Torch process	Oxy-fuel gas welding	OFW
Other processes	Laser beam welding	LBW
	Electron beam welding	EBW
	Electroslag welding	ESW
	Furnace braze	FB

We will discuss thermal spray processes in Section 19.12.

Basically the arc processes involve melting and coalescence of a rod, wire, or electrode (the consumable) with the substrate. There is dilution of at least the first layer of the hardfacing with melted substrate, and the degree of dilution depends on the deposition process (oxy-fuel gives the lowest dilution). It is common practice to apply two layers to minimize substrate dilution in the outer layer. The types of consumables that can be applied range from relatively expensive cobalt-based alloys to relatively low-cost alloy steel *buildup* materials. Hardfacing deposits have no technical thickness

Figure 19–19
Knife edges rebuilt by OFW hardfacing with a hard alloy (RCo–Cr–A)

limit, but two layers with a finished thickness of 1/8 in. minimum (3 mm) is common. Deposits can be left as-applied on earthmoving machinery, and the like, and in these cases the deposits can be 1/4 in. (6 mm) or more. Deposits are hard as applied, and the hardfacing process is extremely useful for imparting wear and possibly corrosion resistance to selected areas of machinery (Figure 19–19). A significant consideration in the use of fusion hardfacing processes is the weldability of the substrate and sufficient mass in the part to minimize distortion. The major classes of alloys applied by fusion hardfacing are

Tool steel	For the same types of applications as wrought tool steels
Manganese steel	For mineral extraction and gouging wear applications
Copper-based alloys	For rebuilding worn machinery-steel parts
Composites (steel or WC)	For high-stress abrasion applications
Cobalt-based alloys	For galling resistance for applications where wear and corrosions are conjoint
Nickel-based alloys	For metal-to-metal applications
Iron chromium alloys	For high-stress abrasive wear and gouging

Suitable substrates are materials that have good weldability with these alloys. Low-carbon steels are the most common substrates, and the two most popular hardfacings are the iron/chromium alloys and nickel-based alloys. Hard, thick deposits are prone to *checking*, small cracks form from welding shrinkage stresses. This can be a selection concern.

19.12 Thermal Spraying

Thermal spraying has been highly available in the United States and worldwide since the 1990s. It has become a standard tool for improving surfaces in most industries. *Thermal spraying* is the application of a material (the consumable) to a substrate by melting the material into droplets and impinging the softened or molten droplets on a substrate to form a continuous coating. The molten droplets splat cool on impact with the substrate, and the coating is essentially formed by overlapping splats. The coatings can contain porosity, unmelted material, and oxides. The latter are considered to be coating defects, but porosity is inherent in all the nonfusion processes. Some processes produce less porosity than others; the high-velocity processes produce the lowest amount of porosity (Figure 19–20). The mechanism of bonding to the surface is the same as in plating: some atomic interaction, some mechanical interlocking. Most thermal spray processes require abrasive blasting or a bondcoat to optimize the coating adhesion. Bond strengths in shear are about 10 ksi on good deposits. There is no technical thickness limit, but the more expensive coatings are usually only a few mils thick; low-cost buildup materials applied with high-deposition-rate guns may be economical up to thicknesses of 0.1 in. (2.5 mm).

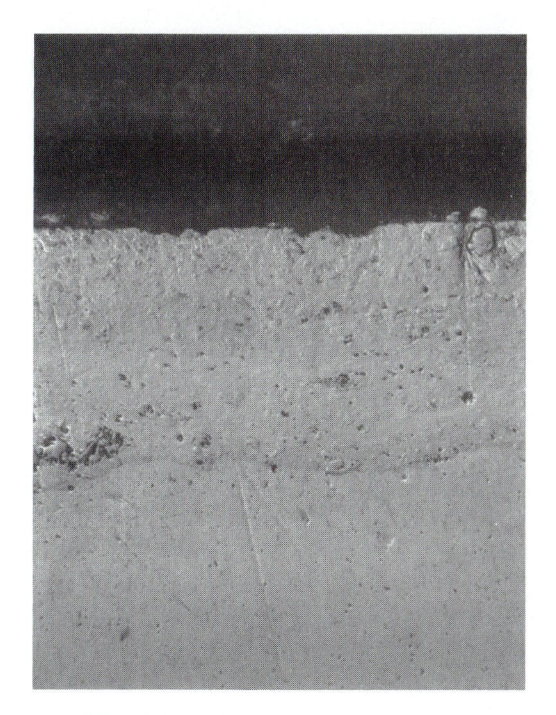

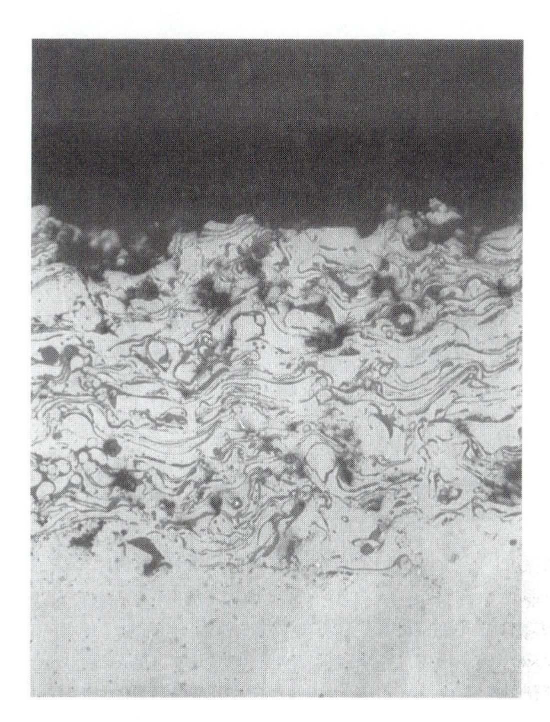

Figure 19–20
Comparison of detonation gun coating microstructure on a WC/Co coating (left) to WC/Co sprayed by a plasma torch (right). (Photos are ×400 magnification.)

The following are the most common thermal spray processes:

Process	*Description*
Flame spraying (FLSP) Gas temperature 5430°F (3000°C)	Powder, rod, or wire is melted by an oxy-fuel flame.
Electric arc spraying (EASP)	Wires are motor driven at each other and melted by an electric arc. The melted droplets are propelled by a gas at the substrate.
Plasma arc spraying (PSP) 25,000°F (14,000°C)	Powder is melted by an arc-generated plasma within the gun.
Detonation gun (d-Gun) 5430°F (3000°C)	Powder is melted in a gun by spark ignition of explosive gas. Particle velocity is 9850 ft/s (3000 m/s).
High-velocity oxy-fuel (HVOF) 4500°F (2500°C)	Powder is melted in a combustion chamber containing oxygen, hydrogen, and a fuel gas (methane, etc.). [Particle velocity is 2250 ft/s (750 m/s)].
Cold spray	Particles do not melt and adhere from high-velocity impact.

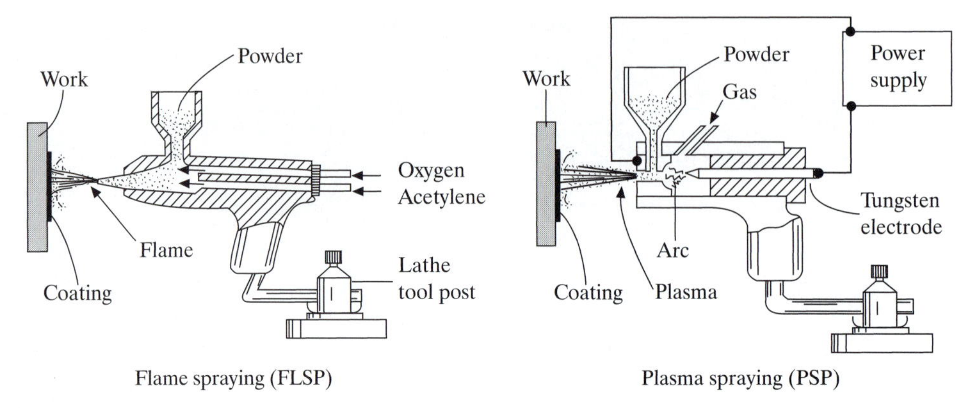

Figure 19–21
Nonfusion processes: flame and plasma spraying

Flame spraying and plasma "torches" are illustrated in Figure 19–21. *Flame spraying* was the first thermal spray process. The concept is the same for plasma spraying, only the source of heat is a plasma with a temperature of over 25,000°F (14,000°C) compared with only about 5000°F (2760°C) for an oxyacetylene flame.

Plasma arc spraying produces denser coatings, less porosity, and better adhesion; it is more suitable for spraying ceramics. Many types of metals and ceramics can be applied to almost any metal substrate. There is very little heating of the substrate; part temperatures rarely exceed 300°F (149°C). Thermosetting plastics can even be ceramic coated.

Thermal spraying of metals and ceramics is useful for part buildups and for producing wear-resistant surfaces. The porosity of these coatings makes them suited to impregnation of lubricants. Bond strengths are usually in the range from 10 to 20 ksi in shear (69 to 138 MPa).

High-velocity thermal spray processes such as HVOF and detonation gun coating serve the same function as the plasma processes, but they usually have better bond strength and lower porosity. They are usually used for applying wear-resistant coatings like carbides or ceramics.

Thermal spray processes sometimes compete with platings for atmospheric corrosion control. Tanks and large pieces of equipment can be plated with zinc, nickel, or other platings for rust protection. Flame spraying of anodic coatings such as zinc or aluminum (with wire-fed arc guns) can be used to protect very large structures from atmospheric rusting. Water tanks, TV towers, and even large bridges can be sprayed. A thermal-sprayed zinc coating 5 mils (125 μm) thick can keep a bridge rust-free for 30 years or so. This process is still somewhat more expensive than painting in the United States, and its use is not as prevalent as it is in, for example, Japan. There it has been used for bridges with spans of 1000 m (3280 ft) or more.

The *cold spray* process is still in the development mode, but it is a variation of the high-velocity processes that does not melt spray particles at all. Particles are heated a small amount (<50°C) and then accelerated at the substrate by expanding a high-pressure gas (usually helium at several hundred psi) through a diverging nozzle. Particle velocities can be mach 3 or more and the particles (10–30 μm in diameter) adhere to the surface as they deform to a splat thickness of about 20% of their diameter. Coating thickness can be more than an inch (25 mm), which allows for the use of this process for making shapes (parts, prototypes, tools). The drawback is that in 2004, this process only works on ductile metals that can deform into a "cold splat" on impact. Currently, it is mostly used to apply corrosion protection coatings of zinc and aluminum or for electrical conductor coatings of pure copper for EMI shielding and the like.

19.13 High-Energy Processes

In 2004 in the United States, there were a variety of high-energy surface treatments being used to alter properties of surfaces; sometimes these processes were performed without adding dimension to the surface. The common systems in use were ion implantation, laser treatments, and electron beam treatments. *Ion implantation* is the process of impinging ions of elements on the surface of a material with sufficient energy (velocity) that the ions embed into the atomic lattice of the substrate (Figure 19–22). This process is performed in a vacuum chamber. Ions are usually produced by an ion gun. There are many variations of equipment, but one way to produce ions is to pass a gas through an electron beam or plasma. The gas atoms become ions from collisions with the electron beam or species in the plasma. When ions are generated, they are not magnetically neutral. They can be accelerated and focused by magnetic coils and impinged on the work surface. When ions implant into a surface of a metal, they allegedly create atomic defects and misfits (and some alloying) that harden and strengthen the surface. The depth of implantation is usually only about 4 μin. (0.1 μm), but the "sphere of influence" of the implantation allegedly is much deeper, maybe a micrometer or so (0.00004 in.).

Various atomic species can be implanted, but the common species are nitrogen, carbon, boron, and chromium. Commercial implantation companies who perform these surface treatments claim significant improvements in tribological and mechanical properties, and alteration of chemical and sometimes physical properties of surfaces. Implantation is not normally performed on soft materials; it is more often applied to hard materials such as tool steel to improve wear or related properties. The parts to be treated must fit into a vacuum chamber. Part heating can be minimal, and most parts never get hotter than 300°F (148°C). The benefit of implantation varies with the application. Implanting a tool would not guarantee improved performance. Development work may be necessary in species and implant dosage. Implantation is in wide commercial use for doping semiconductor materials. Its use for treating tools is not as widespread in the United States in 2004.

Laser and electron beam treatments of surfaces are performed to alter surface properties by rapid heating and cooling. Glazing with laser

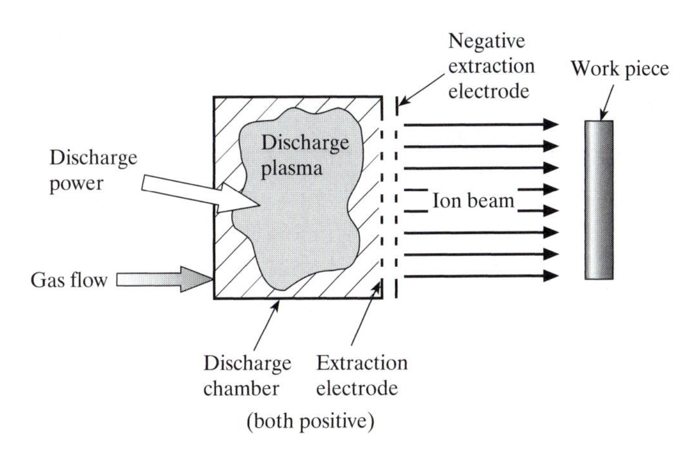

Figure 19–22
Schematic of ion implantation (courtesy Prof. P. Wilbur, Colorado State University)

and EB involves locally heating a surface to slightly below the melting temperature and allowing the mass of the material to rapidly quench the surface. If cooling rates are above about 10^6 °C per second, an amorphous structure can be obtained in some materials. The heated depth is usually on the order of about 0.004 in. (100 μm). The surface properties of these glazed surfaces can be significantly different from the properties of the bulk material. This process is not widely used, but it can be used to make amorphous surfaces if such a surface will offer some property advantage.

A more recent adaptation of this process is to use a laser or electron beam to produce surface alloying. This is really a form of hardfacing. An example of this process is to fuse aluminum oxide powder to the surface of zirconia. The alloy depth is only a few micrometers, but the self-mated wear properties are improved by a factor of ten. This is a very powerful surface engineering tool, and it shows considerable promise as a way of creating special alloy surfaces.

In summary, ion implantation, laser treatment, and electron beam treatment provide yet more ways to engineer surfaces. The result of these treatments is not always known, but with continued research, these processes may be the carburizing and nitriding of the future.

19.14 Diffusion Processes

In Chapter 10 we classified diffusion treatments for steels into categories: thin and thick. We also listed typical processes under each category. All these processes—carburizing, nitriding, carbonitriding, and the like—are surface engineering tools. They compete with coatings and other surface treatments for wear applications and other applications that require a hard surface. Carburizing, carbonitriding, and ferritic nitrocarburizing will produce surfaces that are comparable to hardened steels in wear properties. Nitriding can produce surface hardnesses of 70 HRC on nitriding steel substrates. This is the hardest steel surface. Nitrided surfaces on nitriding steel compete with chromium plating in hardness. If an application requires even more abrasion resistance, boronizing and refractory metal diffusion treatments [titanium carbide (TiC) and vanadium carbide (VC)] produce extremely hard layers about 0.0001 in. (2.5 μm) deep. These coatings are significantly harder than even nitrided surfaces. They are extremely abrasion resistant. Titanium carbide diffusion treatments produce a surface growth of about 0.00005 in. (1.25 μm), and nitriding produces a growth that is a function of the depth of case, but for the most part, diffusion processes can be considered to be into the substrate surface. They can be used where a surface buildup, as in a coating, cannot be tolerated.

19.15 Selective Hardening

Selective hardening applies to a variety of steels and cast irons (Table 19–2). As described in Chapter 10, depth of hardening can be as shallow as 0.01 in. (250 μm) from the surface, or all the way through some sections. Flame hardening is the most appropriate process for deep hardening of parts that may be too large to fit into a furnace. Induction hardening is more suited to hardening small parts—the work must fit within a heating coil. This process is suited for production-type parts, rather than one-of-a-kind parts. Flame hardening and induction hardening processes compete with carburizing and other diffusion processes, with coatings, and with through hardening. There is set-up involved in both flame hardening and induction hardening. This must be considered in the economics of using these processes. The big advantage of using selective hardening processes over other surface engineering processes is that they produce a significant depth of hardening, and this hardening does not significantly alter the part dimensions. There is no intentional buildup. The only size change is

Table 19–2
Materials that are commonly flame, induction, electron beam, or laser hardened

Carbon Steels	Hardness (HRC)	Alloy Steels	Hardness (HRC)
AISI 1025 to 1030	(40–45)	AISI 4140	(50–60)
1035 to 1040	(45–50)	3140	(50–60)
1045	(52–55)	4340	(54–60)
1050	(55–61)	6145	(54–62)
1145	(52–55)	52100	(58–62)
1060	(60–63)		
Tool Steels		**Cast Irons**	
AISI O1	(58–60)	Class 30	(45–55)
S1	(50–55)	Class 45	(55–62)
P20	(45–50)	Ductile 80–60–03	(55–62)

due to the volume expansion associated with changing soft structure to hard structure.

19.16 Special Surface Treatments

Figure 19–2 lists three special treatments that penetrate surfaces: cryo, magnetic, and vibration. Actually these processes (if they really work) affect the bulk material, not just the surface. We mention these processes because they are out there and designers will eventually encounter them. During the 30+ years that these processes have been around in the United States, there have been an equal number of users of these processes and of skeptics who refuse to use them. Cryogenic treatment of hardened steels is widely available on a commercial basis in the United States. This process involves slowly cooling hardened steel to about −300°F (−166°C). It supposedly "densifies" the material and increases wear resistance. We discussed the technical basis for cryogenic treatment of steel in Section 10.2. In the author's experience it has not improved service life in the applications tested, but there are many testimonials that it does improve wear life and dimensional stability.

Vibration treatments are supposed to do the same job as a thermal stress relief. There are many who believe that this process works and a similar number who do not. The same situation exists concerning treatments for steels that involve alternating magnetic fields. This process is supposed to improve wear life of high-speed steels and similar tool materials.

Process/Material Repertoire

Figures 19–23 and 19–24 illustrate some of the processes available for modifying surfaces and also list some discriminating characteristics for each process. Of course, this list is an opinion (the authors'), but most purveyors of surface engineering processes will tend to agree that each process has its niche where it works better than the competition. The same situation exists for materials that are applied with these processes. Plasma-sprayed chromium oxide reportedly performs better than other spray processes and consumables for certain types of paper rolls. Chromium electroplate seems to work better than other coatings in resisting abrasion on hydraulic cylinder shafts. Thus our discriminating factors are perceptions based on industrial experience, and they are intended to serve as a guide for designers who may have not had occasion to use a particular process.

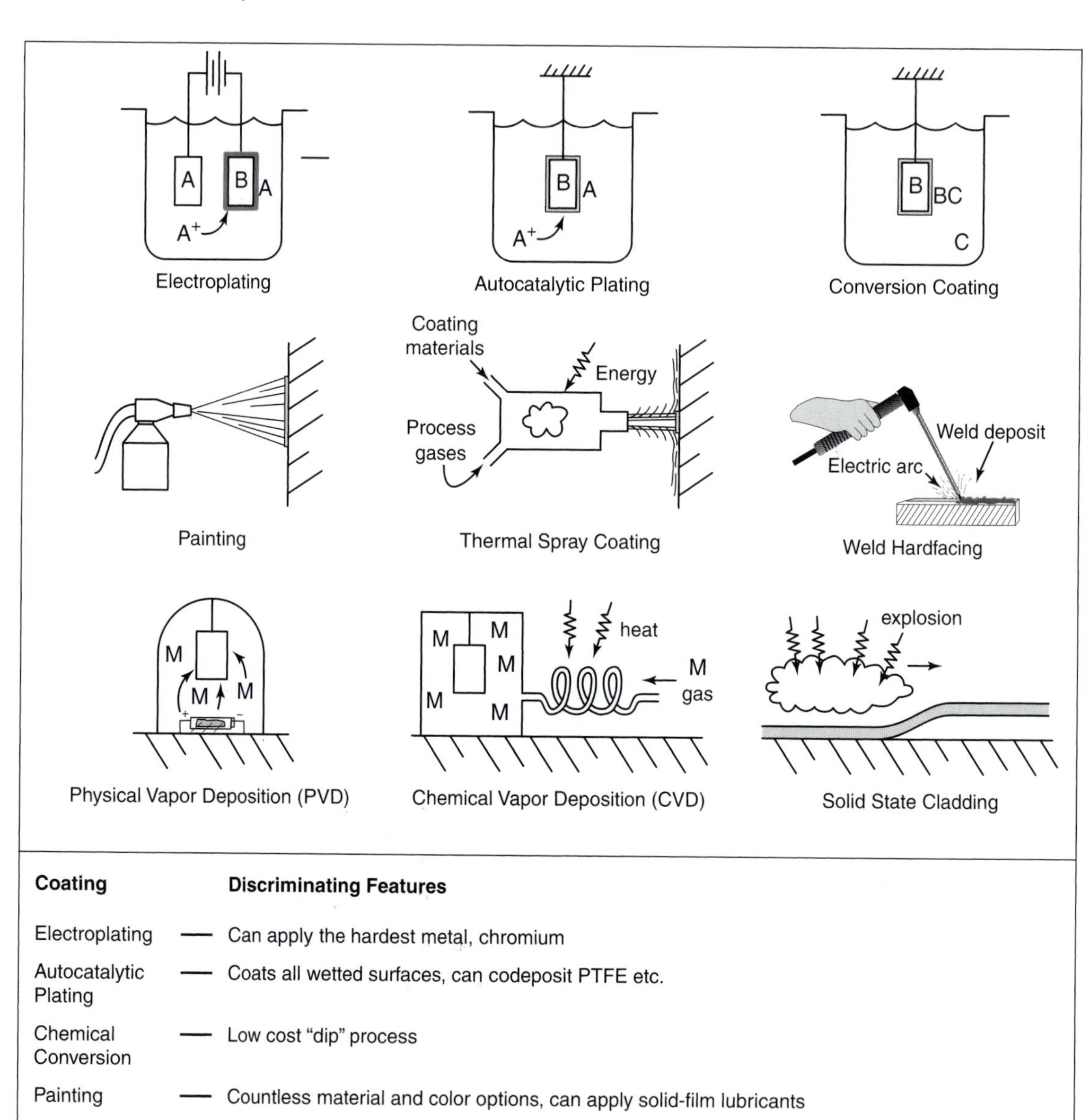

Coating		Discriminating Features
Electroplating	—	Can apply the hardest metal, chromium
Autocatalytic Plating	—	Coats all wetted surfaces, can codeposit PTFE etc.
Chemical Conversion	—	Low cost "dip" process
Painting	—	Countless material and color options, can apply solid-film lubricants
Thermal Spray	—	Many material options, applicable to many substrates
Fusion Hardfacing	—	Many material options, can be very thick
PVD	—	Readily available for application of thin films
CVD	—	Can apply very special materials including diamond
Cladding	—	Can apply special metals to large substrate areas

Figure 19–23
Important surface engineering processes that apply a coating

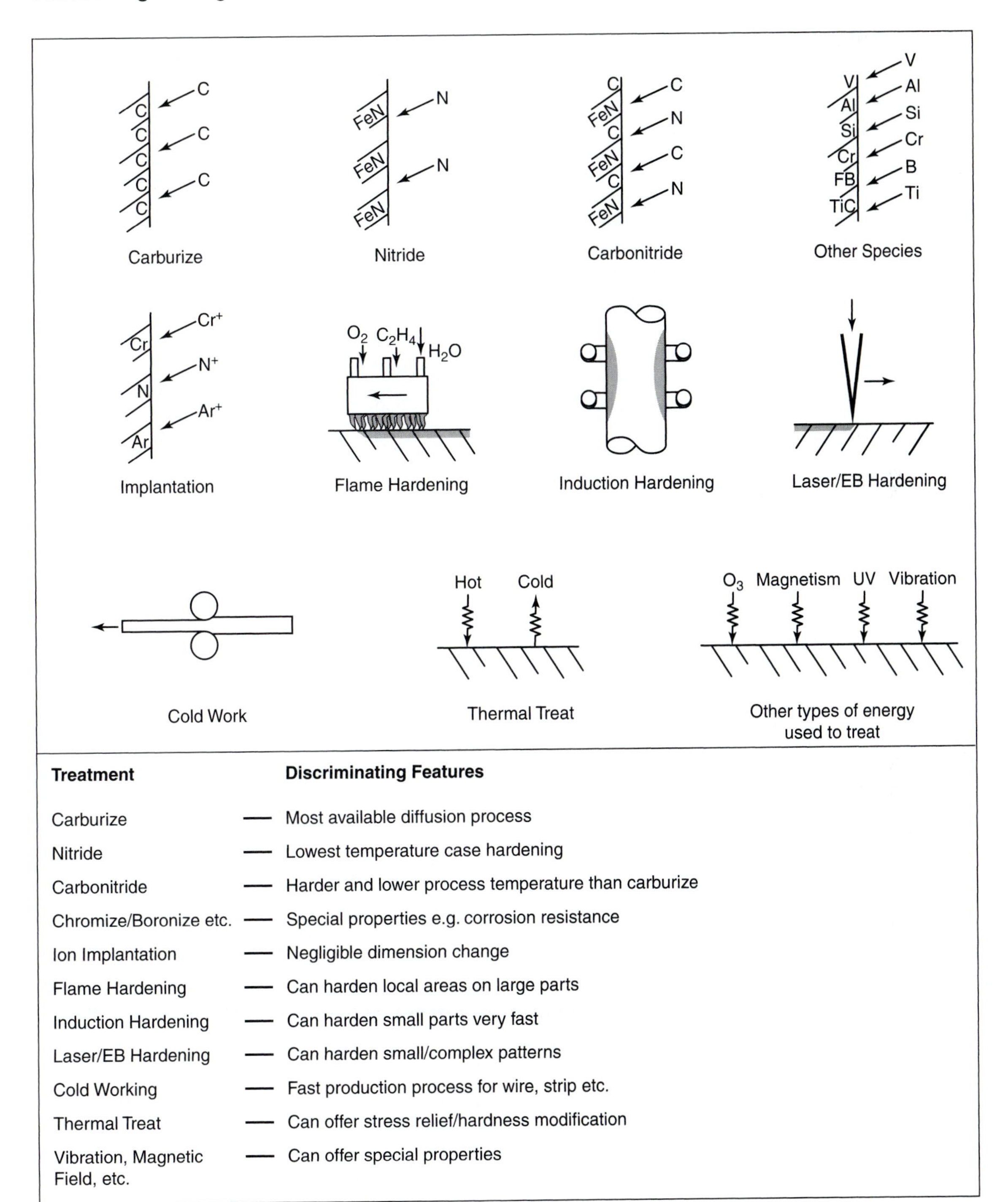

Treatment		Discriminating Features
Carburize	—	Most available diffusion process
Nitride	—	Lowest temperature case hardening
Carbonitride	—	Harder and lower process temperature than carburize
Chromize/Boronize etc.	—	Special properties e.g. corrosion resistance
Ion Implantation	—	Negligible dimension change
Flame Hardening	—	Can harden local areas on large parts
Induction Hardening	—	Can harden small parts very fast
Laser/EB Hardening	—	Can harden small/complex patterns
Cold Working	—	Fast production process for wire, strip etc.
Thermal Treat	—	Can offer stress relief/hardness modification
Vibration, Magnetic Field, etc.	—	Can offer special properties

Figure 19–24
Important surface engineering treatments

Useful Materials

Table 19–3 presents some useful processes and materials for various kinds of application. Figure 19–25 shows how our repertoire might apply to the various purposes for using surface engineering. This illustration shows that some processes (e.g., plating) can be used in all application categories. Thermal spray coatings similarly are used for a number of reasons. This reflects the importance of different processes (Figure 19–25). The remainder of this section contains some comments on the applicability of processes and materials for various types of applications. The following section will couple these processes and materials with a selection methodology.

For appearance: If the goal of a surface engineering process is good appearance, the candidates may include the decorative platings, PVD coatings, black oxide, paints, and others. Anodizing is probably the most-used process for improving the appearance of aluminum surfaces. Anodize coatings can be dyed many colors, and these types of surfaces are widely used on architectural elements. Black oxide is used to make machine parts look better, but it has very little durability. Powder coatings and paint can provide durable and aesthetically pleasing surfaces. PVD titanium nitride is gold colored, and this process has been used to make coatings for jewelry. Impurities in the vacuum system can make the gold titanium nitride coatings iridescent; these coatings on titanium can make attractive earrings, and the like.

For corrosion: Platings and conversion coatings are widely used for corrosion mitigation, but there are precautions to keep in mind. Recalling our discussions of galvanic attack, if a defect is present in a coating that is more noble than the substrate, severe pitting will occur. Even without galvanic action, if a coating is used for protection from a chemical that is aggressive to the basis metal, severe pitting will still occur at coating discontinuities.

Most platings are acceptable for resistance to atmospheric corrosion. Indoors, corrosion of steel will not occur unless the relative humidity is greater than 20%. Because most factories have atmospheres with humidity significantly above this value, machine components need corrosion protection if corrosion can affect serviceability. For outdoor service, the need for protective coatings is obvious. The most important platings for atmospheric corrosion protection are cadmium, chromium, zinc, electroless nickel, anodizing, black oxide, chromating, and phosphating. Cadmium and zinc are both sacrificial to ferrous substrates. Scratches and defects are prevented from rusting by the anodic behavior of the coating. For equal thickness and comparable environments, the corrosion characteristics of these two coatings are about the same. Zinc may be slightly better in industrial environments and cadmium slightly better than zinc in seacoast environments. Coating thicknesses of 0.0005 to 0.001 in. (12 to 25 μm) are usually adequate for indoor use. The use of cadmium in the United States is declining because of environmental restrictions.

Chromium is used for protection of industrial machine parts. The optimum thickness for chromium depends on the applications; steel mill rolls may have a flash of 0.00008 in. (2 μm), and paper mill rolls may have a thickness of 0.005 in. (125 μm). On industrial machines, chromium is not used unless the scratch resistance of the chromium is needed. It resists abuse and handling better than all other platings.

Chromium does not need an underplating of copper, nickel, or both for deposition purposes. Chromium can be applied directly to steel, but all hard chromium plates contain microscopic cracks that can allow corrodents to reach the protected substrate. The use of underplatings reduces the likelihood of a continuous corrodent path through the cracked chromium to the substrate. Thus the use of copper and nickel under chromium is advisable if corrosion protection is the primary purpose of the coating. Without an underplating, a chromium thickness

Table 19–3
Surface engineering processes/materials

	Typical Thickness[a] mils(μm)	Approximate Hardness (kg/mm^2)	Use
Platings			
Chromium	Flash 0.05 to 0.1 (1 to 2)	900	Decorative surfaces wear parts
	Avg. 0.1 to 1 (2 to 25)		
	Heavy 0.5 to 2.5 (12 to 63)		
Copper	Avg. 0.5 to 1 (12 to 25)	150	Conductive surface underplating
	Heavy 2.0 to 5.0 (50 to 127)		
Gold	0.02 to 0.1 (0.8 to 2)	80	Electrical contacts
Nickel (soft)	0.1 to 0.5 (2 to 10)	250	Atmospheric corrosion resistance
(hard)	0.7 to 1.5 (18 to 31)	500	durable surfaces
	Heavy 2 to 50 (50 to 1270)		
Silver	0.1 to 0.3 (2 to 7)	100	Galling resistance
Zinc	0.2 to 0.5 (5 to 12)	100	Rust protection on steel
Electroless nickel + PTFE	0.5 to 1 (12 to 25)	450 as plated	Atmospheric corrosion resistance, wear
	0.5 to 1 (12 to 25)	650 aged	surfaces
Conversion Coatings			
Anodizing	0.4 to 0.6 (10 to 13)	1100	Corrosion protection
Hardcoating	1 to 3 (25 to 75)	1100	Abrasion resistance
Phosphate/chromate	0.001 to 0.4 (0.02 to 10)	—	Improved rust resistance
Black oxide	0.001 to 0.1 (0.02 to 25)	150	Appearance
Thin-Film Coatings			
PVD TiN/TiCN	0.05 to 0.1 (1.2 to 2.5)	800 to 1800	Abrasion resistance
PVD Pure Metals	0.005 to 0.01 (0.1 to 0.25)	—	Decorative
Hardfacing			
ERCoCr–A (cobalt[b] base)	100 to 200 (2500 to 5000)	420	Galling/corrosion
420 stainless steel	100 to 200 (2500 to 5000)	550	Corrosion/wear
RFe Cr–A1 (iron/chromium)	125 to 250 (3150 to 6250)	600	High stress abrasion
Thermal Spray Coatings			
Tungsten Carbide (WC/Co)	3 to 5 (75 to 125)	1500	Wear surface
Chromium Oxide	3 to 5 (75 to 125)	1800	Wear surface
420 Stainless Steel	5 to 10 (125 to 2500)	550	Buildup
Diffusion Treatments			
Carburizing	2 to 60 (50 to 500)	600	Wear surfaces
Nitriding	1 to 30 (25 to 250)	900	Wear surfaces
TiC/VC	0.05 to 0.1 (1.5 to 2.5)	2000	Wear surfaces
Selective Hardening			
Laser/EB	2 to 30 (50 to 250)	700[c]	Thin wear surfaces
Flame Hardening	60 to 250 (1500 to 6250)	600[c]	Wear surfaces on large parts
Induction Hardening	10 to 80 (250 to 1600)	600[c]	Small parts
Other Treatments			
Paints	1 to 3 (25 to 75)		Corrosion protection
Repair Cements	50 to 200 (1250 to 5000)		Corrosive wear
Dry-Film Lubricants	1 to 2 (25 to 50)		Sliding systems
Ion Implantation	0.004 to 0.006 (0.1 to 0.15)		Corrosion protection, wear surfaces
Laser/EB Glazing	5 to 10 (125 to 250)		Special applications
CVD Coatings	0.05 to 10 (0.5 to 250)		Wear surfaces

[a] 1 mil = 2.5 μm
[b] American Welding Society designation per AWS A5.13-80
[c] On 1045 steel

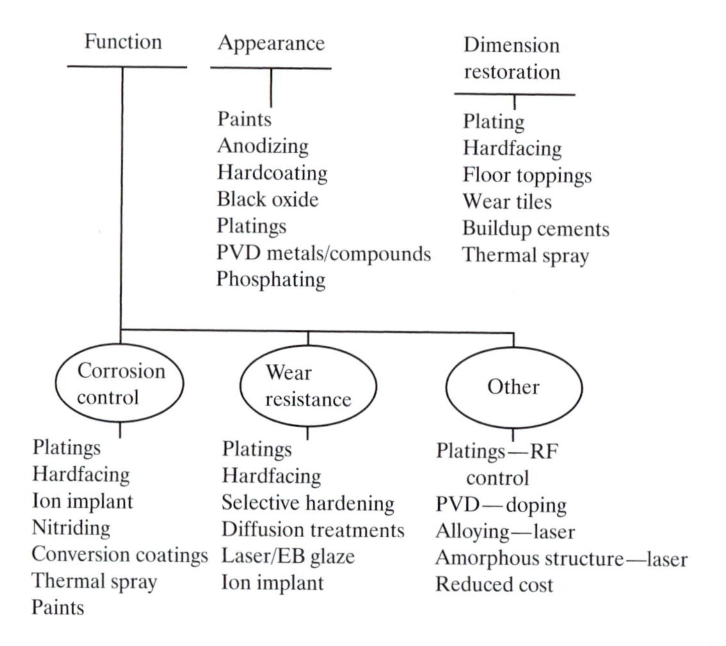

Figure 19–25
Reasons for using surface engineering processes

in the range of 0.0004 to 0.001 in. (6 to 25 μm) may be required for rust resistance.

Autocatalytic nickel in the thickness range of 0.0005 to 0.0007 in. (12.5 to 17.5 μm) is becoming a very popular coating for machine components. It is a noble coating compared with cadmium and zinc, but on the other hand it is more corrosion resistant by itself than these platings. The important advantages of this process are uniformity of the coating and the ability to plate in holes and recesses. The as-deposited hardness of about 43 HRC makes it very scratch resistant and durable.

The merits of anodizing and hardcoating were discussed at length in Chapter 17. In comparison with the other coatings on our list, they are undoubtedly the best coatings to use on aluminum parts for atmospheric corrosion resistance. Chromating is a faster and cheaper process, but it is not as effective. Chromating should be used in place of anodizing only if cost is a more important factor than serviceability.

Black oxide is only marginally effective in preventing even indoor atmospheric corrosion on steels. It should not be considered if rusting will affect serviceability.

Phosphating is not as good as an electroplate for atmospheric corrosion resistance, but a heavy zinc phosphate will work about as well as thin cadmium or zinc plating as long as the parts to be coated are not handled frequently. Phosphate coatings impregnated with oil/wax are fairly effective in preventing rusting of steels (Figure 19–26). They are often used on gears, cams, and

Zinc phosphate coating weight g/m^2	Supplemental coating	Salt spray hours to 5% red rust (ASTM B 119)
11	None	<8
11	Water-born wax	24
11	Protective oil	72
11	Paint	72
>11	Wax/oil emulsion	96–240

Figure 19–26
Corrosion resistance of zinc phosphate conversion coatings on steel

mechanical components where the phosphate aids break-in wear. The oil-impregnated phosphate coating will usually prevent rusting during storage. Once on a machine, the machine oils will prevent rusting over the life of the part.

For wear: There is no question that one of the most important applications of the coatings under discussion is to reduce wear in sliding systems. Almost every sprocket chain used in machines comes from the vendor with a conversion coating on it. Plating of cutting tools to improve their service life is becoming common. Many hand tools are plated to reduce abrasion. High-performance automobile engines have plated engine cylinders and crankshaft throws, and so on. The application of surface engineering processes to reduce wear are many and varied.

The results of metal-to-metal laboratory wear tests on platings are shown in Figure 19–27. In running against a hardened steel, silver outperformed all the other platings in the test. Hard nickel and a softer low-stress nickel from a sulfamate bath had favorable wear characteristics. Chromium in various forms—a heavy plate, a *flash deposit*, and a proprietary deposition—had low net wear, but the coating (chromium) seems to cause more wear on the other member in the sliding couple. Electroless nickel had very bad wear characteristics. The wear life, however, is significantly improved by heat treating.

Hard anodize on aluminum caused severe wear on the mating member. Electrodeposits of black nickel, tin, and cadmium did not wear the mating metal, but they themselves showed very high wear rates.

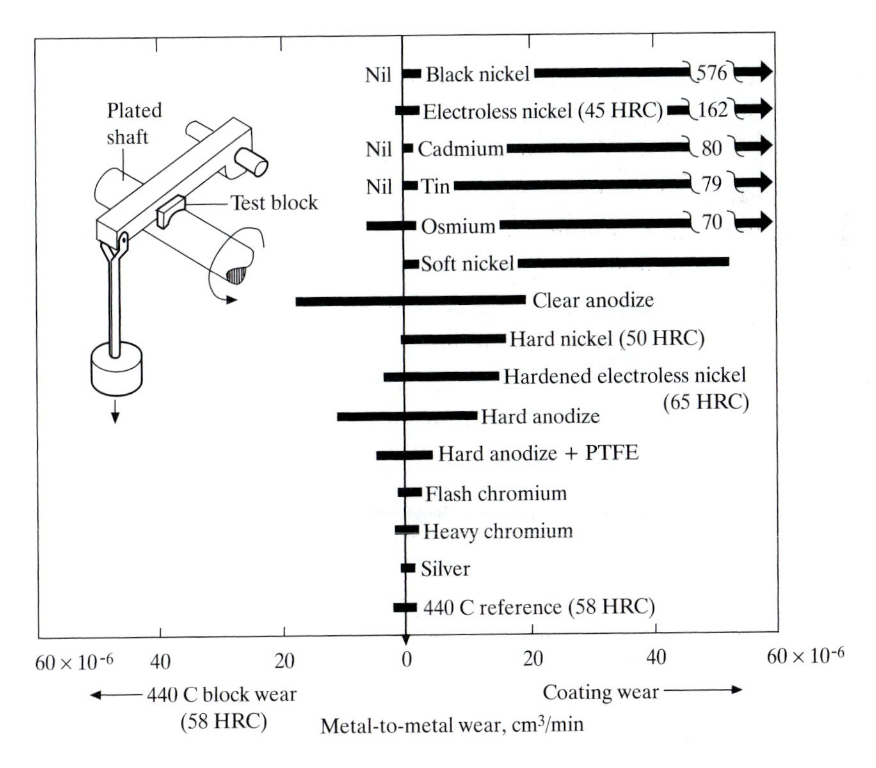

Figure 19–27
Metal-to-metal wear characteristics of various coatings

Although not shown in the plating wear test results, phosphate conversion coatings are very effective in reducing metal-to-metal wear. Chromate coatings and oxides are not normally used in wear systems because their thickness is usually so small that they are quickly worn away in almost any sliding system. For metal-to-metal sliding systems, silver, hard nickel, and chromium are the most desirable coatings to use. Thickness should be at least 2 mils (50 μm). Do not use hard anodize for a bushing or similar device contacting a rotating shaft. It will wear the shaft. It is usually acceptable as a coating for linear bushings and sliding systems. Another precaution to keep in mind when using platings and coatings for wear systems: Do not use coatings for applications involving point or line contact loading, because they may spall.

Some laboratory test data on the abrasion resistance of platings are shown in Figure 19–28. Chromium emerged as the best. The surprising result in this test is that it is far superior in resistance to abrasion from dry sand than is a hard anodized aluminum, which is supposedly harder than chromium. This lower abrasion resistance is no doubt due to the inherent porosity in anodic coatings. In summarizing the abrasion characteristics of platings, test data suggest that chromium is the preferred coating.

The wear resistance of thermal spray coatings matches the wear spectrum of the materials that can be applied. Pure metals such as copper have poor wear characteristics, and hard materials like the ceramics and cermets have exceptional wear characteristics. Wear resistance is a major reason for using thermal spraying. The same thing is true of hardfacings. The wear properties can be that of a tool steel or a steel–carbide composite. Both of these materials can be applied by fusion hardfacing. In general, platings are most practical for wear surfaces with a thickness less than 10 mils (250 μm). Hardfacings are used for deposit thicknesses in the range from about 10 mils (250 μm) to inches thick. The thermal spray processes have no technical limit in thickness, but it is common to use them for wear surfaces between 2 and 5 mils (50 to 250 μm) in finished thickness. HVOF coatings of tungsten carbide (WC/10% Co) and chromium carbide CrC/25% NiCr) have shown to be competitive with chromium plating in heavy

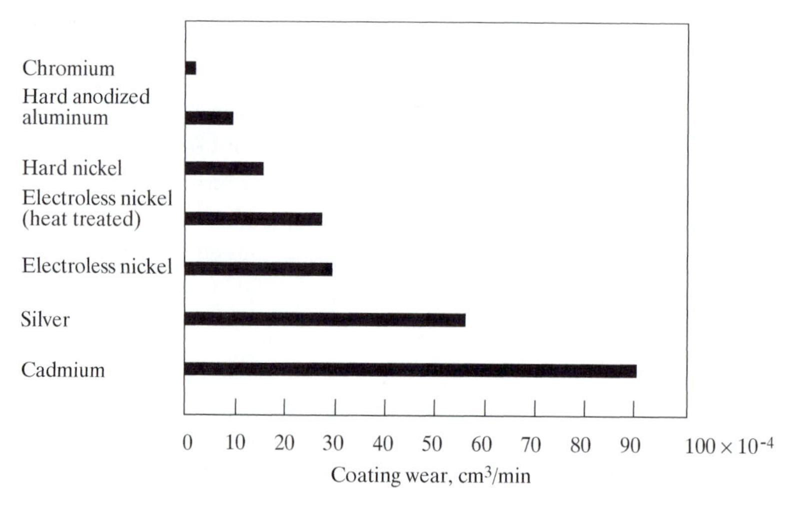

Figure 19–28
Resistance of coatings to abrasion by 60-mesh silica (modified ASTM G 65 test)

thicknesses [>1 mil (25 μm)] on large parts. Thermal spray processes are usually not cost effective for small parts like bolts and shafts that can be racked and batch plated. Thermal spray involves separate handling of each part. This should be kept in mind in process selection.

Diffusion treatments are often the surface treatments of choice when the goal is to improve the durability of a surface (wear and abuse) without increasing part dimensions. Nitriding can make the surface of a nitriding steel 70 HRC. Carburizing can produce surface hardnesses of up to 62 HRC on suitable steels, and the proprietary vanadium and titanium carbide diffusion treatments can produce surface hardnesses greater than 2000 kg/mm^2, harder than cemented carbide. Why not use these processes for all surfaces that need enhanced wear resistance? The design consideration that is significant for these processes as well as for all surface engineering processes is application temperature. As shown in Figure 19–29, diffusion treatments require process temperatures in the range of 1600 to 2000°F (870 to 1100°C). This temperature range is above the critical transformation temperature and the stress relieve temperature of most steels. This means that the treatment could cause distortion (and probably will).

As a rule of thumb, if the surface engineering process requires heating of the substrate to a temperature at, near, or above the stress relieving temperature of the material, there is the likelihood of part distortion. This means only 150°F (65°C) for some plastics, about 300°F (150°C) for aluminums, about 500°F (260°C) for copper-based alloys, and about 1200°F (650°C) for most steels.

Everything that we mentioned about part distortion applies to selective hardening, but it is much better from the distortion standpoint than diffusion processes. The application temperatures are similar, but these processes heat only a small portion of the workpiece, a surface layer. These processes are often used on finish-machined parts.

For dimension restoration: Rebuilding worn parts is a significant part of surface engineering. We already mentioned the use of thermal spray and plating. Fusion hardfacing, applying a weld deposit overlay, is an invaluable process in mining and earthmoving machinery. These kinds of applications would probably spall platings and thermal spray deposits. In addition, they are much thinner than needed. A bulldozer blade might tolerate 10 mm of wear and still work. So when they are rebuilt with hardfacing, they usually receive two or more layers, ending with a thickness of about 10 mm. Laser cladding is being used on rebuilding roll journals and roll faces where the process can be automated.

Iron chromium alloys with an as-deposited hardness in the range of 55 to 60 HCC are popular for the mineral and earthmoving applications. Carbide-containing deposits are used for applications subjected to severe wear but mild impact. Metal matrix carbide deposits are popular in laser cladding, and 420 stainless steel with an as-deposited hardness of about 50 HRC is popular in the United States for rebuilding the faces of steel mill and similar rolls.

19.17 Process/Material Selection

Methodology: Our suggested methodology for implementing a surface engineering selection is:

1. Establish and quantify your objective
2. Define the properties and attributes that you want in the coating or surface treatment
3. List important "wants" and compare process and material candidates on how well they fulfill your "wants"
4. Make a selection

Sometimes it is necessary to perform evaluation tests before finalizing step number four. There are many applications where a failure in service could be very costly or dangerous, or result in warranty changes. Tests may be appropriate for

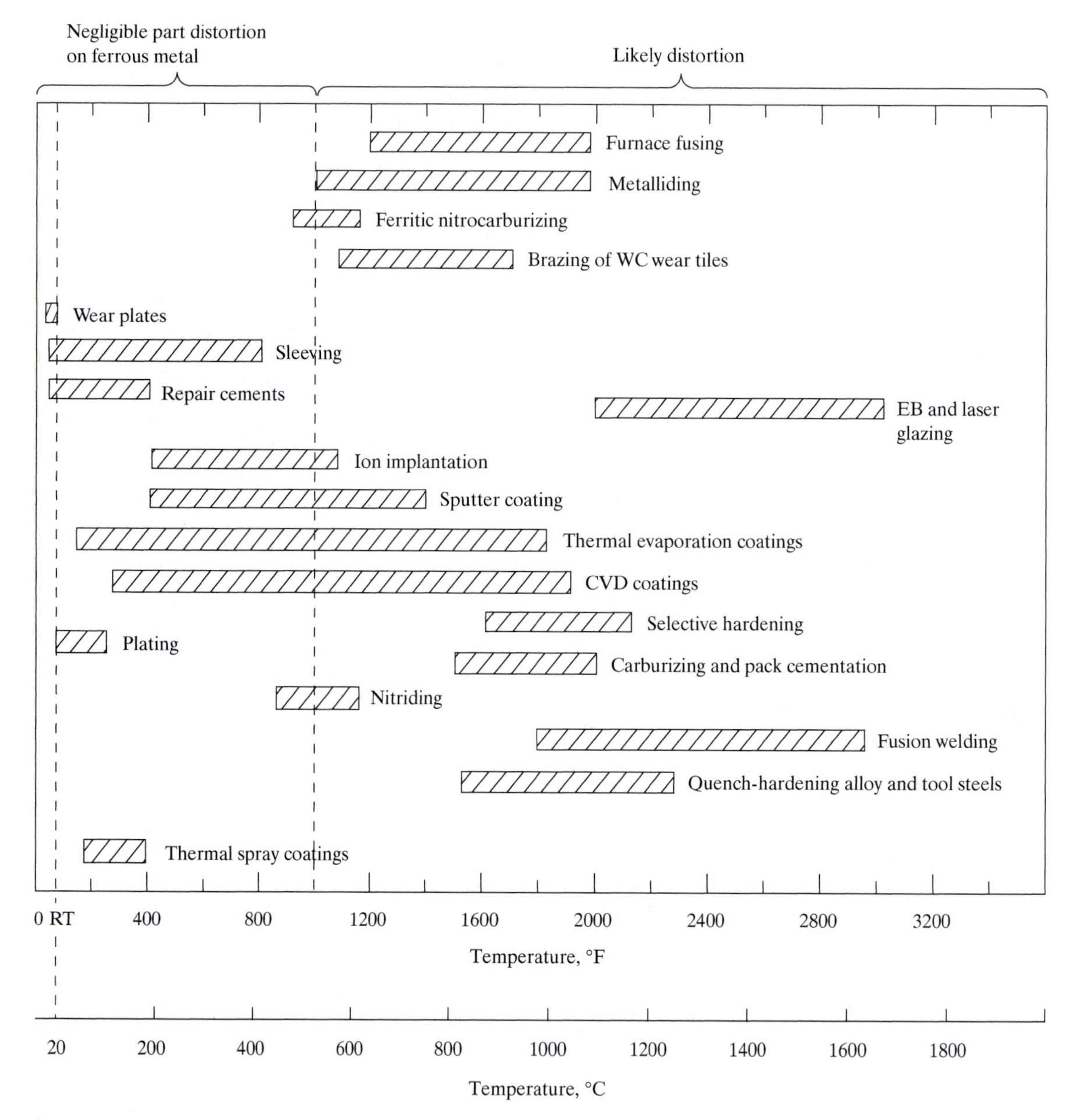

Figure 19–29
Temperatures encountered in various surface engineering processes

1. I want to achieve ____________________
____
____________________with this treatment.
2. My substrate requirements are: _________
____
____
3. My thickness/depth requirements are: ____
____
____
4. My size-change requirements are: _______
____
____
5. My cost, availability and quantity considerations are: ____________________________
____
6. I require these specific properties: ______
____
____
7. My intended service life is: ____________
____
____

Figure 19–30
Surface engineering checklist

evaluating a proposed change in a surface or a surface engineering process. Thus, for step four you may decide to perform evaluating studies and then make a selection.

Figure 19–30 is a checklist that may be helpful in establishing your list of wanted properties and attributes. In fact, later in this chapter we will illustrate how a selection matrix can be used to help maintain selection objectivity with your list of "musts." There will always be a number of surface engineering processes that can solve a specific problem. This matrix can help you finalize a choice.

Also, some surface engineering processes do not require an additional material selection; some do. If you decide to carburize a steel surface, you have already selected the material that will be "applied" to the surface: carbon. However, if you decide to use a thermal spray coating, you must specify not only the type of thermal spray process but also the coating material. There are at least ten process options and hundreds of material options. The selection process can be simplified if you combine the process and material as a candidate. For example, make "HVOF-sprayed WC/Co" a candidate for improving a surface rather than trying to decide on a thermal spray process as well as many consumables. Plating is similar. Make electroplated chromium a candidate rather than reviewing all plating processes and all metals that can be plated. Thus, the first step in the selection methodology is to, wherever possible, combine process and material as a candidate. It helps. The remainder of this section will address each selection step.

Objective: This step in selecting a surface engineering process may seem intuitive. You want to prevent the rusting of a part; you want it to resist wear longer; you want it to be a certain color; you want it to be shiny or dull. However, designers should think thoroughly about an objective and try to quantify that objective. You want to prevent rust—how long do you want the part to be rust-free? If you answer is until the part is purchased, you may want to dip it in oil or use a thin chemical conversion coating. If your answer is: "For 5 years of use and many washings as a kitchen utensil"; the coating must be very different, possibly a heavy organisol or plastisol plastic. One situation where rust protection is needed only while the product is on the shelf is with building consumables and bulk fasteners like nails. Interior-use nails are often sold with a light film of oil left from drawing and forming. If the oil is moisture displacing, it can keep the nails rust-free for years. But not if they get wet. If getting wet is anticipated, you may want to galvanize of zinc plate the nails. Wear resistance is not as simple as it sounds. You must think about the type of wear and anticipated wear life. No sliding part lasts forever. You need to establish a reasonable, anticipated service life for any surface engineering process.

Substrate considerations: From the practical standpoint, anodizing only applies to aluminum and not all alloys accept hardcoat. Titanium and aluminum cannot be electroplated with conventional techniques. There are many similar substrate limitations. Almost every substrate has temperature limitations. If you exceed a particular temperature, you could degrade substrate properties (Figure 19–29). If you have an aluminum substrate, all of the processes that have process temperatures above 1200°F (650°C) are immediately ruled out as candidates. They will melt the aluminum. These films are commonly applied to tool steels. Some thin-film deposition processes occur at temperatures that will soften hardened tool steels. Often, the choice of substrate is fixed, so every process that is incompatible with the chosen substrate is eliminated from the selection process. Sometimes substrate size will be a factor. If you want to improve the wear life of a bulldozer blade, you probably will not want to use a PVD or CVD process that is done in a vacuum chamber or retort. Many vessels for these processes are less than a cubic meter in volume.

Thickness/depth: Coating thickness or depth of surface penetration may be the most critical decision in selecting a surface engineering process. As shown in Figure 19–31, each process has a normal thickness/depth range. Commercial thin-film coatings are almost always applied in the range of 1 to 2 μm. Conversely, it is difficult to apply a weld overlay less than 3 μm thick with shielded metal arc welding. Designers need to firmly establish their coating thickness limits or the depth requirements on treatments that penetrate the original surface.

Size change: A complimentary decision is needed on how much distortion, growth, shrink, or process variation is permissible. Fusion welding will almost always create measurable distortion. Hardening

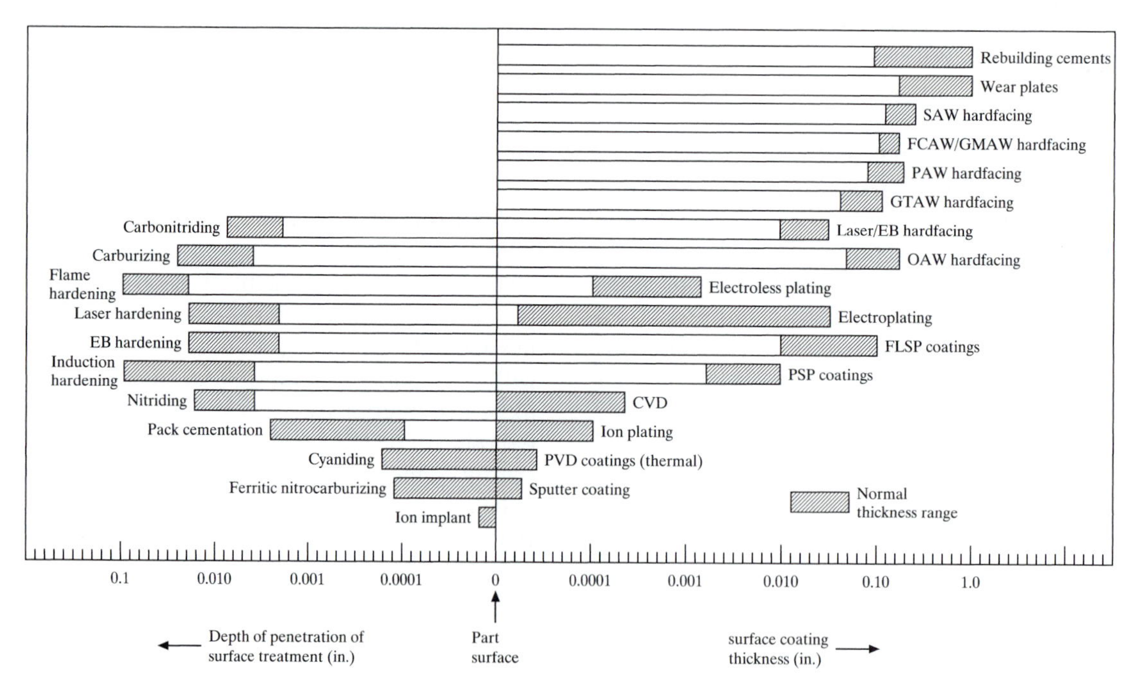

Figure 19–31
Normal thickness/depth ranges for various surface engineering processes

steels creates mostly growth, but occasionally shrink. How much can you tolerate? Flame hardening can produce a hard layer depth of 4 mm in some substrates, but it is likely to vary 3 to 5 mm. Is this acceptable? Any of the processes that involve heating to the red-heat range are likely to produce some size change in a substrate. How much can be tolerated? Some case hardening processes produce less size change than others. Nitriding is the lowest-temperature process and usually produces less size change than carburizing and carbonitriding. Designers need to decide on what is permissible.

Business concerns: There almost always are some critical business concerns. Quantity is always foremost. If you must apply a surface engineering process to one million parts per month, the process selected may be significantly different from that chosen if you only have one special gear to treat.

Most of the processes that we discussed can be made suitable for treating large quantities of small parts, but big parts become a problem for processes that require individual racking or handling. Plating or painting a million automobile bumpers per month requires a huge amount of capital equipment, whereas plating or painting one million screws is very inexpensive—batch processes can be used. Similarly, applying PVD coatings to small drills and milling cutters is low in cost, whereas applying the same coating to wrenches may be cost prohibitive. Drills are very easy to fixture and rotate; wrenches are not. PVD is a line-of-sight process, and all coated surfaces must be exposed to the coating source.

Some proprietary surface engineering processes are costly because of the specialized equipment needed or the complicated processing steps. For example, some computer chip manufacturers apply 250 surface engineering processes to a disk that will be made into individual chips. All of these processes are performed in clean room conditions. This makes it expensive. Some diamond and diamond-like carbon (DLC) coatings are unique and thus their purveyors charge extra for their distinguishing features.

Finally, availability affects any service or product. If there is only one company in the world that can perform hyperbaric nitriding on titanium, they are likely to charge a lot for the process. On the other extreme, probably every commercial heat treater in the world offers carburizing of steels. This makes it a "commodity" process and lowers the cost for most parts.

Business concerns may also include environmental factors. Some manufacturers elect to avoid the use of electroplatings because solution disposal can be on environmental concern. Some processes such as sandblasting present potential health risks (e.g., silicosis) and fines. Many thermal spray processes are extremely noisy and require soundproof enclosures. Every design situation is different, but there will probably always be some business concerns in selecting a surface engineering process and these need to be considered.

Properties: Sometimes a special property is needed that in turn steers the selection process. For example, you may want to produce a hard surface on austenitic stainless but need the surface to remain nonferromagnetic. This rules out the nitriding processes because they tend to make the nitrided layer ferromagnetic. If you want to impart hardness to the polished surface of a hardened gage block with no dimensional change, you may be out of luck. Maybe ion implantation would help, but most other processes would produce measurable size change. One of the most significant property concerns with any coating is coating adhesion. The coatings that rely on a mechanical bond with the surface always have a potential for debonding. The adhesion characteristics of surface engineering coatings or treatments depend on the mechanisms involved in their formation. Figure 19–32 categorizes common surface engineering processes by the mechanism by which they adhere or are produced on a surface. We

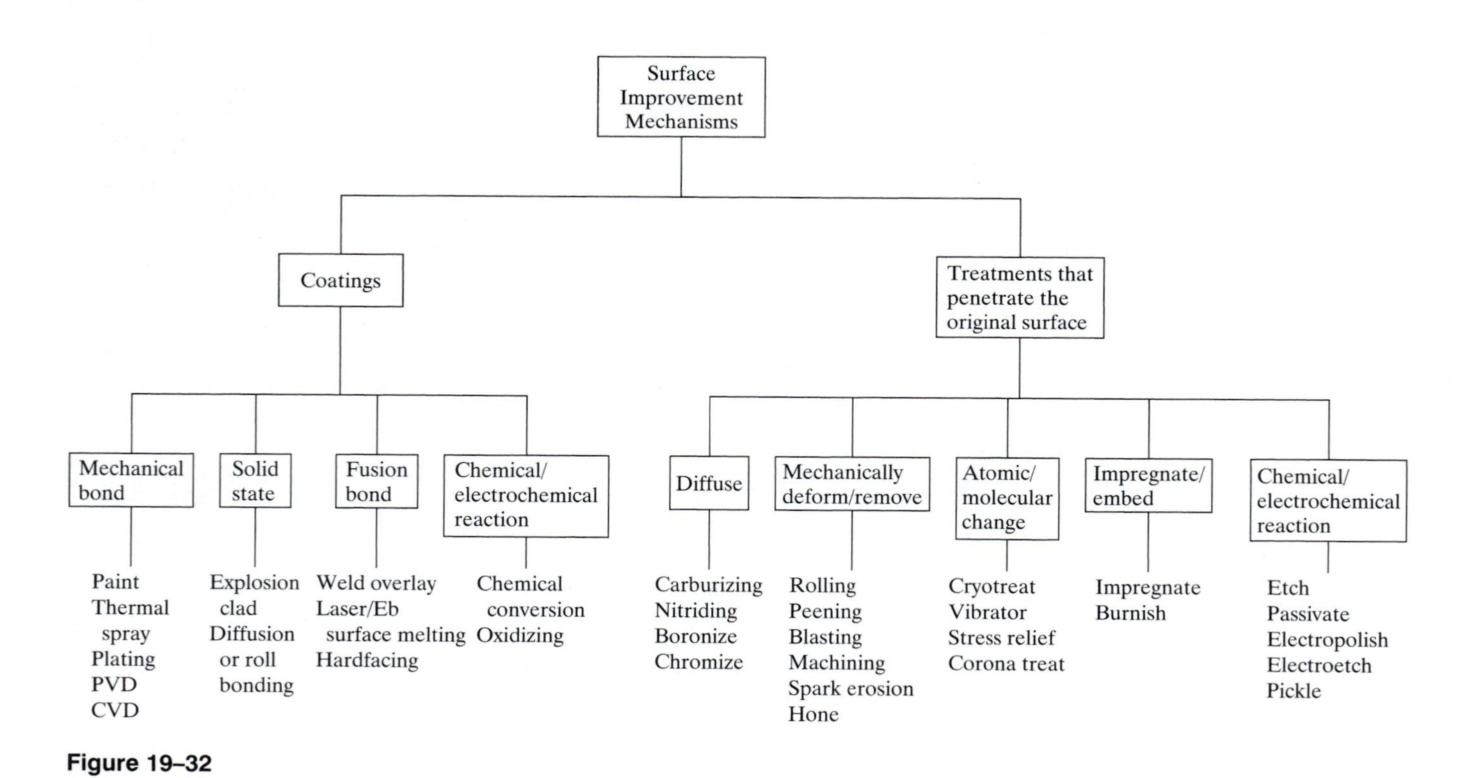

Figure 19–32
Common surface engineering processes categorized by the mechanism of their formation

mentioned that most of the common coatings have mechanical bonds with the surface. The strength of the best mechanical bonds is less than that of a fusion bond, which is considered to be the best bond. Some call it a metallurgical bond. A weld overlay is fusion bonded. A solid-state bond such as cladding is also a metallurgical bond, but it could have defects that preclude bonding. For example, if a stainless steel is roll clad onto carbon steel, there may not be a good bond where a contaminant was present on the surface member before rolling. Coatings that are formed by chemical or electrochemical reaction with a surface are likely to never disbond unless they become thick. If you put a piece of steel in a furnace at 1000°F (538°C) for several hours, it will develop a very adherent oxide that will have an excellent bond with the surface. However, if the same piece of steel is heated for 1000 h at 1000°F (538°C), it will develop a very thick oxide scale that will likely spall under moderate mechanical action. Coatings such as chromate conversion coatings and electrochemical conversion coatings such as anodizing seldom lose their bond to the substrate if they are applied in the normal thickness range. However, whenever a coating is used for improving the properties of a surface, the designer must consider the risk of the coating coming off while in service. Bond strength becomes a critical selection factor.

Most surface treatments that do not involve coating do not have adhesion concerns. The treated depth will probably never spall on a selective hardened wheel, but platings may be removed in the first use. Usually, mechanical bonded coatings do not perform well in Hertzian loading; their use for these applications should be avoided. Some chromium-plated ball bearings are commercially available, but it took years to develop a suitable process.

Some diffusion treatments such as nitriding can produce an unwanted layer (white layer) or growth on top of the hardened layer. Ordinarily, these need to be removed because they can spall in service—especially with Hertzian contact.

In any case, designers need to be aware of the mechanism of formation of an altered layer in the surface and decide it there are any adhesion concerns. These concerns in turn should be made a selection factor.

Serviceability issues: Finally, our checklists ask you to consider how a treated or coated surface may fail in service. Are you most concerned with wear, corrosion, spalling, or even color change? Whatever might cause a warranty problem, part rejection, or failure of a larger machine should be noted as a selection factor. Some machines in the auto industry are only designed for use during a model run, which may only be three years. This would mean that long wear and corrosion life is not a factor. Tool steel tooling instead of carbide tooling may be in order.

Using a selection matrix: Unfortunately, machine designers cannot list a repertoire of surface engineering processes and materials on a drawing, only one process and one material. This can be a daunting task because of the number of choices available. We have tried to simplify selection by recommending a few processes and materials, and we have given you some results of industrial experience. However, we have one more aid to suggest: use of a selection matrix (Figure 19–33). Simply list your "must" properties/attributes on one axis and reasonable process/material candidates on the other axis. Then rate each candidate on a scale from 1 to 10 on how well it meets each requirement on your list.

For example, you are designing a spindle for a test machine. The spindle face will be subject to abrasion from rubbing the edge of the polyester belt. You want to make the spindle from a free-machining steel (probably leaded low-carbon steel) because it has many diameters that need to be precisely controlled. You want it to be black to match the rest of the machine. The tolerance on critical diameters is $\pm$ 0.0005 in. (25 μm), and you will make fewer than ten of these spindles.

	(10 is best; 5 is neutral)		Process/material		Candidates
Needed Features	Black paint	Black oxide	Black nitride	Black nickel plate	Black chromium plate
Rust resistance	8	2	7	9	10
Wear resistance	2	1	10	4	5
Free-machining steel substrate	5	5	5	5	5
No dimension change	1	10	9	8	8
Thickness <1 mil	5	5	5	5	5
Cost <$20	4	10	8	2	2
Total Score	25	33	44	33	35

Figure 19–33
Selection matrix for comparing surface engineering candidates with a list of desired features/benefits

The critical considerations are:

1. Goal: service life of ten years
2. Substrate consideration: make from leaded low-carbon steel
3. Thickness requirement: less than 1 mil (50 μm) hardened surface layer is adequate
4. Tolerance considerations: the spindle is stout and can take elevated temperature heat treatment, but the run-out or the finished part must be less than 0.0005 in. (25 μm) with diameter tolerances of $\pm$ 0.001 in. (50 μm)
5. Business considerations: cost less than $20.00 to harden
6. Property requirements: no wear on the spindle face
   - 55 HRC minimum preferred
   - must be black

Figure 19–33 shows these required features in the first column of a selection matrix and the top row lists five candidate choices. The must-be-black requirement limited the selection process significantly. There are not that many coatings or surface treatments that are black. A number between 1 and 10 is assigned in each box corresponding to the perceived merit of each candidate in each feature area. The merit numbers are summed, and the process/material with the highest merit is the primary candidate. Sometimes you will take the top two or three candidates and run laboratory or field evaluations, but the matrix process is usually helpful in making surface engineering selections. Actually, it works well for the selection of any material. You can weight certain properties if you want, and you can even use hard numbers for properties in the boxes. Instead of using 1 to 10 for hardness, you could insert candidates' hardness in GPa. You can put in euros per kg for a cost factor.

This process almost always helps designers to carefully consider their "wants" versus the candidate properties. We recommend its use.

19.18 Specifications

The selection guidelines and selection matrix presented should answer the question of what process/material to use. A factor remaining is how to specify the desired treatment on an engineering drawing. To be complete, a plating specification should include the following items:

1. Type of coating/treatment
2. Desired thickness (give a range)
3. Areas to be covered/treated
4. Hardness (if various hardnesses are available)
5. Postprocessing treatments (if any)

A typical plating specification might read

> Plate all over with 0.0003- to 0.0005-in. hard chromium at 65 to 70 HRC. Heat treat for 4 h at 400°F immediately after plating to prevent hydrogen embrittlement.

A postplating heat treatment is necessary on most hardened steels (those over 32 HRC). Plating tends to charge steel with hydrogen, and this causes embrittlement. A simple heat treatment like the one described in our sample will remove this embrittlement.

A PVD or CVD coating usually requires no postdeposition treatment, and the hardness is not variable; you get what you get. A typical specification might read:

> Coat all over with cathodic arc sputtered titanium nitride, 1 to 2 μm thick.

Thermal spray process specifications require naming the process as well as items 1 to 5 (from earlier list):

> Coat journal areas with 5 to 8 mils of chromium oxide deposited with plasma arc spraying. Finish grind to drawing dimension. Minimum thickness after grinding 0.005 in.

Hardfacing with fusion welding requires naming the alloy and the process, as well as the joint detail. Often a drawing is required:

> Machine journal to 1.820 in. Hardface with two layers of RCoCrA 0.1-in. min. thickness. Grind to 2.000- to 2.002-in. diameter. Minimum hardness to be 43 HRC.

Some alloys can tolerate a postwelding stress relieve without softening; some cannot.

Paints and powder coatings should be specified by dry film or cured thickness:

> Coat all over with Valspun 1030; dry film thickness shall be 0.0002 to 0.0004 in.

Paints often require abrasive blasting of the substrate for the best adhesion. Substrate blasting may affect the film roughness and texture. Specify gloss and final texture Ra, and so on, if this is a concern. Dry-film lubricant coatings require rigorous adherence to the manufacturers' application specification.

> Coat area "A" with 25 to 40 μm of Slip–140 (Bud Labs, Rochester, NY, USA) using Mil-403 application specification.

> Diffusion treatments require items 1 to 5:

> *Material:* 4615 steel; carburize all over 0.010 to 0.015 in. deep; harden and temper to 58–61 HRC surface hardness.

Selective hardening requires items 1 to 5:

> *Material:* 1095 steel; flame harden gear teeth 0.060 to 0.080 in. deep as shown (show desired hardening profile); temper to 58 to 60 HRC.

Induction hardening would have a similar drawing specification.

Ion implantation has somewhat different ground rules than most other surface treatments. You need to identify the substrate and state the desired implant species, the area to be treated, and the dosage or beam flux:

> Ion implant areas shown with chromium, 10^{17} atoms/cm^2.

You may want a treatment depth of 2 mm, but in 2004 the current technology limited the depth to about 0.1 μm.

Conversion coatings may or may not require the designer to specify thickness. Usually, it is ignored if only a fixture or a tool is to be treated. If the coating will be applied to a million production parts per month, then a coating thickness (weight) is specified.

> *One time:* Black oxide coat all over
>
> *Production:* Coat all over with chromate conversion coating 10 mg/m^2

Anodizing thickness should always be specified.

> *Anodize:* Anodize all over 0.0005 to 0.001 in. thick, dye black
>
> *Hardcoat:* Hardcoat all over 0.001/0.002 in. thick

Mechanical surface treatments alter surface texture, and normal surface finish specification techniques apply.

> *Abrasive blasting:* Abrasive blast with 120-grit aluminum oxide, 100% strike density, final surface roughness 50 to 55 μm Ra
>
> *Lapping:* Lap surface flat to one light band, final surface roughness 0.05 to .1 μm
>
> *Grinding:* G .08–.12 μm

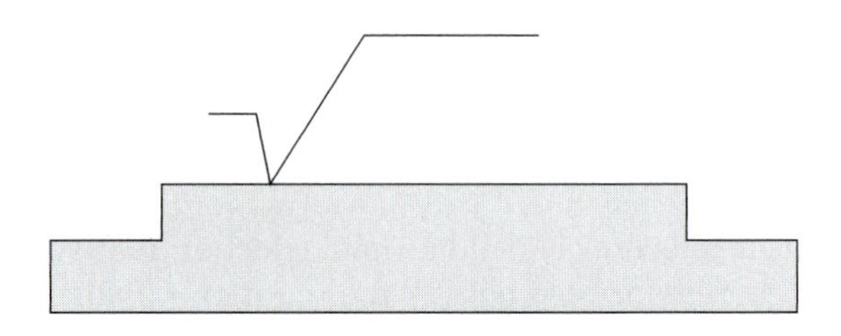

> Electrical discharge machine (EDM): EDM textured areas ("A" and "C") to have a final surface roughness of 4 to 8 μin. Ra.

Specifying fluorocarbon imbedding and other "spray can" surface treatments requires details on how to perform the operation.

> Treat surface A with M21 PTFE spray (Bud Labs, Rochester, NY, USA). Apply thin coat; allow to dry; burnish into surface with cotton flannel cloth. Repeat three times more.

Summary

Surface engineering can be used to create use properties that cannot be obtained from the substrate. In some processes, the only thing that is important to the function of a component is the chemical makeup and properties of what is on the outer few layers of atom. Some electronic components are like this. All the digital information on compact discs is obtained from tiny spots burned in a 100-Å (0.0000004-in.)–thick coating. The point is that coatings and surface treatments are a critical part of modern materials engineering, and design engineers must know how to use them.

Some summary comments on surface engineering are as follows:

- Thermal spray deposits are usually porous—they must be sealed to be impervious.

- Electroplating, PVD coatings, and thermal spray deposits are not fused to the substrate surface, and coating bond strength is a design consideration.
- Electroplating does not go on uniformly; it is heavier on edges. This must be considered in part dimensions.
- Electroplating does not go into holes and recesses without special techniques.
- Some metals, such as aluminum, titanium, and magnesium, cannot be electroplated with conventional techniques.
- Autocatalytic platings cover all wetted surfaces; they will go into holes.
- Areas to be coated by PVD must be essentially in line of sight with the evaporant source in the coating chamber.
- Heat-sensitive materials such as hardened steels can be softened by many PVD and CVD processes. Know the application temperature before choosing a process.
- DLC/AHC coatings may be affected by moisture, such as high relative humidity (%RH).
- PVD coatings can contain nodules that make them abrasive to contacting surfaces.
- Avoid finishing thin PVD coatings; they may be removed.
- Consider coating buildups in part dimensions; specify part dimensions before or after coating.
- For something to be coated with a PVD process or ion implantation, it must fit in a vacuum or treatment vessel.
- The same material can have various properties, depending on the thermal spray process used for application.
- Fusion hardfacing deposits are preferred over mechanical bond coatings for heavy-duty battering applications.
- Conversion coatings are probably the lowest-cost coatings.
- Anodizing is an electrochemical conversion coating, and it is applicable to aluminum, magnesium, and titanium.
- Ion implantation may require experimentation (species, dose, etc.).
- Always consider surface treatment application temperature and substrate compatibility.
- Not all commercially available surface engineering processes work on all substrates. Users need to establish compatibility.
- If chromium plating does not offer sufficient abrasion resistance, consider harder PVD or CVD and diffusion treatments.
- Check paint adhesion. There are many factors that can have a negative effect.
- Many paints cannot withstand long-term UV and IR radiation. Acrylic and cellulosic neat polymers have demonstrated ability to survive outdoors.
- Thermal spray coatings may chip when used for a cutting edge. Their use is usually not recommended for this application.
- Adhesion, process roughness, and application temperature should be concerns whenever PVD/CVD coatings are used.
- Lubricious PVD, CVD, and solid-film lubricant spray coatings are useful for producing self-lubricating surfaces.
- PVD coatings like TiN have utility as cosmetic surfaces.
- Welding hardfacing usually fits a need for wear surfaces that can tolerate millimeters of wear.
- Harder (>60 HRC) arc-welded hardfacings develop small cracks (checks) and this may be objectionable in some applications.
- Laser and EB hardfacings are usually applied less than 1 mm thick.
- The benefits of laser and EB surface remelting/glazing vary with the application. This is not a process with guaranteed benefit.

- Electropolishing can remove significant amounts of metal. Make sure that parts can tolerate the dimensional loss.
- Anodizing and hardcoating are the best processes for improving the durability of aluminum, but not all alloys will accept hardcoat.
- Passivation is necessary on 300 series, ferritic and PH stainless steels as well as other corrosion-resistant alloys.
- Martensitic stainless steels are not normally passivated.
- Carburizing is usually the simplest and lowest-cost process for hardening low-carbon steel surfaces.
- Nitriding is a low-distortion way of adding hardness to alloy steel surfaces.
- EDM is a very useful way of producing a matte surface texture.
- Abrasive blasting can texture surfaces, but it is not as controllable as EDM.
- Impregnated surfaces may not be impregnated deeper than the surface texture unless the substrate contains porosity that goes to the surface.

Critical Concepts

- Surface treatments can be applied to the original surface and some can alter properties below the original surface.
- Adhesion of surface treatments can vary from mechanical bond to metallurgical bond.
- Each surface treatment has a "niche" where it will outperform others.
- Selection should include all treatments as candidates, and relative properties should be weighed when making a choice.
- Users of coatings and surface treatments must establish suitability of surface engineering coatings/treatments for their application. Testing is often necessary.

Terms You Should Remember

electroplating
conversion coatings
physical vapor deposition
electroless plating
phosphate coatings
chromate coating
flash deposit
deposition rate
microcracked
Faraday's law
hydrogen embrittlement
throwing power
corner buildup
local plating
rinsability
hardfacing
ion plating
plasma
sputtering
substrate
thermal spraying
flame spraying
arc spraying
ion implantation
thermal evaporation
EDX
XRD
SEM
AES
SIMS
ESCA
FTIR
macros
buildup
checking

Case History

Extended Razor Blade Life

Machines used to spool silver halide film on cores and insert the spooled film in magazines require fast-acting cutting blades to cut the film to the desired length. Cutting was accomplished by commercially available hardened steel razor blades. They "crush cut" the film against a steel anvil. The razor blades dulled frequently, and the change frequency became a cost concern. The department engineers tried to improve knife life by applying hard coatings. They tried chromium, titanium nitride, DLC, titanium carbonitride, and a few experimental thin-film coatings. None significantly improved knife life.

At that point, the production department requested materials laboratory studies in order to find a solution. Metallographic examination of used coated and uncoated knives showed that the extremely sharp razor blade edge was not dulling by wear, but by plastic deformation. The razor blade cutting edge "mushroomed" on repetitive striking of the steel anvil. The compression stresses in the tip were above the compression yield strength. The solution proposed by the materials engineers was to intentionally dull the edge to produce a controlled compression stress below the compression yield strength. Grinding a flat on the edge was tried, but this also produced a burr that interfered with cutting. Electropolishing was then recommended to dull the edge to a controlled radius. This was the surface engineering process that worked. Knife life was significantly improved. This is an example of how surface engineering is a tool that can be employed to improve existing designs. However, an engineered approach of discriminating among available processes is recommended.

Questions

Section 19.3

1. What makes plating bond to a surface?
2. Calculate the time required to plate 1 mil of zinc on steel, assuming 100% bath efficiency. The normal valence of zinc is 2 and the atomic weight is 65.
3. What is the reason for corner buildup in electroplating?
4. Why is titanium difficult to plate?
5. What makes electrodeposits hard?
6. What is the hardest plating?

Section 19.4

7. What is the mechanism of age hardening in electroless nickel deposits?
8. What is the composition of electroless nickel?
9. Can you mask areas on a part when electroless plating?

Section 19.7

10. What is the difference between a black oxide coating and a phosphate coating?
11. Cite three places where chromate conversion coatings are used.
12. What is the DIP process?

Section 19.8

13. What is the difference between PVD and ion plating?
14. What evaporant is used to produce titanium nitride coatings?
15. What are the two most important limitations in using PVD coatings?
16. Explain ion implantation.

Section 19.9

17. What are suitable substrates for fusion hard-facing?
18. What is the thickness limit of hardfacing?

Section 19.10

19. Describe three property differences between plasma spray coatings and electrodeposited coatings.
20. What is HVOF? How does it differ from plasma spray?
21. How can thermal spray processes be used for corrosion control?

Section 19.11

22. What is the depth of the analyzed layer with SIMS?
23. What process would you use to identify surface microconstituents in an aluminum alloy?

Section 19.16

24. What is a flash plating? When should it be used?

25. You apply a 1-mil hardcoat to a 3-in.-ID bore. How much will the bore change in size?
26. You wore 0.020 in. from the diameter of a 10-in. journal on a roll. What process would you use for repair? Explain.
27. What is the most abrasion-resistant plating?
28. Name three platings that provide cathodic protection on steel.

Section 19.18

29. Specify a plating to repair a worn cylinder in an automobile engine.
30. Specify a PVD coating for a punch press die.

To Dig Deeper

ASM Handbook, Volume 5, Surface Engineering. Materials Park, OH: ASM International, 1994.

Budinski, K. G. *Surface Engineering for Wear Resistance*. Upper Saddle River, NJ: Prentice Hall, 1988.

Dehotre, N., Ed. *Lasers in Surface Engineering, Vol. 1*. Materials Park, OH: ASM International, 1998.

Davis, J. R., Ed. *Surface Engineering for Corrosion and Wear Resistance*. Materials Park, OH: ASM International, 2001.

Davis, J. R. *Surface Hardening of Steels*. Materials Park, OH: ASM International.

Dennis, J. K., and T. E. Such. *Nickel and Chromium Plating*, 3rd ed. Materials Park, OH: ASM International/Woodhead Publishing Ltd., 1993.

Dowson, D. and others. *Thin Films in Tribology*. New York: Elsevier Science, 1993.

Edwards, J. *Coating and Surface Treatment Systems for Metals*. Materials Park, OH: ASM International, 1997.

Graham, A. K. *Electroplating Engineering Handbook*. New York: Van Nostrand Reinhold Co., 1971.

Hochman, R. F. L., Ed. *Ion Plating and Implantation*. Materials Park, OH: ASM International, 1986.

Holmberg, K., and A. Matthews. *Coatings Tribology*. New York: Elsevier Science, 1994.

Lindsay, J. H. *Coatings and Coating Processes for Metals*. Materials Park, OH: ASM International, 1993.

Mattox, D. M. *Handbook of Physical Vapor Deposition*. New York: Noyes Publications, 1998.

Metals Handbook, 9th ed., Volume 5, Surface Cleaning, Finishing, and Coating. Materials Park, OH: ASM International, 1982.

Parthasaradhy, N. V. *Practical Electroplating Handbook*. Upper Saddle River, NJ: Prentice Hall, 1989.

Pierson, H. O. *Handbook of Chemical Vapor Deposition*. New York: Noyes Publications, 1999.

Riedel, W. *Electroless Nickle Plating*. Materials Park, OH: ASM International, 1991.

Stern, K. H., Ed. *Metallurgical and Ceramic Protective Coatings*. Materials Park, OH: ASM International, 1996.

Swarcy, P., Ed. *Surface Coatings, Science and Technology*, 2nd ed. New York: John Wiley & Sons, 1996.

Wierzchon, T., and T. Burakowski. *Surface Engineering of Metals*. Boca Raton, FL: CRC Press LLC, 1999.

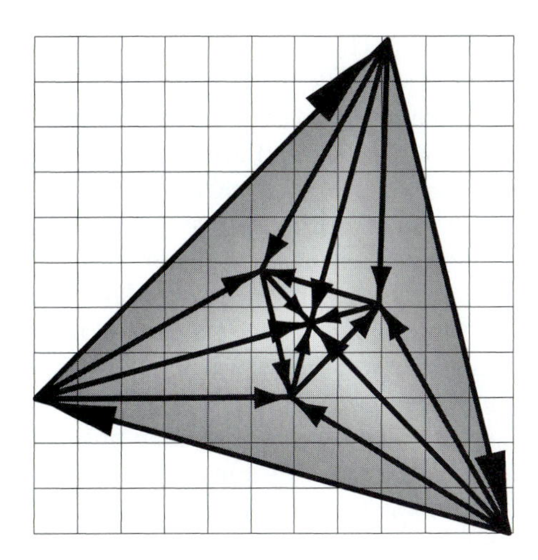

CHAPTER 20

The Selection Process

Chapter Goals

1. An understanding of the roles of cost, availability, and properties in material selection.
2. Guidelines on how to screen candidate materials and arrive at the proper choice.

Rationale

The material used for a design is just as important as the design itself, maybe more so. If on testing a design faults are discovered, then the design can be modified and retested until it works. However, if the wrong material was selected and this does not surface in prototype testing, it often shows up "down the road" as a failure in service in a customer's hands. This chapter deals with the methodology of reviewing acquired knowledge and reference information on material properties, defining material requirements for a design, and then comparing needs with properties to arrive at a proper material selection.

This is the bottom line of materials engineering, and it is why we have written this book. We want designers to select a material for a design that will perform as intended. This chapter is about making the right engineering decisions—the goal of all engineers.

A major objective of this text is to give the designer an understanding of material systems as well as a familiarization with enough specific engineering materials to allow their effective use in daily assignments.

In each chapter a few materials from a particular material system were selected to build a repertoire. This listing does not imply exclusion of other materials. The main point is that a designer should become familiar with a few plastics, a few ceramics, a few tool steels, and so on; it will be found that these will satisfy about 90% of design needs. Save the remaining 10% of material problems for the metallurgist or materials engineer. If none is available, use the materials engineering literature for help.

In this chapter we shall summarize some of the things that we have discussed in the preceding chapters and illustrate how material properties and other selection factors are used to make a material selection for given design situations.

Our recommended approach to material selection is not novel or complicated. It is simply a compilation of the various steps used by most experienced designers.

20.1 The Design Process

Most designers and engineers establish part designs as one aspect of their daily regimen. Civil engineers design highways and buildings; chemical engineers design process equipment; electrical engineers design machine controls, utilities, and electronic devices. Mechanical engineers design machines and products. This listing of activities could go on for other types of engineering disciplines. The common thread throughout engineering is design. Engineers and designers create value for their employers by designing machines to make products and by designing new products, processes, and facilities. The marriage of design and material selection should start early in the design process. This section is intended to show how this should happen.

Engineering/design assignments come in many ways depending on the company, but most times they come from an existing or perceived need. The paint department is getting scratches on fascias as they are transported to the drying ovens. A designer/engineer is asked to solve this problem. An audio/video company senses that there is a market for a digital image display device. There have been several accidents at a particular highway crossing within the past year, and a civil engineer is asked to design a remedy. A chemical engineer is asked to increase the throughput of a powder chemical; process improvements must be designed.

The individual steps in the design process are illustrated in Figure 20–1. In the United States, teams are used in many industries instead of a single person. The teams often include manufacturing engineers, customer representatives, a statistician, or a financial analyst. This type of team composition allows concurrent engineering to occur. The manufacturing engineer starts to investigate how to build each component in a product as it is designed. The financial person delves into costs and *availability* issues; the systems person develops software and controls. Design tools often involve computers. Expert systems guide selection processes; Internet searches can be used for design-related information; finite-element models can be used to investigate stresses, flow characteristics, heat transfer, and the like. Catalogs on CD-ROMs are searched to select components for a product under design. Materials of construction decisions must be made for each design detail. If you purchase a handle for a device, you must specify the material for the handle as well as the fasteners to assemble the handle. Material selection is part of the design process.

The first step in the design process is to sit and think about the task at hand (for appearances, this step should be done with a pencil in hand or in front of a CAD screen containing a few random shapes). Ideas should be produced, which are pursued with sketches, literature searches, models, discussions with the customer, and the like.

If the thinking step is productive, in the second step a design concept may take shape. Sizes are established. Drive and actuation systems are envisioned. Questions must be asked. Are all the desired functions performed? Is the probable cost within the project limits? Are the desired speeds or production rates possible? The device should start to gel in the mind of the designer.

Step three is to make an assembly drawing to find out if the design concept works on paper or on the computer. If it does not, it may be necessary to return to the beginning. Once an assembly drawing is complete and a motion analysis indicates that parts move as they are supposed to and do not hit each other, it is time to get into hard-core engineering: stress, vibration, acceleration analysis, and the like. It is not too early to think of such things as wear mode, lubrication, operating environment, and the effect of component failures.

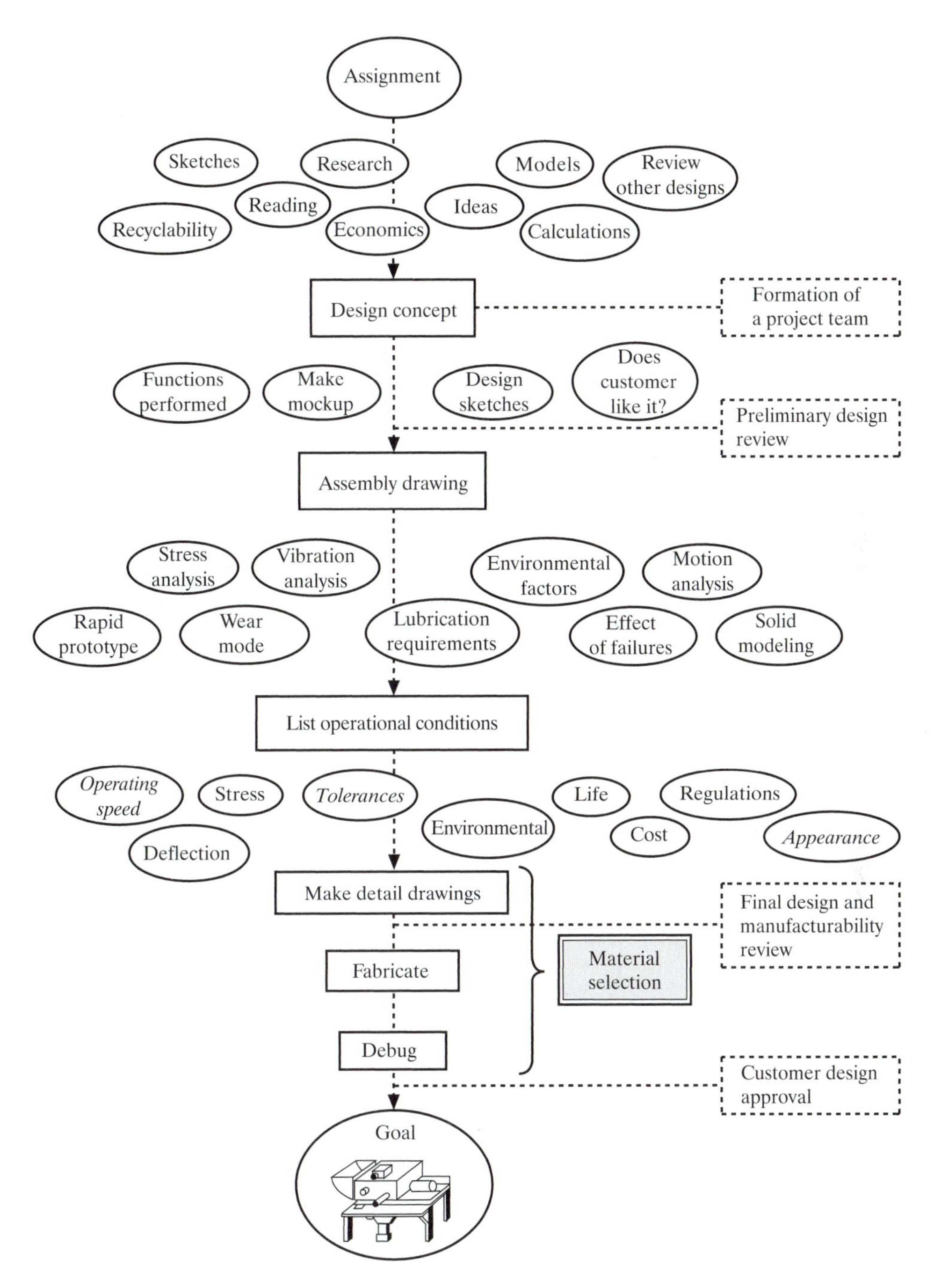

Figure 20–1
Role of material selection in the design process. The dashed boxes show how teams and concurrent engineering fit into the design process.

The next step is to establish all the operational requirements of each component in the device. Which parts need to be hard? How much deflection can be tolerated in this shaft? What flatness is needed on the base plate? Is corrosion a problem in the loading mechanism? Can this arm tolerate any wear? What is the *allowable stress* (bending) in this shaft? These are the types of factors that must be established.

Once the designer has firmed up in his or her mind the requirements of each part, it is time to make detail drawings. Material selection is a key part of this step in the design process. A set of drawings must be produced that shows the shape, dimensions, material of construction, and applicable treatments or special sequences.

The last steps in the process are to fabricate, debug, and put in service. Materials enter into these steps in fabricability and substitution of new materials for the parts that did not make it through debugging. The major points are that the designer should start to think about materials of construction quite early, and, more important, that effective material selection is predicated on knowing the operational requirements of every part. You simply cannot select a material for a part without knowing what that part must do in service.

If an engineering assignment requires the purchase of a piece of machinery or subcontracting of a structure or facility, some of the steps in the design process (Figure 20–1) will be different. Instead of assembly and detail drawings, the designer will probably write design specifications. Material selection should be a part of these specifications. When vendor proposals are obtained, they are matched to the established specifications and once again material selection becomes important. The vendor, for example, wants to substitute gray cast iron for the pump housing that you specified to be CF-8 cast stainless steel. Should you permit it?

Thus material selection is still a part of the engineering process whether you design the machine or somebody else does. All the factors that would go into your own design should also be considered when evaluating someone else's design if it is your responsibility to make the piece of equipment function. If the gray cast iron corrodes through in 6 months, it is your fault, not the pump manufacturer's. You bought the equipment, and it is the engineer's responsibility to buy something compatible with the intended service environment.

20.2 Selection Factors

There are literally hundreds of different properties of materials. The most important to consider when selecting a material for a given part are those that are essential to the function of the part. We shall go into more detail on properties, but most designers put equal importance on properties, availability, and economics. In the United States business issues have arisen in the past decade that also influence material selection. There is no sense in specifying the use of a particular material if it cannot be obtained within the time constraints of a project. Similarly, a material cannot be considered for use if it costs more than the project can bear.

Finally, it is unwise to design a throwaway container from polystyrene foam when this material is known to produce a disposal problem and the foaming agents may use ozone-depleting substances.

Figure 20–2 lists the major selection factors—properties, availability, economics, and business issues—along with some pertinent subfactors.

Availability

One of the first things that many designers ask when initially considering the use of a particular material is whether the material is on hand. A "no" answer will provoke a second question: Can we get it in 1 week? 2 weeks? and so on. If this answer is acceptable, the next question is, Do we have to order a minimum quantity? There are more than 15,000 plastics that are commercially available, but only a dozen or so are available in

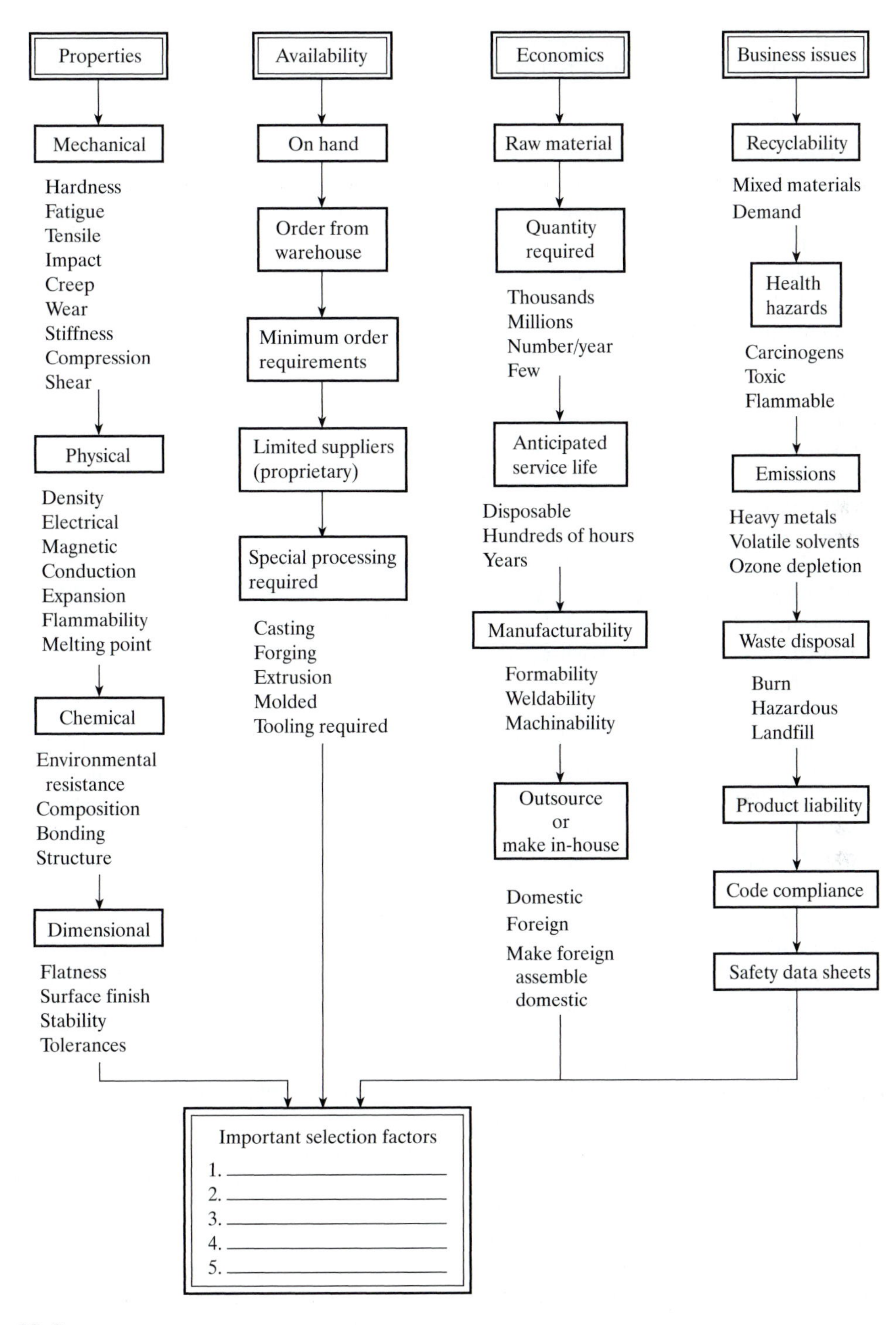

Figure 20–2
Material selection checklist

standard shapes from warehouses. How available is the plastic under consideration? Is it available only in molding pellets?

Another pertinent question: Is the material available from more than one supplier? It is not advisable to use a proprietary material that is available from only one supplier if it is likely that the parts in question will be made again 5 years hence. The same thing holds true if many parts are required. If you are a captive buyer, you are at the mercy of the supplier on cost and delivery.

Sometimes a raw material fabrication technique may limit availability. As an example, it is not uncommon to get delivery times of 40 weeks on special forgings. In lean times, castings may be obtained in 4 weeks; other times, delivery can take 12 weeks. These are all important selection factors under the category of availability. It is the designer's responsibility to establish a time line for procurement of materials, and if a desired material cannot be obtained within the constraints of this schedule, another material will have to be substituted.

It is also advisable to select materials that are known to be readily available. As shown in Figure 20–3, in 2004 about one-third of all the stainless steel consumed in the United States was type 304. This suggests that there is a higher likelihood of finding 1.5-mm-thick 304 stainless steel with a 2-B finish than there would be of finding the same item in type 321 stainless steel. The more popular alloys have greater availability. The materials in our suggested repertoire are usually the more readily available materials.

The driving force for availability studies is often time to market. If a new product is being designed, the design will be "leaked"; if the product shows potential, it will be copied by competitors. The company who beats its competitors to the marketplace will get the initial sales burst that is necessary to pay for development costs. Being the first out often helps continuing business. Using readily available materials can help time to market.

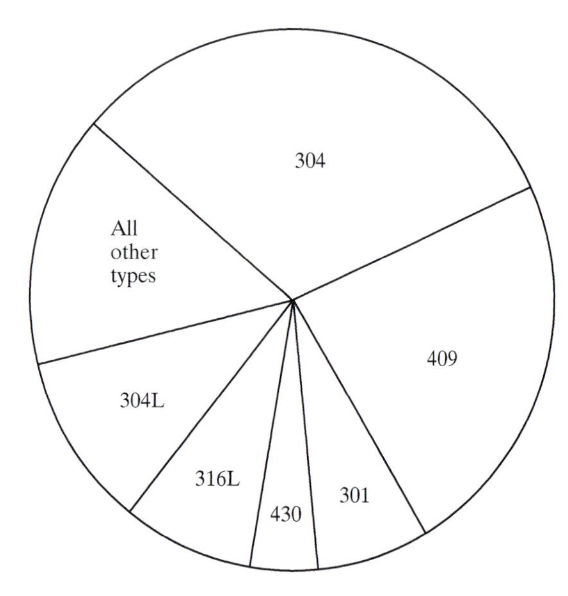

Figure 20–3
Relative usage of stainless steels in the United States in 2004 (various sources)

Economics

Assuming that an established availability criterion can be met, the designer should look into the economics of a material. Consciously or unconsciously, every designer uses the cost of materials as a prime selection factor. If a design calls for a soft, ductile component, a designer would probably never even consider gold or silver because of the belief that these are very expensive metals, too costly to be used in a common machine. However, sometimes materials that are in themselves very expensive are justified because they offer a unique property advantage or because they are cheaper to use than other, lower-cost materials; for example, a design might be simplified and thus made at lower cost.

What are the comparative costs of the materials used in design? No one country has a ready supply of all metals, ceramic compounds, or the raw stocks for polymers. The supply of these commodities depends on everything from droughts to wars. The demand similarly varies with the economic climate, and thus prices fluctuate daily in response to the supply and the demand. It is not possible to put an exact price on a quantity of a material, but in 2003 the relative costs in the United States of some of the materials that we have discussed looked approximately as shown in Figure 20–4.

The most common plastics and metals cost between $0.25 and $1/lb. Some of the new exotic polymers can cost as much as $300/lb, and the precious metals cost as much as $10,000/lb. These are the extremes in cost, but, as was pointed out previously, sometimes the special properties of even very expensive materials can make them the best choice for a particular design. Beryllium costs as much as $900/lb in special sheet form, but it has a stiffness much greater than that of steel, and it is very light. These properties make it the most economical choice for certain applications. Tantalum at $150/lb is used for severe corrosion applications. A small washer of it can be used for patching glass-lined vessels handling strong chemicals. It is the cheapest way to repair these expensive vessels.

Ceramics and composites are noticeably absent from our cost chart. The reason is that these materials are not normally available in standard shapes, such as plates, bars, and sheets. Ceramics must be molded into a specific shape. Composites can be made as sheets and other shapes for fabrication by conventional machining, but only the simpler composites, such as glass-reinforced thermoplastics (RTP) and fabric-reinforced thermosets, are commonly available. The more sophisticated composites, such as carbon-reinforced epoxides and polyesters, are normally made into the desired shape by lay-up or special techniques.

The cost of boron and carbon filaments for high-strength composites is still in the range of $15 to $200/lb; ceramics can be as cheap as $0.60/lb for aluminum oxide grit or as much as $400 for a small molded part of the same material weighing only 100 g. The same thing is true of carbon products. Some small-diameter carbon bushings sell for $0.10 each, whereas the same bushing made from another grade of carbon may cost $20. If ceramics or composites are to be considered for a particular job, there is little alternative other than to obtain a quotation from a suitable supplier. There simply are no good general cost per pound guidelines on these materials.

Getting back to our cost chart, the cost estimates presented are based on bulk quantities, such as molding pellets in plastics and heats of metals. Any engineering material that is bought in small quantities from a warehouse has many extras added on. As an example, 1030 steel may sell for $0.30/lb if you buy a heat, but that same material in a 1000-lb quantity and vacuum melted costs $6/lb. One of the lowest-cost tool steels, W1, may cost only $1.50/lb as a hot-rolled round, but as a centerless ground and polished round the cost becomes around $6/lb. Thus, if you are considering small quantities of materials, you must expect many warehouse and handling extras.

If you are designing a part for large-scale production and it appears that a polymeric material, ceramic, or metal may do the job, one helpful way of comparing material costs is to look at the *cost per unit volume*. Thirty dollars per pound sounds like a lot of money for a plastic, but if that plastic has a specific gravity of 1.5 compared with 7.8 for steel, you get more volume in a pound of plastic than you get in a pound of steel; so the $30 becomes much smaller when divided by its volume. Figure 20–5 shows the costs per unit volume of some common engineering materials. It can be seen that the cost ranking changes on a number of materials. Even beryllium looks less expensive. Similarly, many plastics look less expensive. The lowest-cost material of construction is a plastic. Traditionally, cast iron has always been the lowest-cost metal.

The main point in this discussion of raw material costs is that cost should not be the sole selection factor in making a choice of materials for a particular design. If only a few machine components are to be made, material cost may be almost inconsequential. If a large quantity of material is

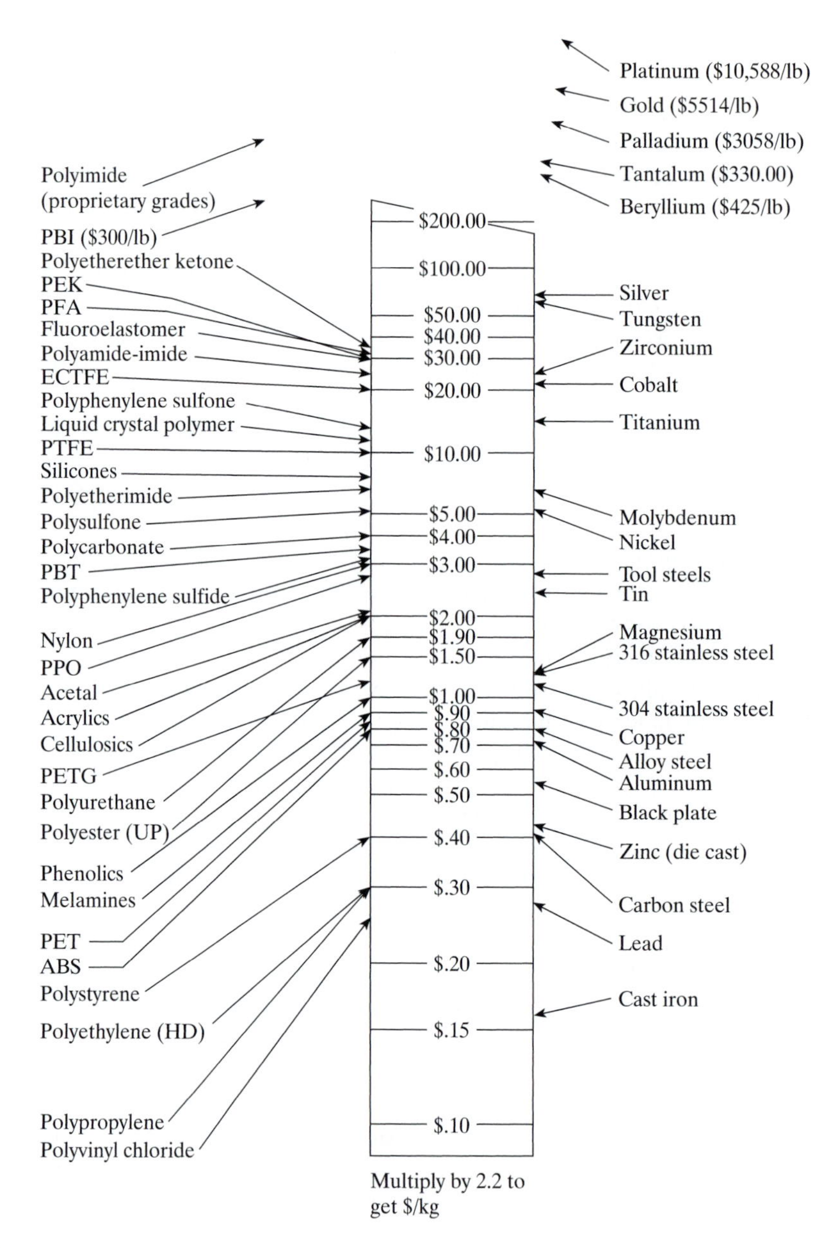

Figure 20–4
Comparative costs of engineering materials per pound in December 2003 in the United States. Plastics are in the form of unfilled molding resin; metals are in the form of mill products in bulk heat quantities.

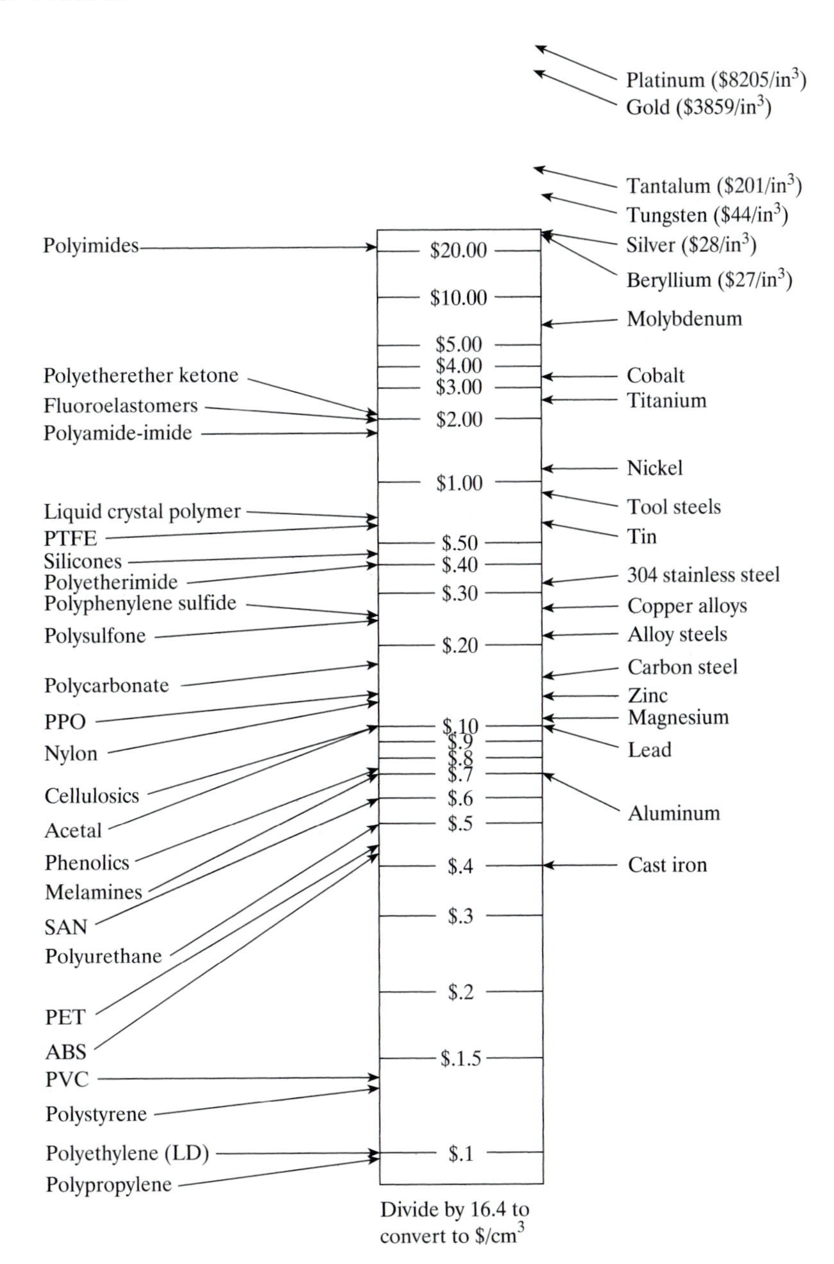

Figure 20–5
Cost of some engineering materials per cubic inch in December 2003; these change frequently. Use this for rough screening, then get the current price.

involved, the answer to the material–cost question should be obtained by a thorough cost study using current material costs.

Other pertinent considerations on the subject of economics are anticipated *service life (serviceability)* and fabricability. If a machine is being designed to last forever, it may be economical to use materials with high raw material costs if these materials have superior strength, wear, or other properties that will contribute to the machine's lasting indefinitely. Some examples of designing for indefinite life are equipment associated with the distribution of utilities to a manufacturing operation. Steam turbines put in service in the nineteenth century are still producing power in many old buildings. At the other extreme, machines used in the manufacture of dashboards on this year's car need last only a year if a model change is anticipated. Thus the designer must use judgment on how much extra to pay for long service life in selecting a material.

Another significant economic factor that we will discuss is *manufacturability*, or how easy it is to convert a raw material to a desired shape. If a material has poor machinability, it will cost more to fabricate than a material with good *machinability*. Some materials such as the cemented carbides cannot be machined at all. They can only be shaped at manufacture or by techniques such as grinding, electrical discharge machining, or electrochemical machining. Most ceramic materials fit into this same category. They cannot even be electrical discharge machined.

Titanium, magnesium, and some other metals in sheet or strip form cannot be formed without special tooling or special hot-forming techniques. These factors make these metals more expensive to use than ductile low-carbon steels or aluminum alloys.

Manufacturing costs can also be a factor with polymeric materials. Only a limited number of plastics are available in standard shapes such as rod and plate. If only a few parts are needed of a special plastic, it may be necessary to pay molding costs.

In some cases, even easy-to-machine plastics involve high machining costs because of their tendency to distort during machining. Some of the acetals fall into this category. If accurate dimensions are required, the machining sequence may involve several intermediate stress-relief heat treatments, thus resulting in high machining costs.

Weldability as an economic factor simply means that some materials require special joining techniques that increase costs. Titanium cannot be welded in air. Thus an inert gas welding chamber or special shielding is required. Inertia welding and explosive bonding are the only common techniques used to join aluminum to steels. These processes involve expensive equipment and special techniques. A common example of weldability increasing costs is welding of low-alloy steels, high-strength steels, and tool steels. Special procedures and heat treating equipment are required. Thus if a part under design requires welding, the weldability of candidate materials can become a selection factor.

The last consideration in the economics column of Figure 20–2 is to outsource or not. *Outsourcing* is a business term that came into vogue in the United States in the 1990s. It means to contract outside of your organization to make an item or perform a service. In 2003, this business concept was widely practiced in the United States to various degrees in different organizations. Some billion-dollar (in sales) consumer product companies had absolutely no manufacturing facilities. They outsourced everything but the business aspects.

Usually outsourcing decisions are made at levels other than the engineer/designer level. However, engineering materials still need to remain a consideration for the responsible designer/engineer. For example, plastic injection molding operations are very material sensitive. If an existing molding operation is outsourced to another continent, it is quite likely that you may not be able to buy "good molding" plastic in that country and a significant quality or production capability problem may arise. The same types of problems occur with steels and other materials of construction. Some countries do not manufacture titanium, some countries do not make super austenitic stainless steels, some countries do not have vacuum carburizing facilities. Designers and engineers often must perform materials investigations when manufacturing operations are outsourced. Designers are ultimately responsible for the overall product or service, no matter where it is made and feedstock differences can affect product success.

In summary of cost factors, the best basis for comparison of candidate materials for a part is the price per piece after manufacturing. This number will include material cost per unit (cost/volume) plus the fabrication cost. Table 20–1 shows some machining cost factors from one fabrication shop. These types of fabrication cost factors are used in estimating job and part costs. These can be used in conjunction with Figure 20–5 to compare finished part costs from various materials.

Table 20–1
Machining cost factors for various metals*

Material	*Machining Cost Index*
Hot-rolled 1020 steel	1.5
Finished 1020 steel	1.0
Cold-rolled 1213 steel	1.0
Cold-rolled 12L13 steel	0.7
Finished 1040 steel	2.0
Prehardened 4150 steel (30 HRC)	3.4
Hot-rolled 4140 steel	2.3
Nitriding alloy steel	3.4
O1 tool steel	2.3
Decarb-free D2/A2 tool steel	3.4
Hot-rolled M2 tool steel	3.4
Class 25 gray cast iron	1.0
Class 45 gray cast iron	2.3
Ductile iron grade 120–90–02	3.4
304/316 stainless steel	3.4
317 stainless steel	5.1
303 stainless steel	1.5
420 stainless steel	2.3
440C stainless steel	3.4
17–4 stainless steel	5.1
CF 8M cast stainless steel	5.1
Brasses	0.7
C95300 bronze	1.5
Monel 400 N04400	2.3
Monel 405 N04405	1.5
Nickel N02200	2.3
Nickel castings (ASTM A 743/CZ100)	3.4
Aluminum alloys (nonhard)	0.7
Aluminum alloys T6	0.4
Magnesium and its alloys	0.4
Pure titanium (4 grades)	3.4
Titanium 6A1–4V	5.1

*If it costs $1.00 to machine a part from 1020 steel; it will cost $5.80 to machine the same part from titanium 6A1–4V. This is the opinion of one machine shop; other shops may have different opinions.

Properties

An entire chapter was spent on the description of material properties (Chapter 2), and we have discussed the relative properties of the various material systems in subsequent chapters. The properties listed in Figure 20–2 can be used as a guide in establishing a list of the material properties that will affect the performance of a part under design. Think about your operational conditions; then ask which are the most important mechanical properties. If a device is cyclic loaded, you may want a material with a high fatigue strength. If you are designing a support device, compressive strength will be an important property. Shock-loaded devices need good impact strength; shafts must have a shear strength adequate to handle the torque transmitted. Stiffness (modulus of elasticity) is important in springs; creep is important in devices used at elevated temperatures. Finally, wear can affect many different types of devices. This latter property should also be qualified as to mode: abrasion, metal-to-metal wear, erosion, or surface fatigue.

Our list of mechanical properties is not complete, but the point is that a list of applicable mechanical properties should be based on the stresses, motion, and forces applied to the part under design.

This same type of procedure applies to the other classes of material properties; the physical, chemical, and dimensional. The checklist of physical properties could conceivably be several pages long. Sometimes an important physical property may be surface conductivity. Devices handling radio-frequency currents require good surface conductivity. Often it is necessary that parts be ferromagnetic or nonmagnetic. Sometimes plastic parts may require a specific degree of self-extinguishability (flammability). The coefficient of thermal expansion is an often-forgotten physical property, but this property is important whenever dissimilar materials are to be fastened together and heated. There are really no bounds on the number of physical properties that should be considered in material selection; sometimes a relatively obscure property may become important when this class of properties is given thoughtful consideration.

One of the most important chemical properties is environmental resistance. Is corrosion by chemicals important? Will rust interfere with the function of a part? If corrosion is determined to be important, it may be advisable to establish an acceptable corrosion rate. Can a rate of $250\,\mu$m/yr be tolerated? Such information is helpful in consulting corrosion data surveys.

Chemical composition may be an important selection factor for reasons other than corrosion resistance. Metal alloys containing cobalt cannot be used in parts of nuclear reactors because radiation can cause the cobalt to form a radioactive isotope with a significant half-life, thus creating a disposal problem if the part ever needs replacement. Sometimes the chemical nature of a surface can be a selection factor. Copper, zinc, and lead alloys resist adhesive bonding. The same thing is true with many plastics.

The crystal structure of some materials can be important. It was previously mentioned that boron nitride in the hexagonal form is soft and weak; it is hard in the cubic form. Carbon in the cubic form is diamond; it is graphite in the hexagonal form, and amorphous carbon has properties that are different from diamond and graphite.

The final class of properties to be considered—dimensional—applies to all components. The designer must set tolerances and surface-finish requirements. If it is determined that abnormally close tolerances are required, this becomes a selection factor. Some engineering materials are more stable in machining than others. If a plastic is being considered, it will be desirable to select a plastic that is not susceptible to movement by moisture absorption. If a long, slender part is under design, it will be necessary to stay away from cold-finished metals. They are notorious for bowing when locally machined. If a part requires a mirror finish, this becomes a material selection factor because some materials can never be made to the low surface roughness required for a mirror finish. Gray cast irons fit into this category; the graphite inclusions preclude the possibility of a mirror finish. The same can be said of fabric-reinforced plastics, porous ceramics, and some coarse-grained cast metals.

Business Issues

In the United States and some European and Asian countries, environmental and regulatory issues can be of equal importance with, or even of more importance than, economic factors. The same can be said for other selection factors that we elect to categorize as business issues. Some of the business issues that most designers will have to consider in selection of materials are illustrated in Figure 20–2. It may not be possible to use a particular material unless it is on the approved list of some regulatory agency. In the United States, the Food and Drug Administration regulates the materials of construction that can be used for prosthetic devices. There are insurance codes that dictate the materials of construction allowed in some types of pressure vessels. Machines sold in Europe must be wired and certified to a European Union (EU) code. We call these selection items "business issues" because businesses must deal with them, like it or not. To ignore these factors can be costly in various ways. If a designer or manufacturer ignores a code, a fine or even imprisonment can result; if environmental concerns are ignored, a company's reputation and sales may be affected. If a product is defective, a product liability lawsuit could bankrupt the manufacturer. We will comment on some of the more important business issues in the following discussion.

Recyclability is becoming more important each year in the United States and most European countries. Countries with many consumer goods have waste streams that are getting out of control. Localities are running out of places to landfill trash. Recycling is an attempt to reduce the waste stream, and it becomes a material selection factor in that some materials are easier to recycle than others. For example, thermoplastic materials are easier to recycle than thermosetting materials. Mixing plastics in an assembly therefore creates a recycling problem. Decorating with gold inks and the like can make a plastic unrecyclable. Consider recycling factors on most designs. It makes good business sense.

Health issues enter material selection in a number of ways. Some materials use *carcinogens* in their manufacture. Some materials are toxic, and this *toxicity* must be considered in designing a consumer product. Materials that involve use of *volatile organic compounds* (*VOCs*) should be avoided where possible. VOCs are regulated by many governments. Some materials produce health risks in handling. Some materials are more flammable than others. If an item can be exposed to open flames or some other occasion of high heat, it is better to select a material with low flammability, reducing a hazard. It is a useful exercise in design of products to try to visualize all the possible ways that a particular material could cause a health hazard or *hazardous waste*. If some concerns surface, this will become a selection factor. There are handbooks of industrial toxicology that show health hazards of most engineering materials.

In the United States today, emissions from manufacturing processes and machinery are significant factors to be dealt with. If you are designing a part that can be stamped or cast to shape, you should consider how you would clean the oil from the sheet metal stampings. If the cleaning process will involve handling chlorinated solvents, it may be more cost-effective to use a casting if it does not have to be cleaned of oil. Handling and disposal of organic solvents has become a formidable task in the United States. If you are considering the use of a material that will require salt-bath hardening, you may want to consider how you will dispose of the salt if the process is to be performed in-house. In some countries, arc welding has severe restrictions because of the metal fumes emitted. It may be simpler to use a mechanical fastened assembly than to deal with the welding emissions.

There is a growing trend toward "smokeless" factories. When you select a material you should consider whether use of that material will add emissions to the atmosphere.

Waste disposal may not be an obvious material selection factor, but it can be related. For example, if you are designing parts that need corrosion protection, you may select chromium plating. In some localities it is almost impossible to get regulatory agency approval to discharge depleted plating solutions through normal sewage systems. Cadmium plating has almost disappeared in the United States because of disposal and related problems. All manufacturing operations produce scrap. It may be difficult to dispose of the scrap from a particular material. For example, polyethylene is widely used in laminating materials. Thin layers of molten poly are used to join materials to make composites. It is very difficult to dispose of the polyethylene "logs" that are produced in drool carts placed under the poly extruders at startup and shutdown. This is a selection concern; maybe another plastic would be recyclable. Sometimes such seemingly innocuous things as waste adhesives are considered to be hazardous materials. Always consider disposal in material selection: Will the use of a particular material generate waste that is costly to dispose of?

Product liability is a business issue of high significance in the United States. In 1965 legislation was passed in Congress aimed at increasing the safety of consumer products. As a direct but unintended result, this legislation has developed into a multibillion dollar industry for lawyers. If any accident involves a consumer product, lawyers often sue the manufacturer of the product as well as the manufacturers of all the components in the product. Sometimes even designers can be sued. The plaintiff will try to prove that there was a design or material defect in the product when it left the manufacturer's company, or there was a fault in marketing the product.

A design defect can be as simple as the designer's failure to check the low-temperature impact strength of a material. "Marketing defect" is a legal term alleging that assembly, cleaning, or use instructions were inadequate, or that not all hazards attendant on use of the product were identified. A classic example is the football-helmet manufacturer that was sued by a player who received a neck injury in a game. The manufacturer had not put a warning sign on the helmet stating that it would not prevent neck injuries. Selecting materials with product liability in mind requires careful consideration of all properties that could be related to failures or hazards. A design review with a safety engineer is recommended for parts that could produce consumer injuries if the part or product failed. Have the safety engineer also recommend appropriate hazard and use instructions.

Our final business issue is code compliance. This simply means that there may be regulatory codes that dictate what material can be used in an application. In the United States and Canada, an international joint commission that controls boundary waters has banned the element chlorine as an industrial feedstock. This could mean no vinyl (PVC) plastics or paints. At present this ban is not enforced, but it should be considered in material selection. Overall, business issues are becoming more important each year. For the foreseeable future these issues need to be considered along with the more engineering-oriented factors such as properties and economics.

Summary

Figure 20–2 may give the impression that material selection involves the consideration of too many factors, but despite the awe the illustration may inspire, it is not that complicated. The following questions are the most frequently asked by materials engineers when they are confronted by a drawing presented by a designer who perceives that he or she has a potential material problem.

1. What does this part/machine do?
2. How many are you going to make?
3. Is corrosion a concern?
4. Does the part rub on anything; is wear or friction a concern?
5. Does it see shock loading?
6. How long do you want it to last?
7. When do you need it?

This short list illustrates the more common selection issues. Essentially we are asking five questions:

1. What are the important properties ("must have" and "want")?
2. When must the part be completed (delivery)?
3. How many are needed?
4. How long does the part need to last?
5. How important is cost?

The wear, corrosion, and impact questions are aimed at the three most common causes of failure.

We have tried to show that the most important factors are properties, availability, economics, and business issues. The relative importance of each of these factors depends on a particular design situation. If you are designing a part to correct a machine shutdown, availability may be most important. If you are designing a missile launcher, properties may be most important. If you are designing an infant seat for an automobile, product liability and fail-safe design will be prime concerns.

Economic factors are most important if your budget is low or if many units are required. Whatever the case, the designer must establish priorities and then develop a list of the important selection factors that apply to the part or device under design. This list does not have to be formal or even written. Sometimes a mental list is adequate. In any case, the establishment of selection factors is an essential part of the selection process.

20.3 A Materials Repertoire

We have reviewed the design process and selection factors, and at this point the designer has the drawing of the device he or she wants to make or to buy, a list of the operational conditions for the device, and a formulated list of the most important selection factors. To cite a simple example, let us assume that the designer wants to make a shaft for a hypothetical device, a wiebol plotter. The drawing for this part is shown in Figure 20–6. The list of operational conditions is as follows:

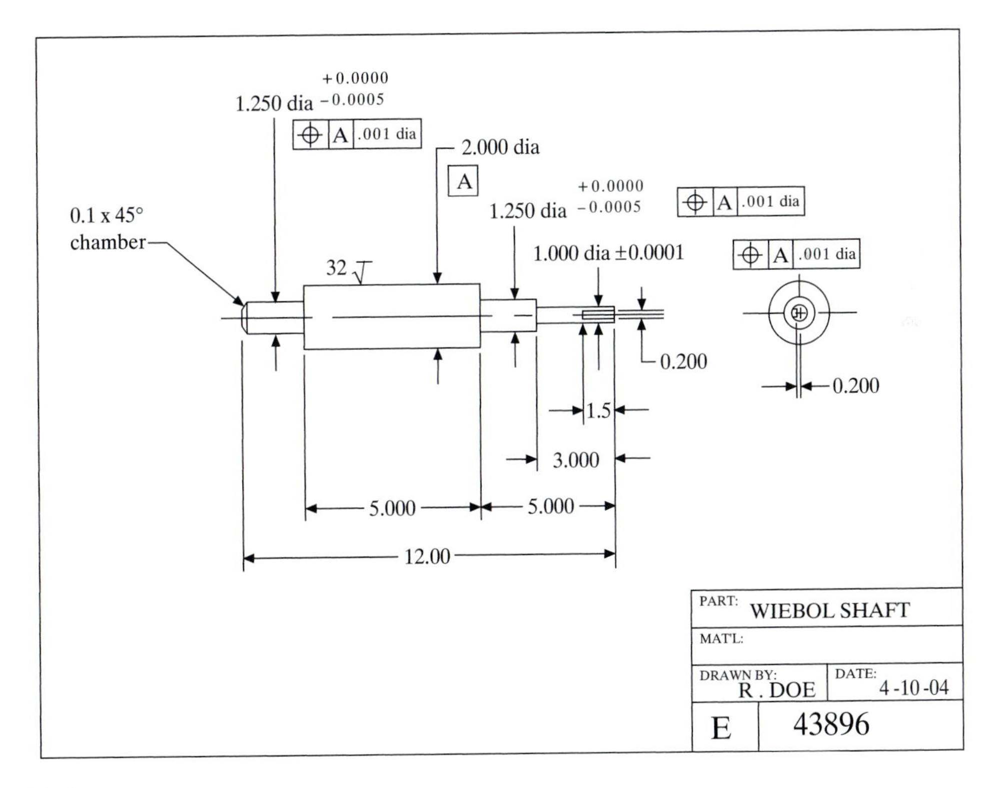

Figure 20–6
Detail drawing for typical machine part

Operational Conditions on Wiebol Shaft

1. 1.250 diameters must fit with X522 ball bearing.
2. Smallest diameter must carry torque of 130 J, shear stress of 10 ksi (68.9 MPa).
3. Possible shock load of 40 J.
4. Small end must resist damage from frequent removal of keyed gear.
5. No inertial requirements.
6. Surface roughness to be 32 μin. Ra max. (0.8 μm).
7. Diameters must be concentric within 0.001 in.

Based on these operational conditions and the cost and time constraints of the project, a list of important selection factors is formulated.

Important Selection Factors

1. A hardness of at least 30 HRC is needed.
2. Fatigue strength must be at least 30 ksi (207 MPa).
3. Impact strength must be high.
4. Stiffness must be as high as possible.
5. Diameter (A) cannot rust in 50% RH room air.
6. Must be stable and straight within 0.001 in.
7. Part is needed in one week.
8. Three units are required.
9. Maximum part cost: $30 each.
10. Expected service life: 5 years.

The selection factor list was obtained by using Figure 20–2 as a checklist. The designer simply goes down the list and notes the properties that apply. Additional properties can be added, or a designer may want to establish his or her own property checklist. Experienced designers may not use a written checklist, but you can be sure that they use one committed to memory.

Proceeding on to the next step in the selection process, it is now time to select a material system. Do the important selection factors dictate a plastic, a metal, a ceramic, or even a composite? The answer to this question is obtained by performing a mental scan of the relative properties of various material systems. We have just reviewed relative material costs (Figures 20–4 and 20–5), and in appropriate sections of the text there are tabulations of significant properties on all important material systems. Figure 20–7 is a comparison of a few material properties, and the same type of comparison chart can be made for other material properties. Many material handbooks and databases have extensive comparative property tabulations that can be used. By this point, the reader should have developed some awareness of the relative properties of the more common material systems. Assuming that this is the case, the designers now can review their repertoire of engineering materials and select a specific material. The materials that we have recommended as useful for most design situations are shown in Figure 20–8. We list about 70 materials, which may seem like too many, but this list is much briefer than the list of metal alloys that are commercially available in the United States, which has some 65,000 entries; the plastics databases, which list hundreds of basic polymers and any number of modifications to each in the form of fillers, plasticizers, and the like; and the ceramic journals, which list hundreds of ceramics that are used for various applications. The reader could go down the list of materials and state some facts about each. It is not a difficult task.

The designer picks a material. The appropriate material specification is then put in the material box on the drawing. This puts the designer in the finish box in the flow chart shown in Figure 20–9. Unfortunately, we are not quite finished yet. More often than not a material may require some treatment to get the desired properties. It

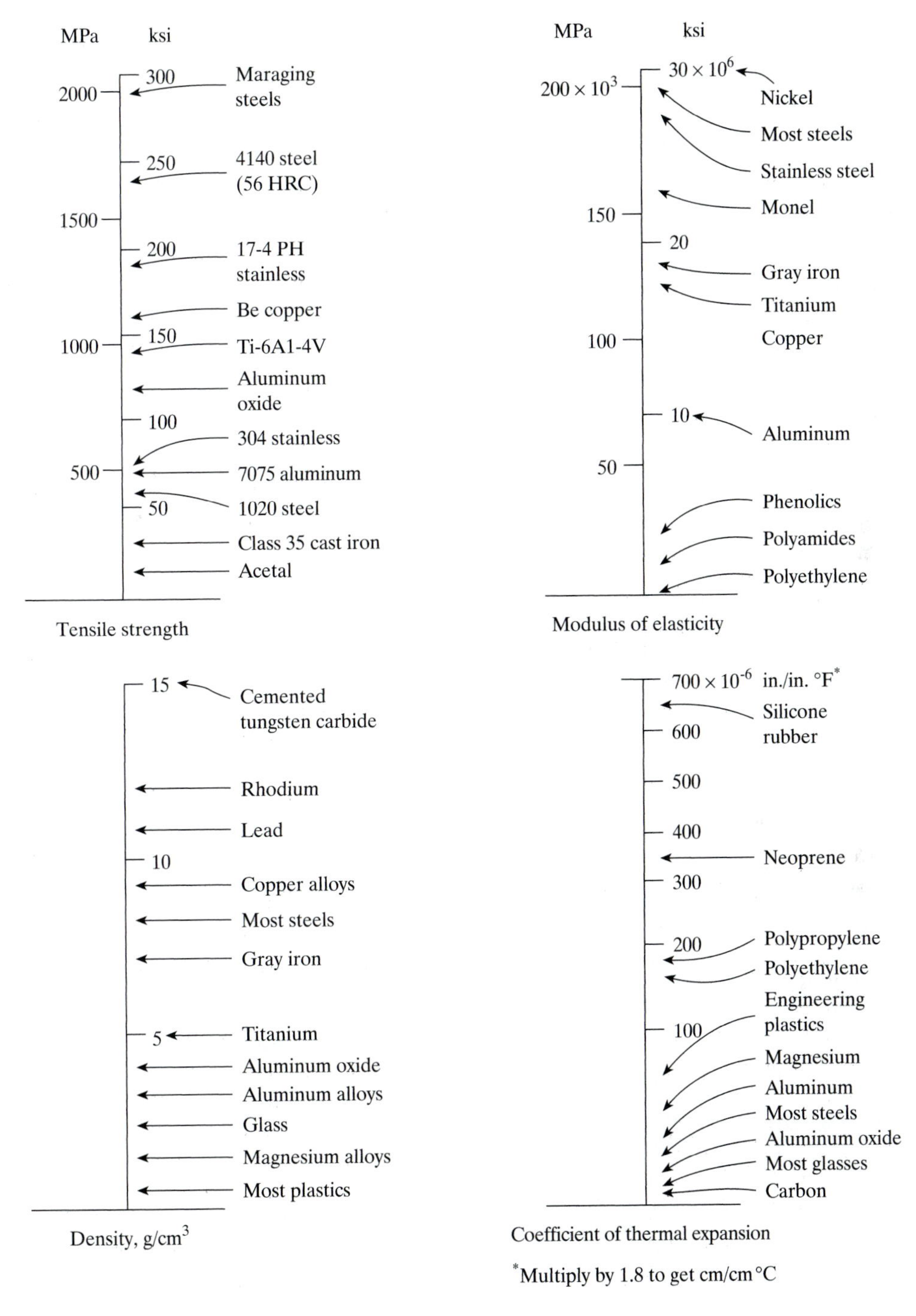

Figure 20–7
Property comparisons of various material systems at room temperature

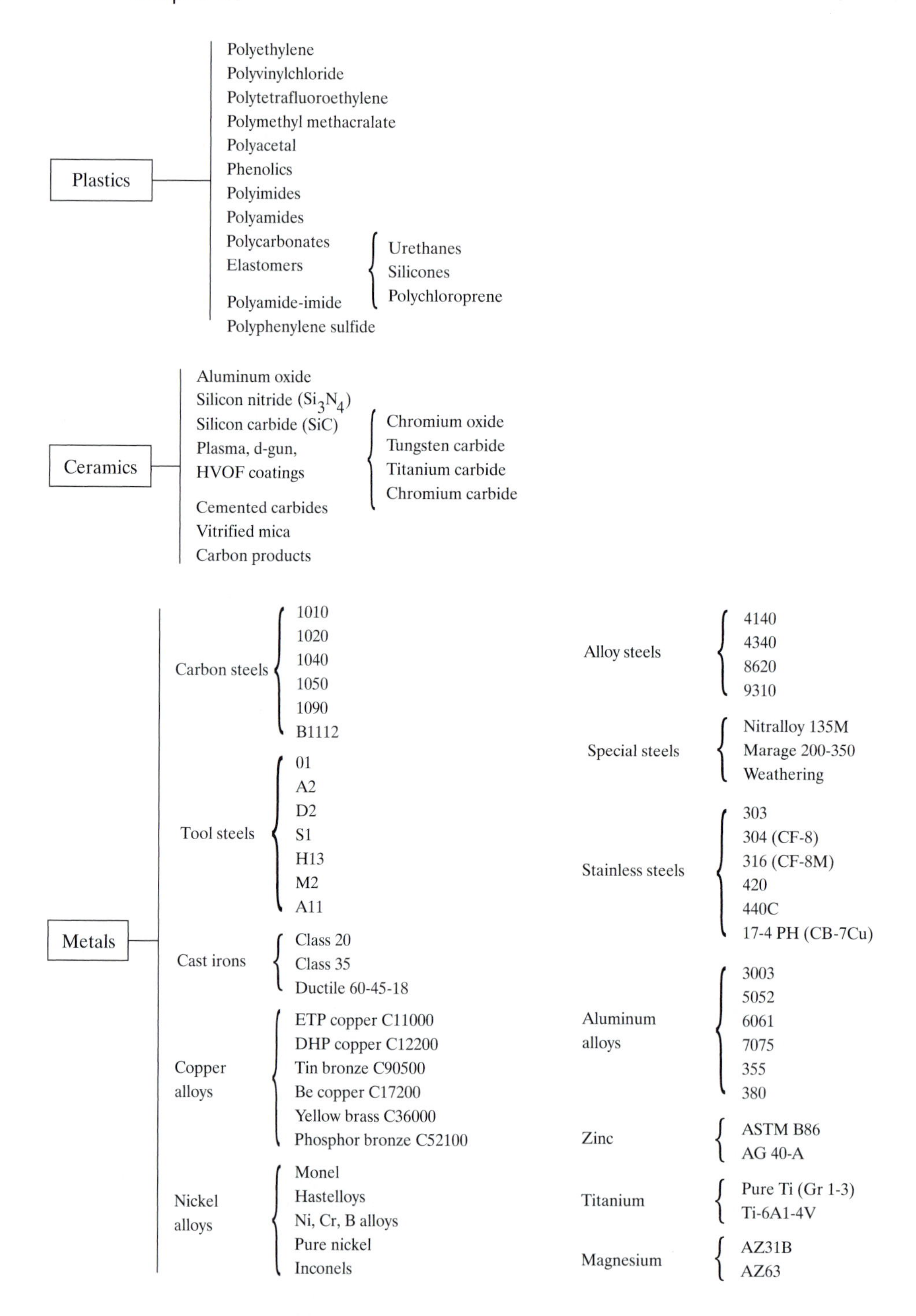

Figure 20–8
Possible designer's repertoire of engineering materials

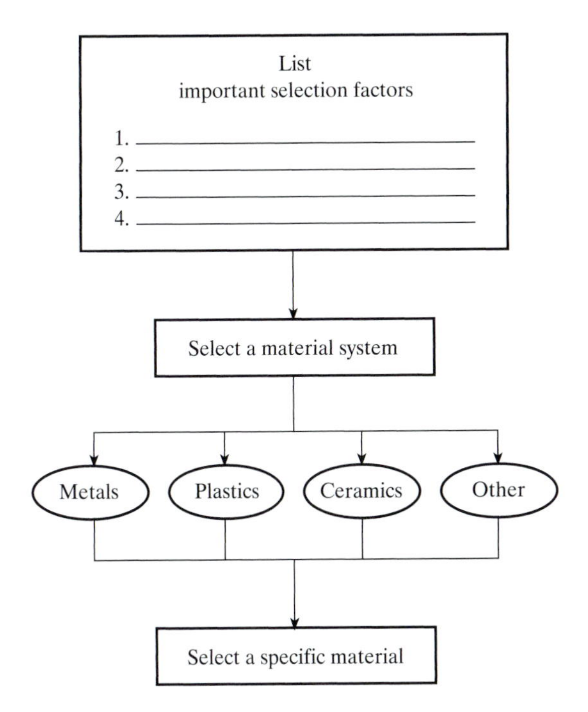

Figure 20–9
Steps involved in material selection

is also the designer's responsibility to specify applicable treatments. Figure 20–10 lists some of the material treatments that we have previously discussed. This can serve as a treatment checklist. At last we can complete an engineering drawing. The accepted name of the selected material is listed along with the appropriate designation. Plastics and ceramics often need a trade name and manufacturer for designation. Desired treatments are listed as a footnote. A proper material designation might look like those illustrated in Figure 20–11.

Let us check out this system on the example of the wiebol shaft. The list of selection factors shows that high stiffness is required. Immediately, this puts us into the metal or ceramic sections of our materials repertoire. The stiffness parameter, the modulus of elasticity, of plastics is very low. The requirement of good impact strength eliminates the brittle ceramics and metals. The hardness and stiffness requirements put us into the category of hardenable steels. Which steels meet the remaining cost, availability, and property factors? There are probably five steels that could work (for about the same cost). This will almost always be the case. There are a number of right answers, but experience in previous use would direct many designers to select AISI 4140 steel for the material of construction for our wiebol shaft. The only selection factor that this material does not meet is resistance to rusting in conditioned air. Thus we add a material treatment, a 0.0005 to 0.001-in. (12.5 to 25-μm)–thick electroless nickel plate on diameter A. The other treatment required is to harden and temper to 28 to 32 HRC. Now we are done, and the engineering drawing can be completed:

> *Material*: AISI 4140 steel prehardened to 28–32 HRC, Electroless nickel plate diameter A 0.0005/0.001-in. thick (12 to 25 μm). Mask all other areas.

Putting the selection process and a material repertoire into writing makes it seem complicated even to someone who has been using it for many years. This is simply because most designers store their property, availability, and cost information in their heads. With experience, these selection steps become intuitive, and as one's knowledge of materials increases, it is not necessary to scour the literature to look up comparative properties. The whole process becomes almost automatic.

20.4 Materials for Typical Machine Components

In our example of the hypothetical wiebol shaft, we outlined a procedure for an in-depth material selection study. However, most machines contain components that are common to other machines. For example, many machines use shafts of one type or another. The same is true

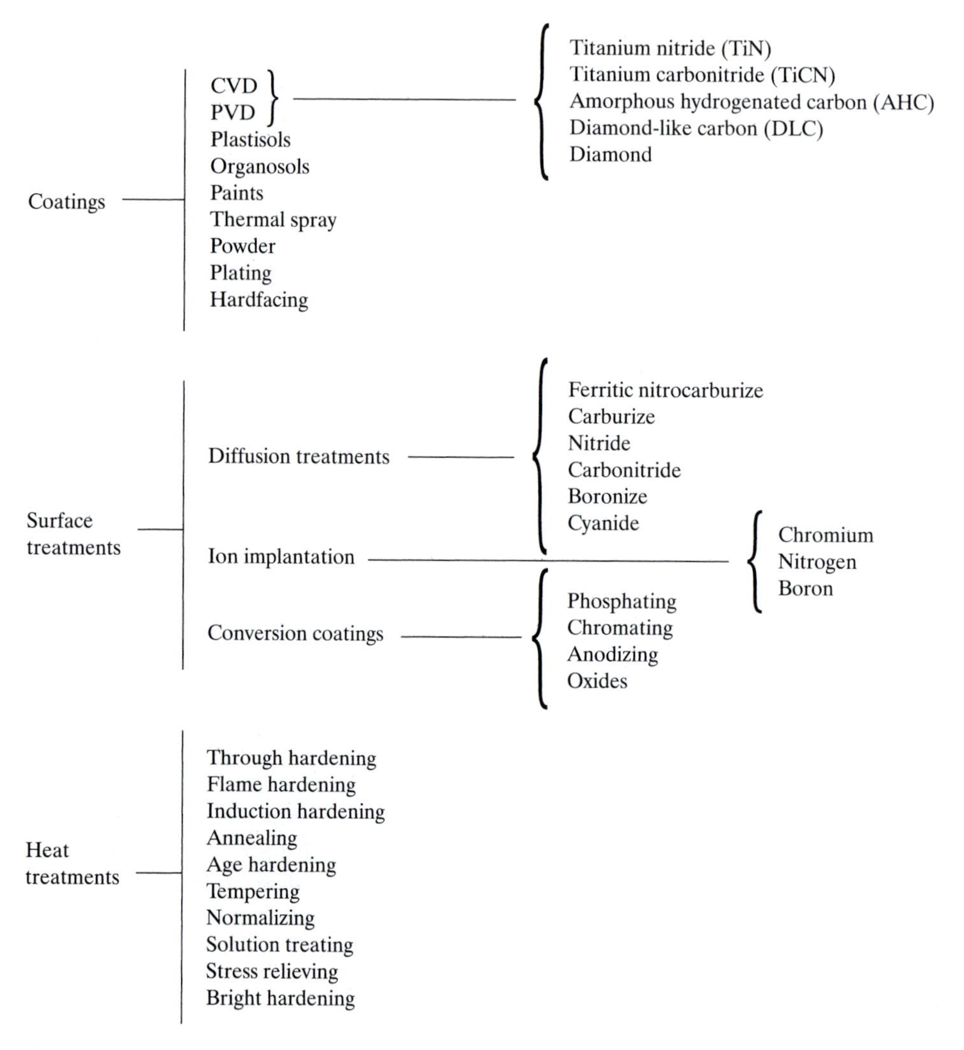

Figure 20–10
Material treatments

of gears, bushings, fasteners, and other parts. A few materials are frequently used for these types of parts. In Table 20–2 we have tabulated a number of these common machine components, and we have listed several candidate materials that can be used in making these components. It is the purpose of this table to give the designer a starting point in selecting materials for a part under design that may be included in the table.

Several materials are listed for each component because in all selection problems there is always more than one material that will work satisfactorily. The first candidate material listed is usually the lowest in cost or the least wear resistant. The other materials are listed in order

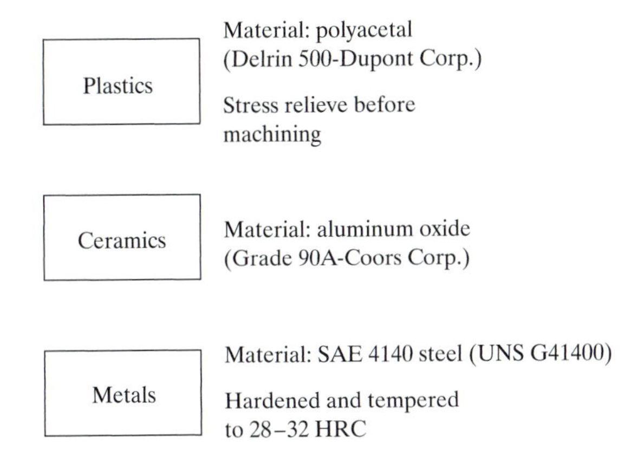

Figure 20–11
Drawing specifications for various engineering materials

of increasing hardness, strength, or wear resistance. The improved serviceability of the latter materials is usually coincident with higher costs in material, fabricability, or both. Thus the designer must still weigh relative merits before making a final selection. As an example, under the subject of springs, four materials are listed: phosphor bronze, 1080 steel, and types 301 and 17-7 stainless steel. Phosphor bronze has the lowest strength, but it has excellent fabricability and does not rust. The 1080 steel, music wire, is used for probably 90% of all wire springs. All wire springs should be made from this material unless there is a reason why it will not work. The 301 stainless steel is used for applications where corrosion resistance is important. The 17-7 wire is used where operating stresses are higher than the 301 can tolerate and elevated-temperature resistance or corrosion is a selection parameter.

20.5 Selection Case Histories

The following are some case histories of selection problems and the recommended solutions. They are intended to further illustrate material selection methodology.

Problem

Materials for a precision gage device (large rectangular frame: overall size 50 × 200 × 400 mm).

Requirements:

1. *Must properties*: Must be stable with time and resistant to damage when used for measurement; will be used to locate tools within 1 μm. The device will be used in a controlled atmosphere, but the temperature may vary by ±5°F, maximum service temperature 80°F; minimum 70°F.
2. *Want properties*: Stiffness comparable to steel, corrosion and wear resistance not necessary, significant machining required, including small tapped holes.
3. *Quantity*: Only one required.
4. *Delivery*: Within 2 months.
5. *Life*: Must last for the anticipated product life of 10 years.
6. *Cost*: Expect to pay several thousand dollars.

The requirement of being stable enough to locate items within 1 μm suggests that thermal expansion should be very low. If the device is made from steel, the 400-mm dimension will change 200 μm when the temperature varies 10° in the use environment. This requirement and the machining requirement make Invar, an iron/nickel low-expansion alloy, the recommended material. There are grades of Invar that have essentially zero expansion within the use temperature limits of the device. Invar meets the must properties, but it is a bit low on the stiffness requirement. It machines like carbon steel, so it meets the machining requirement. The elastic modulus of the lowest expansion grade is only 22×10^6 psi (1.52 GPa), but this certainly should be adequate for a measuring device. This is one of those situations where very few materials can meet the "must" requirement, no expansion with temperature.

Table 20–2
Materials for machine components (*selection database*)

Component or Tool	Candidate Materials*
Base plates	Aluminum jig plate, 1020 steel, gray iron
Battering tools	S1, S7 tool steels
Bolts	Acetal, 1020 steel, 1040 steel, 4140 steel
Boring tools	M2, M3-2, T15 tool steels, cemented carbide
Broaches	M2, M3-2 tool steels
Bushings	PTFE filled acetal, cloth-reinforced phenolic, PTFE fabric, leaded tin bronze, P/M bronze, P/M iron, polyimide
Cavities, injection mold	P20, H13, A2 tool steels, 420 stainless steel
Chisels	S1, S7 tool steels
Choppers	A2, D2, M2 tool steels
Chutes	PVC, UHMWPE, 1020 steel, 304 stainless steel
Die-casting dies (zinc, aluminum)	H13 tool steel
Dies, blanking	O1, A2, D2, A11 tool steels, cemented carbide
Dies, drawing	O1, A2, D2 tool steels, cemented carbide
Dies, drawing, wire	D2, M2 tool steels, cemented carbide
Dies, forging, hot	4340 steel, H13, H21 tool steels
Dies, forming	O1, A2, D2 tool steels
Dies, press brake	4140 steel, S7 tool steel
Dies, threading	M2, M3-2 tool steels
Dowels	W1, A2 tool steels
Drills	M42, M3-2, T15 tool steels
Drill bushings	O1, A2 tool steels
Fixtures	Filled epoxy, 6061 T6 aluminum, O1, A2 tool steel
Gears	MoS_2-filled nylon, acetal, cloth-reinforced phenolic, gray iron, ductile iron, 1020 steel, carburized 4615 steel, flame-hardened 1045 steel, 4340 steel, carburized 9310 steel
Gages	Acetal, A2 tool steel, cemented carbide
Guards	PVC, acrylic, polycarbonate, expanded metal
Hammers	1080 steel, S7 tool steel
Holding blocks (cavity)	P20 tool steel
Hobs	M2, M2-3, T15 tool steels
Keys, drive	Acetal, yellow brass, 1020 steel, 4140 steel
Knives, steel slitting	A2, D2, M2 tool steels, cemented carbide
Knives, paper	A2, D2, M2-3 tool steels, cemented carbide
Knives, corrosive materials	440 C stainless steel, cemented carbide, zirconia
Knives, steel shear	S1, S7 tool steel
Knives, square cutter	Flame-hardened 1045 steel, D2 tool steel
Lathe tools	M2, M2-3, T15 tool steels, cemented carbides
Machine bases	1020, A36 steel, gray iron, ductile iron
Pinions	Carburized 8620 steel, A2 tool steel
Pneumatic tools	S1, S7 tool steels
Punches	W1, A2, D2 tool steels
Reamers	M2, T15 tool steels

Table 20–2
continued

Component or Tool	Candidate Materials*
Rolls, bending	4340 steel
Rolls, calendering	Chill cast iron
Rolls, conveying	6061 T6 aluminum, 1020 steel, polyurethane coated steel
Rolls, engraving	D2 tool steel
Screwdrivers	S1, S7 tool steels
Seals, face	Graphite versus alumina, polyimide versus 440C stainless
Seals, shaft	PTFE versus alumina, graphite versus tin bronze, graphite versus alumina
Shafting	1020, 4140, 4340 steels, 17-4 PH stainless steel, nitrided Nitralloy
Springs, flat	Phosphor bronze, beryllium copper, 1080 steel, 6150 steel, 17-7 PH stainless steel
Springs, wire	Phosphor bronze, 1080 steel (music wire), cold-drawn 301 stainless steel, 17-7 PH stainless steel
Worm wheels	Phosphor bronze, aluminum bronze
Wrenches	1050 steel, S1, S7 tool steels

*Hardenable metals must be heat treated to their recommended working hardness.

Material: Super Invar, hot-rolled plate (Cardec Industrial Metals Co., Redding PA).

Problem

Material for precision spur gears (6-in. pitch diameter, 1-in. face width).

Requirements:

1. *Must properties:* No scoring and very low wear with repeated cycling from rest to 3000 rpm every 3 min. The tooth load is very low; it is less than 50 lb on a 36-pitch tooth.
2. *Want properties:* The material should be capable of low surface roughness (good grindability); corrosion and wear are not a problem. The gears are submerged in an inhibited oil bath. Need very good tooth profile accuracy.
3. *Quantity:* Only two sets are needed. They are for a special testing machine.
4. *Delivery:* There is a 6-month time period allowed for construction of the machine.
5. *Service life:* The gears need to run up to 4 h per day for 5 years.
6. *Cost:* The precision machining required on the teeth will make these gears cost $300 each.

There are only a few materials that are used for precision gears: tool steels, case hardened alloy steel, cast irons, bronzes, and occasionally some plastics. The no-scoring requirement suggests the need for a steel with a high hardness (over 60 HRC). Tool steels will meet this requirement; so will case-hardened alloy steels like carburized 9310. We selected type D2 tool steel because it is very stable in hardening and it is very galling resistant (scoring) when it is self-mated with both members at 60 HRC. The 9310 case-hardened steel would be more impact resistant, but the low tooth load suggests that impact will not be a problem. The D2 tool steel will be more stable in heat treating than the case-hardened steel. They require a liquid quench. D2 is air hardening.

Material: D2 tool steel, hardened and tempered to 59/61 HRC.

Problem

Material for a large welded machine base (1 × 2 × 4 m) to be made from square tubing.

Requirements:

1. *Must properties:* Must have the strength and stiffness to hold drive motors, transport mechanism, and work station.
2. *Want properties:* Would like the weldment to resist rusting in room air.
3. *Quantity:* Only one of these units will be made.
4. *Delivery:* There is a time limit of 3 months for fabrication of the device by an outside supplier.
5. *Service life:* Must last 20+ years.
6. *Cost:* Project has a tight budget, should be low cost (<$2000).

There are a limited number of materials commercially available as square tubing: low-carbon steel, stainless steel, aluminum, and brass. Polyester/glass pultrusions are available, but it would be difficult to form them into a complicated shape. It can only be done by bonding, not welding. We selected carbon steel from the metals. It is stiff, strong, and weldable, and lower in cost than the stainless, the other material with comparable strength and stiffness. The nonferrous metals are higher cost and lower stiffness, and they do not weld as easily. The no-rust requirement can be met by application of an electroless nickel plating to the part.

> *Material:* 1020 carbon steel square tubing, 2″ × 2″ × ⅛″ wall, cold finished seal weld and electroless nickel plate 0.0003/0.0008 in. thick after fabrication. Plug all holes during plating.

Problem

Injection-molded polystyrene (PS) cores for yarn are wearing on spindles at knitting mills. The cores are only loosely fitted on 300 series stainless steel spindles during unwinding; they move slightly with respect to the spindle in a fretting motion. The wear debris create a contamination problem during yarn dying. The selection goal is a core material that is not susceptible to fretting damage in contact with 304 stainless steel.

Requirements:

1. *Must properties:* Injection moldable, same shrink rate as polystyrene (PS), better fretting resistance than polystyrene (PS).
2. *Want properties:* White color, high impact strength.
3. *Quantity:* Five million per year.
4. *Delivery:* Need 500,000 units per month for 10 months.
5. *Service life:* Cores are reusable, but it is assumed that they will be discarded after about four reuses (2 years).
6. *Cost:* Must not cost in excess of 30 cents per unit (U.S.). The part weight is 32 g.

The must requirement of shrinkage characteristics like PS severely limits selection. This was made a requirement to save scrapping of a multiple cavity core mold worth $300,000. The cost is another very restricting selection requirement. Polystyrene is one of the lowest-cost plastics. The only commodity plastics that are occasionally cheaper are the olefins such as polyethylene and polypropylene. Both of these plastics have a shrink rate that is significantly different from that of PS. The proposed solution was polystyrene molding resin containing 15% by weight of a fluorocarbon (PTFE). This raised the molding pellet costs by 20%, but this is still cheaper than building a new mold.

> *Material:* High-impact polystyrene with 15% by volume PTFE; shall be Dor Inc. grade DSC203.

Problem

Small stainless steel transport rollers (10 mm in diameter, 6 mm long) are corroding and scratching a contacting web product. The rollers rotate on 2-mm-diameter pins, and the rotation is produced by friction with the web product. The selection goal is a material that does not scratch the web and that freely rotates on a hardened steel (60 HRC) dowel pin.

Requirements:

1. *Must properties:* Must not corrode in contact with chloride and nitrate salts; cannot be copper or aluminum alloys because of contamination reasons; cannot seize in 24 h/day use.
2. *Want properties:* Low wear, low cost, made by screw machine or injection molding. Would like the material to not require lubrication. The sliding speed can be as high as 3000 rpm (on a 2-mm-diameter pin).
3. *Quantity:* Need 15,000 per year at quarterly intervals.
4. *Delivery:* Need 3750 per quarter throughout the year.
5. *Currently:* These rollers last about a year. When the corrosion gets progressive, the rollers scratch product. This requires their removal. Would like to double life.
6. *Cost:* Would like cost to be less than $1 each.

The "want" of no lubrication suggests that only self-lubricating materials need apply. Steel, ceramics, and many other tool materials cannot slide on a pin for a year without generating copious amounts of wear debris (which can lead to seizure). Self-lubricating materials include plastics, carbon–graphites, and composites made from plastic blends that contain a lubricant. P/M porous bushings could be considered, but the lubricant can migrate out and this would cause contamination. We selected a self-lubricating polyimide containing 20% graphite. This material will not corrode in the subject environments and the graphite makes the plastic self-lubricating. Carbon–graphites were not used because of availability problems. The polyimide molding resin selected cost $9 per pound, but there are about 400 parts per pound. The material cost per roller is only about $0.02.

> *Material:* Injection molding polyimide + 20% graphite; shall be ARUN grade 407. Poly Plastics, Henrietta, NY.

Problem

Flat springs are needed to hold plastic film samples against a platen in a test rig. The springs are 0.5 mm thick, 10 mm wide, and 50 mm long. A material is needed for these springs.

Requirements:

1. *Must properties:* Black color, will not rust in 50% RH, 70°F room environment.
2. *Want properties:* Would like 10^7 flexes at a stress level of 3000 psi (20.68 MPa).
3. *Quantity:* Twenty springs.
4. *Delivery:* Springs must be made within 30 days.
5. *Service life:* The 10^7 flexes equate to about a year service life.
6. *Cost:* Should not cost more than $10 each after fabrication.

Availability controls this selection to a significant degree. There are not many materials in warehouses in the form of 0.5-mm-thick sheet or strip. Two common flat spring materials that have good availability are 1080 steel and beryllium copper. We opted for 1080 hardened and tempered spring material. It can be easily blackened with a black oxide treatment. This spring stock is widely used for flat springs, steel rule dies, and pallet banding. The springs can be cut from the hardened strip with a water jet laden with particles or by EDM machining. The beryllium copper would involve more process steps. It would have to be heat treated. Wrought spring materials such as full-hard cold-rolled wrought steel or copper alloys were considered, but heat treated springs last longer and are more likely to deliver the desired 10^7 cycles.

> *Material:* 1080 steel, hardened and tempered to spring condition (47 HRC). Fabricate complete and black oxide coat all over.

Summary

Our discussions of various material systems should have supplied enough information so that the designer can quantitatively compare the candidate materials and know what heat treatments and processing to use when specifying the material on the drawing. Obviously, it is not possible to compile a list of all the components that are found on machines. Similarly, the materials listed are not the only ones that will work, but the examples reflect a wide range of service experience.

The material selection process should be started as early as possible in a machine or structure design. The key to proper material selection is to establish a checklist of the properties required for a serviceable design. Does the part have to be hard? Does it require corrosion resistance? Are any special physical properties required? Once the required use properties are established and prioritized, designers can draw on their repertoire of available engineering materials to select one or more candidate materials and treatments.

Remember to state the function of the part. What does it do? What are the apparent forces on it? What are the potential unforeseen forces? Calculate stresses based on the perceived forces. List potential corrodents; determine if tribological properties are a consideration. Doing these steps should lead to a short list of "must" properties. Must properties are those that cannot be compromised without compromising the function of the part or project.

Next list the "want" properties. You would like to have these properties, but the part would not fail in service if you did not get them. An example is good machinability. Poor machinability usually increases fabrication cost but does not impact mechanical or physical properties that relate to the intended service (unless poor machinability creates stress concentrations of some surface defect that can lower service ability).

It is intuitively obvious that quantity required, availability, and cost affect selection. These factors affect everything that we buy. Sometimes cost and availability become "must" selection factors. High-production parts often have cost as a priority; replacement of failed parts often makes availability a priority. If you need to get a machine back on-line in 24 h, you will probably make availability a priority.

Our selection examples illustrate that the selection procedure that we advocate is to establish the function of a part, list and prioritize required properties, and then scan your repertoire of material properties for a match. Scan mechanical, physical, dimensional, and chemical properties. Expert systems and computer databases are available to help with the task of scanning material properties. Computers can scan faster than we mortals. The precaution that should be taken when using these selection aids is that they usually do not include parochial restraints. The computer does not know for example that 440C stainless is not readily available as 2-mm-thick × 2-in.-wide strip. Similarly, the computer does not know that there are two dozen grades of polycarbonate molding pellets available from a single supplier when it tells you to use polycarbonate for a part. What grade of polycarbonate? Expert systems and searchable databases are not infallible; consider them as useful tools that can give you the wrong answer. If you follow the selection format that we have proposed, you will end up with probably three materials that will meet your part requirements. This is normal. There are usually a number of right answers to a selection problem. This makes it easier to solve availability or cost problems. Your final selection will probably be a compromise of properties, availability, and cost, but in no case should serviceability be compromised.

The following are some important points to remember:

- Consider materials of construction concurrent with design ideas.
- Certain material shapes require long lead times to purchase (die castings, forgings, and others). If this is a problem, consider alternative shapes.
- Many low-volume usage engineering materials are often unavailable from warehouses. Try to use popular alloys, plastics, and ceramics; they are usually available in warehouses.
- Look at overall costs of using a material, not just the cost of the material used. Fabrication costs often outweigh raw material costs.
- Have your customer define expected service life.
- Include heat treatments and coatings in cost comparisons.
- Determine if welding needs to be a selection factor (maybe for repairs).
- Determine if wear, corrosion, and impact are selection considerations.
- Thoroughly consider the properties that are important to a design. Use these as the basis of selection.
- Use complete specifications for materials, treatments, and coatings. Do not use trade names without referring to the company that owns that trade name.
- Consider how you will check materials and parts for *defects*.
- Plastics, ceramics, and metals and composites thereof should compete as equal candidates for designs.
- Calculate stresses on parts and make sure that the selected material can tolerate the stresses.
- When it appears that three materials will match a selection problem, use the one that you are most familiar/experienced with.
- Remember to consider business issues in material selection.

Critical Concepts

- Consider all materials when cost is a key concern.
- Cost per unit mass may not be as important as net material cost to make a part.
- It is ok to change the material of construction as a design evolves and more is learned about operating conditions.
- Your drawing specification determines what material you will receive; specify correctly.

Terms You Should Remember

cost per unit volume
serviceability
manufacturability
machinability
VOCs
allowable stress
tolerances
availability
machining cost index
carcinogens
hazardous waste
recyclability
safety data sheets
toxicity
stability
service life
selection database
operating speed
appearance
defects
product liability

Case History

Material for Tensioning Rollers

The design for a new product called for two flanged rollers to tension a thin (0.5 mm) polyester belt. The roller was 100 mm in diameter; the belt width was 25 mm; the belt tension was 50 N; and the roller rotated at 10 rpm. The materials engineering lab was asked to make recommendations for the roller material and its bearing system. The rollers were driven only by 90° contact with the belt.

Because annual production would only be one hundred units, we would normally recommend buying commercial belt-tensioning roller systems. However, these tensioning devices were tried on a prototype device and failed because of high friction and grease contamination. The machine would be run in clean room conditions, and grease contamination or rust could not be tolerated. The next obvious choice may be to machine the rollers from steel or aluminum and fit the inside diameter with oil-impregnated bronze bushings that could run on a hardened stripper bolt for a shaft. However, a little more thinking about the design yielded a selection of ultra-high-molecular-weight polyethylene (UHMWPE). At the designated speed and load (pressure/velocity), the plastic rollers could rotate on a steel pin unlubricated, eliminating two to four parts. Also, plastic rollers meet the no-rust requirement and UHMWPE would be resistant to abrasion from belt slippage. This material selection was implemented, and it has worked flawlessly, at lower cost than competing materials that would require bearings.

Questions

Section 20.1

1. How can lubrication be dealt with in material selection?
2. Corrosion resistance is a critical requirement for a particular design. How should you start the design process?
3. List the operational conditions for an automobile tire wrench.
4. Cite an example where a federal regulation is a design parameter.
5. You need high damping of vibrations in a design. Name two candidate materials and explain why they are appropriate.
6. You are designing a low-cost child's toy. Name two candidate materials.

Section 20.2

7. Name five materials that would have long lead times for procurement.
8. Cite two specific examples where the use of precious metals is cost effective.
9. Explain the meaning of minimum-order requirements.
10. Cite an example where product liability is a concern. How would you deal with it?
11. Compare the cost of making a part out of stainless steel to the cost of making the same part from nickel-plated carbon steel.
12. Compare the machining costs to produce a 1-μin. surface roughness (0.02 μm Ra) on a part to the cost of machining the same part to a surface roughness of 20 μin. (0.5 μm) Ra.
13. How do you determine the cost of a ceramic part?
14. Compare the cost of making a base plate ¼ × 4 × 4 in. (12.5 × 100 × 100 mm) out of carbon steel or out of PVC.

Section 20.3

15. What counterface hardness is required for a P/M bronze bushing?
16. Low deflection is a service criterion. Name three materials that are candidates for this application. Explain your choices.

17. A part is designed to tolerate a bending stress of 140 ksi (1034 MPa). Name three candidate materials for this part.
18. When would you use a fluoroelastomer in design? Explain your answer.
19. Write a specification for a CVD coating of TiC on a D2 tool steel punch press die.
20. What would it cost to nitride a 25-mm × 25-mm × 4 m-long 4140 steel rail?
21. Name three materials with higher stiffness than steel.
22. You wish to make a new design for an Allen wrench. Specify the material and heat treatment.
23. When would you use a plastic instead of an oil-impregnated P/M bronze bushing for a plain bearing on a 1-in.-diameter (25-mm) shaft that rotates at 100 rpm?
24. You need an electrically insulating structural material to operate at 700°F (370°C). Name two candidate materials and compare their costs.
25. Cite two examples where the use of cemented carbide tooling is cost effective.
26. A cemented carbide punch is abrading rapidly in service. What material would you use to solve this problem?
27. You calculate the operating stress on a part and find it to be 150 ksi. What are three candidate materials for this application?

To Dig Deeper

Ashby, M. F. *Materials Selection in Mechanical Design*. London: Pergamon Press, 1995.

Ashby, M. F., and D. Cebon. *Cambridge Materials Selector*. Cambridge, MA: Granta Design Ltd., 1996.

ASM Handbook Vol. 20: Materials Selection and Design. Materials Park, OH: ASM International, 1997.

Craig, B. D., Ed. *Handbook of Corrosion Data*. Metals Park, OH: ASM International, 1989.

Machine Design, Materials Reference Issue, an annual. Cleveland, OH: Penton IPC.

Modern Plastics Encyclopedia, an annual. New York: McGraw-Hill Book Co.

Schaffer, J. P., and others. *The Science and Design of Engineering Materials*, 2nd ed. Boston: WCB McGraw-Hill, 1999.

Schweitzer, Philip A. *Corrosion Resistance Tables*, 3rd ed. New York: Marcel Dekker, Inc., 1992.

Waterman, N. F., and M. F. Ashby. *The Material Selector*, 2nd ed. Boca Raton, FL: CRC Press LLC, 1997 (also available on CD-Rom).

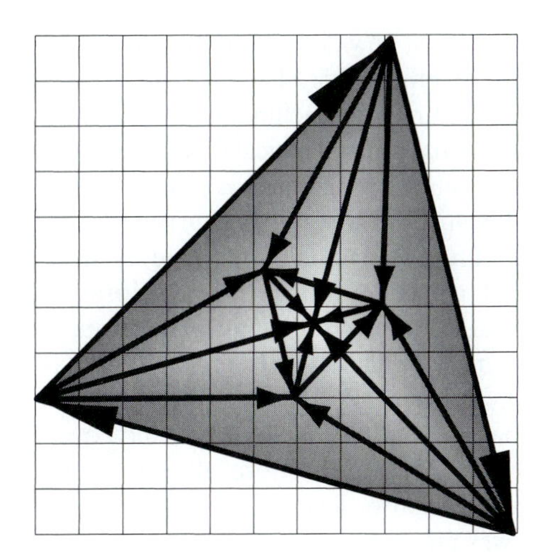

CHAPTER 21

Failure Prevention

Chapter Goals

1. To point out common design faults that cause service failures and to present guidelines on how to prevent them.
2. To show that serviceability requires a sound design, selection of the right material, and inspection to prevent material flaws.
3. To identify some of the more common sources of failures and show how nondestructive evaluation (NDE) can be used to detect them.

Rationale

This chapter should not be necessary. We have discussed the theory of all important material systems; we have given you a repertoire of materials to work with from each important material system; and we have given you our "trade secrets" on how to successfully use these materials. So why do we end this book with a chapter on failure prevention? Because stuff happens! Probably everything that engineers' design eventually fails. We pay a lot of money for new automobiles, even used automobiles, but they all fail. They rust to the point at which they are structurally unsound, or we give up on fixing the mechanical systems as they fail.

We all know that nothing lasts forever; this chapter is about preventing failure during the anticipated service life. Automobiles have an anticipated service life of 50,000 to 100,000 miles in the United States (based on warranties). Water heaters come with a variety of service lives: 5, 10, and 12 years. Most premature failure happens because of either some factor overlooked in the design stage or a defect. This chapter will present guidelines on how to check your designs to make sure that they are not failure-prone and to review inspection and testing techniques that address defects. You need to know how to prevent your designs from failing before completion of their intended service lives.

In the previous chapter, we outlined the steps to be taken in selecting materials for machines or structures that you design or that you are responsible for procuring. Following these steps should result in a material selection that yields a serviceable design. Experience has shown that in spite of intelligent material selections on the part of designers, failures still occur. Every industrial plant has a maintenance department, as do most commercial buildings, apartment houses, utilities, and even the local church. What do these maintenance personnel do? For the most part, they fix things that have failed in service. Were all of these designed with the wrong materials? Some were. Some failed because their design life was exceeded (roofs are good for only 30 years, etc.), but many of these failures were caused by neglect of some commonsense design guidelines or by an unanticipated situation that taxes material properties.

It is the purpose of this chapter to discuss some of the design factors that help to make a correct material selection last as long as intended. We shall concentrate more on design and use conditions than on materials. We have already discussed the important material systems and how to select them. Now, how do you put these materials together so that they work? There are three major causes for failures in machines and structures: wear, corrosion, and mechanical breakage.

Almost any component that fails in service does so because it wore out, it corroded, or it broke (Figure 21–1). Some components suffer two or all of these modes of deterioration. Figure 21–1 estimates the relative percentages of failure modes that occur in metals and plastics. Many years of failure analysis work have shown that most service failures did not occur because of bad material. More often than not, some commonsense design factor was ignored or there was an error in fabrication. In this concluding chapter, we shall discuss prevention of

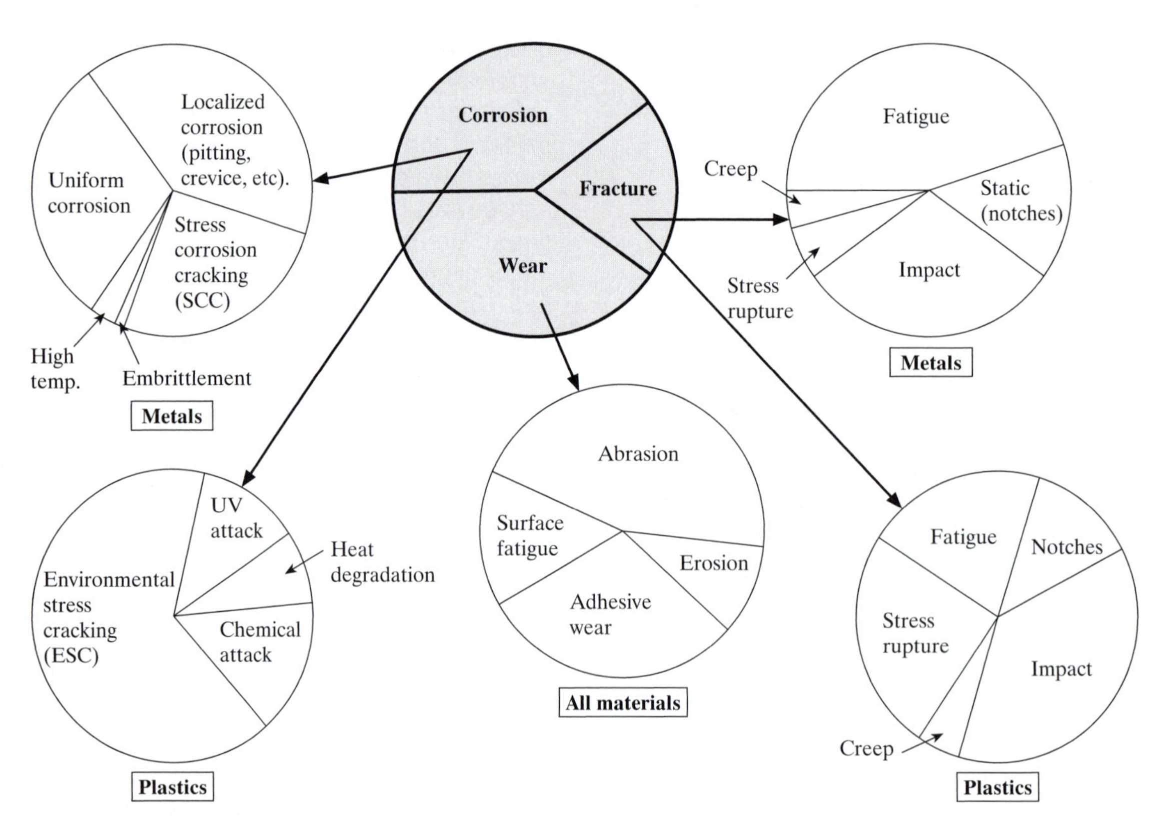

Figure 21–1
Impediments to serviceability

wear, corrosion, and mechanical failures and point out some things that can and should be done to prevent these failures from occurring.

21.1 Preventing Wear Failures

Why do parts wear out? We can answer this question with the following statement: It is their destiny. This may sound facetious, but it really is not. When any contacting solids experience relative motion they will wear; similarly, when a fluid constantly flows over a solid surface, erosion will occur. The wear or erosion can be manifested in a number of forms, but it will occur. The reason why sliding wear must occur is that solid-to-solid sliding wear is due to the same thing that causes friction: interaction between surfaces. When two solids contact, there are bonds between the mating hills and valleys. Some of the bonding is mechanical interlocking; some is due to atomic or molecular forces. Just as there is no way to completely eliminate sliding friction (people have tried for many centuries), there is no way to completely eliminate wear when contact and relative motion exist. There will always be some transfer or attrition from surface asperities when bonds are broken to produce motion. The mechanism of attrition will be different for chemical modes of wear and for some of the other modes of wear, but the eventuality of wear is a fact.

In most machines there are a number of factors that the designer can control to minimize wear and to make the wear that is inevitable at least tolerable. The following are some design hints that apply to the various modes of wear.

Abrasion, Low Stress (Figure 21–2)

1. Eliminate the abrasive. If the source of abrasive is something like airborne particles from an ash pit, install a water spray on the pit to keep particles from becoming airborne. There are often other similarly simple solutions.

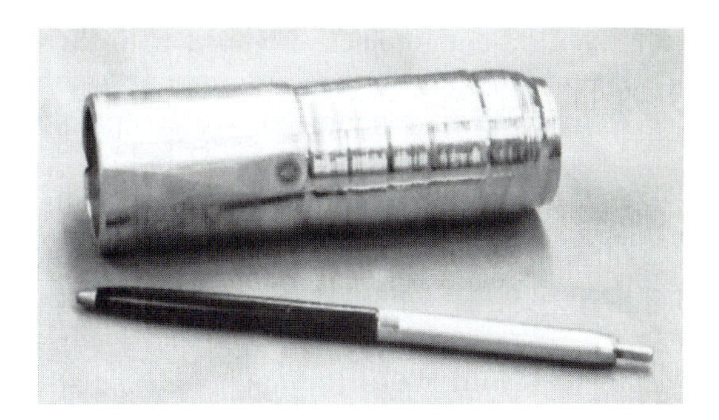

Figure 21–2
Part failure due to abrasive wear. Asbestos packing ran against this stainless steel bushing (type 316).

2. Shield critical surfaces from abrasive substances. Sealed bearings, shaft seals, air purges, and the like are effective ways of keeping abrasive substances from reaching sliding interfaces.
3. Determine the hardness of the substance that is likely to produce abrasion and make the machine surfaces that will see this abrasion harder than the abrading substance. If an abradant has a hardness over about 900 kg/mm^2, only cermets or ceramics are hard enough to resist its cutting action. Some practical high-hardness surfaces are thermal-sprayed ceramics, cemented carbides, and tool steels with high-volume fractions of carbides in their microstructure.
4. Reduce contact stresses between the abrasive and sliding surfaces. If dirt or a similar abrasive substance is likely to find its way into a sliding system, the damaging effect can be minimized by using a mating surface of a polymer or babbitt-type material to provide embeddability of the abrasive substance. This will decrease the contact stress and the cutting action of the abrasive.
5. Use coatings or surface treatments that make the surface harder than or accommodating to the abradant (e.g., rubber).

6. Use ultra-high-molecular-weight polyethylene or 90A polyurethane to resist abrasion in systems where nonmetals are acceptable (pipes, rolls, chutes, etc.).
7. Cemented carbide or aluminum oxide tiles provide abrasion resistance that is superior to most other materials.

Metal-to-Metal Wear

1. Always lubricate. All unlubricated metal-to-metal sliding systems will produce appreciable wear if there is significant sliding. If lubricants are not permitted, try to make one member of the couple from a self-lubricating plastic (e.g., PTFE-filled acetal).
2. A hard–soft metal couple will produce more wear on both members than a hard–hard couple. A hard steel (60 HRC) mated with a soft steel (20 HRC) will not protect the hard member from wear. If one member is to be perishable, mate one member from materials such as plastic, red brass, or gray iron and the other from hardened steel (60 HRC). Ceramics and plastics may need compatibility testing.
3. Avoid soft–soft *sliding couples* in metals. Unhardened metals should be used only for light loads and when service is infrequent or intermittent; they produce very high wear rates.
4. The lowest system metal-to-metal wear and galling resistance will be achieved with a hard–hard couple. Metals with high hardness (> 60 HRC) and fine carbides in their microstructure, self-mated or mixed, are preferred (high-speed steels, carburized alloy steel, P/M tool steels).
5. Avoid austenitic, ferritic, and precipitation-hardened (PH) stainless steels, aluminum alloys (without hardcoat), nickel alloys (without boron), and the pure metals. Most alloys in these categories have poor metal-to-metal wear characteristics self-mated or mated with most other metals.
6. Avoid self-mating plastics, they tend to weld.
7. Ceramics tend to gall when self-mated. A cemented carbide is usually an acceptable counterface.

Surface Fatigue

1. Use materials with high hardness and high compressive strength. Most ball and roller bearings are made from 52100 steel at 60–62 HRC for the races and rolling elements; similar high-hardness steels should be used for rolling applications. Type M50 high-speed steel is also used for high-performance rolling element bearings. Silicon nitride balls are available in high-performance ball bearings (52100 steel races).
2. Thin, case-hardened surfaces and surface coatings have a tendency to spall under Hertzian stresses if the substrate has insufficient strength to support the case. Use only coatings with documented ability to withstand these types of stresses. Run a test if in doubt. Some applications can tolerate chromium-plated ball bearings; some cannot. In 2003 it was the trend to use graded hard coatings. Instead of putting TiN directly on a hard tool steel, first nitride the tool steel, then put on a layer of TiN, followed by a layer of TiCN. This provides increased support and resistance to spalling.
3. Carefully select lubricants. Dry-film lubricants and oils that contain graphite or MoS_2 will promote surface fatigue. The fine particles are pressed into the surface under extraordinary stresses. Similarly, rolling surfaces should be kept free from dirt and abrasives. Dust may reduce bearing life by 50%. In general, it is not advisable to use solid-film lubricants in rolling element bearings. Instead, use oil or grease. Deep surface hardening is preferred for heavily loaded rails and wheels.

Nonmetal Sliding Wear

1. Avoid glass- or mineral-filled plastics mated with metals. These fillers will produce abrasion on even hardened metal.
2. Lubricate plastic sliding systems where possible. If a substance wets the surface, it will lubricate it and reduce the wear rate. Check lubricant compatibility. Will it cause chemical attack? Use self-lubricating plastics, whenever possible.
3. Avoid self-mating of plastics. Many types of plastics will friction weld together. The glass- and mineral-filled grades are more resistant, but compatibility tests may be needed.
4. Keep plastics within their rated PV limit. Most plastic manufacturers supply PV limit data on their bearing materials. Service conditions should be within their recommended limits, but also use common sense. A PV of 5000 theoretically means that the material can withstand a speed of 5000 ft/min with an apparent contact pressure of 1 psi. However, the tests used to develop this PV data did not use speeds this high. This kind of speed will probably melt or char any plastic. Try to determine the conditions of test before applying PV data. Your use conditions should be similar to the test conditions.
5. Use caution in mating ceramics with other ceramics. Some gall when self-mated. Most ceramics cause severe wear to hardened steel counterfaces. Compatibility tests are usually needed to develop suitable sliding couples of ceramics and cermets. All common couples require lubrication. A successful couple that has been used for years on face seals is carbon graphite versus aluminum oxide. Silicon carbide has been successfully used mated with WC/Co in water-lubricated pump bushings (SiC bushing versus WC/Co sleeve on the shaft).

Corrosive Wear/Erosion

1. *Cavitation:* The best cavitation resistance is obtained with materials with high corrosion resistance and tensile strength. The Stellite-type materials and titanium alloys are usually the best. Chromium plate on 300-series stainless steel also shows promise.
2. *Liquid erosion:* Each metal has an erosion velocity limit. If fluid velocities exceed this limit, active/passive metals will lose their protective films and corrosion will occur. Plastics and ceramics do not rely on surface films for corrosion resistance; thus they are preferred for liquid erosion conditions.
3. *Fretting corrosion:* Fretting corrosion and fretting wear are best avoided by preventing the fretting motion. If this is not possible, the faying surfaces should be kept separated by an oil film, dry-film lubricant, plastic film, or soft metal plating (e.g., silver or lead).

Solid Particle Erosion

1. Avoid undesirable *impingement angles*. Soft metals erode fastest at low angles (~20°); brittle materials erode fastest when impingement is normal to the surface.
2. Use elastomers wherever possible. Rubbers are resilient and they have a tendency to elastically deform rather than be cut when hit with high-velocity particles. Try to use low impingement angles or parallel flow.
3. In sand erosion systems, cemented carbides (type C2 or C4) have demonstrated better erosion resistance than most other materials (elastomers, hard steels, and plastics).
4. Many times, replaceable heavy steel impingement plates (wear backs) are the lowest-cost solution to particle erosion problems; get a hole, put on a new plate.

The preceding are some design suggestions that may be helpful in preventing failures due to various modes of wear. As has been stated a number of times in this text, the most important part of designing to prevent wear failures is to first establish the mode of wear that is likely to occur in a particular system. Once this is done, proceed with the selection process using your repertoire of wear-resistant materials and keeping in mind these design hints.

21.2 Preventing Corrosion Failures

In Chapter 13 we concluded with a recommendation that corrosion (Figure 21–3) should be controlled by thoroughly examining the environmental conditions that equipment or structures are likely to see in their lifetime and by then consulting corrosion data to match materials with the conditions. If data are not available, then corrosion tests are essential. This is still our recommendation for preventing corrosion failures, but most corrosion failures are caused by poor designs. Some typical design problems that relate to piping and tanks are shown in Figure 13–25, but some additional design statements can be made to ensure serviceability.

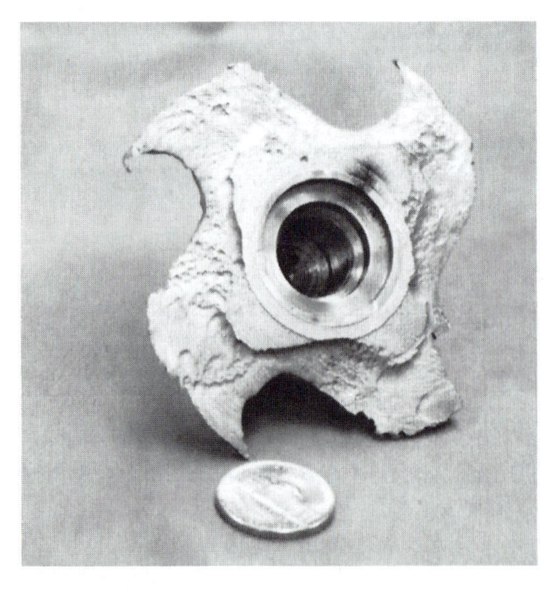

Figure 21–3
Pump impeller that failed due to corrosion

The annual cost of corrosion in the United States is estimated to be in excess of $200 billion. This figure is astronomical because of all of the cars, buses, trucks, and exhaust systems thereof that become useless because of corrosion. Most of these automotive atrocities are due to poor design. Previous editions of this book labeled crevices inside automobile doors as the most costly corrosion problem in the United States. Some U.S. automobile manufacturers seem to have responded with appropriately placed and sized drains to prevent water buildup in door bottoms (no doubt at our urging).

Next we labeled crevices in the spray zone of automobile wheel wells as the source of unnecessary corrosion. Plastic wheel well liners could solve this problem. Some automobile manufactures (at our urging?) have adopted our proposed plastic wheel well liner suggestion. The most costly poor-design corrosion problem in the 7th edition of this book was rebar corrosion in concrete columns and abutments in the snow-prone regions of the United States. Figure 21–4 presents this edition's candidate for very costly and preventable corrosion failure: environmental degradation of plastics and elastomers. The photos in Figure 21–4 illustrate environmental degradation of plastics and elastomers on vehicles, but it is equally costly in other areas:

- Peeling paint on homes and buildings
- Roofing materials that fail in less than a decade
- Vinyl siding that fades within a few years of installation
- Fiber bloom in glass-fiber reinforced plastics
- Elastomer moldings that crack and turn brittle

Failure of paint systems on automobiles [Figure 21–4(a)] was rampant when volatile solvent-based paints were phased out in favor of water-based systems. Cracking and embrittlement of

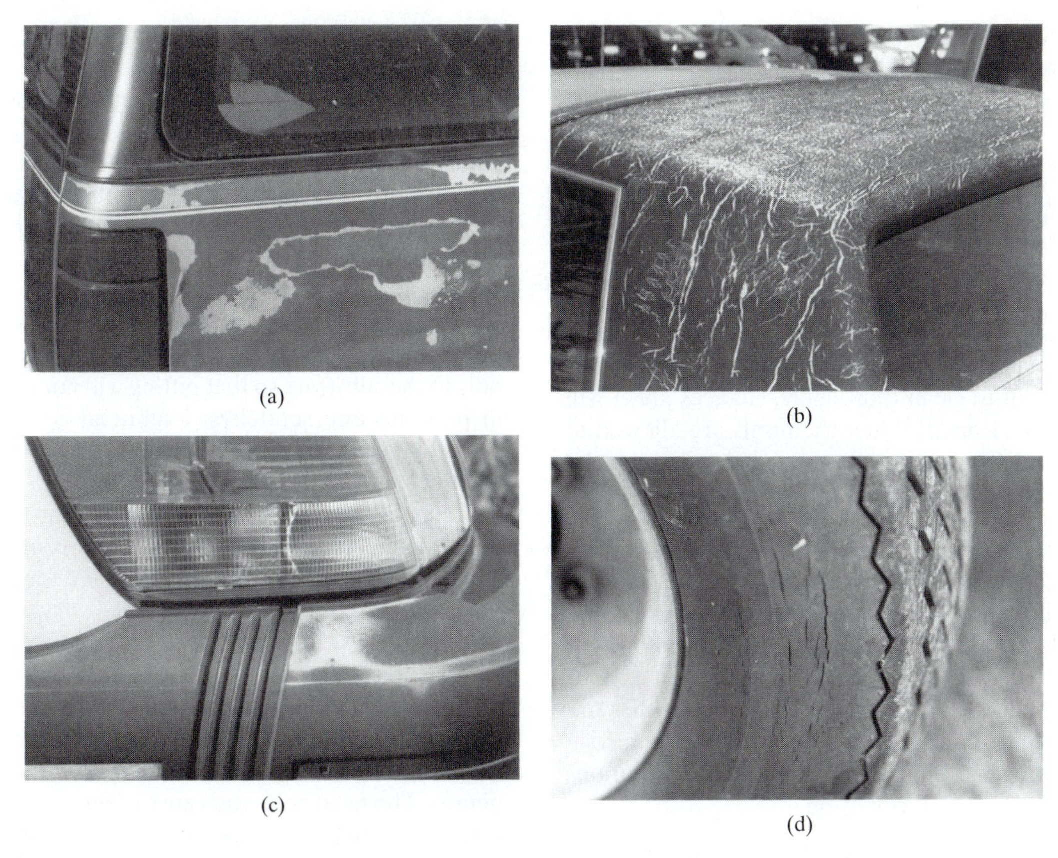

Figure 21–4
Common examples of environmental degradation of plastics and elastomers: (a) peeling paint, (b) vinyl roof cracking, (c) elastomer bumper discoloration, and (d) cracking in tire sidewall

vinyl-covered roofs [Figure 21–4(b)] and dashboards has become epidemic in the south and western parts of the United States. Similarly, elastomer bumpers and fascia that were not painted lost their gloss and color [Figure 21–4(c)] from UV damage. Finally, the cracked tire [Figure 21–4(d)] illustrates a problem that occurs when tires are exposed to UV rays and oxidation.

The common factor in all of these failures is outdoor exposure. Plastics and elastomers are especially susceptible to degradation from UV rays, oxidation, chemicals in the air, and thermal effects. Plastics and elastomers should not be used for outdoor exposure without quantitative data on suitability. There are commercial exposure sites available in different climate areas, and responsible plastic and elastomer manufacturers should use these kinds of facilities to develop long-term weatherability data. If a plastics supplier cannot verify outdoor suitability of their product, then designers should move on to another supplier who has quantitative data on his or her product. In addition, designers should keep in mind that not very many plastics can accommodate long-term outdoor exposure as can the cellulosics, acrylics, polycarbonates, EPDM, and silicone rubbers. These are the safest polymer systems for outdoor use.

Common Problems and Preventive Measures

Underground piping: Check soil conditions; use cathodic protection, plastic- or bitumastic-coated pipe, or polyethylene wrapping if the soil shows high conductivity. This type of soil produces abnormal potential for corrosion of ferous pipes (Figure 21–5).

Underdeposite attack: Design equipment to be cleanable. If chemical handling equipment is difficult to clean thoroughly, chances are it will not be cleaned. When chemicals are allowed to accumulate on surfaces, severe local corrosion can occur under the deposite.

Water coolant systems: Wherever possible use a recirculating system and add proper corrosion inhibitors to the water. Corrosion rates can be reduced by a factor of 10 compared with uninhibited systems. Corrosion protection is also needed in chilled brine or water lines that are subject to condensation. Galvanizing, protective coatings, and insulation are potential solutions to condensation corrosion.

Oil sumps: Use *vapor space inhibitors* in the oil to prevent corrosion in the tank space above the oil level.

Corrosion in storage: Use vapor space inhibitors, inert gas shielding, desiccants, hermetically sealed containers, or moisture-displacing oils on metal surfaces that may be stored for long periods of time before use. Do not store organic solvents in partially filled containers. Water condensation (from air) may form. Many solvents become extremely corrosive when contaminated with water (chlorinated solvents form HCl). Keep storage tanks full whenever possible.

Coatings: When using metallic coatings for corrosion protection, try to use a metal that is anodic to the substrate so that pitting will not occur in pinholes and scratches. Nonmetal coatings should be checked for pinholes with conductive solutions and an electrical continuity test. Thin coatings (<about 125 μm) will never provide long-range corrosion protection. Thick coatings [>0.1 in. (2.5 mm)] are usually inspected for pinholes with a spark test.

Stress corrosion cracking (SCC): Avoid metal–environment combinations that are known to produce SCC. Yellow brasses and some copper alloys are very prone to SCC in ammonia atmospheres. The most common cause of SCC is chlorides in contact with austenitic stainless steels (Figure 21–6). The chloride level to produce SCC can be as low as 0.5 ppm. Most metals have

Figure 21–5
Graphitization damage in a gray-iron pipe that was underground for 8 years

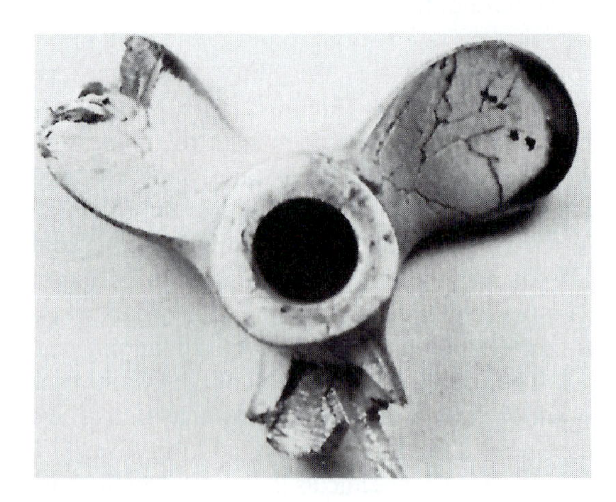

Figure 21–6
Stress corrosion cracking in a stainless steel mixing blade

a threshold temperature below which SCC is not likely to occur (50°C for austenitic SS in chloride solutions, 150°C for carbon steels in caustics). Stay below these *SCC threshold* temperatures. Eliminate residual stresses and try to reduce applied stresses. SCC will not occur if welds are stress relieved or if operating stress levels are kept low by design. Avoid designs that could allow local concentration of chlorides by evaporation in crevices or by intermittent immersion. Many chlorinated solvents cause SCC of plastics. Avoid this type of exposure.

Environmentally assisted cracking: Some plastic failure studies suggest that about 30% of all plastic failures are due to environmental stress cracking (*EAC*) (Figure 21–7). There are chemical environments that will cause plastic to crack within seconds, hours, days, and years. Some corrosion engineers believe that all plastics will eventually develop environmental stress cracks; it is just a matter of time. Plastic manufacturers usually list the environments that cause stress corrosion cracking of their plastics, but these data are usually based on short-term laboratory tests. In long-term applications, environments as seemingly innocuous as water can cause EAC of plastics after 1 or 2 years of exposure. One of the most unique ski boot designs ceased to exist because the polycarbonate ski boot shells developed cracks after a few years. The company went out of business because of EAC.

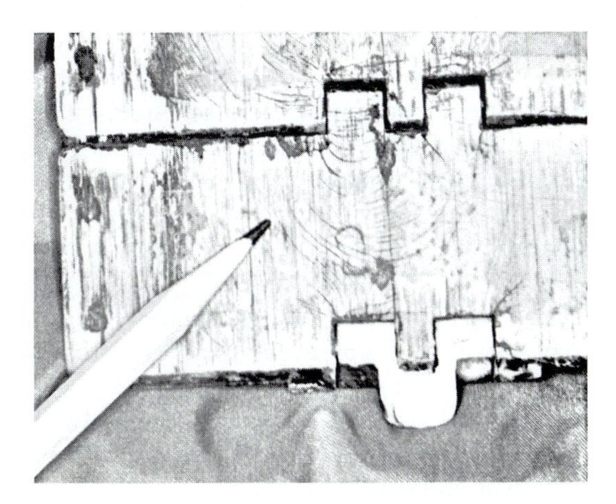

Figure 21–7
Environmental stress cracking in acetal conveyor parts after 1-year of use in silver halides

When long-term environmental resistance data is not available on a plastic, simple no-stress immersion tests can be conducted. If the plastic swells in the test environment within 30 days, it may be subject to EAC in the long term; that environment reacted with the polymer. In general, the amorphous plastics such as ABS, PVC, PES, SAN, PS, PC, and so on are more prone to EAC than the semicrystalline plastics such as nylons, acetals, polyethylene, polypropylene, and PBT. Keep EAC in mind in designs that require plastics to last years in service. The higher the stress and temperature, the greater the susceptibility.

Intergranular corrosion: Use low-carbon or stabilized grades of austenitic stainless steel when welding processes can cause sensitization; check welded tubing and pipe for sensitization (in manufacture) before use (ASTM A 708). Avoid use of stainless steels in the sensitizing range from 800 to 1650°F (427 to 899°C).

High-temperature oxidation: Avoid the use of metals at temperatures in excess of the maximum use temperature in air. Temperatures above the maximum use temperature cause excessive scaling.

Steel	Maximum Operating Temperature in Air, °F (°C)
1 Cr, 0.5 Mo	1050 (565)
410 SS	1300 (205)
430 SS	1550 (845)
446 SS	2000 (1095)
304 SS	1650 (900)
316 SS	1650 (900)
310 SS	2000 (1095)

Galvanic corrosion: Avoid metal couples that are far apart in the galvanic series. Avoid stress cells. Cold-formed elbows welded to annealed pipe can create a galvanic couple. All members should be in the same state of stress. Avoid stagnant areas in vessel and pipe designs to prevent concentration cells. Solutions with differing oxygen levels can cause a solution concentration cell and rapid corrosion in the oxygen-depleted part of the cell.

If dissimilar metal couples are unavoidable, electrically insulate the two or use favorable anode–cathode size ratios. A stainless steel fastener in a large aluminum plate is a favorable ratio. The anode (corroded member) is very large, and the cathode (protected member) is small. The reverse combination would be disastrous.

Dealloying: Avoid systems that are prone to this form of attack. The use of yellow brass in plumbing systems (especially hot water) results in dezincification in almost every instance. The use of cast-iron pipe in corrosive soils will without fail produce graphitization (loss of iron, leaving only the graphite network; see Figure 21–5). Plastic wrapping, cathodic protection (Mg anodes), or plastic pipe should be used.

Welding: Avoid incomplete penetration welds. The unfused part of the joint can become a concentration cell (crevice). Inert gas backing or pickling after welding should be used to prevent weld-contamination corrosion of stainless steels.

Passivating: Always pickle (passivate) stainless or grind stainless steel vessels and equipment to remove iron pickup from fabrication processes. Failure to do so will result in pitting under the iron deposits.

Corrosion fatigue: Avoid cyclic stresses on metals in an environment capable of causing corrosive attack. Surface corrosion causes stress concentrations that can raise the local stress level above the fatigue limit of the material.

Erosion corrosion: When using metals that derive their corrosion resistance from passive films, keep liquid velocities below the threshold levels that tend to remove these films:

Metal	Limiting Velocity for Erosion in Seawater* (m/s)
Copper	1.0
Admiralty brass	1.5
Aluminum	2.0
Aluminum bronze	2.5
Cupronickel	3.5

*Similar tables exist for other liquids.

Aging: UV exposure is a formidable cause for plastic deterioration. If a plastic or rubber is to be used outdoors, make UV resistance a key selection factor. Some transparent plastics that are known to give satisfactory life outdoors are PMMA, PC, and most cellulosics. Opaque plastics can often be filled to obtain acceptable UV resistance.

Handling seawater: The high-copper alloys and bronzes have demonstrated reliability as materials for use in seawater. Monels and cupronickels are higher strength, with similar corrosion resistance. Pure titanium is used for many severe marine applications such as desalinization systems. Conventional austenitic stainless steels have marginal resistance to seawater (at room temperature), but certain grades of superferritic stainless steel provide acceptable seawater resistance.

Corrosion allowances: If a substance is known to be corrosive to a particular metal, a corrosion allowance based on annual corrosion rate and expected life must be added to the vessel wall thickness. Careful measurements of corrosion rates will allow accurate determination of the necessary corrosion allowance and prevent expensive, overly generous corrosion allowances.

There are many other hints, rules of thumb, and precautions that apply to preventing corrosion failures, but the foregoing address areas that are known to be likely areas of concern. Consideration of these hints will help to prevent new car door, wheel well, bridge, and auto weathering problems.

21.3 Preventing Mechanical Failures

When a machine component or structure fails by fracture or by permanent deformation, the mechanism of such a failure is simply a stress that is higher than the material can withstand. Most designers calculate operating stresses on critical components to make sure that they are below the allowable stress for the material used in the design. Why do failures still occur? The answer to this question relates to probably 90% of the mechanical failures that occur in industry: stress concentrations (Figure 21–8). The calculations performed by most designers yield the nominal or apparent stress level. If stress concentrations are present, the local state of stress can be orders of magnitude greater than the perceived stress, and failures usually start in these local areas of high stress. Failures precipitated by stress concentrations are usually conjoint with two other mechanical "happenings," such as impact and fatigue.

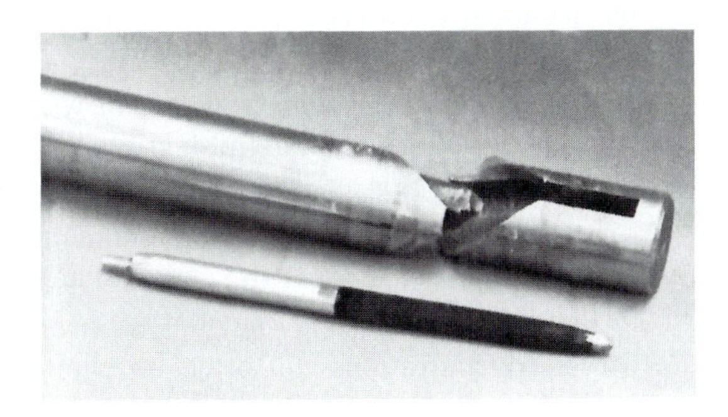

Figure 21–8
Fatigue failure originating in keyway

How does a designer prevent mechanical failures? Nothing can be done from the design standpoint about the mechanical failures that occur when a forklift truck backs into a dial assembly machine, but the machine designer can prevent most of the failures due to stress concentrations by eliminating them on the drawing board. Similarly, some mechanical failures can be prevented by simply having a better understanding of the *fracture resistance* of metals.

In the remainder of this section we shall address these two factors as keys to preventing mechanical failures.

Stress Concentrations

We have discussed stress concentrations many times throughout this text, but we did not address quantitative use in design. When a part is loaded in tension, for example, the stress of the member is simply the load (P) on the member divided by the cross-sectional area (A) of the member, P/A. If the member under load contains a notch, a hole, or certain geometric shape configurations, the stress seen by the member in the area of the notch will be magnified by a *stress concentration factor:*

$$S_{max} = K_f S_n$$

where
S_n = apparent macroscopic stress (Figure 21–9)
S_{max} = local stress in the region of a stress concentration (Figure 21–9)
K_f = the stress concentration factor (for fatigue)

If the stress concentration factor is known, it can be used in design calculations to determine local stress. Allowable stress limits should then be applied to this value (S_{max}), rather than to the apparent stress that is derived from traditional stress calculations for loaded members.

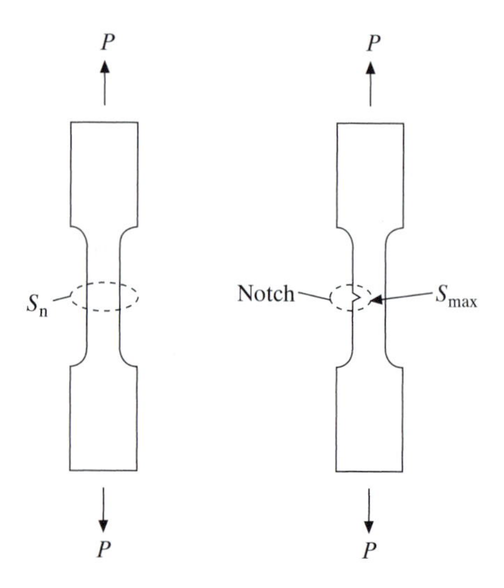

Figure 21–9
Stress concentration in a tensile test specimen

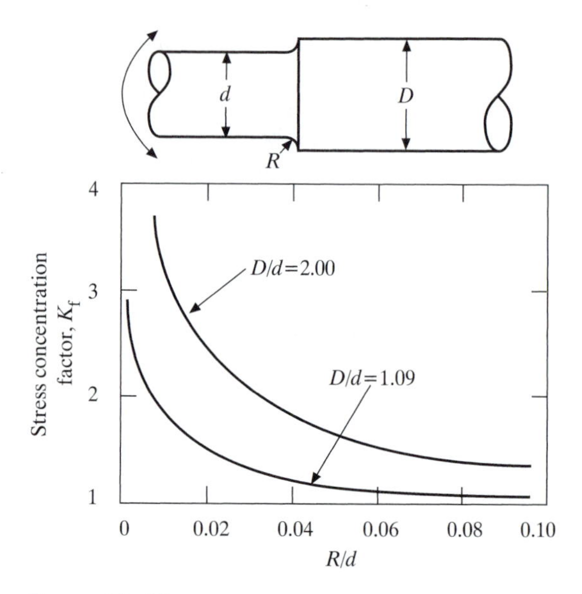

Figure 21–10
Stress concentration factors for bending of a stepped shaft

Stress concentration factors apply to all modes of part loading: tension, bending, and torsion. The mathematical value of a stress concentration factor is a function of the geometry of the loaded member and the mode of loading. In general, stress concentration factors are inversely proportional to the square root of the *notch radius*. The smaller the notch radius, the greater the stress concentration factor. Many handbooks* present stress concentration data similar to those shown in Figure 21–10. If you have a complicated shape that does not appear in handbooks, stress concentration factors can be calculated by computer-assisted finite-element techniques. Figure 21–11 shows some estimated stress concentration factors for various round shaft configurations [assuming a diameter ratio of $D/d = 2$ and notch radii < 0.03 in. (0.75/mm)].

The importance of stress concentrations in preventing mechanical failures cannot be overemphasized. Stress concentrations should be considered in all designs, but their importance increases severalfold when dynamic loads are encountered. Some pertinent design guidelines are as follows:

1. Specify radii as large as permissible on *all* inside corners in highly stressed members.
2. Put transitions in selective hardening or case hardening in low-stress areas. These structure changes can act as notches.
3. Coarse surface roughness can create notches. The lower the surface roughness, the less chance for stress concentrations. Surface lay should be aligned with the applied load.
4. Provide radiused grinding reliefs that will prevent notch formation in finish grinding.
5. Use snap rings judiciously; the notch effect of snap ring grooves cannot be eliminated by radiusing. Put them at the end of shafts or in low-stress areas.
6. Stress concentrations can often be eliminated by increasing the flexibility (lowering the stiffness) of members. Allow flexure in smooth, straight portions of members.

*Peterson, R. E., *Stress Concentration Factors.* New York: John Wiley & Sons, 1974.

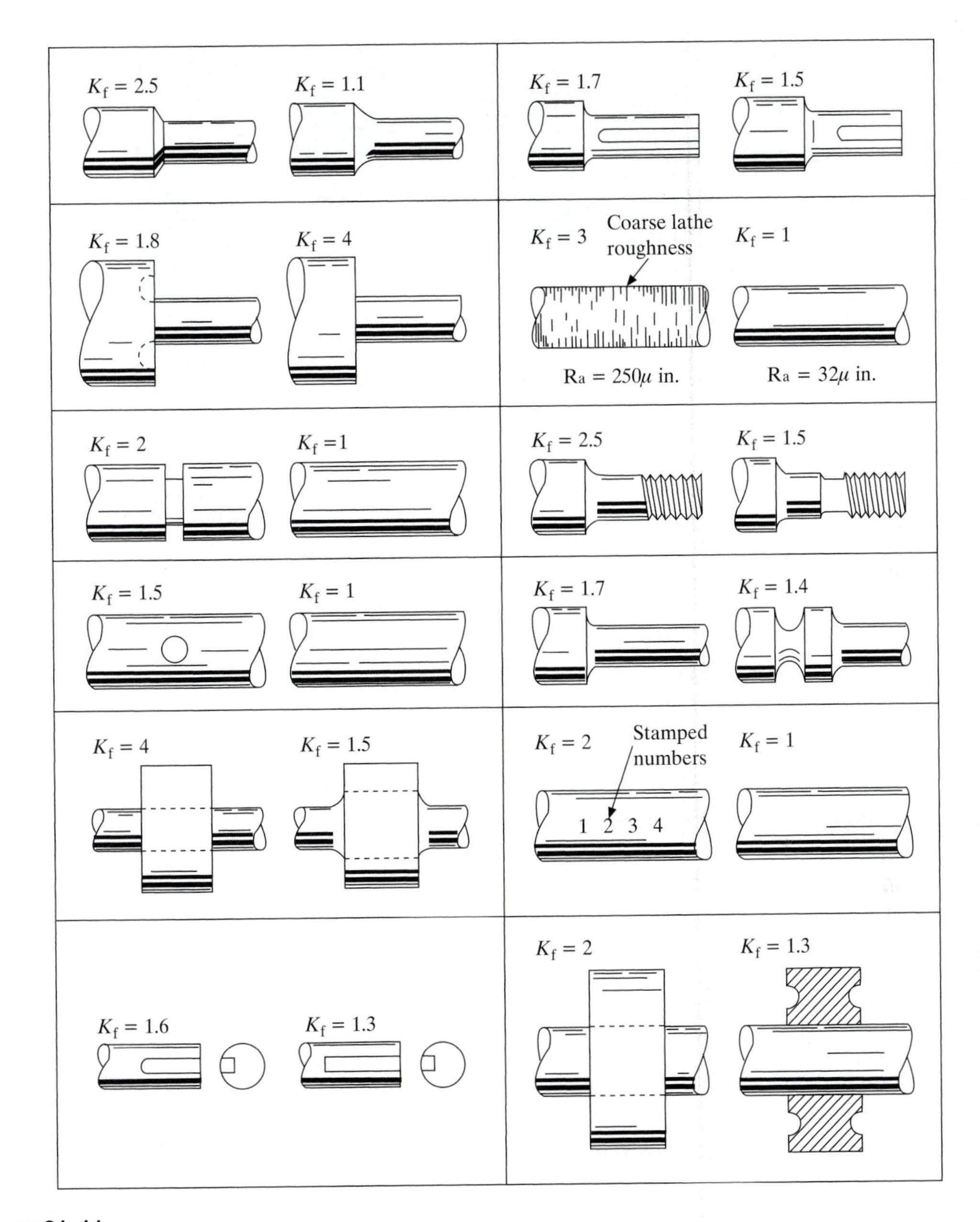

Figure 21–11
Approximate stress concentration factors for various shaft configurations

Figure 21–12
Fatigue crack progressing from a corner nick in a polycarbonate part (×32). Each "circle" is a progression of the crack front.

7. Keep threads in low-stress portions of shafts.
8. Avoid incomplete penetration welds where fatigue is likely.
9. Many plastics are notch sensitive; avoid notches, even surface scratches, on these types of plastics if they are highly stressed. (Figure 21–12).
10. Melt flow junctions (knit lines) in molded plastics often result in stress concentrations. Look for them, and alter gating to eliminate them or move them from high-stress areas.
11. Investigate stress rupture and creep properties of plastics under sustained static loads. Use stress rupture strength or creep data in design rather than yield or flex strength.

Internal defects can produce unseen stress concentrations, but the most ignored stress concentrator is surface grooves produced by machining operations. The lathe is the machine tool that can be blamed for making the most troublesome stress-concentrating notches in shafts. A lathe tool bit with a 0.010-in. (0.25-mm) radius is a wonderful notch generator. The designer, however, can control these machine tool notches by specifying fine finishes in high-stress areas. The prevention of mechanical failures that are caused by stress concentrations can be summarized in one general rule: Use generous radii on changes in section and specify low surface roughness in areas of flexure.

Fracture Resistance

Besides the elimination of stress concentrations, another action that can be taken by the designer to prevent mechanical failures is to select a material that has good fracture resistance. The fracture resistance of an engineering material can be measured in a number of ways, but the simple meaning of the term is that a material has the ability to resist crack propagation. All mechanical failures start with a crack. Fracture occurs when the forces tending to propagate the crack are greater than the forces tending to arrest the crack.

The exact mechanisms of crack propagation are beyond the scope of this discussion, but material properties can be reviewed by a designer to obtain a feeling for the inherent fracture resistance of a material selected for a design. The following is a list of some of these factors:

1. Stress–strain curve
2. Impact strength
3. Fatigue strength
4. *Notched tensile strength*
5. Critical stress intensity factor
6. Stress rupture strength
7. *Fracture toughness*

We cannot discuss these factors in detail, but a few fracture resistance hints can be mentioned in each of these areas.

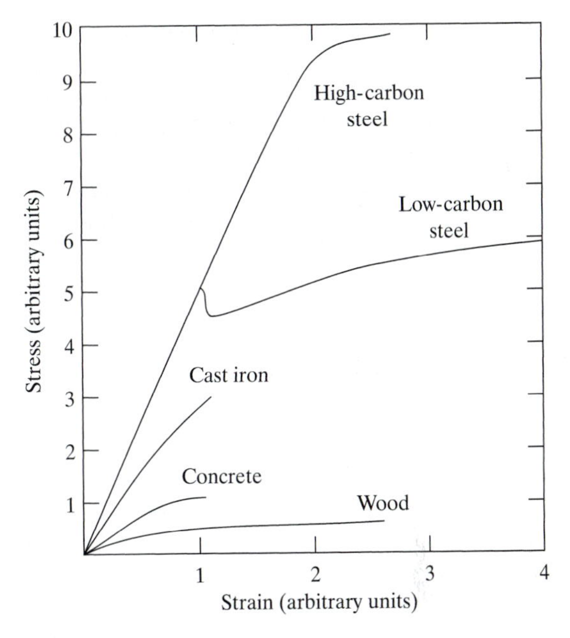

Figure 21–13
Typical stress–strain diagram for various materials

Stress–strain data: As mentioned in Chapter 2, the toughness of a material is measured as the work required to fracture a given volume of material. The simplest measure of the work required to fracture a material is the area under the stress–strain diagram. As shown in Figure 21–13, if we were to integrate the area under each of the stress–strain curves, we would have a quantitative measurement of the amount of work required to break specimens of these different materials. This is a measure of the toughness of a material, and the tougher the material the greater its resistance to crack propagation (fracture resistance).

Most material handbooks do not show complete stress–strain diagrams for various engineering materials, but they almost always tabulate tensile strength and percent of elongation. If a designer is considering two materials for an application and they both have the same tensile strength, but one has a 2% elongation and the other has a 10% elongation, you can almost be certain that the latter has better toughness and fracture resistance. The greater strain that occurred in the latter material before a tensile test failure means that the area under its stress–strain diagram would be larger and its toughness greater. Thus readily available tensile property data can be an aid in evaluating fracture resistance.

Impact strength: Handbook tabulations on the impact strength of materials should be a good indicator of toughness and fracture resistance. Unfortunately, the diversity of test techniques makes it difficult to compare various material systems. Tool steel manufacturers usually publish unnotched Charpy or Izod data, cast-iron foundries often use round bar data, and polymer manufacturers use yet another test. A designer often cannot use these data to compare various candidate material from differing material systems.

These data should be used to compare alloys within a given family of alloys, but not to compare materials from different material families unless the exact test and sample configuration are specified. Assuming a data tabulation meets the "same-test criterion," impact data can then be used to compare fracture resistance. The higher the impact strength is, the greater the fracture resistance.

Fatigue strength: Fatigue strength is a measure of the cyclic stress level that a material is capable of withstanding. When this information is available, it should be used in design calculations as the basis for allowable design stress. When fatigue strength data are not available, it is common practice to use a fraction of the tensile strength (40% for steel, 30% for plastics, 20% for cast iron, etc.). These are acceptable ways to address fatigue, but be cautious in using high-strength materials. The higher the tensile strength, the higher the notch sensitivity. What is good for fatigue (high strength) is bad for toughness. Hardened and tempered steels have very high tensile strength, but it is not advisable to use them for shafting and similar applications that could produce significant bending stresses. Medium hardness steel (<45 HRC), PH stainless steel, titanium 6Al–4V, aluminum/zinc alloys, beryllium copper, ultra-high-strength steel, and many of the nickel-based superalloys are often good for fatigue conditions. Fatigue-loaded shafts are often made from 4140 or 4340 alloy steel hardened and tempered to a hardness of 28 to 32 HRC. Type 17–4 PH stainless steel in condition H1050 (38 HRC) or H925 (43 HRC) is popular for shafting in corrosive atmospheres. Titanium (6A1–4V) is used for applications in which 17–4 stainless steel does not provide adequate corrosion resistance. Low-strength metals such as the pure metals, low-carbon steel, nonhardened austenitic stainless steel, cast irons, and yellow brasses usually have low fatigue resistance.

Fatigue cracks usually originate in tension. Ceramic and cermets in general do not have high tensile strength, and they are not usually loaded in tension. In fact, tensile load is usually not advisable. Ceramics and cermets have high compressive strength, and they can tolerate cyclic compressive loading. WC/Co cermets and alumina ceramics are used as tools in punch press applications.

Plastics with good tensile and impact strength may be poor in fatigue loading. This is particularly true of the amorphous plastics. Applications with known fatigue conditions, even low cycle, need fatigue data in the selection phase. Choose plastics with documented fatigue data that show acceptable behavior in your environment.

Notched tensile strength: Stress concentration data are often obtained by putting a particular shape notch in a particular shape and running a fatigue test. The notch is made sharper and the test is repeated. The difference between the results of these tests is the basis of the numerical value of the stress concentration factor. Developing these data is very expensive and time-consuming. A simpler approach to determining a material's sensitivity to stress concentrations and its fracture resistance is to conduct notched tensile tests. A test bar is prepared in the normal fashion and tested to failure. A second bar is then notched in the gage length and pulled to failure. The notched-to-unnotched tensile strength ratio is a measure of a material's sensitivity to stress concentrations. The higher this number, the greater the fracture resistance.

Unfortunately, good data on this parameter are scarce. If a designer wishes to compare several materials for a critical application, this test could be helpful. One precaution: Notched tensile tests conducted on round tensile samples can be misleading if the application calls for other than round shapes. Round bar tests give data that are high due to geometry effects of a circumferential notch. Flat plate tensile samples

(ASTM A 338) yield results that more closely coincide with material behavior in heavy sections and real shapes.

Fracture Mechanics

As discussed in Chapter 2, *fracture mechanics* is another tool to analyze the fracture resistance of a material. Tabulations of *critical stress intensity factors* can be used to select materials with higher fracture resistance. This parameter will become increasingly important because it is usually the preferred measure of the toughness of ceramics, and ceramics may be the high-strength engineering materials for the new millennium.

Finite-Element Stress Analysis (FEA)

Many computer-assisted design (CAD) programs combine solid modeling and finite-element analysis as part of the software package. This technique allows estimation of stresses, deflections, and strains in shapes that are too complicated for analysis with traditional stress/deflection formulas. It is a powerful engineering tool, but it must be used with discretion. Do not accept every answer from an analysis as fact. The software designers may have made some assumptions that do not apply to your special case. Do your own analysis. Assume that the results are correct; then try to check the model with stress measurements, tests, or additional calculations. FEA is good for calculating stress concentrations on peculiar part geometries. Make it a standard tool to help prevent mechanical failures, but always try to check the results for accuracy. (See Case History at end of this chaper.)

Stress rupture strength: As discussed earlier in Chapters 2 and 6, stress rupture is failure of a part or structure under sustained loading—usually at a stress well below the yield strength at the use temperature. It could happen in 100 h or it may take 10,000 h. Metals are usually not prone to stress rupture failure at temperatures

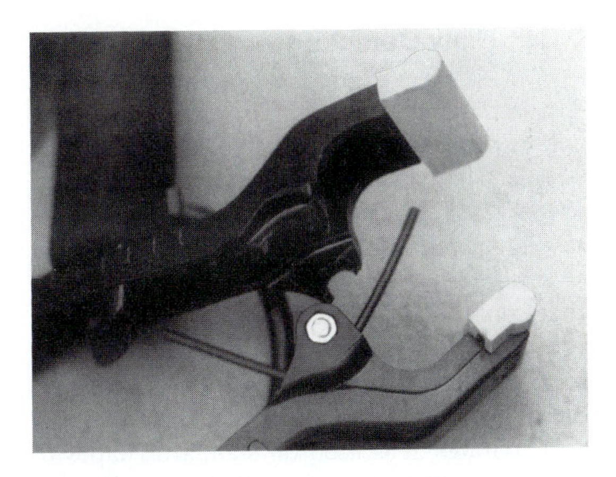

Figure 21–14
Stress rupture failure of a plastic worklight clamp. The clamp failed at first use within 24 h after being clamped to a board.

below 800°F (426°C). It becomes a concern if they are used in the red-heat range. Ceramics are not often used for structural members, so not much effort has been expended on studying their stress rupture characteristics. However, many plastic parts fail by the mechanism of stress rupture. Figure 21–14 shows a clamp for a new halogen worklight that failed with 24 h after it was clamped to a 38-mm-thick piece of wood. Apparently, the designer anticipated that it would only be clamped to something much thinner. It was made from polycarbonate, a plastic that is especially prone to failure by stress rupture. If the room temperature yield strength of a particular polycarbonate grade is 6000 psi (41 MPa) at room temperature, the material should not be used at a design stress higher than about 1500 psi (10 MPa). At 110°F (43°C) that allowable stress should probably be lower than 1000 psi (6.89 MPa). Whenever plastics are used under a significant sustained load, their stress rupture characteristics should be investigated. If they will see elevated temperatures, stress rupture strength at that temperature should be used as a design criterion. Operating stresses must be kept well below the stress rupture strength.

21.4 Flaw Detection

The last failure prevention tool that we will discuss is *nondestructive testing* (*NDT*) or, as some people call it, nondestructive evaluation (NDE). We opt for the former term. Nondestructive testing means to evaluate a piece of material or structure to determine if it contains any flaws that could affect serviceability. The evaluation process must not destroy or alter the material or structure that is being assessed. It is used on metals, plastics, ceramics, composites, cermets, and coatings; it is used on standard shapes of materials as they come from their manufacturer (rods, billets, flats, sheets, bars, and the like); it is used to detect cracks, internal voids, surface cavities, delamination, incomplete or defective welds—any type of flaw that could lead to premature failure. We will briefly discuss the most commonly used NDT techniques and show how these apply to failure prevention.

Table 21–1 lists the more common NDT techniques. There are countless others that are process or material specific. For example, very special tests have been developed to assess the integrity of seams on three-piece tin cans. There are similar special tests to measure the integrity of the coatings on food and beverage cans. These are specific rather than universal NDT tests. NDT tests are almost always used in assessing the integrity of mill shapes. Nondestructive tests are employed to determine if the manufacturing process is out of control and producing defective material. Figure 21–15 is a classic example of the type of defect that manufacturers try to find before the material leaves the mill. This defect is a forging burst in the center of a 7.5-in.-diameter bar of stainless steel. Obviously, a defect this large certainly should have been detected at the mill, but it was not. Thus this is one type of defect that is commonly studied in NDT. A second type of flaw that is assessed by NDT is that caused by service stresses, which start after the part or structure is placed into service.

Referring to Table 21–1, the simplest type of NDT that can be applied is visual inspection. Many flaws can be detected with this technique; visual inspection is particularly effective in detecting poor welds in structures. Many welding

Table 21–1
Commonly used NDT techniques

Technique	Capabilities	Limitations
Visual inspection	Macroscopic surface flaws	Small flaws are difficult to detect; no subsurface flaws
Microscopy (optical/electron)	Small surface flaws	Not applicable to large structures; no subsurface flaws
Radiography (x-ray/gamma rays)	Subsurface flaws	Smallest defect detectable is 2% of the thickness; radiation protection needed
Dye penetrant	Surface flaws	No subsurface flaws; not for porous materials
Ultrasonics	Subsurface flaws	Material must be a good conductor of sound
Magnetic particle	Surface and near-surface flaws	Limited subsurface capability; only for ferromagnetic materials
Eddy current	Surface and near-surface flaws	Difficult to interpret in some applications; only for metals
Acoustic emission	Can analyze entire structures	Difficult to interpret; expensive equipment

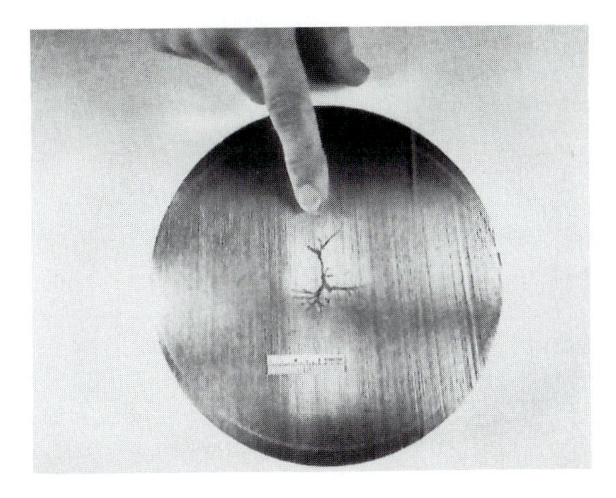

Figure 21–15
Forging burst in the center of a 7½-in.-diameter stainless steel bar

flaws are macroscopic: crater cracking, undercutting, slag inclusions, incomplete penetration welds, cold welds, and the like. Often, visual inspection is suitable for detecting flaws in composite structures and piping of all types. Essentially, visual inspection should be performed the way that one would inspect a new car prior to delivery. The smart buyer pores over every square inch of the car and notes anything that is not as it should be. This same type of inspection should be done on fabrications and components used in fabrications. Some things to look for are the following:

- Wrong materials (rust or color)
- Bad welds or joints
- Missing fasteners or components
- Poor fits
- Wrong dimensions
- Improper surface finish
- Delaminations (coatings and laminated materials)
- Large cracks, cavities, dents
- Indications of mechanical damage (bent parts, scratches, dents, etc.)
- Inadequate size, wrong parts
- Lack of code approval stamps and similar proofs of testing
- Nongraded fasteners

Microscopic examination can be done to look more closely at things that do not look proper in visual inspection. Optical microscopy can be used to determine if suspect areas are really cracked, whether a pit is due to a material defect or to corrosion, whether a die edge is chipped or just dull, if a spot is an inclusion or a surface contaminated with something, if cracks are branched, and the like. Scanning electron microscopy (SEM) is particularly useful for deducing the nature of surface flaws. Most SEMs cannot accommodate a specimen larger than about 1 in.^{3} (25 mm on a side). This limits the use of this NDT technique to tiny parts such as electronic components, but large items can be studied by making silicone or cellulose acetate replicas of the surface in question. Replicas can be PVD coated with metal and then can be studied in the SEM. Often, surface flaws can be accurately diagnosed and assessed for severity by a simple SEM study.

Schematics of some of the most commonly used NDT techniques are presented in Figure 21–16. Radiography has an advantage over some of the other processes in that the radiograph provides a permanent reference for the internal soundness of the object that is radiographed.

The source of x-rays is simply a wire filament emitting electrons. The electrons hit a target and x-rays are emitted due to excitation of the target atoms. The x-rays' ability to penetrate metals is a function of the accelerating voltage in the x-ray tube. For example, a 300-kV machine may penetrate 3 in. (75 mm) of steel. If a void is present in the object being radiographed, more x-rays will pass in that area and the film under the part in turn will have more exposure than in the nonvoid areas. The sensitivity of x-rays is nominally 2% of the material's thickness. Thus,

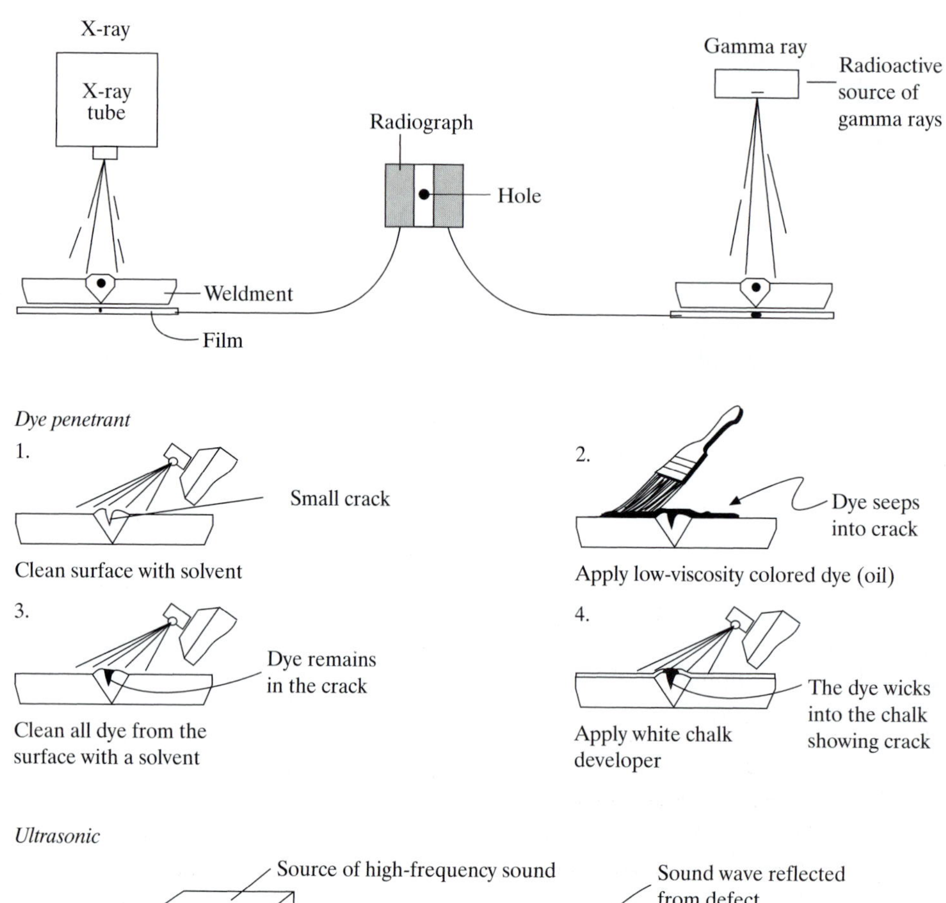

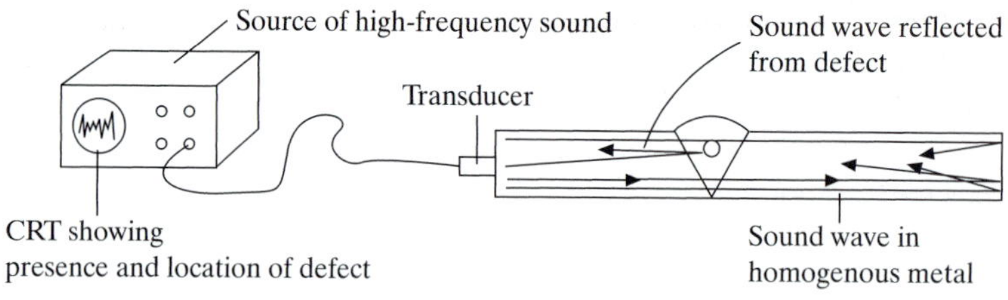

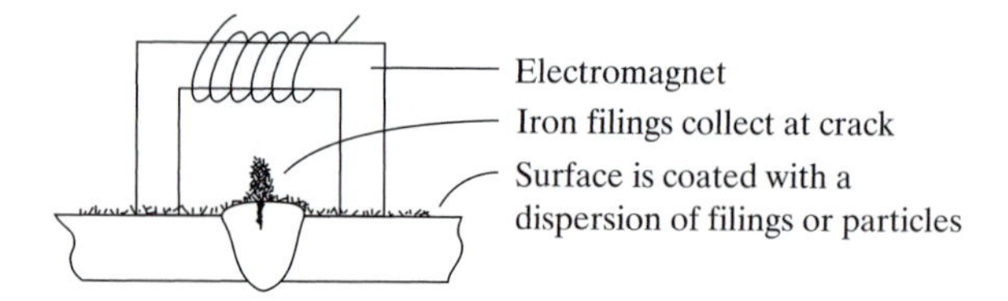

Figure 21–16
Schematic of nondestructive testing techniques

if you were x-raying a piece of steel 1 in. (25 mm) thick, the smallest void that you could detect would be 0.020 in. (0.5 mm) in dimension. For this reason, parts are often radiographed in several different planes. A thin crack would not show up unless the x-rays ran parallel to the plane of the crack. Gamma radiography is identical to x-ray radiography in function. The difference is the source of the penetrating electromagnetic radiation. In gamma radiography, it is a radioactive material such as cobalt 60. Gamma radiography is less popular because of the hazards of handling radioactive materials. An industrial x-ray facility is illustrated in Figure 21–17.

Figure 21–17
X-ray machine being used on a group of small weldments.

The normal sensing medium for radiography is photographic film. X- and gamma rays sensitize films just as light sensitizes conventional photographic film. The more radiation striking the film, the more it is exposed. A crack would appear as a black line on a gray background in normal radiography. Instead of film, the radiation passing through a volume of material can be sensed on a screen that fluoresces when it is irradiated. This is fluoroscopy, and the same process is used in airports for security screening of baggage. This sensing technique is used in industry in places where there is a need to look at many parts quickly. For example, a foundry may set up a fluoroscope on a conveyor or belt carrying small castings. A newer method of sensing radiation through a volume is *computed tomography* (*CT*). This is the process that hospitals often refer to as a CAT scan. In this process, radiation is directed at the subject at many different angles, for example, in 20° increments around 360° in several planes. The penetrating radiation is electronically sensed and a digitized, three-dimensional image can be created on a CRT and printed out in any perspective. The attenuation of the x-rays is a function of the material and its density. CT provides a three-dimensional map of density that indicates cracks, voids, and other defects as well as part shape. The CT image can be used to make a CAD drawing of the part (if one does not exist). This is an extremely powerful tool for finding internal defects, and it is currently being used on a wide variety of parts and machines.

Dye-penetrant facilities are essential for any shop performing welding. The only equipment required is a spray can of cleaner, dye, and developer. You cannot rely on the unaided eye to detect weld cracks. Dye penetrant is very sensitive, and it will detect all cracks and voids that come to the surface. It will not pick up subsurface flaws.

Ultrasonic inspection involves sending a high-frequency sound wave through the part to be inspected. The sound wave travels through the piece and is reflected from the opposing surface. If there is a flaw in the part, the sound wave is reflected before it reaches its destination, and a visual indication is seen on the cathode-ray tube display in the ultrasonic power supply. Many details are involved in the correct interpretation of ultrasonic inspection using conventional CRT display.

Developments in ultrasonic inspection allow pseudo-three-dimensional mapping of defects. A part is immersed in a water couplant and ultrasonic waves are sent and sensed at *xy*-positions on the part. The ultrasonic sending and receiving unit is maintained at a fixed distance from the bottom of the water tank. A computer analyzes the ultrasonic waves reflected from the part (and from defects) and the *z*-dimension is recorded for each *xy*-position. A map is generated that depicts the part and any internal defects within the part. The chief advantage of ultrasonics is that it is extremely sensitive, and there is no real thickness limit on the material that can be penetrated. It can go through 20 ft (6 m) of steel, and there are units that work on plastics and composites.

Magnetic particle inspection is somewhat more involved than is implied by the schematic. In its simplest application, an electromagnet yoke is placed on the surface of the part to be examined, a kerosene–iron filing suspension is poured on the surface (sometimes dry iron filings are used), and the electromagnet is energized. If there is a crack or flaw on the surface of the part, a north and south pole will form at the void. There is a flux loss, and the iron filings will collect and show up in the void. What complicates this form of inspection is that for best sensitivity the lines of magnetic force should be perpendicular to the defect. Thus it is more common to use a machine not unlike a welding power supply to pass current through or around the part. Using the right-hand rule, if current is flowing through a conductor, lines of flux will be formed in the direction of your fingers if you point your right thumb in the direction of the current flow. Magnetization is usually accomplished by placing the part in a special inspection rig and passing current down the axis of the part, as well as looping a coil of current-carrying cable around the part to give axial lines of flux.

Eddy current testing works on the principle that minute electrical currents that are magnetically induced in a material will be affected by the nature of the material. These currents are influenced by voids, cracks, and changes in grain size, as well as by physical separation of the inducing coil from the material. In practice, a probe is placed on the surface of the part to be inspected, and electronic equipment monitors the eddy currents in the work through the same probe. The currents in the workpiece are caused by high-frequency current reversals in the contacting coil. The sensing circuit is part of the sending coil. Pipeline corrosion can be monitored by sending one of these probes through the pipe. A pit will make the material farther from the coil, and this will be sensed. These types of devices are commonly used to test the integrity of resistance-welded piping and tubing. Automatic systems can test welds at speeds up to 100 ft/min (5 m/s). In its simplest form, eddy current gages are commercially available to check coating thickness. Magnetic gages are cheaper, and they work fine for sensing nonferromagnetic coatings on a ferromagnetic substrate, but they do not work on nonferrous metals. Eddy current does.

Acoustic emission is probably the most difficult to interpret of the processes listed in Table 21–1. It is unique in that it can be used to examine a structure as a complete unit. The concept is simple: Stress a structure and listen to the noise it makes. The stress can be to put the structure or vessel into service. Listening for what happens under stress is done by attaching sensitive microphones to the structure. For example, to test a pressure vessel, the vessel is pressurized and acoustic sensors are attached to critical parts of the vessel. Usually, a structure is stressed by applying a range of pressures, and the acoustic responses are recorded and computer analyzed. Essentially, the sensors listen for the sound of crack formation. This inspection process usually requires a specialist in this area for interpretation of the meaning of the vessel's "groans." In its simplest form, acoustic emission devices are commercially available that test the integrity of hardcoatings. Coated specimens are scratched and the acoustic response is recorded to show at what load the coating was removed or spalled from the substrate.

The common denominator for the test techniques that we have discussed is testing flaws or defects without destroying the structure. Another factor that is common to these processes is that they need a criterion for rejection. It is the responsibility of the designer to state the process to be used in an NDT study and what will constitute an unacceptable part. Some of the ultrasonic processes are extremely sensitive. They can pick up conditions that will not affect part serviceability. For example, carbide phases in tool steels can show up as indications, but they are not flaws; they are supposed to be there. Rejection limits can be specified by referring to ASTM or similar inspection standards or by simply stating your desires on the part drawing. A typical drawing notation might read:

> Maximum allowable internal void to be 2 mm in diameter as determined by radiographic inspection.

NDT tools are effective aids in achieving desired serviceability, but the designer must tell the fabricator which process to use and how to use it.

Summary

In this final chapter we have attempted to illustrate that serviceability of parts and structures requires thoughtful material selection as well as consideration of design factors that can lead to failures. When the part is being designed, determine the operating stresses and ask, "Have I done everything possible to prevent a wear failure, a corrosion failure, or a mechanical failure? Does the part contain any harmful stress concentrations? Will rusting affect its function? Will operation in a dirty environment cause abrasion? Does the part contain any flaws? What is the environment? Will it change?"

The keys to serviceability are selecting the right material, giving the material the appropriate treatments (heat treatments, coatings, and so on), and, finally, using a design that is compatible with accepted failure prevention guidelines. We have tried throughout this text to present information on all the materials' considerations that relate to these keys to serviceability. If you adhere to our recipes for material selection and design and still have a failure in service, it can only be blamed on the perversity of matter.

Some concepts to remember are the following:

- Premature wear failures can be minimized by designing parts to resist specific modes of wear.
- Corrosion failures can be minimized by using materials with documented ability to withstand a particular environment.
- Mechanical failures can be minimized by calculating operating stresses and keeping these stresses below the fatigue strength creep, or stress rupture of the materials involved. It is also imperative to prevent the introduction of geometric stress concentrations by careless design. Nondestructive testing should also be used to ensure that flaws from fabrication or service conditions do not exist.
- Nothing has to fail. Roads paved by the Romans 2000 years ago are still in use; many modern highways in the northeast United States last less than 10 years. The difference is, very simply, proper design, the use of appropriate materials, and quality workmanship.

The end

Thank you for your perseverance. You now have been exposed to as much engineering materials knowledge and experience as the publisher's page budget would allow (even though you may have skipped some chapters). We have tried to give you the theory behind the behavior of engineering materials and what you need to know about material selection. The information in this text is based on our cumulative 65-years of experience in material selection in industry.

This book contains much of our "good" data—the property and composition tabulations that need to be consulted on a regular basis. Reviewers of this book have termed it "half-handbook." So be it. That is the way we feel it should be. We look things up in this book on a regular basis, and it is our intention that you will keep this book as a reference to do the same.

Selecting the right material for a part that you have designed is a very important part of having that part work and last as long as you intended. Hopefully this text has given you most of the tools to get the job done. Finally, good luck in your career and thank you again for your attention and interest.

Critical Concepts

- Corrosion (including aging), wear, and fatigue never sleep. They must be dealt with.
- Nondestructive testing (NDT) should be considered on parts that must not fail in service.
- Finite-element modeling and other forms of modeling are useful tools for analysis of stress levels.
- Corrosion wear and erosion protection begins with identification of the wear mode.
- Fatigue failure protection begins with identification of stress concentrations and determination of stress severity.

Terms You Should Remember

impingement angle
vapor space inhibitors
SCC threshold
graphitization
stress concentration factor
notch radius
fracture resistance
fracture toughness
notched tensile strength
sliding couples
critical stress intensity factor
fracture mechanics
radiography
ultrasonics
NDT
magnetic particle
acoustic emission
eddy current
computed tomography (CT)
EAC

Case History

Designing with Computer Models

A project team of experienced engineers was assembled to design a new high-resolution mammography cassette. The key to the new design was to have a curve in the hinged panels that, when closed like a book, produced a spring force high enough to produce intimate contact between the x-ray film and a phosphorous intensifying screen. This intimate contact between film and screen would produce very high resolution images and be a discriminating feature over the competition.

Concepts were resolved in meetings and the cassette was designed on a CAD terminal. All stress and deflections were carefully modeled with finite-element modeling (FEM) techniques using the three-dimensional CAD part drawings. Iterative changes were made to satisfy the FEM analysis and finally the system worked in the model. The solid models were then used to generate injection molding cavities to make the six plastic parts that comprise the cassette. Several dozen of each part were molded and about ten cassettes were assembled to show to the customer. Unfortunately, the cassette did not work. The curve panels buckled when they were closed, and this did not permit intimate screen-to-film contact. Therefore, the design did not work. At this point the team had spent over a million dollars and the customer department canceled the project for fear of spending another million for naught.

Computer models are based on assumptions made by the software writer. Apparently, the software person did not include a buckling factor. This failure could have been prevented by machining prototypes from solid plastic early on—before hundreds of thousand of dollars were spent on injection molding cavities. Then there would have been time and money for designing out the buckling problem. Computer models should be coupled with traditional engineering to avoid such mistakes.

Questions

Section 21.1

1. Name three materials that can resist abrasion by titanium dioxide with a hardness of 1100 kg/mm^2.
2. What is an acceptable counterface for a sliding system with aluminum oxide as one member?
3. What is the allowable PV for a couple of PTFE-filled acetal versus hardened steel?
4. What is the proper hardness for a counterface for a glass-filled PTFE bushing?
5. Name three materials that have good cavitation resistance in water.
6. Name three types of erosion.
7. A 2-in.-diameter has a DN of 80,000. What is its allowable rpm?

Section 21.2

8. Name two examples of dealloying.
9. When should passivating be used?
10. How would you cathodically protect an underground cast iron pipe?
11. You want to make a piping system to carry seawater at 200°F (93°C). Name three candidate materials.
12. How can EAC of plastics be dealt with?
13. Name a metal that can withstand continuous immersion in seawater.

Section 21.3

14. Calculate the stress concentration factor on a round shaft that is reduced in diameter from 1 to 0.5 in. (a) with a step radius of 0.5, and (b) with a step radius of 0.01 in.
15. How can tensile test data be used to deduce the toughness of a steel?
16. Calculate the size flaw that will cause failure in a beryllium rocket case operating at a hoop stress of 10 ksi (assume B to be 0.2).
17. Which material is better to use for a mixer shaft, titanium 6A1–4V or 1020 steel? Explain your answer.
18. Explain the role of crystallinity in the fatigue resistance of plastics.
19. A steel has a tensile strength of 60,000 psi. What is its allowable fatigue strength?

Section 21.4

20. What is the smallest defect that radiography will detect in a 4-in.-thick steel plate?
21. How do you specify allowable defect size using ultrasonic inspection?
22. Name three limitations of magnetic particle inspection.
23. Why is it common to radiograph in multiple directions?
24. What is the most sensitive NDT technique for determining if a flaw exists in an aluminum automobile connecting rod? Explain your choice.

25. What are the five most important material properties that apply to the serviceability of an open-end wrench?
26. When is eddy current testing superior to the other processes listed in Table 21–1?
27. What is CT tomography and how is it used?
28. What are the limitations of dye penetrant inspection?
29. How would you examine a weld to determine if it is serviceable?

To Dig Deeper

Ashby, M. F., and D. R. H. Jones. *Engineering Materials 1,* 2nd ed. Oxford: Butterworth/Heinemann, 1998.

Bolotin, V. *Mechanics of Fatigue*. Boca Raton, FL: CRC Press LLC, 1999.

Brostom, W., and R. D. Corneliussen. *Failure of Plastics.* New York: Hanser Publishers/Macmillan Publishing, 1986.

Characterization and Failure Analysis of Plastics. Materials Park, OH: ASM International, 2003.

Collins, J. A. *Failure of Materials in Mechanical Design: Analysis, Prediction, Prevention*, 2nd ed. New York: John Wiley & Sons, 1993.

Cortz, L. *Nondestructive Testing*. Materials Park, OH: ASM International, 1995.

Dym, J. B. *Product Design with Plastics*. New York: Industrial Press, Inc., 1983.

Entwistle, K. M. *Basic Principles of the Finite Element Method*. London: IOM Communications Ltd., 1999.

McMaster, R. C., Ed. *Nondestructive Testing Handbook*, 2nd ed., vol 1. (Six volumes are available on different subjects.) American Society for Nondestructive Testing. Materials Park, OH: ASM International, 1985.

Nair, N. V., and others. *Fracture Mechanics: Microstructure and Micromechanisms*. Metals Park, OH: ASM International, 1987.

Rueter, W. G., and R. S. Piascik. *Fatigue and Fracture Mechanics*, STP 1417. West Conshohocken, PA: ASTM International, 2002.

Wright, D. *Environmental Stress Cracking of Plastics.* Materials Park, OH: Rapra Technology/ASM International, 1996.

Young, W. C. *Roark's Formulas for Stress and Strain*, 6th ed. New York: McGraw-Hill Book Co., 1989.

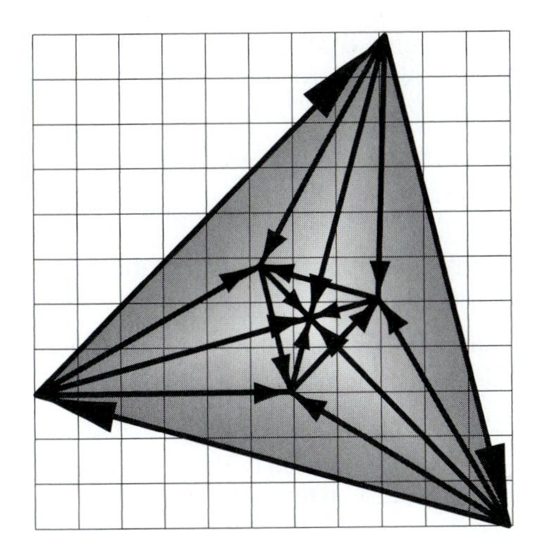

APPENDIX 1

Listing of Selected World Wide Web Sites Relating to Engineering Materials

The following are selected internet references on engineering materials from various sources.

Search Pages	*URL*
Alta Vista	http://www.altavista.com
Google	http://www.google.com
Northern Light	http://www.nlsearch.com
Hotbot	http://www.hotbot.com
Excite	http://www.excite.com
Mamma	http://www.mamma.com
Fast Search	http://www.alltheweb.com
Ingenta	http://www.ingenta.com

Government Sites

The Advanced Materials & Processes Technology Information Analysis Center (AMPTIAC)

- http://amptiac.alionscience.com
- Materials and processing products (books and databooks), technical inquiries, consulting, upcoming conferences, and library services (document location, bibliographies, and referrals)
- Free

National Institute of Standards and Technology (NIST)

- http://nvl.nist.gov
- Online library catalog and other fee-based resource services
- Free online library catalog search; other search services are fee based.

Science and Technical Information Network (STINET)

- http://www.dtic.mil/
- Search of science and technical databases
- Free

U.S. Patent and Trademark Office

- http://patents.uspto.gov/
- Database of front page information from U.S. patents issued from 1/1/76 to present
- Free

National Institute of Standards and Technology (NIST) WebBook

- http://webbook.nist.gov/
- Materials database: thermochemical, heat of reaction, IR spectra, mass spectra, and ion energies data
- Free; might charge in the future.

NIST Databases distributed by the Standard Reference Data Program

- http://www.nist.gov/srd/dblist.htm
- Numerous scientific databases (examples: chemical thermodynamics, corrosion performance, tribomaterials, molten salts, ceramic phase equilibrium database . . .)
- Fee charged

National Institute of Standards and Technology (NIST)–Physics Laboratory

- http://physics.nist.gov/PhysRefData/contents. html
- Materials Database containing physical constants, atomic spectroscopic data, x-ray and gamma data, nuclear physics data, and other NIST data
- Free

Department of Energy (DOE) Reports Bibliographic Database

- http://www.osti.gov/waisgate/gpo.html
- Bibliographic Database
- Free access; reports cost

- Ceramic engine coatings and parts (history, R&D, newsletter, research)
- Free

NIST Atomic Spectra Database

- http://physics.nist.gov/cti-bin/AtData/main_asd
- Database of atomic energy levels, wavelengths, transition probabilities
- Free

Electronic Selected Current Aerospace Notices (E-SCAN)

- http://www.sti.nasa.gov/scan/scan.html
- Searchable database of recently issued report and journal articles dealing with a wide range of engineering and science topics
- Free

European Patents

- http://ep.espacenet.com

U.S. Department of Energy

- http://www.energy.gov

Sandia National Laboratories

- http://www.sandia.gov

National Science Foundation

- http://www.nsf.gov/

CSIRO

- http://www.CSIRO.au

National Sciences and Engineering Research Council of Canada

- http://www.nserc.ca/index.htm

U.S. Environmental Protection Agency (EPA)

- http://www.epa.gov

Commercial Sites

CS ChemFinder Chemical Information Server-Cambridge Soft

- http://chemfinder.camsoft.com/
- Chemical database covering physical properties and 2D chemical structures
- Free

Rapra Technology Ltd.

- http://www.rapra.net/
- Polymer and Rubber Database, polymer industry directory (polymers, additives, machinery, products)
- Cost, depends on service

Metal Suppliers Online

- http://www.suppliersonline.com/
- Metals Database (Search for materials properties, supplier of a specific grade and product, or supplier information.)
- Fee charged

Wiltec Research Co

- http://www.wiltecresearch.com
- Thermodynamic properties

Polymers Dotcom

- http://www.plastics.com
- Basic guides to plastics, description of material properties, links to polymer producers, recyclers, equipment, and tooling
- Free

International Organization for Standardization (ISO)

- http://www.iso.ch/welcome.html
- Information about ISO, meetings, links to worldwide standards organizations
- Free

Mat Web

- http://www.matweb.com
- Material property information
- Free

Ceramic and Industrial Minerals

- http://www.ceramics.com/
- Links to suppliers, manufacturers, and consultants in the ceramic industry
- Free

The Composites Corner

- http://www.advmat.com/links/html
- Links to composite materials sites (academic, associations, consultants, companies, government, suppliers)
- Free

Edinburgh Engineering Virtual Library (EEVL)

- http://www.materials.ac.uk/resources/fullrecord.asp?resourceid=1136
- Databases of engineering/materials resources
- Free searches

Plastics News

- http://www.plasticsnews.com/
- News stories, searchable listings of commercial data, contacts for over 1500 companies (compounders, recyclers, and plastic lumber makers), and historical resin pricing data
- Free, but subscribers get more information

Modern Plastics and Modern Plastics International

- http://www.modplas.com/
- Articles, world tradeshows, trade name directory, plastics and current news
- Free

American Composites Manufacturers Association (ACMA)

- http://www.acmanet.org/index.cfm
- Membership, publications, meetings, seminars, technical services, and links
- Free

Professional Society Sites

ASM International

- http://www.asm-intl.org/
- Free

Society of Automotive Engineers (SAE)

- http://www.sae.org/
- Databases (aerospace specs, fuels and lubricants, highway vehicles, SAE publications and standards), membership information, and meetings
- Free; databases charge

Thermal Spray Society

- http://www.asm-intl.org/tss/
- Thermal spray technology glossary and meetings
- Free

The Copper Page

- http://www.copper.org/
- Bibliographic database, properties, suppliers, market data, standards, and copper links
- Free

The American Ceramic Society (ACerS)

- http://www.acers.org/
- Membership, meetings, and ceramic and glass publications
- Free

The Materials Research Society (MRS)

- http://www.mrs.org/
- Meetings, membership, and materials related links
- Free

NACE International

- http://www.nace.org/
- Corrosion publications, conferences, standards, and corrosion links
- Free

American Welding Society

- http://www.aws.org
- Membership, seminars, publications, events, welding web links
- Free

Association for Iron and Steel (AIST)

- http://www.aistech.org
- Meetings, membership, and publications
- Free

American Society of Testing and Materials (ASTM)

- http://www.astm.org/

- Membership, publications, search for standards, and laboratory directory
- Free

The Minerals, Metals, and Materials Society (TMS)

- http://www.tms.org/
- Meetings, membership, and publications
- Free

The Society of the Plastics Industry (SPI)

- http://www.socplas.org/
- Meetings, plastic links, membership, book orders, and trade shows
- Free

Automotive Steels

- http://www.autosteel.org
- Steelmakers, publications, glossary, events, and related links
- Free

American Iron and Steel Institute—Steel Works

- http://www.steel.org/
- News, steel links, statistics, markets and applications, and publications
- Free

American National Standards Institute (ANSI)

- http://www.ansi.org/
- ANSI information, news, upcoming events, and engineering links
- Free

American Vacuum Society

- http://www.avs.org

United Engineering Foundation

- http://www.engfnd.org/engfnd

Chemical Abstracts

- http://www.cas.org

The European Ceramic Society

- http://www.chem.tue.nl/ecers/
- Members, working groups, conferences, and information on the Journal of the European Ceramic Society
- Free

Aluminum Association

- http://www.aluminum.org
- Information on the aluminum industry
- Free

The Society of Plastics Engineers (SPE)

- http://www.4spe.org/
- Membership, publications, industry events, classified ads, and plastic links
- Free

Polyurethane Foam Association

- http://www.pfa.org/
- Publications, glossary, meetings, and keyword search
- Free

Institute for Scientific Information

- http://www.isinet.com

American Chemical Society

- http://www.acs.org/

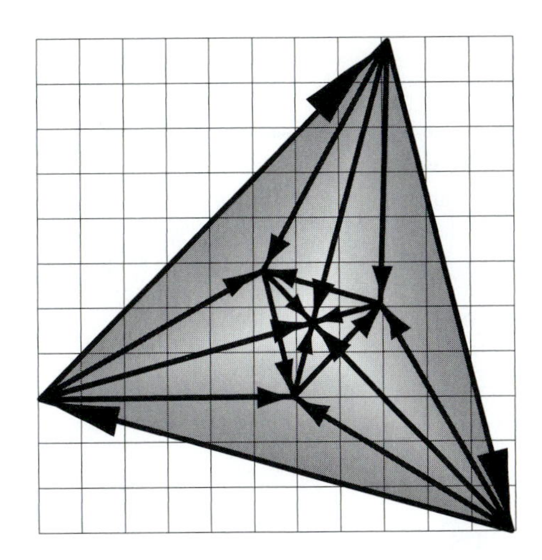

APPENDIX 2

Properties of Selected Engineering Materials

Unfilled/Unreinforced Plastics

Mechanical properties*

Type (Acronym)	Tensile Yield Strength, ksi (MPa)	Elongation %	Flexural Strength, ksi (MPa)	Tensile Modulus of Elasticity, ksi (MPa)	Impact Strength, ft-lb/in. (J/m)	Injection Moldable
Polytetrafluoroethylene (PTFE)	4.5 (31)	300	—	51 (0.35)	3 (88)	No
Polybutylene terephthalate (PBT)	8 (55)	150	12 (83)	—	0.8 (23.6)	Yes
Polysulfone (PSU)	16.2 (70)	75	15.4 (106)	360 (2.48)	1.3 (38.3)	Yes
Polymethylmethacrylate (PMMA)	10.5 (72)	5	16 (110)	425 (2.93)	0.3 (8.8)	Yes
Polyamide-imide (PAI)	26 (179)	15	30 (207)	750 (5.17)	2.5 (73.7)	Yes
Phenolic (PF)	10 (69)	<1	11 (76)	1050 (7.3)	0.35 (10.3)	No
Polyimide (PI)	13 (90)	4	18 (124)	630 (4.3)	0.75 (22)	Most no; some yes
Epoxy (EP)	10.5 (72)	4	16 (110)	450 (3.1)	0.3 (8.8)	No
Polystyrene (PS)	7.5 (51.7)	1.5	12.5 (86)	480 (3.3)	0.3 (8.8)	Yes
Polyethylene (PE)	1.9 (13)	600	—	24 (0.16)	—	Yes
Polyvinylchloride (PVC) (rigid)	6.5 (44.8)	6	13 (89)	375 (2.6)	4 (118)	Yes
Polypropylene (PP)	5 (34)	200	6.5 (44.8)	190 (1.3)	1.3 (38)	Yes
Acrylonitrile butadiene styrene (ABS)	8 (55)	12	11 (76)	335 (2.3)	3 (88)	Yes
Polycarbonate (PC)	9 (62)	110	13 (89.6)	332 (2.28)	15 (442)	Yes
Acetal (POM)	10 (68.9)	35	14 (96)	520 (3.6)	1.5 (44)	Yes
Nylon 6 (PA)	9 (62)	27	10.5 (72.4)	400 (2.75)	1.9 (56)	Yes

Unfilled/Unreinforced Plastics—cont'd

Physical properties*

Type (Acronym)	Density lb/in.^{3} (Sp.Gr.)	Coefficient of Thermal Expansion, 10^{-5} in./in./°F (10^{-5} m/m/°C)	Thermal Conductivity, Btu in./h ft^2 °F (W/m °C)	Volume Electrical Resistivity Ω-cm	Specific Heat, Btu/lb °F (J/kg °C)
Polytetrafluoroethylene (PTFE)	0.08 (2.2)	3.99 (7.2)	1.74 (0.25)	10^{18}	0.22 (926)
Polybutylene terephthalate (PBT)	0.05 (1.31)	4 (4.2)	1.1 (0.15)	20×10^{15}	0.45 (1894)
Polysulfone (PSU)	0.04 (1.24)	3.1 (5.6)	1.8 (0.26)	5×10^{16}	0.24 (1010)
Polymethylmethacrylate (PMMA)	0.043 (1.19)	3.9 (7)	1.44 (0.21)	$>10^{17}$	0.35 (1473)
Polyamide-imide (PAI)	0.05 (1.4)	2 (3.6)	1.7 (0.29)	1.2×10^{17}	—
Phenolic (PF)	0.05 (1.4)	2 (3.6)	0.2 (0.03)	10^{11}	0.37 (1557)
Polyimide (PI)	0.05 (1.43)	2.7 (4.9)	4.6 (0.66)	10^{15}	0.29 (1220)
Epoxy (EP)	0.04 (1.15)	3.3 (5.9)	2.4 (0.34)	6.1×10^{15}	0.45 (1894)
Polystyrene (PS)	0.04 (1.05)	4 (7.2)	0.89 (0.13)	$>10^{16}$	0.325 (1300)
Polyethylene (PE)	0.034 (0.92)	10 (18)	2.28 (0.33)	10^{18}	0.54 (2273)
Polyvinylchloride (PVC)	0.054 (1.44)	3 (5.4)	1.04 (0.15)	10^{15}	—
Polypropylene (PP)	0.034 (0.9)	4.8 (8.6)	1.28 (0.18)	$>10^{17}$	0.45 (1894)
Acrylonitrile butadiene styrene (ABS)	0.04 (1.05)	3.8 (6.8)	1.5 (0.22)	2.7×10^{16}	—
Polycarbonate (PC)	0.045 (1.21)	3.7 (6.7)	1.38 (0.2)	8×10^{16}	0.30 (1263)
Acetal (POM)	0.053 (1.425)	4.5 (8.1)	1.56 (0.22)	1×10^{15}	0.35 (1473)
Nylon 6 (PA)	0.042 (1.10)	5 (9)	1.2 (0.17)	4.5×10^{13}	0.4 (1689)

*Properties vary significantly with manufacturing process. Values shown are room temperature averages from various manufacturers. Do not use for design. Use data from the manufacturer of the specific material that will be used.

Unreinforced Ceramics and Related Materials

Mechanical properties*

Material	Flexural Strength, ksi (MPa)	Compressive Strength, ksi (MPa)	Hardness, HV	Modulus of Elasticity, 10^6 psi (GPa)	Fracture Toughness, K_{IC} MPa$\sqrt{m}$
Aluminum oxide, Al_2O_3	50 (345)	330 (2300)	1900	53 (360)	5
Silicon nitride, Si_3N_4	113 (800)	370 (2620)	1750	45 (310)	5
Silicon carbide, SiC	69 (500)	425 (2950)	2550	60 (405)	3
Toughened zirconia, ZrO_2	150 (1050)	198 (1370)	1300	30 (204)	10
Amorphous glass (silica)	14 (98)	270 (1860)	600	10 (69)	—
Machinable glass ceramic	15 (103)	50 (344)	250	9.3 (64)	—
Carbon–graphite	8 (55)	16 (110)	<100	1.5 (10)	—
Cemented carbide, WC/6%Co	200 (1400)	580 (4000)	1600	90 (612)	15

Physical properties*

Material	Density, lb/in.^{3} (Sp.Gr.)	Thermal Conductivity, Btu ft/h ft^2 °F (W/m °C)	Coefficient of Thermal Expansion 20–100°C, 10^{-6}in./in./°F (10^{-6}m/m/°C)	Electrical Resistivity Ω-m	Maximum Use Temperature no Load °F (C)
Aluminum oxide	0.14 (3.92)	20 (34)	4.1 (7.4)	10^{12}	3000 (1650)
Silicon nitride	0.12 (3.2)	19 (33)	1.7 (3.1)	10^{10}	2300 (1260)
Silicon carbide	0.12 (3.2)	35 (60)	2.5 (4.5)	—	2300 (1260)
Toughened zirconia	0.21 (6)	3.5 (6)	4.7 (8.5)	$>10^8$	—
Amorphous glass	0.08 (2.2)	0.8 (1.4)	0.55 (1)	10^7	—
Machinable glass/ceramic	0.09 (2.5)	0.97 (1.7)	5.2 (9.4)	10^{12}	1800 (980)
Carbon/graphite	0.06 (1.70)	50 (86)	4.2 (9.6)	2.3×10^{-6}	600(AIR) (315)
Cemented carbide	0.62 (17.2)	50 (86)	4.1 (7.4)	6×10^{-8}	—

*Properties vary significantly with manufacturing process. Values shown are room temperature averages from various manufacturers. Do not use for design. Use data from the manufacturer of the specific material that will be used.

Carbon and Alloy Steels

Mechanical properties at indicated hardness (annealed)[a]

Type and (UNS Number)	Tensile Strength, ksi (MPa)	Yield Strength, ksi (MPa)	Elongation %	Hardness, HV	Modulus of Elasticity, 10^6 psi (GPa)	Machinability Index[b] (Annealed)
1010 (G10100)	55 (380)	40 (275)	35	110	30 (206)	100
1020 (G10200)	58 (400)	43 (296)	34	120	30 (206)	150
1040 (G10400)	75 (517)	51 (351)	30	150	30 (206)	150
4140 (G41400)	95 (655)	60 (413)	25	200	30 (206)	230
4340 (G43400)	110 (758)	70 (482)	20	220	30 (206)	230
8620 (G86200)	78 (538)	56 (386)	30	148	30 (206)	330
A36	70 (482)	36 (248)	21	150	30 (206)	150

Physical properties[a]

Type and (UNS Number)	Density lb/in.^{3} (Sp.Gr.)	Thermal Conductivity, Btu ft/h ft^2 °F (W/m °C)	Coefficient of Thermal Expansion, (10^{-6} in./in./°F) (10^{-6} m/m/°C)	Electrical Resistivity 10^{-6} Ω-cm	Ferromagnetic (at Room Temperature)
1010 (G10100)	0.28 (7.87)	27 (47)	6.6 (11.9)	20	Yes
1020 (G10200)	0.28 (7.87)	27 (47)	6.6 (11.9)	20	Yes
1040 (G10400)	0.28 (7.87)	27 (47)	6.3 (11.3)	19	Yes
4140 (G41400)	0.28 (7.87)	27 (47)	6.8 (12.3)	20	Yes
4340 (G43400)	0.28 (7.87)	27 (47)	6.8 (12.3)	20	Yes
8620 (G86200)	0.28 (7.87)	27 (47)	6.6 (11.9)	18	Yes
A36	0.28 (7.87)	27 (47)	6.3 (11.3)	20	Yes

[a] Average room temperature properties from various sources. Do not use for design. Use data from the manufacturer of the specific material that will be used.

[b] SAE 1117 free-machining steel has an index of 100. Higher index numbers mean that some machining operations may be more expensive to perform compared to 1117 steel. This index is based on experience in a particular shop. Other shops may rate machinability of these materials differently.

Tool Steels

Typical mechanical properties at indicated hardness[a]

Type and (UNS Number)	Tensile Strength, ksi (MPa)	Yield Strength, ksi (MPa)	Modulus of Elasticity, 10^6 psi (GPa)	Hardness, HV	Machinability Index[b] (Annealed)
S7 (T41907)	340 (2344)	275 (1896)	30 (207)	560	230
M2 (T11302)	—	—	30 (207)	630	340
H13 (T20813)	258 (1778)	238 (1641)	30 (207)	510	230
A2 (T30102)	252 (1737)	195 (1344)	29.5 (203)	500	320
D2 (T30402)	450 (3012)	220 (1516)	29.5 (203)	610	340
O1 (T31501)	280 (1930)	272 (1875)	31.5 (217)	535	230

Physical properties[a]

Type and (UNS Number)	Density, lb/in.^{3} (Sp.Gr.)	Coefficient of Thermal Expansion, 10^{-6}in./in./°F (10^{-6} m/m/°C)
S7 (T41907)	0.283 (7.83)	6.9 (12.59)
M2 (T11302)	0.293 (8.12)	5.6 (10.08)
H13 (T20813)	0.28 (7.76)	6.8 (12.2)
A2 (T30102)	0.28 (7.87)	7.8 (13.1)
D2 (T30402)	0.28 (7.7)	5.7 (10.2)
O1 (T31501)	0.28 (7.8)	6.0 (10.8)

[a]Average room temperature properties from various sources. Do not use for design. Use data from the manufacturer of the specific material that will be used.

[b]SAE 1117 free-machining steel has an index of 100. Higher index numbers mean that some machining operations may be more expensive to perform compared to 1117 steel. This index is based on experience in a particular shop. Other shops may rate machinability of these materials differently.

Ferrous Castings

Mechanical properties (at indicated hardness)[a]

ASTM Designation	Tensile Strength, ksi (MPa)	Yield Strength, ksi (MPa)	Elongation, % (min)	Hardness, HV	Modulus of Elasticity, 10^6 psi (GPa)	Machinability Index[b] (Annealed)
A48 Class 20 gray iron	20 (138)	20 (38)	Nil	175	12 (82)	100
A48 Class 35 gray iron	35 (241)	35 (241)	Nil	200	16 (110)	120
A48 Class 40 gray iron	40 (275)	40 (275)	Nil	210	18 (124)	140
A536 grade 5 60-45-18 ductile iron	60 (413)	45 (310)	18	175	25 (172)	210
A536 grade 3 80-55-06 ductile iron	80 (551)	60 (413)	6	235	25 (172)	230
A47-grade 32510 malleable iron	50 (344)	32 (220)	10	125	26 (179)	100
A220 grade 60004 malleable iron	80 (551)	60 (413)	3	225	26 (179)	120
A532 grade A white iron	45 (310)	45 (310)	Nil	550	25 (172)	normally not machined
A27 grade 60-30 cast steel	60 413	30 (206)	24	156	30 (206)	120

Physical properties[a]

ASTM Designation	Density, lb/in.^{3} (Sp.Gr.)	Thermal Conductivity Btu ft/h ft^2 °F (W/m °C)	Coefficient of Thermal Expansion, 10^{-6} in./in./°F (m/m °C)	Electrical Resistivity 10^{-6} Ω-cm	Ferromagnetic @ RT	Specific Heat, Btu/lb °F (J/kg °C)
A48 Class 20 gray iron	0.25 (7)	29 (51)	6.0 (10.8)	100	Yes	0.13 (550)
A48 Class 35 gray iron	0.25 (7)	29 (51)	6 (10.8)	100	Yes	—
A48 Class 40 gray iron	0.26 (7.2)	29 (51)	6 (10.8)	80	Yes	—
A536 Grade 5 60-45-18 ductile iron	0.257 (7.2)	20 (34)	6.6 (11.8)	66	Yes	0.13 (550)
A536 Grade 3 80-55-06 ductile iron	0.251 (7)	18 (30)	6.6 (11.8)	68	Yes	—
A47 Grade 32510 malleable iron	0.261 (7.2)	29.5 (50)	5.9 (9)	35.8	Yes	—
A220 Grade 60009 malleable iron	0.267 (7.3)	29.5 (50)	7.5 (13.5)	40.9	Yes	—
A27 Grade 60-30 cast steel	0.283 (7.8)	27 (46)	8.2 (14.7)	15–20	Yes	0.11 (460)

[a] Average room temperature properties from various sources. Do not use for design. Use data from the manufacturer of the specific material that will be used.

[b] SAE 1117 free-machining steel has an index of 100. Higher index numbers mean that some machining operations may be more expensive to perform compared to 1117 steel. This index is based on experience in a particular shop. Other shops may rate machinability of these materials differently.

Stainless Steels

Mechanical properties at indicated hardness (annealed)[a]

Type (UNS Number)	Tensile Strength, ksi (MPa)	Yield Strength, ksi (MPa)	Elongation, %	Hardness, HV	Modulus of Elasticity, 10^6 psi (GPa)	Machinability Index[b] (Annealed)
430 (S43000)	70 (482)	40 (275)	20	260	29 (199)	165
416 (S41600)	75 (517)	40 (275)	22	260	29 (199)	150
420 (S42000)	75 (517)	40 (275)	25	240	29 (199)	300
440C (S44004)	110 (758)	65 (448)	14	260	29 (199)	375
303 (S30300)	75 (517)	30 (206)	35	260	28 (193)	190
304 (S30400)	75 (517)	30 (206)	40	260	28 (193)	340
316 (S31600)	75 (517)	30 (206)	40	260	28 (193)	340
17-4 PH (S17400)	150 (1034)	125 (861)	19	280	28.5 (196)	350
17-7 PH (S17700)	175 (1206)	155 (1068)	12	280	29 (199)	350

Physical properties[a]

Type (UNS Number)	Density, lb/in.^{3} (Sp.Gr.)	Thermal Conductivity Btu ft/h ft^2 °F (W/m °C)	Coefficient of Thermal Expansion, 10^{-6} in./in./°F (10^{-6} m/m/°C)	Electrical Resistivity 10^{-6} Ω-cm	Ferromagnetic @ RT	Specific heat Btu/lb °F (J/kg °C)
430 (S43000)	0.28 (7.8)	15.1 (26)	5.8 (10.5)	60	Yes	0.11 (460)
416 (S41600)	0.28 (7.8)	14.4 (25)	5.5 (9.9)	70	Yes	0.11 (460)
420 (S42000)	0.28 (7.8)	14.4 (25)	5.7 (10.26)	55	Yes	0.11 (460)
440C (S44004)	0.28 (7.8)	14.0 (24)	5.6 (10.1)	60	Yes	0.11 (460)
303 (S30300)	0.29 (8)	9.4 (16)	9.6 (17.3)	72	No	0.12 (500)
304 (S30400)	0.29 (8)	9.4 (16)	9.6 (17.3)	72	No	0.12 (500)
316 (S31600)	0.29 (8)	9.4 (16)	8.9 (16)	74	No	0.12 (500)
17-4 PH (S17400)	0.28 (7.8)	10.4 (18)	6 (10.8)	100	Yes	0.11 (460)
17-7 PH (S17700)	0.276 (7.8)	9.7 (17)	5.6 (10.1)	82	Yes	0.11 (460)

[a] Average room temperature properties from various sources. Do not use for design. Use data from the manufacturer of the specific material that will be used.

[b] SAE 1117 free-machining steel has an index of 100. Higher index numbers mean that some machining operations may be more expensive to perform compared to 1117 steel. This index is based on experience in a particular shop. Other shops may rate machinability of these materials differently.

Aluminum Alloys

Mechanical properties (at temper indicated)[a]

Type (UNS Number)	Tensile Strength, ksi (MPa)	Yield Strength, ksi (MPa)	Elongation %	Hardness, HV	Modulus of Elasticity, 10^6 psi (GPa)	Machinability Index[b] (@ Temper Listed)
1100-H14 (A91100)	18 (124)	17 (117)	9	26	10.0 (69)	180
2024-14 (A92024)	68 (469)	47 (324)	19	120	10.6 (73)	110
3003-H14 (A93003)	22 (152)	21 (145)	8	40	10.0 (69)	160
5052-H34 (A95052)	38 (262)	31 (214)	10	68	10.2 (70)	150
6061-T6 (A96061)	45 (310)	40 (276)	12	95	10.0 (69)	110
6063-T6 (A96063)	35 (241)	31 (214)	12	73	10.0 (69)	110
7075-T6 (A97075)	83 (572)	73 (503)	11	150	10.4 (72)	100
355-T6 (A03550)	32 (2.50)	31 (214)	2	85	10.0 (69)	110
380 (A03800)	42 (302)	40 (276)	2	80	10.0 (69)	130

Physical properties[a]

Type (UNS Number)	Density lb/in.^{3} (Sp.Gr.)	Thermal Conductivity Btu ft/h ft^2 °F (W/m°C)	Coefficient of Thermal Expansion 10^{-6} in./in./°F (10^{-6} m/m/°C)	Electrical Resistivity 10^{-6} Ω-cm	Ferromagnetic @ RT	Specific Heat, Btu/lb°F (J/kg °C)
1100-H14 (A91100)	0.098 (2.71)	125 (210)	13.1 (23.6)	2.8	No	0.22 (920)
2024-T4 (A92024)	0.1 (2.77)	70 (116)	12.9 (23.2)	3.4	No	0.22 (920)
3003-14 (A93003)	0.099 (2.73)	92 (150)	12.9 (23.2)	4.2	No	0.22 (920)
5052-H34 (A95052)	0.097 (2.68)	80 (132)	13.2 (23.8)	4.9	No	0.22 (920)
6061-T6 (A96061)	0.098 (2.70)	116 (190)	13.1 (23.6)	4.0	No	0.23 (960)
6063-T6 (A96063)	0.098 (2.70)	116 (190)	13.0 (23.4)	3.3	No	0.23 (960)
7075-T6 (A97075)	0.101 (2.80)	75 (130)	13.1 (23.6)	5.2	No	0.23 (960)
355 (A03550)	0.1 (2.77)	87 (150)	12.4 (22.3)	—	No	0.22 (920)
380 (A03800)	0.99 (2.73)	58 (100)	11.7 (21.1)	—	No	0.22 (920)

[a]Average room temperature properties from various sources. Do not use for design. Use data from the manufacturer of the specific material that will be used.

[b]SAE 1117 free-machining steel has an index of 100. Higher index numbers mean that some machining operations may be more expensive to perform compared to 1117 steel. This index is based on experience in a particular shop. Other shops may rate machinability of these materials differently.

Copper Alloys

Mechanical properties (for tempers/hardness shown)[a]

Type (UNS Number)	Tensile Strength, ksi (MPa)	Yield Strength, ksi (MPa)	Elongation %	Hardness, HV	Modulus of Elasticity 10^6 psi (GPa)	Machinability Index[b] (Annealed)
ETP copper (C11000)	55 [H08][d] (379)	50 [H08] (344)	4	60	17 (117)	150
DHP copper (C12200)	55 [H08] (379)	50 [H08] (344)	8	60	17 (117)	150
Be copper (C17200) (age hardened)	203 (1400)	170 (1172)	10	410	19 (131)	100
Red brass (C23000)	57 [H02] (393)	49 [H02] (337)	12	65	17 (117)	70
Yellow brass (C26000)	62 [H02] (427)	52 [H02] (358)	25	70	16 (110)	90
Free cutting yellow brass (C36000)	58 [H02] (400)	45 [H02] (310)	25	78	14 (96)	70
Tin bronze (C90700)	45 (310)	21 (145)	20	70	15 (103)	150
Leaded tin bronze (C93700)	35 (241)	18 (124)	20	65	11 (76)	100

Physical properties[a]

Type (UNS Number)	Density, lb/in.^{3} (Sp.Gr.)	Thermal Conductivity, Btu ft/h ft^2 °F (W/m°C)	Coefficient of Thermal Expansion, 10^{-6} in./in./°F (10^{-6} m/m/°C)	Electrical[c] Conductivity, % IACS	Ferromagnetic @ RT	Specific Heat, Btu/lb°F (J/kg °C)
ETP copper (C11000)	0.323 (8.94)	226 (391)	9.8 (17.7)	101	No	0.092 (385)
DHP copper (C12200)	0.323 (8.94)	196 (339)	9.8 (17.7)	85	No	0.092 (385)
Be copper (C17200)	0.298 (8.25)	62 (107)	9.9 (17.8)	22	No	0.10 (425)
Red brass (C23000)	0.316 (8.74)	92 (159)	10.4 (18.7)	37	No	0.09 (380)
Yellow brass (C26000)	0.308 (8.52)	70 (121)	11.1 (20.1)	28	No	0.09 (380)
Free cutting yellow brass (C36000)	0.307 (8.50)	67 (116)	11.4 (20.6)	26	No	0.09 (380)
Tin bronze (C90700)	0.32 (8.78)	40.8 (72)	10.2 (18.4)	9.6	No	0.09 (380)
Leaded tin bronze (C93700)	0.33 (8.95)	27.1 (40)	10.3 (15.5)	10	No	0.09 (380)

[a]Average room temperature properties from various sources. Do not use for design. Use data from the manufacturer of the specific material that will be used.
[b]SAE 1117 free-machining steel has an index of 100. Higher index numbers mean that some machining operations may be more expensive to perform compared to 1117 steel. This index is based on experience in a particular shop. Other shops may rate machinability of these materials differently.
[c]Copper = 3.82×10^{-8} Ω/m.
[d]Temper.

Nickel Alloys

Mechanical properties (at indicated hardness)[a]

Type (UNS Number)	Tensile Strength, ksi (MPa)	Yield Strength, ksi (MPa)	Elongation %	Hardness, HV	Modulus of Elasticity, 10^6 psi (GPa)	Machinability Index[b] (Annealed)
Pure nickel (N02200)	68 (469)	27 (186)	50	170	30 (207)	340
Monel 400 (N04400)	82 (565)	39 (269)	45	137	26 (179)	340
Inconel 600 (N06600)	99 (683)	45 (310)	43	160	31 (213)	300
Inconel 625 (N06625)	130 (896)	70 (483)	50	183	29.8 (205)	300
Inconel X 750 (N07750)	180 (1241)	120 (827)	25	360	31 (213)	360
Incoloy 800 (N08800)	85 (586)	42 (290)	45	155	28.3 (195)	320
Incoloy 825 (N08825)	100 (690)	45 (310)	45	155	29.8 (205)	320
Hastelloy B-2 (N10665)	138 (951)	76 (524)	53	240	31.4 (216)	340
Hastelloy C-276 (N10276)	115 (792)	51 (351)	60	210	29.8 (205)	340

Physical properties[a]

Type (UNS Number)	Density, lb/in.^{3} (Sp.Gr.)	Thermal Conductivity, Btu ft/h ft^2 °F (W/m°C)	Coefficient of Thermal Expansion, 10^{-6} in./in./°F (10^{-6} m/m/°C)	Electrical Resistivity, 10^{-6} Ω-cm	Ferromagnetic @ RT	Specific Heat, Btu/lb°F (J/kg °C)
Pure nickel (N02200)	0.321 (8.89)	40 (70.2)	8.6 (15.5)	11	Yes	0.108 (454)
Monel 400 (N04400)	0.318 (8.8)	26 (46)	9.1 (16.3)	47	Yes (@ rt only)	0.127 (534)
Inconel 600 (N06600)	0.306 (8.5)	8.3 (14.6)	6.8 (12.2)	—	Yes	—
Inconel 625 (412) (N06625)	0.305 (8.44)	5.5 (9.8)	7.1 (12.8)	126	Yes	0.098
Inconel X 750 (N07750)	0.299 (8.28)	6.8 (12)	7 (12.6)	126	Yes	—
Incoloy 800 (N08800)	0.287 (7.44)	6.5 (11.5)	8 (14.4)	91	Yes	0.12 (505)
Incoloy 825 (N08825)	0.294 (8.14)	6.3 (11.1)	7.8 (14)	110	Yes	—
Hastelloy B-2 (N10665)	0.333 (9.24)	7.05 (12.2)	6.2 (11.2)	137	No	0.089 (374)
Hastelloy C-276 (N10276)	0.321 (8.90)	7.5 (12.9)	6.2 (11.2)	130	No	0.102 (429)

[a] Average room temperature properties from various sources. Do not use for design. Use data from the manufacturer of the specific material that will be used.
[b] SAE 1117 free-machining steel has an index of 100. Higher index numbers mean that some machining operations may be more expensive to perform compared to 1117 steel. This index is based on experience in a particular shop. Other shops may rate machinability of these materials differently.

Refractory Metals

Mechanical properties (at indicated hardness)[a]

Type (UNS Number)	Tensile Strength, ksi (MPa)	Yield Strength, ksi (MPa)	Hardness, HV	Modulus of Elasticity, 10^6 psi (GPa)	Machinability Index[b] (Annealed)
Zirconium (R60701)	65 (448)	45 (310)	110	14.4 (99)	320
Molybdenum (R03604)	95 (655)	80 (551)	220	47 (324)	430
Tungsten (R07005)	220 (1516)	220 (1516)	480	59 (407)	520
Tantalum (R05200)	60 (413)	48 (331)	150	27 (186)	320
Hafnium (R02001)	77 (531)	32 (220)	110	20 (137)	500
Columbium/ Niobium (R04200)	50 (345)		60	15 (103)	320

Physical properties[a]

Type (UNS Number)	Density, lb/in.^{3} (Sp.Gr.)	Thermal Conductivity, Btu ft/h ft^2 °F (W/m°C)	Coefficient of Thermal Expansion 10^{-6} in./in./°F (10^{-6} m/m/°C)	Electrical Resistivity, 10^{-6} Ω-cm	Ferromagnetic @ RT	Specific Heat, Btu/lb °F (J/kg °C)
Zirconium (R60701)	0.234 (6.5)	13 (22.4)	3.2 (5.76)	40	No	0.068 (2863)
Molybdenum (R03604)	0.37 (10.28)	84.5 (147)	2.8 (5.04)	5.2	No	0.065 (2736)
Tungsten (R07005)	0.695 (19.32)	96.6 (166)	2.5 (4.5)	—	No	0.034 (1431)
Tantalum (R05200)	0.6 (16.68)	31 (53)	3.6 (6.48)	12.5	No	0.036 (1516)
Hafnium (R02001)	0.47 (13.06)	—	34 (61.20)	30	No	0.035 (1473)
Columbium/ Niobium (R04200)	0.31 (8.5)	25 (43)	4.1 (7.38)	15	No	—

[a] Average room temperature properties from various sources. Do not use for design. Use data from the manufacturer of the specific material that will be used.

[b] SAE 1117 free-machining steel has an index of 100. Higher index numbers mean that some machining operations may be more expensive to perform compared to 1117 steel. This index is based on experience in a particular shop. Other shops may rate machinability of these materials differently.

Miscellaneous Metals

Mechanical properties (at indicated hardness)[a]

Metal and (UNS Number)	Tensile Strength ksi (MPa)	Yield Strength ksi (MPa)	Elongation (%)	Hardness HV	Modulus of Elasticity 10^6 psi (GPa)	Machinability Index[b]
Magnesium AZ31B-H24 (M11311)	38 (262)	26 (179)	12	50	6.5 (44.8)	50
Magnesium AZ91A-F (M11911)	33 (227)	22 (152)	3	63	6.5 (44.8)	50
Beryllium (R19800)	42 (289)	—	3	—	44 (303.4)	520
Titanium unalloyed (R50250)	60 (413)	45 (310)	20	150	15.2 (104)	460
Titanium Ti-6A1-4V (R56401)	160 (1103)	140 (965)	8	360	16.5 (114)	510
Zinc ASTM B86 Gr. AG 40A (Z33520)	41 (283)	32 (221)	10	90	0.012 (0.085)	80
Zinc ASTM B 86 Gr. AC 41A (Z35531)	48 (331)	33 (228)	7	91	>0.012 (>0.085)	80

Physical properties[a]

Metal and (UNS Number)	Density lb/in.^{3} (Sp.Gr.)	Thermal Conductivity Btu ft/h ft^2 °F (W/m°C)	Coefficient of Thermal Expansion 10^{-6} in./in./°F (10^{-6} m/m/°C)	Electrical Resistivity 10^{-6} Ω-cm	Ferromagnetic @RT	Specific Heat Btu/lb °F (J/kg °C)
Magnesium AZ31B-H24 (M11311)	0.064 (1.77)	44 (76)	14 (25.2)	9.2	No	0.245 (1030)
Magnesium AZ91A-F (M11911)	0.065 (1.8)	41 (69)	14.5 (26)	9.3	No	0.25 (1050)
Beryllium (R19800)	0.067 (1.86)	100 (172)	6.3 (11.3)	—	No	—
Titanium unalloyed (R50250)	0.163 (4.5)	9 (15)	5.3 (9.54)	55	No	0.127 (538)
Titanium Ti-6A1-4V (R56401)	0.16 (4.5)	4.2 (7.2)	5.3 (9.54)	177	No	0.135 (560)
Zinc ASTM B86 Gr. AG 40A (Z33520)	0.238 (6.6)	65 (113)	15.2 (27.4)	6.3	No	0.1 (420)
Zinc ASTM B86 Gr AC 41A (Z35531)	0.242 (6.7)	63 (111)	15.2 (27.4)	6.5	No	0.1 (420)

[a] Average room temperature properties from various sources. Do not use for design. Use data from the manufacturer of the specific material that will be used.

[b] SAE 1117 free-machining steel has an index of 100. Higher index numbers mean that some machining operations may be more expensive to perform compared to 1117 steel. This index is based on experience in a particular shop. Other shops may rate machinability of these materials differently.

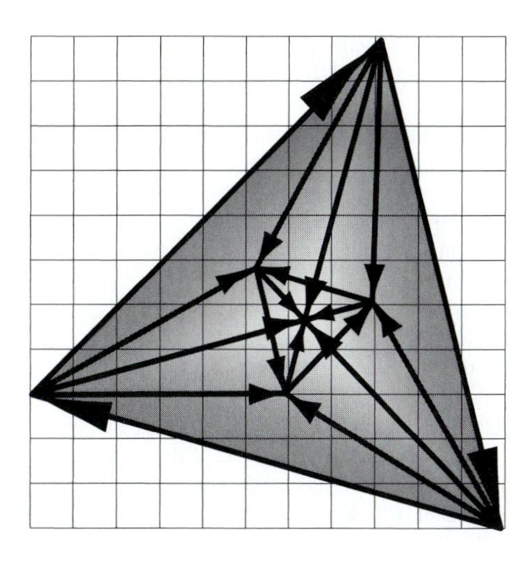

Index